19th International Seminar on High Energy Physics (QUARKS-2016)

EPJ Web of Conferences Volume 125 (2016)

Pushkin, Russia
29 May - 4 June 2016

Editors:

V.A. Andrianov
V.A. Matveev
V.A. Rubakov
V.T. Kim
A.A. Andrianov
M.D. Fitkevich

ISBN: 978-1-5108-3165-0

Printed from e-media with permission by:

Curran Associates, Inc.
57 Morehouse Lane
Red Hook, NY 12571

Some format issues inherent in the e-media version may also appear in this print version.

Printed by Curran Associates, Inc. (2016)

For additional information, please contact EDP Sciences – Web of Conferences at the address below.

EDP Sciences – Web of Conferences
17, Avenue du Hoggar
Parc d'Activité de Courtabœuf
BP 112
F-91944 Les Ulis Cedex A
France

Phone: +33 (0) 1 69 18 75 75
Fax: +33 (0) 1 69 28 84 91

contact-edps@webofconferences.org

Additional copies of this publication are available from:

Curran Associates, Inc.
57 Morehouse Lane
Red Hook, NY 12571 USA
Phone: 845-758-0400
Fax: 845-758-2633
Email: curran@proceedings.com
Web: www.proceedings.com

TABLE OF CONTENTS

PREFACE: "Quarks-2016", 19th International Seminar on High Energy Physics

This publication of Proceedings of "Quarks-2016", the 19th International Seminar on High Energy Physics, is a collection of refereed papers presented at plenary and section sessions. Complete set of written contributions can be found online at EPJ web of Conferences Publishing website.

This Seminar aimed at bringing together people working in various areas of theoretical high energy physics, cosmology, astrophysics and experimental physics. It took place in Pushkin, Saint-Petersburg region (Russia) from May 29 to June 4, 2016 (http://quarks.inr.ac.ru). It was organized by Institute for Nuclear Research of the Russian Academy of Sciences, Joint Institute for Nuclear Research, Dubna; Petersburg Nuclear Physics Institute of NRC "Kurchatov Institute", Gatchina; Saint-Petersburg State University, Peter the Great Saint-Petersburg Polytechnic University and Saint-Petersburg Scientific Center of the Russian Academy of Sciences.

Topics covered at "Quarks-2016" are: physics beyond the Standard Model at colliders and in rare processes; Higgs phenomenology; cosmology and astrophysics as related to particle physics; modern field theory and selected aspects of mathematical physics; quantum chromodynamics and strong interactions; ultra-high energy cosmic rays; selected experimental results.

The organizers and participants of "Quarks-2016" acknowledge the support from the organizing institutions, Russian Foundation for Basic Research and Federal Agency of Scientific Organizations.

Editors: Vladimir Andrianov, co-chairman
Victor Matveev, co-chairman
Valery Rubakov, co-chairman
Alexander Andrianov,
Maxim Fitkevich, scientific secretary
Ivan Kharuk, scientific secretary
Victor Kim,
Sergei Kolevatov

Committees "Quarks2016":

QUARKS-2016 Organizing committee

- V. A. Matveev (JINR, Dubna) Co-Chairman
- V. A. Rubakov (INR, Moscow) Co-Chairman
- V. A. Andrianov (SPbSU, St.Petersburg) Co-Chairman
- M. D. Fitkevich (INR, Moscow) Scientific Secretary
- I. V. Kharuk (INR, Moscow) Scientific Secretary
- A. V. Nevstrueva (INR, Moscow) Secretary
- A. B. Arbuzov (JINR, Dubna)
- A. V. Gladyshev (JINR, Dubna)
- V. K. Novikova (JINR, Dubna)
- A. G. Panin (INR, Moscow)
- G. I. Rubtsov (INR, Moscow)
- P. S. Satunin (INR, Moscow)
- D. S. Gorbunov (INR, Moscow)
- S. V. Troitsky (INR, Moscow)
- M. V. Libanov (INR, Moscow)
- S. V. Demidov (INR, Moscow)
- V. T. Kim (PNPI & SPbPU, St.Petersburg)
- L. N. Lipatov (PNPI & SPbSU, St.Petersburg)
- M.V. Ioffe (SPbSU, St.Petersburg)
- Y. A. Berdnikov (SPbPU, St.Petersburg)
- S. S. Afonin (SPbSU, St.Petersburg)
- S. S. Kolevatov (SPbSU, St.Petersburg)
- O. O. Novikov (SPbSU, St.Petersburg)
- V. N. Kovalenko (SPbSU, St.Petersburg)

QUARKS-2016 Advisory committee

- V. A. Matveev (JINR, Dubna)
- A. A. Andrianov (SPbSU, St.Petersburg)
- A. M. Boyarsky (Leiden U.)
- F. L. Bezrukov (CERN & U. of Connecticut)
- S. L. Dubovsky (New York U.)
- A. P. Isaev (JINR, Dubna)
- D. I. Kazakov (JINR, Dubna & ITEP & MIPT, Moscow)
- V. T. Kim (PNPI & SPbPU, St.Petersburg)
- L. N. Lipatov (PNPI & SPbSU, St.Petersburg)
- A. A. Penin (U. of Alberta, Edmonton & KIT, Karlsruhe)
- D. V. Semikoz (APC, Paris)
- M. E. Shaposhnikov (EPFL, Lausanne)
- S. M. Sibiryakov (CERN & EPFL, Lausanne & INR, Moscow)
- D. T. Son (U. of Chicago)
- P. G. Tinyakov (ULB, Brussels)

Sponsors "Quarks2016":

Russian Foundation for Basic Research and Federal Agency of Scientific Organizations.

List of participants

of 19th International Seminar on High Energy Physics QUARKS-2016
Pushkin, Russia, 29 May - 4 June, 2016.

1. Achasov Nikolay (Sobolev Inst., Novosibirsk)
2. Afonin Sergey (St.Petersburg State University)
3. Ageev Dmitry (Steklov Mathematical Institute of RAS)
4. Akhmedov Emil (ITEP & HSE, Moscow)
5. Akhmedova Valeria (HSE, Moscow)
6. Alberte Lasma (SISSA, Trieste)
7. Alexeev Stanislav (SAI, MSU)
8. Alikhanov Ibragim (Baksan Neutrino Observatory INR RAS)
9. Altsybeev Igor (St.Petersburg State University)
10. Andrianov Vladimir (St.Petersburg State University)
11. Andrianov Alexander (St.Petersburg State University)

12. Antonov Nikolai Viktorovich (St. Petersburg State University)
13. Arbuzov Andrey (JINR, Dubna)
14. Arbuzov Boris (MSU)
15. Arbuzova Elena (Dubna U. & Novosibirsk State U.)
16. Aref'eva Irina (Steklov Math. Institute of RAS)
17. Astapov Konstantin (INR of RAS)
18. Astrakhantsev Nikita Yurievich (ITEP, Moscow)
19. Avdeev Nikita (MSU)
20. Babichev Eugeny (LPT, Orsay)
21. Balitskii Iaroslav (IHEP)
22. Barvinsky Andrei (Lebedev Inst.)
23. Bednyakov Alexander Vadimovich (JINR,Dubna)
24. Bekaert Xavier (Tours U.)
25. Belavin Alexander (Landau Inst.)
26. Belikov Iouri (IPHC)
27. Berezin Victor (INR of RAS,Moscow)
28. Bezrukov Leonid (INR of RAS,Moscow)
29. Bezrukov Fedor (U. of Connecticut)
30. Biryukov Alexandr (Samara State Aerospace University)
31. Blas Diego (CERN)
32. Blatov Alex (IMU)
33. Boos Eduard (SINP, MSU)
34. Braguta Victor Valerievich (IHEP)
35. Braun Mikhail (St.Petersburg State University)
36. Bravina Larissa (University of Oslo)
37. Bykov Dmitry (MPI for Gravitational Physics (Potsdam, Germany),)
38. Chekhov Leonid (Steklov Math. Inst. of RAS)
39. Chetyrkin Konstantin Grigorievich (KIT, Karlsruhe)
40. Chudaykin Anton Sergeevish (INR of RAS)
41. Czarnecki Andrzej (U. of Alberta)
42. Dolgopolov Mikhail (Samara State Aerospace University)
43. Dordevic Milos Predrag (Vinca Institute of Nuclear Sciences)
44. Dorokhov Alexander (JINR, Dubna)
45. Dudas Emilian (Ecole Polytechnique)
46. Dudko Lev (SINP, MSU)
47. Duk Viacheslav (INR RAS & INFN Perugia)
48. Dvornikov Maxim (IZMIRAN & Tomsk State University)
49. Dyadina Polina (SAI, MSU)
50. Echeistov Vselolod (MSU)
51. Feofilov Grigory Alexandrovich (St.-Petersburg State University)
52. Figueroa Daniel (CERN)
53. Fitkevich Maxim (INR of RAS, Moscow)
54. Francia Dario (Scuola Normale Superiore Pisa)
55. Galynskii Mikhail Vladimirovich (Joint Institute for Power and Nuclear Research-Sos)
56. Gelfond Olga (Levedev Inst.)
57. Gladyshev Alexei (JINR, Dubna)
58. Golovnev Alexey (St.-Petersburg State University)

59. Gorbenko Victor (Stanford U.)
60. Gorbunov Dmitry Sergeevich (INR of RAS)
61. Gorsky Alexander (IPPI, Moscow)
62. Grib Andrey Anatoljevich (Russian State Herzen Pedagogical University)
63. Grinyuk Andrey (JINR, Dubna)
64. Gromov Nikolay (Kings College, London)
65. Gubina Nadezhda Valer'evna (MSU)
66. Gulamov Igor Ekhtiramovich (Skobeltsyn Institute of Nuclear Physics (SINP,MSU))
67. Gulitskiy Nikolay Mikhailovich (St.- Petersburg State University)
68. Gurskaya Albina V. (Samara State Aerospace University)
69. Ioffe Mikhail (St.- Petersburg State University)
70. Isaev Alexei (JINR, Dubna)
71. Ivanov Evgeny (JINR, Dubna)
72. Ivanov Mikhail (EPFL & INR)
73. Jenkovszky Laszlo Laszlovich (Bogolyubov ITP, Kiev)
74. Kakin Polina (St.-Petersburg State University)
75. Kalashev Oleg Yevgenyevich (INR of RAS)
76. Kaplan David (Johns Hopkins U.)
77. Karshenboim Savely (Pulkovo observatory & MPQ)
78. Kataev Andrei (INR of RAS, Moscow)
79. Kazakov Dmitry (JINR, Dubna)
80. Kerbikov Boris O (ITEP)
81. Kharuk Ivan (INR of RAS, Moscow)
82. Khmelnitsky Andrei (INR of RAS, Moscow)
83. Khramtsov Mikhail (Steklov Mathematical Institute of RAS)
84. Kim Victor (PNPI)
85. Kirpichnikov Dmitry Viktorovich (INR of RAS)
86. Kirsanov Mikhail (INR of RAS, Moscow)
87. Kiselev Alexey Vladimirovich (Sobolev Institute for Mathematics & Novosibirsk St)
88. Kisselev Alexander (IHEP Protvino, NRC "Kurchatov Institute")
89. Kolevatov Sergey (St.-Petersburg State University)
90. Kolupaeva Liudmila (JINR, Dubna)
91. Komargodski Zohar (Weizmann Inst.)
92. Kompaniets Mikhail (St.-Petersburg State University)
93. Koroteev Peter (Perimeter Institute for Theoretical Physics)
94. Koshelev Alexey (U. da Beira)
95. Kostenko Mariia Mikhaivolna (St.- Petersburg State University)
96. Kotov Andrey Yuryevich (ITEP)
97. Kovalenko Vladimir (St.-Petersburg State University)
98. Kozachuk Anastasia (MSU)
99. Kozhevnikov Arkadii Alexeevich (S.L. Sobolev Institute of Mathematics)
100. Krasnikov Nikolai (INR of RAS, Moscow)
101. Kravtsova Galina (MSU)
102. Krutov Alexander Fedorovich (Samara University)
103. Kudenko Tatiana Grigor'evna (INR of RAS)
104. Kudlis Andrew (St.-Petersburg State University)
105. Kuehn Johann (Karlsruher Insitut fuer Technologie)

106. Kulagin Sergey (INR of RAS)
107. Kuvshinov Viacheslav Ivanovich (BSU, Minsk)
108. Kuznetsov Mikhail (INR of RAS, Moscow)
109. Lakomov Igor (CERN)
110. Larin Sergey (INR of RAS, Moscow)
111. Latosh Boris (Dubna U.)
112. Lawe Matthew David (Lancaster U.)
113. Lebedev Nikita (St.-Petersburg State University)
114. Lee Roman (Budker Inst. of Nuclear Physics Novosibirsk)
115. Levkov Dmitry (INR of RAS, Moscow)
116. Libanov Maxim (INR of RAS, Moscow)
117. Lipatov Lev (PNPI)
118. Liventsev Dmitri (Virginia Tech./KEK)
119. Lobanov Andrey Evgen'evich (MSU)
120. Lopez Nacir Diana Laura (CERN)
121. Lublinsky Michael (Ben-Gurion University of the Negev)
122. Malyshev Mikhail (St.- Petersburg University)
123. Marachevsky Valery (St. Petersburg U.)
124. Marakulin Artur (INR of RAS, Moscow)
125. Martinovic Lubomir (JINR, Dubna)
126. Matveev Victor (JINR, Dubna)
127. Melikhov Dmitri (SINP MSU)
128. Mikhailov Sergey (JINR, Dubna)
129. Mirbabayi Mehrdad (IAS, Princeton)
130. Mironov Andrei (Levedev Inst. & ITEP)
131. Mironov Sergey (INR of RAS, Moscow)
132. Misuna Nikita G. (Lebedev Institute, Moscow)
133. Morozov Alexey (ITEP Moscow)
134. Morozov Andrey (IITP RAS)
135. Muller Daniel (Instituto de FΓsica, Universidade de BrasΓlia)
136. Musaev Edward (Albert Einstein Inst. Postdam)
137. Naumov Dmitry (JINR, Dubna)
138. Nazarova Elizaveta (MSU)
139. Nekrasov Nikita (Stony Brook)
140. Nesterov Dmitry (St.- Petersburg. INP)
141. Nesti Fabrizio (RBI)
142. Nesvizhevsky Valery (ILL Grenoble)
143. Nevzorov Roman (U. of Adelaide & ITEP)
144. Nikiforova Vasilisa (INR of RAS, Moscow)
145. Novikov Oleg (St. Petersburg State University)
146. Novikov Victor alexander (ITEP)
147. Novozhilov Victor Yurievich (St.-Petersburg Centre for Natural Science)
148. Nugaev Emin (INR Moscow)
149. Ospanov Rustem (U. of Manchester)
150. Ostapchenko Sergey (Frankfurt Institute for Advanced Studies (FIAS))
151. Pakhlov Pavel (LPI RAS & MEPhI)
152. Panin Alexander (INR of RAS, RAS)
153. Pankov Alexander (Tech. Univ. of Gomel)

154. Pavlov Yuri Viktorovich (Institute of Problems in Mechanical Engineering of)
155. Penin Alexander (U. of Alberta)
156. Petrov Vladimir (IHEP, Protvino)
157. Petrova Elena Yurievna (SINP MSU)
158. Petti Roberto (U. of South Carolina)
159. Pikelner Andrey Fedorovich (JINR, Dubna)
160. Pismak Yury (St. Petersburg State University)
161. Polezhaev Roman Gennadievich (Samara University)
162. Polukhina Natalia Gennadievna (Lebedev Inst.)
163. Polyarush Alexandr Yurievich (INR of RAS, Moscow)
164. Poppitz Erich Robert (University of Toronto)
165. Postnov Konstantin (SAI MSU)
166. Pozdeeva Ekaterina (SINP MSU)
167. Prokhvatilov Evgeni (St. Petersburg State University)
168. Pshirkov Maxim (SAI MSU)
169. Ptitsina Ksenia (INR of RAS, Moscow)
170. Radovskaya Anna (Lebedev Inst.)
171. Rajendran Surjeet (Berkeley U.)
172. Ramazanov Sabir (Gran Sasso Science Institute)
173. Rebbi Claudio (Boston U.)
174. Rogachevsky Oleg (JINR, Dubna)
175. Romanenkova Evgenia (MEPhI)
176. Romanino Andrea (SISSA)
177. Rubakov Valery (INR of RAS, Moscow)
178. Rubtsov Grigory I. (INR of RAS, RAS)
179. Rumyantsev Dmitry (Yaroslavl State U.)
180. Ryadovikov Vasily (IHEP, Protvino)
181. Samsonov Igor (University of Western Australia)
182. Sapronov Andrey (JINR, Dubna)
183. Satunin Petr (INR of RAS, Moscow)
184. Savvidis Georgios (Demokritos Nat. Research Centre Athens)
185. Schoenning Karin (Uppsala U.)
186. Semenov-Tian-Shansky Kirill (PNPI)
187. Shabanov Arseniy Igorevich (INR of RAS, RAS)
188. Shaparau Valery (Joint Institute for Power and Nuclear Research-Sos)
189. Shevchenko Vladimir (NRC Kurchatov Institute)
190. Shkerin Andrey (EPFL)
191. Shleenkov Mark Alexandrovich (Samara State Aerospace University)
192. Shnir Yakov (JINR, Dubna)
193. Sibiryakov Sergey (CERN & EPFL & INR)
194. Smilga Andrei (U. of Nantes)
195. Smirnov Alexey (INR of RAS,)
196. Smolnikov Anatoly (JINR, Dubna)
197. Smolyakov Mikhail (SINP MSU)
198. Snegirev Timofey (Tomsk SPU & Tomsk Politechnic U.)
199. Sobolev Ivan Vladimirovich (INR of RAS, RAS)
200. Sokolov Aleksandr (St.- Petersburg State University)
201. Sokolova Elena (INR of RAS, RAS)

202. Sokolsky Pierre (U. of Utah)
203. Soloshenko Aleksey Aleksandr (Joint Institute for Nuclear Research)
204. Sosnovikov Artur (INR Moscow)
205. Sotiriou Thomas (University of Nottingham)
206. Starobinsky Alexei (Landau Inst.)
207. Starodubtsev Artem (St.-Petersburg State University)
208. Stavinskiy Alexey Valentinovich (ITEP)
209. Stepanyantz Konstantin (MSU)
210. Studenikin Alexander (MSU)
211. Suetin Daniil Petrovich (Peter the Great St.Petersburg Polytechnic Universi)
212. Taronna Massimo (Brussels U.)
213. Teryaev Oleg (JINR, Dubna)
214. Tetradis Nikolaos (University of Athens, Greece)
215. Tikhanovskaya Maria (NRNU MEPhI)
216. Timiryasov Inar (INR of RAS, Moscow)
217. Tinyakov Peter (ULB)
218. Tlisov Danila Anatol'evich (INR of RAS,)
219. Tokareva Anna (INR of RAS, Moscow)
220. Toporensky Alexey (SAI MSU)
221. Tretyakova Daria (Ural Federal University)
222. Trincherini Enrico (INFN Pisa)
223. Tseytlin Arkady (Imperial College and Lebedev Institute, Moscow)
224. Urban Federico (KBFI, Tallin)
225. Vasiliev Mikhail (Levedev Inst.)
226. Vassilevich Dmitri (St. Petersburg State University)
227. Vazquez Gomez Ricardo (Laboratori Nazionali di Frascati)
228. Vechernin Vladimir Victorovich (St.-Petersburg State University)
229. Vereshagin Vladimir Vladimirovich (St.-Petersburg State University)
230. Verkhodanov Oleg (SAO RAS)
231. Vernov Sergey (SINP MSU)
232. Vikman Alexander (FZU, Prague)
233. Villadoro Giovanni (ICTP)
234. Volkova Victoria (INR of RAS, Moscow)
235. Voloshin Sergei (Wayne State University)
236. Zabzine Maxim (Uppsala U.)
237. Zakharov Valentin (ITEP, Moscow)
238. Zaslavskii Oleg Borisovich (Kharkov V.N. Karazin National University)
239. Zenkevich Yegor (INR of RAS, Moscow)
240. Zhezher Yana Valer'evna (Lomonosov MSU & INR RAS)
241. Zhokhov Roman (IHEP)
242. Zubov Roman (St. Petersburg State University University)

edp sciences
EPJ Web of Conferences

Statement of Peer review

In submitting conference proceedings to EPJ Web of Conferences, I certify to the Publisher that I adhere to the **Policy on Publishing Integrity** of the journal in order to safeguard good scientific practice in publishing.

1. All articles have been subjected to peer review administered by the proceedings editors.
2. Reviews have been conducted by expert referees, who have been requested to provide unbiased and constructive comments aimed, whenever possible, at improving the work.
3. Proceedings editors have taken all reasonable steps to ensure the quality of the materials they publish and their decision to accept or reject a paper for publication has been based only on the merits of the work and the relevance to the journal.

Title, date and place of the conference

19 International Seminar on High Energy Physics,
Pushkin, Saint Petersburg, 29 May - 4 June 2016

Proceedings editor(s):

V.A. Andrianov, V.A. Matveev, V.A. Rubakov, V.T. Kim, A.A. Andrianov,
M.D. Fitkevich, Scientific Secretary, I.V. Kharuk, Scientific Secretary,
S.S. Kolevatov

Date and editor's signature

09.09.2016

EPJ Web of Conferences **125**, 01001 (2016) DOI: 10.1051/epjconf/201612501001
QUARKS-2016

Recent results from CMS

Milos Dordevic[1] on behalf of the CMS Collaboration *

[1] *Vinca Institute of Nuclear Sciences, PO Box 522, 11001 Belgrade, Serbia*

Abstract. The highlights of the most recent CMS results with 13 TeV data will be presented in this overview. The Standard Model precision measurements, including the top quark production, will be shown first. This will be followed by the presentation of Higgs boson studies with the early 13 TeV data. Then the focus will shift to searches for physics beyond the Standard Model, including the searches for several Supersymmetric scenarios, using different analysis techniques. The talk will conclude with searches for the exotic resonances, with an emphasis on studies of the high-mass diphoton production.

1 Introduction

The LHC Run I was very successful for the CMS Collaboration, with the major event being the discovery of the Higgs boson. A large number of studies of the known processes and searches for new phenomena was conducted, all of them confirming the prediction of the Standard Model (SM) with the unprecedented precision. The first studies of 13 TeV data recorded by the CMS experiment in the year of 2015 will be presented in this overview.

2 Standard Model results at 13 TeV

2.1 Studies of two-particle correlations

The two-particle angular correlations for charged particles are studied at 13 TeV [1]. The presence of the ridge-like structure for long range ($|\eta| > 2.0$) near-side $\Delta\phi \approx 0$ pairs is confirmed, as shown in Fig. 1. The magnitude of the correlation is maximal in the range $1.0 < p_T < 2.0$ GeV/c, with an approximately linear increase with the charge particle multiplicity, similar to the one measured at the 7 TeV [2], but now extended to much higher multiplicity values. Strong collision system size dependence is observed when comparing data from pp, pPb and PbPb collisions.

2.2 W and Z boson production

Total inclusive and fiducial W and Z boson cross sections at 13 TeV are measured in electron and muon final states [3]. The total inclusive cross sections times branching fractions are $\sigma(pp \rightarrow W^+X \times B(W^+ \rightarrow l^+\nu) = 11370 \pm 50(stat) \pm 230(syst) \pm 550(lumi)pb$, $\sigma(pp \rightarrow W^-X \times B(W^- \rightarrow l^-\nu) = 8580 \pm 50(stat) \pm 160(syst) \pm 410(lumi)pb$ and $\sigma(pp \rightarrow ZX \times B(Z \rightarrow l^+l^-) = 1910 \pm 10(stat) \pm$

*e-mail: milos.dordevic@cern.ch

$40(syst) \pm 90(lumi)pb$ for the dilepton mass range of 60 to 120 GeV. The measured ratio of the W and Z boson cross section in the electron and muon channels, that is useful also as an additional check of the lepton universality, is shown in Fig. 2. The measurement is found to be in a good agreement with the SM prediction.

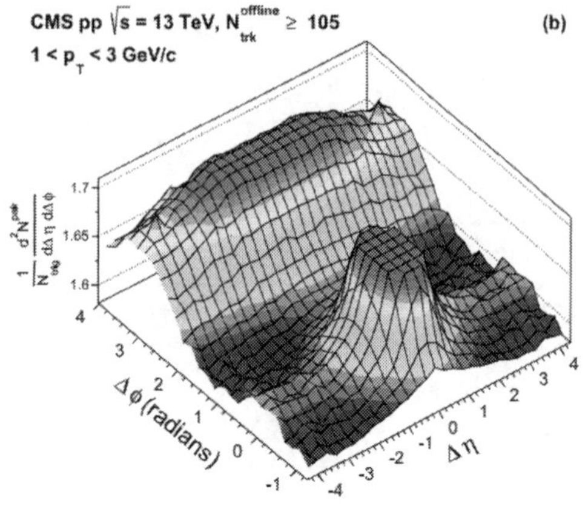

Figure 1. The two-particle correlation function for pairs of charge particles both in range $1 < p_T < 3$ GeV/c, showing the "rigde"-like structure.

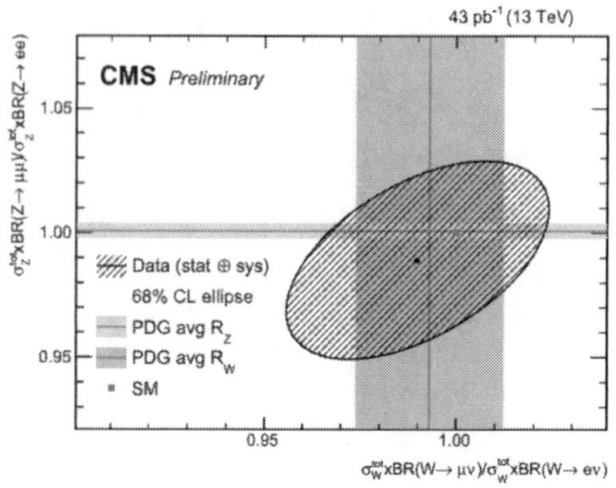

Figure 2. Ratios of W and Z inclusive cross sections in the electron and muon channels, as test of lepton universality and the SM expectation.

2.3 Inclusive jet production

Double-differential inclusive jet cross section, as a function of jet transverse momentum and rapidity, is measured for two jet size parameters R=0.7 and R=0.4 and jets with p_T up to 2 TeV and $|y| = 4.7$ [4] and the results are presented in Fig. 3. The new energy of 13 TeV provides an access to kinematic region of momentum fraction x where the parton density is large. When comparing the data to predictions at NLO accuracy in pQCD, a better agreement is observed for the larger jet size. The NLO prediction matched to parton shower from POWHEG+PYTHIA8 [5, 6] performs equally well for the both jet cone size parameters. The LO predictions of PYTHIA8 and HERWIG++ [7] have significant discrepancies with respect to data, thus stressing the importance of the higher order corrections.

2.4 Single top production

The top quark mass was measured in single top events [8], where the W boson decays to muon and a neutrino. Specific event topology and kinematics, featuring the central production of a light jet and an asymmetry of the initial state, was exploited to suppress the ttbar background. The measured value of $m_t = 172.60 \pm 0.77(stat)^{+0.97}_{-0.93}(syst)$ GeV, extracted from a fit to the top mass distribution (Fig. 4), is in a good agreement with the world average of $m_t = 173.34 \pm 0.76$ GeV. The t-channel single top cross section at 13 TeV is measured as well [9]. Kinematic variables are combined into a multivariate discriminator that was fit to obtain the inclusive cross section of $\sigma_{t-ch.} = 227.8 \pm 9.1(stat.) \pm 14.0(exp.)^{+28.7}_{-27.7}(theo.) \pm 6.2(lumi.)pb = 227.8^{+33.7}_{-33.0}pb$, found to be in a good agreement with the SM expectation.

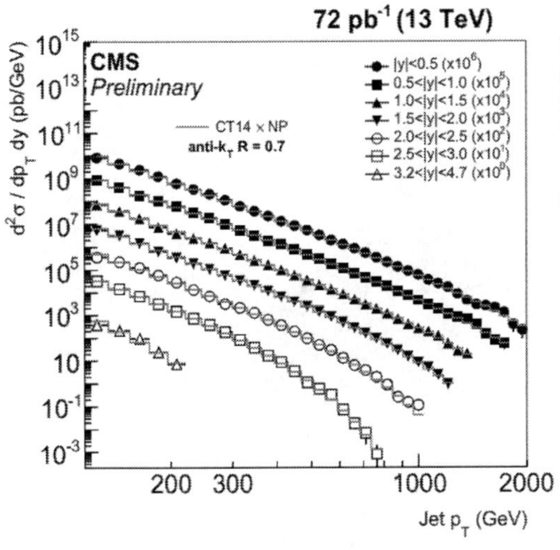

Figure 3. Double-differential inclusive jet cross section, shown as a function of the jet transverse momentum.

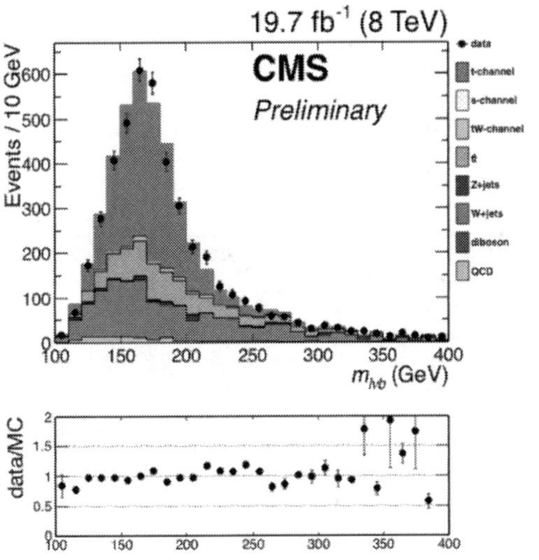

Figure 4. The top quark mass after the final selection including also the charge and pseudorapidity cuts.

3 Higgs studies at the 13 TeV

3.1 ttH production

A search for SM Higgs boson in association with a top quark pair at 13 TeV is performed. The analysis targets the bbar, $\gamma\gamma$ and multilepton decays of the Higgs boson. The multilepton analysis selects final states with two same-sign leptons or more than three leptons produced in association with b jets [10]. The signal extraction is based on the two-dimensional multivariate discriminator trained against each of the ttbar and ttV backgrounds, as shown in Fig. 5. The signal strenght is measured to be the $0.6^{+1.4}_{-1.1}$ times the SM expectation and the observed (expected) 95% CL limits are 3.3 (2.6) times the SM expectation. For the Higgs boson decays to a pair of b quarks, the lepton plus jets and dileptonic decays of the ttbar system are considered [11]. A combination of the matrix element method and multivariate approach is performed. A combined fit of the multivariate discriminant templates to data yielded with the signal strength of $-2.0^{+1.8}_{-1.8}$ and the observed (expected) 95% CL limits are 2.6 (3.6) times the SM expectation. The Higgs to diphotons analysis, although statistically limited at this point, was performed with the inclusion of the two categories with leptonic and hadronic tags of the ttbar pair [12]. The signal strength for the ttH production in this channel is $-3.8^{+4.5}_{-3.6}$, with the two tags corresponding to the ttH process combined statistically.

3.2 Higgs decays to bosons

The diphoton channel remains one of the best to test Higgs properties. It was performed in the gg→H, VBF and ttH categories [12]. Small deficit that is observed is driven by the untagged categories (gg→H). The observed (expected) significance at 125.09 GeV is $1.7(2.7)\sigma$ and the signal strength of $0.69^{+0.47}_{-0.42}$ is consistent with the SM expectation, as shown in Fig. 7. The H→ZZ→4l analysis, based on the matrix element kinematic discriminant between signal and background, is performed [13]. The distribution of the four lepton invariant mass is shown in Fig. 8. The observed(expected) significance

 DOI: 10.1051/epjconf/201612501001

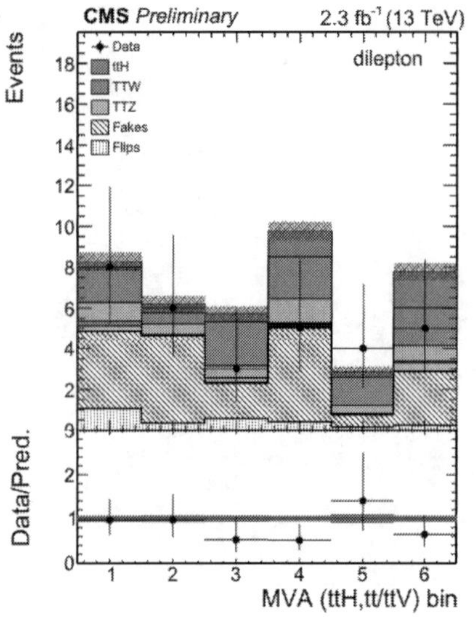

Figure 5. BDT kinematic discriminant for signal extraction in two same-sign lepton category.

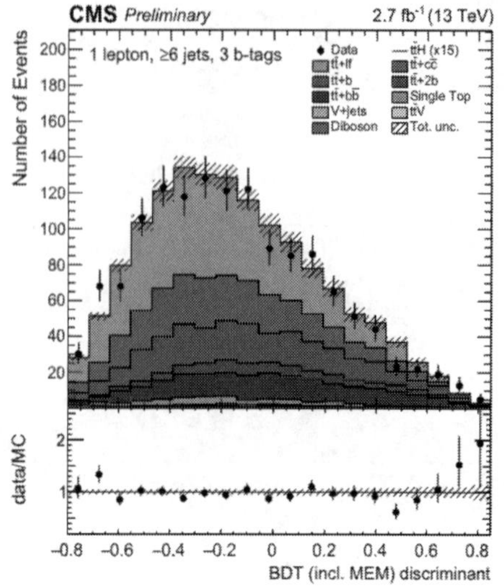

Figure 6. Final discriminator shape of the BDT, including also the MEM discriminator in the training.

at 125.09 GeV of 2.5(3.4)σ and the signal strength of $0.82^{+0.57}_{-0.43}$ are consistent with the SM within the uncertainties.

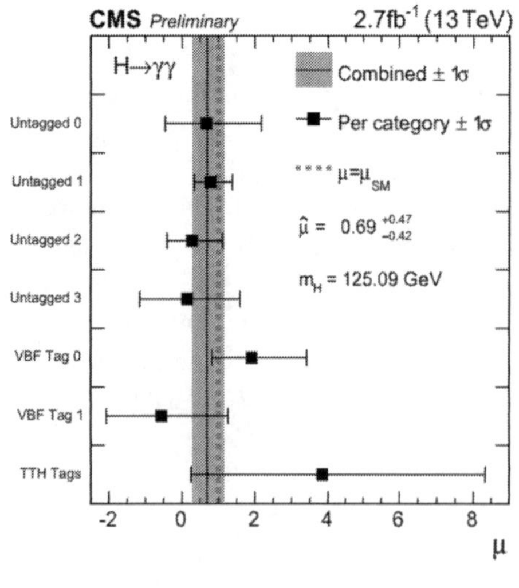

Figure 7. Signal strength modifier for each category in H→ $\gamma\gamma$ channel, compared to the overall signal strength and to the SM expectation.

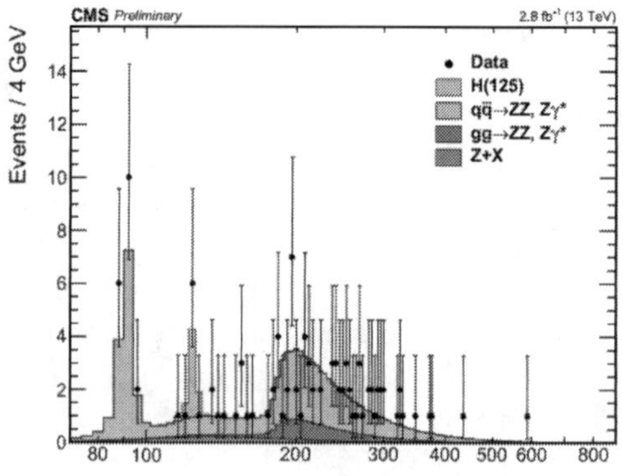

Figure 8. The four lepton invariant mass distribution m_{4l}, shown in the full studied mass range.

4 Supersymmetry searches at the 13 TeV

4.1 SUSY with zero leptons

The searches for gluino pair production with the LSP and zero leptons in the final state are performed [14]. The results of the analysis using the HT variables, as shown in Fig. 9, provides some of the world leading limits on T1bbbb, T1tttt and T1qqqq of 1600, 1530 and 1440 GeV. Similar search was performed using the stransverse mass variable M_{T2}, which is a measure of the transverse momentum imbalance in an event [15]. No excess above the SM background is observed and gluino(neutralino) mass was probed up to 1725(1100) GeV, as shown in Fig. 10, extending the Run 1 sensitivity by more than 300 GeV [16].

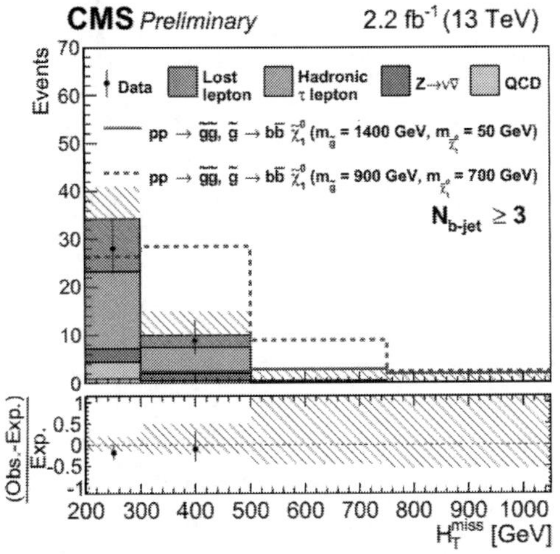

Figure 9. Observed number of events and SM background predictions for one of the search regions sensitive to the T1bbbb scenarios.

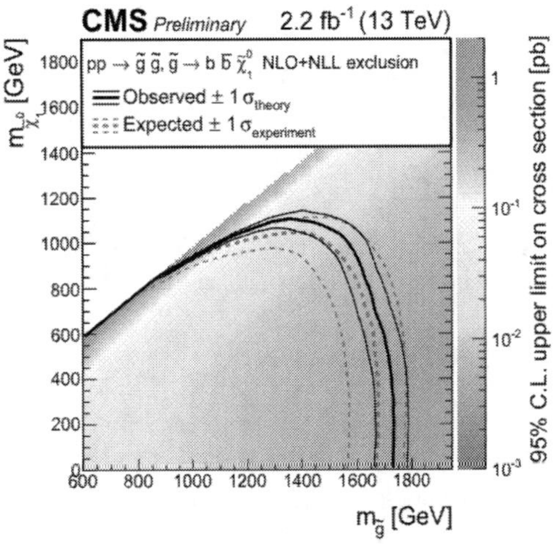

Figure 10. Exclusion limit at 95% CL for gluino mediated bottom-squark production.

4.2 SUSY with one lepton

The zero lepton SUSY search using the Razor variables is extended with an additional one lepton category [17]. The events are further categorised by the number of b-tagged jets in the final state. For the 100 GeV neutralino mass and gluino decaying to bb pairs, the limits are set to 1650 GeV, while for the decays to top(light) quarks this particular SUSY scenario is excluded up to 1600(1350) GeV, as shown in Fig. 11. The search for SUSY in final state with one lepton, one b-tagged jet and large missing transverse energy (MET) was performed using the sum of masses of large radius jets [18]. A key feature of this analysis is that the dominant background ttbar is nearly uncorrelated with the transverse mass of the lepton and MET. The distribution of the M_J variable is shown in Fig. 11. Within the T1tttt scenario the gluino mass is excluded up to 1575 GeV.

5 Exotica searches at the 13 TeV

5.1 Search for black holes

The search for black holes is performed [19] in the framework of ADD model with n-extra dimensions [20–22]. The semiclassical and quantum black holes are considered. The SM background, dominated

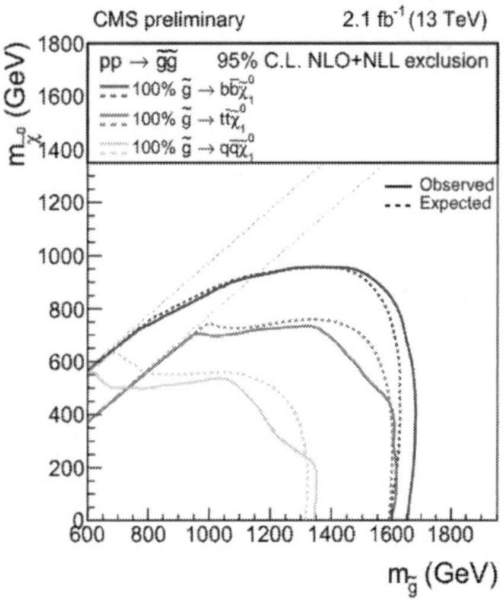

Figure 11. Summary of the exclusion contours at 95% CL, with gluino decays at 100% branching fraction to LSP and bb, tt and qq pairs.

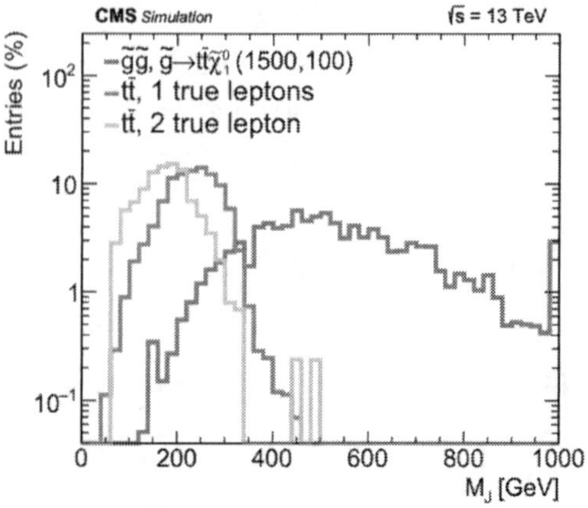

Figure 12. Distribution of M_J from simulated events with a small amount of initial state radiation (ISR).

by the QCD multijet events, is estimated from data. The distribution of the discriminating variable S_T, representing a scalar sum of the transverse energies of all energetic objects in the event, is shown in Fig. 13. No significant excess has been observed and the 95% CL limits in the semiclassical(quantum) black hole interpretation of 8.7(8.0) TeV are established. These limits are surpassing the 8 TeV results of 5.5 - 6.0 TeV [23].

5.2 Search for dark matter

The search for dark matter in the final state with large MET and at least one jet with high transverse momentum is performed [24]. The dark matter is considered to be produced in pairs, via the decay of a vector mediator, and the signal is extracted from a fit to the observed MET distribution in data. The mediator mass up to 1.3 TeV is excluded at 90% CL, as it is shown in Fig. 14 where also the results from the other, non-collider experiments, are overlaid together with this measurement, for the comparison.

5.3 Search for diphoton resonances

The search for diphoton resonances, under the 2HDM (spin 0) or ADD/RS (spin 2) hypothesis, is performed [25]. The high diphoton mass region between the 500 GeV and 4.5 TeV is explored, with the relative width up to 5.6×10^{-2}. The 13 TeV results are combined with similar searches at the 8 TeV [26, 27]. The largest excess is observed at 750 GeV and resonance width of 1.4×10^{-4}, with the local p-value of 3.4σ, reduced to 1.6σ when taking into account of searching for the several signal hypotheses. The corresponding diphoton mass distribution and p-value are shown in Fig. 15 and Fig. 16, respectively.

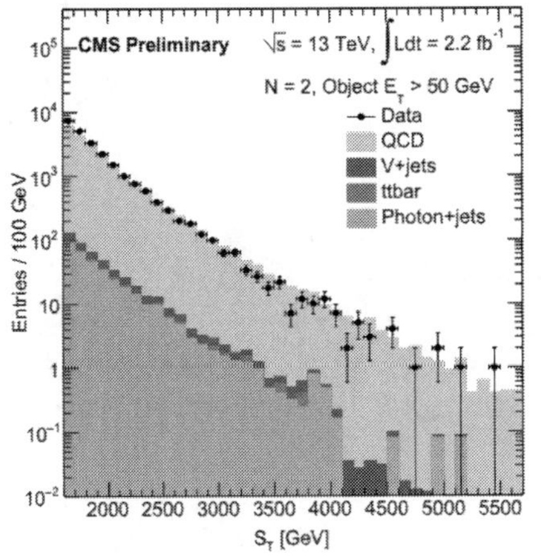

Figure 13. The S_T variable distribution, with the main contribution from QCD multijet background.

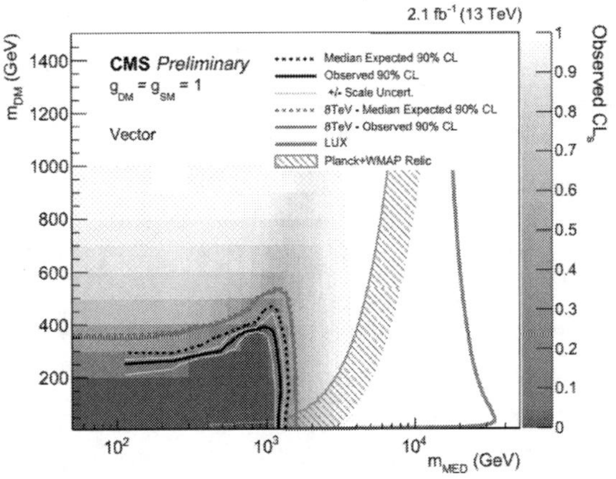

Figure 14. The dark matter exclusion contours in the $m_{med} - m_{DM}$ plane assuming a vector-type interaction.

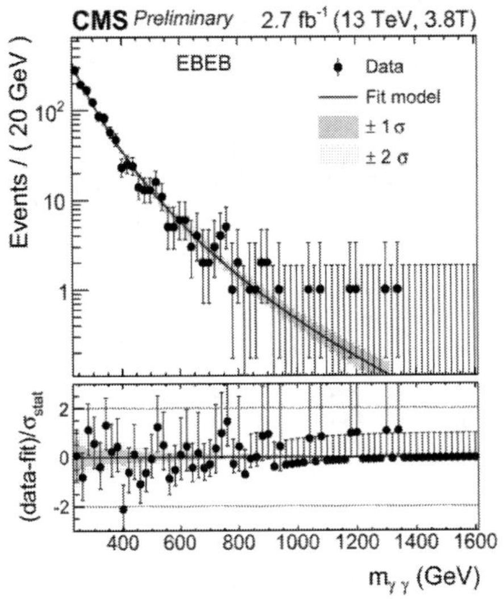

Figure 15. The observed invariant mass spectra for the EBEB channel and using the 3.8T dataset.

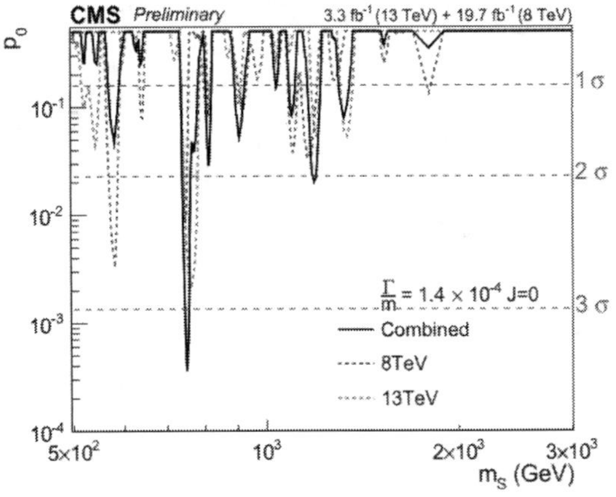

Figure 16. The observed background only p-values for spin 0 and the most narrow width hypothesis, obtained from the combination of the 8 and 13 TeV results.

6 Conclusion

The 2015 has marked a start of CMS taking the 13 TeV data. It was the first operation under the 25ns conditions and physics objects are well commissioned. The SM results are generally in an agreement with the prediction, many new physics results have an improved sensitivity and an excess in the diphoton invariant mass is observed at 750 GeV, that requires more data for a precise scrutiny of its existence.

References

[1] The CMS Collaboration, Phys. Rev. Lett. 116, 172302, (2016)
[2] The CMS Collaboration, JHEP **09** 091, (2016)
[3] The CMS Collaboration, CMS-PAS-SMP-15-004, (2016)
[4] The CMS Collaboration, CMS-PAS-SMP-15-007, (2016)
[5] T. Sjostrand, S. Mrenna and P. Skands, JHEP **05** 026, (2006)
[6] T. Sjostrand, S. Mrenna and P. Skands, Comput. Phys. Comm. **178** 852, (2008)
[7] G. Corcella et al., JHEP **0101** 010, (2001)
[8] The CMS Collaboration, CMS-PAS-TOP-15-001, (2016)
[9] The CMS Collaboration, CMS-PAS-TOP-16-003, (2016)
[10] The CMS Collaboration, CMS-PAS-HIG-15-008 (2016)
[11] The CMS Collaboration, CMS-PAS-HIG-16-004, (2016)
[12] The CMS Collaboration, CMS-PAS-HIG-15-005, (2016)
[13] The CMS Collaboration, CMS-PAS-HIG-15-004, (2016)
[14] The CMS Collaboration, CMS-PAS-SUS-15-002 (2016)
[15] The CMS Collaboration, CMS-PAS-SUS-15-003, (2016)
[16] The CMS Collaboration, JHEP **05** 078, (2015)
[17] The CMS Collaboration, CMS-PAS-SUS-15-004, (2016)
[18] The CMS Collaboration, CMS-PAS-SUS-15-007, (2016)
[19] The CMS Collaboration, CMS-PAS-EXO-15-007, (2016)
[20] N. Arkani-Hamed, S. Dimopoulos, G.R. Dvali, Phys. Lett. B **429** 264, (1988)
[21] N. Arkani-Hamed, S. Dimopoulos, G.R. Dvali, Phys. Lett. B **436** 257, (1988)
[22] N. Arkani-Hamed, S. Dimopoulos, G.R. Dvali, Phys. Rev. D **59** 086004, (1999)
[23] G. Landsberg, Fundam. Theor. Phys., **178** 267 (2015)
[24] The CMS Collaboration, CMS-PAS-EXO-15-003, (2016)
[25] The CMS Collaboration, CMS-PAS-EXO-16-018, (2016)
[26] The CMS Collaboration, Phys. Lett **750** 494-419, (2016)
[27] The CMS Collaboration, CMS-PAS-EXO-12-045, (2016)

EPJ Web of Conferences **125**, 01002 (2016) DOI: 10.1051/epjconf/201612501002
QUARKS-2016

Current results of the NOvA experiment

Liudmila Kolupaeva[1,2,*] on behalf of the NOvA Collaboration

[1]*Joint Institute for Nuclear Research*
[2]*Moscow State University*

Abstract. NOvA is a long baseline neutrino oscillation experiment at Fermi National Accelerator Laboratory, USA. It is one of the so-called "off-beam-axis" experiments with two highly active detectors separated by 810 km of Earth's crust. Both detectors are identical and consist of PVC cells filled with a liquid scintillator. The NOvA collaboration has already published its first results for muon neutrino disappearance [1] and electron neutrino appearance [2] mode with data collected between July 2013 and March 2015. This talk is devoted to a discussion on NOvA's first results.

1 Introduction

NOvA (NuMI Off-Axis ν_e-Appearance) is a long baseline neutrino oscillation experiment consisting of two functionally identical detectors, which have been placed 14 mrad off-axis. The NuMI (Neutrinos at the Main Injector) facility at the Fermi National Accelerator Laboratory (USA) provides a muon (anti-)neutrino beam for the experiment. The neutrino beam, with a peak energy of 2 GeV, travels 810 km through the Earth's crust to the Far Detector at Ash River (Minnesota). Muon neutrinos may oscillate to electron or tau neutrinos as they propagate. The NOvA detectors can measure the transition probabilities $P(\nu_\mu \rightarrow \nu_e)$ and $P(\nu_\mu \rightarrow \nu_\mu)$. From these measurements NOvA is able to extract oscillation parameters: Δm^2_{32}, mixing angle θ_{23}, CP violating phase δ_{CP} and neutrino mass hierarchy.

NOvA published its first results [1], [2] this year. These results are based on data collected between July 2013 and March 2015. More data and results will be presented at the Neutrino 2016 summer conference.

2 The NOvA experiment

Neutrinos are created by colliding 120 GeV protons with a carbon target. The π and K mesons produced will decay into a neutrino and an associated lepton. The 4.2m $\times$ 4.2m$\times$ 15m near detector is located at a distance of about 1km from the target and has mass about 300 t. It is used for measuring the neutrino flux composition before oscillations. The far detector is located at a distance 810 km. It has dimensions of 15.6m $\times$ 15.6m $\times$ 60m with a mass of 14 kt. It is responsible for measuring neutrino oscillations. Both detectors are placed 14 mrad off of the neutrino beam axis. Such a scheme helps to obtain narrow energy spectra near the first oscillation maximum and suppresses the high energy tail.

*e-mail: ldkolupaeva@yandex.ru

The NOvA detectors are huge tracking calorimeters with excellent segmentation. They consist of many PVC cells (6 cm × 4 cm) filled with mineral oil liquid scintillator with 5% admixture of pseudocumene. These cells are constructed in alternate planes, thus NOvA can measure the vertical and horizontal position of an event. The number of cells in the far and near detector are 344 054 and 21 192 respectively. Each cell contains a wavelength-shifting fiber loop with 32-pixel avalanche photodiods used for readout.

The NOvA detectors were under construction during the first analysis data taking period, from July 2013 to March 2015. During this period 3.45×10^{20} POT were delivered to the carbon target. Because the NOvA Far detector size was not constant during this period the NOvA data should be matched to 2.74×10^{20} POT-14-ton-equivalent.

3 The ν_μ Disappearance Results

The ν_μ Disappearance mode is sensitive to θ_{23} and Δm^2_{23} measurements. During the first analysis period NOvA detected 33 ν_μ CC events instead of the 211.8 ± 12.5 events expected under non-oscillation hypothesis. It is a clear observation of muon neutrino oscillations. The oscillation parameter extraction procedure is based on measured spectra in both detectors. Spectra from the Near Detector (Figure 1) is extrapolated to the Far detector and this prediction is compared with far detector measurements. The extrapolation procedure is discribed in Figure 2. Figure 3 shows the final measured far detector neutrino spectra.

Expected background events for the NOvA disappearance analysis are 2 neutral current and 1.4 cosmic ray events. Main systematic contributions are due to Calorimetric Energy Calibration, Muon Energy Scale etc. Total systematic uncertainty for $\sin^2\theta_{23}$ is 6.8 % and Δm^2_{32} is 3.7 %.

The best fit oscillation parameters are:

$$\Delta m^2_{23} = 2.52^{+0.20}_{-0.18} \times 10^{-3}\mathrm{eV}^2 \qquad 0.38 < \sin^2\theta_{23} < 0.65\ (68\%\ \mathrm{CL}) \quad \text{(for normal hierarchy)}$$

$$\Delta m^2_{23} = -2.56 \pm 0.19 \times 10^{-3}\mathrm{eV}^2 \qquad 0.37 < \sin^2\theta_{23} < 0.64\ (68\%\ \mathrm{CL}) \quad \text{(for inverted hierarchy)}$$

Allowed regions for oscillation parameters in comparison with other experiments measurement are shown in the Figure 4.

4 The ν_e Appearance Results

The ν_e appearance channel is sensitive to the mass hierarchy and δ_{CP} phase. The NOvA collaboration developed two selection algorithms for the ν_e analysis. The official analysis LID (Likelihood Identification) is based on dE/dx likelihood calculations independently for transverse and longitudinal slices under different particle hypotheses. Another algorithm is LEM (Library Event Matching) [7]. Information about energy deposition is compared to the huge library of Monte Carlo events.

The extrapolation procedure is also used for far detector prediction (Figure 5 shows near detector spectra). The expected number of background events and their composition is shown in table 1. The systematic uncertainty for the ν_e analysis is 17.6 %. The main contribution to systematics comes from calibration, interaction uncertainties and scintillator saturation.

NOvA observed 6 events selected by LID and 11 events selected by the LEM algorithm (Figure 6). Figure 7 shows the allowed regions for $\sin^2 2\theta_{13}$ for all values of δ_{CP} and both hierarchies. For fixed $\sin^2 2\theta_{13} = 0.086 \pm 0.005$ [5], [6] figure 8 shows the significance at witch different hypotheses are disfavored for all values of δ_{CP} and both hierarchies, with fixed $\sin^2\theta_{23} = 0.5$.

LID analysis disfavors the $0 < \delta_{CP} < 0.8\pi$ range for inverted hierarchy at 90% C.L. LEM analysis disfavors inverted hierarchy for all values of δ_{CP} and $0 < \delta_{CP} < \pi$ for normal hierarchy at 90% C.L.

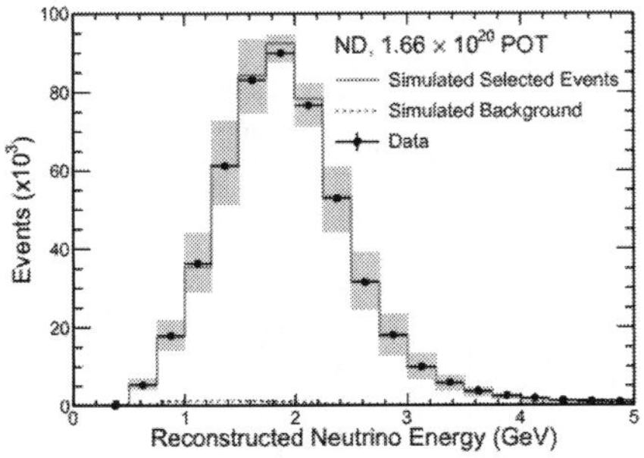

Figure 1. Near Detector spectra, which is used in extrapolation and prediction spectra in Far Detector.

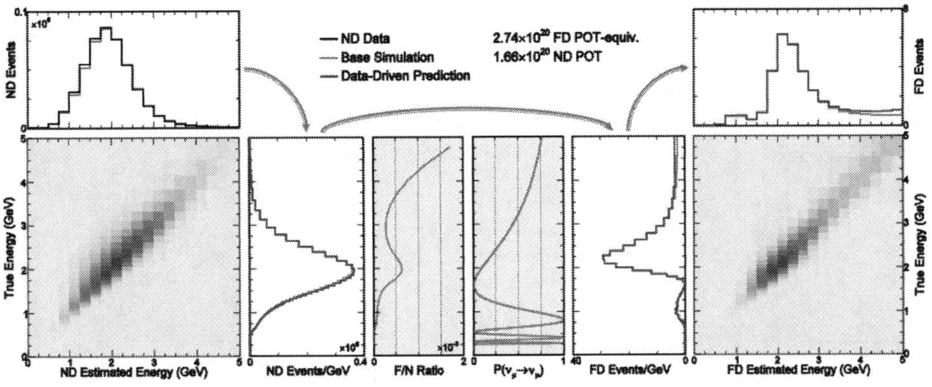

Figure 2. Extrapolation procedure with measured near detector spectra. Measured near detector spectra converted to true energy is multiplied by far-to-near detector ratio and oscillation probability. Calculated far detector true energy spectra are converted to reconstructed energy spectrum.

References

[1] P. Adamson *et al.* [NOvA Collaboration], Phys. Rev. D **93**, no. 5, 051104 (2016)
[2] P. Adamson *et al.* [NOvA Collaboration], Phys. Rev. Lett. **116**, no. 15, 151806 (2016)
[3] P. Adamason *et al.*, Phys. Rev. Lett. **112**, 191801 (2014)
[4] K. Abe *et al.*, Phys. Rev. D **91**, 072010 (2015)
[5] F. P. An *et al.*, Phys. Rev. Lett. **115**, 111802 (2015)
[6] Y. Abe *et al.*, JHEP **1410** 86 (2014)
[7] C. Backhouse and R. B. Patterson, Nucl. Inst. Meth. A **778** (2015)

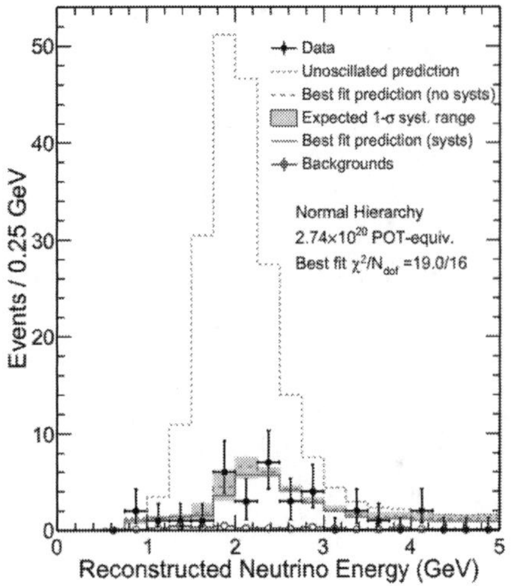

Figure 3. This figure shows the Far Detector ν_μ energy spectra. The green line represents the non-oscillated energy spectra with an expected 211.8 events. NOvA observed 33 ν_μ CC events.

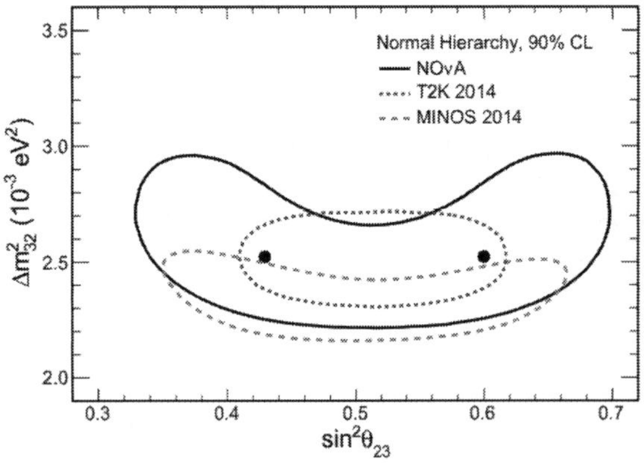

Figure 4. 90% C.L. contour for normal hierarchy on the surface $\sin^2\theta_{23}$ and Δm^2_{32} along with MINOS [3] and T2K [4] results. This plot shows good agreement.

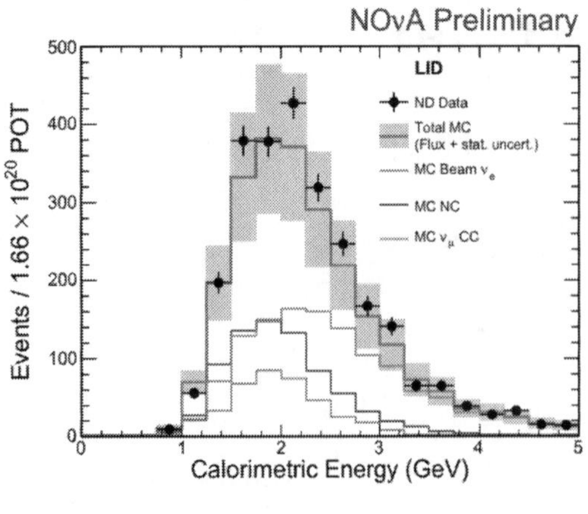

Figure 5. Near Detector LID spectra used for ν_e analysis prediction in Far Detector.

 DOI: 10.1051/epjconf/201612501002

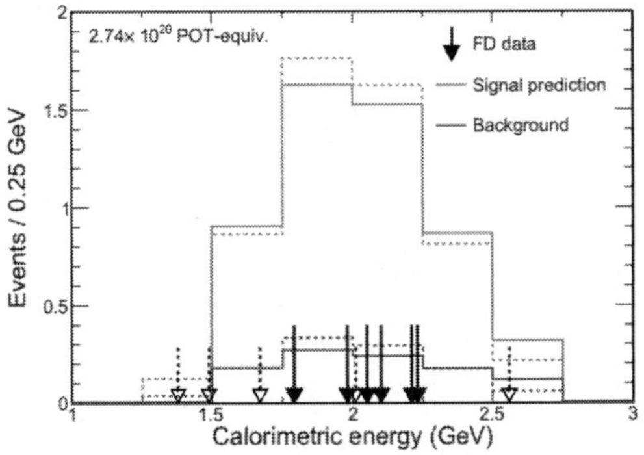

Figure 6. This figure shows both LID and LEM far detector ν_e CC energy spectrum. LID selector (solid) identified 6 events, LEM found 11 events, all LID events were selected with additional 5 events (dashed).

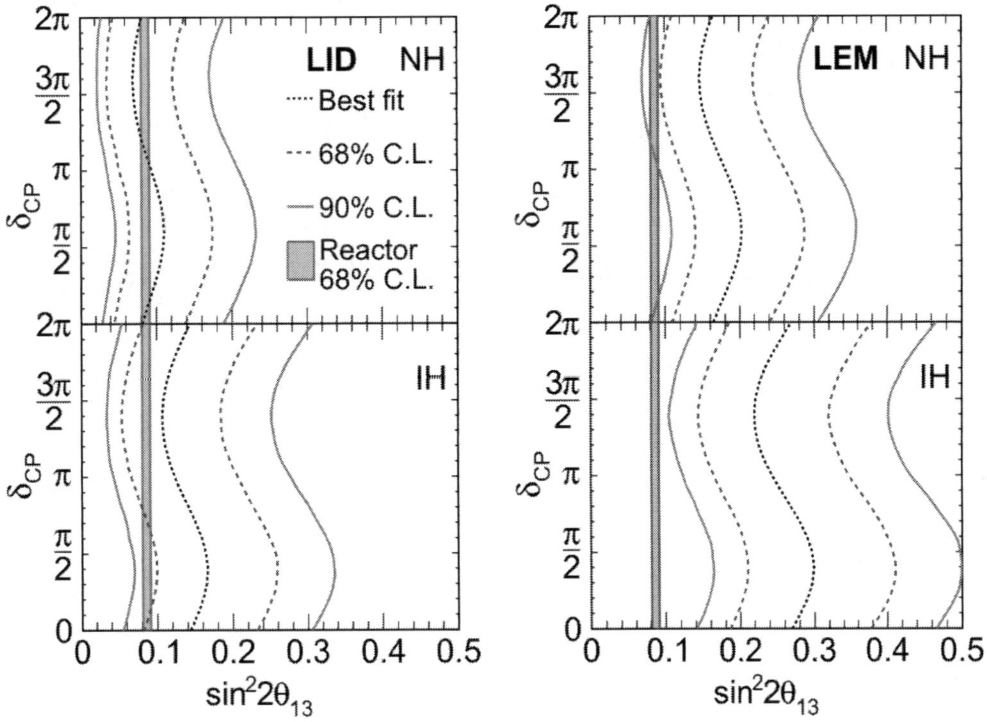

Figure 7. 68% and 90% C.L. regions for LID and LEM selectors in $\sin^2 2\theta_{13}$ and δ_{CP} surface in comparison with rector neutrino experiment data. The top panel is normal and the bottom is inverted hierarchy. $\sin^2 \theta_{23} = 0.5$ remain fixed.

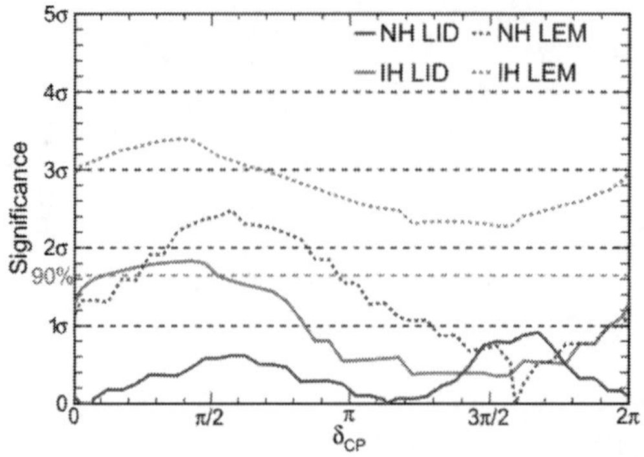

Figure 8. Significance for normal (blue) and inverted (red) neutrino mass hierarchy for all values of δ_{CP}. Both LEM and LID selectors are represented.

Table 1. Expected background for ν_e analysis

	Total bkg.	Beam ν_e	NC	ν_μ CC	ν_τ CC	Cosmic
LID	0.99	0.5	0.37	0.05	0.02	0.06
LEM	1.07	0.5	0.43	0.07	0.02	0.06

EPJ Web of Conferences **125**, 01003 (2016) DOI: 10.1051/epjconf/201612501003
QUARKS-2016

Recent results from the T2K experiment

M. Lawe[1,*] *for the* T2K Collaboration

[1] *Lancaster University, Bailrigg, Lancaster, United Kingdom, LA1 4YW*

Abstract. T2K is a long-baseline neutrino oscillation experiment based in Japan. It was the first experiment to observe ν_e appearance in a ν_μ beam, and has more recently produced the highest precision measurement of the oscillation parameter θ_{23} and has begun to probe the charge-parity violation phase δ. Using a ν_μ beam disappearance sample, T2K measures $\sin^2\theta_{23} = 0.514^{+0.055}_{-0.056}$ and $\Delta m^2_{32} = (2.51 \pm 0.10)\times 10^{-3}$ eV^2/c^4 assuming the normal mass hierarchy, and $\sin^2\theta_{23} = 0.511 \pm 0.055$ and $\Delta m^2_{13} = (2.48 \pm 0.10)\times 10^{-3}$ eV^2/c^4 assuming the inverted mass hierarchy. A joint fit between ν_μ beam appearance and disappearance samples produces consistent results for $\sin^2\theta_{23}$ and Δm^2_{32} whilst also excluding at 90% confidence level the regions $\delta = [0.15, 0.83]\pi$ for the normal hierarchy and $\delta = [-0.08, 1.09]\pi$. Using a $\bar{\nu}_\mu$ beam disappearance sample, $\sin^2\bar{\theta}_{23} = 0.45^{+0.29}_{-0.12}$ and $|\Delta\bar{m}^2_{32}| = (2.51 \pm 0.28) \times 10^{-3}$ eV^2/c^4, consistent with previous measurements and the T2K ν_μ sample.

1 Neutrino Oscillation Theory

The Pontecorvo-Maki-Nakagawa-Sakata (PMNS) formalism for neutrino oscillations describes the relation between the flavour eigenstates, ν_α, through which they interact, and the mass eigenstates, ν_i, through which they propagate [1]

$$\underbrace{\begin{pmatrix} \nu_e \\ \nu_\mu \\ \nu_\tau \end{pmatrix}}_{\text{Flavour state}} = \underbrace{\begin{pmatrix} U_{e1} & U_{e2} & U_{e3} \\ U_{\mu 1} & U_{\mu 2} & U_{\mu 3} \\ U_{\tau 1} & U_{\tau 2} & U_{\tau 3} \end{pmatrix}}_{\text{Mixing matrix}} \underbrace{\begin{pmatrix} \nu_1 \\ \nu_2 \\ \nu_3 \end{pmatrix}}_{\text{Mass state}}. \tag{1}$$

The PMNS mixing matrix, U, describes the 3D rotations between the mass and flavour eigenstates and is traditionally expressed as the product of three 2D rotation matrices

$$U = \begin{pmatrix} 1 & 0 & 0 \\ 0 & c_{23} & s_{23} \\ 0 & -s_{23} & c_{23} \end{pmatrix} \begin{pmatrix} c_{13} & 0 & s_{13}e^{-i\delta} \\ 0 & 1 & 0 \\ -s_{13}e^{i\delta} & 0 & c_{13} \end{pmatrix} \begin{pmatrix} c_{12} & s_{12} & 0 \\ -s_{12} & c_{12} & 0 \\ 0 & 0 & 1 \end{pmatrix}, \tag{2}$$

where $c_{ij} \equiv \cos\theta_{ij}$ and $s_{ij} \equiv \sin\theta_{ij}$, the cosine and sine of the three mixing angles, θ_{12}, θ_{23} and θ_{13}, and δ is the Dirac Charge-Parity (CP) violating phase.

*e-mail: m.lawe@lancaster.ac.uk

The probability that a neutrino produced in flavour state ν_α, with energy E, is detected as flavour state ν_β after propagating a distance L is given by

$$P_{\alpha\to\beta} = \left|\langle \nu_\alpha | \nu_\beta (t) \rangle\right|^2 = \left|\sum_{ij} U^*_{\alpha i} U_{\beta i} U_{\alpha j} U^*_{\beta j} e^{-i \frac{\Delta m^2_{ij}}{2} \frac{L}{E}}\right|, \tag{3}$$

where $\Delta m^2_{ij} = m^2_i - m^2_j$, the mass squared difference between the mass eigenstates.

Three flavour neutrino oscillations are therefore dependent upon six measurable parameters, the three mixing angles θ_{12}, θ_{23} and θ_{13}, the CP-violating phase δ, and two mass squared differences Δm^2_{32} and Δm^2_{21}.

The dependence on $\frac{L}{E}$ in Equation 3 allows experiments to tune their baseline and neutrino energy to maximise the oscillation probability, and therefore their sensitivity to measurements of the oscillation parameters.

There are currently three open questions remaining within the PMNS framework,

- What is the sign of Δm^2_{32}? Is $\Delta m^2_{32} > 0$, known as the "normal" neutrino mass ordering, or is $\Delta m^2_{32} < 0$, known as the "inverted" neutrino mass ordering.
- What is the precise value of θ_{23}? Is it maximal, $\theta_{23} = 45°$, or does it lie in the first, $\theta_{23} < 45°$, or second, $\theta_{23} > 45°$, octant?
- Is there Charge-Parity violation in the lepton sector? Requires $\sin\delta \neq 0$.

By measuring $\nu_\mu \left(\bar{\nu}_\mu\right)$-disappearance and $\nu_e\,(\bar{\nu}_e)$-appearance within an initially produced $\nu_\mu \left(\bar{\nu}_\mu\right)$ (anti)neutrino beam, the Tōkai-to-Kamioka (T2K) experiment can address these questions.

2 The T2K Experiment

The T2K experiment [2] is a long-baseline neutrino oscillation experiment based in Japan. The experiment produces a neutrino beam at the J-PARC research facility in Tōkai village on Japan's east coast. The beam properties are measured with a suite of near detectors at the J-PARC site, and 295 km away with the Super-Kamiokande (SK) detector in Kamioka. T2K was the first long-baseline neutrino oscillation experiment to use an off-axis design to tune the $\frac{L}{E}$ oscillation dependence to the oscillation maximum at SK.

2.1 Beamline

The T2K neutrino beam [3] is produced by impinging a 30 GeV proton beam upon a graphite target. The interaction produces a large flux of mesons, predominately charged pions, which are charge selected and focused by a series of three magnetic horns. The mesons then enter a decay volume where they decay in-flight to produce neutrinos; the dominant process being $\pi^+ \to \mu^+ + \nu_\mu$. By reversing the polarity of the magnetic horns an antineutrino beam can be produced via $\pi^- \to \mu^- + \bar{\nu}_\mu$.

2.2 Off-Axis Design

The T2K neutrino beam centre is directed 2.5° away from SK. Doing so tunes the beam profile into a narrowband neutrino flux spectrum, peaking at 0.6 GeV. This energy and baseline combination maximises the oscillation probability at SK, as seen in Figure 1, whilst reducing backgrounds.

DOI: 10.1051/epjconf/201612501003

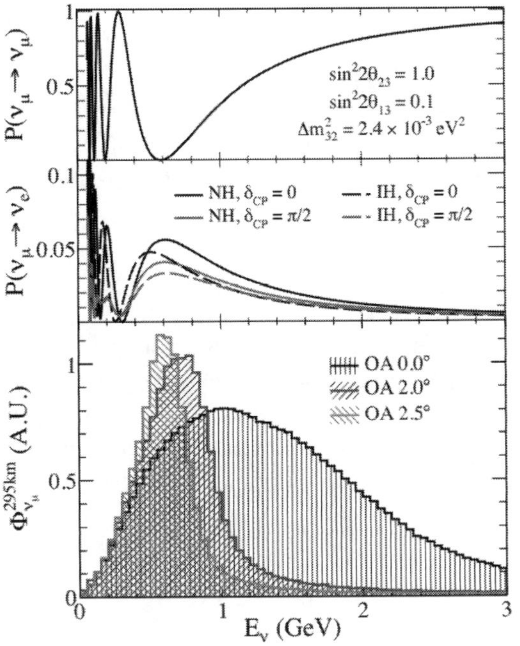

Figure 1. T2K neutrino flux spectrum for three different off-axis angles (lower panel), compared to ν_μ survival probability (upper panel) and ν_e appearance probability (central panel) as a function of neutrino energy at the T2K baseline of 295 km.

2.3 Near Detectors

T2K uses two near detectors, the on-axis Interactive Neutrino GRID (INGRID) and the off-axis Near Detector at 280 m (ND280). Both detectors are positioned within a pit 280 m downstream of the proton target at the J-PARC facility.

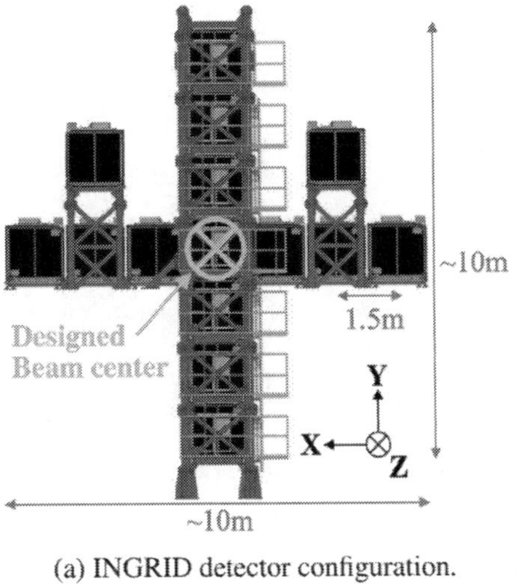

(a) INGRID detector configuration.

(b) Exploded diagram of ND280 configuration. Beam passes from left to right.

Figure 2: T2K near detectors.

2.3.1 INGRID

INGRID [4] is comprised of 14 individual modular detectors placed in a cross configuration, as shown in Figure 2a, which is primarily designed to monitor the neutrino beam direction and profile. Two additional off-axis modules measure the beam asymmetry. Each INGRID module is built up of planes of iron and scintillator, their high mass, and therefore high interaction rate, allows for the neutrino beam to be monitored at a per beam-spill level.

2.3.2 ND280

ND280 is designed to characterise the off-axis beam composition, prior to oscillations having occurred. This requires measuring the unoscillated ν_μ spectrum, along with the inherent ν_e and wrong sign ($\bar{\nu}_\mu$ and $\bar{\nu}_e$) beam contamination (for neutrino operation, particles and antiparticles are reversed for antineutrino operation). The ND280 is a complex detector design, see Figure 2b, featuring both carbon and water target regions within fine-grained scintillator detectors. It has a 0.2 T magnetic field to allow for charge discrimination, momentum measurements and particle identification of charged tracks within time projection chambers.

2.4 Far Detector, Super-Kamiokande

The T2K far detector is Super-Kamiokande [5]; located 295 km from J-PARC it measures the oscillated ν_μ and ν_e spectra. SK is a 50 kT (22.5 kT fiducial volume) water Cherenkov detector, readout by over 11,000 20 inch photo-multiplier tubes. SK is capable of distinguishing between μ and e interactions, based upon the different Cherenkov ring profiles, with greater than 99% accuracy.

2.5 Datasets

T2K has been operating since 2009 and the results presented here are based on the delivery of 6.57×10^{20} Protons-On-Target (POT) in neutrino mode operation, and 4.01×10^{20} POT in antineutrino mode operation.

3 Oscillation Analysis Methodology

The T2K oscillation analysis works by extrapolating the measured, unoscillated, neutrino spectrum at the near detectors, to produce an event rate prediction at SK. The prediction is based on parametrisation of the flux, cross section and detector simulation models. The flux and cross section parameters are initially constrained by external measurements and T2K beamline monitoring, which provide an initial event rate prediction at the ND280. These parameters are then tuned via a fit to ND280 data samples, allowing production of an unoscillated event rate prediction at SK. A further fit is then performed to reconcile the unoscillated prediction with the oscillated measurement at SK, allowing the extraction of the oscillation parameters within the PMNS framework.

3.1 Flux Model

FLUKA [6] is used to model the interaction of the protons with the T2K target, which is tuned to external hadron production data and NA61/SHINE [7] measurements of pion and kaon production data on a carbon target. GEANT3 [8] with GCALOR [9] is then used to simulate the subsequent particle propagation and decay. INGRID and beam monitors provide additional constraint on the initial unoscillated neutrino flux prediction as shown in Figure 3. Prior to any ND280 constraint the flux uncertainty is about 10%, and the covariance between ND280 and SK allows for this uncertainty to be reduced by the ND280 fit.

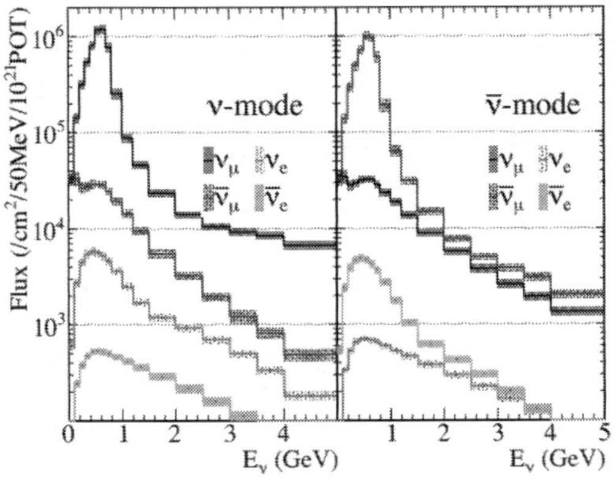

Figure 3. T2K flux prediction at SK in neutrino and antineutrino mode operation, the bands represent the flux uncertainty [10].

3.2 Cross Section Model

Neutrino interactions are modelled using NEUT [11], with the oscillation signal extracted from charged-current quasi-elastic (CCQE) interactions. Unfortunately there is significant uncertainty on current measurements of CCQE interaction cross sections, as seen in Figure 4, and non-CCQE interactions can mimic the CCQE signal, for example via pion absorption or multi-nucleon interactions. Therefore it is crucial to include all relevant physics models within the NEUT interaction generator and they are correctly parametrised. T2K tunes the NEUT model parameters to external cross section measurements, such as those from MiniBooNE [12] and MINERνA [13], along with pion-nucleus scattering data to model pion transportation within the nuclear environment.

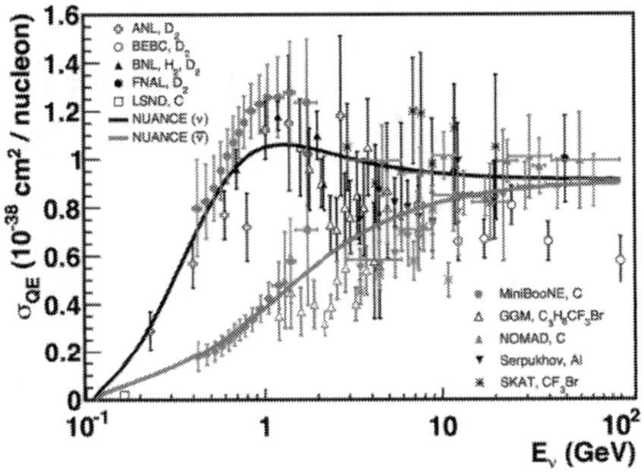

Figure 4. Neutrino (black) and antineutrino (red) CCQE cross section measurements showing significant spread and large uncertainties [1]

3.3 ND280 Fit

The T2K flux and cross section models, in combination with the ND280 GEANT4 [14] simulation, allow for an initial prediction of interaction rates in the ND280. Event samples are then selected by

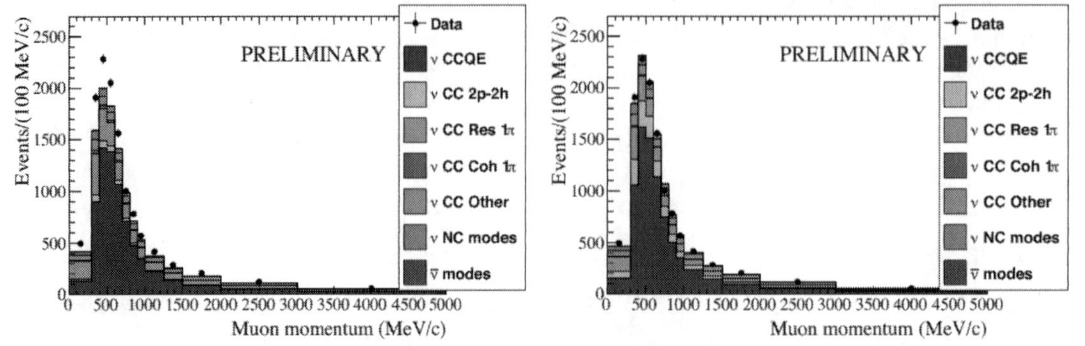

(a) Prefit. (b) Postfit.

Figure 5: Neutrino beam mode, 0π sample, muon momentum distribution pre (left) and post (right) ND280 flux and cross section fit.

the ND280, and these are used to create a new fit to the cross section and flux parameters to better constrain them.

3.3.1 ND280 Event Samples

Seven independent event samples are selected for the ND280 fit, three in neutrino beam and four in antineutrino beam mode. All samples require the selection of the highest momentum, muon-like track, and then use charge discrimination to determine whether it came from a neutrino or antineutrino interaction. For neutrino beam data the three samples are partitioned based upon the number of pions observed during the interactions, providing, 0π, 1π and $N\pi$ charged-current samples. For the antineutrino beam operation the samples are split based on track multiplicity into 1-Track and N-Track charge current samples, but due to the large neutrino contamination within the antineutrino beam, these samples are produced with both a positive and negative muon candidate to give four samples in total.

3.3.2 Parameter Fit

All seven samples are used to simultaneously fit the flux and cross section model parameters, taking into account the uncertainties in the near detector measurements. This improves the data - Monte Carlo agreement, as shown in Figure 5, and reduces the total flux and cross section uncertainty, as shown in Figure 6 when the new parameters are applied to the SK prediction.

3.4 Oscillation Fit

The combination of the ND280 fit and independent cross section models (those not currently constrained by the ND280 fit) are combined with the GEANT3 SK detector model to produce a final event rate prediction at SK.

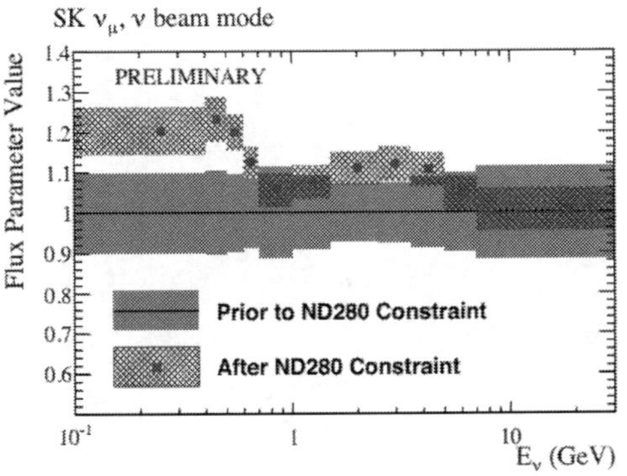

(a) Flux parameters.

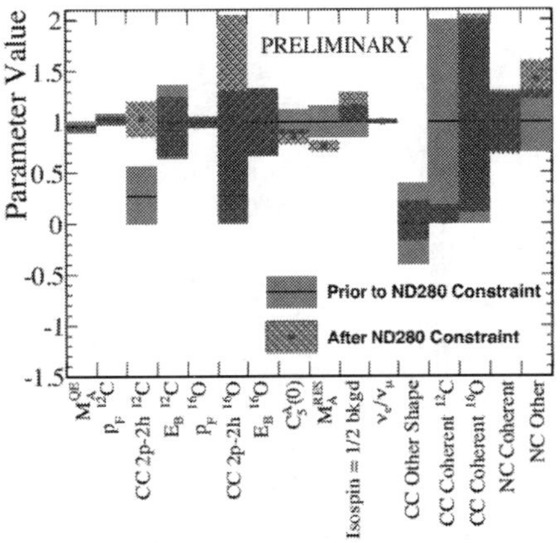

(b) Cross section parameters.

Figure 6: Change in the value and associated uncertainty on the flux (left) and cross section (right) parameters prior to (red) and after (blue) the ND280 fit.

3.4.1 SK Event Samples

All event selection requires single, fully contained Cherenkov rings, with a vertex reconstructed within the SK fiducial volume and in temporal coincidence with the J-PARC neutrino beam. These rings are then characterised as μ-like or e-like based on a likelihood fit.

If a ring is found to be μ-like it is also required to have a reconstructed momentum greater than 200 MeV/c (to reject charged pions and misidentified electrons), and that there are less than two decay electrons reconstructed (to reject events with unobserved muons or pions).

For e-like rings the additional requirements are that the visible energy is greater than 100 MeV (to remove low energy neutral-current interactions and the decay electrons from unobserved parent particles), there are no decay electrons, the reconstructed neutrino energy is less that 1.25 GeV (as above this threshold the intrinsic ν_e beam contamination becomes dominant), and that the event is not consistent with the reconstruction of a π^0. The resultant distributions in data and Monte Carlo (with nominal oscillation parameters) are shown in Figure 7.

3.4.2 Analyses

To extract the oscillation parameters a likelihood fit is performed between the selected SK events and the predicted event distributions which are varied across the full oscillation parameter phase-space. The solar oscillation parameters to which the experiment is insensitive $\left(\theta_{12}, \Delta m^2_{12}\right)$ are either held at, or Gaussian drawn from, the world best fit value, while the remaining oscillation parameters (either all, or a subset of them) fitted depending on the specific analysis being undertaken, as will be described in Section 4. The analyses that will be described are:

- A single fit to the neutrino beam ν_μ disappearance sample;
- A joint fit between the neutrino beam ν_μ disappearance and ν_e appearance samples;
- A single fit to the antineutrino beam $\bar{\nu}_\mu$ disappearance sample; and
- A test for consistency with PMNS oscillations on the antineutrino beam $\bar{\nu}_e$ appearance sample.

DOI: 10.1051/epjconf/201612501003

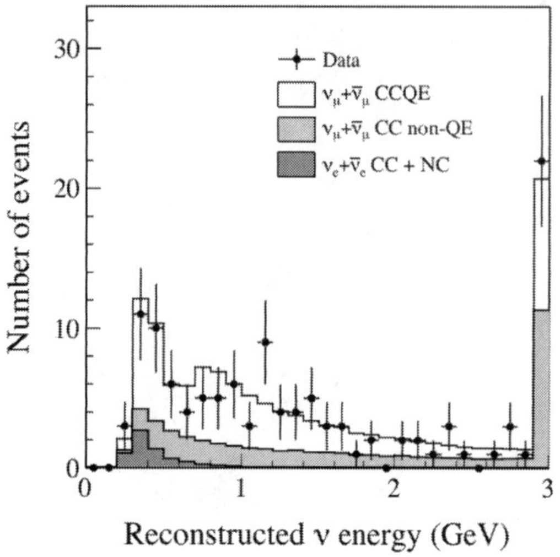

(a) ν_μ disappearance sample.

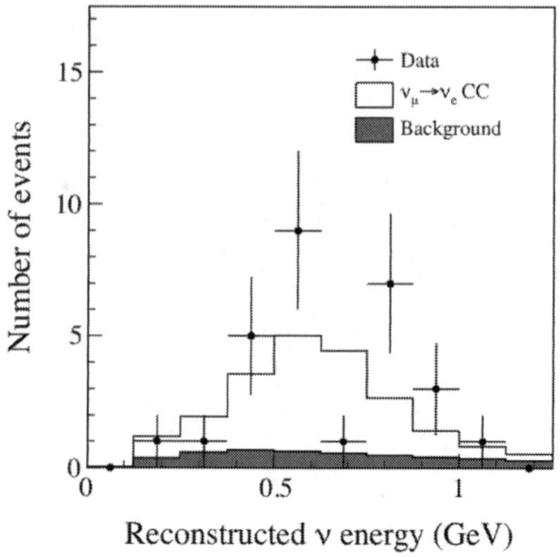

(b) ν_e appearance sample.

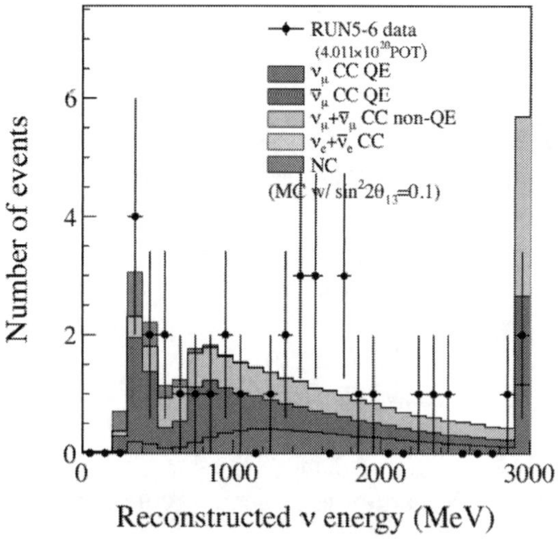

(c) $\bar{\nu}_\mu$ disappearance sample.

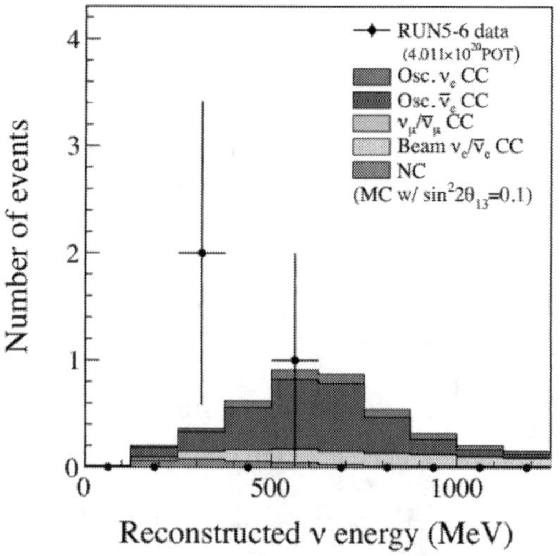

(d) $\bar{\nu}_e$ disappearance sample.

Figure 7: Reconstructed neutrino energy for the four SK event samples.

4 Results

4.1 ν_μ Disappearance Fit

The likelihood fit is performed to the oscillation parameters θ_{23} and $\Delta m^2_{32}\left(\Delta m^2_{13}\right)$ assuming the normal (inverted) hierarchy. The parameters θ_{12}, θ_{13} and Δm^2_{12} are Gaussian thrown from the global best fit with errors and the full range of possible values for δ are sampled ($-\pi \rightarrow +\pi$). The measurement is sensitive to the θ_{23} octant, and the results prefer maximal disappearance and the second octant for both the normal and inverted hierarchy assumptions as shown in Figure 8 [15].

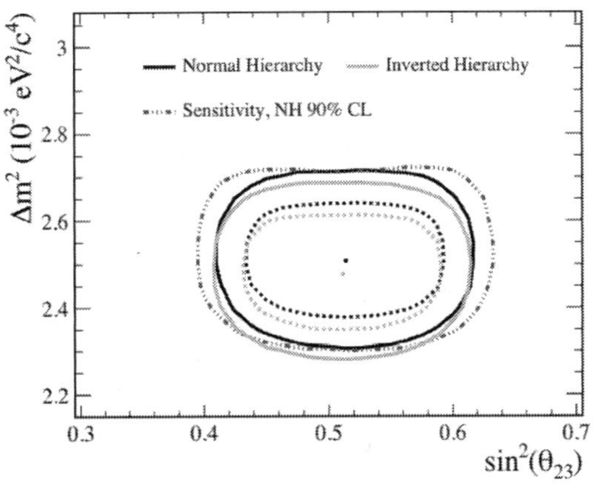

Figure 8. 68% (dashed) and 90% (solid) confidence level intervals for θ_{23} and Δm^2_{32} $\left(\Delta m^2_{13}\right)$ assuming the normal (inverted) hierarchy. The projected 90% confidence level sensitivity contour is also shown assuming the normal hierarchy.

4.2 ν_μ Disappearance and ν_e Appearance Joint Fit

The likelihood fit is performed to the oscillation parameters θ_{23}, θ_{13}, $\left|\Delta m^2\right|$ and δ, where $\left|\Delta m^2\right|$ is derived from a simultaneous fit over both mass hierarchies. The parameters θ_{12} and Δm^2_{12} are fixed to the global best fit values. The initial fit showed a very weak preference for the inverted hierarchy, the second octant with maximal mixing, and a large value for θ_{13} but with a significant error envelope that is consistent with the smaller measurements made by reactor experiments. The fit was insensitive to the charge-parity violating phase δ.

T2K is less sensitive to θ_{13} than the reactor measurements, so this can be used as a constraint within the joint fit by including it as a χ^2 penalty term. By including the reactor constraint the fit subsequently favoured the normal hierarchy and the second octant with maximal mixing as shown in Figure 9. The data also begins to show sensitivity to δ, favouring a value around $\delta \approx -\frac{\pi}{2}$, and rejecting ~35% (~60%) of the possible range of δ at the 90% confidence level within the normal (inverted) hierarchy, as shown in Figure 10 [15].

4.3 $\bar{\nu}_\mu$ Disappearance Fit

Similar to the neutrino mode measurement, this fits $\bar{\theta}_{23}$ and $\Delta\bar{m}^2_{32}$ to the $\bar{\nu}_\mu$ contribution of the $\bar{\nu}_\mu$ disappearance sample. The wrong-sign ν_μ background contribution to the $\bar{\nu}_\mu$ disappearance sample is also used to simultaneously fit θ_{23} and Δm^2_{32} as independent parameters. The parameters θ_{12}, θ_{13} and Δm^2_{12} are fixed to the global best fit values (and are assumed to be the same for neutrinos and antineutrinos), whilst δ is fixed to -1.55 radians from the previously described joint fit analysis. The hierarchy is assumed to be normal due to the low statistics preventing any sensitivity to this parameter. This measurement weakly prefers the first octant, but is consistent with other experiments and the previous T2K neutrino beam results as shown in Figure 11, and is currently statistics limited [10].

4.4 $\bar{\nu}_e$ Appearance Test

There are insufficient statistics within the $\bar{\nu}_e$ sample to perform a full analysis fit, instead a discrete parameter β is defined to modify the PMNS appearance probability such that

$$P\left(\bar{\nu}_\mu \to \bar{\nu}_e\right) = \beta \times P_{\text{PMNS}}\left(\bar{\nu}_\mu \to \bar{\nu}_e\right). \tag{4}$$

Toy experiments are performed with two values of β, $\beta = 0$ where no oscillations occur, and $\beta = 1$ where oscillations occurred in line with the PMNS expectation. Each toy experiment throws the

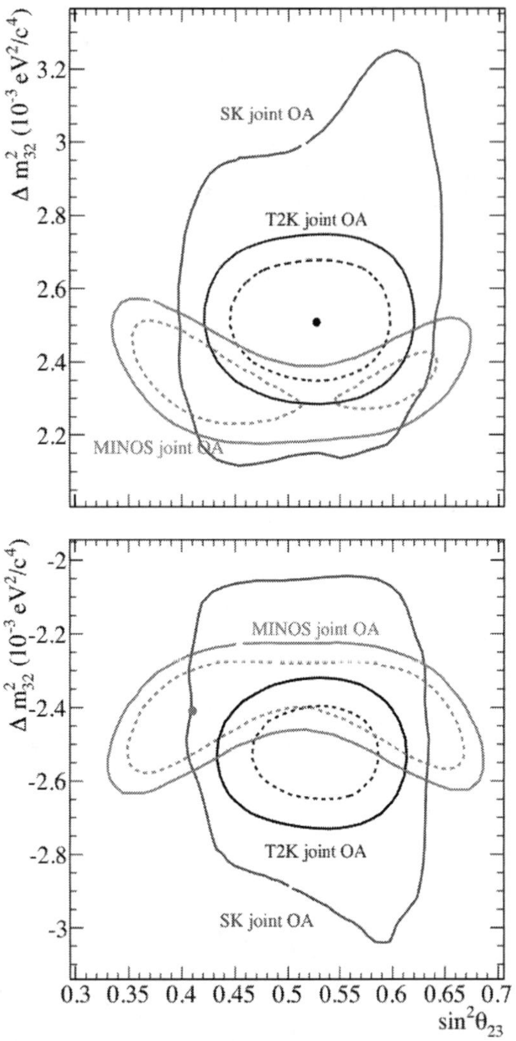

Figure 9. 68% (dashed) and 90% (solid) confidence level intervals for θ_{23} and Δm^2_{32} showing the distribution across the normal (upper) and inverted (lower) hierarchies. Results from the MINOS [16] and SK [17] experiments are also shown for comparison.

parameters θ_{13}, θ_{23}, Δm^2_{32} and δ whilst fixing θ_{12} and Δm^2_{12} the global best fit values. The distribution in expected number of events from each toy throw is shown in Figure 12 for both β hypotheses. As can be seen the average expectation for the no oscillation hypothesis, $\beta = 0$, was for ~1.6 events, and under the PMNS oscillation hypothesis, $\beta = 1$, was for ~4 events. With 3 events recorded in data there is no way to distinguish between these two hypotheses and this will require more data to be taken.

5 Future Sensitivity

Subsequent results will utilise a joint fit between all four SK event samples. In doing so T2K will be able to begin to probe δ at high precision. Figure 13 shows the potential sensitivity of T2K to δ as a function of θ_{13} assuming true $\delta = -\frac{\pi}{2}$ and a 50% neutrino beam and 50% antineutrino beam mix delivering 7.8×10^{21} POT [20]. The systematic uncertainties used are taken from the 2012 T2K oscillation analyses and are ~15%, a figure that has already been improved upon within the most

DOI: 10.1051/epjconf/201612501003

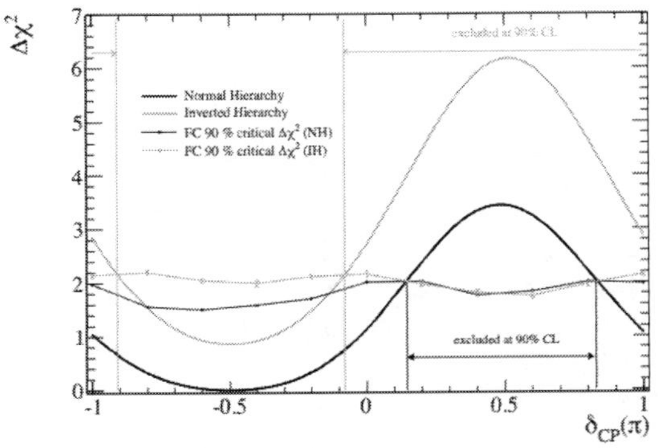

Figure 10. Profiled $\Delta\chi^2$ distribution as a function of δ for the normal (black) and inverted (blue) hierarchy, with the 90% confidence level exclusion critical values displayed.

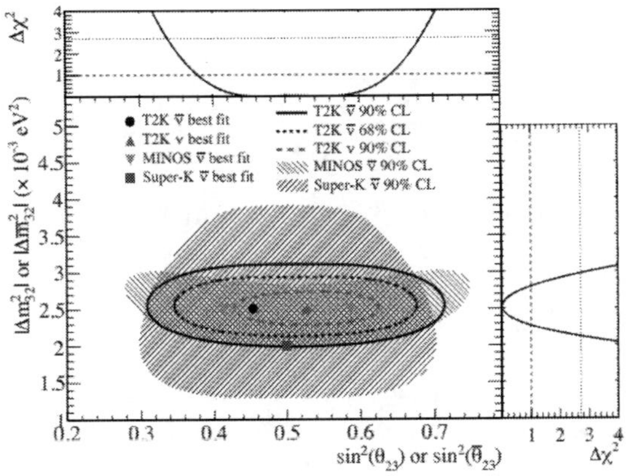

Figure 11. 68% (dashed) and 90% (solid) confidence level intervals for $\bar{\theta}_{23}$ and $\Delta\bar{m}^2_{32}$. Results from the MINOS [18] and SK [19] experiments are also shown for comparison, along with the T2K [15] neutrino beam result.

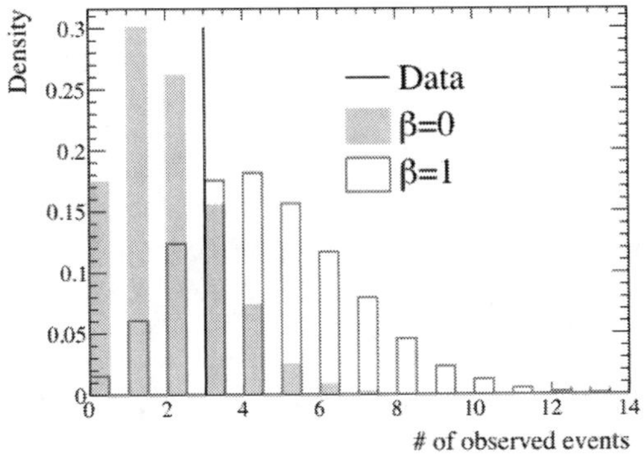

Figure 12. Distribution for the predicted number of events from the $\beta = 0$ and $\beta = 1$ toy experiments along with the data result.

recent analyses. Further constraint from the reactor experiments along with measurements from the NOνA experiment [21][22] will further improve upon this future sensitivity.

6 Conclusion

With current data, about ~14% of the planned total, T2K has achieved world leading measurements of the oscillation parameter θ_{23} and has begun to probe δ. Future updates will include higher statistics, an improved near detector fit (with water target data), full four SK sample joint fit, and the inclusion of additional datasets from the SK charged-current $1\pi^{\pm}$ samples. T2K is also producing a large number

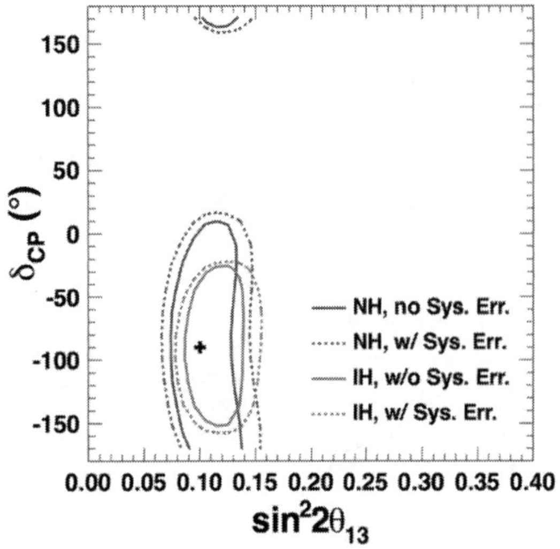

Figure 13. Future sensitivity of T2K to δ and θ_{13} displaying 90% confidence level contours with and without systemic uncertainty for both the normal and inverted hierarchy.

of cross section measurements, and is beginning work on proposed future extensions to the experiment such as T2K-II and Hyper-Kamiokande.

References

[1] K.A. Olive et al. (Particle Data Group), Chin. Phys. **C38**, 090001 (2014)
[2] K. Abe et al., Nucl. Instrum. Meth. **659**, 106 (2011)
[3] K. Abe et al. (T2K Collaboration), Phys. Rev. D **87**, 012001 (2013)
[4] K. Abe et al., Nucl. Instrum. Meth. **694**, 211 (2012)
[5] S. Fukuda et al., Nucl. Instrum. Meth. **501**, 418 (2003)
[6] A. Ferrari et al., Tech. rep. (2005)
[7] N. Abgrall et al., Nucl. Instrum. Meth. **701**, 99 (2013)
[8] R. Brun et al., Tech. rep., CERN DD/EE/84-1 (1987)
[9] C. Zeitnitz, T.A. Gabriel, Nucl. Instrum. Meth. **A349**, 106 (1994)
[10] K. Abe et al. (T2K Collaboration), Phys. Rev. Lett. **116**, 181801 (2016)
[11] Y. Hayato, Acta Phys. Polon. **B40**, 2477 (2009)
[12] A. Aguilar-Arevalo et al., Nucl. Instrum. Meth. **599**, 28 (2009)
[13] L. Aliaga et al., Nucl. Instrum. Meth. **743**, 130 (2014)
[14] S. Agostinelli et al., Nucl. Instrum. Meth. **506**, 250 (2003)
[15] K. Abe et al. (T2K Collaboration), Phys. Rev. D **91**, 072010 (2015)
[16] P. Adamson et al., Phys. Rev. Lett. **112**, 191801 (2014)
[17] A. Himmel (The Super-Kamiokande Collaboration), AIP Conf. Proc. **1604**, 345 (2014)
[18] P. Adamson et al. (The MINOS Collaboration), Phys. Rev. Lett. **108**, 191801 (2012)
[19] K. Abe et al. (Super-Kamiokande Collaboration), Phys. Rev. Lett. **107**, 241801 (2011)
[20] K. Abe et al., Prog. Theor. Exp. Phys. **2015**, 043C01 (2015)
[21] P. Adamson et al. (NOvA Collaboration), Phys. Rev. D **93**, 051104 (2016)
[22] P. Adamson et al. (NOvA Collaboration), Phys. Rev. Lett. **116**, 151806 (2016)

EPJ Web of Conferences **125**, 01004 (2016) DOI: 10.1051/epjconf/201612501004
QUARKS-2016

Upgrade of the ALICE Inner Tracking System

Iouri Belikov[1,⋆] for the ALICE Collaboration

[1] *IPHC, Université de Strasbourg, CNRS-IN2P3, 23 rue du Loess, BP28, 67037 Strasbourg cedex 2*

Abstract. A Large Ion Collider Experiment (ALICE) is built to study the properties of the strongly interacting matter created in heavy-ion collisions at the LHC. With the upgrade of its Inner Tracking System (ITS), the ALICE experiment is going to increase the rate of data taking by almost two orders of magnitude. At the same time, the precision of secondary vertex reconstruction will become by at least a factor 3 better than it currently is. In this talk, we briefly show some selected physics results motivating the upgrade of the ITS, describe the design goals and the layout of the new detector, and highlight a few important measurements that will be realized after the completion of this upgrade.

1 ALICE experiment at CERN

A Large Ion Collider Experiment (ALICE) [1–3] is one of the four major experiments currently operating on the beams of the Large Hadron Collider (LHC) at the European Organization for Nuclear Research (CERN). The aim of this experiment is to study the physics of strongly interacting matter at extreme energy densities, where the formation of a new phase of matter, the quark-gluon plasma, is expected. For this purpose, the collaboration is carrying out series of comprehensive measurements of hadrons, electrons, muons and photons produced in the collisions of heavy nuclei. In addition, ALICE is studying proton-proton interactions both as a reference for the heavy-ion collisions and also in physics areas where ALICE is competitive with other LHC experiments.

2 Selected (Run I) results and their limitations

An important part of the ALICE physics program is dedicated to high precision measurements of charm and beauty production in heavy-ion collisions. The ultimate goals of the heavy-flavour studies include:

- Thermalization of heavy quarks in the produced medium (by measuring baryon-to-meson ratios for charm and beauty particles, ratios of yields of heavy-flavour particles with strange content, and azimuthal flow anisotropy for as many as possible heavy-flavour species).
- Parton mass and colour-charge dependence of in-medium energy losses (by measuring the momentum-dependent nuclear modification factors for B and D mesons, and comparing them with those for light-flavour particles).

⋆e-mail: iouri.belikov@iphc.cnrs.fr

 DOI: 10.1051/epjconf/201612501004

Using the data recorded during the LHC Run I (2009–2013), the ALICE Collaboration has already published several important results on the D-meson production in Pb–Pb collisions. For example, the left panel of Figure 1 shows the p_T-dependent nuclear modification factor for D-mesons (average of D^0, D^+ and D^{*+}) compared to that of pions and charged particles in the 0–10 % centrality class. There is an indication that production of D-mesons in Pb–Pb collisions at lower p_T is less suppressed than that of light-flavour particles. The ALICE measurement of the second-order azimuthal flow harmonic v_2 is presented on the right panel of the Figure. The data suggest that the v_2 for D-mesons is compatible with that of charged particles and is larger than 0 over a wide momentum range. But, in both cases, the current statistical and systematic uncertainties do not allow for firm conclusions. In addition, the read-out rate capabilities and space-point precision of the present Inner Tracking System (ITS) are not sufficient to perform similar measurements with beauty particles (B-mesons), which are important to prove the existence of quark-mass ordering: $R_{AA}^{B} > R_{AA}^{D} > R_{AA}^{\text{charged}}$ and $v_2^{B} < v_2^{D} < v_2^{\text{charged}}$.

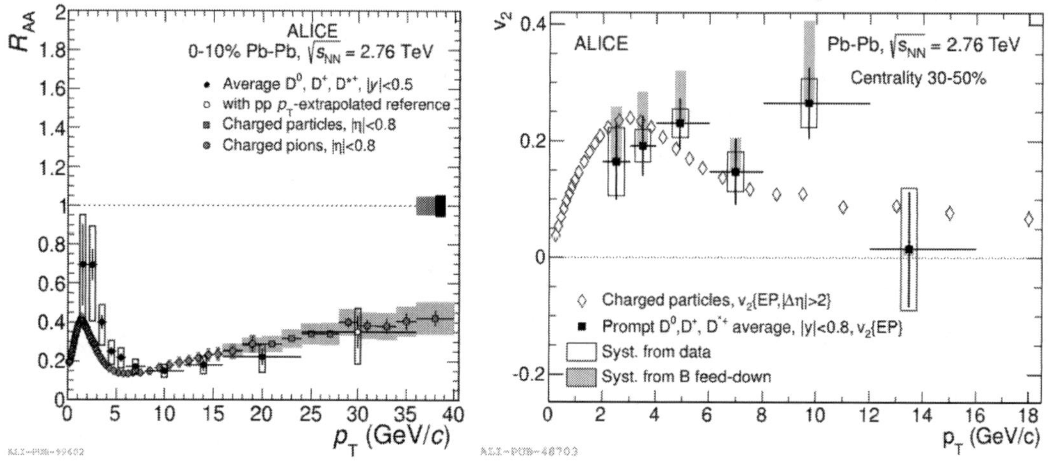

Figure 1. Left panel: Nuclear modification factor R_{AA} for D-mesons, pions and charged particles [4]. Right panel: Amplitude of the second-order flow harmonic v_2 for D-mesons and charged particles [5]. Statistical and systematic uncertainties are represented by bars and boxes respectively.

At the same time, even though ALICE has already measured several baryon-to-meson ratios with light-flavour particles [6, 7], the extension of these results towards the heavy-flavous sector is currently not possible. The read-out rates and precision of track reconstruction provided by the present ITS do not allow for detecting the heavy-flavous baryons (Λ_c and Λ_b) in high-track-multiplicity environment of Pb–Pb collisions at the LHC.

3 Upgrade of the Inner Tracking System

The limitations of the present ITS will be radically reduced with the planned upgrade [8]. The general layout of the upgraded ITS is shown in Figure 2. The detector will be 1.5 m long, with the outer radius of 40 cm, and the overall surface of about 10 m^2 will be covered with silicon pixel sensors. This will be a "12.6 Giga-pixel camera", with an estimated cost of about 13.6 million Swiss Francs.

The detector will consist of seven layers, with the first layer as close to the interaction point as 22.4 mm. The material budget of the three innermost layers will be as low as 0.3 % of a radiation length. Combined with the space-point precision of about 5 μm, the expected track impact-parameter

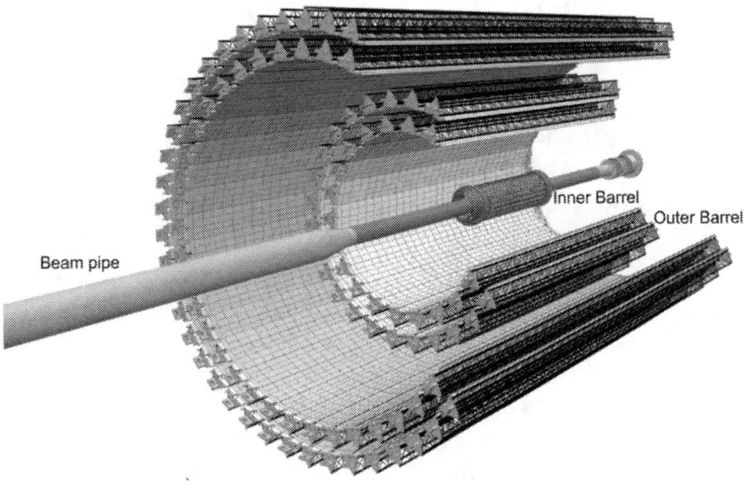

Figure 2. The general layout of the upgraded ITS detector.

resolution at $p_T \sim 1$ GeV/c will be about 20 μm, which is a factor 3 (in the transverse plane) and 5 (in the beam direction) better as compared with the current situation.

The new detector will also allow for an efficient track reconstruction down to very low p_T. The simulations show that the tracking efficiency at $p_T \sim 0.1$ GeV/c will be a factor about 6 higher than it is now. This will be of a great importance for planned di-electron measurements (see Table 1) and the detection of D^* and D_S mesons.

All seven layers of the upgraded ITS will be equipped with the ALPIDE chip [8]. Taking advantages of the 0.18 μm CMOS technology by TowerJazz [9], this chip combines the sensitive part and the read-out electronics within the same piece of silicon. The main characteristics of the chip are:

- overall chip size: 15×30 mm^2;
- pixel size: 29×27 μm^2;
- detection efficiency: > 99 %, with a noise probability of $< 10^{-6}$;
- time resolution: about 2 μs;
- in-pixel discriminators and in-matrix address encoder with asynchronous sparsified readout;
- power consumption: 40 mW/cm^2;
- high radiation tolerance: tested up to 2.7 Mrad (TID), and 1.7×10^{13} 1 MeV n_{eq}/cm^2 (NIEL)

The upgraded ITS will be almost two orders of magnitude faster than now. The read-out electronics of the detector will be able to register events with a typical rate of 50 kHz and a few 100 kHz for minimum bias Pb–Pb and pp collisions respectively.

The parts of the detector will have to be aligned with the precision of better than 5 μm by means of dedicated software. The data coming out of the detector will be reconstructed quasi-online, using a dedicated Online-Offline (O^2) computer farm [10].

Overall, the upgrade of the Inner Tracking System will be accomplished by the end of the LHC Long Shutdown 2 (2021).

4 Physics with the upgraded ITS

The expected physics reach for various observables is summarized in Table 1, in terms of minimum accessible p_T and of statistical uncertainties. We consider a scenario with an integrated luminosity of $10\,nb^{-1}$, fully used for minimum-bias Pb–Pb data collection, and a low-magnetic-field run with $3\,nb^{-1}$ of integrated luminosity for the low-mass di-electron studies. The case of the programme up to Long Shutdown 2 is shown for comparison. In this case, a delivered luminosity of $1\,nb^{-1}$ is assumed, out of which 10% is recorded with a minimum-bias trigger.

In the upgrade case, the systematic uncertainties will be reduced as well. This is because for many measurement, due to the mentioned improvements in the track impact-parameter resolution, the level of the combinatorial background will significantly be be reduced, and the feed-down contribution from beauty to charm measurements will be estimated directly, from the data themselves.

There are also several observables, like those involving the heavy-flavour baryons and the low-momentum di-electrons, that were never accessible before, but will become well within the experimental reach thanks to the unique capabilities of the upgraded ALICE Inner Tracking System.

Table 1. Summary of the physics reach: minimum accessible p_T and relative statistical uncertainty in Pb–Pb collisions for an integrated luminosity of $10\,nb^{-1}$. For heavy flavour, the statistical uncertainties are given at the maximum between $p_T = 2\,GeV/c$ and p_T^{min}. For elliptic flow measurements, the value of v_2 used to calculate the relative statistical uncertainty σ_{v_2}/v_2 is given in parenthesis. The programme up to Long Shutdown 2, with an integrated luminosity of $0.1\,nb^{-1}$ collected with minimum-bias trigger, is shown for comparison.

Observable	Current, $0.1\,nb^{-1}$		Upgrade, $10\,nb^{-1}$	
	p_T^{min} (GeV/c)	statistical uncertainty	p_T^{min} (GeV/c)	statistical uncertainty
Heavy Flavour				
D meson R_{AA}	1	10 %	0	0.3 %
D_s meson R_{AA}	4	15 %	< 2	3 %
D meson from B R_{AA}	3	30 %	2	1 %
J/ψ from B R_{AA}	1.5	15 % (p_T-int.)	1	5 %
B^+ yield	not accessible		2	10 %
Λ_c R_{AA}	not accessible		2	15 %
Λ_c/D^0 ratio	not accessible		2	15 %
Λ_b yield	not accessible		7	20 %
D meson v_2 ($v_2 = 0.2$)	1	10 %	0	0.2 %
D_s meson v_2 ($v_2 = 0.2$)	not accessible		< 2	8 %
D from B v_2 ($v_2 = 0.05$)	not accessible		2	8 %
J/ψ from B v_2 ($v_2 = 0.05$)	not accessible		1	60 %
Λ_c v_2 ($v_2 = 0.15$)	not accessible		3	20 %
Di-electrons				
Temperature (intermediate mass)	not accessible			10 %
Elliptic flow ($v_2 = 0.1$) [11]	not accessible			10 %
Low-mass spectral function [11]	not accessible		0.3	20 %
Hypernuclei				
${}^3_\Lambda H$ yield	2	18 %	2	1.7 %

References

[1] F. Carminati *et al.* [ALICE Collaboration], J. Phys. G **30**, 1517 (2004).
[2] G. Alessandro *et al.* [ALICE Collaboration], J. Phys. G **32**, 1295 (2006).
[3] K. Aamodt *et al.* [ALICE Collaboration], JINST **3**, S08002 (2008)
[4] J. Adam *et al.* [ALICE Collaboration], JHEP **03**, 081 (2016)
[5] B. Abelev *et al.* [ALICE Collaboration], Phys. Rev. Lett. **111**, 102301 (2013)
[6] B. Abelev *et al.* [ALICE Collaboration], Phys. Rev. Lett. **111**, 222301 (2013)
[7] J. Adam *et al.* [ALICE Collaboration], Phys. Rev. C **93**, 034913 (2016)
[8] B. Abelev *et al.* [ALICE Collaboration], J. Phys. G **41**, 087002 (2014).
[9] http://towerjazz.com
[10] ALICE Collaboration, CERN-LHCC-2015-006, (2015)
[11] ALICE Collaboration, CERN-LHCC-2012-012, LHCC-I-022, (2012)

Overview of recent ALICE results

Igor Lakomov (for the ALICE Collaboration)[1,⋆]

[1] *European Organization for Nuclear Research (CERN), CH-1211, Geneva 23, Switzerland*

Abstract. ALICE is one of the four largest LHC experiments. It is dedicated to the study of the properties of the deconfined state of matter formed at large energy densities in heavy-ion collisions — the Quark-Gluon Plasma. The ALICE Collaboration also participated in the pp and p–Pb data-taking periods at the LHC. An overview of recent ALICE results is presented for three collision systems: pp, p–Pb and Pb–Pb.

1 Introduction

The high energy densities reached in heavy-ion collisions at LHC energies lead to the formation of a deconfined state of matter — the Quark-Gluon Plasma (QGP). The ALICE detector allows one to study the properties of this state at the LHC. Since the lifetime of the QGP is very small it cannot be observed directly. Measurement of the final-state particle yields is one of the strongest tools used for such studies. These yields in heavy-ion collisions are then compared to the expected yields from a superposition of incoherent pp collisions scaled by the number of elementary binary collisions. The ratio of the particle yield in heavy-ion collisions to the corresponding scaled yield in pp collisions is called the Nuclear Modification Factor (R_{AA}). In the presence of any hot nuclear matter effects related to the QGP formation, R_{AA} is not equal to unity. However heavy-ion collisions are more complicated, and there are some Cold Nuclear Matter (CNM) effects playing a role. The latter also might affect the R_{AA}. In order to disentangle hot and CNM effects in heavy-ion collisions, the study of pA collisions is necessary. In pA collisions the energy densities are lower than in nucleus-nucleus collisions, and the formation of the QGP is much less probable. Thus, pA collisions allow one to study pure CNM effects. It is also usually done in terms of the Nuclear Modification Factor, which for pA collisions is referred to as R_{pA}. In addition to global properties of the QGP, the geometry of the collision system can be studied using other observables, like elliptic flow. For a more detailed description of the system, differential measurements in transverse momentum (p_T), rapidity (y) and centrality are usually performed. This article presents an overview of the latest ALICE results for different collision systems.

2 Charged particle and open HF production

ALICE recently published one of its first results for Pb–Pb collisions at $\sqrt{s_{NN}} = 5.02$ TeV, the pseudorapidity density of charged particles [1]. The left panel of Fig. 1 shows its dependence on the

⋆e-mail: igor.lakomov@cern.ch

DOI: 10.1051/epjconf/201612501005

collision energy compared to other experimental measurements. The rising trends established by the low-energy data for minimum bias pp, p(d)–A collisions as well as for central AA collisions are confirmed by the ALICE measurements. Moreover, the fact that p–Pb and d–Au measurements lay on the curve for pp collisions leads to the conclusion that the strong rise seen in nucleus-nucleus collisions is not solely related to the multiple interactions undergone by the participants since the proton in pA collisions also encounters multiple nucleons in the nucleus. In the right plot of Fig. 1 the centrality dependence of the pseudorapidity density of charged particles is shown. The results for Pb–Pb at 5.02 TeV are compared to those from pp and Pb–Pb both at 2.76 TeV and to p–Pb measurements at 5.02 TeV. A strong dependence on the average number of participants $N_{\rm part}$ is seen: $\langle \frac{dN}{d\eta} \rangle / \left(\langle N_{\rm part} \rangle / 2 \right)$ is increasing with the increase of $N_{\rm part}$, i.e. from peripheral to central collisions. The data clearly show the scaling of the centrality dependence with energy and collision system . In order to quantify the

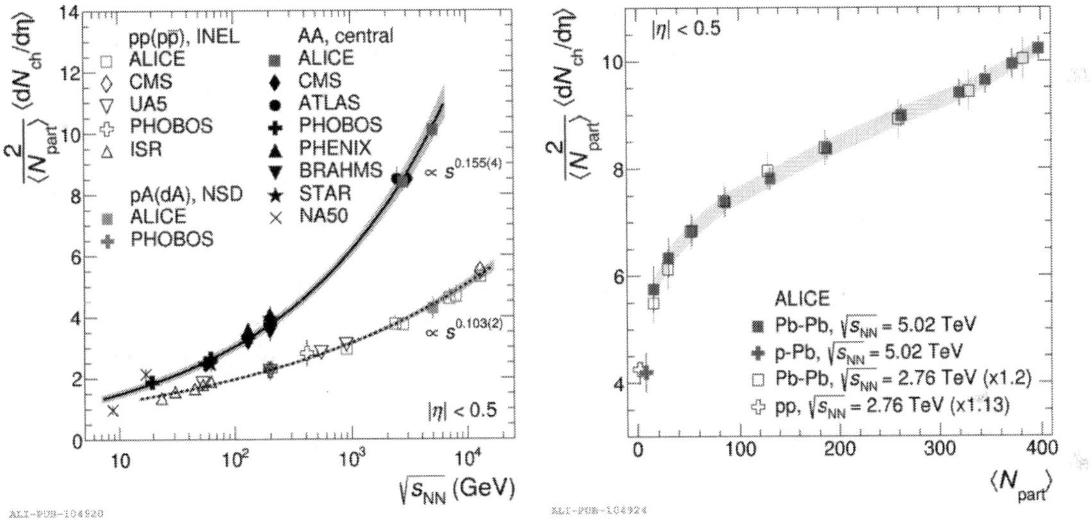

Figure 1: Pseudorapidity density of charged particles $\langle \frac{dN}{d\eta} \rangle / \left(\langle N_{\rm part} \rangle / 2 \right)$ as a function of average number of participants ($\langle N_{\rm part} \rangle$) for central (left) and 0-80% central (right) collisions. In the left panel the results from other experiments for various collision systems and energies are also shown. In the right panel pp and Pb-Pb results are scaled by 1.13 and 1.2, respectively.

nuclear matter effects in p–Pb and Pb–Pb collisions ALICE measured the nuclear modification factors for identified particles. In the left panel of Fig. 2 the $R_{\rm pPb}$ measurements are presented for charged hadrons, pions, kaons and protons [2]. At high $p_{\rm T}$, $R_{\rm pPb}$ is consistent with unity for all species, while at intermediate $p_{\rm T}$, in the Cronin effect region, a mass ordering is seen with an indication of proton enhancement. The right panel of Fig. 2 shows the $R_{\rm PbPb}$ for D-mesons which is compared to previous $R_{\rm pPb}$ measurements for D-mesons [3]. For both central and semi-central Pb–Pb collisions D-meson production is suppressed, up to a factor of 5–6 and 3, respectively, at $p_{\rm T} \sim 10$ GeV/c. Results for p–Pb collisions in both panels of Fig. 2 are indicative of the hot nuclear matter origin of the corresponding suppression seen in Pb–Pb collisions.

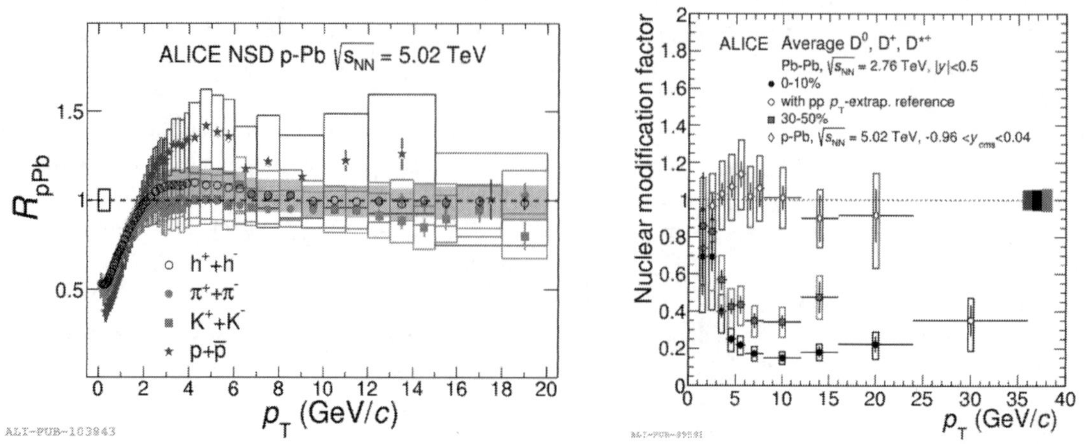

Figure 2: Nuclear modification factors of charged particles in p–Pb collisions (left), and of D-mesons in Pb–Pb collisions (right). In the left panel results are shown separately for different particle species: charged hadrons, pions, kaons and protons. In the right panel the results are shown for central (0 – 10%) and semi-central (30 – 50%) Pb–Pb collisions, and compared to the ALICE p–Pb measurements.

3 Jets and flow

For a deeper understanding of the initial collision geometry, transport mechanisms etc., the momentum anisotropy of the final state particles is usually studied. This is done experimentally in terms of the anisotropic flow coefficients representing the coefficients of the Fourier decomposition of the particle angular distribution. ALICE measured v_2, v_3 and v_4 of the charged particles angular distribution in Pb–Pb collisions at 5.02 TeV and compared them to the corresponding results at lower energy (2.76 TeV) [4]. As can be seen from the left panel of Fig. 3, the values of v_2, v_3 and v_4 increase with $\sqrt{s_{NN}}$ from 2.76 to 5.02 TeV. In the 0 – 50% centrality range they increase by $(3.0 \pm 0.6)\%$, $(4.3 \pm 1.4)\%$ and $(10.2 \pm 3.8)\%$, respectively. Moreover, this increase seems to be independent of centrality as seen from the corresponding ratios. Hydrodynamical models describe these results fairly well. In the right top and bottom panels of Fig. 3, the ALICE v_2 measurements for charged jets in central and semi-central Pb–Pb collisions at 2.76 TeV, respectively, are shown [5]. With a significance higher than 3σ the semi-central Pb–Pb measurements indicate a non-zero $v_2^{\text{ch jet}}$. The observed behavior indicates a path-length dependence of the in-medium parton energy loss. Larger uncertainties on the measurements for central Pb–Pb collisions preclude drawing a strong conclusion on the magnitude of $v_2^{\text{ch jet}}$. Moreover, for central Pb–Pb collisions, the fluctuations in the initial state may contribute to the measured non-zero $v_2^{\text{ch jet}}$. The $v_2^{\text{ch jet}}$ results are compared to single particle v_2 from ALICE and CMS and $v_2^{\text{calo jet}}$ measured by ATLAS. While they cannot be directly quantitatively compared given the difference in the energy scale, the results agree fairly well showing a positive v_2 for both charged particles and jets to high p_T in central and semi-central Pb–Pb collisions. This observation indicates that the sensitivity to the collision geometry persists up to high p_T.

4 Quarkonium production

The study of the production of quarkonia, bound states of a charm (beauty) quark and its antiquark, is of particular interest. Heavy and tightly bound, the heavy quark pairs are produced in the initial hard

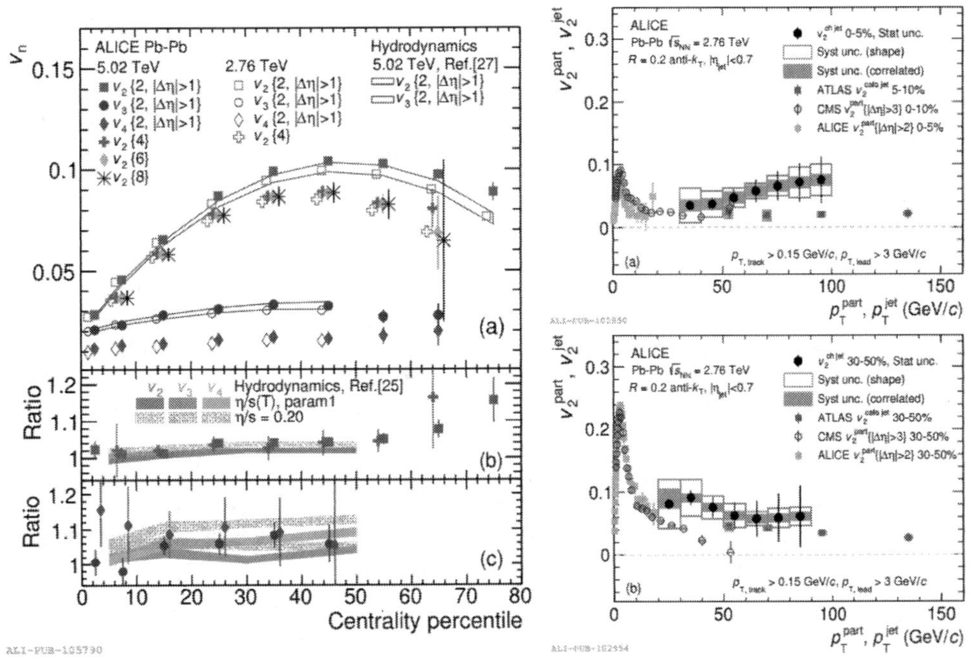

Figure 3: Measurements of the anisotropic flow of charged hadrons (left) and of charged jets (right) in Pb–Pb collisions. In the left panel ALICE results for 5.02 and 2.76 TeV as a function of centrality are compared. In the right part of the figure, results are shown for central (0 – 5%, top) and semi-central (30 – 50%, bottom) collisions as a function of p_T. In both plots on the right, the corresponding measurements from ATLAS and CMS are also shown.

partonic collisions. Sequential suppression of quarkonium states with the increase of the temperature of the collision may be used as a QGP thermometer [6, 7], therefore thorough study of quarkonium production is crucial for understanding the physics in heavy-ion collisions. Figure 4 shows the ALICE measurements of the J/ψ suppression in Pb–Pb collisions at 2.76 TeV as a function of centrality [8] (left panel) and of p_T [9] (right panel). The J/ψ production is suppressed in Pb–Pb collisions. The suppression increases with increasing centrality and for higher p_T. A comparison to the results from PHENIX for Au–Au collisions at 0.2 TeV in the left panel of Fig. 4 indicates a larger J/ψ suppression at RHIC energies compared to that seen at about 10 times higher collision energies at the LHC. This is the effect of recombination which is expected to be significant at LHC energies while at RHIC energies it is almost negligible. At low p_T its effect is expected to be higher, thus high-p_T J/ψ are more suppressed. However it was not expected to explain the strong low-p_T J/ψ excess at low N_{part} (corresponding to peripheral collisions) clearly seen in the right panel of Fig. 4. After more detailed study this excess was understood as an effect from coherent photo-production of J/ψ in ultra-peripheral collisions [9]. After the extensive analysis of the J/ψ production in p–Pb collisions where the CNM effects were quantified as a function of p_T, rapidity and centrality [10–12], and the corresponding measurement of the $\psi(2S)$ production in p–Pb collisions integrated over centrality [13], ALICE recently published the results on the $\psi(2S)$ production in p–Pb collisions as a function of centrality [14]. Figure 5 shows the double ratio of the $\psi(2S)$ and J/ψ cross sections in p–Pb collisions with respect to pp collisions: $\left[\sigma_{\psi(2S)}/\sigma_{J/\psi}\right]_{pPb} / \left[\sigma_{\psi(2S)}/\sigma_{J/\psi}\right]_{pp}$ as a function of the number of binary collisions

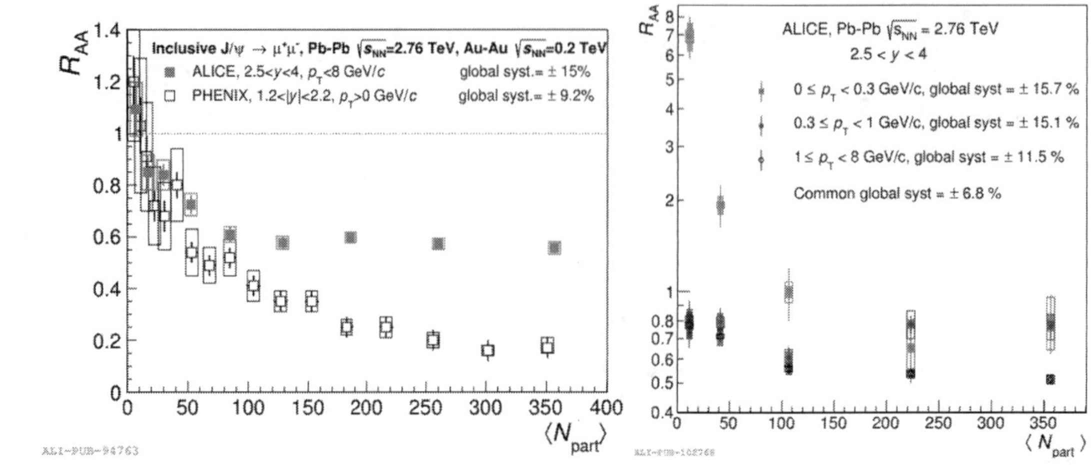

Figure 4: Nuclear modification factor for J/ψ in Pb–Pb collisions as a function of N_{part} integrated over p_{T} (left) and in different p_{T} intervals (right). In the left panel the ALICE data are compared to the corresponding results from PHENIX for Au–Au collisions at 0.2 TeV.

N_{coll} (left panel) and of the crossing time $\langle \tau_{\mathrm{c}} \rangle$ of the $c\bar{c}$ pair in the nuclear matter (right panel). The measured double ratio indicates a decreasing trend with increasing centrality of the collisions (higher N_{coll} values) for forward and backward rapidity intervals, which agrees fairly well with theoretical model calculations including final-state effects. However large uncertainties for the forward-y measurements are also consistent with no centrality dependence. Surprisingly, the ALICE measurements show similar behaviour to the PHENIX d–Au results in the mid-y interval and with about 25 times lower collision energy. The crossing time was varied by studying different p_{T} and centrality intervals. At forward rapidity, the τ_{c} values are much smaller than the resonance formation time τ_{f}, excluding a significant role of final-state interactions with nuclear matter. Effects occurring at later times, such as the break-up by co-movers in the hadronic gas, are suitable candidates for the explanation of the observed $\psi(2S)$ suppression. Significantly larger values of τ_{c} at backward rapidity are still smaller than τ_{f}. The observed τ_{c}-scaling of the double ratios might indicate an effect at least partly related to a dissociation of the fully-formed resonance in nuclear matter.

5 Conclusions

Some recent ALICE results are presented in this overview for different collision systems (pp, p–Pb, Pb–Pb), including both measurements from Run I and Run II of the data-taking period at the LHC. Comparisons to theoretical models and results from other experiments lead to a deeper understanding of the physics of high-energy particle collisions and give additional constraints to theoreticians to tune their models. Many other ALICE results will appear soon based on the collected and coming data.

References

[1] J. Adam et al. (ALICE), Phys. Rev. Lett. **116**, 222302 (2016), 1512.06104
[2] J. Adam et al. (ALICE) (2016), 1601.03658
[3] J. Adam et al. (ALICE), JHEP **03**, 081 (2016), 1509.06888

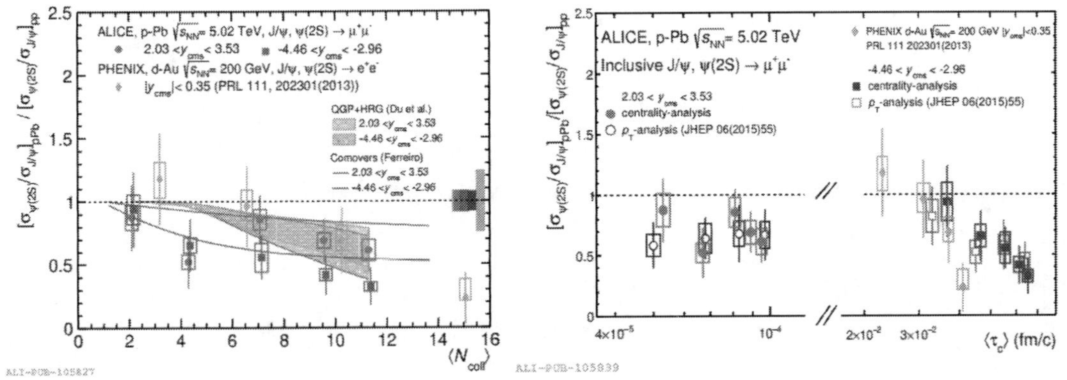

Figure 5: Double ratio of the ψ(2S) and J/ψ cross sections in p–Pb collisions with respect to pp collisions: $\left[\sigma_{\psi(2S)}/\sigma_{J/\psi}\right]_{pPb} / \left[\sigma_{\psi(2S)}/\sigma_{J/\psi}\right]_{pp}$ as a function of the number of binary collisions N_{coll} (left panel) and of the crossing time $\langle\tau_c\rangle$ (right panel) of the $c\bar{c}$ pair in the nuclear matter. ALICE results are also compared to the PHENIX measurements for d–Au collisions at 0.2 TeV.

[4] J. Adam et al. (ALICE), Phys. Rev. Lett. **116**, 132302 (2016), 1602.01119
[5] J. Adam et al. (ALICE), Phys. Lett. **B753**, 511 (2016), 1509.07334
[6] T. Matsui, H. Satz, Phys. Lett. **B178**, 416 (1986)
[7] H. Satz (2006), hep-ph/0602245
[8] J. Adam et al. (ALICE), JHEP **05**, 179 (2016), 1506.08804
[9] J. Adam et al. (ALICE), Phys. Rev. Lett. **116**, 222301 (2016), 1509.08802
[10] B.B. Abelev et al. (ALICE), JHEP **02**, 073 (2014), 1308.6726
[11] J. Adam et al. (ALICE), JHEP **06**, 055 (2015), 1503.07179
[12] J. Adam et al. (ALICE), JHEP **11**, 127 (2015), 1506.08808
[13] B.B. Abelev et al. (ALICE), JHEP **12**, 073 (2014), 1405.3796
[14] J. Adam et al. (ALICE), JHEP **06**, 050 (2016), 1603.02816

EPJ Web of Conferences **125**, 01006 (2016) DOI: 10.1051/epjconf/201612501006
QUARKS-2016

Strong Interaction Studies with PANDA at FAIR

Karin Schönning[1,⋆] for the PANDA collaboration

[1] *Dept. of Physics and Astronomy, Uppsala Universitet, Box 516, S-75120 Uppsala, Sweden*

Abstract. The Facility for Antiproton and Ion Research (FAIR) in Darmstadt, Germany, provides unique possibilities for a new generation of nuclear-, hadron- and atomic physics experiments. The future PANDA experiment at FAIR will offer a broad physics programme with emphasis on different aspects of hadron physics. Understanding the strong interaction in the perturbative regime remains one of the greatest challenges in contemporary physics and hadrons provide several important keys. In these proceedings, PANDA will be presented along with some high-lights of the planned physics programme.

1 Introduction

The Standard Model (SM) of particle physics has been proven successful in describing elementary particles and their interactions. Strong interactions of quarks and gluons are described by the theory of Quantum ChromoDynamics (QCD) [1]. At high energies, the coupling α_s is sufficiently weak to enable a perturbative treatment (pQCD) that has been rigorously and successfully tested [2]. However, at low and intermediate energies, the strong coupling α_s grows and pQCD breaks down. Strongly interacting particles at this scale therefore emerge as hadrons where the quarks and gluons are confined within a radius of ≈1 fm. Understanding the strong interaction at energies in the *confinement domain* up to pQCD is one of the most challenging questions in contemporary physics. More specifically, this can be considered a puzzle containing many pieces, where some of the most important are

- Why and how are quarks confined into hadrons?
- How is mass generated by the strong interaction? Only 1% of the proton mass originates from the current quark masses.
- What is the origin of the so-called *spin crisis*? Only 1/3 of the proton spin originates from the current quarks.
- Are there exotic hadrons, *i.e.* non-$q\bar{q}$ or non-qqq states? Which forces act between their constituents?
- What is the role of strangeness and charm in systems with larger degrees of complexity, *e.g.* nuclei or even neutron stars?

Hadrons are excellent probes of the strong interaction. The scale of the interaction within a hadron is given by the mass of its quarks. For strange hadrons, the scale is governed by the mass of the strange quark, *i.e.* m_s ≈100 MeV. This is close to the QCD cut-off, Λ_{QCD}, which is the scale at which

⋆e-mail: karin.schonning@physics.uu.se

quarks confine into hadrons. Strange hadrons therefore provide a key to confinement. The charm scale is more than ten times larger since the mass of the charm quark is $m_c \approx$1300 MeV. Here, the strong coupling is $\alpha_s \approx 0.3$ which is just barely outside the region where pQCD applies. Charmed hadrons therefore provide a bridge between perturbative and non-perturbative QCD. To summarize, strange and charmed hadron physics probe the strong interaction at two different energy scales and a systematic comparison could be very illuminating.

Furthermore, the study of hadrons in the precision frontier shed light on other open questions within the Standard Model, for example: i) Are there baryon decays that violate CP conservation? ii) What are the hadronic contributions to SM precision observables, *e.g.* the anomalous magnetic moment of the muon, $g-2$?

The future PANDA experiment at FAIR in Darmstadt, Germany, will be a unique facility to study a wide range of aspects of the strong interaction. PANDA will be optimized for strange and charmed hadron physics. In the following, we will present the design and outline its broad physics programme.

2 The PANDA experiment

The PANDA (anti-Proton ANnihilation in DArmstadt) detector will be an integral part of the HESR (High Energy Storage Ring). The antiproton beam will impinge on a fixed target which will provide a high luminosity environment crucial for the physics programme. The HESR will provide anti-protons with momenta from of 1.5 GeV/c up to 15 GeV/c, corresponding to centre-of-mass energies of 2.0 up to 5.5 GeV. The HESR will have two operation modes: the *high resolution mode* and the *high luminosity mode*. The high resolution mode will provide a beam momentum precision up to $5 \cdot 10^{-5}$ thanks to stochastic and electronic cooling. The luminosity will be $\approx 10^{31} cm^{-2} s^{-1}$. The high luminosity mode offers a 20 times higher luminosity, $\approx 2 \cdot 10^{32} cm^{-2} s^{-1}$ and a beam momentum resolution of 10^{-4} from stochastic cooling. During the first few years of PANDA operation, the so-called *Modularized Start Version* (MSV) will be available. It will provide a luminosity of $\approx 10^{32} cm^{-2} s^{-1}$ and a beam momentum resolution of 10^{-4}. The target will be either of cluster-jet type or frozen hydrogen pellet type for the $\bar{p}p$ annihilation programme. For $\bar{p}A$ studies, target foils will be used. PANDA will cover a solid angle of almost 4π and consist of a *Target Spectrometer* (TS) and a *Forward Spectrometer* (FS). The TS will detect particles emitted at polar angles larger than 10° in the horisontal plane and larger than 5° in the vertical plane. The MicroVertex Detector (MVD) will surround the target region and provide precise vertex position measurements. The vertex space resolution will be of the order 50 μm in x and y and 100 μm in z. Gas Electro Multiplier (GEM) planes and a Straw Tube Tracker (STT) will provide precise tracking ($\Delta p_T / p_T$ = 1.2% when combined with the MVD) of charged particles and a 2T solenoid magnet. A barrel Detection of Internally Reflected Cherenkov Light (DIRC) and a Time-of-Flight (TOF) detector will discriminate between pions, kaons and protons and an Electro Magnetic Calorimeter (EMC) equipped with 17200 $PbWO_4$ crystals will provide state-of-the-art photon reconstruction and separation between electrons and pions. Muon detectors surrounding the solenoid magnet will provide identification of muons. The FS will cover angles below 10° in the horizontal plane and below 5° in the vertical plane. It will comprise a Forward Tracking System (FTS) with three pairs of straw tube planes. A dipole magnet bend the charged particle trajectories. Furthermore, an Aerogel Ring Imaging Cherenkov Counter (RICH) and a forward time-of-flight (FTOF) will enable particle identification and a Forward ElectroMagnetic Calorimeter (FEMC) photon reconstruction and electron/pion separation. Also in the FS there will be a muon detector. The PANDA experiment will utilize a new, time-based data acquisition that reads out the data continously from the stream. This put high demands on the readout techniques and the foreseen software trigger that will be used for this purpose, but it also opens up new possibilities regarding readout rates and data storage. The PANDA detector, shown in Figure 2, is described in detail in Ref. [3].

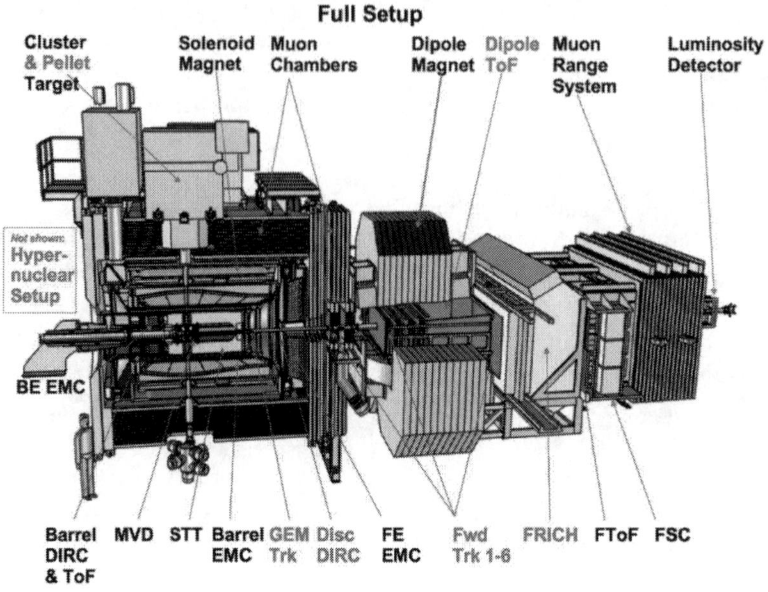

Figure 1. View of the future PANDA detector.

3 Nucleon structure

The proton was discovered almost 100 years ago and together with the neutron, it constitutes the major part of the visible mass of the universe. However, a complete understanding of its structure is yet to be found. As mentioned before, only 1% of its mass and 1/3 of its spin can be explained by the simple quark model. The quark structure is revealed by reactions where absorbtion, radiation or exchange of real or virtual photons. Such processes dominate in reactions involving charged leptons in the inital and/or final state. A multitude of experimental efforts devoted to study different aspects of nucleon structure are ongoing world-wide. The majority are these (*e.g.* at COMPASS, HERMES, JLAB) utilize electromagnetic probes and thereby they explore the space-like (SL) region of the nucleon structure observables. PANDA, on the other hand, will utilize $\overline{p}p$ annihilations which means that time-like (TL, positive momentum transfer q^2) observables will be explored: i) Generalized Distribution Amplitudes (GDAs), which are accessible by wide angle Compton scattering in hard exclusive processes like *e.g.* $\overline{p}p \to \gamma^*\gamma$ and $\overline{p}p \to \gamma^*\pi^0$, ii) Transistion Distribution Amplitudes (TDAs) from *e.g.* $\overline{p}p \to \gamma^* M \to e^+e^- M$ or $\overline{p}p \to \gamma^* M \to J/\Psi M$ where $M = \pi^0, \eta, \omega, \rho^0 ...$ [4], iii) Transverse Momentum distributions (TMDs) in Drell-Yan processes, *i.e* $\overline{p}p \to \gamma^* X \to \mu^+\mu^- X$ and iv) Electro Magnetic Form Factors (EMFFs) in $\overline{p}p \to \gamma^* \to l^+l^-$ and $\overline{p}p \to \gamma^*\pi^0 \to l^+l^-\pi^0$, where $l = e, \mu$.

In the following we will discuss the proton EMFFs. For protons, the differential cross section of the $\overline{p}p \to\to l^+l^-, l = e, \mu$ process can be parameterized in terms of the electric G_E and the magnetic G_M form factors assuming one-photon exchange (see for example [5]). G_E and G_M are complex function of the momentum transfer squared q^2 and the moduli $|G_E|$ and $|G_M|$ can be extracted from the total and differential cross sections. Questions that are addressed in nucleon EMFF studies are in general: i) How does the ratio R = G_E/G_M depend on q^2? Existing data from PS170 [6] differ from BaBar [7] and BES III [8] at low q^2. ii) Do the EMFFs agree with the predictions from pQCD in the high q^2 limit? iii) Down to which q^2 is the pQCD picture valid? iv) Do TL and SL EMFFs agree in the high q^2 limit, as predicted by pQCD and analyticity [9]? Existing high q^2 data from BaBar [7] indicates a difference in the TL and SL magnetic FFs by a factor of 2.

 DOI: 10.1051/epjconf/201612501006

Recently, simulation studies of the $\overline{p}p \to e^+e^-$ reaction with PANDA have been carried out [10]. When the high luminosity mode of the HESR is available, the EMFFs can be determined with an unprecedented precision within $5.4 < q^2 < 13.9$ (GeV$/c)^2$). Furthermore, the so-called effective form factor (see e.g. [7]) can be measured up to $q^2 < 29(\mathrm{GeV}/c)^2$. The very large expected background from *e.g.* $\overline{p}p \to \pi^+\pi^-$ could be suppressed by a factor $\sim 10^8$. Another on-going simulation study show good prospects also for form factor measurement using the $\overline{p}p \to \mu^+\mu^-$ reaction [11].

Since the form factors G_E and G_M are complex in the TL region, they have a relative phase. The phase introduces polarization effects [12]. It is presently being investigated whether an upgraded version of PANDA could comprise a polarized target. This would enable the very first measurement of the phase of the proton EMFFs.

4 Hyperon dynamics

The key question in hyperon physics is "What happens if you replace one (or several) light quark(s) in the nucleon with one (or several) heavier one(s)?" The strange quark is sufficiently light to relate the knowledge about hyperons to nucleons and vice versa. In other words, hyperon physics provide a new angle to increase our understanding of the structure of the nucleon. At the same time the strange quark is heavy enough to be traced in the reactions. In that way one can obtain new information about the structure of baryons which is not accessible in the pure light (u and d) sector. The clearest situation is achieved in exclusive hyperon-antihyperon production.

The scale probed in such reactions is governed by the strange-quark mass, i.e. $m_s \approx 100$ MeV. Since this corresponds to the confinement scale, the relevant degrees of freedom are unclear — quarks and gluons, or hadrons? The production of strange hyperons therefore probes QCD in the intermediate region between the perturbative and non-perturbative regime. Besides effective field theories, which can account for the universal aspects of the long-range forces, phenomenological models are needed, which require guidance by experimental data.

In particular, spin observables have been proven decisive in testing different models for polarizations and spin correlations [13–15]. Spin observables, *e.g.* the polarization, are straight-forward to measure for ground-state hyperons thanks to their weak, self-analyzing decays. This means that the decay products are preferentially emitted along the direction of spin of the parent hadron. In the case of $\Lambda \to p\pi^-$, the angular distribution of the proton is related to the Λ polarization by $I(\cos\theta_p) = \frac{1}{4\pi}(1 + \alpha_\Lambda P_n \cos\theta_p)$, where $\alpha_\Lambda = 0.64$ [2] is the asymmetry parameter. In production of spin 1/2 hyperons Y in $\overline{p}p \to \overline{Y}Y$ with unpolarized beam and target, the polarization of the hyperon and the anti-hyperon and the spin correlation between them are accessible.

Some hyperons decay into other hyperons, *e.g.* the Ξ. In the $\Xi^- \to \Lambda\pi^-, \Lambda \to p\pi^-$ process, the additional asymmetry parameters β and γ of the Ξ^- hyperon are accessible *via* the joint angular distribution of the Λ hyperons and the protons [16].

For spin $\frac{3}{2}$ hyperons, *e.g.* the Ω^-, the spin structure is more complicated. Here, we consider the polarization parameters of individual spin $\frac{3}{2}$ hyperons only. Spin 3/2 hyperons produced in strong processes like $\overline{p}p \to \overline{\Omega^+}\Omega^-$ have seven non-zero polarization parameters. Three of these can be extracted from the angular distribution of the decay baryon, *e.g.* $\Omega^- \to \Lambda K^-$. The remaining four can be obtained by studying the joint angular distribution $I(\theta_\Lambda, \phi_\Lambda, \theta_p, \phi_p)$ of the Λ hyperons from the Ω^- decay and the protons from the subsequent Λ decay.

The PS185 collaboration have provided a large set of high-quality data on single-strange hyperons [17], which triggered various theoretical investigations [13–15]. The ambition is to survey the spin structure of $\overline{p}p \to \overline{Y}Y$ for all ground-state multi-strange and single-charmed hyperons.

Table 1. Results from simulation studies of the various production reactions of ground state hyperons as well as the excited $\Xi^{-*}(1820)$. The efficiencies are exclusive, *i.e.* all final state particles are reconstructed.

$p_{\bar{p}}$ (GeV/c)	Reaction	σ (μb)	Eff (%)	Decay	Rate at 10^{31}cm^{-2}s^{-1}
1.64	$\bar{p}p \to \bar{\Lambda}\Lambda$	64	10	$\Lambda \to p\pi^-$	28 s^{-1}
4	$\bar{p}p \to \bar{\Lambda}\Sigma^0$	≈40	30	$\Sigma^0 \to \Lambda\gamma$	30 s^{-1}
4	$\bar{p}p \to \bar{\Xi}^+\Xi^-$	≈2	20	$\Xi^- \to \Lambda\pi^-$	2 s^{-1}
12	$\bar{p}p \to \bar{\Omega}^+\Omega^-$	≈0.002*	≈30	$\Omega \to \Lambda K^-$	≈4 h^{-1}
12	$\bar{p}p \to \bar{\Lambda}_c\Lambda_c^+$	≈0.1*	≈30	$\Lambda_c \to \Lambda\pi^+$	≈2 d^{-1}
4.6	$\bar{p}p \to \bar{\Xi}^+\Xi^{-*}(1820)$	1	5	$\Xi^{-*} \to \Lambda K^-, \Lambda \to p\pi^-$	15000 d^{-1}

Simulation studies of exclusive hyperon production have been performed using a simplified Monte Carlo framework [18–20]. The estimated signal rates are given in Table 1. They are calculated assuming the Day One luminosity of 10^{31}cm^{-2}s^{-1}. The only well known cross section is $\bar{p}p \to \bar{\Lambda}\Lambda$ at 1.64 GeV/c; the other single- and double-strange cross sections are obtained by extrapolating data collected at other energies. For Ω and Λ_c production in $\bar{p}p$ annihilations, no data exist. The quoted values for $\sigma(\bar{p}p \to \bar{\Omega}^+\Omega^-)$ and $\sigma(\bar{p}p \to \bar{\Lambda}_c\Lambda_c^+)$ are theoretical predictions [21]. From the simulations we conclude that already with the Day One setup, the spin observables of multi-strange baryons can be thoroughly explored for the first time. The full setup will extend these studies to charmed hyperons. As the FAIR facility matures, studies of CP violation in baryon decays will become possible, for instance in double-strange decays such as $\Xi^- \to \Lambda\pi^-$ [19].

5 Baryon spectroscopy

Hadron spectroscopy provides another laboratory in which to explore the dynamics of strongly interacting quarks and gluons confined into hadrons. In baryon spectroscopy, the focus lies on multi-strange baryons. In light and strange baryon spectroscopy we search for answers the following questions: i) to which extent do the excitation spectra of baryons consisting of *u, d, s* follow the systematics of SU(3) flavour symmetry? ii) which degrees of freedom are relevant for the excitation modes of baryons? iii) how important is the dynamics in baryon-meson systems? iv) are there exotic baryon states, *e.g.* pentaquarks or dibaryons? Baryon spectroscopy has been decisive in the development of our understanding of the microscopic world, the best example being the formulation of the quark model. In the dawn of particle physics during 1950's and 1960's, the concept of quarks was not yet known. Thanks to extensive experiments, a plethora of new particles were observed and it soon became clear that they could not all be elementary. Many were found to have a finite life-time and were classified as *strange*. The new-found particles could be organized according to "the eight-fold way" or SU(3) symmetry and as a result, the Ω^- hyperon was predicted. The discovery of the Ω^- baryon [22] demonstrated the success of the eight-fold way and shortly after, the quark model was formulated by Gell-Mann and Zweig [23]. The quark model, where three equivalent valence quarks interacting flavour-independently and color-magnetically with one-gluon exchange, has been proven successful in classifying ground-state baryons and describing some of their static properties. The intense worldwide effort in baryon spectroscopy has been focused on N^* and Δ resonances. Most of the known baryon resonances have masses smaller than 2 GeV/c^2 and were discovered in elastic πN scattering experiments. In recent years, many laboratories worldwide (JLab, ELSA, MAMI, GRAAL, Spring-8 etc) have studied these resonances in electromagnetically induced reactions ([24, 25]). The observed spectra show many features that cannot be explained within the quark model. For example, the predicted so-called *level ordering* of baryonic states with positive and negative parity disagree with the experimental findings. Another problem found by the experiments is that of *missing resonances*: predicted baryonic states that haven't been observed by experiment. One possible explanation is that the states haven't been found because they couple weakly to πN and γN. They would as a consequence be unlikely to be produced in pion- or photoproduction. Searches using different probes, *e.g.* anti-protons, could shed light on this puzzle.

Another explanation of the missing states is that they are simply not there, which in turn would indicate that our understanding of baryon structure is incomplete. It would be very illuminating to see whether the features of the light baryon spectra are carried over to the strange sector, *i.e.* the excited single-, double- and triple-strange hyperon spectra. That would enable a systematic comparison of systems containing different strangeness. The existing data bank is however scarce, in particular regarding double- and triple-strange hyperons. Only one excited Ξ state and zero excited Ω states are considered well established with the PDG classification system. Ongoing efforts at JPARC and JLAB could open up for studies of light double strange hyperons, *i.e.* Ξ^*. Excited baryons can also be studied in charmonium och bottom baryon decays, but the detectors are not optimized for exclusive multi-strange hyperon studies. Thus, the prospects with currently running facilities for heavier double- and triple-strange remain scarce. This is in contrast to the spectroscopy of bottom (b) and charmed (c) baryons is a very active field of research today thanks to numerous B-factories like BaBar, BELLE and LHCb. To summarize, current experimental facilities worldwide study either light baryons or charmed and bottom hyperons while leaving a gap in the strange sector.

In order to investigate the prospect of PANDA being able to fill this gap, a simulation study of the $\overline{p}p \to \overline{\Xi}^{+}\Xi^{-*}(1820) + c.c.$ has been carried out [26]. The $\Xi^{-*}(1820) \to \Lambda K^{-}$ was considered and a BR of 100% was assumed. For a cross section of $\approx 1\ \mu b$ and the Day One luminosity $10^{31}\mathrm{cm}^{-2}\mathrm{s}^{-1}$, an exclusive detection rate of $\approx$ 15000 events per day is expected, as shown in Table 1. This demonstrates that PANDA indeed has the potential of filling the gap in the strange sector from Day One.

6 Charmonium spectroscopy

A long-standing question in hadron physics concerns the existence of so-called exotics, *i.e.* states which are inconsistent with the meson ($q\overline{q}$) or baryon (qqq) picture. QCD allows for *e.g.* multiquarks ($qq\overline{q}\overline{q}$ or $qqqq\overline{q}$), glueballs (gg or ggg) or hybrids ($q\overline{q}g$) but until recently, the evidence have been scarce. In the past, many efforts were made in the light sector (for reviews see for example Refs. [27, 28]) where unambiguous identification is often difficult due to the states being broad, largely overlapping and mixing. The charm sector has proven more straight-forward in the sense that the observed states are relatively narrow and well separated. Recently, the discovery of the first charged charmonium state $Z_c^{\pm}(3900)$ by BES III [29] and and confirmed by Belle [30] implied a major breakthrough and led to an increased activity in the field. Another interesting example is the $J^{PC} = 1^{++}$ state $X(3872)$, that have puzzled the scientific community for more than a decade [31, 32]. The nature of the $X(3872)$ is disputed: is it a multiquark or a $D\overline{D}^*$ molecule, or maybe an more conventioal $\chi_{c1}(2P)$ candidate, with some $D\overline{D}^*$ threshold features? In order to solve this puzzle, a precision measurement of its width and line-shape is required. However, most running or planned facilities have clear disadvantages that limit their scopes. Most of them are either i) e^+e^- colliders (BES III and Belle II), where direct formation of any state that is not $J^{PC} = 1^{--}$ is suppressed, or ii) high energy pp experiments (LHCb), where any charmonium-like state needs to be formed in production together with several tagged or untagged recoils. In both cases, one relies on the detector resolution when determining the properties of the $X(3872)$. With the PANDA experiment, the antiproton beam enables direct formation of any $q\overline{q}$-like state X in $\overline{p}p \to X$. The width and the line-shape of any such state can therefore be measured in an energy scan where the precision is only limited by the precision of the beam momentum, which will be around 10^{-4} with the high luminosity mode and $5 \cdot 10^{-5}$ with the high precision mode. A recent simulation study [33] show that the width can be measured with a sensitivity of $\Gamma/\Delta\Gamma > 5$ for widths between 90 and 120 keV. Furthermore, a simulation of the molecular line-shape show that it will be possible to distinguish between a virtual and a bound state. These measurements can be carried out also using the MSV with 80 days of data taking.

7 Other physics topics

The scope of PANDA is very broad thanks to the antiproton beam and the versatile detector. Important physics topics that will be studied, but were not discussed here, are for instance open charm physics, hypernuclear physics and hadrons in nuclear matter.

8 Summary

PANDA will be a unique experiment with an excellent potential to shed light on many of the most important puzzles of contemporary physics. In the following, high-lights of the scope are outlined as a function of time.

Day One

Already during the start-up year of PANDA, with a reduced HESR luminosity ($10^{31}\text{cm}^{-2}\text{s}^{-1}$) and a staged version of the detector, PANDA can provide unprecedented precision of the $X(3872)$ width and a first measurement of the $X(3872)$ line-shape. Furthermore, PANDA can collect uprecedented data samples on spin observables of single-strange hyperons and the first measurement ever of double-strange spin observables in $\bar{p}p$ annihilations. PANDA can access the full spectrum of single- and double-strange excited hyperons with a unique statistical precision.

First Years

During the following years, the experimental situation will be the same as the first year but since more data will be ackumulated over time, the physics scope widens. PANDA will for the first time measure the seven accessible polarization parameters of the Ω^- hyperon and provide full triple-strange excited hyperon spectrum. PANDA will further explore the precision frontier of charmonium spectroscopy. In addition, the modular, versatile design of PANDA opens up for unique contributions to hypernuclear physics.

Long-term plans

The design luminosity of the HESR ($2{\cdot}10^{32}\text{cm}^{-2}\text{s}^{-1}$) and the complete PANDA detector setup will enable high precision studies of nucleon structure, *e.g.* proton EMFFs. Furthermore, open charm can be studied with world-leading statistical precision. The expected unprecedented hyperon samples will enable searches for CP violation in baryon decays.

References

[1] Fritzsch H *et al.*, Phys. Lett. **B47** 365 (1971); Politzer H D, Phys. Rev. Lett. **30** 1346 (1973); Gross D J and Wilczek F, Phys. Rev. Lett. **30** 1343 (1973); Weinberg S, Phys. Rev. Lett. **31** 494 (1973).

[2] Olive K A *et al.* (Particle Data Group) Chin. Phys. C**38** 090001(2014).

[3] The PANDA collaboration, Technical Progress Report (2005).

[4] The PANDA collaboration, Eur. Phys. J. A **51** 107 (2015).

[5] Zichichi A *et al.*, Nuovo Cim. **24** 170 (1962).

[6] Bardin G *et al.* Nucl. Phys. B **411** 3 (1994).
[7] Lees J P *et al.* Phys. Rev. D **88** 7:072009 (2013).
[8] Ablikim M *et al.* Phys. Rev. D **91** 11:112004 (2015).
[9] S. Pacetti *et al.* Phys. Rep. **550-551** 1 (2015).
[10] The PANDA collaboration, *"Feasibility studies of time-like proton electromagnetic form factors at PANDA at FAIR"* submitted 2016 (arXiv:1606.01118 hep-ex).
[11] Zimmermann I, Talk at the ECT* workshop *Probing transverse nucleon structure at high momentum transfer* (2016)
[12] Dubnickova A *et al.* Nuovo Cim. **109** 241 (1996).
[13] Kohno M and Weise W, Phys. Lett. B **179** 15 (1986); Rubinstein H R and Snellman H Phys. Lett. B **165** 187 (1985); Furui S and Faessler A, Nucl. Phys. A **486** 669 (1987); Burkardt M and Dillig M Phys. Rev. C **37** 1362 (1988); Alberg M A *et al.* Z. Phys. A **331** 207 (1988).
[14] Tabakin F and Eisenstein R A Phys. Rev. C **31** 1857 (1985); Kohno M and Weise W, Phys. Lett. B. **179** 15 (1985); La France P *et al.* Phys. Lett. B **214** 317 (1988); Timmermans R G E *et al.* 1992 Phys. Rev. D **45** 2288 (1992); Haidenbauer J *et al.* Phys. Rev. C **46** 2516 (1992).
[15] Ortega P G *et al.* Phys. Lett. B. **696** 352 (2011).
[16] Koch W *Analysis of scattering and decay* ed. Nikolic M (New York-London-Paris:Gordon and Breach, 1968).
[17] Johansson T 2003, proceedings of *8th Int. Conf. on Low Energy Antiproton Physics* 95 (2003).
[18] The PANDA collaboration, Physics Performance Report (2009).
[19] Thomé E, *"Multi-Strange and Charmed Antihyperon-Hyperon Physics for PANDA"* (Ph. D. Thesis, Uppsala University, 2012).
[20] Grape S, *"Studies of PWO Crystals and Simulations of the* $\bar{p}p \to \bar{\Lambda}\Lambda, \bar{\Lambda}\Sigma^0$ *Reactions for the PANDA experiment"* (Ph.D. Thesis, Uppsala University, 2009).
[21] Kaidalov A B and Volkovitsky P E Z, Phys. C **63** 51 (1994); Titov A I and Kämpfer B, Phys. Rev. C **78** 025201 (2008); Gornitschnig A T *et al.* Eur. Phys. J. A **42** 43 (2009); He J *et al.* Phys. Rev. D **84** 114010 (2011); Khodjamirian A *et al.* Eur. Phys. J. A **48** 31 (2012).
[22] Barnes V E *et al.*, Phys. Rev. Lett. **12** 204 (1964).
[23] Gell-Mann M, Phys. Lett. **8** 214 (1964); Zweig G, CERN-preprint CERN-TH-401 402 (1964).
[24] Klempt E and Richard J M, Rep. Prog. Phys. **82** 1095 (2010).
[25] Crede V and Roberts W, Rep. Prog. Phys. **76** 076301 (2013).
[26] Pütz J, Proceedings of FAIRNESS 2016, Garmisch-Partenkirchen, Germany (2016).
[27] Crede V and Meyer C A, Prog. Part. Nucl. Phys. **63** 74 (2009).
[28] Meyer C A and Swanson E, Prog. Part. Nucl. Phys. **82** 21 (2015).
[29] Ablikim M *et al.*, Phys. Rev. Lett. **110** 252001 (2013).
[30] Liu Z Q *et al.* Phys. Rev. Lett. **110** 252002 (2013).
[31] Abe K *et al.* Phys. Rev. Lett. **91** 262001 (2003).
[32] Aaij *et al.* Phys. Rev. Lett. **110** 222001 (2013).
[33] Nerling F, talk at the *11th International Workshop on Heavy Quarkonium* (2016)

Multiplicity and theremalization time in heavy-ions collisions

Irina Aref'eva[1,*]

[1] *Steklov Mathematical Institute, Russian Academy of Sciences,
Gubkina str. 8, 119991, Moscow, Russia*

Abstract. We present a concise review of quark-gluon plasma formation in heavy-ions collisions within the holographic approach. In particular, we discuss how to get the total multiplicity in heavy ions collision to fit the recent experimental data. We also discuss theoretical estimations of time formation of QGP in heavy ions collision and show that different observables can give the different times of QGP formation.

1 Introduction

It is my pleasure to give a talk here. I would like to thanks the organizers for invitation.

The subject of my talk is holography description of quark-gluon plasma (QGP) formation in heavy-ions collisions (HIC). The plan of the talk is the following.

- I'll start from the physical picture of formation of QGP in HIC.
- Then I'll remind the raisons to use dual approach to physical problems related with QGP. As to holographic approach to QGP there are two options:
  - so called the top-down approach,
  - as well as the bottom-up approach.

  The meaning is obvious from the names. The top theory is of course, the string theory :). There is considerable progress in applications of holography to static properties of QGP [1]
- The main part of the talk will be devoted to discussion of the holographic description of the process of formation of QGP. The goal will be to fit experimental date related with multiplicity and thermalization time in heavy ions collisions. By experimental data I mean data from RHIC and more recent data from LHC. One can find the holographic description of QGP formation in [2, 3].

1.1 Physical picture of QGP formation in heavy ions collisions

Let me start from the physical picture of QGP formation in heavy ions collisions.

In QCD we have quark confinement and asymptotic freedom. It is known (mainly due to lattice calculations) that if the temperature T increases, or density nuclear matter increases, then the nuclear matter undergoes to a phase transition to the deconfined phase. The quark gluon plasma is a macroscopic matter formed from the quarks and gluons at high temperature.

*e-mail: arefeva@mi.ras.ru

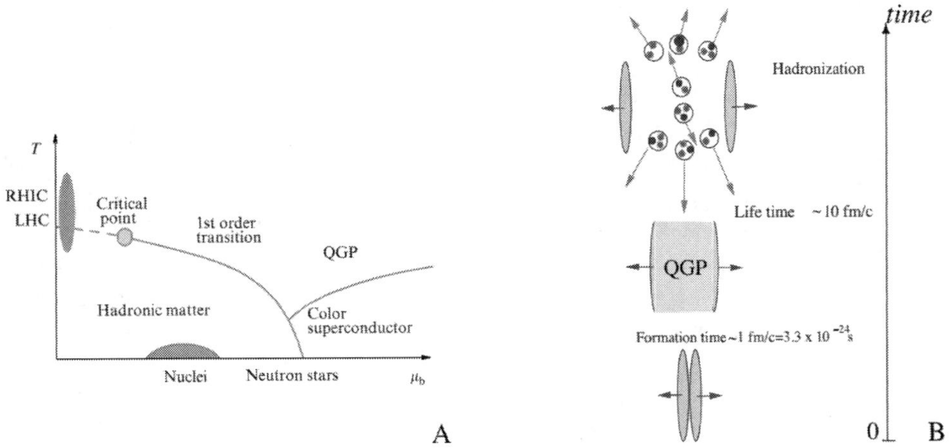

Figure 1. (color on line) A. The QCD phase diagram. B. The schematic picture of heavy ions collision

The phase diagram of QCD is not well known either experimentally or theoretically. A commonly conjectured form of the phase diagram, temperature T vs quark chemical potential μ, is shown in Fig.1.A. The chemical potential μ is a measure of the imbalance between quarks and antiquarks in the system. The phase transition is not sharp and it is supposed to be the 1-st order.

Ordinary nuclear matter in this diagram is at μ = 310 MeV and T close to zero. If we increase the quark density, i.e. increase μ, keeping the temperature low, we go into a phase of more and more compressed nuclear matter neutron stars. Above the (blue on the on-line version of the paper) smeared line there is a transition to the quark-gluon plasma. At ultra-high densities one expects to find the phase of color-superconducting quark matter. In ultra-relativistic heavy ion collisions one studies this matter in the regime of extreme energy density. In the figure the typical values of μ and T in heavy-ion collisions are shown by a filled (blue) region near the T-axis.

Heavy-ions collisions are studied in several experiments. Below I list them with indication of the energy per nucleon.

- At the Brookhaven Alternating Gradient Synchrotron (AGS), started in the 1990's, the the energy per one nucleon is $\sqrt{s_{NN}} = 4.75GeV$,
- At the CERN Super Proton Synchrotron (SPS), $\sqrt{s_{NN}} = 17.2GeV$.
- At the Brookhaven Relativistic Heavy-Ion Collider (RHIC), $\sqrt{s_{NN}} = 200GeV$
- At the LHC collider at CERN, $\sqrt{s_{NN}} = 2.76TeV$ (2011), $\sqrt{s_{NN}} = 5.02TeV$ (2015).

At RHIC the ions of Au are collided and at LHC the ions of Pb are collided. Schematically the picture of heavy ion collisions is presented in Fig.1.B.

There are strong experimental evidences that RHIC or LHC have created some medium which behaves collectively. They are the following

- modification of particle spectra (compared to p+p)

- high p_T -suppression of hadrons
- jet quenching
- elliptic flow
- suppression of quarkonium production and etc.

Study of this medium is related with one of the fundamental questions in physics. This problem is what happens to matter at extreme densities and temperatures as may have existed in the first microseconds 10^{-5}, $T \sim 10^{12}$ K after the Big Bang. One can say that the aim of heavy-ion physics is to collide nuclei at very high energies and create such a state of matter in the lab (Little Big Bang).

As to the theory, there are macroscopic and microscopic approaches to description of QGP formation in HIC. Macroscopic approaches include thermodynamics, hydrodynamics, kinetic theory, etc. The microscopic approach is suppose to be based on QCD, or just on some effective theories. The list of macroscopic approaches is started from the names of Fermi, Landau and Pomeranchuk, who proposed in 50-s the hydrodynamical model to high energy collisions of protons [21–23].

1.2 Holographic approach to static QGP

In the last more than 10 years an other approach has been developed. This approach is based on *holography*, and it is also known as *AdS/CFT* approach. An application to this approach to formation of QGP in HIC is the subject of the present talk.

A remarkable conclusion from the RHIC and LHC experiments is that the QGP does not behave as a weakly coupled gas of quarks and gluons, but rather as a strongly coupled fluid. This makes perturbative methods inapplicable. The lattice formulation of QCD does not work, since we have to study real-time phenomena. This has provided a strong motivation for study the dynamics of strongly coupled quark gluon plasma through the gauge/string duality.

The Gauge/Gravity duality gives an correspondence between the 4-dim physical space where the gauge theory lives and the 5-dimensional space where the supergravity (weak curvature) approximation of the 10-dimensional string theory is valid. Or in others words, the properties of the gauge theory in (physical) Minkowski space in 3+1 dimensions are in one-to-one relation with properties of the bulk 4+1 dimensional theory.

The best known example of such theories is N = 4 super Yang-Mills, a superconformal field theory with matter in the adjoint representation of the gauge group $SU(N_c)$. However, there is not yet exist a gravity dual construction for QCD. Differences between N = 4 SYM and QCD are less significant, when quarks and gluons are in the deconfined phase. Lattice calculations show that QCD exhibits a quasi-conformal behavior at temperatures T >300 MeV and the equation of state can be approximated by e = 3 p (a traceless conformal energy-momentum tensor), see Fig.2.A. where the comparison of the calculations carried out using the highly improved staggered quark (HISQ) and the stout-smeared staggered action (stout) for the trace anomaly, the pressure, and the entropy density are presented. These observations strongly support to use the AdS/CFT correspondence as a tool to get non-perturbative dynamics of quark gluon plasma. The AdS/CFT or holography states the relation between generation functions for correlators of QFT in the 4-dimensional Minkowski space-time and classical solutions in the 5-dimensional gravity [5], see Fig.2.B. In some sense the duality claims a kind of auto-model behaviour of strong correlator systems. There is an similarity with the auto-model behaviour of the parton model [6].

There is the considerable success in description of the static QGP. The main results are collected in the book [1].

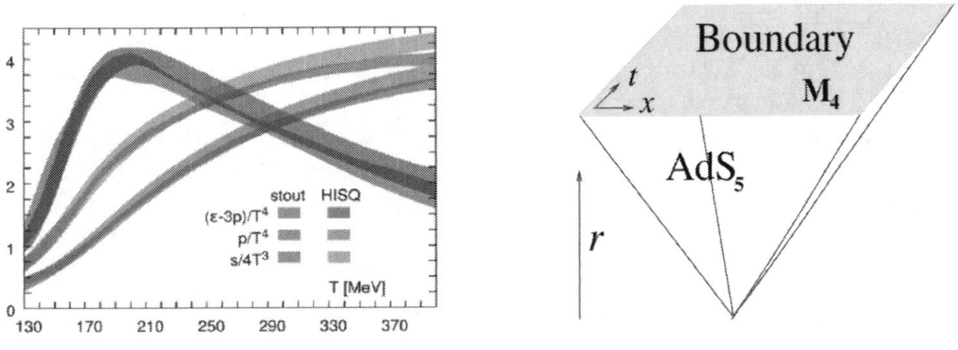

A B

Figure 2. (color on line) A. The QCD equation of state from the lattice. The figure taken from Ref.[4]. B. The schematic picture for holographic calculations ($r = 1/z$ in (1))

1.3 Holographic approach to QGP formation in heavy ions collisions

I am going to discuss the formation of quark gluon plasma within AdS CFT setup. The main conjectures are the following:

- thermal d-dimensional QFT can be described by d+1 background with a black hole
- thermalization of 4-dimensional QFT, i.e. formation of 4-dimensional quark gluon plasma, corresponds to a black hole formation in AdS_5.

Let me now say few words about black holes formation in AdS. To initiate the process of BH formation one has to perturb the initial AdS metric. According to AdS/CFT correspondence this deformation should be related with the deformation of the energy-momentum tensor of the boundary theory. So the idea is to make some perturbation of AdS metric that near the boundary mimics the heavy ions collisions and see what happens.

One can consider several deformations of AdS metric:

- a deformation by adding colliding gravitational shock waves.
- we can drop of a shell of matter with vanishing rest mass ("null dust"),
- we can study the toy 3-dim model with colliding ultra relativistic particles

It seems natural to think about an ultrarelativistic nucleus as a shock wave in 4-dimensional space-time. One can also assume that this shock has a profile function. According AdS CFT correspondence to this energy-momentum tensor corresponds the metric of a shock wave in AdS$_5$ of the following form

$$ds^2 = L^2 \frac{-du\,dv + dx_\perp^2 + \phi(x_\perp, z, L)\delta(u)\,du^2 + dz^2}{z^2}, \tag{1}$$

here u, v are light-cone variables, $u = x_0 + x_\parallel$, $v = x_0 - x_\parallel$, $\phi(x_\perp, z, L)$ is the profile function, L is the scale parameter. For $\phi(x_\perp, z, L) = 0$ the metric (1) is nothing but the Poincare metric for AdS$_5$. For two colliding nuclei corresponds the following metric in AdS$_5$

$$ds^2 = L^2 \frac{-du\,dv + dx_\perp^2 + \phi_1(x_\perp, z, L)\delta(u)\,du^2 + \phi_2(x_\perp, z, L)\delta(v)\,dv^2 + dz^2}{z^2}. \tag{2}$$

DOI: 10.1051/epjconf/201612501007

It is interesting to note that gaussian regularization of the point shock wave profile function $\phi(x_\perp, 0)$ exactly corresponds to the Woods-Saxon profile for the nuclear density. This correspondence give us L equal to 4.3 fermi for gold and L equal to 4.4 fermi for lead.

It is natural to ask the question: can we guaranty the black hole formation under collisions of these shock waves. If the answer is "yes", then I'll take the entropy of this formed black hole as a multiplicity of particles production. This idea has been explored in several papers [7–16]. One can also study the problem numerically and a lot of efforts are performed in this direction [17, 18].

2 Analytical estimation of the entropy of the BH produced in collision of two schock waves

I am going to speak about what can be done analytically. As I have already mentioned in Introduction the main goal is to estimate multiplicity and thermalization time.

2.1 Multiplicity

The lead-lead collisions show a power form of the multiplicity dependence on the energy, and the power is about 0.15, [19, 20]

$$\mathcal{M} \sim s^{0.15}, \tag{3}$$

here s is the Mandelstam variable, that is equal to the square of the energy in the central mass frame, see Fig.3.

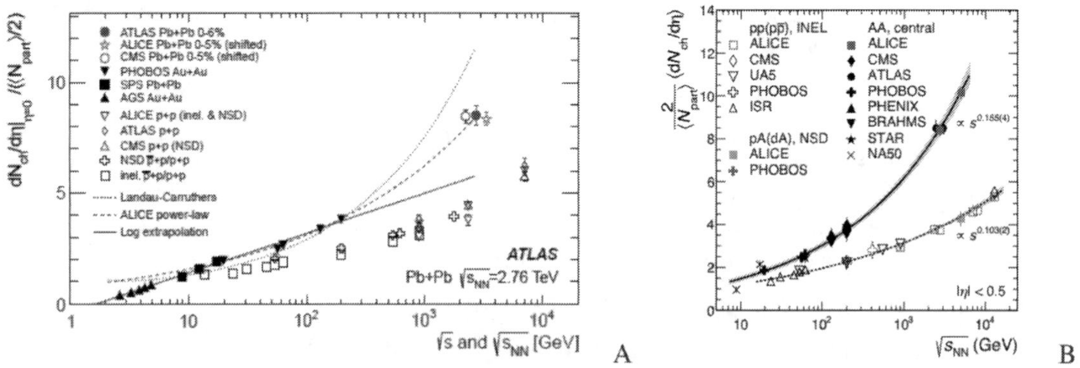

Figure 3. Plots of the total number of charged particles versus energy $\sqrt{s}$ for protons and $\sqrt{s_{NN}}$ per pair of nucleons for ions collisions. A. Plot from ATLAS collaboration [19]. B. Plot from ALICE collaboration [20].

The idea to relate multiplicities with the thermodynamic characteristic of the media produced as a result of a collision of particles comes back to Landau and Fermi [21–24].

The idea to estimate multiplicity by the entropy of BH produced in 5-dim under perturbation of the initial AdS have proposed in [7] and elaborated in [9, 11–14, 16].

The entropy is assumed to be estimated by the area of the trapped surface that forms in collision of two shock waves if the energy of colliding shocks is above a threshold. This estimation is related with the Penrose theorem that claims that black hole formation is equivalent to a trapped surface formation and the area of horizon of the formed black hole can be estimated by the trapped surface area. It worth mention that the Penrose theorem deals with asymptotically flat spaces. But we apply here this theorem to the case of the AdS space-time. This point requires more treatments.

Now let me tell what we get in the holographic estimations of the multiplicity. The simplest holographic model gives [7, 9, 13]

$$\mathcal{M} \sim s^{1/3} \tag{4}$$

This is faraway from the experimental results [19, 20], the power is twice more. The natural idea arises to try to modify the holographic model. The simplest option is to modify the background. We can do this in several ways:

- removed parts of AdS assuming some cuts [7]
- assuming existence of some others fields and change the vacuum solution from AdS to some modification. The simplest possibility is to deal with dilaton-gravity [16, 26, 27].

The more prominent dilation-gravity model is the so called improved holographic QCD [25]. There exist the dilation potential such that the model reproduces 2-loop beta function as well as glueball spectrum. The shock waves collisions in this background have been calculated in [16]. They have improved a little the power (they also got the log corrections), but still the power is not as shown in Fig.3. In [26] it has been show that one can get the required power around $s^{0.15}$, but in this case one has to assume that the dilaton is in fact a phantom. In [27] we have shown that one can get $\mathcal{M} \sim s^{0.15}$ for s bounded from above.

Since in all these considerations more or less general backgrounds reproducing AdS at UV and confinement at IR have been used, one can think that just general assumptions about the background prevent to reproduce the suitable behavior in UV and IR and in the same time give the correct energy dependence of multiplicity at high energy. In particular, we can think that a default assumption of the isotropic form of background metrics is responsible for this discrepancy and for a more realistic description of the entropy production in holographic models one has to consider anisotropic backgrounds.

2.2 Anisotropic holographic backgrounds

In favor of anisotropic holographic backgrounds there are also additional arguments. Up to couple years ego, it was believed that just after heavy-ion collisions, a pre-equilibrium period exists for up to 1 fm/c and then the QGP appears and this QGP is isotropic. However now there is a belief that the QGP is created after a very short time after the collision, $\tau_{therm} \sim 0.1 fm/c$, and it is anisotropic ("anisotropic" means to a spatially anisotropy) for a short time τ after the collision, $0 < \tau_{therm} < \tau < \tau_{iso}$, and the time of locally isotropization is about $\tau_{iso} \sim 2fm/c$.

In the holographic version of this setup it is suitable to consider a black hole formation in a spatially anisotropic background. We consider one possible version of the anisotropic background, namely the so-called Lifshitz-type background

$$ds^2 = L^2\left[\frac{\left(-dt^2 + dx^2\right)}{z^2} + \frac{\left(dy_1^2 + dy_2^2\right)}{z^{2/\nu}} + \frac{dz^2}{z^2}\right]. \tag{5}$$

It has been shown in [30] that for the wall-on-wall collision in the 5d Lifshitz-like background with the critical exponent $\nu = 4$, the dependence of multiplicity on the energy is desirable, i.e. behaves as $s^{1/6}$. $\nu = 1$ corresponds to the AdS case.

This metric is a solution [30, 35] to a 5d model governed by the action

$$S = \frac{1}{16\pi G_5}\int d^5x\sqrt{|g|}\left(R + \Lambda - \frac{1}{6}\left(H_{(3)}^2 + m_0^2 B_{(2)}^2\right)\right), \tag{6}$$

where m_0 and Λ are constant and the 3-form $H_{(3)}$ and the 2-form $B_{(2)}$ are related by $H_{(3)} = dB_{(2)}$. There is a problem with action (6) since up to now we do not know the blacking version of the metric (5) as a solution to the model (6). However there is a possibility to support the metric (5) by another action and to find the corresponding blacking solution[32]. This action has the form

$$S = \int d^5x \sqrt{|g|} \left(R[g] + \Lambda - \frac{1}{2}(\partial\phi)^2 - \frac{1}{4} e^{\lambda\phi} F^2_{(2)} \right), \tag{7}$$

where the 2-form $F_{(2)}$ is the gauge field with $F^2_{(2)} = F_{mn}F^{mn}$, ϕ is the dilaton scalar field, λ is a dilatonic coupling constant and Λ is the cosmological constant. The model (7) can be considered as a truncated supergravity IIA in the style of [35]. Another possible underlying theory is the 5d $SO(6)$ gauged supergravity [36].

The metric of the black brane solution in the theory (7) reads

$$ds^2 = e^{2\nu r}\left(-f(r)dt^2 + dx^2\right) + e^{2r}\left(dy_1^2 + dy_2^2\right) + \frac{dr^2}{f(r)}, \tag{8}$$

with the blackening function given by

$$f(r) = 1 - me^{-(2\nu+2)r}. \tag{9}$$

and the dilaton and Maxwell fields:

$$\phi = \phi(r), \quad e^{\lambda\phi} = \mu e^{4r}, \quad F_{(2)} = \frac{1}{2} q\, dy_1 \wedge dy_2, \tag{10}$$

where μ and q are two constants. For the particular case $\nu = 4$, the metric (8)-(9) along with the ansatz (10) solves the field equations if the constants take the following values:

$$\lambda = \pm\frac{2}{\sqrt{3}}, \quad \Lambda = 90, \quad \mu q^2 = 240. \tag{11}$$

See Appendix in [32] for details.

If the dilaton is constant and the Maxwell field vanishes, the metric (8) with $\nu = 1$ turns out to be the black brane solution in the AdS background:

$$ds^2 = e^{2r}\left(-f(r)dt^2 + dx^2\right) + e^{2r}\left(dy_1^2 + dy_2^2\right) + \frac{dr^2}{f(r)}, \quad f(r) = 1 - me^{-4r} \tag{12}$$

2.3 The Vaidya-Lifshitz geometry

To study the thermalization process we need to use the infalling shell approach based on the Vaidya solution. First, we introduce the coordinate $z = e^{-\nu r}$, which, after the rescaling $(t, x, y_1, y_2) \mapsto \nu^{-1}(t, x, y_1, y_2)$, allows one to rewrite the metric (8) in the form

$$ds^2 = z^{-2}\left(-f(z)dt^2 + dx^2\right) + z^{-2/\nu}(dy_1^2 + dy_2^2) + \frac{dz^2}{z^2 f(z)}, \qquad f = 1 - mz^{2+2/\nu}. \tag{13}$$

To write down the Vaidya-Lifshitz solution, one should consider the ingoing null geodesics

$$dt + \frac{dz}{f(z)} = 0 \tag{14}$$

and introduce the Eddington-Finkelstein coordinate system (v, x, y_1, y_2, z) via

$$dv = dt + \frac{dz}{f(z)}. \tag{15}$$

Owing to (15) we can represent (13) in the following form

$$ds^2 = -z^{-2}f(v,z)dv^2 - 2z^{-2}dvdz + z^{-2}dx^2 + z^{-2/\nu}(dy_1^2 + dy_2^2), \qquad f(v,z) = 1 - m(v)z^{2+2/\nu}, \tag{16}$$

where the mass function $m(v)$ determines the thickness of the shell falling along $v = 0$ and captures the information about the black hole formation. For the infinite thin shell $m(v)$ has the form

$$m(v) = M\theta(v), \tag{17}$$

where M is a constant and $\theta(v)$ is the Heaviside function. One can also consider a smooth function $m(v)$ and get, for instance,

$$f(v,z) = 1 - \frac{m}{2}\left(1 + \tanh\frac{v}{\alpha}\right)z^{2+2/\nu}, \tag{18}$$

where m and α are two constants.

The solution (16) interpolates between the vacuum Lifshitz-type solution (5) inside the shell ($v < 0$) and the Lifshitz black brane geometry (13) outside the shell ($v > 0$).

3 The thermalization process

In [30] we have shown that there is a trapped surface, which forms in the collision of two shock waves in the background (5), controlled by boundary points z_a and z_b, with $z_a < z_b$. This trapped surface defines the location of the horizon for (16).

We can consider the thermalization time $t_{\rm therm}$ of several physical quantities, for example the two-point correlation functions, Wilson loops or the entanglement entropy [31–33]. All these quantities we can find in the holographically within geodesics approximations. The simple case is the the thermalization of the two-point correlation function.

3.1 Thermalization times of two-point correlators

Calculations of the thermalization time $t_{\rm therm}$ of two-point correlators at the scale ℓ is based on finding geodesics with endpoints located at the distance ℓ for a bulk particle. Then, the thermalization time $t_{\rm therm}$ is the time when this geodesic covered by the shell (16).

The general case for the Lagrangian of the point-like probe has the form

$$\mathcal{L} = \sqrt{-z^{-2}f(z)\dot{v}^2 - 2z^{-2}\dot{v}\dot{z} + z^{-2}\dot{x}^2 + z^{-2/\nu}\left(\dot{y}_1^2 + \dot{y}_2^2\right)}, \tag{19}$$

where τ is a parameter and $\dot{x} = \partial_\tau x$, etc. Here we have two possibilities for the choice of τ with respect to the transverse directions.

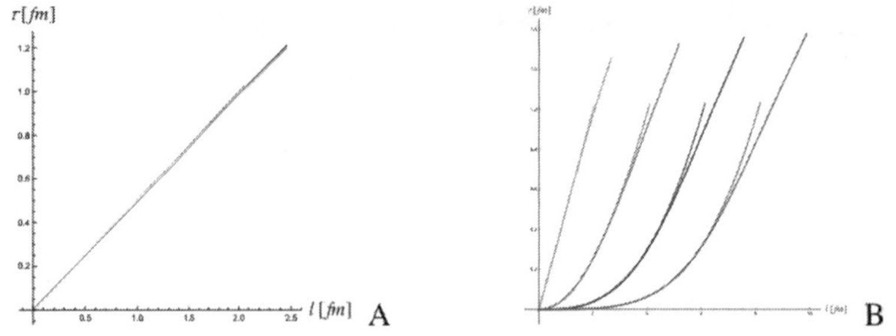

Figure 4. The thermalization time τ as a function of ℓ for the 5-dimensional Lifshitz metric (13)-(**??**) for $\nu = 2, 3, 4$. **A:** Thermalization along the longitudinal direction with $m = 0.5$ and $m = 0.1$. All lines coincide. **B:** Thermalization along the transversal direction, $\nu = 1$ (orange), $\nu = 2$ (brown), $\nu = 3$ (blue) and $\nu = 4$ (gray). The solid and dotted curves correspond to $m = 0.5$ and $m = 0.1$, respectively.

3.1.1 Thermalization along the longitudinal direction

For the case $\tau = x$ (x is a longitudinal direction) we have

$$\mathcal{L} = \frac{\sqrt{\mathcal{R}_x}}{z}, \quad \mathcal{R} = 1 - f(z)(v_x^{'})^2 - 2v_x^{'}z_x^{'}. \tag{20}$$

The integrals of motion corresponding to (20) are

$$\mathcal{J} = -\frac{1}{z\sqrt{\mathcal{R}_x}}, \qquad \mathcal{I} = \frac{f(z)v_x^{'} + z_x^{'}}{z\sqrt{\mathcal{R}}}. \tag{21}$$

We take $\mathcal{I} = 0$ and get

$$z_x^{'} = \pm\sqrt{f(z)\left(\frac{1}{z^2\mathcal{J}^2} - 1\right)}. \tag{22}$$

The turning point z_*, the point there $z_{x^*}^{'} = 0$, is related with $\mathcal{J}$ as $\mathcal{J}^2 = 1/z_*^2$.

For the distance ℓ between the ends of the geodesic and the thermalization time one gets

$$\ell = 2z_*\int_0^1 \frac{wdw}{\sqrt{f(z_*w)(1-w^2)}}, \tag{23}$$

$$t_{\text{therm}} = z_*\int_0^1 \frac{dw}{f(z_*w)}. \tag{24}$$

Note that here we assume that the turning point lies above the horizon, i.e. $z_h > z_*$. The behavior of the thermalization time as a function of the distances for (23)-(24) is represented in Fig.4.A. We see that the thermalization time behaves linearly with ℓ. The results match to those for modified AdS models from [31] and coincide for all values of the dynamical exponent ν.

3.1.2 Thermalization along the transversal direction

In this case $\tau = y_1$. From (19) we have

$$\mathcal{L} = \frac{\sqrt{\mathcal{R}}}{z}, \qquad \mathcal{R} = z^{2-2/\nu} - f(z)(\dot{v}_y)^2 - 2\dot{v}_y\dot{z}_y. \tag{25}$$

The integrals of motion corresponding to (25) are

$$\mathcal{J} = -\frac{z^{1-2/\nu}}{\sqrt{\mathcal{R}}}, \qquad \mathcal{I} = \frac{f(z)\dot{v}_y + \dot{z}_y}{z\sqrt{\mathcal{R}}}. \tag{26}$$

For $\mathcal{I} = 0$ we have the following relation between the turning point z_* and $\mathcal{J}$, $\mathcal{J}^2 = 1/z_*^{2/n}$ and we get

$$\dot{z}_y = \pm\frac{z^{1-\frac{2}{\nu}}\sqrt{f\,(1-\mathcal{J}^2 z^{2/\nu})}}{\mathcal{J}}. \tag{27}$$

From (27) one gets the relation between the ends of geodesic and the thermalization time

$$\ell = 2z_*^{1/\nu}\int_0^1 \frac{w^{-1+2/\nu}dw}{\sqrt{f(wz_*)\,(1-w^{2/\nu})}}, \tag{28}$$

$$t_{\text{therm}} = z_*\int_0^1 \frac{dw}{f(z_*w)}. \tag{29}$$

Here we remove the regularization since $\nu > 0$. The dependence (29) on (28) is given in Fig. 4.B. We see that the thermalization time in the transversal direction depends on the anizotropic parameter ν. In particular, for $\nu = 2$ the thermalization process is more then twice faster as compared to the longitudinal direction. By increasing ν we make the thermalization in the transversal direction faster. We also see that for larger values of ν the dependence on the mass m becomes more essential.

3.2 Thermalization time of Wilson loops

In this section, we compare the thermalization time for spatial Wilson loops with different orientations and its dependence on the value of the dynamical exponent ν. To simplify these estimations we consider the thin shell limit. We are interested in the value of the boundary time t_{therm} when the string profile is totally covered by the thin shell, i.e.:

$$t_{therm}(\ell) = \int_0^{z_*(\ell)} \frac{dz}{f(z)}, \tag{30}$$

where ℓ is the length between the string endpoints on the boundary.

In Fig.5 we plot the dependence on ℓ of the thermalization time for two configurations in the xy_1-plane. In Fig.6.A the behavior of the thermalization time as a function of ℓ for the configuration in the transverse y_1y_2-plane is presented. One can see that the thermalization time decreases with increasing ν for all cases plotted in Fig. 5 and Fig. 6. The dependence on the length ℓ for the loop in the xy_1-plane with the short extent in the x-direction is linear. At the same time, the dependence for the loop in the same plane, but with the short extent in the y_1-direction, as well as for the loop in y_1y_2-plane, is not linear for small ℓ asymptoting to the linear dependence only for large ℓ. It should be noted that the deviation from linearity strengthens with increasing ν. We also see that for the configuration in the transverse plane, the deviation of the thermalization time for the anisotropic cases from the isotropic one the thermalization time is much stronger than for the other orientations. In Fig.6.B the comparison of thermalization times for different orientations in the case $\nu = 4$ is plotted. This plot shows that the dependence on the orientation is crucial, varying the orientation we change the order of thermalization time. This means that characteristic scale depends on the orientation.

Figure 5. The thermalization time for the Wilson loop in the xy_1-plane with a short extent in the x- and y-directions ((a) and (b), respectively). Different curves correspond to different values of $\nu = 1, 2, 3, 4$ (from top to down for each plot).

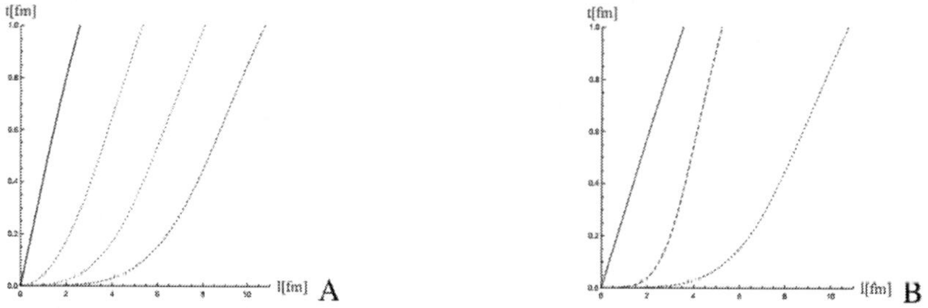

Figure 6. A. The thermalization time as a function of ℓ for the Wilson loop in the y_1y_2-plane, different curves correspond to different values of $\nu = 1, 2, 3, 4$ (from left to right). B. The thermalization time as a function of ℓ for $\nu = 4$, for Wilson loops with a short extent in the x- and y- directions lying in the xy_1-plane and for Wilson loop in the y_1y_2-plane (from left to right).

4 Conclusion

As a conclusion let me summarize. Using conjecture that black hole production in AdS 5 gives quark gluon plasma formation in 4-dim QCD we have got:

- Multiplicity, that fits the recent experimental data [19, 20].
- The thermalization time on ℓ of two-point correlators as well as Wilson loops depends on orientation. It is not clear yet how we can check experimentally this prediction.

As to further directions of application of holography to quark gluon plasma formation in heavy ion collisions, I would like to mention the AdS/CFT approach to dealing with QGP formation with non-zero chemical potential [14]. In the context of the future project NICA this becomes really very actual.

Acknowledgments

I would like to thank the Organizers of Quarks2016 for the kind invitation and warm hospitality. This work is supported by the Russian Science Foundation (project 14-50-00005).

References

[1] J. Casalderrey-Solana, H. Liu, D. Mateos, K. Rajagopal, U. A. Wiedemann, *Gauge/String Duality, Hot QCD and Heavy Ion Collisions*, Cambridge University Press 2014
[2] I. Ya. Aref'eva, *Phys. Usp.* **57** (2014) 527.
[3] O. DeWolfe, S. S. Gubser, C. Rosen and D. Teaney, *Prog. Part. Nucl. Phys.* **75**, 86 (2014).
[4] A. Bazavov *et al.* Phys. Rev. D **90**, 094503 (2014)
[5] J. M. Maldacena, Adv. Theor. Math. Phys. **2**, 231-252 (1998) [hep-th/9711200].
[6] V. Matveev, R. Muradyan and A. Tavkhelidze, Lett. Nuovo Cim. 1972. V.5. P.712
[7] S. S. Gubser, S. S. Pufu and A. Yarom, Phys. Rev. D **78** (2008) 066014; JHEP **0911** (2009) 050
[8] D. Grumiller and P. Romatschke, JHEP **0808** (2008) 027 [arXiv:0803.3226 [hep-th]].
[9] L. Alvarez-Gaume et al., JHEP **0902** (2009) 009
[10] J. L. Albacete, Y. V. Kovchegov and A. Taliotis, JHEP **0807** (2008) 100
[11] S. Lin and E. Shuryak, Phys. Rev. D **79** (2009) 124015
[12] J. L. Albacete, Y. V. Kovchegov and A. Taliotis, JHEP **0905** (2009) 060
[13] I. Y. Aref'eva, A. A. Bagrov and E. A. Guseva, JHEP **0912**, 009 (2009)
[14] I. Y. Aref'eva, A. A. Bagrov and L. V. Joukovskaya, JHEP **1003** , 002, (2010)
[15] Y. V. Kovchegov and S. Lin, JHEP **1003** (2010) 057;
Y. V. Kovchegov, Prog. Theor. Phys. Suppl. **187** (2011) 96
[16] E. Kiritsis and A. Taliotis, JHEP **1204**, 065 (2012)
[17] P. M. Chesler and L. G. Yaffe, Phys. Rev. Lett. **106** (2011) 021601
[18] W. van der Schee, "Gravitational collisions and the quark-gluon plasma," arXiv:1407.1849.
[19] G. Aad *et al.* [ATLAS Collaboration], Phys. Lett. B **710**, 363 (2012)
[20] J. Adam *et al.* [ALICE Collaboration], Phys. Rev. Lett. **116**, no. 22, 222302 (2016)
[21] L.D. Landau, *Izvestya AN. Ser. phys.* **17** 51 (1953)
[22] Fermi E *Progr. Theoret. Phys.* **5** 570 (1950)
[23] I.Ya. Pomeranchuk, *DAN, SSSR* **78** 884 1951
[24] E.L. Feinberg, *Phys. Usp.* **168** (1998) 697.
[25] U. Gursoy, E. Kiritsis, L. Mazzanti and F. Nitti, *JHEP* **05**, 033 (2009);
[26] I. Ya. Aref'eva, E. O. Pozdeeva and T. O. Pozdeeva, Teor. Mat. Fiz. **180**, 35 (2014).
[27] D. S. Ageev and I. Ya. Aref'eva, J. Exp. Theor. Phys. **120**, no. 3, 436 (2015)
[28] M. Strickland, Thermalization and isotropization in heavy-ion collisions, *Pramana* **84**, (2015) 671.
[29] D. Giataganas, Observables in Strongly Coupled Anisotropic Theories, *PoS Corfu* **2012**, 122 (2013).
[30] I. Ya. Aref'eva and A. A. Golubtsova, *JHEP* **1504** (2015) (011).
[31] I. Ya. Aref'eva, *Theor. Math. Phys.*, 184:3 (2015), 1239-1255, arXiv:1503.02185.
[32] I. Y. Aref'eva, A. A. Golubtsova and E. Gourgoulhon, Analytic black branes in Lifshitz-like backgrounds and thermalization, arXiv: 1601.06046
[33] D. S. Ageev, I. Y. Aref'eva, A. A. Golubtsova and E. Gourgoulhon, "Holographic Wilson loops in Lifshitz-like backgrounds," arXiv:1606.03995.
[34] M. Taylor, Non-relativistic holography, arXiv:0812.0530.
[35] T. Azeyanagi, W. Li and T. Takayanagi, *JHEP* **06** (2009) 084.
[36] A. Donos, J. P. Gauntlett and C. Pantelidou, *JHEP* **1201** (2012) 061

Recent results from the Belle experiment

Dmitri Liventsev[1,2,⋆,⋆⋆]

[1] *Virginia Polytechnic Institute and State University, Blacksburg, Virginia 24061*
[2] *High Energy Accelerator Research Organization (KEK), Tsukuba 305-0801*

Abstract. We review recent results obtained using the data recorded with the Belle detector at the KEKB asymmetric-energy e^+e^- collider in KEK, Japan.

1 Introduction

The Belle detector is a large-solid-angle magnetic spectrometer that consists of a silicon vertex detector (SVD), a 50-layer central drift chamber (CDC), an array of aerogel threshold Cherenkov counters (ACC), a barrel-like arrangement of time-of-flight scintillation counters (TOF), and an electromagnetic calorimeter comprised of CsI(Tl) crystals (ECL) located inside a super-conducting solenoid coil that provides a 1.5 T magnetic field. An iron flux-return located outside of the coil is instrumented to detect K_L^0 mesons and to identify muons (KLM). The detector is described in detail elsewhere [1].

The Belle experiment successfully operated for more than a decade until 2010 at the asymmetric-energy e^+e^- collider KEKB [2] in various $\Upsilon(nS)$ resonances, having collected a world-record sample of data over 1 ab^{-1}.

Here we present a review of recent results from Belle based on its full statistics.

2 Angular Analysis of $B^0 \to K^*(892)^0\ell^+\ell^-$

Rare decays of B mesons are an ideal probe to search beyond the Standard Model (SM) of particle physics, since contributions from new particles lead to effects that are of similar size as the SM predictions. The rare decay $B^0 \to K^*(892)^0\ell^+\ell^-$, where $\ell^+\ell^-$ is either e^+e^- or $\mu^+\mu^-$, involves the quark transition $b \to s\ell^+\ell^-$, a flavor changing neutral current that is forbidden at tree level in the SM. Higher order SM processes such as penguin or W^+W^- box diagrams allow for such transitions, leading to branching fractions of less than one in a million. Various extensions to the SM predict contributions from new physics, which can interfere with the SM amplitudes and lead to enhanced or suppressed branching fractions or modified angular distributions of the decay products.

Belle presented an angular analysis [3], using the decay modes $B^0 \to K^*(892)^0e^+e^-$ and $B^0 \to K^*(892)^0\mu^+\mu^-$, in a data sample recorded with the Belle detector. The LHCb collaboration reported a discrepancy in the angular distribution of the decay $B^0 \to K^*(892)^0\mu^+\mu^-$, corresponding to a 3.4$\sigma$ deviation from the SM prediction [4]. In contrast to the LHCb measurement the di-electron channel is also used in this analysis.

⋆e-mail: dmitri.liventsev@kek.jp
⋆⋆On behalf of the Belle collaboration

Table 1. Fitted yields and statistical error for signal (n_{sig}) and background (n_{bkg}) events in the binning of q^2 for both the combined electron and muon channel.

Bin	q^2 range in GeV^2/c^4	n_{sig}	n_{bkg}
0	1.00 – 6.00	49.5 ± 8.4	30.3 ± 5.5
1	0.10 – 4.00	30.9 ± 7.4	26.4 ± 5.1
2	4.00 – 8.00	49.8 ± 9.3	35.6 ± 6.0
3	10.09 – 12.90	39.6 ± 8.0	19.3 ± 4.4
4	14.18 – 19.00	56.5 ± 8.7	16.0 ± 4.0

K^* candidates are formed in the channel $K^{*0} \to K^+\pi^-$ and are combined with oppositely charged lepton pairs to form B meson candidates. The large combinatoric background is suppressed by applying requirements on kinematic variables. Two independent variables can be constructed using constraints that in $\Upsilon(4S)$ decays B mesons are produced pairwise and each carries half the center–of–mass (CM) frame beam energy, E_{Beam}. These variables are the beam constrained mass, M_{bc}, and the energy difference, ΔE, in which signal features a distinct distribution that can discriminate against background. The variables are defined in the $\Upsilon(4S)$ rest frame as

$$M_{\text{bc}} \equiv \sqrt{E_{\text{Beam}}^2/c^4 - |\vec{p}_B|^2/c^2} \text{ and} \tag{1}$$

$$\Delta E \equiv E_B - E_{\text{Beam}}, \tag{2}$$

where E_B and $|\vec{p}_B|$ are the energy and momentum of the reconstructed candidate, respectively. Correctly reconstructed candidates are located around the nominal B mass in M_{bc} and feature ΔE of around zero. Candidates are selected satisfying $5.22 < M_{\text{bc}} < 5.3\ \text{GeV}/c^2$ and $-0.10\ (-0.05) < \Delta E < 0.05$ GeV for $\ell = e\ (\ell = \mu)$.

Large irreducible background contributions arise from charmonium decays $B \to K^{(*)}J/\psi$ and $B \to K^{(*)}\psi(2S)$, in which the $c\bar{c}$ state decays into two leptons. These decays have the same signature as the desired signal and are vetoed with the following requirements on $q^2 = M_{\ell^+\ell^-}$, the invariant mass of the lepton pair: $-0.25\ \text{GeV}/c^2 < M_{ee(\gamma)} - m_{J/\psi} < 0.08\ \text{GeV}/c^2$, $-0.15\ \text{GeV}/c^2 < M_{\mu\mu} - m_{J/\psi} < 0.08\ \text{GeV}/c^2$, $-0.20\ \text{GeV}/c^2 < M_{ee(\gamma)} - m_{\psi(2S)} < 0.08\ \text{GeV}/c^2$ and $-0.10\ \text{GeV}/c^2 < M_{\mu\mu} - m_{\psi(2S)} < 0.08\ \text{GeV}/c^2$.

For the angular analysis the number of signal events n_{sig} and background events n_{bkg} in the signal region $M_{\text{bc}} > 5.27\ \text{GeV}/c^2$ are obtained by a fit to M_{bc} in bins of q^2. The extracted yields and the definition of the bin ranges are presented in Table 1.

We perform an angular analysis of $B^0 \to K^*(892)^0\ell^+\ell^-$ including the electron and muon modes. The decay is kinematically described by three angles θ_ℓ, θ_K and ϕ and the invariant mass squared of the lepton pair q^2. Definitions of the angles and the full angular distribution follow Ref. [5]. The binning in q^2 is detailed in Table 1 together with the measured signal and background yields. Uncovered regions in the q^2 spectrum arise from vetoes against backgrounds of the charmonium resonances $J/\psi \to \ell^+\ell^-$ and $\psi(2S) \to \ell^+\ell^-$ and vetos against π^0 Dalitz decays and photon conversion.

The observables $P'_{i=4,5,6,8}$, introduced in Ref. [6] are functions of Wilson coefficients, containing information about the short-distance effects and can be affected by new physics and are considered to be largely free from form-factor uncertainties [7]. The statistics in this analysis are not sufficient to perform an eight-dimensional fit, therefore a folding technique is used explained in more detail in Refs. [8].

The signal and background fractions are derived from a fit to beforehand, where the yields are listed in Table 1. The M_{bc} variable is split into a signal (upper) and sideband (lower) region at

5.27 GeV/c^2. In the second step the shape of the background for the angular observables is estimated on the M_{bc} sideband. This is possible as the angular observables have shown to be uncorrelated to M_{bc} in the background sample.

All observables $P'_{i=4,5,6,8}$ are extracted from the data in the signal region using three-dimensional unbinned maximum likelihood fits in four bins of q^2 and the additional zeroth bin using the folded signal PDF, fixed background shapes and a fixed number of signal events. Each $P'_{i=4,5,6,8}$ is fitted with the K^* longitudinal polarization F_L and the transverse polarization asymmetry $A_T^{(2)}$. Counting also the zeroth bin, which exhibits overlap with the range of the first and second bin, 20 measurements are performed.

The measurements are compared with SM predictions based upon different theoretical calculations. The result for P'_5 is shown in Fig. 1 together with available SM prediction and LHCb measurements. A deviation with respect to the SM prediction is observed with a significance of 2.1σ in the q^2 range $4.0 < q^2 < 8.0$ GeV^2/c^4. This deviation is into the same direction and in the same q^2 region where the LHCb collaboration reported the so-called P'_5 anomaly [4, 8].

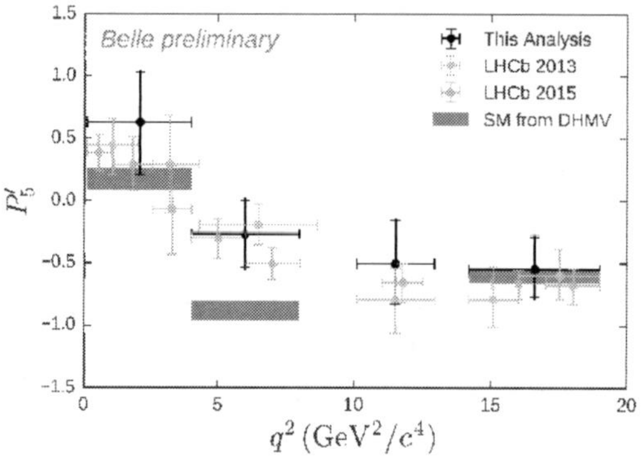

Figure 1. Result for the P'_5 observable compared to SM predictions from various sources. Results from LHCb [4, 8] are shown for comparison.

3 Measurement of the branching ratio of $\bar{B}^0 \to D^{*+}\tau^-\bar{\nu}_\tau$

Semitauonic B meson decays of the type $b \to c\tau\nu_\tau$ are sensitive probes to search for physics beyond the Standard Model (SM). Charged Higgs bosons, which appear in supersymmetry and other models with at least two Higgs doublets, may contribute to the decay to due to large mass of the τ lepton and induce measurable effects in the branching fraction. Similarly, leptoquarks, which carry both baryon number and lepton number, may also contribute to this process. The ratio of branching fractions

$$\mathcal{R}(D^{(*)}) = \frac{\mathcal{B}(\bar{B} \to D^{(*)}\tau^-\bar{\nu}_\tau)}{\mathcal{B}(\bar{B} \to D^{(*)}\ell^-\bar{\nu}_\ell)} \quad (\ell = e, \mu), \tag{3}$$

is typically used instead of the absolute branching fraction of $\bar{B} \to D^{(*)+}\tau^-\bar{\nu}_\tau$, to reduce several systematic uncertainties such as those on the experimental efficiency, the CKM matrix elements $|V_{cb}|$, and on the form factors. The SM calculations on these ratios predict $\mathcal{R}(D^*) = 0.252 \pm 0.003$ [9] and $\mathcal{R}(D) = 0.297 \pm 0.017$ [10, 14] with precision of better than 2% and 6% for $\mathcal{R}(D^*)$ and $\mathcal{R}(D)$, respectively. Exclusive semitauonic B decays were first observed by the Belle Collaboration [12], with subsequent studies reported by Belle [13, 14], *BABAR* [11], and LHCb [15] Collaborations. All results are consistent with each other, and the average values of Refs. [11, 14, 15] have been found to be $\mathcal{R}(D^*) = 0.322 \pm 0.018 \pm 0.012$ and $\mathcal{R}(D) = 0.391 \pm 0.041 \pm 0.028$ [16], which exceed the SM predictions for $\mathcal{R}(D^*)$ and $\mathcal{R}(D)$ by 3.0σ and 1.7σ, respectively. The combined analysis of $\mathcal{R}(D^*)$ and $\mathcal{R}(D)$, taking into account measurement correlations, finds that the deviation is 3.9σ from the SM prediction.

In the paper [17] Belle reported the first measurement of $\mathcal{R}(D^*)$ using the semileptonic tagging method. Signal $B^0\bar{B}^0$ events are reconstructed in modes where one B decays semi-tauonically $B^0 \to D^{*+}\tau^-\bar{\nu}_\tau$ where $\tau^- \to \ell^-\bar{\nu}_\ell\nu_\tau$, (referred to hereafter as $B_{\rm sig}$) and the the other B decays in a semileptonic channel $\bar{B}^0 \to D^{*+}\ell^-\bar{\nu}_\ell$ (referred to hereafter as $B_{\rm tag}$). To reconstruct normalization $B^0\bar{B}^0$ events, which correspond to the denominator in $\mathcal{R}(D^*)$, both B mesons are reconstructed decaying to semileptonic decay modes $D^{*\pm}\ell^{\mp}\bar{\nu}_\ell$.

To tag semileptonic B decays, we combine D^{*+} meson and lepton candidates of opposite electric charge and calculate the cosine of the angle between the momentum of the B meson and the $D^*\ell$ system in the $\Upsilon(4S)$ rest frame, under the assumption that only one massless particle is not reconstructed:

$$\cos\theta_{B\text{-}D^*\ell} \equiv \frac{2E_{\rm beam}E_{D^*\ell} - m_B^2 - M_{D^*\ell}^2}{2|\vec{p}_B|\cdot|\vec{p}_{D^*\ell}|}, \tag{4}$$

where $E_{\rm beam}$ is the energy of the beam, and $E_{D^*\ell}$, $\vec{p}_{D^*\ell}$ and $M_{D^*\ell}$ are the energy, momentum, and mass of the $D^*\ell$ system, respectively. The variable m_B is the nominal B meson mass [18], and $\vec{p}_B$ is the nominal B meson momentum. All variables are defined in the $\Upsilon(4S)$ rest frame. Correctly reconstructed B candidates in the tag and normalization mode $D^*\ell\nu_\ell$ are expected to have a value of $\cos\theta_{B\text{-}D^*\ell}$ between -1 and $+1$. On the other hand, correctly reconstructed B candidates in the signal decay mode $D^*\tau\nu_\tau$ or falsely reconstructed B candidates would tend to have values of $\cos\theta_{B\text{-}D^*\ell}$ below the physical region due to contributions from additional particles and a large negative correlation with missing mass squared, $M^2_{\rm miss} = (2E_{\rm beam} - \sum_i E_i)^2/c^4 - |\sum_i \vec{p}_i|^2/c^2$, where $(\vec{p}_i, E_i)$ is four-momentum of the particles in the $\Upsilon(4S)$ rest frame.

In each event we require two tagged B candidates that are opposite in flavor. Signal events may have the same flavor due to the $B\bar{B}$ mixing, however we veto such events as they lead to ambiguous $D^*\ell$ pair assignment and larger combinatorial background. We require that at most one B meson is reconstructed in a D^+ mode, in order to avoid large background from fake neutral pions when forming D^* candidates. In each signal event we assign the candidate with the lowest value of $\cos\theta_{B\text{-}D^*\ell}$ (referred to hereafter as $\cos\theta^{\rm sig}_{B\text{-}D^*\ell}$) as $B_{\rm sig}$.

To separate reconstructed signal and normalization events, we employ a neural network approach. The variables used as inputs to the network are (i) $\cos\theta^{\rm sig}_{B\text{-}D^*\ell}$, (ii) missing mass squared, $M^2_{\rm miss}$, and (iii) visible energy $E_{\rm vis} = \sum_i E_i$, where E_i is energies of the particles in the $\Upsilon(4S)$ rest frame. To separate signal and normalization events from background processes, we use the extra energy, $E_{\rm ECL}$, which is defined as the sum of the energies of neutral clusters detected in the ECL that are not associated with reconstructed particles.

We extract the signal and normalization yields using a two-dimensional extended maximum-likelihood fit in NN and $E_{\rm ECL}$. The projection of the fitted distributions are shown in Figure 2. The yields of signal and normalization events are measured to be 231 ± 23(stat) and 2800 ± 57(stat), respectively. The ratio $\mathcal{R}(D^*)$ is therefore found to be

$$\mathcal{R}(D^*) \quad = \quad 0.302 \pm 0.030 \pm 0.011, \tag{5}$$

where the first and second errors correspond to statistical and systematic uncertainties, respectively.

We calculate the statistical significance of the signal as $\sqrt{-2\ln(\mathcal{L}_0/\mathcal{L}_{\rm max})}$, where $\mathcal{L}_{\rm max}$ and $\mathcal{L}_0$ are the maximum likelihood and the likelihood obtained assuming zero signal yield, respectively. We

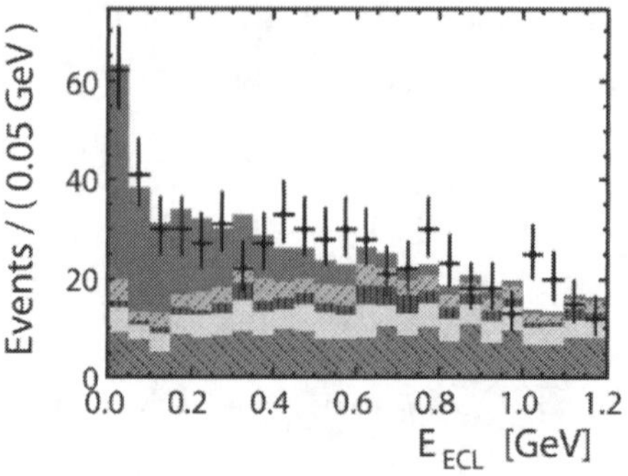

Figure 2. Projections of the fit results with data points overlaid of E_{ECL} distribution with signal-enhanced *NN* region (*NN* > 0.8. The background categories are described in detail in the text, where "others" refers to predominantly $B \to X_c D^*$ decays.

obtain a statistical significance of 13.8σ. We also estimate the compatibility of the measured value of $\mathcal{R}(D^*)$ and the SM prediction. The effect of systematic uncertainties are included by convolving the likelihood function with a Gaussian distribution. We obtain that our result is larger than the SM prediction by 1.6σ.

We investigated the compatibility of the data samples with type II two-Higgs-doublet model (2HDM) and leptoquark models. We find our data is compatible with the SM and type II 2HDM with $\tan\beta/m_{H^+} = 0.7\ \mathrm{GeV}^{-1}$, while the R_2 type leptoquark model with $C_T = +0.36$ is disfavored.

4 Observation of the decay $B_s^0 \to K^0\bar{K}^0$

The two-body decays $B_s^0 \to h^+h'^-$, where $h^{(\prime)}$ is either a pion or kaon, have now all been observed [18]. In contrast, the neutral-daughter decays $B_s^0 \to h^0h'^0$ have yet to be observed. The decay $B_s^0 \to K^0\bar{K}^0$ is of particular interest because the branching fraction is predicted to be relatively large. In the Standard model, the decay proceeds mainly via a $b \to s$ loop (or "penguin") transition and the branching fraction is predicted to be in the range $(16-27)\times10^{-6}$ [20]. The presence of non-SM particles or couplings could enhance this value [21]. It has been pointed out that CP asymmetries in $B_s^0 \to K^0\bar{K}^0$ decays are promising observables in which to search for new physics [22].

The current upper limit on the branching fraction, $\mathcal{B}(B_s^0 \to K^0\bar{K}^0) < 6.6\times10^{-5}$ at 90% confidence level, was set by the Belle Collaboration using 23.6 fb^{-1} of data recorded at the $\Upsilon(5S)$ resonance [23]. In paper [19] Belle updates this result using the full data set of 121.4 fb^{-1} recorded at the $\Upsilon(5S)$. The analysis presented here uses improved tracking, K^0 reconstruction, and continuum suppression algorithms. The data set corresponds to $(6.53\pm0.66)\times10^6$ $B_s^0\bar{B}_s^0$ pairs [24] produced in three $\Upsilon(5S)$ decay channels: $B_s^0\bar{B}_s^0$, $B_s^{*0}\bar{B}_s^0$ or $B_s^0\bar{B}_s^{*0}$, and $B_0^{*0}\bar{B}_s^{*0}$. The latter two channels dominate, with production fractions of $f_{B_s^{*0}\bar{B}_s^0} = (7.3\pm1.4)\%$ and $f_{B_s^{*0}\bar{B}_s^{*0}} = (87.0\pm1.7)\%$ [25].

Candidate K^0 mesons are reconstructed via the decay $K_s \to \pi^+\pi^-$ using a neural network (NN) technique. To identify $B_s^0 \to K_sK_s$ candidates, we define two variables: the beam-energy-constrained mass $M_{\mathrm{bc}} = \sqrt{E_{\mathrm{beam}}^2 - |\vec{p}_B|^2c^2}/c^2$; and the energy difference $\Delta E = E_B - E_{\mathrm{beam}}$, where E_{beam} is the beam energy and E_B and $\vec{p}_B$ are the energy and momentum, respectively, of the B_s^0 candidate. These quantities are evaluated in the e^+e^- center-of-mass frame. We require that events satisfy $M_{\mathrm{bc}} > 5.34\ \mathrm{GeV}/c^2$ and $-0.20\ \mathrm{GeV} < \Delta E < 0.10\ \mathrm{GeV}$. To suppress background arising from continuum $e^+e^- \to q\bar{q}$ $(q = u, d, s, c)$ production, we use a second NN that distinguishes jetlike continuum events from more spherical $B_s^{(*)0}\bar{B}_s^{(*)0}$ events. The NN has a single output variable ($C_{\rm NN}$) that ranges from -1 for backgroundlike events to $+1$ for signal-like events. We require $C_{\rm NN} > -0.1$, which rejects approximately 85% of $q\bar{q}$ background while retaining 83% of signal decays. We subsequently translate $C_{\rm NN}$ to a new variable

$$C'_{\rm NN} = \ln\left(\frac{C_{\rm NN} - C_{\rm NN}^{\rm min}}{C_{\rm NN}^{\rm max} - C_{\rm NN}}\right), \tag{6}$$

where $C_{\rm NN}^{\rm min} = -0.1$ and $C_{\rm NN}^{\rm max}$ is the maximum value of $C_{\rm NN}$ obtained from a large sample of signal MC decays. The distribution of $C'_{\rm NN}$ is well modeled by a Gaussian function. Backgrounds arising from other B_s^0 and non-B_s^0 decays were studied using MC simulation and found to be negligible.

After applying all selection criteria, approximately 1.0% of events have multiple B_s^0 candidates. For these events, we retain the candidate having the smallest value of χ^2 obtained from the deviations of the reconstructed K_s masses from their nominal values [18]. According to MC simulation, this criterion selects the correct B_s^0 candidate > 99% of the time.

We measure the signal yield by performing an unbinned extended maximum likelihood fit to the variables $M_{\rm bc}$, ΔE, and $C'_{\rm NN}$.

The results of the fit are $29.0^{+8.5}_{-7.6}$ signal events and $1095.0^{+33.9}_{-33.4}$ continuum background events. Projections of the fit are shown in Fig. 3. The branching fraction is calculated via

$$\mathcal{B}(B_s^0 \to K^0\bar{K}^0) = \frac{Y_s}{2N_{B_s^0\bar{B}_s^0}(0.50)\mathcal{B}_{K^0}^2\varepsilon}, \tag{7}$$

where Y_s is the fitted signal yield; $N_{B_s^0\bar{B}_s^0} = (6.53 \pm 0.66) \times 10^6$ is the number of $B_s^0\bar{B}_s^0$ events; $\mathcal{B}_{K^0} = (69.20 \pm 0.05)\%$ is the branching fraction for $K_s \to \pi^+\pi^-$ [18]; and $\varepsilon = (46.3 \pm 0.1)\%$ is the signal efficiency as determined from MC simulation. The factor 0.50 accounts for the 50% probability for $K^0\bar{K}^0 \to K_sK_s$ (since $K^0\bar{K}^0$ is CP even). Inserting these values gives $\mathcal{B}(B_s^0 \to K^0\bar{K}^0) = (19.6^{+5.8}_{-5.1} \pm 1.0 \pm 2.0) \times 10^{-6}$, where the first uncertainty is statistical, the second is systematic, and the third reflects the uncertainty due to the total number of $B_s^0\bar{B}_s^0$ pairs. This value is in good agreement with the SM predictions [20], and it implies that the Belle II experiment [26] will reconstruct over 1000 of these decays. Such a sample would allow for a much higher sensitivity search for new physics in this $b \to s$ penguin-dominated decay.

The signal significance is calculated as $\sqrt{-2\ln(\mathcal{L}_0/\mathcal{L}_{\rm max})}$, where $\mathcal{L}_0$ is the likelihood value when the signal yield is fixed to zero, and $\mathcal{L}_{\rm max}$ is the likelihood value of the nominal fit. We include systematic uncertainties in the significance by convolving the likelihood function with a Gaussian function whose width is equal to that part of the systematic uncertainty that affects the signal yield. We obtain a signal significance of 5.1 standard deviations; thus, our measurement constitutes the first observation of this decay.

5 Study of $e^+e^- \to B^{(*)}\bar{B}^{(*)}\pi^\pm$ at $\sqrt{s} = 10.866$ GeV

Two new charged bottomonium-like resonances, $Z_b(10610)$ and $Z_b(10650)$, have been observed recently by the Belle Collaboration in $e^+e^- \to \Upsilon(n{\rm S})\pi^+\pi^-$, $n = 1, 2, 3$ and $e^+e^- \to h_b(m{\rm P})\pi^+\pi^-$, $m = 1, 2$ [28, 29]. Analysis of the quark composition of the initial and final states reveals that these hadronic objects have an exotic nature: Z_b should be comprised of (at least) four quarks including a $b\bar{b}$ pair. Several models [30] have been proposed to describe the internal structure of these states. In Ref. [31], it was suggested that $Z_b(10610)$ and $Z_b(10650)$ states might be loosely bound $B\bar{B}^*$ and $B^*\bar{B}^*$ systems, respectively. If so, it is natural to expect the Z_b states to decay to final states with $B^{(*)}$ mesons at substantial rates.

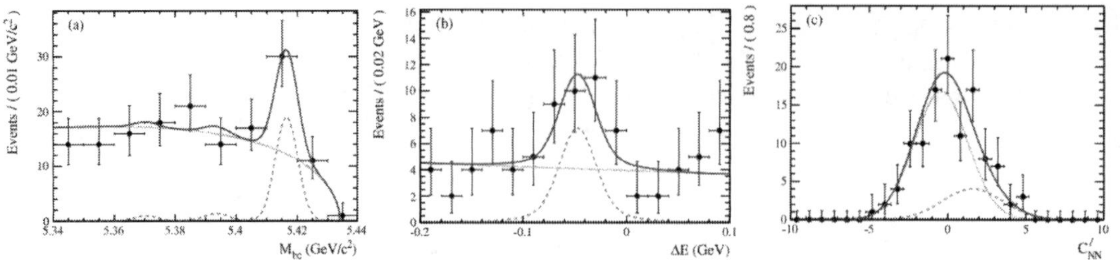

Figure 3. Projections of the 3D fit to the real data: (a) M_{bc} in -0.11 GeV $< \Delta E <$ 0.02 GeV and $C'_{NN} > 0.5$; (b) ΔE in 5.405 GeV/c^2 $< M_{bc} <$ 5.427 GeV/c^2 and $C'_{NN} > 0.5$; and (c) C'_{NN} in 5.405 GeV/c^2 $< M_{bc} <$ 5.427 GeV/c^2 and -0.11 GeV $< \Delta E <$ 0.02 GeV. The points with error bars are data, the (green) dashed curves show the signal, (magenta) dotted curves show the continuum background, and (blue) solid curves show the total. The χ^2/(number of bins) values of these fit projections are 0.30, 0.43, and 0.26, respectively, which indicate that the fit gives a good description of the data. The three peaks in M_{bc} arise from $\Upsilon(5S) \to B^0_s\bar{B}^0_s$, $B^{*0}_s\bar{B}^0_s + B^0_s\bar{B}^{*0}_s$, and $B^{*0}_s\bar{B}^{*0}_s$ decays.

Evidence for the three-body $\Upsilon(10860) \to B\bar{B}^*\pi$ decay has been reported previously by Belle, based on a data sample of 23.6 fb^{-1} [32]. In the analysis [27] Belle uses a data sample with an integrated luminosity of 121.4 fb^{-1} collected near the peak of the $\Upsilon(10860)$ resonance ($\sqrt{s}$ = 10.866 GeV) with the Belle detector. Note that we reconstruct only three-body $B^{(*)}\bar{B}^{(*)}\pi$ combinations with a charged primary pion. For brevity, we adopt the following notations: the set of $B^+\bar{B}^0\pi^-$ and $B^-B^0\pi^+$ final states is referred to as $BB\pi$; the set of $B^+\bar{B}^{*0}\pi^-$, $B^-B^{*0}\pi^+$, $B^0B^{*-}\pi^+$ and $\bar{B}^0B^{*+}\pi^-$ final states is referred to as $BB^*\pi$; and the set of $B^{*+}\bar{B}^{*0}\pi^-$ and $B^{*-}B^{*0}\pi^+$ final states is denoted as $B^*B^*\pi$.

B mesons are reconstructed in the following decay channels: $B^+ \to J/\psi K^{(*)+}$, $B^+ \to \bar{D}^{(*)0}\pi^+$, $B^0 \to J/\psi K^{(*)0}$, $B^0 \to D^{(*)-}\pi^+$ (eighteen in total). The dominant background comes from $e^+e^- \to c\bar{c}$ continuum events, where true D mesons produced in e^+e^- annihilation are combined with random particles to form a B candidate. This type of background is suppressed using variables that characterize the event topology. We identify B candidates by their reconstructed invariant mass $M(B)$ and momentum $P(B)$ in the center-of-mass (c.m.) frame. We require $P(B) <$ 1.35 GeV/c to retain B mesons produced in both two-body and multibody processes. The $M(B)$ distribution for B candidates is shown in Fig. 4(a). We perform a binned maximum likelihood fit of the $M(B)$ distribution to the sum of a signal component parameterized by a Gaussian function and two background components: one related to other decay modes of B mesons and one due to continuum $e^+e^- \to q\bar{q}$ processes, where $q = u, d, s, c$. We find 12263 ± 168 fully reconstructed B mesons. The B signal region is defined by requiring $M(B)$ to be within 30 to 40 MeV/c^2 (depending on the B decay mode) of the nominal B mass.

Reconstructed B^+ or $\bar{B}^0$ candidates are combined with π^-'s — the right-sign (RS) combination — and the missing mass, $M_{\rm miss}(B\pi)$, is calculated as $M_{\rm miss}(B\pi) = \sqrt{(\sqrt{s} - E_{B\pi})^2/c^4 - P^2_{B\pi}/c^2}$, where $E_{B\pi}$ and $P_{B\pi}$ are the measured energy and momentum of the reconstructed $B\pi$ combination. Signal $e^+e^- \to BB^*\pi$ events produce a narrow peak in the $M_{\rm miss}(B\pi)$ spectrum around the nominal B^* mass while $e^+e^- \to B^*B^*\pi$ events produce a peak at $m_{B^*} + \Delta m_{B^*}$, where $\Delta m_{B^*} = m_{B^*} - m_B$, due to the missed photon from the $B^* \to B\gamma$ decay. It is important to note here that, according to signal MC, $BB^*\pi$ events, where the reconstructed B is the one from the B^*, produce a peak in the $M_{\rm miss}(B\pi)$ distribution at virtually the same position as $BB^*\pi$ events, where the reconstructed B is the primary one. To remove the correlation between $M_{\rm miss}(B\pi)$ and $M(B)$ and to improve the resolution, we use $M^*_{\rm miss} = M_{\rm miss}(B\pi) + M(B) - m_B$ instead of $M_{\rm miss}(B\pi)$. The $M^*_{\rm miss}$ distribution for the RS combinations is shown

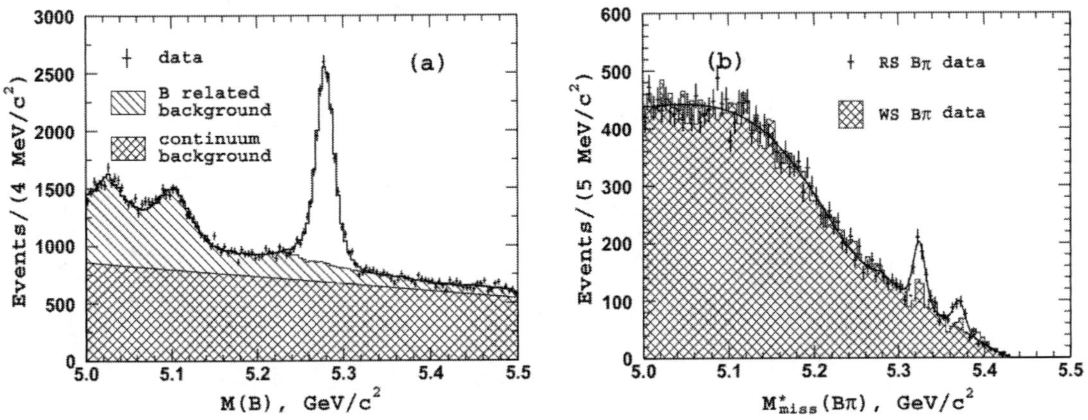

Figure 4. The (a) invariant mass and (b) $M^*_{\rm miss}(B\pi)$ distribution for B candidates in the B signal region. Points with error bars represent the data. The open histogram in (a) shows the result of the fit to data. The solid line in (b) shows the result of the fit to the RS $B\pi$ data; the dashed line represents the background level.

in Fig. 4(b), where peaks corresponding to the $BB^*\pi$ and $B^*B^*\pi$ signals are evident. Combinations with π^+ — the wrong sign (WS) combinations — are used to evaluate the shape of the combinatorial background. There is also a hint for a peaking structure in the WS $M^*_{\rm miss}$ distribution, shown as a hatched histogram in Fig. 4(b). Due to $B^0 - \bar{B}^0$ oscillations, we expect a fraction of the produced B^0 mesons to decay as $\bar{B}^0$ given by $0.5x_d^2/(1 + x_d^2) = 0.1861 \pm 0.0024$, where x_d is the B^0 mixing parameter [18].

A binned maximum likelihood fit is performed to fit the $M^*_{\rm miss}$ distribution. ISR events produce an $M^*_{\rm miss}$ distribution similar to that for $q\bar{q}$ events; these two components are modeled by a single threshold function. The resolution of the signal peaks in Fig. 4(c) is dominated by the c.m. energy spread and is fixed at 6.5 MeV/c^2 as determined from the signal MC. The fit to the RS spectrum yields $N_{BB\pi} = 13 \pm 25$, $N_{BB^*\pi} = 357 \pm 30$ and $N_{B^*B^*\pi} = 161 \pm 21$ signal events. The statistical significance of the observed $BB^*\pi$ and $B^*B^*\pi$ signal is 9.3σ and 8.1σ, respectively. The statistical significance is calculated as $\sqrt{-2\ln(\mathcal{L}_0/\mathcal{L}_{\rm sig})}$, where $\mathcal{L}_{\rm sig}$ and $\mathcal{L}_0$ denote the likelihood values obtained with the nominal fit and with the signal yield fixed at zero, respectively.

For the subsequent analysis, we require $|M^*_{\rm miss} - m_{B^*}| < 15$ MeV/c^2 to select $BB^*\pi$ signal events and $|M^*_{\rm miss} - (m_{B^*} + \Delta m_B)| < 12$ MeV/c^2, where $\Delta m_B = m_{B^*} - m_B$, to select $B^*B^*\pi$ events. For the selected $B^*B^{(*)}\pi$ candidates, we calculate $M_{\rm miss}(\pi) = \sqrt{(\sqrt{s} - E_\pi)^2/c^4 - P_\pi^2/c^2}$, where E_π and P_π are the reconstructed energy and momentum, respectively, of the charged pion in the c.m. frame. We perform a simultaneous binned maximum likelihood fit to the RS and WS samples, assuming the same number and distribution of background events in both samples and known fraction of signal events in the RS sample that leaks to the WS sample due to mixing. To fit the $M_{\rm miss}(\pi)$ spectrum, we use the function

$$F(m) = [f_{\rm sig}S(m) + B(m)]\epsilon(m)F_{\rm PHSP}(m), \tag{8}$$

where $m \equiv M_{\rm miss}(\pi)$; $f_{\rm sig} = 1.0$ (0.1105 ± 0.0016, [33]) for the RS (WS) sample; $S(m)$ and $B(m)$ are the signal and background PDFs, respectively; and $F_{\rm PHSP}(m)$ is the phase space function. To account for the instrumental resolution, we smear the function $F(m)$ with a Gaussian function.

Table 2. Summary of fit results to the $M_{miss}(\pi)$ distributions for the three-body $BB^*\pi$ and $B^*B^*\pi$ final states.

Mode	Parameter	Model-0	Model-1		Model-2		Model-3
			Solution 1	Solution 2	Solution 1	Solution 2	
$BB^*\pi$	$f_{Z_b(10610)}$	1.0	1.45 ± 0.24	0.64 ± 0.15	1.01 ± 0.13	1.18 ± 0.15	–
	$f_{Z_b(10650)}$	–	–	–	0.05 ± 0.04	0.24 ± 0.11	–
	$\phi_{Z_b(10650)}$, rad.	–	–	–	-0.26 ± 0.68	-1.63 ± 0.14	–
	f_{nr}	–	0.48 ± 0.23	0.41 ± 0.17	–	–	1.0
	ϕ_{nr}, rad.	–	-1.21 ± 0.19	0.95 ± 0.32	–	–	–
	$-2\log\mathcal{L}$	−304.7	−300.6	−300.5	−301.4	−301.4	−344.5
$B^*B^*\pi$	$f_{Z_b(10650)}$	1.0	1.04 ± 0.15	0.77 ± 0.22			–
	f_{nr}	–	0.02 ± 0.04	0.24 ± 0.18			1.0
	ϕ_{nr}, rad.	–	0.29 ± 1.01	1.10 ± 0.44			–
	$-2\log\mathcal{L}$	−182.4	−182.4	−182.4			−209.7

We first analyze of the $BB^*\pi$ $[B^*B^*\pi]$ data with the simplest hypothesis, Model-0, that includes only the $Z_b(10610)$ $[Z_b(10650)]$ amplitude. Results of the fit are shown in Fig. 5; the numerical results are summarized in Table 2. The fraction f_X of the total three-body signal attributed to a particular quasi-two-body intermediate state is calculated as $f_X = \int |\mathcal{A}_X|^2\, dm / \int S(m)\, dm$, where $\mathcal{A}_X$ is the amplitude for a particular component X of the three-body amplitude. Next, we extend the hypothesis to include a possible non-resonant component, Model-1, and then the $BB^*\pi$ data is fit to a combination of two Z_b amplitudes, Model-2. In both cases, we do not get a statistically significant improvement in the data description: the likelihood value is only marginally improved (see Table 2). The addition of extra components to the amplitude also produces multiple maxima in the likelihood function. As a result, we use Model-0 as our nominal hypothesis. Finally, we fit both samples to a pure non-resonant amplitude (Model-3). In this case, the fit is significantly worse.

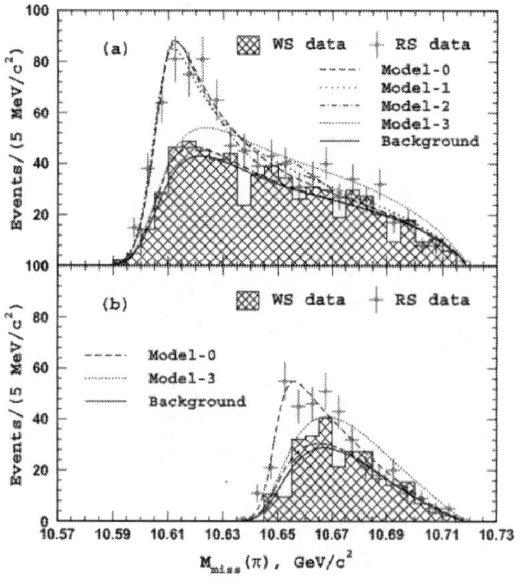

Figure 5. The $M_{miss}(\pi)$ distribution for the (a) $BB^*\pi$ and (b) $B^*B^*\pi$ candidate events.

If the parameters of the Z_b resonances are allowed to float, the fit to the $BB^*\pi$ data with Model-0 gives 10605 ± 6 MeV/c^2 and 25 ± 7 MeV for the $Z_b(10610)$ mass and width, respectively, and the

Table 3. $\mathcal{B}$ branching fractions for the $Z_b^+(10610)$ and $Z_b^+(10650)$ decays. The first quoted uncertainty is statistical; the second is systematic.

Channel	Fraction, %	
	$Z_b(10610)$	$Z_b(10650)$
$\Upsilon(1S)\pi^+$	$0.60 \pm 0.17 \pm 0.07$	$0.17 \pm 0.06 \pm 0.02$
$\Upsilon(2S)\pi^+$	$4.05 \pm 0.81 \pm 0.58$	$1.38 \pm 0.45 \pm 0.21$
$\Upsilon(3S)\pi^+$	$2.40 \pm 0.58 \pm 0.36$	$1.62 \pm 0.50 \pm 0.24$
$h_b(1P)\pi^+$	$4.26 \pm 1.28 \pm 1.10$	$9.23 \pm 2.88 \pm 2.28$
$h_b(2P)\pi^+$	$6.08 \pm 2.15 \pm 1.63$	$17.0 \pm 3.74 \pm 4.1$
$B^+\bar{B}^{*0} + \bar{B}^0B^{*+}$	$82.6 \pm 2.9 \pm 2.3$	–
$B^{*+}\bar{B}^{*0}$	–	$70.6 \pm 4.9 \pm 4.4$

fit to the $B^*B^*\pi$ data gives 10648 ± 13 MeV/c^2 and 23 ± 8 MeV for the $Z_b(10650)$ mass and width, respectively. The large errors here reflect the strong correlation between the resonance parameters.

Using the results of the fit to the $M_{\rm miss}(\pi)$ spectra with the nominal model (Model-0 in Table 2) and the results of the analyses of $e^+e^- \to \Upsilon(nS)\pi^+\pi^-$ [28] and $e^+e^- \to h_b(mP)\pi^+\pi^-$ [34, 35], we calculate the ratio of the branching fractions $\mathcal{B}(Z_b(10610) \to B\bar{B}^* + c.c.)/\mathcal{B}(Z_b(10610) \to \text{bottomonium}) = 4.76 \pm 0.64 \pm 0.75$ and $\mathcal{B}(Z_b(10650) \to B^*\bar{B}^*)/\mathcal{B}(Z_b(10650) \to \text{bottomonium}) = 2.40 \pm 0.44 \pm 0.50$.

We calculate the relative fractions for Z_b decays, assuming that they are saturated by the already observed $\Upsilon(nS)\pi$, $h_b(mP)\pi$, and $B^*B^{(*)}$ channels. The results are summarized in Table 3.

In conclusion, we report the first observations of the three-body $e^+e^- \to BB^*\pi$ and $e^+e^- \to B^*B^*\pi$ processes with a statistical significance above 8σ. The analysis of the $B^{(*)}B^*$ mass spectra indicates that the total three-body rates are dominated by the intermediate $e^+e^- \to Z_b(10610)^{\mp}\pi^{\pm}$ and $e^+e^- \to Z_b(10650)^{\mp}\pi^{\pm}$ transitions for the $BB^*\pi$ and $B^*B^*\pi$ final states, respectively.

6 Observation of the decay $\Lambda_c^+ \to pK^+\pi^-$

Several doubly Cabibbo-suppressed (DCS) decays of charmed mesons have been observed. Their measured branching ratios with respect to corresponding Cabibbo-favored (CF) decays play an important role in constraining models of the decay of charmed hadrons and in the study of flavor-$SU(3)$ symmetry [37]. Because of the smaller production cross sections for charmed baryons, DCS decays of charmed baryons have not yet been observed, and only an upper limit, $\mathcal{B}(\Lambda_c^+ \to pK^+\pi^-)/\mathcal{B}(\Lambda_c^+ \to pK^-\pi^+) < 0.46\%$ with 90% confidence level, has been reported by the FOCUS Collaboration [38]. Theoretical calculations of DCS decays of charmed baryons have been limited to two-body decay modes [39, 40].

Recently Belle reported the first observation of the DCS decay $\Lambda_c^+ \to pK^+\pi^-$ and the measurement of its branching ratio with respect to its counterpart CF decay $\Lambda_c^+ \to pK^-\pi^+$. In the letter [36], Belle reports the first observation of the DCS decay $\Lambda_c^+ \to pK^+\pi^-$ and the measurement of its branching ratio with respect to its counterpart CF decay $\Lambda_c^+ \to pK^-\pi^+$. Unlike charmed meson decays, internal W emission and W exchange are not suppressed for charmed baryon decays. In previous studies of CF or singly Cabibbo-suppressed (SCS) decays of Λ_c^+ and Ξ_c^0, direct evidence of W exchange and internal W emission has been observed [41]. When we consider that W exchange is prohibited in $\Lambda_c^+ \to pK^+\pi^-$ but allowed in $\Lambda_c^+ \to pK^-\pi^+$, the contribution of W exchange to Λ_c^+ decays can be estimated by comparing the measured branching ratio with the naïve expectation [38], $\tan^4\theta_c$ (0.285%), where θ_c is the Cabibbo mixing angle [42] and $\sin\theta_c = 0.225 \pm 0.001$ [18]. This approach does not take into account effects of flavor-$SU(3)$ symmetry breaking.

A Λ_c^+ candidate is reconstructed from the three charged hadrons. To suppress combinatorial backgrounds, especially from B meson decays, we place a requirement on the scaled momentum: $x_p > 0.53$, where x_p is defined as $p^* / \sqrt{E_{\rm cm}^2/4 - M^2}$; here, $E_{\rm cm}$ is the total center-of-mass energy, p^* is the momentum in the center-of-mass frame, and M is the mass of the Λ_c^+ candidate. In addition, the χ^2 value from the common vertex fit of the charged tracks must be less than 40.

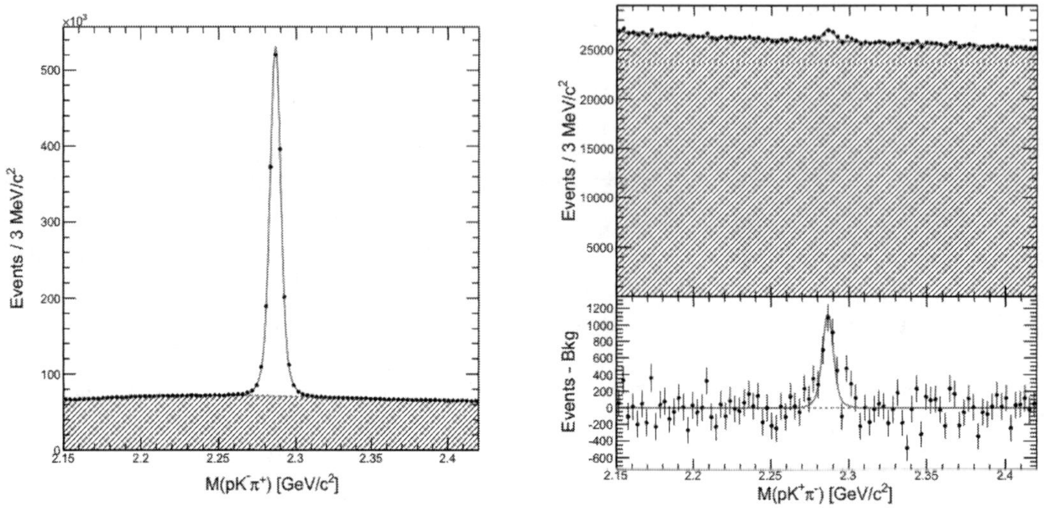

Figure 6. Distribution of $M(pK^-\pi^+)$ (left), $M(pK^+\pi^-)$ (top right) and residuals of data with respect to the fitted combinatorial background (bottom right). The curves indicate the fit result: the full fit mode 1 (solid) and the combinatoric background only (dashed).

Figures 6, *a)* and 6, *b)* show invariant mass distributions, $M(pK^-\pi^+)$ (CF) and $M(pK^+\pi^-)$ (DCS), with the final selection criteria. DCS decay events are clearly observed in $M(pK^+\pi^-)$. We perform a binned least-χ^2 fit to the two distributions from 2.15 GeV$/c^2$ to 2.42 GeV$/c^2$ with 0.01 MeV$/c^2$ bin width, and the figures are drawn with merged bins. The DCS decay has a peaking background from the SCS decay $\Lambda_c^+ \to \Lambda K^+$ with $\Lambda \to p\pi^-$, which has the same final state topology. However, because of the long Λ lifetime, many of the Λ vertexes are displaced by several centimeters from the main vertex so the geometrical requirements suppress most of this background. The remaining SCS-decay yield is included in the signal yield of $\Lambda_c^+ \to pK^+\pi^-$ decay and is estimated via the relation

$$N(SCS;\Lambda \to p\pi^-) = \frac{\epsilon(SCS;\Lambda \to p\pi^-)}{\epsilon(CF)} \frac{\mathcal{B}(SCS;\Lambda \to p\pi^-)}{\mathcal{B}(CF)} N(CF), \tag{9}$$

where $N(CF)$ is the signal yield of the CF decay, $\mathcal{B}(SCS;\Lambda \to p\pi^-)/\mathcal{B}(CF) = (0.61 \pm 0.13)\%$ is the branching ratio [18], and $\epsilon(SCS;\Lambda \to p\pi^-)/\epsilon(CF) = 0.023$ is the relative efficiency found using MC samples. After subtraction of this SCS component, the signal yield of the DCS decay is $3379 \pm 380 \pm 78$, where the first uncertainty is statistical and the second is systematic due to this subtraction. To estimate the statistical significance of the DCS signal, we exclude the SCS signal by vetoing events with 1.1127 GeV$/c^2$ $< M(p\pi^-) <$ 1.1187 GeV$/c^2$. The significance is estimated as $\sqrt{-2\ln(\mathcal{L}_0/\mathcal{L})}$, where $\mathcal{L}_0$ and $\mathcal{L}$ are the maximum likelihood values from binned maximum likelihood fits with the signal yield fixed to zero and allowed to float, respectively. The calculated significance corresponds to 9.4σ.

The branching ratio, $\mathcal{B}(\Lambda_c^+ \to pK^+\pi^-)/\mathcal{B}(\Lambda_c^+ \to pK^-\pi^+)$, is $(2.35 \pm 0.27 \pm 0.21) \times 10^{-3}$, where the uncertainties are statistical and systematic, respectively. The branching fraction of the CF decay, $(6.84 \pm 0.24^{+0.21}_{-0.27}) \times 10^{-2}$, was already well-measured in a previous Belle analysis [43]. Combining that with our measurement, we determine the absolute branching fraction of the DCS decay to $(1.61 \pm 0.23^{+0.07}_{-0.08}) \times 10^{-4}$, where the first uncertainty is the total uncertainty of the branching ratio and the second is uncertainty of the branching fraction of CF decay. This measured branching ratio corresponds to $(0.82 \pm 0.12) \tan^4 \theta_c$, where the uncertainty is the total.

References

[1] A. Abashian *et al.* (Belle Collab.), Nucl. Instr. and Meth. A **479**, 117 (2002).
[2] S. Kurokawa and E. Kikutani, Nucl. Instrum. Methods Phys. Res., Sect. A **499**, 1 (2003), and other papers included in this volume; T. Abe *et al.*, Prog. Theor. Exp. Phys. (2013) 03A001 and following articles up to 03A011.
[3] A. Abdesselam *et al.* (Belle Collaboration), arXiv:1604.04042.
[4] R. Aaij *et al.* JHEP **02**, 104 (2016).
[5] R. Aaij *et al.* JHEP **08**, 131 (2013).
[6] S. Descotes-Genon, J. Matias, M. Ramon, and J. Virto, JHEP **01**, 048 (2013).
[7] S. Descotes-Genon, T. Hurth, J. Matias, and J. Virto. JHEP **1305**, 137 (2013).
[8] R. Aaij *et al.* Phys. Rev. Lett. **111**, 191801 (2013).
[9] S. Fajfer, J.F.Kamenik, and I. Nisandzic, Phys. Rev. D **85**, 094025 (2012).
[10] J.F. Kamenik, and F. Mescia, Phys. Rev. D **78**, 014003 (2008).
[11] J.P. Lees *et al.* (*BABAR* Collaboration), Phys. Rev. Lett. **109**, 101802 (2012); J.P. Lees *et al.* (*BABAR* Collaboration), Phys. Rev. D **88**, 072012 (2013);
[12] A. Matyja *et al.* (Belle Collaboration), Phys. Rev. Lett. **99**, 191807 (2007).
[13] A. Bozek *et al.* (Belle Collaboration), Phys. Rev. D **82**, 072005 (2010).
[14] M. Huschle *et al.* (Belle Collaboration), Phys. Rev. D **92**, 072014 (2015).
[15] R. Aaij *et al.* (LHCb Collaboration), Phys. Rev. Lett. **115**, 111803 (2015).
[16] Y. Amhis *et al.*, arXiv:1412.7515 and online update at http://www.slac.stanford.edu/xorg/hfag/
[17] A. Abdesselam *et al.* (Belle collaboration), arXiv:1603.06711.
[18] K. A. Olive *et al.* (Particle Data Group), Chin. Phys. C **38**, 090001 (2014).
[19] B. Pal *et al.* (Belle Collaboration), Phys. Rev. Lett. **116**, 161801 (2016).
[20] C. H. Chen, Phys. Lett. B **520**, 33 (2001); A. R. Williamson and J. Zupan, Phys. Rev. D **74**, 014003 (2006); A. Ali *et al.*, Phys. Rev. D **76**, 074018 (2007); C. K. Chua, Phys. Rev. D **78**, 076002 (2008); K. Wang and G. Zhu, Phys. Rev. D **88**, 014043 (2013); J. J. Wang, D. T. Lin, W. Sun, Z. J. Ji, S. Cheng and Z. J. Xiao, Phys. Rev. D **89**, 074046 (2014); Q. Chang, J. Sun, Y. Yang and X. Li, Phys. Lett. B **740**, 56 (2015); H. Y. Cheng, C. W. Chiang and A. L. Kuo, Phys. Rev. D **91**, 014011 (2015).
[21] Q. Chang, X. Q. Li and Y. D. Yang, J. Phys. G **41**, 105002 (2014).
[22] S. Baek, D. London, J. Matias and J. Virto, J. High Energy Phys. **12**, (2006) 019; A. Hayakawa, Y. Shimizu, M. Tanimoto and K. Yamamoto, Prog. Theor. Exp. Phys. **2014**, 023B04 (2014).
[23] C.-C. Peng *et al.* (Belle Collaboration), Phys. Rev. D **82**, 072007 (2010).
[24] C. Oswald *et al.* (Belle Collaboration), Phys. Rev. D **92**, 072013 (2015)
[25] S. Esen *et al.* (Belle Collaboration), Phys. Rev. D **87**, 031101(R) (2013).
[26] T. Abe *et al.* (Belle II Collaboration), arXiv:1011.0352.
[27] A. Garmash *et al.* (Belle Collaboration), Phys. Rev. Lett. **116**, 212001 (2016).

[28] A. Bondar *et al.* (Belle Collaboration), Phys. Rev. Lett. **108**, 122001 (2012).

[29] A. Garmash *et al.* (Belle Collaboration), Phys. Rev. D **91**, 072003 (2015).

[30] D.-Y. Chen and X. Liu, Phys. Rev. D **84**, 094003 (2011); A. Ali, C. Hambrock and W. Wang, Phys. Rev. D **85**, 054011 (2012); I. V. Danilkin, V. D. Orlovsky and Y. A. Simonov, Phys. Rev. D **85**, 034012 (2012); S. Ohkoda, Y. Yamaguchi, S. Yasui, K. Sudoh and A. Hosaka, Phys. Rev. D **86**, 014004 (2012); E. Braaten, C. Langmack and D. Hudson Smith, Phys. Rev. D **90**, 014044 (2014).

[31] A. Bondar, A. Garmash, A.I. Milstein, R. Mizuk, and M.B. Voloshin, Phys. Rev. D **84**, 054010 (2011).

[32] A. Drutskoy *et al.* (Belle Collaboration), Phys. Rev. D **81**, 112003 (2010).

[33] Determined using mixing parameter x_d and the ratio of charged to neutral B yields measured in data from two-body $e^+e^- \to B^{(*)}B^{(*)}$ processes.

[34] I. Adachi *et al.* (Belle Collaboration), Phys. Rev. Lett. **108**, 032001 (2012).

[35] The fits to the $M_{\rm miss}(\pi)$ distributions performed in Ref. [28] give fractions of $f_{Z(10610)^\pm\pi^\mp} = 0.423^{+0.05+0.007}_{-0.127-0.009}$ $(0.352^{+0.156+0.001}_{-0.004-0.134})$ and $f_{Z(10650)^\pm\pi^\mp} = 0.602^{+0.103+0.11}_{-0.211-0.038}$ $(0.648^{+0.152+0.067}_{-0.114-0.155})$ for the $h_b(1P)$ $(h_b(2P))$ decays.

[36] S.B. Yang *et al.* (Belle collaboration), Phys. Rev. Lett. **117**, 011801 (2016).

[37] B.R. Ko *et al.* (Belle Collaboration), Phys. Rev. Lett. **102**, 221802 (2009).

[38] J.M. Link *et al.* (FOCUS Collaboration), Phys. Lett. B **624**, 166 (2005).

[39] K.K. Sharma and R.C. Verma, Phys. Rev. D **55**, 7067 (1997).

[40] T. Uppal, R.C. Verma, and M.P. Khanna, Phys. Rev. D **49**, 3417 (1994).

[41] S. Henderson *et al.* (CLEO Collaboration), Phys. Lett. B **283**, 161 (1992); P. Avery *et al.* (CLEO Collaboration), Phys. Rev. Lett. **71**, 2391 (1993); K. Abe *et al.* (Belle Collaboration), Phys. Lett. B **524**, 33 (2002); B. Aubert *et al.* (*BABAR* Collaboration), Phys. Rev. Lett. **95**, 142003 (2005).; R. Chistov *et al.* (Belle Collaboration), Phys. Rev. D **88**, 071103(R) (2013).

[42] N. Cabibbo, Phys. Rev. Lett. **10**, 531 (1963).

[43] A. Zupanc *et al.* (Belle Collaboration), Phys. Rev. Lett. **113**, 042002 (2014).

EPJ Web of Conferences **125**, 01009 (2016) DOI: 10.1051/epjconf/201612501009
QUARKS-2016

Chiral Imbalance in QCD and its consequences

Alexander Andrianov[1,2,⋆] *Vladimir* Andrianov[1,⋆⋆] and *Domenec* Espriu[2,⋆⋆⋆]

[1]*Faculty of Physics, Saint Petersburg State University, Universitetskaya nab. 7/9, Saint Petersburg 199034, Russia, and*
[2]*Departament de Física Quàntica i Astrofísica and Institut de Ciéncies del Cosmos (ICCUB),Universitat de Barcelona, Martí i Franqués 1, 08028 Barcelona, Spain*

Abstract. Under extreme conditions of high temperature and/or large quark (baryon) density, the vacuum of QCD changes its properties, and deconfinement, chiral symmetry restoration as well as chiral symmetry breaking take place. These transitions (phases) are accompanied by the rapid change in the rate and nature of topological transitions connecting different topological sectors. The heavy ion collisions (HIC) program opens a possibility to study these phenomena in so-called non-Abelian Quark-gluon plasma (QGP). In these phases the currents of light quarks (vector and axial-vector) can be independently examined for right-handed (RH) and left-handed (LH) quarks. To describe such a quark matter chiral chemical potential can be introduced to quantify the presence of chirality imbalance (ChI) i.e. the difference between the average numbers of RH and LH quarks in the fireball after HIC. In this review talk we will focus our attention on the discussion of the ChI related developments in heavy ion physics at central collisions and the plans for the future experiments aimed at establishing (or falsifying) the presence of Local spacial Parity Breaking (LPB) in heavy ion data. We describe some of experimental observables in detecting the signal of LPB. A number of measurements is proposed that allow to reach a definite conclusion on the occurrence of LPB effects in non-Abelian QGP produced in central heavy ion collisions and its simulation within a number of QCD-inspired models is outlined. Based on the effective meson theory in the presence of Chern-Simons interaction it is found that the spectrum of massive vector mesons splits into three polarization components with different effective masses. Moreover a resonance broadening occurs that leads to an increase of spectral contribution to the dilepton production as compared to the vacuum state. The asymmetry in production of longitudinally and transversely polarized states of ρ and ω mesons for various values of the dilepton invariant mass can serve as a characteristic indication of the LPB in CERES, HADES, PHENIX, STAR, NA60 and ALICE experiments.

1 Introduction and Outline

Nowadays Quantum Chromodynamics (QCD) represents a theory with known symmetries and well established elementary constituents. However their interaction leading to the observed hadron states

⋆e-mail: andrianov@icc.ub.edu
⋆⋆e-mail: v.andriano@rambler.ru
⋆⋆⋆e-mail: espriu@icc.ub.edu

still needs in the adequate description beyond the perturbation theory, especially, at low and intermediate energies.

It is widely believed that such a non-perturbative dynamics occurs due to the nontrivial topology of QCD vacuum as a consequence of the compactness of non-Abelian color gauge group $SU(3)$. Its compactness allows for topologically nontrivial configurations of the gluon field. These configurations essentially determine the QCD vacuum structure. The latter is composed of the superposition of an infinite set of topologically distinct states connected by tunneling instanton-like transitions [1] and becomes a "θ-vacuum" of the theory [2, 3]. It is likely that topological effects in QCD are responsible for the chiral symmetry breaking (see [4, 5] for a review).

With the light current quark masses negligible compared to relevant energy scales, the QCD Lagrangian has an approximate chiral symmetry. However this symmetry is not realized in the ground state due to formation of quark-anti-quark pair condensate in the QCD vacuum. This is of course the well-known phenomenon of spontaneously broken chiral symmetry which is a fundamental property of QCD. On the other hand, lattice QCD computations have shown that with increasing temperature such a vacuum condensate will be melting with the chiral symmetry being restored at sufficiently high temperature $T > T_c \sim 170MeV$ [6, 7]. Therefore, chiral symmetry restoration at high temperatures is also a basic prediction of QCD. With such a symmetry in light quark sector, the currents (vector and axial-vector) in quark-gluon phase (QGP) can be independently examined for right-handed (RH) and left-handed (LH) quarks. To describe such a matter a chiral chemical potential can be introduced in order that to quantify the presence of chirality imbalance i.e. the difference between the numbers of RH and LH chiral fermions in the system. Quarks with specific chirality have their momentum direction correlated with spin orientation.

There remains however the crucial question of how to achieve a chiral QGP with chirality imbalance in the first place. The process of creating nonzero chirality pertains to the famous "chiral anomaly" [8, 9]. This anomaly implies the breaking of axial current conservation by quantum loop effects. Correspondingly, locally, chirality imbalance can be generated through nonorthogonal electric and magnetic field strength. Such fields can be either QED EM fields or QCD chromo-EM ones. In QCD the gluon configurations generating chiral imbalance globally exist: they are the topological solitons such as instantons and sphalerons. These topological objects play crucial role for non-perturbative dynamics in QCD vacuum [10] as well as in hot QGP [11, 12]. Such configurations have their global topological winding number Q_w, and when coupled with chiral fermions, can generate an amount of global chirality imbalance via chiral anomaly: $N_R - N_L = 2Q_w$ for each flavor of light quarks. In short, the chirality imbalance is a direct manifestation of the QCD topological fluctuations and could become observable through the asymmetry in polarizations of dilepton pairs [13] in central heavy ion collisions (HIC) and the Chiral magnetic wave (CMW) in non-central HIC [14]. Summarizing one can say that the Chiral imbalance (ChI) of quarks is a remarkable phenomenon that stems from highly nontrivial interplay of chiral symmetry of QCD, axial anomaly, and the topology of gluon configurations.

In this talk we focus the reader's attention on the discussion of ChI related developments in heavy ion physics at *central* collisions and the plans for the future experiments aimed at establishing (or falsifying) the presence of Local parity breaking (LPB) in HIC data.

We describe:

- Where and how does chiral imbalance appear: Phenomenology of parity breaking in finite volume.
- Chiral imbalance in quark-gluon phase: topological charge vs. axial (chiral) charge: topological vs. chiral chemical potentials.
- Chiral imbalance in hadronic phase: scalar, pseudoscalar, vector meson spectra[13].

▸ Observables sensitive to chiral imbalance - manifestation of LPB in heavy ion collisions (HIC): dilepton polarization asymmetry [13].

2 Phenomenology of parity breaking in finite volume

Parity is a well established global symmetry of strong interactions. Are there any reasons to believe it may be broken in a finite volume?! Local large fluctuations in the topological charge presumably exist in a hot environment and are responsible for LPB.

- Quantum fluctuations of θ parameter: *P*-odd bubbles [14], their manifestation in Chiral Magnetic Effect (CME)[15]) (an earlier proposal: [16]).
- New QCD phase characterized by a spontaneous parity breaking due to formation of neutral pion-like background [17, 18] .
- High energy production of pseudoscalar gluelumps $\Leftrightarrow$ of parity-odd bunches of gluon jets $\Rightarrow$ then a PB background remains inside a hot dense nuclear fireball in HIC !?

Observable effects:

- For *peripheral* heavy ion collisions they lead to the Chiral Magnetic Effect (CME): Large $\vec{B}$ $\Rightarrow$ large $\vec{E}$ $\Rightarrow$ charge separation[14].
- For *central* collisions (and light quarks) they correspond to a phase with axial chemical potential $\mu_5 \neq 0$ located in "fluctons" of few-Fermi size [13].

The two effects are complementary!

3 Topological charge and topological chemical potential

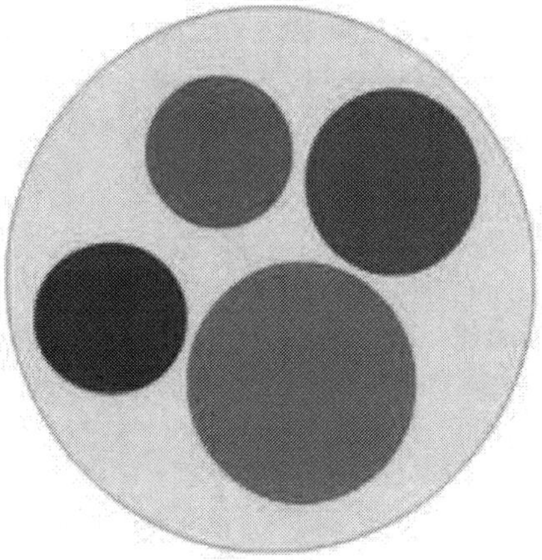

Figure 1. Interplay between domains with positive (in red) and negative (in blue) topological charge in the fireball interior

In hot QCD (in heavy ion collisions) a metastable topological charge $\langle T_5\rangle$ may arise in a finite volume due to sphaleron-like transitions [19].

$$T_5 = \frac{1}{8\pi^2}\int_{\mathrm{vol.}} d^3x\varepsilon_{jkl}\mathrm{Tr}\left(G^j\partial^k G^l - i\frac{2}{3}G^jG^kG^l\right)$$

it may survive for a sizeable lifetime in a heavy-ion fireball

$$\langle\Delta T_5\rangle \neq 0 \quad \text{for} \quad \Delta t \simeq \tau_{\text{fireball}} \simeq 5 \div 10 \text{ fm/c};$$

$$(7 \pm 1 \text{ fm/c} \ \ [20])$$

For the fireball lifetime one can trigger the value of $\langle\Delta T_5\rangle \neq 0$ introducing into the QCD Lagrangian a topological chemical potential μ_θ in a gauge invariant way via $\Delta\mathcal{L}_{top} = \mu_\theta\Delta T_5$, where

$$\Delta T_5 = T_5(t_f) - T_5(0) = \frac{1}{8\pi^2}\int_0^{t_f} dt \int_{\mathrm{vol.}} d^3x \mathrm{Tr}\left(G^{\mu\nu}\widetilde{G}_{\mu\nu}\right).$$

We notice that in finite 3 + 1 dimensional volume the so-called index theorem does not work and the jump of topological charge is not necessarily integer.

4 Axial baryon charge and axial chemical potential

The exact operator law in QCD, the partial conservation of axial current (broken by current quark masses m_q for light quarks $q = u, d, s$ and gluon anomaly)

$$\partial_\mu J_5^\mu - 2im_qJ_5 = \frac{1}{8\pi^2}\mathrm{Tr}\left(G^{\mu\nu}\widetilde{G}_{\mu\nu}\right)$$

predicts the rigid relations between changes in topology of gauge configurations and the induced axial charge (for small quark masses $m_q \simeq 0$). If the gluon and quark flows across the volume boundary of a fireball are reasonably small (a fireball is metastable) then,

$$\frac{d}{dt}\left(Q_5 - 2N_fT_5\right) \simeq 0, \quad Q_5 = \int_{\mathrm{vol.}} d^3x \sum_{u,d,s} q^\dagger\gamma_5 q = \langle N_L - N_R\rangle,$$

and the chiral charge generated in a heavy ion collision is approximately conserved during $\tau_{fireball}$.

The characteristic left-right oscillation time is governed by inverse quark masses. Indeed if an excess of left(or right)-handed quark states $q_L(x_0 = 0)$ arises in the fireball in the initial moment of central collision then their further evolution in the rest frame can be approximately characterized by the following equation,

$$q(x_0) \simeq \left(\cos(m_qx_0) - i\gamma_0\sin(m_qx_0)\right)q_L(0). \tag{1}$$

Therefore after a period $\Delta x_0 \sim \pi/2m_q$ left-handed quark states become essentially right-handed.

- For u, d quarks $1/m_q \sim 1/5 \text{ MeV}^{-1} \sim 40 \text{ fm} \gg \tau_{\text{fireball}}$ and the left-right quark mixing can be neglected during fireball life time.
- For s quark $1/m_s \sim 1/150 \text{ MeV}^{-1} \sim 1 \text{ fm} \ll \tau_{\text{fireball}}$ and the chiral charge averaged over the fireball volume and its lifetime vanishes $\langle Q_5^s\rangle \simeq 0$ due to left-right oscillations.

Correspondingly for u, d quarks in a background with the jump of topological charge $\langle \Delta T_5 \rangle \neq 0$ or with the chiral (axial) charge $\langle Q_5 \rangle \neq 0$ QCD can be equally described at the Lagrangian level by introducing a topological chemical potential μ_θ or an axial chemical potential μ_5

$$\langle \Delta T_5 \rangle \simeq \frac{1}{2N_f} \langle Q_5 \rangle \Longleftrightarrow \mu_5 \simeq \frac{1}{2N_f} \mu_\theta,$$

$$\Delta \mathcal{L}_{top} = \mu_\theta \Delta T_5 \Longleftrightarrow \Delta \mathcal{L}_q = \mu_5 Q_5$$

Thus one can think about a generation of chiral imbalance from non-trivial topological charge or visa versa inducing a topologically non-trivial gluon configuration by imbalance of left- and right-handed quark densities. As both properties are related locally one may get a parity violation generated by either charge in a finite volume of fireball and visible during fireball lifetime.

How does μ_5 affect the hadronic phenomenology?

5 Effective meson theory in a medium with LPB

In the environment with chiral chemical potential the scalar sector can be described by using the spurion technique in the chiral Lagrangian with an isosinglet μ_5 [21],

$$D_\nu \Longrightarrow D_\nu - i\{\mathbf{I}_q \mu_5 \delta_{0\nu}, \cdot\} = D_\nu - 2i\mathbf{I}_q \mu_5 \delta_{0\nu}, \tag{2}$$

where for three light flavors $\mathbf{I}_q \equiv \mathrm{diag}[1, 1, 0]$ in accordance with the preceding discussion. An effective Lagrangian is required to include the lightest isoscalar degrees of freedom such as σ and $a_0(980)$, which will be mixed with their pseudoscalar partners η, η' and π, respectively.

In the lowest order of chiral expansion such an effective lagrangian reads,

$$\mathcal{L} = \frac{1}{4} \mathrm{Tr}\left(D_\mu H D^\mu H^\dagger\right) + \frac{B}{2} \mathrm{Tr}\left[\hat{m}(H + H^\dagger)\right] + \frac{M^2}{2} \mathrm{Tr}\left(HH^\dagger\right)$$
$$-\frac{\lambda_1}{2} \mathrm{Tr}\left[(HH^\dagger)^2\right] - \frac{\lambda_2}{4}\left[\mathrm{Tr}\left(HH^\dagger\right)\right]^2 + \frac{c}{2}(\det H + \det H^\dagger)$$
$$+\frac{d_1}{2} \mathrm{Tr}\left[\hat{m}(HH^\dagger H + H^\dagger H H^\dagger)\right] + \frac{d_2}{2} \mathrm{Tr}\left[\hat{m}(H + H^\dagger)\right] \mathrm{Tr}\left(HH^\dagger\right) \tag{3}$$

where the $SU(3)_L \times SU(3)_R$ chirally invariant part is parameterized by constants $M, \lambda_{1,2}, c$ and soft chiral symmetry breaking is realized with the help of vertices containing $\hat{m} = \mathrm{diag}[m_u, m_d, m_s]$ and constants $B, d_{1,2}$. The convenient splitting of scalar fields is given by

$$H = \xi \Sigma \xi, \quad \xi = \exp\left(i\frac{\Phi}{2f}\right), \quad \Phi = \lambda^a \phi^a, \quad \Sigma = \lambda^b \sigma^b$$

The v.e.v. of the neutral scalars are defined as $v_i = \langle \Sigma_{ii} \rangle$ where $i = u, d, s$, and, in the chiral limit $\hat{m} = 0$, satisfy the following gap equations:

$$(2\mu_5^2 + M^2) - 2(\lambda_1 + \lambda_2)v_q^2 - \lambda_2 v_s^2 + c v_s = 0;$$

$$(-2\lambda_2 + \frac{c}{v_s})v_q^2 + M^2 - (2\lambda_1 + \lambda_2)v_s^2 = 0,$$

where for non-vanishing isosinglet μ_5 we find our solutions to be $v_u = v_d \equiv v_q \neq v_s$.

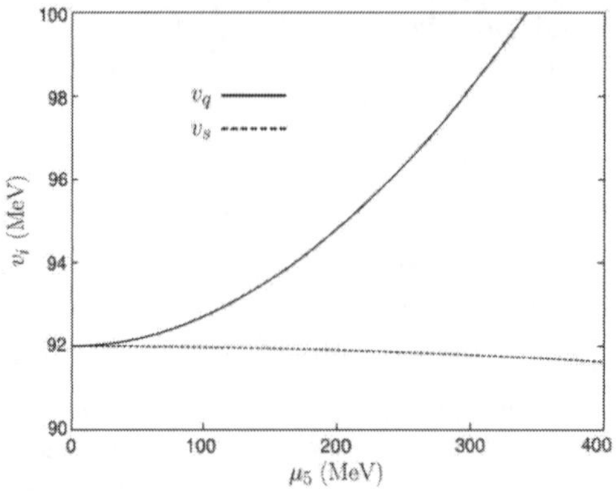

Figure 2. Vacuum condensates in the presence of chiral imbalance

Evidently for large μ_5 the correct solution has the following asymptotics,

$$v_q^2 = \frac{2\mu_5^2 + M^2 + \frac{c^2}{4\lambda^2}}{2(\lambda_1 + \lambda_2)} + O(1/\mu_5^2), \qquad v_s = \frac{c}{2\lambda_2} + O(1/\mu_5^2).$$

These solutions are depicted in the Fig.2.

We stress that the consistent solution with a finite limit for strange quark constituent mass $v_s \ll v_q$ exists if the term breaking $U(1)_A$ symmetry is in action. Meantime the constituent masses for light u, d flavours is growing with chiral chemical potential, first as $\sim \mu_5^2$ and next as $\sim \mu_5$. Moreover even if the spontaneous chiral symmetry breaking is absent for $M^2 < 0$ in parity-even environment this CSB will be triggered by chiral imbalance.

The lattice computations perfectly confirm [22] this behavior against chiral chemical potential.

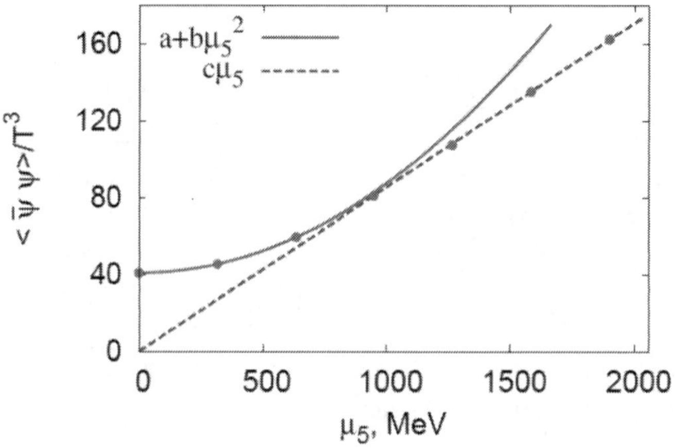

Figure 3. Chiral condensate in two-color QCD lattice as a function of μ_5 in the confinement phase (T = 158 MeV) .

For further purposes we need the non-strange meson sector and η_s

$$\Phi = \begin{pmatrix} \eta_q + \pi^0 & \sqrt{2}\pi^+ & 0 \\ \sqrt{2}\pi^- & \eta_q - \pi^0 & 0 \\ 0 & 0 & \sqrt{2}\eta_s \end{pmatrix}, \Sigma = \begin{pmatrix} v_u + \sigma + a_0^0 & \sqrt{2}a_0^+ & 0 \\ \sqrt{2}a_0^- & v_d + \sigma - a_0^0 & 0 \\ 0 & 0 & v_s \end{pmatrix}$$

$$\begin{pmatrix} \eta_q \\ \eta_s \end{pmatrix} = \begin{pmatrix} \cos\psi & \sin\psi \\ -\sin\psi & \cos\psi \end{pmatrix} \begin{pmatrix} \eta \\ \eta' \end{pmatrix}$$

For $\mu_5 = 0$, in the chiral limit $\hat{m} = 0$ we adopt $v_u = v_d = v_s = v_0 \equiv f_\pi \approx 92$ MeV. The coupling constants (in units of MeV) are fitted to phenomenology assuming isospin symmetry with the help of χ^2 minimization (MINUIT):

$$b = -3510100/m,\ M^2 = 1255600,\ c = 1252.2,\ \lambda_1 = 67.007,$$

$$\lambda_2 = 9.3126,\ d_1 = -1051.7/m,\ d_2 = 523.21/m,$$

where $m \equiv m_q = (m_u + m_d)/2$ and $m/m_s \simeq 1/25$.

One, of course, should not expect the range of applicability of dim-4 meson lagrangian for chemical potentials larger than few hundreds of MeV. Nevertheless the predictions for lightest states, pions can be well controlled within the chiral perturbation theory.

5.1 New eigenstates of strong interactions with LPB (isotriplet)

We present a simple case of mixing due to LPB in the isotriplet sector with π and a_0. The kinetic and mixing terms in the Lagrangian are given by

$$\mathcal{L} = \frac{1}{2}(\partial a_0)^2 + \frac{1}{2}(\partial\pi)^2 - \frac{1}{2}m_1^2 a_0^2 - \frac{1}{2}m_2^2\pi^2 - 4\mu_5 a_0\dot{\pi},$$

where

$$\begin{aligned} m_1^2 =& -2[M^2 - 2(3\lambda_1 + \lambda_2)v_q^2 - \lambda_2 v_s^2 \\ & - cv_s + 2(3d_1 + 2d_2)mv_q + 2d_2 m_s v_s + 2\mu_5^2] \\ m_2^2 =& \frac{2m}{v_q}\left[b + (d_1 + 2d_2)v_q^2 + d_2 v_s^2\right] \end{aligned}$$

After diagonalization in the momentum representation, the new (momentum-dependent) eigenstates are defined $\tilde{\pi}$ and $\tilde{a}_0$.

$\tilde{a}_0$ mass shows an enhancement, but μ_5 has to be understood as a perturbatively small parameter. A better treatment of $\tilde{a}_0$ would require heavier degrees of freedom.

5.2 New eigenstates of strong interactions with LPB (isosinglet)

Two new processes are likely to appear inside the fireball: the decays $\eta, \eta' \to \pi\pi$ that are strictly forbidden in QCD on parity grounds.

In a medium where parity is broken: are these processes relevant within the fireball? Can these particles reach thermal equilibrium?!

There are no $O(p^2)$ terms in the chiral Lagrangian involving vertices $\eta\pi\pi$.

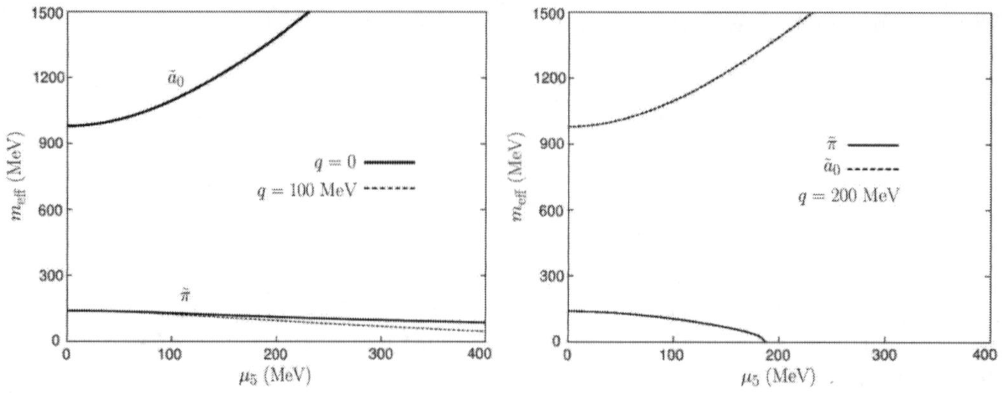

Figure 4. For high energies $k_0, |\vec{k}| > m_1 m_2/(4\mu_5) \equiv k^c_{\tilde{\pi}}$, in-medium $\tilde{\pi}$ goes tachyonic. Nevertheless, energies are always positive (no vacuum instabilities).

$O(p^4)$ terms lead to couplings such as

$$\mathcal{L}_{\eta\pi\pi} \sim \frac{16\mu_5}{F_\eta f_\pi^2} L\,\partial\eta\,\partial\pi\partial\pi,$$

where $L \sim 10^{-3}$. A rough estimate of the partial width shows a strong dependence on μ_5 as $\Gamma_{\eta\to\pi\pi} \propto \mu_5^2$ and gives values similar or higher than $\Gamma_{\rho\to\pi\pi} = 150$ MeV.

In the isosinglet sector, we show the mixing of η, σ and η'. The kinetic and mixing terms in the Lagrangian are given by

$$\mathcal{L} = \frac{1}{2}[(\partial\sigma)^2 + (\partial\eta_q)^2 + (\partial\eta_s)^2] - \frac{1}{2}m_3^2\sigma^2 - \frac{1}{2}m_4^2\eta_q^2 - \frac{1}{2}m_5^2\eta_s^2$$
$$- 4\mu_5\sigma\dot{\eta}_q - 2\sqrt{2}cv_q\eta_q\eta_s,$$

where

$$m_3^2 = -2(M^2 - 6(\lambda_1 + \lambda_2)v_q^2 - \lambda_2 v_s^2 + cv_s$$
$$+ 6(d_1 + 2d_2)mv_q + 2d_2 m_s v_s + 2\mu_5^2),$$
$$m_4^2 = \frac{2m}{v_q}\left[b + (d_1 + 2d_2)v_q^2 + d_2 v_s^2\right] + 2cv_s,$$
$$m_5^2 = \frac{2m_s}{v_s}[b + 2d_2 v_q^2 + (d_1 + d_2)v_s^2] + \frac{cv_q^2}{v_s}.$$

After diagonalization, the new eigenstates are $\tilde{\sigma}$, $\tilde{\eta}$ and $\tilde{\eta}'$.

Again, for high energies $k_0, |\vec{k}| > k^c_{\tilde{\eta}}$ with $k^c_{\tilde{\eta}} \equiv \frac{m_3}{4\mu_5 m_5}\sqrt{m_4^2 m_5^2 - 8c^2 v_q^2}$, in-medium $\tilde{\eta}$ goes tachyonic. $\tilde{\eta}'$ mass also shows an enhancement and a better treatment would require the inclusion of heavier degrees of freedom.

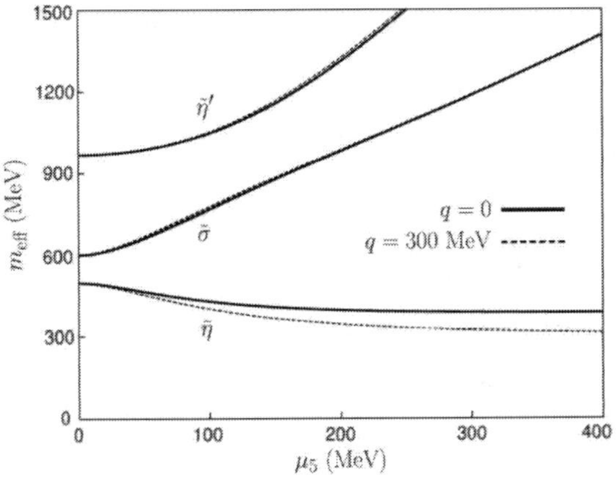

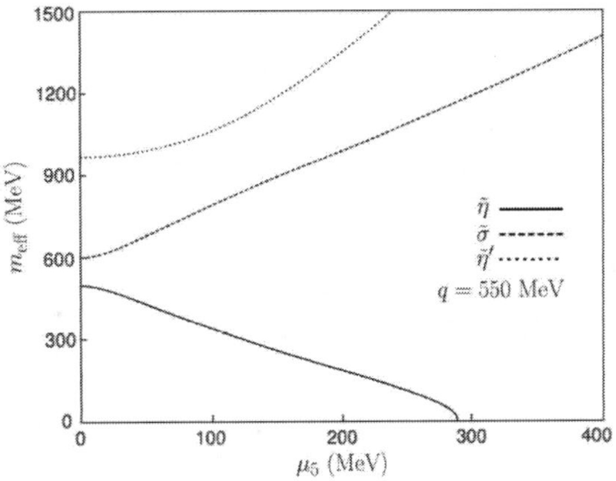

The cubic couplings used to calculate the widths $\tilde{\eta}, \tilde{\sigma}, \tilde{\eta}' \to \tilde{\pi}\tilde{\pi}$ from the Lagrangian are given by

$$\mathcal{L}_{\sigma aa} = 2[(3d_1 + 2d_2)m - 2(3\lambda_1 + \lambda_2)v_q]\sigma\vec{a}_0^2,$$

$$\mathcal{L}_{\sigma\pi\pi} = \frac{1}{v_q^2}\left[(\partial\vec{\pi})^2 v_q - (b + 3(d_1 + 2d_2)v_q^2 + d_2 v_s^2)m\vec{\pi}^2\right]\sigma,$$

$$\mathcal{L}_{\eta a\pi} = \frac{2}{v_q^2}\vec{a}_0[\partial\eta_q\partial\vec{\pi}v_q - (b + (3d_1 + 2d_2)v_q^2 + d_2 v_s^2)m\eta_q\vec{\pi}],$$

$$\mathcal{L}_{\sigma a\pi} = -\frac{4\mu_5}{v_q}\sigma\vec{a}_0\dot{\vec{\pi}}, \quad \mathcal{L}_{\eta aa} = -\frac{2\mu_5}{v_q}\dot{\eta}_q\vec{a}_0^2, \quad \mathcal{L}_{\eta\pi\pi} = 0.$$

After diagonalization, one replaces the initial $\{\eta_q, \eta_s, \sigma\}$ to $\{\tilde{\eta}, \tilde{\sigma}, \tilde{\eta}'\}$ and $\{\pi, a_0\}$ to $\{\tilde{\pi}, \tilde{a}_0\}$.

The widths are firstly computed at the rest frame of the decaying particle and secondly with a boosted particle.

$\tilde{\eta}$ exhibits a smooth behaviour with $\langle\Gamma_{\tilde{\eta}}\rangle \sim 60$ MeV $\leftrightarrow$ mean free path ~ 3 fm $\lesssim L_{\rm fireball} \sim 5 \div 10$ fm. Possible thermalization! Down to $\mu_5 \sim 100$ MeV, $\tilde{\sigma}$ width decreases and becomes stable. The visible bumps in these two channels seem to reflect the tachyonic nature of the decaying $\tilde{\pi}$. $\tilde{\eta}'$ width grows up to the GeV scale (violation of unitarity). More degrees of freedom are needed.

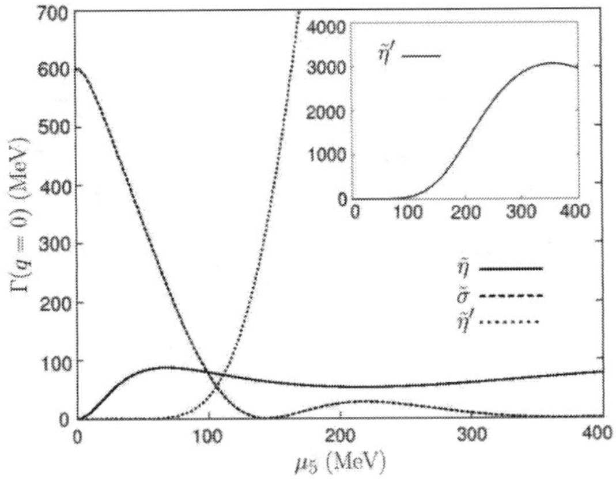

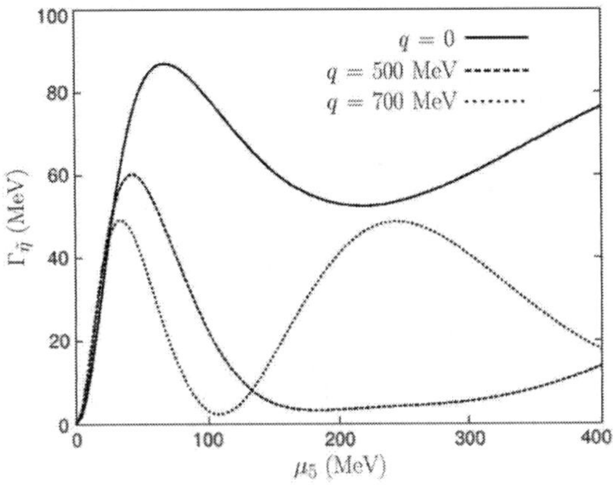

Small variations at low 3-momenta: 2 initial bumps slowly separate as one increases q. $\Gamma_{\tilde{\eta}}(\mu_5, q)$ exhibits a saddle point at $\mu_5^* \sim 240$ MeV and $q^* \sim 500$ MeV. For large 3-momenta, a third interme-

 DOI: 10.1051/epjconf/201612501009

diate bump appears (creation of 2 tachyons) and grows fast as one increases q becoming the global maximum when $q \gtrsim 700$ MeV.

6 Vector mesons in a medium with LPB

LPB can be investigated in e.m. production of leptons and photons in hot/dense nuclear matter via heavy ion collisions.

- Thus e.m. interaction extends the axial charge by e.m. anomaly,

$$Q_5^q \to \tilde{Q}_5 = Q_5^q - T_5^{\text{em}}, \quad T_5^{\text{em}} = \frac{N_c}{8\pi^2} \int_{\text{vol.}} d^3x\, \varepsilon_{jkl} \operatorname{Tr}\left(\hat{A}^j \partial^k \hat{A}^l\right).$$

- Now chiral chemical potential μ_5 is conjugated to (nearly) conserved $\tilde{Q}_5$.
- In the hadron phase QCD bosonization of total $\tilde{Q}_5$ can be realized [13] with the help of meson Lagrangian and the Vector Meson Dominance providing a good precision in coefficients.

$$\mathcal{L}_{\text{int}} = \bar{q}\gamma_\mu \hat{V}^\mu q; \quad \hat{V}_\mu \equiv -eA_\mu Q + \frac{1}{2} g_\omega \omega_\mu \mathbb{I} + \frac{1}{2} g_\rho \rho^0_\mu \tau_3,$$
$$(V_{\mu,a}) \equiv \left(A_\mu,\, \omega_\mu,\, \rho^0_\mu\right)$$

where $Q = \frac{\tau_3}{2} + \frac{1}{6}$, $g_\omega \simeq g_\rho \equiv g \simeq 6$.

In the parity-odd effective lagrangian for vector mesons the following term is generated

$$\Delta\mathcal{L} \simeq \varepsilon^{\mu\nu\rho\sigma} \operatorname{Tr}\left[\zeta_\mu V_\nu V_{\rho\sigma}\right]$$

with $\zeta_\mu = \zeta\delta_{\mu 0}$ for a spatially homogeneous and isotropic background and $\zeta \propto \mu_5$.

Two different cases of isospin structure for μ_5:

► Isosinglet pseudoscalar background ($T \gg \mu$) [RHIC, LHC]

► Pion-like (isotriplet) background ($\mu \gg T$) [FAIR, NICA]

Massive MCS electrodynamics for vector mesons is described by the following lagrangian

$$\mathcal{L}_{MCS} = -\tfrac{1}{4} F^{\alpha\beta}(x) F_{\alpha\beta}(x) + \tfrac{1}{2} m^2 A_\nu(x) A^\nu(x) + \tfrac{1}{2} \zeta_\mu A_\nu(x) \tilde{F}^{\mu\nu}(x) + \text{g.f.}$$

In momentum space wave Eqs.

$$\left[g^{\lambda\nu}\left(k^2 - m^2\right) - k^\lambda k^\nu + i\,\varepsilon^{\lambda\nu\alpha\beta} \zeta_\alpha k_\beta\right] \mathbf{a}_\lambda(k) = 0; \quad k^\lambda\, \mathbf{a}_\lambda(k) = 0. \tag{4}$$

Energy spectrum for transversal polarizations is governed by

$$\omega_{\mathbf{k},\pm} = \sqrt{\mathbf{k}^2 + m^2 \pm \zeta|\,\mathbf{k}\,|}, \tag{5}$$

and for longitudinal polarization it remains intact,

$$\omega_{\mathbf{k},L} = \sqrt{\mathbf{k}^2 + m^2} \tag{6}$$

After diagonalization of mass matrix the mass spectrum is given by

$$m^2_{V,\epsilon} = m^2_V - \epsilon\zeta|\vec{k}|, \tag{7}$$

where $\epsilon = 0, \pm 1$ is the type of meson polarization. Thus $m^2_{V,+} < m^2_{V,L} < m^2_{V,-}$.

If $\hat{\zeta}$ is an isosinglet the photon itself happens to be unaffected by CS interaction.

The position of the poles for $\pm$ polarized vector mesons is moving with wave vector $|\vec{k}|$.

This splitting unambiguously signifies LPB. Can it be measured?

$\rightarrow$ Dilepton production in HIC from the decays $\rho, \omega \rightarrow e^+e^-$ potentially contains sufficient information on LPB.

6.1 Manifestation of LPB in heavy ion collisions in dilepton production

$L, \pm$ contribution for vector mesons before acceptance corrections:

$$\frac{dN^\epsilon_{ee}}{dM} \simeq c_V \frac{\alpha^2 \Gamma_V m^2_V}{3\pi^2 g^2 M^2} \left(\frac{M^2 - n^2_V m^2_\pi}{m^2_V - n^2_V m^2_\pi} \right)^{3/2}$$

$$\times \sum_\epsilon \int_M^\infty dk_0 \frac{\sqrt{k^2_0 - M^2}}{e^{k_0/T} - 1} \frac{m^4_{V,\epsilon}}{\left(M^2 - m^2_{V,\epsilon}\right)^2 + m^4_{V,\epsilon} \frac{\Gamma^2_V}{m^2_V}},$$

where $n_V = 2, 0$; $|\vec{k}| = \sqrt{k^2_0 - M^2}$ and $M^2 > n^2_V m^2_\pi$. c_V absorbs combinatorial factors different for ρ and ω, μ, finite volume suppression.

Simulations of distorted spectral functions are implemented with PHENIX acceptance: $|y_{ee}| < 0.35$, $|\vec{p}^e_t| > 200$ MeV, gaussian M_{ee} smearing (width=10 MeV)

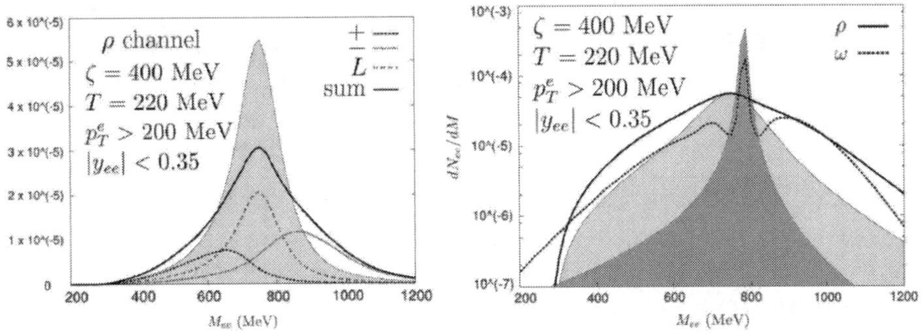

Figure 5. On the left: polarization splitting in ρ spectral function for LPB $\zeta = 400$ MeV ($\mu_5 = 290$ MeV) compared with $\zeta = 0$ (shaded region). On the right: in-medium ρ and ω channels (solid and dashed line) and their vacuum contributions (light and dark shaded regions) for $\zeta = 400$ MeV. In-medium ρ is enhanced by a factor 1.8 due to $\pi\pi$ regeneration into ρ.

DOI: 10.1051/epjconf/201612501009

The PHENIX/STAR anomaly: an abnormal e^+e^- yield is observed in central HIC.

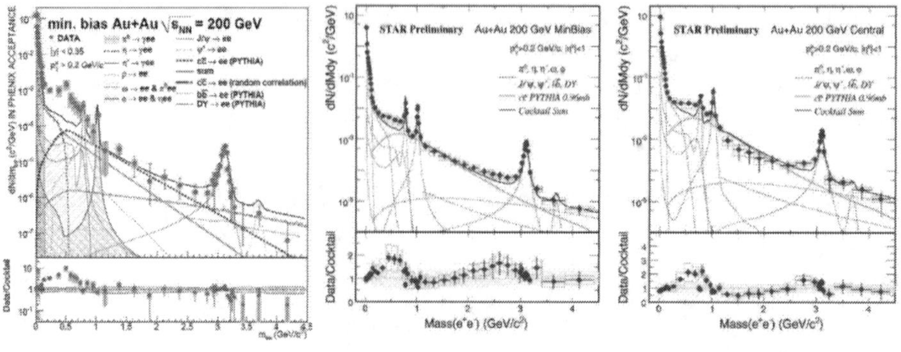

Figure 6. Abnormal excess of dileptons: on the left PHENIX data[23] and on the right STAR data[24]

It may be explained partially by the local parity breaking shown in Fig.5

6.2 Polarization asymmetry: dilepton angular distributions

In [25] it was suggested to detect a splitting of dilepton polarizations averaged over angular sectors. There are two convenient choices of this splitting.

Case A: the angle θ_A between the two outgoing leptons in the laboratory frame.

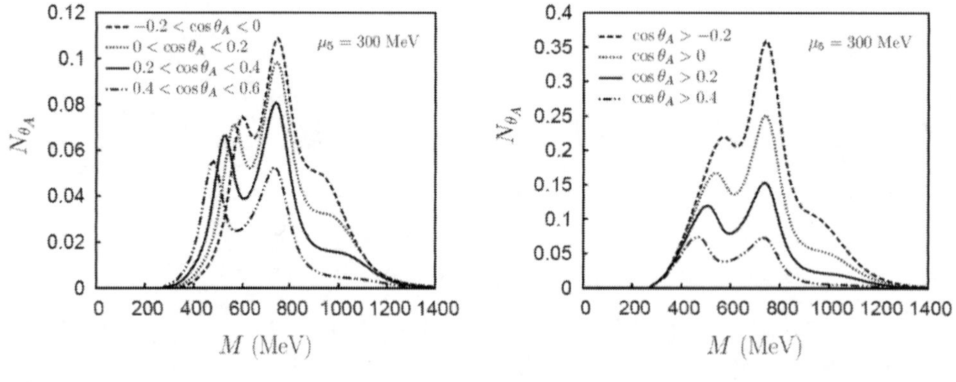

Figure 7. The ρ spectral function is presented for fixed $\mu_5 = 300$ MeV. $\cos\theta_A \in [-0.2, 0], [0, 0.2], [0.2, 0.4], [0.4, 0.6]$ and $[0.6, 0.8]$ in the left panel, and $\cos\theta_A \geq -0.2, 0, 0.2, 0.4$ in the right one. Results are corrected for PHENIX experimental acceptance.

Case B: the angle θ_B between one of the two outgoing leptons in the laboratory frame and the same lepton in the dilepton rest frame

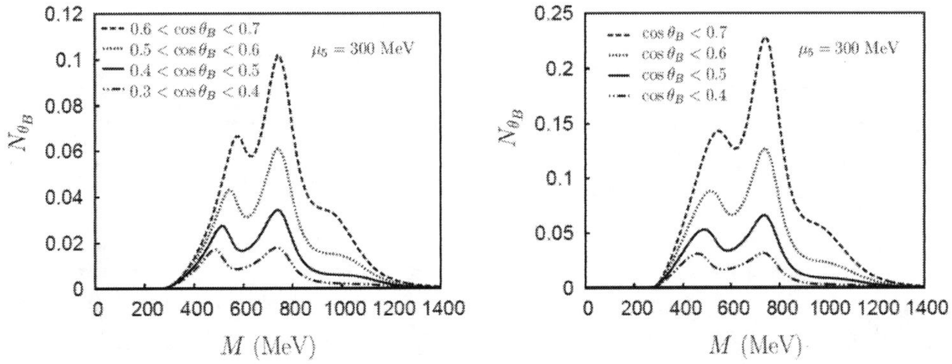

Figure 8. The ρ spectral function is presented for fixed $\mu_5 = 300$ MeV. $\cos\theta_B \in [0.3, 0.4], [0.4, 0.5], [0.5, 0.6]$ and $[0.6, 0.7]$ in the left panel, and $\cos\theta_B \leq 0.4, 0.5, 0.6, 0.7$ in the right one. Results are corrected for PHENIX experimental acceptance.

7 Conclusions and Outlook

In the report a medium (a fireball) is regarded with chiral imbalance occurring in heavy ion collisions at high energies. To describe such a quark (hadron) matter chiral chemical potential can be introduced to quantify the presence of chirality imbalance i.e. the difference between the numbers of Right-(Left-)handed quarks in the fireball after HIC at central collisions. In such a medium (in hadron phase) the possibility of phase formation with a Local special Parity Breaking is described. Based on the effective meson theory in the presence of Chern-Simons interaction it is found that the spectrum of massive vector mesons splits into three polarization components with different effective masses. Accordingly a resonance broadening occurs that leads to an increase of spectral contribution to the dilepton production as compared to the vacuum state. In particular, in a hadron phase some of experimental observables are presented for detecting the signal of LPB. The asymmetry in production of longitudinally and transversely polarized states of ρ and ω mesons for various values of the dilepton invariant mass can serve as a characteristic indication of the LPB in CERES, HADES, PHENIX, STAR, NA60 and ALICE experiments. Also QCD-motivated models are considered from whose properties the phenomenon of LPB could be found. The results are compared with lattice calculations for the investigated phases.

Acknowledgements

This work has been supported through grants FPA2013-46570, 2014-SGR-104 and Consolider CPAN. Funding was also partially provided by the Spanish MINECO under project MDM-2014-0369 of ICCUB (Unidad de Excelencia 'Maria de Maeztu'). A.A. and V.A. were supported by Grant RFBR projects 16-02-00348 and also got a financial support of SPbSU, in particular, by projects 11.42.1049.2016, 11.42.1051.2016.

References

[1] A.A. Belavin, A.M. Polyakov, A.S. Schwartz and Y. S. Tyupkin, Phys. Lett. B **59**, 85 (1975).
[2] C. G. Callan, Jr., R. F. Dashen and D. J. Gross, Phys. Lett. B **63**, 334 (1976).

[3] G. 't Hooft, Phys. Rev. D **14**, 3432 (1976) [Erratum-ibid. D **18**,2199 (1978)].
[4] D.I. Dyakonov and V.Y. Petrov, Phys. Lett. B **147**, 351 (1984); Sov. Phys. JETP **62**, 204 (1985); ibid. **62**, 431 (1985); Nucl. Phys. B **272**, 457 (1986).
[5] T. Schaefer and E. V. Shuryak, Rev. Mod. Phys. **70**, 323 (1998); [hep-ph/9610451]
[6] S. Borsanyi et al. [Wuppertal-Budapest Collaboration], JHEP. **073**, 1009, (2010).
[7] A. Bazavov et al., Phys. Rev. D **85**, 054503 (2012).
[8] S. L. Adler, Phys. Rev. **177**, 2426 (1969).
[9] J. S. Bell and R. Jackiw, Nuovo Cim. A **60**, 47 (1969).
[10] G. 't Hooft, hep-th/0010225; T. Schaefer and E. V. Shuryak, Rev. Mod. Phys. **70**, 323 (1998).
[11] E. Shuryak, arXiv:1412.8393 [hep-ph]; Prog. Part. Nucl. Phys. **62**, 48 (2009).
[12] J. Xu, J. Liao and M. Gyulassy, Chin. Phys. Lett. 32, no. 9, 092501 (2015); arXiv:1508.00552 [hep-ph].
[13] A.A. Andrianov, V.A. Andrianov, D. Espriu, X. Planells, Physics Letters B **710**,230 (2012); Phys. Rev. D **90**, 034024 (2014).
[14] D.E. Kharzeev, R.D. Pisarski, M.H.G. Tytgat, Phys. Rev. Lett. **81**, 512 (1998); K. Buckley, T. Fugleberg, and A. Zhitnitsky, Phys. Rev. Lett. **84**, 4814 (2000); D.E. Kharzeev, Phys. Lett. B **633**, 260 (2006); D.E. Kharzeev, Ann. Phys. (NY) **325**, 205 (2010); D.E. Kharzeev, A. Zhitnitsky, Nucl. Phys. A **797**, 67 (2007); D.E. Kharzeev, L.D. McLerran, H.J. Warringa, Nucl. Phys. A **803**, 227 (2008); K. Fukushima, D.E. Kharzeev, H.J. Warringa, Phys. Rev. D **78**, 074033 (2008); K. Fukushima, D.E. Kharzeev, H.J. Warringa, Nucl. Phys. A **836**, 311 (2010).
[15] T. D. Lee and G. C. Wick Phys. Rev. D **9**, 2291 (1974).
[16] A.Vilenkin, Phys.Rev. D **22**, 3080 (1980).
[17] A.A.Anselm, Phys.Lett. B **217**,169 (1989); A.A.Anselm, M.G.Ryskin, A.G.Shuvaev, Surv. in HEP **11**, 27 (1997.
[18] A.A. Andrianov, D. Espriu, Phys. Lett. B **663**,450 (2008); A.A. Andrianov, V.A. Andrianov, D. Espriu, Phys. Lett. B **678**,416 (2009).
[19] F. R. Klinkhamer and N. S. Manton, Phys. Rev. D **30**, 2212 (1984); V. A. Kuzmin, V. A. Rubakov, and M. E. Shaposhnikov, Phys. Lett. B **155**, 36 (1985); L. McLerran, E. Mottola, and M. E. Shaposhnikov, Phys. Rev. D **43**, 2027 (1991); G. D. Moore and K. Rummukainen, Phys. Rev. D **61**, 105008 (2000); E. Shuryak and I. Zahed, Phys. Rev. D **67**, 014006 (2003).
[20] R. Rapp and J. Wambach "Chiral symmetry restoration and dileptons in relativistic heavy-ion collisions," in: Advances in Nuclear Physics (Adv. Phys. Part. Nucl. (APPN), Vol. 25, J. W. Negele and E. Vogt, eds.), Kluwer, New York (2002), pp. 1–205; W. Liu and R. Rapp, Nucl. Phys. A **796**, 101 (2007); H. van Hees and R. Rapp, Nucl. Phys. A **806**, 339 (2008).
[21] A. A. Andrianov, D. Espriu & X. Planells, Eur. Phys. J. C **73**:2294 (2013).
[22] V. V. Braguta and A. Y. Kotov, Phys. Rev. D **93**, no. 10, 105025 (2016); V. V. Braguta, V. A. Goy, E. M. Ilgenfritz, A. Yu. Kotov, A. V. Molochkov, M. Muller-Preussker, and B. Petersson. JHEP, 06:094, 2015.
[23] A. Adare et al. [PHENIX Collaboration], Phys. Rev. C 81 (2010) 034911, arXiv:0912.0244 [nucl-ex].
[24] L. Adamczyk et al. [STAR Collaboration], Phys. Rev. Lett. **113**, no. 2, 022301 (2014) Addendum: [Phys. Rev. Lett. **113**, no. 4, 049903 (2014)] arXiv:1312.7397 [hep-ex].
[25] A.A. Andrianov, V.A. Andrianov, D. Espriu, X. Planells, Phys. Rev. D **90**, 034024 (2014).

EPJ Web of Conferences **125**, 01010 (2016) DOI: 10.1051/epjconf/201612501010
QUARKS-2016

Euler-Lagrange equations for high energy actions in QCD and in gravity *

L. N. Lipatov[1,2,**]

[1] *St. Petersburg State University*
[2] *Petersburg Nuclear Physics Institute*

Abstract. The high energy scattering in QCD and gravity can be described in terms of reggeized gluons and gravitons, respectively. At $N = 4$ SUSY the BFKL Pomeron is dual to the reggeized graviton living in the 10-dimensional anti-de-Sitter space. We discuss the corresponding effective actions for reggeized gluon and graviton interactions. The Euler-Lagrange equations for these effective theories are constructed with a variational approach and by using an invariance under the gauge and general coordinate transformations. We discuss their solutions and applications to the calculation of effective Reggeon vertices and trajectories.

1 Introduction

The hadron scattering amplitudes $A(s,t)$ at high energies $\sqrt{s}$ and fixed momentum transfers $q = \sqrt{-t}$ can be presented as sums of the amplitudes A^p with definite signatures $p = \pm 1$

$$A^p(s,t) = \int_{a-i\infty}^{a+i\infty} \frac{d\,j}{2\pi} \left((-s)^j + p\,s^j\right) f_j^p(t)\,. \tag{1}$$

The t-channel partial waves $f_j^p(t)$ have the poles with their positions depending on t

$$f_j^p(t) \sim \frac{1}{j - j(t)}\,. \tag{2}$$

They lead to the Regge behavior of the amplitudes

$$A^p(s,t) \sim (-s)^{j(t)} + p s^{j(t)}\,. \tag{3}$$

Pomeron is a special Regge pole with $p = 1$ and $j(0) \approx 1$, which is responsible for a slowly growing behavior of total cross sections σ_t at large energies and for the fulfilment of the Pomeranchuk theorem for the particle-particle and particle-anti-particle cross sections.

The exchange of two or more reggeons generates more complicated Mandelstam singularities of $f_j(t)$ in the j-plane [1]. To take into account all possible Pomeron contributions V.N. Gribov constructed the Reggeon calculus based on the 2+1 field theory of a complex scalar field [2].

*The investigation is supported by the St.Petersburg State University grant and RFBI grant 14-2200281.
**e-mail: lipatov@thd.pnpi.spb.ru

DOI: 10.1051/epjconf/201612501010

In the Quantum Chromodynamics (QCD) the scattering amplitudes with the gluon quantum numbers in the t-channel have the Regge form in the so-called leading logarithmic approximation (LLA) [3]. Pomeron is a colorless composite state of the reggeized gluons. Its wave function in QCD and in other field models with the gauge group $SU(N_c)$ satisfies the BFKL equation [3]. It is remarkable, that the equations for the composite states of several reggeized gluons with the color singlet and octet quantum numbers are integrable at large N_c [4, 5]. Thus, it looks natural to reformulate QCD and other gauge models in terms of the effective degrees of freedom corresponding to the reggeized gluons. The gluon Regge trajectory and various reggeon couplings in upper orders of perturbation theory can be calculated from the effective action presented in Ref. [6, 7]. In the $N = 4$ SUSY the Pomeron is dual to the reggeized graviton in the anti-de-Sitter space due to AdS/CFT correspondence [8–10]. The effective action for the high energy scattering in gravity was derived in ref. [11]. Below we consider the Euler-Lagrange equations for the effective actions describing interactions of reggeized gluons and gravitons.

2 Effective action for high energy processes in QCD

The effective action in QCD is written for a cluster of usual quarks, gluons and reggeized gluons having their rapidities y in the interval η around its central value y_0.

Apart from the anti-hermitian matrix $N_c \times N_c$ for the gluon field v_μ we introduce also the fields $A_\pm = A_0 \pm A_3$ describing the production and annihilation of the reggeized gluons

$$v_\mu(x) = -iT^a v_\mu^a(x)\,,\ A_\pm(x) = -iT^a A_\pm^a(x)\,,\ [T^a, T^b] = if^{abc}T_c\,. \tag{4}$$

The operators T^a are generators of the gauge group $SU(N_c)$ in the fundamental representation.

The fields $A^\mp = A_\pm$ are invariant under local gauge transformations of the gluon fields v_μ

$$\delta v_\mu = \frac{1}{g}\,[D_\mu, \chi(x)]\,,\ D_\mu = \partial_\mu + g v_\mu\,,\ \delta A_\pm = 0 \tag{5}$$

with the parameters $\chi(x)$ vanishing at $x \to \infty$. For the case of the global SU_n rotations with constant χ the Reggeon fields $A^\pm$ are transformed as usual gluon fields v_μ. They satisfy also the kinematical constraints

$$\partial_- A^- = \partial_+ A^+ = 0\,,\ \partial_\pm = \frac{\partial}{\partial x^0} \pm \frac{\partial}{\partial x^3} \tag{6}$$

corresponding to the fact, that the Sudakov components α_i, β_i of the cluster momenta k_i are strongly ordered $\beta_i \gg \beta_{i+1}$, $\alpha_i \ll \alpha_{i+1}$ in the multi-Regge kinematics.

The effective action for a cluster of real and virtual particles with their rapidities belonging to the small rapidity interval η has the form [6]

$$S_{eff} = \int d^4x \left(L_{QCD} + Tr\,\partial_\mu A^+ \partial_\mu A^-\right) + S_{ind}\,,\ S_{ind} = Tr \int d^4x\,(V_+ \partial_\mu^2 A^+ + V_- \partial_\mu^2 A^-)\,, \tag{7}$$

where L_{QCD} is the usual QCD lagrangian

$$L_{QCD} = i\bar{\psi}\hat{\partial}\psi + \frac{1}{2} Tr\, G_{\mu\nu}^2\,,\ G_{\mu\nu} = \frac{1}{g}\,[D_\mu, D_\nu]\,. \tag{8}$$

The anti-hermitian operators $V_\pm$ in the induced action S_{ind} are expressed in terms of eikonal amplitudes for a massless particle scattered off the external gluon field v. By chousing an appropriate normalization of the fields $A^\pm$ we can present $V_\pm$ in the form [6]

$$V_\pm = \frac{1}{g}\,\partial_\pm \frac{1}{D_\pm}\,\overleftarrow{\partial}_\pm \equiv -\frac{\partial_\pm}{g} + v_\pm - v_\pm \frac{g}{\partial_\pm} v_\pm + v_\pm \frac{g}{\partial_\pm} v_\pm \frac{g}{\partial_\pm} v_\pm + \ldots\,, \tag{9}$$

where by definition the derivative $\overleftarrow{\partial}$ acts on functions situated to the left from it. The contribution $-\partial_\pm/g$ is negligible in S_{int}, because the fields $A^\pm$ do not depend on $x^\pm$. It is natural to define the action of the integral operator $1/\partial_\pm$ in the symmetric form

$$\frac{1}{\partial_\pm} f(x^\pm) = \frac{1}{2}\left(\frac{1}{\partial_0} \mp \frac{1}{\partial_3}\right) f(x^\pm) = \frac{1}{4}\left(\int_{-\infty}^{x^\pm} dx'^\pm f(x'^\pm) - \int_{x^\pm}^{\infty} dx'^\pm f(x'^\pm)\right). \tag{10}$$

The invariance of the action under gauge transformations is a consequence of the relation

$$V'_\pm = \frac{1}{g}\partial_\pm e^{-\chi}\frac{1}{D_\pm}e^{\chi}\overleftarrow{\partial}_\pm \to V_\pm \tag{11}$$

valid after its integration over $x^\pm$ due to the vanishing of the gauge parameter $\chi(x)$ at $x \to \infty$.

The matrix $V_\pm$ can be written in one of two factorized forms

$$V_\pm = -\frac{\partial_\pm}{g}\, O(x^\pm) = O^+(x^\pm)\,\frac{\overleftarrow{\partial}_\pm}{g}, \tag{12}$$

where the operators $O(x^\pm)$ and $O^+(x^\pm)$ are defined below

$$O(x^\pm) \equiv -\frac{1}{D_\pm}\overleftarrow{\partial}_\pm, \quad O^+(x^\pm) = \partial_\pm\frac{1}{D_\pm}. \tag{13}$$

The poles $1/D_\pm$ in $V_\pm$ appear from propagators of massless particles with other rapidities emitting gluons inside the given kinematical interval η. These particle virtualities $\sim k_\pm$ are large. Hence it looks natural to define the action of the operators $1/\partial_\pm$ as it was done above

$$\frac{1}{\partial_\pm} f(x^\pm) = \frac{1}{4}\int d\overline{x}^\pm\, \epsilon(x^\pm - \overline{x}^\pm)\, f(\overline{x}^\pm), \tag{14}$$

but for all terms of the perturbative expansion of $1/D_\pm$. Such definition of $V_\pm$ is compatible with its anti-hermicity property and leads to the principal value prescription for corresponding poles $1/k_\pm$ in the momentum space.

Integrating S_{ind} over $x^\pm$ we can present it in terms of asymptotic values of functions $O_p(x^\pm)$

$$S_{ind} = -\frac{Tr}{g}\int d^2x_\perp\left(\int_{-\infty}^{\infty} dx^-\, O|_{x^+=-\infty}^{x^+=\infty}\,\partial_\sigma^2 A^+ + \int_{-\infty}^{\infty} dx^+ O|_{x^-=-\infty}^{x^-=\infty}\,\partial_\sigma^2 A^-\right). \tag{15}$$

The induced action is real due to the anti-hermicity relation for $O|_{x^\pm=-\infty}^{x^\pm}$. It can be written also in the four-dimensional form

$$S_{ind} = Tr\int d^4x\left(v_+(x)\,O(x^+)\,\partial_\sigma^2 A^+ + v_-(x)\,O(x^-)\,\partial_\sigma^2 A^-\right). \tag{16}$$

The product of the operators $O^+(x^\pm)$ and $O(x^\pm)$ does not depend on $x^\pm$, but generally $O(x^\pm)$ are not unitary operators. The absence of unitarity is related to our use of the principal value prescription $\theta(|k^\pm| - \epsilon)/k^\pm$ for particle propagators. The infrared cut-off ϵ can be chosen to be proportional to $e^{-\eta}$ where η is a low limit for the relative rapidity of neigbouring clusters described by the effective action.

We can use the representation of $1/D_\pm$ in terms of a modified P-exponents

$$\frac{1}{D_\pm} = P\,\frac{e^{-\frac{g}{4}\int_{-\infty}^{x_\pm} d\overline{x}^\pm v_\pm}}{e^{-\frac{g}{4}\int_{x_\pm}^{\infty} d\overline{x}^\pm v_\pm}}\,\frac{1}{\partial_\pm}\,\bar{P}\,\frac{e^{-\frac{g}{4}\int_{x_\pm}^{\infty} d\overline{x}^\pm v_\pm}}{e^{-\frac{g}{4}\int_{-\infty}^{x_\pm} d\overline{x}^\pm v_\pm}}. \tag{17}$$

It allows to write $O(x^{\pm})$ in the form

$$O(x^{\pm}) = P\,\frac{e^{-\frac{g}{4}\int_{-\infty}^{x_{\pm}} d\tilde{x}^{+} v_{\pm}}}{e^{-\frac{g}{4}\int_{x_{\pm}}^{\infty} d\tilde{x}^{+} v_{\pm}}}\,\frac{1}{2}\left(Pe^{\frac{g}{4}\int_{-\infty}^{\infty} d\tilde{x}^{+} v_{\pm}} + \bar{P}e^{-\frac{g}{4}\int_{-\infty}^{\infty} d\tilde{x}^{+} v_{\pm}}\right). \tag{18}$$

The first factor here is an unitary matrix and second one is an hermitian operator. We obtain a simple representation for the difference of $O(x^{\pm})$ at $x^{\pm} = \pm\infty$

$$O\,|_{x^{\pm}=-\infty}^{x^{\pm}=\infty} = \frac{1}{2}\left(Pe^{-\frac{g}{4}\int_{-\infty}^{\infty} d\tilde{x}^{+} v_{\pm}} - \bar{P}e^{\frac{g}{4}\int_{-\infty}^{\infty} d\tilde{x}^{+} v_{\pm}}\right)\left(Pe^{\frac{g}{4}\int_{-\infty}^{\infty} d\tilde{x}^{+} v_{\pm}} + \bar{P}e^{-\frac{g}{4}\int_{-\infty}^{\infty} d\tilde{x}^{+} v_{\pm}}\right), \tag{19}$$

which leads to an explicit expression for S_{ind} having the reality property.

Note, that effective vertices for the reggeized gluon with the indices ±1, color index c and momentum q, annihilated to $r+1$ gluons with the polarization indices $\mu_0, \mu_1, ..., \mu_r$, color indices $a_0, a_1, ..., a_r$ and momenta $k_0, k_1, ..., k_r$, can be written in the form [6, 7]

$$\Delta^{\mu_0\mu_1...\mu_r\pm}_{a_0a_1...a_rc} = -q^2 \prod_{i=0}^{r} \delta^{\mu_i}_{\pm} \Delta_{a_0a_1...a_rc}(k_0^{\pm}, k_1^{\pm}, ...k_r^{\pm}),\ \Delta_{a_0c} = \delta_{a_0c}\,,\ \sum_{i=0}^{r} k_r^{\pm} = q^{\pm} = 0\,. \tag{20}$$

Here $\Delta_{a_0...a_r}$ has the Bose symmetry and satisfies the recurrent relation (the Ward identity) [6]

$$\Delta_{a_0a_1...a_r}(k_0^{\pm}, k_1^{\pm}, ...k_r^{\pm}) = \frac{1}{k_r^{\pm}} \sum_{t=0}^{r-1} i f_{aa_ta_r}\, \Delta_{a_0...a_{t-1}aa_{t+1}...a_r}(k_0^{\pm}, k_1^{\pm}, ...k_{t-1}^{\pm}, k_t^{\pm} + k_r^{\pm}, k_{t+1}^{\pm}, ...k_r^{\pm}),$$

where f_{abc} is the structure constant of the gauge group. Thus, these vertices for the principal value prescription do not depend on the color representations of the matrices $v_{\pm}(x)$ in S_{ind}. In particular, one can use the adjoint representation for these matrices.

3 Classical equation for the effective action in QCD

The variation of the induced action S_{ind} over $v_{\pm}$ can be written as follows

$$\delta S_{ind} = -Tr \int d^4x \left(\partial_+ \frac{1}{D_+}\,\delta v_+\,\frac{1}{D_+}\,\overleftarrow{\partial}_+\,\partial_\mu^2 A^+ + \partial_- \frac{1}{D_-}\,\delta v_-\,\frac{1}{D_-}\,\overleftarrow{\partial}_-\,\partial_\mu^2 A^-\right). \tag{21}$$

Using the relation

$$\partial_{\pm}\frac{1}{D_{\pm}}\,\delta v_{\pm}\,\frac{1}{D_{\pm}}\,\overleftarrow{\partial}_{\pm} = O^{+}(x^{\pm})\,\delta v_{\pm}\,O(x^{\pm}),$$

we obtain for the variation of S_{ind}

$$\delta S_{ind} = -Tr \int d^4x \left(\delta v_+\, O(x^+)\,\partial_\mu^2 A^+ O^+(x^+) + \delta v_-\, O(x^-)\,\partial_\mu^2 A^- O^+(x^-)\right). \tag{22}$$

Therefore the Euler-Lagrange equations for S_{eff} in a pure glue-dynamics have the form

$$[D_\mu, G^{\mu\nu}]^{\perp} = 0\,,\ [D_\mu, G^{\mu\pm}] = j^{\pm}_{ind}\,, \tag{23}$$

with the anti-hermitian currents

$$j^{\pm}_{ind} = O(x^{\pm})(\partial_\sigma^2 A^{\pm})O^{+}(x^{\pm})\,. \tag{24}$$

 DOI: 10.1051/epjconf/201612501010

They are conserved

$$[D_\pm, j^\pm_{ind}] = 0 \tag{25}$$

due to the relations $D_\pm\, O(x^\pm) = 0$.

In the quasi-elastic kinematics, where, for example, $A^+ = 0$, after the gauge transformation, leading to the constraint $v'_- = 0$, one obtains

$$j'^{\,-}_{ind} = \lambda\, \partial^2_\sigma A^- \lambda^+\,,\ \lambda = \frac{1}{2}\left(Pe^{\frac{g}{4}\int_{-\infty}^{\infty} d\widetilde{x}^- v'_-} + \bar{P}e^{-\frac{g}{4}\int_{-\infty}^{\infty} d\widetilde{x}^- v'_-}\right) = 1\,. \tag{26}$$

In this light cone gauge there is an explicit solution of the Euler-Lagrange equations

$$\overline{v^\sigma} = \delta^\sigma_- A^-. \tag{27}$$

Thus, the reggeon field A^- has the physical interpretation of a classical solution of the Euler-Lagrange equations for $A^+ = 0$ in the light-cone gauge $v_- = 0$. This solution is a superposition of the shock waves proportional to $\delta(x^\mp - x_0^\mp)\ln|x^\perp - x_0^\perp|^2$.

Inserting the solution of the Euler-Lagrange equation for a general kinematics in the effective action, one obtains a generating function for the reggeized gluon vertices in the tree approximation [6, 7]. Note, that even for the quasi-elastic kinematics the Euler-Lagrange equations have other solutions corresponding to more complicated constraints at large times $t \to \pm\infty$. The asymptotic behavior of these contributions is fixed at $t = -\infty$ and $t = \infty$ in terms of two arbitrary functions $v_1(\vec{x})$ and $v_2(\vec{x})$, respectively. The effective action calculated on the general solution depending on $A^\pm$ and $v_{1,2}(\vec{x})$ gives a possibility to find a generating functional in a tree approximation for all possible scattering amplitudes and an arbitrary number of reggeized gluons [6]. Furthermore, by calculating the functional integral over the quantum fluctuations δv_μ around classical solutions we can find various effective vertices with loop corrections [12–14].

4 Effective action for the high energy gravity

According to J. Maldacena the $N = 4$ super-symmetric gauge theory is equivalent to the 10-dimensional super-string model living on the 10-dimensional anti-de-Sitter space [8–10]. As a result, here the BFKL Pomeron is dual to the reggeized graviton (see, for example, [15]) and the Gribov calculus for Pomerons should be generally covariant. Note, that the graviton Regge trajectory and its various couplings in a leading order were calculated many years ago [16].

The generally covariant effective action in gravity was constructed for a cluster of gravitons and reggeized gravitons having their rapidities in an interval around their central value [11]. Apart from the usual Einstein-Hilbert action and a kinetic term for the reggeon fields $A^{\pm\pm}$

$$S = -\frac{1}{2\kappa^2}\int d^4x\left(\sqrt{-g}\,R + \partial_\sigma A^{++}\partial_\sigma A^{--}\right) + S_{ind}\,, \tag{28}$$

it contains the induced term S_{ind}

$$S_{ind} = -\frac{1}{2\kappa^2}\int d^4x\left(\frac{j_{++}}{2}\,\partial^2_\sigma A^{++} + \frac{j_{--}}{2}\,\partial^2_\sigma A^{--}\right) \tag{29}$$

with the currents $j_{\pm\pm}$ being functionals of the metric tensor $g^{\mu\nu}$.

Fields $A^{\pm\pm}$ describing the production and annihilation of reggeized gravitons are invariant under the general coordinate transformations which are reduced to the Poincare group at large distances. They satisfy the kinematical constraints

$$\partial_+ A^{++} = \partial_- A^{--} = 0\,, \tag{30}$$

corresponding to the strong ordering of the Sudakov components for momenta of produced clusters in the multi-Regge kinematics.

The gravity fields $h_{\mu\nu}$ are introduced as fluctuations of $g_{\mu\nu}$ around the Minkowsky metric tensor $\eta_{\mu\nu}$ having the diagonal structure $(1,-1,-1,-1)$. One can expand the Hilbert-Einstein lagrangian in h and obtain the solution of the effective Euler-Lagrange equation at small $A_{\pm\pm}$ in the form $\widetilde{h}_{\pm\pm} \approx A_{\pm\pm}$, providing that the currents $j_{\pm\pm}$ can be expanded in h as follows

$$j_{\pm\pm} = h_{\pm\pm} + O(h^2)\,. \tag{31}$$

The functional form of the current $j_{\pm\pm} = \partial_\pm j^\mp$ is fixed by the general covariance of the action. Due to this constraint the new current $j^\mp = 2x^\mp - \omega^\mp$ satisfies the Hamilton-Jacobi (HJ) equation [11]

$$g^{\rho\sigma}\,(\partial_\rho\omega^\mp)(\partial_\sigma\omega^\mp) = 0\,. \tag{32}$$

The function $\omega^\mp$ describes the light-front shock wave moving in the gravitational field.

If we search the solution of the HJ equation in the form $\omega^\mp = 2x^\mp - 2x^\mp(x^\pm, x^\perp)$, the quantities $x'^\mp = \omega^\mp/2$ can be considered as light-cone components of the coordinate transformation $x'(x)$ to the systems with the global light-cone times $x^\mp$. In these systems the light-cone component of $g^{\rho\sigma}$ is zero

$$g'^{\mp\mp} = g^{\mu\nu}\partial_\mu x'^\mp\,\partial_\nu x'^\mp = 0\,. \tag{33}$$

This HJ equation does not fix completely the global light-cone time systems. It is naturally to impose on $g'^{\rho\sigma}$ more restrictive constraints expressed in terms of the Minkowsky tensor

$$g'^{\rho\mp} = \eta^{\rho\mp}\,,\ \eta^{\mp\mp} = \eta_\perp^{\rho\mp} = 0\,,\ \eta^{\pm\mp} = 1\,. \tag{34}$$

They correspond to the global light-cone time inertial systems where we have in particular

$$g'^{\rho\mp}\,\frac{\partial}{\partial x'^\rho} = \frac{\partial}{\partial x'^\pm}\,. \tag{35}$$

Note, that $j^\mp$ in the effective action may be substituted by $-\omega^\mp$ because the term $2x^\mp$ gives a vanishing contribution. After its integration over $x^\pm$ the induced action can be expressed only in terms of the Hamilton-Jacobi functions $\omega^\mp$ at $x^\pm = \infty$ where we have the Minkowsky metric

$$S_{ind} = \frac{1}{2\kappa^2}\int d^2x^\perp\left(\int_{-\infty}^{\infty}\frac{dx^-}{4}\,\omega^-\,\partial_{\sigma\perp}^2 A^{++}\,|_{x^+=-\infty}^{x^+=\infty} + \int_{-\infty}^{\infty}\frac{dx^+}{4}\,\omega^+\,\partial_{\sigma\perp}^2 A^{--}\,|_{x^-=-\infty}^{x^-=\infty}\right). \tag{36}$$

5 Classical equations for the effective gravity

Variations $\delta g^{\mu\nu}$ and $\delta\omega^\mp$ along particle trajectories are not independent due to the HJ equation

$$\partial_\mu\omega^\mp\,\partial_\nu\omega^\mp\,\delta_\tau g^{\mu\nu} + 2\,g^{\rho\sigma}\,\partial_\rho\omega^\mp\,\delta_\tau\partial_\sigma\omega^\mp = 0 \tag{37}$$

and expressed through the corresponding infinitesimal shifts in the proper time τ

$$\delta_\tau g^{\mu\nu} = d\tau\,(\partial_\sigma g^{\mu\nu})\,\frac{d}{d\tau}x^\sigma\,,\ \delta_\tau\partial_\sigma\omega^\mp = d\tau\frac{d}{d\tau}\,\partial_\sigma\omega^\mp\,. \tag{38}$$

They can be calculated with the help of the Hamilton equations

$$\frac{d}{d\tau}x^\sigma = g^{\sigma\rho}\,\partial_\rho\omega^\mp\,,\ \frac{d}{d\tau}\,\partial_\sigma\omega^\mp = -\frac{1}{2}\,(\partial_\sigma g^{\mu\nu})\,\partial_\mu\omega^\mp\partial_\nu\omega^\mp\,. \tag{39}$$

Because $x'^{\mp} = \omega^{\mp}/2$ are light-cone components of $x'(x)$ in systems with $g'^{\mp\mp} = 0$ we have

$$2g^{\rho\sigma}\,\partial_\rho\omega^{\mp}\,\delta_\tau\partial_\sigma\omega^{\mp} = -2d\tau\, g'^{\chi\mp}\frac{\partial g^{\mu\nu}}{\partial x'^{\chi}}\,\partial_\mu\omega^{\mp}\partial_\nu\omega^{\mp} = -2d\tau\,\frac{\partial g^{\mu\nu}}{\partial x'^{\pm}}\,\partial_\mu\omega^{\mp}\partial_\nu\omega^{\mp} = \delta_\tau\frac{\partial\omega^{\mp}}{\partial x'^{\pm}}\,, \tag{40}$$

where the property $g'^{\chi\mp} = \eta^{\chi\mp}$ of the global light-cone time inertial systems was used. By integrating these equalities over the invariant phase space it is possible to verify the relations

$$\int d^4x\,\sqrt{-g}\,2g^{\rho\sigma}\,\partial_\rho\omega^{\mp}\,\delta_\tau(\partial_\sigma\omega^{\mp}) = \int d^4x'\,\sqrt{-g'}\,\delta_\tau\frac{\partial\omega^{\mp}}{\partial x'^{\pm}} = \int d^2x^{\perp}\int_{-\infty}^{\infty}\frac{dx^{\mp}}{4}\,\delta_\tau\omega^{\mp}\,|_{x^{\pm}=-\infty}^{x^{\pm}=\infty}\,,$$

where we took into account, that in the global light-cone inertial system g' does not depend on $x'^{\pm}$, which allows to integrate over this coordinate and neglect the factor $\sqrt{-g'}$ at $x'^{\pm} = \infty$.

The last expression enters in the variation of the induced action

$$\delta S_{ind} = \frac{1}{2\kappa^2}\int d^2x^{\perp}\left(\int_{-\infty}^{\infty}\frac{dx^-}{4}\delta\omega^-\,\partial^2_{\sigma\perp}A^{++}|_{x^+=-\infty}^{x^+=\infty} + \int_{-\infty}^{\infty}\frac{dx^+}{4}\delta\omega^+\,\partial^2_{\sigma\perp}A^{--}|_{x^-=-\infty}^{x^-=\infty}\right). \tag{41}$$

Therefore, taking into consideration also the HJ relation between $\delta_\tau g^{\mu\nu}$ and $\delta_\tau\omega^{\mp}$ on particle trajectories, one can present the variation of S_{ind} over $g^{\mu\nu}$ in a simple 4-dimensional form

$$\delta S_{ind} = -\frac{1}{2\kappa^2}\int d^4x\,\sqrt{-g}\,\delta g^{\mu\nu}\left(\partial_\mu\omega^-\,\partial_\nu\omega^-\,\partial^2_{\sigma\perp}A^{++} + \partial_\mu\omega^+\,\partial_\nu\omega^+\partial^2_{\sigma\perp}\,A^{--}\right). \tag{42}$$

The integrand contains the factor $\sqrt{-g}$ as a result of our transformation from the special coordinate system with the global light-cone time $x'^{\mp}$ to a general system with coordinates x.

The above expression for δS_{ind} allows one to calculate easily the induced energy-momentum tensor $\theta_{\mu\nu}$ in the Euler-Lagrange equations

$$R_{\mu\nu} - \frac{1}{2}g_{\mu\nu}R = -\theta_{\mu\nu}\,,\;\; \theta_{\mu\nu} = \partial_\mu x'^-\partial_\nu x'^-\partial^2_\sigma A^{++} + \partial_\mu x'^+\partial_\nu x'^+\partial^2_\sigma A^{--}, \tag{43}$$

where $x'^{\pm} = \omega^{\pm}$ are coordinates in the corresponding light-cone global systems. This result can be found independently from considerations related to the general covariance of the Euler-Lagrange equations.

In a quasi-elastic kinematics, where $A^{++} = 0$, it is convenient to work in the inertial system with the metric tensor obeying the constraints $g'^{\rho+} = \eta^{\rho+}$ with the global light-cone time x'^+, because here the energy-momentum tensor is $T_{\mu\nu} \sim \delta^+_\mu\delta^+_\nu$. The classical equations in this system have the simple solution

$$g'^{\rho\sigma} = \eta^{\rho\sigma} + \delta^\rho_-\delta^\sigma_- A^{--}\,. \tag{44}$$

It is a superposition of the plane-wave solutions of Aichelburg and Sexl with the gravitation centers situated at $x^+ = z^+$, $x^{\perp} = z^{\perp}$ and distributed with the weight function $\partial^2_{\tau\perp}A^{--}(z^+, z^{\perp})$. The coordinate transformation $x' = x'(x)$ to this global light-cone time inertial system satisfies the equations

$$g^{\mu\nu}\frac{\partial x'^\rho}{\partial x^\mu}\frac{\partial x'^+}{\partial x^\nu} = \eta^{\rho+} \tag{45}$$

and the tensors $\theta_{\mu\nu}$ and $T_{\rho\sigma}$ in these systems are related as follow

$$\theta_{\mu\nu}\frac{\partial x^\mu}{\partial x'^\rho}\frac{\partial x^\nu}{\partial x'^\sigma} = T_{\rho\sigma}\,. \tag{46}$$

Thus, the covariant energy-momentum tensor for the quasi-elastic kinematics is

$$\theta_{\mu\nu} = \partial_\mu x'^+ \partial_\nu x'^+ \partial_\sigma^2 A^{--} \,, \tag{47}$$

where partial derivatives $\partial_\mu x'^+$ are found from the solution of the Hamilton-Jacobi equation

$$g^{\mu\nu} \partial_\mu x'^+ \partial_\nu x'^+ = 0 \,. \tag{48}$$

The covariant energy-momentum tensor for a general kinematics is conserved

$$D_\mu \theta^{\mu\nu} = 0 \tag{49}$$

due to the kinematical constraints $\partial_\pm A^{\pm\pm} = 0$. As a consequence of the Hamilton-Jacobi equation for $x'^\pm$ the tensor $\theta_{\mu\nu}$ is traceless

$$g^{\mu\nu} \theta_{\mu\nu} = 0 \,. \tag{50}$$

Thus, we constructed the generally covariant Euler-Lagrange equations for the effective action in the high energy gravity. In a quasi-elastic kinematics with $A^{++} = 0$ one of their solutions at an inertial coordinate system with the global light-cone time x'^+ has a simple form $g'^{--} = A^{--}$. There are also many solutions which are parameterized by their asymptotic behavior at $t \to -\infty$ and $t \to \infty$, respectively. The effective action calculated on these solutions allows to construct a generating functional for the effective multi-graviton scattering amplitudes in a tree approximation and to reproduce independently the known results [16]. The integration over fluctuations around the classical solutions gives a possibility to calculate loop corrections to reggeized graviton interactions and to effective graviton vertices. In particular, using the one-loop Regge trajectory obtained in Refs [11, 16], the graviton scattering amplitude in the double-logarithmic approximation was found [17]. A possible generalization of our approach to the super-gravity in the 10-dimensional AdS space will be interesting for the construction of the Gribov Pomeron calculus in N=4 SUSY.

References

[1] S. Mandelstam, *Nuovo Cim.* **30**, 1148 (1963).

[2] V. N. Gribov, *Sov. Phys. JETP* **26**, 414 (1968).

[3] L. N. Lipatov, *Sov. J. Nucl. Phys.* **23** 338 (1976);
V. S. Fadin, E. A. Kuraev, L. N. Lipatov, *Phys. Lett.* **B 60** 50 (1975);
E. A. Kuraev, L. N. Lipatov, V. S. Fadin, Sov. Phys. JETP **44** 443 (1976);
Ya. Ya. Balitsky, L. N. Lipatov, *Sov. J. Nucl. Phys.* **28** (1978) 822.

[4] L. N. Lipatov *High energy asymptotics of multi-colour QCD and exactly solvable lattice models*, hep-th/9311037, unpublished.

[5] L. N. Lipatov, *J. Phys.* **A 42**:304020 (2009).

[6] L. N. Lipatov, *Nucl. Phys.* **B 452**, 369 (1995); *Phys. Rept.* **286**, 131 (1997).

[7] E. N. Antonov, L. N. Lipatov, E. A. Kuraev, I. O. Cherednikov, *Nucl. Phys.* **B 721**, 111 (2005).

[8] J. M. Maldacena, *Adv. Theor. Math. Phys.* **2** 231 (1998).

[9] S. S. Gubser, I. R. Klebanov, A. M. Polyakov, *Phys. Lett.* **B 428** 105 (1998).

[10] E. Witten, *Adv. Theor. Math. Phys.* **2** 253 (1998).

[11] L. N. Lipatov, Phys. Part. Nucl. **44** 391 (2013), arXiv: 1105.31277 [hep-ph].

[12] V. S. Fadin, L. N. Lipatov, *Phys. Lett.* **B 429** 127 (1998);
M. Ciafaloni and G. Camici, *Phys. Lett.* **B 430** 349 (1998).

[13] A. V. Kotikov, L. N. Lipatov, *Nucl. Phys.* **B 582** 19 (2000).
[14] A. V. Kotikov, L. N. Lipatov, *Nucl. Phys.* **B 661** 19 (2003).
[15] A. V. Kotikov, L. N. Lipatov, Nucl. Phys. **B 874** 889 (2013).
[16] L. N. Lipatov, Phys. Lett. **116** (1982) 411.
[17] J. Bartels, L. N. Lipatov, A. Sabio Vera, J. High Energy Phys. **07**(2014) 056.

EPJ Web of Conferences **125**, 01011 (2016) DOI: 10.1051/epjconf/201612501011
QUARKS-2016

Trajectories of bright stars at the Galactic Center as a tool to evaluate a graviton mass

Alexander Zakharov[1,2,3,4,⋆], *Predrag* Jovanović[5,⋆⋆], *Dusko* Borka[6,⋆⋆⋆], and *Vesna* Borka Jovanović[6,⋆⋆⋆⋆]

[1] *Institute of Theoretical and Experimental Physics, 117259 Moscow, Russia*
[2] *Bogoliubov Laboratory for Theoretical Physics, JINR, 141980 Dubna, Russia*
[3] *National Research Nuclear University MEPhI (Moscow Engineering Physics Institute), 115409, Moscow, Russia*
[4] *North Carolina Central University, Durham, NC 27707, USA*
[5] *Astronomical Observatory, Volgina 7, 11060 Belgrade, Serbia*
[6] *Atomic Physics Laboratory (040), Vinča Institute of Nuclear Sciences, University of Belgrade, P.O. Box 522, 11001 Belgrade, Serbia*

Abstract. Scientists worked in Saint-Petersburg (Petrograd, Leningrad) played the extremely important role in creation of scientific school and development of general relativity in Russia. Very recently LIGO collaboration discovered gravitational waves [1] predicted 100 years ago by A. Einstein. In the papers reporting about this discovery, the joint LIGO & VIRGO team presented an upper limit on graviton mass such as $m_g < 1.2 \times 10^{-22} eV$ [1, 2]. The authors concluded that their observational data do not show violations of classical general relativity because the graviton mass limit is very small. We show that an analysis of bright star trajectories could bound graviton mass with a comparable accuracy with accuracies reached with gravitational wave interferometers and expected with forthcoming pulsar timing observations for gravitational wave detection. This analysis gives an opportunity to treat observations of bright stars near the Galactic Center as a tool for an evaluation specific parameters of the black hole and also to obtain constraints on the fundamental gravity law such as a modifications of Newton gravity law in a weak field approximation. In that way, based on a potential reconstruction at the Galactic Center we give a bounds on a graviton mass.

1 Brief history of GR development in Russia

It is well-known that general relativity (GR) was discovered in November 1915 [3, 4] (see also [5–9] for a more detailed historical description of the issue). The gravitational waves in the framework of GR have been discussed in A. Einstein paper [10] published on June 22, 1916. The Quarks-2016 was organized in Pushkin near Saint-Petersburg. Four Saint-Petersburg institutions organized the seminar, namely, Petersburg Nuclear Physics Institute of NRC "Kurchatov Institute", Saint-Petersburg

⋆e-mail: zakharov@itep.ru
⋆⋆e-mail: pjovanovic@aob.rs
⋆⋆⋆e-mail: dusborka@vin.bg.ac.rs
⋆⋆⋆⋆e-mail: vborka@vinca.rs

State University, Peter the Great Saint-Petersburg Polytechnic University and Saint-Petersburg Scientific Center of RAS. Saint-Petersburg is practically the birthplace of Soviet (Russian) GR school and perhaps the school started to work almost immediately after the revolution and World War I, when Vsevolod Frederiks returned to Russia in 1918 (in the same year the main Soviet review journal Uspekhi Fizicheskikh Nauk has been established).[1]. In 1918 Soviet Russia was in ruins after revolutions and World War I, financial and human losses were gigantic, but in spite of that a scientific life was going on. In 1921 Frederiks published three papers Physics – Uspekhi journal including a paper [22], where he described basic concepts of GR. Frederiks and Friedmann planned to write a book "Foundations of General Relativity", but unfortunately Friedmann died after severe typhoid in 1925 when he was only 37 and only the first chapter "Tensor calculus" has been written and published [23]. As V.A. Fock wrote [24]: "Alexander Alexandrovich Friedmann and Vsevolod Konstantinovich Frederiks were professors of Petrograd University (now Leningrad University) and they were the first scientists who started to teach Russian physicists (working in Petrograd) GR...Main speakers were V.K. Frederiks and A.A. Friedmann. Styles of their talks were rather different. Frederiks clearly understood physical aspects of theory[2] and disliked to show mathematical calculations, while Friedmann placed emphasis on mathematics but not on physics. He aimed at mathematical rigor and paid attention to complete and precise form ulations of initial assumptions. Discussions between Frederiks and Friedmann were very interesting." Now it is clear that it was not strange that one of the most important solutions of Einstein equations has been found by A. A. Friedmann [25, 26] and now we live in the Friedmann universe. The hot Universe model (the Big Bang model) has been introduced by G. Gamov [27] who was A. Friedmann's student [28]. Conferences on general relativity and gravitation

[1]The crucial role of Frederiks in development of Russian science has been pointed out in the literature, for instance, V. P. Vizgin and V. Ya. Frenkel named their article as "Vsevolod Konstantinovich Frederiks as a pioneer of relativity and liquid crystal physics in USSR" [11]. After finishing high school Vsevolod Kostantinovich Frederiks continued his education in Europe as it was rather usual for young people from noble families if they decided to work in science or education [12, 13]. He selected the physical department of the Geneva University. There were two reasons to select namely the Geneva University, first, the Geneva location in a center of Western Europe is very convenient to visit other European scientific centers, second, the Geneva University had a very strong school of experimental solid state physics and electromagnetism and a famous scientist Charles-Eugene Guye was a professor and Director of Physical Institute at the Geneva University. Guye was Frederiks's supervisor. Earlier, Guye was a Privatdozent at Zurich Polytechnic where he was a teacher of A. Einstein. Starting in 1907 Guye and his assistants conducted wonderful experiments where they showed that electron masses depend on their speed [14–16] and experimentally confirmed predictions of special relativity. After defence of PhD thesis in Geneva Frederiks went to Göttingen to work since 1911 as an assistant of Woldemar Voigt who was Director of Physical Institute of the Göttingen University (Woldemar Voigt had a number of scientific titles, in particular, in 1913 he was elected a Honored Member of Imperial Saint-Petersburg University). Voigt was the first scientist who assumed universal speed of light in 1887 and tried to find transformations which keep an invariance of the wave equations [17] (see also [18, 19] for a more detailed discussion of the issue, in particular, differences between Voigt and Lorentz transformations). In summer 1914 W. Voigt left the Physical Institute Director position. In August 1914 World War I started and it was unexpected for many Russians staying in Europe at the time, after that Russian people (including V. Frederiks) had a status of civilian prisoners-of-war. It means that they have number of restrictions, in particular, Russian people have no right to have a job with wages from governmental institutions. In these circumstances, Frederiks has been resigned from the Physical Institute. However, Voigt asked authorities to permit Frederiks to work in laboratory. As a result of their joint studies in 1915 Voigt and Frederiks published an important paper on piezoelectricity [20], where the authors proved that Voigt model of piezoelectricity is more correct than a model proposed by W. C. Röntgen [21]. Since 1912 Hilbert was interested in mathematical aspects of theoretical physics and he had assistants as consultants on actual problems of physics. Duties of Hilbert's assistant were to report every day about the most important results in branches of physics which were interesting for Hilbert. Since it was impossible to obtain an official position and official wages from the Göttingen University for Frederiks as a citizen of enemy country, Hilbert offered Frederiks to be an assistant on physics and he paid an assistant salary from his own personal funds. The Hilbert's action had an important impact on development of general relativity, geometry and mathematical aspects of physics in Russia.

[2]"When I asked question: "How do you formulate a law of motion of many bodies in GR?" he immediately answered: "Motion of bodies is determined by motion of singular points of metric tensor". It was said many years before derivations of equations of motion for system of bodies (as it was done in Einstein et al. and my papers in 1938 – 1939)" [24].

play very important role in a development of the branch of science. One of the first conferenc[3] was organized under supervision of W. Pauli in 1955 and it was a celebration of forty years since a creation of GR and fifty years since the discovery of special relativity. Two participants A.D. Alexandrov and V.A. Fock represented USSR where Alexandrov presented an invited plenary talk on mathematical aspects of GR. In 1955 A.D. Alexandrov was a rector of Leningrad University and corresponding member of Academy of Sciences of USSR. He was only 42 but he delivered a talk not because he had a high administrative position but because of his high scientific reputation. Unfortunately, now it is not common practice when young leading scientists are rectors of top universities. Academician A.D. Alexandrov had many outstanding students including famous G. Perelman who was last Alexandrov's student and a recipient of a number of the many prizes (including Fields medal and the Millennium prize) but he rejected to accept these prizes.

2 Introduction

Very recently long-term efforts of theorists and experimentalists have been led to a remarkable discovery of gravitational waves [1]. In June 2016 LIGO collaboration reported about the detection of the second gravitational wave event and one suspected case [29] (LIGO people reported about the result one week before the century jubilee since a publication of the first Einstein paper on gravitational waves [10]).[4] This discovery represents a brilliant confirmation of general relativity and gives also a confirmation of black hole existence in binary systems. Different alternative theories of gravity, including massive graviton theories introduced by Fierz and Pauli [30] are a subject of intensive studies. Gravitational wave observations could constrain a graviton mass since in the case of massive graviton a gravitational wave signal is different from a signal of general relativity [31, 32]. Analyzing and comparing observations with massive and massless models one could obtain a graviton mass constraint. Additionally, in the case if source of gravitational radiation signal is known one could detect a time delay of gravitational wave signal in respect to electromagnetic and neutrino signals. A slow progress in understanding of dark matter (DM) and dark energy (DE) problems stimulates a growth of interest to alternative theories of gravity. Nowadays, there are many alternative theories of gravity which have been proposed. In spite of the evident success of the conventional general relativity (GR) since after 100 years of its development there are necessity to cure its shortcomings. Also, perhaps there are deviations from Newtonian gravity at the Solar system scales [33, 34] in spite of a model of the thermal origin of the anomaly [35]. Some other alternative theories of Lorentz invariant massive gravity has been introduced [30], but scieniest found a number of problems with such theories such as existence of ghosts, vDVZ discontinuity [36–39]. Some of these problems can be overcame [39–48]. We have to note that some of alternative theories do not have the Newtonian limit in a weak gravitational field approximation. For example, Yukawa-like corrections have been obtained, as a general feature, in the framework of $f(R)$ theories of gravity [49]. A number of gravity theories have a Yukawa limit for a weak gravitational field. In this investigation, we will discuss observational consequences of massive gravity where one can expect a Yukawa type of weak gravitational field limit instead of Newtonian one. Experimental and observational ways to bound a graviton mass can be very different. Will considered an opportunity to evaluate a graviton mass analyzing a time delay in electromagnetic waves such as supernova or gamma-ray burst [32], or from from gravitational wave signal alone [31]. Some

[3]Now it is called GR0 conference according to the standard scientific chronology while the first GR conference (GR1) has been organized by Bryce De Witt in Chapel Hill (NC) in 1957.

[4]As one could see the LIGO collaboration reported about the discoveries of gravitational wave events in the time interval between one century jubilee since the GR discovery in November 1915 and one century since the first Einstein paper on gravitational waves published in June 1916.

other suggestions are given in following references [50–55]. It was shown that speed of gravity practically coincides with speed of light [56–60], and it can be used to get constraints of a graviton mass. Constraints on the range λ of Yukawa type force in Solar system has been considered [61] and one can derive the bound $\lambda > 2.8 \times 10^{12}$ km [32] from these constraints assuming a natural modification of the Newtonian potential [32, 41]:

$$V(r) = -\frac{GM}{(1+\delta)r}\left[1 + \delta e^{-\left(\frac{r}{\lambda}\right)}\right], \tag{1}$$

where δ is a universal constant. In our previous investigation [62] we found constraints on parameters of Yukawa gravity.

One more way that can be used to evaluate a graviton mass [63] is pulsar timing. An idea to use pulsar timing for gravitational wave detection has been proposed many years ago [64]. An analysis of the cross-correlation function between pulsar timing residuals of pulsar pairs could give an opportunity to detect gravitational waves [65, 66]. With 90% probability, massless gravitons can be distinguished from gravitons heavier than 3×10^{-22} eV (Compton wavelength $\lambda_g = 4.1 \times 10^{12}$ km) [63]. Predictions show that if bi-weekly observation of 60 pulsars is performed for 5 years with a pulsar rms timing accuracy of 100 ns and if 10 year observation of 300 pulsars with 100 ns timing accuracy would probe graviton masses down to 3×10^{-23} eV ($\lambda_g = 4.1 \times 10^{13}$ km).

Aim of our investigation is that our previous results [62] about the constraints on parameters of Yukawa gravity from observations of trajectories of bright stars near the Galactic Center use to obtain a graviton mass bounds.

3 Graviton mass estimates from S2 star orbit

Monitoring bright stars at the Galactic Center is given in the following references [67–75]. The astrometric observations of S2 star were given in the references [76]. Recently, they were used to evaluate parameters of black hole and to test and constrain several models of modified gravity at mpc scales [77–81]. Comparisons of results obtained in Yukawa gravity model with the NTT/VLT astrometric observations of S2 star [76] resulted with the constraints on the range of Yukawa interaction λ [62]. Comparison showed that λ is most likely on the order of several thousand astronomical units $\lambda \sim 5000 - 6000$ AU. This result can be used to constrain the lower bound for Compton wavelength λ_g of the graviton, and thus the upper bound for its mass $m_g = hc/\lambda_g$, assuming Yukawa gravitational potential of a form $\propto r^{-1}\exp(-r/\lambda_g)$ [see e.g. 31]. We find these constraints using chi-square test of goodness of the S2 star orbit fits by Yukawa gravity potential (1).

In order to obtain the reliable statistical criterion,we used the following expression:

$$\chi^2 = \sum_{i=1}^{n}\left[\frac{\left(x_i^o - x_i^c\right)^2}{\sigma_{xi}^2 + \sigma_{int}^2} + \frac{\left(y_i^o - y_i^c\right)^2}{\sigma_{yi}^2 + \sigma_{int}^2}\right], \tag{2}$$

where (x_i^o, y_i^o) is the i-th observed position, (x_i^c, y_i^c) is the corresponding calculated position, n is the number of observations, σ_{xi} and σ_{yi} are uncertainties of observed positions, while σ_{int} accounts for the "intrisic dispersion" of the data [84]. σ_{int} is usually introduced whenever the observed uncertainties are not mutually consistent, like it is the case with SN Ia data in cosmology (see e.g. [82, 83] and references therein).

We performed the fits of $n = 70$ positions of S2 star observed by NTT/VLT [76] with its simulated orbits in Yukawa gravity potential (1), by varying λ between 1500 and 20000 AU. We assume value

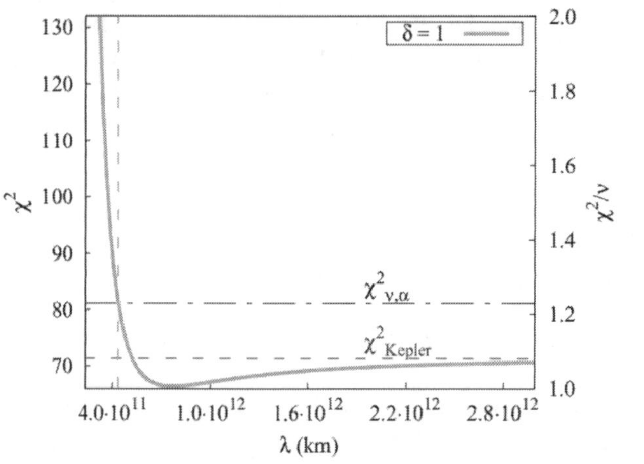

Figure 1. χ^2 (solid lines) as a function of Yukawa range of interaction λ, i.e. the graviton Compton wavelength λ_g, obtained from the fits of NTT/VLT observations of S2 star [76] using the gravity potential (1) for $\delta = 1$. For comparison, the value of the Keplerian fit $\chi^2_{\text{Kepler}} = 71.34$ is also denoted by the horizontal dashed line. The critical value for $\nu = 66$ degrees of freedom and $\alpha = 0.1$ significance level, $\chi^2_{\nu,\alpha} = 81.08$, is presented by the horizontal dash-dotted line, and the upper bounds λ_x of the corresponding exclusion regions for λ_g by the vertical dotted lines. The values $\lambda_g < \lambda_x$ can be excluded with 90% probability.

of $\delta = 1$. In total 4 parameters were fitted: two components of initial position and two components of initial velocity in orbital plane. The number of degrees of freedom is $\nu = 70 - 4 = 66$ [62]. Fitting is performed by minimization of χ^2 (see 2), where σ_{int} is estimated from the requirement that the global minimum of reduced χ^2 over the whole range of λ is $\chi^2/\nu = 1$ (see Fig. 1). We obtained that $\sigma_{int} = 1.13$ mas satisfies this requirement.

The resulting value of χ^2 for $\delta = 1$ as functions of λ is presented in Fig. 1. We can see that χ^2 asymptotically approaches to the corresponding value of the Keplerian fit $\chi^2_{\text{Kepler}} = 71.34$ (horizontal dashed line in Fig. 1), for the large values of λ. Also, χ^2 has global minimum at $\lambda = 5100 \pm 50$ AU $\approx 7.6 \times 10^{11}$ km in the case of $\delta = 1$. We can conclude that the obtained results for λ are consistent with those from [62].

Further, we use the obtained values of χ^2 to test the null hypothesis that λ should be at least on the order of 10^3 AU. In order to test this hypothesis we assumed the significance level $\alpha = 0.1$ and we calculated the critical value $\chi^2_{\nu,\alpha} = 81.08$ for $\nu = 66$ degrees of freedom (horizontal dash-dotted line in Fig. 1). As it can be seen from Fig. 1, there is an exclusion range of λ with some upper bound λ_x where $\chi^2 > \chi^2_{\nu,\alpha}$, so the cases $\lambda < \lambda_x$ can be excluded with very high probability of $1 - \alpha = 90\%$. This upper exclusion bound is $\lambda_x = 2900 \pm 50$ AU $\approx 4.3 \times 10^{11}$ km. Since the null hypothesis can be considered as valid for $\lambda > \lambda_x$, this value can be taken as the lower allowed bound for the graviton Compton wavelength $\lambda_g > \lambda_x$. Our results are in agreement with Solar System and LIGO constraints on λ_g (see Fig. 8 from [2]).

We obtain that the corresponding upper bound for graviton mass, $m_g = hc/\lambda_x$, is $m_g = 2.9 \times 10^{-21}$ eV in the case of $\delta = 1$ [84]. These constraints are consistent with those obtained from a gravitation wave signal GW150914 [2]. It significantly exceeds 1.2×10^{-22} eV which represents 90% probability limit for distinguishing massless gravitons from massive gravitons using pulsar timing experiments [see e.g. 63].

4 Conclusions

We consider phenomenological consequences of massive gravity and show that an analysis of bright star trajectories could bound the graviton mass. We perform simulations of the S2 star orbit around the supermassive black hole at the Galactic Center in Yukawa gravity [62] and we compare our results with the NTT/VLT astrometric observations of S2 star [76]. We obtain the constraints on the range of Yukawa interaction $\lambda > 4.3 \times 10^{11}$ km and corresponding most likely upper bound for graviton mass $m_g < 2.9 \times 10^{-21}$ eV.

Our estimate of a graviton mass is slightly greater than the estimate obtained by the LIGO collaboration with the first detection of gravitational waves from the binary black hole system. We can conclude that a) our estimate is consistent with the LIGO estimations; b) our estimation is based on trajectories analysis of bright stars near the Galactic Center; c) in the future our estimate may be improved with forthcoming observational facilities like GRAVITY [85], E-ELT [86] and TMT [87]. We believe that in the future such facilities may improve the discussed estimates of graviton mass.

A.F.Z. thanks the organizers of Quarks-2016 for their kind and invitation to present the talk at the seminar. A.F.Z. appreciates also NSF (HRD-0833184) and NASA (NNX09AV07A) at NASA CADRE and NSF CREST Centers (NCCU, Durham, NC, USA) for a partial support. P.J., D.B.and V.B.J. wish to acknowledge the support by the Ministry of Education, Science and Technological Development of the Republic of Serbia through the project 176003 "Gravitation and the large scale structure of the Universe".

References

[1] B. P. Abbott *et al.*, (LIGO Scientific Collaboration and Virgo Collaboration), Phys. Rev. Lett. **116**, 061102 (2016).

[2] B. P. Abbott *et al.*, (LIGO Scientific Collaboration and Virgo Collaboration), LIGO Document P1500213-v27, arXiv:1602.03841 (2016).

[3] A. Einstein, Sitzungsber. K. Preuss. Akad. Wiss. 844 (1915).

[4] D. Hilbert, Göttingen Nachr. **3**, 395 (1915).

[5] J. Mehra, *Einstein, Hilbert and the Theory of Gravitation*, (Dordrecht, Boston, 1974).

[6] J. Earman, C. Glymour Arch. Hist. and Exact. Sci. **19**, 291 (1978).

[7] V.P. Vizgin, Ya.A. Smorodinskii, Physics – Uspekhi **22**, 489 (1979).

[8] V.P. Vizgin, Physics – Uspekhi **44**, 1283 (2001).

[9] A.A. Logunov, M.A. Mestvirishvili, V.A. Petrov, Physics – Uspekhi **47**, 607 (2004).

[10] A. Einstein, Sitzungsber. K. Preuss. Akad. Wiss. **1**, 688 (1916).

[11] V. P. Vizgin, V. Ya. Frenkel, in *Einstein's collection 1983– 1984 (Einteinovskij sbornik 1983–1984)*, 106 (1988) (in Russian).

[12] A. S. Sonin, V. Ya. Frenkel, Priroda **10**, 86 (1994) (in Russian).

[13] A. S. Sonin, V. Ya. Frenkel, *Vsevolod Konstantinovich Frederiks*, (Nauka, Moscow, 1995) (in Russian).

[14] C.E. Guye, S. Ratnowsky, Comptes Rendus de l' Académie **150** 326 (1910).

[15] C.E. Guye, C. Lavanchy, Comptes Rendus de l' Académie **161** 52 (1915).

[16] C.E. Guye, C. Lavanchy, Comptes Rendus de l'Académie **42** 286 (1916).

[17] W. Voigt, Nachrichten von der Königl. Gesellschaft der Wissenschaften und der Georg-Augusts-Universität zu Göttingen, 41 (1887).

[18] A. Ernst, J.-P. Hsu, Chinese J. Phys. **39**, 211 (2001).

[19] A. Ernst, J.-P. Hsu, in *Lorentz and Poincare invariance*, ed. J.-P. Hsu (World Scientific, Singapore, 2001), p. 4.
[20] W. Voigt, V. Freedericksz, Annalen der Physik **353**, 145 (1915).
[21] W. C. Röntgen, Wied. Ann. **89** 16 (1889).
[22] V. K. Frederiks, Uspekhi Fiz. Nauk **2**, 162 (1921) (in Russian); reprinted in Phys. – Usp. **42**, 1235 (1999).
[23] V. K. Frederiks, A.A. Friedmann, *Tensor calculus* (Leningrad, Academia, 1921) (in Russian).
[24] V. A. Fock, Sov. Phys. Usp. **6**, 473 (1964).
[25] A.A. Friedmann, Zeitschrift für Physik **10**, 377 (1922); English translation in General Relativity and Gravitation **31**, 1991 (1999).
[26] A.A. Friedmann, Zeitschrift für Physik **21**, 326 (1924); English translation in General Relativity and Gravitation **31**, 2001 (1999)
[27] G. Gamow, Phys. Rev. **70**, 572 (1946).
[28] E. A. Tropp, V. Ya. Frenkel, A. D. Chernin, *Alexander A. Friedmann: The man who made the Universe expand*, (Cambridge University Press, 2006).
[29] B. P. Abbott *et al.*, (LIGO Scientific Collaboration and Virgo Collaboration), Phys. Rev. Lett. **116**, 241103 (2016).
[30] M. Fierz and W. Pauli, Proc. of the Royal Society of London, Series A **173**, 211 (1939).
[31] C. Will, Phys. Rev. D **57** 2061 (1998); [gr-qc/9709011].
[32] C. Will, Living Reviews in Relativity **17**, 4 (2014); arXiv:1403.7377v1[gr-qc].
[33] J. D. Anderson, P. A. Laing, E. L. Lau, A. S. Liu, M. M. Nieto and S. G. Turyshev, Phys. Rev. Lett. **81**, 2858 (1998).
[34] J. D. Anderson, J. K. Campbell, J. E. Ekelund, J. Ellis and J. F. Jordan, Phys. Rev. Lett. **100**, 091102 (2008).
[35] S. G. Turyshev, V. T. Toth, G. Kinsella, S. C. Lee, S. M. Lok and J. Ellis, Phys. Rev. Lett. **108**, 241101 (2012).
[36] V. I. Zakharov, JETP Letters **12**, 447 (1970).
[37] H. van Dam and M. Veltman, Nucl. Phys. B **22**, 397 (1970) .
[38] Y. Iwasaki, Phys. Rev. D **2**, 2255 (1970).
[39] V. A. Rubakov and P. G. Tinyakov, Physics – Uspekhi **51**, 759 (2008).
[40] A. I. Vainshtein, Phys. Lett. B **39**, 393 (1972).
[41] M. Visser, Gen. Rel. Grav. **30**, 1717 (1998).
[42] G. Dvali, G. Gabadadze and M. Porrati, Phys. Lett. B **485**, 112 (2000).
[43] I. I. Kogan, S. Mouslopoulos, A. Papazoglou and G. G. Ross, Nucl. Phys. B **595**, 225 (2001).
[44] C. Deffayet, G. Dvali, G. Gabadadze and A. Vainshtein, Phys. Rev. D **65**, 044026 (2002).
[45] L. S. Finn and P. J. Sutton, Phys. Rev. D **65**, 044022 (2002).
[46] T. Damour, I. I. Kogan and A. Papazoglou, Phys. Rev. D **67**, 064009 (2003).
[47] M. Maggiorre, *Gravitational Waves*, Vol. VI, (Oxford University Press, 2008).
[48] K. Hinterbichler, Rev. Mod. Phys. **84**, 671 (2012).
[49] S. Capozziello, A. Stabile and A. Troisi, Mod. Phys. Lett. A **24**, 659 (2009).
[50] A. S. Goldhaber and M. M. Nieto, Phys. Rev. D **9**, 1119 (1974).
[51] P. G. O. Freund, A. Maneshwari and E. Schonberg, Astrophys. J. **157**, 857 (1969).
[52] A. A. Logunov and M. A. Mestvirishvili, Theor. Math. Phys. **61**, 1170 (1984).
[53] S. S. Gershtein, A. A. Logunov and M. A. Mestvirishvili, Physics – Uspekhi **49**, 1179 (2006).

[54] S. S. Gershtein, A. A. Logunov and M. A. Mestvirishvili, Phys. Part. Nucl. **39**, 1 (2008).
[55] A. S. Goldhaber and M. M. Nieto, Rev. Mod. Phys. **82**, 939 (2010).
[56] S. M. Kopeikin, Astrophys. J. Lett. **556**, L1 (2001); [gr-qc/0105060].
[57] E. B. Fomalont and S. M. Kopeikin, Astrophys. J. **598**, 704 (2003); [astro-ph/0302294].
[58] S. M. Kopeikin, Class. Quant. Grav. **21**, 3251 (2004), [gr-qc/0310059].
[59] S. M. Kopeikin, Class. Quantum Grav. **22**, 5181 (2005); [gr-qc/0501048].
[60] S. M. Kopeikin, Int. J. Mod. Phys. D **15**, 273 (2006); [gr-qc/0501001].
[61] C. Talmadge, J.-P. Berthias, R. W. Hellings and E. M. Standish, Phys. Rev. Lett. **61**, 1159 (1988).
[62] D. Borka, P. Jovanović, V. Borka Jovanović, A. F. Zakharov, J. Cosm. Astropart. Phys. **11**, 050 (2013).
[63] K. J. Lee, F. A. Jenet, R. H. Price, N. Wex and M. Kramer, Astrophys. J. **722**, 1589 (2010).
[64] M. V. Sazhin, Sov. Astron. **22**, 36 (1978).
[65] F. A. Jenet, G. B. Hobbs, K. J. Lee and R. N. Manchester, Astrophys. J. **625**, L123 (2005).
[66] K. J. Lee, F. A. Jenet and R. H. Price, Astrophys. J. **685**, 1304 (2008).
[67] A. M. Ghez, M. Morris, E. E. Becklin, A. Tanner and T. Kremenek,Nature **407**, 349 (2000).
[68] G. F. Rubilar and A. Eckart, Astron. Astrophys. **374**, 95 (2001).
[69] R. Schödel, T. Ott, R. Genzel *et al.*, Nature **419**, 694 (2002).
[70] A. M. Ghez, G. Duchêne, K. Matthews *et al.*, Astrophys. J. Lett. **586**, L127 (2003).
[71] A. M. Ghez, S. A. Wright, K. Matthews *et al.*, Astrophys. J. Lett. **601**, L159 (2004).
[72] A. M. Ghez, S. Salim, N. N. Weinberg *et al.*, Astrophys. J. **689**, 1044 (2008).
[73] S. Gillessen, F. Eisenhauer, S. Trippe *et al.*, Astrophys. J. **692**, 1075 (2009).
[74] S. Gillessen, R. Genzel, T. K. Fritz *et al.*, Nature, **481**, 51 (2012).
[75] L. Meyer, A. M. Ghez, R. Schödel *et al.*, Science, **338**, 84 (2012).
[76] S. Gillessen, F. Eisenhauer, T. K. Fritz *et al.*, Astrophys. J. **707**, L114 (2009).
[77] A. F. Zakharov, A. A. Nucita, F. De Paolis and G. Ingrosso, Phys. Rev. D **76**, 062001 (2007).
[78] A. F. Zakharov, S. Capozziello, F. De Paolis, G. Ingrosso and A. A. Nucita, Space Sci. Rev. **48**, 301 (2009).
[79] D. Borka, P. Jovanović, V. Borka Jovanović and A. F. Zakharov, Phys. Rev. D **85**, 124004 (2012).
[80] A. F. Zakharov, J. Astrophys. Astron. **36**, 539 (2015).
[81] A. F. Zakharov, in *Gravitation, Astrophysics, and Cosmology - Proc. Twelfth Asia-Pacific International Conference*, edited by J.-P. Hsu and V.N. Melnikov (World Scientific Publ., Singapore, 2016) p. 176.
[82] P. Astier, J. Guy, N. Regnault *et al.*, Astron. Astrophys. **447**, 31 (2006).
[83] A. F. Zakharov and V. N. Pervushin, Phys. Atom. Nucl. **75**, 1418 (2012).
[84] A. F. Zakharov, P. Jovanović, D. Borka, V. Borka Jovanović, J. Cosm. Astropart. Phys. **05**, 045 (2016).
[85] N. Blind, F. Eisenhauer, S. Gillessen *et al.*, arXiv:1503.07303 [astro-ph.IM].
[86] A. Ardeberg, J. Bergeron, A. Cimatti *et al.*, *An Expanded View of the Universe Science with the European Extremely Large Telescope*, ed. by M. Lyubenova and M. Kissler-Patig, ESO, Munich.
[87] W. Skidmore *et al.*, *Thirty Meter Telescope Detailed Science Case: 2015 (TMT Observatory Corporation).*

EPJ Web of Conferences **125**, 01012 (2016) DOI: 10.1051/epjconf/201612501012
QUARKS-2016

Quantum corrections in higher spin theory

Arkady Tseytlin[1,⋆]

[1] *Blackett Laboratory, Imperial College London, U.K. and Lebedev Institute, Moscow, Russia*

Abstract. We review some recent results on computation of quantum corrections in conformal and massless higher spin theory.

1 Introduction

Below we shall review some recent results of the study of quantum corrections in theories containing infinite number of massless or conformal higher spin states [1–7].

There is a long history of theoretical studies of higher spin theories. Massive HS fields appear in string theory and may become massless in a certain "tensionless" limit (not well defined in flat space). Massless HS theories are known to exist in AdS space and play a natural role in vectorial AdS/CFT duality. Theories dual to free adjoint $U(N)$ CFT's contain both massless tower of states plus infinite set of massive fields in AdS. There is also closely associated theory of conformal higher spins (with non-unitary higher derivative kinetic terms) that admits a consistent in flat-space expansion. Hypothetic theory of massless HS fields in flat space that should have infinite dimensional gauge and global symmetries should have "trivial" S-matrix to comply with analogs of Coleman-Mandula theorem and Weinberg soft theorem.

Theories of infinite number of fields require specific definition at quantum level that should be consistent with their symmetries and provide a prescription of how to sum over all spins. Similar ambiguities in string field theory are fixed by the existence of world-sheet prescription for computation of observables, but in the case of HS theories such a "first-quantized" formulation is not know. To make first steps towards understanding these issues one would like to compute some simple partition functions and scattering amplitudes with higher spin exchanges in massless higher spin (MHS) theory and conformal higher spin (CHS) theory.

In the context of vectorial AdS_{d+1}/CFT_d duality one starts with a free complex scalar in fundamental representation of $U(N)$, i.e. $\int d^d x\, \partial_m \Phi_i^* \partial_m \Phi_i$ $(i = 1, ..., N)$ with a tower of on-shell conserved higher spin currents $J_s \sim \Phi_i^* \partial_{(m_1} ... \partial_{m_s)} \Phi_i + ...$ $(s = 1, 2, ...)$ which are $U(N)$ invariant CFT primaries of dimension $\Delta = s + d - 2$ which are dual to massless HS fields ϕ_s field in AdS_{d+1}. $J_0 = \Phi_i^* \Phi_i$ with $\Delta = d-2$ is dual to massive scalar with $\Delta(\Delta-d) = m^2 = -2$. This is the same spectrum as in Vasiliev's massless HS theory in AdS_{d+1}. This MHS theory in AdS "summarizes" correlators of $U(N)$ singlet primaries J_s in free CFT: they are reproduced by HS interactions in AdS_{d+1} with coupling $\sim 1/N$, i.e. the classical action for Fronsdal fields ϕ_s in AdS should be

$$S = N \int d^{d+1}x \Big[\sum_s \phi_s(-\nabla^2 + m_s^2)\phi_s + \sum C_{s_1 s_2 s_3}(\nabla)\, \phi_{s_1} \phi_{s_2} \phi_{s_3} + ... \Big] \equiv N\bar{S} \quad (1)$$

⋆e-mail: tseytlin@imperial.ac.uk

Quantum corrections are described by an effective action $\Gamma = N\bar{S} + \Gamma_1 + N^{-1}\Gamma_2 + ...$ On the CFT side we may consider the generating functional for correlators of currents $J_s(\Phi) = \Phi_i^* \mathcal{J}_s \Phi_i$, $\mathcal{J}_s \sim \partial^s$ introducing source fields h_s. These will have the interpretation of conformal higher spin (CHS) fields that are gauge fields for the infinite dimensional algebra of symmetries of the free scalar theory. Their linearized gauge transformations $\delta h_{m_1\cdots m_s} = \partial_{(m_1}\varepsilon_{m_2\cdots m_s)} + \eta_{(m_1 m_2}\alpha_{m_3\cdots m_s)}$ are given by linearized differential and algebraic symmetries that generalize diffeomorphism and Weyl symmetry of the conformal C^2 gravity. Then the generating functional for connected correlators of currents is found by integrating over free fields Φ_i

$$\Gamma[h] = N \log\det(-\partial^2 + \textstyle\sum_s h_s \mathcal{J}_s) \tag{2}$$

From the AdS/CFT point of view $\Gamma[h]$ should follow from the massless HS theory in AdS$_{d+1}$ upon integrating over AdS$_{d+1}$ fields ϕ_s with Dirichlet b.c.

$$e^{-\Gamma[h]} = \int_{\phi_s|_{\partial \rm AdS}=h_s} [d\phi_s] \exp(-N\bar{S}[\phi]) \tag{3}$$

This matching suggests that all quantum corrections to AdS effective action $\Gamma - N\bar{S} = \Gamma_1 + N^{-1}\Gamma_2 + ...$ should vanish, i.e. the quantum MHS theory in AdS should be defined so that to satisfy this. Summing up infinite sets of (integrated) correlators suggests also further conjecture: partition function of the CFT on a non-trivial background should match the MHS partition function in (asymptotically) AdS with the corresponding boundary. Then all quantum corrections $(N^0, N^{-1}, ...)$ to the AdS partition function should vanish to all orders. In particular, matching at 1-loop order requires $\Gamma_1 = 0$.

This suggests that massless HS theories should have hidden simplicity due to large underlying symmetry (i.e. in a sense are similar to "topological" or "integrable" models). Below we shall discuss some examples of quantum computations in massless and conformal HS theories that demonstrate this.

2 Free massless HS theory in flat space and in AdS

One may consider a collection of free massless HS $s = 0, 1, 2, 3, ..., \infty$ fields with gauge-invariance $\delta\phi_{m_1...m_s} = \partial_{(m_1}\epsilon_{m_2...m_s)}$ that are described by the Fronsdal action $S = \int d^4x\, \partial^n\phi^{m_1...m_s}\partial_n\phi_{m_1...m_s} + ...$ Like massless vector, graviton, etc. for $s > 0$ they have 2 dynamical d.o.f. in 4d. One then observes a curious fact: the total number of of d.o.f. is zero if each spin enters once and te sum is defined using zeta-function regularization: $1 + \sum_{s=1}^{\infty} 2 = 1 + 2\zeta(0) = 0$. Equivalent statement is true for free partition function. For each spin s we have

$$Z_{{\rm MHS},s} = \Big[\frac{\det\Delta_{s-1\perp}}{\det\Delta_{s\perp}}\Big]^{1/2} = \Big[\frac{(\det\Delta_{s-1})^2}{\det\Delta_s \det\Delta_{s-2}}\Big]^{1/2} = \Big(\frac{1}{\sqrt{\det(-\partial^2)}}\Big)^2 \tag{4}$$

where $\Delta_s = -\partial^2$ defined on symmetric rank s traceless tensor. Then the total partition function is [3]

$$Z_{\rm MHS} = \prod_{s=0}^{\infty} Z_{{\rm MHS},s} = \Big[\frac{1}{\det\Delta_0}\Big]^{1/2}\Big[\frac{\det\Delta_0}{\det\Delta_{1\perp}}\Big]^{1/2}\Big[\frac{\det\Delta_{1\perp}}{\det\Delta_{2\perp}}\Big]^{1/2}\Big[\frac{\det\Delta_{2\perp}}{\det\Delta_{3\perp}}\Big]^{1/2} ... = 1 \tag{5}$$

Here we see cancellation between the physical spin s determinant and the ghost determinant for spin $s + 1$ giving the total $Z = 1$. This is similar to what happens in a supersymmetric or topological theory. Here this cancellation reflects the large gauge symmetry of the theory. The cancellation of an infinite number of factors is of course a priori ambiguous: it assumes a particular regularization that should be consistent with the underlying symmetry. The formal cancellation of factors we assumed is, in fact, consistent with the above zeta-function prescription for count of d.o.f: $Z_{\rm tot} = (Z_0)^{\nu_{\rm tot}}$, $\nu_{\rm tot} =$

$1 + \sum_{s=1}^{\infty} 2 = 0$. This zeta-function prescription is similar to the one used in computation of vacuum energy in bosonic string theory which is consistent with target space symmetries of the theory (the massless vector in $d = 26$, etc.). Equivalent to zeta-function is the prescription where one introduces a particular exponential cutoff in the sum over spins and then drops all singular terms. In general dimension d this reads

$$\nu_{\rm tot} = 1 + \sum_{s=1}^{\infty} \nu_s\, e^{-\epsilon[s+\frac{1}{2}({\rm d}-4)]}\Big|_{\epsilon\to 0,\ {\rm fin.}} = 0 \tag{6}$$

where in d = 4 we have $\nu_s = 2$.

The fact of triviality of the partition function applies also to the MHS theory in $AdS_{\rm d}$ dual to a free scalar CFT in d dimensions ($\mathrm{d} = d + 1$), i.e. $Z = 1$ holds in AdS vacuum of Vasiliev theory. The 1-loop partition function is given by the product of the contributions of spin s Fronsdal fields in AdS

$$Z_{\rm MHS} = \prod_{s=0}^{\infty} Z_{{\rm MHS},s}\,, \qquad Z_{{\rm MHS},s} = Z_{s,s-1} = \Big[\frac{\det \Delta_{s-1\,\perp}(M^2_{s-1,s})}{\det \Delta_{s\,\perp}(M^2_{s,s-1})}\Big]^{1/2}, \tag{7}$$

$$\Delta_s(M^2_{s,k}) \equiv -\nabla^2_s + M^2_{s,k}\,, \qquad M^2_{s,k} = -s + (k-1)(k+{\rm d}-2)\,, \qquad k = 0,1,\ldots,s-1\,. \tag{8}$$

One may define the determinants using spectral zeta-function so that (Λ=UV cutoff, r=AdS radius)

$$\ln\det\Delta_s = -\zeta_{\Delta_s}(0)\ln(\Lambda^2 r^2) - \zeta'_{\Delta_s}(0) \tag{9}$$

The natural prescription where one first sums over s for finite argument z of the zeta-function and then takes $z \to 0$ appears to be consistent with HS symmetries and one finds [3, 8]

$$\zeta_{\rm tot}(z) = \sum_{s=0}^{\infty} \zeta_{\Delta_s}(z)\,, \qquad \zeta_{\rm tot}(z\to 0) = 0 + 0\times z + O(z^2)\,, \qquad Z_{\rm MHS}({\rm AdS_d}) = 1 \tag{10}$$

An equivalent regularization prescription is same as in (6)

$$\ln Z_{\rm MHS}({\rm AdS_d}) = \sum_{s=0}^{\infty} \ln Z_{{\rm MHS},s}\, e^{-\epsilon[s+\frac{1}{2}({\rm d}-4)]}\Big|_{\epsilon\to 0,\ {\rm fin.}} = 0 \tag{11}$$

A natural conjecture is that the same triviality applies to all all-loop vacuum partition function: $Z_{\rm MHS}({\rm AdS_d}) = 1$: this is required for consistency with the vectorial AdS/CFT duality to free $U(N)$ scalar theory at the boundary.

3 Conformal higher spin theory

Conformal higher spins generalize Maxwell ($s = 1$) and Weyl ($s = 2$) theories and describe pure spin s states off shell – they have maximal gauge symmetry consistent with locality. That requires higher-derivative kinetic terms. The 4d CHS theory being a consistent local gauge theory with infinite dimensional higher spin symmetry which is closely associated to massless HS theory in AdS_5 is of interest as a model regardless the issue of its apparent non-unitarity. Free CHS field in flat 4d has the action $S_s = \int d^4x\, h_s P_s \partial^{2s} h_s$ where P_s is projector to transverse traceless totally symmetric rank s field. The CHS field h_s thus has dimension $2-s$ which is the same as of sources or "shadow" operators for spin s conserved currents $J_s(\Phi)$ that have dimension $2+s$. h_s also play the role of boundary values for the MHS fields in AdS_{d+1} dual to J_s.

The interacting CHS theory may be defined as "induced" one starting with $I = \int d^4x[\partial\Phi^*\partial\Phi + \sum_s \Phi^* \mathcal{J}_s \Phi\, h_s]$ and keeping the local log UV divergent part of the generating functional in (2)

$$S_{\rm CHS} = \log\det(-\partial^2 + \sum_s \mathcal{J}_s h_s)\Big|_{\log\Lambda} \tag{12}$$

The corresponding partition function in flat $d = 4$ space counting effective d.o.f. is given by

$$Z_{{\rm CHS},s} = (Z_0)^{\nu_s} = (\det\Delta_0)^{-\nu_s/2}\,, \qquad \nu_s = s(s+1) = 2, 6, ... \tag{13}$$

$$Z_{\rm CHS} = \prod_{s=0}^{\infty}(Z_0)^{\nu_s} = (Z_0)^{\nu_{\rm tot}}\,, \qquad \nu_{\rm tot} = \sum_{s=0}^{\infty}\nu_s \tag{14}$$

Using again the same regularization

$$\sum_{s=0}^{\infty} F(s) \;\rightarrow\; \sum_{s=0}^{\infty} F(s)\, e^{-\epsilon(s+\frac{d-3}{2})}\Big|_{\epsilon\to 0,\ {\rm fin.}} = 0 \tag{15}$$

we find $\nu_{\rm tot} = 0$ and thus $Z_{\rm CHS} = 1$. This regularization is the same as implied by the relation of MHS theory in AdS_{d+1} and CHS theory at the boundary. In fact, the 1-loop $Z_{\rm CHS}(S^4)$ in same regularization is again found to be equal to 1. This is consistent with the relation to the ratio of MHS partition functions in AdS_5 with two choices (D/N) of boundary conditions [5]

$$Z_{{\rm CHS},s}(S^4) = \frac{Z^-_{{\rm MHS},s}(AdS_5)}{Z^+_{{\rm MHS},s}(AdS_5)} \tag{16}$$

Then $Z_{\rm CHS}(S^4) = 1$ is directly related to $Z^{\pm}_{\rm MHS}(AdS_5)_{\rm tot} = 1$. This regularization is consistent with symmetries of the CHS theory: it leads also to the vanishing of conformal anomaly after summation over all spins [7]. For each spin s field one finds

$$T^m_m = -{\rm a}\,R^*R^* + {\rm c}\,C^2\,, \quad {\rm a}_s = \tfrac{1}{720}\nu_s^2(14\nu_s+3)\,, \quad {\rm c}_s - {\rm a}_s = \tfrac{1}{720}\nu_s(15\nu_s^2 - 45\nu_s + 4)\,, \quad \nu_s = s(s+1) \tag{17}$$

and then using the same regularization (15) one gets

$$\sum_{s=1}^{\infty}({\rm c}_s - {\rm a}_s) = 0\,, \qquad \sum_{s=1}^{\infty}{\rm a}_s = 0 \tag{18}$$

Thus the total conformal anomaly vanishes providing a check of 1-loop quantum consistency of the CHS theory (where Weyl symmetry is one of the gauge symmetries). The summation over infinite set of spins spins using this particular prescription provides a novel mechanism of getting UV finiteness in a bosonic theory (not completely dissimilar to the one in string theory).

4 Scalar scattering via conformal higher spin exchange

The issue of regularization of the sum over spins appears again in the context of S-matrix. One example is the scattering of external conformal scalars interacting through exchange of higher spin particles (here conformal, compared to massive tower in string theory). One may start with

$$S[\Phi, h] = \int d^4x[\partial\Phi^*\partial\Phi + \sum_s h_s J_s(\Phi)] + S[h]\,, \qquad S[h] = \kappa\sum_s \int h_s\, P_s \partial^{2s} h_s + O(h^3) \tag{19}$$

Then four-scalar tree-level scattering amplitude is found to be [2]

$$A_s^{(t)}(\mathsf{s},\mathsf{t},\mathsf{u}) = P_s(\tfrac{\mathsf{s}-\mathsf{u}}{\mathsf{s}+\mathsf{u}}) \tag{20}$$

where $\mathsf{s},\mathsf{t},\mathsf{u}$ are Mandelstam variables: $\mathsf{s}+\mathsf{t}+\mathsf{u}=0$ and $P_s(x)$ is the Legendre polynomial (which is replaced by the Gegenbauer polynomial $C_s^{(\frac{d-3}{2})}$ in d dimensions). The amplitude is manifestly scale-invariant. Summed over spins the t-channel amplitude is

$$A^{(t)}(\mathsf{s},\mathsf{t},\mathsf{u}) = \kappa^{-1}\ F_4(-\tfrac{\mathsf{s}-\mathsf{u}}{\mathsf{s}+\mathsf{u}}), \qquad F_4(z) = \sum_{s=0}^{\infty}(s+\alpha_d)P_s(z), \qquad \alpha_d \equiv \tfrac{d-3}{2} = \tfrac{1}{2} \tag{21}$$

For generic z the sum over spins diverges and requires a regularization. The same regularization as in (15) gives a delta-function: $F_4^{\rm reg}(z) = \delta(z-1)$. Then the total amplitude for $\Phi\,\Phi \to \Phi\,\Phi$ scattering given by the sum of t-channel plus u-channel contributions) is

$$A_{\Phi\Phi\to\Phi\Phi} = \kappa^{-1}\ [\ \delta(\tfrac{\mathsf{s}}{\mathsf{t}}) + \delta(\tfrac{\mathsf{s}}{\mathsf{u}})\ .\ ]$$

It actually vanishes for physical momenta: in c.o.m. frame $\vec{p}_1 + \vec{p}_2 = 0 = \vec{p}_3 + \vec{p}_4$ where in terms of the scattering angle $\frac{\mathsf{s}}{\mathsf{t}} = -(\sin^2\frac{\theta}{2})^{-1}$, $\frac{\mathsf{s}}{\mathsf{u}} = -(\cos^2\frac{\theta}{2})^{-1}$ the arguments of delta-functions never vanish for real θ, i.e. $A_{\Phi\Phi\to\Phi\Phi} = 0$. Similarly,

$$A_{\Phi\Phi^*\to\Phi\Phi^*} = \tfrac{\kappa^{-1}}{2}[\ \delta(\tfrac{\mathsf{u}}{\mathsf{t}}) + \delta(\tfrac{\mathsf{u}}{\mathsf{s}})\ ] = \tfrac{\kappa^{-1}}{2}[\ \delta(\cot^2\tfrac{\theta}{2}) - \delta(\cos^2\tfrac{\theta}{2})\ ] \tag{22}$$

so here the t-channel and s-channel contributions cancel each other and again $A_{\Phi\Phi^*\to\Phi\Phi^*} = 0$. Thus while the individual spin s exchange contributions are nontrivial the total amplitude vanishes in particular summation prescription.

The reason is in the large underlying global HS symmetry that constrains the S-matrix (like in the case of hidden conserved charges in 2d theories). The global part of the CHS symmetry contains the conformal generators plus other higher spin generators, e.g., "hyper-translations" $\delta\Phi = \varepsilon^{\mu_1\ldots\mu_r}\partial_{\mu_1}\ldots\partial_{\mu_r}\Phi$, For example, this fixes the $A_{\Phi\Phi\to\Phi\Phi}$ amplitude to be $A_{\Phi\Phi\to\Phi\Phi}(\mathsf{s},\mathsf{t},\mathsf{u}) = k_1(\mathsf{t},\mathsf{u})\,\delta(\mathsf{s}) + k_2(\mathsf{s},\mathsf{u})\,\delta(\mathsf{t}) + k_3(\mathsf{t},\mathsf{s})\,\delta(\mathsf{u})$ In addition, the invariance under dilatations $p \to \lambda p$ $A_{\Phi\Phi\to\Phi\Phi}(\lambda^2\mathsf{s},\lambda^2\mathsf{t},\lambda^2\mathsf{u}) = A_{\Phi\Phi\to\Phi\Phi}(\mathsf{s},\mathsf{t},\mathsf{u})$ then implies that the only solution consistent with crossing and scaling symmetry is

$$A_{\Phi\Phi\to\Phi\Phi}(\mathsf{s},\mathsf{t},\mathsf{u}) = 0\,. \tag{23}$$

We conclude that the regularization in which the tree-level scalar amplitude vanishes is thus consistent with the CHS symmetry. This conclusion generalizes also to scattering amplitudes involving CHS fields on external lines [11].

5 Scalar scattering in massless HS theory in flat space

Massless 2-derivative higher spins define a unitary free theory ($\int d^4x\ \partial\phi_s\partial\phi_s$, $\delta\phi_s = \partial\epsilon_{s-1}$) but it is an open question if there is a consistent interacting theory also in flat space. Such theory may be of interest if there is a flat space limit of the consistent MHS theory in AdS (given by Vasiliev's equations); it might be related to a tensionless limit of string theory, if such limit can be properly defined in flat space. It is not clear (in contrast to CHS case) which could be the underlying HS symmetry but one expects again an infinite tower of HS conserved charges. These should then constrain the S-matrix. To explore such theory one may use the known cubic couplings to compute some simple scattering amplitudes.

Despite various $s > 2$ "no-go theorems" prohibiting, in particular, minimal interactions (i.e. higher spin long-range forces) one may look for a consistent theory that contains (i) infinite tower of all spins $s = 0, 1, 2, 3, ..., \infty$; (ii) higher derivative (non-minimal) cubic interactions $\partial^n \phi_{s_1}\phi_{s_2}\phi_{s_3}$, $s_2 + s_3 - s_1 \leq n \leq s_2 + s_3 + s_1$ $(s_1 \leq s_2 \leq s_3)$ (see a review in [1]). Using Noether procedure one may then deform the transformations $\delta\phi_s = \partial\epsilon_{s-1} + ...$, and attempt to determine the 4-point and higher interaction vertices. This should fix 3-point coupling constants like was done in the l.c. gauge in [9] $g_{s_1 s_2 s_3} = g\,\frac{\ell^{s_1+s_2+s_3-1}}{(s_1+s_2+s_3-1)!}$ The resulting theory should have two parameters: g= dimensionless and ℓ= length

$$S = \frac{1}{g^2}\int d^4x\,\Big[\sum_s \partial\phi_s\partial\phi_s + \sum \ell^{n-1}\partial^n\phi_{s_1}\phi_{s_2}\phi_{s_3} + \sum \ell^{k-2}\partial^k\phi^4 + ...\Big] \tag{24}$$

where ϕ^3 terms contain two covariant structures $\partial^{s_1+s_2+s_3}$ and $\partial^{s_2+s_3-s_1}$. Whether there exists a consistent and local ϕ^4 vertex remains an open question. The theory is in any case effectively non-local: the number of derivatives grows with s and with number of fields n in ϕ^n vertex. It may still have hidden simplicity under a particular summation over spins prescription.

It is useful to describe symmetric higher spin tensors by $\phi_s(x, u) = \phi^{a_1...a_s}(x)\, u_{a_1} \dots u_{a_s}$ where u^a is a constant vector. Then the free Fronsdal action is

$$S^{(2)}[\phi_s] = \int d^dx\, [\phi_s(x, \partial_u)\, K\, \phi_s(x, u)]_{u=0}\ , \tag{25}$$

$$K = (1 - \tfrac{1}{4}u^2\partial_u^2)[\partial_x^2 - (u\cdot\partial_x)(\partial_x\cdot\partial_u) - \tfrac{1}{2}(u\cdot\partial_x)\partial_u^2] \tag{26}$$

The off-shell field ϕ_s double-traceless $(\partial_u^2)^2\phi_s(x, u) = 0$ and linearized gauge transformations are $\delta_s^{(0)}\phi_s(x, u) = (u\cdot\partial_x)\varepsilon_{s-1}(x, u)$. In de Donder gauge $\partial^{a_1}\phi_{a_1...a_s} + ... = 0$ the equations of motion become $\Box\phi_s(x, u) = 0$.

To discuss scattering of spin 0 particles: we need cubic interaction vertex with $(s_1 = 0, s_2, s_3)$ which reads $(\partial_{x_{ij}} \equiv \partial_{x_i} - \partial_{x_j})$

$$S^{(3)}[\phi_0, \phi_{s_2}, \phi_{s_3}] = g_{0 s_2 s_3}\int d^dx\Big[(\partial_{u_2}\cdot\partial_{x_{31}})^{s_2}(\partial_{u_3}\cdot\partial_{x_{12}})^{s_3}\phi_0(x_1)\phi_{s_2}(x_2, u_2)\phi_{s_3}(x_3, u_3)\Big]_{\substack{u_i=0\\ x_i=x}} \tag{27}$$

The spin s propagator is in $d = 4$

$$\mathcal{D}_s(u, u'; p) = -\tfrac{i}{p^2}\,\mathcal{P}_s(u, u')\ , \qquad \mathcal{P}_s(u, u') = \tfrac{1}{(s!)^2}\left(\sqrt{u^2u'^2}\right)^s T_s\!\left(\tfrac{u\cdot u'}{\sqrt{u^2u'^2}}\right), \tag{28}$$

$$T_s(z) \equiv \tfrac{s}{2}\sum_{k=0}^{[s/2]}\tfrac{(-1)^k(s-k-1)!}{k!(s-2k)!}(2z)^{s-2k} = \tfrac{1}{2}\Big[(z + \sqrt{z^2-1})^s + (z - \sqrt{z^2-1})^s\Big] \tag{29}$$

where T_s is the Chebyshev polynomial of first kind. Following [10] we may then compute the scattering of spin 0 fields via echange of the tower of massless HS fields with the above cubic interactions and making a conjecture about possible quartic coupling. One finds in s-channel [1]

$$\mathcal{A}_{exch}(\mathrm{s,t,u}) = \sum_{s=0,2,4,...}^{\infty}\mathcal{A}^s_{exch}(\mathrm{s,t})\ , \qquad \mathcal{A}^s_{exch}(\mathrm{s,t,u}) = -\tfrac{ig^2_{00s}}{\mathrm{s}}\,(\mathrm{t+u})^s\ T_s\!\left(\tfrac{\mathrm{t-u}}{\mathrm{t+u}}\right), \tag{30}$$

$$\mathcal{A}_{exch}(\mathrm{s,t,u}) = -\frac{i}{\mathrm{s}}\Big[F(\sqrt{\mathrm{s+t}} + \sqrt{\mathrm{t}}) + F(\sqrt{\mathrm{s+t}} - \sqrt{\mathrm{t}})\Big], \tag{31}$$

$$F(z) \equiv \sum_{s=0,2,4,...}^{\infty} g^2_{00s}\left(\tfrac{z^2}{4}\right)^s = \tfrac{1}{8}g^2\,(\ell z)^2[I_0(\ell z) - J_0(\ell z)]\ , \tag{32}$$

DOI: 10.1051/epjconf/201612501012

with the total result being

$$\hat{\mathcal{A}}_{exch}(\mathrm{s},\mathrm{t},\mathrm{u}) = \mathcal{A}_{exch}(\mathrm{s},\mathrm{t},\mathrm{u}) + \mathcal{A}_{exch}(\mathrm{t},\mathrm{s},\mathrm{u}) + \mathcal{A}_{exch}(\mathrm{u},\mathrm{t},\mathrm{s})\,.$$

The sum over spins here is convergent giving Bessel functions with non-trivial dependence on the Mandelstam variables and the scale ℓ. In the Regge limit $\mathrm{t} \to \infty$, s=fixed

$$\hat{\mathcal{A}}_{exch}(\mathrm{s},\mathrm{t},\mathrm{u}) \sim -\frac{ig^2}{\mathrm{s}}\ell^2 \mathrm{t}\, I_0(\ell\sqrt{8\mathrm{t}}) \sim -\frac{ig^2}{\mathrm{s}}(\ell^2\mathrm{t})^{3/4}\, e^{\ell\sqrt{8\mathrm{t}}} \tag{33}$$

while in the fixed angle UV limit when $\mathrm{s},\mathrm{t},\mathrm{u} \to \infty$, $\frac{\mathrm{t}}{\mathrm{s}} = -\sin^2\frac{\theta}{2}$, $\frac{\mathrm{u}}{\mathrm{s}} = -\cos^2\frac{\theta}{2}$ we get

$$\hat{\mathcal{A}}_{exch}(\mathrm{s},\mathrm{t},\mathrm{u}) \sim ig^2|\mathrm{s}|^{3/4}\, e^{\ell\sqrt{|\mathrm{s}|}\, f(\theta)} \to \infty\,, \qquad f(\theta) > 0 \tag{34}$$

This amplitude has exponential UV growth. This may be compared to soft UV limit of the Shapiro-Virasoro amplitude in string theory:

$$A_4 = g^2\frac{\Gamma(-1-\frac{1}{4}\alpha'\mathrm{s})\Gamma(-1-\frac{1}{4}\alpha'\mathrm{s})\Gamma(-1-\frac{1}{4}\alpha'\mathrm{s})}{\Gamma(2+\frac{1}{4}\alpha'\mathrm{s})\Gamma(2+\frac{1}{4}\alpha'\mathrm{s})\Gamma(2+\frac{1}{4}\alpha'\mathrm{s})} \tag{35}$$

$$A_4 \to g^2|\mathrm{s}|^{-6}(\sin\theta)^{-6}e^{-\alpha'|\mathrm{s}|\,h(\theta)} \to 0\,, \qquad h(\theta) = -\tfrac{1}{4}\big(\sin^2\tfrac{\theta}{2}\log\sin^2\tfrac{\theta}{2} - \cos^2\tfrac{\theta}{2}\log\cos^2\tfrac{\theta}{2}\big) > 0$$

To get the full 4-point amplitude we need to add the contribution of 0-0-0-0 vertex. It is expected to be effectively "non-local" with infinite series in ∂^n; this may "soften" the UV behaviour of the exchange contribution. One may try to guess the 4-scalar vertex in flat-space HS action from its form in AdS action reconstructed using AdS/CFT in [12]. This leads to

$$S^{(4)}[\phi_0] = g^2\int d^4x\Big[\sum_{s=0}^{\infty} f_{2s}(\Delta_{x_{34}})\,(\partial_{x_{12}}\cdot\partial_{x_{34}})^{2s}\,\phi_0(x_1)\phi_0(x_2)\phi_0(x_3)\phi_0(x_4)\Big]_{x_i=x}$$

where $\Delta_{x_{34}} \equiv (\partial_{x_3}+\partial_{x_4})^2$, $\quad\partial_{x_{12}} \equiv \partial_{x_1} - \partial_{x_2}$ and $f_{2s}(z)$ is an infinite series in z, regular at $z=0$:

$$z \to \infty: \qquad f_{2s}(z) \to c_{2s}\frac{\ell^{4s-2}}{z}\,, \qquad c_{2s} = \frac{1}{[(2s-1)!!]^2}$$

Then the asymptotic contribution to the 4-scalar amplitude is

$$\sum_{s=0}^{\infty} f_{2s}(\mathrm{s})\,(\mathrm{t}-\mathrm{u})^{2s} = \tfrac{2\mathrm{t}+\mathrm{s}}{2\mathrm{s}}\Big[I_0(2\ell\sqrt{2\mathrm{t}+\mathrm{s}}) - J_0(2\ell\sqrt{2\mathrm{t}+\mathrm{s}})\Big]$$

This is similar to the above exchange contribution so some cancellation between the two remains a possibility. Constraints of gauge invariance on other amplitudes with HS particles on external lines suggest that quartic vertices should be non-local [13]. Assuming the existence of a consistent gauge-invariant action for massless HS theory in flat space it may then turn out that the resulting S-matrix will be nearly trivial due to the cancellation between the exchange and 4-vertex contributions.

To conclude, we are beginning to learn how to do quantum computations in theories with infinite number of massless higher spin fields. This brings up the importance of proper definition of quantum theory (sum over spins, UV regularization) consistent with symmetries of the theory. The remarkable large symmetry of higher spin theories implies various simplifications (1-loop $Z = 1$, zero effective number of d.o.f.) and simplification of S-matrix (at least in conformal HS theory).

This work was supported by the ERC Advanced grant No.290456 and by the Russian Science Foundation grant 14-42-00047 associated with Lebedev Institute.

References

[1] D. Ponomarev and A. A. Tseytlin, "On quantum corrections in higher-spin theory in flat space," JHEP **1605**, 184 (2016) [arXiv:1603.06273].

[2] E. Joung, S. Nakach and A. A. Tseytlin, "Scalar scattering via conformal higher spin exchange," JHEP **1602**, 125 (2016) [arXiv:1512.08896].

[3] M. Beccaria and A. A. Tseytlin, "On higher spin partition functions," J. Phys. A **48** (2015) no.27, 275401 [arXiv:1503.08143].

[4] M. Beccaria and A. A. Tseytlin, "Higher spins in AdS_5 at one loop: vacuum energy, boundary conformal anomalies and AdS/CFT," JHEP **1411**, 114 (2014) [arXiv:1410.3273].

[5] M. Beccaria, X. Bekaert and A. A. Tseytlin, "Partition function of free conformal higher spin theory," JHEP **1408**, 113 (2014) [arXiv:1406.3542].

[6] S. Giombi, I. R. Klebanov and A. A. Tseytlin, "Partition functions and Casimir energies in higher spin AdS_{d+1}/CFT_d," Phys. Rev. D **90**, no. 2, 024048 (2014) [arXiv:1402.5396].

[7] A. A. Tseytlin, Nucl. Phys. B **877**, 598 (2013) [arXiv:1309.0785].

[8] S. Giombi, I. R. Klebanov and B. R. Safdi, "Higher Spin AdS_{d+1}/CFT_d at One Loop," Phys. Rev. D **89**, no. 8, 084004 (2014) [arXiv:1401.0825].

[9] R. R. Metsaev, "Poincare invariant dynamics of massless higher spins: Fourth order analysis on mass shell," Mod. Phys. Lett. A **6**, 359 (1991).

[10] X. Bekaert, E. Joung and J. Mourad, "On higher spin interactions with matter," JHEP **0905**, 126 (2009) [arXiv:0903.3338].

[11] M. Beccaria, S. Nakach and A. A. Tseytlin, "On triviality of S-matrix in conformal higher spin theory," arXiv:1607.06379 [hep-th].

[12] X. Bekaert, J. Erdmenger, D. Ponomarev and C. Sleight, "Quartic AdS Interactions in Higher-Spin Gravity from Conformal Field Theory," JHEP **1511**, 149 (2015) [arXiv:1508.04292].

[13] M. Taronna, "Higher-Spin Interactions: four-point functions and beyond," JHEP **1204**, 029 (2012) [arXiv:1107.5843].

EPJ Web of Conferences **125**, 02001 (2016) DOI: 10.1051/epjconf/201612502001
QUARKS-2016

The SM extensions with additional light scalar singlet, nonrenormalizable Yukawa interactions and $(g-2)_\mu$

S.N.Gninenko[1] and N.V.Krasnikov[1,2]

[1] *INR RAS, 117312 Moscow*

[2] *Joint Institute for Nuclear Research, 141980 Dubna*

Abstract. We consider the SM extension with additional light real singlet scalar, right-handed neutrino and nonrenormalizable Yukawa interaction for the first two generations. We show that the proposed model can explain the observed $(g-2)$ muon anomaly. Phenomenological consequenses as flavour violating decays $\tau \to \mu\mu\mu, \mu\mu e, \mu ee$ are briefly discussed. We also propose the $U_R(1)$ gauge generalization of the SM with complex scalar singlet and nonzero right-handed charges for the first two generations.

1 Introduction

The discovery of the neutrino oscillations [1, 2] means that at least two neutrino have nonzero masses. The minimal extension of the SM with nonzero neutrino masses is the νMSM [3, 4]. In this model one adds to the SM three additional massive Majorana(right-handed) fermions ν_{Ri}, $i = 1, 2, 3$. Due to seesaw mechanism [3, 5] after the spontaneous $SU_L(2) \otimes U(1)$ electroweak symmetry breaking the neutrinos acquire masses $m_{\nu_i} = \frac{m^2_{Di}}{M_{Ri}}$. Here m_{Di} are the Dirac neutrino masses and M_{Ri} are the masses of the ν_{Ri} neutrinos. The νMSM has a candidate - the lightest Majorana neutrino with a mass $M_{\nu_R} \leq O(50)\ KeV$ - for dark matter. Besides, the model with light Majorana neutrino can solve the problem of the baryon asymmetry in our Universe [4].

In this report which is based mainly on Refs.[6] we consider the extension of the νMSM with additional scalar field and nonrenormalizable Yukawa interaction for the first two generations. We show that the SM extension with additional light real singlet field can explain the $(g-2)$ muon anomaly. Phenomenological consequences of the proposed model as flavour violating decays $\tau \to \mu\mu\mu, \mu ee$ are brifly discussed. We also propose the $U_R(1)$ gauge generalization of the SM with complex scalar singlet and nonzero righthanded charges for the first two generations.

2 The νMSM extension with additional real scalar isosinglet and nonrenormalizable Yukawa interaction

In this section we consider the extension of the νMSM with additional scalar field and nonrenormalizable Yukawa interaction for the first two generations [6]. The Lagrangian of the model has the form

$$L_{tot} = L_{SM} + L_{Qd\phi} + L_{Qu\phi} + L_{Le\phi} + L_{\phi} + L_{\nu_R} \,. \tag{1}$$

Here

$$L_{Qd\phi} = -\frac{h_{Qd\phi,ik}}{M}\bar{Q}_{Li}\bar{H}\phi d_{Rk} + \, H.c. \,, \tag{2}$$

$$L_{Qu\phi} = -\frac{h_{Qu\phi,ik}}{M}\bar{Q}_{Li}H\phi u_{Rk} + \, H.c. \,, \tag{3}$$

$$L_{Le\phi} = -\frac{h_{Le\phi,ik}}{M}\bar{L}_{Li}\bar{H}\phi e_{Rk} + \, H.c. \,, \tag{4}$$

$$L_{\phi} = \frac{1}{2}\partial^{\mu}\phi\partial_{\mu}\phi - \frac{M_{\phi}^2\phi^2}{2} - \lambda\phi^4 \,, \tag{5}$$

$$L_{\nu_R} = i\bar{\nu}_{Rj}\hat{\partial}\nu_{Rj} - (\frac{M_{\nu_{Rj}}}{2}\nu_{Rj}\nu_{Rj} + h_{Lij}\bar{L}_i H\nu_{Rj} + H.c.) \,, \tag{6}$$

where L_{SM} is the SM Lagrangian, ν_{Rj} - are the Majorana neutrinos and $L_1 = (\nu_{eL}, e_L)$, $L_2 = (\nu_{\mu L}, \mu_L)$, $L_3 = (\nu_{\tau L}, \tau_L)$, $e_{R1} = e_R$, $e_{R2} = \mu_R$, $e_{R3} = \tau_R$, $Q_{L1} = (u_L, d_L)$, $Q_{L2} = (c_L, s_L)$, $Q_{L3} = (t_L, b_L)$, $d_{R1} = d_R$, $d_{R2} = s_R$, $d_{R3} = b_R$, $u_{R1} = u_R$, $u_{R2} = c_R$, $u_{R3} = t_R$, $\bar{H} = (-(H^-)^*, (H^0)^*)^1$. The main peculiarity of the model (1-6) is the use of nonrenormalizable Yukawa interactions (2-4)[2]. Here we consider the particular case of the general model (1 - 6) with nonzero renormalizable Yukawa interaction only for

[1] Here $H = (H^0, H^-)$ is the SM Higgs doublet and M is some high energy scale .

[2] The nonrenormalizable Yukawa interactions have been considered in Refs.[7].

the third fermion generation. We assume that the masses of the first two light generations arise due to nonrenormalizable interactions (2-4). We impose the discrete symmetry

$$\phi \to -\phi \,, \tag{7}$$

$$e_{Rk} \to -e_{Rk} \;\; (k = 1, 2) \,, \tag{8}$$

$$d_{Rk} \to -d_{Rk} \;\; (k = 1, 2) \,. \tag{9}$$

The discrete symmetry (7 - 9) restricts the form of nonrenormalizable interactions (2-4), namely

$$h_{Qd\phi,i3} = h_{Qu\phi,i3} = 0 \,, \tag{10}$$

$$h_{Le\phi,i3} = 0 \,, \tag{11}$$

As a consequence of the (7-9) the renormalizable SM Yukawa interaction with the first two quark and lepton generations vanishes and the fermions of the first two generations acquire masses only due to nonrenormalizable interactions (2-4).

We can consider the nonrenormalizable interactions (2 - 4) as some effective interactions arising from renormalizable interactions. For instance, the interaction (2) can be realized in renormalizable extension of the SM with additional scalar field ϕ and new massive quark $SU(2)_L$ singlet fields D_R, D_L with a mass M_D and $U(1)$ hypercharges $Y_{D_L} = Y_{D_R} = -\frac{1}{3}$. The interaction of new quark fields D_R, D_L with ordinary quarks and the neutral scalar field ϕ is

$$L_{qD\phi} = -c_i \bar{Q}_{Li} \bar{H} D_R - k_j \bar{D}_L d_{Rj} \phi + H.c.\,. \tag{12}$$

In the heavy D-quark mass limit $M_D \to \infty$ we obtain the effective interaction (2) with

$$\frac{h_{Qd\phi,ij}}{M} = \frac{c_i k_j}{M_D} \,. \tag{13}$$

Analogously we can consider nonrenormalizable interaction (4) as an effective interaction which arises in renormalizable extension of the SM with additional scalar field and new massive lepton $SU(2)_L$ singlet fields E_R, E_L with $Y_{E_L} = Y_{E_R} = 1$ The interaction of new lepton fields E_R, E_L with ordinary quarks and the neutral scalar field ϕ is

$$L_{eE\phi} = -d_i \bar{L}_{Li} \bar{H} E_R - f_j \bar{E}_L e_{Rj} \phi + H.c.\,. \tag{14}$$

In the heavy E-lepton mass limit $M_E \to \infty$ we obtain the effective interaction (4) with

$$\frac{h_{Le\phi,ij}}{M} = \frac{d_i f_j}{M_E}\,. \tag{15}$$

After the spontaneous $SU(2)_L \otimes U(1)$ electroweak symmetry breaking the Yukawa interaction of the scalar field with charged leptons takes the form

$$L_{ll\phi} = -\bar{h}_{Le,ik}\bar{e}_{Li}\phi e_{Rk} + H.c\,, \tag{16}$$

where

$$\bar{h}_{Le,ik} = h_{Le\phi,ik}\frac{< H >}{M}\,, \tag{17}$$

$i = 1,2,3$, $k = 1,2$ and $e_{L1} = e_L$, $e_{L2} = \mu_L$, $e_{L3} = \tau_L$, $< H >= 174\ GeV$. Nonzero vacuum expectation value for the real field ϕ generates nonzero lepton masses for electrons and muons while the mass of τ-lepton arises due to renormalizable Yukawa coupling. The lepton mass matrix and the Yukawa lepton $\phi^{'} = \phi - < \phi >$ interactions are

$$L_{ll} = -\bar{h}_{Le,ik} < \phi > \bar{e}_{Li}e_{Rk} - m_\tau \bar{e}_{L3}e_{R3} + H.c\,, \tag{18}$$

$$L_{ll\phi} = -\bar{h}_{Le,ik}\bar{e}_{Li}\phi^{'} e_{Rk} + H.c\,, \tag{19}$$

where $i = 1,2,3$ and $k = 1,2$. The mass terms (18) and the interaction (19) have different flavour structure that leads to the tree level flavour changing transitions like $\tau \to \mu + \phi$, $\tau \to e + \phi$ and as a consequence to the flavour violating decays like

$$\tau^- \to \mu^- + \phi^* \to \mu^-\mu^+\mu^-\,, \tag{20}$$

$$\tau^- \to \mu^- + \phi^* \to \mu^- e^+ e^-\,, \tag{21}$$

$$\tau \to e^- + \phi^* \to e^-\mu^+\mu^-\,, \tag{22}$$

$$\tau \to e^- + \phi^* \to e^- e^+ e^-\,. \tag{23}$$

At present state of art we can't predict the value of flavour violating Yukawa couplings $\bar{h}_{Le31}, \bar{h}_{Le32}$.

The interaction (19) leads, in particular, to the additional one loop contribution to muon magnetic moment due to $\phi^{'}$ scalar exchange, namely [8]

$$\Delta a_\mu = \frac{h^2_{Le,22}}{8\pi^2}\frac{m^2_\mu}{M^2_\phi}\int_0^1 \frac{x^2(2-x)}{(1-x)(1-\lambda^2 x) + \lambda^2 x}\,, \tag{24}$$

where $\lambda = \frac{m_\mu}{m_\phi}$. In the limit $M_\phi >> m_\mu$

$$\Delta a_\mu = \frac{1}{4\pi^2}\frac{m_\mu^2}{M_\phi^2}\bar{h}^2_{Le,22}[ln(\frac{M_\phi}{m_\mu}) - \frac{7}{12}]. \tag{25}$$

The precise measurement of the anomalous magnetic moment of the positive muon from the Brookhaven AGS experiment [9] gives a result which is 3.6σ higher than the Standard Model (SM) prediction

$$a_\mu^{exp} - a_\mu^{SM} = (288 \pm 80)\cdot 10^{-11}, \tag{26}$$

where $a_\mu \equiv \frac{g_\mu - 2}{2}$. Using the formulae (25, 26) we find that for $m_\Phi = (100, 10, 1, 0.5)\ GeV$ the muon $g-2$ anomaly can be explained for

$$\bar{h}^2_{Le,22} = (1.6 \pm 0.5)\cdot 10^{-2} \quad for \quad m_\phi = 100\, GeV, \tag{27}$$

$$\bar{h}^2_{Le,22} = (2.6 \pm 0.8)\cdot 10^{-4} \quad for \quad m_\phi = 10\, GeV, \tag{28}$$

$$\bar{h}^2_{Le,22} = (6.2 \pm 1.9)\cdot 10^{-6} \quad for \quad m_\phi = 1\, GeV. \tag{29}$$

$$\bar{h}^2_{Le,22} = (2.6 \pm 0.8)\cdot 10^{-6} \quad for \quad m_\phi = 0.5\, GeV. \tag{30}$$

For the opposite limit $m_\mu \gg m_\phi$

$$\Delta a_\mu = \frac{3h^2_{Le,22}}{16\pi^2} \tag{31}$$

and as a consequence of (26, 31) we find that

$$\bar{h}^2_{Le,22} = (1.5 \pm 0.5)\cdot 10^{-7}. \tag{32}$$

As in the SM the Yukawa couplings $\bar{h}_{Le.ii}$ are proportional to the lepton masses. It means that the interaction of the ϕ scalar with electrons is weaker than the interaction of the ϕ scalar with muons by factor $m_\mu/m_e \approx 200$ and the contribution of the ϕ scalar to the electron magnetic moment is suppressed at least by factor $(m_e/m_\mu)^2$ in comparison with the muon magnetic moment even for superlight $m_\phi \ll m_e$ scalar. For instance, for $m_\phi = 1\ GeV$ the contribution of the ϕ scalar to the electron magnetic moment is

$$(\Delta a_e)_\phi = (0.16 \pm 0.05)\cdot 10^{-17} \tag{33}$$

that is much smaller the bound from a_e [10]

$$\Delta a_e = a_e^{exp} - a_e^{SM} = (-1.06 \pm 0.82) \cdot 10^{-12} \tag{34}$$

Due to the suppression factor $\frac{m_e}{m_\mu}$ for electon Yukawa coupling in comparison with muon Yukawa coupling the search for light ϕ scalar in electron fixed target experiments or e^+e^- experiments is very problematic but not hopeless.[3] The search for very light ϕ scalar in $\pi \rightarrow (\phi \rightarrow e^+e^-)\gamma$ decay is possible but again we have additional suppression factor $(\frac{m_u - m_d}{m_\mu})^2 \sim O(10^{-2})$. Light scalar particle ϕ with a mass $m_\phi \lesssim 1\ GeV$ decaying into muon pair can be searched for at CERN SPS secondary muon beam in full analogy with the search for new light vector boson Z' [11].

3 The $U(1)$ gauge generalization of the model with real scalar field

Here we outline one of possible generalizations of the model (1-6). In the proposed generalization instead of real scalar ϕ we use complex scalar Φ and new abelian gauge groop $U_R(1)$[4] with nonzero charges for right handed fermions of the first and second generations, namely

$$Q_X(u_R) = Q_X(c_R) = Q_X(\nu_{eR}) = Q_X(\nu_{\mu R}) = -Q_X(d_R) = -Q_X(s_R) = -Q_X(e_R) = -Q_X(\mu_R)\,. \tag{35}$$

The nonrenormalizable Yukawa interactions of the first and second generation fermions in full analogy with the (2-4) interactions take the form

$$L_{Qd\Phi} = -\frac{h_{Qd\Phi,ik}}{M}\bar{Q}_{Li}\bar{H}\Phi d_{Rk} + H.c.\ , \tag{36}$$

$$L_{Qu\Phi} = -\frac{h_{Qu\Phi,ik}}{M}\bar{Q}_{Li}H\Phi^* u_{Rk} + H.c.\ , \tag{37}$$

$$L_{Le\Phi} = -\frac{h_{Le\Phi,ik}}{M}\bar{L}_{Li}\bar{H}\Phi e_{Rk} + H.c.\ , \tag{38}$$

$$L_{L\nu\Phi} = -\frac{h_{L\nu\Phi,ik}}{M}\bar{L}_{Li}\bar{H}\Phi^* \nu_{Rk} + H.c.\ , \tag{39}$$

[3] Roughly speaking we have to improve the discovery potential by 3-4 orders of magnitude.

[4] In Refs. [12] new light vector boson interacting with the $L_\mu - L_\tau$ current has been proposed for $(g-2)_\mu$ anomaly explanation, see also Ref.[13] where the model with new light gauge boson interacting with the SM electromagnetic current has been proposed for the $(g-2)_\mu$ anomaly explanation and Ref.[14] where the interaction of light gauge boson with $(B-L)+xY$ current has been considered.

Note that proposed model is free from γ_5-anomalies and we can consider the origin of the $U_R(1)$ gauge group as a result of the gauge symetry breaking $SU_L(2) \otimes SU_R(2) \otimes U(1) \to SU_L(2) \otimes U_R(1) \otimes U(1)$. We assume that in the considered model $<\Phi> \neq 0$ that leads to nonzero X gauge boson mass and nonzero fermion masses for the first and second generations. In the unitaire gauge $\Phi = \phi + <\Phi>$, where $\phi = \phi^*$ is real scalar field as in previous section plus we have massive vector boson X. So in this model after $U_R(1)$ gauge symmetry breaking in addition to the νMSM spectrum we have both scalar and vector particles. The one loop contribution to the anomalous muon(electron) magnetic moment due to the ϕ and X exchanges is

$$\Delta a_\mu = \Delta a_\mu(\phi) + \Delta a_\mu(X), \tag{40}$$

where the $\Delta a_\mu(\phi)$ contribution is given by the formulae (24,25,31) and the vector X boson contribution for $(V + A)$ right-handed coupling[5] with fermions is [8]

$$\Delta a_\mu = \frac{g_X^2}{8\pi^2}\frac{m_\mu^2}{M_{X_\mu}^2}\int_0^1 \frac{2x^2(2-x) + 2x(1-x)(x-4) - 4\lambda^2 x^3}{(1-x)(1-\lambda^2 x) + \lambda^2 x}, \tag{41}$$

$$\Delta a_\mu = -\frac{g_X^2}{3\pi^2}\frac{m_\mu^2}{M_X^2} \quad (for\ M_X \gg m_\mu). \tag{42}$$

The X boson contribution (41) to the $(g-2)$ is negative. For instance, for $\alpha_X = \frac{g_X^2}{4\pi} = 10^{-8}$ and $M_X = 500\ MeV$ the X-boson contribution to Δa_μ is $\Delta a_\mu(X) = -1.8 \cdot 10^{-10}$. The positive contribution due to ϕ boson exchange is positive and cancels the negative contribution from the X boson exchange. The most perspective experiments for the search for light vector X-boson with the electron coupling constant $\alpha_X = O(10^{-8})$ are the electron fixed target experiments or e^+e^- experiments. The X boson can decay into electron-positron or muon-antimuon pairs, also invisible decays of the X boson into light sterile neutrino are possible. The experiment NA64 [15] at CERN will be able to search for both invisible and visible X boson decay modes with the $\alpha_X \geq O(10^{-12})$ [16].

4 Conclusion

The νMSM with additional scalar field and nonrenormalizable interaction for the first two generations can explain the observed muon $(g-2)$ anomaly. The model predicts the existence of flavour violating

[5]The interaction Lagrangian of the vector X field with muons is $L_{X\bar{\mu}\mu} = g_X X^\nu \bar{\mu}\gamma_\nu(1+\gamma_5)\mu$.

quark and lepton decays like $\tau \to \mu\mu\mu, \mu\mu e, \mu ee$. Besides the $U(1)$ gauge generalization of the model with real isosinglet scalar field is also able to explain muon $(g-2)$ anomaly.

We are indebted to colleagues from INR theoretical department for discussions and comments.

References

[1] As a recent review, see for example:
Y.Wang and Z.Xing, arXiv:1310.7478[hep-ex];
M.Laveder and C.Giundi, arXiv:1310.7478[hep-ex].

[2] K.A. Olive et al. (Particle Data Group), Chin. Phys. C **38** (2014) 090001.

[3] P.Minkowski, Phys.Lett. **B67** (1977) 421.

[4] As a review, see for example: A.Boyarsky, O.Ruchayskiy and M.Shaposhnikov, Ann.Rev.Nucl.Part.Sci. **59** (2009 191.

[5] T.Yanagida, in Proceedings of the Workshop
on Unified Theory and the Baryon Number of the Universe,
edited by O.Sawada and A.Sugamoto (KEK, Tsukuba, 1979), p.95:
M.Gell-Mann, P.Ramond and R.Slansky,
in Supergravity, edited by P. van Nieuwenhauzen and D.Freedman,
(North Holland, Amsterdam, 1979).

[6] S.N.Gninenko and N.V.Krasnikov, Phys.Rev. **D92** (2015) 034009;
S.N.Gninenko and N.V.Krasnikov, arXiv:1602.03548[hep-ph].

[7] See for example:
C.D.Froggat and H.B.Nielsen, Nucl.Phys. **B147** (1979) 277;
N.V.Krasnikov, Phys.Lett., **B140** (1984) 337;
W.Buchmuller and D.Wyler, Nucl.Phys. **B268** (1986) 621;
V.Barger et al., Phys.Rev. **D67**,(2003) 115001.

[8] As a recent review, see for example:
F.Jegelehner and A.Nyffeler, Phys.Repts. **477** (2009) 1.

[9] G.W.Benett et al., [Muon G-2 Collaboration], Phys.Rev. **D73** (2006) 072003.

[10] H.Davvoudiasl, H.S.Lee and W.J.Marciano, Phys.Rev. **D86** (2001) 095009;
M.Endo, K.Hamaguchi and G.Mishima, Phys.Rev. **D86** (2012) 095029;
H.S.Lee, Phys.Rev. **D90** (2014) 091702.

[11] S.N.Gninenko, N.V.Krasnikov and V.A.Matveev, Phys.Rev. **D91** (2015) 095015.

[12] S.N.Gninenko and N.V.Krasnikov, Phys.Lett. **B513** (2001) 119;
S.Baek et al., Phys.Rev. **D64** (2001) 055006;
E.Ma, D.P.Roy and S.Roy, Phys.Lett. **B525** (2002) 101.

[13] M.Pospelov, Phys.Rev. **D80** (2009) 095002.

[14] H.S.Lee and S.Yan, arXiv:1604.0213[hep-ph].

[15] S.Andreas et al., arXiv:1312.3309[hep-ph].

[16] S.N.Gninenko, N.V.Krasnikov, M.M.Kirsanov and D.V.Kirpichnikov, arXiv:1604.08432[hep-ph].

EPJ Web of Conferences **125**, 02002 (2016) DOI: 10.1051/epjconf/201612502002
QUARKS-2016

Upgraded LHC experiments as a check of non-perturbative effects of the Electro-Weak Interaction

Boris A. Arbuzov[1,*] and *Ivan V.* Zaitsev[1,**]

[1] *M.V. Lomonosov Moscow State University*

Abstract. Recently reported diphoton excesse at LHC is interpreted to be connected with heavy WW zero spin resonances. The resonances appears due to the wouldbe anomalous triple interaction of the weak bosons, which is defined by coupling constant λ. The $\gamma\gamma$ $750\,GeV$ anomaly is considered to correspond to weak isotopic spin 0 pseudoscalar state. We obtain estimates for the effect, which qualitatively agree with ATLAS data. Effects are predicted in a production of W^+W^-, $(Z,\gamma)(Z,\gamma)$ via resonance X_{PS} with $M_{PS} \simeq 750\,GeV$, which could be reliably checked at the upgraded LHC at $\sqrt{s} = 13\,TeV$. In coupling constant of the triple anomalous interaction is estimated to be $\lambda = -0.010 \pm 0.005$ in an agreement with existing restrictions. Specific predictions of the hypothesis are significant effects in decay channels $X_{PS} \to \gamma\, l^+ l^-$, $X_{PS} \to l^+ l^- l^+ l^-$ $(l = e, \mu)$.

1 Introduction

In experiments [1] indications for excesses in the production of boson pairs WW, WZ, ZZ were observed at invariant mass $M_R \simeq 2\,TeV$. Data for these processes are also present in works [2, 3]. Despite the fact that the wouldbe effect is not finally established yet, the publication causes numerous proposals for an interpretation mostly in terms of theories beyond the Standard Model (see, *e.g.* [4]).

In work [5] we have considered interpretation of the effect in terms of a weak isotopic spin 2 scalar $W\,W$ state. Indeed, pair of triplets W^a could form a resonance state, the so-called W-ball. Of course the well known gauge interaction of these bosons with coupling $g(M_W) = 0.65$ can not bind them in the resonance state with mass being of a TeV scale. However, there might exist also an additional effective interaction, *e.g.* the anomalous triple boson interaction [7, 8], which increases with increasing energy scale. In case the interaction becomes sufficiently strong at a TeV scale, it might lead to a formation of a resonance under discussion. Note, that the most recent data at $\sqrt{s} = 13\,TeV$ [6] do not contradict estimates of [5].

There are recent data on $\gamma\gamma$ anomaly at $M(\gamma\gamma) \simeq 750\,GeV$ [11, 12], which also caused numerous proposals for an interpretation [13]. Could this anomaly also be prescribed to analogous states? We would consider this problem in the present talk.

*e-mail: arbuzov@theory.sinp.msu.ru
**e-mail: zaytsev_iv@mail.ru

2 A model for the WW resonance

Now let us consider a possibility of a heavy resonance in case of an existence of the anomalous three-boson interaction, which in conventional notations [7, 8] looks like

$$-\frac{G}{3!}\,F\,\epsilon_{abc}\,W^a_{\mu\nu}\,W^b_{\nu\rho}\,W^c_{\rho\mu}\,;\quad G\;=\;-\,\frac{g\,\lambda}{M_W^2} \tag{1}$$

$$W^a_{\mu\nu}\;=\;\partial_\mu W^a_\nu - \partial_\nu W^a_\mu\;+g\,\epsilon_{abc}W^b_\mu W^c_\nu\,;$$

where $g \simeq 0.65$ is the electro-weak coupling. The best limitations for parameter λ read [14]

$$\lambda_\gamma = -\,0.022 \pm 0.019\,;\quad \lambda_Z = -\,0.09 \pm 0.06\,; \tag{2}$$

where a subscript denote a neutral boson being involved in the experimental definition of λ. Let us emphasize that $F \equiv F(p_i)$ in definition (1) denotes a form-factor, which is either postulated as in original works [7, 8] or it is just uniquely defined as in works on a spontaneous generation of effective interaction (1) [9, 10]. In any case the form-factor guarantees the effective interaction to act in a limited region of the momentum space. That is it vanishes for momenta exceeding scale Λ. Formfactor F is explicitly presented *e.g.* in work [5]. Calculations were done in the framework of an approximate scheme, which accuracy was estimated to be $\simeq (10-15)\%$ [15]. Would-be existence of effective interaction (1) leads to important non-perturbative effects in the electro-weak interaction.

In particular, one might expect resonances to appear in the system of two W^a-bosons. A possibility of an appearance of such states (W-balls) was already discussed, *e.g.* in works [10, 16]. In the previous work [5] we studied the 2000 GeV anomaly and appeared to the conclusion, that data can be described in terms of weak isospin 2 scalar resonance. The effect is due to anomalous interaction (1) and we come to a conclusion [5], that there is a possibility to describe data [1] with

$$\lambda\;=\;-0.015 \pm 0.005. \tag{3}$$

There are recent indications for existence of the other effect: the $\gamma\gamma$ enhancement at invariant mass $M_{\gamma\gamma} \simeq 750\,GeV$. Let us consider this effect being explained by existence of zero weak isotopic spin pseudoscalar state X_{PS}, which interaction with electroweak bosons is described by the following expression

$$L_{eff}\;=\;\frac{G_{PS}}{4}\,\delta_{ab}\,\epsilon^{\mu\nu\rho\sigma}W^a_{\mu\nu}\,W^b_{\rho\sigma}\,X_{PS}\,; \tag{4}$$

Let us consider a Bethe-Salpeter equation for a pseudoscalar resonance consisting of two W. We consider such state corresponding to value of the weak isospin: $I = 0$. With interaction (1) we have the following Bethe-Salpeter equation for state X_{PS} in correspondence to diagrams presented in Fig 1 under assumption of existence of interaction (4).

$$\Psi_{PS} = G_{PS} + \frac{G^2}{8\,\pi^2}\Big(\frac{1}{6x}\int_0^x \Psi_{PS}(y)y^2dy - \frac{1}{2}\int_0^x \Psi_{PS}(y)ydy -$$

$$\frac{x}{2}\int_x^\infty \Psi_{PS}(y)dy + \frac{x^2}{6}\int_x^\infty \frac{\Psi_{PS}(y)}{y}dy\Big); \tag{5}$$

where coupling constant G_{PS} is defined by (4). Here in view of large value $M_X \simeq 0.75\,TeV$ of the wouldbe resonance we neglect W mass.

$$z = \frac{G^2\,x^2}{128\,\pi^2};\; t = \frac{G^2\,y^2}{128\,\pi^2}; \tag{6}$$

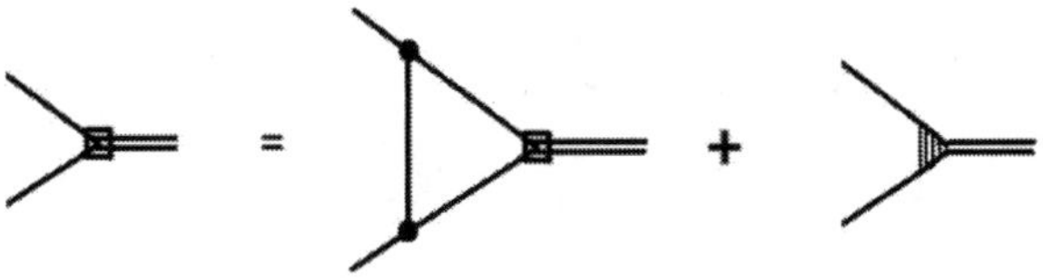

Figure 1. Diagram form of equation (5). Simple lines represent W-s, a double line corresponds to the resonance, black circles correspond to interaction (1), squares correspond to BS wave function. The triangle corresponds to vertex (4).

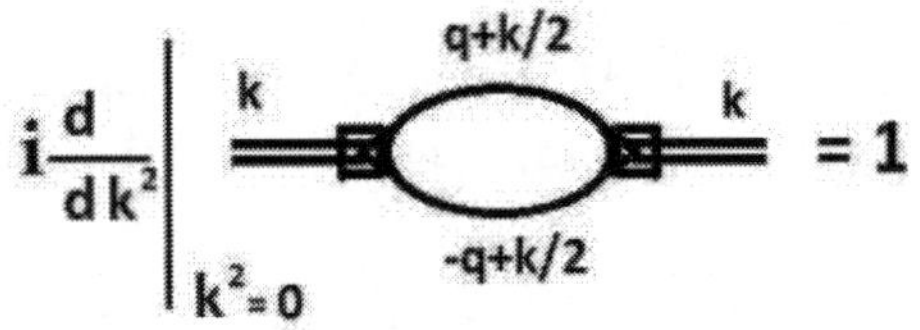

Figure 2. Diagram representation of normalization condition for coupling G_{PS}. Simple lines represent W-s, a double line corresponds to the resonance, squares correspond to BS wave function.

we come to the following equation

$$\Psi_{PS}(z) = G_{PS} + \frac{4}{3\sqrt{z}}\int_0^z \Psi_R(t)\sqrt{t}dt - 4\int_0^z \Psi_{PS}(t)dt - 4\sqrt{z}\int_z^\infty \frac{\Psi_{PS}(t)}{\sqrt{t}}dt + \frac{4z}{3}\int_z^\infty \frac{\Psi_{PS}(t)}{t}dt. \tag{7}$$

Equation (7) satisfies condition

$$\Psi_{PS}(0) = G_{PS}. \tag{8}$$

By successive differentiations of equation (7) we obtain a Meijer differential equation for function $\Psi_{PS}(z)$

$$\left[\left(z\frac{d}{dz}+\frac{1}{2}\right)\left(z\frac{d}{dz}\right)\left(z\frac{d}{dz}-\frac{1}{2}\right)\left(z\frac{d}{dz}-1\right)+z\right]\Psi_{PS}(z) = 0. \tag{9}$$

Then the solution, which fulfill boundary condition both at zero and at the infinity (see *e.g.* [18]) is the following

$$\Psi_{PS}(z) = \frac{G_{PS}}{2}G_{04}^{30}(z\,|\,_{0,1/2,1,-1/2}). \tag{10}$$

The normalization condition for Bethe-Salpeter wave function (10) give according to diagram Fig. 2 with account of definition (6) the following relation

$$\frac{9}{64\,\pi^2}\int_0^\infty dy\,\Psi_{PS}(y)^2 = \frac{9\,\sqrt{2}\,G_{PS}^2}{16\pi\,G}\,I = 1; \tag{11}$$
$$I = \int_0^\infty \frac{G_{04}^{30}\big(t\,|\,_{0,1/2,1,-1/2}\big)^2}{2\,\sqrt{t}}dt = \frac{\pi}{8}.$$

With values I (11), $g = 0.65$ and with account of (1) we obtain coupling G_{PS}

$$G_{PS} = \frac{8}{3\,M_W}\sqrt{g\,|\lambda|\,\sqrt{2}} = 0.00318\,\frac{1}{GeV}; \tag{12}$$

where numerical value corresponds to $\lambda = \lambda_0 = -0.01$, that is safely inside restrictions (2) and estimates (3).Value (12) corresponds to scale $\Lambda \simeq 0$. We take value (12) for estimates of effects, bearing in mind, that for other values of scale Λ coupling G_{PS} is defined by solution (10), namely

$$G_{PS}(\Lambda) = \frac{G_{PS}}{2}G_{04}^{30}(z_\Lambda|\,_{0,1/2,1,-1/2});\; z_\Lambda = \frac{G^2\,\Lambda^4}{128\,\pi^2}. \tag{13}$$

We evaluate the pseudoscalar resonance decay probabilities with $\Lambda_D = M_{PS} = 750\,GeV$. For estimations of cross sections we take Λ in correspondence to maxima of structure functions. That is $\Lambda(\sqrt{s}) \simeq \sqrt{s}/7$. Then for the decay and for two values $\sqrt{s} = 8\,TeV$ and $\sqrt{s} = 13\,TeV$ we have the following values for effective coupling G_{PS}

$$G_{PS}(\Lambda_D) = G_{PS}(0.75\,TeV) = 0.00303; \tag{14}$$
$$G_{PS}(8\,TeV/7) = 0.00285;\; G_{PS}(13\,TeV/7) = 0.00242.$$

Thus we have interaction (4) with parameters G_{PS} (12,14) and $M_{PS} = 750\,GeV$. We use well-known relation

$$W^0 = \cos\theta_W\,Z + \sin\theta_W\,A; \tag{15}$$

and obtain for partial decay widths of the pseudoscalar X_{PS}

$$\Gamma(W^+\,W^-) = 31.06\,GeV\,(43.5\%);\quad \Gamma(Z\,Z) = 9.01\,GeV\,(12.6\%);$$
$$\Gamma(Z\,\gamma) = 4.72\,GeV\,(6.6\%);\quad \Gamma(\gamma\,\gamma) = 0.91\,GeV\,(1.2\%); \tag{16}$$
$$\Gamma(W^+\,W^-\,Z) = 21.34\,GeV\;29.9\%);\quad \Gamma(W^+\,W^-\,\gamma) = 4.45\,GeV\,(6.2\%) \tag{17}$$
$$\Gamma_t(X_{PS}) = 71.49\,GeV.$$

We would present also probabilities for the following specific channels, where l means light lepton (μ, e)

$$\Gamma(l^+\,l^-\gamma) = 0.404\,GeV\,(0.64\%);\; \Gamma(l^+\,l^-\,l^+\,l^-) = 0.0411\,GeV(0.065\%). \tag{18}$$

Then we calculate cross sections for X_{PS} production in $p\,p$ collisions for $\sqrt{s} = 8\,TeV$ and for $\sqrt{s} = 13\,TeV$. In doing this we use CompHEP package [19].

Thus we consider possible pseudoscalar neutral resonance with mass $\approx 750\,TeV$, which decay into

$$W^+W^-;\quad ZZ;\quad Z\gamma,\quad \gamma\gamma. \tag{19}$$

Table 1. Results for cross-sections of X_{PS} production in $p + p$ collisions at $\sqrt{s} = 8\,TeV$ and $\sqrt{s} = 13\,TeV$ with $\lambda = -0.01$ ($G_{PS} = 0.00318\,GeV^{-1}$).

$p + p \to X+$	$\sigma\, fb,\, 8\,TeV$	$\sigma\, fb,\, 13\,TeV$
$W^+ X_{PS}$	12.96	23.10
$W^- X_{PS}$	3.70	7.89
$Z X_{PS}$	5.87	11.31
γX_{PS}	1.85	3.54
$q(\bar{q})\, jets\, X_{PS}$	81.68	131.10
$W^+ W^- X_{PS}$	36.70	289.87
$W^+ Z X_{PS}$	27.36	208.05
$W^- Z X_{PS}$	6.60	53.45
$W^+ \gamma X_{PS}$	1.61	4.47
$W^- \gamma X_{PS}$	0.40	1.10
$\sigma(X_{PS})_{tot}$	178.74	734.11

According to Table 1 the cross-section of the resonance production at $\sqrt{s} = 8\,TeV$ is four times less than at $\sqrt{s} = 13\,TeV$. Available $\sqrt{s} = 8\,TeV$ data [20–22] do not contradict to our estimates with account of branching ratios (16). Namely, for $\sqrt{s} = 8\,TeV$ we have

$$\sigma(p\,p \to X_{PS}) \cdot BR\,(X_{PS} \to \gamma\,\gamma) = 178.8 \cdot 0.012 = 2.15\,fb; \tag{20}$$

that do not contradict the most recent limitations [23]. Limitations for $W\,W$ and $Z\,Z$ decay modes [24, 25] also do not contradict the present results. For example, CMS data [22] give for $750\,GeV$ resonance with width $\simeq 100\,GeV$ limitation $\sigma BR\,(X_{PS} \to \gamma\gamma) < 40\,fb$ with prediction $5.2\,fb$ from Table 1. Let us note, that our result for effect in channel $X_{PS} \to ZZ$ gives for l^+l^- ($l = e, \mu$) decays of Z-bosons with integral luminosity $L = 5.3\,fb^{-1}$ [20] the following estimate for an event number

$$\sigma(X_{PS}, 8\,TeV) \cdot BR\,(X_{PS} \to l^+l^-l^+l^-) \cdot L = \\ 178.74 \cdot 0.00065 \cdot 5.3 = 0.62\,. \tag{21}$$

It is worth mentioning, that in experimental results at $\sqrt{s} = 8\,TeV$ [20] there is one event just at $M(l^+l^-l^+l^-) = 750\,GeV$ and no other events for $M(l^+l^-l^+l^-) > 600\,GeV$. Of course this coincidence proves nothing due to the poor statistics, we may only state, that results [20] do not contradict our estimates.

Now what for $\sqrt{s} = 13\,TeV$? First of all let us estimate an effect in channel $\gamma\,\gamma$. We have for possible number of events with (16) and data from Table 1

$$N_{\gamma\gamma} = \sigma(p\,p \to X_{PS}) \cdot BR\,(X_{PS} \to \gamma\,\gamma) \cdot L = 8.81 \cdot L(fb^{-1}). \tag{22}$$

Thus we have for $L \simeq 3\,fb^{-1}$ few tens events, that agrees observations [11, 12].

It may be advisable to study effect not only in channel $X_{PS} \to \gamma\,\gamma$ but also in channel $X_{PS} \to \gamma\,l^+\,l^-$. According to (16) we have

$$\frac{BR\,(X_{PS} \to \gamma\,l^+\,l^-)}{BR\,(X_{PS} \to \gamma\,\gamma)} = \frac{0.445}{1.0} = 0.445\,; \tag{23}$$

that is actually only two times smaller than already observed effect in 2γ.

Let us also calculate the effect for $l^+ l^- l^+ l^-$ at $\sqrt{s} = 13\,GeV$ in the resonance region $\simeq 750\,GeV$

$$N(l^+ l^- l^+ l^-) = \sigma(X_{PS}, 13\,TeV) \cdot BR\,(X_{PS} \to l^+ l^- l^+ l^-) = 0.48\,L\,(fb^{-1}). \qquad (24)$$

So for $L \sim 20\,fb^{-1}$ the effect in four leptons channel may become noticeable. The more so as for this channel background conditions are favorable [20]. Effects $X_{PS} \to \gamma\, l^+ l^-$; $X_{PS} \to l^+ l^- l^+ l^-$ with intensities (23,24) would confirm definitely the interpretation of the $750\,GeV$ state being W-ball. Note, that existing limitations on a possible extra contribution of decay $X_{PS} \to \gamma Z$ with invisible decay $Z \to \bar{\nu}\nu$ [26] do not contradict our estimates.

There is one more suspicious point in data by ATLAS [27], where process

$$p + p \to jets + Z(\to l^+ l^-) + missingE_T; \qquad (25)$$

was studied at $\sqrt{s} = 8\,TeV$. SM calculations for the number of events for integrated luminosity $L = 20.3\,fb^{-1}$ give $N = 9.8 \pm 2.9 \pm 1.4$. Experiment [27] gives $N = 29$ events.

However with our premises we have for process $p + p \to X_{PS} + everything$ with decay $X_{PS} \to Z(\to l^+ l^-) + Z(\to \bar{\nu}\nu)$ also missing E_T due to the invisible Z-decay. Thus we have for the additional contribution

$$\Delta N = \sigma(pp \to X_{PS}) \cdot L \cdot BR\,(X_{PS} \to ZZ) \cdot 2 \cdot BR\,(Z \to l^+ l^-)\,BR(Z \to \bar{\nu}\nu) = 12.3; \qquad (26)$$

for results in tables. Contribution (26) corresponds to the discrepancy within error bars. In any case the data do not contradict the W-ball interpretation of $X(750)$.

Let us remind, that all the estimates were made with $\lambda = -0.01$. Calculations for another value of λ are straightforward with the prescriptions presented above.

3 Conclusion

Existence of W-balls would testify for anomalous gauge interaction (1), which would be due to non-perturbative effects in the electroweak interaction. Thus we could come to important conclusion, that non-perturbative contributions are appropriate not only to QCD, but to the electroweak interaction as well. In this case the anomalies in the electroweak boson pair production do not contradict the Standard Model and do not need extra efforts for a choice of a theory beyond the SM.

Data on effects under discussion might give information on a value of parameter λ. According to our considerations it could be expected in range $\lambda = -0.010 \pm 0.005$. Of course experiments on direct measurement of λ, *e.g.* in processes of $W^+ W^-$, $W^\pm Z(\gamma)$ production are also quite desirable.

Acknowledgments

The work is supported in part by grant NSh-7989.2016.2 of the President of Russian Federation.

References

[1] G. Aad *et.al.* (ATLAS Collaboration), J. High Energy Phys., **1512**, 055 (2015).
[2] V. Khachatryan *et.al.* (CMS Collaboration), J. High Energy Phys., **1408**, 173 (2014).
[3] V. Khachatryan *et.al.* (CMS Collaboration), J High Energy Phys., **1408**, 174 (2014).
[4] D. Gonsalves, F. Krauss and M. Spannovsky, Phys. Rev. D **92** (2015) Article 053010.

[5] B. A. Arbuzov and I. V. Zaitsev, Int. J. Mod. Phys. A **30** (2015) Article 1550221.
[6] G. Aad *et.al.* (ATLAS Collaboration), arXiv: 1606.04833 [hep-ex] (2016).
[7] K. Hagiwara, R.D. Peccei, D. Zeppenfeld and K. Hikasa, Nucl. Phys. B **282**, 253 (1987).
[8] K. Hagiwara, S. Ishihara, R. Szalapski and D. Zeppenfeld, Phys. Rev. D **48**, 2182 (1993).
[9] B. A. Arbuzov, Eur. Phys. J. C **61**, 51 (2009).
[10] B. A. Arbuzov and I. V. Zaitsev, Phys. Rev. D **85** (2012) Article 093001.
[11] G. Aad *et.al.* (ATLAS Collaboration), arXiv: 1606.03833 [hep-ex] (2016).
[12] V. Khachatryan *et al.* (CMS Collaboration), arXiv: 1606.04093 [hep-ex] (2016).
[13] J. Ellis *et al.*, arXiv: 1512.05327 [hep-ph].
[14] K. A. Olive *et al.* (Particle Data Group), *Review of particle physics*, Chin. Phys. C **38** (2014) Article 090001.
[15] B. A. Arbuzov, Theor. Math. Phys. **140**, 1205 (2004).
[16] B. A. Arbuzov and I. V. Zaitsev, Int. J. Mod. Phys. A **27** (2012) Article 1250012.
[17] H. Bateman and A. Erdélyi, *Higher Transcendental Functions V.1*, McGraw-Hill, New-York, Toronto, London, 1953.
[18] B. A. Arbuzov, *Non-perturbative Effective Interactions in the Standard Model*, De Gruyter, Berlin, 2014.
[19] E. E. Boos *et al.* (CompHEP Collaboration), Nucl. Instr. Meth. A **534**, 250 (2004).
[20] V. Khachatryan *et al.* (CMS Collaboration), Phys. Lett. B **721**, 190 (2013).
[21] G. Aad *et al.* (ATLAS Collaboration), Phys. Rev. Lett **113** (2015) Article 71801.
[22] V. Khachatryan *et al.* (CMS Collaboration), Phys. Lett. B **750**, 494 (2015).
[23] G. Aad *et al.* (ATLAS Collaboration), Phys. Rev. D **92** (2015) Article 032004.
[24] G. Aad *et al.* (ATLAS Collaboration), Eur. Phys. J. C, **75**, 209 (2015).
[25] G. Aad *et al.* (ATLAS Collaboration), arXiv: 1512.05099 [hep-ex].
[26] V. Khachatryan *et al.* (CMS Collaboration), Phys. Lett. B **755**, 102 (2016).
[27] G. Aad *et al.* (ATLAS Collaboration), Eur. Phys. J. C **75**, 318 (2015).

Sgoldstino rate estimates in the SHiP

Konstantin Astapov[1,2,⋆] and *Dmitry* Gorbunov[1,3,⋆⋆]

[1] *Institute for Nuclear Research of the Russian Academy of Sciences, Moscow 117312, Russia*
[2] *Moscow Institute of Physics and Technology, Dolgoprudny 141700, Russia*
[3] *Physics Department, Moscow State University, Vorobievy Gory, Moscow 119991, Russia*

Abstract. For supersymmetric extensions of the Standard Model with light sgoldstinos—scalar and pseudoscalar superpartners of goldstinos—we estimate the signal rate anticipated at the recently proposed fixed target experiment SHiP utilizing a CERN Super Proton Synchrotron beam of 400 GeV protons.

1 Introduction

Low energy supersymmetry (SUSY) is perhaps the most developed extension of the Standard Model of (SM) of particle physics [1, 2]. While, inherent in supersymmetry, a technically natural solution to the gauge hierarchy problem implies the SM superpartners are at or below the TeV energy scale, other and much lighter new particles can exist as well. In particular, if supersymmetry is spontaneously broken at not very high energy scales, the particles from the SUSY breaking sector may show up at quite low energies. Their effective couplings to the SM particles are anticipated to be rather weak; therefore a high intensity beam is required to test the model via production of the new particles. The CERN Super Proton Synchrotron (SPS) provides us with a high intensity beam of 400 GeV protons, and the recently proposed beam-dump Search for Hidden Particles (SHiP) experiment can perform the task. We estimated the signal rate expected at the SHiP experiment in supersymmetric models with sufficiently light particles of the Goldstino supermultiplet. The latter contains Goldstino (the Nambu–Goldstone field, fermion) and its superpartners, scalar and pseudoscalar sgoldstinos. While goldstinos are R odd, sgoldstinos are R even and hence can be singly produced in scatterings of the SM particles and can subsequently decay into the SM particles. Interaction of the Goldstino supermultiplet with other fields is suppressed by the parameter F. To the leading order in $1/F$, sgoldstino coupling to SM gauge fields [photons $F_{\mu\nu}$, gluons $G^a_{\mu\nu}$, where the index $a = 1, ...8$ runs $SU(3)$ color group] and matter fields (leptons f_L, up and down quarks f_U and f_D) at the mass scale above Λ_{QCD} but below electroweak symmetry breaking reads

⋆e-mail: astapov@ms2.inr.ac.ru
⋆⋆e-mail: gorby@ms2.inr.ac.ru

$$\mathcal{L}_{eff} = -\frac{1}{2\sqrt{2}F}\left(m_S^2 S\tilde{\bar{G}}\tilde{G} + i m_P^2 P\tilde{\bar{G}}\gamma_5\tilde{G}\right) - \frac{1}{2\sqrt{2}}\frac{M_{\gamma\gamma}}{F} S F^{\mu\nu}F_{\mu\nu} + \frac{1}{4\sqrt{2}}\frac{M_{\gamma\gamma}}{F} P\epsilon^{\mu\nu\rho\sigma}F_{\mu\nu}F_{\rho\sigma} -$$
$$- \frac{1}{2\sqrt{2}}\frac{M_3}{F} S G^{\mu\nu\,a}G^a_{\mu\nu} + \frac{1}{4\sqrt{2}}\frac{M_3}{F} P\epsilon^{\mu\nu\rho\sigma}G^a_{\mu\nu}G^a_{\rho\sigma} - \frac{\tilde{m}^{LR\,2}_{D_{ij}}}{\sqrt{2}F} S\bar{f}_{D_i}f_{D_j} - i\frac{\tilde{m}^{LR\,2}_{D_{ij}}}{\sqrt{2}F} P\bar{f}_{D_i}\gamma_5 f_{D_j} -$$
$$- \frac{\tilde{m}^{LR\,2}_{U_{ij}}}{\sqrt{2}F} S\bar{f}_{U_i}f_{U_j} - i\frac{\tilde{m}^{LR\,2}_{U_{ij}}}{\sqrt{2}F} P\bar{f}_{U_i}\gamma_5 f_{U_j} - \frac{\tilde{m}^{LR\,2}_{L_{ij}}}{\sqrt{2}F} S\bar{f}_{L_i}f_{L_j} - i\frac{\tilde{m}^{LR\,2}_{L_{ij}}}{\sqrt{2}F} P\bar{f}_{L_i}\gamma_5 f_{L_j}\,. \quad (1)$$

Here M_3 is the gluino mass, $M_{\gamma\gamma} = M_1 \sin^2\theta_W + M_2\cos^2\theta_W$ with M_1 and M_2 being $U(1)_Y$- and $SU(2)_W$-gaugino masses and θ_W the weak mixing angle, and $\tilde{m}^{LR\,2}_{U_{ij}}$ and $\tilde{m}^{LR\,2}_{D_{ij}}$ are left-right up- and down- squark soft mass terms. Sgoldstino decays into two electrically charged SM particles yield the signature well recognizable at SHiP: two charged tracks from a single vertex supplemented with a peak in the invariant mass of outgoing particles. Sgoldstino couplings to the SM fields are inversely proportional to the parameter of the order of the squared scale of SUSY breaking in the whole model. This unique feature of the Goldstino supermultiplet allows us to probe the SUSY breaking scale by hunting for the light sgoldstinos.

To illustrate the sensitivity of the SHiP experiment to sgoldstino couplings, in the next sections we present numerical results for the set of values of MSSM parameters (the benchmark point in the MSSM parameter space) shown in Table 1. It is an arbitrary choice, except we suppose that all the

M_1, GeV	M_2, GeV	M_3, GeV	μ, GeV	$\tan\beta$
100	250	1500	1000	6
m_A, GeV	A_l, GeV	m_l, GeV	A_Q, GeV	m_Q, GeV
1000	2800	1000	2800	1000

Table 1. MSSM benchmark point.

model parameters take experimentally allowed values and the lightest Higgs boson mass is 125 GeV. Trilinear soft supersymmetry breaking parameters $A_{l,Q}$ are defined by the relations $\tilde{m}^{LR\,2}_{D_{ii}} \equiv m_{D_i}A_Q$, $\tilde{m}^{LR\,2}_{U_{ii}} \equiv m_{U_i}A_Q$, $\tilde{m}^{LR\,2}_{L_{ii}} \equiv m_{L_i}A_l$, where we use SM fermion masses m_{D_i,U_i,L_i}; m_A, m_Q, m_l are *CP*-odd Higgs boson, squark and slepton masses, correspondingly.

2 Scalar sgoldstino

In this section we consider two different production mechanisms of the scalar sgoldstino relevant for the SHiP setup. The first one is the direct production via hard gluon fusion in proton scatterings off the target material. The second one is the production in decays of mesons emerging due to the proton scattering.

2.1 Gluon fusion

If the sgoldstino is much heavier than the QCD energy scale of 100 MeV, it can be produced directly via gluon fusion. The relevant parts of the sgoldstino interaction Lagrangians (1), read

$$\mathcal{L}_{Sgg} = \left(\theta\, g^{one-loop}_{hgg}(m_S) - \frac{\alpha_s(m_S)\beta(\alpha_s(M_3))}{\beta(\alpha_s(m_S))\alpha_s(M_3)}\frac{M_3}{2\sqrt{2}F}\right) S\, G^{\mu\nu\,a}G^a_{\mu\nu}, \quad (2)$$

where the first term, associated with Higgs-sgoldstino mixing (**??**), is proportional to the Higgs effective coupling to gluons (appearing at one-loop level via virtual quark exchanges) [3]. Note that both terms in Eq. (2) are inversely proportional to supersymmetry breaking parameter F. To obtain a reliable estimate of the direct scalar sgoldstino cross section $\sigma_{pp\to S}$, we properly rescale the results of Ref. [4], where a coupling similar to Eq. (2) is responsible for the light inflaton production at the fixed target experiment with a 400 GeV proton beam. For the MSSM parameters from Table 1 we find the following numerical approximation to the cross section as a function of the sgoldstino mass and supersymmetry breaking parameter,

$$\log_{10}(\frac{\sigma_{pp\to S}}{\sigma_{pp,\text{total}}}) = -15.8666 - 0.93934\times\left(\frac{m_S}{1\text{GeV}}\right) + 0.02025\times\left(\frac{m_S}{1\text{GeV}}\right)^2 + 0.00052\times\left(\frac{m_S}{1\text{GeV}}\right)^3 - \\ - 4\log_{10}\left(\frac{\sqrt{F}}{100\text{TeV}}\right), \quad (3)$$

where $\sigma_{pp,\text{total}}$ is the total pp cross section for the 400 GeV proton beam. This approximation is illustrated in Fig. 1 (left) for two reference values of $\sqrt{F}$. Given the number of protons on target

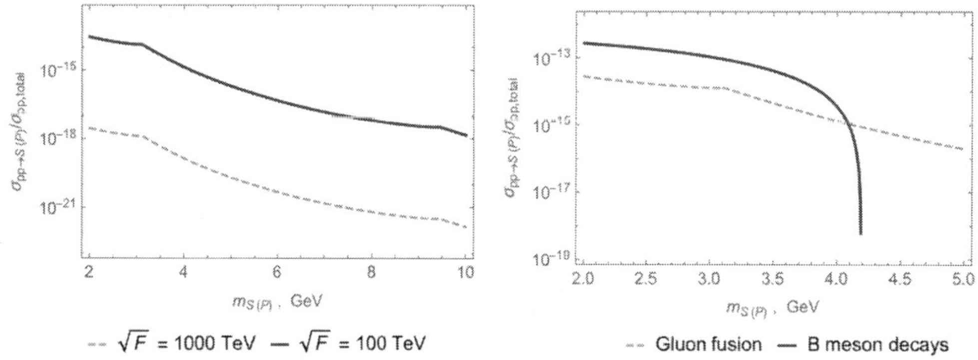

Figure 1. Cross sections of scalar and pseudoscalar sgoldstino production in gluon fusion as functions of sgoldstino mass. No difference between scalar and pseudoscalar cases is expected (on the left panel). And cross sections of scalar sgoldstino production in gluon fusion and in B-meson decays for the model with $\sqrt{F} = 100\,\text{TeV}$ (on the right panel).

expected at SHiP, about 2×10^{20} [5], one concludes from Fig. 1 (left) that the direct production can provide us with sgoldstinos only in the models with a supersymmetry breaking scale below 1000 TeV, if the MSSM superpartner scale is in the TeV range.

2.2 B meson decays

Scalar sgoldstinos of masses in the GeV range are dominantly produced by decays of heavy mesons appearing by proton scattering off target. In the context of the SHiP experiment the main source of sgoldstinos is decays of B mesons. Decays of charmed mesons are suppressed as compared to beauty meson decays due to the smallness of the CKM (Cabibbo – Kobayashi – Maskawa) matrix element in the corresponding amplitude. The process is described by the triangle diagram with a t quark and W bosons running in the loop. The sgoldstino is emitted by the virtual t quark through sgoldstino-top-top coupling (1) and sgoldstino-Higgs mixing (the latter dominates for the values shown in Table 1).

Adopting the same logic as used in [4] for the light inflaton, we calculate the branching ratio of B-meson decay into S,

$$\mathrm{Br}(B \to X_s S) = 0.3 \times (\frac{m_t}{m_W})^4 \left(1 - \frac{m_S^2}{m_b^2}\right)^2 (A_Q v + F\theta)^2 \left(\frac{100\ \mathrm{TeV}}{\sqrt{F}}\right)^4 , \tag{4}$$

where X_s stands for the strange meson channel mostly saturated by a sum of pseudoscalar and vector kaons; m_b, m_t and m_W stand for b, t quarks and $W^\pm$ boson masses, correspondingly. The scalar sgoldstino production cross section is then a product of the branching ratio above and the beauty cross section evaluated at the SHiP energy scale as $1.6 \times 10^{-7} \times \sigma_{pp,\mathrm{total}}$ [5]. In Fig. 1 (right) we compare the sgoldstino production cross sections provided by the direct and the indirect mechanisms for the same value of the supersymmetry breaking parameter, $\sqrt{F} = 100$ TeV. One can observe from Eqs. (3) and (4) that both cross sections scale as $\propto 1/F^2$. Thus, we conclude from the plot in Fig. 1 (right) that the meson channel dominates sgoldstino production when the kinematics allows. In what follows we concentrate on this case and comment on prospects of searches for the heavier sgoldstinos, $M_{S(P)} \gtrsim 4$ GeV, (available only via the direct production) in due course.

2.3 Sgoldstino decay pattern

The sgoldstino is R even and can decay into pairs of SM particles, if it is kinematically allowed. For the sgoldstino of the (sub-)GeV mass-range, the main decay channels are $\gamma\gamma$, e^+e^-, $\mu^+\mu^-$, $\pi^0\pi^0$, $\pi^+\pi^-$, K^+K^-, $K^0\bar{K}^0$ (see Ref. [6] for details).

Between these mass ranges, where $m_s \approx 1.2 - 4$ GeV, neither description method is reliable where some heavy mesons, as well as multimeson final states, enter the game. However, comparing contributions for relevant hadronic modes calculated within these two approaches, we observe that deviations do not exceed an order of magnitude. Thus, we conclude that order-of-magnitude estimates of the sgoldstino lifetime for the mass interval 1.2 – 4 GeV can be obtained by some extrapolation of the chiral theory approach.

The branching ratios of the scalar sgoldstino decays for the values of MSSM soft parameters chosen in Table 1 are shown in Fig. 2 (left). Hadronic channels, $\pi\pi$, KK, naturally dominate (if kine-

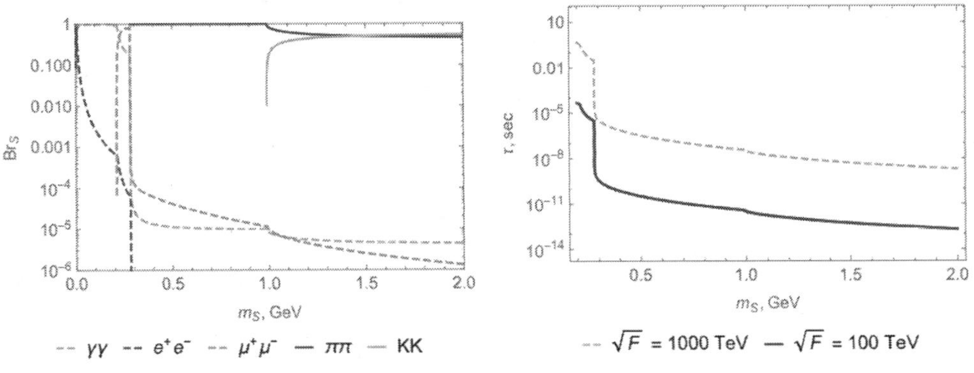

Figure 2. Branching ratios of a scalar sgoldstino (on the left panel). And Lifetime of a scalar sgoldstino as a function of its mass (on the right panel).

matically allowed), while $\gamma\gamma$ and $\mu^+\mu^-$ give small but noticeable contributions.

The sgoldstino lifetime for a set of values of F is presented in Fig. 2 (right). To reach the main detector of the SHiP experiment, the sgoldstino has to cover a distance of about 100 meters [5]. The results in Fig. 2 (right) suggest that SHiP can be sensitive mostly to the models with supersymmetry breaking scale of about 100 TeV and higher. The lifetime scales as $\tau \propto F^2$ and as $\tau \propto 1/M_3^2$ since the hadron channel dominates.

2.4 Sgoldstino signal event rate at SHiP

Now we collect all the ingredients required to achieve the main goal of this article, the estimate of the number of sgoldstino decay events inside the fiducial volume of the SHiP experiment. The SHiP construction is outlined in Ref. [5]. The 400 GeV proton beam fueled by the SPS hits the target and produces bunches of mesons, which can decay into new particles (sgoldstinos in the case at hand). The latter can also appear directly from the proton-proton collisions (through gluon fusion). The detector is placed at a distance of $l_{sh} = 63.8$ m from the target. The vacuum vessel length is about $l_{det} = 60$ m. It forms a cylinder along the beam axis with an elliptical base of 5 m×10 m. The trajectories of electrically charged particles emerging from the new particle decays can be traced in the detector volume and their energies and types can be determined by the registration system utilizing devices arranged at the far end of the detector.

Number of signal events as

$$N_{\text{signal}} = \frac{N_{\text{POT}}}{\sigma_{pp,\text{total}}} \int w_{det} \frac{d\sigma_{pp\to S(P)}}{dp d\theta_p d\phi_p} d^3\vec{p}\,, \tag{5}$$

where the expected number of protons on the target is $N_{\text{POT}} = 2 \times 10^{20}$ [7] and w_{det} denotes the probability for the sgoldstino to decay inside the fiducial volume of the detector,

$$w_{det}(E_{S(P)}, m_{S(P)}, \sqrt{F}) = \exp(-l_{sh}/\gamma c\tau_{S(P)}) \left[1 - \exp(-l_{det}/\gamma(E_{S(P)}) c\tau_{S(P)})\right], \tag{6}$$

with the sgoldstino gamma factor $\gamma(E_{S(P)}) = E_{S(P)}/m_{S(P)}$.

In Fig. 3 we indicate the region in the model parameter space $(m_{S(P)}, 1/\sqrt{F})$, where the number of sgoldstino decays inside the SHiP fiducial volume exceeds 3, $N_{\text{signal}} > 3$. That is, if no events were observed (the background for the two-body decays into charged SM particles is zero [5]) the region is excluded at the confidence level of 95%, in accordance with the Poisson statistics. The upper boundary in Fig. 3 is the region where the sgoldstino coupling constants $\propto 1/F$ are large enough to initiate very fast decay of the sgoldstino before it reaches the detector. The lower boundary in Fig. 3 is the region where the couplings are so small that sgoldstinos escape from the detector without decay. The number of signal events here scales with the model parameters as $N_{\text{signal}} \propto M_3^2\mu^6/F^4$. The region in Fig. 3 of the heaviest sgoldstino reachable at SHiP, $m_S \approx 3.6$ GeV, is the meeting point of the lower and the upper boundaries. Here, the sgoldstino decay length is about 100 m, which is the scale of both the SHiP detector length and the distance from the target, $\gamma c\tau \sim l_{det} \sim l_{sh}$. In this case the number of signal events scales as $N_{\text{signal}} \propto \mu^6/F^2$. The scalings of the signal events imply that models with a higher (as compared to that presented in Fig. 3) scale of supersymmetry breaking can be tested if MSSM parameters μ, M_3 are appropriately bigger (as compared to those presented in Table 1).

Sgoldstinos of masses 3.6–4.2 GeV, which can be produced through B-meson decays (see Figs. 1 (right) and 3), seem to be beyond the SHiP's grip for our choice of MSSM parameters presented in Table 1. However, the signal scaling with model parameters explained above suggests that sgoldstinos of masses above 3.6 GeV can be tested at SHiP in models with a higher scale of SM superpartners. Finally, from the results presented in Figs. 1 (right) and 2 (right) one can conclude that sgoldstinos

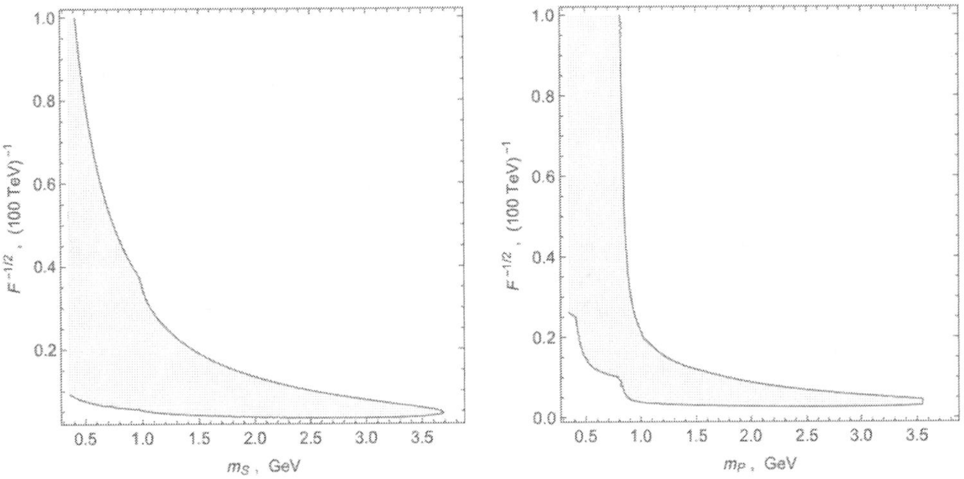

Figure 3. The shaded region will be probed at the SHiP experiment. Left plot is for probing scalar sgoldstino and right is for pseudoscalar.

of masses above 4 GeV, which can appear only via direct production, cannot be tested at SHiP. Both sgoldstino production and decay are governed by the same ratio M_3^2/F^2, and the sgoldstino lifetime $\tau \propto 1/m_S^3$ is too short for the reasonably high sgoldstino production. One needs much higher intensity of the proton beam at SPS to probe this region of model parameter space.

3 Pseudoscalar sgoldstino

If parity is (strongly) violated in the sfermion sector of the MSSM, sgoldstino couplings to the SM fermions violate it, too. Then, scalar and pseudoscalar sgoldstinos are very similar as regards the SHiP phenomenology. However, if sgoldstino couplings conserve parity (that takes place, e.g., in left-right extensions of the MSSM), the phenomenology of pseudoscalar and scalar sgoldstins is different in some aspects (see Refs. [6, 8] for details). In this section we investigate SHiP sensitivity to the pseudoscalar sgoldstino couplings.

The pseudoscalar sgoldstino P can be directly produced via the gluon fusion with the same cross section as the scalar sgoldstino; hence, Fig. 1 is valid for both cases. However, its production through the meson decays is somewhat different from that of the scalar sgoldstino because of the absence of mixing with the light MSSM Higgs.[1]

The light pseudoscalar can be produced in B-meson decays. The corresponding one-loop diagram is very similar to that in the case of the scalar sgoldstino discussed in Sec. 2.2, but only the sgoldstino-top-top pseudoscalar coupling (1) contributes. Given the pseudoscalar nature of P, for the two-body decay it is accompanied by the vector kaon K^*. The decay rate can be obtained by properly replacing the coupling constants in the result presented in Ref. [9] for the case of decay into the axion. In contrast to the scalar S, the pseudoscalar sgoldstino P does not decay into a meson pair. However, it mixes with pseudoscalar mesons π and η, as explained in Ref. [6]. Since mesons exhibit four-meson coupling, the pseudoscalar sgoldstino can decay into three mesons through the virtual meson state, $P \to \pi^{0^*}/\eta^* \to 3$ mesons.

[1]There is a mixing with CP-odd Higgs A^0, which is negligibly small for our choice of the benchmark point in Table 1.

The pseudoscalar sgoldstino lifetime and its relevant decay branching ratios are presented in Fig. 4

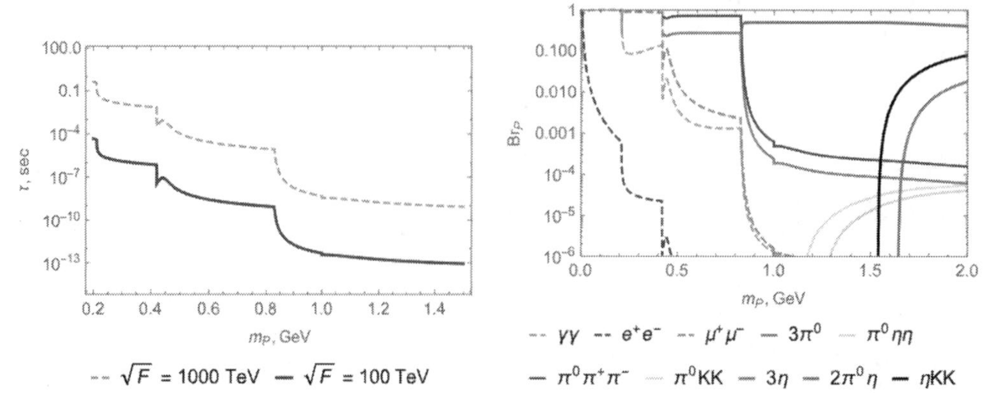

Figure 4. Pseudoscalar sgoldstino lifetime for $\sqrt{F}$ = 100, 1000 TeV (lines from bottom to top) (on the left pannel). And branchings ratios of the pseudoscalar sgoldstino (on the right panel).

Hadronic channels, mainly 3π, ηKK, $\eta\pi\pi$ and 3η, dominate sgoldstino decay. Given its geometry, the SHiP experiment is sensitive to the supersymmetry breaking scales of about $\sqrt{F} \sim 100\,\text{TeV}$ and above.

Performing for the pseudoscalar case the same procedure as that adopted in Sec. 2.4 for the scalar sgoldstino, we estimate the SHiP sensitivity to the pseudoscalar sgoldstino interaction. In Fig. 3 (right) we present the plot displaying the region in the parameter space $(F^{-1/2}, m_S)$ expected to be explored with the SHiP experiment. One observes that in the case of the pseudoscalar sgoldstino the SHiP is sensitive to models with a lower SUSY breaking scale $\sqrt{F}$ as compared to the scalar sgoldstino (cf. Fig.3).

References

[1] H. E. Haber and G. L. Kane, "The Search for Supersymmetry: Probing physics beyond the standard model," Phys. Rep. **117**, 75 (1985).

[2] S. P. Martin, "A Supersymmetry primer, In *Kane, G.L. (ed.): Perspectives on supersymmetry II* 1-153 [hep-ph/9709356].

[3] J. R. Ellis, M. K. Gaillard, D. V. Nanopoulos, and C. T. Sachrajda, Phys. Lett. B **83**, 339 (1979).

[4] F. Bezrukov and D. Gorbunov, "Light inflaton Hunter's Guide," JHEP **1005** (2010) 010 [arXiv:0912.0390 [hep-ph]].

[5] M. Anelli *et al.* [SHiP Collaboration], "A facility to Search for Hidden Particles (SHiP) at the CERN SPS," arXiv:1504.04956 [physics.ins-det].

[6] D. S. Gorbunov, "Light sgoldstino: Precision measurements versus collider searches," Nucl. Phys. B **602**, 213 (2001) [hep-ph/0007325].

[7] S. Alekhin *et al.*, "A facility to Search for Hidden Particles at the CERN SPS: the SHiP physics case," arXiv:1504.04855 [hep-ph].

[8] D. S. Gorbunov and V. A. Rubakov, "Kaon physics with light sgoldstinos and parity conservation," Phys. Rev. D **64**, 054008 (2001) [hep-ph/0012033].

[9] M. Freytsis, Z. Ligeti and J. Thaler, "Constraining the Axion Portal with B —> K l+ l-," Phys. Rev. D **81**, 034001 (2010) [arXiv:0911.5355 [hep-ph]].

On Geoneutrinos

L. B. Bezrukov[1,*], A. S. Kurlovich[1], B. K. Lubsandorzhiev[1], V. V. Sinev[1], V. P. Zavarzina[1], and V. P. Morgalyuk[2]

[1]*Institute for Nuclear Researches of Russian Academy of Sciences, Prospekt 60-letia Oktyabrya 7a, Moscow, 115409, Russia*
[2]*A. N. Nesmeyanov Institute of Organoelement Compounds of Russian Academy of Sciences, Vavilova 10, Moscow, 115409, Russia*

Abstract. Experimental data on geoneutrinos allow to admit that masses of U, Th and K in the Earth can be up to $m_U = 1.7 \cdot 10^{17}$ kg, $m_{Th} = 6.7 \cdot 10^{17}$ kg and $m_K/m_{Earth} \sim 2\%$. These values correspond to intrinsic Earth heat flux in ~300 TW. The most part of this flux goes up in rift zones as a heated gases. Argo Project results and the measurements of the Moon intrinsic heat flux support the existence of such a big flux. So large of U, Th, K abundances were predicted by Adjusted Hydridic Earth model.

Introduction

The connection of geoneutrinos and Earth thermal flux was known for a long time. Detecting of geoneutrinos on the surface can give the exact value of radiogenic part of all heat produced inside the Earth [1]. Thermal flux is obtained by measuring the temperature gradient in the crust. The most exact value 47 ± 2 TW is obtained by the averaging of experimentally measured points. This method includes only part of the energy passing through the crust by conductivity way and does not take into account other possible ways of energy transfer. For example, the formation of gases could take some heat and decrease the conductivity termal flux. Gas may pass the crust and release the energy somewhere in the ocean.

We considered a number of possible Earth thermal flux values and concluded that value of order 300 TW is preferable. This value may be easily explained if to admit the abudances of potassium and uranium with thorium in larger amounts inside of the Earth core. We proposed here the Adjusted Hydridic Earth model which can explain as the value of thermal flux in 300 TW and also the measured number of events in ongoing geoneutrino experiments.

1 Earth models

The Bulk Silicat Earth (BSE) model [2] gives:

$$\begin{aligned} m_{BSE}(\mathrm{U}) &= 0.81 \cdot 10^{17}\mathrm{kg}, \\ m_{BSE}(\mathrm{Th}) &= 3.16 \cdot 10^{17}\mathrm{kg}, \\ m_{BSE}(^{40}\mathrm{K}) &= 0.58 \cdot 10^{17}\mathrm{kg}. \end{aligned} \tag{1}$$

*e-mail: bezrukov@inr.ac.ru

Table 1. Geochemical model of modern Hydridic Earth

Geosphere	Depth range, km	Composition
Lithosphere	0 - 150	CaO, MgO, Al_2O_3, SiO_2, Na_2O, FeO, H_2O...
Astenosphere	150	Thin layer of Metalsphere with high hydrogen concentration
Metalsphere	150 - 2900	$Mg_2Si : Si : FeSi = 6 : 3 : 1$
External core	2900 - 5000	$MgH_{0.1}$, $SiH_{0.1}$, $FeH_{0.1}$ $+MgH_n$, SiH_n, FeH_n (n = 7)
Inernal core	5000 - 6371	MgH_n, SiH_n, FeH_n (n = 7)

This amount distributed only in the Crust and Upper Mantle in framework of BSE model.

There is an alternative Earth model [3], [4] named Hydridic Earth (HE) model which predicts the primordial chemical elements composition of the Earth.

Vladimir Larin [3] used the idea that the separation of the chemical elements in the solar system (chemical differentiation) was originated from the magnetic field of the Protosun. He found the law:

$$(\frac{X_M}{X_{\rm Si}})_{Planet} = (\frac{X_M}{X_{\rm Si}})_{Sun} \cdot F(R_{Planet}, E_{IP}(M)) \tag{2}$$

where X_M is the mass fraction in the planet mass of the chemical element with atomic number M, $E_{IP}(M)$ is the ionization potential of the chemical element with atomic number M in eV, R_{Planet} is the Planet distance to the Sun. In the work [4] the function F was proposed for the Earth in the form:

$$F(R_{Earth}, E_{IP}(M)) = 12100 \cdot e^{-1.1537 \cdot E_{IP}(M)}. \tag{3}$$

We obtained from (2) and (3):

$$\begin{aligned} m_{HE}(\mathrm{U}) &= 3.18 \cdot 10^{17}\mathrm{kg}, \\ m_{HE}(\mathrm{Th}) &= 1.05 \cdot 10^{18}\mathrm{kg}, \\ m_{HE}(^{40}\mathrm{K}) &= 2.63 \cdot 10^{19}\mathrm{kg}. \end{aligned} \tag{4}$$

by using experimental data of $(\frac{X_M}{X_{\rm Si}})_{Sun}$ from [5] for the Sun and the data for E_{IP} from [6]. Function F from (3) is the theoretical one and can be adjusted.

The 18.3% of the Earth primordial mass is predicted to be Hydrogen [4]. The inner Earth would have been and still could be hydrogen rich. The most part of primordial hydrogen have escaped to atmosphere and space through the degassing of the mantle. Model suggests that large amounts of hydrogen are still located in the core.

Hydridic Earth model predicts huge amount of ^{40}K in the Earth.

2 Experimentally measured uranium and thorium geoneutrinos

Geoneutrinos were detected with two experimental setups for the moment. They are: Borexino detector [7] constructed for the ^7Be solar neutrinos registartion placed in Gran Sasso Laboratory and KamLAND detector [8] for testing neutrino oscillation parameters in solar region placed in Kamioka Observatory. Both detectors discovered geoneutrino flux that are in accordance with predicted values. Prediction is based on BSE model. On figure 1 antineutrino spectra are shown for three main isotopes emitting geoneutrinos. It is seen that only part of spectrum can be detected for ^{238}U and ^{232}Th using

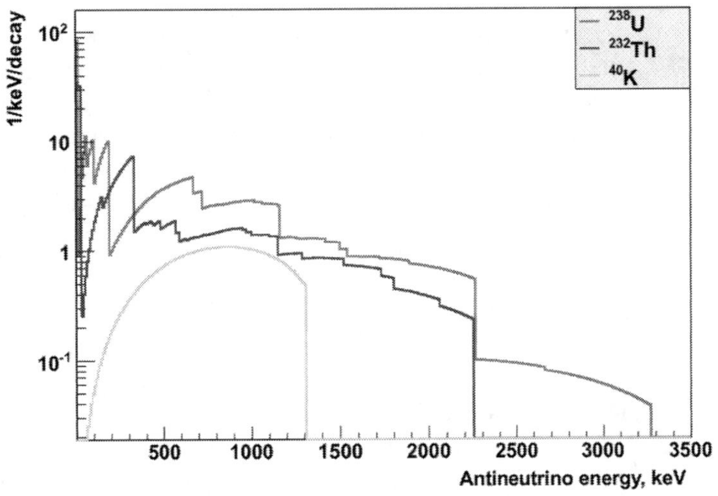

Figure 1. Antineutrino spectra from ^{238}U, ^{232}Th and ^{40}K per decay. On X axis the antineutrino energy and on Y axis the number of neutrinos emitted per keV energy bin in one decay are shown.

Table 2. Counting rates of Borexino and KamLAND

Detector	T_{meas}, days	N_p	N_{event}	R, TNU*
Borexino	2056	$(0.977 \pm 0.05) \times 10^{31}$	$23.7^{+6.5(st)+0.8}_{-5.7(st)-0.6}$	43.5
KamLAND	2991	$(5.98 \pm 0.13) \times 10^{31}$	116^{+28}_{-27}	23.7

*TNU = 1 event per 10^{32} protons per year

the inverse beta-decay reaction on proton. This reaction has a threshold 1.8 MeV. ^{40}K spectrum is totally under this threshold.

Counting rates of geoneutrinos are shown in table 2. From the table one can see that counting rate of geoneutrinos is extremely low for both detectors. It does not contradict to any Earth model, better to say it agrees with all possible models. To check models and distinguish between them one needs to accumulate sufficient statistics (4000-5000 events). Easily to estimate that Borexino needs to be operatable at least 1000 years to achieve this statistics. The only one way for investigation of geoneutrino fluxes is to develope a special detector having target mass no less than 10 kt and low background.

3 Earth thermal flux

Thermal flux of the Earth is known for a long time. It is considered that the flux is produced by decays of long lived radioactive elements such as ^{238}U, ^{232}Th and ^{40}K. According to BSE model (1) the total radiogenic thermal flux is about 15 TW. Other part of the flux is considered to be produced by a number of not radioactive reasons [2]. Let consider the possible values of thermal flux that are known today.

3.1 Conductive part of heat flux

The value of Earth thermal flux was measured by installing of thermosensors in the drilled holes all over the world in cotinental and ocean crust [9]. The data were averaged and some methods of data interpolation were used. As a result the value 47 ± 2 TW was obtained. It is nesessary to stress that this value was changing from 30 to 47 TW during 2000-2010 years with changing of the interpolation methods.

This method does not account the possible ways of thermal transmission other than thermoconductivity.

3.2 Heat transfer in oceans

Energy imbalance in oceans was observed by use of monitoring temperature profile down to 2 km of depth in a numerous points. It is so called the ARGO project[10]. By monitoring of temperature during 6 years and using about 4000 special devices floating in different parts of world ocean they found that the accumulated energy in 2 km layer of ocean corresponds to $\sim 5\times10^{22}$J. This fact can be interpreted that ocean was heated this period mostly by intrinsic Earth source of heat. The value of thermal flux from this source must be 300 TW.

3.3 Earth thermal flux recalculated from the measured one on the Moon

Interpretation of surface heat flow is considerably less ambiguous on the Moon becauce is thought to be directly related to the abundance of radioactive elements. If the Moon and the Earth consisted of the same elements the thermal flux of the Earth is proportionally larger than the one of the Moon with coefficient of masses ratio. For the moment we know three measurements of Moon thermal flux: measurements by Apollo 15 and 17 missions [11], registering of radio waves from the Moon of 10 cm and 20 cm wavelengths [12] and measuring of surface temperature at the permanent dark places on Moon surface on the Moon south pole region [13], [14]. According Apollo measurements the Earth flux is appeared 43-65 TW. Second method gives the value about 170 TW. Third method gives the largest value 420 TW.

4 Potassium abundance problem

BSE model does not give the large value of potassium abundance in total Earth mass. It is considered that all potassium is concentrated in the crust. Potassium abundance in the crust is about 2% in weight. But the absence of potassium inside of Earth is based on the fact that meteorites have low abundance of potassium. In reality we have no information about potassium abundance inside the Earth below the crust. If its abundance in the Earth is about the same as in the crust the thermal flux could be much more larger than predicted BSE model. The only way to obtain information on the potassium inside the Earth is to measure antineutrino and possibly neutrino fluxes from ^{40}K which content in natural potassium is 1.17×10^{-4}.

Calculated heat flux from ^{40}K is ten times larger than one from ^{238}U and ^{232}Th in sum in case of its abundance on the level of 2% all over the Earth. From previous section where we discussed possible values of thermal flux we showed that Earth termal flux might be up to 300 TW.

Thermal flux from ^{40}K can be nondirectly checked by registering antineutrino and neutrino fluxes produced by ^{40}K. In [15],[16] the spectrum of recoiled electrons from ^{40}K was compared with spectra from other neutrino sources in Borexino detector.

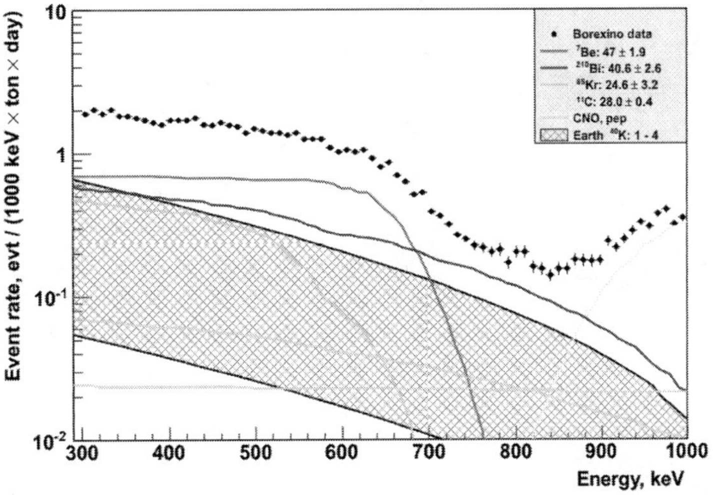

Figure 2. Area of ^{40}K spectrum and solar neutrino spectra (pep and CNO). Main background spectra are also shown. On X axis the energy released per single event in the detector and on Y axis the event number per 1000 keV energy range per day in one ton of a target are shown. Numbers show the count rate per day from a source.

5 Adjusted Hydridc Earth model.

Can we explain the results of geoneutrino flux measurements and the value of heat flux derived from Argo experiment in the frame of HE model?

We propose the following adjusted HE (AHE) model. We took the function F from (3) about twice less in the region of small E_{IP}. It gives us m_{AHE}(U) = $1.7 \cdot 10^{17}$ kg, m_{AHE}(Th) = $6.7 \cdot 10^{17}$ kg, $m_{AHE}(^{40}\mathrm{K}) = 1.2 \cdot 10^{19}$ kg.

The equation relating masses and heat production is

$$H = m \cdot \frac{N_{Avog}}{A} \cdot \frac{E_{release}}{\tau} \cdot \alpha, \quad (5)$$

where N_{Avog}– Avogadro number, A– atomic number, $E_{release}$– energy release in decay chain in MeV, $\tau = t_{1/2}/ln2$– mean lifetime of isotope, α– the conversion factor 1 MeV = $1.602 \cdot 10^{-13}$ J.

$$H(\mathrm{U}) = 1.7 \cdot 10^{20}[\mathrm{g}]\frac{6 \cdot 10^{23}}{238}[\mathrm{g}^{-1}] \cdot \frac{47.7[\mathrm{MeV}]}{6.45 \cdot 10^{9} \cdot 3.15 \cdot 10^{7}[\mathrm{s}]} \cdot 1.602 \cdot 10^{-13}[\mathrm{J}] = 16.1\ \mathrm{TW}, \quad (6)$$

$$H(\mathrm{Th}) = 10.5 \cdot 10^{20}[\mathrm{g}]\frac{6 \cdot 10^{23}}{232}[\mathrm{g}^{-1}] \cdot \frac{40.4[\mathrm{MeV}]}{2.03 \cdot 10^{10} \cdot 3.15 \cdot 10^{7}[\mathrm{s}]} \cdot 1.602 \cdot 10^{-13}[\mathrm{J}] = 18.0\ \mathrm{TW}, \quad (7)$$

Table 3. Number of events in modern detectors for $1kt \cdot year$ exposition

Site	Experimental count rate	Calculated one
Borexino	43.5 ± 12.5	38.1
KamLand	23.67 ± 5.61	31.8

$$H(^{40}\mathrm{K}) = 1.2 \cdot 10^{22}[\mathrm{g}]\frac{6 \cdot 10^{23}}{40}[\mathrm{g}^{-1}] \cdot \frac{0,89 \cdot 0.598[\mathrm{MeV}]}{1.8 \cdot 10^{9} \cdot 3.15 \cdot 10^{7}[\mathrm{s}]} \cdot 1.602 \cdot 10^{-13}[\mathrm{J}] = 270\ \mathrm{TW}. \quad (8)$$

$$H = H(\mathrm{U}) + \mathrm{H(Th)} + \mathrm{H}(^{40}\mathrm{K}) = 304\ \mathrm{TW}. \quad (9)$$

This heat value is produced mostly by ^{40}K decay and is consistent with Argo experiment results [10].

The experimental signal of the geoneutrino detectors depends not only on m(U) and m(Th) but also on the distribution of U and Th in the Earth's interior. The calculations show that decays in the crust matter surrounding of the detector give the main contribution to the experimental signal [2]. The Hydridic Earth model predicts the existence of the Earth hydrogen degassation process. Hydrogen appears on the surface of the Earth core and goes up to the cosmic space through the long chain of processes. During this way the hydrogen flow purifies the metallic mantel volume up to lithosphere. So U and Th must locate in the present-day Earth mostly in the core and the lithosphere. The Internal core of present-day Earth still could be primordial and could have the primordial U and Th densities. The existence in the present day Earth core of the large amount of U and Th is natural in the frame of the Hydridic Earth model.

We took for the crust U and Th distribution according to BSE model and for mantel and core the concentrations of U and Th were chosen to satisfy the observed counting rates (Borexino and KamLAND). The results of calculation of geoneutrino detector counting rate for Borexino and KamLAND in the frame of AHE model are given in Table 3.

We demonstrated that Borexino and KamLAND results and Argo experiment result can be understood in the frame AHE model.

Conclusion

1. We found that the total Earth thermal flux might be larger than one measured using the temperature gradient.

2. We proposed Adjusted Hydridic Earth model with $m_{AHE}(\mathrm{U}) = 1.7 \cdot 10^{17}$ kg, $m_{AHE}(\mathrm{Th}) = 6.7 \cdot 10^{17}$ kg, $m_{AHE}(^{40}\mathrm{K}) = 1.2 \cdot 10^{19}$ kg. This model can explain the observed geoneutrino fluxes and the high thermal flux.

3. The detector with a target 10 kt is needed to accumulate at least 4000–5000 geoneutrino events in 10–20 years of operation time. Directional analysis could clarify the question: Has the core some radioactivity or not? Extra pure scintillator allows to measure ^{40}K spectrum and estimate the potassium amount in the whole Earth. This will make understandable the total Earth thermal flux. For better separation of ^{40}K spectrum and neutrinos from CNO cycle it is very desirable to have independent measurement of CNO neutrinos.

Acknowledgements

We are grateful to Vladimir Larin and Stefan Schoenert for useful discussions. This work was supported by grant of Russian Foundation of Basic Research No 13-02-92440.

References

[1] Livia Ludhova (Ed.), *Geo-neutrinos* (Open Academic Press, Berlin, Germany, 2016)
[2] Gianni Fiorentini, Marcello Lissia and Fabio Mantovani, Phys. Rept. **453**, 117 (2007); arXiv:0707.3203 [physics.geo-ph]; Fabio Mantovani, Luigi Garmignani, Gianni Fiorentini and Marcello Lissia, arXiv:hep-ph/030913v2
[3] V. N. Larin (ed. C. Warren Hunt), *Hydridic Earth: the New Geology of Our Primordially Hydrogen-Rich Planet* (Polar Publishing, Calgary, Alberta, Canada, 1993) 247.
[4] Herve Toulhoat, Valerie Beaumont, Viacheslav Zgonnik, Nikolay Larin and Vladimir N. Larin, arXiv:1208.2909 [astro-ph.EP]
[5] Martin Asplund, Nicolas Grevesse and Jacques Sauval, Nucl.Phys. A **777**, 1 (2006); arXiv:astro-ph/0410214v2
[6] David R. Lich (ed), *CRC Handbook of Chemistry and Physics* (84th Edition. CRC Press. Boca Raton, Florida, 2003); Section 10, Atomic, Molecula and Optical Physics; Ionizatoin Potentials of Atoms and Atomic Ions
[7] M. Agostini, S. Appel, G. Bellini *et al.* (Borexino Collaboration), Phys. Rev. D **92**, 031101 (2015); arXiv:1506.04610 [hep-ex]
[8] A. Gando, Y. Gando, H. Hanakago *et al.* (KamLAND Collaboration), Phys. Rev. D **88**, 033001 (2013); arXiv:1303.4667v2
[9] J. H. Davies and D. R. Davies, Solid Earth **1**, 5 (2010)
[10] J. Hansen, M. Sato, P. Kharecha and K. von Schuckmann, Atmos. Chem. Phys. **11**, 13421 (2011)
[11] Langseth, M. G., S. P. Clark Jr., J. L. Chute, S. J. Keihm and A. E. Wechsler, The Moon **4**, 390 (1972)
[12] V. D. Krotikov and V. S. Troitsky, Soviet J. Phys. Uspekhi **6**, 841 (1964)
[13] Paige, D. A., *et al.*, Space Sci. Rev. **150**, 125 (2010)
[14] Watson, K., J. Geophys. Rev. **72**, 331 (1967)
[15] V. V. Sinev, L. B. Bezrukov *et al.*, Phys. Particles and Nuclei **46**, 186 (2015)
[16] V. V. Sinev, L. B. Bezrukov, E. A. Litvinovich, I. M. Machulin, M. D. Skorokhvatov, S. V. Sukhotin, Phys. Part. Nucl. **46**, 186 (2015)

EPJ Web of Conferences **125**, 02005 (2016) DOI: 10.1051/epjconf/201612502005
QUARKS-2016

Higgs-radion phenomenology in stabilized RS models

Eduard Boos[1,⋆], *Viacheslav* Bunichev[1], *Sergey* Keizerov[1], *Maxim* Perfilov[1], *Eduard* Rakhmetov[1], *Mikhail* Smolyakov[1], *Kseniia* Svirina[1], and *Igor* Volobuev[1]

[1] *D.V. Skobeltsyn Institute of Nuclear Physics, M.V. Lomonosov Moscow State University*

Abstract. An important general prediction of stabilized brane world models is the existence of a bulk scalar radion field, whose lowest Kaluza-Klein (KK) mode is the scalar particle called the radion. This field comes from the fluctuations of the metric in the extra dimension and the radion mass can be smaller than that of all the massive KK modes of the other particles propagating in the multidimensional bulk. Due to its origin, the radion and its KK tower couple to the trace of the energy-momentum tensor of the Standard Model. These fields have the same quantum numbers as the neutral Higgs field and can mix with the latter, if they are coupled. We present a short review of some aspects of Higgs-radion phenomenology in stabilized brane-world models. In particular, we discuss the possibility of explaining the 750 GeV excess by the production of a radion-dominated state.

1 Introduction to the model

The discovery of the 125 GeV scalar boson [1], rather accurate measurements of its mass [2], still not very precise but compatible with the Standard Model (SM) expectations for the Higgs boson measurements of its properties (see [3]), and the recent observation by the ATLAS and CMS experiments of the 750 GeV diphoton excess [4] attracted much attention to the scalar states predicted by various extensions of the SM. One of such SM extensions that naturally gives rise to extra scalar fields is the Randall-Sundrum model with two branes stabilized by a bulk scalar field [5, 6], which is necessary for the model to be phenomenologically acceptable.

A stabilized brane world model in five-dimensional space-time $E = M_4 \times S^1/Z_2$ with coordinates $\{x^M\} \equiv \{x^\mu, y\}$, $M = 0, 1, 2, 3, 4$, $\mu = 0, 1, 2, 3$, the coordinate $x^4 \equiv y$, $-L \leq y \leq L$ parameterizing the fifth dimension, is defined by an action that is the sum of the action of five-dimensional gravity $g_{MN}(x, y)$ and the action of the stabilizing five-dimensional scalar field $\phi(x, y)$, which also includes potentials on the branes.

The space of extra dimension is the orbifold S^1/Z_2, which is realized as the circle of circumference $2L$ with the points y and $-y$ identified. Correspondingly, the metric g_{MN} and the scalar field ϕ satisfy the standard orbifold symmetry conditions. The branes are located at the fixed points of the orbifold, $y = 0$ and $y = L$, and it is assumed that the SM fields with Lagrangian L_{SM} live on the brane at $y = L$.

This action gives rise to a system of nonlinear differential equations for the metric and the scalar field, which has a background solutions with anti-de Sitter metric and the scalar field depending

⋆e-mail: boos@theory.sinp.msu.ru

only on the coordinate of extra dimension. The background (vacuum) solution for the Higgs field is standard, all the other SM fields being equal to zero.

Suppose we have a solution for the background metric and the scalar field for an appropriate choice of the parameters of the potentials such that the interbrane distance is stabilized and equal to L. It means that the vacuum energy of the scalar field has a minimum for this value of the interbrane distance. Now the linearized theory is obtained by taking the fluctuations $h_{MN}(x, y)$ of the gravitational field and the fluctuations $f(x, y)$ the scalar field above this background [6].

There are three scalar fields in the linearized theory: the fields $f(x, y)$, $h_{44}(x, y)$ and $\gamma^{\mu\nu}h_{\mu\nu}(x, y)$. However, these fields are not independent, because they are connected by the equations of motion of the linearized theory and a gauge condition [6, 7]. For this reason we can use any one of them to describe the scalar states. Usually, similar to the case of the unstabilized Randall-Sundrum model, the field $h_{44}(x, y)$ is taken and expanded in KK modes, the lowest mode being called the radion field and the higher modes belonging to its KK tower. Thus, one can say that the radion field is the collection of the spin-0 degrees of freedom, coming from the scalar fluctuations of the metric in extra dimension and of the stabilizing scalar field [6, 7]. In a number of studies it was argued that the radion can be significantly lighter than all the other massive KK excitations [8, 9].

The phenomenology of the radion follows from a simple interaction Lagrangian, which is the product of the trace of the SM energy-momentum tensor and the radion field $r(x)$, divided by a dimensional scale parameter:

$$L_r = -\frac{r(x)}{\Lambda_r} T^{\mu}_{\mu}, \tag{1}$$

where the trace includes not only the kinetic terms, but also the interaction terms of the SM fields [10], as well as the terms corresponding to the conformal anomaly of massless gluon and photon fields. Thus, the radion couples to the massless photon and gluon fields via the well-known conformal anomaly in the trace of the energy momentum tensor since the radion also plays the role of the dilaton. When the massive SM fields, as supposed in most of the studies, are taken on the mass-shell, the radion interaction with them, at the lowest order in the SM couplings, is the same as that of the Higgs boson with the replacement of the radion scale by the Higgs vacuum expectation value. Thus, the interaction vertices of the radion with the SM fields are similar to those of the SM Higgs boson except for the anomaly enhanced interactions with gluons and photons. This leads to the corresponding relative enhancement in gluon and photon decay modes and the gluon-gluon fusion production channel of the radion. Various aspects of the decay and production properties of the radion have been touched upon in a number of studies, including a possible mixing of the radion with the Higgs boson, which is important for the radion and Higgs phenomenology [7, 11, 12].

However, if the massive SM fields are not on the mass shell, there appear additional terms in the radion interaction Lagrangian, which contribute, in particular, to the processes with tree level diagrams involving the radion and off-shell fermion lines. The extra terms in the Lagrangian lead to a momentum dependence in the vertices potentially giving additional contributions if virtual fermions participate in the process. For the main radion decay processes the fermions are on-shell and these additional terms in the vertices vanish, but this might not be the case for the radion production.

In the present short review we discuss various aspects of the Higgs-radion phenomenology. First we consider the phenomenology of the single radion production for the cases of on- and off-shell fermions, as well as the associated Higgs boson-radion production. Next we discuss the Higgs-radion mixing due to the merging of the mechanisms of spontaneous symmetry breaking on the brane and of the gravitational bulk stabilization and study the phenomenology of the Higgs-dominated and radion-dominated states arising due to this mixing. Based on these results, we discuss the possibility of explaining the 750 GeV excess by the production of a radion-dominated state.

2 Single radion and associated Higgs boson-radion production

In all the main radion production processes, there are contributing Feynman diagrams involving off-shell fermions. The radion can be emitted from various fermion currents containing off-shell fermion propagators in various production processes in e^+e^- collisions, as well as in production processes in hadron collisions at the LHC. The off-shell fermions run in the fermion loops of the diagrams of the gluon-gluon fusion production process at the LHC and the gg, $\gamma\gamma$, γZ decay modes of the radion. In all these processes there are additional to the Higgs boson case non-trivial vertices (even for massless fermions) of fermion-radion interactions that follow from the structure of the trace of the gauge invariant SM energy-momentum tensor. One can show that all the contributions to the perturbative amplitudes of physical processes with a single radion are cancelled out due to the corresponding additional terms for both massless and massive off-shell fermions.

For a simple example, the radion production in association with Z boson in e^+e^- collisions, one can show that the matrix element squared is exactly equal to that for the Higgs boson with the replacement $\Lambda_r \to v$ and $m_r \to m_h$. Thus, all the additional contributions are cancelled out. The same property of the cancellation of additional to the Higgs-like contributions takes place for the associated radion and $W^\pm$ boson production, for example, $u\bar{d} \to rW^+$.

The property of the cancellation of all the contributions additional to the Higgs-like part also takes place in the case, where an arbitrary number, say N, of gauge bosons is emitted from the line of a massive fermion in association with the radion as shown in Fig. 1. One can show [10] that the sum of

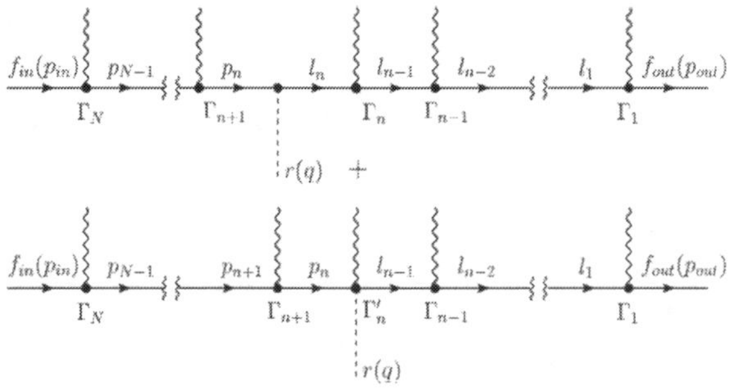

Figure 1. Fermion current radiating the radion and N SM vector gauge bosons in a three-point vertex (upper graph) and in a four-point vertex (lower graph).

all the contributions to the amplitude leads to only the Higgs-like type of the contribution and all the other parts are cancelled out explicitly.

This result can be easily generalized to the loop case. One can find more details in [10].

It turns out that the Higgs boson-radion similarity mentioned above takes place also in the processes of the associated Higgs boson-radion production, although in this case it is more complicated. Namely, it is shown that the perturbative amplitudes of the associated Higgs boson-radion production that do not include the anomalous vertices can be transformed into the corresponding amplitudes of the Higgs boson pair production by a replacement of the radion mass by the Higgs mass and the radion coupling constant by the Higgs field vacuum expectation value, as well as a rescaling of the triple Higgs vertex coupling [13].

In other words, the explicit amplitude of the associated Higgs boson-radion production process, which contains additional non-standard Feynman graphs and has a complicated structure of the vertices, can be obtained from the well studied result for the pair Higgs boson production by a simple rescaling of the parameters. This property follows from the structure of any massive fermion current emitting the radion and an arbitrary number of any SM gauge bosons. One can find in [13] simple examples of the associated Higgs boson-radion production in fermion-antifermion annihilation and in gluon fusion with only two bosons (*rh* or *hh*) in the final state demonstrating the explicit cancellation. One can show that a similar general property takes place in the case of the associated Higgs boson-radion production by considering a fermion current (or a fermion loop) with the emission of an arbitrary number of SM bosons with all possible permutations. A rigorous proof of this property can be found in paper [13].

3 Higgs-radion mixing and the phenomenology of Higgs-dominated and radion-dominated states

As we have already mentioned, the radion has the same quantum numbers as the neutral Higgs boson. The same is valid for the radion field and its excitations. Thus, these fields can mix with the Higgs field, if they are coupled.

A Higgs-radion coupling was first considered in the unstabilized Randall-Sundrum model by introducing a Higgs-curvature term on the brane that was done in [11]. Then such a coupling and the resulting Higgs-radion mixing in the case of the stabilized model were discussed in paper [7] without taking into account the KK tower of higher scalar excitations. The phenomenology of the Higgs-radion mixing originating from the Higgs-curvature term was also considered in view of the discovery of the Higgs-like boson at the LHC [1], various assumptions about the masses and the mixings of the scalar states having been analyzed in numerous papers. In particular, it was shown that the light radion-dominated state with mass below or above the observed 125 GeV boson is still not completely excluded by all the EW precision constrains and the LHC data.

Below we consider a model, where a Higgs-radion coupling naturally arises due to a mechanism of spontaneous symmetry breaking on the brane involving the stabilizing scalar field. This approach takes into account the influence of the KK tower of higher scalar excitations on the parameters of the Higgs-radion mixing, which turns out to be of importance. It also has the advantage that it modifies only the scalar sector of the model and leaves intact the masses and the coupling constants of the graviton KK excitations.

In brane-world models, the vacuum solution for gravity and the stabilizing scalar field and the vacuum solution for the SM fields are independent. If we consider a brane world model to be an indivisible theory, it is reasonable to believe that there should be a common interconnected vacuum solution for all these fields. To this end we modify the action of the model so that it contains a quartic interaction of the stabilizing scalar field and of the Higgs field, which would connect the stabilization of the extra dimension size and the SM spontaneous symmetry breaking. This can be achieved by replacing the SM Higgs potential by the interaction Lagrangian

$$L_{int}(\phi, H) = -\lambda\left(|H|^2 - \frac{\xi}{M}\phi^2\right)^2, \tag{2}$$

M denoting the fundamental energy scale and ξ being a positive dimensionless parameter. The equations for the background configuration following from such a modified action give a relation between the vacuum value of the Higgs field and the value of the field ϕ on the brane at $y = L$

$$\phi^2(L) = \frac{Mv^2}{2\xi}. \tag{3}$$

This means that in such a scenario the Higgs field vacuum expectation value, being proportional to the value of the stabilizing scalar field on the TeV brane, arises dynamically as a result of the gravitational bulk stabilization.

In this scenario, the linearized theory is obtained by expanding the metric and the scalar field in the standard way and the Higgs field in the unitary gauge about the background solution as

$$H(x) = \begin{pmatrix} 0 \\ \frac{v+\sigma(x)}{\sqrt{2}} \end{pmatrix}. \tag{4}$$

After substituting this representation into the action of the model and keeping the terms of the second order in h_{MN}, f and σ one gets the Lagrangian of this action which is the standard free Lagrangian of the SM (i.e. the masses of all the SM fields are expressed in the same way as usually in terms of the vacuum value of the Higgs field and the coupling constants) together with the standard second variation Lagrangian of the stabilized RS model [6] supplemented by an interaction term of the scalar fields f and σ on the brane coming from interaction Lagrangian (2) [14].

Besides the fields f and σ, there two more scalar fields in the linearized theory, – the fields $h_{44}(x, y)$ and $\gamma^{\mu\nu}h_{\mu\nu}(x, y)$. As we have already mentioned in the Introduction, the fields $f(x, y)$, $h_{44}(x, y)$ and $\gamma^{\mu\nu}h_{\mu\nu}(x, y)$ are not independent and are connected by the equations of motion of the linearized theory and a gauge condition [6, 7].

The field $h_{44}(x, y)$ can be expanded in KK modes, which induces the corresponding expansion of the bulk scalar field $f(x, y)$. Substituting the latter expansion into the second variation Lagrangian and integrating over the extra dimension coordinate, one gets an effective four-dimensional Lagrangian. In case the Higgs and the radion masses are much smaller than the masses of the radion excitations one can pass to a low energy approximation in the four-dimensional Lagrangian by integrating out the radion excitation fields. This gives an effective Lagrangian for the interaction of the Higgs field $\sigma(x)$ and the radion field, which we will denote here $\phi_1(x)$, with the SM fields. However, due to the Higgs-radion mixing terms the fields $\sigma(x)$ and $\phi_1(x)$ are not mass eigenstates.

The physical mass eigenstate fields $h(x)$, $r(x)$ are, as usually, obtained by a rotation diagonalizing the mass matrix

$$\begin{aligned} h(x) &= \cos\theta\,\sigma(x) + \sin\theta\,\phi_1(x) \\ r(x) &= -\sin\theta\,\sigma(x) + \cos\theta\,\phi_1(x). \end{aligned} \tag{5}$$

The field $h(x)$ is called the Higgs-dominated field and the field $r(x)$ is called the radion-dominated field. Finally, one gets the effective interaction Lagrangian of the physical scalar fields $h(x)$ and $r(x)$ with the Standard Model fields in the following form [14]:

$$\begin{aligned} L_{h-r} &= \frac{1}{2}\partial_\mu h(x)\partial^\mu h(x) - \frac{1}{2}m_h^2 h^2(x) + \frac{1}{2}\partial_\mu r(x)\partial^\mu r(x) - \frac{1}{2}\mu_r^2 r^2(x) \\ &- \frac{(c\cos\theta + \sin\theta)}{\Lambda_r}h(x)(T_\mu^\mu + \Delta T_\mu^\mu) + \frac{(c\sin\theta - \cos\theta)}{\Lambda_r}r(x)(T_\mu^\mu + \Delta T_\mu^\mu) - \\ &- \sum_f \frac{m_f}{v}\bar\psi_f\psi_f(\cos\theta\, h(x) - \sin\theta\, r(x)) + \frac{2M_W^2}{v}(W_\mu^- W^{\mu+})(\cos\theta\, h(x) - \sin\theta\, r(x)) + \\ &+ \frac{M_Z^2}{v}(Z_\mu Z^\mu)(\cos\theta\, h(x) - \sin\theta\, r(x)) + \frac{M_W^2}{v^2}(W_\mu^- W^{\mu+})(\cos\theta\, h(x) - \sin\theta\, r(x))^2 + \\ &+ \frac{M_Z^2}{2v^2}(Z_\mu Z^\mu)(\cos\theta\, h(x) - \sin\theta\, r(x))^2. \end{aligned} \tag{6}$$

The effective four-dimensional interaction Lagrangian (6) expressed in terms of the physical Higgs-dominated $h(x)$ and radion-dominated $r(x)$ fields contains their interactions with the SM fields and involves only five parameters in addition to those of the SM: the masses of the Higgs-dominated and radion-dominated fields m_h and m_r, the mixing angle θ, the (inverse) coupling constant of the radion to the trace of the energy-momentum tensor of the SM fields Λ_r and the parameter c that accommodates the contributions of the integrated out heavy scalar modes and is constrained by the other parameters as follows: $0 < c < c_{max} = \frac{(m_r^2 - m_h^2)\sin 2\theta}{m_r^2\cos^2\theta + m_h^2\sin^2\theta}$.

There are two possible scenarios, where the observed 125 GeV boson is either a Higgs-dominated state or a radion-dominated state. A detailed statistical analysis carried out in paper [14] shows that the Higgs-dominated state is the strongly preferred scenario.

4 A radion-dominated state as an explanation of the 750 GeV excess

The statistical analysis admits the existence of a heavy radion-dominated state with mass 750 GeV. In order to understand whether the 750 GeV observed excess can be interpreted as a radion-dominated state let us consider the main decay and production properties of such a state that follow from effective Lagrangian.

Practically for all the values of the mixing parameter $\sin\theta$ the branching ratios are distributed between the modes close to those for the SM Higgs boson, if it had had mass 750 GeV. The NLO corrections are included following the HDECAY code [15]. The dominating decay modes are the decays to heavy SM particles W^+W^-, ZZ boson pairs and the top quark pair. However, for some rather small values of the parameter $\sin\theta$ close to approximately v/Λ_r all the branching ratios are significantly decreased, and the dominating decay mode becomes the mode to two gluons. Also in this region of the parameter space the branching to two photons is significantly increased. Such a property could be easily understood from the structure of the interaction vertices of the radion-dominated state and the SM fermions and gauge bosons (see the Feynman rules in [14]). Indeed all the vertices for the fermions and massive gauge bosons contain the factor $\frac{\cos\theta - c\cdot\sin\theta}{\Lambda_r} - \frac{\sin\theta}{v}$, which becomes very small for $\sin\theta$ close to v/Λ_r. This occurs due to the cancellation of the contributions to the vertices coming from the SM type part of the interactions and the part coming from the trace of the energy-momentum tensor. In contrast, the interaction vertices of the radion-dominated state and the massless gluons and photons have anomaly enhanced contributions and the mentioned cancellation does not take place for small values of the parameter $\sin\theta$ close to v/Λ_r. The corresponding cancellation occurs for the gluon-gluon and photon-photon vertices at much larger values of the parameter $\sin\theta$.

A crucial point for the possible interpretation is the production cross section, which should be in the range from a few to 10 fb. The production cross section for the radion-dominated state has been calculated including the contributions of all the production channels for the radion-dominated state (ggF, VBF, rV, rtt) with the decay to two photons. We have also included the NNLO K-factors taken from the Higgs cross section working group web page [16, 17]. It turned out that, as expected, the gluon-gluon fusion dominates the production cross section in the most interesting region of the parameter space, where the cross section has a maximum. The maximum occurs for the same values of parameters, for which the corresponding gg and $\gamma\gamma$ decay branching ratios have the maximum. As for the branching ratios in the range close to the maximum, the cross section depends very weakly on the the parameter c.

One should note that the parameter region allowed by the signal strength measurements for 125 GeV boson at 8 TeV and 13 TeV LHC energy and presented in Fig. 2 includes the above mentioned region, where the two-photon cross section has a maximum.

However, the interpretation of the observed excess as the radion-dominated state is very problematic or even impossible in the simplest variant of the discussed brane-world models, where only the gravitational degrees of freedom are allowed to propagate in the bulk. Indeed, on can show that the

DOI: 10.1051/epjconf/201612502005

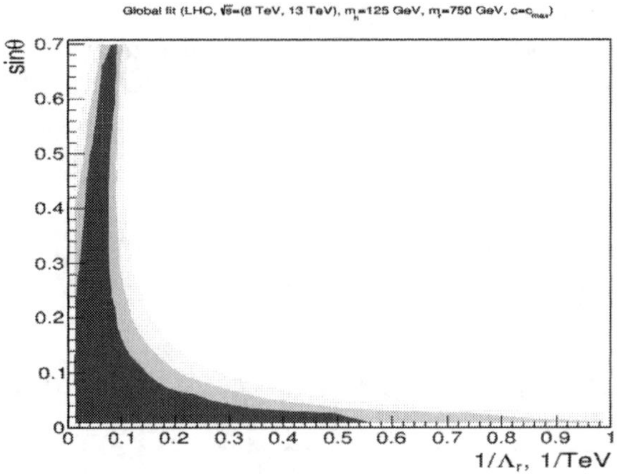

Figure 2. Exclusion contours for the global χ^2 fit in the $(\sin\theta, 1/\Lambda_r)$ plane for the LHC at $\sqrt{s}$ =8 and 13 TeV and m_h=125 GeV, m_r=750 GeV, $c = c_{max}$. The dark, medium and light shaded areas correspond to CL of the fit 65%, 90% and 99% respectively.

cross section has a maximum of about 0.14 fb, which is by a factor of $50 \div 100$ smaller than what is needed to achieve the observed level of the cross section for the 750 GeV excess. One can easily check that in the other areas of the parameter space the production cross section gets even smaller.

Thus, one has to conclude that in the simplest variant of the model under consideration, where only the gravitational and stabilizing scalar fields propagate in the bulk, the observed excess cannot be understood as a radion-dominated resonance. More details are given in [18].

In order to increase the two-photon signal rate some other heavy particles should propagate in the gluon-gluon-scalar and photon-photon-scalar loop vertices, the scalar in our case being the Higgs-dominated or the radion-dominated states. Such heavy particles in brane-world models could be the excited KK states of those SM fields, which are also allowed to propagate in the multidimensional bulk. However, the calculation of the contributions of the excited states in stabilized brane-world models is not straightforward: the energy-momentum tensor of an excited state of the bulk fields has a component corresponding to the extra dimension, to which the radion couples two times stronger than to the trace of the four-dimensional components [12, 19] thus modifying the overall coupling to the excited state. Moreover, the radion couplings to the energy-momentum tensors of the excited states depend on the radion wave function in the extra dimension as well as on the wave functions of the excited states, which are model dependent. The same problem with the wave functions arises, if one considers the coupling of the bulk Higgs field to the excited states. Thus, taking into account the contributions of the excited KK states of the bulk fields is a complicated problem that needs a special thorough investigation.

5 Conclusion

In this short review we have presented the results of a systematic study of the radion in stabilized brane-world models. We have shown that, despite the differences in the Higgs boson and radion couplings to off-shell fermions, there is a Higgs boson-radion similarity in the processes involving off-shell fermions and an arbitrary number of gauge bosons that is present at the tree and loop levels.

We have also discussed a new mechanism of Higgs-radion mixing immanent in stabilized brane-world models and based on unifying the spontaneous symmetry breaking on the brane and the gravitational stabilization of the extra dimension size. In the framework of this approach, we have derived the corresponding effective Lagrangian of the interaction of the Higgs-dominated and radion-dominated states that includes a cumulative effect of higher excited scalar states. For special values of the model parameters, it reproduces a number of SM extensions by a singlet scalar boson.

We have shown that the interpretation of the 125 GeV scalar state as the Higgs-dominated state is strongly preferred. However, a rather wide range of the parameter space for the radion-dominated state is allowed by the present data, the allowed regions somewhat growing with the growth of the radion (inverse) coupling constant Λ_r.

We have also studied the possibility of interpreting the 750 GeV excess as a radion-dominated state in stabilized brane-world models and found out that the radion-dominated state mass of 750 GeV is allowed by the 125 Higgs signal strength measurements. But the production cross section of this state is too small to explain the LHC 750 GeV excess.

Acknowledgements

The work was supported by grant 14-12-00363 of the Russian Science Foundation.

References

[1] G. Aad *et al.* [ATLAS Collaboration], Phys. Lett. B **716**, 1 (2012) ,
S. Chatrchyan *et al.* [CMS Collaboration], Phys. Lett. B **716**, 30 (2012)
[2] G. Aad *et al.* [ATLAS and CMS Collaborations], Phys. Rev. Lett. **114**, 191803 (2015)
[3] G. Aad *et al.* [ATLAS and CMS Collaborations], arXiv:1606.02266 [hep-ex]
[4] G. Aad *et al.* [ATLAS Collaboration], *Search for high-mass diphoton resonances in pp collisions at* $\sqrt{s} = 13TeV$ *with the ATLAS detector*, ATLAS-CONF-2015-081,
S. Chatrchyan *et al.* [CMS Collaboration], *Search for High-Mass Diphoton Resonances in pp Collisions at* $\sqrt{s} = 13TeV$ *with the CMS Detector*, CMS PAS EXO-15-004
[5] W. D. Goldberger and M. B. Wise, Phys. Rev. Lett. **83**, 4922 (1999)
[6] E. E. Boos, Y. S. Mikhailov, M. N. Smolyakov and I. P. Volobuev, Mod. Phys. Lett. A **21**, 1431 (2006)
[7] C. Csaki, M. L. Graesser and G. D. Kribs, Phys. Rev. D **63**, 065002 (2001)
[8] C. Csaki, M. Graesser, L. Randall and J. Terning, Phys. Rev. D **62**, 045015 (2000)
[9] C. Charmousis, R. Gregory and V. A. Rubakov, Phys. Rev. D **62**, 067505 (2000)
[10] E. Boos, S. Keizerov, E. Rakhmetov, K. Svirina, Phys. Rev. D **90**, 095026 (2014)
[11] G. F. Giudice, R. Rattazzi and J. D. Wells, Nucl. Phys. B **595**, 250 (2001)
[12] C. Csaki, J. Hubisz and S. J. Lee, Phys. Rev. D **76**, 125015 (2007)
[13] E. Boos, S. Keizerov, E. Rakhmetov and K. Svirina, arXiv:1604.00998 [hep-ph]
[14] E. E. Boos, V. E. Bunichev, M. A. Perfilov, M. N. Smolyakov and I. P. Volobuev, Phys. Rev. D **92**, 095010 (2015)
[15] A. Djouadi, J. Kalinowski and M. Spira, Comput.Phys.Commun, **108**, 56 (1998)
[16] S. Dittmaier *et al.* [LHC Higgs Cross Section Working Group Collaboration], arXiv:1101.0593
[17] S. Heinemeyer *et al.* [LHC Higgs Cross Section Working Group Collaboration], arXiv:1307.1347
[18] E. E. Boos, V. E. Bunichev and I. P. Volobuev, arXiv:1603.04495 [hep-ph].
[19] T. G. Rizzo, JHEP **0206**, 056 (2002)

Finite-Temperature Higgs Potentials

Dolgopolov M.V.[1,*], Gurskaya A.V.[1,**], and Rykova E.N.[1,***]

[1]*Samara University, Academician Pavlov 1, 443011 Samara, Russia.*

Abstract. In the present article we consider the short description of the "Finite-Temperature Higgs Potentials" program for calculating loop integrals at vanishing external momenta and applications for extended Higgs potentials reconstructions. Here we collect the analytic forms of the relevant loop integrals for our work in reconstruction of the effective Higgs potential parameters in extended models (MSSM, NMSSM and etc.).

1 Algebraic-Derivative recursion method

The smallness of external state masses relative to the masses of loop particles allows us to take the limit where the external particles are massless. These integrals are particularly easy to solve using algebraic recursion relations or residue method (also remember [1]). But it is necessary to keep the external momentum p_μ general (i. e. nonzero) until it can be converted into an external mass in the amplitude, after which point one may take accurate limit $p \to 0$, especially in the cases with equivalent internal masses in loop.

In d dimensions such two-point loop integral B_0 is defined as:

$$\frac{1}{(4\pi)^2} B_0(m^2, M^2) \;=\; \mu^{4-d} \int \frac{d^d k}{(2\pi)^d} \frac{i}{(k^2 - m^2)\,[k^2 - M^2]}. \tag{1}$$

In order to derive asymptotic results we expand $B_0(p^2, m_1^2, m_2^2)$ in powers of the external momentum p^2:

$$B_0(p^2, m_1^2, m_2^2) = -\Delta_\mu - \frac{m_1^2 \log m_1^2 - m_2^2 \log m_2^2}{m_2^2 - m_1^2} + O\left(\frac{p^2}{m_1^2 + m_2^2}\right).$$

The 3- and 4-point loop integrals at vanishing external momenta are defined as:

$$\frac{1}{(4\pi)^2} C_{2n}(m_1^2, m_2^2, m_3^2) \;=\; \mu^{4-d} \int \frac{d^d k}{(2\pi)^d} \frac{i k^{2n}}{\prod_i^3 (k^2 - m_i^2)}, \tag{2}$$

$$\frac{1}{(4\pi)^2} D_{2n}(m_1^2, m_2^2, m_3^2, m_4^2) \;=\; \mu^{4-d} \int \frac{d^d k}{(2\pi)^d} \frac{i k^{2n}}{\prod_i^4 (k^2 - m_i^2)}. \tag{3}$$

We follow the 'Buras' convention of Appendix A.5 in [2], and [3] with some extent, corrections and prove. Note that the above equations are manifestly equivalent under interchange of the arguments

*Dolgopolov Mikhail Vyacheslavovich, e-mail: mvdolg@yandex.ru
**Gurskaya Albina Valentinovna, e-mail: a-gurska@yandex.ru
***Rykova Elza Nurovna, emailelzarykova@rambler.ru

DOI: 10.1051/epjconf/201612502006

and explicit calculations show that one may reshuffle the order of the arguments without changing the result (for functions with unequal arguments it is obvious, and for functions with the equivalent arguments it demands accuracy!).

The two-point loop integral in tensor reduction B_1 can be defined as:

$$\frac{1}{(4\pi)^2} p_\mu B_1(p, m^2, M^2) = \mu^{4-d} \int \frac{d^d k}{(2\pi)^d} \frac{ik_\mu}{(k^2 - m^2)\left[(k+p)^2 - M^2\right]}. \tag{4}$$

Such integrals appear in the self-energy contributions.

It turns out that this is very simple to evaluate integrals (1), (2), (3) with the following algebraic relations:

$$\frac{1}{(k^2 - m_1^2)(k^2 - m_2^2)} = \frac{1}{m_1^2 - m_2^2}\left(\frac{1}{k^2 - m_1^2} - \frac{1}{k^2 - m_2^2}\right), \tag{5}$$

$$\frac{k^2}{k^2 - m^2} = 1 + \frac{m^2}{k^2 - m^2}. \tag{6}$$

Note that denominators of integrals under consideration are spherically symmetric. And for the case of all equivalent masses the particularly well-known general integral can be performed using Γ function:

$$\int \frac{d^n k}{(2\pi)^n} \frac{(k^2)^b}{(k^2 - A^2)^a} = \frac{i}{(4\pi)^{n/2}} (-1)^{a+b} (A^2)^{b-a+n/2} \frac{\Gamma(b + n/2)\Gamma(-b + a - n/2)}{\Gamma(n/2)\Gamma(a)}, \tag{7}$$

where the left-hand side is an integral in n-dimensional Minkowski space. In the private case at $b = 0$

$$\int \frac{d^n k}{(2\pi)^n} \frac{1}{(k^2 - A^2)^a} = \frac{i}{(4\pi)^{n/2}} (-1)^a (A^2)^{-a+n/2} \frac{\Gamma(a - n/2)}{\Gamma(a)}. \tag{8}$$

1.1 Results for one-loop scalar integrals

The explicit formulae at $d \to 4$ are listed below (we give also expressions for some 3-point functions proportional to higher momenta powers, may be useful in contributions for effective Higgs potential calculations [4–6]) and may be represented only in terms of B_0 integral, or A_0 (8):

$$\frac{1}{(4\pi)^2}A_0(m^2) = \mu^{4-d}\int\frac{d^dk}{(2\pi)^d}\frac{i}{k^2-m^2} = i\frac{\mu^{4-d}i}{(4\pi)^{d/2}}(-1)(m^2)^{-1+d/2}\Gamma(1-d/2) = \tag{9}$$

$$= \frac{-ii}{16\pi^2}\left(1+\frac{\varepsilon}{2}\log(4\pi\mu^2)+O(\varepsilon^2)\right)m^2\left(1-\frac{\varepsilon}{2}\log(m^2)+O(\varepsilon^2)\right)\times$$

$$\times (-1)\left(\frac{2}{\varepsilon}+1-\gamma_E+O(\varepsilon)\right) \Rightarrow$$

$$\Rightarrow A_0(x) = x(-\Delta_\mu+\log x), \tag{10}$$

$$B_0(x,y) = \frac{1}{x-y}\left[A_0(x)-A_0(y)\right] = \frac{-1}{x-y}\left[x\Delta_\mu - y\Delta_\mu - x\log x + y\log y\right] =$$

$$= -\Delta_\mu - \frac{-x\log x + y\log x - y\log x + y\log y}{x-y} =$$

$$= -\Delta + \log\frac{x}{\mu^2} - \frac{y}{x-y}\log\frac{y}{x} = \tag{11}$$

$$= B_0(y,x) = -\Delta + \log\frac{y}{\mu^2} - \frac{x}{y-x}\log\frac{x}{y}, \tag{12}$$

$$C_0(x,y,z) = \frac{1}{y-z}\left[B_0(x,y)-B_0(x,z)\right] = \frac{y\log\frac{y}{x}}{(x-y)(z-y)} + \frac{z\log\frac{z}{x}}{(x-z)(y-z)}, \tag{13}$$

$$C_0(x,y,y) = \frac{\partial}{\partial y}B_0(x,y) = \frac{1}{y-x}\left[1+\frac{x\log\frac{x}{y}}{y-x}\right], \tag{14}$$

$$C_2(x,y,z) = B_0(x,y) + zC_0(x,y,z) =$$

$$= -\Delta - \log\frac{\mu^2}{x} + \frac{y^2\log\frac{y}{x}}{(x-y)(z-y)} + \frac{z^2\log\frac{z}{x}}{(x-z)(y-z)}, \tag{15}$$

$$C_2(x,y,y) = B_0(y,x) + yC_0(x,y,y) =$$

$$= -\Delta_\mu + \log y - \frac{x}{y-x}\log\frac{x}{y} + \frac{y}{y-x}\left[1+\frac{x\log\frac{x}{y}}{y-x}\right] =$$

$$= -\Delta - \log\frac{\mu^2}{y} - \frac{x}{x-y}\left[\frac{y}{x} - \frac{x\log\frac{x}{y}}{x-y}\right] = \tag{16}$$

$$= B_0(x,y) + yC_0(x,y,y) =$$

$$= -\Delta - \log\frac{\mu^2}{x} + \frac{y}{y-x}\left[1-\frac{(y-2x)\log\frac{x}{y}}{y-x}\right], \tag{17}$$

$$= \frac{\partial}{\partial y}A_0(y) + xC_0(x,y,y) =$$

$$= -\Delta - \log\frac{\mu^2}{y} - \frac{x}{x-y}\left[\frac{y}{x} - \frac{x\log\frac{x}{y}}{x-y}\right]. \tag{18}$$

The answers (16), (18), and (17) are identical indeed, check by Mathematica is done. Check for last equations:

$$\begin{aligned}\frac{1}{(4\pi)^2}\frac{\partial}{\partial m^2}A_0(m^2) &= \mu^{4-d}\int\frac{d^dk}{(2\pi)^d}\frac{i}{(k^2-m^2)^2} = i\,\frac{\mu^{4-d}i}{(4\pi)^{d/2}}(-1)^2(m^2)^{-2+d/2}\Gamma(2-d/2) = (19)\\ &= \frac{i\,i}{16\pi^2}\left(1+\frac{\varepsilon}{2}\log(4\pi\mu^2)+O(\varepsilon^2)\right)\left(1-\frac{\varepsilon}{2}\log(m^2)+O(\varepsilon^2)\right)\times\\ &\times\left(\frac{2}{\varepsilon}-\gamma_E+O(\varepsilon)\right) = (-\Delta_\mu+1)+\log x\\ yes\quad &\frac{\partial}{\partial x}A_0(x) = -\Delta_\mu+\log x+1. \qquad (20)\end{aligned}$$

Equal arguments correspond to equivalent masses. The divergent part is

$$\Delta_{\{\mu;\,1\}} = \frac{2}{4-d}+\log(4\pi\{\mu^2;\,1\})-\gamma_{\mathrm{E}}+1$$

where μ is the renormalization scale.

And

$$C_{11}(x,y) = -\frac{x-3y}{4(x-y)^2}+\frac{y^2}{2(x-y)^3}\log\frac{y}{x}, \qquad (21)$$

$$C_{12}(x,y) = -\frac{x+y}{2(x-y)^2}-\frac{xy}{3(x-y)^3}\log\frac{y}{x}. \qquad (22)$$

The explicit formulae for the four-point integrals at vanishing external momenta are (possible cross-checks are provided):

$$\begin{aligned}D_0(x,y,z,t) = &\frac{1}{z-t}\left(C_0(x,y,z)-C_0(x,y,t)\right) =\\ = &\frac{y\log\frac{y}{x}}{(y-x)(y-z)(y-t)}+\frac{z\log\frac{z}{x}}{(z-x)(z-y)(z-t)}+\frac{t\log\frac{t}{x}}{(t-x)(t-y)(t-z)}, \qquad (23)\end{aligned}$$

$$\begin{aligned}D_0(x,y,z,z) &= \frac{\partial}{\partial z}C_0(z,y,x) = \frac{\partial}{\partial z}\left(\frac{y\log\frac{y}{z}}{(z-y)(x-y)}+\frac{x\log\frac{x}{z}}{(z-x)(y-x)}\right) =\\ &= \frac{1}{(x-z)(y-z)}+\frac{x\log\frac{x}{z}}{(x-y)(x-z)^2}+\frac{y\log\frac{y}{z}}{(y-x)(y-z)^2}, \qquad (24)\end{aligned}$$

$$\begin{aligned}D_2(x,y,z,t) &= C_0(x,y,z)+tD_0(x,y,z,t) =\\ &= \frac{y^2\log\frac{y}{x}}{(y-x)(y-z)(y-t)}+\frac{z^2\log\frac{z}{x}}{(z-x)(z-y)(z-t)}+\frac{t^2\log\frac{t}{x}}{(t-x)(t-y)(t-z)}, \qquad (25)\end{aligned}$$

 DOI: 10.1051/epjconf/201612502006

$$D_2(0,x,y,z) = \frac{1}{y-x}(B_0(z,y)-B_0(z,x)) = -\frac{1}{y-x}\left(\frac{x\log\frac{x}{z}}{x-z}-\frac{y\log\frac{y}{z}}{y-z}\right), \tag{26}$$

$$D_2(0,x,x,y) = C_0(y,x,x) = -\frac{1}{y-x}\left(1-\frac{y\log\frac{y}{x}}{y-x}\right), \tag{27}$$

$$D_2(x,y,z,z) = \frac{\partial}{\partial z}C_2(z,y,x) = \frac{z}{(x-z)(y-z)}+\frac{x^2\log\frac{x}{z}}{(x-y)(x-z)^2}+\frac{y^2\log\frac{y}{z}}{(y-x)(y-z)^2}. \tag{28}$$

A null argument represents the negligible mass of the virtual particle.

The case with two pairs of equivalent masses in D_0 can be reduced with the following algebraic relation:

$$\frac{1}{(k^2-A^2)^2(k^2-B^2)^2} = \frac{1}{(A^2-B^2)^2}\left(\frac{1}{(k^2-A^2)^2}+\frac{1}{(k^2-B^2)^2}\right)-\frac{2}{(A^2-B^2)^3}\left(\frac{1}{k^2-A^2}-\frac{1}{k^2-B^2}\right), \tag{29}$$

or by derivatives:

$$D_0(x,x,y,y) = \frac{\partial^2}{\partial x\,\partial y}B_0(x,y) = \frac{\partial}{\partial x}C_0(x,y,y) = \frac{2}{(x-y)^2}+\frac{x+y}{(y-x)^3}\log\frac{x}{y}. \tag{30}$$

The explicit formula for the 2-point loop integral B_1 at vanishing external momentum is:

$$B_1(0,x,y) = \frac{1}{2}B_0(0,x,y)-\frac{1}{4}\frac{x+y}{x-y}+\frac{1}{2}\frac{xy}{(x-y)^2}\log\frac{x}{y} = \frac{1}{4}+\frac{1}{2}C_2(x,y,y), \tag{31}$$

where $C_2(x,y,y)$ is given in eq. (16) or (17) or equivalently (18).

And some additional private checks:

$$B_0(m^2) = \frac{\partial}{\partial m^2}A_0(m^2) = -\Delta_\mu + \log m^2 + 1, \tag{32}$$

$$C_0(m^2) = \frac{1}{2}\frac{\partial}{\partial m^2}B_0(m^2) = \frac{1}{2m^2}, \tag{33}$$

$$D_0(m^2) = \frac{1}{3}\frac{\partial}{\partial m^2}C_0(m^2) = -\frac{1}{6m^4}. \tag{34}$$

2 Finite temperature one-loop integrals

In the finite temperature field theory Feynman diagrams with boson propagators, containing Matsubara frequencies $\omega_n = 2\pi nT$ ($n = 0, \pm1, \pm2, ...$), lead to structures of the form

$$I[m_1, m_2, ..., m_b] = T\sum_{n=-\infty}^{\infty}\int\frac{d\mathbf{k}}{(2\pi)^3}\prod_{i=1}^{b}\frac{(-1)^b}{(\mathbf{k}^2+\omega_n^2+m_j^2)}, \tag{35}$$

Here $\mathbf{k}$ is the three-dimensional momentum in a system with the temperature T. In the following calculations first we perform integration with respect to $\mathbf{k}$ and then take the sum, using the reduction to three-dimensional theory in the high-temperature limit for zero frequencies. At $n \neq 0$ the result is [7, 8]

$$I[m_1, m_2, ..., m_b] = 2T\,(2\pi T)^{3-2b}\,\frac{(-1)^b \pi^{3/2}}{(2\pi)^3}\,\frac{\Gamma(b-3/2)}{\Gamma(b)}\,S(M, b-3/2), \tag{36}$$

where

$$S(M, b-3/2) = \int \{dx\} \sum_{n=1}^{\infty} \frac{1}{(n^2 + M^2)^{b-3/2}}, \qquad M^2 \equiv \left(\frac{m}{2\pi T}\right)^2. \tag{37}$$

For $b > 1$ the parameter m^2 is a linear function dependent on m_i^2 and the variables $\{dx\}$ of Feynman parametrization, which are the integration variables in (37). At the integer values of b the integrand in (37) is a generalized Hurwitz zeta-function [9, 10]. Note that for the leading threshold corrections to effective parameters of the two-doublet potential $b > 2$, so the wave-function renormalization appears in connection with the divergence at $b = 2$ (which is suppressed by vertex factors, see [11, 12]).

A number of integrals can be easily calculated. We calculate the integral

$$J_0[a_1, a_2] = \int \frac{d\mathbf{k}}{(2\pi)^3} \frac{1}{(\mathbf{k}^2 + a_1^2)(\mathbf{k}^2 + a_2^2)} = \frac{1}{4\pi(a_1 + a_2)}, \tag{38}$$

taking a residue in the spherical coordinate system. Here $a_{1;2}^2$ are the sums of squared frequency and squared mass, see (35). Derivatives of J_0 with respect to a_1 and a_2 can be used for calculation of integrals

$$J_1[a_1, a_2] = \int \frac{d\mathbf{k}}{(2\pi)^3} \frac{1}{(\mathbf{k}^2 + a_1^2)^2(\mathbf{k}^2 + a_2^2)} = -\frac{1}{2a_1}\frac{\partial J_0}{\partial a_1} = \frac{1}{8\pi a_1 (a_1 + a_2)^2}, \tag{39}$$

$$J_2[a_1, a_2] = \int \frac{d\mathbf{k}}{(2\pi)^3} \frac{1}{(\mathbf{k}^2 + a_1^2)^2(\mathbf{k}^2 + a_2^2)^2} = \frac{1}{4a_1 a_2}\frac{\partial^2 J_0}{\partial a_1 \partial a_2} = \frac{1}{8\pi a_1 a_2 (a_1 + a_2)^3}. \tag{40}$$

and [1]

$$J_3[a_1, a_2, a_3] = \int \frac{d\mathbf{k}}{(2\pi)^3} \frac{1}{(\mathbf{k}^2 + a_1^2)(\mathbf{k}^2 + a_2^2)(\mathbf{k}^2 + a_3^2)} = \frac{1}{4\pi(a_1 + a_2)(a_1 + a_3)(a_2 + a_3)}, \tag{41}$$

$$J_4[a_1, a_2, a_3] = \int \frac{d\mathbf{k}}{(2\pi)^3} \frac{1}{(\mathbf{k}^2 + a_1^2)^2(\mathbf{k}^2 + a_2^2)(\mathbf{k}^2 + a_3^2)} = \frac{2a_1 + a_2 + a_3}{8\pi a_1 (a_1 + a_2)^2 (a_1 + a_3)^2 (a_2 + a_3)}, \tag{42}$$

Thus, the procedure of Feynman parametrization is not used. Substituting $a_1^2 = 4\pi^2 n^2 T^2 + m_1^2$ and $a_2^2 = 4\pi^2 n^2 T^2 + m_2^2$ to (38) and taking the sum over Matsubara frequencies after the integration we get

$$I_0[m_1, m_2] = \sum_{n=-\infty, n\neq 0}^{\infty} J_0^n[m_1, m_2] = \sum_{n=-\infty, n\neq 0}^{\infty} \frac{1}{4\pi(\sqrt{4\pi^2 n^2 T^2 + m_1^2} + \sqrt{4\pi^2 n^2 T^2 + m_2^2})}. \tag{43}$$

or, after redefinition of mass parameters $M_{1;2} = m_{1;2}/2\pi T$ the temperature corrections to effective potential are expressed by summed integrals

$$I_1[M_1, M_2] = -\frac{1}{64\pi^4 T^2} \sum_{n=-\infty, n\neq 0}^{\infty} \frac{1}{\sqrt{M_1^2 + n^2}(\sqrt{M_1^2 + n^2} + \sqrt{M_2^2 + n^2})^2}, \tag{44}$$

[1] The same results for J_3 and J_4 can be found in [14] and [15, 16], where they appear in the context of high temperature dimensional reduction.

$$I_2[M_1, M_2] = \frac{1}{256\pi^5 T^4} \sum_{n=-\infty, n\neq 0}^{\infty} \frac{1}{\sqrt{M_1^2+n^2}\sqrt{M_2^2+n^2}(\sqrt{M_1^2+n^2}+\sqrt{M_2^2+n^2})^3}. \tag{45}$$

Note that the series (43) are divergent, but the derivatives (44) and (45) are convergent for all $M_{1;2}$. In the following it will be convenient to keep separately the terms for zero and nonzero modes in the sum. Both terms will be temperature-dependent since the zero-mode integrals coincide with (38)-(40), where $a_i^2 = m_i^2$ and the factor T should be accounted for. Numerical check of the zero temperature limiting case $T \to 0$ demonstrates that the non-temperature field theory results are successfully reproduced. In the high-temperature limit the zero mode gives dominant contribution in agreement with a known suppression of quantum effects at increasing temperatures.

The sum of integrals (44) and (45) can be expressed by means of the generalized zeta-function. Such forms can be derived if we introduce Feynman parameters in the integrand of (38)

$$\frac{1}{[\mathbf{k}^2+m_a^2][\mathbf{k}^2+m_b^2]} = \int_0^1 \frac{dx}{([\mathbf{k}^2+m_a^2]x + [\mathbf{k}^2+m_b^2](1-x))^2}, \tag{46}$$

and redefine $\mathbf{k} \longrightarrow \mathbf{p} = \mathbf{k}/2\pi T$, $M^2(M_a, M_b, x) = (M_a^2 - M_b^2)x + M_b^2$. Then we get

$$\frac{1}{[\mathbf{k}^2+m_a^2][\mathbf{k}^2+m_b^2]} = \frac{1}{(2\pi T)^4} \int_0^1 \frac{dx}{(\mathbf{p}^2+n^2+M^2)^2}. \tag{47}$$

and divergent series for (38) ($d\mathbf{k} = (2\pi T)^3 d\mathbf{p}$)

$$I_0[M_a, M_b] = \frac{1}{2\pi T} \int_0^1 dx \sum_{n=-\infty, n\neq 0}^{\infty} \int \frac{d\mathbf{p}}{(2\pi)^3} \frac{1}{(\mathbf{p}^2+n^2+M^2)^2}, \tag{48}$$

With the help of dimensional regularization or differentiating the integral

$$\int \frac{d\mathbf{p}}{(2\pi)^3} \frac{1}{(\mathbf{p}^2+M^2)} = -\frac{M}{4\pi} + O(\frac{M^2}{T^2}) \tag{49}$$

over the parameter M, the equation (48) can be reduced to

$$I_0[M_a, M_b] = \frac{1}{16\pi^2 T} \int_0^1 dx\, \zeta(2, \frac{1}{2}, M^2), \tag{50}$$

where $\zeta(u, s, t)$ is the generalized Hurwitz zeta-function [9, 10]: [2]

$$\zeta(u, s, t) = \sum_{n=1}^{\infty} \frac{1}{(n^u + t)^s}. \tag{51}$$

So in the case under consideration the sums of integrals (44) and (45) can be calculated by differentiation of (50) with respect to mass parameters participating in $M = M(M_a, M_b, x)$. Differentiation increases the power s in the denominator of (50) giving convergent integrals

$$I_1[M_a, M_b] = \frac{T}{2M_a} \frac{\partial}{\partial M_a} I_0 = -\frac{1}{64\pi^4 T^2} \int_0^1 dx\, x\, \zeta[2, \frac{3}{2}, M^2(x)], \tag{52}$$

$$I_2[M_a, M_b] = -\frac{1}{2M_b} \frac{\partial}{\partial M_b}(-I_1) = \frac{3}{256\pi^6 T^4} \int_0^1 dx\, x\,(1-x)\, \zeta[2, \frac{5}{2}, M^2(x)]. \tag{53}$$

The integrals (52) and (53) are equal to the series (44) and (45), respectively.

[2] Note that (non-generalized) Hurwitz zeta-function is defined by $\zeta(s, t) = \sum_{n=0}^{\infty} \frac{1}{[n+t]^s}$.

3 Examples. Results with one-loop contributions

Two pairs of different virtual masses

Four-point function with pairs of equivalent virtual masses:

$$\underline{J_2(m_1,m_2)} = i\mathrm{Tr}\int \frac{d^4k}{(2\pi)^4}\frac{i^4}{(\hat{k}-m_1)(\hat{k}-m_2)(\hat{k}-m_1)(\hat{k}-m_2)} = \tag{54}$$

$$= 4\frac{1}{16\pi^2}B_0(m_1^2,m_1^2)+4(m_1^2+3m_2^2+4m_1m_2)\frac{1}{16\pi^2}C_0(m_2^2,m_1^2,m_1^2)+8m_2^2(m_2+m_1)^2\frac{1}{16\pi^2}D_0(m_1^2,m_1^2,m_2^2,m_2^2).$$

This result (3) have to be symmetric under interchange $m_1 \rightleftarrows m_2$. Check for equivalent virtual masses:

$$\lim_{m_2\to m_1} J_2(m_1,m_2) = J_2(m) = \frac{1}{16\pi^2}\left(-4\Delta_\mu + 4\log m^2 + 4 + 16 - \frac{16}{3}\right) \tag{55}$$

coincides.

Three different virtual masses

$$\underline{J_3(m_1,m_2,m_3)} = i\mathrm{Tr}\int \frac{d^4k}{(2\pi)^4}\frac{i^4}{(\hat{k}-m_1)(\hat{k}-m_2)(\hat{k}-m_1)(\hat{k}-m_3)} =$$

$$= 4\frac{1}{16\pi^2}B_0(m_1^2,m_1^2) + 4(3m_1^2+m_2m_3+2m_1(m_2+m_3))\frac{1}{16\pi^2}C_0(m_1^2,m_2^2,m_3^2)+$$

$$+8m_1^2(m_1^2+m_1(m_2+m_3)+m_2m_3)\frac{1}{16\pi^2}D_0(m_1^2,m_1^2,m_2^2,m_3^2). \tag{56}$$

The selfcheck for last is $\lim\limits_{m_3\to m_2} J_3(m_1,m_2,m_3) = J_2(m_1,m_2)$ is fulfilled.

Four different virtual masses

$$\underline{J_4(m_1,m_2,m_3,m_4)} = i\mathrm{Tr}\int \frac{d^4k}{(2\pi)^4}\frac{i^4}{(\hat{k}-m_1)(\hat{k}-m_2)(\hat{k}-m_3)(\hat{k}-m_4)} =$$

$$= 4\frac{1}{16\pi^2}B_0(m_3^2,m_4^2) + 4m_1^2\frac{1}{16\pi^2}C_0(m_1^2,m_3^2,m_4^2)+$$

$$+4(m_2^2+m_1m_2+(m_1+m_2)(m_3+m_4)+m_3m_4)\frac{1}{16\pi^2}C_0(m_2^2,m_3^2,m_4^2)+$$

$$+4\{m_2^2[m_1^2+m_1m_2+(m_1+m_2)(m_3+m_4)+m_3m_4)]+m_1m_2m_3m_4\}\frac{1}{16\pi^2}D_0(m_1^2,m_2^2,m_3^2,m_4^2). \tag{57}$$

The self-checks are interchange symmetry between parameters and again $\lim\limits_{m_4\to m_3} J_4(m_1,m_2,m_3,m_4) = J_3(m_1,m_2,m_3)$ and etc.

The common principal check is also the contraction of fermion and boson loops diverging parts.

Note, that the integrals J_2, J_3 can be rewritten in terms of J_4, which is expressed in terms of one integral A_0 (or B_0) with the help of Mathematica program.

Four equivalent virtual masses

$$\underline{J_1(m)} = i\mathrm{Tr}\int \frac{d^4k}{(2\pi)^4}\frac{i^4}{(\hat{k}-m)(\hat{k}-m)(\hat{k}-m)(\hat{k}-m)} = \frac{1}{16\pi^2}\left(-4\Delta + 4\log\frac{m^2}{\mu^2} + \frac{44}{3}\right). \tag{58}$$

3.1 One-loop gaugino-higgsino contributions to λ_1

The couplings to be taken into account are the following:

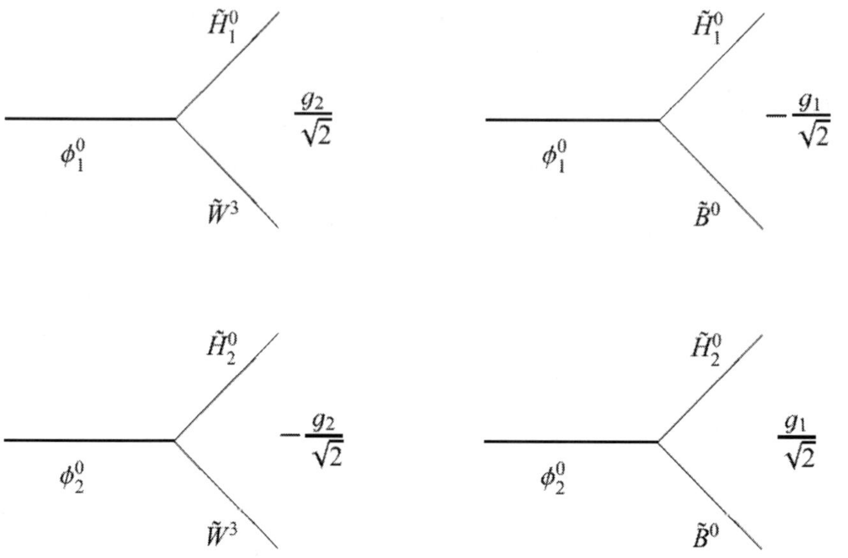

The general scheme of one-loop calculation is illustrated by the following example of λ_1 evaluation:

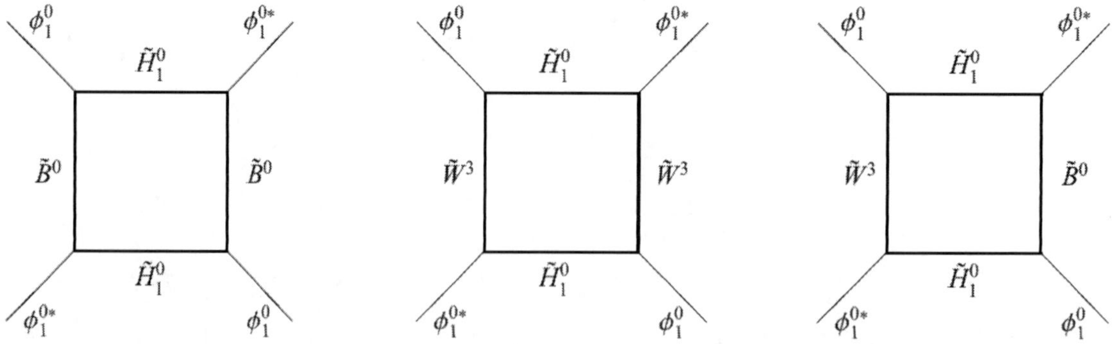

Calculation of all corrections at the one-loop gives the following result for the shift $\Delta\lambda_i$ (see λ_i normalization in [12, 13]) of the effective parameters

$$-\frac{1}{2}\Delta\lambda_1 = -\left(\frac{g_1^4}{4}J_2(m_{\tilde{H}_1^0}, m_{\tilde{B}^0}) + \frac{g_2^4}{4}J_2(m_{\tilde{H}_1^0}, m_{\tilde{W}^3}) + \frac{g_1^2 g_2^2}{4}J_3(m_{\tilde{H}_1^0}, m_{\tilde{W}^3}, m_{\tilde{B}^0})\right), \tag{59}$$

where we have to deal with the above mentioned integrals over fermion loops.

4 Summary and Outlook

Within the framework of the program the calculation technique is developed for one- and two-loop integrals in the finite-temperature field theory. The main object is the one-loop finite-temperature contribution to the mass state, had the same view in the form with the expression for the bosons ideal gas free energy. Principal object is the sum over Matsubara frequencies for relevant integrals allows representation in the form of generalized Hurwitz function. When masses are nondegenerated should further use operations with decomposition of rational fractions in the sum of (by analogy with Passarino–Veltman calculations methods), resulting in an arbitrary one-loop temperature diagram with several propagators expressed as derivatives of the basic object and rational fractions with itself. For the primary object there are well-known and well defined high-and low-temperature decompositions and renormalizability. Explicit submission as an infinite series or through the generalized Hurwitz function facilitate study of arbitrary graphs representing one-loop effective contributions to the potential. Conditions for the effective potential global and local minimums existence are included in the criteria for the validity of models with an extended Higgs sector [17, 18].

References

[1] R.P.Feynman, Phys.Rev.**76**:769-789,1949.

[2] A. J. Buras, P. H. Chankowski, J. Rosiek and L. Slawianowska, Nucl. Phys. B **659**, 3 (2003) [arXiv:hep-ph/0210145].

[3] Athanasios Dedes, Janusz Rosiek, and Philip Tanedo, Phys.Rev. D **79**, 055006 (2009).

[4] Dolgopolov M.V., Dubinin M.N., Rykova E.N. Two-Higgs doublet potential of the MSSM at finite temperature and Higgs boson masses // Quarks. 2008. V. 1.

[5] E. Akhmetzyanova, M. Dolgopolov, M. Dubinin, Self-couplings of Higgs bosons in scenarios with mixing of CP-even/CP-odd states, Workshop on CP Studies and Non-Standard Higgs Physics, CERN-2006-009 (May 2004 - December 2005), eds. S. Kraml, G. Azuelos, D. Dominici, J. Ellis, G. Grenier, H.,E. Haber, J. S. Lee, D. J. Miller, A., CERN, Geneva, 2006, 133-138, arXiv: hep-ph/0608079.

[6] M. Dolgopolov, M. Dubinin, E. Rykova, J. Modern Phys., 2:5 (2011), 301-322.

[7] L. Vergara, Evaluating one loop integrals at finite temperature. Journal of Physics A**30** (1997) 6977–6980

[8] P. Amore, Non perturbative regularization of one loop integrals at finite temperature. Journal of Physics A**38** (2005) 6463–6472 (hep-th/0503142)

[9] P. Amore, Convergence acceleration of series through a variational approach. Journal of Mathematical Analysis and Applications **323** (2006) 63–77

[10] M. Abramowitz and I. Stegun, Handbook of Mathematical Functions, Dover Publications, NY, 1964

[11] E. Akhmetzyanova, M. Dolgopolov, M. Dubinin, Higgs bosons in the two-doublet model with CP violation. Phys.Rev. D**71** (2005) 075008 (hep-ph/0405264)

[12] E. Akhmetzyanova, M. Dolgopolov, M. Dubinin, Violation of CP Invariance in the Two-Doublet Higgs Sector of the MSSM. Phys.Part.Nucl. **37** (2006) 677–734

[13] M.Dolgopolov, E.Rykova, M. Dubinin, I Erofeev Threshold corrections to the MSSM effective Higgs potential: gaugino and higgsino contributions // 20th International Workshop High Energy Physics and Quantum Field Theory, QFTHEP 2011, 24 September - 1 October 2011, Sochi 2011, P. 1–8.

[14] M. Laine, Effective theories of MSSM at high temperature. Nucl.Phys. B**481** (1996) 43–84 (hep-ph/9605283); Erratum-ibid. B548: 637–638, 1999
[15] G.R. Farrar, M. Losada, SUSY and the electroweak phase transition. Phys.Lett. B**406** (1997) 60–65 (hep-ph/9612346)
[16] M. Losada, High temperature dimensional reduction of the MSSM and other multiscalar models. Phys.Rev. D**56** (1997) 2893–2913 (hep-ph/9605266)
[17] M. V. Dolgopolov, S. P. Zavodov, E. Yu. Petrova, Bifurcation sets of extended Higgs potential, Vestn. Samar. Gos. Tekhn. Univ. Ser. Fiz.-Mat. Nauki, 4(33) (2013), 173–183.
[18] M. N. Dubinin, E. Yu. Petrova, High-temperature Higgs potential of the two-doublet model in catastrophe theory, Teoret. Mat. Fiz., 184:2 (2015), 315–337.

The muon g-2: retrospective and future

A. E. Dorokhov[1,2,3,⋆], A. E. Radzhabov[4,⋆⋆], and A. S. Zhevlakov[3,4,⋆⋆⋆]

[1] *Bogoliubov Laboratory of Theoretical Physics, JINR, 141980 Dubna, Russia*
[2] *N.N.Bogoliubov Institute of Theoretical Problems of Microworld, M.V.Lomonosov Moscow State University, Moscow 119991, Russia*
[3] *Department of Physics, Tomsk State University, Lenin ave. 36, 634050 Tomsk, Russia*
[4] *Institute for System Dynamics and Control Theory SB RAS, 664033 Irkutsk, Russia*

Abstract. Soon, new experiments at FNAL and J-PARC will measure the muon anomalous magnetic moments with better accuracy than before. From theoretical side, the uncertainty of the standard model prediction is dominated by the hadronic contributions. Current status of the experimental data and theoretical calculations are briefly discussed.

1 Introduction

Cosmology tell us that about 95% of matter is not detected by modern measurements. We think that the dark matter surround us, however, we don't see it. There are two strategies to search for the physics beyond the standard matter: high energy and low energy experiments. In the first case, due to high energy we attempt to excite the heavy degrees of freedom. There are no firm evidences on the deviation of measured cross sections from the predictions of the standard model. In the case of the low energy experiments, it is possible to reach very high precision of the measured quantities because of huge statistics. And within this kind of experiments there are some rough redges of the standard model. The most famous deviation is observed for the muon anomalous magnetic moment and it remains stable for many years.

The anomalous magnetic moment (AMM) of charged leptons ($l = e, \mu, \tau$) is defined as

$$a_l = \frac{g_l - 2}{2}, \tag{1}$$

with the gyromagnetic ratio g_l of the lepton magnetic moment to its spin, in Bohr magneton units. The Dirac equation $g = 2$ predicts for a free point-like fermion with spin $1/2$ and thus there is no anomaly at tree level (Fig. 1a). However, deviations appear when taking into account the interactions leading to fermion substructure and thus to nonzero a_l. In the standard model it appears from the radiative corrections to the tree fermion-photon vertex (Fig. 1) due to the coupling of the lepton spin to virtual fields, which in the SM are induced by QED, weak and strong (hadronic) interactions (Fig. 1)

$$a^{\rm SM} = a^{\rm QED} + a^{\rm weak} + a^{\rm hadr}. \tag{2}$$

⋆e-mail: dorokhov@theor.jinr.ru
⋆⋆e-mail: aradzh@icc.ru
⋆⋆⋆e-mail: zhevlakov1@gmail.com

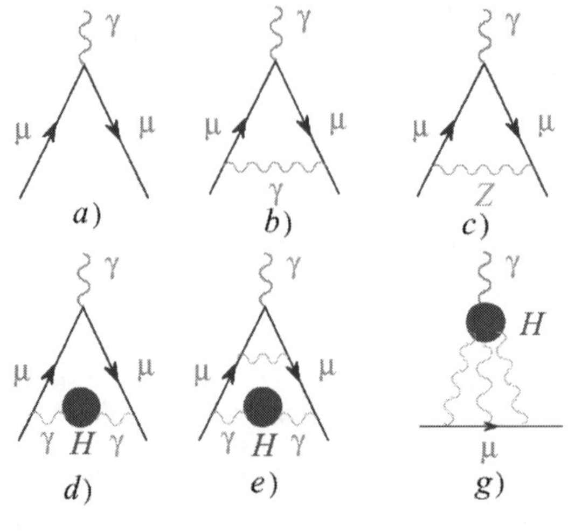

Figure 1. Representative diagrams for the SM contributions to a_μ. Here, H is for the hadronic block.

2 The electron $g-2$ and the fine coupling constant

The electron AMM was experimentally discovered by Kush and Foley from Columbia University, New York (USA) in 1947 [1, 2] with the result [3]

$$a_e^{\text{Columbia}} = 0.00118 \pm 0.00003. \quad (3)$$

This was immediately confirmed by Schwinger in the framework of quantum electrodynamics as a result of the radiative correction (Fig. 1b)

$$a_e^{\text{Schwinger}} = \frac{\alpha}{2\pi} = 0.001161. \quad (4)$$

Since that time, enormous progress has been reached in experiment and theory. The latest measurement by the Gabrielse' group from Harvard university (USA) provides the result with fantastic accuracy [4]

$$a_e^{\text{Harvard}} = 1\ 159\ 652\ 180.73\ (0.28) \times 10^{-12} \quad [0.24\ \text{ppb}]. \quad (5)$$

Within the standard mode (1), the electron AMM is almost completely dominated by QED contribution calculated with 5-loop accuracy

$$a_l^{\text{QED}} = \sum_{n=1}^{5} C_{2n}^l \left(\frac{\alpha}{\pi}\right)^n + ..., \quad (6)$$

where the value of coefficients are given in Table 1.

The result (6) allows to determine the fine structure constant α with the extraordinary precision [6, 7]

$$\alpha^{-1}(a_e) = 137.035\ 999\ 1570\ (29)(27)(18)(331) \quad [0.25\ \text{ppb}], \quad (7)$$

where the first two uncertainties are due to errors in $C_{8,10}^l$, the third one is uncertainty from hadronic and weak corrections, and the last one is due to experimental error in the measurement of a_e^{Harvard}.

DOI: 10.1051/epjconf/201612502007

Table 1. The coefficients in the QED contribution to the lepton AMM (6). The first three coefficients are known analytically. The errors in $C^{\mu}_{4,6}$ are due to the experimental uncertainties in m_μ/m_e ratio. The errors in $C^{l}_{8,10}$ are due to numerical simulations.

	l=e	l=μ	source
C^l_2	0.5	0.5	Schwinger[3]
C^l_4	-0.328 478 444 00...	0.765 857 425(17)	Laporta, Remiddi [5]
C^l_6	1.181 234 017...	24.050 509 96(32)	Laporta, Remiddi [5]
C^l_8	-1.912 06(84)	130.879 6(63)	Kinoshita, Nio, et.al. [6]
C^l_{10}	7.791(336)	753.29(1.04)	Kinoshita, Nio, et.al. [6]

This determination became possible after the complete QED contribution to the electron AMM up to tenth order in the coupling constant were achieved numerically by the Prof. T. Kinoshita group [6] (for recent review see [7]).

In 2010, the direct determination of the fine coupling constant became possible from measurement of the ratio $\hbar/m_{Rb}$ [8]

$$\alpha^{-1}(Rb) = 137.035\,999\,049\,(90) \quad [0.66 \text{ ppb}]. \tag{8}$$

With this α, the the SM prediction for the electron AMM becomes

$$a_e^{\rm SM,Rb} = 1\,159\,652\,181.643\,(764) \times 10^{-12} \quad [0.24 \text{ ppb}]. \tag{9}$$

Both determination of the electron AMM (5) and (9) are consistent within the errors.

$$a_e^{\rm exp} - a_e^{\rm SM,Rb} = -0.91\,(81) \times 10^{-12}. \tag{10}$$

Thus the experimental and SM results for the electron AMM are in perfect agreement.

3 The muon $g-2$: experiment vs standard model. Electroweak contributions.

In 2006, there were published the results obtained by the E821 collaboration at Brookhaven National Laboratory [9] on measurements of the muon AMMa_μ

$$a_\mu^{\rm BNL} = 116\,592\,08.0\,(6.3) \times 10^{-10} \quad [0.54 \text{ ppm}]. \tag{11}$$

Later on, this value was corrected [10, 11] for a small shift in the ratio of the magnetic moments of the muon and the proton as

$$a_\mu^{\rm BNL,CODATA} = 116\,592\,09.1\,(6.3) \times 10^{-10}. \tag{12}$$

It is well-known that the effect of the second-order contribution, due to exchange by the particle with mass M, to the AMM of the lepton with mass m_l is proportional to $a_l \propto (m_l/M)^2$. It means, that sensitivity for the muon to the interaction with scale M is by factor $m_\mu^2/m_e^2 \propto 40000$ higher than for the electron. This fact compensates a less experimental accuracy of the muon AMM measurements (11) relatively to the electron one (5), and make the study of the muon AMM more perspective in search for new physics.

Another exciting point is that soon the new data on the muon AMM will be available from experiments proposed at Fermilab (USA) [12] and J-PARC (Japan) [13]. These experiments plan to reduce the present experimental error by factor 4, to a precision of 0.14 ppm.

In SM, the contributions to the muon AMM from QED (Fig. 1b) [14] and weak (Fig. 1c) [15, 16] interactions (Fig. 1c) are known with high accuracy

$$a_\mu^{\mathrm{QED,Rb}} = 11\,658\,471.8951\,(0.0080) \times 10^{-10}, \tag{13}$$

$$a_\mu^{\mathrm{QED,a_e}} = 11\,658\,471.8846\,(0.0037) \times 10^{-10}, \tag{14}$$

$$a_\mu^{\mathrm{weak}} = 15.36\ (0.10) \times 10^{-10}, \tag{15}$$

The most important feature of new estimate for the weak sector, that significantly increases the theoretical precision, is to use precise Higgs-boson mass value measured at LHC. The remaining theory error comes from unknown three-loop contributions and dominantly from light hadronic uncertainties in the second-order electroweak diagrams with quark triangle loops. The accuracy of these calculations is enough for any planed experiments in new future.

Subtracting from the experimental result the well-defined contributions from QED and weak interactions one gets

$$a_\mu^{\mathrm{BNL}} - a_\mu^{\mathrm{QED,Rb}} - a_\mu^{\mathrm{weak}} = 721.65\ (6.38) \times 10^{-10}, \tag{16}$$

where the error is only due to the experiment. We can treat this number as an experimental result for the rest contributions, i.e. of the strong interaction of SM and of the hypothetical interactions beyond SM.

4 Hadronic contributions to the muon $g-2$. Vacuum polarization effect.

Strong (hadronic) interaction produces relatively small contributions to a_μ, however they are known with an accuracy comparable to the experimental uncertainty in (11). In the leading in α orders, these contributions can be separated into three terms

$$a_\mu^{\mathrm{hadr}} = a_\mu^{\mathrm{HVP,LO}} + a_\mu^{\mathrm{HVP,HO}} + a_\mu^{\mathrm{HLbL}}. \tag{17}$$

In (17), a_μ^{HVP} is the leading in α contribution due to the hadron vacuum polarization (HVP) effect in the internal photon propagator of the one-loop diagram (Fig. 1d), a_μ^{ho} is the next-to-leading order contribution related to iteration of HVP (Fig. 1e). The last term is not reduced to HVP iteration and it is due to the hadronic light-by-light (HLbL) scattering mechanism (Fig. 1g).

Hadronic contributions in (17) are determined by effects dominated by long distance dynamics, the region where the methods of perturbation theory of Quantum Chromodynamics (QCD) do not applicable and one must use less reliable nonperturbative approaches. However, in case of HVP, using analyticity and unitarity (the optical theorem) a_μ^{HVP} can be expressed as the spectral representation integral [17, 18]

$$a_\mu^{\mathrm{HVP}} = \frac{\alpha}{\pi} \int_{4m_\pi^2}^{\infty} \frac{dt}{t} K(t) \rho_{\mathrm{V}}^{(\mathrm{H})}(t), \tag{18}$$

which is a convolution of the hadronic spectral function

$$\rho_V^{(\mathrm{H})}(t) = \frac{1}{\pi} \mathrm{Im}\Pi^{(\mathrm{H})}(t) \tag{19}$$

with the known from QED kinematical factor

$$K(t) = \int_0^1 dx \frac{x^2(1-x)}{x^2 + (1-x)t/m_\mu^2}, \tag{20}$$

where m_μ is the muon mass. The QED factor is sharply peaked at low invariant masses t and decreases monotonically with increasing t. Thus, the integral defining a_μ^{HVP} is sensitive to the details of the spectral function $\rho_V^{(\text{H})}(t)$ at low t. At present there is no direct theoretical tools that allow to calculate the spectral function at low t with required accuracy. Fortunately, $\rho_V^{(\text{H})}(t)$ is related to the total $e^+e^- \to \gamma^* \to$ hadrons cross-section $\sigma(t)$ at center-of-mass energy squared t by

$$\sigma^{e^+e^-\to\text{hadrons}}(t) = \frac{4\pi\alpha}{t}\rho_V^{(\text{H})}(t), \tag{21}$$

and this fact is used to get quite accurate estimate of a_μ^{HVP}. The most precise recent phenomenological evaluations of a_μ^{HVP}, using recent $e^+e^- \to$ hadrons data, provide the results

$$a_\mu^{\text{HVP,LO},e^+e^-} = \begin{cases} 692.3\ (4.2)\times 10^{-10}, & [19] \\ 694.91\ (4.27)\times 10^{-10}. & [20] \end{cases} \tag{22}$$

In addition, data on inclusive decays of the τ-lepton into hadrons are used to replace the e^+e^- data in certain energy regions. This is possible, since the vector current conservation law relates the $I = 1$ part of the electromagnetic spectral function to the charged current vector spectral function measured in $\tau \to \nu$ +non-strange hadrons (see, i.e. [21]). All these allows to reach during the last decade a substantial improvement in the accuracy of the contribution from the HVP.

Similar dispersion relation approach and the same phenomenological input lead to the estimate of the next-to-leading hadronic contribution (Fig. 1e) [20, 22]

$$a_\mu^{\text{HVP,NLO}} = -9.84\ (0.07)\times 10^{-10}, \tag{23}$$

$$a_\mu^{\text{HVP,NNLO}} = 1.24\ (0.01)\times 10^{-10} \tag{24}$$

Thus, the HVP and next-to-leading order contribution related to HVP are known with an accuracy better than 1 per cent.

In near future it is expected that new and precise measurements from CMD3 and SND at VEPP-2000 in Novosibirsk, BES III in Beijing and KLOE-2 at DAFNE in Frascati allow to significantly increase accuracy of predictions for a_μ^{HVP} and a_μ^{ho} and resolve some inconsistency problems between different set of data.

Subtracting from the experimental result the contributions from electroweak interaction and hadronic vacuum effect one gets

$$a_\mu^{\text{BNL}} - a_\mu^{\text{QED,Rb}} - a_\mu^{\text{weak}} - a_\mu^{\text{HVP}} = 37.95\ (7.64)\times 10^{-10}, \tag{25}$$

where one can treat this number as an experimental result for the rest contributions, i.e. of the strong interaction of due to the light-by-light mechanism and of the hypothetical interactions beyond SM.

5 Hadronic contributions to the muon $g-2$. Light-by-light scattering mechanism.

The basic element for calculations of the hadronic LbL contribution to the muon AMM (Fig. 1g) is the fourth-rank light quark hadronic vacuum polarization tensor

$$\Pi_{\mu\nu\lambda\rho}(q_1,q_2,q_3) = \int d^4x_1 \int d^4x_2 \int d^4x_3 e^{i(q_1x_1+q_2x_2+q_3x_3)}\times \\ \times \left\langle 0|T(j_\mu(x_1)j_\nu(x_2)j_\lambda(x_3)j_\rho(0))|0\right\rangle, \tag{26}$$

where $j_\mu(x)$ are light quark electromagnetic currents and $|0\rangle$ is the QCD vacuum state.

The muon AMM can be extracted by using the projection [23]

$$a_\mu^{\rm LbL} = \frac{1}{48 m_\mu} {\rm Tr}\left((\hat{p} + m_\mu)[\gamma^\rho, \gamma^\sigma](\hat{p} + m_\mu)\Pi_{\rho\sigma}(p, p)\right),$$

where

$$\begin{aligned}
\Pi_{\rho\sigma}(p', p) = -ie^6 \int \frac{d^4 q_1}{(2\pi)^4} \int \frac{d^4 q_2}{(2\pi)^4} \frac{1}{q_1^2 q_2^2 (q_1 + q_2 - k)^2} \times \\
\times \gamma^\mu \frac{\hat{p}' - \hat{q}_1 + m_\mu}{(p' - q_1)^2 - m_\mu^2} \gamma^\nu \frac{\hat{p} - \hat{q}_1 - \hat{q}_2 + m_\mu}{(p - q_1 - q_2)^2 - m_\mu^2} \gamma^\lambda \times \\
\times \frac{\partial}{\partial k^\rho} \Pi_{\mu\nu\lambda\sigma}(q_1, q_2, k - q_1 - q_2),
\end{aligned} \tag{27}$$

m_μ is the muon mass, $k_\mu = (p' - p)_\mu$ and it is necessary to consider the limit $k_\mu \to 0$.

In general, the HLbL scattering amplitude is a complicated object for calculations. It is a sum of different diagrams including the dynamical quark loop, the meson exchanges, the meson loops and the iterations of these processes. Fortunately, already in the first papers devoted to the calculation of the HLbL contributions [24–26], it has been recognized that these numerous terms show a hierarchy. This is related to existence of two small parameters: the inverse number of colors $1/N_c$ and the ratio of the characteristic internal momentum to the chiral symmetry parameter $m_\mu/(4\pi f_\pi) \sim 0.1$. The former suppresses the multiloop contributions, so that the leading contribution is due to the quark loop diagram and the two-loop diagrams with mesons in the intermediate state. In latter case, the contribution of the diagram with intermediate pion is enhanced by small pion mass in the meson propagator. The leading in $1/N_c$ diagrams are drawn in Fig. 2. They are the box diagram with dynamical quarks (Fig. 3) and the meson exchange diagrams in pseudoscalar, scalar and axial-vector channels.

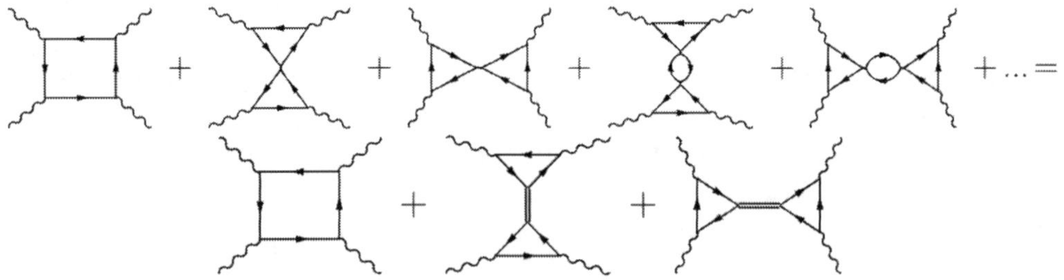

Figure 2. A schematic illustration for the diagrams contributing to the four-rank polarization tensor to the leading in $1/N_c$ order. The four-fermion interaction is introduced through (28). The nonlocal multi-photon vertices are not shown for simplicity, see Fig. 3.

For explicit calculations of the hadronic contributions to the muon AMM due to the light-by-light scattering mechanism [27–31] we use the $SU(3)\times SU(3)$ chiral quark model with nonlocal Lagrangian (NχQM)

$$\begin{aligned}
\mathcal{L} = \bar{q}(x)(i\hat{\partial} - m_c)q(x) + \frac{G}{2}[J_S^a(x)J_S^a(x) + J_{PS}^a(x)J_{PS}^a(x)] \\
- \frac{H}{4} T_{abc}[J_S^a(x)J_S^b(x)J_S^c(x) - 3J_S^a(x)J_{PS}^b(x)J_{PS}^c(x)],
\end{aligned} \tag{28}$$

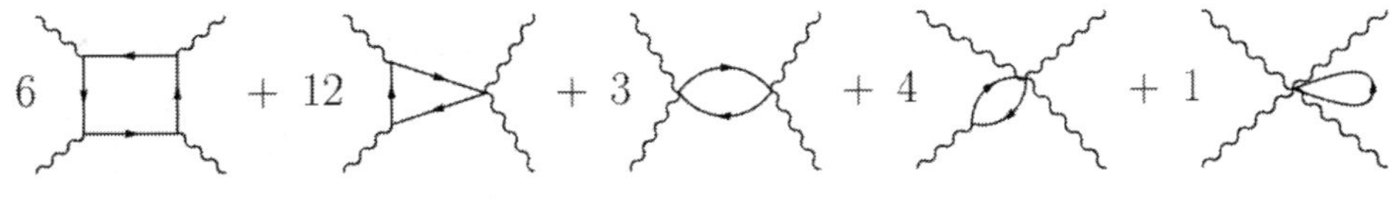

Figure 3. Contact terms which are gave contribution to $\Pi_{\mu\nu\lambda\rho}(q_1, q_2, q_3)$. Numbers in front of diagrams are the degeneracy factors.

where $q(x)$ are the quark fields, m_c ($m_u = m_d \neq m_s$) is the diagonal matrix of the quark current masses, G and H are the four- and six-quark coupling constants. Second line in the Lagrangian represents the Kobayashi–Maskawa–t'Hooft determinant vertex with the structural constant

$$T_{abc} = \frac{1}{6}\epsilon_{ijk}\epsilon_{mnl}(\lambda_a)_{im}(\lambda_b)_{jn}(\lambda_c)_{kl},$$

where λ_a are the Gell-Mann matrices for $a = 1, .., 8$ and $\lambda_0 = \sqrt{2/3}I$. The nonlocal structure of the model is introduced via the nonlocal quark currents

$$J_M^a(x) = \int d^4x_1 d^4x_2\, f(x_1) f(x_2)\, \bar{Q}(x - x_1, x)\, \Gamma_M^a Q(x, x + x_2), \tag{29}$$

where $M = S$ for the scalar and $M = PS$ for the pseudoscalar channels, $\Gamma_S^a = \lambda^a$, $\Gamma_{PS}^a = i\gamma^5\lambda^a$ and $f(x)$ is a form factor with the nonlocality parameter Λ reflecting the nonlocal properties of the QCD vacuum. In (29), the gauge-invariant interaction with an external photon field V_μ^a is introduced through the Schwinger phase factor

$$Q(x, y) = \mathcal{P}\exp\left\{i\int_x^y dz^\mu V_\mu^a(z)\, T^a\right\} q(y). \tag{30}$$

In order to guarantee the Ward-Takahashi identities, it induces the quark-antiquark–n-photon vertices. Additionally, there appear the meson–quark-anti-quark–n-photon vertices.

The numerical results for the value of a_μ^{HLbL} are presented in Fig. 4 for the $SU(2)$ and $SU(3)$ models. The estimates for the partial contributions to a_μ^{HLbL} (in 10^{-10}) are the π^0 contribution 5.01(0.37) [28], the sum of the contributions from π^0, η and η' mesons 5.85(0.87) [28], the scalar σ, $a_0(980)$ and $f_0(980)$ mesons contribution 0.34(0.48) [29, 30], and the quark loop contribution is 11.0(0.9) [30]. The total contribution obtained in the leading order in the $1/N_c$ expansion is (see also [30])

$$a_\mu^{\mathrm{HLbL,N\chi QM}} = 16.8(1.25)\cdot 10^{-10}. \tag{31}$$

The error bar accounts for the spread of the results depending on the model parameterizations. Comparing with other model calculations, we conclude that our results are quite close to the recent results obtained in [32, 33].

If we add the result (31) to all other known contributions of the standard model to a_μ, (14),(22) and (23), we get that the difference between experiment (12) and theory is

$$a_\mu^{\mathrm{BNL,CODATA}} - a_\mu^{\mathrm{SM}} = 18.73 \times 10^{-10}, \tag{32}$$

which corresponds to 2.43σ. If one uses the hadronic vacuum polarization contribution from the τ hadronic decays instead of e^+e^- data

$$a_\mu^{\mathrm{HVP,LO-\tau}} = 701.5(4.7)\times 10^{-10} \qquad [19], \tag{33}$$

the difference decreases to 12.14×10^{-10} (1.53σ) in the NχQM model (31).

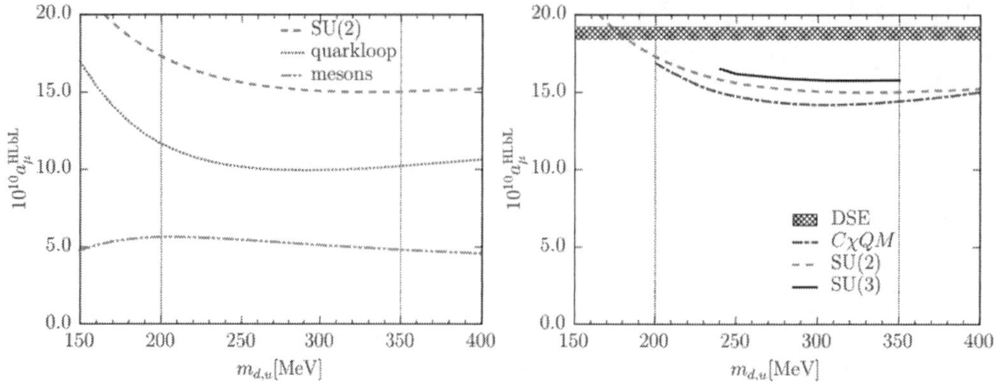

Figure 4. (Left) The results for $a_\mu^{\rm HLbL}$ in the $SU(2)$ model: the red dashed line is the total result, the green dotted line is the quark loop contribution and the magenta dash-dot-dot line is the $\pi + \sigma$ contribution. Thin vertical line indicates the region for estimation of $a_\mu^{\rm HLbL}$ error band. (Right) The results for $a_\mu^{\rm HLbL}$: the black solid line is the $SU(3)$-result, the red dash line corresponds to the $SU(2)$-result, the blue dash-dotted line is the $C\chi QM$ result [32], hatched region correspond to DSE result [33].

6 Conclusions

Clearly, a further reduction of both the experimental and theoretical uncertainties is necessary. On the theoretical side, the calculation of the still badly known hadronic light-by-light contributions in the next-to-leading order in the $1/N_c$ expansion (the pion and kaon loops) and extension of the model by including heavier vector and axial-vector mesons is the next goal. The contribution of these effects and the model error induced by them are not included in the result (31). Preliminary studies [25, 34] show that these contributions are one order smaller than the pseudoscalar exchanges and the quark loop contributions. However, the interesting point that inclusion of vector channel can strongly suppress contribution from the quark loop due photon–vector meson exchange which lead to appearance in each photon vertex additional VMD-like factor. This was found in local NJL model [26] and should be carefully investigated in the nonlocal one.

New experiments at FNAL and J-PARC have to resolve the muon $g-2$ problem, increasing effect or leading to its disappearance.

Numerical calculations are performed on computing cluster "Ac. V.M. Matrosov". The work is supported by Russian Science Foundation grant (RSCF 15-12-10009).

References

[1] P. Kush, H.M. Foley, Phys.Rev. **72**, 1256 (1947)
[2] H.M. Foley, P. Kush, Phys.Rev. **73**, 412 (1948)
[3] J.S. Schwinger, Phys.Rev. **73**, 416 (1948)
[4] D. Hanneke, S. Fogwell, G. Gabrielse, Phys. Rev. Lett. **100**, 120801 (2008), 0801.1134
[5] S. Laporta, E. Remiddi, Phys. Lett. **B379**, 283 (1996), hep-ph/9602417
[6] T. Aoyama, M. Hayakawa, T. Kinoshita, M. Nio, Phys. Rev. Lett. **109**, 111807 (2012), 1205.5368

[7] T. Kinoshita, Int. J. Mod. Phys. **A29**, 1430003 (2014)
[8] R. Bouchendira, P. Clade, S. Guellati-Khelifa, F. Nez, F. Biraben, Phys. Rev. Lett. **106**, 080801 (2011), 1012.3627
[9] G. Bennett, others (Muon g 2 collaboration) (Muon g-2), Phys.Rev. **D73**, 072003 (2006), hep-ex/0602035
[10] P.J. Mohr, B.N. Taylor, D.B. Newell, Rev. Mod. Phys. **84**, 1527 (2012), 1203.5425
[11] K.A. Olive et al. (Particle Data Group), Chin. Phys. **C38**, 090001 (2014)
[12] G. Venanzoni, J. Phys. Conf. Ser. **349**, 012008 (2012)
[13] N. Saito (J-PARC g-'2/EDM), AIP Conf. Proc. **1467**, 45 (2012)
[14] T. Aoyama, M. Hayakawa, T. Kinoshita, M. Nio, Phys. Rev. Lett. **109**, 111808 (2012), 1205.5370
[15] A. Czarnecki, W.J. Marciano, A. Vainshtein, Phys. Rev. **D67**, 073006 (2003), [Erratum: Phys. Rev.D73,119901(2006)], hep-ph/0212229
[16] C. Gnendiger, D. Stöckinger, H. Stöckinger-Kim, Phys. Rev. **D88**, 053005 (2013), 1306.5546
[17] C. Bouchiat, L. Michel, J. Phys. Radium **22**, 121 (1961)
[18] L. Durand, Phys. Rev. **128**, 441 (1962)
[19] M. Davier, A. Hoecker, B. Malaescu, Z. Zhang, Eur.Phys.J. **C71**, 1515 (2011), 1010.4180
[20] K. Hagiwara, R. Liao, A.D. Martin, D. Nomura, T. Teubner, J. Phys. **G38**, 085003 (2011), 1105.3149
[21] F. Jegerlehner, R. Szafron, Eur. Phys. J. **C71**, 1632 (2011), 1101.2872
[22] A. Kurz, T. Liu, P. Marquard, M. Steinhauser, Nucl. Phys. **B879**, 1 (2014), 1311.2471
[23] S.J. Brodsky, E. De Rafael, Phys. Rev. **168**, 1620 (1968)
[24] E. de Rafael, Phys.Lett. **B322**, 239 (1994), hep-ph/9311316
[25] M. Hayakawa, T. Kinoshita, A.I. Sanda, Phys. Rev. Lett. **75**, 790 (1995), hep-ph/9503463
[26] J. Bijnens, E. Pallante, J. Prades, Phys. Rev. Lett. **75**, 1447 (1995), [Erratum: Phys. Rev. Lett.75,3781(1995)], hep-ph/9505251
[27] A.E. Dorokhov, W. Broniowski, Phys. Rev. **D78**, 073011 (2008), 0805.0760
[28] A.E. Dorokhov, A.E. Radzhabov, A.S. Zhevlakov, Eur. Phys. J. **C71**, 1702 (2011), 1103.2042
[29] A.E. Dorokhov, A. Radzhabov, A.S. Zhevlakov, Eur.Phys.J. **C72**, 2227 (2012), 1204.3729
[30] A.E. Dorokhov, A.E. Radzhabov, A.S. Zhevlakov, JETP Lett. **100**, 133 (2014), 1406.1019
[31] A.E. Dorokhov, A.E. Radzhabov, A.S. Zhevlakov, Eur. Phys. J. **C75**, 417 (2015), 1502.04487
[32] D. Greynat, E. de Rafael, JHEP **07**, 020 (2012), 1204.3029
[33] C.S. Fischer, T. Goecke, R. Williams, Eur. Phys. J. **A47**, 28 (2011), 1009.5297
[34] K. Melnikov, A. Vainshtein, Phys. Rev. **D70**, 113006 (2004), hep-ph/0312226

New Physics search in the processes of the electroweak top quark production

Alexey Baskakov[1], *Edward* Boos[1], *Vycheslav* Bunichev[1], *Mikhail* Dubinin[1], *Lev* Dudko[1], *Igor* Myagkov[1], *Maxim* Perfilov[1], and *Georgy* Vorotnikov[1]

[1] *Skobeltsyn Institute of Nuclear Physics, MSU, 119992 Moscow, Russia*

Abstract. The processes involving single top quark production have an electroweak nature with unique properties. They are very interesting from both theoretical and experimental points of view. The article contains a brief overview of possible "New Physics" manifestations in the single top quark production processes.

1 Introduction

In various aspects the top quark (t) is a very unique object. The top quark is the heaviest elementary particle found so far with a mass slightly less than the mass of the gold nucleus. In comparison with other quarks the difference comes from two experimental facts, namely, a very large top quark mass compared to masses of all other quarks and very small mixing to quarks of the first and second generations. The value of the top quark mass measured is known now with a precision of 0.28% $M_t = 172.44 \pm 0.48$ GeV [1] being the most precisely known quark mass. Top Yukawa coupling $\lambda_t = 2^{3/4} G_F^{1/2} m_t$ is very close to unity. The quark mixing in the Standard Model (SM) is encoded in matrix elements of the Cabibbo-Kobayashi-Maskawa matrix. The matrix element V_{tb} is close to one while the elements V_{ts} and V_{td} are significantly smaller than one. These two experimental facts, large mass and small mixing, lead to the conclusion that in SM the top quark decays to W-boson and b-quark with a probability close to 100%. The width of the top quark being calculated in SM at the NLO level [2] is about 1.4 GeV. From one side, the top width is much smaller than its mass, and therefore the top quark is a narrow resonance. From the other side the top width is significantly larger than a typical QCD scale $\Lambda = 200$ MeV. As a result the top quark life time ($\tau_t \approx 5 \times 10^{-25}$s) as predicted by SM is much smaller than a typical time for formation of QCD bound states ($\tau_{QCD} \approx 1/\Lambda_{QCD} \approx 3 \times 10^{-24}$s). Therefore, the top quark decays long before it can hadronize, so top quark containing hadrons does not exist [3]. Thus the top quark provides a very clean source for fundamental information.

Since the top quark decays before hadronization its spin properties are not spoiled. Therefore the spin correlation in top production and decays is an interesting issue of the top quark physics.

Three mechanisms of the single top production are distinguished by the virtuality Q_W^2 of the W-boson involved: *t*-channel ($Q_W^2 < 0$), *s*-channel ($Q_W^2 > 0$), associated tW ($Q_W^2 = M_W^2$). The single top quark production at hadron colliders was considered for the first time in Ref. [4] and later in Ref. [5]–[15]. The authors of Ref. [12, 15] studied the most complete tree-level set of processes in the SM that contribute to the single top quark production. QCD NLO and NNLO corrections to various single top production processes have been calculated in several papers [16]–[29]. In particular, NLO corrections to kinematic distributions were presented [20]. The influence of NLO corrections not only to the production but also to the subsequent top quark decay has been studied in Ref. [22, 25]. Potentially important corrections at the threshold region have been resummed up to NNLL accuracy [27]. Monte-Carlo (MC) analyses of the production processes of the single top quark allowing to extract it from main backgrounds were performed in Ref. [15, 19].

In the single top quark t-channel and s-channel processes in the SM top quark is produced through the left-handed current interaction. The production is very similar to the top decay process turned backward in time. For the polarized top decay, it is well-known that the charged lepton tends to point along the direction of top spin [30]. In the production process this is the direction of the initial $\bar{d}$-quark for the s-channel, and the dominant direction of the final $\bar{d}$-quark for the t-channel. Therefore, in the top quark rest frame there is strong correlation in the angle of produced lepton with respect to one of the above directions [31, 32]: $\frac{1}{\sigma}\frac{d\sigma}{d\cos\theta_\ell^*} = \frac{1}{2}(1 + \cos\theta_\ell^*)$. Spin properties in the tW production process are more involved. Here, in order to maximize spin correlations, one can find a kinematic region in which top quarks are produced with the polarization vector preferentially close to the direction of the charged lepton or the d,s-quark momentum from the associated W decay. In this kinematic region, the direction of the produced charged lepton or the d,s-quark should be as close as possible to the direction of the initial gluon beam in the top quark rest frame [32].

Due to unique properties physics of the single top quark will be an important part of research programs for all future hadron and lepton colliders and will include studies of top quark properties, various new physics channels, and kinematical characteristics of top quark events as significant backgrounds to a number of other processes.

Many details of theoretical studies and experimental analysis of single top production and decay properties could be found in a number of review papers Ref. [33]–[39].

2 Manifestation of "New Physics" in single top quark production processes.

Since the top is so heavy and point-like at the same time one might expect possible deviations from the SM predictions more likely occur in the top sector.

Single top processes are extremely interesting for different types of possible manifestations of "New Physics". In general, two main situations are possible depending on relations between characteristic collision energy and thresholds of possible new states. If the collision energy is smaller than the production threshold $E_{\text{collisions}} < E_{\text{threshold}}$ new states can not be produced directly and "New Physics" may manifest as deviations in production cross sections and kinematic distributions due to possible anomalous couplings or the interference of new resonances with the SM contribution below thresholds. In the case of single top, the anomalous couplings could be anomalous Wtb and/or FCNC couplings. If the collision energy is greater than the production threshold $E_{\text{collisions}} > E_{\text{threshold}}$ new states can be produced directly in the production or in the decay of the top quark. Such states are predicted in many extensions of the SM and could include additional heavy vector (W′) or heavy scalar (H^+) bosons decaing to top and bottom quarks or rather light not yet excluded charged scalars appearing in the top quark decay.

3 Anomalous Wtb couplings

Generic parametrisation of the anomalous couplings follows from the effective Lagrangian approach [40]. There are number of dimension 6 effective operators which preserve the SM gauge invariance [41, 42] and lead to modifications of Wtb vertex. The effective Wtb vertex has the following form [43]:

$$\mathcal{L} = \frac{g}{\sqrt{2}} V_{tb} \bar{b} \gamma^{\mu} (f_V^L P_L + f_V^R P_R) t W_{\mu}^{-} - \frac{g}{\sqrt{2}} \bar{b} \frac{V_{tb} \sigma^{\mu\nu} \partial_\nu W_{\mu}^{-}}{M_W} (f_T^L P_L + f_T^R P_R) t + h.c. \quad (1)$$

where M_W is the mass of the W boson, V_{tb} is the Cabibbo-Kobayashi-Maskawa matrix element, $P_{L,R} = \frac{1 \mp \gamma_5}{2}$, $\sigma_{\mu\nu} = \frac{i}{2}(\gamma_\mu \gamma_\nu - \gamma_\nu \gamma_\mu)$, g is the coupling constant of the weak interaction, the form factor f_V^L (f_V^R) represents the left-handed (right-handed) vector coupling, and f_T^L (f_T^R) represents the left-handed (right-handed) tensor coupling. The anomalous couplings $f_V^L, f_V^R, f_T^L, f_T^R$ are related to the constants in front of the effective operators [41] in the following way:

$$|f_V^L| = 1 + |C_{\phi q}^{(3,3+3)}| \frac{v^2}{V_{tb}\Lambda^2}, \quad |f_V^R| = \frac{1}{2} |C_{\phi\phi}^{33}| \frac{v^2}{V_{tb}\Lambda^2}, \quad |f_T^L| = \sqrt{2} |C_{dW}^{33}| \frac{v^2}{V_{tb}\Lambda^2}, \quad |f_T^R| = \sqrt{2} |C_{uW}^{33}| \frac{v^2}{V_{tb}\Lambda^2}. \quad (2)$$

Note that the couplings C are naturally of the order of unity. Therefore, one may expect the natural value of f couplings of the order of $\frac{v^2}{\Lambda^2}$ which is about 0.05 or less. The theoretical estimations performed in Refs. [44–47] have shown the Tevatron and the LHC collider potentials showing the natural size of the parameters which could be achieved. The expected limits could not be much better because uncertainties are dominated by systematics. Expected bounds could be improved by a factor of $2 \div 3$ at a Linear Collider esspecially if the $e\gamma$-collision mode will be realised [48, 49].

In the single top processes the Wtb anomalous couplings contribute to the production and to the subsequent decay of produced top quark, and also change the total width of the top quark. The resulting dependence of the signal process on the anomalous parameters is more complicated than a simple polinomial structure. Such a dependence affects spin correlations between production and decay and can be exploited in experimental analysis. For example, anomalous operators are detectable in the W boson helicity fractions in the decay of top quarks [50, 51]. This property is used in many LHC measurements of W boson helicity fractions [52–54] The tightest limits coming from W boson helicity fractions are obtained by the ATLAS Collboration [52] at 7 TeV: $-0.20 < f_V^R < 0.23$, $-0.14 < f_T^L < 0.11$, $-0.08 < f_T^R < 0.04$, and by the CMS Collaboration [55]: $-0.05 < f_T^R < 0.03$

The recent and most tight direct limits[1] on the anomalous couplings f_V^R and f_T^L in Wtb vertex are achieved by the CMS Collaboration in the single top quark production process (pp collisions at $\sqrt{s} = 7$ TeV) [57]. New method of modelling of the anomalous interactions in this analysis is performed using auxillary vector fields [58]. The resulting CMS limits are: $|f_V^L| > 0.90$, $|f_V^R| < 0.34$, $|f_T^L| < 0.09$.

The best direct limit on the f_T^R coupling in Wtb vertex is obtained from combination of measurements in the single and the pair top quark production processes on the Tevatron by the D0 Collaboration [59]: $|f_T^R| < 0.35$. There are also results for f_V^R and f_T^L couplings: $|f_V^R| < 0.55$, $|f_T^L| < 0.22$. Assuming the scale $\Lambda = 1$ TeV these limits translate to the corresponding limits on the anomalous operators to be $|C_{\phi q}^{(3,3+3)}| < 14.7$, $|C_{\phi\phi}^{33}| < 18.0$, $|C_{dW}^{33}| < 2.5$, and $|C_{uW}^{33}| < 4.1$.

3.1 FCNC top quark anomalous couplings

FCNC top quark anomalous couplings can be probed in the production or in the rare decays of top quark [33]. In particular, tug and tcg FCNC couplings may affect the single top production rate,

[1]Indirect limits on the anomalous Wtb couplings follow from measurements of the $b \to s\gamma$ decays[56], however, the limits are obtained under the assumption that there are no other BSM contributions in the loop.

what was exploited at the Tevatron and the LHC to set the corresponding limits. The gauge invariant effective Lagrangian describing the top FCNC interections has the following form:

$$\mathcal{L}_{\rm FCNC} = \frac{1}{\Lambda}\left[\kappa_{\rm tq}^{\gamma,Z} e\bar{\rm t}\sigma_{\mu\nu}{\rm q}F^{\mu\nu}_{\gamma,Z} + \kappa_{\rm tq}^{g} g_s\bar{\rm t}\sigma^{\mu\nu}\frac{\lambda^i}{2}{\rm q}G^i_{\mu\nu}\right] + h.c. \quad (3)$$

where $q = u$ or c, with u, c and t representing the quark fields; $\kappa_{\rm tq}^{\gamma,Z}$ and $\kappa_{\rm tq}^{g}$ define the strength of couplings with the photon, Z and gluon, respectively; g_s and λ^i are the strong coupling constant and color matrices; $G^i_{\mu\nu}$ is the gauge field tensor of the gluon. From the above effective Lagrangian one easily obtains the partial rare decay widths: $\Gamma({\rm t}\to{\rm qg}) = \left(\frac{\kappa^g_{\rm tq}}{\Lambda}\right)^2 \frac{8}{3}\alpha_s m_{\rm t}^3$. The partial width is directly proportional to the ratio $\left(\frac{\kappa^g_{\rm tq}}{\Lambda}\right)^2$, therefore various authors and collaborations present limits either in terms of constants κ or in terms of the corresponding branching ratio. The representative diagrams contributing to various processes [60–62] of single top production due to presence of FCNC interactions are shown in Fig. 1. The model independent analysis based on the signal diagrams in

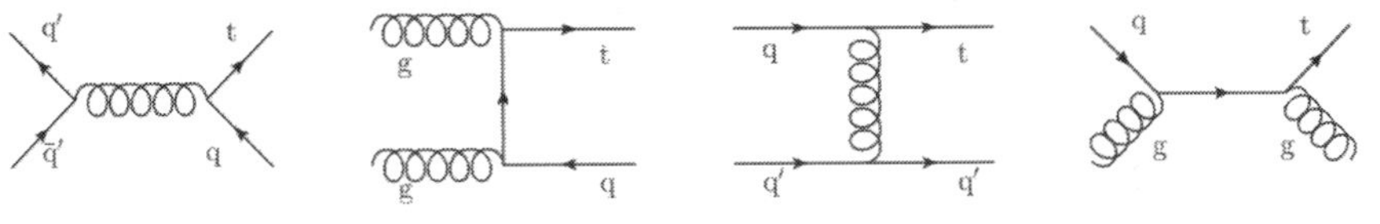

Figure 1. The representative leading order Feynman diagrams with FCNC tqg couplings.

Fig. 1 corresponds to the case when an extra hard jet is detected in association with the top quark. The most recent results on FCNC anomalous coupling limits in terms of both couplings and branchings are presented in Table 1. The accumulated luminosities for each of the analysis are listed explicitly in the table. One should stress the best current limits are far above very small SM value for the decay branching ratio $BR^{SM}({\rm t}\to{\rm cg}) \approx 5\times10^{-11}$, however, being in the interesting range for some extensions of the SM [33].

Table 1. Observed 95% C.L. upper limits on anomalous FCNC couplings and corresponding branching ratios.

	κ^g_{tu}/Λ, TeV^{-1}	κ^g_{tc}/Λ, TeV^{-1}	$Br(t\to ug)$	$Br(t\to cg)$
D0 (1.96 TeV, 2.3 fb^{-1})[63]	1.3×10^{-2}	5.7×10^{-2}	2.0×10^{-4}	3.9×10^{-4}
CDF (1.96 TeV, 2.2 fb^{-1})[64]			3.9×10^{-4}	5.7×10^{-3}
ATLAS (7 TeV, 2.05 fb^{-1})[65]	6.9×10^{-3}	1.6×10^{-2}	5.7×10^{-5}	2.7×10^{-4}
ATLAS (8 TeV, 20 fb^{-1})[66]	5.8×10^{-3}	1.3×10^{-2}	4.0×10^{-5}	1.7×10^{-4}
CMS (7 TeV, 5 fb^{-1})[57]	1.8×10^{-2}	5.6×10^{-2}	3.6×10^{-4}	3.4×10^{-3}

3.2 Top quark in the decay of new charged vector boson W′

Single top could be produced in the decay of new charged vector (W′) and observed as s−channel resonance decaying to the top and bottom quarks. Many extensions of the SM, in particular, composite models, modern variants of Technicolor models, superstrings inspired and Grand Unification models, models with extra dimensions, Left-Right symmetric models etc., predict an existence of such new vector boson (bosons) W′. In searches for W′ in a model independent way, the effective Lagrangian approach is used. Corresponding effective Lagrangians have the following form, neglecting higher-dimensional structures:

$$\mathcal{L} = \frac{V_{q_iq_j}}{2\sqrt{2}} g_w \bar{\mathrm{q}}_i \gamma_\mu (a^R_{q_iq_j}(1+\gamma^5) + a^L_{q_iq_j}(1-\gamma^5)) \mathrm{W}'^\mu \mathrm{q}_j + \mathrm{H.c.} \tag{4}$$

where $a^R_{q_iq_j}, a^L_{q_iq_j}$ - left and right couplings of W′ to quarks, $g_w = e/(s_w)$ is the SM weak coupling constant and $V_{q_iq_j}$ is the SM CKM matrix element. The notations are taken such that for so-called SM-like W′ $a^L_{q_iq_j} = 1$ and $a^R_{q_iq_j} = 0$. For the W′ searches the limits on W′ mass depend on both left and right couplings a^L, a^R, and for the case of right-handed interaction of W′ they depend on a relation between the W′ mass and the mass of possible right-handed neutrino ν_R. If the decay of W′ to the right-handed neutrino is kinematically allowed, the branching ratio of W′ decay to top and bottom quarks is smaller and corresponding mass limits are expected to be worse. The NLO corrections in the s−channel W′ production with subsequent decay were computed in Ref. [67] and an influence of W′–W^{SM} interference was demonstrated in Ref. [68]. In the case when the W′ is a first KK state in models with extra deminsions, an additional interference with the rest of KK tower should be taken into account Ref. [69]. The 95% C.L. limits obtained by the Tevatron at $\sqrt{s} = 1.96$ TeV and the LHC collaborations at $\sqrt{s} = 7, 8$ and 13 TeV are collected in the Table 2 for the cases when the couplings a^L, a^R are equal to 1, 0 (Left), 0, 1 (Right) and 1, 1 (Left, Right).

Table 2. Observed 95% C.L. limits on the lowest bound of the mass of W′ boson.

Scenario	$M_{W'}$(Left)	$M_{W'}$(Right), $M_{W'} > M_{\nu_R}$	$M_{W'}$(Right), $M_{W'} < M_{\nu_R}$	$M_{W'}$(Left, Right), $M_{W'} > M_{\nu_R}$
D0 (1.96 TeV)[70]	863 GeV	885 GeV	890 GeV	916 GeV
CDF (1.96 TeV)[71]		800 GeV	825 GeV	
CMS (8 TeV)[72]	1.92 TeV	2.15 TeV	2.02 TeV	2.15 TeV
ATLAS (8 TeV)[73]	1.70 TeV		1.92 TeV	

3.3 Charged Higgs boson and top quark

Charged Higgs bosons appear if the Higgs sector includes two (or more) isodoublets. The most known is the two-Higgs-doublet model (THDM) [?] and its special case, the effective Higgs potential of the Minimal Supersymmetric Standard Model (MSSM) [?] with Yukawa sector of Type-II. Five physical scalar fields appear in the two-doublet models: two CP-even Higgs bosons h and H, a CP-odd Higgs boson A and two charged Higgs bosons $H^\pm$. If Yukawa sector of the model is Type-II, which is the case of MSSM, the couplings of the charged Higgs boson to u-type quarks are proportional to $\cot\beta$ and the couplings to the d-type quarks and charged leptons are proportinal to $\tan\beta$ ($\tan\beta = v_2/v_1$, $\sqrt{v_1^2 + v_2^2} = v = 246$ GeV). Charged Higgs interaction Lagrangian can be written as

$$\mathcal{L} = \frac{g_w V_{q_iq_j}}{2\sqrt{2}} \mathrm{H}^+ \bar{\mathrm{q}}_i \left[g_L^{ij}\left(\frac{1-\gamma^5}{2}\right) + g_R^{ij}\left(\frac{1+\gamma^5}{2}\right)\right] \mathrm{q_j} \tag{5}$$

The LEP2 experiments defined a model-independent lower limit of about 80 GeV on the charged Higgs mass [74] at a 95% CL.

If the top quark can decay through the channel $\mathrm{t} \to \mathrm{H^+b}$, so the mass of the charged Higgs boson is smaller than $m_\mathrm{t} - m_\mathrm{b}$, main single charged Higgs boson production mechanism is $\mathrm{t\bar{t}} \to \mathrm{H^+W^-b\bar{b}}$

when the top decays to $H^+\bar{b}$ and the antitop decays to W^-b. In the MSSM scenarios frequently considered [75] for moderate $\tan\beta$ the charged Higgs decays to $\tau^+\nu_\tau$. The ATLAS and CMS experiments determined an upper limits on $BR(t \to H^+b)$ with the following decay $H^+ \to \tau^+\nu_\tau$, τ^+ and W^- decay hadronically. 95% CL upper limits on $BR(t \to H^+b) \times BR(H^+ \to \tau^+\nu_\tau)$ in the mass range $m_{H^\pm} = 80 \div 160$ GeV on the level 0.2÷1.2% have been set by ATLAS experiment [76] using $\sqrt{s} = 8$ TeV data. For the $e\tau$ and $\mu\tau$ final states 95% CL upper limit of 3÷9% for $BR(t \to H^+b)$ have been found by ATLAS and CMS experiments in the mass range $m_{H^\pm} = 80 \div 160$ GeV assuming $BR(H^+ \to \tau^+\nu_\tau) = 1$ at the energy $\sqrt{s} = 7$ TeV [77]. In the hadronic decay mode $H^+ \to c\bar{s}$ which is dominant for small $\tan\beta$ the upper limit on the level of 1÷5% (95% CL) have been set by ATLAS experiment on $BR(t \to H^+b)$ assuming that $BR(H^+ \to c\bar{s}) = 1$ at the energy $\sqrt{s} = 7$ TeV [78].

If the charged Higgs boson mass exceeds $m_t - m_b$, the main production mechanism is the bottom-top fusion (for the four-flavor scheme and for the five-flavor scheme). In the four-flavor scheme there are no b-quark PDF's causing a double-counting removal procedure [79] similar to the procedure for single top quark production. The contributions of the four- and the five-flavor schemes are calculated at NLO and combined using socalled "Santander matching scheme" [80]. In a number of popular MSSM scenarios the $H^+ \to \tau^+\nu_\tau$ mode dominates at moderate masses of $m_{H^\pm} \sim 200 \div 220$ GeV. For large values of $m_{H^\pm}$ and large $\tan\beta$ the channel $H^\pm \to tb$ becomes dominant. However, the contribution of $H^+ \to \tau^+\nu_\tau$ is non-negligible allowing one to set an upper limit of 0.004÷0.8 pb on $\sigma(pp \to t\bar{b}H^+) \times BR(H^+ \to \tau^+\nu_\tau)$ for $m_{H^\pm} = 200 \div 1000$ GeV using ATLAS data at the energy $\sqrt{s} = 8$ TeV [76]. First results of the CMS experiment on the direct charged Higgs boson production ($m_{H^\pm} > m_t - m_b$) in the $H^+ \to t\bar{b}$ at the energy $\sqrt{s} = 8$ TeV can be found in [81].

The available experimental limits for the charged scalar Higgs particle decaying to top and bottom quarks are published by D0 collaboration [82]. The mass limits are given in the mass range greater than 180 GeV ($m_{H^+} > m_t + m_b$) since lower mass range is covered with better precision from the top pair production with the top decay to the charged Higgs and b-quark. The limits strongly depend on the ratio of two Higgs vacuum expectation values $\tan\beta$ and the type of 2HDMs (Type I, Type II and Type III).

4 Conclusions and outlook

The top quark, being the heaviest elementary particle discovered so far, has been observed first in the pair-production process and later in the single production mode. Single top quark registration at the Tevatron and then at the LHC is an important confirmation of the structure of the electroweak part of the Standard Model. The single top production cross sections measured at the Tevatron and the LHC are in a reasonable agreement with the SM computations including NLO and in several cases NNLO corrections. The single top production is a powerful tool to search for a delicate deviations from the SM. No such deviations have been observed yet but the limits on the anomalous Wtb and FCNC top quark couplings, as well as on the parameters of new vector and scalar bosons appearing in the top quark production or its subsequent decays have been reliably extracted. The mass limits for new resonances are expected to be much higher with increasing of accumulated statistics at nominal LHC energy and in future LHC runs with higher luminosity and energy. While the limits on anomalous couplings will be dominated by systematic uncertainties and therefore might be improved only by a factor of two-three, much tighter limits at a percentage level of accuracy could be achieved at a future Linear Collider.

Acknowledgments

The authors are grateful to numerous colleagues from theory groups and numerous colleagues from the single top groups of the D0 and the CMS collaborations for clarifying discussions and joined studies. The work was supported by grant 16-12-10280 of the Russian Science Foundation.

References

[1] V. Khachatryan *et al.* [CMS Collaboration], Phys. Rev. D **93**, no. 7, 072004 (2016)
[2] M. Jezabek and J. H. Kuhn, Phys. Rev. D **48**, 1910 (1993) [Erratum-ibid. D **49**, 4970 (1994)]; M. Jezabek and J. H. Kuhn, Nucl. Phys. B **314**, 1 (1989).
[3] I. I. Bigi, Y. L. Dokshitzer, V. A. Khoze, J. H. Kuhn and P. M. Zerwas, Phys. Lett. B **181** (1986) 157.
[4] S. S. D. Willenbrock and D. A. Dicus, Phys. Rev. D 34, 155 (1986).
[5] C. P. Yuan, Phys. Rev. D 41, 42 (1990).
[6] G. V. Dzhikia and S. R. Slabospitsky, Sov. J. Nucl. Phys. 55, 1387 (1992); Phys. Lett. B 295, 136 (1992).
[7] S. Cortese and R. Petronzio, Phys. Lett. B 253, 494 (1991).
[8] R. K. Ellis and S. J. Parke, Phys. Rev. D 46, 3785 (1992).
[9] G. Bordes and B. van Eijk, Z. Phys. C 57, 81 (1993).
[10] D. O. Carlson, E. Malkawi, and C. P. Yuan, Phys. Lett. B 337, 145 (1994).
[11] T. Stelzer and S. Willenbrock, Phys. Lett. B 357, 125 (1995).
[12] A. P. Heinson, A. S. Belyaev, and E. E. Boos, Phys. Rev. D 56, 3114 (1997).
[13] S. Bar-Shalom, G. Eilam, A. Soni, and J. Wudka, Phys. Rev. D 57, 2957 (1998).
[14] T. Tait and C. P. Yuan, hep-ph/9710372.
[15] A. S. Belyaev, E. E. Boos, and L. V. Dudko, Phys. Rev. D 59, 075001 (1999)
[16] G. Bordes and B. van Eijk, Nucl. Phys. B 435, 23 (1995)
[17] R. Pittau, Phys. Lett. B 386, 397 (1996)
[18] M. C. Smith and S. Willenbrock, Phys. Rev. D 54, 6696 (1996); T. Stelzer, Z. Sullivan, and S. Willenbrock, Phys. Rev. D 56, 5919 (1997);
[19] T. Stelzer, Z. Sullivan, and S. Willenbrock, Phys. Rev. D 58, 094021 (1998).
[20] B. W. Harris *et al.*, Phys. Rev. D 66, 054024 (2002).
[21] Z. Sullivan, Phys. Rev. D 70, 114012 (2004).
[22] J. Campbell, R. K. Ellis, and F. Tramontano, Phys. Rev. D 70, 094012 (2004).
[23] J. M. Campbell and F. Tramontano, Nucl. Phys. B **726**, 109 (2005); J. M. Campbell, R. Frederix, F. Maltoni and F. Tramontano, Phys. Rev. Lett. **102**, 182003 (2009).
[24] Q. H. Cao, R. Schwienhorst, and C. P. Yuan, Phys. Rev. D 71, 054023 (2005); Q. -H. Cao, R. Schwienhorst, J. A. Benitez, R. Brock, C. -P. Yuan, Phys. Rev. **D72**, 094027 (2005); S. Heim, Q. -H. Cao, R. Schwienhorst, C. -P. Yuan, Phys. Rev. **D81**, 034005 (2010);
[25] R. Schwienhorst, C. -P. Yuan, C. Mueller and Q. -H. Cao, Phys. Rev. D **83**, 034019 (2011).
[26] N. Kidonakis, Phys. Rev. D **74**, 114012 (2006); N. Kidonakis, Phys. Rev. D **75**, 071501 (2007); N. Kidonakis, Nucl. Phys. A **827**, 448C (2009); N. Kidonakis, Phys. Rev. Lett. **102**, 232003 (2009); N. Kidonakis, Phys. Rev. D **81**, 054028 (2010); N. Kidonakis, Phys. Rev. D **82**, 054018 (2010); N. Kidonakis, Phys. Rev. D **83**, 091503 (2011);
[27] N. Kidonakis, Phys. Part. Nucl. **45**, no. 4, 714 (2014).
[28] M. Brucherseifer, F. Caola and K. Melnikov, Phys. Lett. B **736**, 58 (2014)

[29] E. L. Berger, J. Gao, C.-P. Yuan and H. X. Zhu, arXiv:1606.08463 [hep-ph].
[30] M. Jezabek and J. H. Kuhn, Phys. Lett. B **329**, 317 (1994).
[31] G. Mahlon and S. J. Parke, Phys. Rev. D 55, 7249 (1997); G. Mahlon and S. J. Parke, Phys. Lett. B 476, 323 (2000).
[32] E. E. Boos and A. V. Sherstnev, Phys. Lett. B 534, 97 (2002).
[33] M. Beneke *et al.*, "Top quark physics," arXiv:hep-ph/0003033.
[34] T. M. P. Tait and C. P. P. Yuan, Phys. Rev. D **63**, 014018 (2000)
[35] S. Willenbrock, hep-ph/0211067.
[36] R. Kehoe, M. Narain and A. Kumar, Int. J. Mod. Phys. A **23**, 353 (2008)
[37] A. Heinson and T. R. Junk, Ann. Rev. Nucl. Part. Sci. **61**, 171 (2011)
[38] E. Boos and L. Dudko, Int. J. Mod. Phys. A **27**, 1230026 (2012)
[39] E. Boos, O. Brandt, D. Denisov, S. Denisov and P. Grannis, Phys. Usp. **58**, no. 12, 1133 (2015)
[40] W. Buchmuller and D. Wyler, Nucl. Phys. B **268**, 621 (1986).
[41] J. A. Aguilar-Saavedra, Nucl. Phys. B **812** (2009) 181
[42] C. Zhang, N. Greiner, and S. Willenbrock, arXiv:1201.6670 [hep-ph]; C. Zhang and S. Willenbrock, Phys. Rev. D **83**, 034006 (2011).
[43] G. L. Kane, G. A. Ladinsky and C. P. Yuan, Phys. Rev. D **45**, 124 (1992).
[44] E. Boos, L. Dudko, and T. Ohl, Eur. Phys. J. C 11, 473 (1999)
[45] S. Tsuno, I. Nakano, Y. Sumino and R. Tanaka, Phys. Rev. D **73**, 054011 (2006)
[46] J. A. Aguilar-Saavedra, Nucl. Phys. B **804**, 160 (2008)
[47] W. Bernreuther, P. Gonzalez and M. Wiebusch, Eur. Phys. J. C **60**, 197 (2009)
[48] E. Boos, M. Dubinin, A. Pukhov, M. Sachwitz, H. J. Schreiber, Eur. Phys. J. **C21**, 81-91 (2001).
[49] E. Boos, A. Pukhov, M. Sachwitz and H. J. Schreiber, Phys. Lett. B **404**, 119 (1997)
[50] J. A. Aguilar-Saavedra, J. Carvalho, N. F. Castro, F. Veloso and A. Onofre, Eur. Phys. J. C **50** (2007) 519
[51] C. -R. Chen, F. Larios and C. -P. Yuan, Phys. Lett. B **631**, 126 (2005) [AIP Conf. Proc. **792**, 591 (2005)]
[52] G. Aad *et al.* [ATLAS Collaboration], JHEP **1206** (2012) 088
[53] V. Khachatryan *et al.* [CMS Collaboration], JHEP **1501** (2015) 053
[54] [CMS Collaboration], CMS-PAS-TOP-12-025.
[55] S. Chatrchyan *et al.* [CMS Collaboration], JHEP **1310** (2013) 167
[56] J. Drobnak, S. Fajfer and J. F. Kamenik, Nucl. Phys. B **855**, 82 (2012);
[57] CMS Collaboration [CMS Collaboration], CMS-PAS-TOP-14-007.
[58] E. Boos, V. Bunichev, L. Dudko and M. Perfilov, arXiv:1607.00505 [hep-ph].
[59] V. M. Abazov *et al.* [D0 Collaboration], Phys. Lett. B **713**, 165 (2012)
[60] T. Han, M. Hosch, K. Whisnant, B. -L. Young and X. Zhang, Phys. Rev. D **58**, 073008 (1998)
[61] J. J. Liu, C. S. Li, L. L. Yang and L. G. Jin, Phys. Rev. D **72**, 074018 (2005)
[62] J. Gao, C. S. Li, J. J. Zhang, H. X. Zhu, Phys. Rev. **D80**, 114017 (2009);
[63] V. M. Abazov *et al.* [D0 Collaboration], Phys. Lett. B **693**, 81 (2010)
[64] T. Aaltonen *et al.* [CDF Collaboration], Phys. Rev. Lett. **102**, 151801 (2009)
[65] G. Aad *et al.* [ATLAS Collaboration], Phys. Lett. B **712**, 351 (2012)
[66] G. Aad *et al.* [ATLAS Collaboration], Eur. Phys. J. C **76**, no. 2, 55 (2016)
[67] Z. Sullivan, Phys. Rev. D **66**, 075011 (2002)
[68] E. Boos, V. Bunichev, L. Dudko and M. Perfilov, Phys. Lett. B **655**, 245 (2007)

[69] E. E. Boos, I. P. Volobuev, M. A. Perfilov and M. N. Smolyakov, Theor. Math. Phys. **170**, 90 (2012)

[70] V. M. Abazov *et al.* [D0 Collaboration], Phys. Lett. B **699**, 145 (2011)

[71] T. Aaltonen *et al.* [CDF Collaboration], Phys. Rev. Lett. **103**, 041801 (2009)

[72] V. Khachatryan *et al.* [CMS Collaboration], JHEP **1602**, 122 (2016)

[73] G. Aad *et al.* [ATLAS Collaboration], Phys. Lett. B **743** (2015) 235

[74] ALEPH Collaboration, Phys. Lett. B **543**, 1 (2002); DELPHI Collaboration, Eur. Phys. J. C **34**, 399 (2004); L3 Collaboration, Phys. Lett. B **575**, 208 (2003); OPAL Collaboration, Eur. Phys. J. C **72**, 2076 (2012).

[75] S. Heinemeyer, W. Hollik, and G. Weiglein, Comput. Phys. Commun. **124**, 76 (2000)

[76] ATLAS Collaboration, JHEP **03**, 088 (2015)

[77] ATLAS Collaboration, JHEP **03**, 076 (2013); CMS Collaboration, JHEP **07**, 143 (2012).

[78] ATLAS Collaboration, Eur. Phys. J. C **73**, 2465 (2013).

[79] LHC Higgs Cross Section Working Group Collaboration, arXiv:1307.1347[hep-ph]

[80] R. Harlander, M. Kramer, and M.Schumacher, CERN Technical Report PH-TH/2011-134, 2011 (arXiv:1112.3478[hep-ph])

[81] CMS Collaboration, JHEP **1511**, 018 (2015)

[82] V. M. Abazov *et al.* [D0 Collaboration], Phys. Rev. Lett. **102**, 191802 (2009)

EPJ Web of Conferences **125**, 02009 (2016) DOI: 10.1051/epjconf/201612502009
QUARKS-2016

New Physics at 1TeV?

S.I. Godunov[1,2,⋆]

[1] *Institute for Theoretical and Experimental Physics, 117218, Moscow, Russia*
[2] *Novosibirsk State University, 630090, Novosibirsk, Russia*

Abstract. We consider the extension of the Standard Model with an extra scalar S which decays can be responsible for the diphoton excess with invariant mass $\sim$ 750 GeV observed at the 13 TeV LHC run. Two scenarios of S production are considered: gluon fusion through a loop of heavy isosinglet quark(s) and photon fusion through a loop of heavy isosinglet leptons. In the second case many heavy leptons are needed or/ and they should have large electric charges in order to reproduce experimental data on $\sigma_{pp\to SX} \cdot \mathrm{Br}(S \to \gamma\gamma)$.

1 Introduction

ATLAS and CMS collaborations recently announced a small enhancement over smooth background of two photon events with invariant mass 750 GeV [1, 2]. Though statistical significance of this enhancement is not large (within 3 standard deviations), it induced a whole bunch of theoretical papers devoted to its interpretation. The reason for this explosive activity is clear: maybe the Standard Model is changed at one TeV scale, and we are witnessing the first sign of New Physics.

Let us suppose that the observed enhancement is due to the $\gamma\gamma$ decay of a new particle. Then it should be a boson with spin different from one: the simplest possibility is a scalar particle S with $m_S = 750$ GeV. Since it decays to photons, it should be neutral and colorless, therefore in pp-collisions it can be produced in gluon-gluon fusion through the loop of colored particles and in photon-photon fusion through the loop of charged particles. Let us suppose that S decays to these particles are kinematically forbidden, otherwise $\mathrm{Br}(S \to \gamma\gamma)$ reduces significantly which makes $S \to \gamma\gamma$ decays unobservable at the LHC.

We suppose that the particles propagating in the loop are Dirac fermions, so they have tree level masses, and that they are $SU(2)_\mathrm{L}$ singlets. Nonzero hypercharges provide couplings of these particles with photon and Z-boson. These particles can be quark(s) (color triplets) T_i or lepton(s) (color singlets) L_i. They couple with S by Yukawa interactions with coupling constants λ^i_T and λ^i_L correspondingly.

This talk is based on the paper [3] written in collaboration with A.N. Rozanov, M.I. Vysotsky, and E.V. Zhemchugov.

⋆e-mail: sgodunov@itep.ru

2 Quarkophilic S

In the case of one heavy quark T the following terms should be added to the Standard Model lagrangian:

$$\Delta\mathcal{L} = \tfrac{1}{2}(\partial_\mu S)^2 - \tfrac{1}{2}m_S^2 S^2 + \bar{T}\gamma_\mu(\partial_\mu - \tfrac{i}{2}g_s A_\mu^i \lambda_i - ig'\tfrac{Y_T}{2}B_\mu)T + m_T\bar{T}T + \lambda_T\bar{T}TS, \tag{1}$$

where A_μ^i and B_μ are gluon and $U(1)$ gauge fields respectively, and λ_i are Gell-Mann matrices. S coupling with gluons is generated by the T-quark loop:

$$M_{gg} = \frac{\alpha_s}{6\pi}\frac{\lambda_T}{m_T}F(\beta)G_{\mu\nu}^{(1)}G_{\mu\nu}^{(2)}S, \tag{2}$$

$$\beta = \left(\frac{2m_T}{m_S}\right)^2, \quad F(\beta) = \frac{3}{2}\beta\left[1 - (\beta - 1)\arctan^2\frac{1}{\sqrt{\beta - 1}}\right]. \tag{3}$$

Inclusive cross section of S production in pp collision at the LHC through gluon fusion is given by:

$$\sigma_{pp\to SX} = \frac{\alpha_s^2}{576\pi}\left(\frac{\lambda_T}{m_T}\right)^2 |F(\beta)|^2 m_S^2 \left.\frac{dL_{gg}}{d\hat{s}}\right|_{\hat{s}=m_S^2}, \tag{4}$$

where the so-called gluon-gluon luminosity is given by the integral over gluon distributions:

$$\frac{dL_{gg}}{d\hat{s}} = \frac{1}{s}\int\limits_{\ln\sqrt{\tau_0}}^{-\ln\sqrt{\tau_0}} g(\sqrt{\tau_0}e^y, Q^2)g(\sqrt{\tau_0}e^{-y}, Q^2)dy, \tag{5}$$

$\tau_0 = \hat{s}/s$, $s = (13\text{ TeV})^2$, and we use $Q^2 = m_S^2$. Integrating gluon distributions from [4] for $\sqrt{\hat{s}} =$ 750 GeV, $\sqrt{s} = 13$ TeV, we get $dL_{gg}/d\hat{s} \approx 4.0$ nb, $m_S^2\, dL_{gg}/d\hat{s} \approx (1/0.69\text{ nb}) \cdot 4.0\text{ nb} \approx 5.8$. In order to take into account gluon loop corrections, (4) should be multiplied by the so-called K-factor which is close to 2 for $\sqrt{s} = 13$ TeV, according to [5] (see also Fig. 2 in [6]).

In this way for $m_T = m_S$ and $\lambda_T = 1$, substituting $\alpha_s(m_S) = 0.090$, we obtain $\sigma_{pp\to SX} \approx 41$ fb, which should be multiplied by $\text{Br}(S \to \gamma\gamma)$ in order to be compared with experimental observations [1, 2]. Total width of S is dominated by the $S \to gg$ decay, and from (2) we get:

$$\Gamma_{S\to gg} = \left(\frac{\alpha_s}{6\pi}\right)^2 \cdot 8\frac{m_S^3\lambda_T^2}{16\pi m_T^2}|F(\beta)|^2 \approx 3.1\text{ MeV}. \tag{6}$$

Thus for the models we consider, the width of S should be much smaller than 45 GeV which is preferred by the ATLAS data. Let us note that CMS data prefer narrow S; see also [7].

T-quark loop contributes to $S \to \gamma\gamma$ decay as well. The corresponding matrix element equals

$$M_{\gamma\gamma} = \frac{\alpha}{3\pi}\frac{\lambda_T}{m_T}F(\beta)F_{\mu\nu}^{(1)}F_{\mu\nu}^{(2)} \cdot 3_c Q_T^2, \tag{7}$$

where the factor 3_c corresponds to the three colors, and Q_T is the T-quark electric charge. For $\gamma\gamma$ width we get:

$$\Gamma_{S\to\gamma\gamma} = \left(\frac{\alpha}{3\pi}\right)^2 (3_c Q_T^2)^2 \frac{m_S^3\lambda_T^2}{16\pi m_T^2}|F(\beta)|^2 \approx 22\text{ keV}, \tag{8}$$

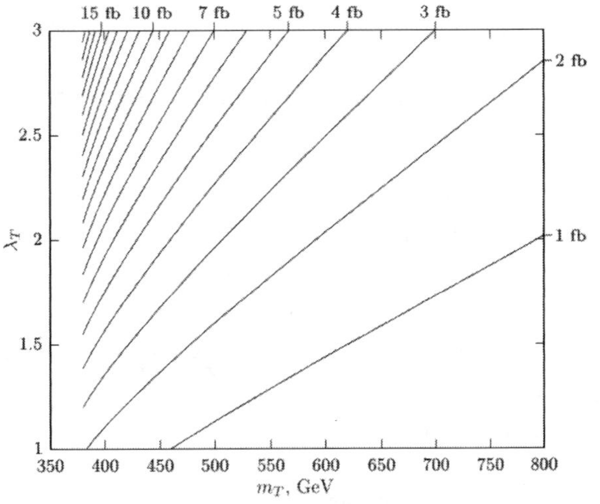

Figure 1. Contour plot of $\sigma_{pp\to SX} \cdot \mathrm{Br}(S \to \gamma\gamma)$.

$$\mathrm{Br}(S \to \gamma\gamma) \approx \left(\frac{\alpha}{\alpha_s}\right)^2 \frac{(3_c Q_T^2)^2}{2} \approx 0.0070, \tag{9}$$

where we substituted $Q_T = 2/3$ and $\alpha = 1/125$.[1] Finally, we obtain:

$$\sigma_{pp\to SX} \cdot \mathrm{Br}(S \to \gamma\gamma) \approx 0.28 \text{ fb}. \tag{10}$$

Experimental data provides a value approximately 36 times larger:

$$[\sigma_{pp\to SX} \cdot \mathrm{Br}(S \to \gamma\gamma)]_{\mathrm{exp}} \approx 10 \text{ fb}. \tag{11}$$

This value follows from the the fact that both collaborations see about 15 events with 3 fb^{-1} luminosity collected by at 13 TeV and effectivity of $\gamma\gamma$ registration $\varepsilon \approx 0.5$ [1].

In order to reproduce experimental result (11) we should suppose that six T-quarks exist. In this case $\mathrm{Br}(S \to \gamma\gamma)$ remains the same, while the cross section of S production should be multiplied by the same factor 36, and (11) is reproduced.[2]

However, unappealing multiplication of the number of T-quarks can be avoided by diminishing T-quark mass and increasing the value of λ_T. The experimental result is reproduced for $m_T = 400$ GeV and $\lambda_T = 2.5$.[3] Curves of constant $\sigma_{pp\to SX} \cdot \mathrm{Br}(S \to \gamma\gamma)$ are shown in Figure 1 on (λ_T, m_T) plot.

In the rest of this section we consider the model with one additional quark T and $m_T = 400$ GeV, $\lambda_T = 2.5$. As it was shown above, this model can explain the diphoton excess but let us check that other existing bounds are not violated.

It is natural to suppose that T-quark mixes with u-, c-, and t-quark which makes it unstable. To avoid LHC Run 1 bounds on m_T following from the search of the decays $T \to Wb$, $T \to Zt$ and $T \to Ht$ [8–10] which exclude T-quark with mass below 700 GeV, we suppose that $T - t$ mixture is

[1] Fine structure constant is substituted by its running value at $q^2 = m_S^2$.

[2] The experimental result (11) will be reproduced if at 1 TeV scale we have a "mirror image" of the Standard Model with three vector-like generations.

[3] As far as $\lambda_T^2/4\pi$ is a parameter of perturbation theory, this value of λ_T is close to the maximum allowed value in order for the perturbation theory to make sense.

small, and T-quark mixing with u- and c-quarks dominates [11]. Though since we want the mass of T-quark to be just above the threshold of $S \to TT$ decay the mixing with the first two generations should not be too large in order to prevent the diminishing of $\mathrm{Br}(S \to \gamma\gamma)$ due to $S \to TT^* \to TWq$ decays.

Since at $\sqrt{s} = 8$ TeV the gluon-gluon luminosity is 4.6 times smaller that at $\sqrt{s} = 13$ TeV, the CMS bound from Run 1 $[\sigma_{pp\to SX}\mathrm{Br}(S \to \gamma\gamma)]_{8\ \mathrm{TeV}} < 1.5$ fb [12] is (almost) not violated.

Concerning S decays, let us note that the dominant $S \to gg$ decay is hidden by the two jets background produced by strong interactions. At 8 TeV LHC energy the following upper bound was obtained [13]:

$$[\sigma_{pp\to SX} \cdot \mathrm{Br}(S \to gg)]^{\mathrm{exp}}_{8\ \mathrm{TeV}} < 30\ \mathrm{pb}. \tag{12}$$

while in the considered model $[\sigma_{pp\to SX}]^{\mathrm{theor}}_{8\ \mathrm{TeV}} \approx 0.39$pb.

Three modes of S decays to neutral vector bosons do exist and have the following hierarchy $\Gamma_{S\to\gamma\gamma} : \Gamma_{S\to Z\gamma} : \Gamma_{S\to ZZ} = 1 : 2(s_W/c_W)^2 : (s_W/c_W)^4$, where $s_W(c_W)$ is the sine (cosine) of electroweak mixing angle (here we suppose that mixing of S with Higgs doublet is negligible). Thus if the existence of S will be confirmed by new data, $S \to \gamma Z$ and $S \to ZZ$ decays should be also looked for.

Scalar S can mix with the Standard Model Higgs boson due to renormalizable interaction term $\mu\Phi^\dagger\Phi S$, where Φ is the Higgs isodoublet. Such an extension of the Standard Model was studied in our recent paper [14]. Due to doublet admixture there are tree level decays $S \to WW, ZZ, t\bar{t}$ and hh, where h is the 125 GeV Higgs boson. According to Eqs. (16)–(20) from [14], the sum of these widths equals approximately $\sin^2\alpha \cdot m_S^3/8\pi v_\Phi^2 \approx \sin^2\alpha \cdot 300$ GeV, where α is the mixing angle, and $v_\Phi = 246$ GeV is the Higgs vev. Ratio of partial widths at small α is $\Gamma_{S\to WW} : \Gamma_{S\to ZZ} : \Gamma_{S\to hh} \approx 2 : 1 : 1$. As a result, S width grows and $\mathrm{Br}(S \to \gamma\gamma)$ diminishes correspondingly.

To reproduce experimental result (11) and in order not to violate bounds from searches in these modes, the mixing angle α should be small enough. For example, for $\sin\alpha < 1/150$ we obtain at most 12 MeV (or 11%) increase of the width of S, which is acceptable. It corresponds $|\mu| < 15$ GeV which is not unnaturally small.

Let us check if for such mixing angle $S \to ZZ$ decays (due to doublet admixture) do not exceed experimental bounds obtained at the LHC. In the considered model

$$[\sigma_{pp\to SX} \cdot \mathrm{Br}(S \to ZZ)]_{13\ \mathrm{TeV}} < 33\ \mathrm{fb}, \tag{13}$$

well below experimental upper bound which, according to Fig. 11 from [15], equals 4 fb/$(\mathrm{Br}(Z \to 4\ell))^2 = 400$ fb at 2σ (see also [16]).

For 8 TeV (taking into account energy dependence of the K-factor)

$$[\sigma_{pp\to SX} \cdot \mathrm{Br}(S \to ZZ)]_{8\ \mathrm{TeV}} < 9.0\ \mathrm{fb}, \tag{14}$$

while the experimental upper bound is 60 fb (Fig. 12 from [17]).

More stringent upper bound comes from the search of $S \to hh$ decays [18] and equals 40 fb, while in our case the cross section does not exceed 10 fb.

Let us say a few words about future prospects. If the existence of S will be confirmed with larger statistics at the LHC, then it can be studied at e^+e^--colliders as well. For the cross section of S production in photon fusion, according to [19, Eq. (48.47)], [20], we have:

$$\sigma_{ee\to eeS}(s) = \frac{8\alpha^2}{m_S^3}\Gamma_{S\to\gamma\gamma}\left[f\left(\frac{m_S^2}{s}\right)\left(\ln\left(\frac{m_T^2 s}{m_e^2 m_S^2}\right) - 1\right)^2 - \frac{1}{3}\ln^3\left(\frac{s}{m_S^2}\right)\right], \tag{15}$$

$$f(z) = \left(1 + \tfrac{1}{2}z\right)^2 \ln\tfrac{1}{z} - \tfrac{1}{2}(1-z)(3+z), \tag{16}$$

and $\Gamma_{S\to\gamma\gamma}$ is given in Eq. (8). At e^+e^- collider CLIC with $s = (3\,\text{TeV})^2$ for $\lambda_T = 2.5$ and $m_T = 400$ GeV we obtain

$$\sigma^{\text{CLIC}}_{ee\to eeS} \approx 0.46\ \text{fb}. \tag{17}$$

With projected CLIC luminosity $L = 6\cdot 10^{34}/(\text{cm}^2\cdot\text{sec})$ [19, p. 393], during one accelerator year ($t = 10^7$ sec) about 300 S resonances should be produced.

3 Leptophilic S

S production in $\gamma\gamma$ fusion will be analyzed in this section (see also [21–23]).

Let us suppose that heavy leptons L_i which couple to S have electric charges Q_L, and there are N such degenerate leptons. The lagrangian is similar to that of the heavy quarks case (1):

$$\Delta\mathcal{L} = \tfrac{1}{2}(\partial_\mu S)^2 - \tfrac{1}{2}m_S^2 S^2 + \bar{L}_i\gamma_\mu(\partial_\mu - ig'\tfrac{Y_L}{2}B_\mu)L_i + m_L\bar{L}_iL_i + \lambda_L\bar{L}_iL_iS, \tag{18}$$

where we assume equal lepton masses and S couplings. For $S\to\gamma\gamma$ width we obtain:

$$\Gamma_{S\to\gamma\gamma} = \left(\frac{\alpha}{3\pi}\right)^2 (NQ_L^2)^2 \frac{m_S^3\lambda_L^2}{16\pi m_L^2}|F(\beta)|^2,\ \beta = \left(\frac{2m_L}{m_S}\right)^2. \tag{19}$$

Production of S at the LHC occurs through fusion of two virtual photons emitted by quarks which reside in the colliding protons. Let us estimate the production cross section. For the partonic cross section we get:

$$\sigma^{(\gamma\gamma)}_{q_1q_2\to q_1q_2S}(\hat{s}) = \frac{8\alpha^2}{m_S^3}e_1^2e_2^2\Gamma_{S\to\gamma\gamma}\left[f\left(\frac{m_S^2}{\hat{s}}\right)\left(\ln\left(\frac{m_L^2\hat{s}}{\Lambda^2_{\text{QCD}}m_S^2}\right)-1\right)^2 - \frac{1}{3}\ln^3\left(\frac{\hat{s}}{m_S^2}\right)\right], \tag{20}$$

where e_1, e_2 are charges of the quarks, $\hat{s} = x_1x_2s \equiv \tau s$, and $f(z)$ is given by (16). We should multiply (20) by PDFs and integrate over x_1 and x_2:

$$\sigma^{(\gamma\gamma)}_{pp\to SX}(s) = \sum_{q_1,q_2}\int\limits_{m_S^2/s}^{1}\sigma^{(\gamma\gamma)}_{q_1q_2\to q_1q_2S}(\tau s)d\tau\cdot s\cdot\frac{dL_{q_1q_2}}{d\hat{s}}(Q^2,\tau). \tag{21}$$

$$\frac{dL_{q_1q_2}}{d\hat{s}}(Q^2,\tau) = \frac{1}{s}\int\limits_{\ln\sqrt{\tau}}^{-\ln\sqrt{\tau}} q_1(x_1,Q^2)q_2(x_2,Q^2)dy, \tag{22}$$

$x_1 = \sqrt{\tau}e^y$, $x_2 = \sqrt{\tau}e^{-y}$. We take $Q^2 = m_S^2$ and use PDFs from [4].

Cross sections in the case of one heavy lepton with charge $Q_L = 1$, Yukawa coupling constant $\lambda_L = 2$ and mass $m_L = 400$ GeV are shown in Table 1. For $\Lambda_{\text{QCD}} = 300$ MeV and $\sqrt{s} = 13$ TeV we get $\sigma^{(\gamma\gamma)}_{pp\to SX} \approx 11$ ab,[4] while the experimental result (11) is three orders of magnitude larger. We come to the conclusion that $\sum NQ_L^2 \approx 30$ is needed: we need either 30 leptons with unit charges, or one lepton with charge 6, or several multicharged leptons.[5]

[4] According to Eq. (12) from the recent paper [24], this cross section equals 25 ab.

[5] If $\sigma^{(\gamma\gamma)}_{pp\to SX} = 25$ ab, then 30 should be replaced with 20.

Table 1. Cross sections (in ab) for double photon production in the leptophilic model for different values of Λ_{QCD} and proton collision energies.

		Λ_{QCD}, GeV		
		0.1	0.3	1.0
$\sqrt{s}$, TeV	7	2.5	1.9	1.3
	8	3.8	2.9	2.0
	13	15	11	7.8

It is natural to suppose that leptons with charge one mix with the Standard Model leptons and become unstable. Search for such particles was performed at the LHC, and the lower bound $m_L >$ 170 GeV was obtained [25]. See also [26], where bounds on masses and mixings of L are discussed. For masses above 200 GeV the existence of L is still relatively unconstrained.

Cross section for quasielastic S production can be estimated with the help of the following equation:

$$\sigma_{pp\to ppS} = \frac{8\alpha^2}{m_S^3}\Gamma_{S\to\gamma\gamma}\left[f\left(\frac{m_S^2}{s}\right)\left(\ln\left(\frac{s}{m_S^2}\right)-1\right)^2 - \frac{1}{3}\ln^3\left(\frac{s}{m_S^2}\right)\right]. \qquad (23)$$

For $\sqrt{s} = 13$ TeV, $\lambda_L = 2$ and $m_L = 400$ GeV it equals 4.1 ab.[6]

4 Conclusions

We analyze the possibility that the enhancement at 750 GeV diphoton invariant mass observed by ATLAS and CMS is due to decays of a new scalar S. We found that production of S in gluon fusion in a minimal model with one additional heavy Dirac quark T can have value of $\sigma_{pp\to SX}\cdot \mathrm{Br}(S\to\gamma\gamma)$ compatible with data. An upper bound on the mixing of S with $h(125)$ is obtained. If heavy leptons L are introduced instead of T, then S can be produced at LHC in photon fusion, however, in order to reproduce experimental data many leptons L_i are needed and/or they should be multicharged.

S. G. is partially supported by MK-4234.2015.2 and by RFBR under grants 16-32-60115, 14-02-00995, and 16-02-00342. S. G. is also grateful to Dynasty Foundation for the support.

References

[1] *The ATLAS collaboration*, ATLAS-CONF-2015-081 (2015)
[2] *The CMS collaboration*, CMS-PAS-EXO-15-004 (2015)
[3] Godunov S.I., Rozanov A.N., Vysotsky M.I., Zhemchugov E.V., JETP Lett. **103**, 635 (2016)
[4] Harland-Lang L. A., Martin A. D., Motylinkski P., Thorne R. S., Eur. Phys. J. C **75**, 204 (2015)
[5] Baglio J., Djouadi A., JHEP **1103**, 055 (2011)
[6] Harlander R. V., Kilgol W., Phys. Rev. Lett. **88**, 201801 (2002)
[7] Buckley M. R., Eur. Phys. J. C **76**, 345 (2016)
[8] *The ATLAS collaboration*, JHEP **1510**, 150 (2015)
[9] *The ATLAS collaboration*, JHEP **1508**, 105 (2015)
[10] *The CMS collaboration*, Phys. Rev. D **93**, 012003 (2016)
[11] Buchkremer M., Proc. of 49th Rencontres de Moriond on Electroweak Interactions and Unified Theories, 519 (2014)

[6]According to Eq. (24) from [24], quasielastic cross section is two times smaller.

[12] *The CMS collaboration*, Phys. Lett. B **750**, 494 (2015)
[13] *The ATLAS collaboration*, Phys. Rev. D **91**, 052007 (2015)
[14] Godunov S. I., Rozanov A. N., Vysotsky M. I., Zhemchugov E. V., Eur. Phys. J. C 76, 1 (2016)
[15] *The ATLAS collaboration*, ATLAS-CONF-2015-59.
[16] Franceschini R. et. al., JHEP **1603**, 144 (2016)
[17] *The ATLAS collaboration*, Eur. Phys. J. C **76**, 45 (2016)
[18] *The ATLAS collaboration*, Phys. Rev. D **92**, 092004 (2015)
[19] Olive K. A. et al. (Particle Data Group), Chin. Phys. C **38**, 090001 (2014)
[20] Budnev V. M., Ginzburg I. F., Meledin G. V., Serbo V. G., Phys. Reports **15**, 181 (1975)
[21] Fichet S., vob Gersdorff G., and Royon C., Phys. Rev. D **93**, 075031 (2016)
[22] Csaki C., Hubisz J., and Terning J., Phys. Rev. D **93**, 035002 (2016)
[23] Pilaftis A., Phys. Rev. D **93**, 015017 (2016)
[24] Harland-Lang L. A., Khoze V. A., Ryskin M. G., JHEP **1603**, 182 (2016)
[25] *The ATLAS collaboration*, JHEP **1509**, 108 (2015)
[26] A. Djouadi, J. Ellis, R. Godbole, J. Quevillon, arXiv:1601.03696.

EPJ Web of Conferences **125**, 02010 (2016) DOI: 10.1051/epjconf/201612502010
QUARKS-2016

New Physics beyond the 750 GeV diphotons events

Dmitry Gorbunov[1,⋆]

[1] *Institute for Nuclear Research of Russian Academy of Sciences, 60-th October Anniversary pr. 7a, 117312 Moscow, Russia*

Abstract. the 750 GeV diphoton anomaly observed by ATLAS and CMS collaborations in 2015, and the models suggested by theorists to explain it. The talk has been presented in the special session devoted to this anomaly and possible (new) physics behind it. Since all the new physics models proposed by the beginning of the Conference predict many new particles accessible for direct study at present and next stages of LHC operation, which is very exciting, I chose the preposition "beyond" in the title.

1 Reports by ATLAS and CMS on December 15th, 2015

The first results obtained at the new for LHC center-of-mass energy of 13 TeV have been presented at CERN Seminar on December 15th [1]. The first talk was delivered by a CMS representative, who announces the results to be published in 6+17 new preprints based on the data collected from June 2015 to November 2015. Provided by the higher center-of-mass energy the data are of a special interest to the new physics models (17 new papers), where the direct production of new particles gets amplified as compared to the previous LHC run at 8 TeV. The most reliable new data sample refers to $L = 2.2\,\mathrm{fb}^{-1}$ integral luminosity, and the models under discussion predict the production cross sections σ obeying the inequality

$$(\sigma \times L)_{13\,\mathrm{TeV},\,2.2\,\mathrm{fb}^{-1}} > (\sigma \times L)_{8\,\mathrm{TeV},\,19.8\,\mathrm{fb}^{-1}} \tag{1}$$

where integral luminosity $L = 19.8\,\mathrm{fb}^{-1}$ stands for most reliable data collected at 8 TeV by CMS. Models with sufficiently heavy new particles may naturally satisfy (1). In particular, production cross section of gluons and squarks with masses of 1.5 TeV is enhanced at 13 TeV by a factor of 35 with respect to 8 TeV. CMS new data gave no any support of the previously announced signal at 1.8-2 TeV in diboson channels WW, WZ, ZZ with weak bosons reconstructed by jets. However, the CMS collaboration observed a new signal in $\gamma\gamma$ final state at about 750 GeV, while searching for massive gravitons expected within the Randall–Sundrum models of extra spatial dimensions. The anomaly is recognized only for photons collected in the barrel part of electromagnetic calorimeter: both photons carry significant transverse-to-beam momentum. This feature remain unexplained and may point at some errors in reconstruction, at a fake signal or at a real signal which confirmation requires more statistics.

⋆e-mail: gorby@ms2.inr.ac.ru
[1]CERN seminar, https://indico.cern.ch/event/442432/

The second talk was by ATLAS, which prepared 32 preprints with new results. This number noticeably exceeds that of the CMS, because of noticeably higher integral luminosity, 3.5-3.2 fb^{-1}, collected by ATLAS. The experiment CMS suffered from technical problems, one of the widely discussed is with magnetic field of the muon solenoid. At the new energy frontier ATLAS investigated 44 new signal regions in supersymmetric extensions of the Standard Model of particle physics (SM), extensions with heavy massive Z'- and W'-bosons, models with blackholes, models of dark matter, etc. The new data happened to be in some tension with previously found anomalies in diboson modes at 1.8-2 TeV and in the mode with missing energy and Z-boson decaying into electrons. A discouraging news was the lack of the Higgs bosons in the new data: a combined analysis of $\gamma\gamma$- and ZZ^*-modes revealed 1.4 σ signal while 3.4 σ was expected. At the same time ATLAS observed an anomaly in $\gamma\gamma$-channel with invariant mass of about $m_{\gamma\gamma} = 750$ GeV, which looked very much like the anomalous signal at CMS.

The CMS and ATLAS measurements of event distribution over the diphoton invariant mass $m_{\gamma\gamma}$ distribution are presented in Fig. 1. The signal in the ATLAS presentation (right panel) was much

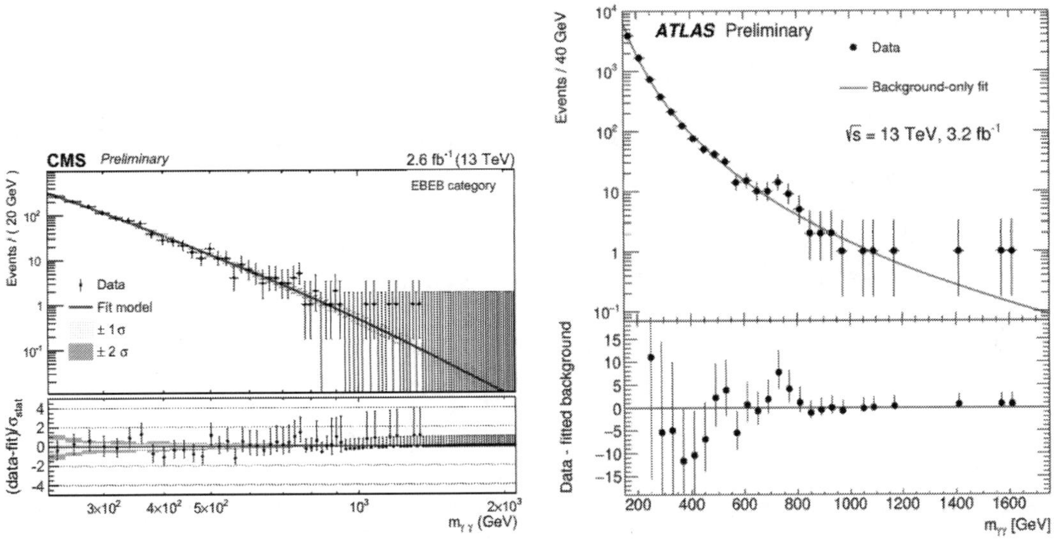

Figure 1. Diphoton invariant mass distribution measured by CMS [1] (left panel) and ATLAS [2] (right panel) collaborations. The anomalous signal was observed in both experiments at $m_{\gamma\gamma} \simeq 750$ GeV.

more pronounced than that of CMS (left panel), which might be caused by higher statistics (3.2 fb^{-1} vs 2.6 fb^{-1}). The ATLAS exhibited higher sensitivity to the signal. When fitted by zero-width resonance (i.e. when the diphoton invariant mass resolution $\Delta m_{\gamma\gamma}$ exceeds the resonance width $\Gamma_{\gamma\gamma}$), the local p-value characterizing the signal significance was $p = 3.6\,\sigma$. It became 3.9 σ for wide-width approximation, the best fit value $\Gamma_{\gamma\gamma}/m_{\gamma\gamma} \approx 0.06$ corresponds to $\Gamma_{\gamma\gamma} \approx 46$ GeV. Thus, the data slightly favored the wide resonance. Note, that the p-values get dropped to 2-2.3 σ, respectively, when look elsewhere effect is properly accounted for. The penalty is taken for searching over 0.2-2 TeV diphoton invariant mass interval and also for scanning over the resonance width at $\Gamma_{\gamma\gamma}/m_{\gamma\gamma} < 0.1$. The signal significance of CMS signal was even lower (3.0 σ turned into 1.7 σ) but the fact that both experiment point at the same mass $m_{\gamma\gamma} \simeq 750$ GeV was reassuring.

2 Presented this Spring results (13 TeV and 8 TeV)

Being inspired by these results both collaborations proceeded to elaborate on the data, and later, at the Moriond Conference[2], have presented the extended analyses. A significant improvement was achieved by CMS [3], which included the data with zero solenoid magnetic field. The problems with solenoid seemed absolutely irrelevant for photon reconstruction in the electromagnetic calorimeter. These new data added 0.6 fb^{-1} to the CMS data used in the preliminary analysis and allowed CMS to achieve the sensitivity in the diphoton mode similar to that of ATLAS. Both data sets exhibit the signal at about 750 GeV, while the absence of the anomalous events with one or both photons flying along the beam axis (and to be registered in the end-cup calorimeters) has been confirmed. As a result, the local p-value jumped up to 2.9(2.8) σ for spin-0(2) resonance; thus CMS 13 TeV data give no hint of the resonance spin. Further, CMS had analyzed the data of 8 TeV run where some indication of the signal was observed. The combined fit favors the narrow resonance $\Gamma/M_{\gamma\gamma} = 1.4 \times 10^{-4}$ with local p-value of 3.4 σ decreasing to 1.6 σ due to the look elsewhere effect.

It is instructive to compare these results with the preliminary results of the combined analysis also shown at the CERN Seminar in December. To illustrate the situation, I plot in Fig. 2 the results

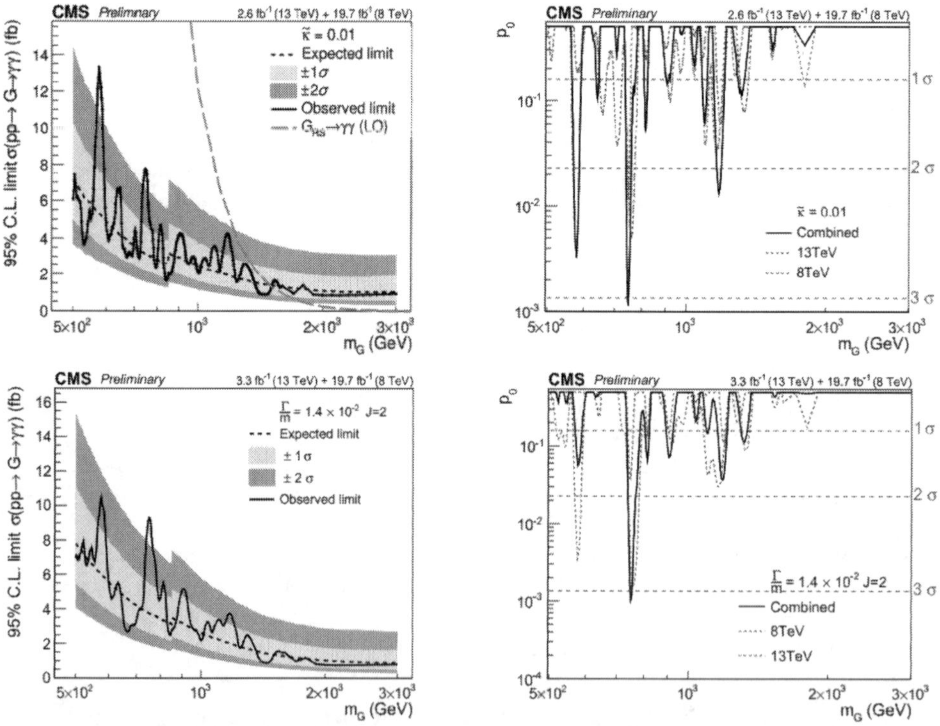

Figure 2. An example of combined CMS results: preliminary results presented at the CERN Seminar (top panel), results from the Moriond Conference (bottom panel).

obtained assuming $J = 2$ resonance (Randall–Sundrum models) with a particular width $\Gamma/M_{\gamma\gamma} = 1.4 \times 10^{-2}$ (it refers to $\tilde{k} = 0.01$ in the alternative notations). One can observe, first, that the new data of 0.6 fb^{-1} obtained with zero-magnetic field helped to get rid of the two peaks at $M_G \simeq 0.6$ TeV and $M_G \simeq 1.1$ TeV, seen in the combined 8+13 TeV data. Second, the preliminary analysis has been

[2] 51st Rencontres de Moriond, EW 2016, https://indico.in2p3.fr/event/12279/

Table 1. Summary of the diphoton anomaly [3, 4]; [QUARKS'16] indicate somewhat different numbers presented at the Conference by ATLAS and CMS speakers.

	ATLAS	CMS
p-values, loc(glob)	$3.9(2.3)\,\sigma$, $(2.0)\,\sigma$ [QUARKS'16]	$2.9(1.4)\,\sigma$, $(<1)\,\sigma$ [QUARKS'16]
resonance mass	$M_{\gamma\gamma} = 750\,\text{GeV}$	$M_{\gamma\gamma} = 760\,\text{GeV}$
resonance width	$\Gamma_{\gamma\gamma} \approx 45\,\text{GeV}$	$\Gamma_{\gamma\gamma} \approx 0\,\text{GeV}$
production at 13 TeV	$\sigma_{pp\to X} \times \text{Br}_{X\to\gamma\gamma} = (10 \pm 3)\,\text{fb}$	$\sigma_{pp\to X} \times \text{Br}_{X\to\gamma\gamma} = (3.7 \pm 1.4)\,\text{fb}$
integral luminosity	$3.3\,\text{fb}^{-1}$	$3.2\,\text{fb}^{-1}$
production at 8 TeV	$\sigma_{pp\to X} \times \text{Br}_{X\to\gamma\gamma} = (0.4 \pm 0.8)\,\text{fb}$	$\sigma_{pp\to X} \times \text{Br}_{X\to\gamma\gamma} = (0.5 \pm 0.6)\,\text{fb}$
integral luminosity	$20.3\,\text{fb}^{-1}$	$19.7\,\text{fb}^{-1}$
spin	certainly (?) not $J = 2$	no preference

checked and, apparently, some errors have been corrected including the data of 8 TeV run: compare the red dotted curve on the top right plot and blue dotted curve on the bottom right plot of Fig. 2 and find they are different.

ATLAS also added 8 TeV run data of $20.3\,\text{fb}^{-1}$ but arrived at qualitatively the opposite result [4]: the old data show (if any) a weaker signal. Namely, for the best-fit value $\Gamma/M_{\gamma\gamma} = 0.06$ and spin-0 resonance the data exhibit an excess of $1.9\,\sigma$, while one would expect a $1.2(2.1)\,\sigma$ stronger excess for the resonance produced in $gg(q\bar{q})$ scattering[3]. For spin-2 resonance no hint was recognized in the old data: the discrepancies between the observed signal end expected are $2.7(3.3)\,\sigma$ for the the gluon(quark) production channel.

The results of the both experiments are summarized in Table 1. One can see that, as compared to ATLAS, CMS measures 2.5 times smaller production cross section at 13 TeV, but observes a hint of signal at 8 TeV and has no preferences for the resonance spin.

It is worth to add that neither ATLAS no CMS reported any anomaly in $Z\gamma$, ZZ, WW, gg or hh channels in the interesting region of the corresponding invariant masses. Recall also, that the SM Higgs boson signal at 13 TeV is observed at much lower level than expected. Having all these in mind, still it is tempting to consider the anomaly observed by ATLAS and CMS as a hint of new physics: the direct production of the a heavy neutral boson. The numerous discrepancies between the anomalous results obtained at different energies and reported by different collaborations may be simply due to a lack of statistics.

3 Response from theorists

Theorists have immediately reacted to the announcement of the possible signal of new physics in $\gamma\gamma$-mode: 8 preprints have been submitted to the Cornell arXiv on the day of the CERN Seminar. By the date of this talk, May 31st, 365 preprint with references to the ATLAS preprint [2] have been issued in the arXiv. It is a desperate response of the overheated "gas of theorist" awaited for any signal of new physics in the LHC data. To say the truth, most probably several groups of LHC-related physicists were aware of details of the anomaly some time before the CERN Seminar. It would explain the submission of the very first paper [5] in 30 minutes after the Seminar. Likewise, the 8 papers of the first day after the Seminar are of 6, 23, 6, 19, 45, 6, 21, 3 pages, and it is rather difficult to believe that 20-page paper with a properly performed new physical analysis can be prepared in one evening.

[3]The difference is because of different evolution of gluon and quark parton distribution functions from 8 TeV to 13 TeV energy scale.

Anyway, the mostly cited paper (286 references[4] by May 31st) "What is the resonance at 750 GeV?" [7] was submitted on the first day and written by 10 authors; and 7 out of 10 are from CERN, which is also very elucidative. It is a paper of 45 pages, which "contains everything", that is all the information from experiments needed to theorist who is looking for a new beyond-the-SM (BSM) explanation of the observed anomaly. The violent reaction by theorists is naturally attributed to the fact that many well developed BSM theories contain (or can be straightforwardly supplemented with) a candidate for 750 GeV resonance. Indeed, the authors of 8 preprints issued (long) before the CERN Seminar later placed the references on the ATLAS and CMS results because the models they studied might be potentially relevant for the explanation of the anomalous events.

This activity have been recognized by a broader scientific community, and there is a paper [8] issued in March with predictions of time-evolution of the total number of papers referring to the ATLAS and CMS results, see Fig. 3. The paper presents a statistical analysis of the cumulative number

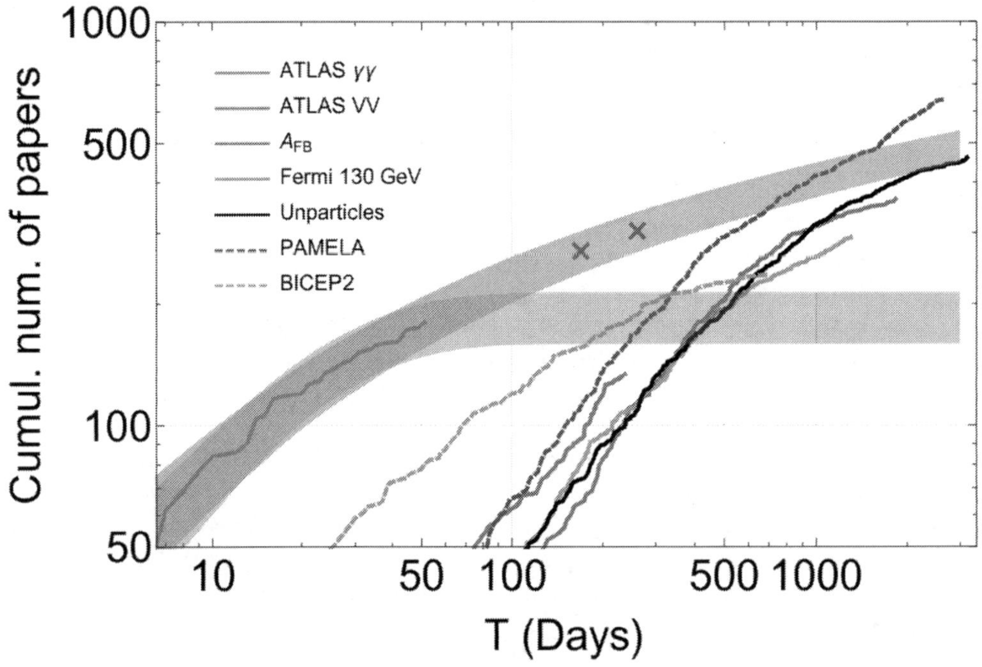

Figure 3. Time-distribution of the cumulative number of papers citing the papers which announced the particular anomalous results in particle physics, see Ref. [8] for details. Red crosses indicate the predictions for the papers referring to the ATLAS paper [2] about 750 GeV diphoton events.

of papers submitted in response to several well-known anomalies (relic gravity waves by BICEP2, cosmic positrons by PAMELA, γ-rays by Fermi, etc) which might point at the BSM physics but later have been explained within the SM. The conclusion was that the evolution follows the dilogarithm function and the number of papers referring to the ATLAS paper [2] must not exceed 310 by the 1st of June. From the number presented above one sees that the prediction was wrong: the theorist turned out to be more active and productive than expected. There are two reasons: reanalysis with new data presented at the Moriond Conference, and the fact that LHC just started to collect data promising to multiply the statistics by a factor of ten in a year.

[4]The next-to-it Ref. [6] was cited 208 times.

There are many reviews and books solely devoted to possible physics at LHC, mostly BSM physics. In particular, looking through the outlines of the books [9, 10], which may be treated as the LHC Bible and New Testament, respectively, one can find many hypothetical particles—nonstandard Higgs bosons, additional scalars, sgoldstino, sneutrino, technipions, massive gravitons—proposed in literature as the 750 GeV resonance, and many BSM theories—extra dimensions, supersymmetry, supersymmetry with R-parity violation, heavy quarks, dark matter, Z'- and W'-bosons, compositeness, heavy Majorana neutrino—which can be responsible for the resonance production at LHC. These are candidates for the soloist and its backing in this LHC play.

One can ask, given so many reliable candidates known for many years, did anybody predict a neutral particle of exactly 750 GeV mass? As far as I know, certainly not inside the particle physics community, though one may argue it can be related to the general scale of new physics taming the quadratic corrections to the Higgs boson mass squared. However, the almost exact number has been predicted by authors of the very popular cartoon series "Simpson" in 1998, when the main character Homer Simpson wrote a formula

$$M(H^0) = \pi\left(\frac{1}{137}\right)^8 \sqrt{\frac{hc}{G}}$$

presumably for the Higgs boson mass; here h, c and G stand for the Planck constant, speed of light in vacuum and the gravitational constant, respectively. Calculating this value one finds 775 GeV !! Hence, one can name the resonance under discussion as the Homer boson, which nicely fits to the notation H^0 from the cartoon. It is obviously an ad hoc prediction, however there are examples of predictions in particle physics which looked as ad hoc for decades. One is the "Greisen equation" in Cosmic ray physics dated to 1956,

$$\Pi(t, E_0/\epsilon_c) \approx \frac{0.31}{\sqrt{\beta_0}} \times \exp\left[t\left(1 - \frac{3}{2}\log s\right)\right],$$

which describes the number of particles in the shower at depth t, given the energy of initial high-energy cosmic ray particle E_0 and critical energy ϵ_c; here t is in units of the radiation depth, and I introduced the so-called shower age $s \equiv 3t/(t + 2\beta_0)$ and $\beta_0 = \log(E_0/\epsilon_c)$. Though the Greisen equation looks simple it works strikingly well. There was no deviation of this formula in the Greisen works or further papers, review, books on the subject, till 2007, when the detailed analysis has been performed in public [11].

4 Models, Problems, Predictions...

Parameters of any models proposed to explain the anomaly must be adjusted in a way to fit into the estimates of the resonance production cross section, mass and width shown in Table 1. The production channel and the decay branching ratio into two photons is not determined from the data, and the total width is either fixed (by ATLAS) or limited from above (by CMS). Even the spin of the resonance is not definitely fixed. Examples of the favored parameter space are presented in Fig. 4, assuming gluon fusion (left panel) and b-quark fusion (right panel) as the dominant production mechanisms.

The first definite statement about the resonance in diphoton mode is that it is either spin-0 or spin-2 (though the exotic Chern–Simons terms may make spin-1 viable as well). Now, from the ATLAS data spin-2 is disfavored; if not, the Kaluza-Klein gravitons may be suggested as possible candidates for the observed resonance. Since the new particle X is neutral, its interactions with gluons and photons are effective, like we have for the SM Higgs boson, $XG_{\mu\nu}G^{\mu\nu}$ and $XF_{\mu\nu}F^{\mu\nu}$, but then the new charged particles are needed, which couple both to X and the SM gauge bosons. They must be light and in a

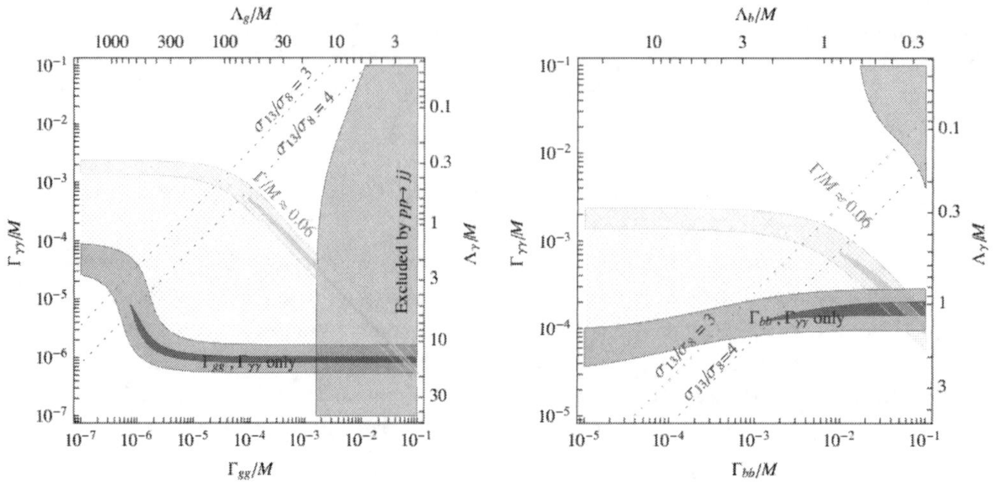

Figure 4. The regions in the resonance model parameter space (M is its mass, $\Gamma_{\gamma\gamma}$, Γ_{gg} and Γ_{bb} are its decay rates into two photons, two gluons and two b-quarks, respectively, and Λ_γ, Λ_g, Λ_b are new physics energy scales possibly responsible for the effective resonance coupling to photons, gluons and b-quarks, respectively). The resonance production goes through gg (left panel) or bb (right panel) fusions, correspondingly, see Ref. [7] for details. The gray regions are experimentally excluded, the green regions and blue regions show the models with wide and narrow resonances, in the latter case only $\gamma\gamma$ and the production channel contribute to the total resonance width. The red dotted lines are lines of the constant ratio of the production cross sections at 13 TeV and 8 TeV indicated by the label σ_{13}/σ_8. The green regions, blue regions and the yellow regions between the former two are favored by the observation of the 750 GeV resonance.

large number to ensure the sufficiently strong coupling constants in front of these interaction terms, because the constants scale with new particle masses M_f (say fermions) and numbers N_f as

$$\alpha_{Xgg} \propto \alpha_s \frac{N_f}{M_f}, \quad \alpha_{X\gamma\gamma} \propto \alpha \frac{e_f^2 N_f}{M_f}.$$

The most natural production mechanism is the gluon fusion; the quark fusion and photon fusion are rather difficult to accommodate, though in models with extra dimension Kaluza–Klein modes may help.

The simplest option for the new light particles is vector-like (to avoid gauge anomalies) quarks or leptons. Actually, the new bosons are somewhat preferred to new fermions, since they can make the Electroweak vacuum stable, which is needed for any high-energy scale inflation and especially for the Higgs-inflation. Note that mixing of the new resonance with the SM Higgs boson is rather constrained. Other options include sneutrinos in supersymmetric extensions with broken R-parity and sbino in models with Dirac-type bino.

The new boson may belong to the extended Higgs sector, it can be heavy CP-even H or axial A Higgs of 2 or 3 Higgs doublet models, H, A or additional singlet scalar n from (non)minimal supersymmetric extensions of the SM, extensions with inert Higgs doublets, scalar bosons from extended gauge sectors, like those in $U(1)_{B-L}$, $SU(2)_L \times SU(2)_R$, $SU(5)$, $SO(10)$, E_6, or those involved into seesaw type II mechanism, or radiation mechanism giving masses to active neutrinos etc. As a rule, always new charged particles are required.

The new resonance may have the Goldstone nature, like pseudoscalar mesons of QCD, and its decay into photons may originate from a quantum anomaly, like $\pi^0 \to \gamma\gamma$. Usually, it is not alone, but with mates (like charged pions in QCD), and examples are models with heavy axion, coloron, dilaton, radion, branons, sgoldstinos. Similar to other cases, in this type of models one must introduce new particles at ~TeV scale.

In most models, given the $U(1)_Y$ gauge symmetry, one has to observe the resonance in γZ and ZZ channels as well, with the rate determined by the gauge symmetry. In many models the resonance must also show up in WW and gg channels.

There are many models nicely fitted to the resonance, if it is narrow, as CMS favors. The wide resonance is difficult to accommodate in many otherwise theoretically consistent and phenomenologically viable models. The width-to-mass ratio favored by ATLAS is very similar to what we have for many hadrons of QCD, which may suggest some technicolor model, or models with technicolor in a hidden sector. It might be that an invisible decay mode (e.g. into dark matter particles) dominates the resonance width, but the LHC searches for a single photon and a single jet events with missing energy severely constrain this option. Alternatively, one may consider decays of several different almost degenerate in mass particle mimicking the wide resonance. Other options are a three body decay into a photon pair and an elusive particle and decay into a couple of light pion-like states subsequently decaying into photon pairs with kinematics like two single photons.

5 Prospects

There are many models capable of explaining the diphoton anomaly. In most cases (with a few exceptions) these are extensions of earlier suggested BSM theories supplemented with new relatively light particles. All the models predict new charged with respect to the SM gauge group particles, light enough to be searched for at the LHC in the nearest future. Thus one must know by the end of the year, whether the Homer family will be invited to Stockholm. If the signal is real, the LHC physics must be fabulously rich, and we only start probing the heavy particle sector.

The work is supported by the RSF grant 14-12-01430.

References

[1] CMS Collaboration (CMS), CMS-PAS-EXO-15-004 (2015)
[2] The ATLAS collaboration, ATLAS-CONF-2015-081 (2015)
[3] CMS Collaboration (CMS), CMS-PAS-EXO-16-018 (2016)
[4] The ATLAS collaboration, ATLAS-CONF-2016-018 (2016)
[5] K. Harigaya, Y. Nomura, Phys. Lett. **B754**, 151 (2016), 1512.04850
[6] J. Ellis, S.A.R. Ellis, J. Quevillon, V. Sanz, T. You, JHEP **03**, 176 (2016), 1512.05327
[7] R. Franceschini, G.F. Giudice, J.F. Kamenik, M. McCullough, A. Pomarol, R. Rattazzi, M. Redi, F. Riva, A. Strumia, R. Torre, JHEP **03**, 144 (2016), 1512.04933
[8] M. Backović (2016), 1603.01204
[9] N.V. Krasnikov, V.A. Matveev, *New Physics at Large Hadron Collider, in Russian* (2014), http://urss.ru/cgi-bin/db.pl?lang=RU&blang=ru&page=Book&id=187848
[10] N.V. Krasnikov, V.A. Matveev, *Discovery of the Higgs Boson and Search for New Physics at Large Hadron Collider at energies 7 and 8 TeV, in Russian* (2015), http://urss.ru/cgi-bin/db.pl?lang=RU&blang=ru&page=Book&id=191169
[11] R.W. Schiel, J.P. Ralston, Phys. Rev. **D75**, 016005 (2007), hep-ph/0607248

EPJ Web of Conferences **125**, 02011 (2016) DOI: 10.1051/epjconf/201612502011
QUARKS-2016

The Higgs bosons decays in the NMSSM with CP-violation

A.V. Gurskaya[1,⋆] and M.V. Dolgopolov[1,⋆⋆]

[1] *Samara University, Academician Pavlov 1, 443011 Samara, Russia.*

Abstract. We investigate the Higgs bosons decay width in the framework of NMSSM in the one-loop approximation. The Higgs sector under consideration contains CP-violation in the general case due to explicit and spontaneous violations of the CP-symmetry. The local minimum conditions for the Higgs bosons physical states are taken into account as necessary requirements. So, calculation of the decay width reduces to the calculation of corresponding processes amplitudes. The strategy of calculation is based on the perturbation theory.

Introduction

The Higgs sector is the main object of our research. The Higgs boson with mass 125 GeV is observed in the experiments at the LHC. There are no signals of other particles that could be Higgs bosons with smaller masses. Why do we extend the Higgs sector?

First, the Higgs sector is responsible for the masses of fundamental particles. But there is a discrepancy between the SM predictions and neutrino oscillation experiments. We know that different types of neutrino transit into each other [1, 2]. Therefore, the neutrino has mass. Since the SM does not explain this problem, then the Higgs sector of this model is still under consideration in this sense.

Secondly, there is the vacuum stability problem. There are researches [3, 4], in which it is considered that the vacuum is the metastable state for the Higgs boson with mass of order 125 GeV. Then, we need to understand where the "island of stability" is and for what values of physical parameters it exists?

The third problem is the baryon asymmetry justification. There is a hypothesis that the baryon asymmetry appear already at an early stage of the Universe formation in the framework of process, so-called, baryogenesis. It is the formation of hadrons from quarks and gluons. One of the condition of baryon asymmetry is combined violation of C- and P- parity (CP-violation) [5]. The SM contains the complex CP-violating-phase in the fundamental fermions mix matrix. However, this phase is not enough to explain of the baryon asymmetry. May be, there are additional CP-violation sources in the Higgs sector. It is reason for we need to extend the Higgs sector [6, 7].

Note also that the supersymmetric models contain extended Higgs sector naturally. There are requirements of supermultiplets chirality. It is leading to the replacing the hermitian conjugate doublet in the new doublet. We can not realize supersymmetry in SM because the degrees of freedom for bosons and fermions states are not the same. The simplest susy-model is MSSM, but it has so-called

⋆e-mail: a-gurska@yandex.ru
⋆⋆e-mail: mvdolg@yandex.ru

μ-problem. The solution of this problem is the introduction of the additional Superpole chiral singlet in the Higgs sector. This model is called Next-to-MSSM or NMSSM [8, 9].

1 Higgs bosons in NMSSM with CP-violation

In order to introduce the CP-violation in our model, we need to rewrite the complex structure in Lagrangian (complex couplings or complex vacuum expectation values parameters). Note also, the supersymmetry is broken, otherwise superparticles would have the same masses as their partners in the SM. Terms, which violate the supersymmetry, also violate CP-symmetry.

The CP-violation case in the NMSSM was considered by author in earlier work [10]. There is also a similar work dedicated to the potential of the Higgs NMSSM with explicit CP-violation [11]. In general case the explicit CP-violation and spontaneous violation of CP-symmetry take place. In order to calculate physical states of the Higgs bosons, we have to find the local minimum conditions of Higgs potential. This procedure is necessary for finding stable vacuum states. So, for it, we found the first and second derivations of potential on the fields in the minimum. We need to diagonalize of the 5x5 mass squared matrix of neutral Higgs bosons, and does not have viable analytical solutions. The eigenstates considered to be positive in local minimum.

The local minimum requirements fix all parameters ranges and set fixed parameters like valid scenarios. Several scenarios compare with CP-conserving case were considered in [10]. the scenario with the first Higgs boson mass less then 125 GeV is of the principal interest. We are interested in this scenario in the context of the baryon asymmetry problem and baryogenesis. There is the hypothesis of electroweak baryogenesis, in which a strong phase transition of the first order is required. And this phase transition is realized for the Higgs bosom with the mass less than 50 GeV. The considered scenario satisfies this hypothesis.

2 Higgs bosons decay width calculation

We will consider $1 \to 2$ process. General formula for this case is well known:

$$\Gamma(H \to X\overline{X}) = \frac{|A_{H\to X\overline{X}}(\varphi)|^2}{16\pi M_H} \sqrt{1 - \frac{4m_X^2}{M_H^2}}, \tag{1}$$

where M_H – Higgs boson mass, m_X – final particle mass, $A_{H\to X\overline{X}}(\varphi)$ – the process amplitude, depending on CP-violation phase.

And we consider decay into vector bosons and into fermions (see Fig.1).

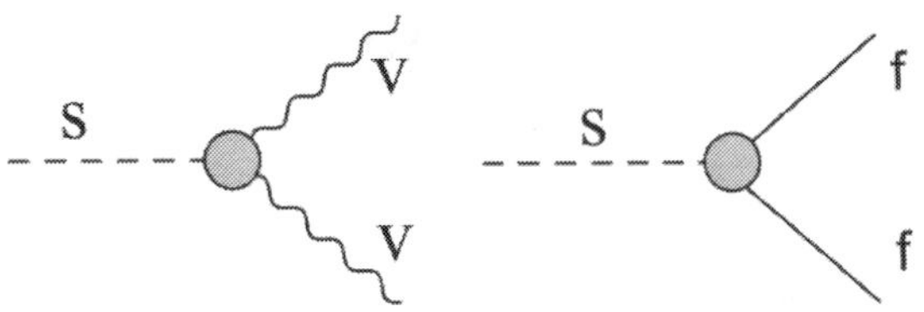

Fig. 1. Two types for decay of scalar particle without definition of loop topology.

Choosing the method of calculating the next perturbation order, we can represent the results of calculations close to the full set of loop contributions. The advantage of this approach is that the computation reduces to the minimum set of scalar integrals, and those, in turn, are expressed through known functions of the Veltman–Passarino A0, B0, C0 [12, 13]. The calculation of the vertex functions in the appropriate order of perturbation theory in the framework of this approach should be limited only to the numerical calculation of the reduced analytical results for the select set of generalized structures that contributed to this vertex function, bypassing the stage tensor, dimensional and algebraic reduction [13, 14].

Diagrams on Fig.1 present general case for decay of scalar particle. In particular, we have five types of diagrams for final states (see Fig. 2, 3).

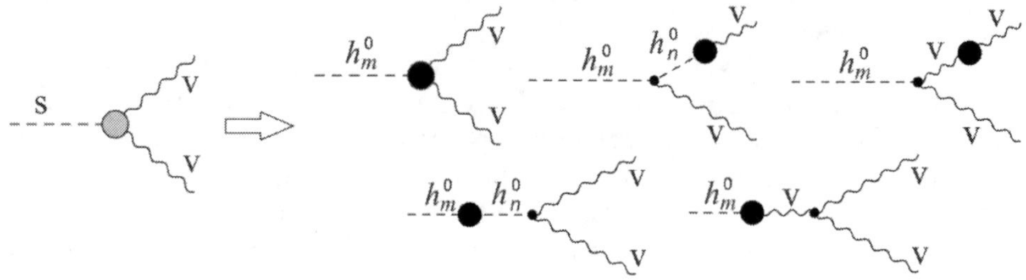

Fig.2. Diagrams of the neutral Higgs bosons decay to the gauge.

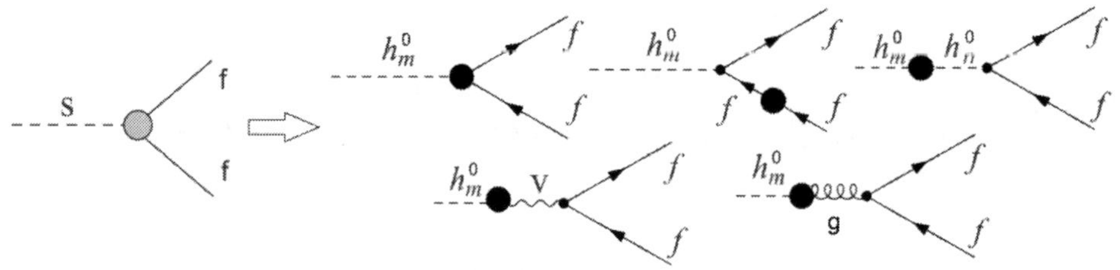

Fig 3. Diagrams of the neutral Higgs bosons decay to the fermions.

For example, there is the diagram type, when the Higgs boson has decay into any vector particles. We may define it like this (Fig 4):

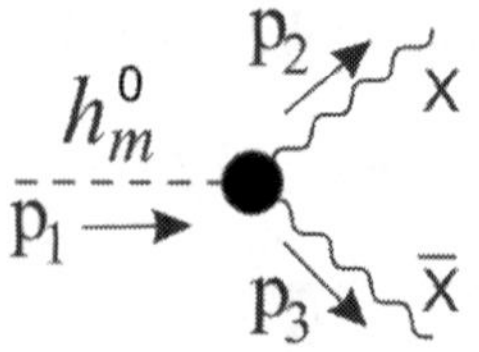

Fig.4. Diagram of the scalar particle decay into the vector particles.

Formula for this one-loop correction is given by

$$\left. \begin{aligned} T^{svv} = \epsilon^{\mu\nu\sigma\rho} p_{2\sigma} p_{3\rho} G^{(1)}_{h_m X\overline{X}} + g^{\mu\nu} G^{(2)}_{h_m X\overline{X}} + p_2^{\mu} p_2^{\nu} G^{(3)}_{h_m X\overline{X}} \\ + p_3^{\mu} p_3^{\nu} G^{(4)}_{h_m X\overline{X}} + p_2^{\mu} p_3^{\nu} G^{(5)}_{h_m X\overline{X}} + p_3^{\mu} p_2^{\nu} G^{(6)}_{h_m X\overline{X}} \end{aligned} \right\} \quad (2)$$

Each factor has analytical expression. For example, first $G^{(1)}$:

$$G^{(1)}_{h_m X\overline{X}} = -\frac{iN_c K}{4\pi^2}[F^{(1)}m_1 C_0 + (F^{(1)}m_1 - F^{(2)}m_2)C_1 + (F^{(1)}m_1 + F^{(3)}m_3)C_2],$$

where $F^{(1)}$, $F^{(2)}$, $F^{(3)}$ are defined from $s + p\gamma^5$ and $\upsilon\gamma^\mu + a\gamma^\mu\gamma^5$ as its combinations.

In the CP-violating case in the Higgs sector the scalar (s) and pseudoscalar (p) Yukawa couplings are following:

$$\overline{u}uh_j : \ -\frac{gM_u\left(\mathrm{A}_{1j}s_\beta - i\gamma^5\mathrm{A}_{2j}c_\beta + \mathrm{A}_{3j}c_\beta\right)}{2m_W s_\beta} \tag{3}$$

$$\overline{d}dh_j : \ \frac{gM_d\left(-\mathrm{A}_{1j}c_\beta + s_\beta\left(\mathrm{A}_{3j} + i\gamma^5\mathrm{A}_{2j}\right)\right)}{2c_\beta m_W}, \tag{4}$$

where A_{ij} – squared mass matrix components of CP-add and CP-even mixing between physical states.

We take into account the mixing between CP-states in two stages. The parametrization is given by:

$$H_1 = \begin{pmatrix} \frac{1}{\sqrt{2}}(\upsilon_1 + \phi_1^0 + i\chi_1) \\ \phi_1^- \end{pmatrix}, \quad H_2 = e^{i\theta}\begin{pmatrix} \phi_2^+ \\ \frac{1}{\sqrt{2}}(\upsilon_2 + \phi_2^0 + i\chi_2) \end{pmatrix} \tag{5}$$

$$S = \frac{1}{\sqrt{2}}e^{i\varphi}(\upsilon_3 + \phi_3^0 + i\chi_3),$$

where υ_i – the vacuum expectation of Higgs field; θ, φ – spontaneous CP-violation phases.

First, we get the basis of fields with defined CP-parity $\left(H, h, A^0, G^0\right)$:

$$\begin{pmatrix} \phi_1^0 \\ \phi_2^0 \end{pmatrix} = \begin{pmatrix} cos\beta & -sin\beta \\ sin\beta & cos\beta \end{pmatrix}\begin{pmatrix} H \\ h \end{pmatrix}; \begin{pmatrix} \chi_1 \\ \chi_2 \end{pmatrix} = \begin{pmatrix} cos\beta & -sin\beta \\ sin\beta & cos\beta \end{pmatrix}\begin{pmatrix} G^0 \\ A^0 \end{pmatrix}$$

And then the second twist is following:

$$\begin{pmatrix} H \\ A^0 \\ h \\ \phi_3^0 \\ \chi_3 \end{pmatrix} = A_{ij}\begin{pmatrix} h_1 \\ h_2 \\ h_3 \\ h_4 \\ h_5 \end{pmatrix}, \tag{6}$$

where h_i – physical states without defined CP-parity.

3 Summary and Outlook

In the presented note we deal with the one-loop corrections to Higgs bosons decay widths in the framework of NMSSM. We considered combined CP-violation due to explicit and spontaneous violations in the Higgs sector. We conclude that the local minimum conditions of potential have the crucial role for the Higgs bosons physical states existence in agreement with allowed ranges of NMSSM parameters.

Authors are going to consider the temperature evolution of the Higgs potential in the next upcoming work. It will allow the range selection for free and control parameters of the model. Some analysis in the study of this problem was developed by methods of the catastrophes theory [15, 16].

References

[1] Fukuda Y et al. (Super-Kamiokande Collab.) Phys. Rev. Lett. 81 1562 (1998); 82 2644 (1999); Fukuda S et al. (Super-Kamiokande Collab.) Phys. Rev. Lett. 85 3999 (2000)

[2] Hampel W et al. (GALLEX Collab.) Phys. Lett. B 447 127 (1999); Altmann M et al. (GNO Collab.) Phys. Lett. B 490 16 (2000); Bellotti E (for the GNO Collab.) Nucl. Phys. B: Proc. Suppl. 91 44 (2001).

[3] Elias-Miro J. Higgs mass implications on the stability of the electroweak vacuum / J. Elias-Miro [et al.] // e-print: arXiv:1112.3022v1.

[4] Degrassi G. Higgs mass and vacuum stability in the Standard Model at NNLO / G. Degrassi [et al.] // e-print: arXiv:1205.6497v2.

[5] Sakharov A.D. Violation of CP invariance, C-asymmetry and baryon asymmetry of the Universe // Pisma v ZhETF, 1967, Vol. 5, Issue 1, 32–35.

[6] Dolgopolov M.V., Rykova E.N. Constraints on electroweak baryogenesis in models involving an extended Higgs sector // Physics of Atomic Nuclei Volume 72, Issue 1, Pages 173-177.

[7] Dolgopolov M.V., Dubinin M.N., Rykova E.N. Two-Higgs doublet potential of the MSSM at finite temperature and Higgs boson masses // Quarks. 2008. V. 1.

[8] Higgs bosons in a nonminimal supersymmetric model / J. Ellis [et al.] // Physics Review D. 1989. V.39. P.845–869.

[9] Maniatis, M. The Next-to-minimal supersymmetric extension of the standard model reviewed // IJMP A. 2010. V.25. P.3505–2602.

[10] Gurskaya A. Higgs bosons in NMSSM with CP-violation / Proceedings of the XXII International Workshop QFTHEP'2015 – MSU, SINP.

[11] T. V. Volkova, M. V. Dolgopolov, M. N. Dubinin, E. N. Rykova, Effective Higgs potential in Next-to-Minimal Supersymmetric Standard Model, Vestn. Samar. Gos. Tekhn. Univ. Ser. Fiz.-Mat. Nauki, 2(31) (2013), 233242.

[12] 't Hooft G. Scalar one loop integrals / G. 't Hooft, M.Veltman // Nucl.Phys.B. 1979. V.153. 365-401.

[13] Passarino G. One loop correction for e^+e^- annihilation into $\mu^+\mu^-$ in the weinberg model / G.Passarino, M.Veltman // Nucl.Phys.B. 1979. V.160. P.151.

[14] Philippov, Y.P. The new presentation of the algebraic reduction results for B0, C0 scalar integrals // Teoreticheskaya psizika. 2005. V.6. 86–97.

[15] M. N. Dubinin, E. Yu. Petrova, High-temperature Higgs potential of the two-doublet model in catastrophe theory, Teoret. Mat. Fiz., 184:2 (2015), 315-337.

[16] M. V. Dolgopolov, S. P. Zavodov, E. Yu. Petrova, Bifurcation sets of extended Higgs potential, Vestn. Samar. Gos. Tekhn. Univ. Ser. Fiz.-Mat. Nauki, 4(33) (2013), 173-183.

EPJ Web of Conferences **125**, 02012 (2016) DOI: 10.1051/epjconf/201612502012
QUARKS-2016

Constraining Dark Matter and Ultra-High Energy Cosmic Ray Sources with Fermi-LAT Diffuse Gamma Ray Background

Oleg Kalashev[1,⋆]

[1] *Institute for Nuclear Research of the Russian Academy of Sciences, Moscow 117312, Russia*

Abstract. We use the recent measurement of the isotropic γ–ray background (IGRB) by Fermi LAT and analysis of the contribution of unresolved point γ–ray sources to IGRB to build constraints on the models of ultra-high cosmic rays (UHECR) origin. We also calculate the minimal expected diffuse γ–ray flux produced by UHECR interactions with an interstellar photon background. Finally, for the subclass of dark matter (DM) models with decaying weakly interacting massive particles (WIMP), we build constraints on the particle decay time using minimal expected contributions to the IGRB from unresolved point γ–ray sources and UHECR.

1 Introduction

The Fermi LAT Collaboration [1] has presented their measurement of the IGRB within an unprecedentedly wide energy range up to 820 GeV based on 50 months of observations at Galactic latitude (b) $|b| > 20$. The IGRB spectrum can be well described over nearly four decades in energy by a power law with exponential cut-off, having a spectral index of 2.3 and a break energy of about 0.3 TeV. The origin of IGRB is not fully understood. However, it has been demonstrated recently [2] that unresolved point γ–ray sources, such as Active Galactic Nuclei or Star Forming Galaxies may account for up to 100% of IGRB flux and at least 86% of integral isotropic flux above 50 GeV. This leaves little space for other possible contributions, such as the electromagnetic cascades produced by UHECR interactions and products of Dark Matter (DM) decay (or annihilation). UHECR existence has been confirmed experimentally and therefore the guaranteed minimal contribution to the IGRB from UHECR must exist. On the contrary, the DM-originated flux of γ–rays is strongly model dependent and not limited from below. In this work, we discuss both mechanisms in detail. We start from building constraints imposed by the IGRB measurement on UHECR source models in section 2 and calculate minimal expected cascade γ–ray flux, originated from UHECR interactions with the interstellar photon background. In section 3, we consider decaying WIMP DM models and use the IGRB spectrum along with minimal contributions to isotropic γ–ray flux from UHECR and unresolved point sources to build conservative constraints on WIMP decay time, assuming various decay channels. Indirect DM detection methods and their connection to the IGRB have been studied in a number of works (for recent review see [3]), as well as constraints on UHECR origin models (see e.g. [4] and references therein). To our knowledge however, no study of DM models has been performed taking into account all guaranteed contributions to the IGRB.

⋆e-mail: kalashev@inr.ac.ru

2 UHECR and diffuse γ–ray background

Despite the impressive progress in the UHECR experimental study, the chemical composition of cosmic rays at highest energy is still a subject of debate. All modern cosmic ray experiments extract information on the UHECR spectrum and composition from observations of extensive air showers (EAS), initiated by cosmic ray particles. The method involves simulation of EAS development and suffers from unavoidable systematic uncertainties of primary particle energy determination at a level of 15 – 20% and even larger uncertainties in UHECR composition, since showers initiated by protons and compound nuclei are quite similar. Pierre Auger Observatory (PAO) [5], Telescope Array (TA) [6] and HiRes [7] observe light element composition in the energy range $1-4EeV$, however PAO reports transfer to heavier composition at higher energies, while TA and HiRes data is compatible with pure proton composition above $1EeV$. Observation of accompanying flux of secondary particles from UHECR interactions in intergalactic space can serve as an independent method of the UHECR composition study. In this paper, we focus on the secondary diffuse γ–ray flux. Protons and heavier nuclei produce e^+e^- pairs on the Cosmic Microwave Background (CMB) and infrared-ultraviolet extragalactic background light (EBL)

$$N + \gamma_{\rm CMB,EBL} \to e^+ + e^- + N. \tag{1}$$

$e^\pm$ then initiate electromagnetic (EM) cascades driven by the chain of inverse Compton scattering of electrons

$$e^\pm + \gamma_{\rm CMB} \to e^\pm + \gamma, \tag{2}$$

and pair production by high energy photons on the CMB and EBL

$$\gamma + \gamma_{\rm CMB,EBL} \to e^+ + e^- \tag{3}$$

Relatively fast EM cascade development leading to energy loss and exponential growth of the number of particles proceeds until photons reach the threshold energy for e^+e^-–pair production on EBL $E_{th,\gamma} \simeq$ TeV. The universe becomes essentially transparent for photons with $E_\gamma \lesssim 0.1$TeV. For this reason, the energy injected to the EM cascade by UHECR interactions, or by any other mechanisms is accumulated in the form of diffuse γ-radiation.

The threshold nucleus energy for the process (1) on a background photon with energy ε

$$E_{th} = \frac{m_e(m_N + m_e)}{\varepsilon} \simeq 0.5 \times A\left(\frac{\varepsilon}{10^{-3}eV}\right)^{-1} EeV, \tag{4}$$

is roughly proportional to nucleus mass number A. For this reason, compound nuclei are less efficient in $e^\pm$ pair production. Therefore, below we focus on the UHECR source models assuming pure proton composition. We shall see below that IGRB data imposes strict constraints on this class of models.

Also, UHE protons and light nuclei may contribute to the EM cascade through the GZK mechanism [**?**]

$$N + \gamma_{\rm CMB} \to \pi^{\pm,0} + N', \tag{5}$$

via products of $\pi^{\pm,0}$ decay, however since the GZK energy threshold for protons is about 50 times higher than $e^\pm$–pair production threshold energy (4), its contribution to cascade radiation is subdominant unless a very hard primary proton energy spectrum is assumed.

2.1 Conservative constraints on proton UHECR sources models

Below we use a simple phenomenological model for the UHECR source. We assume homogeneously distributed sources, emitting protons with power-law generation spectra

$$Q_p(E,z) \propto (1+z)^3 H(z)\left(\frac{E}{E_0}\right)^{-p}, \quad E \in [E_{\rm min}, E_{\rm max}], \tag{6}$$

where E_0 is an arbitrary normalization energy. Below, unless we state explicitly, we cut the injection spectrum below $E_{min} = 0.1$ EeV and above $E_{max} = 10^{2.5}$ EeV without loss of generality. Indeed the main contribution to the EM cascade comes from protons with energies in the interval from 1 EeV to a few EeV unless the injection spectrum is too flat ($\gamma_g \leq 2$) which is forbidden anyway as it will be demonstrated below. Also note, that models with $E_{max} > 10^{2.5}$ EeV don't substantially improve UHECR fit but may overproduce secondary ν-flux. We also introduce the evolution of source luminosity density with redshift z given by the term $H(z)$ in comoving volume, assuming that the source spectrum shape does not depend on z. For the evolution term we use general form

$$H(z) = H(0)(1+z)^m \quad \text{for } 0 \leq z \leq z_{max}. \tag{7}$$

We also consider two specific cases of source density proportional to the star formation rate (SFR) [8]

$$H_{SFR}(z) \propto \begin{cases} (1+z)^{3.4}, & z \leq 1 \\ (1+z)^{-0.3}, & 1 < z \leq 4 \\ (1+z)^{-3.5}, & z > 4. \end{cases} \tag{8}$$

and evolution of BL Lac/FR I sources [9]. For the latter, we derived the following analytical parametrisation

$$H_{BLLac}(z) \propto \frac{(a+z)^l}{(b+z)^k} exp(-z/z_c), \quad z < 6 \tag{9}$$

$a = 0.000283$; $l = 0.850383$; $b = 0.7662$; $k = 6$; $z_c = 10.006$

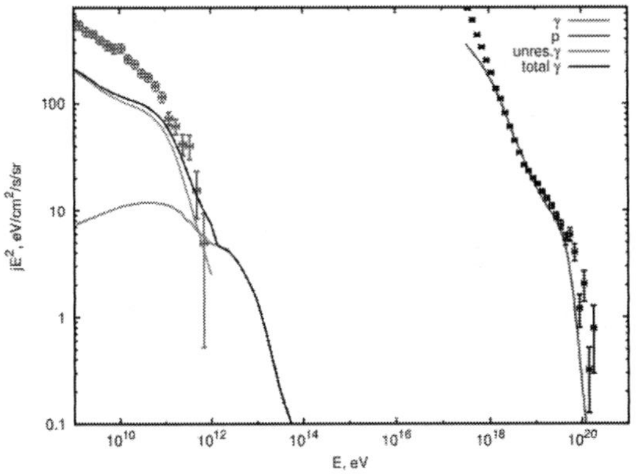

(a) $p = 2.6$, BL Lac/FR I sources

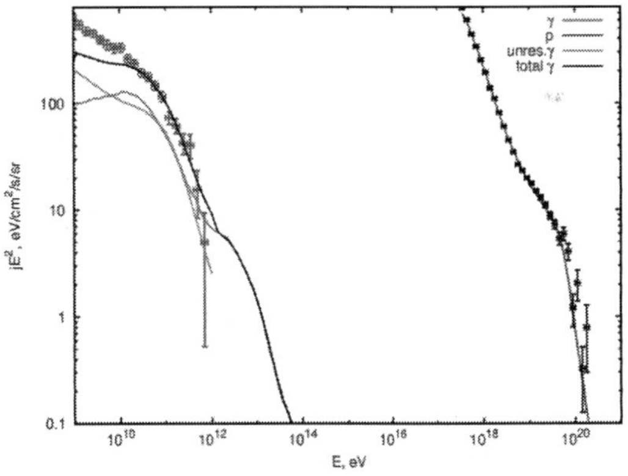

(b) $p = 2.44$, SFR evolution

Figure 1: The energy spectra of protons (blue curve) and cascade photons (red curve) from sources emitting protons normalized on the TA spectrum [10]. Also, the Fermi IGRB measurements [1] are shown assuming minimal galactic foreground allowed by systematic uncertainty, and minimal contribution to IGRB from unresolved γ–ray sources [2] (green curve). Total cascade + minimal unresolved source flux is shown by the black solid curve.

To obtain the UHECR spectrum and the spectrum of secondary γ after propagation through intergalactic space, we use the numerical code [11] that solves transport equations with terms describing

UHECR and electromagnetic cascade interactions with CMB and EBL (for the latter, we use the recent estimate [12]). Following Ref. [4], to derive conservative constraints on proton source models, we don't require a perfect UHECR spectrum fit in terms of χ^2. Such a requirement would lead to preference for strong evolving models with a relatively hard injection spectrum. These models are strongly constrained by Fermi IGRB and by Ice Cube UHECR ν flux limit [13]. Instead, for every m in the physically motivated range $-1 \leq m \leq 7$ and several values of $z_{\max}$ we find optimal value of spectrum power index p, providing a reasonable fit. We can do this safely for the following reasons. First of all, the systematic uncertainty of energy spectrum measurement is much higher than the statistical uncertainty in all present experiments. Secondly, the toy effective spectrum model (6) that we use may not precisely describe the actual picture, e.g. the actual spectrum may be different from power law or may be distorted by local sources, etc.

In Fig. 1 we show sample calculations of proton and secondary cascade γ made assuming SFR and BL Lac/FR I evolution. One can see (Fig. 1b) that cascade γ flux, when added to the minimal expected contribution from unresolved sources, may easily exceed the IGRB flux measured by Fermi LAT.

To define the formal conservative criterion of model consistency, we use maximal true isotropic flux

$$\Phi^{\rm iso}_{\max}(E_\gamma) \equiv \Phi^{\rm IGRB}_{\max}(E_\gamma) - \Phi^{\rm astro}_{\min}(E_\gamma),$$

which is calculated by subtraction of minimal unresolved source flux [2] from the maximal IGRB flux allowed by systematic galactic foreground uncertainty. We compare it with the cascade γ flux $\Phi_{\rm cas}(E_i)$ in each IGRB energy bin i with statistical error σ_i by constructing $\chi^2_{\rm cas}$:

$$\chi^2_{\rm cas} \equiv \sum_{\Phi_{\rm cas}(E_i)>\Phi^{\rm iso}_{\max}(E_i)} \frac{(\Phi_{\rm cas}(E_i) - \Phi^{\rm iso}_{\max}(E_i))^2}{\sigma_i^2} \qquad (10)$$

and require $\chi^2_{\rm cas} < 4$ or $\chi^2_{\rm cas} < 9$ to obtain an approximate estimate of 2-σ or 3-σ constraint respectively. The dependence of $\chi^2_{\rm cas}$ on generic model parameters m and $z_{\max}$ for models roughly fitting TA

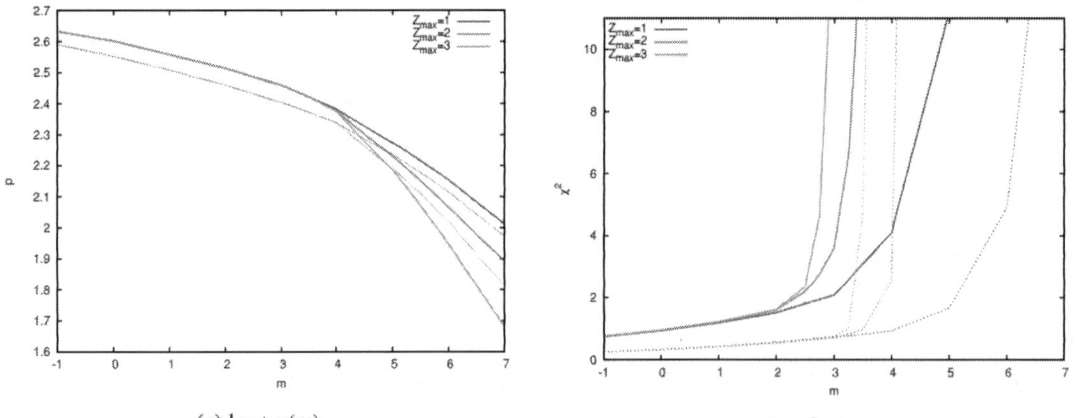

(a) best $p(m)$ (b) $\chi^2_{\rm cas}(m)$

Figure 2: Optimal model parameter $p(m)$ (Fig. a) and $\chi^2_{\rm cas}(m)$ (Fig. b) obtained for models fitting unshifted (solid lines) and shifted ($\delta E/E = -20\%$, dotted lines) TA energy spectrum for 3 values of $z_{\max}$

energy spectrum is illustrated by Fig. 2b. One can see that depending on the assumed z_{zmax} models with $m \gtrsim 2.5-4$ tend to overproduce diffuse γ. The constraint weakens if we shift the TA energy scale by 20% towards lower energies. In this case, models with $m \gtrsim 6$ are rejected for $z_{max} \geq 1$. Models with stronger evolution would require even smaller z_{max}.

2.2 Minimal diffuse γ–ray flux from UHECR sources

As mentioned above, PAO UHECR composition analysis indicates that at least part of UHECR events is caused by compound nuclei. In Ref. [5] the fraction of protons was estimated in the total UHECR flux measured by PAO at various energies. To make a realistic estimate of the minimal UHECR contribution to the diffuse γ–ray flux, we perform the fit of the proton only component of the PAO UHECR spectrum and calculate the secondary γ flux disregarding possible contribution of the heavier nuclei. Also, to minimize the cascade γ–ray flux, we assume negative though realistic source luminosity density evolution (9). The result of this calculation is presented in Fig. 3. In this case the cascade γ flux is

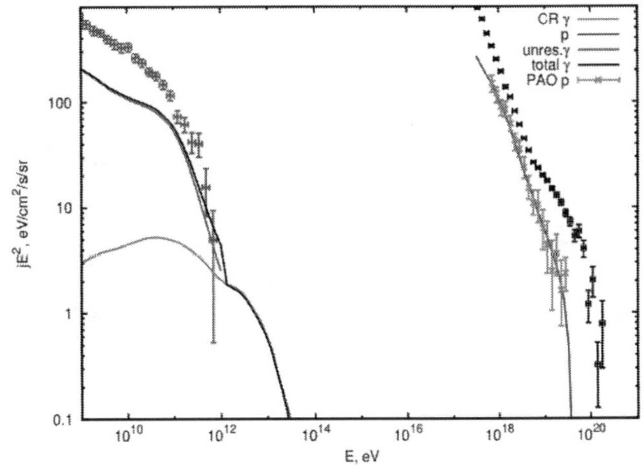

Figure 3: Energy spectra of protons (blue curve) and cascade photons (red curve) from BL Lac/FR I sources emitting protons with spectrum $E^{-2.79}$, $E < 4$EeV normalized on PAO proton spectrum.

well below the IGRB level everywhere except the last energy bin, where it becomes compatible with the flux of unresolved γ sources. This result is used in the next section.

3 WIMP Dark Matter decay and diffuse γ–ray background

In this section, we obtain conservative constraints on WIMP dark matter decay time, based on the observations of the IGRB, taking into account the minimal contribution to the IGRB from unresolved γ sources and from UHECR. We use the same method as in the previous section to derive the conservative constraints. Namely, we construct the quantity

$$\chi^2_{DM} \equiv \sum_{\Phi^{DM}_\gamma(E_i) > \Phi^{DM}_{max}(E_i)} \frac{(\Phi^{DM}_\gamma(E_i) - \Phi^{DM}_{max}(E_i))^2}{\sigma_i^2}, \tag{11}$$

where the term $\Phi^{\rm DM}_{\rm max}(E_\gamma)$ stands for the maximal isotropic γ–ray flux, which could be attributed to dark matter:

$$\Phi^{\rm DM}_{\rm max}(E_\gamma) \equiv \Phi^{\rm IGRB}_{\rm max}(E_\gamma) - \Phi^{\rm astro}_{\rm min}(E_\gamma) - \Phi^{\rm cas}_{\rm min}(E_\gamma)$$

In fact, the signal from WIMP dark matter is not expected to be isotropic since the DM particles

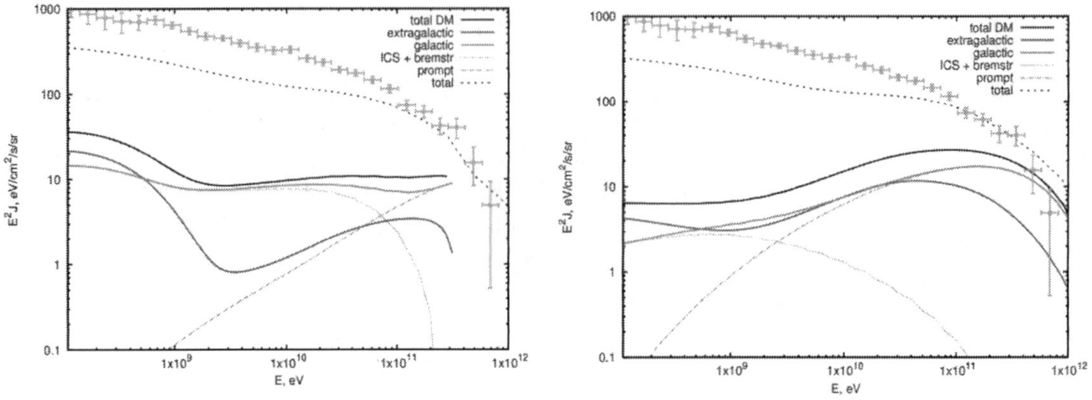

(a) $\mu^+\mu^-$: $\tau = 9\times10^{26}s$, $M_{DM} = 800GeV$

(b) $u\bar{u}$: $\tau = 3\times10^{27}s$, $M_{DM} = 5TeV$

Figure 4: Sample spectra of γ–rays from DM decay.

should be concentrated in halos. In our analysis we therefore use the minimal flux, which comes from the direction to the Milky Way anti-centre. We consider decay channels through e, μ, τ, u, b, W with γ and $e^\pm$ final states and calculate fluxes of prompt and secondary γ from $e^\pm$ interactions. It is produced by the DM in the Milky Way (MW) halo and remote galaxies (EG)

$$\Phi_\gamma = \Phi^{\rm MW}_\gamma + \Phi^{\rm EG}_\gamma \qquad (12)$$

The prompt galactic γ–ray flux from the anti-centre direction in the decay channel f with rate Γ_f is given by

$$\frac{d\Phi_\gamma}{dEd\Omega} = \frac{\Gamma_f}{4\pi M_{\rm DM}\tau_{\rm DM}}\frac{dN^f_\gamma}{dE}\int_{r_\odot}^{\infty}\rho(r)dr, \qquad (13)$$

where $M_{\rm DM}$ and $\tau_{\rm DM}$ are DM particle mass and decay time and $\rho(r)$ is the DM energy density profile, for which we use the Navarro, Frenk and White parametrisation [14]. We calculate the injection spectra $dN^f_{\gamma,e}/dE$ and the flux of secondary γ–rays from bremsstrahlung and inverse Compton scattering of electrons in halo using the public PPPC 4 DM code [15]. Note that secondary γ flux from $e^\pm$ interactions in halo has some uncertainty related to insufficient knowledge of galactic media and magnetic field. We use the "medium" interaction model implemented in [15].

To calculate the second term in (12) we approximate the extragalactic sources by continuous distribution and simulate the EM cascade propagation taking into account interactions of $e^\pm$ and γ with the CMB and EBL using numerical code [11] with the source density

$$\frac{dQ_{e,\gamma}}{dE} = \frac{\bar\rho_0}{M_{\rm DM}\tau_{\rm DM}}(1+z)^3\Gamma_f\frac{dN^f_{e,\gamma}}{dE}, \qquad (14)$$

where $\bar\rho_0 = 1.15\times10^{-6}\,{\rm GeV\,cm^{-3}}$ is the average dark matter energy density at $z = 0$.

In Fig. 4 we show the example of these calculations for the decay channels, $M_{\rm DM}$ and $\tau_{\rm DM}$ indicated in captions. Extragalactic and Milky Way halo contributions are shown separately to illustrate that the fluxes are comparable. Note that model shown in Fig. 4a also provides a good fit for AMS-02 positron-electron data [16].

By imposing requirement $\chi^2_{\rm DM} \leq 9$ we constrain the minimal decay time of DM particles in particular channels. These constraints are presented in Fig. 5. For comparison, we also show the limits obtained disregarding minimal contribution of UHECR interaction products and unresolved γ–ray sources (green curves) or disregarding just UHECR products (blue curves). The limits on $\tau_{\rm DM}$ for $M_{\rm DM} \gtrsim$ few TeV obtained in this work are on average 3-5 times stronger than the conservative constraints presented in the most recent analysis [16] due to more precise calculation of extragalactic contribution and account for unresolved astrophysical source and UHECR cascade γ.

Acknowledgments

Section 2 is a continuation of the original work [4] initiated in collaboration with V. Berezinsky and A. Gazizov. The author thanks M. Cirelli for help with the PPPC 4 DM package. The work was supported by the Russian Science Foundation, grant 14-12-01430.

References

[1] M. Ackermann et al. (Fermi-LAT), Astrophys. J. **799**, 86 (2015), 1410.3696
[2] M. Di Mauro (Fermi-LAT), *The origin of the Fermi-LAT γ-ray background*, in *14th Marcel Grossmann Meeting on Recent Developments in Theoretical and Experimental General Relativity, Astrophysics, and Relativistic Field Theories (MG14) Rome, Italy, July 12-18, 2015* (2016), 1601.04323, http://inspirehep.net/record/1415711/files/arXiv:1601.04323.pdf
[3] J.M. Gaskins (2016), 1604.00014
[4] V. Berezinsky, A. Gazizov, O. Kalashev (2016), 1606.09293
[5] A. Aab et al. (Pierre Auger), Phys. Rev. **D90**, 122006 (2014), 1409.5083
[6] R.U. Abbasi et al., Astropart. Phys. **64**, 49 (2015), 1408.1726
[7] Y. Fedorova (HiRes), *HiRes Stereo Cosmic Rays Composition Measurements*, in *Proceedings, 30th International Cosmic Ray Conference (ICRC 2007)* (2007), Vol. 4, pp. 463–466, http://indico.nucleares.unam.mx/contributionDisplay.py?contribId=1236&confId=4
[8] H. Yuksel, M.D. Kistler, J.F. Beacom, A.M. Hopkins, Astrophys. J. **683**, L5 (2008), 0804.4008
[9] G. Giacinti, M. Kachelrieß, O. Kalashev, A. Neronov, D.V. Semikoz, Phys. Rev. **D92**, 083016 (2015), 1507.07534
[10] R.U. Abbasi et al. (Telescope Array), Astropart. Phys. **80**, 131 (2016), 1511.07510
[11] O.E. Kalashev, E. Kido, J. Exp. Theor. Phys. **120**, 790 (2015), 1406.0735
[12] Y. Inoue, S. Inoue, M.A.R. Kobayashi, R. Makiya, Y. Niino, T. Totani, Astrophys. J. **768**, 197 (2013), 1212.1683
[13] J. Heinze, D. Boncioli, M. Bustamante, W. Winter (2015), 1512.05988
[14] J.F. Navarro, C.S. Frenk, S.D.M. White, Astrophys. J. **462**, 563 (1996), astro-ph/9508025
[15] M. Cirelli, G. Corcella, A. Hektor, G. Hutsi, M. Kadastik, P. Panci, M. Raidal, F. Sala, A. Strumia, JCAP **1103**, 051 (2011), [Erratum: JCAP1210,E01(2012)], 1012.4515
[16] W. Liu, X.J. Bi, S.J. Lin, P.F. Yin (2016), 1602.01012

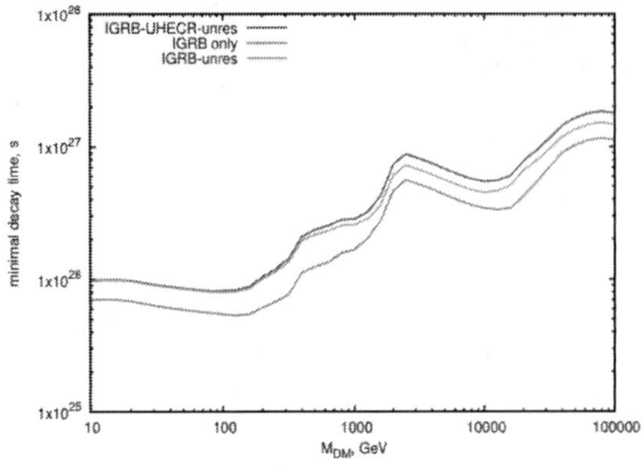

(a) $\mu^+\mu^-$

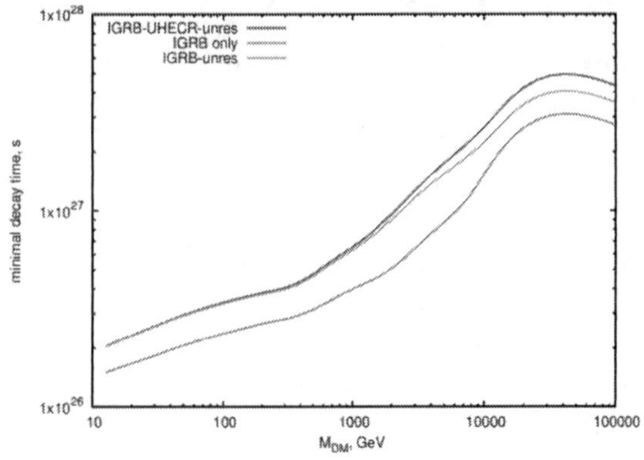

(b) $b\bar{b}$

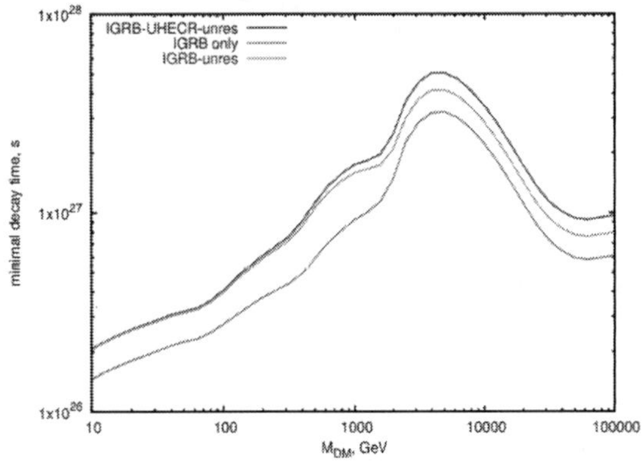

(c) $\tau^+\tau^-$

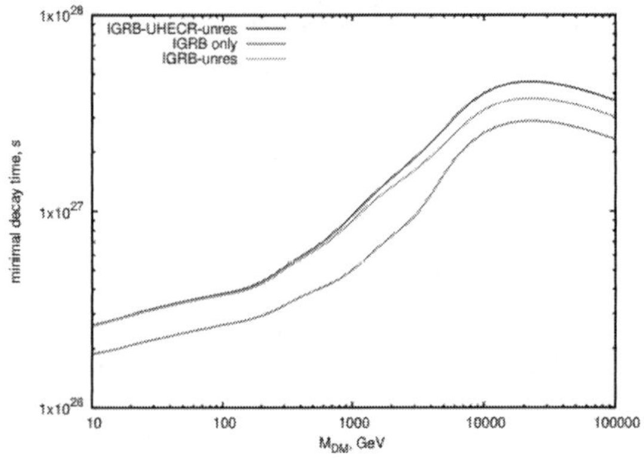

(d) $u\bar{u}$

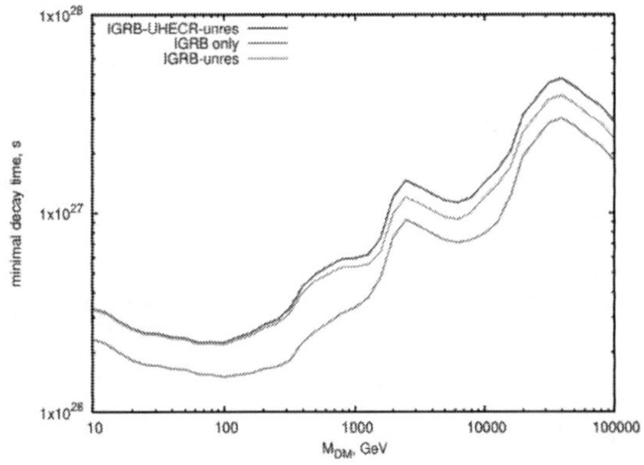

(e) e^+e^-

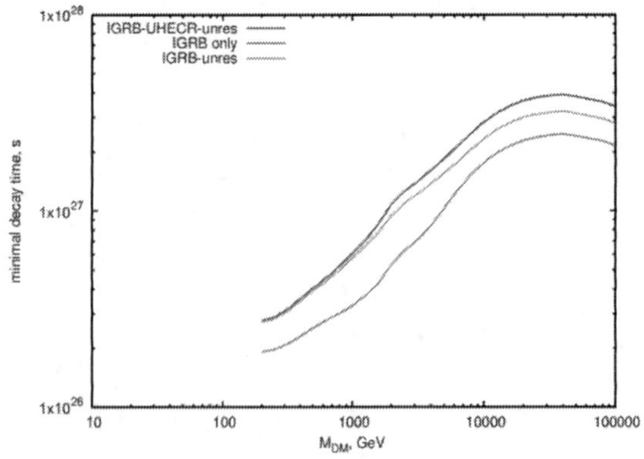

(f) W^+W^-

Figure 5: Minimal DM decay time for various decay channels.

EPJ Web of Conferences **125**, 02013 (2016) DOI: 10.1051/epjconf/201612502013
QUARKS-2016

Exotica searches at 13 TeV by CMS

Mikhail Kirsanov[1,*]

[1] *INR RAS Moscow*

Abstract. Results from searches for exotica signals in proton-proton collisions at $\sqrt{s}$ = 13 TeV in the CMS detector at the LHC are presented. The analysed data are collected during the first year of the LHC Run II in 2015 and correspond to integrated luminosities from 2.2 to 3.3 fb^{-1}.

1 Introduction

The Standard Model (SM) [1], [2] of particle physics has been very successful in explaining most of experimental data over the last half-century, and the recently discovered boson with a mass of 125 GeV [3], [4] could be the final particle required in this theory, the Higgs boson. However, the SM is not without its shortcomings, for instance one requires fine-tuned cancellations of large quantum corrections in order for the Higgs boson to have a mass at the electroweak symmetry breaking scale. This is otherwise known as the hierarchy problem. Other experimental data, such as neutrino masses (very small as compared to other particles) and oscillations, the large quantity of dark matter in the universe, look not quite natural or are difficult to fit into the SM. For this reason the searches for particles beyond the SM continue. In this paper the main searches for such particles in the CMS detector are described. Most of the searches for for particles that arise in various supersymmetric models are beyond the scope of this paper, although some overlap exists.

2 The CMS detector

A detailed description of the CMS detector can be found elsewhere [5]. The main features and those most pertinent to this analysis are described below. The central feature is a superconducting solenoid, 13 m in length and 6 m in diameter, which provides an axial magnetic field of 3.8 T. The bore of the solenoid is instrumented with particle detection systems. The steel return yoke outside the solenoid is instrumented with gas detectors used to identify muons. Charged particle trajectories are measured by the silicon pixel and strip tracker, with full azimuthal coverage within $|\eta|$ <2.5, where the pseudorapidity η is defined as $\eta = -ln[tan(\Theta/2)]$, with Θ being the polar angle of the trajectory of the particle with respect to the counterclockwise beam direction. A lead-tungstate crystal electromagnetic calorimeter (ECAL) and a brass/scintillator hadron calorimeter (HCAL) surround the tracking volume and cover the region $|\eta|$ <3. The ECAL barrel extends to $|\eta|$ =1.48. A lead/silicon-strip preshower detector is located in front of the ECAL endcap. A steel/quartz-fibre Cherenkov forward calorimeter extends the

*e-mail: Mikhail.Kirsanov@cern.ch

calorimetric coverage to $|\eta|$ <5.0. In the region $|\eta|$ <1.74, the HCAL cells have widths of 0.087 in both pseudorapidity and azimuth (ϕ). In the (η, ϕ) plane, and for $|\eta|$ <1.48, the HCAL cells map on to 5x5 ECAL crystal arrays to form calorimeter towers projecting radially outwards from points slightly offset from the nominal interaction point. In the endcap, the ECAL arrays matching the HCAL cells contain fewer crystals. Calibration of the ECAL uses π^0s, $W \rightarrow e\nu$, and $Z \rightarrow ee$. Deterioration of transparency of the ECAL crystals due to irradiation during the LHC running periods and their subsequent recovery is monitored continuously and corrected for using light injected from a laser and LED system.

The sophisticated reconstruction algoriths are used to reconstruct various physical objects in CMS, such as charged particles, photons, muons, jets etc. The "Particle flow" algorithm is used to calculate the missing E_T of the event, i.e. the information from fifferent subdetectors is analysed and combined to create "particles" before summing them up. Particle - based isolation is used in the reconstruction of leptons and photons. Multivariate algoriths are used to identify τ and to reconstruct momentum of leptons and photons.

3 Run II in 2015

Run II of the LHC with the collision energy of 13 TeV started in 2015. The start of the run for CMS was not without problems. Because of them different analyses had to use data samples corresponding to rather different integrated luminosity. The data sample "good for all" corresponds to 2.2 fb^{-1}, the data sample for the analyses that don't use missing transverse energy (MET) corresponds to 2.6 fb^{-1}.

As a rule, exotic particles that are searched for are rather heavy because for smaller masses they are already excluded. In general, the reach of experiment to such particles is expressed in terms of effective luminosity, i.e. the luminosity of quarks and gluons that can produce an object with a certain mass. The comparison of Run I and Run II in this sense is shown in Figure 1. According to this figure, it is expected that the sensitivity of the first part of Run II should exceed the sensitivity of Run II for particles heavyer than 2.5 TeV.

4 Searches for resonances

The largest class of searches for exotic particles is the searches for resonances. They are summarized in Table 1.

Most interesting results are obtained in the search for diphoton resonances: the excess of events at about 750 GeV. This excess slightly increases when combined with Run I data. In Run I the excess at this mass was not very significant, but this is expected because of smaller collision energy. To increase statistics the analysis of data taken with solenoid off (0.6 fb^{-1}) was performed. The $M_{\gamma\gamma}$ distribution in Run II data is shown in Figure 2. The p-values of the combination of all available data are shown in Figure 3.

5 Dark matter searches

The interest to searches of Dark Matter iscreased significantly during the last few years. In CMS Dark Matter can manifest itself as monojets with large MET. Dark matter particle pairs could be produced through a vector mediator and escape from the detector. The results of the corresponding analysis using the first part of Run II are published in EXO-16-013 [15]. Some results can be seen as limits on a two-dimensional plot ($M_{DM} vs M_{mediator}$ in Figure 4.

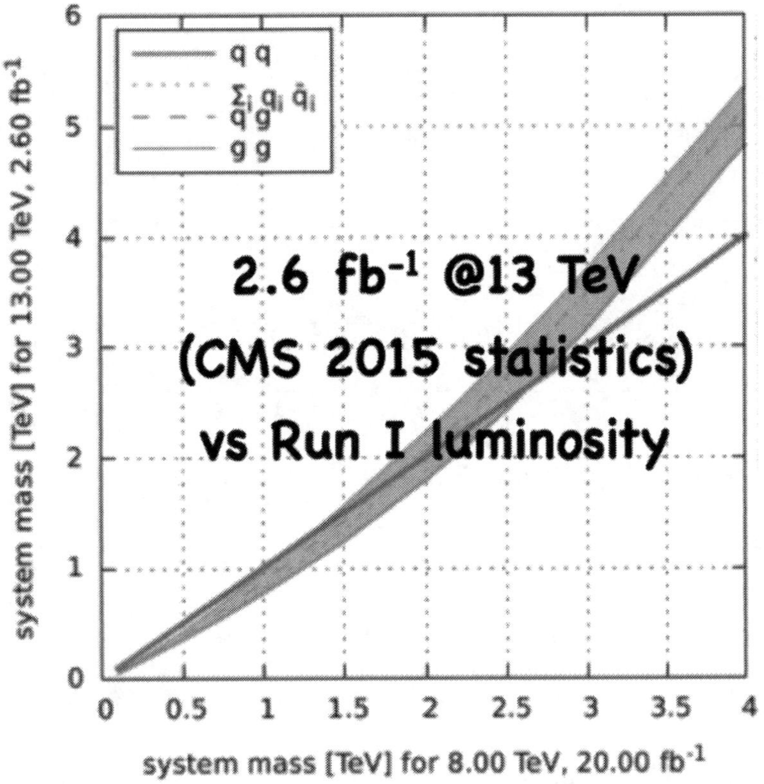

Figure 1. The Run I - Run II comparison of the effective luminosity for high mass objects production.

6 Searches for new long-lived particles

In CMS there is an extensive program of searches for long-lived particles. The following manifestations of their existence are searched for:

- Heavy, long-lived, charged particles, can produce high ionization due to slow speed. Time of flight can be measured.
- Displaced vertices of electrons, muons or jets
- Delayed decays
- Visible decays in flight

Heavy stable charged particles search is presented in EXO-15-010 [16]. The mass limits are set from 800 to 1600 GeV depending on the nature of such particles (most stringent limits are for gluino production).

7 Searches for Black Holes

The search for microscopic Black Holes is performed using the variable S_T, the scalar sum of E_T of all particles in the event. To reduce background, it is also required that the event MET exceeds 50 GeV.

Table 1. Summary of Higgs analyses and documentation

Analysis	Data used, fb^{-1}	Mass limit, TeV	References
Z', neutral spin 1 particle to ee, $\mu\mu$	2.6 - 2.8	2.6 - 3.15	EXO-15-005 [6]
High mass $e\mu$ resonances e.g. QBH production	2.7	2.5 - 4.5	EXO-16-001 [7]
$W' \rightarrow e, \mu + MET$ (SSM)	2.2	4.4	EXO-15-006 [8]
$W' \rightarrow tb \rightarrow bl\nu$	2.2	2.38	B2G-15-004 [9]
$W' \rightarrow \tau\nu$	2.3	3.3	EXO-16-006 [10]
Dijet narrow resonances	2.3	2 - 7	PRL [11]
Diphotons, scalar and spin 2 e.g. for spin 2 graviton	3.3	Limit on σB up to 3 TeV	EXO-16-016 EXO-16-018 [12]
$Z\gamma$	2.7	Limit on σ B up to 2 TeV	EXO-16-019 [13]
Diboson $WW \rightarrow l\nu qq$ e.g graviton	2.3	Limit on σ B up to 1 TeV	B2G-16-004
tt (boosted tops)	2.6	3 - 3.5	B2G-15-002 [14]

The results are reported in EXO-15-007 [17]. In the existing models the Black Hole cross section production is rather large. For this reason the limits on the Black Hole mass are rather stringent, of the order of 8 - 9 TeV. For the same reason they are much more stringent than the Run I limits.

8 Searches for Type III SeeSaw fermions.

The seesaw mechanism could explain the existence and smallness of neutrino masses, one of the fenomena that do not neturally fit into the Standard Model. The type III seesaw mechanism assumes the existence of heavy fermions that can be produced at the LHC. The analysis EXO-16-002 [18] searches for pairs of such fermions that decay to ordinaly leptons, charged and neutrinos. The final state with at least three charged leptons (electrons or muons) and MET is the signature of this analysis. The main background is WZ production. Heavy fermions with masses up to 450 GeV are excluded. At these masses the effective luminosity for Run II 2015 data is significantly smaller than for Run I data, however, due to a number of improvements of the analysis, the sensitivity of EXO-16-002 is better than in the similar Run I analysis.

9 Conclusion.

The exploration of initial 13 TeV data of Run II has given some interesting sesulta and allowed to improve the sensitivity of meny searches for exotic particles. Mor data is a key to understanding if the diphoton excess is real or just a fluctuation. In the nearest future at least ten times more 13 TeV data are expected.

References

[1] S. L. Glashow, "Partial-symmetries of weak interactions" *Nucl. Phys.* **22** (1961) 579.

[2] S. Weinberg, "A model of leptons" *Phys. Rev. Lett.* **19** (1967) 1264.

[3] CMS Collaboration, "Observation of a new boson at a mass of 125 GeV with the CMS experiment at the LHC", *Phys. Lett. B* (2012), arXiv:1207.7235.

DOI: 10.1051/epjconf/201612502013

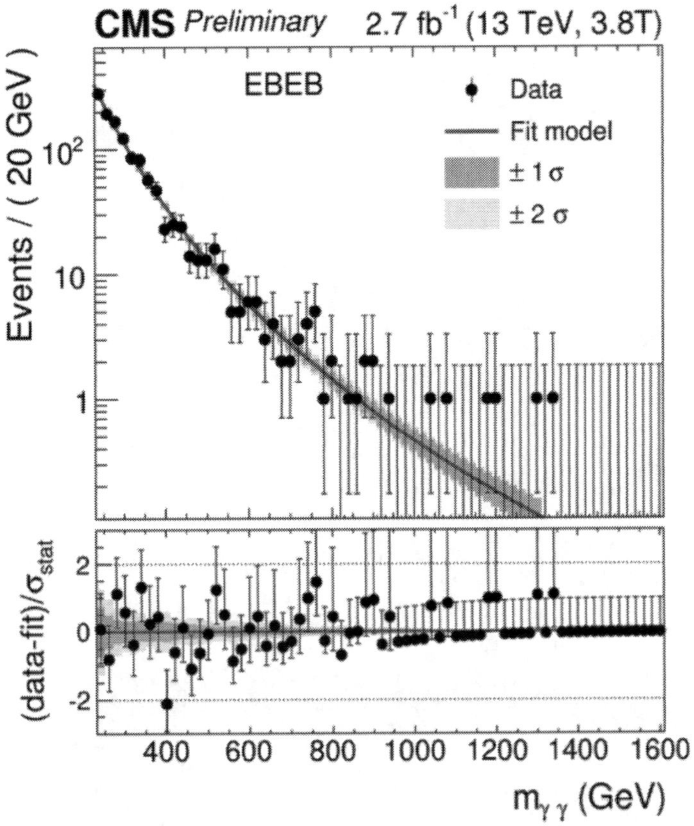

Figure 2. $M_{\gamma\gamma}$ distribution for the category EBEB (both gammas in the barrel)

[4] ATLAS Collaboration, "Observation of a new particle in the search for the Standard Model Higgs boson with the ATLAS detector at the LHC", *Phys. Lett. B* (2012), arXiv:1207.7214.

[5] CMS Collaboration, "The CMS experiment at the CERN LHC" *JINST* **3** (2008) S08004.

[6] CMS Collaboration, "Search for a Narrow Resonance Produced in 13 TeV pp Collisions Decaying to Electron Pair or Muon Pair Final States", CMS PAS EXO-15-005.

[7] CMS Collaboration, "Search for high-mass resonances and quantum black holes in the $e\mu$ final state in proton-proton collisions at $\sqrt{s}$ = 13 TeV", CMS PAS EXO-16-001.

[8] CMS Collaboration, "Search for SSM W' production, in the lepton+MET final state at a center-of-mass energy of 13 TeV", CMS PAS EXO-15-006.

[9] CMS Collaboration, "Search for W' boson resonances decaying into a top quark and a bottom quark in the leptonic final state at $\sqrt{s}$ = 13 TeV", CMS PAS B2G-15-004.

[10] CMS Collaboration, "Search for W' decaying to tau lepton and neutrino in proton-proton collisions at $\sqrt{s}$=13 TeV", CMS PAS EXO-16-006.

[11] CMS Collaboration, "Search for narrow resonances decaying to dijets in proton-proton collisions at $\sqrt{s}$ = 13 TeV", PRL **116**, 071801 (2016).

[12] CMS Collaboration, "Search for new physics in high mass diphoton events in 3.3 fb^{-1} of proton-proton collisions at $\sqrt{s}$ = 13 TeV and combined interpretation of searches at 8 TeV and 13 TeV", CMS PAS EXO-16-018.

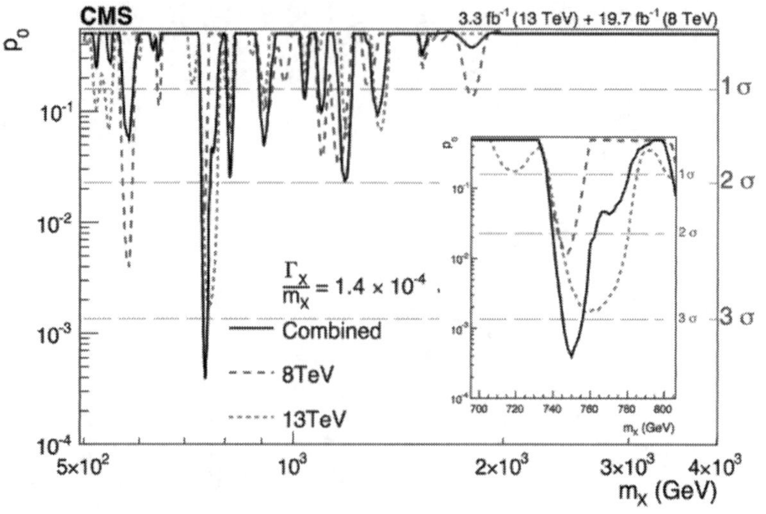

Figure 3. p-values of the combination of all available diphoton data, for the hypothesis of narrow scalar resonance

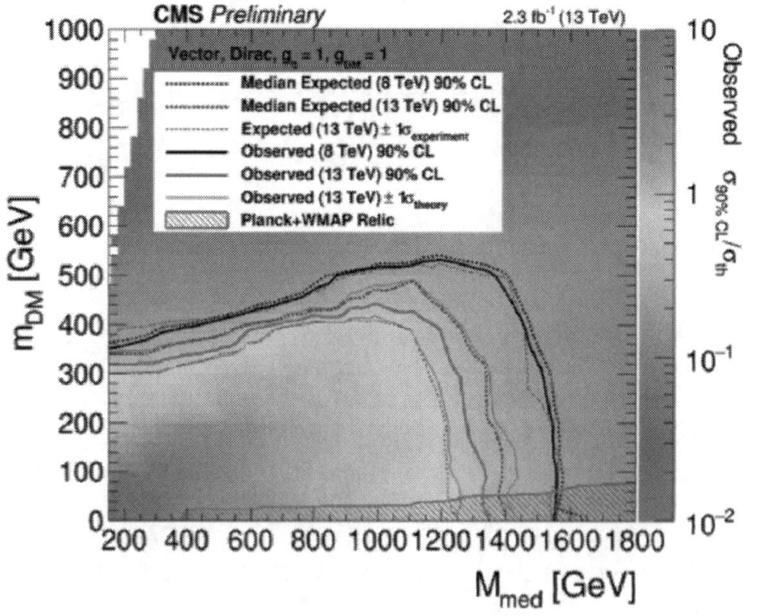

Figure 4. Limits on the Dark Matter particle masses

[13] CMS Collaboration, "Search for high-mass resonances in $Z\gamma \to e^+e^-\gamma/\mu^+\mu^-\gamma$ final states in proton-proton collisions at $\sqrt{s} = 13$ TeV", CMS PAS EXO-16-019.

[14] CMS Collaboration, "Search for $t\bar{t}$ resonances in boosted semileptonic final states in pp collisions at $\sqrt{s} = 13$ TeV", CMS PAS B2G-15-002.

[15] CMS Collaboration, "Search for dark matter production in association with jets, or hadronically decaying W or Z boson at $\sqrt{s} = 13$ TeV", CMS PAS EXO-16-013.

[16] CMS Collaboration, "Searches for Long-lived Charged Particles in Proton-Proton Collisions at $\sqrt{s} = 13$ TeV", CMS PAS EXO-15-010.
[17] CMS Collaboration, "Search for Black Holes with Early Run 2 Data", CMS PAS EXO-15-007.
[18] CMS Collaboration, "Search for Type-III Seesaw Heavy Fermions with Multilepton Final States using 2.3 fb^{-1} of $sqrts = 13$ TeV proton–proton Collision Data", CMS PAS EXO-16-002.

Warped extra dimension and inclined events at Pierre Auger Observatory

Alexander Kisselev[1,*] and *Liliya* Shkalina[2,**]

[1] *Institute for High Energy Physics, NRC "Kurchatov Institute", 142281 Protvino, Russian Federation*
[2] *Department of Physics, Moscow State University, 119991, Moscow, Russian Federation*

Abstract. The generalized solution for the warp factor of the Randall-Sundrum metric is presented which is symmetric with respect to both branes and explicitly periodic in extra variable. Given that the curvature of the 5-dimensional space-time is small, the expected rate of neutrino-induced inclined events at the Surface Detector of the Pierre Auger Observatory is calculated. Both the "downward-going" (DG) and "Earth-skimming" (ES) neutrinos are considered. By comparing the expected event rate with the recent Auger data on searching for neutrino candidates, the lower bound on the fundamental gravity scale M_5 is obtained. The ratio of the number of the ES air showers to the number of the DG showers is estimated as a function of M_5.

1 Introduction

Ultra high energy (UHE) cosmic neutrinos play a key role in the determination of the composition of the ultra high energy cosmic rays (UHECRs) and their origin. UHE neutrinos are expected to be produced in astrophysical sources in the decays of charged pions created in the interactions of UHECRs with matter or radiation. They can be also produced via interaction of the UHECRs with the cosmic microwave background during propagation to the Earth (cosmogenic neutrinos). UHE cosmic neutrinos are not deviated by magnetic fields and could point back to their sources.

Recently, three neutrinos of energy 1-2 PeV, as well as tens of neutrinos above 10 TeV were detected with the IceCube experiment [1]. The cosmic neutrinos with energies near 1 EeV are detectable with the Surface Detector (SD) of the Pierre Auger Observatory (PAO) [2]. In order to isolate neutrino-induced events at the SD of the PAO, it is necessary to look for deeply penetrating quasi-horizontal (inclined) air showers [3]-[5]. The PAO can efficiently search for two types of neutrino-induced inclined air showers (see fig. 1):

1. Downward-going (DG) neutrino-induced showers. They are initiated by neutrinos moving with large zenith angle θ which interact in the atmosphere close to the SD. At the PAO the search is restricted to showers with $\theta > 60°$ [6]. Note that the background from hadronic showers above 10^{17} eV is O(1) in 20 years, and it is negligible above 10^{19} eV [7].

*e-mail: alexandre.kisselev@ihep.ru
**e-mail: la.shkalina@physics.msu.ru

2. Earth-skimming (ES) showers induced by upward tau neutrinos. They can interact in the Earth's crust producing tau leptons. The latter are efficiently produced at zenith angles $90° < \theta < 95°$ [6]. The tau leptons escape the Earth and decay in the atmosphere close to the SD [8]-[9].

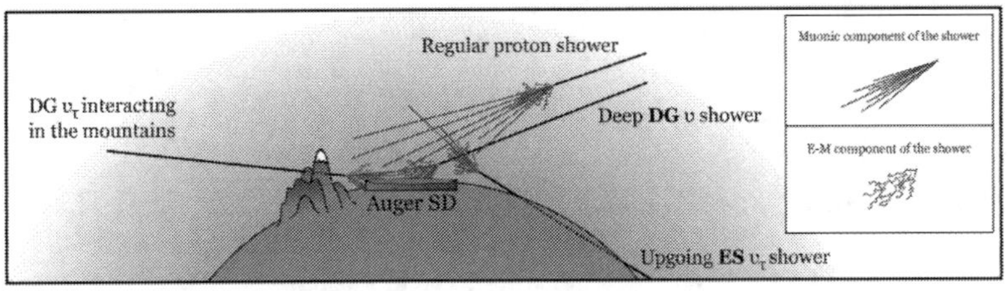

Figure 1. Different types of showers induced by DG and ES neutrinos (fig. 1 in [10]).

Recently, the Auger Collaboration reported on searches for DG neutrinos in the zenith angle bins $60° - 75°$ and $75° - 90°$, as well as for ES neutrinos. The data were collected by the SD of the PAO from 1 January 2004 until 20 June 2013 [6].[1] No neutrino candidates were found. Assuming the diffuse flux of UHE neutrinos to be $dN/dE_\nu = kE_\nu^{-2}$ in the energy range 1.0×10^{17} eV - 2.5×10^{19} eV, the stringent limit was obtained:

$$k < 6.4 \times 10^{-9} \text{ GeV cm}^2 \text{ s}^{-1} \text{ sr}^{-1} . \tag{1}$$

This Auger limit is a factor 3.64 below the Waxman-Bachall bound on neutrino production in optically thin astrophysical sources [11]:

$$E_\nu^2 \frac{dN}{dE_\nu} = 2.33 \times 10^{-8} \text{ GeV cm}^2 \text{ s}^{-1} \text{ sr}^{-1} . \tag{2}$$

In the Standard Model (SM) neutrino-nucleon cross sections are expected to be very small in comparison with hadronic cross sections even at UHEs [12]. That is why, the UHE cosmic neutrinos can be regarded as unique probes of new interactions. In the present paper a theory with an extra dimension (ED) is considered to be a "new physics" theory. We will see that effects coming from the ED can be significant or even dominant in the νN-scattering at UHEs.

2 General solution for the Randall-Sundrum metric

In ref. [13] the 5-dimensional space-time with non-factorizable geometry was suggested as an alternative to the models with flat extra dimensions. The Randall-Sundrum (RS) model [13] predicts an existence of heavy Kaluza-Klein (KK) excitations (massive gravitons). These massive KK resonances are intensively searched for at the LHC.

The RS scenario is described by the following background warped metric

$$ds^2 = e^{-2\sigma(y)} \, \eta_{\mu\nu} \, dx^\mu \, dx^\nu - dy^2 , \tag{3}$$

where $\eta_{\mu\nu}$ is the Minkowski tensor with the signature $(+, -, -, -)$, and y is an extra coordinate. The periodicity condition $y = y + 2\pi r_c$ is imposed, and the points (x_μ, y) and $(x_\mu, -y)$ are identified. Thus, one has a model of gravity in the AdS_5 space-time compactified to the orbifold S^1/Z_2. The orbifold

[1] This search period is equivalent of 6.4 years of a complete Auger SD working continuously [6].

has two fixed points, $y = 0$ and $y = \pi r_c$. It is assumed that there are two branes located at these points. All the SM fields live on one of these branes.

The classical action of the RS scenario is given by [13]

$$S = \int d^4x \int_{-\pi r_c}^{\pi r_c} dy\, \sqrt{G}\,(2\bar{M}_5^3 \mathcal{R} - \Lambda) + \int d^4x \sqrt{|g^{(1)}|}\,(\mathcal{L}_1 - \Lambda_1) + \int d^4x \sqrt{|g^{(2)}|}\,(\mathcal{L}_2 - \Lambda_2)\,, \tag{4}$$

where $G_{MN}(x, y)$ is the 5-dimensional metric, with $M, N = 0, 1, 2, 3, 4$, $\mu = 0, 1, 2, 3$. The quantities

$$g^{(1)}_{\mu\nu}(x) = G_{\mu\nu}(x, y = 0)\,, \quad g^{(2)}_{\mu\nu}(x) = G_{\mu\nu}(x, y = \pi r_c) \tag{5}$$

are induced metrics on the branes, $\mathcal{L}_1$ and $\mathcal{L}_2$ are brane Lagrangians, $G = \det(G_{MN})$, $g^{(i)} = \det(g^{(i)}_{\mu\nu})$. $\bar{M}_5$ is the *reduced* 5-dimensional Planck scale. The quantity Λ is a 5-dimensional cosmological constant, while $\Lambda_{1,2}$ are brane tensions.

For the first time, the function $\sigma(y)$ was obtained in [13] (see also the expression at the bottom of fig. 2):

$$\sigma_0(y) = \kappa|y| - C_0\,, \tag{6}$$

where κ is a parameter with a dimension of mass, which defines the curvature of the 5-dimensional space-time. This solution is consistent with the orbifold symmetry $y \to -y$. However, it is attributed

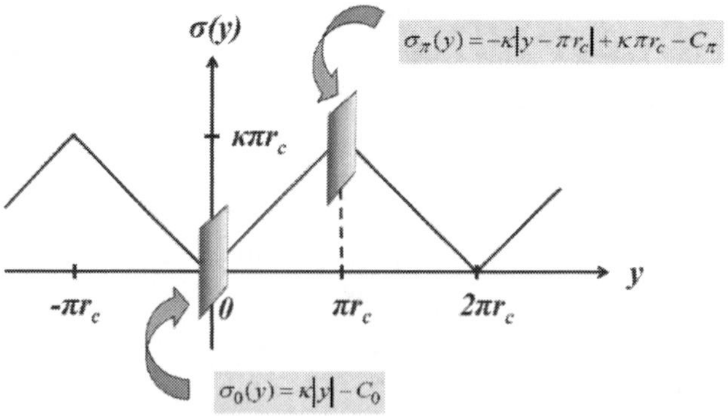

Figure 2. Two equivalent solutions of Einstein-Hilbert's equations for the function $\sigma(y)$ related to the different branes located at $y = 0$ and $y = \pi r_c$.

to the brane $y = 0$ (see fig. 2), and it is not symmetric with respect to the branes.[2]

Instead of (6), one can use another *equivalent* solution related to the brane $y = \pi r_c$ (see the expression at the top of fig. 2):

$$\sigma_\pi(y) = -\kappa|y - \pi r_c| + |\kappa|\pi r_c - C_\pi\,, \tag{7}$$

In order to get an expression symmetric with respect to both branes, we take half the sum of expressions (6) and (7). As a result, we come to the solution:

$$\sigma(y) = \frac{\kappa}{2}(|y| - |y - \pi r_c|) + \frac{|\kappa|\pi r_c}{2} - C\,, \tag{8}$$

[2]Note that in contrast to (6), the RS1 solution [13] has no constant term.

where $C = (C_0 + C_\pi)/2$. The constant terms in (8) are chosen in such a way that one has $\sigma(y) = \kappa y - C$ within the interval $0 < y < \pi r_c$.

Note that neither original RS1 solution nor generalized RS-like solution (8) obeys periodicity in y *explicitly*. One has to keep the periodicity condition in "mind". Fortunately, a new solution was recently proposed [14] which is both symmetric with respect to the branes and *periodic* function of y:

$$\sigma(y) = \frac{\kappa r_c}{2} \left[\left| \mathrm{Arccos}\left(\cos\frac{y}{r_c}\right) \right| - \left| \pi - \mathrm{Arccos}\left(\cos\frac{y}{r_c}\right) \right| \right] + \frac{\pi |\kappa| r_c}{2} - C \,, \tag{9}$$

with the fine tuning relations:

$$\Lambda = -24 \bar{M}_5^3 \kappa^2 \,, \quad \Lambda_1 = -\Lambda_2 = 24 \bar{M}_5^3 \kappa \,. \tag{10}$$

Here Arccos(z) is a *principal value* of the multivalued inverse trigonometric function arccos(z). It is define by the inequality (see, for instance, [15]):

$$0 \leqslant \mathrm{Arccos}(z) \leqslant \pi \,, \quad -1 \leqslant z \leqslant 1 \,. \tag{11}$$

It means that [15]

$$\mathrm{Arccos}(\cos x) = \begin{cases} x - 2n\pi, & 2n\pi \leqslant x \leqslant (2n+1)\pi \,, \\ -x + 2(n+1)\pi, & (2n+1)\pi \leqslant x \leqslant 2(n+1)\pi \,, \end{cases} \tag{12}$$

where $n = 0, 1, \ldots$ It follows from (9), (12) that $\sigma(y) + C = \kappa y$ for $0 < y < \pi r_c$, $\sigma(y) + C = \kappa(2\pi r_c - y)$ for $\pi r_c < y < 2\pi r_c$. and so on (see fig. 2).

Our solution (9)

- is symmetric with respect to the branes: $\sigma(y)$ remains unchanged if $y \to \pi r_c - y$, $k \to -k$;[3]
- obeys the orbifold Z_2-symmetry: $\sigma(y) = \sigma(-y)$;
- reproduces the jumps of $\sigma'(y)$ on both branes: $\sigma''(y) = 2\kappa[\delta(y) - \delta(y - \pi r_c)]$, $0 \leqslant y \leqslant \pi r_c$;
- is the periodic function of the extra coordinate: $\sigma(y + 2\pi r_c n) = \sigma(y)$, $n = \pm 1, \pm 2, \ldots$ Note that the warped function $\sigma(y)$ depends on the constant C.

3 RS-like scenario with the small curvature

By taking different values of C in eq. (9), we come to quite *diverse physical scenarios* [14], [16]. One of them ($C = 0$) is in fact the RS1 model [13]. Another scheme ($C = \kappa \pi r_c$) describes a geometry with a small curvature of five-dimensional space-time (RSSC model) [17]-[19]. In the RSSC model the hierarchy relation looks like (we assume that $\kappa \pi r_c \gg 1$)

$$\bar{M}_{\mathrm{Pl}}^2 = \frac{\bar{M}_5^3}{\kappa} \, e^{2\kappa\pi r_c} \,. \tag{13}$$

So, one can take, for instance, $\kappa \sim 1$ GeV, $M_5 \sim 1$ TeV. Let us underline that the original RS scenario does not admit the parameters to lie in this region, since $\kappa \sim \bar{M}_5 \sim M_{\mathrm{Pl}}$ TeV in it [13].

The masses of the KK gravitons are proportional to the curvature parameter κ:

$$m_n = x_n \kappa \,, \tag{14}$$

[3]Under the replacement $y \to \pi r_c - y$ the positions of the branes are interchanged ($y = 0 \rightleftarrows y = \pi r_c$), while under the replacement $k \to -k$ their tensions are interchanged ($\Lambda_1 \rightleftarrows \Lambda_2$).

where x_n are zeros of the Bessel function $J_1(x)$. The coupling of the massive gravitons to the SM fields is given by the constant:

$$\Lambda_\pi = \bar{M}_5 \left(\frac{\bar{M}_5}{\kappa} \right)^{1/2} . \tag{15}$$

Thus, the RSSC scenario predicts a spectrum of the KK gravitons similar to that of the ADD model [20]-[21]. For the LHC phenomenology of the RSSC model, see [22], [23].

4 Neutrino-nucleon amplitude in the RSSC model

For the UHE cosmic neutrinos, the scattering takes place in the trans-Planckian kinematical region:

$$\sqrt{s} \gtrsim \bar{M}_5 \gg -t , \tag{16}$$

with $\sqrt{s}$ being the colliding energy and $t = -q_\perp^2$ four-dimensional momentum transfer. Remember that the inequality $\kappa \ll \bar{M}_5$ is also satisfied. In the eikonal approximation, which is valid in the kinematical region (16), the neutrino-proton amplitude is of the form:

$$A_{\nu p}(s,t) = 4\pi i \, s \int_0^\infty db b J_0(b \, q_\perp) \left\{ 1 - \exp[i\chi(s,b)] \right\} . \tag{17}$$

In its turn, the eikonal is given by the Fourier-Bessel transformation of the eikonal:

$$\chi(s,b) = \frac{1}{4\pi s} \int_0^\infty dq_\perp q_\perp J_0(q_\perp b) A^{\mathrm{B}}_{\nu p}(s,t) . \tag{18}$$

The Born amplitude is given by the sum of gravi-Reggeons, i.e. reggeized gravitons in the t-channel. Because of a presence of the ED, the Regge trajectory of the graviton is splitting into an infinite sequence of trajectories enumerated by the KK number n [24]:

$$\alpha_n(t) = 2 + \alpha'_g t - \alpha'_g m_n^2, \quad n = 0, 1, \ldots . \tag{19}$$

In string theories, the slope of the gravi-Reggeons is universal, and $\alpha'_g = M_s^{-2}$, where M_s is the string scale. As a result, in the RSSC model the gravity Born amplitude for the neutrino scattering off *a point-like* particle looks like [18], [25]:

$$A^{\mathrm{B}}_{\mathrm{grav}}(s,t) = \frac{\pi \alpha'_g s^2}{2\Lambda_\pi^2} \sum_{n \neq 0} \left[i - \cot \frac{\pi \alpha_n(t)}{2} \right] \left(\frac{s}{\bar{M}_5} \right)^{\alpha_n(t)-2} . \tag{20}$$

The *hadronic* Born amplitude in (18) is defined by a convolution of the gravity amplitude (20) and skewed (t-dependent) PDFs $f_i(x,\mu^2;t)$:

$$A^{\mathrm{B}}_{\nu p}(s,t) = \sum_{i=q,\bar{q},g} \int_0^1 dx A^{\mathrm{B}}_{\mathrm{grav}}(xs,t) \, f_i(x,\mu^2;t) . \tag{21}$$

We take the t-dependent PDFs in a factorized form:

$$f_i(x,\mu^2;t) = f_i(x,\mu^2) D(t) . \tag{22}$$

For the PDFs $f_i(x, \mu^2)$ we use the CT14 parametrization [26] and put $\mu^2 = |t|$. As for a suppression factor $D(t)$, at hadron scales it was evaluated to be $D(t) = \exp(tr_0^2)$, with $r_0^2 = 0.62$ GeV^{-2} [27]. For larger $|t|$, we assume $D(t)$ to have a power-like behavior. As a result, we come to the parametrization:

$$D(t) = \begin{cases} \exp(tr_0^2)\,, & |t| \leqslant q_0^2\,, \\ \left(e|t|/q_0^2\right)^{-q_0^2 r_0^2}\,, & |t| > q_0^2\,, \end{cases} \tag{23}$$

where $q_0 = m_\rho$ is the ρ-meson mass, and $\ln e = 1$.

The total cross sections with the account of the contribution from the KK gravitons are shown in fig. 3. We adopted the SM νN cross sections from [12]. One can see, the total cross section noticeably exceeds the SM one at energies higher than 10^{18} or 10^{19} eV, depending on the reduced 5-dimensional Planck scale $\bar{M}_5$.

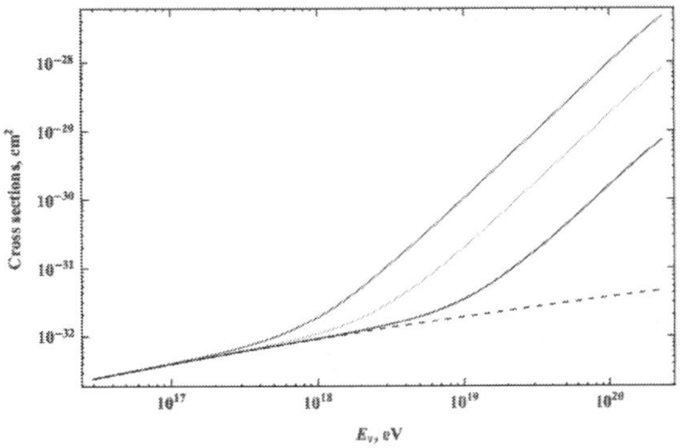

Figure 3. Solid lines (from above): the neutrino total cross sections for $\bar{M}_5 = 1.5$, 2.0 and 3.0 TeV. Dashed line: neutrino CC total cross section.

5 Expected number of neutrino events at the PAO

In [7] the following functional dependence of the DG event rate on the new physics cross section $\sigma_{\rm NP}$ was proposed for a bin in neutrino energy ($10^{18.5}$ eV $< E_\nu < 10^{19.5}$ eV):

$$N_{\rm BSM}^{\rm DG} = N_{\rm SM}^{\rm DG}\,\frac{\sigma_{\rm CC} + \sigma_{\rm NP}}{\sigma_{\rm CC}}\,, \tag{24}$$

where $N_{\rm SM}^{\rm DG}$ is the number of DG events in the absence of new physics, $\sigma_{\rm CC}$ is a charged current (CC) total cross section. $N_{\rm BSM}^{\rm DG}$ is the number of DG events with the account of interactions beyond the SM. We propose a more general formula:

$$\mathcal{E}_{\rm BSM}^{\rm DG}(E_\nu) = \mathcal{E}_{\rm SM}^{\rm DG}(E_\nu)\,\frac{\sigma_{\rm SM}^{\rm eff}(E_\nu) + \sigma_{\rm NP}(E_\nu)}{\sigma_{\rm SM}^{\rm eff}(E_\nu)}\,, \tag{25}$$

where $\mathcal{E}_{\rm BSM}^{\rm DG}$ ($\mathcal{E}_{\rm SM}^{\rm DG}$) is the *exposure* of the SD of the PAO with (without) account of the new interaction. In addition, instead of $\sigma_{\rm CC}$, an effective SM cross section $\sigma_{\rm SM}^{\rm eff}$ is introduce in (25):

$$\sigma_{\rm SM}^{\rm eff} = \sigma_{\rm CC} \sum_{i=e,\mu,\tau} m_{\rm CC}^i + 3\sigma_{\rm NC}\, m_{\rm NC} + \sigma_{\rm CC}\, m_{\rm mount}\,. \tag{26}$$

Here $m^i_{\rm CC}$ and $m_{\rm NC}$ are relative mass apertures for charged current and neutral current (NC) interactions of the DG neutrinos at the PAO. The mass aperture $m_{\rm mount}$ corresponds to the CC interaction of a τ neutrino within the mountains around the PAO (see fig. 1). The relative mass apertures as functions of the neutrino energy where calculated using data from Table I of ref. [10].[4] Note that $\sum_{i=e,\mu,\tau} m^i_{\rm CC} + 3m_{\rm NC} + m_{\rm mount} = 1$.

In contrast to the DG neutrino exposure, the exposure of the ES neutrinos *decreases* with the rise of the neutrino total cross section:[5]

$$\mathcal{E}^{\rm ES}_{\rm BSM}(E_\nu) = \mathcal{E}^{\rm ES}_{\rm SM}(E_\nu)\,\frac{\sigma^2_{\rm CC}(E_\nu)}{[\sigma_{\rm CC}(E_\nu) + \sigma_{\rm NP}(E_\nu)]^2}\,. \tag{27}$$

The formulas (25) and (27) allowed us to calculate expected exposures of the SD of the PAO for the period 1 January 2004 - 20 June 2013. The Auger data on exposures for the SM neutrino interactions in the region from $\log(E_\nu/{\rm eV}) = 17$ to 20.5 in steps of 0.5 were used (see also fig. 3 from [6]). The results of our calculations are presented in fig. 4.

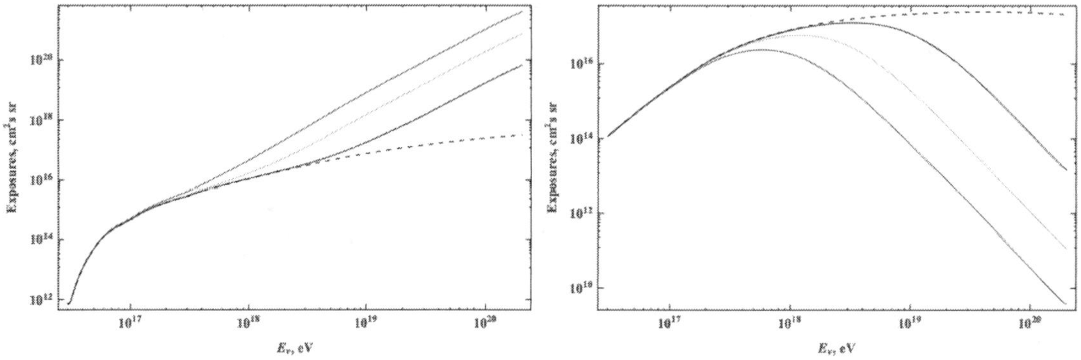

Figure 4. Left panel: expected Auger exposures for the DG neutrinos with zenith angle $75° < \theta < 90°$. Red, green and blue line corresponds to $\bar{M}_5$ = 1.5, 2.0 and 3.0 TeV, respectively. Dashed line: SM exposure taken from [6]. Right panel: the same but for the ES neutrinos.

The number of neutrino events is given by the formula:

$$N_{\rm ev} = \int \frac{dN_\nu}{dE_\nu}(E_\nu)\,\mathcal{E}(E_\nu)\,dE_\nu\,. \tag{28}$$

We assume differential neutrino flux $dN/dE_\nu \sim E_\nu^{-2}$, as well as a ratio $\nu_e : \nu_\mu : \nu_\tau = 1 : 1 : 1$. In fig. 5 we compare the number of neutrino events predicted in the RSSC model with the expected number of neutrino events in the absence of a new physics. Both the DG events with zenith angles $60° < \theta < 90°$ and ES events was taken into account in the full energy region of the sensitivity of the SD ($1.0\times10^{17} - 2.5\times10^{19}$ eV). As was already mentioned above, no neutrino events were seen at the SD of the PAO. Thus, from fig. 5 we obtain the lower bound on 5-dimensional *reduced* Planck scale:

$$\bar{M}_5 > 2.75 \text{ TeV}\,. \tag{29}$$

Correspondingly, for the 5-dimensional Planck scale we find $M_5 > 2.75\times(2\pi)^{1/3} = 5.07$ TeV.

We have also calculated an expected ratio of the ES neutrino showers to the DG neutrino showers with the zenith angles $75° < \theta < 90°$ (see fig. 6). We predict $N_{\rm ES}/N_{\rm DG}$ to be 2.8 (5.7) for $\bar{M}_5$ = 2.8

[4]In particular, $\sum_{i=e,\mu,\tau} m^i_{\rm CC}$ is equal to 0.69 (0.65) at 10^{18} (10^{19}) eV.

[5]Is is a generalization of the formula for the rate for ES showers in the range $10^{18.5}$ eV $< E_\nu < 10^{19.5}$ eV [7].

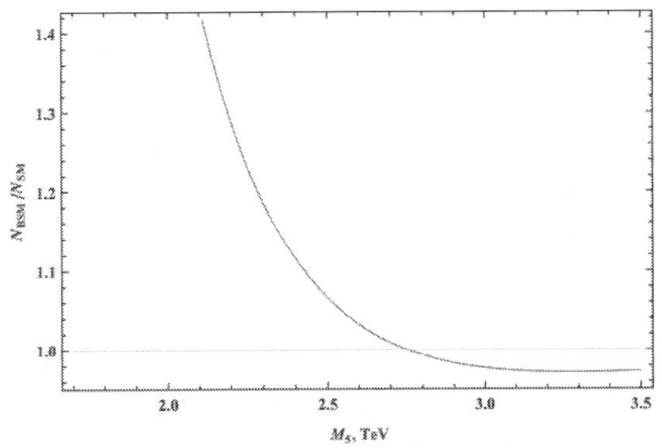

Figure 5. The ratio of the neutrino events with and without effects from EDs as a function of the gravity scale $\bar{M}_5$, expected at the SD of the PAO. The event numbers correspond to the energy range $1.0 \times 10^{17} - 2.5 \times 10^{19}$ eV.

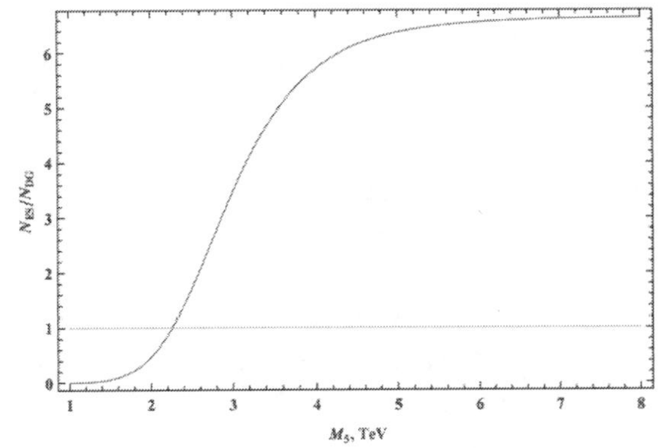

Figure 6. The expected ratio of the ES neutrinos to the DG neutrinos (with zenith angle $75° < \theta < 90°$) at the SD of the PAO as a function of the gravity scale M_5.

(4.0) TeV. For the SM interactions, this ratio was estimated by the Auger Collaboration to be equal to 6.0 [6].

6 Conclusions

In the present paper we have studied the neutrino-induced inclined (quasi-horizontal) events at the Surface Detector of the Pierre Auger Observatory in the Randall-Sundrum scenario with the extra dimension and warped metric. We have presented the general solution for the metric (9) which is symmetric with respect to both branes and explicitly periodic in extra variable. In the framework of the RS-like model with the small curvature of the 5-dimensional space-time, the exposures of the SD of the PAO for the downward going and Earth-skimming neutrinos are estimated (fig. 4). The lower bound on the fundamental gravity scale $\bar{M}_5$ is obtained (29). The ratio of the number of the ES air showers to the number of the DG showers, N_{ES}/N_{DG}, is calculated as a function of $\bar{M}_5$ (fig. 6).

Acknowledgements

The authors are indebted to J. Alvarez-Muñiz for sending numerical data on the Auger exposures shown in fig. 3 from ref. [6].

References

[1] M.G. Aartsen *et al.* (IceCube Collab.), Phys. Rev. Lett. **113**, 101101 (2014)
[2] J. Abraham *et al.* (Auger Collab.), Nucl. Instrum. Meth. A **523**, 50 (2004)
[3] V.S. Berezinsky and G.T. Zatsepin, Phys. Lett. B **28**, 423 (1969); V.S. Berezinsky and A.Yu. Smirnov, Astrophys. Space Sience **32**, 461 (1975)
[4] K.S. Capelle, J.W. Cronin, G. Parente and E. Zas, Astropart. Phys. **8**, 321 (1998)
[5] E. Zas, New J. Phys., **7**, 130 (2005)
[6] A. Aab *et al.* (Auger Collab.), Phys. Rev. D **91**, 092008 (2015)
[7] L. Anchordoqui, H. Goldberg, D. Góra *et al.*, Phys. Rev. D **82**, 043001 (2010)
[8] X. Bertou et al., Astropart. Phys. **17**, 183 (2002)
[9] J.L. Feng, P. Fisher, F. Wilczek and T.M. Yu, Phys. Rev. Lett. **88**, 161102 (2002)
[10] P. Abreu *et al.* (Auger Collab.), Phys. Rev. D **84**, 122005 (2011)
[11] E. Waxman and J.N. Bachall, Phys. Rev. D **64**, 023002 (2001)
[12] A. Cooper-Sarkar and S. Sarkar, JHEP **0801**, 075 (2008)
[13] L. Randall and R. Sundrum, Phys. Rev. Lett. **83**, 3370 (1999)
[14] A.V. Kisselev, Proc. of the Third Annual Large Hadron Collider Physics Conference (LHCP2015), August 31-September 2, 2016, St. Petersburg, Russia
[15] I.S. Gradshteyn and I.M. Ryzhik, *Table of Integrals, Series, and Products, Eighth Edition* (Eds. D. Zwillinger and V. Moll, Academic Press, US, 2014) 55-56.
[16] A.V. Kisselev, *Generalization of the Randall-Sundrum solution*, arXiv:1512.01091, Nucl. Phys. B **909** (2016), to appear.
[17] G. F. Giudice, T. Plehn and A. Strumia, Nucl. Phys. B **706**, 455 (2005)
[18] A.V. Kisselev and V.A. Petrov, Phys. Rev. D **71**, 124032 (2005)
[19] A.V. Kisselev, Phys. Rev. **D 73**, 024007 (2006)
[20] N. Arkani-Hamed, S. Dimopoulos and G. Dvali, Phys. Lett. B **429**, 263 (1998); Phys. Rev. D **59**, 086004 (1999)
[21] I. Antoniadis, N. Arkani-Hamed, S. Dimopoulos and G. Dvali, Phys. Lett. B **436**, 257 (1998)
[22] A.V. Kisselev, JHEP **0809**, 039 (2008)
[23] A.V. Kisselev, JHEP **1304**, 025 (2013)
[24] A.V. Kisselev and V.A. Petrov, Eur. Phys. J. C **36**, 103 (2004); *ibid* **37**, 241 (2004)
[25] A.V. Kisselev, Open Astron. J. **2**, 12 (2009)
[26] S. Dulat *et al.*, Phys. Rev. D **93**, 033006 (2016)
[27] V.A. Petrov and A.V. Prokudin, Eur. Phys. J. C, **23**, 135 (2002)

EPJ Web of Conferences **125**, 02015 (2016) DOI: 10.1051/epjconf/201612502015
QUARKS-2016

Rare radiative leptonic B-decays

Anastasiia Kozachuk[1,2], *Dmitri* Melikhov[1,3,4], and *Nikolai* Nikitin[1,2,5]

[1] *D. V. Skobeltsyn Institute of Nuclear Physics, M. V. Lomonosov Moscow State University, 119991, Moscow, Russia*
[2] *M. V. Lomonosov Moscow State University, Physical Facutly, 119991, Moscow, Russia*
[3] *Institue for High-Energy Physics, Austrian Academy of Sciences, Nikolsdorfergasse 18, A-1050 Vienna, Austria*
[4] *Faculty of Physics, University of Vienna, Boltzmanngasse 5, A-1090 Vienna, Austria*
[5] *A. I. Alikhanov Institute for Theoretical and Experimental Physics, 117218 Moscow, Russia*

Abstract. We obtain predictions for $B_{(s)} \to e^+e^-\gamma$ and $B_{(s)} \to \mu^+\mu^-\gamma$ decays. All the contributions containing long-distance QCD effects are calculated in the framework of relativistic quark model. The contributions of the light vector-meson resonances related to the virtual photon emission from valence quarks of the B-meson are included. The highest branching ratios for the radiative leptonic B-decays are $\mathcal{B}(\bar{B}_s^0 \to e^+e^-\gamma) = 18.8 \times 10^{-9}$ and $\mathcal{B}(B_s^0 \to \mu^+\mu^-\gamma) = 12.2 \times 10^{-9}$. We also give the distribution of the foward-backward asymmetry.

1 Introduction

Rare radiative leptonic $B_{(s)} \to \ell^+\ell^-\gamma$ decays are induced by the flavour-changing neutral current transitions $b \to s, d$ which are forbidden at tree level in the Standard Model and are described by penguin and box diagrams, leading to small probabilities of the order of $10^{-8} - 10^{-10}$ (see e.g. [1]). Due to such small values this sector of heavy flavours represented by rare semi-leptonic and radiative decays is expected to be senscitive to contributions of New Physics. Many of the decays have already been studied theoretically and experimentally measured by LHCb, Belle and Babar collaborations, and no considerable descrepancies between experimental results and predictions of the Standard Model have been found. Nevetherless there are several tensions of the order of $2 - 3\sigma$ (see discussion in [2–4]: The first one is that the ratio $\mathcal{R}_K = \mathcal{B}(B^+ \to K^+\mu^+\mu^-)/\mathcal{B}(B^+ \to K^+e^+e^-)$ is 25% lower than the SM prediction at 2.6σ [5–8]. In an independent measurement, the braching ratio of $B^+ \to K^+\mu^+\mu^-$ is itself 30% lower that the SM value at 2σ [9–13]. Another inconsistency is probably related to the decay $B_s \to \mu^+\mu^-$. The joint CMS and LHCb measurement of its branching ratio [14] also gives the value which is 25% lower than the SM prediction, but here it is 1σ effect only. For the branching ratios of $B_s \to \mu^+\mu^-$ and $B_s \to \mu^+\mu^-\gamma$ decays the following relation takes place

$$\frac{\mathcal{B}(B_s \to \ell^+\ell^-\gamma)}{\mathcal{B}(B_s \to \ell^+\ell^-)} \sim \left(\frac{M_{B^0}}{m_\ell}\right)^2 \frac{\alpha_{em}}{4\pi}, \tag{1}$$

were the squared ratio of masses $(M_{B^0}/m_\ell)^2$ means that the radiative decay $B_s \to \mu^+\mu^-\gamma$ does not have the chirality constraint, α_{em} comes from the photon emission and 4π in the denominator is the difference between three- and two-particle phase space. For muons one can easily get the estimation $(M_{B^0}/m_\mu)^2 \sim 2.5 \times 10^3 \sim 4\pi/\alpha_{em}$, which means that the branching ratios are approximately equal $\mathcal{B}(B_s \to \mu^+\mu^-\gamma) \sim \mathcal{B}(B_s \to \mu^+\mu^-)$. In fact, $\mathcal{B}(B_s \to \mu^+\mu^-\gamma)$ is a little bit larger due to additional dynamical effects, such as resonant contributions.

The paper is organized as follows: In Section 2 we discuss the contributions to the decay amplitude $\langle\gamma\ell^+\ell^-|H_{\rm eff}(b \to q\ell^+\ell^-)|B\rangle$. In Section 3 we calculate the transition form-factors via dispersion approach based on constituent quark picture. In Section 4 we give numerical predictions for the branching ratios, differential distributions for the decay rates and forward-backward asymmetry.

2 The effective Hamiltonian and the amplitude

The effective Hamiltonian describing the $b \to q$ ($q = d, s$) weak transition has the form ([15, 16])

$$\mathcal{H}_{\rm eff}^{b\to q} = \frac{G_F}{\sqrt{2}} V_{tb}V_{tq}^* \sum_i C_i(\mu)\, O_i(\mu), \tag{2}$$

where G_F is the Fermi constant, C_i are the scale-dependent set of Wilson coefficients, and O_i are the basis operators. For B decays the scale parameter μ is appriximately equal to 5 GeV. The amplitudes of the basis operators between the initial and final states may be parametrised in terms of the Lorentz-invariant form factors. These form factors contain nonperturbative QCD contributions, and therefore their calculation is one the main problems when considering $B \to \ell^+\ell^-\gamma$ decays.

2.1 Emission of the virtual photon from the penguin

The most important part which contains nonperturbative QCD contributions corresponds to the cases, when the real photon is directly emitted from the valence b or d quarks, and the $\ell^+\ell^-$ pair is coupled to the penguin. The effective lagrangian in this case takes the form[1]

$$\begin{aligned} H_{\rm eff}^{b\to d\ell^+\ell^-} &= \frac{G_F}{\sqrt{2}}\frac{\alpha_{\rm em}}{2\pi} V_{tb}V_{tq}^* \left[-2im_b \frac{C_{7\gamma}(\mu)}{q^2} \cdot \bar{d}\sigma_{\mu\nu}q^\nu(1+\gamma_5)b \cdot \bar{\ell}\gamma^\mu\ell \right. \\ &\left. + C_{9V}^{\rm eff}(\mu, q^2)\cdot \bar{d}\gamma_\mu(1-\gamma_5)b\cdot\bar{\ell}\gamma^\mu\ell + C_{10A}(\mu)\cdot\bar{d}\gamma_\mu(1-\gamma_5)b\cdot\bar{\ell}\gamma^\mu\gamma_5\ell \right], \end{aligned} \tag{3}$$

and the corresponding diagrams are shown in Fig. 1. The coefficient $C_{9V}^{\rm eff}(\mu, q^2)$ includes long-distance

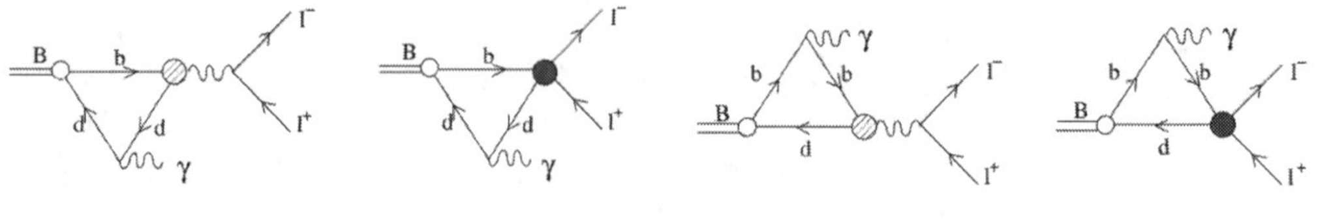

Figure 1. Diagrams contributing to $B \to \ell^+\ell^-\gamma$ discussed in section 2.1. Dashed circles denote the $b \to d\gamma$ operator $O_{7\gamma}$. Solid circles denote the $b \to d\ell^+\ell^-$ operators O_{9V} and O_{10AV}.

effects related to $\bar{c}c$ resonances in the q^2-channel, where q^2 is the invariant mass of the $\ell^+\ell^-$ pair [20–

[1]Our notations and conventions are: $\gamma^5 = i\gamma^0\gamma^1\gamma^2\gamma^3$, $\sigma_{\mu\nu} = \frac{i}{2}[\gamma_\mu, \gamma_\nu]$, $\varepsilon^{0123} = -1$, $\epsilon_{abcd} \equiv \epsilon_{\alpha\beta\mu\nu}a^\alpha b^\beta c^\mu d^\nu$, $e = \sqrt{4\pi\alpha_{\rm em}}$.

 DOI: 10.1051/epjconf/201612502015

22]. The $B \to \gamma$ transition form factors of the basis operators in (3) are defined according to [19]

$$\begin{aligned}
\langle \gamma(k,\,\epsilon)|\bar{d}\gamma_\mu\gamma_5 b|B(p)\rangle &= i\,e\,\epsilon^*_\alpha\left(g_{\mu\alpha}\,pk - p_\alpha k_\mu\right)\frac{F_A(q^2)}{M_B},\\
\langle \gamma(k,\,\epsilon)|\bar{d}\gamma_\mu b|B(p)\rangle &= e\,\epsilon^*_\alpha\,\epsilon_{\mu\alpha\xi\eta}p_\xi k_\eta\,\frac{F_V(q^2)}{M_B}, \qquad (4)\\
\langle \gamma(k,\,\epsilon)|\bar{d}\sigma_{\mu\nu}\gamma_5 b|B(p)\rangle\,(p-k)^\nu &= e\,\epsilon^*_\alpha\left[g_{\mu\alpha}\,pk - p_\alpha k_\mu\right]F_{TA}(q^2,0),\\
\langle \gamma(k,\,\epsilon)|\bar{d}\sigma_{\mu\nu} b|B(p)\rangle\,(p-k)^\nu &= i\,e\,\epsilon^*_\alpha\epsilon_{\mu\alpha\xi\eta}p_\xi k_\eta\,F_{TV}(q^2,0).
\end{aligned}$$

The penguin form factors $F_{TV,TA}(q_1^2, q_2^2)$ are defined as functions of two variables: q_1 is the momentum of the photon emitted from the penguin, and q_2 is the momentum of the photon emitted from the valence quark of the B meson. We calculate the form factors in the framework of the dispersion approach based on constituent quark picture, the details are presented in Section 3.

2.2 Emission of the virtual photon from B-meson valence quarks

Another process contributing to the amplitude is that with the real photon emitted from the penguin, whereas one of the valence quarks directly emits the virtual photon which then goes into the final $\ell^+\ell^-$ pair. This process is described by the diagrams of Fig. 2. The corresponding amplitude has the same

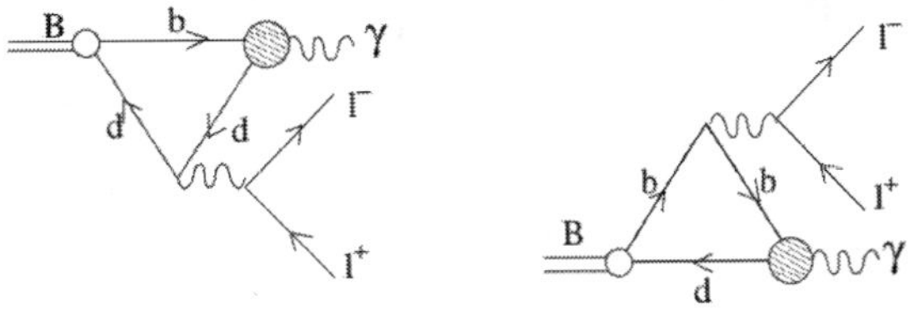

Figure 2. Diagrams contributing to $B_d \to \ell^+\ell^-\gamma$ discussed in section 2.2. Dashed circles denote the $b \to d\gamma$ operator $O_{7\gamma}$.

structure as the $C_{7\gamma}$ part of the amplitude in 2.1 with $F_{TA,TV}(q^2,0)$ replaced by $F_{TA,TV}(0,q^2)$. The form factors $F_{TA,TV}(0,q^2)$ at the necessary timelike momentum transfers are not known. The difficulty is connected with appearance of neutral light vector meson resonances, ρ^0 and ω for B-decays and ϕ for B_s-decays, in the physical $B \to \gamma\ell^+\ell^-$ decay region. We calculate the form factors $F_{TA,TV}(0,q^2)$ for $q^2 > 0$ with the use of gauge-invariant version [23, 24] of the vector meson dominance [25–27]

$$F_{TV,TA}(0,q^2) = F_{TV,TA}(0,0) - \sum_V 2\,f_V g_+^{B\to V}(0)\frac{q^2/M_V}{q^2 - M_V^2 + iM_V\Gamma_V}, \qquad (5)$$

where M_V and Γ_V are the mass and the width of the vector meson resonance, $g_+^{B\to V}(0)$ are the $B \to V$ transition form factors, defined according to the relations

$$\langle V(q,\varepsilon)|\bar{d}\sigma_{\mu\nu}b|B(p)\rangle = i\varepsilon^{*\alpha}\,\epsilon_{\mu\nu\beta\gamma}\left[g_+^{B\to V}(k^2)g_{\alpha\beta}(p+q)^\gamma + g_-^{B\to V}(k^2)g_{\alpha\beta}k^\gamma + g_0^{B\to V}(k^2)p_\alpha p^\beta q^\gamma\right] \qquad (6)$$

and calculated in [28, 29] via relativistic quark model. The leptonic decay constant of a vector meson is given by

$$\langle 0|\bar{d}\gamma_\mu d|V(\varepsilon,p)\rangle = \varepsilon_\mu M_V f_V. \qquad (7)$$

2.3 Bremsstrahlung

Fig. 3 gives diagrams describing Bremsstrahlung. The corresponding contribution to the $B \to \ell^+\ell^-\gamma$ amplitude reads

$$A_\mu^{\rm Brems} = -i\, e \frac{G_F}{\sqrt{2}} \frac{\alpha_{\rm em}}{2\pi} V^*_{td} V_{tb} \frac{f_{B_q}}{M_B} 2\hat{m}_\ell\, C_{10A}(\mu)\, \bar{\ell}(p_2) \left[\frac{(\gamma\epsilon^*)(\gamma p)}{\hat{t} - \hat{m}_\ell^2} - \frac{(\gamma p)(\gamma\epsilon^*)}{\hat{u} - \hat{m}_\ell^2} \right] \gamma_5\, \ell(-p_1), \quad (8)$$

$f_B > 0$.

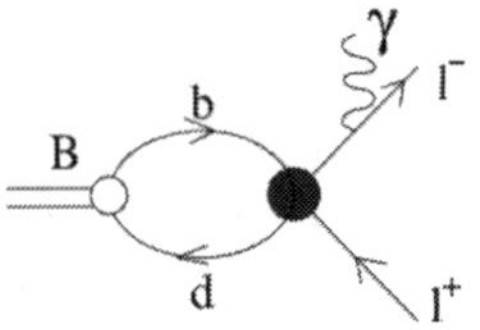

Figure 3. Diagrams describing photon Bremsstrahlung. Solid circles denote the operator O_{10A}.

2.4 Weak annihilation contribution

The weak annihilation contribution is given by triangle diagrams of Fig 3. One should take into

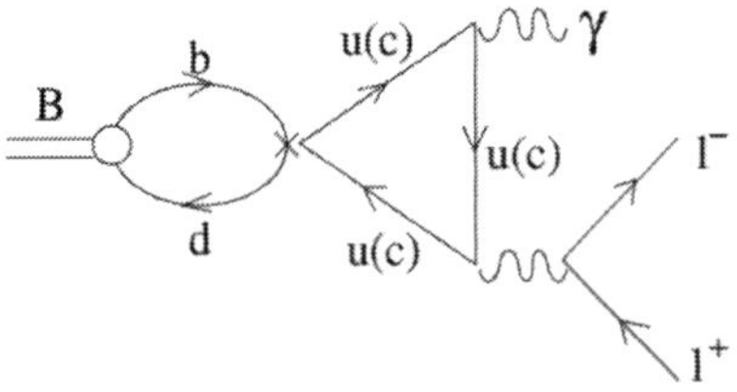

Figure 4. Weak annihilation diagrams contributing to the process $B_s \to \mu^+\mu^-$.

account u and c quarks in the loop. The vertex describing the $\bar{b}d \to \bar{Q}Q$ transition ($Q = u, c$) reads

$$H_{\rm eff}^{B\to\bar{Q}Q} = -\frac{G_F}{\sqrt{2}} a_1\, V_{Qb} V^*_{Qd}\, \bar{d}\gamma_\mu(1-\gamma_5) b\, \bar{Q}\gamma_\mu(1-\gamma_5) Q, \quad (9)$$

with $a_1 = C_1 + C_2/N_c$, N_c number of colors [30]. For $N_c = 3$ one finds $C_1 = PUT$, $C_2 = PUT$, $a_1 = PUT$.

3 Transition form factors

We calculate the transition form factors in the framework of relativistic quark model, which is a dispersion approach based on constituent quark picture [28, 29]. All hardron observables are given by dispersion representations in terms of hardronic relativistic wave functions and spectral densities of corresponding feynman diagrams with constituent quarks in the loops. For the wave functions we use a Gaussian parametrization $\phi(s) = A(s,\beta)e^{-k^2(s)/(2\beta^2)}$. The simpliest relation can be obtained for a pseudoscalar or vector meson decays constant

$$f_M = \int ds\phi_M(s)\rho(s), \tag{10}$$

were $\phi(s)$ is the meson relativistic wave function and $\rho(s)$ is the spectral dencity. The latter is obtained as a direct result of feynman rules for the corresponding feynman diagram. The example of the diagram for a B-meson decay constant is given in Fig. 5. In this work we consider meson-to-photon

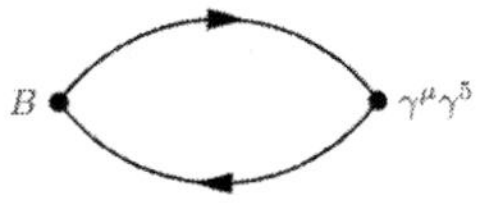

Figure 5. Feynman diagram corresponding to dispersion representation of B-meson decay constant.

transitions; the corresponding form factors $F_{V,A,TV,TA}$ may be obtained in the form of the spectral represenation

$$F(q_1,q_2) = \int ds\phi(s)\frac{ds'\Delta(s,s',q_2^2)}{s'-q_1^2}, \tag{11}$$

were q_1 and q_2 are momenta of the emitted photons. The form factors $F_{V,A}$ were calculated in [31]. We now perform the calculation of the form factors $F_{TV,TA}$. Each of these form factors contains two contributions corresponding to the diagrams of Fig. 2:

$$\begin{aligned} F_{TV} &= Q_d F^{(1)}_{TV}(m_d,m_b) + Q_b F^{(1)}_{TV}(m_b,m_d), \\ F_{TA} &= Q_d F^{(1)}_{TA}(m_d,m_b) + Q_b F^{(1)}_{TA}(m_b,m_d). \end{aligned} \tag{12}$$

The spectral representations for the form factors in (12) have the form

$$F^{(1)}_{TV}(s) = -\int\limits_{(m_1+m_2)^2}^{\infty} dsg_2(s,m_1,m_2) - \frac{M_B^2+q^2}{M_B^2-q^2}\int\limits_{(m_1+m_2)^2}^{\infty} dsg_1(s,m_1,m_2), \tag{13}$$

$$F^{(1)}_{TA}(s) = -\int\limits_{(m_1+m_2)^2}^{\infty} dsg_2(s,m_1,m_2) - \int\limits_{(m_1+m_2)^2}^{\infty} dsg_1(s,m_1,m_2), \tag{14}$$

where m_1 is the mass of the quark, which emits the photon, m_2 is the mass of the spectator, and

$$\begin{aligned} g_1(s,m_1,m_2) &= \phi_B(s,m_1,m_2)\frac{M_B^2-q^2}{(s-q^2)^2}\left(\frac{s+m_1^2-m_2^2}{2s}\sqrt{\lambda(s,m_1,m_2)}-\right. \\ &\left. -m_1^2\log\frac{s+m_1^2-m_2^2+\sqrt{\lambda(s,m_1,m_2)}}{s+m_1^2-m_2^2-\sqrt{\lambda(s,m_1,m_2)}}\right), \end{aligned} \tag{15}$$

$$\begin{aligned} g_2(s,m_1,m_2) &= \phi_B(s,m_1,m_2)\frac{1}{s-q^2}\left(\sqrt{\lambda(s,m_1,m_2)}-\right. \\ &\left. m_1(m_2-m_1)\log\frac{s+m_1^2-m_2^2+\sqrt{\lambda(s,m_1,m_2)}}{s+m_1^2-m_2^2-\sqrt{\lambda(s,m_1,m_2)}}\right). \end{aligned} \tag{16}$$

The model contains only few parameters such as the constituent quark masses and the parameter of the wave function β. These parameters were fixed in [31] using relations (10) for meson decay constants so that our results reproduce the predictions from QCD sum rules and lattice QCD.

4 Numerical results

4.1 Branching ratios

For numerical estimates we use the following values of Wilson coefficients at $\mu = 5$ GeV: $C_1(5\,GeV) = 0.235$, $C_2(5\,GeV) = -1.1$, $C_{7\gamma}(5\,GeV) = 0.308$, $C_{10A}(5\,GeV) = 4.63$. The $C_{9V}^{eff}(\mu, s)$ evolution including cc-resonances is taken from [20–22]. We obtained several distributions for the differential branching fractions, they are shown in Fig.6 and 7. We obtained results for light leptons

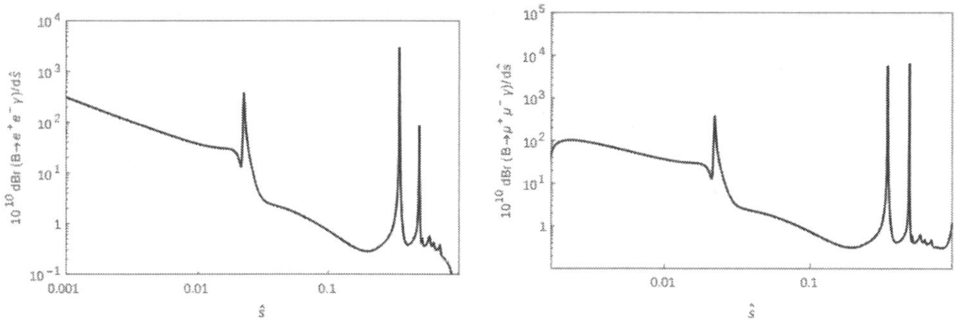

Figure 6. Differential branching fractions for $B \to e^+e^-\gamma$ (left) and $B \to \mu^+\mu^-\gamma$ (right) decays

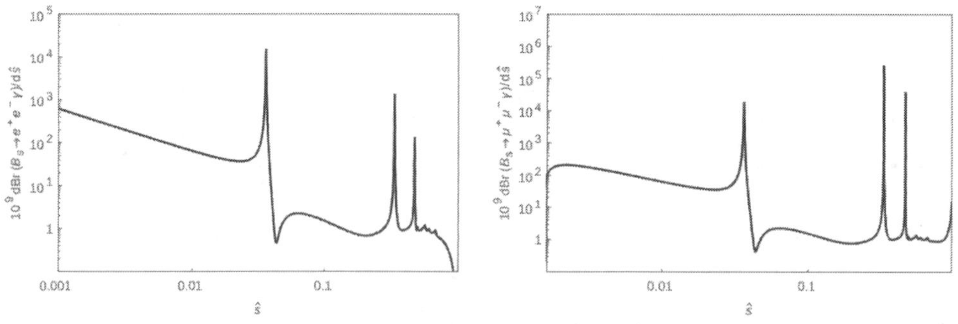

Figure 7. Differential branching fractions for $B_s \to e^+e^-\gamma$ (left) and $B_s \to \mu^+\mu^-\gamma$ (right) decays

for the value of the photon energy cut (the minimal photon energy in the B-meson rest frame) $E_{min}^{\gamma} =$ 80 MeV. This particular choice of E_{min}^{γ} is connected with the energy resolution of the LHCb detector [32, 33]. Our results for the branching ratios are presented in Table 1.

 DOI: 10.1051/epjconf/201612502015

	this work	[34]	[35]	[17]	[18]
$Br(B \to e^+e^-\gamma) \times 10^{10}$	4.84	3.95	5.8	1.01	–
$Br(B \to \mu^+\mu^-\gamma) \times 10^{10}$	1.60	1.31	5.8	0.61	–
$Br(B_s \to e^+e^-\gamma) \times 10^9$	18.8	24.6	16.24	3.29	20
$Br(B_s \to \mu^+\mu^-\gamma) \times 10^9$	12.2	18.8	16.24	2.00	12

Table 1. Numerical estimates for the branching ratios of $B_{(s)} \to \ell^+\ell^-\gamma$ decays

4.2 Forward-backward asymmetry

We obtained distribution for the forward-backward asymmetry, defined by the relation

$$A_{FB}(\hat{s}) = \frac{\int\limits_0^1 d\cos\theta \, \frac{d^2\Gamma(B_{(s)}\to\ell^+\ell^-\gamma)}{d\hat{s}\, d\cos\theta} - \int\limits_{-1}^0 d\cos\theta \, \frac{d^2\Gamma(B_{(s)}\to\ell^+\ell^-\gamma)}{d\hat{s}\, d\cos\theta}}{\frac{d\Gamma(B_{(s)}\to\ell^+\ell^-\gamma)}{d\hat{s}}}, \tag{17}$$

where $\hat{s} = q^2/M_B^2$, θ is the angle between $\vec{p}$ and $\vec{p}_2$. The distribution is presented in Fig.8.

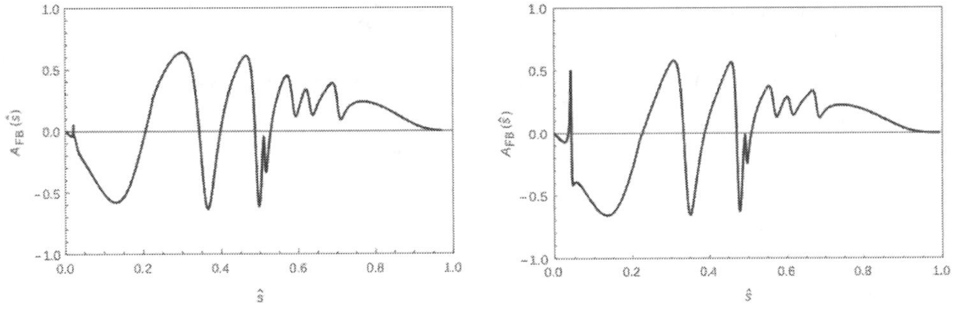

Figure 8. Forward-backward asymmetry for $B \to \mu^+\mu^-\gamma$ (left) and $B_s \to \mu^+\mu^-\gamma$ (right) decays

The decay rates and the forward-backward asymmetry were previosly calculated in several works [17–19, 34–36]. In [17] and [18] not all the contributions were taken into account and in [17–19, 34] the transition form factors were estimated from symmetry considerations coming from LEET. We made direct calculation of the form factors in the framework of the relativistic quark model. Our results agree nicely with [17, 18]. The results [17, 35] are based on not fully consistent models for the form factos and therefore do not seem to us convincing.

5 Conclusions

We obtained predictions for the differential distributions and the branching ratios for the $B_{(s)} \to e^+e^-\gamma$ and $B_{(s)} \to \mu^+\mu^-\gamma$ decays taking into account the following contributions to the amplitude of the process: the photon emission from the $d(s)$ and the b valence quarks of the B-meson, the weak annihilation, and the bremsstrahlung. The corresponding form factors were calculated in the framework of the relativistic quark model.

Acknowledgements

The work was supported by grant 16-12-10280 of the Russian Science Foundation.

References

[1] A. Ali, hep-ph/9606324
[2] S. L. Glashow, D. Guadagnoli and K. Lane, Phys. Rev. Lett. **114**, 091801 (2015)
[3] D. Guadagnoli and K. Lane, Phys. Lett. B **751**, 54 (2015)
[4] D. Guadagnoli, D. Melikhov and M. Reboud, Phys. Lett. B **760**, 442 (2016)
[5] R. Aaij *et al.* [LHCb Collaboration], Phys. Rev. Lett. **113**, 151601 (2014)
[6] C. Bobeth, G. Hiller and G. Piranishvili, JHEP **0712**, 040 (2007)
[7] C. Bouchard *et al.* [HPQCD Collaboration], Phys. Rev. Lett. **111**, no. 16, 162002 (2013) Erratum: [Phys. Rev. Lett. **112**, no. 14, 149902 (2014)]
[8] G. Hiller and F. Kruger, Phys. Rev. D **69**, 074020 (2004)
[9–13]
[9] R. Aaij *et al.* [LHCb Collaboration], JHEP **1406**, 133 (2014)
[10] R. Aaij *et al.* [LHCb Collaboration], JHEP **1302**, 105 (2013)
[11] C. Bobeth, G. Hiller and D. van Dyk, JHEP **1107**, 067 (2011)
[12] C. Bobeth, G. Hiller, D. van Dyk and C. Wacker, JHEP **1201**, 107 (2012)
[13] C. Bobeth, G. Hiller and D. van Dyk, Phys. Rev. D **87**, no. 3, 034016 (2013) [Phys. Rev. D **87**, 034016 (2013)]
[14] V. Khachatryan *et al.* [CMS and LHCb Collaborations], Nature **522**, 68 (2015)
[15] B. Grinstein, M. J. Savage and M. B. Wise, Nucl. Phys. B **319** 271 (1989)
[16] A. J. Buras and M. Munz, Phys. Rev. D **52**, 186 (1995)
[17] C. Q. Geng, C. C. Lih and W. M. Zhang, Phys. Rev. D **62**, 074017 (2000)
[18] Y. Dincer and L. M. Sehgal, Phys. Lett. B **521**, 7 (2001)
[19] F. Kruger and D. Melikhov, Phys. Rev. D **67**, 034002 (2003)
[20] F. Kruger and L. M. Sehgal, Phys. Rev. D **55**, 2799 (1997)
[21] D. Melikhov, N. Nikitin and S. Simula, Phys. Lett. B **430**, 332 (1998)
[22] D. Melikhov, N. Nikitin and S. Simula, Phys. Rev. D **57**, 6814 (1998)
[23] D. Melikhov, Phys. Lett. B **516**, 61 (2001)
[24] D. Melikhov, O. Nachtmann, V. Nikonov and T. Paulus, Eur. Phys. J. C **34**, 345 (2004)
[25] J. J. Sakurai, Annals Phys. **11**, 1 (1960).
[26] M. Gell-Mann and F. Zachariasen, Phys. Rev. **124**, 953 (1961).
[27] G. J. Gounaris and J. J. Sakurai, Phys. Rev. Lett. **21**, 244 (1968).
[28] D. Melikhov, Phys. Rev. D **53**, 2460 (1996)
[29] D. Melikhov, Phys. Rev. D **56**, 7089 (1997)
[30] M. Neubert and B. Stech, Adv. Ser. Direct. High Energy Phys. **15**, 294 (1998)
[31] A. Kozachuk, D. Melikhov and N. Nikitin, Phys. Rev. D **93**, no. 1, 014015 (2016)
[32] [LHCb Collaboration], "LHCb calorimeters: Technical design report," CERN-LHCC-2000-036.
[33] R. Voss and A. Breskin, "The CERN Large Hadron Collider, accelerator and experiments,"
[34] D. Melikhov and N. Nikitin, Phys. Rev. D **70**, 114028 (2004)
[35] W. Y. Wang, Z. H. Xiong and S. H. Zhou, Chin. Phys. Lett. **30**, 111202 (2013)
[36] I. Balakireva, D. Melikhov, N. Nikitin and D. Tlisov, Phys. Rev. D **81**, 054024 (2010)

EPJ Web of Conferences **125**, 02016 (2016) DOI: 10.1051/epjconf/201612502016
QUARKS-2016

Charmed particles production in pA-interactions at 70 GeV

Vasily Ryadovikov[1,*]
(on behalf of the SVD Collaboration)

[1] *Institute for High Energy Physics, Sq. Nauki 1, Protvino, Moscow region 142281, Russia*

Abstract. The results of the SERP-E-184 experiment at the U-70 accelerator (IHEP, Protvino) are presented. Interactions of the 70 GeV proton beam with carbon, silicon and lead targets were studied to detect decays of charmed D^0, $\bar{D}^0$, D^+, D^- mesons and Λ_c^+ baryon near their production threshold. Measurements of lifetimes and masses have shown a good agreement with PDG data. The inclusive cross sections of charm production and their A-dependencies have been obtained. The yields of these particles are compared with the theoretical predictions and the data of other experiments. The measured cross section of the total open charm production $\sigma(c\bar{c}) = 7.1 \pm 2.3(stat) \pm 1.4(syst)$ μb/nucleon at the collision c.m. energy $\sqrt{s} = 11.8$ GeV is well above the QCD model predictions. The contributions of different kinds of charmed particles to the total cross section of the open charm production in proton-nucleus interactions vary with energy.

1 Introduction

We represent the results of data processing of E-184 experiment on studying of charmed particles production mechanisms in pA-interactions at 70 GeV. Experiment is carried out by Cooperation SINP MSU (Moscow) – JINR (Dubna) – IHEP (Protvino) on the U-70 accelerator in Protvino. Importance of work is that:
- The majority of experiments on a charm are executed with electron beams where their main properties are studied (measurements of mass, decay branching etc.).
- Experiments with hadron and heavy ion beams gave an opportunity to study mechanisms of charmed particle production in varying nuclear media from threshold energies to energies of LHC.
- Hadron production of a charm is sensitive to medium modifications. Charmed particles are good probes to investigate properties of this medium, which may show up as quark-gluon plasma.
- More then 20 years ago in IHEP the measurements of cross sections of the charmed particle production at the near-threshold energy region were performed in beam-dump experiment with an absorber of muons [1], in the SCAT bubble chamber experiment [2] and in the experiment with BIS-2 spectrometer [3]. The measured total cross sections in the energy range of the primary beam from 40 to 70 GeV have proved to be much higher than the model predictions based on QCD.
Experiment of E-184 is carried out on the Spectrometer with Vertex Detector setup (SVD).

*e-mail: riadovikov@ihep.ru

2 SVD setup detectors

The main elements of the setup for charm searching [4] are the high-precision micro-strip vertex detector (MSVD) with an active target (AT) and a magnetic spectrometer (MS) of the aperture equal to 1.8 x 1.2 m2 and the field of 1.18 T within the region 3 m long. Active target (AT) (Fig. 1) contains 5 Si-detectors each 300 μm thickness and 1-mm pitch strips, the Pb-plate (220 μm thick) and C-plate (500 μm thick). The micro-strip tracking part of MSVD consists of 10 Si-detectors.

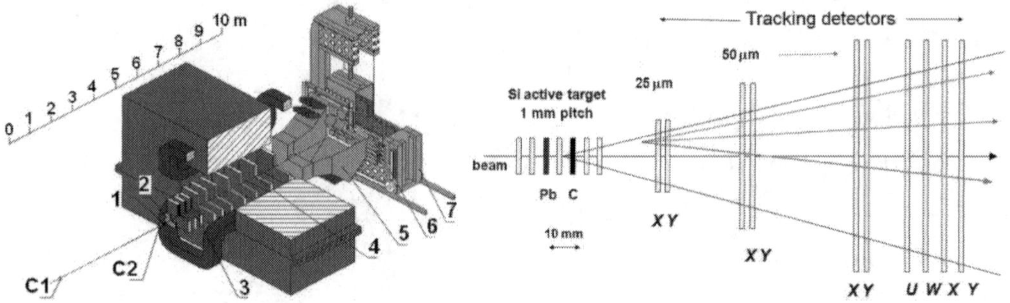

Figure 1. SVD-2 layout and scheme of MSVD. C1,C2 - beam scintillation counters; 1 - Si active target (AT); 2 - microstrip vertex Si-detector (MSVD); 3, 4 - MWPC of magnetic spectrometer (MS); 5 - threshold Cherenkov counter (CC); 6 - scintillation hodoscope (SC); 7 - detector of gamma-quanta (DEGA).

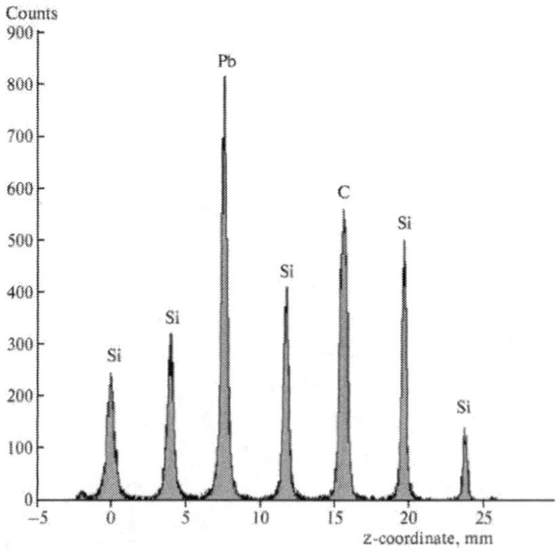

Figure 2. The reconstructed Z-coordinates of the primary vertices in AT.

The vertex detector plays the main role in search of secondary vertexes with possible decays of a charm as the distance between a point of interaction and D meson decay point doesn't exceed 2 mm. We have the accuracy of primary vertex reconstruction along a beam of 100 microns (Fig. 2), secondary vertex – 200 microns. Coordinate accuracy is equal in the cross plane to 10 microns. The spectrometer features allow one to get the effective mass resolution of σ = 4.4 MeV/c^2 for K_s^0 and 1.6 MeV/c^2 for Λ^0 masses, for the particles decayed before the MSVD tracking detectors.

3 Simulations and data processing

Modeling is needed for:
- Optimization of event selection criteria. As a result of modeling selection criteria of events with the minimum background were taken.
- Detection efficiency was defined for charm.
- The confidence that we have really charm particles. Comparison of parameters (decay length, etc.) for MC and experimental systems of charm particles decays.

Proton-nucleus interactions have been simulated by means of FRITIOF7.02 code [5]. The Fermi motion of nucleons, the deformation of nucleus and multiple re-scattering are taken into account. The nucleon distribution density in a nucleus is described in our case by the Woods–Saxon potential $\rho(r) = \rho(0)/(1 + exp[(r - r(0) \cdot A^{1/3}]/c)$, with $r(0) = 1.16(1 - 1.16A^{-2/3})$ fm and c = 0.5 fm. Hadronization processes described by the Lund scheme using the fragmentation function $f(z) \sim z^{-1}(1 - z)^{\alpha} \cdot exp(-bm_t^2/z)$. The parameters of this function were chosen as a = 0.18 and b = 0.34 GeV^{-2} in accordance with the results of the e^+e^- experiments OPAL [6] and [7] where the parameters were adjusted by the measured spectra of D and D^* mesons. Default values were used for the remaining parameters in the FRITIOF code.

GEANT3.21 package [8] was used to simulate registration of pA-interactions. The geometry of active and passive elements of the setup have been defined by means of metrological measurements and corrected with results of the "straight tracks" software alignment. The measured grid map of the magnetic field was applied. Charge spreading over the MSVD strips, noises and cutoff amplitudes were introduced channel-by-channel in accordance with experimentally measured values and data acquisition parameters. The actual efficiencies obtained experimentally for the proportional wire chambers were used for the magnetic spectrometer.

The simulated events have been processed by really using data handling system of the SERP-E-184 experiment. The common data processing procedure started with the filtration of MSVD data and reconstruction of tracks and primary vertices [9]. Next, we selected events with the secondary vertices close to the interaction points as the candidates for charm-production events. For this purpose the method of an analysis in the space of track parameters [10] was used. In this space each track is presented with a point and all points for the tracks from the same vertex are located on a straight line. The results of simulation were compared with the experimental data. A good agreement was found for the numbers of minimum bias events in each of AT plates, for multiplicities of charged particles in primary vertices and for their momenta.

In case of $D^0(\bar{D}^0) \rightarrow K\pi$ process the detection of $K_s^0 \rightarrow \pi\pi$ decay in the MSVD can serve as a reference procedure, because the well-known kaon production cross section is many times larger than for charmed particles. It has been used to estimate the detection efficiency of V^0 decay near the primary vertex and validate the data processing algorithms for $D^0(\bar{D}^0)$ [11].

Additionally, the background minimum bias events were simulated without charm production. This procedure was necessary to estimate the background conditions. The characteristics of three-prong systems ($K\pi\pi$ and $pK\pi$) for MC events were compared to the experimental data [12], [13]. There is a good agreement between the simulated and experimental distributions for a path length, momentum and x_F variable.

4 Selection of events with charmed particles

We have 52 million inelastic events in AT (C, Si, Pb).

4.1 $D^0 \to K^-\pi^+$, $\bar{D}^0 \to K^+\pi^-$ decays

Candidates for events with D^0 or $\bar{D}^0$ particle and its decay into $K\pi$ system were selected using the following criteria [11]:

(i) The distance between the primary vertex and the V^0 vertex is more than 0.5 mm.

(ii) The decay tracks of the V^0 particle have non-zero impact parameters with respect to the primary vertex, and the track of V^0 particle points to the primary vertex.

(iii) The effective mass of the $K\pi$ system differs from the world-average value of the D^0 mass (1.865 GeV/c^2) by less than 0.5 GeV/c^2.

(iv) The momentum of the $K\pi$ system is higher than 10 GeV/c.

(v) The transverse momentum of the decay particle with respect to the direction of motion of the $K\pi$ system is higher than 0.3 GeV/c. It follows from the analysis of the Armenteros–Podolansky criterion and suppresses the background from neutral kaons and Λ^0 hyperons decays (Fig. 3 left).

(vi) The candidate for the ($D^0/\bar{D}^0$) particle survives the visual inspection of events by physicist, by means of specially designed graphic package on the PC screen with high resolution. Having rather small number of events, we used opportunity to check visually all of them. The analysis used full information on event including, for example, hits not used in any track. It helped to discard events possibly having extra beam tracks, secondary interactions in the targets, tracks possibly attributed to more than one vertex, ghost tracks and alike.

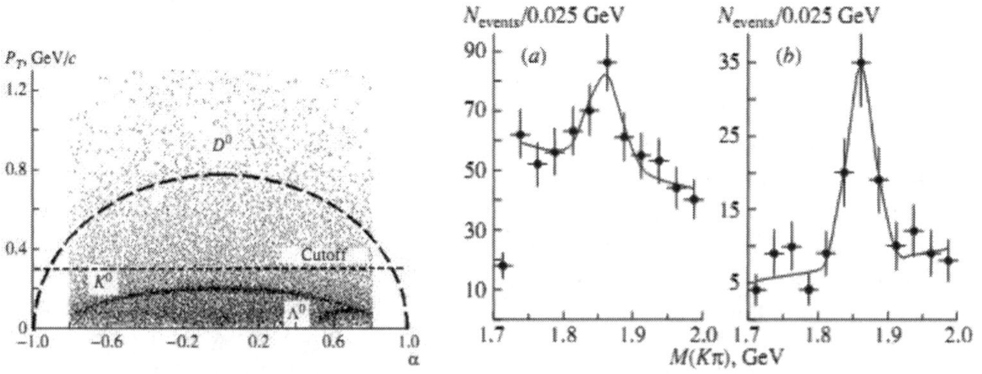

Figure 3. Left: Armenteros–Podolansky plot for K^0, Λ^0 and D^0; $\alpha = (P_L^+ - P_L^-)/(P_L^+ + P_L^-)$. Right: Effective mass spectra of the $K\pi$ system (a) before and (b) after visual inspection.

The effective mass spectra of the $K\pi$ system after applying the criteria (i-v) and visual inspection of the events are presented in Figs. 3a and 3b (right). The mass interval for the visual inspection has been limited to the region from 1.7 to 2.0 GeV/c^2. The visual inspection significantly reduces the background in the region of interest, but the charm + anti-charm ($D^0/\bar{D}^0$) signal decreased only on 20%. The data fit by the sum of the straight line and the Gaussian function is shown in Fig. 3b (right). It gives 1861 MeV/c^2 for ($D^0/\bar{D}^0$) mass with a standard deviation σ = 21 MeV/c^2 (χ^2/NDF = 5.5/6) and the signal to noise ratio of (51 ± 17)/(38 ± 13). The detection efficiency of ($D^0/\bar{D}^0$) particles with efficiency of visual inspection taken into account is equal to $\varepsilon(D^0/\bar{D}^0) = 0.036$.

4.2 $D^+ \to K^-\pi^+\pi^+$, $D^- \to K^+\pi^-\pi^-$ decays

For charged mesons we analyzed the $(K\pi\pi)$ systems: $D^+ \to K^-\pi^+\pi^+$, $D^- \to K^+\pi^-\pi^-$. The charged charmed mesons were found by analyzing the events with a three-prong secondary vertex. The selection procedure of events went as follows (more details in [12]):

(i) Search for the third track to be associated with a two-prong secondary vertex taking into account charges and kinematical correspondence to interaction vertex.

(ii) Selection of the $K\pi\pi$ systems with momentum P>7 GeV/c.

(iii) Cutting out non-physical regions and those with the highest background using Dalitz plot for M($K\pi_1$) and M($K\pi_2$) variables.

(iv) Within three tracks systems, two possible hypotheses for K^0 have shown a significant peak in the mass distribution. Most of background was cut out after applying the condition M$(\pi^+\pi^-)_1$ + M$(\pi^+\pi^-)_2$ < 1.2MeV/c^2.

(v) Hypotheses for $(K^-K^+\pi)$ system (possible D_S) were discarded with the M$(K^-K^+\pi)$ > 1.93 MeV/c^2 cut.

(vi) Simulations have shown concentration of background events at the smallest decay lengths, the condition L > 0.12 mm was applied to eliminate it (L = L_{vis}*M/P).

The mass distributions for D^+ and D^- that was obtained with primary selection (i, ii) are shown in Fig. 4. Thay have significant background contributions, eliminated with the rest of cuts (see Fig. 5).

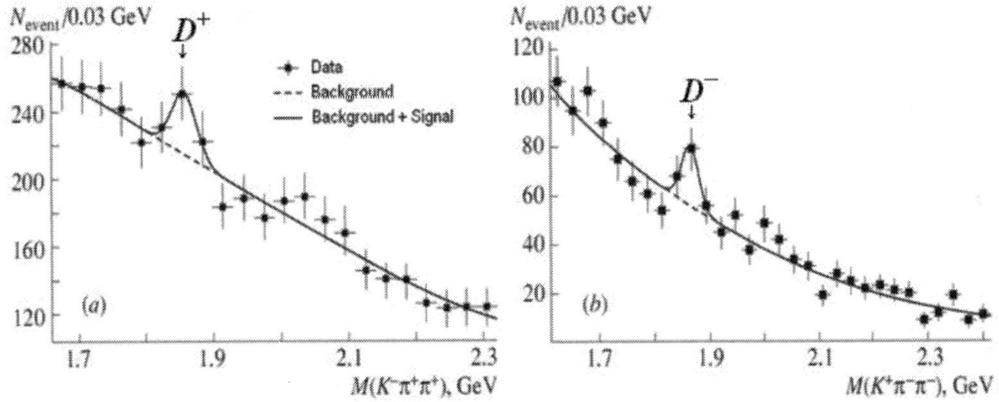

Figure 4. Effective mass spectra of the $K^-\pi^+\pi^+$ (a) and $K^+\pi^-\pi^-$ (b) systems after applying the conditions (i, ii). Data fitted with the sum of a Gaussian function and a sixth order polynomial.

After parameterizing the spectrum in Fig. 5 (left) in terms of the sum of the Gaussian function and a fifth order polynomial (χ^2/NDF = 13.5/30), we got 15.5 ± 5.6 signal events from the D^+ meson decay over the background of 16.6 ± 6.0 events. The measured D^+-meson mass is 1874 ± 5 MeV/c^2 (the world-average value is 1869.6 MeV/c^2) with a standard deviation of 11.5 MeV/c^2. The detection efficiency $\varepsilon(D^+)$ = 0.014 was defined from simulations. The same procedure has been applied to search for the D^- meson signal (Fig. 5 right) in the mass spectrum of the $K^+\pi^-\pi^-$. The spectrum was parameterized as a sum of the Gaussian function and the second order polynomial (χ^2/NDF = 3.6/20). The number of events in the signal was 15.0 ± 4.7 over the background of 8.7 ± 2.7 events. The D^--meson mass was 1864 ± 8 MeV/c^2, with the standard deviation of 22 MeV/c^2. The detection efficiency obtained through the simulation for the D^- meson signal was equal to $\varepsilon(D^-)$ = 0.008.

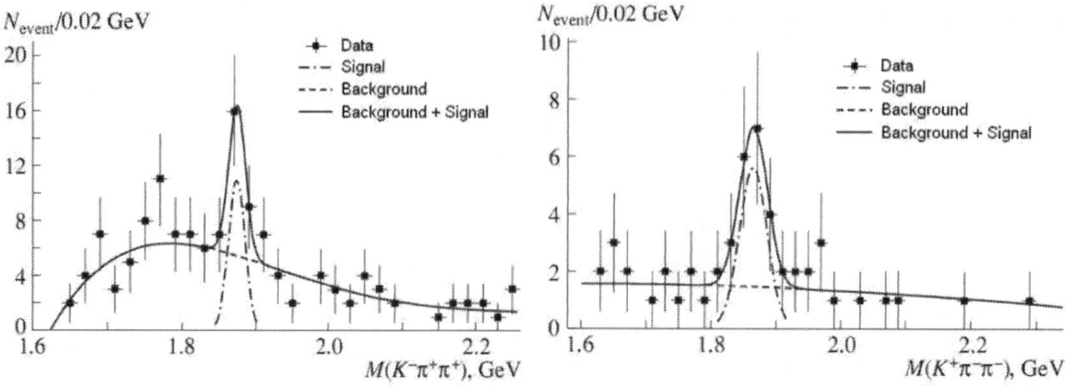

Figure 5. Effective mass spectra of the $K^-\pi^+\pi^+$ (left) and $K^+\pi^-\pi^-$ (right) systems after using all selection criteria.

4.3 $\Lambda_c^+ \to pK^-\pi^+$ decays

The charmed Λ_c^+ baryon was analyzed with the three-prong decay $\Lambda_c^+ \to pK^-\pi^+$. The primary procedures were similar to those for the $D^\pm$ mesons decays (i, ii). At the absence of particle identification, we had two hypotheses for positively charged particles of the $(pK^-\pi^+)$ system in the effective mass spectrum (Fig. 6 left). The next selection procedure of the events was carried out as follows (more details in [13]):

(iii) Dalitz plot for MC events with M($K^-\pi^+$) and M(K^-p) variables were used in order to cut out the experimental data background.

(iv) To eliminate the K^0 background, for the three-particle system M(3π) it was required to have > 1.2 GeV/c^2. According to simulations, it cuts out about 85% of the background and only 12% of the signal.

(v) The $pK^-\pi^+$ system momentum met the following requirement: 25 < P < 60 GeV/c.

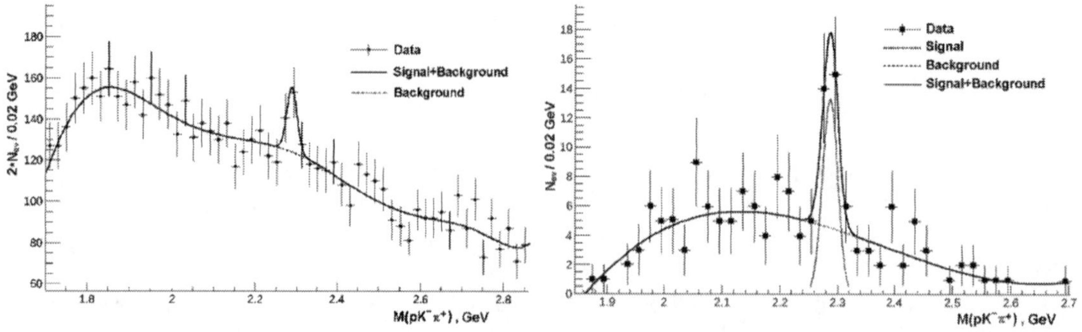

Figure 6. Effective mass spectra of the $pK^-\pi^+$ system (two hypotheses) after the primary selection of events (left) and after using all selection criteria (right).

Application of all the criteria resulted in the effective mass spectrum is shown in Fig. 6 right, with Λ_c^+ signal of 21.6 ± 6.0 events over the background of 16 ± 4 events, the mass 2287 ± 4 MeV/c^2 and σ = 13.1 MeV/c^2. In case when both of hypotheses turned out to be in the peak region, they are taken with weights of 0.5 (there were 5% of such events in simulations and no one in the experimental data).

Table 1. The results of effective mass spectra fitting.

Particle decay	Signal events	Background events	χ^2/NDF	Detection efficiency	Mass, GeV/c^2 (*PDGmass*)
$D^0 \to K\pi$	51±17	38±13	5.5/6	0.036	1861±7 (*1864.8)*
$D^+ \to K^-\pi^+\pi^+$	15.5±5.6	16.6±6.0	13.5/30	0.014	1874±5 (*1869.6)*
$D^- \to K^+\pi^-\pi^-$	15.0±4.7	8.7±2.7	3.6/20	0.008	1864±8 (*1869.6)*
$\Lambda_c^+ \to pK^-\pi^+$	21.6±6.0	16±4	12.7/33	0.011	2287±4 (*2286.5)*

The effective mass spectrum of the $pK^-\pi^+$ system has been parameterized by the sum of the Gaussian function and the polynomial of the third degree (χ^2/NDF = 12.7/33). The detection efficiency for $\Lambda_c^+ \to pK^-\pi^+$ was equal $\varepsilon = 0.011$. The summary results of fitting the effective mass charmed particles spectra are presented in Table 1.

5 Cross sections for charmed particle production and their A-dependence

The following relation has been used to calculate the inclusive cross sections for a given charmed particle:

$$N_s(i) = [N_0 \times (\sigma(i) \times A^{\alpha})/(\sigma_{pp} \times A^{0.7})] \times [(B(i) \times \varepsilon(i))/K_{tr}],$$

where $i = D^0, \bar{D}^0, D^+, D^-$ or Λ_c^+; $N_s(i)$ is the number of events in the signal for (i) type of the charmed particle produced in the given target; N_0 is the number of inelastic interactions in this target; $\sigma(i)$ is the cross section for charmed particle (i) production at a single nucleon of the target; A is the atomic mass number of AT material (C, Si or Pb); $\alpha(i)$ is an exponent parameter in A-dependence of the charm cross section; σ_{pp} is the cross section of the inelastic proton-proton interaction at 70 GeV (= 31440 μb); $B(i)$ is the branching ratio for the charmed particle decay ($B(D^0/\bar{D}^0) \to K\pi$) = 0.038, $B(D^\pm \to K\pi\pi) = 0.094$, $B(\Lambda_c^+ \to pK^-\pi^+) = 0.05$); $\varepsilon(i)$ is the detection efficiency of the charmed particle from Table 1 and $K_{tr} = 0.57$ is the trigger efficiency of registration of inelastic events in our experiment.

Substituting $C(i) = [N_0/(\sigma_{pp} \times A^{0.7})] \times [B(i) \times \varepsilon(i)/K_{tr}]$, the relation takes the following form: $N_s(i) = C(i) \times \sigma(i) \times A^{\alpha}$ or $\ln(N_s(i)) = \alpha\times\ln(A)+\ln(\sigma(i))$. Figure 7 shows A-dependences of the charmed particles production in AT.

The parameter α obtained from the straight linear approximations in Fig. 7 for each particle is presented in Table 2 together with inclusive cross section and path length. The α-parameters are close to 1 for all charmed particles, as it was found earlier for the hidden charm (J/ψ and ψ') production cross sections in the proton-nucleus interactions [14], [15], [16], [17].

On the basis of results presented in Table 2 and using the relation [18]

$$\sigma_{tot}(c\bar{c}) = 0.5\cdot(\sigma_{D+} + \sigma_{D0} + \sigma_{D-} + \sigma_{\bar{D}0} + \sigma_{\Lambda c+} + \sigma_{Ds} + \sigma_{\bar{D}s})$$

total cross section of the charmed particles production in proton-nucleon interactions at 70 GeV can be estimated as $\sigma(c\bar{c}) = 7.1 \pm 2.3(stat) \pm 1.4(syst)$ μb/nucleon. The charmed particles yields measured in our experiment are given in Table 3 and in Fig. 8 along with the data from other experiments and theoretical predictions.

The contributions of charmed particles to the total cross section vary with energy. For example, the particles (D^0 and D^+) contributions go down as the interaction energy decreases to 70 GeV, while the antiparticles ($\bar{D}^0$, D^- and Λ_c^+) contributions grow. A large difference of the charmed particle and

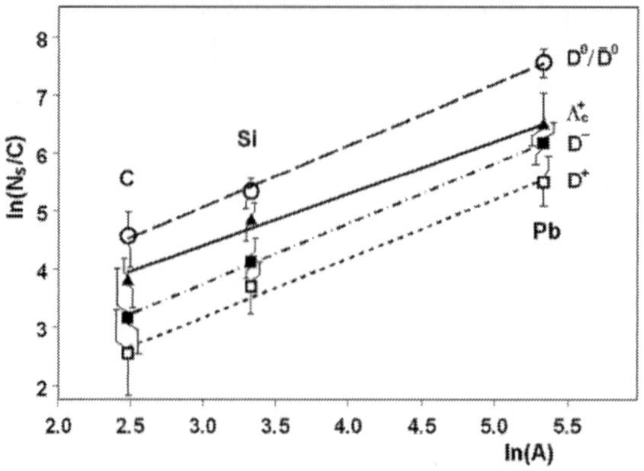

Figure 7. The A-dependence of cross sections for the charmed particles production in pA-interactions.

Table 2. Characteristics of the charmed particles production. The first error is statistical, the second – the systematical one.

Num	Type of charmed particle	Inclusive cross section for all x_F (μb/nucleon	α-parameter	$c\tau$, mm	
				SVD-2	PDG
1	D^+	1.2±0.4±0.2	1.02±0.26	0.291±0.075	0.311
2	D^-	1.9±0.6±0.4	1.04±0.27	0.341±0.088	0.311
3	D^0	2.5±0.8±0.5	1.08±0.12	0.123±0.024	0.124
4	$\bar{D}^0$	4.6±1.6±0.9			
5	$D^0/\bar{D}^0$	7.1±1.8±0.9			
6	Λ_c^+	4.0±1.6±1.2	0.9±0.2	0.051±0.011	0.059

Table 3. The charmed particles yields, $\sigma(i)/\sigma_{tot}(c\bar{c})$.

Yields \ particle	PYTHIA pp	FRITIOF pA	SVD-2 pA	Other experiments	
				NA-27 [23]	HERA-B [24]
D^0	0.28	0.51	0.35±0.16	0.57±0.08	0.44±0.18
$\bar{D}^0$	0.74	0.59	0.65±0.31	0.43±0.09	0.54±0.23
D^+	0.13	0.29	0.16±0.07	0.31±0.06	0.19±0.08
D^-	0.24	0.27	0.27±0.17	0.34±0.06	0.25±0.11
Λ_c^+	0.55	0.36	0.56±0.27	0.52±0.35 | 0.42±0.13 | 0.18±0.01 BIS-2 [25] | E-769 [26] | SELEX-2 [27]	

antiparticle yields was firstly observed experimentally in neutron–nucleus interactions at the average neutron-beam energy of 43 GeV in the BIS-2 experiment [19]. Only antiparticle ($\bar{D}^0$ and D^-) decays were observed there, while particle (D^0 and D^+) decays were not found (their cross sections proved to be below the sensitivity threshold). The results shown in Fig. 8 are compatible with the predictions of the statistical hadronization model [18], [20].

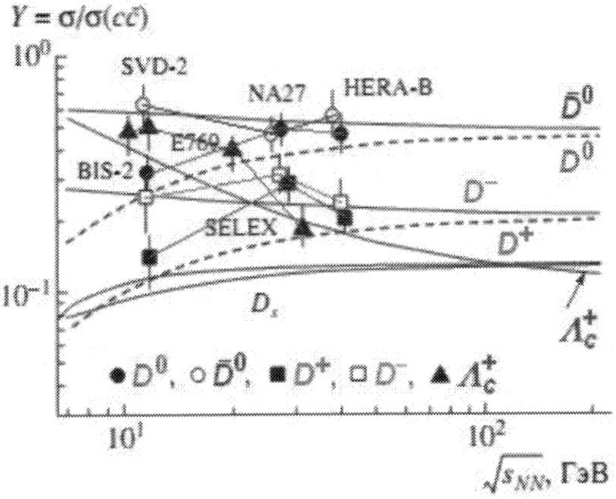

Figure 8. Relative yields of charmed particles. The experimental points are taken from Table 3, the theoretical curves (with designation of a particle) are taken from [18].

6 Results

We present total cross section for open charm production and cross sections for the inclusive production of D mesons and Λ_c^+ baryon.

- The total open charm production cross-section at the collision c.m. energy $\sqrt{s}$ = 11.8 GeV is well above QCD models predictions and inclusive cross sections are close to the prediction of QGSM for D mesons and for Λ_c^+ baryon at this energy (Fig. 9 left).

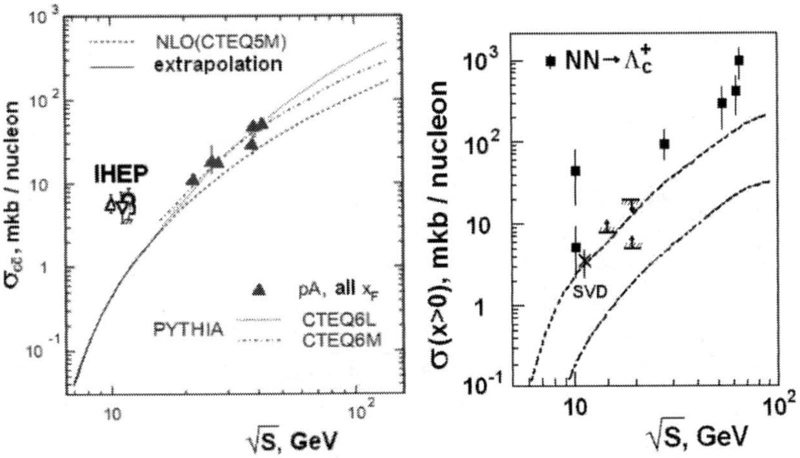

Figure 9. Left: The total cross section for charm production in pA-interactions from [1] - [3], [18]. Extrapolation (solid line) was carried out without points from IHEP: ○ - SVD-2 experiment (see above), ▽ - beam-dump experiment with muon absorber [1], ⊔ - SCAT bubble chamber experiment [2] and △ -the experiment with BIS-2 spectrometer [3]. Other lines are from various models [18]. Right: The inclusive cross section for Λ_c^+ baryon production at $x_F > 0$. The experimental data are from [25], [28] - [29] (CERN, FNAL, BIS-2), point (X) is from our experiment. Dashed lines are for predictions of two versions of QCD model [25].

- The cross sections for Λ_c^+ production (Fig. 9 right) at the collision c.m. energy $\sqrt{s} > 30$ GeV contradict the total cross sections for the open charm production (Fig. 9 left). The experimental cross sections for Λ_c^+ are extraordinarily large in this area.
- The contributions of inclusive charmed particle cross-sections into the total one vary at lower collision energies.

Acknowledgements

We appreciate IHEP (Protvino) staff for active and long-term joint work in the experiment at U-70 accelerator dedicated to the charmed particles production at low energies.

References

[1] A.E. Astratyan et al, Phys. Lett. B **79**, 497 (1978).
[2] V.V. Ammosov et al, Physics of Atomic Nuclei, **53**, 999 (1991).
[3] A.N. Aleev et al, Physics of Atomic Nuclei, **56**, 147 (1993).
[4] V.V. Avdeichikov et al, Instruments and Experimental Techniques, **56**, 9 (2013).
[5] H.Pi, Computer Physic Communication, **71**, 173 (1992).
[6] G. Alexander et al. OPAL collaboration, Zeit. fur Physik C **72**, 16 (1996).
[7] G.S. Huang et al, Phys. Rev. Letters, **94**, 011802 (2005)
[8] GEANT 3.21, CERN Program Library Long Write Up W5013.
[9] A.A. Kiriakov et al, Instruments and Experimental Techniques, **47**, 611 (2004).
[10] A.P. Vorobiev et al, IHEP Preprint 2008-17, Protvino, 2008 (in russian). http://web.ihep.su/library/pubs/prep2008/ps/2008-17.pdf
[11] V.N. Ryadovikov (on behalf of the SVD-2 Collab.), Physics of Atomic Nuclei, **73**, 1539 (2010).
[12] V.N. Ryadovikov (on behalf of the SVD-2 Collab.), Physics of Atomic Nuclei, **77**, 716 (2014).
[13] V.N. Ryadovikov (on behalf of the SVD-2 Collab.), Physics of Atomic Nuclei, **79**, 144 (2016).
[14] J. Badier et al, Z.Phys. C - Particle and Fields **20**, 101 (1983).
[15] D.M. Alde et al, Phys. Rev. Letters, **66**, 133 (1991).
[16] M.C. Abreu et al, Physics Letters, B **410**, 327 (1997).
[17] M.J. Leitch et al, Nuclear Physics, A **544**, 197 (1992).
[18] A. Andronic, P. Braun-Munzinger, K. Redlich, J. Stachel, Phys. Lett. B **659**, 149 (2008). arxiv:0708.1488
[19] A. Aleev et al, Z.Phys. C - Particle and Fields **37**, 243 (1988).
[20] A. Andronic, F. Beutler, P. Braun-Munzinger et al., arxiv:0904.1368v2
[21] R.Vogt, Phys. Rev. C **61**, 035203 (2000).
[22] A.N. Aleev et al, SVD-2 Collaboration, Physics of Atomic Nuclei, **74**, 324 (2011).
[23] M. Aguilar - Benitez, Na-27 Collaboration, Phys. Letters, B **189**, 476 (1987).
[24] I. Abt et al, Eur. Phys. J. C **52**, 531 2007. http://arxiv.org/abs/0708.1443v1
[25] A. N. Aleev et al., BIS-2 Collaboration, Z.Phys. C - Particle and Fields, **23**, 333 (1984). E. N. Chudakov, Master's Thesis (1987), JINR Preprint 1-87-183, 1987 (in russian)
[26] J. Engelfried, www.ifisica.uaslp.mx/ jurgen/public/dpc99-talk.ps.gz, 1999.
[27] F.G. Garcia et al., SELEX Collaboration, Phys. Lett. B **528**, 49 (2002). http://arxiv.org/pdf/hep-ex/0109017.pdf
[28] Bailay R. et al., Nucl. Phys. B **239**, 15 (1984).

 DOI: 10.1051/epjconf/201612502016

[29] M. Basile et al., Lett. Nuovo Cimento, **30**, 487 (1981). M. Basile et al., Lett. Nuovo Cimento, **30**, 481 (1981). F. Muller, Proceedings of IV Warsaw Symposium, 1981 G. Bari et al., Nuovo Cimento, A **104**, 571 (1991).

Sgoldstino physics and flavor-violating Higgs boson decays

Sergei Demidov[1,⋆] and *Ivan* Sobolev[1,2,⋆⋆]

[1] *Institute for Nuclear Research of the Russian Academy of Sciences, 60th October Anniversary prospect 7a, Moscow 117312, Russia*

[2] *Department of Particle Physics and Cosmology, Physics Faculty, M. V. Lomonosov Moscow State University, Vorobjevy Gory, 119991, Moscow, Russia*

Abstract. In this talk we study effects of Higgs-sgoldstino mixing on flavour-violation in Higgs sector and in particular on $h \to \mu\tau$ decay.

1 Introduction

One of the most interesting anomalies in the LHC run-1 was a hint of Higgs flavour-violating (HFV) decay. CMS collaboration observed small excess of signal events which can be interpreted as $h \to \mu\tau$ decay with $\mathrm{Br(h} \to \mu\tau) = \left(8.4^{+3.9}_{-3.7}\right) \times 10^{-3}$ [1]. In the Standard Model (SM) in the absence of neutrino masses this process is forbidden since in this case muon and tau lepton numbers are conserved individually. When extending the SM to include neutrino masses, such a processes can occur via 1-loop diagrams with virtual neutrinos but their rate is suppressed by powers of $\frac{m_\nu^2}{M_W^2}$, so their effect is far too small to affect Higgs decay in practice. A lot of ideas was proposed in an attempt to explain the mentioned excess. [3–14]. In this contribution we briefly present results of our paper [15] and make some new comments and remarks on the results obtained.

2 Model description

2.1 Lepton nubmer violation in the Minimal Supersymmetric Standard Model (MSSM)

The SM is minimally flavour violating: the only sources of flavor violation are the Yukawa matrices. In this way, if assuming that the neutrino mass is zero, the SM posseses accidental $U(1)_e \times U(1)_\mu \times U(1)_\tau$ symmetry. The same holds for purely supersymmetric part of the MSSM Lagrangian (i.e. for Kähler potential and superpotential). However this is not the case for SUSY breaking soft terms, and in particular for slepton mass terms which in general can violate this accidental lepton number symmetry. The quadratic part of slepton lagrangian can be written as

$$\mathcal{L}_{slepton} \supset -\tilde{l}^\dagger_{Li} M_{\tilde{l}LLij} \tilde{l}_{Lj} - \tilde{l}^\dagger_{Ri} M_{\tilde{l}RRij} \tilde{l}_{Rj} - \tilde{l}^\dagger_{Li} M_{\tilde{l}LRij} \tilde{l}_{Rj} - \tilde{l}^\dagger_{Ri} M_{\tilde{l}RLij} \tilde{l}_{Lj}, \quad (1)$$

⋆e-mail: demidov@ms2.inr.ac.ru
⋆⋆e-mail: sobolev.ivan@physics.msu.ru

where

$$\tilde{l} = (\tilde{e}, \tilde{\mu}, \tilde{\tau}) \tag{2}$$

and $M^2_{\tilde{l}LL}$,$M^2_{\tilde{l}RR}$,$M^2_{\tilde{l}RL}$,$M^2_{\tilde{l}LR}$ are 3×3 matrices. Obviously any non-diagonal term in either of these matrices leads to flavour-violation in charged lepton sector. Several comments here are in order. Fist, HFV decays exclusively in the scope of the MSSM have been studied in [3, 16]. It was found that parameter space capable of explaining CMS excess is very narrow. In other words, one needs significant fine-tuning of the MSSM parameters to reach Br(h $\to \tau\mu$) of order of $\sim 1\%$. Second, we assume that non-diagonal terms in $M^2_{\tilde{l}LL}\left(M^2_{\tilde{l}RR}\right)$ are suppressed and thus can be neglected. We will elaborate on this assumption in the next section.

2.2 Models with low SUSY breaking scale

If supersymmetry is actually realized in Nature it should be spontaneously broken at low energies. In this fashion phenomenologically viable supersymmetric theory should possess some sector [17] (which is usually called "hidden" in contradistinction to "visible" or MSSM sector) where symmetry breaking occurs and then is transmitted to "visible" sector by means of some mediation mechanism. According to the analogue of Goldstone theorem, a massless fermion – goldstino – should appear in the spectrum of a model. We utilize the simplest case of a "hidden" sector described by the following lagrangian

$$\mathcal{L}_\Phi = \int d^2\theta\, d^2\bar{\theta}\left(\Phi^\dagger\Phi + \tilde{K}(\Phi^\dagger, \Phi)\right) - \left(\int d^2\theta F\Phi + \text{h.c.}\right), \tag{3}$$

where $\Phi = \phi + \sqrt{2}\theta\psi + \theta^2 F_\phi$ is a chiral superfield and $\tilde{K}$ can be chosen in the form

$$\tilde{K}(\Phi^\dagger, \Phi) = -\frac{m_s^2 + m_p^2}{8F^2}(\Phi^\dagger\Phi)^2 - \frac{m_s^2 - m_p^2}{12F}((\Phi^\dagger\Phi^\dagger\Phi) + (\Phi\Phi^\dagger\Phi)) \tag{4}$$

Chiral superfield is a singlet under the action of the SM gauge group and it contains massless godstino ψ and its superpartners – scalar (s) and pseudoscalar (p) sgoldstinos which form complex field ϕ.The last linear term in the superpotential of Eq. (3) forces the auxiliary field F_ϕ to acquire non-zero vacuum expectation value $\langle F_\phi \rangle = F + O\left(\frac{1}{F}\right)$ and hence triggers spontaneous supersymmetry breaking. The value $\sqrt{F}$ can be regarded as supersymmetry breaking scale. When $\sqrt{F}$ is not far away from the electroweak scale (say it is of order of $\sim O$(TeV)) "hidden" sector becomes not so hidden and sgoldstino interactions with MSSM fields should be included into the low-energy theory. Here we briefly recall a minimal set of interactions of goldstino multiplet with MSSM fields. It can be obtained by making use of spurion method [18]. We will restrict ourselves only to the simplest set of operators which reproduce soft-terms when auxiliary field F_ϕ acquires its vacuum expectation value. Furthermore we assume R-parity conservation and presume that all of the free parameters of the model are real and hence neglect any possible CP-violation.

$$\mathcal{L}_{\text{Kähler}} = -\int d^2\theta\, d^2\bar{\theta} \sum_k \frac{m_k^2}{F^2}\Phi^\dagger\Phi\Phi_k^\dagger e^{g_1V_1+g_2V_2+g_3V_3}\Phi_k, \tag{5}$$

where k runs over all chiral MSSM superfields.

$$\mathcal{L}_{\text{superpotential}} = \int d^2\theta \left\{\epsilon_{ij}\left(-\frac{B}{F}\Phi H_d^i H_u^j + \frac{A_{ab}^L}{F}\Phi L_a^j E_b^c H_d^i\right) + \sum_\alpha \frac{M_\alpha}{2F}\Phi \text{Tr}\, W^\alpha W^\alpha\right\} + \text{h.c.} \tag{6}$$

Several important remarks here are in order. First, lagrangians (4)–(6) contain operators with dimension higher than four. Thereby this theory should be regarded as *effective theory* valid under some cutoff scale. Possible values of the cutoff Λ in the low scale supersymmetry breaking models have been discussed some time ago in Refs. [18, 20, 21]. It has been found that the cutoff for this model can lie somewhere between the level of soft masses of matter scalars $\tilde{m}$ (the largest of which can not exceed $\sqrt{F}$ (see table 1)) and the value $\Lambda^2 = 16\pi F^2/\tilde{m}^2$ (which is the scale of perturbative violation of unitarity of the effective theory). We choose the latter for our numerical estimates. Second, non-renormalizable interactions cause divergences in loop diagrams which involve particles from the hidden sector. These divergences turn out to be of two types: logarithmic and quadratic. The latter are caused by box diagrams in $\tau \to 3\mu$ process and are proportional to non-diagonal entries in $M^2_{\tilde{l}LL}(M^2_{\tilde{l}RR})$ blocks [20]. This is the reason why we assume that this entries are suppressed (in our analysis we set them to zero). Logarithmic divergences take place in $\tau \to \mu\gamma$ [1] due to sgoldstino-Higgs mixing. That is why they are proportional to $\sim \sin^2\frac{\theta}{2}\log\frac{\Lambda}{m_S}$, where θ is the mixing angle, and in general are suppressed (for more detailed discussion see [15]).

2.3 Higgs-sgoldstino mixing

From (5)-(6) one can derive the following Higgs-sgoldstino mixing term

$$\mathcal{L}_{s-h} = -\frac{X}{F} s \cdot h, \tag{7}$$

with [2]

$$X = 2\mu^3 v \sin 2\beta + \frac{v^3}{2}\left(g_1^2 M_1 + g_2^2 M_2\right)\cos^2 2\beta \tag{8}$$

which was obtained in so-called "decoupling limit" (i.e. when all the MSSM Higgs bosons are considerably heavier than the lightest one) and after the EWSB. The mixing angle can be written as

$$\tan 2\theta = \frac{2X}{F(m_s^2 - m_h^2)}. \tag{9}$$

We stress that m_s and m_h in this expression are *non-physical* masses of scalar sgoldstino and Higgs respectively. This remark is important since corresponding *physical* masses can differ drastically from the mentioned ones in case of large of mixing angles and this will be the case for most of the interesting points in the parameter space of our model. We will denote mass states as $\tilde{s}$ and $\tilde{h}$ keeping notations s and h for non-physical states. Next, the second term in (7) is suppressed compared to the first one. In this way the sign of the mixing angle is dictated by the sign of μ parameter. Unlike Higgs boson sgoldstino has flavour-violating couplings to the SM fermions

$$\mathcal{L}_{s-fermions} \supset \frac{A^L_{ab}}{\sqrt{2}F}\bar{e}_b\, l_a\, v_d\, s + \text{h.c.}. \tag{10}$$

So, small admixture of sgoldstino to the lightest Higgs boson generates flavor violating couplings of the latter

$$Y^{\tilde{h}}_{\mu\tau(\tau\mu)} = \frac{v\, A_{\mu\tau(\tau\mu)}\cos\beta\sin\theta}{\sqrt{2}F}. \tag{11}$$

[1]Logarithmic divergences are also present in triangle diagrams contributing to $\tau \to 3\mu$ decay but here they are suppressed compared to the leading tree-level diagrams, see [15].

[2]Here parameters $\beta, M_{1,2}, \mu$ are conventional MSSM parameters (see [17] for details), $v = 174$ GeV, $g_{1,2}$ are $U(1)$ and $SU(2)$ coupling constants respectively.

The corresponding decay width reads

$$\Gamma(\tilde{h} \to \mu\tau) \equiv \Gamma(\tilde{h} \to \bar{\tau}\mu) + \Gamma(\tilde{h} \to \bar{\mu}\tau) = \frac{m_{\tilde{h}}}{8\pi}\left(\left|Y^{\tilde{h}}_{\mu\tau}\right|^2 + \left|Y^{\tilde{h}}_{\tau\mu}\right|^2\right). \quad (12)$$

3 Model analysis and results

We perform scan over the following parameter space (see Table 1). Recall that the consistency of the effective field theory approach to the model (3)–(6) requires that the parameters which become soft terms after the spontaneous supersymmetry breaking should be smaller than $\sqrt{F}$. In what follows we fix this value to 8 TeV. Values of $\sqrt{F}$ smaller than about 8 TeV turn out to be disfavored the

$\tan\beta$	$1.5\ldots50.5$
$\lvert\mu\rvert$	$100\ldots2000$ GeV
M_1	$100\ldots2000$ GeV
M_2	$200\ldots2000$ GeV
M_3	$1.5\ldots4.0$ TeV
$A_{\mu\tau}, A_{\tau\mu}$	$0.1\sqrt{F}\ldots\sqrt{F}$

Table 1. Parameter space used in the analysis

results of direct searches. Namely, at smaller $\sqrt{F}$ the coupling constants of sgoldstino to the SM particles increase and such sgoldstino is phenomenologically unacceptable. In this case, very light sgoldstino, which decays mostly to $b\bar{b}$ due to large mixing with the Higgs boson, becomes excluded by the LEP [22] and TeVatron [23] results. Heavier sgoldstino with $\sqrt{F}$ smaller than about 8 TeV is excluded by the results of the ATLAS and CMS searches for diboson resonances. If we enlarge our parameter space by increasing, in particular, the upper bound on μ in the Table 1, we expect that somewhat lower values of $\sqrt{F}$ and larger values of the sgoldstino mass will be allowed. We find that for the most interesting cases the mass parameter of the scalar sgoldstino m_s should not be very heavy or very small. In the case of heavy sgoldstino, the mixing angle (9) is small and, as a consequence, the width of $\tilde{h} \to \mu\tau$ decay is suppressed. On the other hand, very light sgoldstinos with large

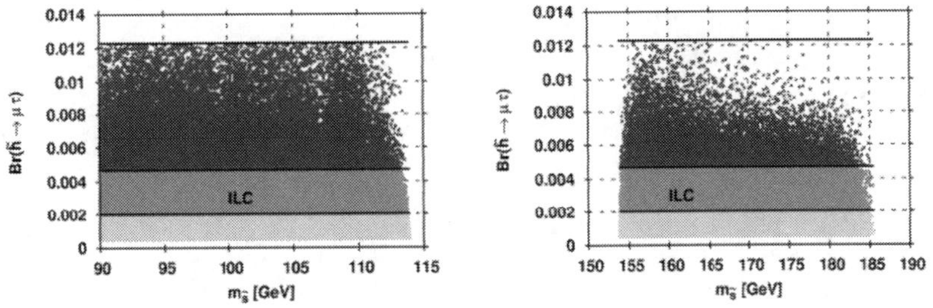

Figure 1. Scatter plots in plane $(m_s, \text{Br}(h \to \tau\mu))$ for $\sqrt{F} = 8$ TeV and sgoldstinos lighter (left) and heavier (right) than Higgs.

Higgs boson admixture are phenomenologically unacceptable due to results from the LEP [22] and

Tevatron [23] experiments. So, we limit ourselves to the regimes in which the scalar sgoldstino mass parameter is somewhat smaller (90–114 GeV) or larger (150–200 GeV) than the Higgs boson mass. Some parameters, which are not of primary importance for the analysis, were fixed to reasonable benchmark values. For example, soft trilinear constant of b-quark A_{bb} is fixed to $A_{bb} = 0.5\sqrt{F}$ and the mass of the pseudoscalar sgoldstino is fixed to $m_p = 200$ GeV. For each point in the mentioned parameter space we calculate signal strength

$$\mu_f = \frac{\sigma(pp \to \tilde{h}) \times \mathrm{Br}(\tilde{h} \to f)}{\sigma(pp \to h^{SM}) \times \mathrm{Br}(h^{SM} \to f)}, \tag{13}$$

where final state f stands for $b\bar{b}$, W^+W^-, $ZZ, \gamma\gamma, \tau\bar{\tau}$. We compare it with experimental constraints by CMS: [24–30] and ATLAS: [31–35]. Next, we check whether the scalar sgoldstino-like resonance is

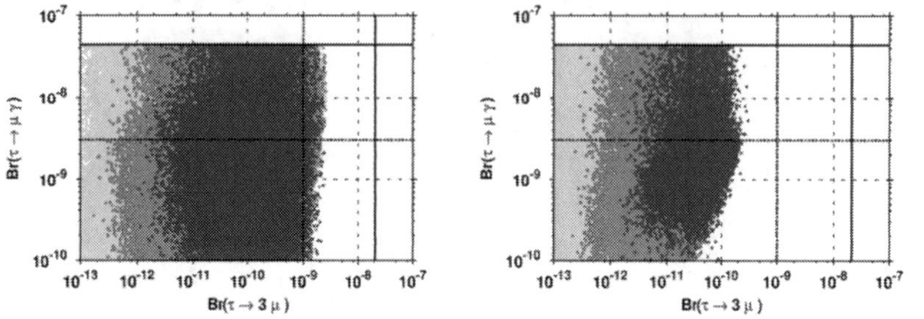

Figure 2. Scatter plots in plane $(\mathrm{Br}(\tau \to \mu\gamma), \mathrm{Br}(\tau \to 3\mu))$ for $\sqrt{F} = 8$ TeV and sgoldstinos lighter (left) and heavier (right) than Higgs. Solid lines correspond to present limits on branching fraction of both decays, whereas dashed line represents expected SuperKEKB sensitivity [46].

allowed by existing experimental results. For the case of the LHC searches for diboson resonances, we use the observables $\sigma(pp \to \tilde{s}) \times \mathrm{Br}(\tilde{s} \to \mathrm{f})$ where f stands for a pair of photons [36], W [37, 38] or Z [39] since these final states are the most constraining for sgoldstinos with discussed masses. Due to large tree level coupling to the gluons M_3, gluon-gluon fusion will be the dominating mechanism for sgoldstino production with cross-section given by the following expression

$$\sigma_{\tilde{s}} = \frac{\pi^2}{8}\frac{\Gamma(\tilde{s} \to gg)}{s m_{\tilde{s}}} \int_{m_{\tilde{s}}^2/s}^{1} \frac{dx}{x} f_{p/g}(x, m_{\tilde{s}^2})\, f_{p,\bar{p}/g}\left(\frac{m_{\tilde{s}}^2}{xs}, m_{\tilde{s}}^2\right), \tag{14}$$

where $\Gamma(\tilde{s} \to gg)$ is a partial width of sgoldstino-like state decaying into two gluons, s is a center-of-mass energy squared and $f_{p/g}(x, Q^2)$ are the parton density functions defined at factorisation scale Q. We use CTEQ6L [40] parametrization for the parton density functions. Finally we make use of constraints on $\mathrm{Br}(\tau \to \mu\gamma)$ [41] and $\mathrm{Br}(\tau \to 3\mu)$ [42] decays. For more detailed discussion on calculation of $\mathrm{Br}(\tau \to \mu\gamma)$ in our model see arguments in section 2.2 and in [15]. The result of the scan over parameter space are presented on Fig.1-5. For illustrating purposes, we present only the models with sufficiently large branching fraction $\mathrm{Br}(\tilde{h} \to \mu\tau) > 5 \cdot 10^{-4}$. By blue color we mark the models which are capable of explaining the CMS excess, by purple color to we mark points which lie somewhat below the CMS excess but still have significant (more than 0.2%). According to the latest studies this level of branching fraction will be reachable at experiments on HL-LHC and ILC; see refs. [43–45]. We also make predictions for flavour-violating decays of τ-lepton (Fig.2),signal strength μ_f for different final states f and Higgs production channels (Fig.3,4) and make predictions for observable $\sigma(pp \to \tilde{s}) \times \mathrm{Br}(\tilde{s} \to \mathrm{W^+W^-}$ or ZZ) for the case of 13 TeV (Figs.5,6). For more results and detailed discussion see [15].

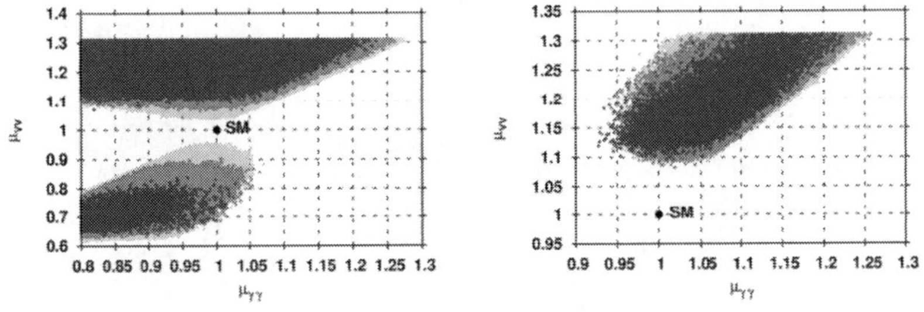

Figure 3. Scatter plots in plane $\left(\mu_{\gamma\gamma}, \mu_{VV}\right)$ for $\sqrt{F} = 8$ TeV and sgoldstinos lighter (left) and heavier (right) than Higgs.

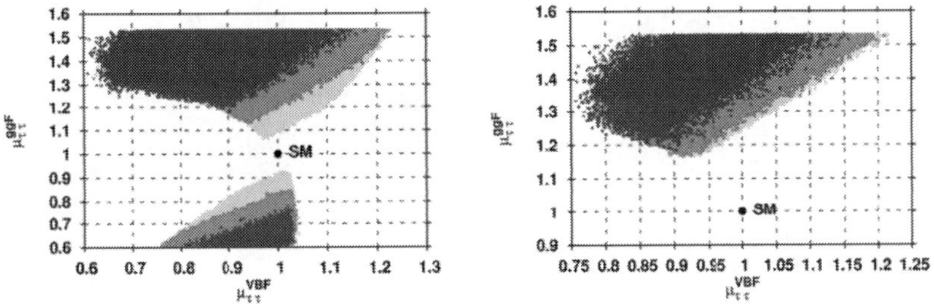

Figure 4. Scatter plots in plane $\left(\mu_{\tau\tau}^{\rm VBF}, \mu_{\tau\tau}^{\rm ggF}\right)$ for $\sqrt{F} = 8$ TeV and sgoldstinos lighter (left) and heavier (right) than Higgs.

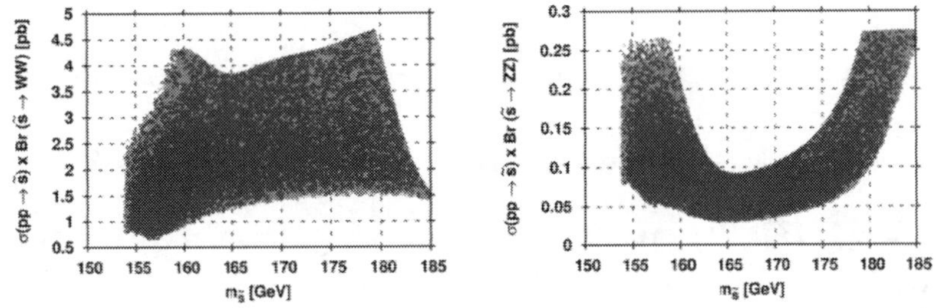

Figure 5. Scatter plots in $\sigma(pp \to \tilde{s}) \times \mathrm{Br}(\tilde{s} \to W^+W^-$ or $ZZ)$-plane for heavy sgoldstino and $\sqrt{s} = 13$ TeV.

4 Conclusions

In this contribution we showed that in models with low scale supersymmetry breaking, the Higgs boson can have considerable branching ratio of $h \to \mu\tau$ decay due to mixing with sgoldstino and demonstrated that the CMS excess in $h \to \mu\tau$ decay can be explained in this framework. We stress, that the features we pointed out in [15] are common in the class of models in question.

 DOI: 10.1051/epjconf/201612502017

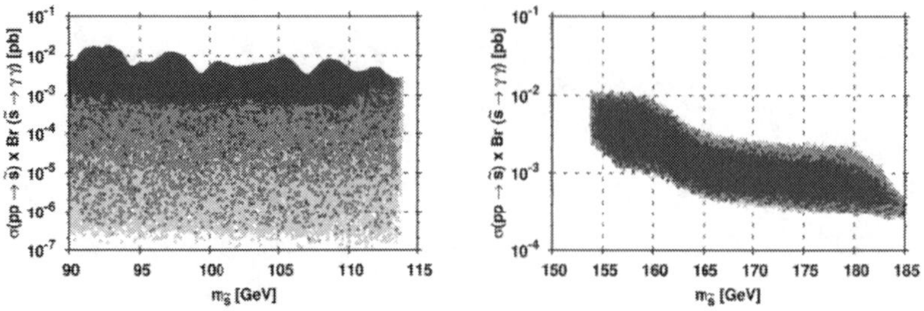

Figure 6. Scatter plots in $\sigma(pp \to \tilde{s}) \times \mathrm{Br}(\tilde{s} \to \gamma\gamma)$-plane for light (left panel) and heavy (right panel) sgoldstino and $\sqrt{s} = 13$ TeV.

Acknowledgements

The work of IS is supported by the RFBR grant 16-32-00894. Numerical calculations have been performed on the Computational cluster of the Theoretical division of INR RAS.

References

[1] V. Khachatryan *et al.* [CMS Collaboration], Phys. Lett. B **749**, 337 (2015) [arXiv:1502.07400 [hep-ex]].
[2] CMS Collaboration [CMS Collaboration], CMS-PAS-HIG-16-005.
[3] A. Dery, A. Efrati, Y. Nir, Y. Soreq and V. Susič, Phys. Rev. D **90**, 115022 (2014) [arXiv:1408.1371 [hep-ph]].
[4] M. D. Campos, A. E. C. Hernández, H. Päs and E. Schumacher, Phys. Rev. D **91**, no. 11, 116011 (2015) [arXiv:1408.1652 [hep-ph]].
[5] D. Aristizabal Sierra and A. Vicente, Phys. Rev. D **90**, no. 11, 115004 (2014) [arXiv:1409.7690 [hep-ph]].
[6] J. Heeck, M. Holthausen, W. Rodejohann and Y. Shimizu, Nucl. Phys. B **896**, 281 (2015) [arXiv:1412.3671 [hep-ph]].
[7] A. Crivellin, G. D'Ambrosio and J. Heeck, Phys. Rev. Lett. **114**, 151801 (2015) [arXiv:1501.00993 [hep-ph]].
[8] I. Doršner, S. Fajfer, A. Greljo, J. F. Kamenik, N. Košnik and I. Nišandžic, JHEP **1506**, 108 (2015) [arXiv:1502.07784 [hep-ph]].
[9] X. G. He, J. Tandean and Y. J. Zheng, JHEP **1509**, 093 (2015) [arXiv:1507.02673 [hep-ph]].
[10] W. Altmannshofer, S. Gori, A. L. Kagan, L. Silvestrini and J. Zupan, Phys. Rev. D **93**, no. 3, 031301 (2016) [arXiv:1507.07927 [hep-ph]].
[11] E. Arganda, M. J. Herrero, X. Marcano and C. Weiland, Phys. Rev. D **93**, no. 5, 055010 (2016) [arXiv:1508.04623 [hep-ph]].
[12] F. J. Botella, G. C. Branco, M. Nebot and M. N. Rebelo, Eur. Phys. J. C **76**, no. 3, 161 (2016) [arXiv:1508.05101 [hep-ph]].
[13] S. Baek and K. Nishiwaki, Phys. Rev. D **93**, no. 1, 015002 (2016) [arXiv:1509.07410 [hep-ph]].
[14] M. Buschmann, J. Kopp, J. Liu and X. P. Wang, arXiv:1601.02616 [hep-ph].
[15] S. V. Demidov and I. V. Sobolev, arXiv:1605.08220 [hep-ph].

[16] E. Arganda, M. J. Herrero, R. Morales and A. Szynkman, JHEP **1603**, 055 (2016) [arXiv:1510.04685 [hep-ph]].
[17] S. P. Martin, Adv. Ser. Direct. High Energy Phys. **21**, 1 (2010) [Adv. Ser. Direct. High Energy Phys. **18**, 1 (1998)] [hep-ph/9709356].
[18] A. Brignole, F. Feruglio and F. Zwirner, Nucl. Phys. B **501**, 332 (1997) doi:10.1016/S0550-3213(97)80767-5 [hep-ph/9703286].
[19] A. Brignole, J. A. Casas, J. R. Espinosa and I. Navarro, Nucl. Phys. B **666**, 105 (2003) doi:10.1016/S0550-3213(03)00539-X [hep-ph/0301121].
[20] A. Brignole and A. Rossi, Nucl. Phys. B **587**, 3 (2000) doi:10.1016/S0550-3213(00)00458-2 [hep-ph/0006036].
[21] A. Brignole, F. Feruglio and F. Zwirner, Phys. Lett. B **438**, 89 (1998) doi:10.1016/S0370-2693(98)00974-5 [hep-ph/9805282].
[22] S. Schael *et al.* [ALEPH and DELPHI and L3 and OPAL and LEP Working Group for Higgs Boson Searches Collaborations], Eur. Phys. J. C **47**, 547 (2006) [hep-ex/0602042].
[23] V. M. Abazov *et al.* [D0 Collaboration], Phys. Rev. D **88**, no. 5, 052011 (2013) [arXiv:1303.0823 [hep-ex]].
[24] S. Chatrchyan *et al.* [CMS Collaboration], Phys. Rev. D **89**, no. 1, 012003 (2014) [arXiv:1310.3687 [hep-ex]].
[25] V. Khachatryan *et al.* [CMS Collaboration], Phys. Rev. D **92**, no. 3, 032008 (2015) [arXiv:1506.01010 [hep-ex]].
[26] S. Chatrchyan *et al.* [CMS Collaboration], Nature Phys. **10**, 557 (2014) [arXiv:1401.6527 [hep-ex]].
[27] S. Chatrchyan *et al.* [CMS Collaboration], JHEP **1405**, 104 (2014) [arXiv:1401.5041 [hep-ex]].
[28] S. Chatrchyan *et al.* [CMS Collaboration], JHEP **1401**, 096 (2014) [arXiv:1312.1129 [hep-ex]].
[29] S. Chatrchyan *et al.* [CMS Collaboration], Phys. Rev. D **89**, no. 9, 092007 (2014) [arXiv:1312.5353 [hep-ex]].
[30] V. Khachatryan *et al.* [CMS Collaboration], Eur. Phys. J. C **74**, no. 10, 3076 (2014) [arXiv:1407.0558 [hep-ex]].
[31] G. Aad *et al.* [ATLAS Collaboration], JHEP **1501**, 069 (2015) [arXiv:1409.6212 [hep-ex]].
[32] G. Aad *et al.* [ATLAS Collaboration], JHEP **1504**, 117 (2015) [arXiv:1501.04943 [hep-ex]].
[33] G. Aad *et al.* [ATLAS Collaboration], Phys. Rev. D **92**, no. 1, 012006 (2015) [arXiv:1412.2641 [hep-ex]].
[34] G. Aad *et al.* [ATLAS Collaboration], Phys. Rev. D **91**, no. 1, 012006 (2015) [arXiv:1408.5191 [hep-ex]].
[35] G. Aad *et al.* [ATLAS Collaboration], Phys. Rev. D **90**, no. 11, 112015 (2014) [arXiv:1408.7084 [hep-ex]].
[36] V. Khachatryan *et al.* [CMS Collaboration], Phys. Lett. B **750**, 494 (2015) [arXiv:1506.02301 [hep-ex]].
[37] G. Aad *et al.* [ATLAS Collaboration], Phys. Lett. B **716**, 62 (2012) [arXiv:1206.0756 [hep-ex]].
[38] V. Khachatryan *et al.* [CMS Collaboration], JHEP **1510**, 144 (2015) [arXiv:1504.00936 [hep-ex]].
[39] G. Aad *et al.* [ATLAS Collaboration], Eur. Phys. J. C **76**, no. 1, 45 (2016) [arXiv:1507.05930 [hep-ex]].
[40] J. Gao *et al.*, Phys. Rev. D **89**, no. 3, 033009 (2014) [arXiv:1302.6246 [hep-ph]].

[41] B. Aubert *et al.* [BaBar Collaboration], Phys. Rev. Lett. **104**, 021802 (2010) [arXiv:0908.2381 [hep-ex]].

[42] K. Hayasaka *et al.*, Phys. Lett. B **687**, 139 (2010) doi:10.1016/j.physletb.2010.03.037 [arXiv:1001.3221 [hep-ex]].

[43] S. Banerjee, B. Bhattacherjee, M. Mitra and M. Spannowsky, arXiv:1603.05952 [hep-ph].

[44] Y. n. Mao and S. h. Zhu, Phys. Rev. D **93**, no. 3, 035014 (2016) [arXiv:1505.07668 [hep-ph]].

[45] I. Chakraborty, A. Datta and A. Kundu, arXiv:1603.06681 [hep-ph].

[46] T. Aushev *et al.*, arXiv:1002.5012 [hep-ex].

EPJ Web of Conferences **125**, 02018 (2016) DOI: 10.1051/epjconf/201612502018
QUARKS-2016

Post-Higgs THDM scenarios associated with the new diphoton resonance data

Mikhail Dubinin[1,2,⋆] and *Elena* Petrova[1,2,⋆⋆]

[1] *Skobeltsyn Institute of Nuclear Physics, Lomonosov Moscow State University, 119991 Moscow, Russia*
[2] *Physics Department, Lomonosov Moscow State University, 119991 Moscow, Russia*

Abstract. The CP-even and CP-odd scalar states of the two-Higgs-doublet model (THDM) and the Higgs sector of the Minimal Supersymmetric Standard Model (MSSM), with properties which could lead to the 750 GeV excess in the $\gamma\gamma$ invariant mass distribution, are studied. It is shown that THDM and MSSM are viable in explaining the observed excess without additional assumptions only when the alignment and decoupling limits break down. If the origin of the excess is not in a Higgs sector, regions of the MSSM parameter space are separated where the lightest CP-even Higgs boson is identical to the Standard Model scalar.

1 Introduction

Most significant deviation in the diphoton invariant mass spectrum at $M(\gamma\gamma) \sim 750$ GeV observed by ATLAS (integrated luminosity 3.2 fb^{-1}) and CMS (2.6 fb^{-1}) collaborations in pp collisions at $\sqrt{s} = 13$ TeV [1] deserves an interpretation in models with an extended Higgs sector. The local excess is about 4σ, however, the 'look elsewhere effect' reduces the global statistical signiffcance by about 1σ [2]. Although the excess may be a statistical fluctuation or misidentification of a hadronic state [3] in an absence of other hints of new physics we will treat it as a real physics signal from a new particle denoted as Φ.

Specific properties of the signal rule out the option that it comes from the decay of a spin-1 particle by virtue of the Landau-Yang theorem [4]. A spin 2 graviton is disfavoured because it couples universally to the conserved energy momentum tensor, but no peak is seen in lepton distributions. Bound states with spin 2 which have nothing to do with gravity can couple differently to different particles, and can have odd parity, different angular distributions, which are as motivated as spin 3 bound states [2]. Furthermore, the resonance should be Higgs-like and couple only very weakly to light quarks as, if produced in $q\bar{q}$ annihilation, it should have already been observed at the first LHC run with $\sqrt{s} = 8$ TeV and 20 fb^{-1} data. We focus here on the more likely possibility of spin zero.

The most natural extension of the Standard Model (SM) is a CP-conserving two-Higgs doublet model (THDM), where two Higgs doublets provide five physical states: two CP-even neutral Higgs bosons h and H, one CP-odd neutral Higgs boson A and two charged scalars $H^{\pm}$, whose masses are free parameters. We suppose that the light CP-even state h is identical to one observed with mass 125

⋆e-mail: dubinin@theory.sinp.msu.ru
⋆⋆e-mail: petrova@theory.sinp.msu.ru

GeV [5]. In order to make in a natural way the lightest *CP*-even state h SM-like, one can invoke the "alignment limit" [6] in which the two mixing angles of scalar states satisfy the condition $\beta - \alpha = \pi/2$. The h couplings to up and down type fermions and gauge bosons, normalized to the SM values, are then close to one, $g_{hVV} \approx g_{huu} \approx g_{hdd} \approx 1$, while the couplings of the *CP*-even H and charged $H^{\pm}$ states reduce to those of the pseudoscalar A, $g_{AVV} = 0$.

A particular case of THDM is the Minimal Supersymmetric Standard Model (MSSM) with Yukawa sector of type II, in which very strong constraints are imposed on model parameters at the tree level. We assume that large radiative corrections from the third generation quarks and quarks superpartners play an important role in generation of the effective two-doublet Higgs potential at the scale below the superpartner mass scale M_S. Besides the alignment limit a favored regime in the MSSM in the context of modern experimental data is the decoupling limit, $\cos(\beta - \alpha) \approx 0$, when $m_A \gg m_Z$, $m_H \approx m_A \approx m_{H^{\pm}}$, in a widely discussed region of low $\tan\beta$ [7] ($\tan\beta = v_2/v_1$, where $\sqrt{v_1^2 + v_2^2} = 246$ GeV) .

In an extended scalar sector different possibilities exist for the Φ particle, which can be either from an isospin singlet state, or from an isospin doublet. Although supersymmetric possibilities include such interpretations of the diphoton resonance as a sgoldstino [8] or a Higgs boson in the Next-to-Minimal SSM (NMSSM) [9] in the following we analyse a possibility of *CP*-even and/or *CP*-odd Higgs boson of the THDM contributing to the Φ signal.

2 Cross section estimates

The excess in the invariant mass distribution at $m_{\gamma\gamma}$ =750 GeV looks like a resonance with the width Γ of about 45 GeV ($0.06 M_{\Phi}$) with the cross section estimates [10]

$$\sigma(pp \to \gamma\gamma) = \begin{cases} 6 \pm 3 \text{ fb} & \text{(CMS)} \\ 10 \pm 3 \text{ fb} & \text{(ATLAS)} \end{cases} \tag{1}$$

In the following we assume that the excess in the s-channel process $pp \to \Phi \to \gamma\gamma$ is produced by gluon fusion and subsequent decay to two photons through t, b quarks and $W^{\pm}$, $H^{\pm}$ boson loops. The possibilities of large $\tan\beta$ values, $\tan\beta \geq 30$, are excluded for M_{Φ}=750 GeV by experimentas searches in the channel $\Phi \to \tau^+\tau^-$ performed at the run I of the LHC [11], so only small and moderate $\tan\beta$ will be considered in the following evaluations.

In the infinitely small width approximation for the cross section times branching fraction

$$\sigma(gg \to \Phi \to \gamma\gamma) = \sigma(gg \to \Phi) \times \mathrm{BR}(\Phi \to \gamma\gamma), \tag{2}$$

where only main decay channels, $t\bar{t}$ and $b\bar{b}$ (see branching fractions in Fig.1), are taken into account

$$\mathrm{BR}(\Phi \to \gamma\gamma) \simeq \frac{\Gamma(\Phi \to \gamma\gamma)}{\sum_{f=t,b} \Gamma(\Phi \to f\bar{f})}. \tag{3}$$

Then partial widths [12]

$$\Gamma(\Phi \to \gamma\gamma) = \frac{G_\mu \alpha^2 M_\Phi^3}{128\sqrt{2}\pi^3} \left| \sum_i A_i(\tau_i) \right|^2, \tag{4}$$

$$\Gamma(\Phi \to f\bar{f}) = \frac{3 G_\mu m_f^2}{4\sqrt{2}\pi} g_{\Phi ff}^2 M_\Phi \beta_f^{p_\Phi}, \tag{5}$$

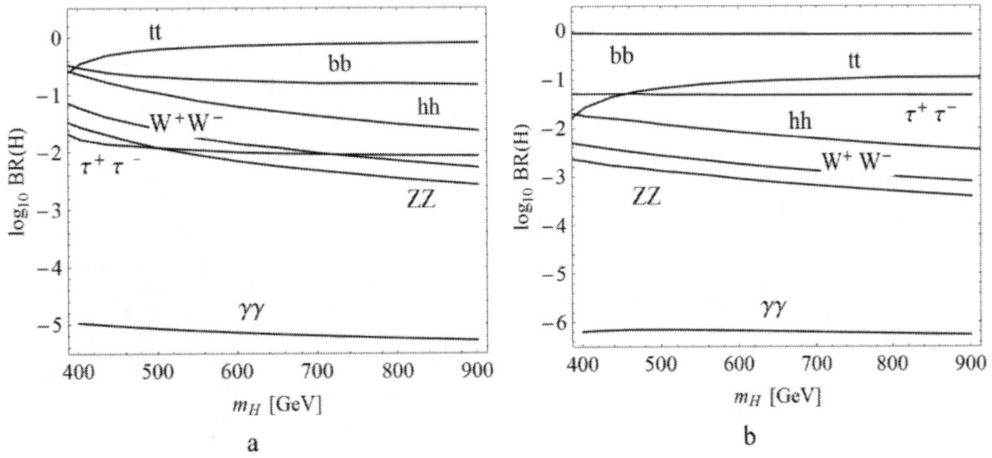

Figure 1. MSSM BR(H) in the decoupling limit for $\tan\beta = 4$ (a) and $\tan\beta = 10$ (b). Here $M_S = 1.5$ TeV, $\mu = 1$ TeV and $\tan\beta = 4$, $A = 2.45M_S$ (a); $\tan\beta = 10$, $A = 1.5M_S$ (b).

where $\beta_f = (1 - 4m_f^2/M_\Phi^2)^{1/2}$ is the quark velocity and p_Φ=3 (1) for the *CP*-even (odd) Higgs boson. The amplitudes of various contributions are

$$A_f = N_c Q_f^2 g_{\Phi ff} F_{1/2}(\tau_f), \qquad A_{W^\pm} = g_{\Phi W^+ W^-} F_1(\tau_{W^\pm}), \qquad A_{H^\pm} = g_{\Phi H^+ H^-} \frac{m_W^2}{m_{H^\pm}^2} F_0(\tau_{H^\pm}) \tag{6}$$

with 1/2, 1, 0 spin formfactor

$$F_{1/2}^H(\tau) = -2[\tau + (\tau - 1) f(\tau)]\tau^{-2}, \qquad F_{1/2}^A(\tau) = 2\tau^{-1} f(\tau), \tag{7}$$

$$F_0^H(\tau) = [\tau - f(\tau)]\tau^{-2}, \qquad F_1^H = [2\tau^2 + 3\tau + 3(2\tau - 1) f(\tau)]\tau^{-2},$$

where $\tau_f = M_\Phi/(4m_f^2)$ and the function $f(\tau)$ above the kinematical threshold is defined as

$$f(\tau) = -\frac{1}{4}\left(\log\frac{1 + \sqrt{1 - \tau^{-1}}}{1 - \sqrt{1 - \tau^{-1}}} - i\pi\right)^2, \qquad \tau > 1. \tag{8}$$

The production cross section at $\sqrt{s} = 13$ TeV for *CP*-even and *CP*-odd Higgs bosons of the mass M_Φ=750 GeV is $\sigma(gg \to H) \approx 0.85 \cot^2\beta$ pb and $\sigma(gg \to A) \approx 1.7 \cot^2\beta$ pb [13]. More precise calculation containing a combination of fixed order at exact NNLO plus approximate N^{3}LO and threshold resummation at N^{3}LL [14] does not change the picture significantly [15].

As a rule the cross section value $\sigma(gg \to H(A) \to \gamma\gamma)$ is too small ($\sim$ 0.015 fb for $\tan\beta = 1$ in the alignment limit), however the appropriate ranges are still viable for $\tan\beta \simeq 1$ and $\alpha \simeq -0.01$, see Fig.2. In the case of the MSSM, additional contributions provided by supersymmetric particles running in the loops, $\tilde{t}_i$ and $\tilde{\chi}_i^\pm$, do not change the results. Then the couplings $g_{hVV} \simeq 0.7$, $g_{huu} \approx 1.4$, $g_{hdd} \simeq 0.01$ are not SM-like any more. So partial breakdown of the alignment (and decoupling) limit leads to observable value of the THDM (MSSM) production cross section.

On the other hand most explanations of a new resonance at 750 GeV in the alignment limit require the *ad-hoc* introduction of new, additional particles to the spectrum [13]. The absence of a strong

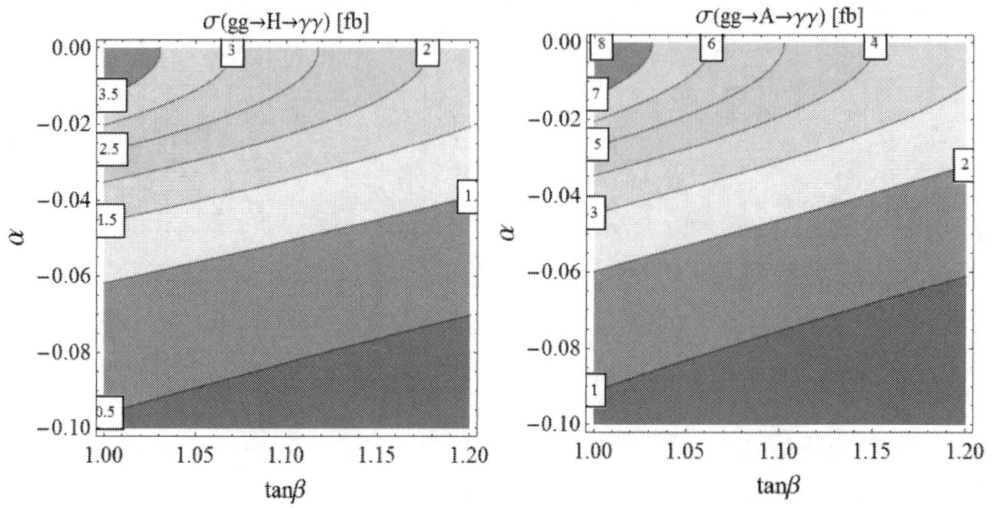

a b

Figure 2. The cross section $\sigma(gg \to \Phi \to \gamma\gamma)$ [fb] for $\Phi = H$ (a) and $\Phi = A$ (b), where m_Φ=750 GeV and $m_{H^\pm}$=1.07 TeV.

signal in the LHC collisions at 8 TeV motivates gluon fusion as the main production mechanism, presumably mediated by massive coloured fermions and/or bosons. Vector-like quarks or leptons strongly enhance the heavy Higgs couplings to photons and eventually gluons, without altering couplings of the already observed 125 GeV state.

3 Restricting the MSSM parameter freedom

In contrast to THDM, the MSSM is strongly constrained. In particular, masses and mixing angles are functions dependent on parameters of radiative corrections. The CP-even MSSM Higgs boson squared masses can be expressed as [16, 17]

$$m^2_{H,h} = \frac{1}{2}(m^2_A + m^2_Z + \Delta\mathcal{M}^2_{11} + \Delta\mathcal{M}^2_{22} \pm \sqrt{m^4_A + m^4_Z - 2m^2_A m^2_Z c_{4\beta} + C}), \tag{9}$$

$$m^2_{H^\pm} = m^2_W + m^2_A - \frac{v^2}{2}(\mathrm{Re}\Delta\lambda_5 - \Delta\lambda_4),$$

where

$$C = 4\Delta\mathcal{M}^4_{12} + (\Delta\mathcal{M}^2_{11} - \Delta\mathcal{M}^2_{22})^2 - 4(m^2_A + m^2_Z)\Delta\mathcal{M}^2_{12}s_{2\beta} - 2(m^2_A - m^2_Z)(\Delta\mathcal{M}^2_{11} - \Delta\mathcal{M}^2_{22})c_{2\beta}$$

and $\Delta\mathcal{M}^2_{ij}$, $\Delta\lambda$ are radiative corrections to the mass matrix elements and self-couplings. The identification of 125 GeV Higgs boson with the CP-even light boson h defines the squared mass of the CP-odd boson m^2_A (see Eq. (9))

$$m^2_A = \frac{m^2_h(C_1 - m^2_h) + m^2_Z(C_2 - C_3) - \Delta\mathcal{M}^2_{11}\Delta\mathcal{M}^2_{22} + \Delta\mathcal{M}^4_{12}}{C_1 - C_2 - C_3 + m^2_Z c^2_{2\beta}}, \tag{10}$$

where

$$C_1 = \Delta\mathcal{M}^2_{11} + \Delta\mathcal{M}^2_{22}, \qquad C_2 = m^2_h - \Delta\mathcal{M}^2_{12}s_{2\beta}, \qquad C_3 = \Delta\mathcal{M}^2_{11}s^2_\beta + \Delta\mathcal{M}^2_{22}c^2_\beta.$$

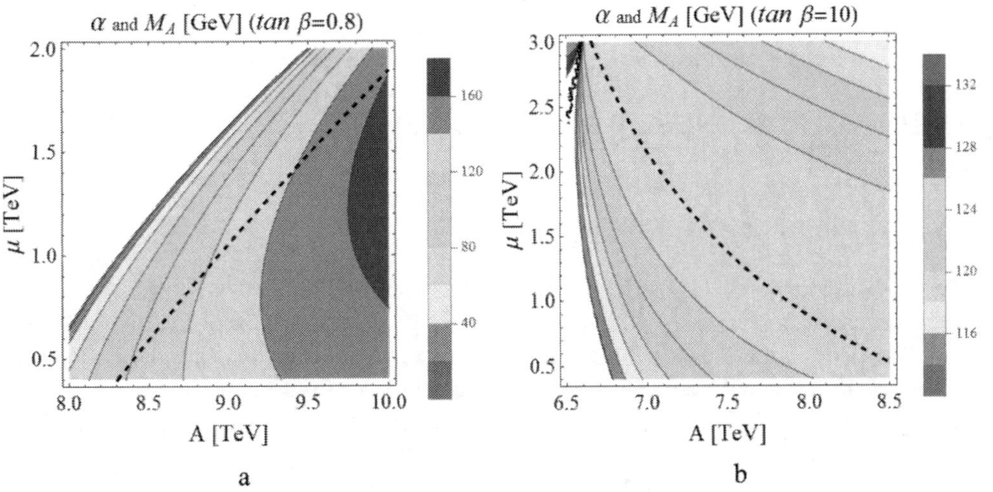

Figure 3. *CP*-odd Higgs mass and a requirement $\alpha \approx \beta - \pi/2$ (dashed black line) in (A, μ)-plane for small (a) and moderate $\tan\beta$ (b), $M_S = 2$ TeV.

Thus all Higgs masses can be unambiguously evaluated in the post-Higgs MSSM. The expression for mixing angle α can be written as [16]

$$\tan 2\alpha = \frac{2\Delta \mathcal{M}_{12}^2 - (m_Z^2 + m_A^2)s_{2\beta}}{(m_Z^2 - m_A^2)c_{2\beta} + \Delta \mathcal{M}_{11}^2 - \Delta \mathcal{M}_{22}^2}. \tag{11}$$

A requirement $\alpha \approx \beta - \pi/2$ and an additional assumption that $m_\Phi \sim 1$ TeV (in particular, 750 GeV) result in a rather significant constraints imposed on the MSSM paramer space. The following analysis is based on the calculation of radiative corrections in the MSSM Higgs sector by means of the effective potential method [18], taking into account carefully the nonleading soft supersymmetry breaking *D*-terms, the wave-function renormalization terms and also including the QCD and weak corrections to Yukawa couplings up to two loops [19]. Only top and bottom quarks and their superpartners in a loop are taken into account. In such a scenario the MSSM parameter space has the dimention four - $\tan\beta$, M_S and soft-breaking SUSY parameters A (in the general case $A_t \neq A_b$, for simplicity we assume $A_t = A_b$) and μ. These parameters respect the conditions

$$2|m_t A| < M_S^2, \qquad 2|m_t \mu| < M_S^2 \tag{12}$$

if we restrict the analysis of effective potential by the dimension-four operators. For numerical calculations we consider parameter sets $(M_S, \tan\beta)$, where $M_S = 1.5, 2, 5, 10$ TeV and $\tan\beta = 0.8, 2, 3, 10, 20$, and check general mass basis conditions for them

$$m_h = 125 \text{ GeV}, \qquad m_{H,A,H^\pm} > 0. \tag{13}$$

It turns out that the alignment limit $\alpha \approx \beta - \pi/2$ is satisfied in rather restricted domain of A and μ. With combination of mass basis conditions it resuts in almost degenerate mass spectrum $m_h = m_H$ and $m_H \approx m_A \approx m_H^\pm$. Characteristic structure is presented in Fig.3 for $M_S = 2$ TeV and $\tan\beta = 0.8$ and 10.

The possibility of $m_\Phi \sim 1$ TeV breaks the alignment limit. In this case the large $\tan\beta$ region for M_S up to 10 TeV is completely excluded. For M_S of about a few TeV the low $\tan\beta$ region is severely restricted. As a rule only two couplings $g_{hVV} \approx g_{huu} \approx 1$ are SM-like. The additional constraint of the σ value, see (1), results in unambiguously defined parametric sets A and μ for fixed M_S. For example, $A = 61.490$ TeV, $\mu = 19.334$ TeV for $M_S = 5$ TeV, and for $\tan\beta$=1 the corresponding values are $m_h = 125$ GeV, $m_H = 750$ GeV, $m_A = 893$ GeV, $m_{H^\pm} = 1073$ GeV, $\alpha = -0.0001$ and $g_{hVV} \simeq 0.7$, $g_{huu} \simeq 1.4$, $g_{hdd} \simeq 0.0001$.

4 Summary

A number of estimates for the post-Higgs THDM and the MSSM effective Higgs sector respecting the restrictions on the model parameter space inspired by the observation of the diphoton excess at an invariant mass of about 750 GeV have been considered. If the signal within these models is real and has a Higgs nature it will result in significant consequences for models parameter space. Either the models include the SM-like state h in the approximate alignment limit with additional charged particles introduced *ad-hoc*, or the alignment limit is broken and the state h is not SM-like fully or partially. In any case the allowed parameter space is strongly restricted.

In the case of minimal supersymmetry it follows from the estimates that the SM-like Higgs properties of the observed boson h in the stringent alignment limit exclude completely masses of the MSSM Higgs bosons of about 1 TeV and lead to close to degenerate mass spectrum $m_h = m_H, m_H \approx m_A \approx m_H^\pm$. Renunciation of this requirement results in satisfactory description of the CP-even h and H (or A) boson mass, width (for h) and cross section for the gg initial state in the case when there are deviations from the alignment limit and properties of the state h are not fully SM-like.

This work was supported by Russian Science Foundation, grant 16-12-10280.

References

[1] ATLAS collaboration, ATLAS-CONF-2015-081; ATLAS-CONF-2016-018

[2] A. Strumia, Interpreting the 750 GeV digamma excess: a review, arXiv:1605.09401[hep-ph]

[3] K. Bondarenko et al., Features in the Standard Model diphoton background, arXiv:1606.09592[hep-ph]

[4] L. Landau, Dokl. Akad. Nauk USSR, Ser.Fiz. **60**, 207 (1948); C.N. Yang, Phys. Rev. **77**, 242 (1950)

[5] G. Aad et al. (ATLAS Collaboration), Phys. Lett. B. **716**, 1 (2012); S. Chatrchyan et al. (CMS Collaboration), Phys. Lett. B.**716**, 30 (2012)

[6] M. Carena et al., Phys. Rev. D **91**, 035003 (2015)

[7] E. Bagnashi et al., LHCHXSWG-INT-2015-004

[8] P. Baratella et al., A closer look to the sgoldstino interpretation of the diphoton excess, arXiv:1603.05682v1[hep-ph]; K. O. Astapov, D. S. Gorbunov, Phys. Rev. D **93**, 035008 (2016)

[9] S. Heinemeyer, The 750 GeV diphoton excess and SUSY, arXiv:1605.08324v1[hep-ph]

[10] R. Franceschini et al., What is the $\gamma\gamma$ resonance at 750 GeV? arXiv:1512.04933[hep-ph]

[11] ATLAS collaboration, JHEP **11**, 056 (2014); CMS collaboration, JHEP **1410**, 160 (2014).

[12] A. Djouadi et al., The coupling of the lightest SUSY Higgs boson to two photons in the decoupling regime, arXiv:9612362v1[hep-ph]; A. Djouadi, Phys. Rept. **457**, 1-216 (2008); A. Djouadi et al., Fully covering the MSSM Higgs sector at the LHC, arXiv:1502.05653v2[hep-ph]

[13] A. Angelescu, A. Djouadi, Gr. Moreau, Scenarii for interpretations of the LHC diphoton excess: two Higgs doublets and vector-like quarks and leptons, arXiv:1512.04921v2[hep-ph]

[14] T. Ahmed et al., Pseudo-scalar Higgs boson production at N^3LO_A+N^3LL', arXiv:1606.00837v1[hep-ph]

[15] C. Anastasiou et al., CP-even scalar boson production via gluon fusion at the LHC, arXiv:1605.05761v1[hep-ph]

[16] Y. Okada, M. Yamaguchi, T. Yanagida, Phys. Lett. B **262**, 54 (1991); J. Ellis, G. Ridolfi, F. Zwirner, Phys. Lett. B **257**, 83 (1991); H.E. Haber, R. Hempfling, Phys. Rev. Lett. **66**, 1815 (1991); R. Barbieri, M. Frigeni, F. Caravaglios, Phys. Lett. B **258**, 167 (1991)

[17] A. Djouadi et al., Eur. Phys. J. C **73**, 2650 (2013); E. Bagnashi et al., LHCHXSWG-INT-2015-004

[18] E. Akhmetzyanova, M. Dolgopolov, M. Dubinin, Phys. Rev. D **71**, 075008 (2005); Fiz. Elem. Chast. Atom. Yadra **37**, 677 (2006)

[19] H. E. Haber, R. Hempfling, A. Hoang, Z. Phys. C **75**, 539 (1997); M. Carena, M. Quiros, C.E.M. Wagner, Nucl. Phys. B **461**, 407 (1996); S. Heinemeyer, W. Hollik, G. Weiglein, Eur. Phys. J. C **9**, 343 (1999); A. Pilaftsis, C.E.M. Wagner, Nucl. Phys.B **553**, 3 (1999)

Hadron Diffraction: Results and Problems

Vladimir Petrov[1,★]

[1] *Institute for High Energy Physics NRC Kurchatov Institute, 142281, Protvino, RF*

Abstract. This is a cursory review of diffractive studies at the LHC with physical interpretation of the experimental data and related problems of theory.

1 Introduction

What is hadron diffraction? We mean under this term a class of processes in hadron-hadron collisions which are characterized by specific angular distribution of the scattered particles similar to diffraction patterns in optics and by the mating feature: large rapidity gaps. The particles to be detected are produced at small angles w.r.t. the colliding beams. Why are such studies worthy of being made? Which fundamental problems of modern physics can be resolved with help of the corresponding data? In terms of characteristic spatio-temporal scales diffractive studies deal with distances and times much larger than those which characterize the majority of the LHC experiments, e.g. the Higgs or hard jets production. Moreover, this difference gets more drastic with the energy growth. When we say "larger" we mean "larger or even significantly larger than 1 fermi". Well, such distances are a routine feature in low-energy nuclear physics.Why expend energy to rediscover them again? The matter is that 1 fm at 100 MeV grossly differs from 1 fm at 1 TeV. In what follows we try to show this exposing the latest results obtained in diffractive studies at the LHC and trying to understand their physical meaning.

2 A Bit of History

One of the most claimed quantities in high-energy physics is total cross-section $\sigma_{\rm tot}$ as function of energy, $\sqrt{s} = 2\sqrt{p^2 + m^2}$ in the center-of-mass frame (p is the c.m.s momentum of the colliding protons and m is the proton mass). This quantity is often loosely described as "an effective area" (transverse to the beam(s)direction) where something happens with colliding hadrons. The formal definition is

$$\sigma_{\rm tot} = \sum_X \sigma_X, \tag{1}$$

where σ_X is the cross-section of any possible process $p + p \longrightarrow X$. "Optical theorem" states that the total cross section is simply related with the imaginary part of the amplitude of elastic scattering $p + p \longrightarrow p + p$:

$$\sigma_{\rm tot} \cdot 2p\sqrt{s} = \lim_{\theta \to 0} T(s,t), \tag{2}$$

★e-mail: Vladimir.Petrov@ihep.ru

where $t = -2p^2(1-\cos\theta) \approx p^2\theta^2$. We carefully use the symbol *lim* to stress that zero-angle scattering doesn't mean that "nothing happened". With no massless exchange particle a non-trivial limit exists. The optical theorem shows that the total cross-section is not only a geometrical quantity but is also defined by the intensity of interaction embodied in the amplitude. Some 50 years ago physicists believed that at very high energies total cross-sections achieve a limiting constant value naturally defined by a "strong interaction radius" $\sim m_\pi^{-1}$. The only dissident was W. Heisenberg who predicted $\sigma_{\rm tot} \sim \ln^2 s$. However his derivation was discarded by fellow theoreticians like Landau et al. for years to come. When in 1971 experiments at the 70 GeV "Serpukhov" proton synchrotron in Protvino showed that the cross-section of interaction of the K^+-meson with the proton *grows* with energy experimentalists decided that this is but a transitory effect and finally the constant value will be achieved. In two years the story had repeated at the CERN ISR , this time with the proton-proton cross-section and at energies five times higher . Physicists were very much embarrassed but desperate attempts of theoreticians to save the asymptotic constancy failed one after another. For the sake of fairness we have to notice that 18 years since the Heisenberg paper two bold theorists T. T. Wu and H. Cheng published a paper where they predicted, in the framework of some specific model, the asymptotic growth of the total cross-sections $\sim \ln^2 s$ though their arguments differed from those by Heisenberg. The experimental discovery of the growth of the total cross-sections was not the discovery of exactly logarithmic growth - other options had (and have) enough room - and it posed a difficult problem before the theorists: why and how do cross-sections grow? We will touch this problem below but now I would like to come directly to the latest available results from the LHC.

3 LHC: Diffractive Products[1].

3.1 Integral Cross-Sections: Total, Elastic and Inelastic .

Since the operation of the ISR no new data on the total cross-sections in pp interactions were available. Due to known similarity of pp and $\bar{p}p$ we relied on the results in $\bar{p}p$ obtained in $S\bar{p}pS$ at CERN and $TEVATRON$ at Fermilab. " For sake of poverty" the cosmic data were also in use in spite of huge errors. These partially surrogate data indicated that cross-sections continue to grow. So, generally it was not a great surprise when first LHC measurements revealed that σ_{tot}^{pp} have grown at energies two orders of magnitude higher than those at the ISR though the growth in cross-sections was only slightly more than two times. Nonetheless I would like to emphasize that this result cannot be considered as a routine one or so because even with account of the $\bar{p}p$ and cosmic ray data σ_{tot}^{pp} could go quite in different ways, up to the flattening. As $\sigma_{tot} = \sigma_{el} + \sigma_{inel}$ the growth of the total cross-section does not necessary lead to the growth of both σ_{el} and σ_{inel}. So the discovery made by the TOTEM Collaboration that both σ_{el} and σ_{inel} (confirmed afterwards by ATLAS and, in regard to inelastic cross-sections, by the CMS and ALICE Collaborations) grow seems quite non-trivial. Moreover, not only σ_{el} grows absolutely but the ratio $\frac{\sigma_{el}}{\sigma_{tot}}$ has also grown from 18 percent at the ISR to 27 percent at the LHC. In simple words, the more is the relative velocity of colliding protons the less they "want" to give up their energy for inelastic processes. If this trend will persist till the full dominance of elastic scattering or it will finally stop or changed for the opposite? This is a very exciting and pressing problem. Certainly, answers of the type "they grow because parameter $\Delta = 0.08$ " is not what we mean.

3.2 Interaction Radius Grows with Energy

Experiments at the ISR revealed the appearance of the dip in $\dfrac{d\sigma^{pp}}{dt}$ at $\approx$1.3 GeV^2 (the onset of it was "felt" already at the Serpukhov synchrotron). The dip (minimum) is a typical diffractive feature inherent to all ondulatory phenomena due to interference. Its position is inversely proportional to the square of something which can be vaguely called "interaction radius". TOTEM has found that the dip moved closer to zero and arrived to $\approx 0.5\ GeV^2$ at 7 TeV. Thus the interaction radius grows with energy. We have to keep in mind that this is not equivalent to the growth of the cross-sections. A similar phenomenon was observed by TOTEM for the so-called forward slope whose growth also signals the growth of the interaction radius. And again both phenomena are not the same.

3.3 $\frac{d\sigma^{pp}}{dt}$ at Small Transferred Momenta: Simple or Complicated?

Gaussian form $\exp Bt$ for $\frac{d\sigma}{dt}$ in many cases appeared quite adequate but already at the ISR energies it was noticed that more complicated form $\exp(Bt + Ct^2)$ is not ruled out. Such a conclusion has been strengthened by the observation made by TOTEM that at the level of 7 σ the t-distribution should be described by $\exp(Bt + Ct^2)$ or even by $\exp(Bt + Ct^2 + Dt^3)$. It is not easy to interpret such a finding in simple non-formal terms. We can only say at the moment that different forms shown above ere related with different forms of the "profiles" in the impact parameter space. Furthermore, it's probable that alternative forms can be invented to describe the data no worse than these exponentials or combinations of them.

3.4 How to Find out the Phase ?

Protons are electrically charged and so it is impossible to get rid off the ubiquitous photons both surrounding the colliding protons and exchanged between them. This leads to a specific phenomenon of the Coulomb-nuclear interference: there exists the region of t where Coulomb and strong ("nuclear") interactions are comparable. At lower transfers the Coulomb part absolutely dominates, at higher ones it dies-off giving the floor to the strong interactions. In the narrow interval of transfers the interference term is directly related to the phases of both participants. As to the Coulomb phase (without strong interactions) it can be calculated theoretically. Another case is the strong interaction phase. At the moment no feasible model exists for it. Those which dominated the "market" during several decades should be ruled out as theoretically inconsistent. So the measurements provided by the TOTEM Collaboration on this subject are very valuable and are directly related to the possibility of the extraction of the phase from the data (even if only in a limited interval of t and with a certain degree of model dependence). The information on the strong interaction phase is extremely important for our understanding of the spatial character of the high-energy collisions : are they more central or more peripheral?

4 Where is QCD?

Up to now all model predictions failed to describe the data on the differential cross-sections measured by TOTEM, especially near the dip and to the right of it. However, the failure of models (some or even all of them) is not a conceptual disaster : after some repair the models try to cope with data again and this process repeats itself already during several decades. Certainly we would like to have something more fundamental to compare with the data, something the failure or success of which would mean much for our world view. LHC experiments successfully test the Standard Model in its electroweak part with a crucial result on the Higgs existence. Strong interaction part of the Standard Model is embodied in quantum chromodynamics(QCD). Can we test QCD in diffraction experiments? Everybody knows that QCD is a weakly coupled theory when applied to the processes characterized by short distances with all pleasures of perturbative calculations. At which distances do occur diffraction processes? Alas, in this case the characteristic spatial scales, as was mentioned in the Introduction, are by no means small. Even worse, they grow with energy. And we are standing before the great problems of QCD at large distances which are very far from being resolved. As realists, we understand that it is quite silly to hope that QCD can give us some exact results on the scattering amplitude of such complicated composite objects like protons. But we can fairly hope to use QCD for obtaining at least a few important parameters, such as the intercepts and the slopes of leading Regge trajectories. This could help us to understand, both quantitatively and qualitatively, the gross features of diffractive processes at high energies.

And at last we, theorists, have to thank our fellow experimentalists for giving us so rich food for our minds.

Acknowledgements

I am grateful to the Organizers of QUARKS-2016 for invitation and vergy good organization of the seminar.

References

[1] TOTEM Collaboration, Europhys. Lett. **101**, 21002, 21004(2013) ;
Phys. Rev. Lett., **111**,012001 (2013);
ATLAS Collaboration, Nucl. Phys. **889B**, 486 (2014);
CMS Collaboration, Physics Letters **B722**, 5 (2013);
ALICE Collaboration, Eur. Phys. J. **C 73**, 2456 (2013) .

Precise determination of $Z - Z'$ mixing in charged gauge boson pair production at the CERN LHC

Alexander Pankov[1,*], *Andrei* Tsytrinov[1,**], and *Vadim* Bednyakov[2,***]

[1]*Abdus Salam ICTP Affiliated Centre and Gomel State Technical University, Gomel 246746, Belarus*
[2]*Laboratory of Nuclear Problems, Joint Institute for Nuclear Research, Dubna 141980 Russia*

Abstract. We quantify the expected sensitivity to Z' boson effects in the W boson pair production at the LHC. The diboson production allows to place stringent constraints on the $Z - Z'$ mixing angle. We find that the present LHC bounds on the mixing angle obtained at the LHC energy of 8 TeV and integrated luminosity of 20 fb^{-1} are of the same order as those derived from the electroweak data. Further improvement on the constraining of this mixing can be achieved from the analysis of data at the LHC with nominal energy and luminosity, 14 TeV and 100 fb^{-1}.

1 Introduction

Various new physics (NP) scenarios beyond the Standard Model (SM) [1], including superstring and left-right-symmetric models, predict the existence of new neutral gauge bosons, which might be light enough to be accessible at current and/or future colliders [2].

The search for these Z' particles is an important aspect of the experimental physics program of current and future high-energy colliders. Present limits from direct production at the LHC and virtual effects at LEP, through interference or mixing with the Z boson, imply that new Z' bosons are rather heavy and mix very little with the Z boson. Depending on the considered theoretical model, Z' masses of the order of 2.8–3.4 TeV [3, 4] and Z-Z' mixing angles at the level of a few per mil are excluded [5, 6]. The size of the mixing angle is strongly constrained by very high precision Z-pole experiments at LEP and the SLC [7]. They contain measurements from the Z line shape, from the leptonic branching ratios normalized to the total hadronic Z decay width and from leptonic forward-backward asymmetries. A Z' boson, if lighter than about 5 TeV, could be discovered at the LHC [8, 9] with $\sqrt{s} = 14$ TeV in the Drell-Yan process

$$pp \to Z' \to \ell^+\ell^- + X \tag{1}$$

with $\ell = e, \mu$. The future e^+e^- International linear collider (ILC) with high c.m. energies and longitudinally polarized beams could indicate the existence of Z' bosons via its interference

*e-mail: pankov@ictp.it
**e-mail: tsytrin@rambler.ru
***e-mail: Vadim.Bednyakov@cern.ch

effects in fermion pair production processes, with masses up to about $6 \times \sqrt{s}$ while Z-Z' mixing will be constrained down to $\sim 10^{-4} - 10^{-3}$ in the process $e^+e^- \to W^+W^-$ [10].

After the discovery of a Z' boson at the LHC via the process (1), some diagnostics of its couplings and Z-Z' mixing needs to be done in order to identify the correct theoretical framework. In this paper we study the potential of the LHC to discover Z-Z' mixing effects in the process

$$pp \to W^+W^- + X \tag{2}$$

and compare it with that expected at the ILC.

The $W^\pm$ boson pair production process (2) is rather important for studying the electroweak gauge symmetry at the LHC. Properties of the weak gauge bosons are closely related to electroweak symmetry breaking and the structure of the gauge sector in general. In addition, the diboson decay modes of Z' directly probe the gauge coupling strength between the new and the standard-model gauge bosons. The coupling strength strongly influences the decay branching ratios and the natural widths of the new gauge bosons. Thus, detailed examination of the process (2) will both test the gauge sector of the SM with the highest accuracy and throw light on NP that may appear beyond the SM.

Direct searches for a heavy WW resonance have been performed by the CDF and D0 collaborations at the Tevatron. The D0 collaboration explored diboson resonant production using the $\ell\nu\ell'\nu'$ and $\ell\nu jj$ final states [11]. The CDF collaboration also searched for resonant WW production in the $e\nu jj$ final state, resulting in a lower limit on the mass of an RS graviton, Z' and W' bosons [5].

The direct WW resonance search by the ATLAS Collaboration using $l\nu l'\nu'$ final-state events in 4.7 fb^{-1} pp collision data at the collider energy of 7 TeV set mass limits on such resonances [12]. Also, the $l\nu jj$ final state allows to reconstruct the invariant mass of the system, under certain assumptions on the neutrino momentum from a W boson decay.

Here, we examine the feasibility of observing a Z' boson in the $W^\pm$ pair production process at the LHC, which in contrast to the Drell-Yan process (1) is not the principal discovery channel, but can help to understand the origin of new gauge bosons.

2 Z' models

There are many theoretical models which predict a Z' with mass possibly in the TeV range. Popular classes of models are represented by E_6-motivated models, the Left-Right Symmetric Model (LR), the Z' in an 'alternative' left-right scenario and the Sequential Standard Model (SSM), which has a heavier boson with couplings like those of the SM Z. Searching for Z' in the above models has been widely studied in the literature [2] and applied at LEP2, the Tevatron and the LHC. For the notation we refer to [10], where also a brief description can be found. The different models considered are: (i) Models related to the breaking of E_6, parametrized by a parameter β, familiar cases are the Z'_χ, Z'_ψ, Z'_η and Z'_I models; (ii) Left-right models, originating from the breaking down of an $SO(10)$ grand-unification symmetry, leading to a $Z'_{\rm LR}$; (iii) The sequential $Z'_{\rm SSM}$, which has couplings to fermions being the same as those of the SM Z.

The mass-squared matrix of the Z and Z' can have non-diagonal entries δM^2, which are related to the vacuum expectation values of the fields of an extended Higgs sector:

$$M^2_{ZZ'} = \begin{pmatrix} M^2_Z & \delta M^2 \\ \delta M^2 & M^2_{Z'} \end{pmatrix}. \tag{3}$$

Here, Z and Z' denote the weak gauge boson eigenstates of $SU(2)_L \times U(1)_Y$ and of the extra $U(1)'$, respectively. The mass eigenstates, Z_1 and Z_2, diagonalizing the matrix (3), are then obtained by the rotation of the fields Z and Z':

$$Z_1 = Z\cos\phi + Z'\sin\phi\,, \tag{4a}$$
$$Z_2 = -Z\sin\phi + Z'\cos\phi\,. \tag{4b}$$

Here, the mixing angle ϕ is expressed in terms of masses as:

$$\tan^2\phi = \frac{M_Z^2 - M_1^2}{M_2^2 - M_Z^2} \simeq \frac{2M_Z\Delta M}{M_2^2}\,, \tag{5}$$

where $\Delta M = M_Z - M_1 > 0$, M_Z being the mass of the Z_1 boson in the absence of mixing, i.e., for $\phi = 0$. Once we assume the mass M_1 to be determined experimentally, the mixing depends on two free parameters, which we identify as ϕ and M_2.

From (4), one obtains the vector and axial-vector couplings of the Z_1 and Z_2 bosons to fermions:

$$v_{1f} = v_f\cos\phi + v'_f\sin\phi\,, \tag{6a}$$
$$a_{1f} = a_f\cos\phi + a'_f\sin\phi\,,$$
$$v_{2f} = v'_f\cos\phi - v_f\sin\phi\,, \tag{6b}$$
$$a_{2f} = a'_f\cos\phi - a_f\sin\phi\,,$$

with $(v_f, a_f) = (g_L^f \pm g_R^f)/2$, and (v'_f, a'_f) similarly defined in terms of the Z' couplings. The fermionic Z' couplings can be found, e.g. in [10].

Analogously, one obtains according to the remarks above:

$$g_{WWZ_1} = \cos\phi\; g_{WWZ}\,, \tag{7a}$$
$$g_{WWZ_2} = -\sin\phi\; g_{WWZ}\,, \tag{7b}$$

where $g_{WWZ} = \cot\theta_W$.

3 Cross section

The cross section for the process (2) from initial quark-antiquark states can be written as

$$\frac{d\sigma_{q\bar{q}}}{dM\,dy\,dz} = K\frac{2M}{s}\sum_q [f_{q|P_1}(\xi_1)f_{\bar{q}|P_2}(\xi_2) + f_{\bar{q}|P_1}(\xi_1)f_{q|P_2}(\xi_2)]\,\frac{d\hat{\sigma}_{q\bar{q}}}{dz}\,. \tag{8}$$

Here, s is the proton-proton center-of-mass energy squared; $z = \cos\theta$ with θ the W^--boson–quark angle in the W^+W^- center-of-mass frame; y is the diboson rapidity; as well as $f_{q|P_1}(\xi_1, M)$ and $f_{\bar{q}|P_2}(\xi_2, M)$ are parton distribution functions in the protons P_1 and P_2, respectively, with $\xi_{1,2} = (M/\sqrt{s})\exp(\pm y)$ the parton fractional momenta; finally, $d\hat{\sigma}_{q\bar{q}}/dz$ are the partonic differential cross sections. In (9), the K factor accounts for next-to-leading order QCD contributions. For simplicity, we will use as an approximation a global flat value $K = 1.2$ both for the SM and Z' boson cases. For numerical computation, we use CTEQ-6L1 parton distributions [13]. Since our estimates will be at the Born level, the factorisation scale

μ_F enters solely through the parton distribution functions, as the parton-level cross section at this order does not depend on μ_F. As regards the scale dependence of the parton distributions we choose for the factorization scale the WW invariant mass, i.e., $\mu_F^2 = M^2 = \hat{s}$, with $\hat{s} = \xi_1 \xi_2 s$ the parton subprocess c.m. energy squared. We have checked that the obtained constraints presented in the following are not significantly modified when μ_F is varied in the interval $\mu_F/2$ to $2\mu_F$.

Taking into account the experimental rapidity cut relevant to the LHC experiments, ($Y_{\rm cut} = 2.5$), one should carry out the integration over the phase space in (9) determined as [14, 15]:

$$|y| \leq Y = \min\left[\ln(\sqrt{s}/M), Y_{\rm cut}\right] = \ln(\sqrt{s}/M), \tag{9}$$

where we do not consider low masses, $\ln(\sqrt{s}/M) < Y_{\rm cut}$. This leads to a cut in the production angle

$$|z| \leq z_{\rm cut} = \min\left[\tanh(Y_{\rm cut} - |y|)/\beta_W, 1\right] , \tag{10}$$

where $\beta_W = \sqrt{1 - 4M_W^2/\hat{s}}$ and M_W is the W boson mass.

The resonant Z' production cross section of process (2) needed in order to estimate the expected number of Z' events, can be derived from (9) by integrating its right-hand-side over z, the rapidity of the $W^\pm$-pair y and invariant mass M around the resonance peak $(M_R - \Delta M/2, M_R + \Delta M/2)$:

$$\sigma(pp \to W^+W^- + X) = \int_{M_R-\Delta M/2}^{M_R+\Delta M/2} dM \times \tag{11}$$
$$\times \int_{-Y}^{Y} dy \int_{-z_{\rm cut}}^{z_{\rm cut}} dz \frac{d\sigma_{q\bar{q}}}{dM\, dy\, dz} .$$

We adopt the parametrization of the experimental mass resolution ΔM in reconstructing the diboson invariant mass of the W^+W^- system, ΔM vs. M. (After integration over y, interference effects vanish.)

The parton level $W^\pm$ boson pair production can be described, within the gauge models discussed here, by the subprocesses

$$q\bar{q} \to \gamma, Z_1, Z_2 \to W^+W^-, \tag{12}$$

as well as t- and u-channel amplitudes.

The differential (unpolarized) cross section of process (12) can be written as:

$$\frac{d\hat{\sigma}_{q\bar{q}}}{dz} = \frac{1}{N_C} \frac{\beta_W}{32\pi\hat{s}} \sum_{\lambda,\lambda',\tau,\tau'} |F_{\lambda\lambda'\tau\tau'}(\hat{s},\theta)|^2 . \tag{13}$$

Here, N_C is the number of quark colors; $\lambda = -\lambda' = \pm 1/2$ are the quark helicities; the helicities of the W^- and W^+ are denoted by $\tau, \tau' = \pm 1, 0$. The helicity amplitudes $F_{\lambda\lambda'\tau\tau'}(\hat{s},\theta)$ are summarized in Ref. [14]. There $\hat{s}$, $\hat{t}$, $\hat{u}$ are the Mandelstam variables defined as $\hat{t} = M_W^2 - \hat{s}(1-\beta_W z)/2$, $\hat{u} = M_W^2 - \hat{s}(1+\beta_W z)/2$; $\Gamma_{1,2}$ are $Z_{1,2}$ boson decay widths; $g_{1,f}^\lambda = v_{1,f} - 2a_{1,f}\lambda$, $g_{2,f}^\lambda = v_{2,f} - 2a_{2,f}\lambda$; and $\gamma_W = \sqrt{\hat{s}}/2M_W$. In the t- and u-channel exchanges we account for the initial $q = u, d, s, c$, only the CKM favoured quarks in the approximation of unity relevant matrix element.

In evaluation of the total width Γ_2 of the Z_2 boson we take into account its decay channels into fermions and $W^{\pm}$ boson pair [16]:

$$\Gamma_2 = \sum_f \Gamma_2^{ff} + \Gamma_2^{WW} . \qquad (14)$$

Further contributions of decays involving Higgs and/or gauge bosons and supersymmetric partners (including sfermions), which are not accounted for in (14), could increase Γ_2 by a model-dependent amount typically as large as 50% [16]. For definiteness the Z_2 width Γ_2 is assumed to scale with the Z_2 mass $\Gamma_2 = (M_2/M_1)\Gamma_1 \approx 0.03\, M_2$. This scaling is what would be expected for the reference model SSM [17]. The $W^{\pm}$-pair invariant mass distribution $(d\sigma/dM)$ is calculated with the same parton distribution functions and event selection criterion as those used in Ref. [18].

4 Constraints on Z'

We focus on the WW production via intermediate Z' and subsequent purely leptonic decay of on-shell $W^{\pm}$, that will be probed at LHC:

$$pp \to WW + X \to l\nu l'\nu' + X \quad (l, l' = e \text{ or } \mu), \qquad (15)$$

and, we follow the analysis given in [14], to evaluate the main backgrounds and possible cuts to enhance the Z' signal to background ratio.

In our analysis, we denote by $N_{\rm SM}$ and $N_{Z'}$ the numbers of 'background' and 'signal' events, and we adopt the criterion $N_{Z'} = 2\sqrt{N_{\rm SM}}$ or 3 events, whichever is larger, as the minimum signal for reach at the 95% C.L.

Another process where one can search for a new diboson resonance such as the Z' is represented by the subsequent WW decay into an $l\nu jj$ final state, i.e., a charged lepton (electron or muon), large missing transverse momentum ($E_T^{\rm miss}$), and at least two jets,

$$pp \to WW + X \to l\nu jj + X. \qquad (16)$$

An advantage of that process is that it has a higher cross section with respect to the pure leptonic final states. Also, the $l\nu jj$ final state allows the reconstruction of the invariant mass of the WW system, under certain assumptions for the neutrino longitudinal momentum from a W boson decay. As a result, a sharper Z' signal can be obtained. On the other hand, this channel has large QCD backgrounds due to the Wjj production, as well as Zjj with Z decaying leptonically and one of the leptons being missed. Also, $t\bar{t}$ production contributes to the background. However, the large QCD background can be reduced by making use of the characteristic harder transverse momenta of the charged lepton and the jets in the Z' signal.

In Table 1, we collect our limits on the Z' parameters for the models listed in Section II. Also shown in Table 1 are the current limits on various Z' boson masses from the LEP2 and Tevatron from studies of diboson W^+W^- pair production. The limits on ϕ and M_2 at the Tevatron assume that no decay channels into exotic fermions or superpartners are open to the Z'. Otherwise, the limits would be moderately weaker. LEP2 constrains virtual and Z-Z' boson mixing effects by the angular distribution of W bosons. Table 1 shows that the limits on ϕ from the EW precision data are generally competitive with and in many cases stronger than those from the colliders, except for the ILC (1 TeV) and LHC (14 TeV) that possess

 DOI: 10.1051/epjconf/201612502020

Table 1. Sensitivity to Z-Z' mixing angle ϕ at 95% C.L.

Collider, process	$\|\phi\|\times$	Z'_χ $\|\phi\|$	Z'_ψ $\|\phi\|$	Z'_η $\|\phi\|$	$Z'_{\rm SSM}$ $\|\phi\|$	$M_{Z'}$ (TeV)
LEP2, $e^+e^- \to W^+W^-$	10^{-2}	6	15	50	7	≥ 1
Tevatron, $p\bar{p} \to W^+W^- + X$	10^{-2}	–	–	–	2	0.4–0.9
electroweak data	10^{-3}	1.6	1.8	4.7	2.6	–
ILC (0.5 TeV), $e^+e^- \to W^+W^-$	10^{-3}	1.5	2.3	1.6	1.2	≥ 3
ILC (1.0 TeV), $e^+e^- \to W^+W^-$	10^{-3}	0.4	0.6	0.5	0.3	≥ 3
LHC (8 TeV), $pp \to W^+W^- \to l\nu l'\nu'$	10^{-3}	–	–	–	5.2	3
LHC (14 TeV), $pp \to W^+W^- \to l\nu l'\nu'$	10^{-4}	4–8	3–6	3–6	5–9	3–4.5

high potential to improve substantially the current bounds on the Z-Z' mixing angle. We stress that these limits are highly complementary.

If a new Z' boson exists in the mass range $\sim$ 3–4.5 TeV, its discovery is possible in the Drell–Yan channel. Moreover, the detection of the $Z' \to W^+W^-$ mode is eminently possible and gives valuable information on the Z-Z' mixing. It might be the only mode other than the dileptonic one, $Z' \to l^+l^-$, that is accessible. Our results demonstrate that it might be possible to detect a new heavy Z' boson from the totally leptonic or semileptonic WW channels at the LHC. The LHC at nominal energy and integrated luminosity provides the best opportunity of studying a new heavy Z' through its WW decay mode and creates the possibility of measuring (or constraining) the Z-Z' mixing, thus providing insight into the pattern of symmetry breaking.

Acknowledgments

This research has been partially supported by the Abdus Salam ICTP (TRIL Programme), the Collaborative Research Center SFB676/1-2006 of the DFG at the Department of Physics of the University of Hamburg and the Belarusian Republican Foundation for Fundamental Research.

References

[1] K. A. Olive *et al.* [Particle Data Group Collaboration], Chin. Phys. C **38**, 090001 (2014).
[2] P. Langacker, Rev. Mod. Phys. **81**, 1199 (2009).
[3] The ATLAS collaboration, ATLAS-CONF-2015-070.
[4] CMS Collaboration, CMS-PAS-EXO-15-005.
[5] T. Aaltonen *et al.* [CDF Collaboration], Phys. Rev. Lett. **104**, 241801 (2010).
[6] V. V. Andreev and A. A. Pankov, Phys. Atom. Nucl. **75**, 76 (2012) [Yad. Fiz. **75**, 67 (2012)].

[7] S. Schael *et al.* [ALEPH and DELPHI and L3 and OPAL and SLD and LEP Electroweak Working Group and SLD Electroweak Group and SLD Heavy Flavour Group Collaborations], Phys. Rept. **427**, 257 (2006).
[8] M. Dittmar, A. S. Nicollerat and A. Djouadi, Phys. Lett. B **583**, 111 (2004).
[9] S. Godfrey and T. Martin, arXiv:1309.1688 [hep-ph].
[10] V. V. Andreev, G. Moortgat-Pick, P. Osland, A. A. Pankov and N. Paver, Eur. Phys. J. C **72**, 2147 (2012).
[11] V. M. Abazov *et al.* [D0 Collaboration], Phys. Rev. Lett. **107**, 011801 (2011).
[12] G. Aad *et al.* [ATLAS Collaboration], Phys. Lett. B **718**, 860 (2013).
[13] J. Pumplin, D. R. Stump, J. Huston, H. L. Lai, P. M. Nadolsky and W. K. Tung, JHEP **0207**, 012 (2002).
[14] V. V. Andreev, P. Osland and A. A. Pankov, Phys. Rev. D **90**, no. 5, 055025 (2014).
[15] P. Osland, A. A. Pankov, N. Paver and A. V. Tsytrinov, Phys. Rev. D **78**, 035008 (2008).
[16] A. A. Pankov and N. Paver, Phys. Rev. D **48**, 63 (1993).
[17] D. Benchekroun, C. Driouichi and A. Hoummada, Eur. Phys. J. direct C **3**, N3 (2001).
[18] N. Agarwal, V. Ravindran, V. K. Tiwari and A. Tripathi, Phys. Lett. B **690**, 390 (2010)

EPJ Web of Conferences **125**, 02021 (2016) DOI: 10.1051/epjconf/201612502021
QUARKS-2016

E_6 inspired composite Higgs model and 750 GeV diphoton excess

Roman Nevzorov[1,2,★] and *Anthony* Thomas[1]

[1]*ARC Centre of Excellence for Particle Physics at the Terascale and CSSM, Department of Physics, The University of Adelaide, Adelaide SA 5005, Australia*
[2]*Institute for Theoretical and Experimental Physics, Moscow 117218, Russia*

Abstract. In the E_6 inspired composite Higgs model (E_6CHM) the strongly interacting sector possesses an SU(6)$\times U(1)_B \times U(1)_L$ global symmetry. Near scale $f \gtrsim 10\,\text{TeV}$ the SU(6) symmetry is broken down to its SU(5) subgroup, that involves the standard model (SM) gauge group. This breakdown of SU(6) leads to a set of pseudo–Nambu–Goldstone bosons (pNGBs) including a SM–like Higgs and a SM singlet pseudoscalar A. Because of the interactions between A and exotic fermions, which ensures the approximate unification of the SM gauge couplings and anomaly cancellation in this model, the couplings of the pseudoscalar A to gauge bosons get induced. As a result, the SM singlet pNGB state A with mass around 750 GeV may give rise to sufficiently large cross section of $pp \to \gamma\gamma$ that can be identified with the recently observed diphoton excess.

1 Introduction

The discovery of the Higgs boson with mass $m_h \simeq 125\,\text{GeV}$ allows one to estimate the values of parameters of the Higgs potential. In the standard model (SM) the Higgs scalar potential is given by

$$V(H) = m_H^2 H^\dagger H + \lambda (H^\dagger H)^2 \,. \tag{1}$$

The 125 GeV Higgs mass corresponds to $m_H^2 \approx -(90\,\text{GeV})^2$ and $\lambda \approx 0.13$. At the moment current data does not permit to distinguish whether Higgs boson is an elementary particle or a composite state. Although the discovered Higgs boson can be composed of more fundamental degrees of freedom, the rather small values of $|m_H^2|$ and the Higgs quartic coupling λ indicate that the Higgs field may emerge as a pseudo–Nambu–Goldstone boson (pNGB) from the spontaneous breaking of an approximate global symmetry of some strongly interacting sector.

The minimal composite Higgs model (MCHM) [1] includes weakly–coupled elementary and strongly coupled composite sectors (for a recent review, see [2]). The weakly–coupled elementary sector involves all SM fermions and gauge bosons. The strongly coupled sector gives rise to a set of bound states that contains Higgs doublet and massive fields with the quantum numbers of all SM particles. These fields are associated with the composite partners of the quarks, leptons and gauge bosons. The elementary states couple to the composite operators of the strongly interacting sector leading to

★e-mail: roman.nevzorov@adelaide.edu.au

mixing between these states and their composite partners. In this framework, which is called partial compositeness, the couplings of the SM states to the composite Higgs are set by the fractions of the compositeness of these states. The observed mass hierarchy in the quark and lepton sectors can be accommodated through partial compositeness if the fractions of compositeness of the first and second generation fermions are quite small. In this case the flavour-changing processes and the modifications of the W and Z couplings associated with the light SM fermions are somewhat suppressed. At the same time, the top quark is so heavy that the right–handed top quark t^c should have sizeable fraction of compositeness.

The strongly interacting sector of the MCHM possesses global SO(5) $\times$ $U(1)_X$ symmetry that contains the $\mathrm{SU}(2)_W \times U(1)_Y$ subgroup. Near the scale f the SO(5) symmetry is broken down to SO(4) so that the SM gauge group remains intact, resulting in four pNGB states which form the Higgs doublet. The custodial global symmetry $\mathrm{SU}(2)_{cust} \subset \mathrm{SO}(4)$ allows one to protect the Peskin–Takeuchi $\hat{T}$ parameter against new physics contributions. Experimental limits on the parameter S imply that $m_\rho = g_\rho f \gtrsim 2.5\,\mathrm{TeV}$, where m_ρ is a scale associated with the masses of the set of spin-1 resonances and g_ρ is a coupling of these ρ–like vector resonances. This set of resonances, in particular, contains composite partners of the SM gauge bosons. Even more stringent bounds on f come from the non–observation of flavor changing neutral currents (FCNCs). In the composite Higgs models, adequate suppression of the non–diagonal flavour transitions can be obtained only if f is larger than 10 TeV. This bound on the scale f can be considerably alleviated in the models with additional flavour symmetries FS. In the models with $\mathrm{FS} = U(2)^3 = U(2)_q \times U(2)_u \times U(2)_d$ symmetry, the bounds that originate from the Kaon and B systems can be satisfied even for $m_\rho \sim 3\,\mathrm{TeV}$. In these models the appropriate suppression of the baryon number violating operators and the Majorana masses of the left–handed neutrino can be achieved if global $U(1)_B$ and $U(1)_L$ symmetries, which ensure the conservation of the baryon and lepton numbers to a very good approximation, are imposed. Thus the composite Higgs models under consideration are based on

$$\mathrm{SU}(3)_C \times \mathrm{SO}(5) \times U(1)_X \times U(1)_B \times U(1)_L \times \mathrm{FS}\,. \tag{2}$$

The couplings of the elementary states to the strongly interacting sector explicitly break the SO(5) global symmetry. As a consequence, the pNGB Higgs potential arises from loops involving elementary states. This leads to the suppression of the effective quartic Higgs coupling λ.

2 E_6 inspired composite Higgs model

In the E_6 inspired composite Higgs model ($\mathrm{E_6CHM}$) the Lagrangian of the strongly coupled sector is invariant under the transformations of an $\mathrm{SU}(6)\times U(1)_B\times U(1)_L$ global symmetry. The $\mathrm{E_6CHM}$ can be embedded into $N = 1$ supersymmetric (SUSY) orbifold Grand Unified Theories (GUTs) in six dimensions which are based on the $E_6 \times G_0$ gauge group [3]. (Different aspects of the E_6 inspired models with low-scale supersymmetry breaking were recently considered in [4]-[19].) Near some high energy scale, M_X, the $E_6\times G_0$ gauge group is broken down to the $\mathrm{SU}(3)_C\times \mathrm{SU}(2)_W\times U(1)_Y\times G$ subgroup where $\mathrm{SU}(3)_C \times \mathrm{SU}(2)_W \times U(1)_Y$ is the SM gauge group. Gauge groups G_0 and G are associated with the strongly interacting sector. Fields belonging to this sector can be charged under both the E_6 and G_0 (G) gauge symmetries. The weakly–coupled sector includes elementary states that participate in the E_6 interactions only. Due to the conservation of the $U(1)_B$ and $U(1)_L$ charges all elementary states with different baryon and/or lepton numbers are components of different bulk 27–plets, whereas all other components of these 27–plets have to acquire masses of the order of M_X. All fields from the strongly interacting sector reside on the brane where E_6 symmetry is broken down to the SU(6) that contains an $\mathrm{SU}(3)_C \times \mathrm{SU}(2)_W \times U(1)_Y$ subgroup. As a result at high energies the

Lagrangian of the composite sector respects SU(6) global symmetry. The SM gauge interactions violate this symmetry. Nevertheless, SU(6) can remain an approximate global symmetry of the strongly coupled sector at low energies if the gauge couplings of this sector are considerably larger than the SM ones.

As in most composite Higgs models, the global SU(6) symmetry in the E_6CHM is expected to be broken below scale f. Here we assume that it gets broken to SU(5) subgroup, so that the SM gauge group is preserved. Since E_6CHM does not possesses any extra custodial or flavour symmetry, the scale f must be much larger than the weak scale, i.e. $v \ll f$. In particular, the adequate suppression of the FCNCs requires $f \gtrsim 10\,\mathrm{TeV}$. The $SU(6)/SU(5)$ coset space includes eleven pNGB states that correspond to the broken generators $T^{\hat{a}}$ of SU(6). These pNGB states can be parameterised by

$$\begin{aligned} \Omega^T &= \Omega_0^T \Sigma^T = e^{i\frac{\phi_0}{\sqrt{15}f}} \left(C\phi_1 \quad C\phi_2 \quad C\phi_3 \quad C\phi_4 \quad C\phi_5 \quad \cos\frac{\tilde{\phi}}{\sqrt{2}f} + \sqrt{\frac{3}{10}} C\phi_0 \right), \\ C &= \frac{i}{\tilde{\phi}} \sin\frac{\tilde{\phi}}{\sqrt{2}f}, \qquad \tilde{\phi} = \sqrt{\frac{3}{10}\phi_0^2 + |\phi_1|^2 + |\phi_2|^2 + |\phi_3|^2 + |\phi_4|^2 + |\phi_5|^2}, \end{aligned} \tag{3}$$

where the SU(6) generators are normalised so that $\mathrm{Tr}T^a T^b = \frac{1}{2}\delta_{ab}$ and

$$\Omega_0^T = (0 \quad 0 \quad 0 \quad 0 \quad 0 \quad 1), \qquad \Sigma = e^{i\Pi/f}, \qquad \Pi = \Pi^{\hat{a}} T^{\hat{a}}.$$

In the leading approximation the Lagrangian, that describes the interactions of the pNGB states, can be written as

$$\mathcal{L}_{pNGB} = \frac{f^2}{2} \left| \mathcal{D}_\mu \Omega \right|^2. \tag{4}$$

The field ϕ_0 is real and does not participate in the $SU(3)_C \times SU(2)_W \times U(1)_Y$ gauge interactions. Five components of vector Ω, i.e $\tilde{H} \sim (\phi_1\ \phi_2\ \phi_3\ \phi_4\ \phi_5)$, form a fundamental representation of the unbroken SU(5) subgroup of SU(6). The components $H \sim (\phi_1\ \phi_2)$ transform as an $SU(2)_W$ doublet. Therefore H corresponds to the SM-like Higgs doublet. Three other components of $\tilde{H}$, i.e. $T \sim (\phi_3\ \phi_4\ \phi_5)$, transform as an $SU(3)_C$ triplet. In the E_6CHM neither H nor T carry baryon and/or lepton number.

The pNGB effective potential $V_{eff}(\tilde{H}, T, \phi_0)$ is induced by the interactions of the elementary states with their composite partners, which break SU(6) global symmetry. The analysis of the structure of this potential including the derivation of quadratic terms $m_H^2|H|^2$ and $m_T^2|T|^2$ in the composite Higgs models, which are similar to the E_6CHM, shows that there is a considerable part of the parameter space where m_H^2 is negative and m_T^2 is positive [20]–[21]. In this parameter region the $SU(2)_W \times U(1)_Y$ gauge symmetry gets broken to $U(1)_{em}$, associated with electromagnetism, while $SU(3)_C$ colour is preserved. Because in the E_6CHM the scale $f \gtrsim 10\,\mathrm{TeV}$, a significant tuning, $\sim 0.01\%$, is required to get the appropriate value of the parameter m_H^2 that results in a 125 GeV Higgs state.

Since in the E_6CHM all states in the strongly interacting sector fill complete SU(5) representations the corresponding fields contribute equally to the beta functions of the $SU(3)_C$, $SU(2)_W$ and $U(1)_Y$ interactions in the one–loop approximation. As a consequence the convergence of the SM gauge couplings is determined by the matter content of the weakly–coupled sector. In this case, approximate gauge coupling unification can be achieved if the right–handed top quark, t^c, is entirely composite and the weakly–coupled elementary sector involves the following set of multiplets (see also [22]):

$$(q_i,\, d_i^c,\, \ell_i,\, e_i^c) + u_\alpha^c + \bar{q} + \bar{d^c} + \bar{\ell} + \bar{e^c} + \eta, \tag{5}$$

where $\alpha = 1, 2$ runs over the first two generations and $i = 1, 2, 3$ runs over all three. In Eq. (5) u_α^c, d_i^c and e_i^c represent the right-handed up- and down-type quarks and charged leptons, q_i and ℓ_i

correspond to the left-handed quark and lepton doublets, whereas $\bar{q}$, $\bar{d}^c$, $\bar{\ell}$ and $\bar{e}^c$ are exotic states which have opposite $SU(3)_C \times SU(2)_W \times U(1)_Y$ quantum numbers to the left-handed quark doublets, right-handed down-type quarks, left-handed lepton doublets and right-handed charged leptons, respectively. An additional SM singlet exotic state, η, with spin $1/2$ is included to ensure the phenomenological viability of the model under consideration. The set of elementary fermion states (5) is chosen so that the weakly–coupled sector contains all SM fermions except right–handed top quark and anomaly cancellation takes place. Using the one–loop renormalisation group equations (RGEs) one can find the exact gauge coupling unification is attained for $\alpha(M_Z) = 1/127.9$, $\sin^2\theta_W = 0.231$ and $\alpha_3(M_Z) \simeq 0.109$. The scale where the unification of the SM gauge couplings takes place is somewhat close to $M_X \sim 10^{15} - 10^{16}$ GeV. This estimation demonstrates that for the phenomenologically acceptable values $\alpha_3(M_Z) \simeq 0.118$ the SM gauge couplings can be reasonably close to each other at very high energies around $M_X \simeq 10^{16}$ GeV.

The scenario under consideration implies that the dynamics of the strongly coupled sector below the scale f leads to the composite $\mathbf{10} + \mathbf{\bar{5}} + \mathbf{1}$ multiplets of SU(5). Because of the conservation of the $U(1)_B$ and $U(1)_L$ charges all components of the 10–plet, i.e. Q, E^c and t^c, carry the same baryon and lepton numbers as the right–handed top quark t^c. The components of $\mathbf{\bar{5}}$ (D^c and L) and $\mathbf{1}$ ($\bar{\eta}$) can have baryon charges $-1/3$ and $+1/3$ [3]. It is expected that the composite multiplets Q, E^c, D^c, L and $\bar{\eta}$ get combined with the elementary exotic states $\bar{q}$, $\bar{e}^c$, $\bar{d}^c$, $\bar{\ell}$ and η, respectively, giving rise to a set of vector–like fermion states. The only exceptions are the components of the 10–plet associated with the composite right–handed top quark which survive down to the electroweak scale.

In the E_6CHM the lightest exotic fermion state has to be stable. Indeed, the baryon number conservation implies that the Lagrangian of the E_6CHM is also invariant under the transformations of the discrete Z_3 symmetry which can be defined as

$$\Psi \longrightarrow e^{2\pi i B_3/3}\Psi, \qquad B_3 = (3B - n_C)_{\text{mod }3}\,. \tag{6}$$

Here B is the baryon number of the given multiplet Ψ and n_C is the number of colour indices ($n_C = 1$ for $\mathbf{3}$ and $n_C = -1$ for $\mathbf{\bar{3}}$). This symmetry is called baryon triality [20]. All states in the SM have $B_3 = 0$. At the same time exotic fermion states carry either $B_3 = 1$ or $B_3 = 2$. As a result the lightest exotic state with non–zero B_3 can not decay into SM particles and should be stable. Since models with stable charged particles are ruled out by various experiments [23]-[24], the lightest exotic fermion in the E_6CHM must be neutral. It is also worth noting that the coupling of this neutral Dirac fermion to the Z–boson have to be extremely suppressed. Otherwise this stable exotic state would scatter on nuclei resulting in unacceptably large spin–independent cross sections. Thus, only a Dirac fermion, which is mostly a superposition of η and $\bar{\eta}$, can be the lightest exotic state in the E_6CHM.

3 750 GeV diphoton resonance

The SM singlet pNGB state ϕ_0 can be identified with the 750 GeV diphoton resonance recently reported by ATLAS and CMS. It is important that no 750 GeV resonance has been observed in other channels like $pp \to t\bar{t}, WW, ZZ, b\bar{b}, \tau\bar{\tau}$ and jj. This may be an indication that the detected signal is just a statistical fluctuation. At the same time, if these observations are confirmed this should set stringent constraints on the new physics models that may lead to such a signature. For example, in the E_6CHM the field ϕ_0 can mix with the Higgs boson which would result in large partial widths of the 750 GeV resonance associated with the decays of this state into pairs of Z-bosons, W-bosons and $t\bar{t}$. The corresponding mixing can be suppressed if invariance under the CP transformation is imposed. Indeed, in this case ϕ_0 manifests itself in the Yukawa interactions with fermions as a pseudoscalar field. In particular, the couplings of the SM singlet pNGB state $\phi_0 = A$ to the top quarks is induced by

$$\mathcal{L}_{AT} = \frac{y_t}{\Lambda_t} A(i\bar{t}_L H^0 t_R + h.c.) . \tag{7}$$

Because of the almost exact CP–conservation the mixing between the Higgs boson and pseudoscalar A is forbidden.

The Lagrangian that describes the interactions between A and exotic fermions can be written in the following form [25]

$$\mathcal{L}_{AF} = A\left(i\kappa_D \bar{d}^c D^c + i\kappa_Q \bar{q} Q + i\lambda_L \bar{\ell} L + i\lambda_E \bar{e}^c E^c + i\lambda_\eta \bar{\eta}\eta + h.c.\right) . \tag{8}$$

In the most general case the couplings κ_i and λ_i in Eq. (8) and the exotic fermion masses μ_i, induced below scale f, i.e.

$$\mathcal{L}_{mass} = \mu_D \bar{d}^c D^c + \mu_Q \bar{q} Q + \mu_L \bar{\ell} L + \mu_E \bar{e}^c E^c + \mu_\eta \bar{\eta}\eta + h.c. , \tag{9}$$

are entirely independent parameters, which are not constrained by the SU(6) and SU(5) symmetries. In order to get the cross section $\sigma(pp \to \gamma\gamma)$, which corresponds to the production and sequential diphoton decays of the pseudoscalar A, of about 5 − 10 fb we assume that μ_D, μ_Q, μ_L, μ_E and μ_η are larger than 375 GeV. As a result the on-shell decays of A into the exotic fermions, that result in the strong suppression of the branching ratios of the decays of this pNGB state into photons, are not kinematically allowed. Integrating out the exotic fermion states one obtains the effective Lagrangian that describes the interactions of the pseudoscalar A with the SM gauge bosons [25]

$$\mathcal{L}^A_{eff} = c_1 A B_{\mu\nu} \widetilde{B}^{\mu\nu} + c_2 A W^a_{\mu\nu} \widetilde{W}^{a\mu\nu} + c_3 A G^\sigma_{\mu\nu} \widetilde{G}^{\sigma\mu\nu} , \tag{10}$$

where $B_{\mu\nu}$, $W^a_{\mu\nu}$, $G^\sigma_{\mu\nu}$ are field strengths for the $U(1)_Y$, $SU(2)_W$ and $SU(3)_C$ gauge interactions, $\widetilde{G}^{\sigma\mu\nu} = \frac{1}{2}\epsilon^{\mu\nu\lambda\rho} G^\sigma_{\lambda\rho}$, $\widetilde{W}^{a\mu\nu} = \frac{1}{2}\epsilon^{\mu\nu\lambda\rho} W^a_{\lambda\rho}$, $\widetilde{B}^{\mu\nu} = \frac{1}{2}\epsilon^{\mu\nu\lambda\rho} B_{\lambda\rho}$, whereas

$$\begin{aligned}
c_1 &= \frac{\alpha_Y}{16\pi}\left[\frac{2\kappa_D}{3\mu_D} B(x_D) + \frac{\kappa_Q}{3\mu_Q} B(x_Q) + \frac{\lambda_L}{\mu_L} B(x_L) + 2\frac{\lambda_E}{\mu_E} B(x_E)\right] , \\
c_2 &= \frac{\alpha_2}{16\pi}\left[3\frac{\kappa_Q}{\mu_Q} B(x_Q) + \frac{\lambda_L}{\mu_L} B(x_L)\right] , \\
c_3 &= \frac{\alpha_3}{16\pi}\left[\frac{\kappa_D}{\mu_D} B(x_D) + 2\frac{\kappa_Q}{\mu_Q} B(x_Q)\right] , \\
B(x) &= 2x \arcsin^2[1/\sqrt{x}] , \qquad \text{for} \qquad x \geq 1 .
\end{aligned} \tag{11}$$

In Eq. (11) $x_D = 4\mu_D^2/m_A^2$, $x_Q = 4\mu_Q^2/m_A^2$, $x_L = 4\mu_L^2/m_A^2$, $x_E = 4\mu_E^2/m_A^2$, $m_A \simeq 750$ GeV is the mass of the SM singlet pNGB state A, $\alpha_Y = 3\alpha_1/5$ while α_1, α_2 and α_3 are (GUT normalised) gauge couplings of $U(1)_Y$, $SU(2)_W$ and $SU(3)_C$ interactions. Using Eqs. (10)–(11) one can obtain an analytical expression for the coupling of the pseudoscalar A to the electromagnetic field $F_{\mu\nu}$

$$\mathcal{L}^{A\gamma\gamma}_{eff} = c_\gamma A F_{\mu\nu} \widetilde{F}^{\mu\nu} , \qquad c_\gamma = c_1 \cos^2\theta_W + c_2 \sin^2\theta_W , \tag{12}$$

where θ_W is the weak mixing (Weinberg) angle and $\widetilde{F}^{\mu\nu} = \frac{1}{2}\epsilon^{\mu\nu\lambda\rho} F_{\lambda\rho}$.

Since at the LHC the pseudoscalar A is predominantly produced through gluon fusion the cross section $\sigma_{\gamma\gamma} = \sigma(pp \to A \to \gamma\gamma)$ can be presented in the following form [26]

$$\sigma_{\gamma\gamma} \simeq \frac{C_{gg}}{m_A s} \frac{\Gamma(A \to gg)\Gamma(A \to \gamma\gamma)}{\Gamma_A} \simeq 7.3\,\text{fb} \times \left(\frac{\Gamma(A \to gg)\Gamma(A \to \gamma\gamma)}{\Gamma_A m_A} \times 10^6\right) , \tag{13}$$

where $C_{gg} \simeq 3163$, $\sqrt{s} \simeq 13\,\text{TeV}$, Γ_A is a total width of the pseudoscalar A while partial decay widths $\Gamma(A \to \gamma\gamma)$ and $\Gamma(A \to gg)$ are given by

$$\Gamma(A \to gg) = \frac{2m_A^3}{\pi}|c_3|^2\,, \qquad \Gamma(A \to \gamma\gamma) = \frac{m_A^3}{4\pi}|c_\gamma|^2\,. \tag{14}$$

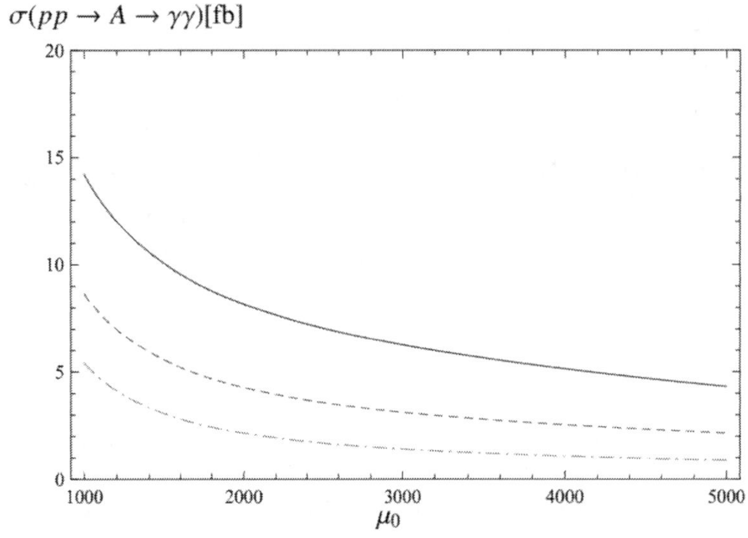

Figure 1. The cross section of $\sigma(pp \to A \to \gamma\gamma)$ is shown as a function of $\mu_Q = \mu_D = \mu_L = \mu_0$ for $\mu_E = 400\,\text{GeV}$ (solid lines), $\mu_E = 500\,\text{GeV}$ (dashed lines) and $\mu_E = 800\,\text{GeV}$ (dashed–dotted lines), for the case $\Lambda_t = 80\,\text{TeV}$ and $\kappa_D = \kappa_Q = \lambda_L = \lambda_E = \sigma = 1.5$.

First of all it is worthwhile to identify the scenario that leads to the suppression of the decay rates $A \to t\bar{t}, WW, ZZ, \gamma Z$ because no indication of the 750 GeV resonance has been observed in the channels associated with these decay modes. The analytical expressions for the corresponding partial decay widths can be presented in the following form

$$\Gamma(A \to t\bar{t}) = \frac{3m_A m_t^2}{8\pi\Lambda_t^2}\sqrt{1 - \frac{4m_t^2}{m_A^2}}\,, \tag{15}$$

$$\Gamma(A \to WW) = \frac{m_A^3}{2\pi}|c_2|^2\left(1 - \frac{4M_W^2}{m_A^2}\right)^{3/2}\,, \tag{16}$$

$$\Gamma(A \to ZZ) = \frac{m_A^3}{4\pi}\left|c_1 \sin^2\theta_W + c_2\cos^2\theta_W\right|^2\left(1 - \frac{4M_Z^2}{m_A^2}\right)^{3/2}\,, \tag{17}$$

$$\Gamma(A \to \gamma Z) = \frac{m_A^3}{8\pi}\sin^2 2\theta_W |c_1 - c_2|^2\left(1 - \frac{M_Z^2}{m_A^2}\right)^3\,. \tag{18}$$

In the model under consideration $\Lambda_t \simeq \sqrt{15}f$ if t^c is mainly a component of $\mathbf{20}'$ of SU(6). Since $f \gtrsim 10\,\text{TeV}$ the partial decay width (15) tends to be rather small, i.e. $\Gamma(A \to t\bar{t}) \lesssim \Gamma(A \to \gamma\gamma)$.

Here we set $\Lambda_t \simeq 80\,\text{TeV}$. The partial decay widths (16)–(18) become substantially smaller than $\Gamma(A \to \gamma\gamma)$ if $|c_2| \ll |c_1|$. The appropriate suppression of $|c_2|$ can be achieved when the exotic fermions that form $SU(2)_W$ doublets are considerably heavier than the $SU(2)_W$ singlet exotic states. On the other hand the non-observation of any new coloured particles with masses below 1 TeV at the LHC implies that the exotic coloured fermions in the E_6CHM should be rather heavy. Thus to simplify our numerical analysis we assume that $\mu_D = \mu_Q = \mu_L = \mu_0 \gtrsim \mu_E$ and $\kappa_D = \kappa_Q = \lambda_L = \lambda_E = \sigma$. For $\mu_0 \gg \mu_E$ the decay rates for $A \to t\bar{t}, WW, ZZ$ and $Z\gamma$ are very suppressed. However μ_0 cannot be too large, otherwise the LHC production cross section of the pseudoscalar A becomes smaller than $5-10\,\text{fb}$. In our numerical analysis we vary μ_0 from 1 TeV to 5 TeV. In this case A mainly decays into a pair of gluons. As a consequence $\Gamma_A \approx \Gamma(A \to gg)$ and the cross section (13) is determined by $\Gamma(A \to \gamma\gamma)$. Then the value of μ_E can be adjusted so that $(\Gamma(A \to \gamma\gamma)/m_A) \sim 10^{-6}$ leading to $\sigma(pp \to A \to \gamma\gamma) \simeq 5-10\,\text{fb}$.

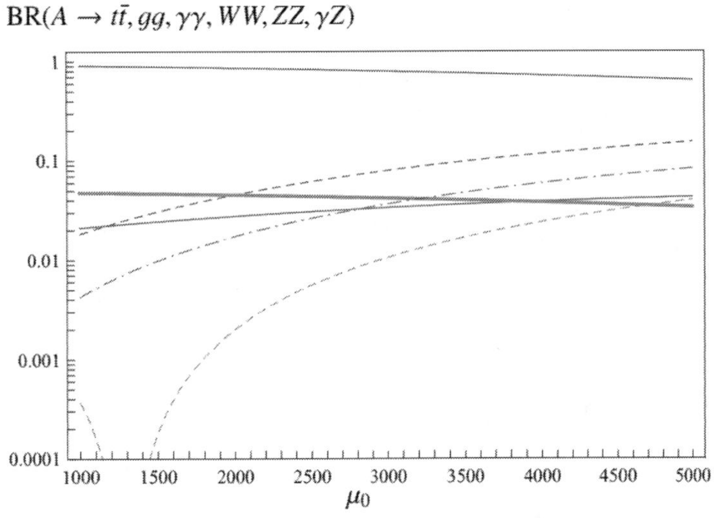

Figure 2. The branching ratios of the decays of the pseudoscalar A into $t\bar{t}$ (dashed–dotted lines), gg (highest solid lines), $\gamma\gamma$ (highest dashed lines), WW (thick solid lines), ZZ (lowest solid lines) and γZ (lowest dashed lines) are presented as a function of $\mu_Q = \mu_D = \mu_L = \mu_0$ for $\mu_E = 400\,\text{GeV}$, $\Lambda_t = 80\,\text{TeV}$ and $\kappa_D = \kappa_Q = \lambda_L = \lambda_E = \sigma = 1.5$.

Fig. 1 demonstrates that $\sigma(pp \to A \to \gamma\gamma)$ decreases very substantially when μ_E increases. For $\mu_E = 400\,\text{GeV}$ and $\sigma \gtrsim 1.5$ the cross section (16) can be about of 5 fb, even when all other exotic fermions are rather heavy $\mu_0 \simeq 5\,\text{TeV}$. At the same time for $\mu_E \simeq 700-800\,\text{GeV}$ the corresponding diphoton production cross section becomes sufficiently large only if $\mu_0 \simeq 1\,\text{TeV}$. When μ_0 changes from 5 TeV to 1 TeV the ratio Γ_A/m_A increases from 10^{-5} to 10^{-4} that corresponds to the variation of the total LHC production cross section of the pseudoscalar A from 100 fb to 1 pb.

In Fig. 2 the dependence of the branching ratios of the pseudoscalar A on μ_0 is explored for $\mu_E \simeq 400\,\text{GeV}$ and $\sigma = 1.5$. From this figure it follows that $A \to gg$ is the dominant decay channel. Its branching fraction is always close to 100%. When $\mu_0 \simeq 5\,\text{TeV}$ the branching ratio $\text{BR}(A \to \gamma\gamma)$ is the second largest one. $\text{BR}(A \to WW)$, $\text{BR}(A \to ZZ)$, $\text{BR}(A \to Z\gamma)$ and $\text{BR}(A \to t\bar{t})$ are substantially smaller than $\text{BR}(A \to \gamma\gamma)$. This might be a reason why the decays $A \to WW, ZZ, Z\gamma, t\bar{t}$ have not been detected yet. The decays $A \to gg$ can be rather problematic to observe because the total LHC production cross section of the pseudoscalar A is quite small. The branching fractions $\text{BR}(A \to ZZ)$ and $\text{BR}(A \to WW)$ increase while $\text{BR}(A \to \gamma\gamma)$ and $\text{BR}(A \to t\bar{t})$ decrease with decreasing μ_0.

When $\mu_0 \simeq 1\,\text{TeV}$ the branching ratios BR($A \to ZZ$) and BR($A \to WW$) are somewhat bigger than BR($A \to \gamma\gamma$). Nevertheless the experimental detection of $A \to ZZ$ and $A \to WW$ can be rather difficult since the W and Z bosons decay predominantly into quarks. In this scenario BR($A \to \gamma Z$) remains the lowest branching fraction and vanishes at some value of μ_0.

This work was supported by the University of Adelaide and the Australian Research Council through the ARC Center of Excellence in Particle Physics at the Terascale (CE 110001004) and through grant LF099 2247 (AWT).

References

[1] K. Agashe, R. Contino, A. Pomarol, Nucl. Phys. B **719**, 165 (2005).
[2] B. Bellazzini, C. Csáki, J. Serra, Eur. Phys. J. C **74**, 2766 (2014).
[3] R. Nevzorov, A. W. Thomas, Phys. Rev. D **92**, 075007 (2015).
[4] S. F. King, S. Moretti, R. Nevzorov, Phys. Rev. D **73**, 035009 (2006).
[5] S. F. King, S. Moretti, R. Nevzorov, Phys. Lett. B **634**, 278 (2006).
[6] S. F. King, S. Moretti, R. Nevzorov, AIP Conf. Proc. **881**, 138 (2007).
[7] S. F. King, S. Moretti, R. Nevzorov, Phys. Lett. B **650**, 57 (2007).
[8] S. F. King, R. Luo, D. J. Miller, R. Nevzorov, JHEP **0812**, 042 (2008).
[9] P. Athron, S. F. King, D. J. Miller, S. Moretti, R. Nevzorov, Phys. Lett. B **681**, 448 (2009).
[10] P. Athron, S. F. King, D. J. Miller, S. Moretti, R. Nevzorov, Phys. Rev. D **80**, 035009 (2009).
[11] J. P. Hall, S. F. King, R. Nevzorov, S. Pakvasa, M. Sher, Phys. Rev. D **83**, 075013 (2011).
[12] P. Athron, S. F. King, D. J. Miller, S. Moretti, R. Nevzorov, Phys. Rev. D **84**, 055006 (2011).
[13] P. Athron, S. F. King, D. J. Miller, S. Moretti, R. Nevzorov, Phys. Rev. D **86**, 095003 (2012).
[14] R. Nevzorov, Phys. Rev. D **87**, 015029 (2013).
[15] R. Nevzorov, S. Pakvasa, Phys. Lett. B **728**, 210 (2014).
[16] R. Nevzorov, Phys. Rev. D **89**, 055010 (2014).
[17] P. Athron, M. Mühlleitner, R. Nevzorov, A. G. Williams, JHEP **1501**, 153 (2015).
[18] P. Athron, D. Harries, R. Nevzorov, A. G. Williams, Phys. Lett. B **760**, 19 (2016).
[19] S. F. King, R. Nevzorov, JHEP **1603**, 139 (2016).
[20] M. Frigerio, J. Serra, A. Varagnolo, JHEP **1106**, 029 (2011).
[21] J. Barnard, T. Gherghetta, T. S. Ray, A. Spray, JHEP **1501**, 067 (2015).
[22] K. Agashe, R. Contino, R. Sundrum, Phys. Rev. Lett. **95**, 171804 (2005).
[23] J. Rich, M. Spiro, J. Lloyd–Owen, Phys. Rept. **151**, 239 (1987).
[24] T. K. Hemmick et al., Phys. Rev. D **41**, 2074 (1990).
[25] R. Nevzorov, A. W. Thomas, arXiv:1605.07313 [hep-ph].
[26] R. Franceschini et al., JHEP **1603**, 144 (2016).

EPJ Web of Conferences **125**, 02022 (2016) DOI: 10.1051/epjconf/201612502022
QUARKS-2016

Experiments on muon radiography with emulsion track detectors

Andrey Aleksandrov[1], *Alexander* Bagulya[1], *Sergei* Baklagin[2], *Mikhail* Chernyavsky[1], *Vladimir* Galkin[2], *Victor* Grachev[3], *Nina* Konovalova[1], *Alexander* Managadze[2], *Natalya* Polukhina[1,3,★], *Tatiana* Roganova[2], *Nikolai* Starkov[1], *Tatiana* Shchedrina[1], *Valeri* Tioukov[1], *Mykhailo* Vladymirov[1], and *Svetlana* Zemskova[1]

[1]*Lebedev Physical Institute, Russian Academy of Sciences, Moscow, 119991 Russia*
[2]*Skobeltsyn Institute of Nuclear Physics, Moscow State University, Moscow, 119991 Russia*
[3]*National Research Nuclear University MEPhI, Moscow, 115409 Russia*

Abstract. Muon radiography is a method of study the internal structure of large natural and industrial objects based on sensing an object with a flux of cosmic muons with their subsequent registration and analysis of the pattern of their dispersion, or conplete (or partial) absorption.The Lebedev Physical Institute of the Russian Academy of Sciences and the Skobeltsyn Institute of Nuclear Physics of Moscow State University have started a series of muon radiography experiments with nuclear emulsion detectors. As a result, the optimal conditions for experiment arrangement have been determined, algorithms of data processing have been worked out, and peculiarities of the method have been ultimately investigated.

1 The muon radiography method applications

Interest to the application of muon radiography (MR) for the purposes of the Earth sciences arose soon after the discovery of cosmic rays and, in particular, the cosmic muons. It is conditioned by the fact that the comic muon flux is sufficiently significant and stable in the cosmic radiation, and in addition, the cross section of the particle interactions at typical energies makes them the perfect agent for the tomography method, as they are able to cross hundreds of meters of rock with the absorption proportional to the traversed substance thickness [1]. Muon radiography applies the same basic principles as medical radiography, namely absorption of the beam (muon instead of an x-ray) crossing matter (rock or constructional material instead of living tissue) and registration with a sensitive device (detector). Advantages of this method are its non-invasiveness, i.e. no damage of the object material, and the natural source of radiation. Since muon is a charged particle, its detection occurs directly and is not associated with additional technical difficulties. Muons of cosmic rays with an energy of 1 TeV has path of 8.6 km water equivalent, what makes muon radiography applicable to the objects even a kilometer in size, with the prerequisite of their location above the detection level, because of the experiment specificity.

★e-mail: poluhina@sci.lebedev.ru

Figure 1. Radiographic image of the volcano Showa-Shinzan (Japan) below a volcanic crater floor. Top – the volcano external appearance; bottom – the density distribution of the breed in a vertical plane passing through the peak. Height of the object is about 1 km; size of detector is 0.4 m^2; exposure time is 60 days; resolution is about 50 mrad [6]

The most actual applications of muon radiography are geosciences [2, 3], geological exploration and mining research [4], alternative method of seismic research [5] and volcanology [6–10] (information about volcanic orifice and lava movement, see Fig. 1).

In addition, the method of muon radiography can be applied for nondestructive testing of massive industrial structures (bridges, dams, blast furnaces, etc.) [11] and for imaging of large archaeological objects. The first archaeological application on the method was performed in the 70-ies of the last century in an attempt to detect a hidden camera inside the body of one of the Egyptian pyramids (pyramid of Khafre) [12] and proved its feasibility and applicability, although the studies didn't show any presence of additional cameras. Using the similar approach with special emulsion photographic plates, in 2016 the international team of scientists received the first "muonographic" picture of the internal structure of the so-called "broken" pyramid of Sneferu (Bent Pyramid), the oldest pyramid of Ancient Egypt, and found in its body previously unknown secondary camera (camera 2 on Fig. 2) [13]. Located at the royal necropolis of Dahshur (around 26 miles from Cairo), this particular structure may be older than even the Great Pyramid, and is believed to be constructed by Old Kingdom Pharaoh Sneferu (circa 2600 BC).

Lately, the capabilities of muon radiography for nuclear reactor monitoring were actively developed [14, 15]. The need for such a research became apparent after the devastating accident at the Fukushima-1 nuclear power plant in Japan [16, 17] (see Fig. 3).

The muon radiography method modification can also be used for control over nuclear materials hidden in luggage or vehicles [18, 19].

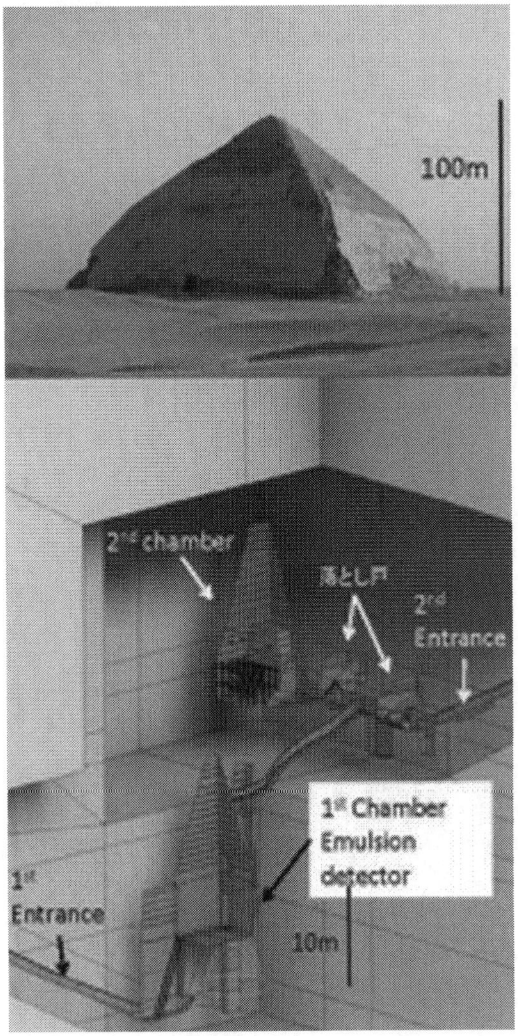

Figure 2. Observation of Bent Pyramid (2015-2016). General view and the scheme of muon detector inside the pyramid

2 Application of the emulsion track detectors in the muon radiography

In the world practice, studies of the internal structure of large objects by the muon radiography method is carried out either with electronic instrument, [21–24], or with track detectors on base of nuclear emulsions [25–27] (see Fig. 4). Electronic devices used for MR are typically rather bulky and difficult to operate. In contrast, emulsion track detectors are of small dimensions and simple construction, and currently have the high spatial (<1 μm) and angular (~ 1 MDA) resolution, large information capacity, ease of transportation and application in complex conditions (such as volcanoes or nuclear reactors). In addition, the most important advantages of nuclear emulsions are their independence from power supply sources and no need for on-line electronic reading systems. The detector size and number of emulsion layers are determined by the desired sensitivity of the experiment depending on the expected angular and linear dimensions of the studied object.

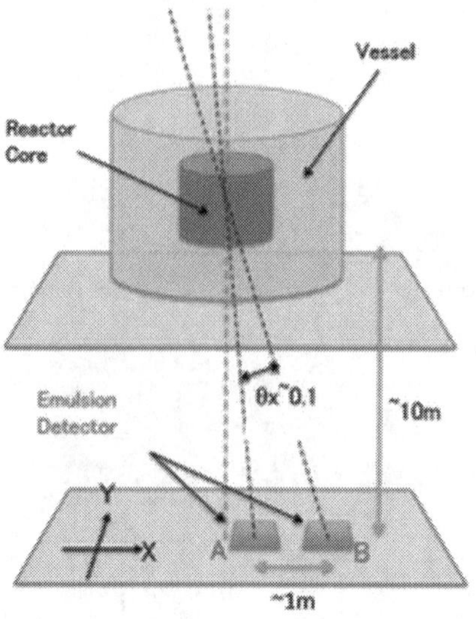

Figure 3. Scheme of the successfully implemented experiment with the "muonographic" inspection of a nuclear reactor Joyo (Japan). The diagram shows the placement of emulsion detectors to inspect the condition of an operational reactor using muon radiography [20]

In previous ten years in many countries (Japan, Italy, Switzerland, Canada, etc.) a significant number of projects for different massive object study with muon radiography method with the use of emulsion track detectors were performed. This approach became possible due to development of automated methods for emulsion data processing and the creation of high-speed scanning apparatus [28–31]. In Russia, the experiments of this type were not carried out up to date, in particular, because of the lack of a reliable manufacturer of nuclear emulsions. However, in the last years, the production of nuclear emulsions with characteristics required for the registration of relativistic particles was established in the Russian company "AVK Slavich" in Pereslavl-Zalessky, Yaroslavl region. It allowed the researches from Lebedev Physical Institute of the Russian Academy of Sciences (LPI) and the Skobeltsyn Institute of Nuclear Physics of Moscow State University (SINP MSU) to carry out the test experiments on the method application.

3 Test experiments in Russia on muon radiography method with the use of emulsion detectors

The LPI and SINP MSU research groups have carried out a number of test experiments with the aim of implementation of muon radiography method on base of emulsion track detectors. The object of the first study was a steel column of 23 tons (magnet yoke of SINP MSU cyclotron) used as a massive absorber of atmospheric muons creating a "shadow" in the particle flux (Fig. 5). In the experiment there were used emulsions of two types: the emulsions produced by Russian company "AVK Slavich" and the ones of Japanese company Fuji Photo Film analogous to those applied for the international experiment OPERA [32]. Detectors installed in the body of the steel column and outcide were exposed for 49 days.

DOI: 10.1051/epjconf/201612502022

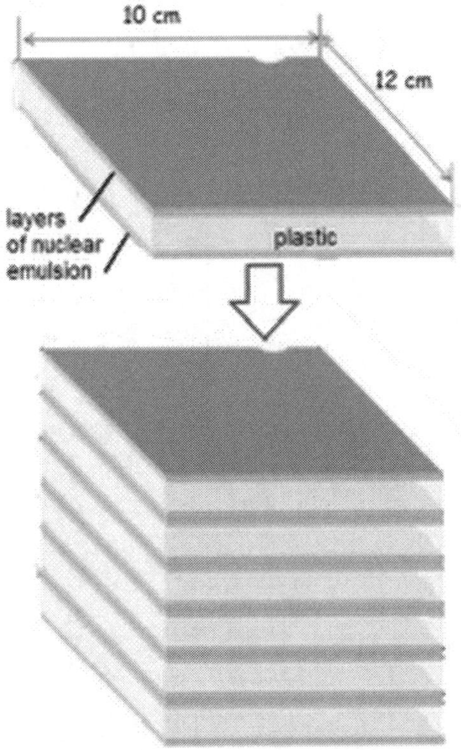

Figure 4. Scheme of a nuclear emulsion track detector

The exposed emulsions were analysed on automated measuring facilities available in LPI [33] and SINP MSU [34] equipped with optical tables of high accuracy and digital high-resolution cameras for recording and digitizing optical images and providing a high scanning speed of nuclear emulsions.

Result of each emulsion plate processing was a data array defining trajectory of each particle registered in the detector by its zenith and azimuthal angles applied in this technique. The data obtained in the automatic processing are subsequently subjected to the procedure of subtraction of the background tracks accumulated in the emulsion storage and transportation. The angular distribution of the registered muons obtained in the first test experiment demonstrated variations of muon fluxes coming to the detector in the different selected angle ranges, corresponding to the different path length in the iron absorber (column).

The second test experiment was held in the Institute of the tire industry in Moscow, the exposition of detectors was 135 days. The monitoring objects were the 40 tone steel wheels of a tire dino rollers (Fig. 6). The scheme of the emulsion track detector assembling was identical to the previous experiment, one detector included three packs with two emulsion layers in each.

Fig. 7 shows the experimental angular distribution of muons obtained in the detector located directly under the steel wheel surface. The result presentation was performed with the tangents of track angle projections

$$t_x = \frac{dx}{dz} = tg(\theta)\cos(\varphi)$$

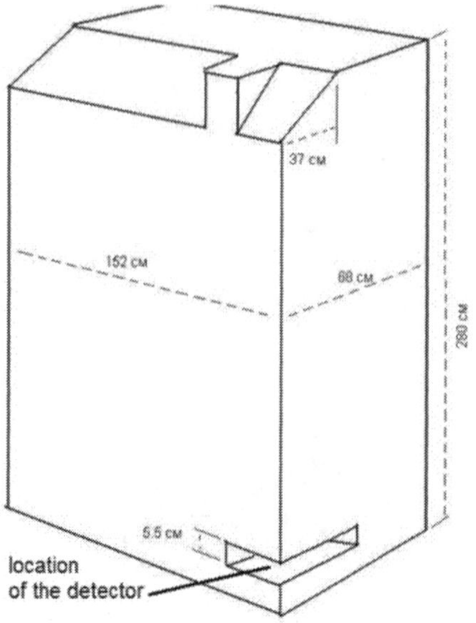

Figure 5. Steel column (part of a cyclotron) as a muon flux absorber. Location of the detector inside the column is shown

Figure 6. The object of study by MR method – a massive steel wheels of a tire dino rollers (40 tonnes, 3m in diameter). Arrows indicate the location of emulsion track detectors

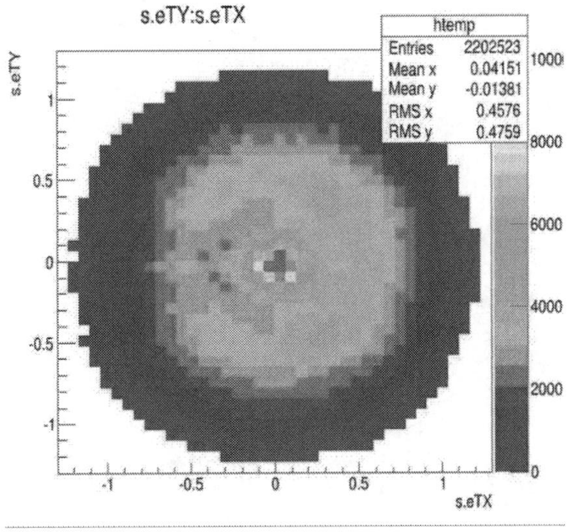

Figure 7. The angular distribution of muons registered in one of the nuclear emulsion detectors after their passage through the object (steel wheels of a tire dino rollers)

and

$$t_y = \frac{dy}{dz} = tg\,(\theta)\sin(\varphi)$$

the on the planes xz and yz, respectively. In this coordinates the obtained picture becomes close of the object photo image in "muon beams". The distribution presented on Fig. 7 clearly shows the "shadow" from the steel wheels.

The third experiment on the muon radiography method was set up at a depth of 30 m in the underground mine located on the territory of the Geophysical service of Russian Academy of Sciences in Obninsk. The experiment was aimed at registration of attenuation of the atmospheric muon flux by comparing the readings of detectors on the earth level and at a depth in the mine after muon passing through the soil layer. In the number of the experiment tasks there was included the possibility of the detection of a hollow in the surrounding soil (vertical elevator shaft) with a help of detector located at a depth of 30 m. In the direction to the elevator shaft (angles $\theta \sim 25°$ and $\varphi \sim 135°$ in the coordinate system indicated in Fig. 8) the absorption of muons is weakened by the presence of the cavity, and hence the flux of muons coming to the detector from this direction should significantly exceed the incoming one from other directions.

In order to determine the optimal exposure time sufficient to obtain the necessary information the detectors were exposed for different periods (2 and 4 months). The obtained experimental distributions of muons traversing the detector exhibited at the depth of 30 m within 4 months is shown in Fig. 9a. For comparison, Fig. 9b shows similar distribution obtained in the model calculations using GEANT4 toolkit [35].

To select signal muon tracks accumulated in the emulsion during the experiment from background ones, the background averaged over the azimuthal angle φ was subtracted from the original distribution (tx, ty). This method gives a better presentation for the difference in track density. The distribution in Fig. 9a clearly shows the signal from the elevator shaft (highlighted with a white frame). As well, the signals from other inhomogeneities in the soil layers above the detector are clearly seen in the picture. Peculiarities in the center of the distribution are associated with the presence of a lense of

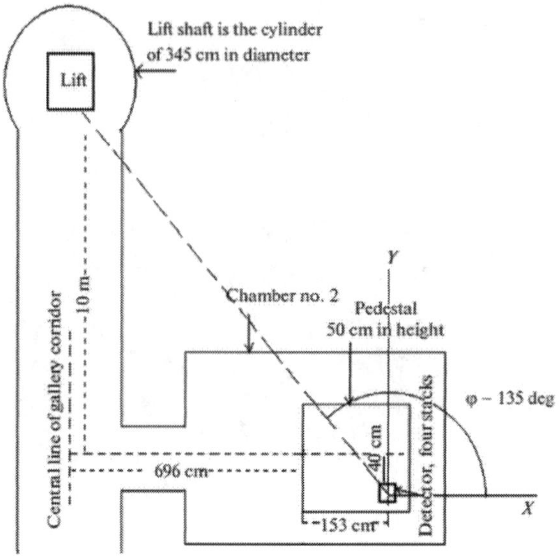

Figure 8. Scheme of relative location of elevator shaft and underground detectors (top view). Orientation of x and y axes in the detector system and the direction to the elevator shaft are shown

marble-like limestone around the geophysical mine. In the model calculations the specificity of the surrounding soil was also taken into account, as sown in Fig. 9b.

4 Conclusions

The analysis of the results obtained in muon radiography test experiments indicates that the emulsion track technique allows to obtain reliable data on the structure of investigated objects. The experimental data are in accordance with the predictions of the model calculations. The possibility of development of muon radiography in Russia is demonstrated using the emulsion track detectors of the proposed design and emulsion data processing facilities available at Russian institutes.

References

[1] K. Nagamine, *Introductory Muon Science*, (Cambridge University Press, Cambridge UK, 2003) 208 pp

[2] J. Marteau, D. Gibert, N. Lesparre, et al., Muons tomography applied to geosciences and volcanology, Nucl. Instr. Meth. Phys. Res. A **695**, 23–28 (2012)

[3] H.K.M. Tanaka, H. Muraoka, Interpreting muon radiographic data in a fault zone: possible application to geothermal reservoir detection and monitoring, Geosci. Instrum. Method. Data Syst. **2**, 145–150 (2013)

[4] D. Bryman, J. Bueno, J. Jansen, Blind Test of Muon Geotomography for Mineral Exploration, ASEG Extended Abstracts 2015(1), pp. 1-3

[5] H.K.M. Tanaka, H. Miyajima, T. Kusagaya, et al., Cosmic muon imaging of hidden seismic fault zones: rainwater permeation into the mechanical fractured zones in Itoigawa–Shizuoka Tectonic Line, Earth and Planetary Science Letters, **306**, 156-162 (2011)

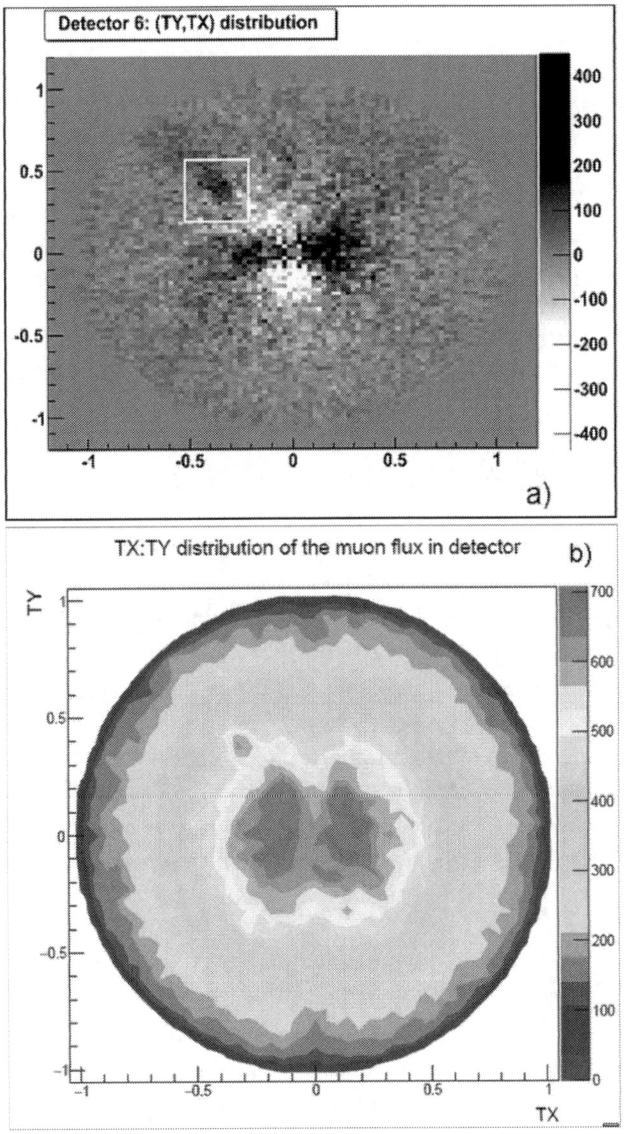

Figure 9. (a) 2D distribution (tx, ty) of muon flux at a depth of 30 m after subtracting of background, averaged by angle φ; the white square frame marks out the imprint of elevator shaft. (b) The same distribution, obtained in model calculation.

[6] H.K.M. Tanaka, T. Nakano, S. Takahashi, et al., Imaging the conduit size of the dome with cosmic-ray muons: The structure beneath Showa-Shinzan Lava Dome, Japan, Geophysical Research Letters, **34**, L22311, (2007)

[7] H.K.M. Tanaka, T. Nakano, S. Takahashi, et al., Radiographic imaging below a volcanic crater floor with cosmic-ray muons, American Journal of Science, **308**, 843–850, (2008)

[8] H.K.M. Tanaka, H. Taira, T. Uchida, et al., Three-dimensional computational axial tomography scan of a volcano with cosmic ray muon radiography, Journal of Geophysical Research, **115**,

B12332 (2010)

[9] F. Fehr, Density imaging of volcanos with atmospheric muons, Proc. ICRC2011, v. 4/12 (0671), p. 321 (2011)

[10] H.K.M. Tanaka, T. Kusagaya & H. Shinohara, Radiographic visualization of magma dynamics in an erupting volcano, Nature Communications, **5**, 3381 (2014)

[11] W.B. Gilboy, P.M. Jenneson, S.J.R. Simons, et al., Muon radiography of large industrial structures, Nucl. Instr. Meth. Phys. Res. B **263**, 317–319 (2007)

[12] L.W. Alvarez, J.A. Anderson, F. El Bredwei, et al., Search for Hidden Chambers in the Pyramids, Science, **167**, 832-839, (1970)

[13] http://www.realmofhistory.com/2016/04/28/egyptian-pyramid-interior-revealed-muon-particles/

[14] J Perry, M Azzouz, J Bacon, et al., Imaging a nuclear reactor using cosmic ray muons, Journal of Applied Physics **113** (18), 184909 (2013)

[15] J. M. Durham, E. Guardincerri, C. L. Morris, et al., Tests of cosmic ray radiography for power industry applications, AIP Advances, **5**, 067111 (2015)

[16] K. Borozdin, S. Greene, Z. Lukić, et al., Cosmic ray radiography of the damaged cores of the Fukushima reactors, Physical Review Letters, **109** (15), 152501 (2012)

[17] H. Miyadera, K.N. Borozdin, S.J. Greene, et al., Imaging Fukushima Daiichi reactors with muons, AIP Advances, **3** (5), 052133 (2013)

[18] C.L. Morris, C.C. Alexander, J.D. Bacon et al., Tomographic imaging with cosmic ray muons, Sci. Global Security, **16**, 37–53 (2008)

[19] M. Alamaniotis, S. Terrill, J. Perry, et al., A multisignal detection of hazardous materials for homeland security, Nuclear Technology & Radiation Protection, **24** (1), 46-55 (2009)

[20] K. Morishima, T. Nakano, N. Naganawa et al., Development on the Cosmic Ray Muon Radiography of the Nuclear Reactor with Nuclear Emulsion and an Appilcation Examination to the Fukushima Daiichi Nuclear Power Plant Accident, report on OPERA collaboration meeting, Nagoya, 28.03.2012

[21] G.V. Valery, High Resolution Radiography with Cosmic-ray Muons, A dissertation submitted to the Department of Physics, University of Surrey, 2010

[22] N. Lesparre, J. Marteau, Y. Declaiset, et al., Design and operation of a field telescope for cosmic ray geophysical tomography, Geosci. Instrum. Method. Data Syst., **1**, 33-42 (2012)

[23] C. Carloganu, V. Niess, S.Bene, et al., Towards a muon radiography of the Puy de Dome, Geosci. Instrum. Method. Data Syst., **2**, 55-60 (2013)

[24] H.K.M. Tanaka, Evaluation of positioning and density profiling accuracy of muon radiography by utillizing a 15-ton steel block, Geosci. Instrum. Method. Data Syst., **2**, 79-83 (2013)

[25] C.F. Powell, P.H. Fowler, D.H. Perkinset, *The study of elementary particles by the photographic method; an account of the principal techniques and discoveries, illustrated by an atlas of photomicrographs*, (Pergamon Press, London, New York, 1959)

[26] K. Morishima, Latest Developments in Nuclear Emulsion Technology, Physics Procedia, **80**, 19 – 24 (2015)

[27] T. Ariga, A. Ariga, K. Kuwabara, et al., Extra-large crystal emulsion detectors towards future large-scale experiments, JINST, **11**, P03003 (2016)

[28] L. Arrabito, E. Barbuto, C. Bozza, et al, Hardware performance of a scanning system for high speed analysis of nuclear emulsions, Nucl. Instrum. Meth. A **568**, 578-587 (2006)

[29] C. Bozza, T. Nakano, *Automatic microscopes for nuclear emulsion readout in high-energy and particle physics Current Microscopy Contributions to Advances in Science and Technology* (A.

Méndez-Vilas, Ed.) 2012

[30] T. Fukuda, S. Fukunaga, H. Ishida, Automatic scanning of nuclear emulsions with wide-angle acceptance for nuclear fragment detection, JINST, **8**, P01023 (2013)

[31] A. Alexandrov, A. Buonaura, L. Consiglio, et al., A new generation scanning system for the high-speed analysis of nuclear emulsions, Journal of Instrumentation, **11** (06), P06002-P06002 (2016)

[32] N. Agafonova, A. Aleksandrov, A. Anokhina, et al. (OPERA Collaboration), Search for oscillations with the OPERA experiment in the CNGS beam, JHEP **07** 004 (2013)

[33] A. Aleksandrov, L. Kashkarov, N. Polukhina, N. Starkov, The Pattern Recognition Software for Automatic Treatment of Track Detector Data at the PAVICOM Completely Automated Measuring Facility, Radiat. Meas., **43**, Suppl. 1, S120-S124 (2008)

[34] A.B. Aleksandrov, A.V. Bagulya, M.M. Chernyavsky, et al., Test Experiments on Muon Radiography with Emulsion Track Detectors in Russia, Physics Procedia, **80**, 78-80 (2015)

[35] S. Agostinelli , J. Allison, K. Amakoeet al., Geant4 - a simulation toolkit, Nucl. Instr. Meth. Phys. Res. A **506** (3), 250–303 (2003)

EPJ Web of Conferences **125**, 02023 (2016) DOI: 10.1051/epjconf/201612502023
QUARKS-2016

Dark matter at the SHiP experiment

Inar Timiryasov[1,2,⋆]

[1] *Institute for Nuclear Research of the Russian Academy of Sciences, 60th October Anniversary Prospect, 7a, 117312 Moscow, Russia*
[2] *Department of Particle Physics and Cosmology, Physics Faculty, Moscow State University, Vorobievy Gory, 119991 Moscow, Russia*

Abstract. We study prospects of dark matter searches in the SHiP experiment. SHiP (Search for Hidden Particles) is the recently proposed fixed target experiment which will exploit the high-intensity beam of 400 GeV protons from the CERN SPS. In addition to the hidden sector detector, SHiP will be equipped with the ν_τ detector, which presumably would be sensitive to dark matter particles. We describe appropriate production and detection channels and estimate SHiP's sensitivity for a scalar dark matter coupled to the Standard model through the vector mediator.

1 Introduction

We address the question of a possibility of dark matter (DM) search in the SHiP experiment. SHiP (Search for Hidden Particles) is the recently proposed [1, 2] fixed target experiment exploiting 400 GeV proton beam from the CERN SPS. The original motivation of the experiment [3] was the search for $O(1)$ GeV sterile neutrinos of the νMSM (see [4] for review of the model). Singlet fermions and active neutrinos are mixed and this mixing is responsible for both production of singlets in decays of heavy mesons (generated by protons on target) and subsequent singlet decays into SM particles (the main signature for the SHiP detector), see [5] for details. The flux of secondary charged particles (mostly muons) is suppressed by the very dense shielding placed downstream. The main idea is to place a large detector ($5 \times 10\,\text{m}^2 \times 50\,\text{m}$ [2]) as close to the target as possible (at a distance of about 60 m [2]) in order to maximize covered solid angle. This setup makes SHiP be a universal tool to probe any models of new physics containing light and long-lived particles which could be produced by protons on target and then decaying into SM particles [1, 6, 7].

The SHiP experiment will be equipped with a tau neutrino detector. In addition to it's main purpose — the first direct observation of $\bar{\nu}_\tau$ — the tau neutrino detector will be capable of observing light DM particles, produced in the beam dump. The possibility of this type searches was briefly discussed in Refs. [1, 2]. In what follows we quote relevant production rates and cross sections and estimate expected sensitivity of the SHiP.

⋆e-mail: timiryasov@inr.ac.ru

2 Dark matter candidate

Models of light dark matter (DM) usually contain light mediator particles. Such mediators are required to provide the realistic value of DM abundance in the Universe [8]. The presence of these mediators provides a possibility to produce a beam of DM particles at fixed target experiments. These DM particles, in turn, could be detected in neutrino detectors. SHiP setup contains all elements required to provide that type of DM search: a high-intensity proton beam, a target designed to suppress neutrino background and ν_τ detector consisting of OPERA-type bricks employing Emulsion Cloud Chamber (ECC) technology (see Ref. [2] for details). DM candidate assumed to be a complex scalar χ. These particles interact with SM particles through the exchange of vector boson A'. We assume that $m_{A'} > 2m_\chi$, so that decay $A' \to \chi\chi^\dagger$ is kinematically allowed and contributes to the invisible decay mode of A'.

The simplest, say, toy Lagrangian of this type with extra $U'(1)$ group reads:

$$\mathcal{L} = \mathcal{L}_{SM} - \frac{1}{4}F'_{\mu\nu}F'^{\mu\nu} + \frac{\epsilon}{2}F'_{\mu\nu}F^{\mu\nu} + \frac{m_{A'}^2}{2}A'_\mu A'^\mu + |D_\mu\chi|^2 - m_\chi^2|\chi|^2 \tag{1}$$

where χ and A' are DM and mediator fields, $F_{\mu\nu} = \partial_\mu A'_\nu - \partial_\nu A'_\mu$, $D_\mu = \partial_\mu + ie'A'_\mu$, e' is the $U'(1)$ coupling constant and ϵ is the parameter of kinematic mixing. In the Eq. (1) we have directly written mass term $m_{A'}$ for the vector mediator, however, we suppose that this mass term stems from the extra-Higgs mechanism in the dark sector. Note that this model, even in its renormalizable form with explicit extra-Higgs mechanism, is not free from anomalies. However, the model is attractive since the whole phenomenology is controlled only by four parameters: $\epsilon, e', m_{A'}, m_\chi$.

We consider scalar DM candidate since it exhibits p-wave annihilation, which is crucial to satisfying cosmological constraints. At the same time, fermion coupled through a massive vector implies s-wave annihilation. Thus it could be only sub-dominant component of a DM.

3 Estimate of dark matter flux

In the SHiP setup, DM particles χ could be produced directly in the proton-proton, or proton-neutron collisions via s-channel exchange of the hidden photon A'. If the hidden photon is sufficiently long-lived and $\Gamma(A') \ll m_{A'}$, then one can use the *narrow width approximation*. In this approximation χ production cross section factorizes,

$$\sigma(pN \to \chi\bar{\chi}) = \sigma(pN \to A')\mathrm{Br}(A' \to \chi\bar{\chi}), \tag{2}$$

and the flux of DM particles $\Phi_\chi = 2\Phi_{A'}\mathrm{Br}(A' \to \chi\bar{\chi})$. Therefore, in this case the problem of DM flux computation reduces to that of A' flux computation.

Despite the popularity of the vector mediator framework, there is no generally accepted method of its production rate computation. The Weizsäcker-Williams approximation was employed in Ref. [9] in order to estimate the flux of A' produced in the process of proton bremsstrahlung. This method deals with the entire proton. Therefore, in order to be sure that the method works properly one should introduce the proton form factors [10]. These form factors sharply drop (as q^{-4}, where q is 4-momentum transfer) for A' heavier than $O(1)$ GeV and lead to underestimation of the flux of heavy vector mediators.

In order to estimate the flux of hidden photons we employ results of Ref. [10], where two different mechanisms of A' production were considered:
1) proton bremsstrahlung

2) radiative decays of secondary mesons (π^0 and η).

The average energy of the DM particles, produced via the proton bremsstrahlung is

$$\langle E_{in} \rangle \simeq (170-190)\left(\frac{m_\chi}{m_{A'}}\right) \text{ GeV}, \quad \text{for } m_{A'} \text{ in range } 10^{-4}-5 \text{ GeV} \tag{3}$$

The production cross section (and the flux) is proportional to $\propto \epsilon^2$. The number of DM particles produced by 10^{20} protons on target via bremsstrahlung is shown in Fig. 1 as function of mediator mass $m_{A'}$ for a fixed value of $\epsilon = 10^{-4}$.

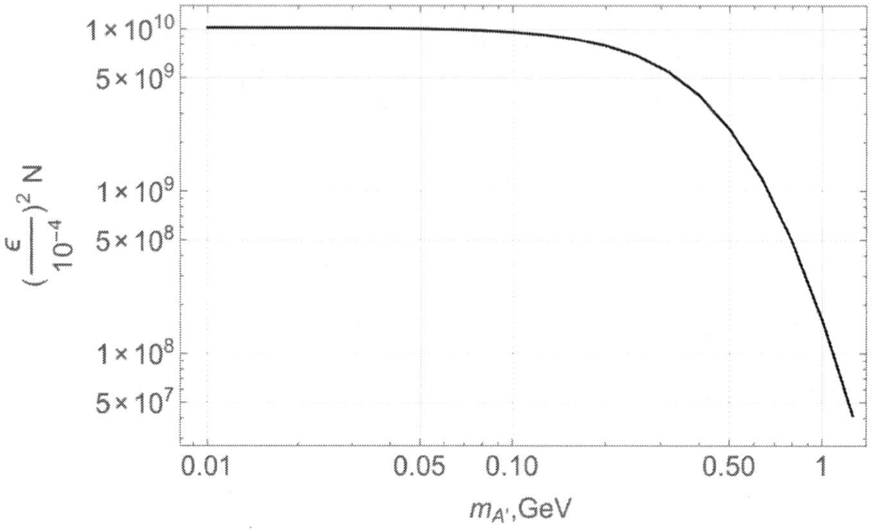

Figure 1. Number of DM particles N_χ which will traverse the detector (i. e. with $\theta < 0.025$, where θ is the angle between the direction of particle's 3-momenta and the beam direction) as function of the mediator mass $m_{A'}$ for $\epsilon = 10^{-4}$. We assume that $\text{Br}(A' \to \chi\bar{\chi}) \simeq 1$.

The mixing ϵ is responsible for A' decays into pairs of charged SM particles, while e' governs decay into DM scalars χ. The partial decay width into a lepton pair is given by

$$\Gamma_{A'}^{l^+l^-} = \frac{1}{3}\alpha\, m_{A'}\epsilon^2 \sqrt{1-\frac{4m_l^2}{m_{A'}^2}}\left(1+\frac{2m_l^2}{m_{A'}^2}\right), \tag{4}$$

where m_l is the mass of the lepton and α is the fine structure constant.

The partial decay width into scalar DM particles is given by

$$\Gamma_{A'}^{\chi\chi^\dagger} = \frac{1}{4}\alpha'\, m_{A'}\left(1-\frac{4m_\chi^2}{m_{A'}^2}\right)^{3/2}, \tag{5}$$

where $\alpha' \equiv e'^2/4\pi$. In what follows assume that $\alpha' \gg \alpha\epsilon^2$.

4 Dark matter scattering cross sections

Dark matter particles could scatter both on electrons and nucleons via the exchange of light mediator.

The elastic DM-electron scattering cross section has the following form [12]:

$$\frac{d\sigma_{\chi e\to\chi e}}{dE_e} = 4\pi\epsilon^2\alpha\alpha' \frac{2m_e E_{in}^2 - (2m_e E_{in} + m_\chi^2)(E_e - m_e)}{(E_{in}^2 - m_\chi^2)(m_{A'}^2 + 2m_e E_e - 2m_e^2)^2}, \tag{6}$$

where E_e is the energy of the recoil electron and E_{in} is the energy of the incident dark matter particle and $\alpha' \equiv e'^2/4\pi$.

The lowest order scalar DM-nucleon cross section was calculated in Ref. [13] and reads (for proton and neutron $N = p, n$, respectively):

$$\frac{d\sigma_{\chi N\to\chi N}}{dQ^2} = 4\pi\epsilon^2\alpha\alpha' \frac{F^2(Q^2)[q_N^2 A(E_{in}, Q^2) - \frac{1}{4}\kappa_N^2 B(E_{in}, Q^2)]}{2m_N(m_{A'} + Q^2)^2(E_{in}^2 - m_\chi^2)}, \tag{7}$$

where $Q^2 = 2m_N(E_{in} - E_\chi)$ is the momentum transfer, E_χ is the energy of the outgoing DM particle, form factor in the simplest form is $F = 1/(1 + Q^2/m_N^2)^2$ with $q_p = 1, q_n = 0, \kappa_p = 1.79$ and $\kappa_n = -1.9$. The functions A and B are defined as:

$$A(E_{in}, Q^2) = 2m_N E_{in}(E_{in} - Q^2/2m_N) - m_\chi^2 Q^2/2m_N,$$
$$B(E_{in}, Q^2) = (Q^2/2m_N)[(2E_{in} - Q^2/2m_N)^2 + Q^2 - 4m_\chi^2].$$

It is known (see, e.g. Fig 4.1 in Ref. [14]) that the neutrino-proton cross section is approximately two orders magnitude greater than the neutrino-electron cross section for neutrino energies greater than hundreds MeV. In order to determine the most relevant interaction pattern for the SHiP setup, we present the total cross sections of two processes described above in Fig. 2. To obtain these total cross sections we integrate the differential cross sections (6) and (7) over the allowed kinematical range (described in Appendix A) and apply an additional cut on the recoil energy of the target particle $E_T > E_{cut} = 1$ GeV.

As one can see from Fig. 2, $\chi - p$ total cross section is more than one order magnitude greater than that of $\chi - e$ scattering. However, one can see from Fig. 3, that the $\chi - p$ differential cross section is steeply falls with the Q^2. Therefore the resulting signal rate of the χ-nucleon scattering will depend on the assumed cut on the nucleons recoil energy E_{cut}.

5 Signal calculation

For the reason described above, we will consider only $\chi - e$ scattering and assume that the energy of the recoil electron $E_e > E_{cut} = 1$ GeV.

We consider production of DM particles χ in decays of hidden photons A': $A' \to \chi\chi^\dagger$. Hidden photons, in turn, could be produced via the proton bremsstrahlung or in decays of neutral mesons. These production channels were studied in Ref. [10] and we employ the flux of A' calculated in this work. We compute energy-angle distribution of χ for $m_\chi = 0.1$ MeV and for $m_{A'}$ in range $(10^{-4} - 1)$ GeV. Using this distribution we determine the signal rate:

$$N_{sig} = \int_{det} d\Omega \int dE_{in}\, n\sigma(E_{in})\, L_{det} \frac{dN_\chi}{dE_{in} d\Omega}, \tag{8}$$

where number of electrons per volume $n = ZN_A\rho/A$, and Z is atomic number, N_A is Avogadro's number, ρ is density of the medium and A is mass of a mole. We assume that the length of the lead

DOI: 10.1051/epjconf/201612502023

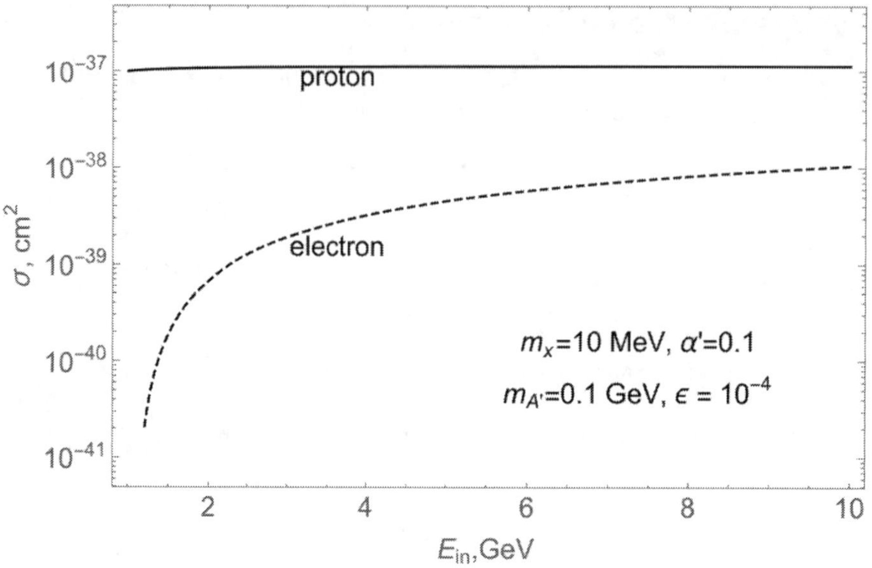

Figure 2. $\sigma_{\chi e\to\chi e}$ and $\sigma_{\chi N\to\chi N}$ as function of the incident χ energy. We have applied the cut on the final energy of the target (electron or proton): $E_T > 1$ GeV.

detector is $L_{det} = 100$ cm. The result for $N_{sig} = 10$ and 10^{20} protons on target is shown in Fig. 4. The dark gray region corresponds to A' produced via the proton bremsstrahlung while light gray region corresponds to A' produced in decays of secondary mesons.

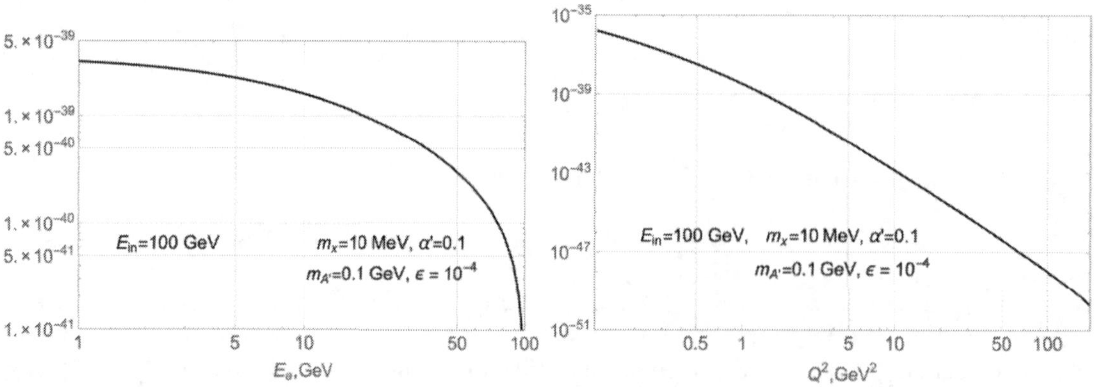

Figure 3. *Left panel*: differential cross section of χ - electron scattering (6) as function of recoil electron energy E_e. *Right panel*: differential cross section of χ - proton scattering (7) as function of the momentum transfer Q^2. The energy of incident DM particle $E_{in} = 100$ GeV. We have applied a cut on the final energy of the target (electron or proton): $E_T > 1$ GeV. Reference values of parameters, which were used in this plots, are not excluded by present experiments.

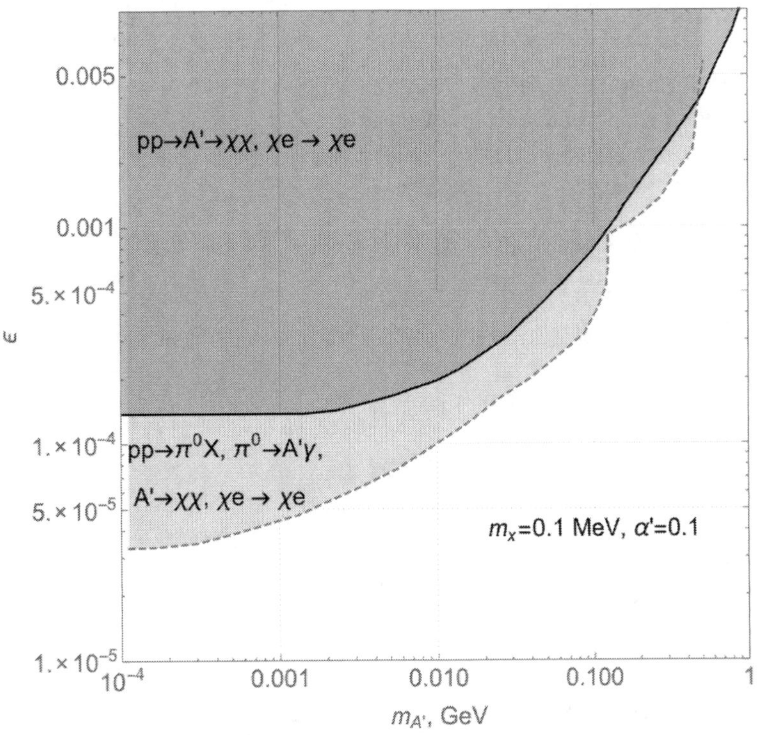

Figure 4. Estimate of SHiP's sensitivity to scalar DM χ (10 signal events, 10^{20} protons on target) in $m_{A'} - \epsilon$ plane. The dark gray region corresponds to A' produced via the proton bremsstrahlung while light gray region corresponds to A' produced in decays of secondary mesons.

6 Conclusions

To summarize, we have estimated the flux of light dark matter particles expected in the SHiP experiment and outlined the region in the model parameter space (see Fig. 4), where 10 χ-e scatterings with $E_e > 1$ GeV are expected for each 10^{20} protons on target. We assert that the SHiP experiment provides a possibility to search for light dark matter particles. The high energy of incident protons together with a huge statistics allows investigating yet not reached regions of the parameter space. A promising channel of dark matter - proton scattering deserves further studies. One needs to estimate, what is the minimal value of transferred momentum Q^2_{min} required to separate the signal from the background induced by elastic neutrino scatterings.

Acknowledgments

The author is indebted to D. S. Gorbunov for useful discussions and inspiration. The author is grateful to N. G. Polukhina and N. I. Starkov for valuable comments and interest to the work. This work has been supported by Russian Science Foundation grant 14-22-00161.

Appendix A: Kinematics

In order to calculate a total cross section of DM scattering one need to know an appropriate integration limits. In this Appendix we present some relevant details.

Consider the process $\chi(p_a) + T(p_b) \to \chi(p_1) + T(p_2)$, where T is a target particle of mass m_T, $T = e, p, n$. Standard Mandelstam variables reads:

$$\begin{aligned} s &= (p_a + p_b)^2 = (p_1 + p_2)^2 = m_\chi^2 + m_T^2 + 2m_T E_a, \\ t &= (p_a - p_1)^2 = (p_b - p_2)^2 = 2m_T^2 - 2m_T E_2, \\ u &= (p_a - p_2)^2 = (p_b - p_1)^2 = m_\chi^2 + m_T^2 - 2m_T E_1. \end{aligned} \tag{9}$$

From (9) and $s + t + u = 2m_\chi^2 + 2m_T^2$ one immediately obtains two useful forms for the momentum transfer:

$$\begin{aligned} Q^2 &= -t = 2m_T(E_a - E_1), \\ Q^2 &= -t = 2m_T(E_2 - m_T). \end{aligned} \tag{10}$$

The integration limits in terms of E_1 or E_2 are defined by the physical region of $2 \to 2$ scattering:

$$-\frac{\lambda(s, m_\chi^2, m_T^2)}{s} \le t \le 0, \tag{11}$$

where $\lambda(x, y, z) = (x - y - z)^2 - 4yz$ is ordinary Källén triangle function.

Requiring $E_2 > E_{cut}$ one also gets $t \le -2m_T(E_{cut} - m_T)$.

Finally,

$$\begin{aligned} E_a - \frac{2m_T}{s}(E_a^2 - m_T^2) \le &E_1 \le E_a + m_T - E_{cut} \\ E_{cut} \le &E_2 \le m_T + \frac{2m_T}{s}(E_a^2 - m_T^2) \end{aligned} \tag{12}$$

References

[1] S. Alekhin *et al.*, arXiv:1504.04855 [hep-ph].

[2] M. Anelli *et al.* [SHiP Collaboration], arXiv:1504.04956 [physics.ins-det].

[3] S. N. Gninenko, D. S. Gorbunov and M. E. Shaposhnikov, Adv. High Energy Phys. **2012** (2012) 718259 doi:10.1155/2012/718259 [arXiv:1301.5516 [hep-ph]].

[4] A. Boyarsky, O. Ruchayskiy and M. Shaposhnikov, Ann. Rev. Nucl. Part. Sci. **59** (2009) 191 doi:10.1146/annurev.nucl.010909.083654 [arXiv:0901.0011 [hep-ph]].

[5] D. Gorbunov and M. Shaposhnikov, JHEP **0710** (2007) 015 Erratum: [JHEP **1311** (2013) 101] doi:10.1007/JHEP11(2013)101, 10.1088/1126-6708/2007/10/015 [arXiv:0705.1729 [hep-ph]].

[6] D. Gorbunov and I. Timiryasov, Phys. Rev. D **92** (2015) no.7, 075015 doi:10.1103/PhysRevD.92.075015 [arXiv:1508.01780 [hep-ph]].

[7] K. O. Astapov and D. S. Gorbunov, Phys. Rev. D **93** (2016) no.3, 035008 doi:10.1103/PhysRevD.93.035008 [arXiv:1511.05403 [hep-ph]].

[8] M. Pospelov, A. Ritz and M. B. Voloshin, Phys. Lett. B **662** (2008) 53 doi:10.1016/j.physletb.2008.02.052 [arXiv:0711.4866 [hep-ph]].

[9] J. Blümlein and J. Brunner, Phys. Lett. B **731** (2014) 320 doi:10.1016/j.physletb.2014.02.029 [arXiv:1311.3870 [hep-ph]].

[10] D. Gorbunov, A. Makarov and I. Timiryasov, Phys. Rev. D **91** (2015) no.3, 035027 doi:10.1103/PhysRevD.91.035027 [arXiv:1411.4007 [hep-ph]].

[11] K. A. Olive *et al.* [Particle Data Group Collaboration], Chin. Phys. C **38** (2014) 090001. doi:10.1088/1674-1137/38/9/090001

[12] B. Batell, R. Essig and Z. Surujon, Phys. Rev. Lett. **113** (2014) no.17, 171802 doi:10.1103/PhysRevLett.113.171802 [arXiv:1406.2698 [hep-ph]].

[13] P. deNiverville, D. McKeen and A. Ritz, Phys. Rev. D **86** (2012) 035022 doi:10.1103/PhysRevD.86.035022 [arXiv:1205.3499 [hep-ph]].

[14] A. Strumia and F. Vissani, hep-ph/0606054.

EPJ Web of Conferences **125**, 03001 (2016) DOI: 10.1051/epjconf/201612503001
QUARKS-2016

Cosmology tests in rare kaon decays

Viacheslav Duk[1,⋆] for the NA62 collaboration⋆⋆

[1] *INFN Perugia (Italy) and INR RAS (Moscow, Russia)*

Abstract.
The Standard Model (SM) of particle physics is an extremely successful theory that effectively describes strong and electroweak interactions up to the energies presently accessible. Still, the SM does not explain the observed parameters of neutrino oscillations, baryon asymmetry of the Universe and Dark Matter (DM), and contains a fine-tuning of 16 orders of magnitude (the gauge hierarchy problem). Various New Physics (NP) models beyond the SM have been developed in order to address the above limitations. This paper concentrates on several models related to cosmology and their tests in rare kaon decays. In particular, recent NA48/2 results on the search for heavy neutrinos, light inflatons and dark photons are presented. Prospects for the ongoing NA62 experiment are discussed.

1 Introduction

For many years kaon physics has been a powerful tool both to explore the flavor sector of the SM and to search for NP. A very high experimental sensitivity (down to $O(10^{-12})$) comes from the possibility to create high-intensity kaon beams, clear decay signatures and low background environment. From the theoretical side, precise predictions are available for various processes.

NP manifestations in kaon sector could be indirect or direct. An example of the former is the deviation from the SM values in the branching ratios (BR) for rare decays (e.g. $K \to \pi\nu\bar{\nu}$, $K \to e\nu$) or form factor distributions (e.g. $K \to e\nu_e\pi^0$, $K \to \mu\nu_\mu\pi^0$). The direct manifestations include peak searches ($K \to eX$, $K \to \mu X$, $K \to \pi X$) where a new particle X escapes from the detector and decay searches where X decays to known particles (leptons, pions) that can be detected.

Among various NP scenarios that can be probed by kaon decays, we will concentrate on those related to cosmology. One of the simplest extensions of the SM is the νMSM model [1]. The model contains three additional leptons N_1, N_2 and N_3 and explains observed parameters of neutrino oscillations, baryon asymmetry of the Universe (BAU) and the existence of DM. The lightest lepton N_1 ($m \sim$ keV) constitutes DM, while N_2 and N_3 (m up to a few GeV) are responsible for neutrino oscillations and BAU and can be searched for in kaon decays $K \to lN (l = e, \mu)$.

Light inflaton models explain inflation in the early Universe and together with the νMSM form a complete extension of the SM [2, 3]. In this class of models an additional scalar particle χ coupling to the SM Higgs boson is introduced and can be searched for in the decay $K \to \pi\chi$.

⋆e-mail: Viacheslav.Duk@cern.ch

⋆⋆Birmingham, Bratislava, Bristol, Bucharest, CERN, Dubna, Fairfax, Ferrara, Florence, Frascati, Glasgow , Liverpool, Louvain, Mainz, Merced, Moscow, Naples, Perugia, Pisa, Prague, Protvino , Rome I, Rome II, San Luis Potosí, SLAC, Sofia, TRIUMF, Turin, Vancouver

Another example of models that can be probed in kaon decays is a class of models with extra U(1) gauge symmetry where the interaction between DM and the SM particles is mediated by a hypothetical gauge boson A' through the kinetic mixing [4]. A possible observation channel of this boson is the decay chain $K \to \pi\pi^0, \pi^0 \to \gamma A'$.

Recently several results on the topic have been obtained by the NA48/2 collaboration which is one of two most recent kaon experiments at CERN, the second one being the NA62 experiment currently running in the CERN North Area.

In the following sections a brief description of the NA48/2 setup is given, then the results on heavy neutrino, light inflation and dark photon search are presented. Finally, prospects for the NA62 experiment are discussed.

2 NA48/2 experiment

The NA48/2 is a fixed-target kaon experiment at CERN SPS that recorded a large data set of kaon decays in 2003–2004. The schematic view of the setup is shown in Fig. 1.

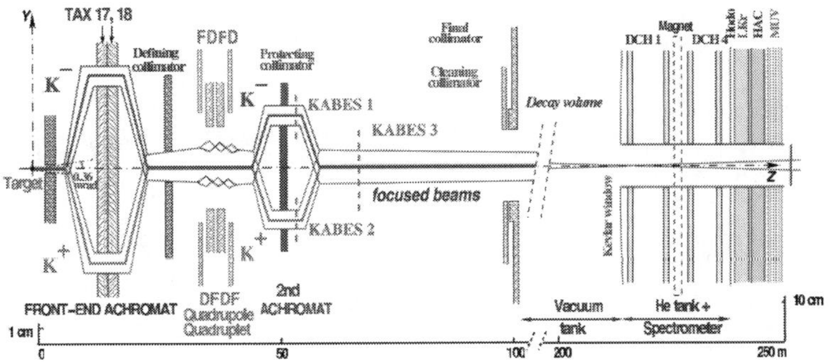

Figure 1. NA48/2 experimental setup.

The experiment used simultaneous beams of K^+ and K^- produced by 400 GeV/c protons from SPS hitting a beryllium target. Charged particles were selected by an achromate system containing four dipole magnets. The kaon momentum was 60±3 GeV/c. After the selection and collimation, the beam particles entered a 114 m long decay volume. The momentum of secondary particles was measured by the magnetic spectrometer located downstream the decay volume. The spectrometer consisted of a dipole magnet, two drift chambers upstream the magnet (DCH1, DCH2) and two chambers downstream it (DCH3, DCH4). The spectrometer was followed by a plastic scintillator hodoscope (Hodo) creating a fast trigger for charged particles passing trough it. The electromagnetic (LKr) and hadronic (HAC) calorimeters were used for the particle identification. They were followed by a muon detector (MUV). The detailed description of the setup can be found in [5].

3 Heavy neutrino search

Searches for heavy neutrinos are performed either by looking for the evidence of their production (a peak in the decays $K \to lN$, l = e, μ) or by detecting their decay products. The νMSM heavy neutral leptons N_2, N_3 can be searched for in the following processes: $K^\pm \to \mu^\pm N, N \to \pi^\pm\mu^\mp$ or $K^\pm \to \mu^\pm N, N \to \pi^\mp\mu^\pm$.

The NA48/2 has recently set limits on the $N_{2,3}$ decays to pions and muons [6]. In this analysis, 3-track decays were selected with the invariant mass close to the kaon one. Then, a peak search was performed in the $\pi^{\pm}\mu^{\mp}$ or $\pi^{\mp}\mu^{\pm}$ invariant mass. The obtained upper limits are shown in Fig. 2. In the first case there is an irreducible background $K^{\pm} \to \mu^{\pm}\pi^{\pm}\mu^{\mp}$, while for the second decay mode the background is much lower, because the decay $K^{\pm} \to \mu^{\pm}\pi^{\mp}\mu^{\pm}$ is prohibited in the SM due to lepton number violation.

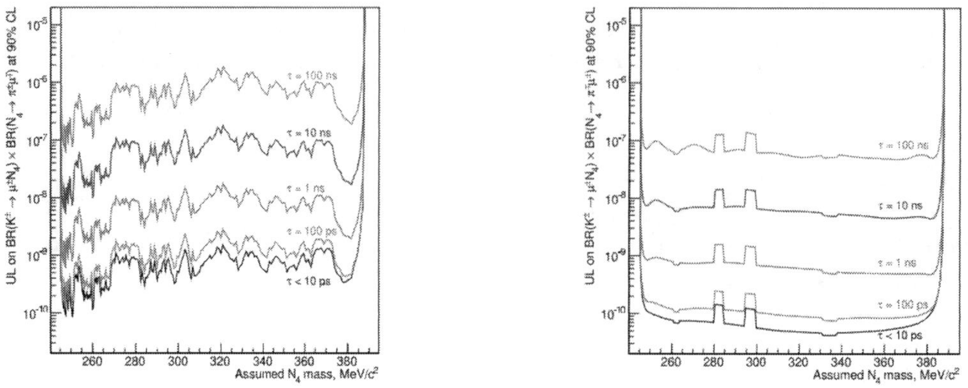

Figure 2. Upper limits on BR($K^{\pm} \to \mu^{\pm}N$, $N \to \pi^{\pm}\mu^{\mp}$) (left) and BR($K^{\pm} \to \mu^{\pm}N$, $N \to \pi^{\mp}\mu^{\pm}$) (right) for different lifetimes of N.

4 Light inflaton search

Light inflaton models contain two free parameters [2, 3]. The parameter values are constrained both by theory and experimental searches [3]. In the context of experimental tests it is straightforward to exploit the following three variables: inflaton mass m_{χ}, its lifetime τ_{χ} and the Higgs-inflaton squared mixing angle θ^2 characterizing the effective inflaton coupling to SM fields (and hence branching ratios of SM particle decays to inflaton). Any two parameters can be considered as free, e.g. for given τ_{χ} and m_{χ}, the value θ^2 is known. The expected inflaton decay modes and current limits on the inflaton phase space are shown in Fig. 3 (left).

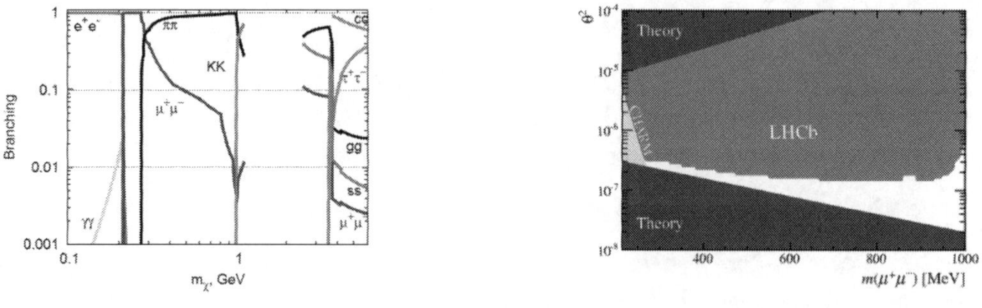

Figure 3. Light inflaton decay modes (left) and current limits on θ^2 (right).

The most promising processes for light inflaton search are the rare meson decays $B \to K\chi$ and $K \to \pi\chi$. The first search in the B sector has recently been published by the LHCb experiment at CERN [7]. The exclusion plot is shown in Fig.3 (right). The mass range 265 MeV< m_χ < 350 MeV is currently outside the experimental reach in the B sector but can be covered in the kaon sector by the decay $K \to \pi\chi, \chi \to \mu^+\mu^-$ ($\chi \to \mu^+\mu^-$ is a dominant decay mode in this mass range, as can be seen in Fig. 3, left). This region contains inflatons with $\tau < 10^{-9}$ s and $\theta^2 = (2 \div 3) \times 10^{-7}$, corresponding to branching ratios $BR(K \to \pi\chi) \times BR(\chi \to \mu^+\mu^-) < 3 \times 10^{-10}$.

The NA48/2 event selection for the inflaton search was the same as for heavy neutrino search [6]. After selecting 3-track candidates, the scan over the the invariant mass of two muons $M_{\mu\mu}$ was performed. The obtained upper limits are shown in Fig. 4.

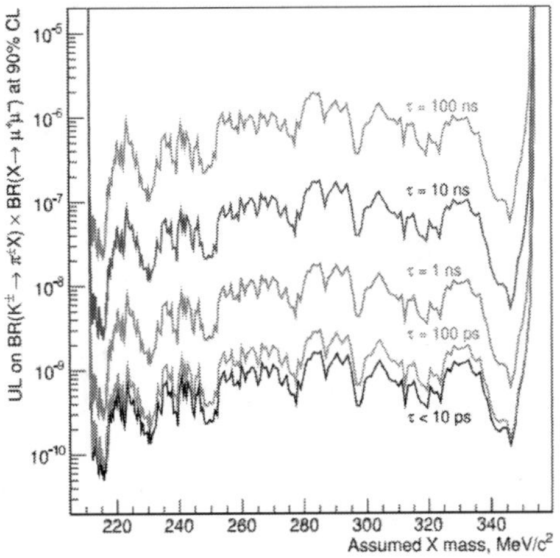

Figure 4. Upper limits on BR($K^\pm \to \pi^\pm\chi$, $\chi \to \mu^+\mu^-$) for different lifetimes of χ.

5 Dark photon search

In 2003–2004 the NA48/2 experiment collected a large set of π^0's tagged from kaon decays $K^\pm \to \pi^\pm\pi^0$ and $K^\pm \to \mu^\pm\nu_\mu\pi^0$ which allows to search for the dark photon (DP)[8]. The BR of the DP production is $BR(\pi^0 \to \gamma A') = 2\epsilon^2\left(1 - \frac{m_{A'}^2}{m_{\pi^0}^2}\right) \times BR(\pi^0 \to \gamma\gamma)$ [9]. Here ϵ is the mixing parameter, $m_{A'}$ is the DP mass. The only allowed decay mode of the DP in the mass range $2m_e < m_{A'} < m_{\pi^0}$ is $A' \to e^+e^-$. Thus, the DP search was performed by selecting Dalitz decays $K^\pm \to \pi^\pm\pi^0_D$ and $K^\pm \to \mu^\pm\nu_\mu\pi^0_D$, where $\pi^0_D \to e^+e^-\gamma$. These decays have the same signature as the DP signal and therefore reproduce irreducible background.

The distribution over the invariant mass M_{ee} for $K^\pm \to \pi^\pm\pi^0_D$ is shown in Fig. 5 (left). A DP signal would correspond to a narrow peak in this distribution. A scan over the mass range 9 MeV/c^2 < $m_{A'}$ < 120 MeV/c^2 was performed. The range limits are due to the limited accuracy of the $\pi^0_D \to e^+e^-\gamma$ simulation (left side) and not competitive sensitivity to the mixing parameter ϵ^2 (right side).

In total 404 mass hypotheses were tested. For none of them the local statistical significance exceeded 3σ. Upper limits at 90% CL on ϵ^2 were calculated for each mass hypothesys. The results compared to other exclusion limits are presented in Fig. 5(right). The world limits on the mixing parameter ϵ with DP are improved in the range 9–70 MeV/c^2.

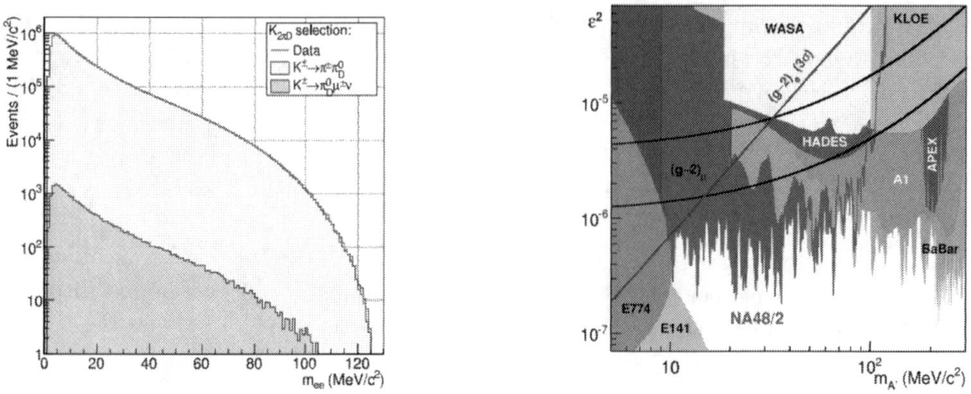

Figure 5. Invariant mass M_{ee} for $K^{\pm} \to \pi^{\pm}\pi^0_D$ (left) and current limits on DP search (right).

6 Prospects for the NA62 experiment

The NA62 experiment is aimed at measuring the BR($K^+ \to \pi^+\nu\bar{\nu}$) with ~10% precision. It is expected to collect ~ 10^{13} K^+ decays (~50 times more than the NA48/2 statistics) in 2016–2018. The schematic view of the setup is shown in Fig. 6. The high-intensity (750 MHz) unseparated (~6% of positively charged kaons) hadron beam with a 75 GeV/c momentum is obtained from the SPS primary beam (protons with 400 GeV/c momentum) impinging on a Be target. Kaons are identified by a differential Cerenkov counter (CEDAR). The momentum and direction of beam particles are measured by a silicon pixel beam spectrometer (Gigatracker). The vacuumed decay region has a length of ~65 m. The momentum of secondary charged particles (produced in kaon decays) is measured by a magnetic spectrometer composed of four stations of straw tubes and a magnet with p_T ~270 MeV/c. The energy of secondary photons and electrons flying at small angles with respect to the beam axis is measured by a liquid krypton electromagnetic calorimeter, while large angle photons are detected by a system of 12 rings of lead-glass blocks (Large angle photon veto). The identification of secondary particles is done in the RICH detector filled with Ne at atmospheric pressure, liquid krypton calorimeter, hadron calorimeter (both calorimeters were previously used in the NA48/2 experiment). A highly segmented muon identification system (Muon veto) is located behind the krypton calorimeter.

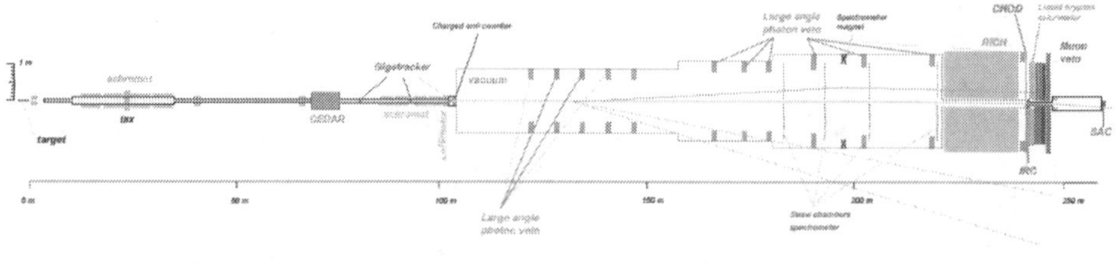

Figure 6. NA62 experimental setup.

The detailed description can be found in [10]. A new spectrometer operated in vacuum will allow to significantly improve the momentum resolution crucial for the searches discussed above (at least by a factor of two). The kaon identification by the KTAG detector will reduce the muon halo contribution. In addition to the decay searches, it is expected to improve the current world limits on the heavy neutrino produciton in $K^+ \to \mu^+ N$ decay, at least for neutrino masses above 300 MeV/c^2. The advantage if this mode is that the production searches are model-independent.

7 Conclusions

Various NP models that allow to solve existing problems in cosmology can be tested in kaon decays, among them the νMSM, light inflaton models and models with dark photons. Recently new upper limits were set by the NA48/2 experiment on the BR of related processes $K^\pm \to \mu^\pm N$, $N \to \mu^\pm \pi^\mp$; $K^\pm \to \mu^\pm N$, $N \to \pi^\pm \mu^\mp$ and $K^\pm \to \pi^\pm \chi$, $\chi \to \mu^+ \mu^-$. The new limits on the mixing parameter ϵ with DP are obtained in the range 9–120 MeV/c^2. The ongoing NA62 experiment should have significantly better sensitivity due to larger statistics and better performance of the detectors.

References

[1] T. Asaka et al., Phys. Lett. **B620** 17 (2005)
[2] F. Bezrukov and D. Gorbunov, JHEP **1005** 010 (2010)
[3] F. Bezrukov and D. Gorbunov, JHEP **1307** 140 (2013)
[4] B. Holdom, Phys. Lett. **B166** 196 (1986)
[5] J. Batley et al., Eur.Phys.J. **C52**, 875–891 (2007)
[6] K. Massri, talk at the Moriond EW 2016 conference
[7] R. Aaij et al., Phys. Rev. Lett. **115** 161802 (2015)
[8] J. Batley et al., Phys. Lett. **B746**, 178 (2015)
[9] B. Batell, M. Pospelov, A. Ritz, Phys. Rev. **D80** 095024 (2009)
[10] F. Hahn et al., NA62 Technical Design Document, **NA62-10-07** (2010)

Kinetics of spontaneous baryogenesis in non-stationary background

Elena Arbuzova[1,2,⋆], *Alexander* Dolgov[1,3,4,⋆⋆], and *Victor* Novikov[3,⋆⋆⋆]

[1] *Novosibirsk State University, Novosibirsk, 630090, Russia*
[2] *Department of Higher Mathematics, Dubna State University, 141980 Dubna, Russia*
[3] *ITEP, Bol. Cheremushkinsaya ul., 25, 117259 Moscow, Russia*
[4] *Dipartimento di Fisica e Scienze della Terra, Università degli Studi di Ferrara Polo Scientifico e Tecnologico - Edificio C, Via Saragat 1, 44122 Ferrara, Italy*

Abstract. Generation of the cosmological baryon asymmetry in frameworks of spontaneous baryogenesis is studied in detail. It is shown that the relation between baryonic chemical potential and the time derivative of the (pseudo)Goldstone field essentially depends upon the representation chosen for the fermionic fields with non-zero baryonic number (quarks). Kinetic equation is modified and numerically solved in equilibrium for the case of time dependent external background or finite integration time to be applicable to the case when energy conservation law is formally violated.

1 Introduction

One of the popular scenarios of baryogenesis is the spontaneous baryogenesis (SBG) proposed in papers [1–3], for reviews see e.g. Refs. [4, 5]. It is assumed that in the unbroken phase the theory is invariant with respect to the global $U(1)$-symmetry, which ensures conservation of baryonic number. This symmetry is spontaneously broken and in the broken phase the Lagrangian density acquires the term

$$\mathcal{L}_{SB} = (\partial_\mu \theta) J^\mu_B \,, \tag{1}$$

where θ is the Goldstone field and J^μ_B is the baryonic current. Due to the spontaneous symmetry breaking (SSB) this current is not conserved. The next step is the statement that the Hamiltonian density corresponding to $\mathcal{L}_{SB}$ is simply the Lagrangian density taken with the opposite sign:

$$\mathcal{H}_{SB} = -\mathcal{L}_{SB} = -(\partial_\mu \theta) J^\mu_B \,. \tag{2}$$

For the spatially homogeneous field $\theta = \theta(t)$ this Hamiltonian is reduced to $\mathcal{H}_{SB} = -\dot\theta n_B$, where $n_B \equiv J^4_B$ is the baryonic number density, so it is tempting to identify $\dot\theta$ with the chemical potential, μ, of the corresponding system. If this is the case, then in thermal equilibrium the baryon asymmetry would evolve to:

$$n_B = \frac{g_S B_Q}{6}\left(\mu T^2 + \frac{\mu^3}{\pi^2}\right) \to \frac{g_S B_Q}{6}\left(\dot\theta T^2 + \frac{\dot\theta^3}{\pi^2}\right), \tag{3}$$

⋆e-mail: arbuzova@uni-dubna.ru
⋆⋆e-mail: dolgov@fe.infn.it
⋆⋆⋆e-mail: novikov@itep.ru

where T is the cosmological plasma temperature, g_S and B_Q are respectively the number of the spin states and the baryonic number of quarks, which are supposed to be the bearers of the baryonic number.

It is interesting that for successful SBG two of the three Sakharov's conditions for the generation of the cosmological baryon asymmetry, namely, breaking of thermal equilibrium and a violation of C and CP symmetries are unnecessary. This scenario is analogous the baryogenesis in absence of CPT invariance, if the masses of particles and antiparticles are different. In the latter case the generation of the cosmological baryon asymmetry can also proceed in thermal equilibrium [6, 7].

In this work the classical version of spontaneous baryogenesis is studied. The talk is organized as follows. In Section 2 the general features of the spontaneous breaking of baryonic $U(1)$-symmetry are described, and the (pseudo)Goldstone mode, its equation of motion, and baryonic chemical potential are introduced. Next, in Sec. 3 the standard kinetic equation in stationary background is presented. In Sec. 4 we derive kinetic equation in time dependent external field and/or for the case when energy is not conserved because of finite limits of integration over time. Several examples, when such kinetic equation is relevant, are presented in Sec. 5. Lastly in Sec. 6 we conclude.

2 Spontaneous symmetry breaking and goldstone mode

Let us consider the theory of complex scalar field Φ interacting with "quarks", Q, and "leptons", L, with the Lagrangian:

$$\mathcal{L}(\Phi) = g^{\mu\nu}\partial_\mu\Phi^*\partial_\nu\Phi - V(\Phi^*\Phi) + \bar{Q}(i\gamma^\mu\partial_\mu - m_Q)\,Q + \bar{L}(i\gamma^\mu\partial_\mu - m_L)L + \mathcal{L}_{int}(\Phi, Q, L)\,, \tag{4}$$

where $\mathcal{L}_{int}$ describes the interaction between Φ and fermionic fields. In the toy model studied below we take it in the form:

$$\mathcal{L}_{int} = \frac{\sqrt{2}}{m_X^2}\frac{\Phi}{f}\,(\bar{L}\gamma_\mu Q)(\bar{Q}^c\gamma_\mu Q) + h.c.\,, \tag{5}$$

where Q^c is charged conjugated quark spinor, m_X is a parameter with dimension of mass, and f is related to the vacuum expectation value of Φ defined below in Eq. (7). Such an interaction can appear e.g. in $SU(5)$ Grand Unified Theory. For simplicity, in our toy model we do not take into account the quark colors.

B-non conserving interaction may have many different forms. The one presented above describes transition of three quark-type fermions into (anti)lepton. There may be transformation of two or three quarks into equal number of antiquarks. Such interaction describes neutron-antineutron oscillations. There even can be a "quark" transition into three "leptons". Depending on the interaction type the relation between $\dot{\theta}$ and the effective chemical potential would have different forms.

Note that Q and L can be any fermions, not necessarily quarks and leptons of the standard model. For example, they can be new heavy fermions. They may possess similar or the same quantum numbers as the quarks and leptons of the standard model and may couple to the ordinary quarks and leptons. In section 4 we consider another model to study kinetics of the baryon asymmetry generation which allows for the transformation $3L \leftrightarrow Q$ or $2Q \leftrightarrow 2\bar{Q}$. They are surely not permitted for the standard quarks. However, the process $3q \leftrightarrow 3\bar{q}$ is permitted and kinetics of this process is essentially the same. We denote by q the fermionic field with the same quantum number as the usual quark.

The theory (4) considered in this section is invariant under the following $U(1)$-transformations:

$$\Phi \to e^{i\alpha}\Phi, \quad Q \to e^{-i\alpha/3}Q, \quad L \to L\,. \tag{6}$$

In the unbroken symmetry phase this invariance leads to the conservation of the total baryonic number which includes the baryonic number of Φ, taken to be unity, and that of quarks, equal to $1/3$. In realistic model the interaction of left- and right-handed fermions may be different but we neglect this possible difference in what follows.

We assume that the global $U(1)$-symmetry is spontaneously broken at the energy scale f in the usual way, e.g. via the potential of the form

$$V(|\Phi|) = \lambda\left(\Phi^*\Phi - f^2/2\right)^2 . \tag{7}$$

The resulting scalar field vacuum expectation value is $\langle\Phi\rangle = fe^{i\phi_0/f}/\sqrt{2}$ with a constant phase ϕ_0.

Below the scale f we can neglect the heavy radial mode of Φ with the mass $m_{radial} = \lambda^{1/2}f$, since being very massive it is frozen out, but this simplification is not necessary and is not essential for the baryogenesis. The remaining light degree of freedom is the variable field ϕ, which is the Goldstone boson of the spontaneously broken $U(1)$. Up to a constant factor the field ϕ is the angle around the bottom of the Mexican hat potential described by Eq. (7). Correspondingly we introduce the dimensionless angular field $\theta \equiv \phi/f$:

$$\Phi = fe^{i\phi/f}\sqrt{2} = fe^{i\theta}/\sqrt{2}. \tag{8}$$

As a result the following effective Lagrangian for θ is obtained:

$$\mathcal{L}_1(\theta) = \frac{f^2}{2}\partial_\mu\theta\,\partial^\mu\theta + \bar{Q}_1(i\gamma^\mu\partial_\mu - m_Q)Q_1 + \bar{L}(i\gamma^\mu\partial_\mu - m_L)L + \left(\frac{e^{i\theta}}{m_X^2}(\bar{L}\gamma_\mu Q_1)(\bar{Q}_1^c\gamma_\mu Q_1) + h.c.\right) - U(\theta). \tag{9}$$

Here we introduced "by hand" potential $U(\theta)$, which may appear due to an explicit symmetry breaking and can lead, in particular, to a nonzero mass of θ. We use the notation Q_1 for the quark field to distinguish it from the phase rotated field Q_2 introduced below in Eq. (11). In a realistic model the quark fields should be (anti)symmetrized with respect to color indices, omitted here for simplicity.

If $U(\theta) = 0$, the theory still remains invariant under the global transformations (i.e. with $\alpha = const$):

$$Q \to e^{-i\alpha/3}Q, \quad L \to L, \quad \theta \to \theta + \alpha . \tag{10}$$

If we only rotate the quark field as above but with coordinate dependent $\alpha = \theta(t, \mathbf{x})$, introducing the new field $Q_1 = e^{-i\theta/3}Q_2$, then the Lagrangian (9) is transformed into:

$$\mathcal{L}_2(\theta) = \frac{f^2}{2}\partial_\mu\theta\partial^\mu\theta + \bar{Q}_2(i\gamma^\mu\partial_\mu - m_Q)Q_2 + \bar{L}(i\gamma^\mu\partial_\mu - m_L)L + \left(\frac{1}{m_X^2}(\bar{Q}_2\gamma_\mu L)(\bar{Q}_2\gamma_\mu Q_2^c) + h.c.\right) + (\partial_\mu\theta)J^\mu - U(\theta), \tag{11}$$

where the quark baryonic current is $J_\mu = (1/3)\bar{Q}\gamma_\mu Q$. Note that the current has the same form in terms of Q_1 and Q_2.

The equation of motion for the quark field Q_1 obtained from Lagrangian (9) has the form:

$$(i\gamma^\mu\partial_\mu - m_Q)Q_1 + \frac{e^{-i\theta}}{m_X^2}\left[\gamma_\mu L(\bar{Q}_1\gamma_\mu Q_1^c) + 2\gamma_\mu Q_1^c(\bar{Q}_1\gamma_\mu L)\right] = 0 . \tag{12}$$

DOI: 10.1051/epjconf/201612503002

Analogously the equation of motion for the phase rotated field Q_2 derived from Lagrangian (11) is:

$$\left(i\gamma^\mu \partial_\mu - m_Q + \frac{1}{3}\gamma^\mu \partial_\mu \theta\right) Q_2 + \frac{1}{m_X^2}\left[\gamma_\mu L(\bar{Q}_2 \gamma_\mu Q_2^c) + 2\gamma_\mu Q_2^c(\bar{Q}_2 \gamma_\mu L)\right] = 0\,. \tag{13}$$

Equations for θ-field derived from these two Lagrangians in flat space-time have respectively the forms:

$$f^2(\partial_t^2 - \Delta)\theta + U'(\theta) + \left[\frac{i\,e^{-i\theta}}{m_X^2}(\bar{Q}_1\gamma_\mu L)(\bar{Q}_1\gamma_\mu Q_1^c) + h.c.\right] = 0 \tag{14}$$

and

$$f^2(\partial_t^2 - \Delta)\theta + U'(\theta) + \partial_\mu J_B^\mu = 0\,, \tag{15}$$

where $U'(\theta) = dU/d\theta$.

Using either the equation of motion (12) or (13) we can check that the baryonic current is not conserved. Indeed, its divergence is:

$$\partial_\mu J_B^\mu = \frac{i\,e^{-i\theta}}{m_X^2}(\bar{Q}_1\gamma_\mu Q_1^c)(\bar{Q}_1\gamma^\mu L) + h.c. \tag{16}$$

(and similarly for Q_2 but without the factor $\exp(-i\theta)$). So the equations of motion for θ in both cases (14) and (15) coincide, as expected.

In the spatially homogeneous case, when $\partial_\mu J_B^\mu = \dot{n}_B$ and $\theta = \theta(t)$, and if $U(\theta) = 0$, equation (15) can be easily integrated giving:

$$f^2\left[\dot\theta(t) - \dot\theta(t_{in})\right] = -n_B(t) + n_B(t_{in})\,. \tag{17}$$

It is usually assumed that the initial baryon asymmetry vanishes, $n_B(t_{in}) = 0$.

The evolution of $n_B(t)$ is governed by the kinetic equation discussed in Sec. 3, which allows to express n_B through $\theta(t)$ and thus to obtain the closed systems of, generally speaking, integro-differential equations. In thermal equilibrium the relation between $\dot\theta$ and n_B may become an algebraic one, but this is true only in the case when the integration over time is sufficiently long and if $\dot\theta$ is constant or slowly varying function of time.

In cosmological Friedmann-Robertson-Walker (FRW) background and space-independent $\theta(t)$ equation (15) is transformed to:

$$f^2(\partial_t + 3H)\dot\theta + U'(\theta) = -(\partial_t + 3H)n_B. \tag{18}$$

We do not include the curvature effects into the Dirac equations because this is not necessary for what follows. Still we are using expression for the current divergence in the form $\mathcal{D}_\mu J^\mu = \dot{n}_B + 3Hn_B$, but not just $\dot{n}_B$.

If particles (fermions) are in thermal equilibrium with respect to baryo-conserving interactions, then their phase space distribution has the form:

$$f_{eq} = \left[1 + \exp(E/T - \xi_B)\right]^{-1}, \tag{19}$$

where dimensionless chemical potential ξ_B has equal magnitude but opposite signs for particles and antiparticles. The baryonic number density, for small ξ_B, is usually given by the expression

$$n_B = g_S B_Q \xi_B T^3/6 \tag{20}$$

(compare to Eq. (3)). However, the relation (20) between the baryonic number density and chemical potential is true only for the normal relation between the energy and three-momentum, $E = \sqrt{p^2 + m^2}$. This is not the case if the dispersion relation has the form

$$E = \sqrt{p^2 + m^2} \pm \dot{\theta}/3, \tag{21}$$

derived from the equation of motion (13), where the signs ± refer to particles or antiparticles respectively, as we see a little below. We should note that the above dispersion relation is derived under assumption of constant or slow varying $\dot{\theta}$. Otherwise the Fourier transformed Dirac equation cannot be reduced to the algebraic one.

If the baryon number is conserved, n_B remains constant in comoving volume and it means in turn that $\xi_B = const$ for massless particles. If and when non-conservation of baryons is switched on, ξ_B evolves according to kinetic equation. Complete thermal equilibrium in the standard theory demands $n_B \to 0$, but a deviation from thermal equilibrium of B-nonconserving interaction leads to generation of non-zero ξ_B and correspondingly to non-zero n_B. As we will see in Sec. 4, SBG allows for generation of nonzero baryonic number in complete thermal equilibrium.

In terms of ξ_B and the new function $\eta = \dot{\theta}/T^3$ equation (18) takes the same form as eq. (17):

$$f^2\left[\eta(t) - \eta(t_{in})\right] = -\frac{g_S B_Q}{6}\left[\xi_B(t) - \xi_B(t_{in})\right] \tag{22}$$

and thus

$$f^2\left[\frac{\dot{\theta}(t)}{T^3(t)} - \frac{\dot{\theta}(t_{in})}{T^3_{in}}\right] = -\frac{g_S B_Q}{6}\left[\xi_B(t) - \xi_B(t_{in})\right]. \tag{23}$$

As we have already mentioned, $n_B(t_{in}) = 0$, so according to Eq. (20) we should also take $\xi_B(t_{in}) = 0$. However this initial condition for chemical potential is not true in the theory with the Lagrangian (11) and the Dirac equation (13) for the quark field, though the condition $n_B(t_{in}) = 0$ is supposed to be always valid. Indeed, the B-nonconserving interaction now conserves energy and thus this process does not split the energies of quarks and antiquarks. However, these energies are split from the very beginning due to relation (21). Correspondingly using eqs. (19) and (21) we find in the massless case:

$$n_B = \int \frac{d^3p}{(2\pi)^3}(f_B - f_{\bar{B}}) = \frac{g_S B_Q}{6}\left(\xi_B - \frac{\dot{\theta}}{3T}\right)T^3. \tag{24}$$

If initially $n_B = 0$, then $\xi_B(t_{in}) = \dot{\theta}_{in}/(3T_{in})$. In the case of conserved baryonic number, n_B remains zero and thus in equilibrium the relation $\xi_B = \dot{\theta}/(3T)$ must be true at any time. When the B-nonconserving interaction is on, the chemical potential would evolve and might evolve even down to zero, leading to generation of non-zero baryonic density, as is discussed in Sec. 4.

In the pseudogoldstone case, when $U(\theta) \neq 0$, equations of motion (15) or (18) cannot be so easily integrated, but in thermal equilibrium the system of equations containing $\theta(t)$ and $\xi_B(t)$ can be reduced to ordinary differential equations which are easily solved numerically. Out of equilibrium one has to solve much more complicated system of the ordinary differential equation of motion for $\theta(t)$ and the integro-differential kinetic equation. It is discussed below in Sec. 3.

3 Kinetic equation for time independent amplitude

The temporal evolution of the distribution function of i-th type particle, $f_i(t, p)$, in an arbitrary process $i + Y \leftrightarrow Z$ in the FRW background, is governed by the kinetic equation:

$$\frac{df_i}{dt} = (\partial_t - H\, p_i \partial_{p_i}) f_i = I_i^{coll}, \tag{25}$$

with the collision integral equal to:

$$I_i^{coll} = \frac{(2\pi)^4}{2E_i} \sum_{Z,Y} \int d\nu_Z \, d\nu_Y \delta^4(p_i + p_Y - p_Z)$$

$$\left[|A(Z \to i + Y)|^2 \prod_Z f \prod_{i+Y} (1 \pm f) - |A(i + Y \to Z)|^2 f_i \prod_Y f \prod_Z (1 \pm f) \right], \tag{26}$$

where $A(a \to b)$ is the amplitude of the transition from state a to state b, Y and Z are arbitrary, generally multi-particle states, $(\prod_Y f)$ is the product of the phase space densities of particles forming the state Y, and

$$d\nu_Y = \prod_Y \overline{dp} \equiv \prod_Y \frac{d^3p}{2E\,(2\pi)^3}. \tag{27}$$

The signs '+' or '−' in $\prod(1 \pm f)$ are chosen for bosons and fermions respectively. We neglect the effects of the space-time curvature in the collision integral which is generally a good approximation.

In the lowest order of perturbation theory the amplitude of transition from an initial state $|in\rangle$ to a final state $|fin\rangle$ is given by the integral of the matrix element of Lagrangian density between these states, integrated over 4-dimensional space d^4x. The quantum field operators are expanded in terms of creation-annihilation operators with a plane wave coefficients: $\sim \exp(-iEt + i\mathbf{px})$.

When the amplitude of the process is time-independent, then the integration of the product of the exponents in infinite integration limits leads to the energy-momentum conservation factors:

$$\int dt d^3x \, e^{-i(E_{in} - E_{fin})t + i(\mathbf{P_{in}} - \mathbf{P_{fin}})\mathbf{x}} = (2\pi)^4 \delta(E_{in} - E_{fin})\, \delta((\mathbf{P_{in}} - \mathbf{P_{fin}}), \tag{28}$$

where E_{in}, E_{fin}, $\mathbf{P_{in}}$, and $\mathbf{P_{in}}$ are the total energies and 3-momenta of the initial and final states respectively. The amplitude squared contains delta-function of zero which is interpreted as the total time duration, t_{max}, of the process and as the total space volume, V. The probability of the process given by the collision integral is normalized per unit time and volume, so it must be divided by V and t_{max}.

We are interested in the evolution of the baryon number density, which is the time component of the baryonic current J^μ: $n_B \equiv J^4$. Due to the quark-lepton transitions the current is non-conserved and its divergence is given by Eq. (16). The similar expression is evidently true in terms of Q_2 but without the factor $\exp(-i\theta)$. Let us first consider the latter case, when the interaction is described by the Lagrangian (11), which contains the product of three "quark" and one "lepton" operators, and take as an example the process $q_1 + q_2 \leftrightarrow \bar{q} + l$.

Since the interaction in this representation does not depend on time, the energy is conserved and the collision integral has the usual form with conserved four-momentum. Quarks are supposed to be in kinetic equilibrium but probably not in equilibrium with respect to B-nonconserving interactions, so their distribution function has the form:

$$f_Q = \exp\left(-\frac{E}{T} + \xi_B\right) \text{ and } f_{\bar{Q}} = \exp\left(-\frac{E}{T} - \xi_B\right). \tag{29}$$

Here and in what follows the Boltzmann statistics is used. Since the dispersion relation for quarks and antiquarks (21) depends upon $\dot{\theta}$, the baryon asymmetry in this case is given by eq. (24) and the kinetic equation takes the form:

$$\frac{g_S B_Q}{6} \frac{d}{dt}\left(\xi_B - \frac{\dot{\theta}}{3T}\right) = -c_1 \Gamma \xi_B, \tag{30}$$

where c_1 is a numerical factor of order unity and Γ is the rate of baryo-nonconserving reactions. If the amplitude of these reactions has the form presented in Eq. (13), then $\Gamma \sim T^5/m_X^4$.

For constant or slow varying temperature the equilibrium solution to this equation is $\xi_B = 0$ and the baryon number density is proportional to $n_B \sim \dot{\theta} T^2$, with $\dot{\theta}$ evolving according Eq. (17) with n_B expressed through $\dot{\theta}$.

Let us check now what happens if the dependence on θ is moved from the quark dispersion relation to the B-nonconserving interaction term (14). The expression for the collision integral (26) is valid only in absence of external field depending on coordinates. In our case, when quarks "live" in the $\theta(t)$-field, the collision integral should be modified in the following way. Now we have an additional factor under the integral (28), namely, $\exp[\pm i\theta(t)]$. In general case this integral cannot be taken analytically, but if we can approximate $\theta(t)$ as $\theta(t) \approx \dot{\theta} t$ with a constant or slowly varying $\dot{\theta}$, the integral is simply taken giving e.g. for the process of two quark transformation into antiquark and lepton, $q_1 + q_2 \leftrightarrow \bar{q} + l$, the energy balance condition imposed by $\delta(E_{q_1} + E_{q_2} - E_{\bar{q}} - E_l - \dot{\theta})$. In other words the energy is non-conserved due to the action of the external field $\theta(t)$. The approximation of linear evolution of θ with time can be valid if the reactions are fast in comparison with the rate of the θ-evolution.

Returning to our case we can see that the collision integral integrated over the three-momentum of the particle under scrutiny (i.e. particle i in eq. (26)) e.g. for process the $q_1 + q_2 \to l + \bar{q}$ turns into:

$$\dot{n}_B + 3Hn_B \sim \int d\tau_{l\bar{q}} d\tau_{q_1 q_2} |A|^2 \delta(E_{q_1} + E_{q_2} - E_l - E_{\bar{q}} - \dot{\theta}) \delta(\mathbf{P}_{in} - \mathbf{P}_{fin}) e^{-E_{in}/T} \left(e^{\xi_L - \xi_B + \dot{\theta}/T} - e^{2\xi_B} \right), \quad (31)$$

where $d\tau_{l,\bar{q}} = d^3 p_l d^3 p_{\bar{q}} / [4 E_l E_{\bar{q}} (2\pi)^6]$. We assumed here that all participating particles are in kinetic equilibrium, i.e. their distribution functions have the form

$$f = 1/[\exp(E/T - \xi) + 1], \quad (32)$$

with $\xi = \mu/T$ being dimensionless chemical potential. In expression (31) ξ_B and ξ_L denote baryonic and leptonic chemical potentials respectively and the effects of quantum statistics are neglected but only for brevity of notations. Their effects are not essential in the sense that they do not change the conclusion. The assumption of kinetic equilibrium is well justified because it is enforced by the very efficient elastic scattering. Another implicit assumption is the usual equilibrium relation between chemical potentials of particles and antiparticles, $\bar{\mu} = -\mu$, imposed e.g. by the fast annihilation of quark-antiquark or lepton-antilepton pairs into two and three photons. Anyhow the assumption of chemical equilibrium is one of the cornerstones of the spontaneous baryogenesis.

The conservation of $(B + L)$ implies the following relation: $\xi_L = -\xi_B/3$. Keeping this in mind, we find

$$\dot{n}_B + 3Hn_B \approx -\left(1 - e^{\dot{\theta}/T - 3\xi_B + \xi_L}\right) I \approx \left(\frac{\dot{\theta}}{T} - \frac{10}{3}\xi_B\right) I, \quad (33)$$

where we assumed that ξ_B and $\dot{\theta}/T$ are small. In relativistic plasma with temperature T the factor I, coming from the collision integral, can be estimated as $I = T^8/m^4$, where m is a numerical constant with dimension of mass. It differs from m_X, introduced in eq. (9), by a numerical coefficient.

The asymmetry between quarks and antiquarks having the distribution (32) with equal by magnitude but opposite by sign chemical potentials and identical dispersion relations is equal to

$$n_B = C_B \xi_B T^3, \quad (34)$$

where C_B is a constant, see Eq. (3) in the limit of $\mu \ll T$, because in the realistic case the baryon asymmetry is quite small.

For a large factor I we expect the equilibrium solution $\xi_B = (3/10)\dot\theta/T$, so $\dot\theta$ up to the different numerical factor seems to be the baryonic chemical potential, as expected in the usually assumed SBG scenario. An emergence of the factor $3/10$ instead of $1/3$ in the equilibrium expression is due to the conservation law $B + L = const$. However, as we have seen above, the baryonic chemical potential is not aways proportional to $\dot\theta(t)$.

4 Kinetic equation for time-varying amplitude

In the case the interaction proceeds in a time dependent field and/or the time duration of the process is finite, then the energy conservation delta-function in (31) does not emerge and the described in Sec. 3 approach becomes invalid, so one has to make the time integration with an account of time-varying background and integrate over the phase space without energy conservation.

In what follows we consider two-body inelastic process with baryonic number non-conservation with the amplitude obtained from the last term in Lagrangian (9). At the moment we will not specify the concrete form of the reaction but only will say that it is the two-body reaction

$$a + b \leftrightarrow c + d, \tag{35}$$

where a, b, c, and d are some quarks and leptons or their antiparticles. The expression for the evolution of the baryonic number density, n_B, follows from eq. (25) after integration of its both sides over $d^3p_i/(2\pi)^3$. Thus we obtain:

$$\dot n_B + 3Hn_B = -\frac{(2\pi)^3}{t_{max}} \int dv_{in}dv_{fin}\, \delta(\mathbf{P}_{in} - \mathbf{P}_{fin})\, |A|^2 \left(f_a f_b - f_c f_d\right), \tag{36}$$

where e.g. $dv_{in} = d^3p_a d^3p_b/[4E_aE_b(2\pi)^6]$ and the amplitude of the process is defined as

$$A = \left(\int_0^{t_{max}} dt\, e^{i[(E_c+E_d-E_a-E_b)t+\theta(t)]}\right) F(p_a, p_b, p_c, p_d), \tag{37}$$

and F is a function of 4-momenta of the participating particles, determined by the concrete form of the interaction Lagrangian. In what follows we consider two possibilities: $F = const$ and $F = \psi^4 m_X^{-2}$, where in the last case ψ^4 symbolically denotes the product of the Dirac spinors of particles a, b, c, and d.

In the case of equilibrium with respect to baryon conserving reactions the distribution functions have the canonical form, $f_a = \exp(-E_a/T + \xi_a)$, where $\xi_a \equiv \mu_a/T$ is the dimensionless chemical potential. So for constant F the product $|A|^2(f_a f_b - f_c f_d)$ depends upon the particle 4-momenta only through E_{in} and E_{fin}, where

$$E_{in} = E_a + E_b, \quad \text{and} \quad E_{fin} = E_c + E_d. \tag{38}$$

To integrate Eq. (36) over the phase space it is convenient to change the integration variables, according to:

$$\frac{d^3p_a}{2E_a}\frac{d^3p_b}{2E_b} = d^4P_{in}\, d^4R_{in}\, \delta(P_{in}^2 + R_{in}^2)\, \delta(P_{in}R_{in}), \tag{39}$$

where $P_{in} = p_a + p_b$ and $R_{in} = p_a - p_b$ and masses of the particles are taken to be zero. Analogous expressions are valid for the final state particles. Evidently the time components of the 4-vectors P are the sum of energies of the incoming and outgoing particles, $P_{in}^{(4)} = E_{in}$ and $P_{fin}^{(4)} = E_{fin}$. Now we can perform almost all (but one) integrations and finally we obtain the kinetic equation in the following form:

$$\dot{n}_B + 3Hn_B = -\frac{T^5}{2^5\pi^6\, t_{max}} \int_0^\infty dy \left[e^{\xi_a+\xi_b}\left(|A_+|^2 + |A_-|^2 e^{-y}\right) - e^{\xi_c+\xi_d}\left(|A_-|^2 + |A_+|^2 e^{-y}\right)\right], \tag{40}$$

where $y = E_-/T$ is the dimensionless energy with E_- being the difference between initial and final energies of the system, $E_- = E_{in} - E_{fin}$, A_+ and A_- are amplitudes taken at positive and negative E_-, respectively. Note, that with the substitution $E_- \to |E_-|$ the only difference between A_+ and A_- is that $A_-(\theta) = A_+(-\theta)$.

The equilibrium is achieved when the integral in Eq. (40) vanishes. Clearly it takes place at

$$\xi_a + \xi_b - \xi_c - \xi_d = \frac{\langle |A_+|^2 e^{-y} + |A_-|^2 \rangle}{\langle |A_+|^2 + |A_-|^2 e^{-y} \rangle} - 1, \tag{41}$$

where the angular brackets mean integration over dy as indicated in Eq. (40).

This results above are obtained for the amplitude which does not depend upon participating particle momenta. The calculations would be be somewhat more complicated if this restriction is not true. For example if the baryon non-conservation takes place in four-fermion interactions, then the amplitude squared can contain the terms of the form $(p_a p_b)^2/m_X^4$ or $(p_a p_c)^2/m_X^4$, etc. The effect of such terms results in a change of the numerical coefficient in Eq. (33) but the latter is unknown anyhow, and what is more important the temperature coefficient in front of the integral in this equation would change from T^5 to T^9/m_X^4.

5 Examples of time-varying θ

5.1 Constant $\dot{\theta}$

This is the case usually considered in the literature and the simplest one. The integral (37) is taken analytically resulting in:

$$|A|^2 \sim \frac{2 - 2\cos[(\dot{\theta} - E_-)t_{max}]}{(\dot{\theta} - E_-)^2}, \tag{42}$$

where E_- is running over the positive semi-axis.

For large t_{max} this expression tends to $\delta(E_- - \dot{\theta})$, so $|A_+|^2 = 2\pi\delta(E_- - \dot{\theta})t_{max}$ and $|A_-|^2 = 2\pi\delta(E_- + \dot{\theta})t_{max} = 0$, if $\dot{\theta} > 0$ and vice versa otherwise. Hence the equilibrium solution is

$$\xi_a + \xi_b - \xi_c - \xi_d - \dot{\theta} = 0, \tag{43}$$

coinciding with the standard result.

The limit of $\dot{\theta} = const$ corresponds to the energy non-conservation by the rise (or drop) of the energy of the final state in reaction (35) exactly by $\dot{\theta}$. However if t_{max} is not sufficiently large, the non-conservation of energy is not equal to $\dot{\theta}$ but somewhat spread out and the equilibrium solution would be different. There is no simple analytical expression in this case, so we have to take the integral (41) over y numerically to find at what values of chemical potentials, ξ_k, it vanishes and this point determines the equilibrium values of the chemical potentials in external $\dot{\theta}$ field.

The results of the calculations are presented in Fig. 1. In the left panel the values of the r.h.s. of Eq. (41) are compared with $\dot{\theta}/T$ (thick line) for two values of the cut-off in time integration $\tau \equiv t_{max}T = 10$ (dashed line) and $\tau = 3$ (dotted line). In the right panel relative differences between the r.h.s. of Eq. (41) and $\dot{\theta}/T$, normalized to $\dot{\theta}/T$, as functions of $\dot{\theta}$ for different maximum time of the

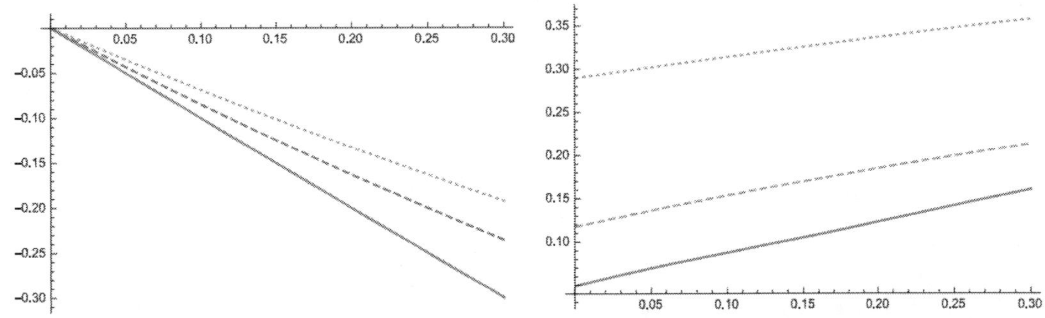

Figure 1. *Left*: $\dot{\theta}/T$ for infinite time integration (thick line). Cut-off in time integration $\tau \equiv t_{max}T = 10$ (dashed); 3 (dotted). *Right*: relative differences between the equilibrium solutions and $\dot{\theta}/T$, normalized to $\dot{\theta}/T$, as functions of $\dot{\theta}$ for different t_{max}: $\tau \equiv t_{max}T = 30$ (thick); 10 (dashed); 3 (dotted).

integration are depicted. We see that for $\tau = 30$ (thick line) the deviations are less than 10%, while for $\tau = 3$ (dotted line) the deviations are about 30%. If we take τ close to unity, the deviations are about 100%. The value of $\dot{\theta}/T$ is bounded from above by 0.3 because at large $\dot{\theta}/T$ the linear expansion, used in our estimates, is invalid.

5.2 Second order Taylor expansion of $\theta(t)$

Here we assume that $\theta(t)$ can be approximated as

$$\theta(t) = \dot{\theta}\, t + \ddot{\theta}\, t^2/2, \tag{44}$$

where $\dot{\theta}$ and $\ddot{\theta}$ are supposed to be constant or slowly varying. In this case the integral over time (37) can also be taken analytically but the result is rather complicated. We need to take the integral

$$\int_0^{t_{max}} dt \exp[i\theta(t)]. \tag{45}$$

Its real and imaginary parts are easily expressed through the Fresnel functions. So the amplitude squared is given by the functions tabulated in Mathematica and the position of the equilibrium point can be calculated, as in the previous case, by numerical calculation of one dimensional integral.

The r.h.s. of Eq. (41) as functions of $\dot{\theta}$ for different values of τ are presented in Fig. 2, left panel. It is interesting that the dependence on τ is not monotonic. This can be explained by that at small τ the effects of $\ddot{\theta}t^2$ are not essential.

To check the dependence on $\ddot{\theta}$ we calculated again the r.h.s. of Eq. (41) but now as functions of $\ddot{\theta}$ presented in the right panel in Fig. 2. We see that the equilibrium point oscillates as a function of $\ddot{\theta}$.

6 Conclusion

We argue that in the standard description $\dot{\theta}$ is not formally the chemical potential, though in thermal equilibrium μ_B tends to $\dot{\theta}$ with numerical, model dependent, coefficient. Moreover, this is not always true but depends upon the chosen representation for the "quark" fields. In the theory described by the Lagrangian (9) which appears "immediately" after the spontaneous symmetry breaking, $\theta(t)$ directly enters the interaction term in this Lagrangian and in equilibrium $\mu_B \sim \dot{\theta}$ indeed. On the other hand,

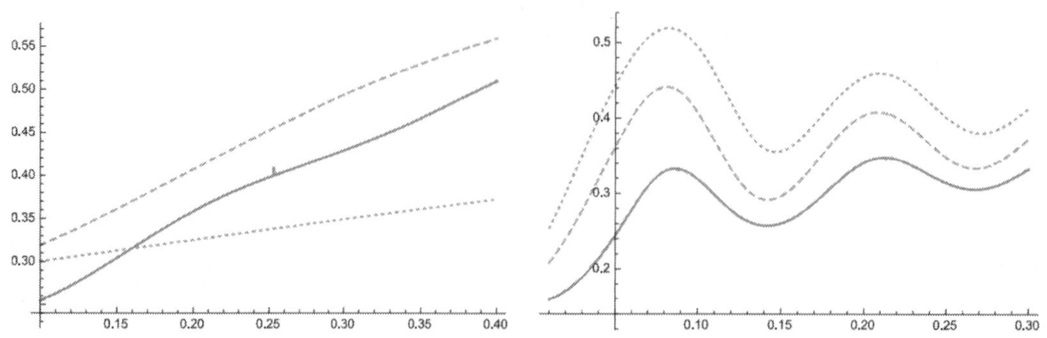

Figure 2. Relative differences between equilibrium solutions and $\dot{\theta}/T$, normalized to $\dot{\theta}/T$. *Left:* as functions of $\dot{\theta}$ for different t_{max}: $\tau \equiv t_{max}T = 30$ (thick); 10 (dashed); 3 (dotted); $\ddot{\theta} = 0.1$. *Right:* as functions of $\ddot{\theta}$ for different $\dot{\theta}$: $\dot{\theta} = 0.1$ (thick); 0.2 (dashed); 0.3 (dotted); $\tau \equiv t_{max}T = 10$.

if we transform the quark field, so that the dependence on θ is shifted to the bilinear product of the quark fields (11), then chemical potential in equilibrium does not tend to $\dot{\theta}$, but to zero. On the other hand, the magnitude of the baryon asymmetry in equilibrium is always proportional to $\dot{\theta}$. It can be seen, according to the equation of motion of the Goldstone field, that $\dot{\theta}/T$ drops down in the course of the cosmological cooling as T^2, so the baryon number density in the comoving volume decreases in the same way. So to avoid complete vanishing of n_B the baryo-violating interaction should switch-off at some non-zero T. This is always the case but the dependence on the interaction strength is non-monotonic.

The assumption of constant or slowly changing $\dot{\theta}$, which is usually done in the SBG scenario, may be not fulfilled and to include the effects of an arbitrary variation of $\theta(t)$ as well as the effects of the finite time integration we transform the kinetic equation in such a way that it becomes operative in the case of non-conserved energy. A shift of the equilibrium value of the baryonic chemical potential due to this effect is numerically calculated.

Acknowledgement. EA and AD thank the support of the Grant of President of Russian Federation for the leading scietific schools of the Russian Federation, NSh-9022.2016.2. VN thanks the support of the Grant RFBR 16-02-00342.

References

[1] A. Cohen, D. Kaplan, Phys. Lett. **B 199**, 251 (1987).
[2] A. Cohen, D. Kaplan, Nucl.Phys. **B308**, 913 (1988).
[3] A. G. Cohen, D.B., A.E. Nelson, Phys.Lett. **B263**, 86-92 (1991).
[4] A.D.Dolgov, Phys. Repts **222** (1992) No. 6;
V.A. Rubakov, M.E. Shaposhnikov, Usp. Fiz. Nauk **166**, 493 (1996);
A. Riotto, M. Trodden, Ann. Rev. Nucl. Part. Sci. **49**, 35 (1999);
M. Dine, A. Kusenko, Rev. Mod. Phys. **76**, 1 (2004).
[5] A.D. Dolgov, Surveys in High Energy Physics **13**, 83 (1998).
[6] A.D. Dolgov, Ya.B. Zeldovich, Uspekhi Fizicheskih Nauk **130**, 559 (1980); Rev. Mod. Phys. **53**, 1-41 (1981).
[7] A.D. Dolgov, Phys. Atom. Nucl. **73**, 588-592 (2010).

EPJ Web of Conferences **125**, 03003 (2016) DOI: 10.1051/epjconf/201612503003
QUARKS-2016

Cosmological particle creation in conformal gravity

Victor Berezin[1,⋆], *Vyacheslav* Dokuchaev[1,2,⋆⋆], and *Yury* Eroshenko[1,⋆⋆⋆]

[1] *Institute for Nuclear Research of the Russian Academy of Sciences, Moscow, Russia*
[2] *National Research Nuclear University MEPhI, Moscow, Russia*

Abstract. We constructed the conformally invariant model for scalar particle creation induced by strong gravitational fields. Starting from the "usual" hydrodynamical description of the particle motion written in the Eulerian coordinates we substituted the particle number conservation law (which enters the formalism) by "the particle creation law", proportional to the square of the Weyl tensor (following the famous result by Ya. B. Zel'dovich and A. A. Starobinsky). Then, demanding the conformal invariance of the whole dynamical system, we have got both the (Weyl)-conformal gravity and the Einstein–Hilbert–dilaton gravity action integral. Thus, we obtained something like the induced gravity suggested first by A. D. Sakharov. It is shown that the resulting system is self-consistent. Some future developments of the theory are discussed in the concluding Chapter.

1 Introduction

The gravitational field causes particle creation by disturbing vacuum fluctuations of scalar, vector, spinor, etc., fields. Owing the results of works by many outstanding scientists [1–6] we know that the renormalization procedure inherent in quantum field theories, requires the adding to the original Einstein–Hilbert gravitational action integral some new terms quadratic in Ricci tensor and scalar curvature (in one loop approximation). The very necessity of these additions supports strongly the idea, first introduced by A. D. Sakharov in 1967 [7] and now known as the induced gravity.

The above mentioned investigations of the particle creation were confined by considering the quantum scalar field on the given background metric (cosmological homogeneous, but anisotropic, space-times). The main obstacle to the accounting for the back reaction is that the rigorous solution of the quantum problem requires knowing the boundary conditions, while the latter can be imposed only after solving the (classical) Einstein equations. Meanwhile the back reaction seems very important because not only the created particles will change the geometry, but the very process of creation, which is purely quantum phenomenon, may violate the well known energy dominance condition and, thus, should be crucial in understanding the global space-time geometry. To avoid this difficulty, we propose to describe the particle creation processes phenomenologically, on the classical level, what should be rather reasonable, when the gravitational fields are strong enough (e. g., in the early universe

⋆e-mail: berezin@inr.ac.ru
⋆⋆e-mail: dokuchaev@inr.ac.ru
⋆⋆⋆e-mail: eroshenko@inr.ac.ru

and inside black holes). The first attempts to the fulfil such a program proved that, indeed, the energy dominance condition may be violated [8, 9].

What would be the "creation law"? Surely, it should be some function of the invariants constructed from the fields responsible for the particle creation. Ya. B. Zel'dovich and A. A. Starobinsky [10] showed that the production rate (e. g., the number of particles created per unit volume per unit time) is proportional to the *square of the Weyl tensor*, the coefficient being depend on the nature of the particles. In our model we adopt this as the first postulate. It will be shown that this leads us inevitably to the Weyl's gravitational integral. If we neglect all the other possible quadratic combinations of Riemann and Ricci tensors as well as scalar curvature, we will arrive at the so called conformal gravity theory, and, at the same time, confine ourselves to the definite choice of the counterparts, arising from the renormalization of the quantum fields.

Thus, we are almost forced to adopt the second postulate — *the conformal invariance* of the field theory. In the present paper we consider only the scalar filed, making it responsible for the particle creation. Demanding the conformal invariance we have to add to the "normal" scalar field Lagrangian the term proportional to the curvature scalar, thus reproducing the Einstein–Hilbert dilaton action integral.

The conformal gravity was invented by H. Weyl in 1918 [11]. His motivation was to construct the unified theory of two (known at the time) fundamental fields: electromagnetic and gravitational ones. Since the electromagnetic field ("identified with the Maxwell equations") is invariant under the conformal transformations, H. Weyl proposed the conformal invariant Lagrangian for the gravitational field. Then. it was recognized that the Weyl's gravity allows only massless particles to exist. On this ground the theory was rejected by H. Weyl himself and by A. Einstein. But, nowadays, this unpleasant feature can be "corrected" by Braut–Englert–Higgs mechanism for the spontaneous symmetry breaking [12]. The vacuum space-time with very high symmetry is a good candidate for the creation of the universe from "nothing" [13]. It can be easily verified that all the homogeneous isotropic space-times have zero Weyl tensor. So. they automatically solve the vacuum equations of the conformal gravity. The idea that the initial state of the universe should be conformal invariant is advocated also by R. Penrose [14, 15] and G. 't Hooft [16].

This paper is devoted to the detailed description of our model for particle creation in the conformal gravity. We will use, in particular, the formalism from our previous papers [17, 18] on the conformal gravity.

Throughout the paper we use the units $\hbar = c = 1$ and the sign convention as in [19], i. e., the signature of the metric tensor is $g_{\mu\nu}$ is $(+, -, -, -)$, the Riemann curvature tensor is defined as

$$R^{\mu}{}_{\nu\lambda\sigma} = \frac{\partial\Gamma^{\mu}_{\nu\sigma}}{\partial x^{\lambda}} - \frac{\partial\Gamma^{\mu}_{\nu\lambda}}{\partial x^{\sigma}} + \Gamma^{\mu}_{\varkappa\lambda}\Gamma^{\varkappa}_{\nu\sigma} - \Gamma^{\mu}_{\varkappa\sigma}\Gamma^{\varkappa}_{\nu\lambda}, \tag{1}$$

while the Ricci tensor is the following convolution

$$R_{\nu\sigma} = R^{\mu}{}_{\nu\mu\sigma}. \tag{2}$$

The scalar curvature is $R = g^{\nu\sigma}R_{\nu\sigma}$, and $\Gamma^{\lambda}_{\mu\nu}$ are the metric connections, i. e., the covariant derivatives of the metric tensor are zero.

2 Phenomenological model

2.1 Hydrodynamics with varying particle number

We start with construction of the hydrodynamical part of our model. In the "classical" hydrodynamics there exist two different sets of dynamical variables, the so called Lagrangian and Eulerian coordinates. The first of them are comoving, i. e., the observer is sitting on some world-line. So, using the least action principle, one has to vary the trajectory of the (quasi)-particles. Since in such a case we cannot take into account the very processes of both creation and annihilation of particles (i. e., trajectories), it is not appropriate for our purposes. Therefore, we need to use the Eulerian description, when the dynamical variables are fields, namely, the particle number density $n(x0)$ and the four-velocities. The action integral in this case is [20]

$$S_{\text{hydro}} = -\int \varepsilon(X,n)\sqrt{-g}\,dx + \int \lambda_0(u^\mu u_\mu - 1)\sqrt{-g}\,dx + \int \lambda_2 X_{,\mu}u^\mu \sqrt{-g}\,dx + \int \lambda_1 (nu^\mu)_{;\mu}\sqrt{-g}\,dx, \quad (3)$$

where $\varepsilon(X,n)$ is the invariant energy density, $n(x)$ — invariant particle number density, $u^\mu(x)$ — four-velocity of the particle flow, $X(x)$ is the auxiliary dynamical variable introduced in order to avoid the identically zero vorticity of particle flow. It enters the action integral with the Lagrange multiplier λ_2, indicating the constraint $X_{,\mu}u^\mu = 0$, i.e., $X(x) = const$ on the trajectories, thus enumerating them. The other two Lagrange multipliers, $\lambda_0(x)$ and $\lambda_1(x)$ are responsible, respectively, for the constraints $u^\mu u_\mu = 1$ (natural normalization of the four-velocities) and $(nu^\mu)_{;\mu} = 0$ — particle number conservation law. The semicolon ";" denotes a covariant derivative with respect to the metric $g_{\mu\nu}$.

Our aim is to incorporate into the formalism the particle "creation law"

$$(nu^\mu)_{;\mu} = \Phi(inv) \neq 0. \quad (4)$$

Evidently, the function Φ should depend on some invariants of the fields causing this particle creation. Here we would like to explore the fundamental result by Ya. B. Zel'dovich and A. A. Starobinsky [10] obtain for the cosmological particle production

$$(nu^\mu)_{;\mu} = \beta C^2. \quad (5)$$

where C^2 is the square of the Weyl tensor $C^\mu_{\ \nu\lambda\sigma}$ (its definition as well as some most important properties see e. g., in [17, 18]) and the coefficient β depends on the type of particles under consideration. We will consider this "creation law" as our first postulate. So, the hydrodynamical part of the action integral now becomes

$$S_{\text{hydro}} = -\int \varepsilon(X,n)\sqrt{-g}\,dx + \int \lambda_0(u^\mu u_\mu - 1)\sqrt{-g}\,dx + \int \lambda_2 X_{,\mu}u^\mu \sqrt{-g}\,dx + \int \lambda_1((nu^\mu)_{;\mu} - \beta C^2)\sqrt{-g}\,dx, \quad (6)$$

Very important note. The Lagrange multiplier λ_1 is, actually, defined up to the additive constant. Indeed, let us replace $\lambda_1 \to \lambda_1 + \gamma_0$, $\gamma_0 = const$, then

$$\gamma_0 \int \left((nu^\mu)_{;\mu} - \beta C^2\right)\sqrt{-g}\,dx = \gamma_0 \int \left((n\sqrt{-g}u^\mu)_{,\mu} - \beta C^2\sqrt{-g}\right)dx \quad (7)$$

Due to the identity $(nu^\mu)_{;\mu}\sqrt{-g} = (n\sqrt{-g}u^\mu)_{,\mu}$, the corresponding volume integral transforms into the surface integral with no effect on the dynamical equations. Thus, we are left with the same "creation law" as before plus the Weyl gravitational action

$$S^{\text{Weyl}}_{\text{grav}} = -\gamma_0\beta \int C^2 \sqrt{-g}\,dx. \quad (8)$$

Thus, the conformal gravity is intrinsically contained in our hydrodynamical part of the total action integral, prior to the introducing the gravitational action itself!

2.2 Scalar field and conformal invariance

By the conformal transformation we will understand the space-time dependent scaling of the metric tensor $g_{\mu\nu}$,

$$ds^2 = g_{\mu\nu}(x)dx^\mu dx^\nu = \Omega^2\hat{g}_{\mu\nu}(x)dx^\mu dx^\nu = \Omega^2(x)d\hat{s}^2. \tag{9}$$

The conformal invariance means

$$\frac{\delta S_{\text{tot}}}{\delta\Omega} = 0. \tag{10}$$

Therefore, we can (and will) consider the conformal factor Ω as a dynamical variable and make variations independently in Ω and in $\hat{g}_{\mu\nu}$ [16].

Let

$$S_{\text{tot}} = S_{\text{grav}} + S_{\text{matter}}. \tag{11}$$

By definition

$$\delta S_{\text{matter}} = \frac{1}{2}\int T_{\mu\nu}\sqrt{-g}\,\delta g^{\mu\nu}dx \quad \text{and} \quad \delta S_{\text{matter}} = \frac{1}{2}\int \hat{T}_{\mu\nu}\sqrt{-\hat{g}}\,\delta\hat{g}^{\mu\nu}dx, \tag{12}$$

where $T_{\mu\nu}(\hat{T}_{\mu\nu})$ is the matter energy-momentum tensor. Consider, first, the following transformation of the metric tensor

$$\delta g^{\mu\nu} = -\frac{2}{\Omega^3}\hat{g}^{\mu\nu}\delta\Omega = -\frac{2}{\Omega}g^{\mu\nu}\delta\Omega. \tag{13}$$

Suppose

$$\frac{\delta S_{\text{grav}}}{\delta\Omega} = 0, \tag{14}$$

then

$$0 = \delta S_{\text{matter}} = -\int T_{\mu\nu}g^{\mu\nu}\frac{\delta\Omega}{\Omega}\sqrt{-g}\,dx, \tag{15}$$

that is, the trace of the energy-momentum tensor should be zero:

$$\mathrm{Tr}\,(T_{\mu\nu}) = \mathrm{Tr}\,(\hat{T}_{\mu\nu}) = 0. \tag{16}$$

If one considers the metric tensor transformation of the kind

$$\delta g^{\mu\nu} = \Omega^2\delta\hat{g}^{\mu\nu}, \tag{17}$$

then, as can be easily seen,

$$\hat{T}_{\mu\nu} = \Omega^2 T_{\mu\nu} \quad (\hat{T}^\mu_\nu = \Omega^4 T^\mu_\nu, \quad \hat{T}^{\mu\nu} = \Omega^6 T^{\mu\nu}). \tag{18}$$

Let us go further on. The question arises: quanta of what kind a field are creating? The most simple choice is the scalar field. And the simplest action integral is

$$S_{\text{scalar}} = \int\left(\frac{1}{2}\chi^\mu\chi_\mu - \frac{1}{2}m^2\chi^2\right)\sqrt{-g}\,dx. \tag{19}$$

Here $\chi_\mu = \chi_{,\mu}$ (comma denotes the partial derivative), $\chi^\mu = g^{\mu\nu}\chi_\nu$ and m is some constant with the dimension of mass. After the "standard" conformal transformations, namely

$$g^{\mu\nu} = \Omega^2\hat{g}^{\mu\nu}, \quad \chi = \frac{1}{\Omega}\hat{\chi}, \tag{20}$$

one gets

$$S_{\text{scalar}} = \int \left(\frac{1}{2}\hat{\chi}^{\mu}\hat{\chi}_{\mu} - \frac{1}{\Omega}\hat{\chi}_{\mu}\Omega^{\mu} + \frac{1}{2}\frac{\hat{\chi}^2}{\Omega^2}\Omega_{\mu}\Omega^{\mu} - \frac{1}{2}m^2\Omega^2\hat{\chi}^2 \right) \sqrt{-\hat{g}}\, dx. \tag{21}$$

Now indices are raising and lowering with the metric $\hat{g}_{\mu\nu}(\hat{g}^{\mu\nu})$. How to make this action conformally covariant? The recipe is well known: one should add into the Lagrangian the term $(R/12)\chi^2$, where R is the scalar curvature, constructing from the metric $g_{\mu\nu}$. The result is

$$\begin{aligned} S_{\text{scalar}} &= \int \left(\frac{1}{2}\chi^{\mu}\chi_{\mu} + \frac{R}{12}\chi^2 - \frac{1}{2}m^2\chi^2 \right) \sqrt{-g}\, dx \\ &= \int \left(\frac{1}{2}\hat{\chi}^{\mu}\hat{\chi}_{\mu} + \frac{\hat{R}}{12}\hat{\chi}^2 - \frac{1}{2}m^2\Omega^2\hat{\chi}^2 \right) \sqrt{-\hat{g}}\, dx - \frac{1}{2}\int \left(\hat{\chi}^2\frac{\Omega^{\lambda}}{\Omega} \right)_{|\lambda} \sqrt{-\hat{g}}\, dx. \end{aligned} \tag{22}$$

Here the vertical line "|" denotes the covariant derivative with respect to the metric $\hat{g}_{\mu\nu}$. The last term can be transformed to the surface integral, it does not effect the dynamics. Remarkably enough, that started with no gravitational action at all, we have got now both the conformal gravity (as a part of the "creation law") and the Einstein-Hilbert-dilaton gravity (as a part of the conformally covariant scalar field Lagrangian).

It seems that if one puts $m = 0$, everything else will be all right. But, it is not so easy, there exists a problem [16]. This problem concerns the signs. With the "correct" sign for the kinetic term $(1/2)\chi^{\mu}\chi_{\nu}$, we have the "wrong" sign for the Einstein-Hilbert-dilaton part, $+(1/12)\hat{R}\hat{\chi}^2$ (with our sign convention there should be "-" instead of "+"), and vice-versa. Our choice is the "correct" sign for $\hat{R}$, i. e., $-(1/12)\hat{R}\hat{\chi}^2$, and the "wrong" sign for the kinetic term, i. e., $-(1/2)\chi^{\mu}\chi_{\nu}$. This requires some explanation. First of all, we do not care about the "correct" sign for the kinetic term, because our scalar χ is not the "genuine" (i. e., fundamental) one. Some part of it we have already "used" as the created particles. The residual part can be viewed as the vacuum fluctuations that consist of virtual particles, including the conformal anomaly, which is responsible for the creation process. Moreover, the "wrong" sign in the kinetic term means the absence of the lower bound for the energy and allows even infinite number of the created particles (let us remember the C-field in the "steady state" cosmological model by F. Hoyle and J. V. Narlikar [21]). Besides, we are not going to consider our field χ as an independent dynamical variable. One more thing. If the scalar field χ is an independent dynamical variable, then, why it "knows" about the conformal transformation $g_{\mu\nu} = \Omega^2\hat{g}_{\mu\nu}$ and adjusts itself properly, i. e., $\hat{\chi} = \Omega\chi$? Only, when this field is a part of it! Fortunately, in our case it is not so, and one can always choose the conformal factor, $\Omega = \varphi$, in such a way that

$$\hat{\chi} = \frac{1}{\ell}\varphi, \tag{23}$$

where ℓ is some factor having dimension of length (it is introduced in order to keep the action integral dimensionless). Then, the action integral for the scalar field takes the form

$$S_{\text{scalar}} = -\frac{1}{\ell^2}\int \left(\frac{1}{2}\varphi^{\mu}\varphi_{\mu} + \frac{\hat{R}}{12}\varphi^2 - \frac{1}{2\ell^2}m^2\varphi^4 \right) \sqrt{-\hat{g}}\, dx \tag{24}$$

There appears the self-interaction term, φ^4. It must be noted that the power 4 in this term is only in the case of the four-dimensional space-time (it depends on the space-time dimensions). Here, two comments are in order. First, the above action is covariant under the conformal transformation, $\varphi(new) = \hat{\Omega}\varphi(old)$, $g_{\mu\nu}(old) = \hat{\Omega}^2 g_{\mu\nu}(new)$, $\sqrt{-g}(old) = \hat{\Omega}\sqrt{-g(new)}$, what can be easily checked. Second, it is now evident, that $3m^2 = \Lambda$ plays the role of the (bare) cosmological term.

To finish this Subsection we write down the energy-momentum tensor $T_{\mu\nu}$ for our (new) scalar field φ, obtained by varying S_{scalar} in $\hat{g}_{\mu\nu}$:

$$\begin{aligned}\hat{T}^{\text{scalar}}_{\mu\nu} &= -\frac{1}{\ell^2}\varphi_{\mu}\varphi_{\mu} + \frac{1}{2\ell^2}\varphi^{\sigma}\varphi_{\sigma}\hat{g}_{\mu\nu} - \frac{1}{2\ell^2}m^2\varphi^4\hat{g}_{\mu\nu} \\ & -\frac{1}{6\ell^2}\left(\varphi^2(\hat{R}_{\mu\nu} - \frac{1}{2}\hat{g}_{\mu\nu}\hat{R}) - 2\left((\varphi\varphi_{\nu})_{|\mu} - (\varphi\varphi^{\sigma})_{|\sigma}\hat{g}_{\mu\nu}\right)\right).\end{aligned} \tag{25}$$

Note the appearance of the second derivatives. The trace of this tensor equals

$$\mathrm{Tr}\,(\hat{T}^{\text{scalar}}_{\mu\nu}) = -\frac{1}{\ell^2}\left(\varphi\varphi^{\sigma}_{|\sigma} - \frac{\hat{R}}{6}\varphi^2 + 2m^2\varphi^4\right). \tag{26}$$

2.3 Hydrodynamics and conformal covariance

Since we consider now the conformal factor φ and transformed metric tensor $\hat{g}_{\mu\nu}$ as the independent dynamical variables, the above written hydrodynamical action integral should be properly "updated". Let us start with analysing the "creation law",

$$0 = \left((nu^{\mu})_{;\mu} - \beta C^2\right)\sqrt{-g} = \left((n\sqrt{-g}u^{\mu})_{,\mu} - \beta C^2\sqrt{-g}\right). \tag{27}$$

It is well known that in the four-dimensional space-time the combination $C^2\sqrt{-g}$ is invariant under conformal transformation, i. e.,

$$C^2\sqrt{-g} = \hat{C}^2\sqrt{-\hat{g}}. \tag{28}$$

So should be the full derivative $(nu^{\mu}\sqrt{-g})_{,\mu}$. The square of the interval ds^2 transforms as

$$ds^2 = g_{\mu\nu}dx^{\mu}dx^{\nu} = \varphi^2\hat{g}_{\mu\nu}dx^{\mu}dx^{\nu} = \varphi^2 d\hat{s}^2, \tag{29}$$

therefore, the four-velocity u^{μ} behaves as

$$u^{\mu} = \frac{dx^{\mu}}{ds} = \frac{1}{\varphi}\frac{dx^{\mu}}{d\hat{s}} = \frac{1}{\varphi}\hat{u}^{\mu}, \tag{30}$$

and, respectively,

$$u_{\mu} = g_{\mu\nu}u^{\nu} = \varphi\hat{g}_{\mu\nu}\hat{u}^{\nu} = \varphi\hat{u}_{\mu}. \tag{31}$$

Thus

$$n\sqrt{-g}u^{\mu} = n\varphi^3\sqrt{-\hat{g}}\hat{u}^{\mu} = \hat{n}\hat{u}^{\mu}, \tag{32}$$

where we introduced the new notation

$$\hat{n} = n\varphi^3\sqrt{-\hat{g}}. \tag{33}$$

It is clear that in the comoving coordinate system $\hat{n}$ is nothing but the particle number per unit spatial coordinate volume and, thus, the conformally invariant quantity. So, the "creation law" does not contain now the conformal factor φ explicitly. Therefore, the hydrodynamical part of the total action integral becomes

$$\begin{aligned}S_{\text{hydro}} &= -\int\varepsilon\left(X, \frac{\hat{n}}{\varphi^3\sqrt{-\hat{g}}}\right)\varphi^4\sqrt{-\hat{g}}\,dx + \int\lambda_0(\hat{u}^{\mu}\hat{u}_{\mu} - 1)\varphi^4\sqrt{-\hat{g}}\,dx \\ &+ \int\lambda_2 X_{,\mu}\hat{u}^{\mu}\varphi^3\sqrt{-\hat{g}}\,dx + \int\lambda_1\left((\hat{n}\hat{u}^{\mu})_{,\mu} - \beta\hat{C}^2\sqrt{-\hat{g}}\right)dx,\end{aligned} \tag{34}$$

and now the hydrodynamical variables are $\hat{n}$, $\hat{u}^\mu$ and X. Let us write down the corresponding equations of motion

$$\frac{\delta S_{\text{hydro}}}{\delta \hat{n}} = -\frac{\partial \epsilon}{\partial n}\frac{1}{\varphi^3\sqrt{-\hat{g}}} - \lambda_{1,\sigma}\hat{u}^\sigma = 0, \tag{35}$$

$$\frac{\delta S_{\text{hydro}}}{\delta \hat{u}^\mu} = 2\lambda_0 \hat{u}_\mu \varphi^4 + \lambda_2\varphi^3 X_{,\mu} - \lambda_{1,\mu}\frac{\hat{n}}{\sqrt{-\hat{g}}} = 0, \tag{36}$$

$$\frac{\delta S_{\text{hydro}}}{\delta X} = -\frac{\partial \epsilon}{\partial X}\varphi^4 - \frac{(\lambda_2\varphi^3\sqrt{-\hat{g}}\hat{u}^\sigma)_{,\sigma}}{\sqrt{-\hat{g}}} = 0. \tag{37}$$

To these we should add, of course, the constraints that follow from variation of the action integral in Lagrange multipliers λ_0, λ_1 and λ_2:

$$\hat{u}^\sigma \hat{u}_\sigma = u^\sigma u_\sigma = 1, \qquad X_\sigma \hat{u}^\sigma = X_\sigma u^\sigma = 0, \qquad (\hat{n}\hat{u}^\mu)_{,\mu} = \beta \hat{C}^2\sqrt{-\hat{g}}, \tag{38}$$

the last of them being equivalent to $(nu^\mu)_{;\mu} = \beta C^2$. The above equations of motion can be also written in terms of the quantities without "hats", namely

$$-\frac{\partial \epsilon}{\partial n} - \lambda_{1,\sigma}u^\sigma = 0, \quad 2\lambda_0 u_\mu + \lambda_2 X_{,\mu} - n\lambda_{1,\mu} = 0, \quad -\frac{\partial \epsilon}{\partial X} - (\lambda_2 u^\sigma)_{,\sigma} = 0. \tag{39}$$

It is not difficult to extract the Lagrange multiplier λ_0 from these equations. Indeed, by making the convolution of the second of the equations with the four-velocity vector u^μ and using the constraints, we get, after comparing the results with the first of the equations, that

$$2\lambda_0 = -n\frac{\partial \epsilon}{\partial n}. \tag{40}$$

Then, introducing the pressure p in the usual way, $p = -\epsilon + n\frac{\partial \epsilon}{\partial n}$, one obtains

$$2\lambda_0 = -(\epsilon + p). \tag{41}$$

The next step is to compute the hydrodynamical part of the total energy-momentum tensor. Omitting the details, we present here the result:

$$\hat{T}^{\text{hydro}}_{\mu\nu} = -\frac{\hat{n}}{\varphi^3\sqrt{-\hat{g}}}\frac{\partial \epsilon}{\partial n}\hat{g}_{\mu\nu}\varphi^4 + \epsilon\varphi^4\hat{g}_{\mu\nu} - 2\lambda_0\varphi^4\hat{u}^\mu\hat{u}^\nu - \lambda_0(\hat{u}^\sigma\hat{u}_\sigma - 1)\varphi^4\hat{g}_{\mu\nu} - \lambda_2 X_{,\sigma}\hat{u}^\sigma\varphi^3\hat{g}_{\mu\nu}$$
$$-4\beta\varphi^2\left((\lambda_1\hat{C}_{\mu\sigma\nu\lambda})^{|\lambda|\sigma} + \frac{1}{2}\lambda_1\hat{C}_{\mu\lambda\nu\sigma}\hat{R}^{\lambda\sigma}\right).$$

or

$$\hat{T}^{\text{hydro}}_{\mu\nu} = (\varepsilon + p)\varphi^4\hat{u}_\mu\hat{u}_\nu - p\varphi^4\hat{g}_{\mu\nu} - 4\beta\varphi^2\left((\lambda_1\hat{C}_{\mu\sigma\nu\lambda})^{|\lambda|\sigma} + \frac{1}{2}\lambda_1\hat{C}_{\mu\lambda\nu\sigma}\hat{R}^{\lambda\sigma}\right). \tag{42}$$

with the trace, equals to

$$\text{Tr}\,(T^{\text{hydro}}_{\mu\nu}) = (\varepsilon - 3p)\varphi^4. \tag{43}$$

The total trace equals

$$\text{Tr}\,(T^{\text{tot}}_{\mu\nu}) = -\frac{1}{\ell^2}\left(\varphi\varphi^{\sigma}_{|\sigma} - \frac{\hat{R}}{6}\varphi^2 + 2m^2\varphi^4\right) + (\varepsilon - 3p)\varphi^4. \tag{44}$$

Finally, let us write the result of the variation of the total action integral in φ, which can be considered as one of the equations of motion as well as the consequence of the postulated conformal invariance. One gets

$$\frac{1}{\ell^2}(\varphi^{\sigma}{}_{|\sigma} - \frac{1}{6}\hat{R}\varphi + 2m^2\varphi^3) + (\varepsilon - 3p)\varphi^3 = 0, \tag{45}$$

as it should be: $\mathrm{Tr}\,(T^{\mathrm{tot}}_{\mu\nu}) = 0$! The relation $\hat{T}_{\mu\nu} = \varphi T_{\mu\nu}$ can be also easily verified. This proves the self-consistency of our model. Note that in no way φ can to be zero value, since this would lead to to the degeneracy of the whole space-time.

3 Conclusion and Discussion

A. Conclusion

We have constructed the self-consistent conformally invariant phenomenological model for particle creation in the presence of strong gravitational fields. The word "phenomenological" means that we adopted classical description both for the created particles (hydrodynamics) and for the "creation law".

This "creation law" enters the action integral with the corresponding Lagrange multiplier and substitute the particle number conservation law in the conventional hydrodynamics. The idea (and our hope) is that such an inclusion of the particle creation law straight into the least action formalism will cause the essential change in the structure of the energy-momentum tensor and will lead to the violation of the energy dominance condition and, thus, will take into account (to some extent) the quantum character of the particle creation process. This idea is not quite new, it was already explored by one of the authors. The new thing is combining of the method with the postulated conformal invariance of the whole theory. This allows to restrict the possible functional form of the "creation law" up to the square of the Weyl tensor.

It appeared, to our surprise, that the above mentioned Lagrange multiplier can be determined only uo to an arbitrary constant. This means that the Weyl gravitational action is, actually, already incorporated into the formalism and does not need to be introduced artificially. The local conformal invariance, taken as the fundamental symmetry, has one more important consequence. In order to make it possible to create particles we need some fields which quanta are these very particles. The simplest is the scalar field. One needs it also because it is the scalar self-interacting field that gives the masses to particles through the Brout–Englert–Higgs mechanism (which also makes the conformal gravity meaningful). If one uses the simplest (again!) form for the scalar field Lagrangian, i.e., "the kinetic term + the mass term", then, in order to make it conformally covariant, it is necessary to introduce also the term proportional to the scalar curvature. Therefore, starting from hydrodynamics, needed for the description of the created particles and introducing the conformally invariant creation law plus the conformally covariant scalar field Lagrangian, we arrived at the conformal gravity theory with the Weyl Lagrangian plus the Einstein–Hilbert–dilaton gravity. This supports the idea, first discussed by A. D. Sakharov, about the induced gravity.

One more thing. In order to have the "correct" sign for the scalar curvature one has to choose the "wrong" sign for the kinetic term in the scalar field Lagrangian. But this causes no conceptual difficulties at all, since our scalar field is not the genuine (fundamental), it is simply the "vacuum residual" part of some entity, the "above-vacuum" part of which is already present in the form of the created particles, and its "conformal anomalous" part is already included into the "creation law". Therefore, it is very "natural" to identify the "vacuum residual" part with the conformal factor of the metric tensor. The mass term now plays the twofold role, it produces the self-interaction and the cosmological term, both initially absent.

B. Discussion

1. The model presented above, is very minimalistic. The matter is not only in that we did not include into consideration the electromagnetic (abelian) and other (nonabelian) gauge fields, causing creation of the particle–antiparticle pairs with opposite charges. Here we restricted ourselves by the specific form of the "creation law", when the rate of particle production is proportional to the square of the Weyl tensor. The absence of other possible terms may be explained by the adopted conformal invariance principle. Indeed, $C^2\sqrt{-g}$ is conformally invariant. It is usually claimed that the latter is the only conformally invariant combination quadratic in Riemann curvature tensor in four dimensions. But, there exists yet another quadratic conformally invariant combination, namely, the so-called Hirzebruch–Pontryagin density $R^{\mu\nu\lambda\sigma}\,{}^*R_{\mu\nu\lambda\sigma}\sqrt{-g}$, where "star" means that ${}^*R_{\mu\nu\lambda\sigma} = \epsilon_{\mu\nu\alpha\beta}R^{\alpha\beta}{}_{\lambda\sigma}$. It is the total derivative and, therefore, when in the action integral, does not alter the equations of motion. In our model, however, it would enter together with the Lagrange multiplier (λ_1) and would have an influence on the whole situation. Of course, it is not a genuine scalar, but pseudoscalar. And may be be, it is good, manifesting the T–violation in the irreversible particle creation processes.

2. Actually, our gravitational equations, when written in terms of quantities without "hats", look as

$$R_{\mu\nu} - \frac{1}{2}g_{\mu\nu}R - \Lambda g_{\mu\nu} = 6l^2 T^{\text{hydro}}_{\mu\nu} \tag{46}$$

(Λ is the cosmological constant). Therefore this is just Einstein gravity, but with the modified energy-momentum tensor, which includes the terms responsible for the particle creation. Nevertheless, the whole theory is conformally invariant on the action integral level, due to the special choice of the dynamical variables. From this it follows immediately that the only solution to the spherically symmetric vacuum equations, free of singularities and stable against the particle creation, is the (anti)-de Sitter or Minkowski space-time. (In the case of spherical symmetry, the vanishing C^2 implies the vanishing C itself, and this singles out the above mentioned metrics.) It sounds not good because the absence of the observers (no particles at all!) would seem to be in favour of the conformally invariant vacuum state. Remember that the existence of such a vacuum was one of the main motivation to reconsider the conformal gravity after the decades of disregarding.

There exists, however, the possibility (may be, exotic one) to switch on the particle creation and still have a vacuum state. To reach this goal, let us allow to create particles of both positive and negative energies (in equal amounts). This resembles the Dirac's sea. Then, we would be much less restrictive in constructing, the already dynamical, vacuum solution. Possible quantum fluctuations will have either positive or negative energies. Those with negative energies will be spreading away due to the self–anti–gravity, while that ones with positive energies will be subject to the usual gravitational instability.

3. The most important problem is how to organize the Braut–Englert–Higgs mechanism for generating particle's rest masses. The conventional line of reasoning is inapplicable here, because the usual (and the most convenient) solution $\varphi = 0$ is impossible in our scheme, since it would mean the conformal factor vanishes and, thus, the very notion if the metrical space–time would become meaningless.

4. Let us imagine that the above problem is already solved. Then, we are able to construct the observers equipped with the clocks and other measurement devices, engines for correcting trajectories and all that. In the self-consistent theory these observers cannot be arbitrary at all. For example, the so called vacuum observers (those, "sitting" outside the matter distribution) should see (= measure) different things, depending on their trajectories: the uniformly accelerated ones must be surrounded by the thermal bath, having the Unruh temperature. The simplest construction of this kind (but without the thermal bath) was done in [22].

5. All these problems are for the future investigations.

Acknowledgments

The reported study was partially supported by RFBR, research project No. 15-02-05038 a. The authors would like to thank Alexey Smirnov for numerous discussions and comments. One of us (V. B) is also grateful to Artyom Starodubtsev, Michail Smolyakov and Igor Gulamov for helpful discussions.

References

[1] L. Parker, Phys. Rev., **183**, 1057–1068 (1969)
[2] A. A. Grib and S. G. Mamaev, Sov. J. Nucl. Phys., **10**, 722–725 (1970)
[3] V. A. Berezin, V. A. Kuzmin and I. I. Tkachev, Phys. Lett. B, **124**, 479–483 (1983)
[4] Ya. B. Zel'dovich, JETP Lett., **9**, 307–312 (1970)
[5] Ya. B. Zel'dovich and A. A. Starobinskii, Sov. Phys. — JETP, **34**, 1159–1165 (1972)
[6] V. N. Lukash, A. A. Starobinskii, Sov. Phys. — JETP, **39**, 742–747 (1974)
[7] A. D. Sakharov, Sov. Phys. Doklady, **12**, 1040–1044 (1968)
[8] V. A. Berezin, Intern, J. Mod. Phys. A, **2**, 1591–1615 (1987)
[9] V. A. Berezin, Proc. 15th Russian Grav. Conf. "RUSGRAV-15", Jun. 30 — Jul. 05, Kazan, Russia, 58–62 (2014); arXiv:1404.3582 [gr-qc]
[10] Ya. B. Zel'dovich and A. A. Starobinskii, JETP Lett., **26**, 252–255 (1977)
[11] H. Weyl, Mathem. Zeit., **2**, 384–411 (1918)
[12] G. 't Hooft, arXiv:1410.6675 [gr-qc].
[13] A. Vilenkin, Phys. Lett. B, **117**, 25–28 (1982)
[14] R. Penrose, *Cycles of time: An extraordinary new view of the Universe* (The Random House, London, 2010)
[15] R. Penrose, Found. Phys, **44**, 873–890 (2014)
[16] G. 't Hooft, arXiv:1511.04427 [gr-qc]
[17] V. A. Berezin, V. I. Dokuchaev, and Yu. N. Eroshenko, JCAP **2016**, 019 (2016); arXiv:1412.2917v1 [gr-qc]
[18] V. A. Berezin, V. I. Dokuchaev, and Yu. N. Eroshenko, *J. Mod. Phys. A*, **31**, 1641004 (2016); arXiv:1509.00212 [gr-qc]
[19] L. D. Landau and E. M. Lifsitz, *The classical theory of fields,“ Course of theoretical physics*, vol 2. Forth edition (Butterworth–Heinemann, Oxford, 2000)
[20] J. R. Ray, J. Math. Phys. **13**, 1451–1453 (1972)
[21] F. Hoyle and J. V. Narlikar, *Proc. Roy. Soc. London. A*, **273**, 1–11, (1963)
[22] V. A. Berezin, arXiv:1106.0670 [gr-qc]

EPJ Web of Conferences **125**, 03004 (2016) DOI: 10.1051/epjconf/201612503004
QUARKS-2016

A dark matter component decaying after recombination: lensing constraints with Planck data

Anton Chudaykin[1,2,⋆], *Dmitry* Gorbunov[1,2,⋆⋆], and *Igor* Tkachev[1,3,⋆⋆⋆]

[1] *Institute for Nuclear Research of the Russian Academy of Sciences, Moscow 117312, Russia*
[2] *Moscow Institute of Physics and Technology, Dolgoprudny 141700, Russia*
[3] *Novosibirsk State University, Novosibirsk 630090, Russia*

Abstract. It was recently proposed [1] that the model with a fraction of decaying cold dark matter is able to reconcile measurements in high redshift (CMB) and low redshift (probes of cluster abundance and the Hubble constant). We check this statement employing the full likelihood of CMB Planck data. We find that the lensing effect calculated from anisotropy spectra measured by Planck imposes the strong constraint on the fraction of unstable dark matter as $F < 8\,\%$ (2σ). However, combining the CMB data with conflicting measurements in low redshift we obtain that the model with $F \approx 2 - 5\,\%$ improves the goodness-of-fit by $1.5 - 2\sigma$ depending on A_s and τ priors in comparison with the concordance ΛCDM model.

1 Introduction

The nature of dark matter and dark energy in the Universe remains mysterious. Order-of-magnitude equality between contributions of baryonic matter and dark matter to the energy density of the Universe today is treated as a chance coincidence. In this situation dark matter may easily consists of several components. To distinguish them we suppose that one of that is unstable and decays into invisible radiation and the other behaviours as Cold Dark Matter (CDM).

In recent paper [1] the authors consider the model with a subdominant fraction of dark matter decaying after recombination and argued that such a model is able to reconcile the emerging tension between the cosmological probes in high redshift and low redshift. Namely, they reveal that the model under consideration explains extremely high value of the Hubble constant $h = 0.738 \pm 0.024$ [2], $h = 0.743 \pm 0.021$ [3] obtained in observations of astronomical Standard Candles in comparison with the corresponding value $h = 0.6727 \pm 0.0066$ extracted from Planck spectra using TT, TE, EE + lowP dataset in notations of the original paper [4]. This model enable us to bring into accord constraints on parameters σ_8, Ω_m imposed by different Large Scale Structure tests with Planck observations as well. In our case, we check the consistency of the model with the galaxy cluster counts measured by Planck via Sunyaev-Zeldovich effect [4] which prefer lower values of σ_8.

However, in Ref. [1] the authors did not exploit the Planck likelihood properly. Instead of this they fixed all primary cosmological parameters relevant at recombination to Planck best-fit values

⋆e-mail: chudy@ms2.inr.ac.ru
⋆⋆e-mail: gorby@ms2.inr.ac.ru
⋆⋆⋆e-mail: tkachev@ms2.inr.ac.ru

– this approach guaranties that anisotropies at the last scattering will be identical in the Decaying Dark Matter (DDM) and ΛCDM models. Furthermore, the authors assume that the angular size of the sound horizon at the last scattering remains the same, or equivalently they put the quantity θ_* to the Planck value – this implies that CMB spectra calculated in DDM almost coincide with those obtained in ΛCDM. A possible deviation from expected quantities can occur only due to gravitational distortions of spectra between last-scattering and the present epoch. Although such distortions are subdominant, the current cosmological data can be sensitive to them.

There are two sources of gravitationally distortions. The first one consists in the rise in the amplitude C_l at low multipoles according to the integrated Sachs-Wolfe effect. This is caused by large values of the cosmological constant Λ in spatially flat Universe since part of dark matter decays. This effect turns out to be insignificantly small giving the correction to C_l below cosmic variance in the relevant range of multipoles (find explanation in [1]).

The second effect which was missed in [1] related to lensing of CMB anisotropies on the compact structure in the late Universe leaves imprints on observable CMB maps as well. The difference in lensing power in DDM and ΛCDM can be used as a smoking gun to resolve which model is better consistent with high quality data and provides better parameter convergence.

To fill the gap we fit all parameters of the DDM model to the current data accounting for the Planck likelihood properly. Our goal is to find out whether DDM is really able to reconcile cosmological measurements in high and low multipoles. We employ the complete Planck 2015 CMB data [5, 6], Hubble measurement [7] and the Planck cluster counts via Sunyaev-Zeldovich effect [8].

2 The model, data sets and procedure

2.1 Decaying Dark Matter model

The two component DDM model has two additional parameters, the fraction of decaying part in the total amount of dark matter , F, and the width of corresponding decay, Γ. We define the parameter F in terms of initial quantities of stable and decaying dark matter density fractions as follows $F \equiv \omega_{ddm}/(\omega_{sdm}+\omega_{ddm})$. Initial means that these densities are measured if $\Gamma = 0$. According to Ref. [1] we assume that unstable dark matter decays into massless invisible particles and the width Γ is measured in units of km/s/Mpc as H_0.

2.2 Cosmological data sets

The main cosmological dataset we always used here is TT,TE,EE which include all CMB spectra measured by Planck in the multipole range $l > 30$ in full compliance with [5]. It already contains gravitational distortions of CMB anisotropies due to lensing on compact structures in the late Universe. Lensing results in smoothing acoustic oscillation in the CMB spectra, so the peaks become lower and the troughs higher.

In some cases we include information about CMB anisotropies at low multipoles including polarisation data and refer to this set as lowP in later according to [5].

Since the lensing effect is the most critical for us we also include the direct measurement of the lensing potential $C_l^{\phi\phi}$ computed by Planck Collaboration using non-Gaussian (connect) part of 4-point correlation functions [6]. We use the notation lens for this measurement.

For the low redshift measurements we employ the constraints on the present expansion rate of the Universe and on abundance of Large Scale Structure today. Namely, we include the current measurement of the Hubble constant with smaller error recently updated by Riess *et. al* [7] and denote it as H_0.

Dataset
(TT, TE, EE + H_0 + CL) + lowP
(TT, TE, EE + H_0 + CL) + lens
(TT, TE, EE + H_0 + CL) + lowP + lens

Table 1. Datasets used in our analysis.

As for structure tests we adopt the galaxy cluster counts measured by Planck via Sunyaev-Zeldovich effect which was used in Ref. [1] as well. We indicate this Planck probe below as CL.

Throughout the paper we always use the data combination (TT, TE, EE + H_0 + CL). However, the measurement of lensing amplitude obtained here is in moderate conflict with one extracted from lens, see [5, 6]. Therefore we decide to separate these probes of the lensing effect at late times and divide all data into three blocks: with lowP, lens and lowP+lens. Our data separation reflects the main difference in lensing power predicted by various cosmological observations.

2.3 Numerical procedure

To find the allowed region in DDM model parameter space we carry out the Markov Chain Monte Carlo analysis using the Monte Python package [9] with the CLASS Boltzman code [10, 11] adopted to the two component DDM model. Eight parameters have been fitted to various observational data: two of them describe intrinsic properties of the DDM model, the fraction F and the width Γ, whereas the remaining six – the angular size of the sound horizon r_s at last-scattering $\theta_* \equiv 100 \times r_s(z_*)/D_A(z_*)$, the baryon density ω_b, initial CDM density $\omega_{cdm} = \omega_{sdm} + \omega_{ddm}$, the optical depth τ, the squared amplitude A_s and tilt n_s of the power spectrum of primordial scalar perturbations – are standard. In our numerical analysis we calculate the perturbations in linear regime only preliminary checked that nonlinear corrections to all relevant quantities are rather small in our approach. For instance, the nonlinear contribution to the matter power spectrum is below 1% at scales relevant for estimation the parameter σ_8. We consider the Universe is spatially flat, neglect possible tensor perturbations in metric and assume normal hierarchy pattern with the sum of the active neutrino masses $m_\nu = 0.06\,\text{eV}$.

3 Constraints on DDM

3.1 Planck data only

Firstly, we emphasize the particular role of the lensing effect which affects the TT power spectrum. For this purpose we depict the difference in this quantity between DDM ($F = 0.1$, $\Gamma = 2000$ km/s/Mpc) and ΛCDM predictions assuming the other parameters are fixed to the best-fit values obtained in TT, TE, EE + lowP analysis. The corresponding difference is shown in the left-hand panel of Fig. 1 by solid red line. The Planck data with the preliminary subtracted prediction of ΛCDM is depicted on the same plot as blue points with error bars as well. We see that the Planck's sensitivity and the difference in lensing power between DDM and ΛCDM models are comparable to each other; this enables the data to feel the peculiar feature of DDM – different lensing power. Although the lensing effect is accounted for in both models, the prediction of the concordance ΛCDM model does not correspond the data points perfectly: the data points oscillate closely to zero. This means that the Planck likelihood prefers a bit stronger lensing effect in comparison with the theoretical prediction of

ΛCDM model; corresponding disagreement is at 2σ confidence level [6]. Although the difference in best-fit predictions of DDM and ΛCDM models (red solid line) is comparable with deviation of the residuals of the data (blue points), it is out of phase. This reflects lensing power even weaker than in the standard cosmology model. As a result of that, proper fitting to TT, TE, EE + lowP likelihood impose the strong constraint on the parameter of DDM, $F < 0.07(2\sigma)$. This observation was missed in the previous work [1].

To emphasize the role of the lensing effect and to reveal the importance of that more explicitly in our approach we show the difference in the lensing power spectrum between DDM ($F = 0.1$, $\Gamma = 2000$ km/s/Mpc) and ΛCDM predictions for the same best-fit parameters obtained in the TT, TE, EE + lowP analysis. The solid red curve in the right panel of Fig. 1 depicts this difference whereas the data

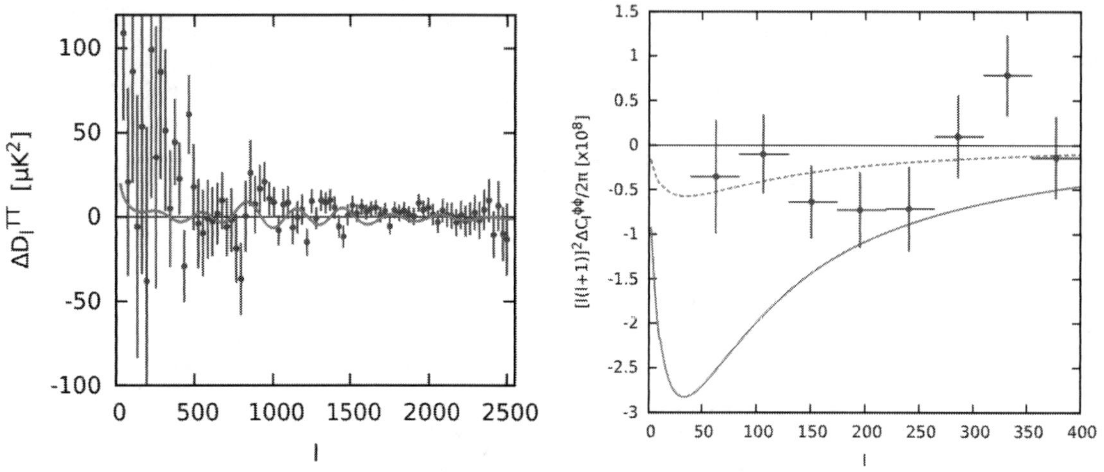

Figure 1. *Left.* Data points with error bars show residuals after subtraction from the measured TT power spectrum the ΛCDM prediction with the best-fit parameters obtained in TT, TE, EE + lowP analysis. Solid curve corresponds to the difference between TT spectra in DDM ($F = 0.1$, $\Gamma = 2000$ km/s/Mpc) and ΛCDM for the same best-fit parameters. *Right.* Data points with error bars show residuals after subtraction from the directly measured lensing power spectrum by Planck the best-fit ΛCDM prediction within TT, TE, EE + lowP analysis. The solid curve depicts the difference in $C_l^{\phi\phi}$ between DDM ($F = 0.1$, $\Gamma = 2000$ km/s/Mpc) and ΛCDM models assuming the same values of standard parameters as in the left panel. The dotted curve corresponds to the best-fit DDM model obtained in the TT, TE, EE + lowP + lens analysis.

points with error bars correspond to the direct measurement of the lensing power spectrum computed from the connected 4-point functions by Planck Collaboration after subtraction the best-fit ΛCDM model from the data. As we can see from the right-hand plot of Fig. 1 lensing power in DDM is actually weaker than in the standard cosmology as mentioned above. However the DDM model pushes the $C_l^{\phi\phi}$ values towards the prediction of the lens dataset. Therefore the model with DDM is worse consistent with Planck data spectrum and better consistent with the direct measurement of the lensing power spectrum than the model with CDM. Because of the first statement the DDM model is tightly constrained by TT, TE, EE likelihood and because of the second one – taking into account lens data provides a more conservative constraint on the parameter F.

The data set lowP does not restrict lensing power directly. However, the situation is a little more tricky and the role of the polarisation measurement at low multipoles coupled to the TT, TE, EE probe turns out to be important. The lack of lensing power in DDM causes higher amplitude values of primordial scalar spectra, A_s, which, in turn, amplifies the amplitude of C_l^{TT}. To compensate the latter growth larger values of optical depth, τ, is needed but the quantity τ is strongly constrained by lowP likelihood. As a result, the parameter F becomes even more restricted in combined TT, TE, EE + lowP dataset, $F < 0.04(2\sigma)$, in comparison with the prediction of TT, TE, EE data only.

The preference of the lens data is opposite. The direct measurement of $C_l^{\phi\phi}$ prefers a little bit weaker lensing power as compared to the ΛCDM prediction [5, 6]. It means that lens likelihood favours DDM consideration rather then ΛCDM one and causes the weaker constraint on the DDM parameter, $F < 0.08(2\sigma)$.

3.2 Planck data and conflicting low z measurements

Now we combine the Planck probes in high redshift with the measurements of H_0, Ω_m and σ_8 in low redshift. Since the DDM is restricted by lensing which acts in opposite directions in different cosmological observations we explore the parameter space in DDM using three different cosmological datasets which are listed in Table 1.

The corresponding constraints on DDM parameters and on the relevant quantities in low redshift measurements are plotted in Figs. 2 - 3. One can see from Fig. 2 that the preferred values $F \simeq 2-5\%$

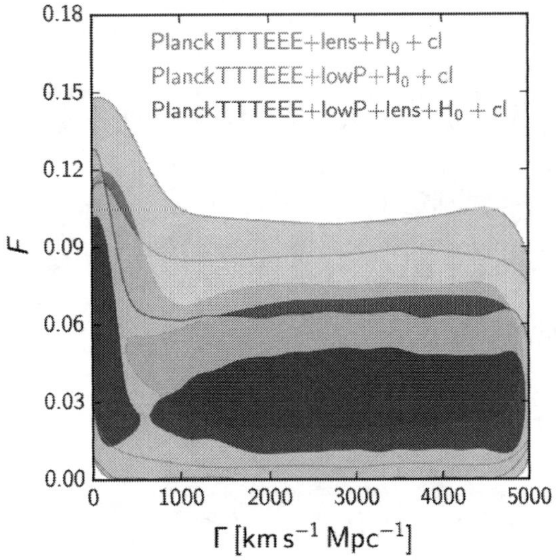

Figure 2. Posterior distributions (1 σ and 2 σ confident regions) of parameters F, Γ in the DDM model.

are almost independent of the parameter Γ in the range $1000\,\mathrm{km/(s\,Mpc)} \lesssim \Gamma \lesssim 5000\,\mathrm{km/(s\,Mpc)}$ which means that all amount of unstable dark matter completely decays after recombination since $\Gamma \gg H_0$. We emphasize that the line $F = 0$ which corresponds to the models with a minuscule fraction of DDM is within 2σ confident region for the (TT, TE, EE + H_0 + CL) + lens and (TT, TE, EE + H_0 + CL) + lowP + lens datasets, but outside for (TT, TE, EE + H_0 + CL) + lowP. The increase in F for lower values of Γ corresponds to the models with non vanishingly small amount of DDM today. In the limit $\Gamma \ll H_0$, DDM at the present is undistinguishable from common CDM. The regions of small Γ is unresolved here and deserve a special study beyond the current research. In the left-hand panel of Fig. 3 the allowed regions at 2σ confidence level with highest values of F and lowest values of H_0 correspond to the range $\Gamma \ll 1000\,\mathrm{km/(s\,Mpc)}$ in Fig. 2 with long-lived DDM particles. In the

dataset	$\Delta\chi^2$	p-value	improvement
(TT, TE, EE + H_0 + CL) + lowP	5.34	0.0692	1.81σ
(TT, TE, EE + H_0 + CL) + lens	4.68	0.0963	1.66σ
(TT, TE, EE + H_0 + CL) + lowP + lens	4.28	0.1176	1.56σ

Table 2. Improvement of DDM over ΛCDM in three datasets considered and listed Tab. 1 in case of 2 extra degrees of freedom in DDM.

right panel of Fig. 3 the regions with highest allowed values of σ_8 and lowest values of Ω_m map to the parameter space 1000 km/(s Mpc) $\lesssim \Gamma \lesssim$ 5000 km/(s Mpc) and highest values of F in Fig. 2.

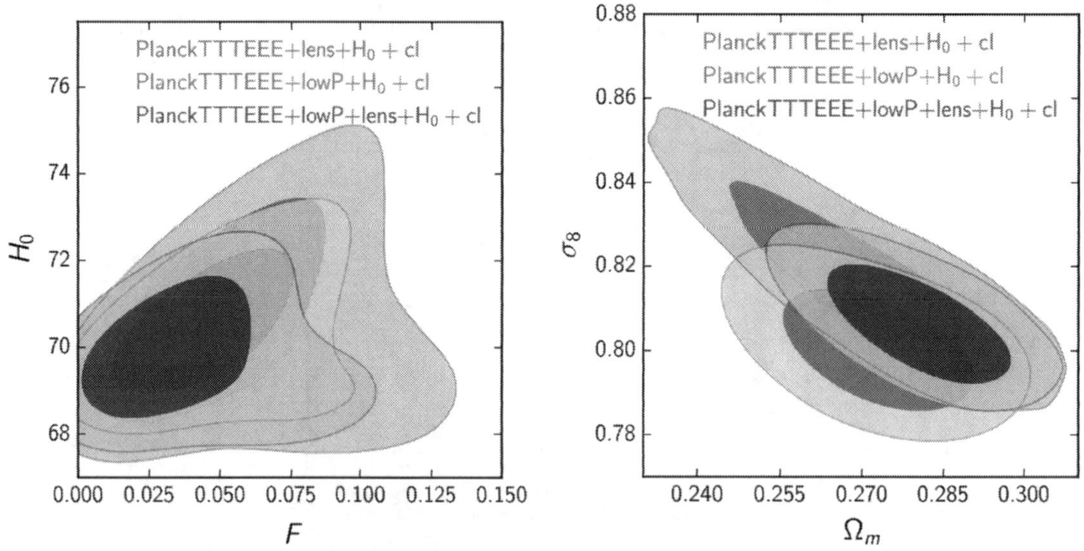

Figure 3. Same as Fig. 2 but for pairs of the parameters H_0-F and σ_8-Ω_m.

We notice that proper fitting gives significantly stronger upper limit on the fraction F which contradicts the result obtained in Ref. [1]. In addition, the allowed region for parameters F, Γ obtained in Ref. [1] is well outside the 2σ contours depicted in Fig. 2. However, the DDM consideration still prefers non-zero values of the parameter F and absence of the decaying part of dark matter here is disfavoured.

To understand which model better describes various observational data we compare logarithmic likelihood $\log L$ calculated for both models in their best-fit points. The difference $2 \cdot \Delta \log L$ between DDM and ΛCDM is distributed as χ^2 with effective number of degree of freedom equalled to the difference in fitting parameters in these two models. In our research the number of fitting parameters equals 2 and corresponds to the DDM parameters, F and Γ. The results of such quantitative analysis are represented in Tab 1.

The DDM consideration indeed improves the goodness-of-fit in comparison with ΛCDM one as suggested in Ref. [1]. However the improvement is not so significant as suggest before because the

Planck data suffers from the lensing conflict represented in Fig. 1. It is entered in all datasets and is worse for DDM than for ΛCDM.

In principle, the corresponding constraint of F is able to rule out DDM completely in future. But we caution the reader while we do not understand the origin of the lensing conflict within one precise measurement we are not able to make any strong conclusion about favour or disfavour of the two component DDM model in comparison with ΛCDM one.

4 Conclusion

We confirm that the DDM consideration provides better description of the current observational data which consists of the precise Planck probe of the Universe at hight redshifts and measurements of the cosmological parameters at low redshifts. However, the favoured values of the fraction F are significantly smaller than ones claimed in the previous work [1].

We explain why lensing information is critical for verification of the two component DDM model using the current cosmological data. Planck measures lensing power in two different ways. Firstly, lensing is extracted from CMB anisotropy maps as smoothing the acoustic structure of the spectra. The second probe of lensing power consists in the direct measurement of the lensing power spectrum $C_l^{\phi\phi}$ using nun-Gaussian parts of all 4-point correlation functions. But both these measurements are conflicting with the ΛCDM prediction which could be also treated as disagreement between low-z and high-z experiments. Namely, the concordance ΛCDM model predicts a bit stronger lensing power as compared to the direct measurements of $C_l^{\phi\phi}$ and significantly weaker one in comparison with the preference of TT, TE, EE + lowP likelihood. Therefore the former data actually favours DDM whereas the latter tightly constrains.

Taking into account the lensing conflict inside the Planck likelihood, we have checked whether the DDM model is able to reconcile measurements at high and low redshifts. The improvement of DDM over ΛCDM is observed in our research but this is not so prominent as suggested before. We feel that the lensing information extracted from the current data is contradictory and requires further investigation. This means that arguments based on present lensing information only should not be taken until the lensing conflict in the Planck data has not been totally resolved.

We extend gratitude to Z. Berezhiani, A. Dolgov and J. Lesgourgues for helpful discussion of the material. The work of A. Chudaykin and D. Gorbunov has been supported by Russian Foundation for Basic Research grant 14-02-00894. All simulations in the paper are carried out with MVS-10P supercomputer of Joint Supercomputer Center of the Russian Academy of Sciences (JSCC RAS).

References

[1] Z. Berezhiani, A.D. Dolgov, I.I. Tkachev, Phys. Rev. **D92**, 061303 (2015), 1505.03644
[2] A.G. Riess, L. Macri, S. Casertano, H. Lampeitl, H.C. Ferguson, A.V. Filippenko, S.W. Jha, W. Li, R. Chornock, Astrophys. J. **730**, 119 (2011), [Erratum: Astrophys. J.732,129(2011)], 1103.2976
[3] W.L. Freedman, B.F. Madore, V. Scowcroft, C. Burns, A. Monson, S.E. Persson, M. Seibert, J. Rigby, Astrophys. J. **758**, 24 (2012), 1208.3281
[4] P.A.R. Ade et al. (Planck) (2015), 1502.01597
[5] P.A.R. Ade et al. (Planck) (2015), 1502.01589
[6] P.A.R. Ade et al. (Planck) (2015), 1502.01591
[7] A.G. Riess et al., Astrophys. J. **826**, 56 (2016), 1604.01424
[8] P.A.R. Ade et al. (Planck), Astron. Astrophys. **571**, A20 (2014), 1303.5080

[9] B. Audren, J. Lesgourgues, K. Benabed, S. Prunet, JCAP **1302**, 001 (2013), 1210.7183
[10] J. Lesgourgues (2011), 1104.2932
[11] D. Blas, J. Lesgourgues, T. Tram, JCAP **1107**, 034 (2011), 1104.2933

EPJ Web of Conferences **125**, 03005 (2016) DOI: 10.1051/epjconf/201612503005
QUARKS-2016

Verification of f(R)-gravity in binary pulsars

Polina Dyadina[1,*], *Stanislav* Alexeyev[1], *Salvatore* Capozziello[2], and *Mariafelicia* De Laurentis[2]

[1] *Sternberg Astronomical Institute, Lomonosov Moscow State University*
[2] *Dipartimento di Scienze Fisiche, Università di Napoli "Federico II"*

Abstract. We develop the parameterized post-Keplerian approach for class of analytic $f(R)$-gravity models. Using the double binary pulsar system PSR J0737-3039 data we obtain restrictions on the parameters of this class of $f(R)$-models and show that $f(R)$-gravity is not ruled out by the observations in strong field regime.

1 Introduction

The theory of general relativity (GR) greatly improved the understanding of our Universe. It allows going beyond the Newtonian picture of the world. Anyway our present understanding has several shortcomings. A set of phenomena exists that we can not directly explain in the framework of GR. For explanation these phenomena many extended gravity models are created. One of the most perspective way to expand GR is to include additional corrections in terms of the Ricci scalar into the Lagrangian. This method underlies $f(R)$-gravity [1], [2]. This theory of gravity could allow to explain Dark Energy (DE) and Dark Matter (DM) by changing the geometry of a space-time and each scale demonstrates its specific geometry. This fact fully agrees with the early spirit of GR that could not act in the same way at all scales [3], [4].

However any theory of gravity should be verifiable. In this work we use the data for the binary pulsar PSR J0737-3039 for test $f(R)$-gravity in the strong field regime[6], [7]. PSR J0737-3039 is the only known double binary pulsar. The extraordinary closeness of the system components, small orbital period and the fact that we see almost edge-on system allow to investigate the manifestation of relativistic effects with the highest available precision.

The structure of the paper is the following. In Section 2 we develop the post-Keplerian formalism for analytic $f(R)$-gravity models and obtain observational constraints arising from PSR J0737-3039. The Section 3 contains our conclusions.

2 $f(R)$-Gravity and Restriction Constraints

2.1 $f(R)$-gravity background

The $f(R)$-gravity models are based on different expansions and enlargements of GR. Higher-order curvature invariants and minimally or non-minimally coupled scalar fields are added into the gravity

* e-mail: guldur.anwo@gmail.com

action. In $f(R)$-gravity one considers a general function $f(R)$ instead of the Hilbert-Einstein Lagrangian R that is linear with respect to the Ricci scalar R. The only assumption that we make is that $f(R)$ is an analytic function. The gravitational action now looks like[3], [4], [8], [9]:

$$S = \int d^4x \sqrt{-g} \left[f(R) + \kappa L_{\rm m} \right], \tag{1}$$

where $\kappa = 16\pi G/c^4$ is the coupling, g is the determinant of the metric and $L_{\rm m}$ is the standard matter contribution.

We assume that the $f(R)$ Lagrangian is analytic (i.e. Taylor expandable) in terms of the Ricci scalar, therefore[8], [9]:

$$f(R) = \sum_n \frac{f^n(R_0)}{n!}(R - R_0)^n \simeq f_0 + f_0' R + \frac{1}{2} f_0'' R^2 + \ldots, \tag{2}$$

where

$$f_0 = const, \quad f_0' = \left.\frac{df(R)}{dR}\right|_{\rm R=0}, \quad f_0'' = \left.\frac{d^2 f(R)}{dR^2}\right|_{\rm R=0}. \tag{3}$$

The flat background is recovered for $R = R_0 \simeq 0$. GR is recovered in the limit $f_0 = 0, f_0' = 4/3, f_0'' = 0$[3], [4]. Hereafter we assume $f_0 = 0$ [3], [4].

2.2 The parameterized post-Keplerian formalism

The parameterized post-Keplerian formalism (PPK) was developed by Damour and Deruelle[10] and further improved by Damour and Taylor [11]. It was created to link the arrival time of pulses and their radiation time in the frame of the pulsar. Damour et al. have showed that all the independent $O(v^2/c^2)$ timing effects can be represented in a simple mathematical way that is generic for a wide class of extended gravities. PPK is a strong-field analogue of the PPN formalism [5]. It includes such effects as the Einstein time delay, Römer time delay, Shapiro time delay and the effects of aberration. The general form of these corrections is model-independent, therefore all possible manifestations of the extended gravity model can be expressed through the 8 PPK parameters $\dot{\omega}$, γ, $\dot{P}_b$, r, s, δ_θ, $\dot{e}$, $\dot{x}$. However, we consider only the most accurately measured parameters, so we do not take into account the last three of them. In this work we obtain the analytical form of four PPK parameters:

$$
\begin{aligned}
\dot{\omega} = &\left(\frac{2\pi}{P_b}\right)^{5/3} \frac{G^{2/3}M_\odot^{2/3}(m_1+m_2)^{2/3}}{c^2(1-e^2)} \times \left(\frac{(f_0'')^2}{2(f_0')+2f_0'R+3(f_0'')^2}\right)^{2/3} \\
&\times\left(\frac{7}{2} - \frac{2(f_0'')^2}{f_0'+f_0'R+2(f_0'')^2} - \frac{1}{16}\frac{(2f_0'+2f_0''R+3(f_0'')^2)}{(f_0'+f_0''R+2(f_0'')^2)^4}\right. \\
&\times\left[\frac{(4(f_0')^3+12(f_0')^2f_0''R+22(f_0')^2(f_0'')^2+12f_0'(f_0'')^2R^2}{(f_0'+f_0''R+2(f_0'')^2)^4}+\right. \\
&\left.\left.+\frac{44f_0'(f_0'')^3R+39f_0'(f_0'')^4+4(f_0'')^3R^3+22(f_0'')^4R^2+39(f_0'')^5R+24(f_0'')^6)}{(f_0'+f_0''R+2(f_0'')^2)^4}\right]\right), \\
\gamma = e&\left(\frac{2\pi}{P_b}\right)^{-1/3}\frac{G^{2/3}M_\odot^{2/3}m_2}{c^2(m_1+m_2)^{4/3}} \times \\
&\times\left(m_1+m_2\left[2+\frac{(f_0'')^2}{2f_0'+2f_0'R+3(f_0'')^2}\right]\right)\times \\
&\times\left(1+\frac{(f_0'')^2}{2f_0'+2f_0'R+3(f_0'')^2}\right)^{-1/3}, \\
r = &\frac{1}{4c^3}GM_\odot m_2, \\
s = &\left(\frac{2\pi}{P_b}\right)^{2/3}\frac{cx(m_1+m_2)^{2/3}}{(GM_\odot)^{1/3}\,m_2}\left(1+\frac{(f_0'')^2}{2f_0'+2f_0'R+3(f_0'')^2}\right)^{-1/3}.
\end{aligned}
\tag{4}
$$

The expression for the last parameter $\dot{P}_b$ was obtained by Capozziello and De Laurentis [3], [4] in the form:

$$
\begin{aligned}
\dot{P}_b = &-\frac{6\pi}{20}\left(\frac{2\pi}{P_b}\right)^{5/3}\frac{(GM_\odot)^{5/3}}{c^5(1-e^2)^{7/2}}\frac{m_1m_2}{(m_1+m_2)^{1/3}} \\
&\times\left((f_0'(37e^4+292e^2+96)-\frac{f_0''\pi^2}{2P_b(1+e^2)^3}\right. \\
&\left.\times\left(891e^8+28016e^6+82736e^4+43520e^2+3072\right)\right).
\end{aligned}
\tag{5}
$$

2.3 Restrictions from PSR J0737-3039

Now we can proceed directly to the method of testing models of gravity [11] and our additions. We plot the curves representing post-Keplerian parameters on the plane, in the Y-axis we put the allowed values of the companion masses m_2 and in the X-axis we put possible values of the pulsar m_1 masses. The region of all curves intersection within the measured accuracy displays the possible range of the pulsar's mass and its companion. For the appropriate gravity model all the curves of post-Keplerian parameters should "meet" at one point (taking into account the accuracy). The measurements of n post-Keplerian parameters give $n-2$ tests. In our case there are $6-2=4$ ones.

Here it is importantly to point out that all the curves for PSR J0737-3039 intersect within the measurement accuracy in the GR case (see Fig. 1).

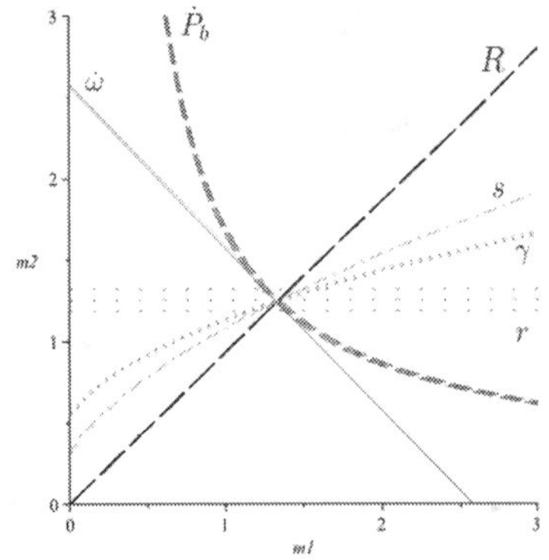

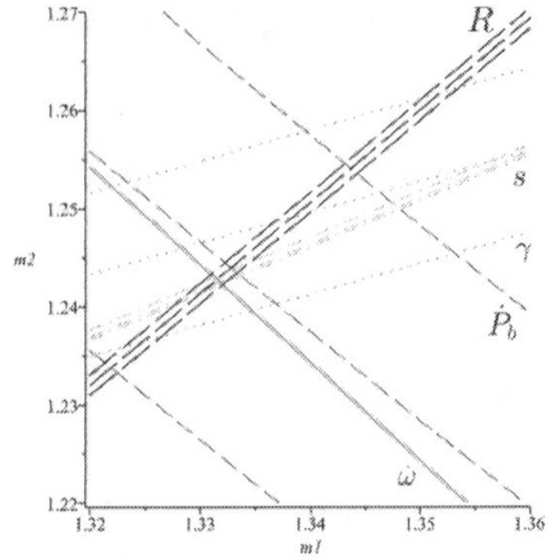

Figure 1. Dependence of the companion mass upon the pulsar one at different scales for GR, $f_0' = 4/3$ and $f_0'' = 0$.

Table 1. Parameters PSR J0737-3039

Parameter	Physical meaning	Experimental value
$P_b(day)$	orbital period	0.10225156248(5)
e	eccentricity	0.0877775(9)
$x(s)$	projected semimajor axis of the pulsar orbit	1.415032(1)
$\dot{\omega}(deg/yr)$	secular advance of the periastron	16.89947(68)
$\tilde{\gamma}(ms)$	time dilation parameter	0.3856(26)
$\dot{P}_b$	secular change of the orbital period	$-1.252(17) \times 10^{-12}$
s	Shapiro delay parameter	0.99974(−39, +16)
$r(\mu s)$	Shapiro delay parameter	6.21(33)
$R = \frac{m_1}{m_2} = \frac{x_2}{x_1}$	mass ratio	1.0714(11)

To obtain restrictions for $f(R)$-gravity we solve the system (4-5) considering observational data within the measurement accuracy for PSR J0737-3039 (see 1). We fix $f_0' = 4/3$ and obtain that the curves intersect only when f_0'' takes the following values (see Fig. 2) :

$$-0.0623 \leq f_0'' \leq 0. \tag{6}$$

This is the maximum limit that we can get for f_0'' when $R << 1$. In this case masses of pulsar and its component take the following values:

$$m_{p1} = 1.3331,$$
$$m_{p2} = 1.2429. \tag{7}$$

Hence the difference between GR and $f(R)$-gravity is 0.5%.

DOI: 10.1051/epjconf/201612503005

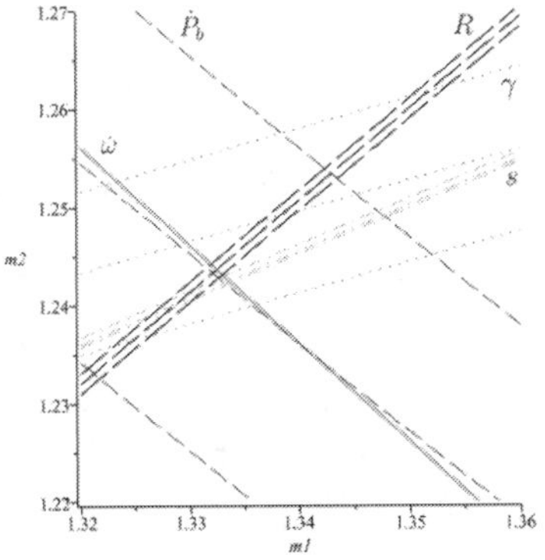

h]

Figure 2. Dependence of the companion mass upon the pulsar one, $f_0' = 4/3$ and $f_0'' = -0.0623$. The last point of intersection $s(m_1, m_2)$, $R(m_1, m_2)$ and $\dot{\omega}(m_1, m_2)$ within the measurement accuracy, when f_0' is fixed.

3 Conclusions

We consider the analytical $f(R)$-gravity in the system of binary pulsars. To find the observational limitations on these models the PPK parameters and other orbital data of the double binary pulsar PSR J0737-3039 [6], [7] are applied. We find the analytical form of the four PPK parameters for the analytic $f(R)$-gravity. It is important that we use the general form of $f(R)$-gravity. The only assumption is that $f(R)$ is supposed to be an analytical function.

In this work we obtained strict limits for free parameter f_0'' (assuming that $f_0' = 4/3$ and $R << 1$):

$$-0.0623 \leq f_0'' \leq 0. \tag{8}$$

This test proves that Extended Theories of Gravity namely the analytic $f(R)$-gravity models are not excluded in strong field regimes. This result gives rise for further studies of $f(R)$-gravity and other models based of such type of GR extension: Palatini $f(R)$-gravity, [13] hybrid metric-Palatini $f(R)$-gravity [14] and also scalar-tensor theories, for example Horndeski gravity [15].

Acknowledgments

The work is supported by the Russian Foundation for Basic Research grant 16-02-00682/16.

References

[1] S. Capozziello and R. de Ritis *Class. Quantum Grav.* **11**, 107 (1994); *Phys. Lett. A* **177**, 1 (1993).

[2] S. Capozziello, M. Demianski, R. de Ritis and C. Rubano *Phys. Rev. D* **52** (1995).

[3] M. De Laurentis and S. Capozziello *Astrop. Phys.* **35**, 257 (2011).

[4] M. De Laurentis and I. de Martino *MNRAS D* **741**, 431 (2013).

[5] C. M. Will, *Theory and Experiment in Gravitational Physics* (Cambridge University Press, New York, 1981); *Living Rev. Relativity* **17**, 4 (2014).

[6] M. Burgay *et al Nature* **426**, 531 (2003).

[7] M. Kramer *et al. Science* **341**, 97 (2006).

[8] S. Capozziello and A. Troisi *Phys. Rev. D* **72**, 044022 (2005).

[9] S. Capozziello, A. Stabile and A. Troisi *Mod. Phys. Lett. A* **21**, 2291 (2006); *Phys. Rev. D* **76**, 104019 (2007); *Class. Quant. Grav.* **25**, 085004 (2008); *Mod. Phys. Lett. A* **24**, 659 (2009); *Int. Jour. of Theor. Phys* **49**, 1251 (2010).

[10] T. Damour and N. Deruelle *Ann. Inst. Henri Poincare A* **43**, 107 (1985); *Ann. Inst. Henri Poincare A* **44**, 263 (1986).

[11] T. Damour and J. H. Taylor *Phys. Rev. D* **45**, 1840 (1992).

[12] T. Faulkner, M. Tegmark, E. F. Bunn and Y. Mao *Phys. Rev. D* **76**, 063505 (2007).

[13] G. J. Olmo *Int. J. Mod. Phys. D* **20**, 413 (2011).

[14] S. Capozziello, T. Harko, T. S. Koivisto, F. S. N. Lobo and G. J. Olmo *JCAP* **1304**, 011 (2013).

[15] G. W. Horndeski, *Int. J. Theor. Phys.* **10**, 363 (1974).

EPJ Web of Conferences **125**, 03006 (2016) DOI: 10.1051/epjconf/201612503006
QUARKS-2016

Systematic treatment of non-linear effects in Baryon Acoustic Oscillations

Mikhail M. Ivanov[1,2,3,⋆]

[1] *FSB/IPHYS/LPPC, École Polytechnique Fédérale de Lausanne, CH-1015, Lausanne, Switzerland*
[2] *Institute for Nuclear Research of the Russian Academy of Sciences, 60th October Anniversary Prospect, 7a, 117312 Moscow, Russia*
[3] *Department of Particle Physics and Cosmology, Physics Faculty, Moscow State University, Vorobjevy Gory, 119991, Moscow, Russia*

Abstract. In this contribution we will discuss the non-linear effects in the baryon acoustic oscillations and present a systematic and controllable way to account for them within *time-sliced perturbation theory*.

1 Introduction

Baryon acoustic oscillations (BAO) are widely used to establish the distance-redshift relation enabling to constrain the expansion history and the composition of the Universe [1, 2]. BAO measurements aim at providing (sub-) percent precision in the near future and thus it is imperative to provide the best possible theoretical control over the BAO in order to fully exploit the potential of future surveys.

The BAO are most prominent in the 2-point correlation function of matter density in position space where they form a peak at $r_{BAO} \sim 110$ Mpc$/h$ in comoving coordinates. It has been observed long ago that the shape of the BAO peak retrieved from N-body simulations differs significantly from the prediction of linear theory (see Fig. 1, left panel). Even though r_{BAO} is significantly larger than the characteristic scale of non-linear clustering $2\pi k_{NL}^{-1} \sim 20$ Mpc$/h$, the leading correction computed in Eulerian standard perturbation theory (SPT) failed to capture the behavior seen in N-body data (see Fig. 1, left panel). This disagreement is caused by large-scale bulk flows whose interaction with short modes is amplified if the distribution of matter has a feature. At leading order the effect of bulk motions is to worsen the correlation between galaxies at separations of order r_{BAO}, which results in the suppression of the BAO peak.

Many approaches have been put forward in order to account for bulk flows. From the numerical side one can mention BAO *reconstruction* [3], which undoes bulk motions directly in the data and yields a sharper BAO peak with the signal-to-noise ratio improved by a factor of 2. From the theoretical side among the most successful approaches we would like to mention renormalized perturbation theory [4, 5], Lagrangian perturbation theory [6] and IR - resummed effective field theory of large scale structures [7, 8]. In this short contribution we will present a new way to systematically describe the non-linear evolution of BAO in the framework of time-sliced perturbation theory (TSPT) [9, 10].

⋆e-mail: mikhail.ivanov@cern.ch

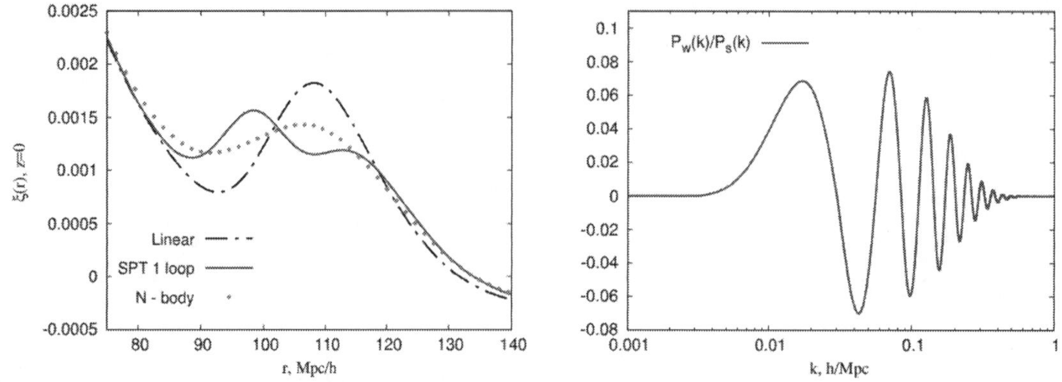

Figure 1. *Left panel*: Two-point matter correlation functions in position space at redshift zero, $\xi(r) \equiv \langle\delta(\mathbf{r})\delta(0)\rangle$: N-body data from the Horizon Run 3 simulation [14] (red points), the prediction of linear theory (black dot-dashed line), and the 1-loop SPT result (blue solid line). *Right panel*: Ratio of oscillatory (wiggly) part P_w of the linear power spectrum to the smooth part P_s. The ΛCDM cosmological parameters have been chosen as in [14].

2 Interaction with bulk flows: a simple physical picture

In this section we present a simple intuitive picture behind the non-linear evolution of the BAO (see Refs [5–12] for more details). We start with the density contrast field in Fourier space, $\delta(\mathbf{k}, \eta) \equiv (\rho(\eta, \mathbf{k}) - \bar{\rho}(\eta))/\bar{\rho}(\eta)$, where η is conformal time, $\rho(\eta, \mathbf{k})$ is the local density and $\bar{\rho}(\eta)$ is the average matter density in the Universe. $\delta(\eta, \mathbf{k})$ is a random stochastic field and one is typically interested in its 2-point correlation function (power spectrum) defined as $P(\eta; k)\delta^{(3)}_{\text{Dirac}}(\mathbf{k} + \mathbf{k}') \equiv \langle\delta(\eta, \mathbf{k})\delta(\eta, \mathbf{k}')\rangle$. The linear power spectrum in our Universe can be decomposed into the smooth and wiggly parts (see Fig. 1, right panel),

$$P_{\text{lin}}(\eta; k) = P_s(\eta; k) + P_w(\eta; k)\,, \tag{1}$$

and these two parts receive non-linear corrections from long-wavelength perturbations (IR modes) in very different ways.

Let us first focus on the situation in which the linear power spectrum has only the smooth part, i.e. is featureless. Consider two galaxies on top of a large scale bulk flow. If the separation between these galaxies is shorter than the long wavelength associated to the flow, then at zeroth order its effect is a uniform acceleration, $\mathbf{a} = -\nabla\Phi_L(x) \simeq \text{const}$ ($\Phi_L(x)$ is a gravitational potential induced by a long-wavelength perturbation). Then, according to the equivalence principle, at a given time the effect of this acceleration can be totally removed by a proper coordinate transformation, so that it cannot affect equal-time physical observables such as the power spectrum. At next-to-leading order one can take into account inhomogeneity in the acceleration, in other words, the fact that the flow is not uniform but rather "converging" or "diverging". In that case two galaxies are likely to be found at separations shorter ("converging" flows) or larger ("diverging" flows) than in linear theory. However, since the smooth power spectrum in the real Universe is close to scale-invariant, the effect of the motions due to long modes should be small. In SPT the latter fact is manifest if one considers the 1-loop correction to the power spectrum in the IR limit ($q \ll k$, where q is the loop momentum),

$$P_{\text{1-loop, SPT}}(k) = P_{\text{lin}}(k) + \left(\frac{569}{735}P_{\text{lin}}(k) - \frac{47}{105}kP'_{\text{lin}}(k) + \frac{1}{10}k^2P''_{\text{lin}}(k)\right)\sigma_S^2\,, \tag{2}$$

where $\sigma_S^2 = \int_0^{q_{\max}\ll k} d^3q P_{\rm lin}(q) \ll 1$ is the large scale variance of the density field. From Eq.(2) it is clear that the interaction between the mode with momentum k and the IR modes is suppressed provided that the derivatives of the linear power spectrum are small, $k^n P_{\rm lin}^{(n)} \sim P_{\rm lin}$. Thus, expanding over the coupling to the IR modes ($\sim \sigma_S^2$) yields a converging perturbative expansion in this case. As the reader might have already guessed, this is not the case for a power spectrum with a feature.

Now consider what happens in the presence of the wiggly component. In that case the effect of non-uniform accelerations is not small anymore. Indeed, the presence of a feature at $r_{BAO} \equiv k_{osc}^{-1}$ means, at the linear level, that two galaxies have bigger probability to be found at this separation. At the non-linear level bulk flows make these galaxies move to shorter ("converging" flows) or larger ("diverging" flows) separations, which results in the degradation of the correlation between them at r_{BAO} once these motions are averaged over long modes. Since matter is collapsing, "converging" flows are more common and two galaxies should be typically found at separations systematically smaller than r_{BAO}. This effect is known as the *physical shift* of the BAO peak [12, 13] and this is a next-to-leading order effect as compared to the broadening due to bulk flows. Assuming $P_w(k) = f_s(k)\sin(k/k_{osc})$ ($f_s(k)$ is a monotonic function) the SPT 1 loop result yields (here we assume $q \le q_{max} \ll k_{osc}$),

$$P_{w,\text{1-loop, SPT}}(k) = P_w(k) + O(1)\times\sigma_S^2 P_w(k) - \frac{k}{k_{osc}}\left(\frac{47}{105} - \frac{kf_s'}{5f_s}\right)\sigma_S^2 f_s \cos(k/k_{osc}) - \frac{k^2}{k_{osc}^2}\frac{\sigma_S^2}{10}P_w\,, \quad \text{(3a)}$$

$$= f_s(k)\left(1 - \frac{k^2}{k_{osc}^2}\frac{\sigma_S^2}{10}\right)\sin\left(\frac{k}{k_{osc}}\left\{1 - \left[\frac{47}{105} - \frac{kf_s'}{5f_s}\right]\sigma_S^2\right\}\right) + O(1)\times\sigma_S^2 P_w(k) + O(\sigma_S^4)\,. \quad \text{(3b)}$$

From (3b) one clearly sees the mentioned effects on acoustic oscillations: the suppression of their amplitude and the shift in their phase w.r.t linear theory. The effective coupling constant to the IR modes $\sim \sigma_S^2(k^2/k_{osc}^2)$ is parametrically enhanced compared to the featureless case and in fact is of order unity for realistic cosmology. Thus, the perturbative expansion as (3b) does not make sense unless one can resumm relevant contributions to all orders in perturbation theory. We will show now how this procedure, called *IR-resummation* can be systematically performed in TSPT.

3 Overview of Time-Sliced Perturbation Theory

Time - sliced perturbation theory is a novel approach to large-scale structure (LSS) formation aimed at studying equal-time correlators in perturbation theory [9]. It is instructive to focus first on the SPT-like approach, in which one studies the following partition function[1],

$$Z[J] = \mathcal{N}^{-1}\int \mathcal{D}\delta_0\, \exp\left\{-\frac{1}{2}\int d^3k \frac{\delta_0(\mathbf{k})\delta_0(-\mathbf{k})}{P_0(k)} + \int d^3k\, \delta_\eta(\mathbf{k})J(-\mathbf{k})\right\}, \quad (4)$$

where the ensemble average is taken over the initial field that has a Gaussian distribution. The time-dependence of the field $\delta(\eta, \mathbf{k})$ is governed by the system of the continuity, Euler and Poisson equations for a perfect pressurless fluid (see, for example, Eq (2.1) of [9]). In SPT one finds the solution to these equations as a series over the initial field δ_0,

$$\delta_\eta(\mathbf{k}) = g(\eta)\delta_0(\mathbf{k}) + g^2(\eta)\int d^3q F_2(\mathbf{q}, \mathbf{k}-\mathbf{q})\delta_0(\mathbf{q})\delta_0(\mathbf{k}-\mathbf{q}) + \dots\,, \quad (5)$$

[1]Typically in LSS one studies the density and velocity divergence fields, however, only one of them is statistically independent and can be an argument of the partition function in Eq.(4). In Ref. [10] we chose this to be the velocity field. In this paper, for presentation reasons, we switched to the density contrast δ. One can show that the resummation procedure in the case of this variable will be completely similar to the one performed in [10].

where $g(\eta)$ is the time propagator (growth factor), F_2 is the SPT integral kernel. Then one can plug the expansion (5) into Eq.(4) and find equal-time correlation functions by taking functional derivatives w.r.t the source J. It is well known that the expansion (5) breaks the invariance under the transformations related to the equivalence principle, which eventually produces spurious IR enhancements in individual SPT diagrams [15, 16]. The latter complicates the analysis of physical IR effects (bulk flows) within SPT.

Since only equal-time correlators typically have practical value, it seems natural to study the partition function for cosmological fields at particular times. To do this, one can transform the measure in (4) $\delta_0 \to \delta_\eta(\delta_0)$, which will define a time-dependent distribution function of δ_η,

$$Z[J] = \tilde{\mathcal{N}}^{-1} \int \mathcal{D}\delta_\eta \, \mathcal{P}[\delta_\eta, \eta] \exp\left\{ \int d^3k \; \delta_\eta(\mathbf{k}) J(-\mathbf{k}) \right\} , \tag{6}$$

The equation that governs the time evolution of $\mathcal{P}[\delta_\eta, \eta]$ is a classical Liouville continuity equation,

$$\frac{\partial}{\partial\eta}\mathcal{P}[\delta_\eta, \eta] + \frac{\partial}{\partial\delta_\eta}\left(\frac{\partial\delta_\eta}{\partial\eta}\mathcal{P}[\delta_\eta, \eta]\right) = 0 \, . \tag{7}$$

One can write $\mathcal{P}[\delta_\eta, \eta] \equiv \exp\left\{-\Gamma[\delta_\eta]\right\}$ and assume, in the spirit of perturbation theory,

$$\Gamma[\delta_\eta] \equiv \sum_{n=1}^{\infty} \frac{1}{n!} \int d^3q_1 ... d^3q_n \; \Gamma_n^{tot}(\eta, \mathbf{q}_1, ..., \mathbf{q}_n) \; \delta_\eta(\mathbf{q}_1) \times ... \times \delta_\eta(\mathbf{q}_n) \, . \tag{8}$$

Then, using the fluid equations for δ_η, one can eliminate $\partial\delta_\eta/\partial\eta$ from Eq.(7) and arrive at the the chain of equations defining the time - evolution of the TSPT vertices Γ_n^{tot}. Remarkably, these equations can be solved *exactly* in the case of Einstein-de Sitter Universe without requiring any truncation of hierarchy. It is convenient to split Γ_n^{tot} into a regular part Γ_n and a singular 'counterterm' C_n,

$$\Gamma_n^{tot} = \Gamma_n + C_n \, . \tag{9}$$

The terms C_n are related to the singular Jacobian that appears as a result of the measure transformation in Eq.(6). The vertices Γ_n represent physical non-linear n-point couplings and are non-local in space. Having imposed Gaussian initial conditions for these vertices one finds that the time evolution factorizes out,

$$\Gamma_n(\eta, \mathbf{k}_1, ..., \mathbf{k}_n) = \frac{1}{g^2(\eta)} \delta^{(3)}_{\text{Dirac}}(\mathbf{k}_1 + ... + \mathbf{k}_n) \bar{\Gamma}_n(\mathbf{k}_1, ..., \mathbf{k}_n), \quad \text{where} \quad g(\eta) = D_+(\eta) \, , \tag{10}$$

and $\bar{\Gamma}_n$ are time-independent. Then one can find cosmological equal-time correlators by performing a perturbative expansion of (6) completely analogous to that of quantum field theory. In this expansion the growth factor $g^2(\eta)$ is treated as a coupling constant, the initial power spectrum $P(k) = \bar{\Gamma}_2^{-1}(\mathbf{k}, -\mathbf{k})$ plays the role of the free propagator, and higher vertices $\bar{\Gamma}_n$ $(n > 2)$ represent connected parts of the amputated tree-level n-point correlation functions.

Remarkably, all TSPT vertices are finite in the soft limit, i.e. when some of their arguments are soft, $\lim_{\epsilon\to 0} \bar{\Gamma}_n(\eta, \mathbf{k}_1, ..., \mathbf{k}_m, \epsilon\mathbf{q}_1, ..., \epsilon\mathbf{q}_{n-m}) = O(\epsilon^0)$. This property is related to the fact that TSPT operates only equal-time quantities, which are protected by the equivalence principle. Thus, each individual TSPT diagram is IR safe in contrast to SPT.

 DOI: 10.1051/epjconf/201612503006

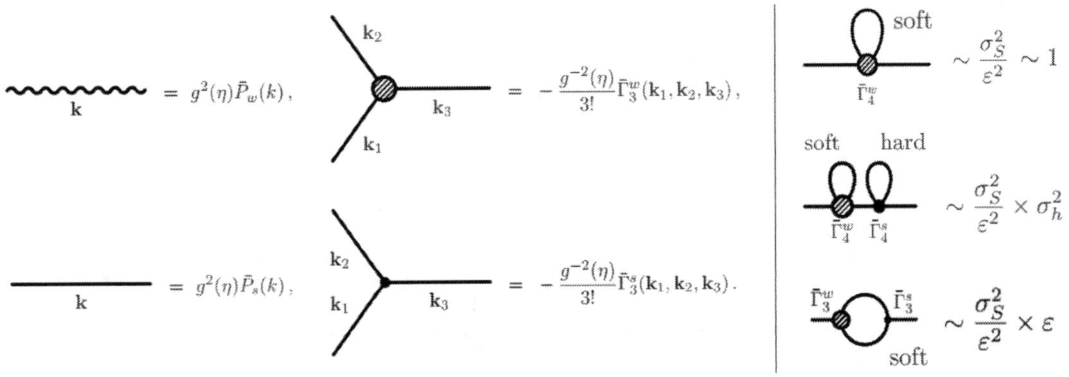

Figure 2. *Left panel:* Examples of Feynman rules for wiggly and smooth elements. *Right panel:* Power-counting rules for some TSPT diagrams.

4 IR - resummation in TSPT

The advantageous IR properties of TSPT can be readily used to account for bulk flows. As we already pointed out, the linear power spectrum in our Universe can be decomposed into the smooth and wiggly parts. Similar decomposition can also be done for the TSPT vertices,

$$\bar{\Gamma}_n(\mathbf{k}_1, ..., \mathbf{k}_n) = \bar{\Gamma}^s_n(\mathbf{k}_1, ..., \mathbf{k}_n) + \bar{\Gamma}^w_n(\mathbf{k}_1, ..., \mathbf{k}_n). \tag{11}$$

Thus, the TSPT loop expansion is split into the smooth and wiggly elements separately (see left panel of Fig. 2), which allows to isolate and resumm the relevant for the BAO contributions. One can find that the wiggly vertices are enhanced in the soft limit, e.g.

$$\lim_{q\to 0} \bar{\Gamma}^w_n(\mathbf{k}_1, ..., \mathbf{k}_m - \sum \mathbf{q}_i, \mathbf{q}_1, ..., \mathbf{q}_{n-m}) \propto \frac{1}{\varepsilon^{n-m}}, \quad \text{where} \quad \varepsilon \sim \frac{k_{osc}}{k} \ll 1, \tag{12}$$

while the smooth vertices do not receive any enhancement in agreement with physical expectations. In order to identify and resumm the enhanced contributions at the diagrammatic level one can establish power-counting rules that determine the order of an arbitrary L - loop diagram. For that one has to set up a separation scale k_S which splits the loop integrals into the soft and hard parts. This scale is a priory arbitrary and any residual dependence on it should be considered a theoretical error. Then, for each diagram containing a wiggly vertex (other diagrams are not IR-enhanced) one has to: (1) Assign each loop to be hard ($q > k_S$) or soft ($q < k_S$). Its value then can be associated with the variance of the density field taken over the corresponding domain,

$$\sigma_S^2 \equiv \int_0^{k_S} d^3q P_s(q), \quad \text{or} \quad \sigma_h^2 \equiv \int_{k_S}^{\Lambda_{UV}} d^3q P_s(q).$$

(2) Count the number of soft legs which are attached to a wiggly vertex. According to Eq.(12) this number is equal to the level of enhancement due to IR modes.

Examples of our power counting rules are shown on the right panel of Fig. 2. It is easy so see that the leading contributions ($\sim (\sigma_S^2/\varepsilon^2)^L$, L-loop order) to the wiggly power spectrum are given by the so-called 'daisy' graphs, whose resummation has a simple diagrammatic representation shown on Fig.3. One can show that the daisy diagrams exponentiate,

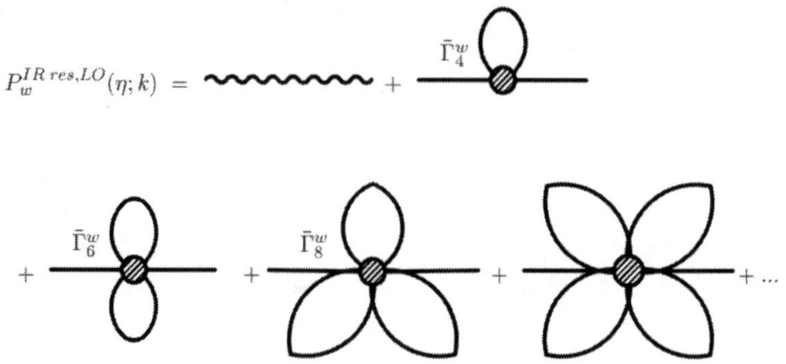

Figure 3. Diagrammatic representation of IR resummation at the leading order.

$$P_w^{IR\ res,\ LO}(\eta;k) = e^{-\Sigma^2 k^2} P_w(\eta;k)\,, \quad \text{where} \quad \Sigma^2(\eta,k_S) \equiv \frac{4\pi}{3}\int_0^{k_S} dq P_s(q)\left[1 - j_0\left(\frac{q}{k_{osc}}\right) + 2 j_2\left(\frac{q}{k_{osc}}\right)\right], \tag{13}$$

and $j_0(x)$, $j_2(x)$ are spherical Bessel functions. Remarkably, at leading order the result is the same for exact dynamics and the Zel'dovich approximation.

At next-to-leading order we resumm the corrections involving a hard loop ($\sim (\sigma_S^2/\varepsilon^2)^L \sigma_h^2$), and the diagrams which contain sub-leading in ε interaction with the soft modes ($\sim (\sigma_S^2/\varepsilon^2)^L \varepsilon$). The result is given by,

$$\begin{aligned} P^{IR\ res,\ NLO}(\eta;k) =& P_s(\eta;k) + e^{-\Sigma^2 k^2} P_w(\eta;k)(1 + \Sigma^2 k^2) \\ &+ P_{1\text{-loop}}[P_s + e^{-\Sigma^2 k^2} P_w] + k^3 \Sigma^2_{\text{NLO,soft}} e^{-\Sigma^2 k^2} \frac{d}{dk} P_w(\eta;k)\,, \end{aligned} \tag{14}$$

where $\Sigma^2_{\text{NLO,soft}} = \Sigma_a^2 + \Sigma_b^2$ has pretty complicated form which can be found in Sec.7 of [10]. Notice that the final NLO IR resummed formula differs quite a lot from the simple form of exponential damping.

5 Comparison with N-body data

In this section we compare our theoretical predictions with the N-body simulations Horizon Run 2 and 3 [14]. For that we evaluate the matter correlation function in position space,

$$\xi(r,z) = \frac{4\pi}{r}\int_0^\infty dk\, k P(z;k)\sin(kr)\,, \tag{15}$$

using the LO and NLO IR - resummed power spectra Eq.(13) and Eq.(14) as an input. The result is shown on Fig. 4. On the left panel we display the LO IR - resummed result. The dependence on the separation scale, which we assume to be a theoretical error, is quite strong in this case. On the right panel of Fig. 4 we show the NLO IR - resummed correlation function. The agreement with the data is significantly improved as compared to LO. Also the dependence on the separation scale is reduced. This is an important observation because any dependence on k_S must vanish, in principle, in the exact result. Thus, the reduction of the theoretical error when going to next-to-leading order ensures the consistency of our resummation scheme. It is useful to compare the TSPT NLO predictions with the

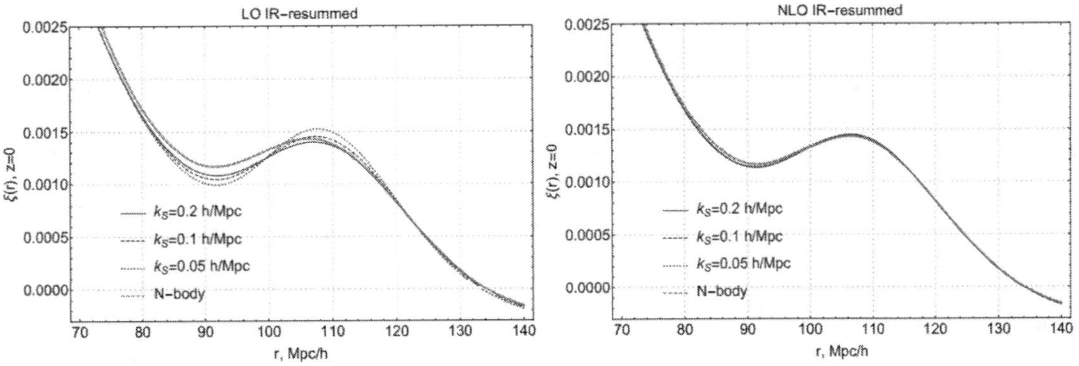

Figure 4. IR-resummed matter correlation function at leading (left panel) and next-to-leading (right panel) order. N-body data are taken from the Horizon Run 3 simulation [14].

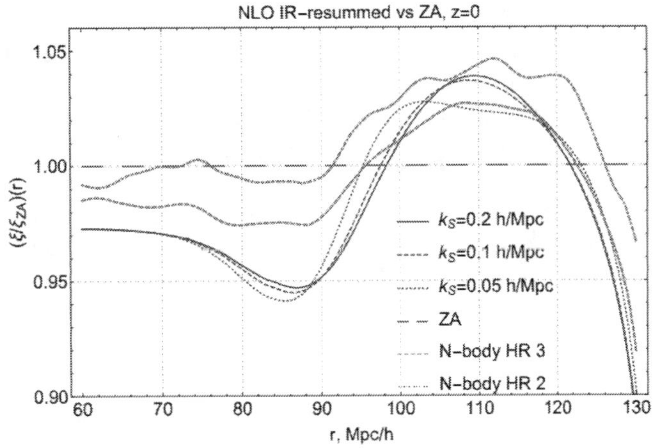

Figure 5. IR-resummed matter correlation function at NLO divided by the Zel'dovich approximation. Also shown are N-body data from the Horizon Run 2 and 3 simulations [14] divided by ZA.

Zel'dovich approximation (ZA), which is known to be a quite accurate model for the BAO feature. On Fig. 5 we show the NLO correlation function divided by ZA, along with the N-body data Horizon Run 2 and 3. The difference between our results and ZA is of order 5% at the BAO peak, and it seems that the NLO result performs slightly better there. We notice, however, that the difference is comparable with the uncertainties in the N-body data in this region[2], thus an apparent improvement over ZA seen in the Horizon Run data should be taken with a grain of salt. Nevertheless, the difference between TSPT NLO and ZA is smaller than the forecasted precision of future surveys. The TSPT framework can also be used to systematically incorporate the NNLO corrections and thus provides a reliable tool for going beyond ZA.

[2]Ref.[14] does not give error bars. An estimate for the statistical variance and the resolution of these simulations suggest the error bars of order few percent at the BAO peak. This level of accuracy is also consistent with the difference between the correlation functions of Horizon Run 2 and Horizon Run 3.

Another non-linear effect on the BAO is the shift of the BAO peak [12, 13]. This effect is quite small $\sim -0.3\%$ for the ΛCDM, but can be measured with future observations [17]. The shift of the BAO peak may be big in modified gravity models [18], which makes it an interesting observable for constraining the deviations from ΛCDM. The shift is induced by terms which are off-phase with the linear BAO oscillations (see (3b)). Those correspond to NLO IR contributions in our power counting (scale as $\sim (\sigma_S^2/\varepsilon^2)^L\varepsilon$), and, in principle, can be explicitly extracted from Eq.(14) to yield $\delta r_{BAO}/r_{BAO} \approx -(0.3 \div 0.5)\%$ which is consistent with values measured in simulations, e.g.[19].

6 Conclusions

In these notes we discussed the non-linear effects in the BAO and sketched the way how one can systematically take them into account within time-sliced perturbation theory. We outlined the physical picture behind the interactions with large scale bulk flows and argued the need for IR resummation if one works within the Eulerian framework. Then we made a short introduction into the TSPT formalism and discussed its key virtues relevant for IR resummation: a clear way to separate the perturbative expansion into the smooth and wiggly components, and manifest IR safety of the TSPT loop integrands. We introduced the power counting rules which were used to identify and resumm relevant sets of diagrams at leading and next-to-leading orders. Finally, we compared our results with N-body data and found good agreement within data errors.

We point out that the TSPT framework can be easily extended to incorporate higher-order corrections due to non-linear clustering, as well as new physics, e.g. the effects of neutrino masses or primordial non-gaussianity.

Acknowledgments Author is grateful to D. Blas, M. Garny and S. Sibiryakov for their contribution to this work. The work is supported by the Swiss National Science Foundation.

References

[1] D. J. Eisenstein, W. Hu and M. Tegmark, Astrophys. J. **504** (1998) L57
[2] D. J. Eisenstein *et al.* [SDSS Collaboration], Astrophys. J. **633** (2005) 560
[3] D. J. Eisenstein, H. j. Seo, E. Sirko and D. Spergel, Astrophys. J. **664** (2007) 675
[4] M. Crocce and R. Scoccimarro, Phys. Rev. D **73** (2006) 063519
[5] M. Crocce and R. Scoccimarro, Phys. Rev. D **77** (2008) 023533
[6] T. Matsubara, Phys. Rev. D **77** (2008) 063530
[7] L. Senatore and M. Zaldarriaga, JCAP **1502**, no. 02, 013 (2015)
[8] T. Baldauf, M. Mirbabayi, M. Simonovic and M. Zaldarriaga, Phys. Rev. D **92** (2015) 4, 043514
[9] D. Blas, M. Garny, M. M. Ivanov and S. Sibiryakov, arXiv:1512.05807 [astro-ph.CO].
[10] D. Blas, M. Garny, M. M. Ivanov and S. Sibiryakov, arXiv:1605.02149 [astro-ph.CO].
[11] S. Tassev, JCAP **1406**, 008 (2014)
[12] B. D. Sherwin and M. Zaldarriaga, Phys. Rev. D **85** (2012) 103523
[13] R. E. Smith, R. Scoccimarro and R. K. Sheth, Phys. Rev. D **77**, 043525 (2008)
[14] J. Kim, C. Park, G. Rossi, S. M. Lee and J. R. Gott, III, J. Korean Astron. Soc. **44**, 217 (2011)
[15] R. Scoccimarro and J. Frieman, Astrophys. J. Suppl. **105**, 37 (1996)
[16] P. Creminelli, J. Noreña, M. Simonović and F. Vernizzi, JCAP **1312**, 025 (2013)
[17] D. H. Weinberg, *et al.* Phys. Rept. **530**, 87 (2013)
[18] E. Bellini and M. Zumalacarregui, Phys. Rev. D **92** (2015) 6, 063522
[19] H. J. Seo *et al.*, Astrophys. J. **720**, 1650 (2010)

EPJ Web of Conferences **125**, 03007 (2016) DOI: 10.1051/epjconf/201612503007
QUARKS-2016

Maeda-Dadhich Solutions as Real Black Holes

S. O. Alexeyev[1,⋆], *A. N.* Petrov[1,⋆⋆], and *B. N.* Latosh[2,3,⋆⋆⋆]

[1] *Sternberg Astronomical Institute, Lomonosov Moscow State University, Universitetsky Prospekt, 13, Moscow 119991, Russia*
[2] *Faculty of Natural and Engineering Science, Dubna International University, Universitetskaya Str., 19, Dubna, Moscow Region, Russia*
[3] *Bogoliubov Laboratory of Theoretical Physics, Joint Institute for Nuclear Research, Joliot-Curie 6, 141980 Dubna, Moscow region, Russia*

Abstract. A four-dimensional static Schwarzschild-like solutions obtained in [4, 5] in the frames of the Einstein-Gauss-Bonnet gravity at the Kaluza-Klein split are analyzed. In such models matter is created by auxiliary dimensions. The main goal of our work is to check that these solutions are physically sensible, and to examine their characteristics, which could be observable.

1 Introduction

String/M-theory [2] is still a perspective candidate for the unified theory of all physical interactions, but a problem of its experimental verification remains unsolved. However, the generic formulation of a string theory is not completed and in some cases it is appropriate to consider the low energy approximation. The second order expansion is the Gauss-Bonnet one. In the framework of Einstein-Gauss-Bonnet (EGB) gravity a set of interesting solutions has been obtained, especially black hole ones [3]. One can use the results of their physical effects to test the string theory and to search for the new physics in gravitation. In the present study we are looking for such effects in the solution obtained in the Maeda and Dadhich (MD) papers [4] devoted to EGB gravity with a cosmological constant. Of course, basic properties of the MD solutions were analyzed carefully by Maeda and Dadhich [4]. On the other hand, properties which could be useful for experimental/observational physics were not considered completely. Thus, it is possible to study physically sensible properties and characteristics, which could be observable.

The main assumption for the MD solutions is the Kaluza-Klein type splitting of the EGB space-time. So, there is a physical (dynamic) space-time (of 2, 3 or 4 dimensions) and a space of auxiliary dimensions. The later one has a constant curvature with the negative sign, whereas in the standard Kaluza-Klein model a space-time of auxiliary dimensions is flat. We concentrate on the 4-dimensional static Schwarzschild-like solution because it seems to be more promising in searching for physically sensible effects. For simplicity we restrict ourselves to the minimal case of 6D EGB space-time that

⋆e-mail: alexeyev@sai.msu.ru
⋆⋆e-mail: alex.petrov55@gmail.com
⋆⋆⋆e-mail: latosh.boris@gmail.com

has been studied separately by Molina and Dadhich [5] and we denote as DM. The EGB solution properties in 6D are similar to those in arbitrary EGB dimensions with $N > 6$.

We check the DM solutions from two points of view. To be physically sensible these solutions have to have, first, acceptable theoretical description, and, second, observable evidences, at least, principally. Theoretical description has to give a non-contradictive definitions of important characteristics. Undoubtedly, for solutions, which pretend to present astrophysical gravitating objects, such a characteristic is a total mass of the system. Not less important theoretical property is a stability of the solution. We notice that a stability is directly connected with a well defined total mass, and independently we state a possibility that the DM solutions to leave stable. Thus, chances to detect related astrophysical objects arise.

Properties, which could be potentially observable and which we study are orbital and thermodynamic effects related to the DM solutions. The last part of the paper is just devoted to these problems. As known, thermodynamic properties are directly connected with a horizon structure of solutions. As it turns out, geometrical structure of the DM solutions is quite non-trivial and we examine it in a significantly more detail than it is given in [4]-[5].

2 Dadhich-Molina solution

The restriction of the EGB gravity in 6D in the DM solution [5] leads to the following form of action:

$$S = \frac{1}{2\kappa_6}\int d^6x\sqrt{-g}\,(R - 2\Lambda + \alpha L_{GB}) + S_{\text{matter}}. \qquad (1)$$

Here and below we use $c = \hbar = 1$ units, κ_6 is the 6-dimensional gravitational (Einstein) constant, $\alpha > 0$ is the Gauss-Bonnet coupling constant; R is a 6-dimensional Ricci scalar, Λ is the cosmological constant and L_{GB} is the Gauss-Bonnet term:

$$L_{GB} = R^2 - 4R_{\mu\nu}R^{\mu\nu} + R_{\mu\nu\rho\sigma}R^{\mu\nu\rho\sigma}\,. \qquad (2)$$

The main MD assumptions are as follows:

- The space-time of EGB gravity (6D in our case) is homeomorphical to $M^4 \times K^2$ where M^4 is a 4-dimensional physical space-time and K^2 is a space of constant curvature with radius r_0 and sign $\bar{k}$ (they are not determined now). It is just the Kaluza-Klein type splitting.
- EGB vacuum case $S_{\text{matter}} = 0$ is considered.
- Model parameters is constrained by the following relation:

$$\bar{k}/r_0^2 = -1/4\alpha = \Lambda\,, \qquad (3)$$

Under such assumption the original solution was obtained [4].

Black hole solution obtained in [4] have the following form

$$ds^2 = f_\pm dt^2 - \frac{dr^2}{f_\pm} - r^2(d\theta^2 - \sin^2\theta d\varphi)\,, \qquad (4)$$

$$f_\pm = 1 + \frac{r^2}{4\alpha}\left[1 \pm \sqrt{\frac{2}{3} + 16\left(\frac{\alpha^{\frac{3}{2}}M}{r^3} - \frac{\alpha^2 q}{r^4}\right)}\right], \qquad (5)$$

where the integration constants M and q appear to be the main solution parameters. Historically, M is treated as a mass of a black hole and q is an additional charge.

3 Total Mass

Energy connected characteristics play crucial role in a description of potentially observable physical objects. So, first of all, we have to define and calculate total mass (energy) of the solution (4)-(5). The mass of the solution may be calculated via superpotential technique. Various forms of superpotentials in $D > 4$ dimensions, applicable in AdS and AdS-like backgrounds, have been constructed both in general relativity (GR) and in EGB theory, see, for example, [6]-[10] and references there in.

Propper calculation were made in paper [1] and the total mass of the solution given by the formula

$$\mathcal{E} = \pm\sqrt{\frac{3}{2}}\alpha^{1/2} M \,. \tag{6}$$

A system is physically sensible if its total energy is not negative $\mathcal{E} \geq 0$, therefore one must set $\pm M_{\text{eff}} \geq 0$. Furthermore, the charge parameter q does not contribute into $\mathcal{E}$, so we have an analogy with the Reisner-Nordsröm-AdS black hole, where $\mathcal{E} = M$, and Q^2 does not contribute into $\mathcal{E}$.

4 Stability

The problem of DM solution stability is important from a theoretical point of view and is crucial in cosmological and astrophysics applications. Usually, for such test one perturbs metric coefficients and checks their further behavior. Technically this is carried out by perturbing initial field equations and solving them. However, to obtain the DM solution special restrictions and conditions have been used. The conditions state that curvature of auxiliary dimensions is constant, perturbations propagating in auxiliary dimensions may violate the condition. Without the condition Maeda-Dadhich metric no longer solve field equations and thus may not be treated as proper background for perturbations. Thus we consider only four-dimensional perturbations of Maeda-Dadhich solution.

It is constructive to consider two cases: when (i) axial symmetry is applied, (ii) there are no symmetries. The case (i) means the existence of perturbations with axial symmetry, corresponding to black holes with small angular momentum. This case is quite important because the MD model excludes rotating black holes as it is stated in [4]-[5]. Further, the instability of the DM metric with axial perturbations means that the solution must be destroyed by a rotation. Such a conclusion could be fatal for cosmological and astrophysics applications where most of objects rotate.

The detail discussion of the task may be found in paper [1]. In the paper it was shown that in both cases (i) and (ii) perturbations admit solution exponentially dumping in time, and, thus, the DM solution has a possibility to conserve a stability.

5 Geometrical Structure

The geometrical structure of the solution is important because solution may describe unphysical objects plagued with singularities. Thus the geometrical structure is crucial research topic. As the DM solution is stable, we analyze its geometrical structure up to $r \to 0$. For the sake of simplicity hereafter we work within $c = \hbar = \alpha = 1$ units. Then $f_\pm$ defined in (5) is rewritten as

$$f_\pm(r) = 1 + \frac{r^2}{4}\left[1 \pm \sqrt{\frac{2}{3} + 16\left(\frac{M}{r^3} - \frac{q}{r^4}\right)}\,\right] . \tag{7}$$

We call f_+ and f_- positive and negative branches of the solution, respectively. We also will consider only objects with positive energy.

In paper [1] more detail analysis is given. In the paper we were able to obtain the diagram describing all possible configurations described by Maeda-Dadhich solution. The diagram is given on Fig. 2. The left hand side of the diagram describe positive branch of the solution, and right hand side – negative branch of the solution. In forbidden regions solution have region with complex metric and the region is not covered by event horizon, thus such regions provide nonphysical solutions. Positive branch of the solution can describe only regular naked singularities and may not provide a description of a black hole. Negative branch, on the other hand, may provide description of bath naked singularity and black hole. It is crucial to highlight that negative branch of the solution have a region with a complex metric for any $q > 0$, but the region is cover by an event horizon.

Thus Maeda-Dadhich solution may provide regular description of a naked singularity and of a black hole, but also describes black holes plagued with singularities.

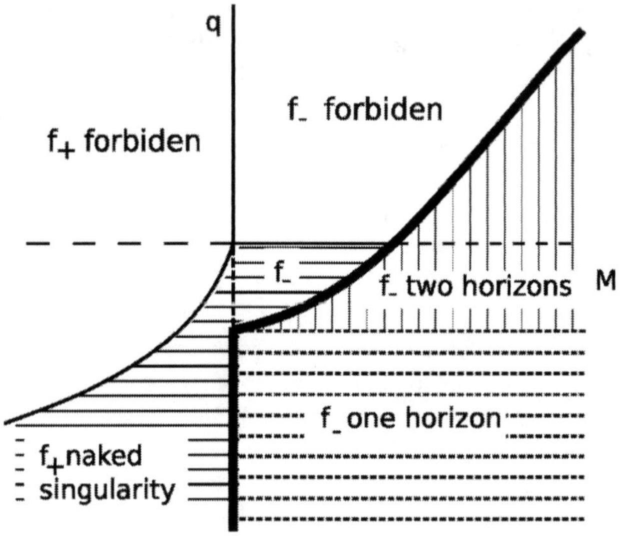

Figure 1. Dadhich-Molina solution phase diagram

6 Orbital effects

Orbital effects play an important role in black hole-like solutions, as they are coupled with an accretion picture that is actually observed [14]. We assume cosmological and astronomical scales and, therefore, can use several approximations. We assume that solution has a mass of stellar range or higher, hence one may work with series expansion of a metric. But that case is reducing to well-know DMPR solution [12]. The solution was studed in series of papers [13, 15–17]

Because the MD metric is static and spherical symmetric the standard method presented in the Chandrasekhar book [11] can be applied effectively to study stable orbits. Already, this method has been used in [13] for examining the Dadhich-Rezania solution [12] that has many similar properties with the DM solution. Therefore, not repeating here the derivation of an application of the Chandrasekhar method, we rather compare the results of its application to both the DM solution and the Dadhich-Rezania solution and the critical dependence between q and M was established

$$|q| \leq \frac{9}{8\sqrt{6}} M^2. \tag{8}$$

If q is bigger than $\frac{9}{8\sqrt{6}}M^2$, test particles orbits change [17] as the quasinormal modes for the scattering [16] and the shadow size of the black hole [15]. Engaging the recent data on M87 galaxy [18], we see that there is no evidence in Dadhich-Rezania model manifestation, so the real difference between the DM solution and the Schwarchild one is negligible small for existing accuracy level.

7 Thermodynamics and Evaporation

Hawking radiation is the other important effect that provides an instrument for model testing at cosmological scales. We use both the Hawking method [19] and the Shankaranarayanan-Padmanabhan-Srinivasan (SPS) one [20, 21]. An interesting particularity of the discussed model is that the temperature grows with M parameter increasing. Similar behavior takes place in Schwarzschild-AdS-like solution [22], but for sufficiently big masses only. If the mass is small, the temperature of the Schwarzschild-AdS-like black hole decreases with increasing mass. Note that there is the temperature area that is unreachable for black holes. However, the DM solution temperature only increases with increasing mass, the temperature can achieve arbitrary values. More detail discussion of the problem may be found in [1].

In [1] it was shown that thermodynamically stable configurations either would be plagued with singularities or would have zero energy. Also it was shown that the evaporation strongly depends upon the initial charge q. Until one estimates the charge of the DM black hole, it is impossible to find the black holes endpoint mass after the evaporation.

8 Conclusions

In this work we studied Maeda-Dadhich solution obtained in the framework of $N > 4$ Einstein-Gauss-Bonnet gravity. We proved the solution stability under linear perturbations, therefore, Maeda-Dadhich metrics is valid for the description of real astrophysical black holes. We investigated the existence of the solution in positive mass region that was not mentioned in the original papers [4, 5]. We found additional limitations on solution parameters. The positive branch is valid only for a naked singularity description, while the negative one could describe a black hole with one or two horizons some of which is plagued with singularities. We established that the black hole temperature is not governed by Hawking evaporation law and this fact causes additional limitations on the Maeda-Dadhich black hole. Unfortunately, the thermodynamically stable configurations either have zero energy or plagued with singularities. Thus we concluding that Maeda-Dadhich solution is not the best candidate for astrophysical black hole.

Acknowledgments

This work was partially supported by individual grants from Dmitry Zimin Foundation "Dynasty" (S.A. & B.L.). We are grateful to Naresh Dadhich for explanation of results in [4, 5]. We also would like to thank Alexander Zakharov and Alexander Shatskiy for useful discussions. At last, we thank Uliana Voznaya and Deepak Baskaran for the help with improving English.

References

[1] S. O. Alexeyev, A. N. Petrov and B. N. Latosh, Phys. Rev. D **92**, no. 10, 104046 (2015) doi:10.1103/PhysRevD.92.104046 [arXiv:1503.06780 [gr-qc]].

[2] John H. Schwarz, String Theory and M-Theory. In "100 Years of Subatomic Physics", Word Scientific, pp. 519-550 (2013).

[3] S.Alexeyev and M.Ponazanov, Phys.Rev. D **55**, 2110 (1997); S. O. Alexeyev, K. A. Rannu, JETP **114**, 406 (2012)

[4] H. Maeda and N. Dadhich, Phys. Rev. D 74, 021501(R) (2006); hep-th/0605031.

[5] A. Molina and N. Dadhich, Int. J. Mod. Phys. D **18**, 599 (2009); arXiv:0804.1194 [gr-qc].

[6] L.F. Abbott and S. Deser, Nucl. Phys. B **195**, 76 (1982).

[7] S. Deser and B. Tekin, Phys. Rev. D **67** 084009 (2003); hep-th/0212292.

[8] N. Deruelle, J. Katz and S. Ogushi, Class. Quantum Grav. **21**, 1971 (2004); gr-qc/0310098.

[9] A.N. Petrov, Class. Quantum Grav. **26**, 135010 (2009); arXiv:0905.3622 [gr-qc].

[10] A.N. Petrov and R.R. Lompay, Gen. Relat. Grav. |bf 45, 545 (2013); arXiv:1211.3268 [gr-qc].

[11] S. Chandrasekhar, *The Mathematical Theory of Black Holes*, Clarendon Press, 1998.

[12] N. Dadhich, R. Maartens, P. Papadopoulos and V. Rezania, Phys. Lett. B **487**, 1 (2000); hep-th/0003061.

[13] S.O. Alexeev, D.A. Starodubceva, JET **138**, 652 (2010).

[14] Shakura, N.I. and Sunyaev, R.A., Astronomy and Astrophysics, **24**, 337-355, 1973

[15] A. F. Zakharov, Phys. Rev. D **90**, no. 6, 062007 (2014) [arXiv:1407.7457 [gr-qc]].

[16] C. Chirenti, A. Saa, J. Skakala, Phys. Rev. D **86**, 124008 (2012).

[17] D. Pugliese, H. Quevedo, R. Ruffini, Phys. Rev. D **83**, 024021 (2011).

[18] S. S. Doeleman et al., Science **338**, 355 (2012).

[19] S.W. Hawking, Commun. Math. Phys. **43**, 199 (1975)

[20] K. Srinivasan and T. Padmanabhan, Phys. Rev. D **60**, 024007 (1999); gr-qc/9812028.

[21] S. Shankaranarayanan, T. Padmanabhan and K. Srinivasan, Class. Quant. Grav. **19**, 2671 (2002); gr-qc/0010042.

[22] J. Crisostomo, R. Troncoso and J. Zanelli, Phys. Rev. D **62**, 084013 (2000); hep-th/0003271.

EPJ Web of Conferences **125**, 03008 (2016) DOI: 10.1051/epjconf/201612503008
QUARKS-2016

Integrable cosmological models with non-minimal coupling and bounce solutions

Ekaterina Pozdeeva[1,*] and *Sergey* Vernov[1,**]

[1]*Skobeltsyn Institute of Nuclear Physics, Lomonosov Moscow State University, Leninskie Gory 1, 119991, Moscow, Russia*

Abstract. We remind the way to obtain integrable models with non-minimally coupled scalar fields. We are interesting to models with bounce solutions and compare bounce solutions in two known integrable models. We show that only one model has a bounce solution that tends to a stable de Sitter solution.

1 Introduction

Models with scalar fields are useful to describe the observable evolution of the Universe as the dynamics of the spatially flat FLRW background with the interval

$$ds^2 = N^2(\tau)d\tau^2 - a^2(\tau)\left(dx_1^2 + dx_2^2 + dx_3^2\right), \tag{1}$$

where $a(\tau)$ is the scale factor, $N(\tau)$ is the lapse function of the parametric time τ, and cosmological perturbations.

The recent high-precision measurements by the Planck telescope [1] confirm predictions of the single-field inflationary models. At the same time, the predictions of the simplest inflationary models with a minimally coupled scalar field lead to sufficiently large values of the tensor-to-scalar ratio of the density perturbations r, and therefore, are ruled out by the Planck data. Such type of inflationary scenarios has the singularity problem. Indeed, the Hawking–Penrose singularity theorems prove that the Universe in models with standard scalar fields minimally coupled to gravity is geodesically past incomplete. At the same time bouncing cosmological scenario with a non-minimally coupled scalar field naturally avoids this singularity problem.

At the bounce point the period of universe contraction changes to a period of universe expansion. Thereby, a bounce point is characterized by two conditions: at this point the Hubble parameter is equal to zero and its cosmic time derivative is positive. Note that in models with standard (not phantom) scalar fields minimally coupled to gravity the Hubble parameter is monotonically decreasing function. The simplest way to get a non-monotonic behavior of the Hubble parameter is to introduce in the model both standard and phantom scalar fields and consider so-called quintom models [2, 3]. In phantom field models the Null

*e-mail: pozdeeva@www-hep.sinp.msu.ru
**e-mail: svernov@theory.sinp.msu.ru

Energy Condition is violate and instability problems arise [4]. To avoid these problems bouncing models with Galileon fields have been constructed [5–10]. Another possibility to get a non-motonic behavior of the Hubble parameter is to consider models with the Ricci scalar multiplied by a function of the standard scalar field [11–17].

In spite of great success of numerical simulations and different approximation schemes for studying of cosmological models, the exact solutions are always useful and allow to catch some qualitative features of differential equations. The use of the FLRW metric essentially simplify the Einstein equations. But, only a few cosmological models with minimally coupled scalar fields are integrable [18–24]. The list of such models has been presented in [22]. In [25] a method for constructing integrable models with non-minimally coupled scalar fields by using the interrelation between the Jordan and Einstein frames has been proposed. Sometimes the integrability of the non-minimally coupled model is more apparent that the integrability of its minimally coupled counterpart [14, 26]. So it is useful to connect integrable models in the Einstein frame and in the Jordan frame [16, 24, 25].

In [14], a flat FLRW cosmological model having the Hilbert–Einstein term, a positive cosmological constant, and a conformally coupled scalar field with a negative quartic potential was investigated. Such a model is exactly integrable and for a large class of initial conditions possesses a bounce, so, it avoids the cosmological singularity. The key property of this model is that the Ricci scalar R is an integral of motion. In [16] it has been shown that this model with a constant R belongs to one-parameter set of integrable models, moreover, some of these integrable models have bounce solutions and polynomial potentials as well. In this short paper we compare the bouncing solutions obtained in different integrable models with non-minimal coupled scalar fields.

2 The simplest integrable model with bounce solutions

Let us consider a cosmological model, described by the following action

$$S=\int d^4x\sqrt{-g}\left[U(\varphi)R-\frac{1}{2}g^{\mu\nu}\varphi_{,\mu}\varphi_{,\nu}-V(\varphi)\right], \tag{2}$$

where $U(\varphi)$ and $V(\varphi)$ are differentiable functions of the scalar field φ, $g=\det(g_{ik})$ is the determinant of the metric tensor g_{ik}, R is the Ricci scalar. Varying action (2) and substituting the FLRW metric (1), one can obtain the following equations [16, 27]:

$$6Uh^2+6U'h\dot{\varphi}=\frac{1}{2}\dot{\varphi}^2+N^2V, \tag{3}$$

$$2U\left[2\dot{h}+3h^2-2h\frac{\dot{N}}{N}\right]+2U'\left[\ddot{\varphi}+2h\dot{\varphi}-\frac{\dot{N}}{N}\dot{\varphi}\right]=N^2V-\left[2U''+\frac{1}{2}\right]\dot{\varphi}^2, \tag{4}$$

$$\ddot{\varphi}+\left(3h-\frac{\dot{N}}{N}\right)\dot{\varphi}-6U'\left[\dot{h}+2h^2-h\frac{\dot{N}}{N}\right]+N^2V'=0, \tag{5}$$

where the function $h=\dot{a}/a$, a "dot" means a derivative with respect to the parametric time τ, and a "prime" means a derivative with respect to the scalar field. If $N(\tau)\equiv 1$, then $h(\tau)$ is the Hubble parameter denoted as $H(t)$, where $t=\tau$ is the cosmic time.

If $U=U_0$ is a constant, then the scalar field minimally coupled to gravity and system (3)–(5) looks simpler, but even in this case the standard way to prove the integrability includes the suitable choice of the function $N(\tau)$ that allows to simplify or even to linearize the equations [22]. At the same time there exists the integrable system with non-minimally coupled scalar field that integrability is obvious and one do not need to guess $N(\tau)$.

As it has been shown in [17] from system (3)–(5) with $N = 1$ one can obtain the following equation

$$2R\left(U + 3U'^2\right) + (6U'' + 1)\dot{\varphi}^2 = 4V + 6V'U', \tag{6}$$

where the Ricci scalar $R = 6(\dot{H} + 2H^2)$.

It is easy to see that a constant R is a solution of this equation if

$$U = U_c = U_0 - \frac{1}{12}\varphi^2, \qquad V = V_c = C_0 + C_4\varphi^4, \tag{7}$$

where U_0, C_0 and C_4 are constants. Indeed, substituting U_c and V_c into Eq. (6), we get

$$R = 2\frac{C_0}{U_0}. \tag{8}$$

To get the standard gravity domain with $U > 0$ we assume $U_0 > 0$.

Formula (8) defines a differential equation for the Hubble parameter:

$$3\left(\dot{H} + 2H^2\right) = \frac{C_0}{U_0}, \qquad \Leftrightarrow \qquad 3\left(\ddot{a}a + \dot{a}^2\right) = \frac{C_0}{U_0}a^2. \tag{9}$$

Therefore, there exists the following integral of motion [17]:

$$I = 3\dot{a}^2a^2 - \frac{C_0}{4U_0}a^4.$$

The considering integrable model is interesting due to bounce solutions. Let us remind that a bounce point t_b is defined by two conditions: the Hubble parameter $H(t_b) = 0$ and $\dot{H}(t_b) > 0$. From system (3)–(5) with $N = 1$ and $U = U_c$, one gets the following conditions on the potential:

$$V(\varphi(t_b)) < 0, \qquad 4V(\varphi(t_b)) - \varphi(t_b)V'(\varphi(t_b)) > 0. \tag{10}$$

Substituting $V = V_c$ in (10), we get $C_0 > 0$ and $C_4 < 0$. Equation (9) with a positive C_0 has two possible real solutions in dependence of the initial conditions:

$$H_1(t) = \sqrt{\frac{C_0}{6U_0}}\tanh\left(\sqrt{\frac{2C_0}{3U_0}}(t - t_0)\right), \qquad H_2(t) = \sqrt{\frac{C_0}{6U_0}}\coth\left(\sqrt{\frac{2C_0}{3U_0}}(t - t_0)\right), \tag{11}$$

where t_0 is an integration constant. The behavior of the Hubble parameter does not depend on the specific dynamics of the scalar field φ, because two-parametric set of functions $\varphi(t)$ corresponds to one-parametric set of $H(t)$. At the same time not all solutions with $H_1(t)$ tend to de Sitter ones, because the scalar field φ may tend to infinity, hence, the function $U(\varphi)$ stands negative.

System (3)–(5) with $N = 1$ and $U = U_c$ can be transformed into the following dynamical system [28]:

$$\begin{cases} \dot{\varphi} = \psi, \\ \dot{\psi} = -3H\psi - \dfrac{\left(12U_0 - \varphi^2\right)V' + 4\varphi V}{12U_0}, \\ \dot{H} = -\dfrac{1}{12U_0}\left(2\varphi^2H^2 + \left[4H\psi - V'\right]\varphi + 2\psi^2\right). \end{cases} \tag{12}$$

Equation (3) with $N = 1$ and $U = U_c$ has the following form

$$6H^2\left(U_0 - \frac{1}{12}\varphi^2\right) - H\varphi\dot{\varphi} - \frac{1}{2}\dot{\varphi}^2 - V = 0\,. \tag{13}$$

If Eq. (13) is satisfied in the initial moment of time, then it is satisfied at any moment of time. By this reason Eq. (13) fixes initial conditions of system (12).

To analyze qualitative behaviour of solutions, in particular, to analyze the stability of de Sitter solutions it is useful to introduce new variables [29], namely, the effective potential

$$V_{eff}(\varphi) = \frac{U_0^2 V(\varphi)}{U(\varphi)^2}\,. \tag{14}$$

and functions

$$P \equiv \frac{H}{\sqrt{U}} + \frac{U'\dot{\varphi}}{2U\sqrt{U}}, \qquad A \equiv \frac{U + 3U'^2}{4U^3}\,. \tag{15}$$

From Eqs. (3) and (4) we get equations that look like the Friedmann equations for models with minimally coupling:

$$3P^2 = A\dot{\varphi}^2 + \frac{1}{2U_0^2}V_{eff}, \qquad \dot{P} = -A\sqrt{U}\,\dot{\varphi}^2. \tag{16}$$

If $U(\varphi) > 0$, then $A(\varphi) > 0$ as well. Therefore, the function P is a monotonically decreasing function at $U > 0$.

De Sitter solutions correspond to extrema of the effective potential: $V'_{eff}(\varphi_{dS}) = 0$. If $U > 0$, then the model has a stable de Sitter solution only if $V_{eff}(\varphi_{dS}) > 0$ and $V''_{eff}(\varphi_{dS}) > 0$ [17]. Let us clarify the condition $V_{eff}(\varphi_{dS}) > 0$. This condition is equivalent to $V(\varphi_{dS}) > 0$. From Eq. (3) we obtain

$$H_{dS} = \pm\sqrt{\frac{V(\varphi_{dS})}{6U(\varphi_{dS})}}\,. \tag{17}$$

If $U(\varphi_{dS}) > 0$, then H_{dS} is a nonzero real number only at $V(\varphi_{dS}) > 0$. In the antigravity region with $U < 0$ one has the following conditions for a stable de Sitter solution: $V_{eff}(\varphi_{dS}) < 0$ and $V''_{eff}(\varphi_{dS}) < 0$.

If the function $U(\varphi)$ is always positive, then one can use the equations in terms of P, A, and V_{eff} instead of the initial Eqs. (3)–(5). However, if the function U changes the sign, then some solutions can be lost. For example, if $U = U_c$, then we get the dynamical system (12). It is easy to see that the point at which $U_c = 0$ is not a singular point of this system, whereas the function P and the potential V_{eff} are singular at $U_c = 0$.

In Fig. 1 we present the effective potential, phase trajectories and the behavior of the Hubble parameter for the integrable system with a constant R. We see that the effective potential has a minimum at $\varphi = 0$ and two maxima. The minimum $\varphi = 0$ corresponds to a stable de Sitter solution. Any bounce point corresponds to a negative value of the effective potential, so the solution should pass via the maximum of V_{eff} to come to zero. The bounce solutions that tends to de Sitter one are denoted by gold and black curves in Fig. 1. Some solutions does not pass the maximum of V_{eff} and move to antigravity domain. Note that the Hubble parameter of the bounce solution is finite in this case as well (see green lines in Fig. 1), but it is not a de Sitter solution. The finiteness of the Hubble parameter that corresponds to infinitely large φ is a characteristic property of this integrable model. The blue curve does not corresponds to de Sitter solution, in this case the initial value of the Hubble parameter is too small and this parameter tends to minus infinity, because $H = H_2(t)$. Colors the Hubble parameter evolutions presented in the right picture of Fig. 1 coincide to the colors of the corresponding phase trajectories in the middle picture. The behavior of bounce solutions in this model has been studied in detail in [14]. In particular, in the conformal time ($N = a$) the function $\varphi(\tau)$ has been found in terms of elliptic functions.

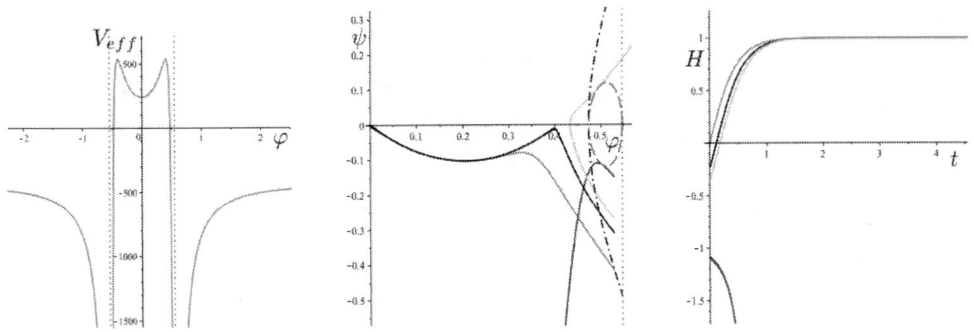

Figure 1. The effective potential (left picture), phase trajectories (middle picture) and the Hubble parameter $H(t)$ (right picture) for $V = C_4\varphi^4 + C_0$ and $U = U_0 - \varphi^2/12$. The parameters are $U_0 = 1/40$, $C_4 = -3$, $C_0 = 0.15$. The initial values are $\varphi_i = 0.53$, and $\psi_i = -0.4164479079$ (gold line), $\psi_i = -0.31$ (black line), $\psi_i = -0.27$ (green line), $\psi_i = -0.15$ (blue line). The black dash curve corresponds to $H = 0$. The blue point lines correspond to $U = 0$. The red dash curve corresponds to $P = 0$.

The red dash curve, defined by the equation $P = 0$, is the boundary of unreachable domain. Any point inside this curve corresponds to a non-real value of the Hubble parameter, calculated by (13). Such domain exists at any model with action (2) that has a bounce solution, because $V(\varphi_b) < 0$. The dynamic of solutions of cosmological models with potentials that are not positive definite has been considered in [17, 30–34]. Solutions can touch the boundary of unreachable domain (see blue curve in Fig. 1). The dynamics of such solutions are similar for integrable and non-integrable models and has been described in [17, 34].

3 Generalizations of the model with a constant R

The considering model with a constant R belongs to one-parametric set of the integrable cosmological models that has been found in [16]. To get this set of integrable models and their general solutions the conformal transformation of the metric and the corresponding model in the Einstein frame have been used [16, 25].

It was shown in [24] that on applying a conformal transformation of the metric $g_{\mu\nu} = \frac{U_0}{U}\tilde{g}_{\mu\nu}$, combined with the use of the new scalar field one arrives to the model with minimally coupled scalar field that has the standard kinetic term and the potential

$$W_c(\phi) = c_1 \cosh^4\left(\frac{\phi}{2\sqrt{3U_0}}\right) + c_2 \sinh^4\left(\frac{\phi}{2\sqrt{3U_0}}\right), \tag{18}$$

where c_1 and c_2 are constants and scalar fields ϕ and φ are connected as follows:

$$\phi = \sqrt{3U_0}\ln\left[\frac{\sqrt{12U_0}+\varphi}{\sqrt{12U_0}-\varphi}\right] \qquad \text{and, inversaly,} \qquad \varphi = \sqrt{12U_0}\tanh\left[\frac{\phi}{\sqrt{12U_0}}\right]. \tag{19}$$

This cosmological model with a minimally coupled scalar field was intensively investigated in [20]. It is known [22] that the model with the potential W_c belongs to the one-parametric set of integrable models with potentials

$$W_\beta(\phi) = c_1\left(\cosh\left[\frac{3\beta\phi}{\sqrt{12U_0}}\right]\right)^{\frac{2(1-\beta)}{\beta}} + c_2\left(\sinh\left[\frac{3\beta\phi}{\sqrt{12U_0}}\right]\right)^{\frac{2(1-\beta)}{\beta}}, \qquad (20)$$

where β is an arbitrary constant. The potential W_c is the potential (20) at $\beta = 1/3$.

Using the inverse conformal transformation, one can get from minimally coupled model with W_β the integrable model with the function U_c and potential

$$V_\beta(\varphi) = \frac{1}{36U_0^2 4^{1/\beta}}\left\{c_1\frac{\left[(\sqrt{12U_0}+\varphi)^{3\beta} + (\sqrt{12U_0}-\varphi)^{3\beta}\right]^{\frac{2(1-\beta)}{\beta}}}{(12U_0-\varphi^2)^{1-3\beta}} + \right.$$
$$\left. + c_2\frac{\left[(\sqrt{12U_0}+\varphi)^{3\beta} - (\sqrt{12U_0}-\varphi)^{3\beta}\right]^{\frac{2(1-\beta)}{\beta}}}{(12U_0-\varphi^2)^{1-3\beta}}\right\}. \qquad (21)$$

For an arbitrary β the general solution in parametric time has the following form [16]:

$$a = \frac{1}{2}\left((\vartheta+\eta)^{1/(3\beta)} + (\vartheta-\eta)^{1/(3\beta)}\right),$$
$$N = \frac{2U_0}{3\beta^2}\left((\vartheta+\eta)^{1/(3\beta)} + (\vartheta-\eta)^{1/(3\beta)}\right)\left(\vartheta^2-\eta^2\right)^{(1-3\beta)/(3\beta)},$$
$$\varphi = \sqrt{12U_0}\frac{(\vartheta+\eta)^{1/(3\beta)} - (\vartheta-\eta)^{1/(3\beta)}}{(\vartheta+\eta)^{1/(3\beta)} + (\vartheta-\eta)^{1/(3\beta)}}, \qquad (22)$$

where ϑ and η are solutions of the following equations:

$$\dot{\vartheta}^2 - 8c_1\frac{U_0}{3\beta^2}\vartheta^{\frac{2(1-\beta)}{\beta}} = E, \qquad \dot{\eta}^2 + 8c_2\frac{U_0}{3\beta^2}\eta^{\frac{2(1-\beta)}{\beta}} = E, \qquad (23)$$

and E is an arbitrary constant.

For some values of a numeric parameter β the potential V_β is polynomial. For example, V_c is equal to V_β at $\beta = 1/3$. It is easy to see that at $\beta = 1/3$ we get $a = \vartheta$.

In the case $\beta = 1$, the potential (21) can be written as

$$V = V_0\left(U_0 - \frac{\varphi^2}{12}\right)^2, \qquad (24)$$

where $V_0 = (c_1+c_2)/U_0^2$. This potential can be negative if and only if the constant $V_0 < 0$. Then, on substituting the potential (24) into the condition (10), we obtain $12U_0 < \varphi^2$. So, the model has no bounce solution with $U_c(\varphi_b) > 0$.

The third case of a polynomial potential has been proposed in [16] and corresponds to $\beta = 2/3$. In this case

$$\tilde{V} = \frac{c_1}{144U_0^2}\left(12U_0 - \varphi^2\right)\left(\varphi^2 + 2V_1\varphi + 12U_0\right), \qquad (25)$$

where $V_1 = 2c_2\sqrt{3U_0}/c_1$.

Substituting $\beta = 2/3$ into (22) and (23), we obtain the general solution in the analytic form:

$$a = \frac{\sqrt{\vartheta+\eta}+\sqrt{\vartheta-\eta}}{2}, \quad N = \frac{3U_0\left[\sqrt{\vartheta+\eta}+\sqrt{\vartheta-\eta}\right]}{2\sqrt{\vartheta^2-\eta^2}}, \quad \varphi = \frac{\sqrt{12U_0}\left[\sqrt{\vartheta+\eta}-\sqrt{\vartheta-\eta}\right]}{\sqrt{\vartheta+\eta}+\sqrt{\vartheta-\eta}},$$

$$\vartheta = \frac{1}{6U_0c_1}\left(9U_0^2c_1^2(\tau-\tau_1)^2 - E\right), \qquad \eta = -\frac{1}{6U_0c_2}\left(9U_0^2c_2^2(\tau-\tau_2)^2 - E\right), \tag{26}$$

where τ_1, τ_2 and E are arbitrary constants.

For some values of parameters [16] the model with potential (25) has a bounce solution with $U_c(\varphi_b) > 0$. We are interesting to a bounce solution that tends to a stable de Sitter ones. Also, we consider only such a solution that $U(\varphi(t)) > 0$ for all $t \geqslant t_b$. Let us check whether such a solution exists for the model with U_c and $\tilde{V}$. To do this we use the effective potential:

$$\tilde{V}_{eff} = \frac{(\varphi^2 + 2V_1\varphi + 12U_0)c_1}{U_0^2(12U_0 - \varphi^2)}. \tag{27}$$

Its first derivative is equal to zero at the points

$$\varphi_{dS} = \frac{2}{V_1}\left(-6U_0 \pm \sqrt{36U_0^2 - 3U_0V_1^2}\right). \tag{28}$$

This result is not valid for $V_1 = 0$, but in this case $\tilde{V} > 0$ at $U_c > 0$, so there is no suitable bounce solutions. By this reason we get the first condition on the parameters of potential: $V_1 \neq 0$. The second condition is that φ_{dS} should be a real number, so $V_1^2 \leqslant 12U_0$. Thus, we get

$$0 < V_1^2 \leqslant 12U_0. \tag{29}$$

A bounce solution corresponds to a negative $\tilde{V}$, whereas de Sitter solution corresponds to a positive $\tilde{V}$, so the necessary condition is that $\tilde{V} = 0$ at some points φ_0 such that $\varphi_0^2 < 12U_0$. The potential $\tilde{V}$ is equal to zero at $\varphi_0 = \pm\sqrt{12U_0}$ and

$$\varphi_\pm = -V_1 \pm \sqrt{V_1^2 - 12U_0}.$$

So, $\varphi_\pm$ are real only if $V_1^2 \geqslant 12U_0$. Comparing this condition with (29) we get $V_1 = \pm\sqrt{12U_0}$, but is this case $\varphi_\pm^2 = 12U_0$. Thus, we come to conclusion that in the case $\beta = 2/3$ no bounce solution tends to de Sitter one.

4 Concluding remarks

In this short paper we remind the way to obtain integrable system with non-minimal coupling and compare bounce solutions in two integrable models with polynomial potentials that belong to the one-parameter set of integrable models.

Note that the dynamics of FLRW Universe can be prolonged smoothly into the region with $U < 0$ (see, for example [28, 35]), however, anisotropic and inhomogeneous corrections are expected to diverge when the function U tends to zero [36, 37]. By this reason it is important to analyse the future dynamics of the obtained bounce solutions.

At $\beta = 1/3$ there exist bounce solutions that tend to a stable de Sitter solution [14]. At $\beta = 2/3$ such solutions do not exist, so in this case bounce solutions tends to antigravity domain, where $U_c < 0$.

Note that the monotonically increasing Hubble parameter $H_1(t)$ also is not suitable for construction of a realistic cosmological scenario. To get a bounce solutions with non-monotonic behaviour of the Hubble parameter that tend to de Sitter ones one can slightly modified functions U_c or V_c [15, 17]. It would be interesting to construct cosmological model with a non-minimally coupled scalar field, a bounce solution of which is suitable for a realistic inflationary scenario.

Acknowledgments. The authors would like to thank Alexander Kamenshchik, Maria Skugoreva, Alexey Toporensky, Alessandro Tronconi, and Giovanni Venturi for the useful discussions. Research of E.P. is supported in part by grant MK-7835.2016.2 of the President of Russian Federation. Research of S.V. is supported in part by grant NSh-7989.2016.2 of the President of Russian Federation. Researches of E.P. and S.V. are supported in part by the RFBR grant 14-01-00707.

References

[1] P.A.R. Ade *et al.* [BICEP2 and Planck Collaborations], Phys. Rev. Lett. **114**, 101301 (2015) [arXiv:1502.00612];
P.A.R. Ade *et. al.* [Planck Collaboration], arXiv:1502.02114

[2] Yi-Fu Cai, E.N. Saridakis, M.R. Setare, and Jun-Qing Xia, Phys. Rept. **493**, 1–60 (2010) [arXiv:0909.2776];
Zong-Kuan Guo, Yun-Song Piao, Xinmin Zhang, Yuan-Zhong Zhang, Phys. Lett. B **608**, 177–182 (2005) [astro-ph/0410654];
R. Lazkoz, G. León, and I. Quiros, Phys. Lett. B **649**, 103–110 (2007) [astro-ph/0701353];
M.R. Setare, E.N. Saridakis, Int.J.Mod.Phys. D **18**, 549–557 (2009) [arXiv:0807.3807];
Yi-Fu Cai, Taotao Qiu, Yun-Song Piao, Mingzhe Li, Xinmin Zhang, JHEP **0710**, 071 (2007) [arXiv:0704.1090]

[3] I.Ya. Aref'eva, A.S. Koshelev, and S.Yu. Vernov, Phys. Rev. D **72**, 064017 (2005) [arXiv:astro-ph/0507067];
S.Yu. Vernov, Theor. Math. Phys. **155**, 544 (2008) [arXiv:astro-ph/0612487];
I.Ya. Aref'eva, N.V. Bulatov, S.Yu. Vernov, Theor. Math. Phys. **163**, 788 (2010) [arXiv:0911.5105]

[4] V.A. Rubakov, Phys. Usp. **57**, 128 (2014) [arXiv:1401.4024]

[5] T. Qiu, J. Evslin, Y. F. Cai, M. Li and X. Zhang, J. Cosmol. Astropart. Phys. **1110**, 036 (2011) [arXiv:1108.0593]

[6] D.A. Easson, I. Sawicki and A. Vikman, J. Cosmol. Astropart. Phys. **1111**, 021 (2011) [arXiv:1109.1047]

[7] M. Osipov and V. Rubakov, J. Cosmol. Astropart. Phys. **1311**, 031 (2013) [arXiv:1303.1221]

[8] M. Koehn, J. L. Lehners and B. A. Ovrut, Phys. Rev. D **90**, 025005 (2014) [arXiv:1310.7577]

[9] M. Libanov, S. Mironov, and V. Rubakov, J. Cosmol. Astropart. Phys. **1608**, 037 (2016) [arXiv:1605.05992]

[10] A. Ijjas and P.J. Steinhardt, arXiv:1606.08880 [gr-qc].

[11] D.A. Tretyakova, A.A. Shatskiy, I.D. Novikov and S. Alexeyev, Phys. Rev. D **85**, 124059 (2012) [arXiv:1112.3770]

[12] A.Yu. Kamenshchik, A. Tronconi, G. Venturi, and S.Yu. Vernov, Phys. Rev. D **87**, 063503 (2013) [arXiv:1211.6272]

[13] E.O. Pozdeeva and S.Yu. Vernov, AIP Conf. Proc. **1606**, 48–58 (2014) [arXiv:1401.7550]
[14] B. Boisseau, H. Giacomini, D. Polarski, and A.A. Starobinsky, J. Cosmol. Astropart. Phys. **1507**, 002 (2015) [arXiv:1504.07927]
[15] B. Boisseau, H. Giacomini and D. Polarski, J. Cosmol. Astropart. Phys. **1605**, 048 (2016) [arXiv:1603.06648]
[16] A.Yu. Kamenshchik, E.O. Pozdeeva, A. Tronconi, G. Venturi and S.Yu. Vernov, Class. Quantum Grav. **33**, 015004 (2016) [arXiv:1509.00590]
[17] E.O. Pozdeeva, M.A. Skugoreva, A.V. Toporensky and S.Yu. Vernov, arXiv:1608.08214
[18] D.S. Salopek and J.R. Bond, Phys. Rev. D **42**, 3936–3962 (1990)
[19] A.J. Maciejewski, M. Przybylska, T. Stachowiak and M. Szydlowski, J. Phys. A **41**, 465101 (2008) [arXiv:0803.2318].
[20] I. Bars and S.H. Chen, Phys. Rev. D **83**, 043522 (2011) [arXiv:1004.0752];
I. Bars, S.H. Chen and N. Turok, Phys. Rev. D **84**, 083513 (2011) [arXiv:1105.3606];
I. Bars, S.H. Chen, P.J. Steinhardt and N. Turok, Phys. Rev. D **86**, 083542 (2012) [arXiv:1207.1940]
[21] A.A. Andrianov, F. Cannata and A.Y. Kamenshchik, J. Cosmol. Astropart. Phys. **1110**, 004 (2011) [arXiv:1105.4515].
[22] P. Fre, A. Sagnotti and A.S. Sorin, Nucl. Phys. B **877**, 1028 (2013) [arXiv:1307.1910].
[23] P. Fre, A. S. Sorin and M. Trigiante, Nucl. Phys. B **881**, 91 (2014) [arXiv:1310.5340]
[24] B. Boisseau, H. Giacomini and D. Polarski, J. Cosmol. Astropart. Phys. **1510**, 033 (2015) [arXiv:1507.00792].
[25] A.Yu. Kamenshchik, E.O. Pozdeeva, A. Tronconi, G. Venturi, and S.Yu. Vernov, Class. Quant. Grav. **31**, 105003 (2014) [arXiv:1312.3540]
[26] A. Borowiec, S. Capozziello, M. De Laurentis, F.S.N. Lobo, A. Paliathanasis, M. Paolella and A. Wojnar, Phys. Rev. D **91**, 023517 (2015) [arXiv:1407.4313]
[27] A.Y. Kamenshchik, I.M. Khalatnikov and A.V. Toporensky, Int. J. Mod. Phys. D **06**, 673 (1997) [gr-qc/9801064]
[28] A.Yu. Kamenshchik, E.O. Pozdeeva, S.Yu. Vernov, A. Tronconi, and G. Venturi, Phys. Rev. D **94**, 063510 (2016) [arXiv:1602.07192]
[29] M.A. Skugoreva, A.V. Toporensky and S.Yu. Vernov, Phys. Rev. D **90**, 064044 (2014) [arXiv:1404.6226]
[30] G.N. Felder, A.V. Frolov, L. Kofman and A.D. Linde, Phys. Rev. D **66**, 023507 (2002) [arXiv:hep-th/0202017]
[31] R. Giambo, J. Miritzis and K. Tzanni, Class. Quant. Grav. **32**, 035009 (2015) [arXiv:1411.0218]
[32] I.Ya. Aref'eva, N.V. Bulatov and R.V. Gorbachev, Theor. Math. Phys. **173**, 1466 (2012) [arXiv:1112.5951]
[33] R. Giambo, J. Miritzis and K. Tzanni, Class. Quant. Grav. **32**, 165017 (2015) [arXiv:1506.08162]
[34] I.Ya. Aref'eva, N.V. Bulatov, R.V. Gorbachev, S.Yu. Vernov, Class. Quant. Grav. **31**, 065007 (2014) [arXiv:1206.2801]
[35] M. Sami, M, Shahalam, M. Skugoreva and A. Toporensky, Phys. Rev. D **86**, 103532 (2012) [arXiv:1207.6691]
[36] A.A. Starobinsky, Sov. Astron. Lett. **7**, 36 (1981)
[37] P. Caputa, S.S. Haque, J. Olson and B. Underwood, Class. Quant. Grav. **30**, 195013 (2013) [arXiv:1306.0012]

EPJ Web of Conferences **125**, 03009 (2016) DOI: 10.1051/epjconf/201612503009
QUARKS-2016

Constraining (pseudo)Conformal Universe and anisotropic inflation with Planck data

Grigory I. Rubtsov[1,⋆] and *Sabir R.* Ramazanov[2,⋆⋆]

[1] *Institute for Nuclear Research of the Russian Academy of Sciences, 60th October Anniversary st. 7a, Moscow 117312, Russia*
[2] *Gran Sasso Science Institute (INFN), viale Francesco Crispi 7, I-67100 L'Aquila, Italy*

Abstract. We review the constraints on the (pseudo)conformal Universe and inflation from the non-observation of statistical anisotropy in the Planck'2013 maps provided at the frequencies 143 GHz and 217 GHz, as well as their cross-correlation.

Statistical isotropy (SI) is one of the cornerstones of the modern cosmology. Deviations from this property – if observed in the cosmic microwave background (CMB) or large scale structure surveys – would set a problem for most of the existing inflationary scenarios. Indeed, according to the Wald's non-hair conjecture, the Universe undergoes a rapid isotropization, what precludes the generation of any subtantial statistical anisotropy (SA). While some counterexamples to the Wald's conjecture are present on the market [1, 2], they either involve ghost instabilities [3] or deal with the fine tuning issues [4]. These problems are absent in some alternatives to inflation, e.g., (pseudo)conformal Universe [5–7], which includes the conformal rolling scenario [5] and the Galilean genesis [6] as the particular realizations.

Statistical anisotropy from (pseudo)Confomal Universe. In this picture, the space-time geometry is effectively Minkowskian at very early times. The state of the Universe is invariant under the transformations of the conformal symmetry group $SO(4|2)$, spontaneously broken down the de Sitter subgroup $SO(4|1)$. The zero-weight conformal field present in the Universe evolves in the symmetry breaking background and its perturbations acquire the Harrison–Zel'dovich spectrum [5–7]. These field perturbations get reprocessed into adiabatic perturbations at much later epoch. The source of non-trivial phenomenology in this setup is the interaction between zero-weight field perturbations and the Goldstone field associated with the symmetry breaking pattern [8–11]. In particular, sufficiently long/short wavelength modes of the Goldstone field give rise to SA [8, 10]/non-Gaussianity [9].

In the conformal rolling scenario and the Galilean genesis, the symmetry breaking pattern $SO(4|2) \rightarrow SO(4|1)$ is achieved by introducing the field ρ characterized by the conformal weight $\Delta = 1$. The homogeneous background for the field ρ is fixed by the dilatation invariance,

$$\rho = \frac{1}{h(t_* - t)} \, . \tag{1}$$

The constant h here is the most important parameter of the conformal rolling scenario and Galilean genesis, as it governs the non-trivial phenomenology, including SA; t_* is the constant of integration,

⋆e-mail: grisha@ms2.inr.ac.ru
⋆⋆e-mail: sabir.ramazanov@gssi.infn.it

which has the meaning of the end-of-roll time. The field θ with the conformal weight $\Delta = 0$ evolves in the background (1) and acquires flat power spectrum at the times close to t_*. On the other hand, long ranged perturbations above the background (1) provoke the small distortion of the power spectrum manifesting as SA.

In the anisotropic Universe, the power spectrum allows for the following generic parametrization,

$$\mathcal{P}_\zeta(\mathbf{k}) \propto \left(1 + \sum_{LM} q_{LM}(k) Y_{LM}(\hat{\mathbf{k}})\right) . \tag{2}$$

Here $Y_{LM}(\hat{\mathbf{k}})$ are the spherical harmonics, $\hat{\mathbf{k}}$ is the unit vector associated with the directions of the cosmological momentum $\mathbf{k}$, i.e., $\hat{\mathbf{k}} = \mathbf{k}/k$, and $q_{LM}(k)$ are the coefficients parametrizing SA.

Though the symmetry breaking pattern $SO(4|2) \to SO(4|1)$ uniquely fixes the shape of SA generated by the end of the roll, the subsequent evolution of the field θ may substantially alter its form. Baring fine-tuning, there are two interesting cases. First, the field θ, frozen at the end of the roll, remains so until the beginning of the hot Big Bang. In this case, the form of SA is fully determined by the symmetry considerations. Another set of predictions takes place, if the field θ follows a long *intermediate* evolution in the Minkowski background after the end of the roll. These two options are referred to as the sub-scenarios A and B, respectively. The associated predictions for SA are summarized below.

Sub-scenario A. In that case, the directional dependence is of the quadrupolar form [8, 11],

$$q_{2M} = \frac{H_0}{k} q'_{2M} + q''_{2M} . \tag{3}$$

Here q'_{2M}'s and q''_{2M}'s encode contributions to SA appearing in the linear and quadratic orders in the parameter h, respectively. Note that the leading order (LO) contribution is characterized by the amplitude decreasing with the wavenumber. Coefficients q'_{2M} obey the Gaussian statistics with the zero mean and variance

$$\langle q'_{2M} q'^*_{2M'} \rangle = \frac{\pi h^2}{25} \delta_{MM'} . \tag{4}$$

The sub-leading order (SLO) contribution is of the axisymmetric form. Namely,

$$q''_{2M} = -\frac{4\pi v^2}{5} Y^*_{2M}(\hat{\mathbf{v}}) , \tag{5}$$

where $\hat{\mathbf{v}} = \mathbf{v}/v$. The components of the vector $\mathbf{v}$ are the Gaussian quantities characterized by the zero means and the dispersions,

$$\langle v_i^2 \rangle = \frac{3h^2}{8\pi} \ln \frac{H_0}{\Lambda} . \tag{6}$$

Here the present Hubble rate H_0 and the scale Λ play a role of the ultraviolet and the infrared cutoffs for the long ranged modes of the perturbations of the field ρ, which serve as the source of SA.

Sub-scenario B. The structure of SA is particularly rich in that case [10]. Namely, all the coefficients q_{LM} with even L are non-zero in (2). They are the Gaussian quantities with zero means and variances given by

$$\langle q_{LM} q^*_{L'M'} \rangle = \frac{3h^2}{\pi(L-1)(L+2)} \delta_{LL'} \delta_{MM'} . \tag{7}$$

Inflation. As it was pointed out in the very beginning, SI is the outcome in the bulk of inflationary scenarios. There are some exceptions, however. The ghost free scenario of inflation capable of generating non-vanishing SA has been put forward in Ref. [2]. There one deals with the Maxwellian field

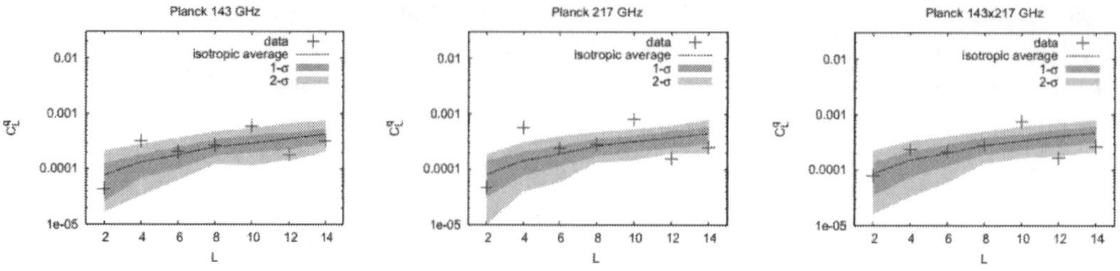

Figure 1. Coefficients C^q_L given by Eq. (11) reconstructed from the Planck data [18]. The scale invariance of the parameters q_{LM} in Eq. (2), i.e., $q_{LM}(k) = q_{LM}$. 68% and 95% C.L. intervals are overlaid with dark grey and light grey, respectively.

non-minimally coupled to the inflaton field. This coupling allows for the prolonged anisotropic evolution of the inflationary Universe, with the preferred direction stretched along the the electric vector **E**. Inflaton perturbations evolving in this anisotropic background, acquire the direction-dependent power spectrum. Modulo the overall factor and the replacement $\mathbf{v} \leftrightarrow \mathbf{E}$, the shape of the resultant SA is exactly the same as in the SLO of the pseudo-conformal Universe. Choosing the z-axis of the cosmic reference frame along the preferred direction (the direction of the vector **E**), one gets for the power spectrum,

$$\mathcal{P}_\zeta \propto \left(1 + g_* \cos^2 \theta\right) . \tag{8}$$

This can be rewritten in the generic form (2), with the only non-zero coeffients q_{2M} related to the amplitude g_* by

$$q_{2M} = \frac{8\pi g_*}{15} Y^*_{2M}(\hat{\mathbf{E}}) . \tag{9}$$

When treating anisotropic inflationary scenarios, one typically constrains the amplitude g_*. This constraint then can be easily converted into the bounds on the intrinsic model parameters.

In the sitiuation, when there are several Maxwellian fields non-minimally coupled to the inflaton [12], the form of the resultant SA is still of the quadrupolar form, but no more possesses the axial symmetry [13]. This case is captured by the following generalization of Eq. (8),

$$\mathcal{P}_\zeta \propto \left(1 + \sqrt{\frac{16\pi}{45}} g_* Y_{20}(\vartheta, \varphi) \cos\chi - \sqrt{\frac{8\pi}{45}} g_* \left(Y_{2,1}(\vartheta, \varphi) - Y_{2,-1}(\vartheta, \varphi)\right) \sin\chi\right) . \tag{10}$$

Here χ is the additional parameter. It has the meaning of the angle between two multipole vectors describing the general quadrupole. When writing Eq. (10), the particular choice of the reference frame is implied: the z-axis is aligned with one of the multipole vectors, while the second multipole vector lies in the Oxz plane.

Constraints. In our paper [18], we used the Planck CMB temperature maps corresponding to the first 15.5 months of observation at the frequencies 143 GHz and 217 GHz [19, 20]. To remove the contamination of the galactic light and point sources we applied the High Frequency Instrument (HFI) power spectrum mask [20, 21], which leaves 43% of the sky unmasked. Averaging over statistically isotropic realizations was performed using 100 Planck simulated multi-frequency CMB maps coadded with the corresponding noise maps and the foreground maps [20, 22]. These maps incorporate the effects of beam asymmetries and complex scanning strategy [23], which are proved to be crucial for the SI studies [17, 22, 24]. The operations with the maps have been performed with the *HEALPix* and *healpy* packages [25].

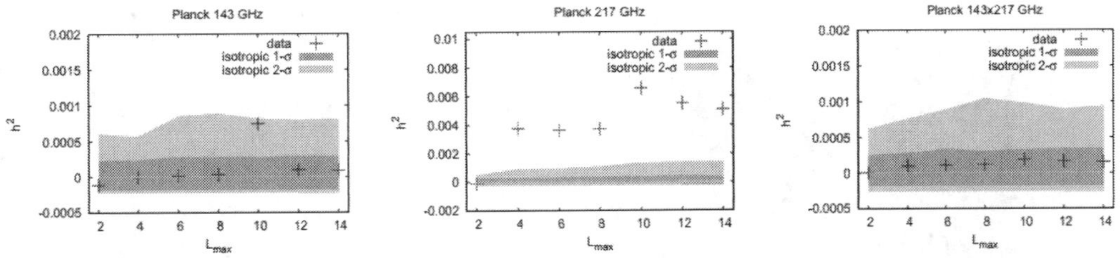

Figure 2. Parameter h^2 of the (pseudo)conformal Universe with long intermediate stage estimated from the Planck data. 68% and 95% C.L. intervals are overlaid with dark grey and light grey, respectively.

We start with constraining sub-scenario A of the (pseudo)conformal Universe. To estimate the parameter h^2, we use the statistics given by the coefficients C_L^q,

$$C_L^q = \frac{1}{2L+1}\sum_M |q_{LM}|^2 , \tag{11}$$

where we set $L = 2$. To reconstruct the parameters q_{2M}, we make use of the quadratic maximum likelihood techniques [14]. The latter has proved to be an efficient tool for searches of SA in the WMAP and Planck data [15–18]. Results are plotted in Fig. 1. Clearly, Planck'2013 data are consistent with the hypothesis of SI. The small excess in the 217 GHz frequency band present at the multipole number $L = 4$ vanishes upon cross-correlating with the 143 GHz frequency band. Hence, it should be considered as the noise/systematical effect rather than the signal of the primordial SA.

For brevity, we discuss only the constraining procedure for the case of the SLO contribution to SA. This is in fact the most relevant contribution, as the LO effect is suppressed by the wavenumber k. First, we make use of Eq. (6) to generate a large number of Gaussian vectors $\mathbf{v}$ at the initial step. Next we use Eq. (5) to convert vectors $\mathbf{v}$ into the sets of coefficients q''_{2M}. For each set, we generate one anisotropic map, and reconstruct the coefficient C_2^q. We demand that no more than 95% of the C_2^q values from simulated maps should exceed the value of the coefficient C_2^q obtained from the data. The final constraints originating from the LO and SLO contributions to SA are presented in Table 1. As it is expected, the SLO effect yields a stronger constraint than the LO suppressed by the wavenumber k [15, 16].

Model/band	143 GHz	217 GHz	143 × 217 GHz	WMAP9, V
Sub-scenario A (LO)	$h^2 < 8.8$	$h^2 < 8.0$	$h^2 < 3.0$	$h^2 < 11$
Sub-scenario A (NLO)	$h^2 \ln \frac{H_0}{\Lambda} < 0.34$	$h^2 \ln \frac{H_0}{\Lambda} < 0.30$	$h^2 \ln \frac{H_0}{\Lambda} < 0.52$	$h^2 \ln \frac{H_0}{\Lambda} < 1.2$
Sub-scenario B	$h^2 < 0.0011$	$h^2 < 0.0090$	$h^2 < 0.0013$	$h^2 < 0.006$
Inflation	$\lvert g_*\rvert < 0.020$	$\lvert g_*\rvert < 0.020$	$\lvert g_*\rvert < 0.026$	$\lvert g_*\rvert < 0.046$

Table 1. Planck 95% C.L. constraints on the parameter h^2 of (pseudo)conformal Universe and on the amplitude of the axisymmetric quadrupole anisotropy g_* [18]. For the comparison purposes, we also present the analogous constraints obtained from the V band of the WMAP9 data [16].

To constrain the parameter h^2 in the sub-scenario B, the C_2^q statistics is not enough, since all the coeffcients q_{LM} with even L are non-zero now. The relevant statistics can be found in Ref. [15], and has a generic form $h^2 = F(C_2^q, C_4^q, ...)$. The results of implementing this estimator for the Planck data

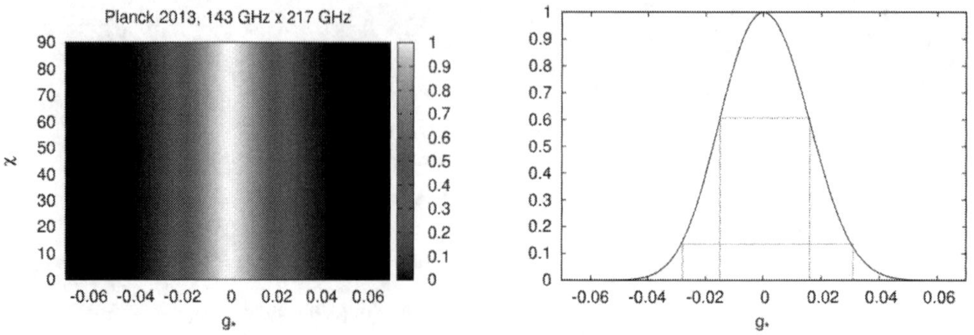

Figure 3. Planck Likelihood plotted as a function of the parameters g_* and χ, whcih describe the general quadrupolar SA (left plot), see Eq. (10). We see that it is independent on the angle χ. The truncated likelihood distribution shown in the right plot implies the 68% limits on the amplitude $g_* = 0.001 \pm 0.015$, essentially the same as the ones derived in Ref. [17].

are shown in Fig. 2. Values of the estimators are plotted for seven ranges of the multipoles starting from the quadrupole $L = 2$ and extending up to $L_{max} = 2, \ldots, 14$. The results for the frequency band 217 GHz clearly exhibit SA, which is again due to the enhancement at $L = 4$, see Fig. 1. The effect vanishes upon cross-correlating the 143 and 217 GHz frequency bands. The rest of the procedure is the same as in the sub-scenario A (see also the papers [15, 16, 18]), but now we use Eq. (7) to generate the sets of coefficients $\{q_{LM}\}$ for some particular value of the parameter h^2. Final constraints are presented in Table 1.

To bound the amplitude g_* relevant for some inflationary scenarios, one folllows the simplified version of the procedure used to constrain the sub-scenario A. Namely, for each fixed value of g_*, we readily construct the set of parameters q_{2M} using Eq. (9), and then generate one anisotropic map for each set. The final bounds are given in Table 1. We finish with discussing the general quadrupolar SA arising in inflation equipped with multiple Maxwellian fields. Our preliminary results are shown in Fig. 3. As it is clearly seen, the Planck'2013 does not allow to distinguish between the cases of the quadrupolar SA with axial symmetry and without. Hence, no constraints for the parameter χ are set at this level, while the bounds on the amplitude g_* remain intact.

We conclude that SA is a significant signature in the sub-scenario B of the (pseudo)conformal Universe. The constraints obtained from the Planck'2013 data demonstrate a factor 5 improvement compared to those derived from WMAP9 maps. On the other hand, SA is a relatively weak signature of the sub-scenario A, as is clearly seen from Table 1. In that case, the advantage of using more sensitive data gives just a factor 2 improvement compared to the previous constraints. The Planck'2013 limits on the amplitude g_* [18] are in a good agreement with the analogous ones [17] derived with the different constraining procedure. For the first time, we address the issue of the general quadrupole anisotropy produced by the inflationary scenarios with multiple vector fields. It appears, however, that the Planck'2013 data is not able to discriminate this case from the standard one. We plan to get back to this issue in the nearest future.

Acknowledgments. The work of GR is supported by the Russian Science Foundation Grant No. 14-12-01430.

References

[1] L. Ackerman, S. M. Carroll and M. B. Wise, Phys. Rev. D **75** (2007) 083502 [Erratum-ibid. D **80** (2009) 069901] [astro-ph/0701357].

[2] M. -a. Watanabe, S. Kanno and J. Soda, Phys. Rev. Lett. **102** (2009) 191302 [arXiv:0902.2833 [hep-th]].

[3] B. Himmetoglu, C. R. Contaldi and M. Peloso, Phys. Rev. D **79** (2009) 063517 [arXiv:0812.1231 [astro-ph]].

[4] N. Bartolo, S. Matarrese, M. Peloso and A. Ricciardone, Phys. Rev. D **87** (2013) 023504 [arXiv:1210.3257 [astro-ph.CO]].

[5] V. A. Rubakov, JCAP **0909** (2009) 030 [arXiv:0906.3693 [hep-th]].

[6] P. Creminelli, A. Nicolis and E. Trincherini, JCAP **1011** (2010) 021 [arXiv:1007.0027 [hep-th]].

[7] K. Hinterbichler and J. Khoury, JCAP **1204** (2012) 023 [arXiv:1106.1428 [hep-th]].

[8] M. Libanov and V. Rubakov, JCAP **1011** (2010) 045 [arXiv:1007.4949 [hep-th]].

[9] M. Libanov, S. Mironov and V. Rubakov, Phys. Rev. D **84** (2011) 083502 doi:10.1103/PhysRevD.84.083502 [arXiv:1105.6230 [astro-ph.CO]].

[10] M. Libanov, S. Ramazanov and V. Rubakov, JCAP **1106** (2011) 010 [arXiv:1102.1390 [hep-th]].

[11] P. Creminelli, A. Joyce, J. Khoury and M. Simonovic, JCAP **1304** (2013) 020 doi:10.1088/1475-7516/2013/04/020 [arXiv:1212.3329].

[12] M. Thorsrud, D. F. Mota and F. R. Urban, Phys. Lett. B **733** (2014) 140 doi:10.1016/j.physletb.2014.04.028 [arXiv:1311.3302 [astro-ph.CO]].

[13] M. Thorsrud, F. R. Urban and D. F. Mota, JCAP **1404** (2014) 010 doi:10.1088/1475-7516/2014/04/010 [arXiv:1312.7491 [astro-ph.CO]].

[14] D. Hanson and A. Lewis, Phys. Rev. D **80** (2009) 063004 [arXiv:0908.0963 [astro-ph.CO]].

[15] S. R. Ramazanov and G. I. Rubtsov, JCAP **1205** (2012) 033 [arXiv:1202.4357 [astro-ph.CO]].

[16] S. R. Ramazanov and G. Rubtsov, Phys. Rev. D **89** (2014) 043517 [arXiv:1311.3272 [astro-ph.CO]].

[17] J. Kim and E. Komatsu, Phys. Rev. D **88** (2013) 101301 [arXiv:1310.1605 [astro-ph.CO]].

[18] G. I. Rubtsov and S. R. Ramazanov, Phys. Rev. D **91** (2015) no.4, 043514 doi:10.1103/PhysRevD.91.043514 [arXiv:1406.7722 [astro-ph.CO]].

[19] P. A. R. Ade *et al.* [Planck Collaboration], Astron. Astrophys. **571**, A1 (2014) doi:10.1051/0004-6361/201321529 [arXiv:1303.5062 [astro-ph.CO]].

[20] http://www.sciops.esa.int/wikiSI/planckpla/index.php

[21] P. A. R. Ade *et al.* [Planck Collaboration], Astron. Astrophys. **571**, A15 (2014) [arXiv:1303.5075 [astro-ph.CO]].

[22] P. A. R. Ade *et al.* [Planck Collaboration], Astron. Astrophys. **571**, A23 (2014) [arXiv:1303.5083 [astro-ph.CO]].

[23] S. Mitra, G. Rocha, K. M. Gorski, K. M. Huffenberger, H. K. Eriksen, M. A. J. Ashdown and C. R. Lawrence, Astrophys. J. Suppl. **193** (2011) 5 [arXiv:1005.1929 [astro-ph.CO]].

[24] D. Hanson, A. Lewis and A. Challinor, Phys. Rev. D **81** (2010) 103003 [arXiv:1003.0198 [astro-ph.CO]].

[25] K. M. Gorski, E. Hivon, A. J. Banday, B. D. Wandelt, F. K. Hansen, M. Reinecke and M. Bartelman, Astrophys. J. **622**, 759 (2005) [astro-ph/0409513].

Fermi bubbles around the M31 galaxy

Maxim Pshirkov[1,2,3,⋆], *Valeriy* Vasiliev[4,⋆⋆], and *Konstantin* Postnov[1,⋆⋆⋆]

[1]*Sternberg Astronomical Institute, Lomonosov Moscow State University, Universitetsky prospekt 13, 119992, Moscow, Russia*
[2]*Institute for Nuclear Research of the Russian Academy of Sciences, 117312, Moscow, Russia*
[3]*Pushchino Radio Astronomy Observatory, 142290 Pushchino,Russia*
[4]*IMPRS Max Planck Institute for Astronomy,D-69117, Heidelberg, Germany*

Abstract. We have searched for an extended high-energy gamma-ray emission around the nearby giant spiral Andromeda galaxy M31 using almost 7 years of Fermi LAT data at energies above 300 MeV. The presence of a diffuse gamma-ray halo with total photon flux $2.6\pm0.6\times10^{-9}$ cm^{-2} s^{-1}, corresponding to a luminosity (0.3-100 GeV) of $(3.2\pm0.6)\times10^{38}$ erg s^{-1} (for a distance of 780 kpc) was found at a 5.3σ confidence level. The best-fit halo template corresponds to two 6-7.5 kpc bubbles symmetrically located perpendicular to the M31 galactic disc, similar to the 'Fermi bubbles' found around the Milky Way centre.

1 Introduction

There are several observational manifestations of extended baryonic coronae around spiral galaxies: soft diffuse X-ray emission extending up to several ten kpc from the central galaxy [1], absorption in O VII line [2, 3], distortions in the shape of gas clouds [4] and stripping of gas in the satellite galaxies by the ram-pressure of the halo gas [5], see [6] and references therein for a review. Such a hot halo around the Milky Way is established by several different methods [7].

Spiral galaxies can also be surrounded by extended cosmic-ray (CR) halos [8, 9]: if there were strong enough magnetic fields (10-100 nG) far away from central regions of the Galaxy, CRs, generated in galaxies, would remain in the magnetized Galactic halo for a cosmological time. These CRs would interact with tenuous ($\sim 10^{-4}$ cm^{-3}) hot plasma producing gamma-rays via pionic channel, the gamma-ray luminosity of such a halo could be around 10^{39} erg s^{-1} at energies above 100 MeV [9]. The size and shape of the halo cannot be established from first principles and depend on the propagation properties of CRs, its predicted size falls in the 20-40 kpc range[9], at smaller scales ($\sim$ 10 kpc), it can be non-uniform as evidenced by the 'Fermi bubbles' (FB) in the Galaxy [10, 11].

It is almost impossible to disentangle such a halo in our Galaxy from the truly extragalactic component. However, they can be searched for around other spiral galaxies. The most natural target is the nearest M31 (Andromeda) galaxy. One can expect the halo size of order of several degrees that can be detected from the Earth-M31 distance of > 700 kpc. The presence of a hot gas around M31, which

⋆e-mail: pshirkov@sai.msu.ru
⋆⋆e-mail: vasilyev@mpia.de
⋆⋆⋆e-mail: kp@sai.msu.ru

 DOI: 10.1051/epjconf/201612503010

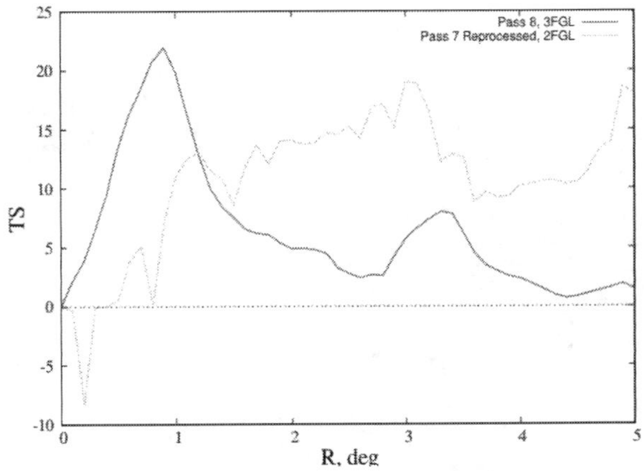

Figure 1. $TS(R_{\text{halo}})$ curves: an earlier version (65 month of data, Pass7 Reprocessed events and 2FGL catalogue) is shown for comparison. The $TS(R_{\text{halo}})$ curve is much smoother when the latest version of event reconstruction and the 3FGL source catalogue are used.

is essential for the gamma-ray emission from the CR halo, was recently demonstrated by the discovery of certain absorption features in UV-spectrum of quasars projected on the sky close to the galaxy [12, 13] and distortions in the observed CMB spectrum in the vicinity of M31 due to interference from the halo gas [14].

2 Data and data analysis

In our analysis we have used 83 months of Fermi LAT data collected since 2008 Aug 04 until 2015 Jul 06. We have selected events that belong to the "SOURCE" class, PASS8_V2 reconstruction and v10r0p5[1] version of the Fermi science tools was used. Only events with energies larger than 300 MeV were selected, because at lower energies the Fermi LAT point spread function (PSF) quickly deteriorates. Usual event quality cut, namely that the zenith angle should be less than 100° (which is sufficient at these energies) has been imposed. For further details of analysis, see [15].

3 Results

3.1 Uniform circle template

Firstly, we searched for diffuse gamma-ray emission from the M31 galaxy. The galaxy was modelled in two different ways: as a point-like source or as an extended object (the IR-based templat, IRAS). The extended template for the M31 galaxy fits the data considerably better than the simple point-like source ($TS_{\text{ext}} = 79$, $TS_{\text{ps}} = 62.3$). The galaxy has a soft spectrum with photon index $\Gamma = 2.40 \pm 0.12$ and the flux $F = (2.6 \pm 0.4) \times 10^{-9}$ph cm^{-2} s^{-1} in the 0.3-100 GeV energy range. The spectrum is even softer if the galaxy is modelled as a point-like source: $\Gamma = 2.64 \pm 0.15$ with the photon flux $F = (1.9 \pm 0.3) \times 10^{-9}$ph cm^{-2} s^{-1}.

The results of fitting with additional halo component are presented in Fig. 1. The fit quality improvement can be easily seen. The highest statistical significance $TS = 22$ was obtained for a

[1]http://fermi.gsfc.nasa.gov/ssc/data/analysis/software/

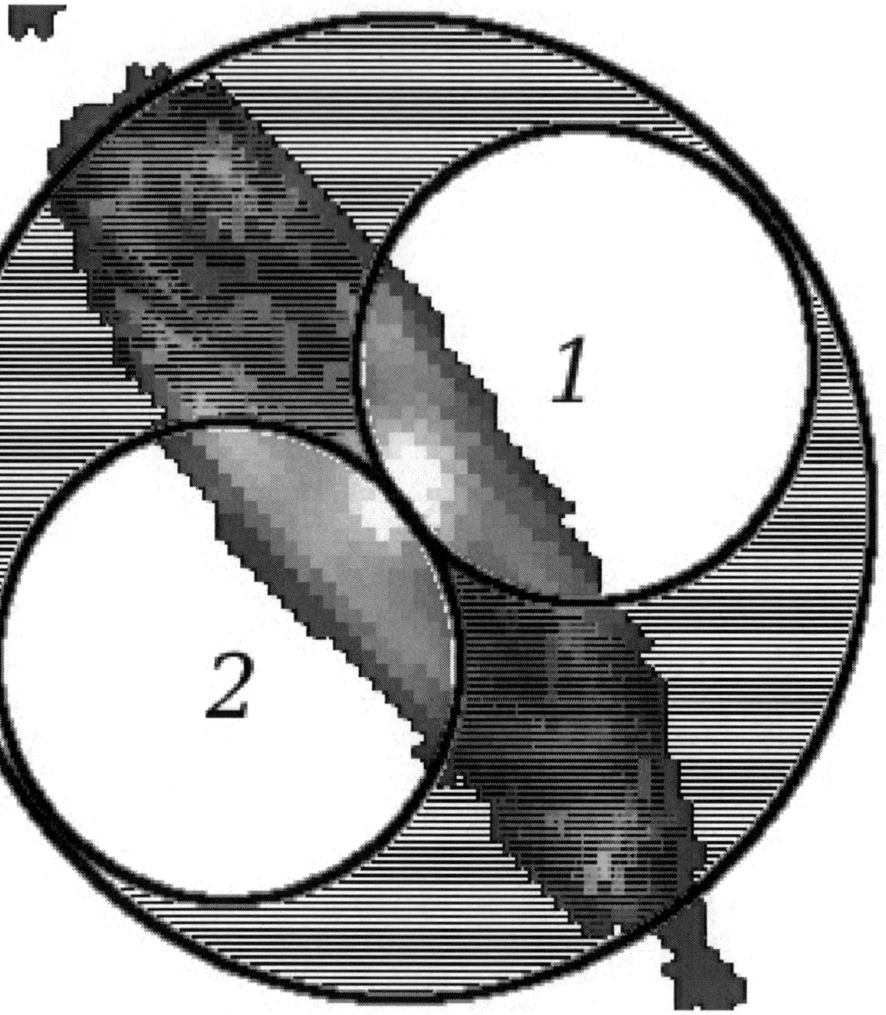

Figure 2. Model templates: IRAS template, 0.9° uniform circle, (1) and (2) – 0.45° bubbles perpendicular to the galactic plane, and region complementary to two bubbles (the shaded area).

halo with radius $R_{\rm halo} = 0.9°$, corresponding to a linear size of ~ 12 kpc. The photon flux from the extended halo and the 0.3-100 GeV luminosity obtained from the fit are ~ $(3.2 \pm 1.0) \times 10^{-9}$ cm^{-2}s^{-1} and $(4.0 \pm 1.5) \times 10^{38}$ erg s^{-1}, respectively, adopting the distance $d = 780$ kpc. The spectral index is found to be rather soft: $\Gamma = 2.30 \pm 0.12$. A marginal improvement ($TS \sim 8$) could be also achieved by adding a 3° halo (~ 35 kpc).

3.2 Bubble template

Despite low statistics (600 – 700 photons), some information about the possible nature of the found extended emission could be inferred from its morphology. The first natural option is that the emission has hadronic origin – the cosmic rays are interacting with the diffuse medium, producing gamma-rays.

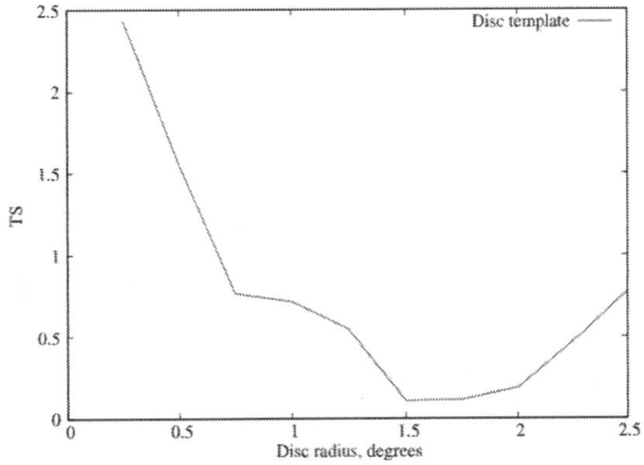

Figure 3. TS curve for an elliptic template with aspect ratio $a/b = 0.22$, representing addition of the HI disc-like emission. Inclusion of these templates does not improve the fit quality.

In this case the emission brightness traces the concentration of targets, i.e., the gas density. Most of the gas mass of the M31 galaxy resides in a flat HI disc (see, e.g [16]) and the expected template should naturally have the shape of an ellipse with aspect ratio $b/a = \cos i \sim 0.22$, where a, b are the semi-major and semi-minor axes, respectively, and $i = 77^\circ$ is the inclination angle. The results are presented in Fig. 3: addition of such a disc-like component does not improve the fit quality. This implies that the hadronic origin of the extended halo due to the cosmic rays interactions with matter is strongly disfavoured.

The Fermi bubbles are almost circular regions of $\sim$ 6 kpc radius located above and below the Galactic centre, their total luminosity is equal to $L_{\mathrm{FB}} \sim 4 \times 10^{37}$ erg s^{-1}. They are believed to stem from the past activity of the central region of the Galaxy, either from the central supermassive black hole (SMBH) or central star-formation burst. We decided to test whether a 'FB-like' morphology can fit the data better than the simple uniform disc template. First of all, we performed fits varying the bubble radius from 0.1 to 0.75 degrees. For each bubble radius, four different models were fitted (see Fig. 2). The results are shown in Fig. 4. There is a clear maximum for the FB 1+2 template with $0.45 - 0.55^\circ$ radius of the bubbles (corresponding to the linear size 6-7.5 kpc), with $TS = 28.2$. The rest of the 0.9° circle (FB$_{\mathrm{compl}}$ template) contributes very little. The parameters of the best-fit model are listed in Table 1. Finally, we have searched for a possible ellipticity of the bubbles. The major semi-axes of both ellipses normal to the disc plane were fixed at $a = 0.45^\circ$, while the aspect ratio b/a was allowed to vary in the 0.1-0.9 range. Fig. 5 shows that although the signal is concentrated around the major semi-axes, it is not completely contained within this narrow jet-like region.

4 Discussion and conclusions

From almost 7 years of the Fermi-LAT observations, we find that the data suggest the presence of a spatially extended diffuse gamma-ray excess around M31. The best-fit morphology of the diffuse emission closely resembles the Fermi bubbles in the Milky Way. The best fit gave $\sim 5.2\sigma$ significance for two 0.45° (6.5 kpc) bubbles with a photon flux of $\sim (2.6 \pm 0.6) \times 10^{-9}$ cm^{-2}s^{-1} and a luminosity of $(3.2 \pm 0.6) \times 10^{38}$ erg s^{-1} in the energy range 0.3–100 GeV. These parameters are fairly close to those of the Fermi bubbles in the Milky Way, though the luminosity of the latter ones is about one order of magnitude lower. The difference in the luminosity can be explained by the presence of a much more

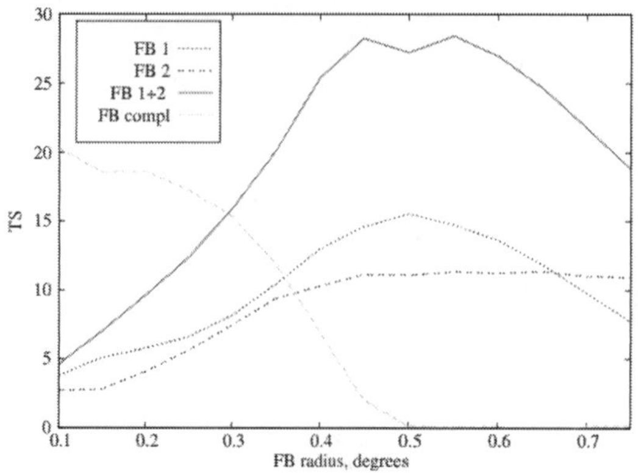

Figure 4. TS curves for different templates. Most of the signal comes from the region corresponding to the FB 1+2 template.

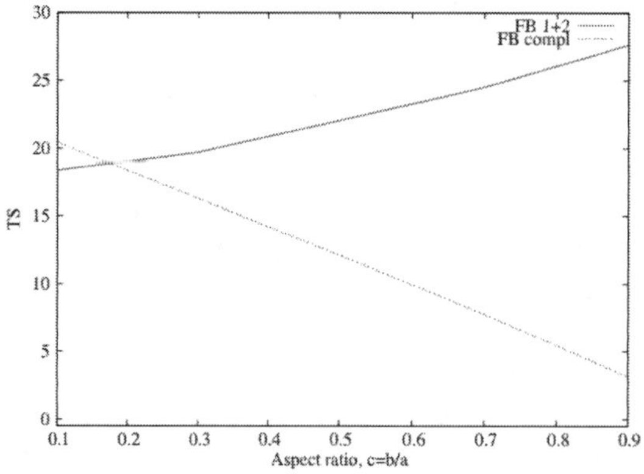

Figure 5. TS curve for elliptic FB templates with different aspect ratio b/a. The semi-major axis is fixed at $a = 0.45°$. The complementary template is defined as above (see Fig. 2)

Table 1. Best-fit parameters for IRAS + different halo templates from Fig. 2. Column 1: model template, column 2: TS value, column 3: integral flux in the 0.3-100 GeV range, column 4: photon spectral index Γ. For comparison, best-fit parameters for uniform 0.9° template are also presented

model	TS	F_{300}, 10^{-9} (cm^{-2} s^{-1})	Γ
IRAS only	–	2.9 ± 0.4	2.4 ± 0.1
FB 1	14.6	1.4 ± 0.5	2.5 ± 0.3
FB 2	11.1	0.72 ± 0.35	2.0 ± 0.2
FB 1+2	28.2	2.6 ± 0.6	2.3 ± 0.1
FB_{compl}	2.0	1.1 ± 0.9	2.5 ± 0.4
0.9° circle	22.0	3.2 ± 1.0	2.3 ± 0.1

DOI: 10.1051/epjconf/201612503010

massive SMBH in the M31 centre. Our findings suggest that the FB phenomenon can be widespread in giant spiral galaxies.

Acknowledgements

The work was supported by the Grant of the President of Russian Federation MK-4167.2015.2 and RFBR grant 14-02-00657.

References

[1] J.T. Li, Z. Li, Q.D. Wang, J.A. Irwin, J. Rossa, MNRAS**390**, 59 (2008), 0807.3587
[2] Q.D. Wang, Y. Yao, T.M. Tripp, T.T. Fang, W. Cui, F. Nicastro, S. Mathur, R.J. Williams, L. Song, R. Croft, ApJ**635**, 386 (2005), astro-ph/0508661
[3] J.N. Bregman, E.J. Lloyd-Davies, ApJ**669**, 990 (2007), 0707.1699
[4] T. Westmeier, R. Braun, D. Thilker, A&A**436**, 101 (2005), astro-ph/0503246
[5] L. Blitz, T. Robishaw, ApJ**541**, 675 (2000), astro-ph/0001142
[6] M.E. Putman, J.E.G. Peek, M.R. Joung, ARA&A**50**, 491 (2012), 1207.4837
[7] M.J. Miller, J.N. Bregman, ApJ**770**, 118 (2013), 1305.2430
[8] F. De Paolis, G. Ingrosso, P. Jetzer, M. Roncadelli, ApJ**510**, L103 (1999), astro-ph/9901033
[9] R. Feldmann, D. Hooper, N.Y. Gnedin, ApJ**763**, 21 (2013), 1205.0249
[10] M. Su, T.R. Slatyer, D.P. Finkbeiner, ApJ**724**, 1044 (2010), 1005.5480
[11] M. Ackermann et al. (Fermi-LAT), Astrophys. J. **793**, 64 (2014), 1407.7905
[12] S.M. Rao, G. Sardane, D.A. Turnshek, D. Thilker, R. Walterbos, D. Vanden Berk, D.G. York, MNRAS**432**, 866 (2013), 1302.7026
[13] N. Lehner, J.C. Howk, B.P. Wakker, ApJ**804**, 79 (2015), 1404.6540
[14] F. De Paolis, V.G. Gurzadyan, A.A. Nucita, G. Ingrosso, A.L. Kashin, H.G. Khachatryan, S. Mirzoyan, E. Poghosian, P. Jetzer, A. Qadir et al., A&A**565**, L3 (2014), 1404.4162
[15] M.S. Pshirkov, V.V. Vasiliev, K.A. Postnov, MNRAS**459**, L76 (2016), 1603.07245
[16] T.R. Cram, M.S. Roberts, R.N. Whitehurst, A&AS**40**, 215 (1980)

EPJ Web of Conferences **125**, 03011 (2016) DOI: 10.1051/epjconf/201612503011
QUARKS-2016

Dense cold matter

Alexey Stavinskiy[1,2,⋆]

[1] *SSC RF ITEP NRC Kurchatov Institute, Moscow, Russia*
[2] *National Research Nuclear University MEPhI, Moscow, Russia*

Abstract. The possibility to study cold nuclear matter with the density of neutron star core and even larger in the laboratory experiment is discussed. Special rare kinematical trigger for relativistic ion-ion collisions is proposed for such study. Expected properties of the matter in such unusual conditions and experimental program for its study is discussed. Possible experimental setup and *R&D* results for position sensitive neutron detector are presented.

1 Introduction

Perturbative aspects of QCD have been tested up to a few percent. In contrast, non-perturbative aspects of QCD (hadronization, confinement, etc.) have barely been tested. The study of the quark matter at different temperature and baryon density is part of effort to consolidate the grand theory of particle physics. A new approach to the laboratory study of extremely dense baryonic matter at low temperature phase diagram region has been proposed by FLINT collaboration [1]. High p_t central rapidity double cumulative trigger for this study was tested experementaly. Wide experimental program with such a trigger has been discussed in A.Stavinskiy's report at WPCF in Kiev [2]. The recent developments of this approach and detectors *R&D* for such a study is discussed here. Special accent is put on *R&D* of the position sensitive neutron detector. Neutron detector is needed for the study of izotopic structure of cold dense nuclear matter and especially for np femtoscopy.

2 Double cumulative trigger for the study of cold matter

Let us consider the light ion collision at NN central mass system energy of the order of few GeV. Extremely high p_t central rapidity trigger selects flucton-flucton interaction [1] (following by D.I.Blokhinzev we will use the term flucton for a few nucleon short range correlation inside nucleus). The higher p_t for trigger photon means the larger minimum number of nucleons involved in the flucton-flucton interaction. Nucleons, involved into flucton-flucton interaction will create multi-baryon system at the final state. This system is the candidate for cold and dense matter droplet and should be studied. Nuclotron energy range is optimal to study of such processes due to two reasons:

1. To identify cold and dense system it should be separated in the phase space from products of ordinary NN interactions with mean $p_t \sim m_\pi$. It means that the beam energy should be sufficiently large (not smaller than 2-3 GeV/nucleon)

⋆e-mail: stavinsk@itep.ru

2. To speak about dense cold matter droplet one needs to obtain as many baryons as possible in the cold and dense final state multibaryon system. Cross sections of the process falls dramatically with increasing of both fluctons masses and pt transfered to multibaryon system. It means that beam energy should be sufficiently small (not larger than 6-10 GeV/nucleon

For the reason of the study of the process with rare trigger measurements at fixed target looks more promising than at collider experiments. But some points of proposed experimental program are not so critical to beam intensity and can be studied with SPD and MPD detectors. Correlation measurements between secondary nucleons and trigger particle provide the location of the cold and dense baryonic system in phase space and its temperature. Once identified, this baryonic system should be studied by femtoscopic (or alternative) method to estimate the size of the interaction volume (density of the baryonic system). It is a nontrivial task for femtoscopy because of both small interaction volume and large 6-dimension density expected for baryonic system produced in selected by trigger flucton-flucton interaction. Methodical problems are planed to be studied at relatively simple MARUSYA-FLINT detector setup at Nuclotron. MARUSYA-FLINT will include all needed detectors (EMCAL for trigger, ToF, monitors, etc.) but in the version with minimal acceptance. The innovation part of the detector setup will be position sensitive neutron detector which will be used for identification of charged baryon (p,d,t,He) and neutrons. Full size measurements within proposed physical program are proposed for BM@N .

One of the most promising measurents are Σ^+ and Σ^- baryon production cross sections and its ratio. The reason for that is quark content of Σ hyperon which is as sensitive to isotopic effects as $(u/d)^2$. It is also important that methodical errors for Σ^+ and Σ^- are similar(in both cases one need to identify one neutron and one charged pion(ising the reverse of the magnetic field). Fast neutrons outside target fragmentation region could be identified by EMCAL with kinetic energy threshold of the order of 100 MeV. For target fragmentation region neutrons kinetic energy of the order of several tenth MeV and we propose to use for neutron identification special neutron detector.

3 Position Sensitive Neutron Detector

3.1 Motivation

Neutrons are one of the main particle species in the energy range of Nuclotron. The only measurable decay mode for strange baryon Σ^- is $n + \pi^-$. For that reasons it is difficult to say about detector impermeability and isotopic ratio measurements in the baryon sector without neutron identification. Scattering experiments, where a proton is knocked-out of the nucleus with high momentum transfer and high missing momentum, show that in C^{12} the neutron-proton fluctons are nearly twenty times as prevalent as proton-proton pairs and, by inference, neutron-neutron pairs. This difference between the types of pairs is due to the nature of the strong force and has implications for understanding cold dense nuclear systems such as neutron stars [3]. We also expect significant difference between space time parameters extracted from femtoscopic analysis of np and pp,nn pairs.

3.2 Design goals

The mean kinetic energy of neutrons in the energy range of Nuclotron is of the order of several hundreds MeV. The neutron momentum accuracy needed for femtoscopic measurements of the order of several tens MeV. To meet these conditions we formulate required features for neutron detector as follows:

1. Neutron energy range from 5 MeV to 200 MeV

2. Accuracy for neutron momentum 10-20 MeV/c
3. Modular structure of detector for correlation measurements
4. Compact installation modules to create large acceptance detector
5. Position resolution ($\sim 1cm$) one order better than module size
6. Time resolution 150-200 psec
7. Gamma-neutron separation
8. Compact module

3.3 Solution and prototype tests

We propose to create for such a study the position sensitive neutron detector which would be next step in neutron identification technique. Except "standard" PMT each detector module will be supplied by several APD in different points of scintillator volume. Scintillator volume is hexagonal parallelepiped with the thickness 200 mm. The distance between opposite side edges is 175 mm. Each side adge is supplied by optic fibers. APDs are at the end of the fibers. The ratio of signal amplitudes for different diodes provides information on neutron interaction point. Detector is equipped by 3 LED in the front faces for monitoring, PMT XP2041 at the back face for ToF measurements, and temperature sensors for monitoring APD gain. The difference between neutron interaction point position and light emission points position within scintillator was estimated within GEANT-4 simulations and proved to be less than 1 cm up to neutron kinetic energy 200 MeV. Position sensitivity of the neutron detector prototype was tested at ITEP cosmic stand. The data analysis is in progress.
This work was supported by the Russian Foundation for Basic Research (project nos.14-02-93108-CNRS-a, 16-02-00637, and 14-02-00896-a).

References

[1] I.G.Alekseev et al., Phys.At.Nucl.,**71**, 1848(2008); Phys.At.Nucl.,**78**, 936(2015)
[2] A.Stavinskiy for the FLINT Collaboration, Proceeding of WPCF-2010, BITP,Kiev,Physics of Particle and Nuclei Letters, **8**, 912(2011)
[3] R.Subedi et al.,Science **320**,1476(2008); arXiv:0908.1514v1[nucl-ex]

EPJ Web of Conferences **125**, 03012 (2016) DOI: 10.1051/epjconf/201612503012
QUARKS-2016

Properties of particles in the ergosphere of black holes

Andrey A. Grib[1,2,⋆] and *Yuri V.* Pavlov[3,4,⋆⋆]

[1] *A. Friedmann Laboratory for Theoretical Physics, St. Petersburg, Russia*
[2] *Theoretical Physics and Astronomy Department, The Herzen University, St. Petersburg, Russia*
[3] *Institute of Problems in Mechanical Engineering, Russian Acad. Sci., St. Petersburg, Russia*
[4] *N.I. Lobachevsky Institute of Mathematics and Mechanics, Kazan Federal University, Kazan, Russia*

Abstract. A new feature of the rotating black holes is the existence in their ergosphere of trajectories of particles with negative and zero energies. Here we analyze general properties of such trajectories comparing them with usual trajectories of particles with positive energy. A comparison with the situation in rotating coordinate frame in Minkowski space-time is made. The possibility of the unbounded growth of the energy of two colliding particles in the centre of mass frame in the ergosphere is analysed.

1 Introduction

As it is known special geodesics on which particles with negative and zero energy (with nonzero angular momentum projection) can move are existing in the ergosphere of rotating black holes [1]–[3]. Such particles surely are not observed in external to the ergosphere region where only particles of positive relative to infinity energy can exist. However it occurs that one can find such geodesics in rotating coordinates frame in Minkowski space-time.

In this paper the systematic analysis of properties of such geodesics in the ergosphere of black hole and comparison with case of rotating coordinates in Minkowski space-time is made and some limitations on values of the energy and angular momentum with any value of the energy are obtained.

As it was shown by us earlier [4] the geodesics of particles with zero energy in the ergosphere of the black hole as well as those for particles with negative energy [5] originate inside the gravitational radius, arrive to the ergosphere and then go back inside the gravitational radius. Estimates for the angular velocities of particles with positive, negative and zero energies are obtained. In the last section the problem of the unlimited growth of energy in the centre of mass frame of two colliding particles for particles with negative and zero energies is investigated.

The system of units with gravitational constant and the light velocity $G = c = 1$ is used in the paper.

⋆e-mail: andrei_grib@mail.ru
⋆⋆e-mail: yuri.pavlov@mail.ru

2 Geodesics in Kerr's metric

The Kerr's metric of the rotating black hole [6] in Boyer-Lindquist coordinates [7] is

$$ds^2 = \frac{\rho^2 \Delta}{\Sigma^2} dt^2 - \frac{\sin^2\theta}{\rho^2} \Sigma^2 (d\varphi - \omega\, dt)^2 - \frac{\rho^2}{\Delta} dr^2 - \rho^2 d\theta^2. \tag{1}$$

where

$$\rho^2 = r^2 + a^2 \cos^2\theta, \quad \Delta = r^2 - 2Mr + a^2, \tag{2}$$

$$\Sigma^2 = (r^2 + a^2)^2 - a^2 \sin^2\theta\, \Delta, \quad \omega = \frac{2Mra}{\Sigma^2}, \tag{3}$$

M is the mass of the black hole, aM is angular momentum. We suppose $0 \le a \le M$. The event horizon for the Kerr's black hole is given by

$$r = r_H \equiv M + \sqrt{M^2 - a^2}. \tag{4}$$

The surface

$$r = r_C \equiv M - \sqrt{M^2 - a^2}. \tag{5}$$

is the Cauchy horizon. The surface of the static limit is defined by

$$r = r_1 \equiv M + \sqrt{M^2 - a^2 \cos^2\theta}. \tag{6}$$

The region of space time between the static limit and the event horizon is called ergosphere [1], [3]. On the frontier of the ergosphere the function

$$S(r, \theta) = r^2 - 2Mr + a^2 \cos^2\theta \tag{7}$$

is zero, inside the ergosphere one has $S(r, \theta) < 0$. Note that

$$\Sigma^2 = (r^2 + a^2)\rho^2 + 2rMa^2 \sin^2\theta > 0, \text{ if } r > 0. \tag{8}$$

Using the relation

$$S\, \Sigma^2 + 4M^2 r^2 a^2 \sin^2\theta = \rho^4 \Delta, \tag{9}$$

one can write the equations of geodesics for the Kerr's metric (1) (see [2], Sec. 62 or [3], Sec. 3.4.1) as

$$\rho^2 \frac{dt}{d\lambda} = \frac{1}{\Delta}\left(\Sigma^2 E - 2MraJ\right), \quad \rho^2 \frac{d\varphi}{d\lambda} = \frac{1}{\Delta}\left(2MraE + \frac{SJ}{\sin^2\theta}\right), \tag{10}$$

$$\rho^2 \frac{dr}{d\lambda} = \sigma_r \sqrt{R}, \quad \rho^2 \frac{d\theta}{d\lambda} = \sigma_\theta \sqrt{\Theta}, \tag{11}$$

$$R = \Sigma^2 E^2 - \frac{SJ^2}{\sin^2\theta} - 4MraEJ - \Delta\left[m^2\rho^2 + \Theta\right], \tag{12}$$

$$\Theta = Q - \cos^2\theta \left[a^2(m^2 - E^2) + \frac{J^2}{\sin^2\theta}\right]. \tag{13}$$

Here E = const is the energy (relative to infinity) of the moving particle, J is the conserved angular momentum projection on the rotation axis, m is the rest mass of the particle, λ — the affine parameter along the geodesic. For the particle with $m \neq 0$ the parameter $\lambda = \tau/m$, where τ is the proper time.

Q is the Carter constant. $Q = 0$ for the movement in the equatorial plane ($\theta = \pi/2$). The constants $\sigma_r, \sigma_\theta = \pm 1$ define the direction of movement on coordinates r, θ.

From (11) follows that the parameters characterizing any geodesic must satisfy the conditions

$$R \geq 0, \quad \Theta \geq 0. \tag{14}$$

For the geodesic being the trajectory of the test particle moving outside the event horizon one must have the condition of movement "forward in time"

$$dt/d\lambda > 0. \tag{15}$$

The conditions (14), (15) lead to inequalities for possible values of the energy E and angular momentum projection J of the test particle at the point with coordinates (r, θ) with fixed value $\Theta \geq 0$ [8].

Outside the ergosphere $S(r, \theta) > 0$,

$$E \geq \frac{1}{\rho^2} \sqrt{(m^2\rho^2 + \Theta)S}, \quad J \in [J_-(r,\theta),\ J_+(r,\theta)], \tag{16}$$

$$J_\pm(r,\theta) = \frac{\sin\theta}{S}\left[-2rMaE\sin\theta \pm \sqrt{\Delta\left(\rho^4E^2 - (m^2\rho^2 + \Theta)S\right)}\right]. \tag{17}$$

On the frontier of the ergosphere (for $\theta \neq 0, \pi$)

$$r = r_1(\theta) \ \Rightarrow\ E \geq 0, \ \ J \leq E\left[\frac{Mr_1(\theta)}{a} + a\sin^2\theta\left(1 - \frac{m^2}{2E^2} - \frac{\Theta}{4Mr_1(\theta)E^2}\right)\right]. \tag{18}$$

The value $E = 0$ is possible on the frontier of the ergosphere when $m = 0, \Theta = 0$. In this case one can have any value of $J < 0$.

Inside the ergosphere $r_H < r < r_1(\theta)$, $S < 0$

$$J \leq \frac{\sin\theta}{-S}\left[2rMaE\sin\theta - \sqrt{\Delta\left(\rho^4E^2 - (m^2\rho^2 + \Theta)S\right)}\right]. \tag{19}$$

and the energy of the particle as it is known can be as positive as negative.

As it is seen from (18), (19) on the frontier and inside the ergosphere the angular momentum projection of particles moving along geodesics can be negative and it can be any large in absolute value number for the fixed value of the energy. This property first found in [8, 9] for the Kerr's metric (as it was shown later in [10]) is valid in the ergosphere of any black hole with the axially symmetric metric.

Note that in the vicinity of horizon from (19) one has

$$J(r) \leq J_H = \frac{2r_HME}{a}, \quad r \to r_H. \tag{20}$$

If the the values of J and $\Theta \geq 0$ are given then from (14), (15) one has at any point outside the horizon

$$E \geq \frac{1}{\Sigma^2}\left[2MraJ + \sqrt{\Delta\left(\frac{\rho^4J^2}{\sin^2\theta} + (m^2\rho^2 + \Theta)\,\Sigma^2\right)}\right]. \tag{21}$$

The lower frontier E corresponds to $R = 0$. As it is seen from (21) negative energies can be only in case of the negative value of J of the angular momentum of the particle and for r such as $2Mra|\sin\theta| > \rho^2\sqrt{\Delta}$, i.e., in accordance with (9), in the ergosphere.

3 General limitations on angular and radial velocities for the metric with rotation

Let us find limitations on the angular velocity of any particle in the metric

$$ds^2 = g_{00}\,dt^2 + 2g_{0\varphi}\,dt d\varphi + g_{\varphi\varphi}\,d\varphi^2 + g_{rr}\,dr^2 + g_{\theta\theta}\,d\theta^2 \tag{22}$$

with $g_{\varphi\varphi} < 0$, $g_{rr} < 0$, $g_{\theta\theta} < 0$ from the condition $ds^2 \geq 0$. Then

$$g_{00}\,dt^2 + 2g_{0\varphi}\,dt\,d\varphi + g_{\varphi\varphi}\,d\varphi^2 \geq 0 \tag{23}$$

and the angular velocity $\Omega = d\varphi/dt$ for any particle is in the limits [1]

$$\Omega_1 \leq \Omega \leq \Omega_2, \quad \Omega_{1,2} = \frac{g_{0\varphi} \mp \sqrt{g_{0\varphi}^2 - g_{00}g_{\varphi\varphi}}}{-g_{\varphi\varphi}}. \tag{24}$$

On the frontier of the ergosphere $g_{00} = 0$ and $\Omega_1 = 0$. Inside the ergosphere $g_{00} < 0$ and $\Omega_{1,2} > 0$ and all particles rotate in the direction of rotation of the black hole [1]–[3].

Putting the components of the metric (1) into (24) one obtains the limiting values of angular velocities for the rotating Kerr black hole

$$\Omega_{1,2} = \frac{2Mra\sin\theta \mp \rho^2\sqrt{\Delta}}{\sin\theta\,\Sigma^2} = \omega \mp \frac{\rho^2\sqrt{\Delta}}{\sin\theta\,\Sigma^2}. \tag{25}$$

Approaching the event horizon one has

$$\lim_{r\to r_H}\Omega_1(r) = \lim_{r\to r_H}\Omega_2(r) = \omega_{\rm Bh} = \frac{a}{2Mr_H}. \tag{26}$$

The value $\omega_{\rm Bh}$ is called angular velocity of rotation of the black hole.

Note that due to nonradial movement of massless particles in ergosphere their radial velocity can have different values (zero for circular orbits).

Let us find limitations on possible values of the radial velocities of particles in the metric (22) from the condition

$$g_{00} + 2g_{0\varphi}\frac{d\varphi}{dt} + g_{\varphi\varphi}\left(\frac{d\varphi}{dt}\right)^2 + g_{rr}\left(\frac{dr}{dt}\right)^2 \geq 0. \tag{27}$$

Taking into account that the sum of first three terms in (27) has the maximal value at

$$\frac{d\varphi}{dt} = \frac{\Omega_2 - \Omega_1}{2}, \tag{28}$$

one obtains

$$\left(\frac{dr}{dt}\right)^2 \leq \frac{g_{\varphi\varphi}}{g_{rr}}\frac{(\Omega_2 - \Omega_1)^2}{4} = \frac{g_{0\varphi}^2 - g_{00}g_{\varphi\varphi}}{g_{rr}g_{\varphi\varphi}}. \tag{29}$$

For the Kerr's metric (1) outside the event horizon $r > r_H$ one obtains

$$\left(\frac{dr}{dt}\right)^2 \leq \frac{\Delta^2}{\Sigma^2}. \tag{30}$$

Close to the event horizon $\Delta(r) \to 0$ and so for any moving particle one has

$$\frac{dr}{dt} \to 0, \text{ if } r \to r_H. \tag{31}$$

For particles moving on geodesics one obtains from Eqs. (10)–(12)

$$\left(\frac{dr}{dt}\right)^2 = \frac{\Delta^2}{\Sigma^2}\left[1 - \frac{\Delta}{(\Sigma^2 E - 2MraJ)^2}\left(\frac{\rho^4 J^2}{\sin^2\theta} + \Sigma^2\left(m^2\rho^2 + \Theta\right)\right)\right]. \tag{32}$$

So the maximal radial velocity $|v_r|_{\max} = \Delta/\Sigma$ is obtained by massless particles with zero projection of the angular momentum moving with fixed value of the angle θ ($\Theta = 0$):

$$\left|\frac{dr}{dt}\right|_{\max} = \frac{\Delta}{\Sigma}, \text{ if } m = 0, \ J = 0, \ Q = -a^2 E^2 \cos^2\theta. \tag{33}$$

4 Particles with zero energy in the Kerr's metric

In literature on black holes there is practically no discussion of properties of particles with zero relative to infinity energy. Here we shall discuss geodesics for such particles.

Values $E = 0$ are possible on the frontier and inside the ergosphere. From the inequality $\Theta \geq 0$ and Eq. (13) one obtains

$$E = 0 \ \Rightarrow \ Q \geq m^2 a^2 \cos^2\theta + J^2 \cot^2\theta \geq 0, \tag{34}$$

i.e. for test particles with zero relative to infinity energy the Carter constant is non negative.

From the inequality (19) one obtains for particles with zero energy inside ergosphere

$$E = 0 \ \Rightarrow \ J \leq -\sin\theta\sqrt{\frac{\Delta\left(m^2\rho^2 + \Theta\right)}{-S}}, \tag{35}$$

i.e. the angular momentum projection of particles with negative energy is negative.

It was shown by us earlier [4] that geodesics of particles with zero energy as well as as for particles with negative [5] energy originate and terminate on r_H. One also obtains

$$E = 0 \ \Rightarrow \ R = -\Delta(m^2\rho^2 + \Theta) - \frac{J^2}{\sin^2\theta} S(r, \theta). \tag{36}$$

So the upper point of the trajectory of massive particles with zero energy is located inside the ergosphere ($r < r_1(\theta)$). Movement along the coordinate θ can be prolonged up to the point with $\Theta(\theta)$ equal to zero. So in case $m = 0$ from (36) one obtains that the upper point of the trajectory of massless particles with zero energy is located on the frontier of the ergosphere $r_1(\theta)$.

The radial velocity of the particle with zero energy is

$$E = 0 \ \Rightarrow \ \frac{dr}{dt} = \frac{\sigma_r \Delta}{2Mra}\sqrt{\frac{-S}{\sin^2\theta} - \Delta\frac{m^2\rho^2 + \Theta}{J^2}} \tag{37}$$

and in general depends on the angular momentum. Indeed in case $m^2\rho^2 + \Theta > 0$ at the point with given r in the ergosphere the radial velocity is the larger in absolute value the lager is $|J|$ and can vary from zero for the maximal value $J = -\sin\theta\sqrt{\Delta(m^2\rho^2 + \Theta)/(-S)}$ to

$$\frac{dr}{dt} = \frac{\sigma_r \Delta\sqrt{-S}}{2Mra\sin\theta}, \quad J \to -\infty. \tag{38}$$

From equations for geodesics it follows that

$$E = 0 \ \Rightarrow \ \frac{d\varphi}{dt} = \frac{-S(r)}{2rMa\sin^2\theta} = \frac{r^2 - 2rM + a^2\cos^2\theta}{-2rMa\sin^2\theta}. \qquad (39)$$

So *the angular velocity of any particle with zero energy does not depend on the value of mass and angular momentum of the particle and is defined by the formula (39).*

If the particle with zero energy approaches the frontier of the ergosphere (then it is necessary that $m = 0$) one has $S(r) = 0$ and

$$r \to r_1(\theta) \ \Rightarrow \ \frac{d\varphi}{dt} \to 0. \qquad (40)$$

Inside the ergosphere the angular velocity of the particle with zero energy as for any particle is positive.

If

$$E = 0, \ \ Q = \cos^2\theta\left[a^2m^2 + \frac{J^2}{\sin^2\theta}\right], \qquad (41)$$

then $\Theta = 0$ and from (11) it follows that movement along geodesic occurs for constant θ. The form of trajectory of such photons does not depend on the value of angular momentum projection as it is seen from (10), (11). The equation of this trajectory can be expressed in elementary functions

$$\begin{aligned}\varphi(r) - \varphi(r_1(\theta)) &= \frac{-\sigma_r}{\sin\theta}\left[\arcsin\frac{r-M}{\sqrt{M^2 - a^2\cos^2\theta}} - \frac{\pi}{2} + \right.\\ &\left. + \frac{a\sin\theta}{\sqrt{M^2-a^2}}\tanh^{-1}\frac{\sqrt{(M^2-a^2)(2rM - r^2 - a^2\cos^2\theta)}}{a(r-M)\sin\theta}\right], \quad a < M, \qquad (42)\end{aligned}$$

and is most simple for $a = M$:

$$\varphi(r) - \varphi(r_1(\theta)) = \frac{-\sigma_r}{\sin\theta}\left[\arcsin\frac{r-M}{\sin\theta} + \frac{\sqrt{2rM - r^2 - M^2\cos^2\theta}}{r-M} - \frac{\pi}{2}\right]. \qquad (43)$$

The corresponding coordinate time of photon movement is

$$\begin{aligned}t(r) - t(r_1(\theta)) &= \frac{-\sigma_r M}{\sqrt{M^2-a^2}}\left[r_H \ln\frac{Mr_H - r\sqrt{M^2-a^2} - a^2\cos^2\theta + a\sin\theta\sqrt{-S(r,\theta)}}{M(r-r_H)} - \right.\\ &\left. - r_C\ln\frac{Mr_C + r\sqrt{M^2-a^2} - a^2\cos^2\theta + a\sin\theta\sqrt{-S(r,\theta)}}{M(r-r_C)}\right], \quad a < M. \qquad (44)\end{aligned}$$

When approaching the event horizon $r \to r_H$ the time of movement diverges logarithmically. Remind that the time of approaching the event horizon by the light ray in Schwarzschild metric also diverges logarithmically [11].

For the extremal rotating black hole

$$t(r) - t(r_1(\theta)) = -\sigma_r 2M\left[\frac{\sqrt{-S(r,\theta)}}{(r-M)\sin\theta} - \ln\frac{r-M}{\sqrt{-S(r,\theta)} + M\sin\theta}\right]. \qquad (45)$$

The time of approaching the event horizon $r \to r_H$ here is divergent as $\sim 1/(r - r_H)$.

Part of the trajectory of the photon moving in equatorial plane on geodesic with zero energy and the time of movement are shown on Fig. 1. The radial coordinate is used r/M. This trajectory as

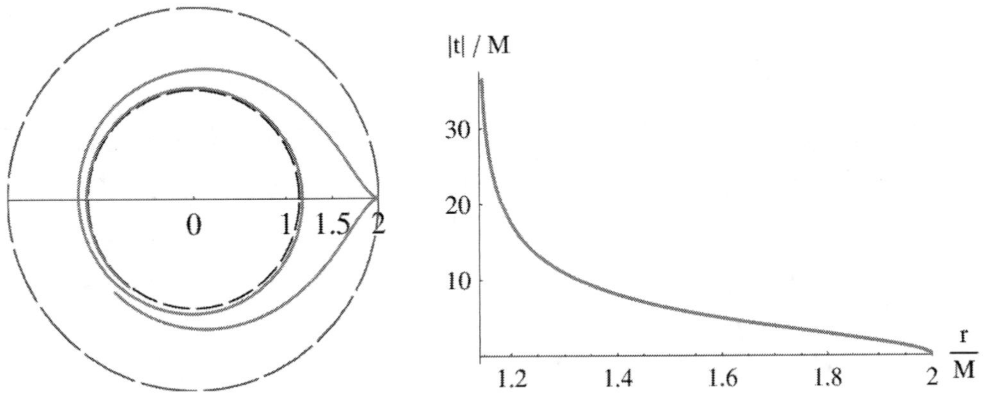

Figure 1. The trajectories (at the left) of photons in the ergosphere of the black hole with $a = 0.99M$ and $E = 0$ and time of movement (at the right).

well as trajectories of any particles intersecting the event horizon of the black hole rotates around the horizon infinitely many times in Boyer-Lindquist coordinate frame. Really from the equations of geodesics (10), (11) it follows that in the vicinity of the horizon

$$\frac{d\varphi}{dr} \sim \frac{1}{\Delta} = \frac{1}{(r - r_H)(r - r_C)}. \tag{46}$$

Let us investigate the problem what simple properties of movement define the difference of particles with negative, zero and positive relative to the infinity energy.

5 Angular velocity of particles with zero and negative energies

From equations of geodesics one obtains for the angular velocity of freely moving particles

$$\frac{d\varphi}{dt} = \frac{2MraE + \frac{SJ}{\sin^2\theta}}{\Sigma^2 E - 2MraJ} = \omega + \frac{\Delta\rho^4 J}{\sin^2\theta\Sigma^2(\Sigma^2 E - 2MraJ)}. \tag{47}$$

From the limitation (19) one obtains for particles with negative energy

$$\frac{J}{EM} \geq \frac{\sin\theta}{-S(r)}\left[2ra\sin\theta + \frac{1}{M}\sqrt{\Delta\left[\rho^4 - \left(\frac{m^2}{E^2}\rho^2 + \frac{\Theta}{M^2E^2}\right)S(r)\right]}\right], \tag{48}$$

and for particles with positive energy in the ergosphere

$$\frac{J}{EM} \leq \frac{\sin\theta}{-S(r)}\left[2ra\sin\theta - \frac{1}{M}\sqrt{\Delta\left[\rho^4 - \left(\frac{m^2}{E^2}\rho^2 + \frac{\Theta}{M^2E^2}\right)S(r)\right]}\right]. \tag{49}$$

Putting boundary values of expressions (48), (49) for $m = 0$ and $\Theta = 0$ into (47) one obtains expressions for Ω_1 and Ω_2 (see (25)). Taking into account that $d\varphi/dt$ due to (47)is growing with the growth of $J/(EM)$ on each interval of continuity $\left(-\infty, \Sigma^2/(2M^2ra)\right)$ and $\left(\Sigma^2/(2M^2ra), +\infty\right)$ one obtains the division of the region of possible angular velocities in the ergosphere on two regions: the

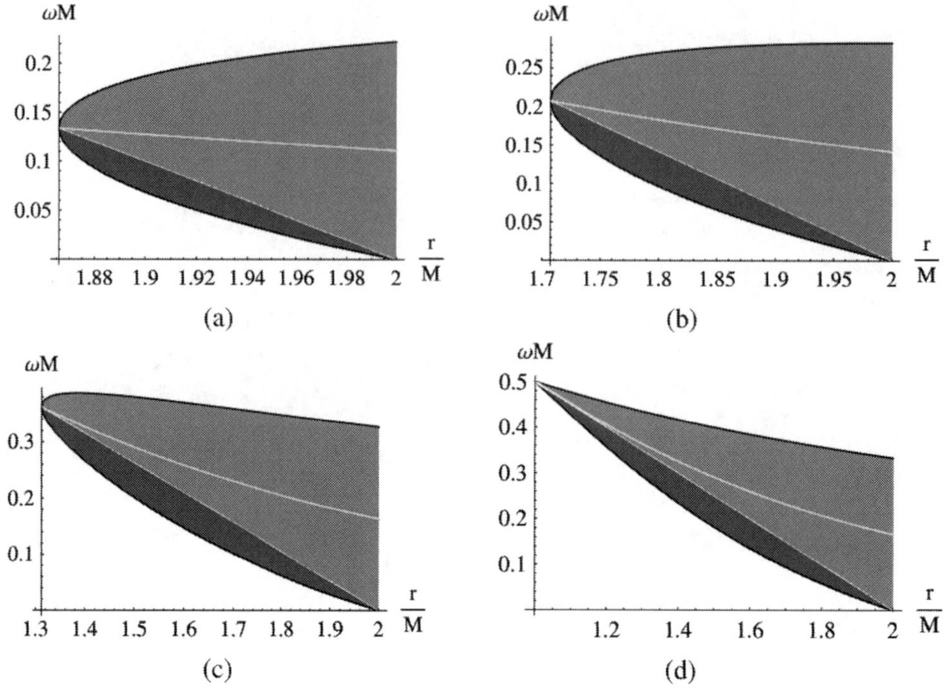

Figure 2. Regions of possible angular velocities of particles in ergosphere for $\theta = \pi/2$ and a) $a/M = 0.5$, b) $a/M = 1/\sqrt{2}$, c) $a/M = 0.95$, d) $a/M = 1$. The red region — angular velocities of particles with positive energy, the blue region — for particles with negative energy, the white line — $E = 0$, the green line — $J = 0$.

lower for particles with negative energy and the upper one for particles with positive energy as it is shown on Fig. 2.

Going in (47) to the limit $|J/(EM)| \to \infty$ one obtains the expression (39). That is why the frontier of regions for particles with positive and negative energies is the line corresponding to the angular velocity of particles with zero energy. In case of rotation in equatorial plane this line is the direct line.

So the value of the angular velocity of the particle in the ergosphere for the given value of the radial coordinate uniquely shows the sign of the energy of the particle relative to the infinity and the value of $J/(EM)$. Particles with zero energy are particles moving in ergosphere with angular velocity (39), corresponding to the white line on Fig. 2. They divide by the value of angular velocity of rotation at given r in the ergosphere particles with negative energies from particles with positive relative to infinity energy. Particles with negative energy are particles rotating in the ergosphere with the angular velocity less than (39)!

The light green line on Fig. 2 corresponds to the angular velocity of free falling particles with zero projection of angular momentum. Note that in this case the energy cannot be negative.

6 The energy of collisions of particles in the centre of mass frame

The energy in the centre of mass frame $E_{\text{c.m.}}$ of two colliding particles with rest masses m_1 and m_2 is found by putting in square the formula

$$(E_{\text{c.m.}}, 0, 0, 0) = p^i_{(1)} + p^i_{(2)}, \tag{50}$$

where $p^i_{(n)}$ — 4-momenta of particles ($n = 1, 2$). Due to $p^i_{(n)} p_{(n)i} = m_n^2$ one obtains

$$E_{\text{c.m.}}^2 = m_1^2 + m_2^2 + 2 p^i_{(1)} p_{(2)i}. \tag{51}$$

For free falling particles with energies E_1 and E_2 (relative to infinity) and angular momenta J_1, J_2 from equations of geodesics one obtains

$$E_{\text{c.m.}}^2 = m_1^2 + m_2^2 - \frac{2}{\rho^2}\sigma_{1\theta}\sigma_{2\theta}\sqrt{\Theta_1\Theta_2} + \\ + \frac{2}{\Delta\rho^2}\left[E_1 E_2 \Sigma^2 - 2Mra(E_1 J_2 + E_2 J_1) - J_1 J_2 \frac{S}{\sin^2\theta} - \sigma_{1r}\sigma_{2r}\sqrt{R_1 R_2} \right]. \tag{52}$$

Let us find the expression of the energy in the centre of mass frame through the relative velocity v_{rel} of particles at the moment of collision [12]. In the reference frame of the first particle one has for the components of 4-velocities of particles at this moment

$$u^i_{(1)} = (1, 0, 0, 0), \quad u^i_{(2)} = \frac{(1, \mathbf{v}_{\text{rel}})}{\sqrt{1 - v_{\text{rel}}^2}}. \tag{53}$$

So $u^i_{(1)} u_{(2)i} = 1 \Big/ \sqrt{1 - v_{\text{rel}}^2}$,

$$v_{\text{rel}} = \sqrt{1 - \left(u^i_{(1)} u_{(2)i}\right)^{-2}}. \tag{54}$$

These expressions evidently don't depend on the coordinate system.

From (51) and (54) one obtains

$$E_{\text{c.m.}}^2 = m_1^2 + m_2^2 + \frac{2 m_1 m_2}{\sqrt{1 - v_{\text{rel}}^2}} \tag{55}$$

and the nonlimited growth of the collision energy in the centre of mass frame occurs due to growth of the relative velocity to the velocity of light [13].

If particles move in radial coordinate in one direction ($\sigma_{1r}\sigma_{2r} = 1$) one obtains in the limit $r \to r_H$ by solving in the last term in (52) the indeterminacy of the form 0/0, for collisions on the horizon

$$E_{\text{c.m.}}^2 = m_1^2 + m_2^2 - \frac{2}{\rho^2}\sigma_{1\theta}\sigma_{2\theta}\sqrt{\Theta_{1H}\Theta_{2H}} + \left(m_1^2 + \frac{\Theta_{1H}}{\rho_H^2}\right)\frac{J_{2H} - J_2}{J_{1H} - J_1} + \\ + \left(m_2^2 + \frac{\Theta_{2H}}{\rho_H^2}\right)\frac{J_{1H} - J_1}{J_{2H} - J_2} + \frac{\rho_H^2}{4M^2 r_H^2 \sin^2\theta}\frac{(J_{1H} J_2 - J_{2H} J_1)^2}{(J_{1H} - J_1)(J_{2H} - J_2)} \tag{56}$$

(see also [14]). If one of the particles has the angular momentum projection $J = J_H$ (the critical particle) and the other particle has $J \neq J_H$ then the energy of collisions is divergent on the horizon. For extremal rotating black holes this was found in the paper of Bañados-Silk-West [15] (BSW-effect). For nonextremal black holes in the vicinity of the horizon $J = J_H$ is not possible however in case of multiple collisions it is possible [16, 17] to have J very close to J_H for $r \to r_H$.

Very large energy of collisions can be obtained if due to multiple collisions or external field the particle has large in absolute value but negative angular momentum projection [8, 9] for the fixed value of the particle energies due to (19). As it was obtained in [8] for the collision in ergosphere one has

$$E_{\text{c.m.}}^2 \approx J_2 \frac{r^2 - 2rM + a^2\cos^2\theta}{\rho^2\Delta\sin^2\theta}\left(\sigma_{1r}\sqrt{J_{1+} - J_1} - \sigma_{2r}\sqrt{J_{1-} - J_1}\right)^2, \quad J_2 \to -\infty. \tag{57}$$

In case $J \to -\infty$ for fixed energy and mass of the particle the trajectory has the form shown in Fig. 1. Unbounded growth of the energy of collision with growing $-J$ due to (55) is conditioned as for BSW effect by the growth of the relative velocity of particles to the velocity of light. Differently from the BSW effect this effect can take place at any point of the ergosphere.

The energy of direct collisions ($\sigma_{1r}\sigma_{2r} = -1$) is divergent on the horizon

$$E_{\rm c.m.}^2 \sim \frac{4a^2}{\Delta\rho^2}(J_{1H} - J_1)(J_{2H} - J_2) \to \infty, \quad r \to r_H, \tag{58}$$

if $J_i \neq J_{iH}$. This way of getting ultrahigh energy is possible for particles one of which moves along white hole geodesic [18, 19].

Formulas (50)–(58) are valid for any colliding particles as with positive as with negative (relative to infinity) energy. For any value of the energies of particles the energy in the centre of mass frame satisfies the inequality

$$E_{\rm c.m.} \geq m_1 + m_2, \tag{59}$$

because colliding particles in the centre of mass frame are moving one to another with some velocities. All three ways of getting unboundedly high energy of collisions are also possible for particles with negative (zero) energy.

The physical importance of these resonances is due to possible conversion of dark matter particles into visible ones in the ergosphere of astrophysical black holes [20, 21].

The results of some calculations of the energy of collisions of two particles with equal masses m one falling from the infinity on the black hole with the other particle with positive, zero and negative energy are given in Ref. [4].

7 Negative and zero energies in rotating coordinate frame

Following [11], § 89, introduce notations for inertial cylindrical space coordinates r', φ', z' and time t. The interval in this coordinate system is

$$ds^2 = c^2dt^2 - dr'^2 - r'^2d\varphi'^2 - dz'^2. \tag{60}$$

Here we use the usual units so that c is the velocity of light. For rotating system let us note the cylindrical coordinates as r, φ, z. Let the rotation axis $z = z'$. Then

$$r' = r, \quad z' = z, \quad \varphi' = \varphi + \Omega t, \tag{61}$$

where Ω is the angular velocity of rotation. Putting (61) into (60) one obtains the expression for the interval in rotational coordinate system

$$ds^2 = (c^2 - \Omega^2 r^2)\,dt^2 - 2\Omega r^2 d\varphi\,dt - dr^2 - r^2 d\varphi^2 - dz^2. \tag{62}$$

Here we use the words "coordinate system" different from the "coordinate frame" because we want to use expression (62) for the interval for any r!

Let the energy and momentum in the inertial system be E', $\mathbf{p}'$. Then the four energy momentum vector due to [11], § 9, in Descartes coordinates is

$$p'^i = \left(\frac{E'}{c}, \mathbf{p}'\right), \quad p'_i = \left(\frac{E'}{c}, -\mathbf{p}'\right). \tag{63}$$

Compute the energy-momentum in rotating coordinate system $p^i = \left(\partial x^i/\partial x'^k\right) p'^k$. Then

$$p^i = \left(\frac{E'}{c},\ p'^r,\ p'^\varphi - \Omega\frac{E'}{c^2},\ p'^z\right), \tag{64}$$

and covariant components are

$$p_i = \left(\frac{E' - \Omega r^2 p'^\varphi}{c},\ -p'^r,\ -r^2 p'^\varphi,\ -p'^z\right) = \left(\frac{E}{c},\ -p'^r,\ -L_z,\ -p'^z\right). \tag{65}$$

So in rotating coordinate system the energy E and component of the angular momentum L_z are

$$E = E' - \Omega L'_z, \quad L_z = L'_z = r^2 p'^\varphi. \tag{66}$$

Surely both these values are constants of motion. It is important that covariant components but not contravariant components of the energy momentum vector occur to be constants of motion. Contravariant components of the momenumt are proportional to the corresponding velocities.

For particles with mass m in cylindrical coordinates of Minkowski space-time one obtains

$$\frac{E'^2}{c^2} - (p'^r)^2 - \frac{1}{r^2}L'^2_z - (p'^z)^2 = m^2c^2. \tag{67}$$

Then we obtain from (66) the possible limits for energy of particle in rotating coordinate system for fixed energy E' in nonrotating system and given value r:

$$E'\left(1 - \frac{\Omega r}{c}\sqrt{1 - \frac{m^2c^4}{E'^2}}\right) \le E \le E'\left(1 + \frac{\Omega r}{c}\sqrt{1 - \frac{m^2c^4}{E'^2}}\right). \tag{68}$$

For $r < c/\Omega$ the energy $E > 0$. For $r = c/\Omega$ the zero energy $E = 0$ can be achieved. For $r > c/\Omega$ the energy $E = 0$ can accept arbitrary values, positive, zero and negative.

This picture is similar to the case of the Kerr metric, when the covariant components of the energy-momentum momentum of particle are

$$p_t = E, \quad p_r = -\sigma_r\frac{\sqrt{R}}{\Delta}, \quad p_\theta = -\sigma_\theta\sqrt{\Theta}, \quad p_\varphi = -J. \tag{69}$$

The negative values of the energy E in both cases can be only in the region of space-time where $g_{00} < 0$ (see the problem 17.9 in the book [22]).

It is well known that for a particle in rotating coordinate frame similar to the Boyer-Lindquist coordinates for Kerr metric there exists region where particle cannot be at rest $r > c/\Omega$ and the frontier $r = c/\Omega$ is the static limit. For the case of the Earth rotating along its axis with a period of one day this distance is $\approx 4.1 \cdot 10^9$ km, which is much farther than Uranus but closer than Neptune.

From formula (68) one sees that negative and zero energy E trajectories are possible in the region $r > c/\Omega$ external to the static limit in rotating coordinate frame. The main difference of the energy of the particle in Kerr metric and in rotating coordinate frame is that in Kerr metric one defines all conserved values relative to space infinity while in rotating coordinates the observer is located at the centre where the projection of angular momentum of any particle is zero so $E = E'$.

Acknowledgements. This work was supported by the Russian Foundation for Basic Research, grant No. 15-02-06818-a and by the Russian Government Program of Competitive Growth of Kazan Federal University.

References

[1] C. W. Misner, K. S. Thorne, J. A. Wheeler, *Gravitation* (Freeman, San Francisco, 1973)
[2] S. Chandrasekhar, *The Mathematical Theory of Black Holes* (Clarendon Press, Oxford University Press, New York, Oxford, 1983)
[3] I. D. Novikov, V. P. Frolov, *Physics of Black Holes* [in Russian] (Nauka, Moscow, 1986); V. P. Frolov, I. D. Novikov, *Black Hole Physics: Basic Concepts and New Developments* (Kluwer Acad. Publ., Dordrecht, 1998)
[4] A. A. Grib, Yu. V. Pavlov, *Black holes and particles with zero and negative energy*, arXiv:1601.02592
[5] A. A. Grib, Yu. V. Pavlov, V. D. Vertogradov, Mod. Phys. Lett. A **29**, 1450110 (2014)
[6] R. P. Kerr, Phys. Rev. Lett. **11**, 237–238 (1963)
[7] R. H. Boyer, R. W. Lindquist, J. Math. Phys. **8**, 265–281 (1967)
[8] A. A. Grib, Yu. V. Pavlov, Europhys. Lett. **101**, 20004 (2013)
[9] A. A. Grib, Yu. V. Pavlov, Teor. Matem. Fiz. **176**, 60–68 (2013) [English transl.: Theor. Math. Phys. **176**, 881–887 (2013)]
[10] O. B. Zaslavskii, Mod. Phys. Lett. A **28**, 1350037 (2013)
[11] L. D. Landau, E. M. Lifshitz, *The Classical Theory of Fields* (Pergamon Press, Oxford, 1983)
[12] M. Banados, B. Hassanain, J. Silk, S. M. West, Phys. Rev. D **83**, 023004 (2011)
[13] O. B. Zaslavskii, Phys. Rev. D **84**, 024007 (2011)
[14] T. Harada, M. Kimura, Phys. Rev. D **83**, 084041 (2011)
[15] M. Bañados, J. Silk, S. M. West, Phys. Rev. Lett. **103**, 111102 (2009)
[16] A. A. Grib, Yu. V. Pavlov, JETP Letters **92**, 125–129 (2010)
[17] A. A. Grib, Yu. V. Pavlov, Astropart. Phys. **34**, 581–586 (2011)
[18] A. A. Grib, Yu. V. Pavlov, Grav. Cosmol. **21**, 13–18 (2015)
[19] A. A. Grib, Yu. V. Pavlov, Teor. Matem. Fiz. **185**, 77–85 (2015) [English transl.: Theor. Math. Phys. **185**, 1425–1432 (2015)]
[20] A. A. Grib, Yu. V. Pavlov, Mod. Phys. Lett. A **23**, 1151–1159 (2008)
[21] A. A. Grib, Yu. V. Pavlov, Grav. Cosmol. **15**, 44–48 (2009)
[22] A. P. Lightman, W. H. Press, R. H. Price, S. A. Teukolsky, *Problem book in relativity and gravitation* (Princeton University Press, New Jersey, 1975)

EPJ Web of Conferences **125**, 03013 (2016) DOI: 10.1051/epjconf/201612503013
QUARKS-2016

Constraints on high energy interaction models from LHC and cosmic ray data

Sergey Ostapchenko[1,2,⋆]

[1] *Frankfurt Institute for Advanced Studies (FIAS), 60438 Frankfurt am Main, Germany*
[2] *D.V. Skobeltsyn Institute of Nuclear Physics, Moscow State University, 119992 Moscow, Russia*

Abstract. Predictions of popular cosmic ray interaction models for some basic characteristics of cosmic ray-induced extensive air showers are analyzed in view of experimental data on proton-proton collisions, obtained at the Large Hadron Collider. The differences between the results are traced down to different approaches for the treatment of hadronic interactions, implemented in those models. Potential measurements by LHC and cosmic ray experiments, which could be able to discriminate between the alternative approaches, are proposed.

1 Introduction

Modeling of high energy hadronic interactions is of considerable importance for experimental studies at the Large Hadron Collider (LHC) and, especially, in high energy cosmic ray (CR) field. In the latter case, one traditionally relies on the extensive air shower (EAS) techniques: the properties of primary CR particles are reconstructed from measured characteristics of nuclear-electro-magnetic cascades induced by CR interactions in the atmosphere. This naturally implies the importance of detailed Monte Carlo simulations of EAS development, particularly, of its backbone – the cascade of nuclear interactions of both the primary particles and of secondary hadrons produced. Thus, the success of the experimental studies depends on the validity of hadronic interaction models used in the analysis.

One usually chooses between two main experimental procedures [1]. In the first case, one deals with the information obtained with scintillation detectors positioned at ground. The energy of the primary particle is reconstructed from the measured lateral density of charged particles (mostly electrons and positrons) while the particle type is inferred from the relative fraction of muons, compared to all charged particles at ground. Alternatively, one may study the longitudinal EAS development by measuring fluorescence light produced by excited air molecules at different heights in the atmosphere, for which purpose dedicated fluorescence telescopes are employed. The primary energy is then related to the total amount of fluorescence light emitted. In turn, the particle type may be determined from the measured position of the shower maximum $X_{\max}$ – the depth in the atmosphere (in g/cm^2) where the number of ionizing particles reaches its maximal value.

Not surprisingly, the observables used to determine the primary particle type – the lateral muon density ρ_μ and the EAS maximum position $X_{\max}$ – appear to be very sensitive to details of high energy hadronic interactions. More precisely, $X_{\max}$ depends strongly on the properties of the primary particle

⋆e-mail: sergei@tf.phys.ntnu.no

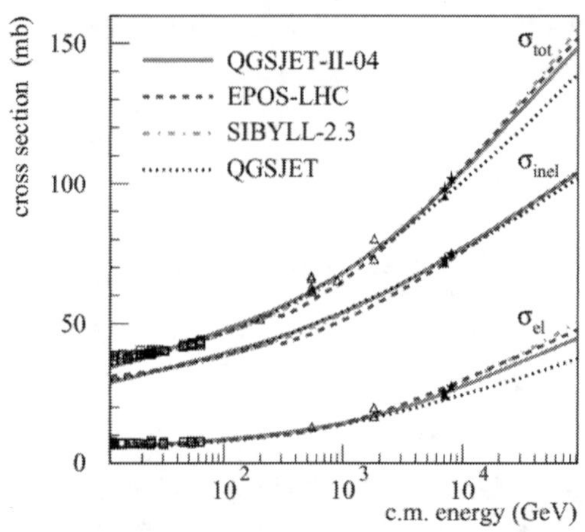

Figure 1. $\sqrt{s}$-dependence of the total, inelastic, and elastic pp cross sections, as calculated using the QGSJET-II-04 [5], EPOS-LHC [3], SIBYLL-2.3 [7], and QGSJET [8] models (solid, dashed, dash-dotted, and dotted lines respectively). Experimental data are from Refs. [10–12].

interaction with air nuclei: the inelastic cross section and the forward spectra of secondary hadrons produced. In turn, the EAS muon content is formed in a multi-step cascade process, driven mostly by interactions of secondary pions and kaons with air. Hence, ρ_μ depends strongly on the properties of pion-air collisions, most importantly, on the multiplicity and spectral shape for charged hadrons produced.

In the following, we shall mainly deal with the most important and best-measured EAS observable – the shower maximum position. The corresponding EAS simulation results are most directly related to the modeling of very high energy proton-proton collisions, hence, are expected to be strongly constrained by experimental studies at LHC. We shall compare the results of $X_{\max}$ calculations, obtained using the most recent versions of the EPOS [2, 3], QGSJET-II [4, 5], and SIBYLL [6, 7] models, which have been updated using experimental data from Run 1 of LHC. Additionally, we shall use the QGSJET model [8] which, though being already outdated physics-wise, demonstrated a generally good agreement with the data from Run 1 of LHC on minimum-bias collisions [9]. Hence, it may be used to study the range of potential variations of model predictions for $X_{\max}$, in view of current LHC data. Our primary goals are to analyze the differences between the model results, to trace their origin to the underlying approaches for the treatment of hadronic collisions, implemented in those models, and to propose potential measurements by LHC and cosmic ray experiments, which could be able to discriminate between the alternative approaches.

2 Model predictions for $X_{\max}$

The uncertainties of model predictions for $X_{\max}$ have been greatly reduced thanks to the LHC studies of very high energy proton-proton collisions. Of particular importance are precise measurements of the total, inelastic, and elastic pp cross sections by the TOTEM and ATLAS experiments [10, 11], which strongly constrained the corresponding model results in the ultra-high energy limit, as illustrated in Fig. 1. In turn, this contributes to the convergence of the model predictions for the inelastic proton-air and nucleus-air cross sections which determine the depth on the first interaction of the primary particle in the atmosphere. It is worth also stressing the importance of the measurements

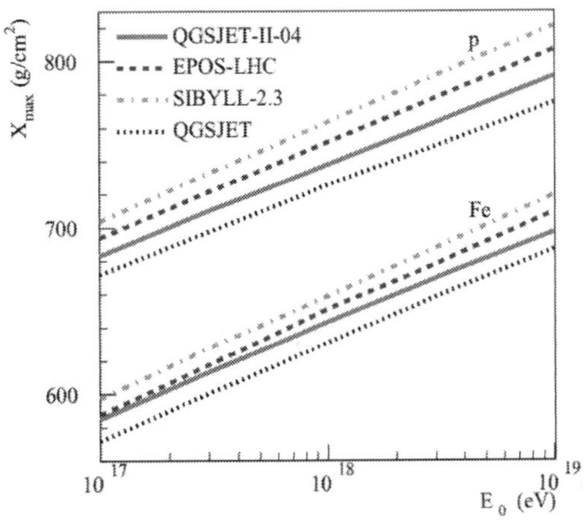

Figure 2. Primary energy dependence of the average shower maximum depth for proton- and iron-initiated vertical EAS, as calculated using the QGSJET-II-04, EPOS-LHC, SIBYLL-2.3, and QGSJET models (solid, dashed, dash-dotted, and dotted lines respectively).

of the differential elastic pp cross section which provides information on the spatial parton distribution in proton. The latter impacts both the calculations of the proton-air cross section and the magnitude of nonlinear interaction effects in proton-proton and proton-nucleus collisions.

Yet there remain considerable differences between the model predictions for $X_{\max}$,[1] as illustrated in Fig. 2. One might relate those to present experimental uncertainties concerning the rate of the inelastic diffraction in high energy pp collisions. Indeed, diffraction has a significant influence on the shape of very forward spectra of secondary hadrons, which in turn makes a strong impact on the longitudinal EAS development. This has been investigated in Ref. [14] in the framework of the QGSJET-II-04 model; the obtained characteristic uncertainty for $X_{\max}$ amounted to 10 g/cm^2, being thus considerably smaller than the differences between the results plotted in Fig. 2. One might then ask if the analysis of Ref. [14] was not general enough or $X_{\max}$ depends on some other characteristics of hadronic interactions, not well constrained by present LHC data.

3 Impact of constituent parton Fock states

An important insight into the problem is provided by recent combined measurements by the CMS and TOTEM experiments of the pseudorapidity η density $dn_{pp}^{\rm ch}/d\eta$ of produced charged hadrons in pp collisions at $\sqrt{s}$ = 8 TeV [15]. In contrast to the EPOS-LHC and QGSJET-II-04 models, hadronic Monte Carlo generators used in the collider field turn to underestimate considerably secondary hadron production at large η. As illustrated in Fig. 3, a similar problem arises for the SIBYLL-2.3 model.

This appears to be related to model assumptions concerning the structure of constituent parton Fock states in hadrons, as discussed in more detail in Ref. [16]. Let us consider a proton-proton interaction in the center of mass (c.m.) frame, which involves multiple binary interactions of partons from two protons' parton clouds formed prior to the collision. We are interested here in the momentum distribution of these multiparton states, which is related to the evolution of parton cascades in hadrons. The picture behind the corresponding treatment of the SIBYLL model and of event generators used

[1] Here and in the following the calculations of extensive air shower development are performed using the CONEX code [13].

DOI: 10.1051/epjconf/201612503013

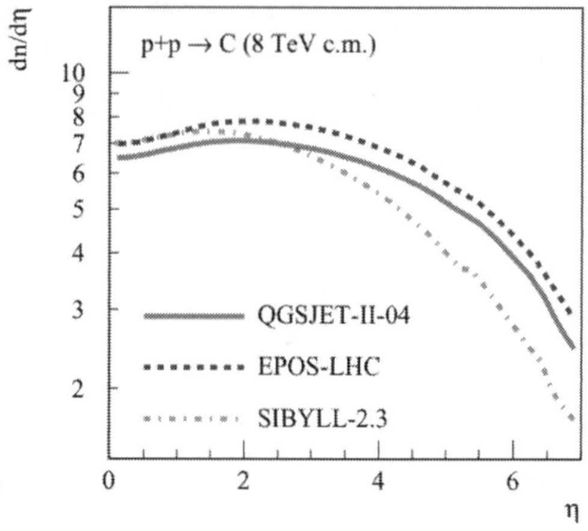

Figure 3. $dn^{\rm ch}_{pp}/d\eta$ for pp collisions at $\sqrt{s}$ = 8 TeV, as calculated using the QGSJET-II-04, EPOS-LHC, and SIBYLL-2.3 models (solid, dashed, and dash-dotted lines respectively) for the nondiffractive event selection of TOTEM: at least one charged hadron produced both at $-6.5 < \eta < -5.3$ and at $5.3 < \eta < 6.5$.

Figure 4. Schematic view of the initial part of the parton cascade in proton. Left: the cascade starts from the same universal parton Fock state; new partons participating in multiple scattering processes emerge from the cascade development, being characterized by $\propto 1/x$ distributions for the momentum fraction. Right: proton is represented by a superposition of Fock states consisting of different numbers of large x constituent partons; the more abundant multiple scattering the larger Fock states get involved in the process.

in the collider field is shown schematically on the left-hand side (lhs) of Fig. 4. At large Feynman-x, one usually starts from the same universal parton Fock state. Additional partons (sea quarks and gluons) giving rise to new branches of the parton cascade, which take part in the multiple scattering processes, result from the evolution of the parton density corresponding to this initial state and their momentum fractions are typically distributed as $\propto 1/x$ in the very high energy limit. Such a picture reflects itself in the hadron production pattern predicted by such models: multiple scattering mostly affects central particle production, while having a weak influence of forward hadron spectra. Indeed, the latter are formed by the hadronization of partons emerging from the initial part of the underlying parton cascade, which starts from the same initial conditions and covers a short rapidity interval, being thus weakly dependent on the further development of the cascade. In particular, the steep fall-down of $dn^{\rm ch}_{pp}/d\eta$ predicted by SIBYLL-2.3 (dash-dotted line in Fig. 3) for large η values reflects the quick decrease of the number of constituent partons when parton momentum fraction increases.

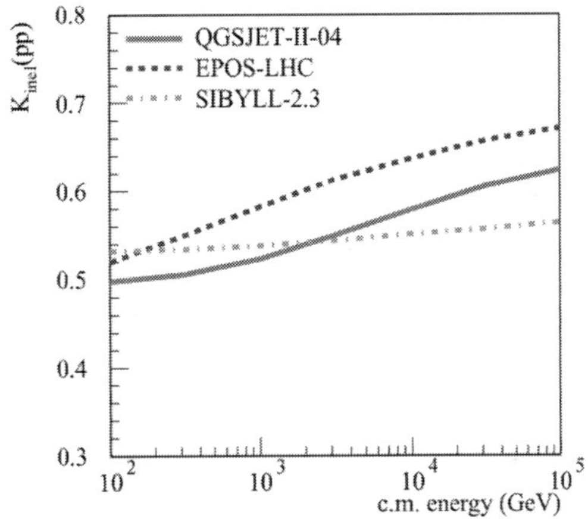

Figure 5. Energy dependence of the inelasticity of leading nucleons in *pp* collisions, as calculated using the QGSJET-II-04, EPOS-LHC, and SIBYLL-2.3 models (solid, dashed, and dash-dotted lines respectively).

In the alternative approach, implemented in the EPOS and QGSJET(-II) models, a proton is represented by a superposition of a number of Fock states containing different numbers of large x constituent partons, as shown schematically on the right-hand side (rhs) of Fig. 4. Further cascading of these partons "dresses" them with low x parton clouds. As the overall parton multiplicity in the central rapidity region is roughly proportional to the number of initial constituent partons, stronger multiple scattering is typically associated with larger Fock states. Thus, there is a strong long-range correlation between central and forward particle production; higher multiplicity in the central rapidity region reflects stronger multiple parton scattering. In turn, this implies that bigger numbers of large x constituent partons are involved in the process, which has a strong impact on forward hadron spectra.

The two approaches give rise to quite different predictions for the energy dependence of the inelasticity $K^{\rm inel}$, i.e. the relative energy loss of leading nucleons, in proton-proton and proton-nucleus collisions. Both schemes predict an enhancement of multiple scattering with the increase of collision energy – as a consequence of the enlarged kinematic space and of an increase of parton densities in the low x limit. However, in the approach corresponding to the schematic picture in the lhs of Fig. 4, this results in a substantial enhancement of secondary particle production in the central rapidity region only, while having a weak impact on forward hadron spectra. Hence, one expects a weak energy dependence of $K^{\rm inel}$ in that case, which we observe indeed for SIBYLL 2.3 in Fig. 5, where its prediction for the inelasticity $K_{pp}^{\rm inel}$ for pp collisions is shown by dash-dotted line. In the alternative approach corresponding to the picture in the rhs of Fig. 4, one obtains a substantial energy-rise of the inelasticity, which is clearly seen in Fig. 5 for QGSJET-II-04 and EPOS-LHC (solid and dashed lines, respectively). The reason for this rise is twofold. First, for any given Fock state, increasing multiple scattering implies that bigger numbers of large x constituent partons are involved in the interaction, thus leaving smaller fractions of the initial proton momentum for spectator partons which finally form the "leading" (most energetic) nucleons. Additionally, Fock states with bigger and bigger numbers of large x constituent partons come into play. The momentum sharing between these partons gives rise to a smaller fraction of the initial proton momentum, possessed by each parton, which thus enhances the energy loss of the leading nucleons. This analysis clearly explains the larger $X_{\rm max}$ values predicted by SIBYLL 2.3 (dash-dotted lines in Fig. 2) – as a smaller inelasticity results in a slower EAS development (see, e.g. Ref. [17]).

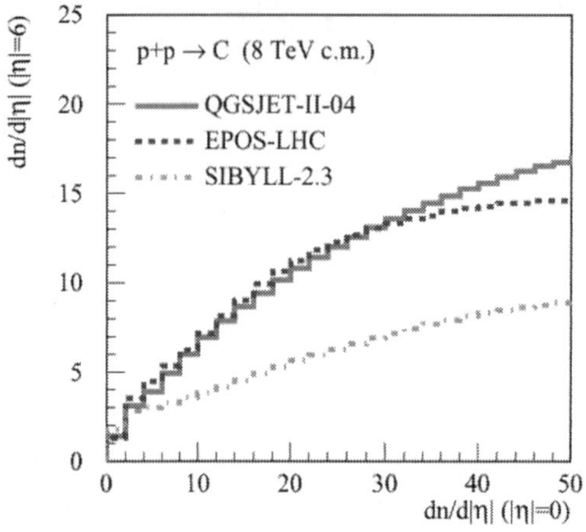

Figure 6. Pseudorapidity density of produced charged hadrons $dn^{\rm ch}_{pp}/d|\eta|$ at $|\eta| = 6$ ($p_{\rm t} > 0$) as a function of $dn^{\rm ch}_{pp}/d|\eta|$ at $|\eta| = 0$ ($p_{\rm t} > 0.1$ GeV) in pp collisions at $\sqrt{s} = 8$ TeV, as calculated using the QGSJET-II-04, EPOS-LHC, and SIBYLL-2.3 models (solid, dashed, and dash-dotted lines respectively).

As is evident from the discussion above, the best way to discriminate between the two theoretical approaches depicted schematically in Fig. 4 is via studies of correlations of the signal strength in central and forward-looking detectors at LHC, which may be performed by the CMS and TOTEM experiments. This is illustrated in Fig. 6, where we plot for $\sqrt{s} = 8$ TeV the η-density of charged hadrons $dn^{\rm ch}_{pp}/d|\eta|$ (for transverse momentum $p_{\rm t} > 0$) at $|\eta| = 6$ (averaged over the interval $5.5 < |\eta| < 6.5$) as a function of the central η-density of charged hadrons of $p_{\rm t} > 0.1$ GeV (averaged over the interval $|\eta| < 1$). Both EPOS-LHC and QGSJET-II-04 predict a strong correlation of the signal strength in CMS and TOTEM. The respective results of the two models practically coincide with each other, apart from the region corresponding to the tails of the multiplicity distributions. In contrast, for SIBYLL 2.3 such a correlation is twice weaker, being thus a "smoking gun" signature for the desirable discrimination.

Another possible way for the model discrimination is via measurements of very forward particle production, e.g. by the LHCf experiment at LHC, when supplemented by triggering different hadronic activities in the central detectors of the ATLAS experiment, as discussed in Ref. [16].

4 Relevance of pion-air interactions

The analysis in Section 3 does not explain the differences in $X_{\rm max}$ predictions of EPOS-LHC, QGSJET-II-04, and QGSJET – as all the three models employ essentially the same treatment of constituent parton Fock states. To reveal the interaction features which are responsible for these differences, we employ the "cocktail" model approach: using QGSJET-II-04 to describe some selected interactions of hadrons in air showers or some particular features of the primary interaction, while treating the rest with one of the other two models (see Ref. [18] for more details). As the first step, we apply QGSJET-II-04 to determine the position of the primary particle interaction in the atmosphere and to describe the production of secondary nucleons in this interaction; all other characteristics of the first proton-air collision and all the subsequent interactions of secondary hadrons in the cascade are treated using EPOS-LHC. This way we check the sensitivity of the calculated shower maximum depth to the model differences concerning the proton-air cross section and the predicted nucleon spectra, which

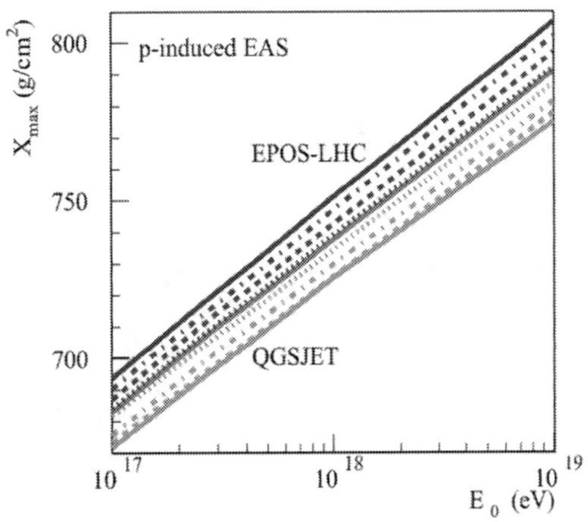

Figure 7. Primary energy dependence of X_{max} for *p*-induced vertical EAS, as calculated using the EPOS-LHC, QGSJET-II-04, and QGSJET models (top blue, middle red, and bottom green solid lines respectively), or applying mixed model descriptions, as explained in the text (dashed, dash-dotted, and dotted lines).

thus comprise the effects of the inelastic diffraction. The calculated X_{max} values shown by the blue dash-dotted line in Fig. 7 differ from the original EPOS-LHC results by not more than 5 g/cm^2, which is well within the uncertainty range obtained in Ref. [14].

Next, we apply QGSJET-II-04 to describe all the characteristics of the primary interaction, while treating the rest of the hadron cascade using EPOS-LHC. The obtained X_{max} shown by the blue dashed line in Fig. 7 is shifted further towards the QGSJET-II-04 results by up to 5 g/cm^2. This additional shift is explained by somewhat harder spectra of secondary mesons, most importantly, of secondary pions in EPOS-LHC, compared to the QGSJET-II-04 model. We also repeat the same calculation describing secondary hadron interactions in the cascade with the QGSJET model, the results being plotted by the green dashed line in Fig. 7. In this case, the difference with the pure QGSJET-based calculation does not exceed 3 g/cm^2, which is due to the fact that forward particle spectra in proton-air collisions are rather similar in QGSJET and QGSJET-II-04.

There remains a large difference between the two dashed lines in Fig. 7 and the results of QGSJET-II-04, which arises from the model treatments of pion-air and kaon-air interactions. In the particular case of QGSJET, this is mainly related to larger pion-air cross section and softer production spectra for secondary mesons, predicted by that model, compared to QGSJET-II-04. The sensitivity of X_{max} to the difference in pion-air cross sections is illustrated by the green dash-dotted line in Fig. 7, obtained by using QGSJET-II-04 results both for the primary interaction and for the inelastic cross sections for all the secondary hadron-air collisions. In turn, applying QGSJET-II-04 to describe in addition pion and kaon spectra in pion-air collisions gives rise to the energy-dependence of X_{max}, shown by the green dotted line in Fig. 7.

In turn, for EPOS-LHC the remaining difference with the QGSJET-II-04 results is due to a copious production of baryon-antibaryon pairs in pion-air and kaon-air collisions, also due to harder (anti-)baryon spectra in EPOS-LHC [19]. This slows down the energy dissipation from the hadronic cascade and thus contributes to the elongation of the shower profile. Indeed, if we apply QGSJET-II-04 to describe both the primary interaction and the production of nucleons and antinucleons in all the secondary pion-air and kaon-air collisions, while treating the rest with EPOS-LHC, the obtained X_{max} shown by the blue dotted line practically coincides with the QGSJET-II-04 results.

5 Relation to muon production depth

As demonstrated in Section 4, a large part of the model uncertainty for the predicted $X_{\max}$ is related to the treatment of pion-air collisions at very high energies. While there exist no experimental data for such interactions above fixed-target energies, the corresponding model treatment may be constrained indirectly by studying other characteristics of cosmic ray-induced air showers. Recently, the Pierre Auger collaboration reported measurements of the maximal muon production depth in air showers, $X^{\mu}_{\max}$ – the depth in the atmosphere (in g/cm^2) where the rate of muon production via decays of pions and kaons reaches its maximal value [20]. In particular, one observed a strong contradiction between the results of EAS simulations with EPOS-LHC and the experimental data: the maximum of the muon production profile was observed substantially higher in the atmosphere than predicted by that model for the heaviest possible primary cosmic ray nuclei.

There are both similarities and differences concerning the relation of $X_{\max}$ and $X^{\mu}_{\max}$ to the properties of hadron-air interactions. Obviously, both characteristics are sensitive to the position X_0 of the primary particle interaction in the atmosphere, which depends on the respective inelastic cross section: fluctuations of X_0 shift the whole cascade upwards and downwards in the atmosphere and thus do so for $X_{\max}$ and $X^{\mu}_{\max}$ for a particular shower. However, in contrast to $X_{\max}$, the maximal muon production depth is much less sensitive to hadron production in the primary interaction. The EAS muon content rather depends on the multi-step hadronic cascade in which the number of pions and kaons increases in an avalanche way until the probability of their decays becomes comparable to the one for interactions. This happens when the energies for most of pions approach the pion critical energy, $E^{\pi^{\pm}}_{\text{crit}} \simeq 80$ GeV [21]. The position of the maximum of the muon production profile is actually close to this turning point.

As a consequence, $X^{\mu}_{\max}$ is very sensitive to the forward spectral shape of secondary mesons in pion-air collisions: producing in each cascade step a meson of a slightly higher energy would mean that a larger number of cascade branching steps is required for reaching the critical energy, with the result that the maximum of the muon production profile will be observed deeper in the atmosphere. A similar effect may be produced by a smaller pion-air cross section as this would increase the pion mean free pass and thereby elongate the muon production profile. In addition, in the particular case of the EPOS model, its predictions for $X^{\mu}_{\max}$ may be influenced by the copious production of (anti-)baryons in pion-air collisions. Indeed, unlike pions and kaons, (anti-)nucleons continue to take part in the hadronic cascade down to the GeV energy range, producing additional generations of secondary hadrons. Muons emerging from decays of secondary pions and kaons created in interactions of low energy nucleons and antinucleons contribute to the elongation of the muon production profile and give rise to larger values of $X^{\mu}_{\max}$.

Performing calculations of $X^{\mu}_{\max}$ for muon energies $E_{\mu} \geq 1$ GeV for proton-induced EAS, we observe substantially larger differences between the results of the different interaction models, compared to the case of $X_{\max}$, as demonstrated in Fig. 8. To reveal the reasons for these differences, we use the same "cocktail" model approach as in Section 4. First, we apply QGSJET-II-04 to describe all the characteristics of the primary interaction, while treating the rest of the hadron cascade using either EPOS-LHC or QGSJET, the results shown respectively by the blue and green dashed lines in Fig. 8. As expected, the obtained $X^{\mu}_{\max}$ values deviate only slightly from the original model calculations – as the differences between the model predictions for $X^{\mu}_{\max}$ are mainly due to secondary (mostly pion-air) interactions in the cascade. For example, the smaller $X^{\mu}_{\max}$ values predicted by QGSJET are mostly due to somewhat larger inelastic pion-air and kaon-air cross sections and softer meson spectra produced by that model, compared to QGSJET-II-04. The first effect is illustrated by the green dash-dotted line in Fig. 8, obtained by applying QGSJET-II-04 to describe both the primary interaction and the inelastic cross sections for all the secondary hadron-air collisions, while treating hadron production

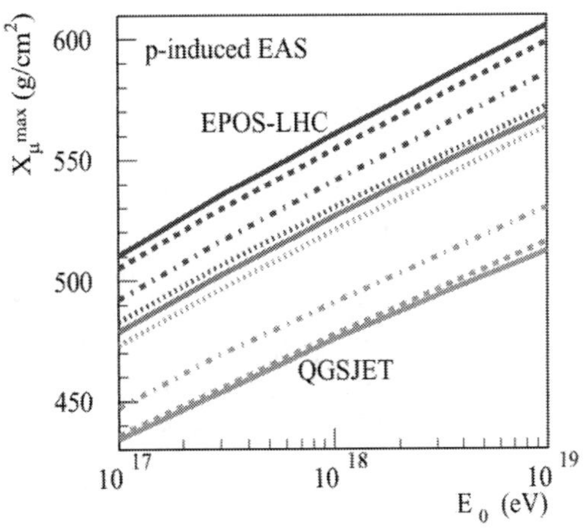

Figure 8. Primary energy dependence of $X^{\mu}_{\rm max}$ ($E_{\mu} \geq 1$ GeV) for p-induced vertical EAS, as calculated using the EPOS-LHC, QGSJET-II-04, and QGSJET models (top blue, middle red, and bottom green solid lines respectively), or applying mixed model descriptions, as explained in the text (dashed, dash-dotted, and dotted lines).

in the secondary interactions with QGSJET. On the other hand, using QGSJET-II-04 results also for the pion and kaon spectra in pion-air collisions results in the energy-dependence of $X^{\mu}_{\rm max}$, shown by the green dotted line in Fig. 8.

In turn, a large part of the difference between EPOS-LHC and QGSJET-II-04 is due to the copious production of baryon-antibaryon pairs in the former model. This is illustrated by the blue dash-dotted line in Fig. 8, obtained by applying QGSJET-II-04 to describe both the primary interaction and the production of nucleons and antinucleons in all the secondary pion-air collisions, while treating the rest with EPOS-LHC. The remaining difference between the two models is due to harder spectra of secondary mesons in EPOS-LHC for pion-air and kaon-air collisions. Indeed, using QGSJET-II-04 results both for the primary interaction and for hadron spectra in secondary pion-air and kaon-air collisions, we get the energy-dependence of $X^{\mu}_{\rm max}$, shown by the blue dotted line in Fig. 8, which is very close to the pure QGSJET-II-04 calculation.

6 Outlook

Experimental studies of proton-proton collisions at LHC furnished valuable information for the CR field, substantially reducing the range of uncertainties for numerical simulations of CR-induced air showers. Nevertheless, there exists yet a significant spread between the results of calculations for the basic EAS observable – the shower maximum position. While a part of this spread may be explained by the present experimental uncertainty concerning the diffraction rate in very high energy pp collisions, a much bigger difference between certain model predictions for $X_{\rm max}$ is related to the model assumptions concerning the momentum structure of constituent parton Fock states in hadrons. The "smoking gun" signature for discriminating between the different theoretical approaches is the correlation of the signal strength between central and forward-looking particle detectors, which may be measured by the CMS and TOTEM experiments at LHC.

Meanwhile, the dominant source of uncertainties for EAS simulations is presently related to the model treatments of very high energy pion-air interactions. This can be constrained indirectly by studying other air shower observables, notably, by measurements of the maximal muon production depth. In particular, the results on $X^{\mu}_{\rm max}$ from the Pierre Auger experiment strongly disfavor the copious production of baryon-antibaryon pairs in pion-air collisions, predicted by the EPOS model.

It is noteworthy that even for QGSJET-II-04 there is a certain tension between the data of the Pierre Auger experiment on $X_{\rm max}$ and $X^{\mu}_{\rm max}$ [20]: the latter point towards a heavier CR composition, compared to the former. According to the analysis presented in this work, one may try to achieve a consistency between the two results by modifying the treatment of pion-air collisions. However, as the potential changes would impact $X^{\mu}_{\rm max}$ much stronger than $X_{\rm max}$, this would imply to aim at higher inelastic cross section and/or softer hadron production spectra for such interactions, which would push one towards a predominantly light, presumably proton-dominated CR composition.

The author acknowledges the support by Deutsche Forschungsgemeinschaft (project OS 481/1).

References

[1] M. Nagano and A. A. Watson, Rev. Mod. Phys. **72**, 689 (2000).
[2] K. Werner, F.-M. Liu and T. Pierog, Phys. Rev. C **74**, 044902 (2006).
[3] T. Pierog, Iu. Karpenko, J. M. Katzy, E. Yatsenko and K. Werner, Phys. Rev. C **92**, 034906 (2015).
[4] S. Ostapchenko, Nucl. Phys. Proc. Suppl. **151**, 143 (2006); Phys. Rev. D **74**, 014026 (2006).
[5] S. Ostapchenko, Phys. Rev. D **83**, 014018 (2011); EPJ Web Conf. **52**, 02001 (2013).
[6] R.S. Fletcher, T. K. Gaisser, P. Lipari and T. Stanev, Phys. Rev. D **50**, 5710 (1994); E.-J. Ahn, R. Engel, T. K. Gaisser, P. Lipari and T. Stanev, *ibid.*, **80**, 094003 (2009).
[7] F. Riehn, R. Engel, A. Fedynitch, T. K. Gaisser and T. Stanev, arXiv:1510.00568 [hep-ph].
[8] N. N. Kalmykov and S. S. Ostapchenko, Phys. Atom. Nucl. **56**, 346 (1993); N. N. Kalmykov, S. S. Ostapchenko and A. I. Pavlov, Nucl. Phys. Proc. Suppl. **52B**, 17 (1997).
[9] D. d'Enterria, R. Engel, T. Pierog, S. Ostapchenko and K. Werner, Astropart. Phys. **35**, 98 (2011); D. d'Enterria, R. Engel, T. Pierog, S. Ostapchenko and K. Werner, Few Body Syst. **53**, 173 (2012).
[10] G. Antchev *et al.* (TOTEM Collaboration), Europhys. Lett. **101**, 21002 (2013); *ibid.*, 21003 (2013); *ibid.*, 21004 (2013); Phys. Rev. Lett. **111**, 012001 (2013).
[11] G. Aad *et al.* (ATLAS Collaboration), Nucl. Phys. B **889**, 486 (2014).
[12] K. Nakamura *et al.* (Particle Data Group), J. Phys. G **37**, 075021 (2010).
[13] T. Bergmann *et al.*, Astropart. Phys. **26**, 420 (2007).
[14] S. Ostapchenko, Phys. Rev. D **89**, 074009 (2014).
[15] S. Chatrchyan *et al.* (CMS and TOTEM Collaborations), Eur. Phys. J. C **74**, 3053 (2014); G. Antchev *et al.* (TOTEM Collaboration), *ibid.* **75**, 126 (2015).
[16] S. Ostapchenko, M. Bleicher, T. Pierog and K. Werner, in preparation.
[17] R. Ulrich, R. Engel and M. Unger, Phys. Rev. D **83**, 054026 (2011).
[18] S. Ostapchenko and M. Bleicher, Phys. Rev. D **93**, 051501 (2016).
[19] T. Pierog and K. Werner, Phys. Rev. Lett. **101**, 171101 (2008).
[20] A. Aab *et al.* (Pierre Auger Collaboration), Phys. Rev. D **90**, 012012 (2014); *ibid.*, **90**, 039904 (2014); *ibid.*, **92**, 019903 (2015).
[21] T. K. Gaisser, *Cosmic Rays and Particle Physics* (Cambridge University Press, Cambridge, England 1990).

EPJ Web of Conferences **125**, 03014 (2016) DOI: 10.1051/epjconf/201612503014
QUARKS-2016

One model of modified gravity with dynamical torsion and its cosmological consequences

Vasilisa Nikiforova[1,*]

[1] *Institute for Nuclear Research of the Russian Academy of Sciences, 60th October Anniversary Prospect 7a, 117312 Moscow, Russia*

Abstract. We consider a model belonging to the class of Poincarè gauge gravities. The model is free of ghosts, tachyons and gradient instabilities about Minkowski and torsionless Einstein backgrounds of sufficiently small curvature. At zero cosmological constant, the model admits a self-accelerating solution with non-Riemannian connection. We study scalar perturbations about the self-accelerating solution and find that the number of scalar modes is the same as in Minkowski background; moreover, in the limit of small effective cosmological constant and below the UV cutoff of the low energy effective theory, the scalar sector does not have pathologies like ghosts or rapid gradient instabilities.

1 Introduction

Modified gravity recently has become of considerable interest mostly in the cosmological context as a viable alternative to General Relativity on cosmological scales. The idea is to replace General Relativity with some other theory which could drive the cosmological acceleration without dark energy.

There are many approaches to modify gravity in IR, see a comprehensive review [1]. These include theories with extra dimensions [2–6] (for a review see Ref. [7]), $f(R)$ and scalar-tensor gravities (for reviews see Refs. [8] and [9]), theories with massive gravitons [10–13] (for a review see Ref. [14]) and theories with explicitly broken Lorentz invariance [15–19] (for a review see Ref. [20]). However, self-accelerating solutions in IR-modified gravities are often plagued with instabilities. In particular, the Minkowski background may be stable, but extra modes may appear in curved backgrounds. They usually have wrong sign kinetic terms and are known as Boulware-Deser modes [21]. Even if the Boulware-Deser modes do not appear, the spectrum can have ghosts or gradient instabilities. One famous example is the DGP model [2], which admits both Minkowski and self-accelerating backgrounds [22, 23]. The former one has only healthy degrees of freedom in the spectrum, while the self-accelerating background has ghost instability [24–28].

Yet another class of theories which are healthy on the Minkowski background [29–32] and may serve as candidates for the consistent infrared modification of gravity are Poincarè gauge gravities. These theories treat the vierbein and connection as independent variables, and the Lagrangians include bilinear terms in the full curvature tensor, as well as bilinear terms in the torsion. The spectrum of these theories about the Minkowski background contains both massless degrees of freedom corresponding to the standard graviton and massive degrees of freedom related to the propagating torsion

*e-mail: nikiforovavas@mail.ru

[31]; the gravitational interaction in these theories is described by both Newtonian and Yukawa terms [30]. The model which we consider in this paper belongs to the class of Poincarè gauge gravities.

This model was studied earlier in Refs. [33–36]. It was shown that ghosts, gradient instabilities and tachyons are absent in the spectrum about the Minkowski background, in de Sitter and anti-de Sitter spaces and in arbitrary torsionless Einstein backgrounds of sufficiently small curvature, provided that certain constraints on the couplings are satisfied [31, 33, 34]. This model admits the self-accelerating solution [36]. It is certainly of interest to understand the properties of perturbations about this solution. Our purpose is to study whether this theory exhibits the Boulware-Deser phenomenon and/or other instabilities in self-accelerating background. The known examples show that possible problems occur in the scalar sector of perturbations. Therefore, we focus in this paper on the scalar sector.

This paper is organized as follows. In Section 2 we present the Lagrangian, write the field equations and derive the generalized Bianchi identities. In Section 3 we present the self-accelerating solution with effective cosmological constant which can be made small by an appropriate choice of parameters. In Section 4 we study perturbations about this solution. We find that there are no extra modes as compared to the Minkowski background. We find also that in the limit of small effective cosmological constant and within the range of validity of the effective low energy theory there are actually no propagating degrees of freedom at all.

2 The Model

2.1 Action and field equations

We make use of the tetrad formalism and consider vierbein (e^i_μ) and connection ($A_{ij\mu} = -A_{ji\mu}$) as independent fields[1]; $\mu = (0,1,2,3)$ is the space-time index, and $i, j = (0,1,2,3)$ are the tangent space indices. In tangent space basis the indices are raised and lowered using the Minkowski metric η_{ij}, so we do not distinguish upper and lower tangent space indices in what follows, if this does not lead to an ambiguity. The signature of metric is $(-,+,+,+)$.

The action of the model is

$$L = \frac{3}{2}(\tilde{\alpha}F - \alpha R) + c_3 F^{ij}F_{ij} + c_4 F^{ij}F_{ji} + c_5 F^2 + c_6(\epsilon \cdot F)^2 \, , \tag{1}$$

where $\alpha, \tilde{\alpha}, c_3, c_4, c_5, c_6$ are the coupling constants, $e \equiv det(e^i_\mu)$; F_{ijkl} is the curvature tensor constructed with the connection $A_{ij\mu}$,

$$F_{ijkl} = e^\mu_k e^\nu_l (\partial_\mu A_{ij\nu} - \partial_\nu A_{ij\mu} + A_{im\mu}A_{mj\nu} - A_{jm\mu}A_{mi\nu}) \, ;$$

$$F_{ij} = \eta^{kl}F_{ikjl} \, , \quad F = \eta^{ij}F_{ij} \, , \quad \epsilon \cdot F \equiv \epsilon^{ijkl}F_{ijkl} \, ;$$

ϵ_{ijkl} is the Levi-Civita symbol defined in such a way that $\epsilon^{0123} = -\epsilon_{0123} = 1$; R_{ijkl} is the Riemann curvature tensor,

$$R_{ijkl} = e^\mu_k e^\nu_l (\partial_\mu \omega_{ij\nu} - \partial_\nu \omega_{ij\mu} + \omega_{im\mu}\omega_{mj\nu} - \omega_{jm\mu}\omega_{mi\nu}) \, ;$$

$$R_{ij} = \eta^{kl}R_{ikjl} \, , \quad R = \eta^{ij}R_{ij} \, ,$$

where $\omega_{ij\mu}$ is the Riemannian spin-connection. The latter is expressed in terms of the vierbein as follows:

$$\omega_{ij\mu} \equiv \omega_{ijk}e^k_\mu = \frac{1}{2}(C_{ijk} - C_{jik} - C_{kij})e^k_\mu \, ,$$

[1]The formalism used in this section is largely the one developed in Ref. [35].

where

$$C_{ijk} = e^{\mu}_{j} e^{\nu}_{k} (\partial_\mu e_{i\nu} - \partial_\nu e_{i\mu}) \, .$$

The connection $A_{ij\mu}$ can be represented as a sum

$$A_{ij\mu} = \omega_{ij\mu} + K_{ij\mu} \, ,$$

where $K_{ij\mu}$ is the contorsion tensor.

It was shown that there are no pathological degrees of freedom in the Minkowski background [31] if one imposes the following conditions on the couplings,

$$c_3 + c_4 = -3c_5 \, , \tag{2}$$

$$\alpha < 0, \quad \tilde{\alpha} > 0, \quad c_5 < 0, \quad c_6 > 0 \, . \tag{3}$$

We consider this case in that follows. In Ref. [34] it was found that there are three propagating modes at the linear level in the Minkowski background: the massless spin-2 mode, the massive spin-2 mode with mass

$$m^2 = \frac{\tilde{\alpha}(\tilde{\alpha} - \alpha)}{2\alpha c_5} \tag{4}$$

and the massive spin-0 mode with mass

$$m_0^2 = \frac{\tilde{\alpha}}{16 c_6} \, . \tag{5}$$

It is worth pointing out that there are two helicity-0 modes, the scalar of mass (5) and longitudinal part of massive spin-2 mode. From the point of view of (3+1)-decomposition of Sec. 4.1, these are scalar perturbations. There are no ghosts or tachyons in the Minkowski background. It was also shown that in the theory equipped with the cosmological constant, the perturbations are healthy in maximally symmetric backgrounds, and in torsionless Einstein backgrounds of sufficiently small curvature as well.

There are two sets of field equations in our model. One consists of the gravitational field equations obtained by varying the action with respect to vierbein,

$$\begin{aligned} \mathcal{G}_{ij} \equiv & \frac{3}{2}\tilde{\alpha}\left(F_{ij} - \frac{1}{2}\eta_{ij}F\right) - \frac{3}{2}\alpha\left(R_{ij} - \frac{1}{2}\eta_{ij}R\right) + c_3\left(F_{ki}F_{kj} + F_{kl}F_{kilj}\right) \\ & + c_4\left(F_{ik}F_{kj} + F_{lk}F_{kilj}\right) + 2c_5 F F_{ij} + 2c_6 \epsilon_{klmi} F_{klmj} (\epsilon \cdot F) - \frac{1}{2}\eta_{ij} L^{(2)} = 0 \, , \end{aligned} \tag{6}$$

where

$$L^{(2)} = c_3 F_{ij} F_{ij} + c_4 F_{ij} F_{ji} + c_5 F^2 + c_6 (\epsilon \cdot F)^2$$

is the part in the Lagrangian which is bilinear in F_{ijkl}. Another set of equations is obtained by varying the action with respect to the connection $A_{ij\mu}$,

$$\begin{aligned} \mathcal{T}_{ijk} \equiv & H_{ijk} + \left\{\left[\eta_{ik}\left(D_m P_{jm} - \frac{2}{3}D_j P\right) - D_i P_{jk}\right] - \left[\eta_{jk}\left(D_m P_{im} - \frac{2}{3}D_i P\right) - D_j P_{ik}\right]\right\} \\ & + 4c_6 \epsilon_{ijkm} D_m (\epsilon \cdot F) + S_{ijk} = 0 \, , \end{aligned} \tag{7}$$

where P_{ij} and P are defined as follows,

$$\begin{aligned} P_{ij} &= c_3 F_{ij} + c_4 F_{ji} \, , \\ P &= \eta^{ij} P_{ij} \, ; \end{aligned}$$

the covariant derivative D_i involves the connection $A_{ij\mu}$,

$$D_i B_j \equiv e_i^\mu D_\mu B_j = e_i^\mu (\partial_\mu B_j - A_{lj\mu} B_l) ;$$

S_{ijk} is defined as

$$S_{ijk} = \frac{2}{3\tilde{\alpha}} H_{mnk} \left(\eta_{im} P_{jn} - \eta_{jm} P_{in} - \frac{2}{3} \eta_{im} \eta_{jn} P + 2c_6 \epsilon_{ijmn} (\epsilon \cdot F) \right)$$

and H_{ijk} is written in terms of contorsion as follows,

$$\frac{2}{3\tilde{\alpha}} H_{ijk} = K_{ikj} - K_{jki} - K_{ill} \eta_{jk} + K_{jll} \eta_{ik} .$$

Because of the invariance of the action under space-time gauge transformations and local Lorentz transformations, not all of field equations are independent.

2.2 Generalized Bianchi identities

The local Lorentz and space-time gauge invariance can be used to find the identities relating the field equations (6), (7) with each other. Let us start with the space-time gauge invariance. Making an infinitesimal gauge transformation, $x'^\mu = x^\mu + \xi^\mu$, we find the variations of e^i_μ and $A_{ij\mu}$:

$$\delta e^i_\mu = -\xi^\lambda \partial_\lambda e^i_\mu - e^i_\lambda \partial_\mu \xi^\lambda = -\xi^\lambda \nabla_\lambda e^i_\mu - e^i_\lambda \nabla_\mu \xi^\lambda , \tag{8}$$

$$\delta A_{ij\mu} = -A_{ij\lambda} \nabla_\mu \xi^\lambda - \xi^\lambda \nabla_\lambda A_{ij\mu} , \tag{9}$$

where ∇_μ is the covariant derivative involving the Christoffel symbols,

$$\nabla_\mu e^i_\nu = \partial_\mu e^i_\nu - \Gamma^{\lambda}_{\cdot\mu\nu} e^i_\lambda .$$

Evaluating the variation of the action with respect to e^i_μ and $A_{ij\mu}$ of the form (8), (9) we find, upon integrating by parts, the following identity:

$$-2\mathcal{G}_{ik} e_{k\nu} \nabla_\mu e^{i\nu} - 2\nabla_\nu (\mathcal{G}_{ik} e_{k\mu} e^{i\nu}) - e^{k\nu} \nabla_\mu A_{ij\nu} \mathcal{T}_{ijk} + \nabla_\nu (e^{k\nu} A_{ij\mu} \mathcal{T}_{ijk}) = 0 \tag{10}$$

(see eqs. (6) and (7) for definitions of $\mathcal{G}_{ij}$ and $\mathcal{T}_{ijk}$).

Another identity comes from the local Lorentz invariance. The model discussed is a gauge theory with Lorentz group as a gauge group. The variations of e^i_μ and $A_{ij\mu}$ under the infinitesimal Lorentz transformation are:

$$\delta e^\mu_i = \omega_{ij} e^\mu_j , \tag{11}$$

$$\delta A_{ij\mu} = -\partial_\mu \omega_{ij} + \omega_{ip} A_{pj\mu} - \omega_{jp} A_{pi\mu} ,$$

where $\omega_{ij} = -\omega_{ji}$ are the parameters of transformation. The invariance of the action under this transformation gives,

$$(\mathcal{G}_{ij} - \mathcal{G}_{ji}) + \nabla_\nu (\mathcal{T}_{ijk} e^{k\nu}) + 2e^{k\nu} (\mathcal{T}_{ipk} A_{jp\nu} - \mathcal{T}_{jpk} A_{ip\nu}) = 0 . \tag{12}$$

The left hand side of eq. (12) contains two free indices, i and j, and is antisymmetric in them.

3 The self-accelerating cosmological solution

The model (1) admits a self-accelerating solution [36],

$$e_0^{\tilde{0}} = 1\,, \quad e_b^{\tilde{a}} = e^{\lambda t}\delta_b^{\tilde{a}}\,, \quad A_{\tilde{0}\tilde{a}\tilde{b}} = f\delta_{\tilde{a}\tilde{b}}\,, \quad A_{\tilde{a}\tilde{b}\tilde{c}} = g\varepsilon_{\tilde{a}\tilde{b}\tilde{c}}\,, \tag{13}$$

with constant λ, f and g, where $a, \tilde{a} = (1, 2, 3)$, tilde denotes tangent space indices, and space-time indices do not have tilde. The parameters f, g and λ are related to couplings as follows,

$$\tilde{\alpha}(f^2 - g^2) - \alpha\lambda^2 - 16c_6 g^2\lambda^2 + 64c_6 f^2 g^2 = 0\,, \tag{14a}$$

$$\tilde{\alpha}(f + \lambda) - 4c_5 g^2\lambda - 32c_6 g^2(\lambda - 2f) = 0\,, \tag{14b}$$

$$\tilde{\alpha}(f^2 - g^2 - \lambda f) - 2\alpha\lambda^2 = 0\,. \tag{14c}$$

One finds that if the following condition is satisfied,

$$c_5 + 16c_6 < 0\,,$$

then there exists the self-accelerating solution with $\lambda > 0$, $f < 0$. The sign of g can be arbitrary, since g is P-odd.

The small value of the effective cosmological constant λ is obtained provided that there is a hierarchy between the dimensionless couplings: for $c_6 = O(\lambda)$, $c_5 = O(\lambda^{-1})$ the parameter λ is the smallest parameter of the problem [36]. It is straightforward to see from eqs. (14a) - (14c) that in this case the parameters f, g and λ are related to couplings as follows:

$$\lambda = \left(-\frac{c_6\tilde{\alpha}^2}{c_5^3}\right)^{1/4}\,,$$

$$f = -\frac{\tilde{\alpha}^{1/2}}{4(-c_5c_6)^{1/4}}\,,$$

$$g = \pm f + O(\lambda)\,.$$

The effective cosmological constant can also be written in terms of the masses (4), (5) of excitations about the Minkowski background:

$$\lambda = m\left(\frac{m}{m_0}\right)^{1/2}\frac{(-\alpha)^{3/4}}{2^{1/4}(\tilde{\alpha} - \alpha)^{3/4}}\,.$$

We point out that the solution (13) does not require any explicit cosmological constant in the Lagrangian (1); the self-acceleration is generated by torsion.

4 Perturbations

Our purpose is to study small perturbations about the self-accelerating background (13). To this end, we work with conformal time

$$\eta = \int e^{-\lambda t}dt = -\frac{1}{\lambda}e^{-\lambda t}\,.$$

Let us also change the notation and denote the vierbein of the original theory by $\hat{e}^i_\mu$. It is convenient to make the following conformal transformation,

$$\hat{e}^i_\mu = \mathrm{e}^{\phi}e^i_\mu\,, \quad \hat{e}^\nu_j = \mathrm{e}^{-\phi}e^\nu_j\,, \tag{15}$$

where

$$e^{\phi} = a(\eta) = -\frac{1}{\lambda\eta} .$$

Note that

$$\phi' = \lambda e^{\phi} ,$$

where prime denotes derivative with respect to conformal time. We do not make any scaling of connection, and keep the notation $A_{ij\mu}$ unchanged. Upon the conformal transformation, the background has flat metric and non-trivial connection,

$$e^{i}_{\mu} = \delta^{i}_{\mu} , \quad A_{\tilde{0}\tilde{a}\tilde{b}} = f\delta_{\tilde{a}\tilde{b}} , \quad A_{\tilde{a}\tilde{b}\tilde{c}} = g\varepsilon_{\tilde{a}\tilde{b}\tilde{c}} . \tag{16}$$

Equation (6) then reads

$$\begin{aligned}
&\frac{3}{2}\tilde{\alpha}e^{-2\phi}\left(F_{ij} - \frac{1}{2}\eta_{ij}F\right) - \frac{3}{2}\alpha e^{-2\phi}\left(R_{ij} - \frac{1}{2}\eta_{ij}R\right) \\
&- \frac{3}{2}\alpha e^{-2\phi}\left[-2e^{\mu}_{i}e^{\nu}_{j}\nabla_{\mu}\nabla_{\nu}\phi - \eta_{ij}g^{\mu\nu}\nabla_{\mu}\nabla_{\nu}\phi + 2e^{\mu}_{i}e^{\nu}_{j}\partial_{\mu}\phi\partial_{\nu}\phi - 2\eta_{ij}g^{\mu\nu}\partial_{\mu}\phi\partial_{\nu}\phi \right.\\
&\left. -\eta_{ij}\left(-3g^{\mu\nu}\nabla_{\mu}\nabla_{\nu}\phi - 3g^{\mu\nu}\partial_{\mu}\phi\partial_{\nu}\phi\right)\right] \\
&+ e^{-4\phi}c_3\left(F_{ki}F_{kj} + F_{kl}F_{kilj}\right) + e^{-4\phi}c_4\left(F_{ik}F_{kj} + F_{lk}F_{kilj}\right) + 2e^{-4\phi}c_5FF_{ij} \\
&+ 2e^{-4\phi}c_6\epsilon_{klmi}F_{klmj}(\epsilon\cdot F) - e^{-4\phi}\frac{1}{2}\eta_{ij}L^{(2)} = 0 ,
\end{aligned}$$

while eq. (7) gives

$$\begin{aligned}
&e^{-\phi}\left(H_{ijk} + 3\tilde{\alpha}\eta_{ik}e^{\mu}_{j}\partial_{\mu}\phi - 3\tilde{\alpha}\eta_{jk}e^{\mu}_{i}\partial_{\mu}\phi\right) \\
&+e^{-3\phi}\left\{\left[\eta_{ik}\left(D_mP_{jm} - \frac{2}{3}D_jP\right) - D_iP_{jk}\right] - \left[\eta_{jk}\left(D_mP_{im} - \frac{2}{3}D_iP\right) - D_jP_{ik}\right]\right\} \\
&+ 4e^{-3\phi}c_6\epsilon_{ijkm}D_m(\epsilon\cdot F) + e^{-3\phi}S_{ijk} = 0 .
\end{aligned}$$

4.1 (3+1)-decomposition

After conformal transformation for the vierbein (15) we have

$$e^{i}_{\mu} = \delta^{i}_{\mu} + \epsilon^{i}_{\mu} , \tag{17}$$

where ϵ^{i}_{μ} is the first order quantity. As mentioned above, the model is invariant under local Lorentz and space-time gauge transformations. The former can be used to make $\epsilon_{i\mu}$ symmetric; this, in particular, means that

$$e^{\mu}_{j} = \delta^{\mu}_{j} - \epsilon_{i\nu}\delta^{\nu}_{j}\delta^{i\mu} ,$$

while the space-time gauge invariance can be used to choose the gauge $e_{\tilde{0}a} = 0$ and the conformal Newtonian gauge.

The full contorsion tensor is written as follows,

$$K_{ijk} = K_{ijk\,(0)} + k_{ijk} ,$$

where $K_{ijk\,(0)} = e^{\mu}_{k}K_{ij\mu\,(0)} = e^{\mu}_{k}(A_{ij\mu\,(0)} - \omega_{ij\mu\,(0)}) = e^{\mu}_{k}A_{ij\mu\,(0)}$ is the background quantity constructed from (16) and k_{ijk} is the first order perturbation.

It is convenient to make the 3-dimensional Fourier decomposition. We use the same notation for the Fourier-transformed variables and replace spatial derivatives $\partial_{\tilde{a}} \equiv e^{\mu}_{\tilde{a}}\partial_\mu$ with $ik_{\tilde{a}}$, where $\mathbf{k}$ is the 3-dimensional momentum. Since the background (16) is invariant under spatial rotations, it is convenient to use (3 + 1)-decomposition of perturbations. This means that we decompose any 3-dimensional tensor into its irreducible components with respect to the small group $SO(2)$ of rotations around the spatial momentum. These irreducible components form sectors with particular values of helicity: scalar sector (helicity-0), vector sector (helicity-1) and tensor sector (helicity-2). Since the field equations are linear, these sectors can be considered separately.

Tensor $k_{ijk} = -k_{jik}$ contains helicity 0, 1, 2 components only. The helicity-2 components form the tensor sector:

$$\begin{aligned} k_{\tilde{0}\tilde{a}\tilde{b}} &= -k_{\tilde{a}\tilde{0}\tilde{b}} = \tau_{\tilde{a}\tilde{b}} \, , \\ k_{\tilde{a}\tilde{b}\tilde{c}} &= k_{\tilde{a}}N_{\tilde{b}\tilde{c}} - k_{\tilde{b}}N_{\tilde{a}\tilde{c}} \, , \end{aligned}$$

where $\tau_{\tilde{a}\tilde{b}}$ and $N_{\tilde{a}\tilde{b}}$ are symmetric, transverse and traceless. All other components of k_{ijk} vanish in the tensor sector.

The non-vanishing helicity-1 components of contorsion are

$$\begin{aligned} k_{\tilde{0}\tilde{a}\tilde{0}} &= -k_{\tilde{a}\tilde{0}\tilde{0}} = \zeta_{\tilde{a}} \, , \\ k_{\tilde{0}\tilde{a}\tilde{b}} &= -k_{\tilde{a}\tilde{0}\tilde{b}} = k_{\tilde{a}}\nu_{\tilde{b}} + k_{\tilde{b}}\mu_{\tilde{a}} \, , \\ k_{\tilde{a}\tilde{b}\tilde{0}} &= k_{\tilde{a}}\kappa_{\tilde{b}} - k_{\tilde{b}}\kappa_{\tilde{a}} \, , \\ k_{\tilde{a}\tilde{b}\tilde{c}} &= k_{\tilde{a}}k_{\tilde{c}}\alpha_{\tilde{b}} - k_{\tilde{b}}k_{\tilde{c}}\alpha_{\tilde{a}} + \eta_{\tilde{a}\tilde{c}}L_{\tilde{b}} - \eta_{\tilde{b}\tilde{c}}L_{\tilde{a}} \, , \end{aligned}$$

all 3-vectors here are transverse. The helicity-0 components form the scalar sector:

$$\begin{aligned} k_{\tilde{0}\tilde{a}\tilde{0}} &= -k_{\tilde{a}\tilde{0}\tilde{0}} = k_{\tilde{a}}\xi \, , \\ k_{\tilde{0}\tilde{a}\tilde{b}} &= -k_{\tilde{a}\tilde{0}\tilde{b}} = k_{\tilde{a}}k_{\tilde{b}}\chi + \delta_{\tilde{a}\tilde{b}}\sigma + \epsilon_{\tilde{a}\tilde{b}\tilde{c}}k_{\tilde{c}}\rho \, , \\ k_{\tilde{a}\tilde{b}\tilde{0}} &= \epsilon_{\tilde{a}\tilde{b}\tilde{c}}k_{\tilde{c}}\theta \, , \\ k_{\tilde{a}\tilde{b}\tilde{c}} &= \epsilon_{\tilde{a}\tilde{b}\tilde{d}}k_{\tilde{c}}k_{\tilde{d}}Q + (k_{\tilde{a}}\epsilon_{\tilde{b}\tilde{c}\tilde{d}} - k_{\tilde{b}}\epsilon_{\tilde{a}\tilde{c}\tilde{d}})k_{\tilde{d}}u + (k_{\tilde{a}}\delta_{\tilde{b}\tilde{c}} - k_{\tilde{b}}\delta_{\tilde{a}\tilde{c}})M \, . \end{aligned}$$

There are altogether 24 components of tensor k_{ijk}. They break up into the 4 tensor components (two components from each of $\tau_{\tilde{a}\tilde{b}}$ and $N_{\tilde{a}\tilde{b}}$), 12 components in vector sector (two components from each of six transverse vectors $\zeta_{\tilde{a}}$, $\nu_{\tilde{a}}$, $\mu_{\tilde{a}}$, $\kappa_{\tilde{a}}$, $\alpha_{\tilde{a}}$ and $L_{\tilde{a}}$) and 8 scalar components ($\xi, \chi, \sigma, \rho, \theta, Q, u$ and M).

In what follows the space-time indices in the first order quantities are raised and lowered with Minkowski metric. The vierbein perturbations (in conformal Newtonian gauge with $\epsilon_{\mu\nu} = \epsilon_{\nu\mu}$ and $e_{\tilde{0}a} = 0$) are:

$$\begin{aligned} \epsilon_{\tilde{0}\tilde{0}} &= -\Phi \, , \\ \epsilon_{\tilde{0}\tilde{a}} &= 0 \, , \\ \epsilon_{\tilde{a}\tilde{b}} &= \Psi\delta_{\tilde{a}\tilde{b}} + i(k_{\tilde{a}}W_{\tilde{b}} + k_{\tilde{b}}W_{\tilde{a}}) + \pi_{\tilde{a}\tilde{b}} \, , \end{aligned}$$

where $W_{\tilde{a}}$ is a transverse vector (vector sector, 2 components), $\pi_{\tilde{a}\tilde{b}}$ is a transverse traceless tensor (tensor sector, 2 components), while Φ and Ψ belong to the scalar sector (2 components).

4.2 Scalar sector

In this paper we study the scalar sector of perturbations. In this sector we have in total 14 scalar field equations; 6 of them are second order in time derivatives and 8 are first order. Not all of these 14 equations are independent, as there are Bianchi identities (10), (12). The left hand side of eq. (10) contains one free index μ and thus can be decomposed into two scalar identities: with $\mu = 0$ and $\mu = a$. Furthermore, the left hand side of eq. (12) is decomposed into two scalar identities, one for $i = a$, $j = 0$ and another for $i = a$, $j = b$. In total we have 4 scalar Bianchi identities relating 14 field equations with each other. Using these identities we can exclude 4 second order equations from consideration. Thus, we have 10 equations. Off hand, two of these equations are second order. However, a sum of each of these second order equations with a linear combination of time-derivatives of the 8 first order equations is a first order equation. Therefore, at this stage we have 10 first order equations for 10 variables characterizing scalar perturbations. This is not the whole story, though.

4.2.1 Number of propagating modes

Let us count the number of propagating degrees of freedom. The variables in the resulting system of 10 first order equations are of two types: the variables of first group,

$$\sigma\,,\ \Phi\,,\ \xi\,,\ \theta\,, \tag{18}$$

enter the linearized equations without time derivatives. The variables of the second group,

$$\Psi\,,\ \chi\,,\ \rho\,,\ Q\,,\ u\,,\ M\,, \tag{19}$$

enter with first time derivatives. Then four of the ten equations can be used to express non-dynamical variables σ, Φ, ξ and θ through the variables (19). After that, there remain 6 first order equations for the 6 variables (19) .

Thus, naively we have 3 degrees of freedom. But this is not the case: two linear combinations of the 6 first order equations actually do not contain time derivatives, i.e., they are algebraic equations. These two equations are used to express two of the variables (19) in terms of the other 4 variables (say, Ψ, χ, Q, u). There remain 4 first order equations for these 4 variables. Thus, we have 2 degrees of freedom, the same number as in the Minkowski background. We concude that the self-accelerating background does not exhibit the Boulevare–Deser phenomenon.

4.2.2 Limit of small λ

Let us now consider the small-λ limit (see Sec. 3). We seek for solutions in the form $F \sim e^{i\int \omega d\eta}$, where F is any of the variables Ψ, χ, Q, u, and

$$\omega \gg \lambda, \quad k^2 \gg \lambda^2, \quad k^2 \equiv k_{\bar{a}} k_{\bar{a}}\ . \tag{20}$$

Then the time derivative is replaced by

$$\partial_0 \equiv i\omega\ .$$

As a result, we arrive at the system of 4 linear homogeneous algebraic equations for Ψ, χ, Q, u, which determines the dispersion relations. Equating the corresponding determinant to zero we get a fourth order equation for ω and then find the 4 dispersion relations. These are:

$$\omega_{1,2} = \frac{ik^2 \pm 2\sqrt{-27f^4 - 3f^2k^2}}{6f^2 + k^2}\lambda\,, \tag{21}$$

$$\omega_3 = -\frac{2if^2}{\lambda}\,, \tag{22}$$

$$\omega_4 = \frac{2if^2(12bf^2 - 2bk^2 + 18f^2 - 3k^2)}{3\lambda(8bf^2 + 12f^2 - k^2)}\,, \tag{23}$$

where $b \equiv c_3/c_5$. We see that

$$\omega_{1,2} \sim \lambda\,,$$
$$\omega_{3,4} \sim \frac{f^2}{\lambda}\,.$$

where $b \equiv c_3/c_5$. These "frequencies" are purely imaginary and naively there are instabilities. However, ω_1 and ω_2 correspond to the processes with characteristic time scale

$$T \sim \frac{1}{\lambda}\,,$$

which means that the instability, if any, does not develop in the Hubble time (in fact, $\omega \sim \lambda$ contradicts our assumption (20), so the result (21) cannot be trusted). The other two "frequencies" are of order

$$\omega_3,\ \omega_4 \sim \frac{f^2}{\lambda}\,.$$

In the limit of small λ they are above a UV cutoff necessarily present in the model. Thus, within the region of validity of the effective low energy theory there are no propagating modes at all.

5 Conclusion

Interestingly, the spectrum of the theory about the self-accelerating background (at least its scalar part) is qualitatively different from the spectrum about the Minkowski background. This is not particularly surprising, since the two backgrounds are not connected by a family of solutions to the field equations, i.e., they belong to two separate branches. It is reassuring that the scalar perturbations about the self-accelerating solution are healthy in the small-λ limit.

We plan to consider tensor and vector sectors in future.

Acknowledgements

The author is indebted to V. Rubakov and S. Randjbar-Daemi for helpful discussions. This work has been supported by Russian Science Foundation grant 14-22-00161.

References

[1] T. Clifton, P. G. Ferreira, A. Padilla and C. Skordis, Phys. Rept. **513**, 1 (2012) [arXiv:1106.2476 [astro-ph.CO]].
[2] G. R. Dvali, G. Gabadadze and M. Porrati, Phys. Lett. B **485**, 208 (2000) [hep-th/0005016].
[3] N. Kaloper and D. Kiley, JHEP **0705**, 045 (2007) [hep-th/0703190].
[4] T. Kobayashi, Phys. Rev. D **78**, 084018 (2008) [arXiv:0806.0924 [hep-th]].
[5] C. de Rham, G. Dvali, S. Hofmann, J. Khoury, O. Pujolas, M. Redi and A. J. Tolley, Phys. Rev. Lett. **100**, 251603 (2008) [arXiv:0711.2072 [hep-th]].
[6] N. Kaloper, Mod. Phys. Lett. A **23**, 781 (2008) [arXiv:0711.3210 [hep-th]].
[7] G. Gabadadze, hep-ph/0308112.
[8] A. De Felice and S. Tsujikawa, Living Rev. Rel. **13**, 3 (2010) [arXiv:1002.4928 [gr-qc]].

[9] Y. Fujii, K. Maeda, *The Scalar-Tensor Theory of Gravitation* (Cambridge University Press, 2003)

[10] S. Deser and A. Waldron, Phys. Rev. Lett. **87**, 031601 (2001) [hep-th/0102166].

[11] S. Deser and A. Waldron, Phys. Lett. B **508**, 347 (2001) [hep-th/0103255].

[12] M. Porrati, JHEP **0204**, 058 (2002) [hep-th/0112166].

[13] C. de Rham, G. Gabadadze and A. J. Tolley, Phys. Rev. Lett. **106**, 231101 (2011) [arXiv:1011.1232 [hep-th]].

[14] K. Hinterbichler, Rev. Mod. Phys. **84**, 671 (2012) [arXiv:1105.3735 [hep-th]].

[15] V. A. Rubakov, hep-th/0407104.

[16] S. L. Dubovsky, JHEP **0410**, 076 (2004) [hep-th/0409124].

[17] S. L. Dubovsky, P. G. Tinyakov and I. I. Tkachev, Phys. Rev. Lett. **94**, 181102 (2005) [hep-th/0411158].

[18] Z. Berezhiani, D. Comelli, F. Nesti and L. Pilo, Phys. Rev. Lett. **99**, 131101 (2007) [hep-th/0703264 [HEP-TH]].

[19] D. Blas, D. Comelli, F. Nesti and L. Pilo, Phys. Rev. D **80**, 044025 (2009) [arXiv:0905.1699 [hep-th]].

[20] V. A. Rubakov and P. G. Tinyakov, Phys. Usp. **51**, 759 (2008) [arXiv:0802.4379 [hep-th]].

[21] D. G. Boulware and S. Deser, Phys. Rev. D **6** (1972) 3368.

[22] C. Deffayet, Phys. Lett. B **502**, 199 (2001) [hep-th/0010186].

[23] C. Deffayet, G. R. Dvali and G. Gabadadze, Phys. Rev. D **65**, 044023 (2002) [astro-ph/0105068].

[24] M. A. Luty, M. Porrati and R. Rattazzi, JHEP **0309**, 029 (2003) [hep-th/0303116].

[25] A. Nicolis and R. Rattazzi, JHEP **0406**, 059 (2004) [hep-th/0404159].

[26] K. Koyama, Phys. Rev. D **72**, 123511 (2005) [hep-th/0503191].

[27] D. Gorbunov, K. Koyama and S. Sibiryakov, Phys. Rev. D **73**, 044016 (2006) [hep-th/0512097].

[28] C. Charmousis, R. Gregory, N. Kaloper and A. Padilla, JHEP **0610**, 066 (2006) [hep-th/0604086].

[29] K. Hayashi and T. Shirafuji, Prog. Theor. Phys. **64**, 866 (1980) Erratum: [Prog. Theor. Phys. **65**, 2079 (1981)].

[30] K. Hayashi and T. Shirafuji, Prog. Theor. Phys. **64**, 1435 (1980) Erratum: [Prog. Theor. Phys. **66**, 741 (1981)].

[31] K. Hayashi and T. Shirafuji, Prog. Theor. Phys. **64**, 2222 (1980).

[32] E. Sezgin and P. van Nieuwenhuizen, Phys. Rev. D **21**, 3269 (1980).

[33] V. P. Nair, S. Randjbar-Daemi and V. Rubakov, Phys. Rev. D **80**, 104031 (2009) [arXiv:0811.3781 [hep-th]].

[34] V. Nikiforova, S. Randjbar-Daemi and V. Rubakov, Phys. Rev. D **80**, 124050 (2009) [arXiv:0905.3732 [hep-th]].

[35] C. Deffayet and S. Randjbar-Daemi, Phys. Rev. D **84**, 044053 (2011) [arXiv:1103.2671 [hep-th]].

[36] V. Nikiforova, S. Randjbar-Daemi and V. Rubakov, arXiv:1606.02565 [hep-th].

EPJ Web of Conferences **125**, 03015 (2016) DOI: 10.1051/epjconf/201612503015
QUARKS-2016

Numerical relativity and the early Universe

Sergey Mironov[1,*]

[1] *Institute for Nuclear Research of the Russian Academy of Sciences, 60th October Anniversary Prospect, 7a, 117312 Moscow, Russia*

Abstract. We consider numerical simulations in general relativity in ADM formalism with cosmological ansatz for the metric. This ansatz is convenient for investigations of the Universe creation in laboratory with Galileons. Here we consider toy model for the software: spherically symmetric scalar field minimally coupled to the gravity with asymmetric double well potential. We studied the dependence of radius of critical bubble on the parameters of the theory. It demonstrates the wide applicability of thin-wall approximation. We did not find any kind of stable bubble solution.

1 Introduction

We study smooth solutions in spherically symmetric 3+1 gravity with additional scalar field. This problem was studied in many aspects, both analytically [1–3] and numerically [4–6] (see [7] for a review). We return to this problem with Galileons in mind. There is an idea, that Galileon field allows one to create a new universe in the laboratory [8]. We just need to create special initial conditions to make some region in our nearly Minkowskian space-time undergo accelerated expansion and create a man-made universe. We want to simulate this process numerically. The most convenient way to do it is to use the following ansatz for the metric

$$ds^2 = \alpha^2 dt^2 + a^2(dr^2 + r^2 d\Omega^2)\,. \quad (1)$$

We need to note, that the whole process is entirely smooth and always asymptotically Minkowskian. Initial configuration is just nearly a homogeneous field (of Galileon-type) in a sufficiently large sphere. The energy density is initially small everywhere, but increases inside the sphere, as the field evolves. While the expanding sphere becomes a new universe, the energy density outside remains small and asymptotes to zero at large distances.

But before proceeding to the numerical relativity with Galileons, we consider the toy model for software debugging: we examine the dynamics of bubbles of domain wall in the theory of standard scalar field minimally coupled to the gravity. The study of transitions in scalar fields from one minimum to another is itself very motivated by cosmology (mostly by early Universe inflationary-like models) and some classical field theory models.

The most popular topic in this context is a critical phenomena, first discovered by Choptuik [5]. Any family of initial configurations will evolve to one of the two end states – flat space (dispersal of

[*] e-mail: mironov@inr.ac.ru

the field to infinity) or a black hole. Critical solution is a transition between these two end states. The non-trivial part is that in spherically symmetric collapse there is a peculiar relation between the mass M of the formed black hole and any parameter p of initial configuration family:

$$M \sim (p - p_*)^\gamma , \tag{2}$$

here p_* is a critical parameter and scaling index γ does not depend on the choice and parametrization of initial family (only on the type of the matter). For a massless scalar field $\gamma \approx 0.37$ (which is a numerical result). The other interesting phenomenon is a self-similarity of the solution. It turns out that near the critical point field exhibits the following scaling relation,

$$\phi(t, x) = \phi(\alpha t, \alpha x) \qquad \alpha = e^{3.44} . \tag{3}$$

It is seen in the simulations, that the same field profile appears on the α times smaller spatial scale, but α times closer to the critical time. Unfortunately, for now this result is not well understood and may be questionable.

But before moving to domain walls in general relativity let us, first, consider standard $\frac{\lambda}{4}(\varphi^2 - v^2)^2$ potential without gravity. There is well-known classical solution: kink in 1+1 dimensions, which generalizes to plain domain wall in 3+1 dimensions

$$\varphi = v \tanh\left(\sqrt{\frac{\lambda}{2}} v(x - x_0)\right) . \tag{4}$$

It is a static solution, we can obtain moving one via boost. Spherically symmetric version of this solution is a bubble (in spherically symmetric situation analytic solution exists only in a thin-wall approximation). It could not be static due to the surface tension. So, we add small asymmetric term to the potential

$$V(\varphi) = \frac{\lambda}{4}(\varphi^2 - v^2)^2 - \epsilon\varphi \tag{5}$$

to obtain static solution (unstable). This solution is exactly a critical one.

The next step is to turn on gravity. First of all, in the thin-wall approximation [2]. In the next section we are going to describe the dynamic of bubbles in this case. In the third section, we will set up a full system of equations and consider numerical simulations. We conclude in the last section.

2 Thin-wall approximation

Thin-wall approximation in gravity-free case is quite trivial, since it is the limit of plain domain wall (4). Still it makes sense to review the dynamics of thin wall in the theory of scalar field with gravity. Let us start with the notations. We use Greek indices for four-dimensional tensors while Latin indices take three values and numerate components of the three-dimensional tensors of the hypersurface (Σ). Superscripts + and − indicate outer and inner four-dimensional volumes from the different sides of the hypersurface. $\sigma_m u^\nu$ is a singular part of stress-energy tensor T_μ^ν on the wall. $K_{ij} = -\Gamma^n_{ij}$ is outer curvature of Σ. Generally dynamic equations for the wall are the consequence of Einstein equations and junction conditions. First junction is a procedure of continuous matching of $g^-_{\mu\nu}$ and $g^+_{\mu\nu}$ on Σ. It allows one to write the metric on the whole manifold

$$ds^2 = \gamma_{ij} dx^i dx^j - dn^2 . \tag{6}$$

The Einstein equations (i, j-components) with the use of K_i^j and σ_i^j could be presented in the following form

$$(\delta_i^j K_m^m - K_i^j)^+ - (\delta_i^j K_m^m - K_i^j)^- = 8\pi\kappa\sigma_i^j\,. \tag{7}$$

In general Einstein equations and junction provide one with a lot of conditions (much more than functions to define) and not always possible to obey. The number of conditions is reduced due to additional symmetries, in our case – spherical symmetry.

$$ds^2 = g_{00}dt^2 + 2g_{01}dtdq + g_{11}dq^2 - r^2 d\Omega^2\,, \tag{8}$$

as usual

$$d\Omega^2 = d\Theta^2 + \sin^2(\Theta)d\varphi^2\,. \tag{9}$$

We require $g_{01} = 0$ for simplicity (gauge freedom). The metric on the shell is

$$ds^2 = d\tau^2 - \rho^2(\tau)d\Omega^2\,. \tag{10}$$

The equations take the following form after imposing spherical symmetry:

$$4\pi\kappa\sigma_0^0 = -K_2^{2+} + K_2^{2-} \tag{11}$$

$$8\pi\kappa\sigma_2^2 = -K_2^{2+} + K_2^{2-} - K_0^{0+} + K_0^{0-} \tag{12}$$

Since we are interested in bubbles of scalar field, we assume stress-energy tensor to be of special type, hence the first of the second junction conditions takes the form

$$\left(1 + \dot\rho^2 - \frac{8\pi\kappa}{3}\epsilon_{in}\rho^2\right)^{1/2} - \left(1 + \dot\rho^2 - \frac{8\pi\kappa}{3}\epsilon_{out}\rho^2 - \frac{2\kappa m}{\rho}\right)^{1/2} = 4\pi\kappa\sigma_0^0\rho \tag{13}$$

here ϵ_{in} is energy density in the inner volume. We will be interested in the case $\epsilon_{out} = 0$, so we omit the subscript in. σ here is a wall tension. Now it is possible to present this equation of motion for the bubble in two different forms

$$\dot\rho^2 = \frac{(\epsilon + 6\pi\kappa\sigma^2)^2}{9\sigma^2}\rho^2 - 1 + \frac{m}{\rho}\left(\kappa - \frac{\epsilon}{6\pi S^2}\right) + \frac{m^2}{16\pi^2 S^2\rho^4} \tag{14}$$

and

$$m = \frac{4\pi}{3}\epsilon\rho^3 + 4\pi\rho^2 S(\dot\rho^2 + 1 - 8\pi\kappa\epsilon\rho^2/3)^{1/2} - 8\pi^2\kappa S^2\rho^3 \tag{15}$$

It is easy to check, that there is no stable equilibrium point for domain wall. Any static solution is unstable (critical).

In other words, the effective potential has no minimum. Though, it is very close to having it, for example, an additional electric charge immediately stabilizes the bubble.

Now let us turn to the numerical simulations and check the presence of stable bubble there and applicability of thin-wall approximation to thick walls in general.

3 Numerical setup

We work in 3+1 ADM formalism [9], it is convenient for numerical simulations because equations are of the first order[1]. Equations of interest are projections of Einstein equations onto the hypersurface, on the normal and mixed projection [10]

$$\left(\partial_t - \mathcal{L}_\beta\right)\gamma_{ij} = -2\alpha K_{ij} \tag{16}$$

[1]we use Planck units $\kappa = 1$

$$\left(\partial_t - \mathcal{L}_\beta\right) K_{ij} = -\nabla_i \nabla_j \alpha +$$
$$+\alpha \left[R_{ij} + K K_{ij} - 2K_{ik} K^k{}_j + \frac{8\pi}{D-2}(S - E)\gamma_{ij} - 8\pi S_{ij} \right] \tag{17}$$

$$R + K^2 - K_{ij} K^{ij} = 16\pi E \tag{18}$$

$$\nabla_j \left(K^{ij} - \gamma^{ij} K \right) = 8\pi j^i \tag{19}$$

For our purposes it is enough to consider spherical symmetry. In this case general expression for the metric is following

$$ds^2 = \left(-\alpha^2 + a^2\beta^2\right) dt^2 + 2a^2\beta dt dr + a^2 dr^2 + r^2 b^2 d\Omega^2. \tag{20}$$

In standard framework a and b are dynamical variables and α and β are gauge parameters. First of all, we impose an orthogonality condition $\beta = 0$, this significantly simplifies all the equations. Among the dynamical variables there are also two extrinsic curvatures K_r^r and K_θ^θ

$$K_i^j = diag(K_r^r, K_\theta^\theta, K_\theta^\theta) \tag{21}$$

and scalar field variables $\Phi \equiv \phi'$ and $\Pi \equiv \frac{a}{\alpha}\left(\dot{\phi} - \beta\phi'\right)$. We consider real scalar field minimally coupled to the gravity.

$$L = \frac{1}{2}\partial_\mu \phi \partial^\mu \phi + V(\phi) - \frac{1}{2}R \tag{22}$$

The projected Einstein equations (16-19) and equations of motion for scalar field can be divided in two groups: dynamical equations (23-28) and constraints (29),(30).

We write them explicitly in the metric $ds^2 = -\alpha^2 dt^2 + a^2 dr^2 + r^2 b^2 d\Omega^2$.

The dynamical equations for the metric components are

$$\dot{a} = -a\alpha K_r^r\,, \tag{23}$$

$$\dot{b} = -\alpha b K_\theta^\theta\,. \tag{24}$$

The evolution equations for the extrinsic curvature components are

$$\dot{K}_r^r = \alpha K_r^r (K_r^r + 2K_\theta^\theta) - \frac{1}{a}\left(\frac{\alpha'}{a}\right)' - \frac{2\alpha}{arb}\left[\frac{(rb)'}{a}\right]' - 8\pi\alpha\left(\frac{\Phi^2}{a^2} + V(\phi)\right), \tag{25}$$

$$\dot{K}_\theta^\theta = \alpha K_\theta^\theta (K_r^r + 2K_\theta^\theta) + \frac{\alpha}{(rb)^2} - \frac{1}{a(rb)^2}\left(\frac{\alpha rb}{a}(rb)'\right)' - 8\pi\alpha V(\phi)\,. \tag{26}$$

The dynamic of scalar field is determined by the following equations

$$\dot{\Phi} = \left(\frac{\alpha}{a}\Pi\right)', \tag{27}$$

$$\dot{\Pi} = \frac{1}{r^2 b^2}\left[r^2 b^2 \left(\frac{\alpha}{a}\Phi\right)\right]' - 2\Pi\frac{\dot{b}}{b} - \alpha a V_\phi(\phi)\,, \tag{28}$$

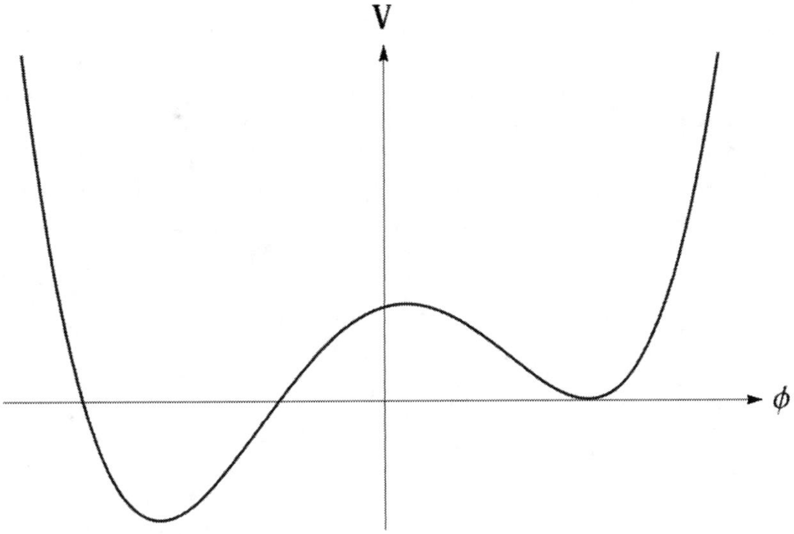

Figure 1. Double well asymmetric potential. We customize the value of the potential in the outside minimum to be zero.

In this ansatz the Hamiltonian constraint reads

$$-\frac{2}{arb}\left[\left(\frac{(rb)'}{a}\right)' + \frac{1}{rb}\left(\left(\frac{rb}{a}(rb)'\right)' - a\right)\right] + 4K_r^r K_\theta^\theta + 2(K_\theta^\theta)^2 = \tag{29}$$

$$= 8\pi\left(\frac{\Phi^2 + \Pi^2}{a^2} + 2V(\phi)\right)$$

and the momentum constraint reads

$$\frac{(rb)'}{rb}\left(K_\theta^\theta - K_r^r\right) - (K_\theta^\theta)' = -4\pi\frac{\Phi\Pi}{a} \tag{30}$$

We consider bubbles of domain wall interpolating between two minima. We customize the value of the potential in the outside minimum to be zero, hence on the spatial infinity metric is Schwarzschild without cosmological constant. Inside the bubble cosmological constant is negative.

$$V(\phi) = \frac{\lambda}{4}(\phi^2 - v^2)^2 + \varepsilon(\phi - v) + c \tag{31}$$

Now, instead of fixing α, we equate a and b while keeping α free[2]. This is a convenient ansatz for our later objectives with Galileons and Universe creation. In other words we customize the gauge

[2]It worth noting that Schwarzschild metric in our ansatz is a bit different:

$$\alpha = \sqrt{1 - \frac{2}{\frac{r}{m} + \frac{m}{4r} + 1}} \qquad a = 1 + \frac{m}{r} + \frac{m^2}{4r^2}$$

condition so that the dynamical variables a and b are equal. This condition combined with (23),(24) provides $K^r_r = K^\theta_\theta \equiv K$. After this procedure we are left with four dynamical equations: one for metric (now (23) and (24) are the same), one for curvature (either (25) or (26)) and still two for scalar field. In addition we got a constraint equation to determine α, it comes from the difference of (25) and (26).

We apply these five equations to evaluate initial configuration, but because of unstable mode we are also forced to keep the constrains (29),(30) satisfied. As for initial conditions, we specify scalar field as domain wall of given radius and thickness and solve the constraint equations to determine α, a and K.

4 Conclusions

We designed and debugged a software for numerical simulations in gravity with scalar fields. The main difference from previous works is the special ansatz for the metric and hence slightly different concept with initial conditions and equations of motion. We studied how gravity affects the dynamics of bubbles in the scalar theory in our ansatz but did not find any kind of stable bubble solution. We see that the thin-wall approximation is applicable for a wide range of bubble configurations, even with fairly thick walls.

Subsequently we plan to gather more data and examine the detailed dependence of the radius of critical bubble on the parameters of the model. Moreover, we naturally are going to simulate the universe creation with Galileon field.

Authors are indebted to Valery Rubakov for useful discussions and inspiration. Authors want to thank Diego Blas and Victoria Volkova for comments and interest to the work. This work was supported by Russian Science Foundation grant 14-22-00161

References

[1] V. Berezin, V. Kuzmin, I. Tkachev, Phys.Lett. B120 (1983) 91-96
[2] V. Berezin, V. Kuzmin, I. Tkachev, Phys.Rev. D36 (1987) 2919
[3] Y. Takamizu, K. Maeda, Phys.Rev. D92 (2015) no.2, 023514, arXiv:1504.05788
[4] R. Marsa, M. Choptuik, Phys.Rev. D54 (1996) 4929-4943, arXiv:gr-qc/9607034
[5] M. Choptuik, Phys.Rev.Lett. 70 (1993) 9-12
[6] K. Clough, E. Lim, arXiv:1602.02568
K. Clough, P. Figueras, H. Finkel, M. Kunesch, E. Lim, S. Tunyasuvunakool, Class.Quant.Grav. 32 (2015) no.24, 245011, Class.Quant.Grav. 32 (2015) 24, arXiv:1503.03436
[7] C. Gundlach, J. Martin-Garcia, Living Rev.Rel. 10 (2007) 5, arXiv:0711.4620
[8] V. Rubakov, Phys.Rev. D88 (2013) 044015, arXiv:1305.2614
[9] R. Arnowitt, S. Deser, C. Misner, Gen.Rel.Grav. 40 (2008) 1997-2027,arXiv: gr-qc/0405109
R. Arnowitt, S. Deser, C. Misner, Phys.Rev. 117 (1960) 1595-1602
[10] M. Zilhao, arXiv:1301.1509

EPJ Web of Conferences **125**, 03016 (2016) DOI: 10.1051/epjconf/201612503016
QUARKS-2016

How to construct a consistent and physically relevant the Fock space of neutrino flavor states?

*A. E.*Lobanov[1,⋆]

[1]*Department of Theoretical Physics, Faculty of Physics, Moscow State University, 119991 Moscow, Russia*

Abstract. We propose a modification of the electroweak theory, where the fermions with the same electroweak quantum numbers are combined in multiplets and are treated as different quantum states of a single particle. Thereby, in describing the electroweak interactions it is possible to use four fundamental fermions only. In this model, the mixing and oscillations of the particles arise as a direct consequence of the general principles of quantum field theory. The developed approach enables one to calculate the probabilities of the processes taking place in the detector at long distances from the particle source. Calculations of higher-order processes including the computation of the contributions due to radiative corrections can be performed in the framework of perturbation theory using the regular diagram technique.

1 Introduction

The Standard Model of electroweak interactions based on the non-Abelian gauge symmetry of the interactions [1], the generation of particle masses due to the spontaneous symmetry breaking mechanism [2], and the philosophy of mixing of particle generations [3] is universally recognized. Its predictions obtained in the framework of perturbation theory are in a very good agreement with the experimental data, and there is no serious reason, at least at the energies available at present, for its main propositions to be revised.

However, in describing such an important and firmly experimentally established phenomenon as neutrino oscillations [4], an essentially phenomenological theory based on the pioneer works by B. Pontecorvo [5] and Z. Maki et al. [6] is used. This theory is well developed and is consistent with the experimental data.

The primary assumption of this theory is that the neutrinos are massive, and moreover, there are three neutrino types with different masses. It is also postulated that the neutrinos produced in reactions are in the states which are superpositions of states with fixed masses, the so-called mass states. These states form the so-called flavor basis. The transformation to this basis from the mass basis is given by a unitary mixing matrix. Initially, the mass basis elements are described by plane waves with the same (three-dimensional) momentum. The time evolution of the flavor states is described by the solution of the corresponding Cauchy problem.

For this reason it is taken for granted that the mass and flavor states can be connected by a unitary transformation. However, this supposition is not correct. This fact is well known. For example, it is

⋆e-mail: lobanov@phys.msu.ru

discussed in paper [7]. So, it is impossible to construct the Fock space for flavor states using the conventional approach. And, as a consequence, it is impossible to calculate the transition probabilities for flavor states in the framework of the perturbation theory.

The purpose of this work is to construct in a consistent and physically relevant manner the Fock space for the Standard Model fermions (leptons and quarks) in such a way that the mixing of the particles and their oscillations would occur automatically.

The basis of the theoretical scheme is as follows. In relativistic quantum field theory a particle is usually associated with an irreducible representation of the Poincaré group [8, 9]. The eigenvalue in this representation of the Casimir operator constructed from the translation operators squared is identified with the observed mass of the particle.

The existence of states which are superpositions of states with different masses contradicts the relativistic invariance of the theory, if the canonical momentum operator is identified with the translation operator. This obviously follows from the fact that the metric is defined on the hyperboloid in the momentum space determined by the value of the particle mass.

To overcome this difficulty it seems natural to associate some set of particles (a multiplet) with an irreducible representation of a wider group. A number of theorems [10] indicates that the only reasonable extension of the symmetry group of the theory is the direct product of the Poincaré group and a group of internal symmetry. However, in this point we encounter the important theorem [11][1].

It seems that due to this theorem the result of extending the symmetry group of the theory will be trivial: the masses of the components of the multiplets will be equal. Just this circumstance is the most important reason of the failure of constructing the Fock space for flavor states. However, it is possible to circumvent this obstacle.

2 Wave function spaces

It is well known (see, for example, [12]) that the derivation algebra of the Poincaré algebra contains not only the translation operators P^μ and generators of the Lorentz group $M^{\mu\nu}$, but also the generator of dilatation D:

$$[P^\mu, D] = P^\mu, \quad [M^{\mu\nu}, D] = 0. \tag{1}$$

So it is possible to construct an external automorphism of Poincaré algebra, which leads to a scaling transformation of the translation generators. For irreducible representations the dilatation changes the value of the Casimir operator which defines the mass of a particle. Hence, such a transformation allows to consider particles with different masses even in the case when the masses were initially equal.

Such considerations conform to the spirit of the Standard Model, where the masses of all particles are generated due to the phenomenon of spontaneous symmetry breaking and are proportional to the vacuum expectation value of the Higgs field.

Let us choose the group $SU(3)$ as the group of internal symmetry. Such an assumption is quite reasonable because the experimental data indicate that there are three generations of fermions. And let us consider the direct product of the Poincaré group and the group $SU(3)$ as an extended symmetry group of the theory.

Let us consider an irreducible representation of this group which is constructed as the direct product of the Dirac representation of the Poincaré group and the fundamental representation of the $SU(3)$

[1]Let Γ be a continuous unitary representation of a finite-dimensional connected Lie group G in a Hilbert space $\mathcal{H}$. Let G contain the inhomogeneous Lorentz group as an analytic subgroup. Let finally the spectrum of the 4-momentum operator P_μ be contained in $\{0\} \bigcup V^+$, V^+ being the future cone in the Minkowski space M^4. If $m_1 > 0$ is an isolated eigenvalue of the mass operator $M = (P_\mu P^\mu)^{1/2}$ then the corresponding eigenspace $\mathcal{H}_1$ is invariant under $\Gamma(G)$.

group. The explicit form of the Lie algebra elements in the present case is obvious. We have the standard realization of the Poincaré group generators multiplied by 3×3 identity matrix $\mathbb{I}$

$$P_\mu = \mathrm{i}\partial_\mu \mathbb{I}, \quad M_{\mu\nu} = \mathrm{i}\big((x_\mu \partial_\nu - x_\nu \partial_\mu) + (\gamma_\mu \gamma_\nu - \gamma_\nu \gamma_\mu)/4\big)\mathbb{I}, \tag{2}$$

and of the Hermitian generators $X_k, k = 1...8$ of the fundamental representation of the $SU(3)$ group defined by the Gell-Mann matrices

$$\begin{aligned}
X_1 &= \left(e^{(1)} \otimes e^{(2)}\right) + \left(e^{(2)} \otimes e^{(1)}\right),\ X_2 = \mathrm{i}\left(e^{(2)} \otimes e^{(1)}\right) - \mathrm{i}\left(e^{(1)} \otimes e^{(2)}\right),\\
X_3 &= \left(e^{(1)} \otimes e^{(1)}\right) - \left(e^{(2)} \otimes e^{(2)}\right),\ X_4 = \left(e^{(1)} \otimes e^{(3)}\right) + \left(e^{(3)} \otimes e^{(1)}\right),\\
X_5 &= \mathrm{i}\left(e^{(3)} \otimes e^{(1)}\right) - \mathrm{i}\left(e^{(1)} \otimes e^{(3)}\right),\ X_6 = \left(e^{(2)} \otimes e^{(3)}\right) + \left(e^{(3)} \otimes e^{(2)}\right),\\
X_7 &= \mathrm{i}\left(e^{(3)} \otimes e^{(2)}\right) - \mathrm{i}\left(e^{(2)} \otimes e^{(3)}\right),\ X_8 = \left(e^{(1)} \otimes e^{(1)}\right) + \left(e^{(2)} \otimes e^{(2)}\right) - 2\left(e^{(3)} \otimes e^{(3)}\right).
\end{aligned} \tag{3}$$

Here the vectors $e^{(l)}$ which, for definiteness, can be chosen as follows

$$e^{(1)} = \begin{pmatrix} 1 \\ 0 \\ 0 \end{pmatrix}, \quad e^{(2)} = \begin{pmatrix} 0 \\ 1 \\ 0 \end{pmatrix}, \quad e^{(3)} = \begin{pmatrix} 0 \\ 0 \\ 1 \end{pmatrix}, \tag{4}$$

are a basis of a three-dimensional vector space over the field of complex numbers.

Let us consider the space $\mathcal{H}_{m,1/2}$ of solutions of the matrix Dirac equation

$$\left(\mathrm{i}\gamma^\mu \partial_\mu \mathbb{I} - m\mathbb{I}\right) \Psi(x) = 0, \tag{5}$$

which describes a multiplet containing three particles. Introduce in this space the scalar product

$$(\Psi, \Phi) = \sum_{s=1}^{3} \int d\mathbf{x}\, \Psi_s^\dagger(\mathbf{x}, t)\, \Phi_s(\mathbf{x}, t) \tag{6}$$

with summation over the coordinates of the vectors $e_s^{(l)}$. Then one can consider this space as the direct sum of the spaces of irreducible unitary representations of the direct product of the Poincaré group and the $SU(3)$ group corresponding to positive and negative frequencies. The action of this group in $\mathcal{H}_{m,1/2}$ is determined by the generators P_μ, $M_{\mu\nu}$, X_k. Obviously, in these representations all the components of the multiplet have equal masses.

A natural basis of multiplet subspace of $\mathcal{H}_{m,1/2}$ with positive frequency can be defined as follows:

$$\Psi_{p,\zeta,l}(x) = \psi_{p,\zeta}(x)\, e^{(l)}. \tag{7}$$

Here $\psi_{p,\zeta}(x)$ are standard plane waves

$$\psi_{p,\zeta}(x) = \frac{1}{\sqrt{2p^0}} u_{p,\zeta} e^{-\mathrm{i}(px)}, \tag{8}$$

where $p^0 = \sqrt{\mathbf{p}^2 + m^2}$, and the spinors $u_{p,\zeta}$ satisfy the equation

$$(\gamma^\mu p_\mu - m) u_{p,\zeta} = 0 \tag{9}$$

and are normalized by the condition

$$\bar{u}_{p,\zeta} u_{p,\zeta'} = 2m\delta_{\zeta,\zeta'}, \quad \bar{u}_{p,\zeta} = u_{p,\zeta}^\dagger \gamma^0. \tag{10}$$

The indices $\zeta, \zeta' = \pm 1$ define the particle polarization. Similarly we can consider the subspace of $\mathcal{H}_{m,1/2}$ with negative frequency.

Let us construct previously mentioned external automorphisms. In the space $\mathcal{H}_{m,1/2}$ the dilatations are determined by the operator

$$D = x_\mu \partial^\mu . \tag{11}$$

The operator of the Dirac equation

$$\left(\mathrm{i}\gamma^\mu \partial_\mu \mathbb{I} - m\mathbb{I}\right) \Psi(x) = 0, \tag{12}$$

commutes with the generators of the extended group. However, for the generator of dilatation we have

$$\left[\left(\mathrm{i}\gamma^\mu \partial_\mu \mathbb{I} - m\mathbb{I}\right), D\right] = \mathrm{i}\gamma^\mu \partial_\mu \mathbb{I}. \tag{13}$$

So, the commutator is equal to zero on solutions for massless particles $m = 0$ only. Therefore, one can construct non-identical spaces $\mathcal{H}^{(i)}_{m,1/2}$, of irreducible representations for multiplets with non-zero masses.

It should be stressed that for massless fermions such a possibility is absent. However, in this case the theory possesses the scale invariance, and, as a consequence, the conformal invariance.

Let us associate the different representation spaces $\mathcal{H}^{(i)}_{m,1/2}$ for multiplets including the neutrinos ($i = \nu$), the charged leptons ($i = e$) as well as the down- ($i = d$) and up- ($i = u$) type quarks with the space $\mathcal{H}_{m,1/2}$ using a unitary intertwining operators $\mathcal{K}^{(i)}$. Then the elements $\Psi^{(i)}(x)$ of the space $\mathcal{H}^{(i)}_{m,1/2}$ and the elements $\Psi(x)$ of the space $\mathcal{H}_{m,1/2}$ become connected by the unitary (with respect to the introduced scalar product (6)) transformation

$$\Psi^{(i)}(x) = \mathcal{K}^{(i)} \Psi(x). \tag{14}$$

The intertwining operators can be constructed as follows. For each multiplet introduce new basis vectors $n^{(l)}(i) \equiv n^{(l)}$, acting by a unitary matrix $\mathbb{V}^{(i)}$ on the old basis vectors $e^{(l)}$:

$$n^{(l)}_s = \sum_{r=1}^{3} V^{(i)}_{sr} e^{(l)}_r . \tag{15}$$

The vectors $n^{(l)}$ are normalized by the conditions (asterisk denotes the complex conjugation)

$$\sum_{s=1}^{3} n^{(l)}_s \overset{*}{n}{}^{(k)}_s = \delta_{kl}, \qquad \sum_{l=1}^{3} n^{(l)}_s \overset{*}{n}{}^{(l)}_r = \delta_{sr}. \tag{16}$$

The operators

$$\mathbb{P}^{(i)}_{(l)} = n^{(l)} \otimes \overset{*}{n}{}^{(l)}, \ \mathbb{P}^{(i)}_{(l)}\mathbb{P}^{(i)}_{(k)} = \delta_{kl}\mathbb{P}^{(i)}_{(l)}, \ \sum_{l=1,2,3} \mathbb{P}^{(i)}_{(l)} = \mathbb{I}. \tag{17}$$

are orthogonal projectors.

We can write the non-trivial intertwining operators in the form

$$\mathcal{K}^{(i)} = \sum_{l=1}^{3} \mathcal{D}^{(i)}_{(l)} \left(n^{(l)} \otimes e^{(l)}\right), \tag{18}$$

where

$$\mathcal{D}^{(i)}_{(l)} = \exp\left((x_\nu \partial^\nu + 3/2) \ln \mu^{(i)}_l\right). \tag{19}$$

Here $\mu_l^{(i)}$ are positive numbers.

The Dirac equations, which are the conditions that the state spaces are irreducible, are now written as follows:

$$\left(\mathrm{i}\gamma^\mu\partial_\mu\mathbb{I} - \mathbb{M}^{(i)}\right)\Psi^{(i)}(x) = 0, \tag{20}$$

where

$$\mathbb{M}^{(i)} = \sum_{l=1}^{3} m_l^{(i)}\mathbb{P}_{(l)}^{(i)}, \quad m_l^{(i)} = m\mu_l^{(i)}. \tag{21}$$

We define bases in the spaces $\mathcal{H}_{m,1/2}^{(i)}$ in the form

$$\Psi^{(i)}_{q,\zeta,\mu_l^{(i)}}(x) = \mathcal{K}^{(i)}\Psi_{p,\zeta,l}(x) = \psi^{(i)}_{q,\zeta,\mu_l^{(i)}}(x)n^{(l)}, \tag{22}$$

where $\psi^{(i)}_{q,\zeta,\mu_l^{(i)}}(x)$ are the plane waves derived from the ordinary plane waves (8) by a dilatation of the coordinates:

$$\psi^{(i)}_{q,\zeta,\mu_l}(x) = \frac{(\mu_l^{(i)})^{3/2}}{\sqrt{2q^0}}u_{q,\zeta}e^{-\mathrm{i}\mu_l^{(i)}(qx)}. \tag{23}$$

Here $q^0 = \sqrt{\mathbf{q}^2 + m^2}$. The spinors $u_{q,\zeta}$ satisfy the equation

$$(\gamma^\mu q_\mu - m)u_{q,\zeta} = 0 \tag{24}$$

and are normalized by the condition

$$\bar{u}_{q,\zeta}u_{q,\zeta'} = 2m\delta_{\zeta,\zeta'}. \tag{25}$$

A direct calculation yields the explicit form of the elements of the Lie algebras of the considered representations. The action of the generators of the $SU(3)$ group

$$X_k^{(i)} = \mathcal{K}^{(i)}X_k\mathcal{K}^{(i)-1} \tag{26}$$

is reduced to the obvious permutations of the basis elements.

The generators of the Lorentz group do not change the form:

$$M_{\mu\nu}^{(i)} = \mathcal{K}^{(i)}M_{\mu\nu}\mathcal{K}^{(i)-1} = \mathrm{i}\big((x_\mu\partial_\nu - x_\nu\partial_\mu) + (\gamma_\mu\gamma_\nu - \gamma_\nu\gamma_\mu)/4\big)\mathbb{I}, \tag{27}$$

since the dilatation operator commutes with the generators of rotations and boosts (see (1)). This fact is quite natural, because the results of the observations, in particular, of the study of particle oscillations cannot depend on the choice of the inertial reference frame used for measurement.

For the translation generators, we have

$$P_\mu^{(i)} = \mathcal{K}^{(i)}P_\mu\mathcal{K}^{(i)-1} = \mathrm{i}\partial_\mu\mathbb{N}^{(i)}, \tag{28}$$

where

$$\mathbb{N}^{(i)} = \sum_{l=1}^{3}\frac{1}{\mu_l^{(i)}}\mathbb{P}_l^{(i)} = \mathbb{M}^{(i)-1}/m. \tag{29}$$

This matrices are proportional to the inverse mass matrices.

We emphasize once again that the eigenvalues of the Casimir operators constructed from the new generators $P_\mu^{(i)}, M_{\mu\nu}^{(i)}, X_k^{(i)}$ take the same values on solutions of the modified Dirac equations as on elements of the initial representation space $\mathcal{H}_{m,1/2}$. In particular, $P^{(i)\mu}P_\mu^{(i)}\Psi^{(i)}(x) = m^2\Psi^{(i)}(x)$, but now the parameter m is not the observed mass, it rather sets the scale of the multiplet masses. In other words, it represents the bare mass of the multiplet. The observed masses are determined by the action of the canonical momentum operator squared on the basis functions $\Psi^{(i)}_{q,\zeta,\mu^{(i)}_l}(x)$ and are equal to $m^{(i)}_l = m\mu^{(i)}_l$ for every component of the multiplet.

The general solutions of the Dirac equations (20) can be expanded in the basis functions that are eigenfunctions of any complete set of operators commuting with the operator of the equation. In contrast to the standard case of the Dirac equation, the complete set for these equations contains five operators.

We now find all complete sets of the operators for which the bases of the solution spaces of the modified Dirac equations (20) consist of plane waves. Let $\mathbb{L}^{(i)}$ be a non-degenerate Hermitian matrix which commutes with $\mathbb{M}^{(i)}$. This matrix can be written as follows:

$$\mathbb{L}^{(i)} = \sum_{l=1}^{3} \lambda^{(i)}_l \mathbb{P}^{(i)}_{(l)}, \quad \lambda^{(i)}_l \neq 0. \tag{30}$$

If we assume that the parameters $\lambda^{(i)}_l$ are pairwise distinct, then, in the space of the solutions of Eq. (20), this matrix defines the action of the operator $\mathcal{M}^{(i)}$ which is a linear combination of $X^{(i)}_3$ (the third projection of the isospin), $X^{(i)}_8$ (the hypercharge) and the Casimir operators of $SU(3)$. (Since the representation is irreducible, then the action of the Casimir operators is given by the matrices that are multiple of the identity matrix). Obviously $\mathcal{M}^{(i)}$ can be chosen as one of the operators of the complete set. This operator is used to isolate the orthogonal subspaces determined by the projectors $\mathbb{P}^{(i)}_{(l)}$. Therefore, the numerical values of the parameters $\lambda^{(i)}_l$ can be arbitrary. It is natural to set $\lambda^{(i)}_l = m^{(i)}_l$, that is, to assume that the action of the operator $\mathcal{M}^{(i)}$ is defined by the matrix $\mathbb{M}^{(i)}$.

For a basis to be a plane-wave one, the complete set must include three operators with continuous spectrum. If we choose the space translation generators, that is, the spatial components of the 4-vector $P^{(i)}_\mu$, as such operators, then any standard spin projector multiplied by $\mathbb{N}^{(i)}$ will determine the spin projection. In this case the complete orthonormal system of solutions of Eq. (20) corresponding to a fixed (positive) frequency delivers the basis $\Psi^{(i)}_{q,\zeta,\mu^{(i)}_l}(x)$ (see Eq. (22)) considered previously.

However, we can take the spatial components of the 4-vector

$$\mathbb{L}^{(i)} P^{(i)}_\mu = \mathrm{i}\partial_\mu \mathbb{L}^{(i)} \mathbb{N}^{(i)} \tag{31}$$

to play the role of the operators with continuous spectrum, where $\mathbb{L}^{(i)}$ is any matrix which satisfies (30). In particular, we can set $\mathbb{L}^{(i)} = (\mathbb{N}^{(i)})^{-1}$. Then the basis functions are the eigenfunctions of the spatial components of the canonical momentum operator $i\partial_\mu$, and

$$\Psi^{(i)}_{p,\zeta,m^{(i)}_l}(x) = \psi^{(i)}_{p,\zeta,m_l}(x) n^{(l)}, \tag{32}$$

where $\psi^{(i)}_{p,\zeta,m^{(i)}_l}(x)$ are the plane waves, describing particles with masses $m^{(i)}_l$:

$$\psi^{(i)}_{p,\zeta,m^{(i)}_l}(x) = \frac{1}{\sqrt{2p^0_l}} u^{(m^{(i)}_l)}_{p,\zeta} e^{-\mathrm{i}(px)}. \tag{33}$$

Here $p^0_l = \sqrt{\mathbf{p}^2 + (m^{(i)}_l)^2}$, the spinors $u^{(m^{(i)}_l)}_{p,\zeta}$ satisfy the equations

$$(\gamma^\mu p_\mu - m^{(i)}_l) u^{(m^{(i)}_l)}_{p,\zeta} = 0 \tag{34}$$

and are normalized by the condition

$$\bar{u}_{p,\zeta}^{(m_l^{(i)})} u_{p,\zeta'}^{(m_l^{(i)})} = 2m_l^{(i)} \delta_{\zeta,\zeta'}. \tag{35}$$

Thus, the matrix $\mathbb{M}^{(i)}$ can be interpreted as the mass matrix of the multiplet, and the parameters $m_l^{(i)} = \mu_l^{(i)} m$ are the observed masses of the particles. The states that are the eigenfunctions of $\mathcal{M}^{(i)}$, can be naturally called the mass states for any choice of other operators of the complete set.

However, we can choose a basis in the representation space $\mathcal{H}^{(i)}_{m,1/2}$ in the form of a superposition of the mass states. Consider an arbitrary unitary matrix $\mathbb{U}$. The functions

$$\Psi^{(i)}_{p,\zeta,\alpha}(x) = \sum_{l=1}^{3} U_{\alpha l} \Psi^{(i)}_{p,\zeta,l}(x) \tag{36}$$

make up a complete orthonormal set in $\mathcal{H}^{(i)}_{m,1/2}$ with the scalar product (6). Here $\Psi^{(\nu)}_{p,\zeta,l}(x)$ are the wave functions of arbitrary mass states. The basis (36) can be obtained as a result of the unitary transformation $\mathcal{U}$ of the space $\mathcal{H}^{(i)}_{m,1/2}$ onto itself:

$$\Psi^{(i)}_{p,\zeta,\alpha}(x) = \mathcal{U}\, \Psi^{(i)}_{p,\zeta,l}(x) = \int \widetilde{K}(x,y)\, \Psi^{(i)}_{p,\zeta,l}(y)\, d\mathbf{y}. \tag{37}$$

The kernel of the transformation is determined by the formula

$$\widetilde{K}(x,y) = \frac{1}{(2\pi)^3} \sum_{\alpha=1}^{3} \sum_{l=1}^{3} \sum_{\zeta=\pm1} \int d\mathbf{q}\, \delta_{\alpha l} \Psi^{(i)}_{p,\zeta,\alpha}(x) \otimes \Psi^{(i)\dagger}_{p,\zeta,l}(y). \tag{38}$$

The elements of this basis no longer are eigenfunctions of the operator $\mathcal{M}^{(i)}$ defined by the mass matrix. The fifth operator from the complete set (denote it as $\mathcal{F}^{(i)}$) is defined now as follows:

$$\mathcal{F}^{(i)} = \mathcal{U}\mathcal{M}^{(i)}\mathcal{U}^{-1}. \tag{39}$$

This operator can explicitly depend on the coordinates of the event space.

It should be emphasized that the form of the causal Green function for the new Dirac equation does not depend on the chosen basis:

$$S_c^{(i)}(x) = \frac{1}{(2\pi)^4} \sum_{l=1}^{3} \mathbb{P}_l^{(i)} \int \frac{(\gamma_\mu p^\mu + m_l^{(i)})\, e^{-\mathrm{i}(px)}}{(m_l^{(i)})^2 - p^2 - \mathrm{i}\epsilon} d^4p. \tag{40}$$

3 Modified model of electroweak interaction

Now we are able to write the electroweak Lagrangian. Once the interaction is carried by the $SU(2) \times U(1)$-gauge fields, the Lagrangian is that of the Standard Model supplemented by the singlets of the right-handed neutrinos.

The Lagrangian for the physical fermion fields in our model is written as follows:

$$\mathcal{L}_f = \mathcal{L}_0 + \mathcal{L}_{int}, \tag{41}$$

where

$$\mathcal{L}_0 = \sum_{i=\nu,e,u,d} \frac{\mathrm{i}}{2}\Big[\big(\bar{\Psi}^{(i)}\gamma^\mu(\partial_\mu \Psi^{(i)})\big) - (\partial_\mu \bar{\Psi}^{(i)})\gamma^\mu \Psi^{(i)}\Big] - \bar{\Psi}^{(i)}\mathbb{M}^{(i)}\Psi^{(i)} \tag{42}$$

is the Lagrangian of free fields and

$$\begin{aligned}\mathcal{L}_{int} = &-\sum_{i=\nu,e,u,d} \bar{\Psi}^{(i)}\mathbb{M}^{(i)}(H/v)\Psi^{(i)} - e\sum_{i=e,u,d} Q^{(i)}\bar{\Psi}^{(i)}\gamma^{\mu}\Psi^{(i)}A_{\mu}\\ &-\frac{g}{2\sqrt{2}}\left(\bar{\Psi}^{(e)}\gamma^{\mu}(1+\gamma^5)\Psi^{(\nu)}W^{-}_{\mu} + \bar{\Psi}^{(\nu)}\gamma^{\mu}(1+\gamma^5)\Psi^{(e)}W^{+}_{\mu}\right)\\ &-\frac{g}{2\sqrt{2}}\left(\bar{\Psi}^{(d)}\gamma^{\mu}(1+\gamma^5)\Psi^{(u)}W^{-}_{\mu} + \bar{\Psi}^{(u)}\gamma^{\mu}(1+\gamma^5)\Psi^{(d)}W^{+}_{\mu}\right)\\ &-\frac{g}{2\cos\theta_{\mathrm{W}}}\sum_{i=\nu,e,u,d}\bar{\Psi}^{(i)}\gamma^{\mu}\left(T^{(i)} - 2Q^{(i)}\sin^2\theta_{\mathrm{W}} + T^{(i)}\gamma^5\right)\Psi^{(i)}Z_{\mu}.\end{aligned} \quad (43)$$

is the interaction Lagrangian between the fermion fields, the vector boson fields $W^{\pm}_{\mu}, Z_{\mu}, A_{\mu}$, and the Higgs field H. Here θ_{W} is the Weinberg angle, $e = g\sin\theta_{\mathrm{W}}$ is the positron electric charge, $T^{(i)}$ is the weak isospin ($T^{(\nu)} = T^{(u)} = 1/2$, $T^{(e)} = T^{(d)} = -1/2$), $Q^{(i)}$ is the electric charge of the multiplet in the units of e. The value v is the vacuum expectation of the Higgs field.

Thus, this Lagrangian formally coincides with the Lagrangian of the Standard Model. However, the wave functions $\Psi^{(i)}$ describe not the individual particles, but the multiplets as a whole. It is possible, so for each multiplet the action defined by the Lagrangian of free fields (42) is explicitly invariant with respect to $SU(3)$ transformations generated by $X^{(i)}_k$ (see (26)).

Therefore, when quantizing the model, the multiplet can be considered as a single particle. The one-particle states in the Fock space are defined as usual, the creation and annihilation operators satisfy the canonical commutation relations. However, these operators carry an additional discrete quantum number that is associated with the mass of the state. The multiplet can be either in one of the three mass states, or in a pure quantum state that is a superposition of the states with fixed masses. In a certain sense we may say that there exist only four fundamental fermions in this model.

We will discuss all this in detail, using the approach described in [13]. Consider the basis (22). We can write the components of the field functions as

$$\begin{aligned}\Psi^{(i)}_s(x) &= \frac{1}{(2\pi)^{3/2}}\sum_{l=1}^{3}\sum_{\zeta=\pm1}\int\frac{d\mathbf{q}}{\sqrt{2q^0}}\,(\mu^{(i)}_l)^{3/2}n^{(l)}_s\left[e^{-\mathrm{i}\mu^{(i)}_l(qx)}u_{q,\zeta}\mathfrak{a}^{-}_{l,\zeta,(i)}(\mathbf{q}) + e^{\mathrm{i}\mu^{(i)}_l(qx)}u_{-q,\zeta}\mathfrak{a}^{+}_{l,\zeta,(i)}(\mathbf{q})\right],\\ \bar{\Psi}^{(i)}_s(x) &= \frac{1}{(2\pi)^{3/2}}\sum_{l=1}^{3}\sum_{\zeta=\pm1}\int\frac{d\mathbf{q}}{\sqrt{2q^0}}\,(\mu^{(i)}_l)^{3/2}\mathring{n}^{(l)}_s\left[e^{-\mathrm{i}\mu^{(i)}_l(qx)}\bar{u}_{-q,\zeta}\mathring{\mathfrak{a}}^{-}_{l,\zeta,(i)}(\mathbf{q}) + e^{\mathrm{i}\mu^{(i)}_l(qx)}\bar{u}_{q,\zeta}\mathring{\mathfrak{a}}^{+}_{l,\zeta,(i)}(\mathbf{q})\right].\end{aligned} \quad (44)$$

Using Noether's theorem we can find the integrals of motion for the fields. The analogue of the energy-momentum tensor associated with the translational symmetry is determined by the relation

$$T^{(i)}_{\alpha\beta} = \frac{1}{2}\sum_{s=1}^{3}\left[\bar{\Psi}^{(i)}_s(x)\gamma_{\alpha}\left(P^{(i)}_{\beta}\Psi^{(i)}_s(x)\right) + \left(P^{(i)*}_{\beta}\bar{\Psi}^{(i)}_s(x)\right)\gamma_{\alpha}\Psi^{(i)}_s(x)\right]. \quad (45)$$

It ensures the existence of the conserved vector

$$\mathcal{P}^{(i)}_{\beta} = \sum_{l=1}^{3}\sum_{\zeta=\pm1}\int q_{\beta}d\mathbf{q}\left[\mathring{\mathfrak{a}}^{+}_{l,\zeta,(i)}(\mathbf{q})\mathfrak{a}^{-}_{l,\zeta,(i)}(\mathbf{q}) - \mathring{\mathfrak{a}}^{-}_{l,\zeta,(i)}(\mathbf{q})\mathfrak{a}^{+}_{l,\zeta,(i)}(\mathbf{q})\right]. \quad (46)$$

The current vector associated with the global gauge symmetry of the Lagrangian is defined by the relation

$$J^{(i)}_{\alpha} = \sum_{s=1}^{3}\bar{\Psi}^{(i)}_s(x)\gamma_{\alpha}\Psi^{(i)}_s(x). \quad (47)$$

It ensures the conservation of the total field charge:

$$Q^{(i)} = \sum_{l=1}^{3} \sum_{\zeta=\pm 1} \int d\mathbf{q} \left[\mathring{a}^{+}_{l,\zeta,(i)}(\mathbf{q}) a^{-}_{l,\zeta,(i)}(\mathbf{q}) + \mathring{a}^{-}_{l,\zeta,(i)}(\mathbf{q}) a^{+}_{l,\zeta,(i)}(\mathbf{q}) \right]. \tag{48}$$

The tensor associated with the $SU(3)$-symmetry of the fields is determined by the relation

$$S^{(i)}_{k\alpha} = \frac{1}{2} \sum_{s=1}^{3} \left[\bar{\Psi}^{(i)}_s(x) \gamma_\alpha (X^{(i)}_k \Psi^{(i)}(x))_s + (\bar{\Psi}^{(i)}(x) X^{(i)}_k)_s \gamma_\alpha \Psi^{(i)}_s(x) \right], \quad k = 1 \ldots 8. \tag{49}$$

It ensures the existence of eight integrals of motion. Using these integrals of motion and the total field charge it is possible to construct nine linear combinations

$$X^{(i)}_{lk} = \sum_{\zeta=\pm 1} \int d\mathbf{q} \left[\mathring{a}^{+}_{l,\zeta,(i)}(\mathbf{q}) a^{-}_{k,\zeta,(i)}(\mathbf{q}) + \mathring{a}^{-}_{l,\zeta,(i)}(\mathbf{q}) a^{+}_{k,\zeta,(i)}(\mathbf{q}) \right], \quad k, l = 1, 2, 3, \tag{50}$$

and three of them are diagonal in the indices l, k.

The integrals of motion for the fields are associated with the generators of the Lie algebra of the symmetry group of the theory as follows:

$$Z\Psi^{(i)}(x) = \left[\Psi^{(i)}(x), \mathcal{Z} \right]. \tag{51}$$

Here $Z = \left\{ P^{(i)}_\beta, Q^{(i)}, X^{(i)}_k \ldots \right\}$ and $\mathcal{Z} = \left\{ \mathcal{P}^{(i)}_\beta, \mathcal{Q}^{(i)}, \mathcal{X}^{(i)}_k \ldots \right\}$. These relations enable one to interpret $\mathring{a}^{+}_{l,\zeta,(i)}(\mathbf{q})$ and $a^{-}_{l,\zeta,(i)}(\mathbf{q})$ as the operators of particle creation and annihilation in the state l with the kinetic momentum $\mathbf{q}$ and the polarization ζ. Accordingly, $a^{+}_{l,\zeta,(i)}(\mathbf{q})$ and $\mathring{a}^{-}_{l,\zeta,(i)}(\mathbf{q})$ are the operators of antiparticle creation and annihilation in the state l with the kinetic momentum $\mathbf{q}$ and the polarization ζ. Equation (51) for $\mathcal{P}^{(i)}_\beta$ and the invariance under the change of particles to antiparticles yields the canonical commutation relations

$$\left[a^{-}_{l,\zeta,(i)}(\mathbf{q}), \mathring{a}^{+}_{k,\zeta',(i)}(\mathbf{q}') \right]_+ = \delta_{lk}\delta_{\zeta\zeta'}\delta(\mathbf{q}-\mathbf{q}'), \quad \left[\mathring{a}^{-}_{l,\zeta,(i)}(\mathbf{q}), a^{+}_{k,\zeta',(i)}(\mathbf{q}') \right]_+ = \delta_{lk}\delta_{\zeta\zeta'}\delta(\mathbf{q}-\mathbf{q}'). \tag{52}$$

Let us now consider linear combinations of these operators:

$$a^{\pm}_{\alpha,\zeta,(i)}(\mathbf{q}) = \sum_{l=1}^{3} U_{\alpha l} a^{\pm}_{l,\zeta,(i)}(\mathbf{q}), \quad \mathring{a}^{\pm}_{\alpha,\zeta,(i)}(\mathbf{q}) = \sum_{l=1}^{3} U^{*}_{\alpha l} \mathring{a}^{\pm}_{l,\zeta,(i)}(\mathbf{q}). \tag{53}$$

Commutation relations for $a^{\pm}_{\alpha,\zeta,(i)}(\mathbf{q})$, $\mathring{a}^{\pm}_{\alpha,\zeta,(i)}(\mathbf{q})$ are canonical. When expressed in terms of these operators, only $\mathcal{P}^{(i)}_\beta$ and $\mathcal{Q}^{(i)}$ are diagonal. However, using $\mathcal{Q}^{(i)}$ and the remaining integrals of motion, we can always construct three linear combinations of the form

$$X'^{(i)}_{\alpha\alpha} = \int d\mathbf{q} \left[\mathring{a}^{+}_{\alpha,\zeta,(i)}(\mathbf{q}) a^{-}_{\alpha,\zeta,(i)}(\mathbf{q}) + \mathring{a}^{-}_{\alpha,\zeta,(i)}(\mathbf{q}) a^{+}_{\alpha,\zeta,(i)}(\mathbf{q}) \right], \quad \alpha = 1, 2, 3. \tag{54}$$

So, the operators $a^{\pm}_{\alpha,\zeta,(i)}(\mathbf{q})$, $\mathring{a}^{\pm}_{\alpha,\zeta,(i)}(\mathbf{q})$ lead to well-defined states in the Fock space. The difference between the number of particles and the number of antiparticles of each type α with the same kinetic momentum $\mathbf{q}$ is an integral of motion.

Consider now the basis (32). We write the components of the field functions as

$$\Psi_s^{(i)}(x) = \frac{1}{(2\pi)^{3/2}} \sum_{l=1}^{3} \sum_{\zeta=\pm 1} \int \frac{d\mathbf{p}}{\sqrt{2p_l^0}} n_s^{(l)} \left[e^{-i(px)} u_{p,\zeta}^{(m_l^{(i)})} \mathfrak{a}_{l,\zeta,(i)}^{-}(\mathbf{p}) + e^{i(px)} u_{-p,\zeta}^{(m_l^{(i)})} \mathfrak{a}_{l,\zeta,(i)}^{+}(\mathbf{p}) \right],$$
$$\bar{\Psi}_s^{(i)}(x) = \frac{1}{(2\pi)^{3/2}} \sum_{l=1}^{3} \sum_{\zeta=\pm 1} \int \frac{d\mathbf{p}}{\sqrt{2p_l^0}} \overset{*}{n}_s^{(l)} \left[e^{-i(px)} \bar{u}_{-p,\zeta}^{(m_l^{(i)})} \overset{*}{\mathfrak{a}}{}^{-}_{l,\zeta,(i)}(\mathbf{p}) + e^{i(px)} \bar{u}_{p,\zeta}^{(m_l^{(i)})} \overset{*}{\mathfrak{a}}{}^{+}_{l,\zeta,(i)}(\mathbf{p}) \right]. \tag{55}$$

The operators $\mathfrak{a}_{l,\zeta,(i)}^{\pm}(\mathbf{p})$, $\overset{*}{\mathfrak{a}}{}^{\pm}_{l,\zeta,(i)}(\mathbf{p})$ arise as a result of scaling transformation

$$\mathfrak{a}_{l,\zeta,(i)}^{\pm}(\mathbf{p}) = (\mu_l^{(i)})^{-3/2} \mathfrak{a}_{l,\zeta,(i)}^{\pm}(\mathbf{q}),\ \overset{*}{\mathfrak{a}}{}^{\pm}_{l,\zeta,(i)}(\mathbf{p}) = (\mu_l^{(i)})^{-3/2} \overset{*}{\mathfrak{a}}{}^{\pm}_{l,\zeta}(\mathbf{q});\ \mathbf{q} = \mathbf{p}/\mu_l^{(i)}. \tag{56}$$

Therefore, using Eq. (51) for $\mathcal{P}_\beta^{(i)}$ and the invariance condition with respect to the change of particles to antiparticles we get that these operators as well as their linear combinations

$$\mathfrak{a}_{\alpha,\zeta,(i)}^{\pm}(\mathbf{p}) = \sum_{l=1}^{3} U_{\alpha l} \mathfrak{a}_{l,\zeta,(i)}^{\pm}(\mathbf{p}),\quad \overset{*}{\mathfrak{a}}{}^{\pm}_{\alpha,\zeta,(i)}(\mathbf{p}) = \sum_{l=1}^{3} U_{\alpha l}^{*} \overset{*}{\mathfrak{a}}{}^{\pm}_{l,\zeta,(i)}(\mathbf{p}) \tag{57}$$

satisfy the canonical commutation relations (52).

A similar reasoning shows that the state described by the operators $\mathfrak{a}_{l,\zeta,(i)}^{\pm}(\mathbf{p})$, $\overset{*}{\mathfrak{a}}{}^{\pm}_{l,\zeta,(i)}(\mathbf{p})$ and $\mathfrak{a}_{\alpha,\zeta,(i)}^{\pm}(\mathbf{p})$, $\overset{*}{\mathfrak{a}}{}^{\pm}_{\alpha,\zeta,(i)}(\mathbf{p})$ are well-defined in the Fock space too. However, there is an important difference. The integral of motion $\mathcal{P}_\beta^{(i)}$ expressed in terms of the operators $\mathfrak{a}_{\alpha,\zeta,(i)}^{\pm}(\mathbf{p})$, $\overset{*}{\mathfrak{a}}{}^{\pm}_{\alpha,\zeta,(i)}(\mathbf{p})$ is non-diagonal. This situation is quite expected. The integral of motion $\mathcal{P}_\beta^{(i)}$ is not the canonical momentum of the field, but the "kinetic momentum". Since the superpositions of the mass states are non-stationary states, the non-diagonal form of $\mathcal{P}_\beta^{(i)}$ reflects the fact of a possible energy transfer from one state to another.

The $SU(3)$-symmetry of the action defined by the interaction Lagrangian (43) is broken even if we will admit that the eigenvectors of the matrices $\mathbb{V}^{(i)}$ are equal. Therefore, the processes of the type $\mu^{\pm} \to e^{\pm} + \gamma$ are forbidden.

If the eigenvectors of the matrices $\mathbb{V}^{(i)}$ are distinct, the terms describing the charged currents automatically generate a phenomenon, which is known as the mixing of generations. The matrix of the mixing coefficients for quarks is an analogue to the Cabibbo–Kobayashi–Maskawa (CKM) matrix

$$\mathbb{U}^{\mathrm{CKM}} = \mathbb{V}^{(u)\dagger} \mathbb{V}^{(d)}, \tag{58}$$

and for leptons it is an analogue to the Pontecorvo–Maki–Nakagawa–Sakata (PMNS) matrix

$$\mathbb{U}^{\mathrm{PMNS}} = \mathbb{V}^{(\nu)\dagger} \mathbb{V}^{(e)}. \tag{59}$$

In experiment, as it is well known, the generation mixing occurs for quarks, while the transfer of energy from one neutrino state to another is seen as the oscillation phenomenon.

4 Conclusion

We put forward a modification of the electroweak interaction theory, in which the fermions with the same electroweak quantum numbers are placed in fermion multiplets and are treated as different quantum states of a single particle. That is, in describing the electroweak interactions it is possible to use four fundamental fermions only. In this model, the mixing and oscillations of the particles arise as a direct consequence of the general principles of quantum field theory.

The developed approach enables one to calculate the probabilities of the processes taking place in the detector at long distances from the particle source. Calculations of higher order processes including the computation of the contributions due to radiative corrections can be performed in the framework of perturbation theory using the regular diagram technique.

The author is grateful to A.B. Arbuzov, A.V. Bolsinov, A.V. Borisov, A.D. Dolgov, D.V. Gal'tsov, A.V. Kartavtsev, E.M. Murchikova, A.A. Polosin, I.P. Volobuev, A.E. Shabad and V.Ch. Zhukovsky for fruitful discussions.

References

[1] S. L. Glashow, Nucl. Phys. **22**, 579 (1961); S. Weinberg, Phys. Rev. Lett. **19**, 1264 (1967); A. Salam, *Elementary Particle Theory*, ed. N. Svartholm, (Almqvist and Wiksells, Stockholm, 1968) p. 137

[2] F. Englert, R. Brout, Phys. Rev. Lett. **13**, 321 (1964); P. W. Higgs, Phys. Rev. Lett. **13**, 508 (1964); G. S. Guralnik, C. R. Hagen, T. W. B. Kibble, Phys. Rev. Lett. **13**, 585 (1964)

[3] N. Cabibbo, Phys. Rev. Lett. **10**, 531 (1963); M. Kobayashi, T. Maskawa, Prog. Theor. Phys. **49**, 652 (1973)

[4] K. Olive *et al.* (Particle Data Group), Chin. Phys. C **38**, 090001 (2014)

[5] B. Pontecorvo, Zh. Eksp. Teor. Fiz. **33**, 549(1957) [Sov. Phys. JETP **6**, 429(1958)]

[6] Z. Maki, M. Nakagawa, S. Sakata, Prog. Theor. Phys. **28**, 870 (1962)

[7] Chiu Man Ho, JHEP **12**, 022 (2012).

[8] E. P. Wigner, Ann. Math. (2) **40**, 149 (1939)

[9] N.N. Bogolubov, A.A. Logunov, A.I. Oksak, I.T. Todorov, *General Principlels of Quantum Field Theory* (Kluwer Academic Publishers, Dordrecht, 1990)

[10] L. Michel, Phys. Rev. **137**, B405 (1965); S. Coleman, Phys. Rev. **138**, B1262 (1965); S. Weinberg, Phys. Rev. **139**, B597 (1965)

[11] R. Jost, Helv. Phys. Acta **39**, 369 (1966); L. O'Raifeartaigh, Phys. Rev. Lett. **14**, 575 (1965); Phys. Rev. **161**, 1571 (1967)

[12] A.O. Barut, R. Raczka, *Theory of Group Representations and Applications* (World Scientific, Singapore, 1986)

[13] N.N. Bogoliubov, D.V. Shirkov, *Introduction to Theory of Quantized Fields* (John Wiley, New York, 1979)

Relaxation of the chiral imbalance in dense matter of a neutron star

Maxim Dvornikov[1,2,3,★]

[1] *Pushkov Institute of Terrestrial Magnetism, Ionosphere and Radiowave Propagation (IZMIRAN), Kaluzhskoe HWY 4, 142190 Troitsk, Moscow, Russia*
[2] *Physics Faculty, National Research Tomsk State University, 36 Lenin Avenue, 634050 Tomsk, Russia*
[3] *University of Hamburg, II Institute for Theoretical Physics, 149 Luruper Chaussee, D-22761 Hamburg, Germany*

Abstract. Using the quantum field theory methods, we calculate the helicity flip of an electron scattering off protons in dense matter of a neutron star. The influence of the electroweak interaction between electrons and background nucleons on the helicity flip is examined. We also derive the kinetic equation for the chiral imbalance. The derived kinetic equation is compared with the results obtained by other authors.

1 Introduction

Some neutron stars (NS) can possess extremely strong magnetic fields $B \gtrsim 10^{15}$ G. These NSs are called magnetars [1]. Despite long observational history of magnetars and numerous theoretical models for the generation of their magnetic fields, nowadays there is no commonly accepted mechanism explaining the origin of magnetic fields in these compact stars.

In Refs. [2–4] we developed the new model for the generation of magnetic fields in magnetars. In this model a large-scale magnetic field in NS can be amplified to the strenth predicted in magnetars owing to the magnetic field instability driven by the electron-nucleon (*eN*) electroweak interaction. Using the proposed model we could account for some properties of magnetars, e.g., ages of these compact stars. Despite the plausibility of the model in Refs. [2–4], some of its ingredients should be substantiated by more detailed calculations based on reliable methods of the quantum field theory (QFT). This work is devoted to the further development of the proposed description of the magnetic fields generation in magnetars. In particular, we will be interested in the evolution of the chemical potentials of the electron gas in NS in the presence of background nucleons.

2 Electron-proton collisions in dense plasma

In NS, the helicity of a massive electron can be changed in electron-proton (*ep*) and electron-electron electromagnetic scatterings as well as in the interaction of an electron with the anomalous magnetic moment of a neutron. As found in Ref. [5], the rate of the former reaction in dense matter of NS is higher than that of the later ones. Therefore, in our analysis, we shall account for only *ep* collisions.

★e-mail: maxdvo@izmiran.ru

2.1 Helicity flip rate in ep collisions

The matrix element for the ep collision, due to the electromagnetic interaction, has the form,

$$\mathcal{M} = \frac{\mathrm{i}e^2}{(k_1 - k_2)^2}\bar{e}(p_2)\gamma^\mu e(p_1) \cdot \bar{p}(k_2)\gamma_\mu p(k_1), \tag{1}$$

where $e > 0$ is the absolute value of the electron charge, $\gamma^\mu = \left(\gamma^0, \boldsymbol{\gamma}\right)$ are the Dirac matrices, $p^\mu_{1,2} = (E_{1,2}, \mathbf{p}_{1,2})$ and $k^\mu_{1,2} = (\mathcal{E}_{1,2}, \mathbf{k}_{1,2})$ are the four momenta of electrons and protons. The momenta of incoming particles are marked with the label 1 and that of the outgoing particles with the label 2.

We shall consider a process when a proton is in the unpolarized states before and after the collision whereas an electron changes its polarization in the scattering. The square of the matrix element in Eq. (1) reads

$$|\mathcal{M}|^2 = \frac{e^4}{(k_1 - k_2)^4}\mathrm{tr}\left[\rho_p(k_1)\gamma^\mu\rho_p(k_2)\gamma^\nu\right] \cdot \mathrm{tr}\left[\rho_e(p_1)\gamma_\mu\rho_e(p_2)\gamma_\nu\right], \tag{2}$$

where the spin density matrices of protons $\rho_p(k_{1,2})$ and electrons $\rho_e(p_{1,2})$ are [6, pp. 106–111]

$$\rho_p(k_{1,2}) = \frac{1}{2}\left(\gamma_\mu k^\mu_{1,2} + M\right), \quad \rho_e(p_{1,2}) = \frac{1}{2}\left(\gamma_\mu p^\mu_{1,2} + m\right)\left(1 + \gamma^5\gamma_\mu a^\mu_{1,2}\right), \tag{3}$$

where M and m are the masses of a proton and an electron, and $\gamma^5 = \mathrm{i}\gamma^0\gamma^1\gamma^2\gamma^3$. The polarization vector of an electron a^μ in Eq. (3) has the form,

$$a^\mu = \left(\frac{(\boldsymbol{\zeta} \cdot \mathbf{p})}{m}, \boldsymbol{\zeta} + \frac{\mathbf{p}\,(\boldsymbol{\zeta} \cdot \mathbf{p})}{m(E + m)}\right). \tag{4}$$

where $\boldsymbol{\zeta}$ is the invariant three vector of the particle polarization, which describes the polarization in the particle rest frame. We shall suppose that electrons are in pure spin states with $\zeta^2 = 1$.

Let us fix the polarizations of an electron. If we study the $R \to L$ transition, i.e. we take that $\boldsymbol{\zeta}_1 = \mathbf{n}_1$ and $\boldsymbol{\zeta}_2 = -\mathbf{n}_2$, one gets that $a^\mu_{1,2} = (\pm p_{1,2}, \pm E_{1,2}\mathbf{n}_{1,2})/m$, where $\mathbf{n}_{1,2} = \mathbf{p}_{1,2}/p_{1,2}$ are the unit vectors towards the electron momenta and $p_{1,2} = |\mathbf{p}_{1,2}|$. Using the approximation of the elastic ep scattering, i.e. assuming that $E_1 = E_2$, and keeping only the leading term in the electron mass m in the computation of the traces of the Dirac matrices in Eq. (2), one gets the matrix element squared in the form,

$$|\mathcal{M}|^2 = 2e^4m^2\left[1 - (\mathbf{n}_1 \cdot \mathbf{n}_2)\right]\frac{\left[\mathcal{E}_1\mathcal{E}_2 + M^2 + (\mathbf{k}_1 \cdot \mathbf{k}_2)\right]}{\left[(\mathcal{E}_1 - \mathcal{E}_2)^2 - (\mathbf{k}_1 - \mathbf{k}_2)^2\right]^2}, \tag{5}$$

which will be used in the following calculations.

The total probability of the ep collision in the ep plasma has the form [6, pp. 247–252],

$$\begin{aligned} W = &\frac{V}{8(2\pi)^8}\int\frac{\mathrm{d}^3p_1\mathrm{d}^3p_2\mathrm{d}^3k_1\mathrm{d}^3k_2}{E_1E_2\mathcal{E}_1\mathcal{E}_2}\delta^4\left(p_1 + k_1 - p_2 - k_2\right)|\mathcal{M}|^2 \\ &\times f_e(E_1 - \mu_\mathrm{R})\left[1 - f_e(E_2 - \mu_\mathrm{L})\right]f_p(\mathcal{E}_1 - \mu_p)\left[1 - f_p(\mathcal{E}_2 - \mu_p)\right], \end{aligned} \tag{6}$$

where we summed over the polarizations of the outgoing proton. Here $f_{e,p}(E) = \left[\exp(\beta E) + 1\right]^{-1}$ are the Fermi-Dirac distributions of electrons and protons, μ_p is the chemical potential of protons, and V is the normalization volume. In Eq. (6) we assume that incoming and outgoing electrons have different chemical potentials: μ_R and μ_L respectively. Protons and electrons are taken to be in the thermal equilibrium having the same temperature T.

Assuming that electrons are highly degenerate and ultrarelativistic, we can calculate the integrals over the electron momenta,

$$I_e = \int \frac{d^3p_1 d^3p_2}{2E_1 2E_2} \frac{(p_1 \cdot p_2)}{E_1 E_2} \delta^4 (p_1 - p_2 - q) \theta(\mu_R - E_1) \theta(E_2 - \mu_L), \tag{7}$$

where $q^\mu = k_2^\mu - k_1^\mu$ and and $(p_1 \cdot p_2) = E_1 E_2 [1 - (\mathbf{n}_1 \cdot \mathbf{n}_2)]$. Using the identity

$$\int \frac{d^3p}{2E} = \int d^4 p \theta(p_0) \delta(p^2), \tag{8}$$

and integrating over p_2 we can obtain that

$$I_e = -\int \frac{d^3p}{2E} \frac{Eq_0 - (\mathbf{p} \cdot \mathbf{q})}{E(E - q_0)} \delta\left(2Eq_0 - 2(\mathbf{p} \cdot \mathbf{q}) - q^2\right) \theta(\mu_R - E) \theta(E - q_0 - \mu_L). \tag{9}$$

Here we recall that we use the approximation of the elastic scattering, i.e. $E_1 = E_2$. Hence we should set $q_0 = \mathcal{E}_2 - \mathcal{E}_1 = 0$ in Eq. (9). Then we represent $d^3p = 2\pi d\cos\theta E^2 dE$ and integrate over $\cos\theta$ using the remaining delta function

$$\delta\left(2E|\mathbf{q}|\cos\theta - \mathbf{q}^2\right) = \frac{1}{2E|\mathbf{q}|} \delta\left(\cos\theta - \frac{|\mathbf{q}|}{2E}\right). \tag{10}$$

Finally, we obtain

$$I_e = \frac{\pi|\mathbf{q}|}{4} \int_0^\infty \frac{dE}{E^2} \theta(\mu_R - E) \theta(E - \mu_L) \approx \frac{\pi|\mathbf{q}|}{4} \frac{\mu_R - \mu_L}{\mu_e^2}, \tag{11}$$

where we take that $\mu_e \approx \mu_R \approx \mu_L$ is the mean chemical potential of the electron gas. Moreover we assume that $\mu_R > \mu_L$ for the total probability to be positive.

Using Eqs. (5) and (11), one gets that the total probability in Eq. (6) takes the form,

$$\begin{aligned} W = & \frac{V\pi e^4 m^2}{4(2\pi)^8} \frac{\mu_R - \mu_L}{\mu_e^2} \int \frac{d^3k_1 d^3k_2}{\mathcal{E}_1 \mathcal{E}_2} \frac{\mathcal{E}_1 \mathcal{E}_2 + M^2 + (\mathbf{k}_1 \cdot \mathbf{k}_2)}{\left[(\mathbf{k}_1 - \mathbf{k}_2)^2 + \omega_p^2\right]^{3/2}} \\ & \times f_p(\mathcal{E}_1 - \mu_p)\left[1 - f_p(\mathcal{E}_2 - \mu_p)\right], \end{aligned} \tag{12}$$

Here, in the denominator, we replace $(\mathbf{k}_1 - \mathbf{k}_2)^2 \to (\mathbf{k}_1 - \mathbf{k}_2)^2 + \omega_p^2$, where ω_p is the plasma frequency in the ep plasma, to avoid the infrared divergencies. Now one can compute the integrals over the proton momenta,

$$\begin{aligned} I_p = & \int \frac{d^3k_1 d^3k_2}{\mathcal{E}_1 \mathcal{E}_2} \frac{\mathcal{E}_1 \mathcal{E}_2 + M^2 + (\mathbf{k}_1 \cdot \mathbf{k}_2)}{\left[(\mathbf{k}_1 - \mathbf{k}_2)^2 + \omega_p^2\right]^{3/2}} f_p(\mathcal{E}_1 - \mu_p)\left[1 - f_p(\mathcal{E}_2 - \mu_p)\right] \\ \approx & \, 32\sqrt{2\tilde{\mu}_p} M^{3/2} \pi^2 T J, \end{aligned} \tag{13}$$

where

$$J = \int_0^1 dx \frac{x^2\sqrt{1 - x^2}}{(x^2 + L^{-2})^{3/2}} = \frac{1}{\sqrt{1 + L^2}} \boldsymbol{K}\left(\frac{L}{\sqrt{1 + L^2}}\right)\left[L + \frac{2}{L}\right] - \frac{2}{L} \boldsymbol{E}\left(\frac{L}{\sqrt{1 + L^2}}\right)\sqrt{1 + L^2}. \tag{14}$$

Here $L^2 = 8M\tilde{\mu}_p/\omega_p^2$, $\boldsymbol{K}(z)$ and $\boldsymbol{E}(z)$ are the complete elliptic integrals, and $\tilde{\mu}_p = \mu_p - M$ is the nonrelativistic part of the protons chemical potential. To derive Eq. (13) we assume that protons are nonrelativistic and the scattering is elastic. Moreover, the proton gas is supposed to have a small nonzero temperature.

Assuming that in a degenerate plasma one has $\omega_p^2 = 4\alpha_{\mathrm{em}}\mu_e^2/3\pi$ (see, e.g., Ref. [7]), where $\alpha_{\mathrm{em}} = e^2/4\pi$ is the fine structure constant, and $\tilde{\mu}_p \approx \mu_e^2/2M$ due to the electroneutrality of the NS matter, we get that $L^2 = 3\pi/\alpha_{\mathrm{em}} \approx 1291 \gg 1$. Decomposing J in Eq. (14) in this limit, one obtains

$$J = \ln 4L - 2 + O\left(L^{-1}\right) = \frac{1}{2}\left[\ln\left(\frac{48\pi}{\alpha_{\mathrm{em}}}\right) - 4\right] > 0. \tag{15}$$

Finally, using Eqs. (12)-(15), we get the total probability for the helicity flip in an ep collision,

$$W = W_0\left(\mu_{\mathrm{R}} - \mu_{\mathrm{L}}\right)\theta\left(\mu_{\mathrm{R}} - \mu_{\mathrm{L}}\right), \quad W_0 = \frac{Ve^4}{32\pi^5}\frac{m^2 M}{\mu_e} T\left[\ln(16L^2) - 4\right]. \tag{16}$$

Note that the total probability is always positive. That is why we introduce the step function $\theta\left(\mu_{\mathrm{R}} - \mu_{\mathrm{L}}\right)$ in Eq. (16). In the next section we shall compare Eq. (16) with the analogous expression accounting for the electroweak interaction between electrons and neutrons.

2.2 Helicity flip rate of electrons, electroweakly interacting with nuclear matter

If we take into account the electroweak interaction between electrons and nucleons, the matrix element for the ep scattering has the same form as in Eq. (1). However, instead of the electron spinors in vacuum, we should use the exact solutions of the Dirac equation for an electron, interacting with a background matter, found in Appendix A.

We shall start with the analysis of $R \to L$ transitions. According to Eq. (1) it is necessary to compute the following quantity:

$$J^{\mu} = (J_0, \boldsymbol{J}) = \bar{u}_{-}(p_2)\gamma^{\mu}u_{+}(p_1). \tag{17}$$

Using Eqs. (31) and (33) one gets

$$J_0 = -\frac{mP_0\left[p_1 + p_2 + E_{+}(p_1) + E_{-}(p_2) - 2\bar{V}\right]}{2\sqrt{E_{0+}(p_1)E_{0-}(p_2)\left[E_{-}(p_2) + p_2 - V_{\mathrm{R}}\right]\left[E_{+}(p_1) + p_1 - V_{\mathrm{L}}\right]}},$$
$$\boldsymbol{J} = -\frac{m\mathbf{P}\left[p_1 - p_2 + E_{+}(p_1) - E_{-}(p_2) - 2V_5\right]}{2\sqrt{E_{0+}(p_1)E_{0-}(p_2)\left[E_{-}(p_2) + p_2 - V_{\mathrm{R}}\right]\left[E_{+}(p_1) + p_1 - V_{\mathrm{L}}\right]}}, \tag{18}$$

where

$$P_0 = w_{-}^{\dagger}(\mathbf{p}_2)w_{+}(\mathbf{p}_1), \quad \mathbf{P} = w_{-}^{\dagger}(\mathbf{p}_2)\boldsymbol{\sigma}w_{+}(\mathbf{p}_1). \tag{19}$$

Here $w_{\pm}$ are the two component spinors corresponding to different helicities, which are defined in Ref. [6, p. 86], $\boldsymbol{\sigma}$ are the Pauli matrices, $V_{\mathrm{L,R}}$ are the effective potentials of the interaction of left and right chiral projections of an electron field with the background matter, defined in Eq. (29), $\bar{V} = \left(V_{\mathrm{L}} + V_{\mathrm{R}}\right)/2$, and $V_5 = \left(V_{\mathrm{L}} - V_{\mathrm{R}}\right)/2$. To obtain Eq. (18) we use the Dirac matrices in the chiral representation [8, pp. 691–696].

For ultrarelativistic electrons one obtains from Eq. (30) that $E_{\pm}(p_{1,2}) = p_{1,2} + V_{\mathrm{R,L}}$. As in Sec. 2.1, here we also study the elastic scattering, in which colliding particles do not transfer to exited states, i.e. $E_{+}(p_1) = E_{-}(p_2)$ and $\mathcal{E}_1 = \mathcal{E}_2$. The equality of energies of incoming and outgoing ultrarelativistic electrons is equivalent to $p_1 - p_2 = 2V_5$. Thus, one obtains that $J = 0$. The computation of P_0 in Eq. (19) gives $|P_0|^2 = [1 - (\mathbf{n}_1 \cdot \mathbf{n}_2)]/2$. Thus, using Eq. (18), one gets that the square of matrix element in Eq. (1) has the form,

$$|\mathcal{M}|^2 = e^4 m^2 \frac{(p_1 + p_2)^2 [1 - (\mathbf{n}_1 \cdot \mathbf{n}_2)]}{8(p_1 - V_5)^2 (p_2 + V_5)^2} \frac{\left[\mathcal{E}_1 \mathcal{E}_2 + M^2 + (\mathbf{k}_1 \cdot \mathbf{k}_2)\right]}{\left[(\mathcal{E}_1 - \mathcal{E}_2)^2 - (\mathbf{k}_1 - \mathbf{k}_2)^2\right]^2}, \tag{20}$$

where we keep only the leading term in the electron mass.

The total probability of the process has the form,

$$\begin{aligned} W = & \frac{V}{2(2\pi)^8} \int \frac{\mathrm{d}^3 p_1 \mathrm{d}^3 p_2 \mathrm{d}^3 k_1 \mathrm{d}^3 k_2}{\mathcal{E}_1 \mathcal{E}_2} \delta^4 (p_1 + k_1 - p_2 - k_2) |\mathcal{M}|^2 \\ & \times f_e(E_1 - \mu_\mathrm{R}) [1 - f_e(E_2 - \mu_\mathrm{L})] f_p(\mathcal{E}_1 - \mu_p) \left[1 - f_p(\mathcal{E}_2 - \mu_p)\right]. \end{aligned} \tag{21}$$

Note that Eq. (21) is slightly different from Eq. (6). It can be accounted for by the different way to normalize electron basis spinors; cf. Eq. (32).

As in Sec. 2.1, first, in Eq. (21) we compute the integrals over the electron momenta,

$$\begin{aligned} I_e = & \int \frac{\mathrm{d}^3 p_1 \mathrm{d}^3 p_2 (p_1 + p_2)^2 [1 - (\mathbf{n}_1 \cdot \mathbf{n}_2)]}{16 (p_1 - V_5)^2 (p_2 + V_5)^2} \\ & \times \delta^4 (p_1 - p_2 - q) \theta (\mu_\mathrm{R} - p_1 - V_\mathrm{R}) \theta (p_2 + V_\mathrm{L} - \mu_\mathrm{L}), \end{aligned} \tag{22}$$

where we suppose that incoming and outgoing electrons are ultrarelativistic, highly degenerate, and have different chemical potentials. We remind that we study $R \to L$ transitions. After the integration over $\mathbf{p}_2$ with help of the delta function, one obtains that

$$\begin{aligned} I_e = & \int \frac{\mathrm{d}^3 p (p + |\mathbf{p} - \mathbf{q}|)^2}{16 (p - V_5)^2 (|\mathbf{p} - \mathbf{q}| + V_5)^2} \left[1 - \frac{(\mathbf{p} \cdot \mathbf{p} - \mathbf{q})}{p |\mathbf{p} - \mathbf{q}|}\right] \\ & \times \delta (p - 2V_5 - |\mathbf{p} - \mathbf{q}|) \theta (\mu_\mathrm{R} - p - V_\mathrm{R}) \theta (|\mathbf{p} - \mathbf{q}| + V_\mathrm{L} - \mu_\mathrm{L}). \end{aligned} \tag{23}$$

Then we represent $\mathrm{d}^3 p = 2\pi \mathrm{d} \cos\theta p^2 \mathrm{d}p$ and integrate over $\cos\theta$ using the remaining delta function

$$\delta (p - 2V_5 - |\mathbf{p} - \mathbf{q}|) = \frac{p - 2V_5}{p |\mathbf{q}|} \delta \left(\cos\theta - \frac{|\mathbf{q}|^2 + 4pV_5 - V_5^2}{2p |\mathbf{q}|} \right). \tag{24}$$

Finally we get the following expression for I_e:

$$\begin{aligned} I_e = & \pi \frac{|\mathbf{q}|^2 - V_5^2}{4|\mathbf{q}|} \int_{\mu_\mathrm{L} - V_\mathrm{R}}^{\mu_\mathrm{R} - V_\mathrm{R}} \frac{\mathrm{d}p}{(p - V_5)^2} = \pi \frac{|\mathbf{q}|^2 - V_5^2}{4|\mathbf{q}|} \frac{(\mu_\mathrm{R} - \mu_\mathrm{L})}{\left(\mu_\mathrm{R} - \bar{V}\right)\left(\mu_\mathrm{L} - \bar{V}\right)} \\ \approx & \frac{\pi |\mathbf{q}|}{4} \frac{(\mu_\mathrm{R} - \mu_\mathrm{L})}{\mu_e^2}, \end{aligned} \tag{25}$$

where we suppose that $\mu_\mathrm{R,L} \gg \bar{V}$, $\mu_\mathrm{R} \approx \mu_\mathrm{L} \approx \mu_e$, and $|\mathbf{q}| \gg V_5$. The most important consequence of Eq. (25) consists in the fact that the eN interaction does not contribute to the difference of the chemical potentials in the numerator.

Eventually, on the basis of Eqs. (21) and (25), one gets the expression for the total probability which coincides with that in Eq. (12). Performing the integration over the protons momenta as in

Sec. 2.1, we obtain that the total probability for the helicity flip in an ep collision has the same form as in Eq. (16). If we study $L \to R$ transitions, making analogous calculations as in the $R \to L$ case, we derive the expression for W, which also coincides with that in Eq. (16). For the sake of brevity we omit these computations.

At the end of this section we shall analyze the approximations made to derive the total probability in the presence of the electroweak eN interaction. First, we assume that ep collisions are elastic. As found in Ref. [9, pp. 205–208], the inelastic contributions to the relativistic collision integral in case of the scattering due to the long range Coulomb interaction are suppressed compared to elastic ones. This situation takes place in our work.

The differential probability of the helicity flip (the analogue of the differential cross section) is $\mathrm{d}W \sim \mathrm{d}\chi/\chi$, where χ is the scattering angle. The maximal probability corresponds to $\chi \sim 0$. That is why we introduce $\omega_p \neq 0$ in Eq. (12) to regularize this infrared divergence. One can estimate $\chi_{\min}$ using Eq. (15) as $\chi_{\min} \sim \alpha_{\mathrm{em}}/48\pi = 4.8 \times 10^{-5}$. Note that the Coulomb cut-off parameter $L_C = \ln(\chi_{\min}^{-1}) \sim 10$ is typical for relativistic plasmas. The energy of an electron can change by $\Delta E = |E_1 - E_2| \sim \chi^2 \langle E \rangle$ in a collision, where $\langle E \rangle \sim 10^2\,\mathrm{MeV}$ is the mean electron energy. If we define χ_5 corresponding to $\Delta E \sim 2V_5$, i.e. when the inelastic effects become comparable with the electroweak interaction contribution, it reads, $\chi_5 = 3.2 \times 10^{-4}$. Thus, one gets that $\chi_5 \gg \chi_{\min}$, i.e. there will be many particles with a big contribution to $\mathrm{d}W$, for which the energy change in a collision is less than $2V_5$. Hence the elasticity is a good approximation even in the presence of the eN interaction. Moreover, the approximation of the elastic ep scattering was also used in Ref. [10], where Γ_f was computed in the situation when protons are nondegenerate. If inelastic effects are taken into account, it can result in some dependence of W_0 in Eq. (16) on $V_{\mathrm{L,R}}$. However, V_5 still will not contribute to the factor $\mu_{\mathrm{R}} - \mu_{\mathrm{L}}$ in Eq. (16).

Second, the proton energies in Eq. (22) are taken as in vacuum: $\mathcal{E}_{1,2} = \sqrt{k_{1,2}^2 + M^2}$, whereas for electrons we exactly account for the contribution of the electroweak interaction to the electron wave function in Eqs. (17) and (18) and and to the electron energies in Eq. (23). However, protons like electrons can electroweakly interact with background neutrons. To justify our choice of the proton energies, we define the analogues of $V_{\mathrm{L,R}}$ for protons and denote them as $V_{\mathrm{L,R}}^{(p)}$. Using Eq. (30), we can estimate the contribution of the electroweak pN interaction to the proton energies as $|\Delta(\mathcal{E}_{1,2})_{\mathrm{EW}}| \sim \bar{V}_p \mp k_{1,2} V_5^{(p)}/M$, where $\bar{V}_p = [V_{\mathrm{L}}^{(p)} + V_{\mathrm{R}}^{(p)}]/2$ and $V_5^{(p)} = [V_{\mathrm{L}}^{(p)} - V_{\mathrm{R}}^{(p)}]/2$. Now, in the energy conservation delta function in Eq. (22), one has $|\Delta(\mathcal{E}_1 - \mathcal{E}_2)_{\mathrm{EW}}| \lesssim p_{\mathrm{F_p}} V_5^{(p)}/M \sim 0.1 V_5$ since $p_{\mathrm{F_p}} \sim 10^2\,\mathrm{MeV}$ is the Fermi momentum for protons in NS, $M \sim 1\,\mathrm{GeV}$, and $V_5^{(p)} \sim V_5$. Analogously for electrons we have $|\Delta(E_1 - E_2)_{\mathrm{EW}}| = 2V_5$. Thus the contribution of the electroweak pN interaction to the conservation of energy is negligible compared to that of the eN interaction: $|\Delta(\mathcal{E}_1 - \mathcal{E}_2)_{\mathrm{EW}}| \ll |\Delta(E_1 - E_2)_{\mathrm{EW}}|$.

2.3 Kinetics of the chiral imbalance

Basing on Eq. (16), one gets the kinetic equations for the total numbers of right and left electrons $N_{\mathrm{R,L}}$ as

$$
\begin{aligned}
\frac{\mathrm{d}N_{\mathrm{R}}}{\mathrm{d}t} &= - W(R \to L) + W(L \to R) = -W_0 \left(\mu_{\mathrm{R}} - \mu_{\mathrm{L}}\right), \\
\frac{\mathrm{d}N_{\mathrm{L}}}{\mathrm{d}t} &= - W(L \to R) + W(R \to L) = -W_0 \left(\mu_{\mathrm{L}} - \mu_{\mathrm{R}}\right),
\end{aligned}
\tag{26}
$$

where $\mu_{\mathrm{R,L}}$ are the chemical potentials of right and left electrons. Let us introduce the number densities as $n_{\mathrm{R,L}} = N_{\mathrm{R,L}}/V$. Using the standard relation between $n_{\mathrm{R,L}}$ and $\mu_{\mathrm{R,L}}$ [2], we get that $\mathrm{d}\left(n_{\mathrm{R}} - n_{\mathrm{L}}\right)/\mathrm{d}t \approx$

$2\dot{\mu}_5\mu_e^2/\pi^2$, where $\mu_5 = (\mu_\mathrm{R} - \mu_\mathrm{L})/2$ is the chiral imbalance. Finally, we get the kinetic equation for μ_5,

$$\frac{\mathrm{d}\mu_5}{\mathrm{d}t} = -\Gamma_f \mu_5, \quad \Gamma_f = \frac{\alpha_\mathrm{em}^2}{\pi}\left[\ln\left(\frac{48\pi}{\alpha_\mathrm{em}}\right) - 4\right]\left(\frac{m}{\mu_e}\right)^2\left(\frac{M}{\mu_e}\right)T, \tag{27}$$

where we use Eqs. (16) and (26).

One can see in Eq. (27) that the helicity flipping term contains μ_5 rather than $\mu_5 + V_5$ as recently suggested in Ref. [11]. Note that this our result is based on the explicit QFT calculation of the total probability of the *ep* scattering, where we exactly account for the *eN* electroweak interaction.

We also mention that, in Eq. (27), we corrected the value of Γ_f used in Refs. [2–4]. The reason for the discrepancy of Γ_f consists in the fact that in Ref. [2] we relied on the results of Ref. [5], where the scattering of unpolarized electrons off protons was studied. However, in our case it is essential to have the fixed opposite polarizations of incoming and outgoing electrons. Hence, the matrix elements used in the present work and in Ref. [5] are different. This fact, explains, e.g., that Γ_f in Eq. (27) is linear in T whereas that used in Refs. [2–4] is proportional to T^2.

We also mention that Γ_f was recently calculated in Ref. [10]. The value of Γ_f obtained in Ref. [10] is independent of T since it was assumed that protons are nondegenerate. This assumption is valid when the early stages of the NS evolution are considered. In Refs. [2–4], we studied the magnetic field generation in a thermally relaxed NS at $t \gtrsim 10^2$ yr after the onset of the supernova collapse. At this time, the proton component of the NS matter should be taken as degenerate. Note that $\Gamma_f \sim \alpha_\mathrm{em}^2$ in Eq. (27) as in Ref. [10].

3 Discussion

The detailed computation of the helicity flip rate in *ep* collisions was the last missing ingredient of the new model, initially proposed in Ref. [2], for the generation of magnetic fields in magnetars driven by the parity violating *eN* electroweak interaction. Note that, in Refs. [2, 3], we estimated the helicity flip rate Γ_f basing on the classical plasma physics. However, the particle spin is a purely quantum object and its evolution should be treated appropriately. In the present work we used the QFT methods to compute the helicity flip rate. It can explain the discrepancy of our results from those in Refs. [2, 3].

The second important result obtained in the present work was the analysis of the influence of the electroweak interaction of colliding electrons with background nucleons on the helicity flip process. Using the approximation of the elastic scattering and assuming that electrons are ultrarelativistic, we have found that the kinetic equation for the chiral imbalance should have the same form as in Refs. [2, 3], see Eq. (27), contrary to the claim in Ref. [11]. It should be also noted that the computation of the anomalous current along the magnetic field made in Refs. [2, 3] and the calculation of the electron helicity flip in the present work imply the same definition of the chemical potentials $\mu_\mathrm{L,R}$; see, e.g., the distribution functions of left and right electrons in Eq. (21)-(23) and in Eq. (7) in Ref. [2]. Thus the results of the present work are consistent with Refs. [2–4].

I am thankful to the organizers of Quarks 2016 for the inivitation, to the Tomsk State University Competitiveness Improvement Program, as well as to RFBR (research project No. 15-02-00293) and DAAD (grant No. 91610946) for partial support.

A Solution of the Dirac equation for an electron, electroweakly interacting with nuclear matter

Let us consider the electroneutral matter in NS consisting of neutrons, protons, and electrons. This matter is supposed to be at rest and unpolarized. The Dirac equation for a test electron, described by the bispinor wave function ψ, electroweakly interacting with neutrons and protons, has the form,

$$\left[i\gamma^{\mu}\partial_{\mu} - m - \gamma^{0}\left(V_{\mathrm{L}}P_{\mathrm{L}} + V_{\mathrm{R}}P_{\mathrm{R}}\right)\right]\psi = 0, \tag{28}$$

where

$$V_{\mathrm{L}} = \frac{G_{\mathrm{F}}}{\sqrt{2}}\left[n_n - n_p(1-4\xi)\right](1-2\xi), \quad V_{\mathrm{R}} = -\frac{G_{\mathrm{F}}}{\sqrt{2}}\left[n_n - n_p(1-4\xi)\right]2\xi, \tag{29}$$

are the effective potentials of the interaction of left and right chiral projections, $n_{n,p}$ are the constant and uniform densities of neutrons and protons, $\xi = \sin^2\theta_{\mathrm{W}} \approx 0.23$ is the Weinberg parameter, and $P_{\mathrm{L,R}} = (1 \mp \gamma^5)/2$ are the chiral projectors.

We shall look for the solution of Eq. (28) in the form, $\psi = e^{-iEt+i\mathbf{pr}}u/\sqrt{V}$, where u is the basis spinor. The energy levels of an electron have the form,

$$E = \bar{V} + E_0, \quad \bar{V} = \frac{V_{\mathrm{L}} + V_{\mathrm{R}}}{2}, \quad E_0^2 = (p - sV_5)^2 + m^2, \quad V_5 = \frac{V_{\mathrm{L}} - V_{\mathrm{R}}}{2}, \tag{30}$$

where $s = \pm 1$. If electrons are ultrarelativistic, one gets from Eq. (30) that $E_{\pm} = p + V_{\mathrm{R,L}}$. Note that the energy levels in Eq. (30) coincide with those found in Ref. [12].

The basis spinors $u_{\pm}$ for different helicities can be also obtained in the explicit form,

$$u_{+} = N_{+}\begin{pmatrix} w_{+} \\ -\frac{m}{E_{+}+p-V_{\mathrm{L}}}w_{+} \end{pmatrix}, \quad u_{-} = N_{-}\begin{pmatrix} -\frac{m}{E_{-}+p-V_{\mathrm{R}}}w_{-} \\ w_{-} \end{pmatrix}, \tag{31}$$

where $w_{\pm}$ are the basis two component spinors which are given in Ref. [6, p. 86]. The spinors $w_{\pm}$ are the eigenvectors of the helicity operator: $(\boldsymbol{\sigma}\cdot\mathbf{p})\,w_{\pm} = \pm|\mathbf{p}|w_{\pm}$. Eq. (31) implies the chiral representation for the Dirac matrices [8, pp. 691–696].

If we normalize the electron wave function as

$$\int d^3x\psi^{\dagger}\psi = 1, \tag{32}$$

we can find the normalization constants $N_{\pm}$ in Eq. (31) in the form,

$$N_{\pm} = \sqrt{\frac{E_{\pm} + p - V_{\mathrm{L,R}}}{2E_{0\pm}}}, \tag{33}$$

where $E_{\pm}$ and $E_{0\pm}$ are given in Eq. (30).

References

[1] R. Turolla, S. Zane, and A. L. Watts, Rep. Prog. Phys. **78**, 116901 (2015).
[2] M. Dvornikov and V. B. Semikoz, Phys. Rev. D **91**, 061301 (2015).
[3] M. Dvornikov and V. B. Semikoz, J. Cosmol. Astropart. Phys. 05 (2015) 032.
[4] M. Dvornikov and V. B. Semikoz, Phys. Rev. D **92**, 083007 (2015).
[5] D. C. Kelly, Astrophys. J. **179**, 599 (1973).
[6] V. B. Berestetskii, E. M. Lifschitz, and L. P. Pitaevskii, *Quantum Electrodynamics*, 2nd ed. (Pergamon, Oxford, 1982).
[7] E. Braaten and D. Segel, Phys. Rev. D **48**, 1478 (1993).
[8] C. Itzykson and J.-B. Zuber, *Quantum field theory* (McGraw-Hill, New York, 1980).
[9] E. M. Lifschitz and L. P. Pitaevskii, *Physical Kinetics* (Pergamon, Oxford, 1981).
[10] D. Grabowska, D. B. Kaplan, and S. Reddy, Phys. Rev. D **92**, 085035 (2015).
[11] G. Sigl and N. Leite, J. Cosmol. Astropart. Phys. 01 (2016) 025.
[12] A. Grigoriev, et al., Grav. Cosmol. **14**, 248 (2008).

Effective description of dark matter as a viscous fluid

Stefan Floerchinger[1], *Mathias* Garny[2], *Nikolaos* Tetradis[2,3,⋆], and *Urs Achim* Wiedemann[2]

[1] *Institut für Theoretische Physik, Universität Heidelberg, 69120 Heidelberg, Germany*
[2] *Theoretical Physics Department, CERN, CH-1211 Genève 23, Switzerland*
[3] *Department of Physics, University of Athens, Zographou 157 84, Greece*

Abstract. Treating dark matter at large scales as an effectively viscous fluid provides an improved framework for the calculation of the density and velocity power spectra compared to the standard assumption of an ideal pressureless fluid. We discuss how this framework can be made concrete through an appropriate coarse-graining procedure. We also review results that demonstrate that it improves the convergence of cosmological perturbation theory.

1 Introduction

The calculation of the power spectrum of cosmological fluctuations for stochastic initial conditions is an important and challenging task. It provides the basis for analyzing data on the large-scale structure of the universe and can lead to constraints on the parameters of the cosmological model. The spectrum can be computed through N-body simulations. However, these are CPU-intensive, while they do not provide an intuitive understanding of the underlying physical processes. The alternative option is an analytical treatment, which amounts to solving the collisionless Vlasov-Poisson equation perturbatively in the density contrast for appropriate classes of initial conditions [1]. The use of perturbative techniques in this context is complicated by the growth of non-linearities at short length scales. Within the strongly non-linear regime, velocity dispersion and virialization becomes important.

The inability of analytical methods to describe the short scales reliably suggests to reformulate cosmological perturbation theory based on an effective description, applicable only above some length scale. The description should contain effective parameters which absorb the effect of short-scale perturbations that are 'integrated out' [2–4]. In the conventional description of dark matter as a pressureless ideal fluid, only the lowest moments of the phase-space distribution are taken into account. Enlarging this framework by including higher moments results in viscous transport coefficients that parametrize a non-ideal stress tensor for the dark matter. In the following we present heuristic arguments that support the suggestion that an efficient fluid description of dark matter must allow for such non-ideal terms. The role of the effective viscosity and sound velocity is to account for the interaction of large-scale fluctuations (with wavenumbers $k < k_m$) with the short-scale ones ($k > k_m$) that are not followed explicitly in a fluid description limited to $k < k_m$. In our analysis the UV contributions are incorporated in the large-scale theory at the level of a one-loop approximation. Going beyond one-loop can be achieved through renormalization-group techniques [7, 8]. The approximation that

⋆e-mail: ntetrad@phys.uoa.gr

we review here provides an intuitive and rather simple framework, while it still resolves the main deficiency of standard perturbation theory (SPT), namely the strong dependence on the short-scale dynamics that are out of the reach of analytical techniques. We review the calculation of the density power spectrum presented in ref. [6] in the context of the coarse-grained theory. We also present results for the velocity and cross spectra within the same framework.

2 The relevant scales

Our approach to the problem of cosmological matter perturbations assumes the presence of two scales: the momentum scale k_Λ at which a fluid description becomes feasible, and the scale $k_m < k_\Lambda$ at which the description can be based on a small number of effective parameters. In practice we can take $k_\Lambda \sim 1-3\ h/\text{Mpc}$, corresponding to length scales $\sim 3-10$ Mpc, and $k_m \sim 0.5-1\ h/\text{Mpc}$, corresponding to length scales $\sim 10-20$ Mpc. At the scale k_m the density contrast is of order 1, while above the scale k_Λ it is much larger than 1 and the dynamics involves shell crossing, making a fluid description unfeasible. The description at k_m must include viscosity terms arising either through the coarse-graining of the scales $k > k_m$ (effective viscosity), or fundamental dark matter interactions (fundamental viscosity). The scales $k > k_\Lambda$ correspond to virialized structures, which are expected to have a negligible effect on long-distance dynamics [5]. It seems reasonable then to expect that the effective viscosity results mainly from the integration of the modes with $k_m < k < k_\Lambda$. The form of the linear power spectrum (which falls off roughly as k^{-3}) is consistent with this expectation: the effective viscosity at k_m is dominated by the modes with k slightly above this value, while the scales $k > k_\Lambda$ give a negligible contribution. This efficient decoupling of deep UV modes is also supported by numerical evidence from N-body simulations as well as analytical arguments based on non-perturbative relations of response functions [9, 10].

There are several reasons for introducing the scale k_m and distinguishing it from k_Λ:

- In the absence of dark-matter interactions, the form of the effective terms beyond the perfect-fluid description (such as effective viscosity) at the scale k_m is in a certain sense universal. They do not depend strongly on the initial conditions at k_Λ, but only on k_m. If fundamental interactions exist, the fundamental viscosity should vary only little between k_Λ and k_m, so that it can be estimated at k_m without loss of accuracy.
- The main role of the scale k_m is to act as an UV cutoff for perturbative corrections in the large-scale theory, thus eliminating contributions of dubious validity, arising from the modes near and above k_Λ. This effect is seen clearly in the good convergence of our results, in contrast to standard perturbation theory [6].
- The scale k_m can be lowered even further than the range that we are considering, if one would like to focus on the effective theory for length scales much larger than the range of baryon acoustic oscillations (BAO). In such a case, the renormalization-group improvement of the effective description is crucial [8].

3 Estimate of the effective viscosity

The validity of the fluid description for cold dark matter cannot be established similarly to systems close to thermal equilibrium. The crucial elements supporting this description are the small dark-matter velocity and the finite age of the universe, during which dark matter particles can drift over a finite distance, much smaller than the Hubble radius. A more quantitative discussion of this point is given in ref. [2]. In the absence of dark-matter interactions, one starts from the Vlasov-Poisson equations and writes the particle phase-space density as

$$f(\mathbf{x}, \mathbf{p}, \tau) = f_0(p)[1 + \delta_f(\mathbf{x}, p, \hat{\mathbf{p}}, \tau)], \tag{1}$$

with τ the conformal time. The Fourier modes of the perturbation are then expanded in terms of Legendre polynomials P_n:

$$\delta_f(\mathbf{k}, p, \hat{\mathbf{p}}, \tau) = \sum_{n=0}^{\infty}(-i)^n(2n+1)\,\delta_f^{[n]}(\mathbf{k}, p, \tau)\,P_n(\hat{\mathbf{k}} \cdot \hat{\mathbf{p}}), \tag{2}$$

with unit vectors $\hat{\mathbf{k}}$, $\hat{\mathbf{p}}$. The Vlasov-Boltzmann equation implies a hierarchy of evolution equations for the higher moments:

$$\frac{d\delta_f^{[n]}}{d\tau} = kv_p\left[\frac{n+1}{2n+1}\delta_f^{[n+1]} - \frac{n}{2n+1}\delta_f^{[n-1]}\right], \qquad n \geq 2, \tag{3}$$

with $v_p = p/am$ the particle velocity. If the typical value of v_p is sufficiently small, the higher moments can be neglected.

A quantitative constraint can be obtained if one assumes that the particle velocity is of the order of the fluid velocity v at small length scales. We emphasize that this assumption is valid only for cold dark matter, for which the particle velocity is induced by the gravitational fields generated by the growth of large-scale structure. The assumption is supported by the estimated typical velocities of dark matter particles in galaxy haloes.[1] At the comoving momentum scale k we expect $[\theta/\mathcal{H}]^2 \sim k^3P^L(k)$, with $P^L(k)$ the linear power spectrum evaluated on the growing mode, $\theta = \vec{\mathbf{k}} \cdot \vec{\mathbf{v}}$ the velocity divergence, and $\mathcal{H} = (1/a)(da/d\tau)$. The linear spectrum scales roughly as k^{-3} above a scale $\sim k_m$, so that $k^3P^L(k)$ is roughly constant, with a value of order 1 today. Its time dependence is given by D_L^2, with D_L the linear growth factor. We expect then that the maximal particle velocity can be identified with the fluid velocity at the scale k_m, and is roughly

$$v_p \sim \frac{\mathcal{H}}{k_m}D_L. \tag{4}$$

The time τ available for the higher moments to grow is $\sim 1/\mathcal{H}$. Therefore, the dimensionless factor characterizing the growth of higher moments is $kv_p/\mathcal{H} \sim D_L\,k/k_m$. Scales with $k \gg k_m/D_L$ require the use of the whole Boltzmann hierarchy and cannot be treated through a fluid description that takes into account only the lowest moments. In practice, the validity of the fluid description extends beyond k_m at all times, as the initial values of the higher moments are much smaller than 1, while $D_L \leq 1$. Therefore, on large scales, cold dark matter can be described as a cosmological fluid, not because microscopic interaction rates are sufficiently large to maintain local thermal equilibrium, but because the lifetime of the universe is too short for dark matter to deviate strongly from local equilibrium.

The above picture can lead to an estimate of the effective viscosity η that must be attributed to the dark-matter fluid because of the particle velocities that deviate from the uniform motion. In general, we expect $\eta/(\rho + p) \sim l_{\rm free}v_p$, with $l_{\rm free}$ the mean free path. For the effective viscosity we can estimate $l_{\rm free} \sim v_p/H$, with $H = \mathcal{H}/a$. In this way we obtain for the kinematic viscosity of the dark-matter fluid

$$\nu_{\rm eff}\mathcal{H} = \frac{\eta_{\rm eff}}{\rho_m\,a}\mathcal{H} \sim l_{\rm free}v_pH \sim \frac{\mathcal{H}^2}{k_m^2}D_L^2, \tag{5}$$

[1] Due to the cosmological expansion, the massive particles that constitute the cold dark matter have negligible velocity originating in their thermal motion before freeze-out. Warm dark-matter particles would have much larger velocities, which makes the application of the fluid description to this case problematic.

where we have neglected the pressure. The numerical value is small, of the order of 10^{-6} today, and even smaller at higher redshifts. However, viscosity plays a significant role in the context of the coarse-grained theory.

We can also estimate the viscosity induced by possible fundamental interactions between the dark-matter particles. For interacting dark matter, with number density n, mass m and cross section σ, the mean free path is $l_{\rm free} \sim 1/(n\sigma)$. We expect then fundamental viscosity

$$\nu_{\rm fund}\mathcal{H} = \frac{\eta_{\rm fund}}{\rho_m\, a}\mathcal{H} \sim l_{\rm free} v_p H \sim \frac{1}{n\sigma}\frac{H^2}{k_m} a D_L \sim \frac{m}{\sigma}\frac{8\pi G_N}{3k_m}\frac{aD_L}{\Omega_m}. \tag{6}$$

The time dependence, given by the last factor, indicates that the fundamental viscosity drops very quickly at early times. We point out that the above estimate is valid only for cold dark matter, for which the typical particle velocity is induced by the late-time growth of structure. For warm dark matter, the typical velocity and the resulting viscosity would evolve differently with time.

We note that that the fundamental viscosity of cold dark matter does not grow indefinitely for decreasing σ. The mean free path of dark matter particles cannot exceed $v_p/H \sim aD_L/k_m$. As a result, for values of σ such that $1/(n\sigma)$ exceeds aD_L/k_m, the particle scattering has no effect. This implies that the fundamental viscosity cannot take values larger than roughly $(\mathcal{H}^2/k_m^2)D_L^2$, and is always subleading or comparable to the effective viscosity.

As a final remark, we would like to emphasize that the constraints derived in refs. [11–13] for the non-ideal properties of dark matter are not applicable to the case of cold dark matter that we are considering. In these works the viscosity and sound velocity are assumed to stay constant or even grow with redshift, in contrast to the redshift-dependence indicated by our eqs. (5) and (6). In our picture, the particle velocity is induced by the growth of structure and is a late-time phenomenon. The assumptions of refs. [11–13] can by justified only within a warm-dark-matter scenario, in which the particles have velocities originating in their primordial motion.

4 One-loop determination of the effective viscosity

For the construction of the coarse-grained theory we follow the Wilsonian approach. A detailed description, including a discussion of the renormalization-group flow of the coarse-grained theory, is given in ref. [8]. The main idea is that the effective parameters of the large-scale theory result from the integration of the UV modes. For the problem at hand, the effective viscosity and sound velocity of the dark-matter fluid are given by one-loop expressions that depend on the dimensionful scale

$$\sigma^2_{dm}(\tilde\eta) = \frac{4\pi}{3}\int_{k_m}^{\infty} dq\, P^L(q,\tilde\eta) = \frac{4\pi}{3}D_L^2(\tilde\eta)\int_{k_m}^{\infty} dq\, P^L(q,0), \tag{7}$$

with $\tilde\eta = \ln a = -\ln(1+z)$ and $D_L(\tilde\eta)$ the linear growth factor: $\delta(\mathbf{k},\tilde\eta) = D_L(\tilde\eta)\,\delta(\mathbf{k},0)$ on the growing mode. For a spectrum that scales as $1/k^3$, the integral is dominated by the region near k_m.

A precise definition of the effective couplings can be given through the propagator of theory. One first defines the doublet

$$\begin{pmatrix} \varphi_1(\mathbf{k},\tilde\eta) \\ \varphi_2(\mathbf{k},\tilde\eta) \end{pmatrix} = \begin{pmatrix} \delta_{\mathbf{k}}(\tilde\eta) \\ -\dfrac{\theta_{\mathbf{k}}(\tilde\eta)}{\mathcal{H}} \end{pmatrix}, \tag{8}$$

and the propagator

$$G_{ab}(\mathbf{k},\tilde\eta,\tilde\eta')\,\delta^{(3)}(\mathbf{k}-\mathbf{k}') = \left\langle \frac{\delta\varphi_a(\mathbf{k},\tilde\eta)}{\delta\varphi_b(\mathbf{k}',\tilde\eta')} \right\rangle. \tag{9}$$

The power spectrum is defined as

$$\langle \varphi_a(\mathbf{k}, \tilde{\eta}) \varphi_b(\mathbf{k}', \tilde{\eta}) \rangle = \delta^{(3)} (\mathbf{k} + \mathbf{k}') P_{ab}(\mathbf{k}, \tilde{\eta}). \tag{10}$$

We concentrate on an Einstein-de Sitter (EdS) background, for which $D_L(\tilde{\eta}) = \exp(\tilde{\eta})$ and the linear propagator for the growing mode in the perfect-fluid theory is

$$g_{ab}(\tilde{\eta} - \tilde{\eta}') = \frac{e^{\tilde{\eta}-\tilde{\eta}'}}{5} \begin{pmatrix} 3 & 2 \\ 3 & 2 \end{pmatrix}. \tag{11}$$

The *one-loop* contribution to the low-energy propagator with wavenumber $k \to 0$, coming from modes with $q \geq k_m$, is given by

$$\delta g_{ab}(\mathbf{k}, \tilde{\eta}, \tilde{\eta}') = -k^2 e^{\tilde{\eta}-\tilde{\eta}'} \sigma^2_{dm}(\tilde{\eta}) \begin{pmatrix} \frac{61}{350} & \frac{61}{525} \\ \frac{27}{50} & \frac{9}{25} \end{pmatrix} \simeq -k^2 e^{\eta} \sigma^2_{dm}(\tilde{\eta}) \frac{3}{50} \begin{pmatrix} 3 \cdot 0.968 & 2 \cdot 0.968 \\ 9 & 6 \end{pmatrix}, \tag{12}$$

where we have taken the limit $\tilde{\eta} \gg \tilde{\eta}'$ for the second equality. On the other hand, the *linear* propagator of a theory with small kinematic viscosity and sound velocity of the form

$$\nu \mathcal{H} = \frac{\eta}{\rho_m a} \mathcal{H} = \beta_\nu e^{2\tilde{\eta}} \frac{\mathcal{H}^2}{k_m^2}, \qquad c_s^2 = \frac{\delta p}{\delta \rho} = \frac{3}{4} \beta_s e^{2\tilde{\eta}} \frac{\mathcal{H}^2}{k_m^2}, \tag{13}$$

contains a contribution

$$\delta g_{ab}(\mathbf{k}, \tilde{\eta}) = -\frac{k^2}{k_m^2} (\beta_\nu + \beta_s) \frac{e^{3\tilde{\eta}}}{45} \begin{pmatrix} 3 & 2 \\ 9 & 6 \end{pmatrix} \tag{14}$$

in addition to the perfect-fluid term (11), where we have kept only the leading contribution for $\eta \gg \eta'$.

We can now derive precise expressions for the effective parameters of the large-scale theory, by identifying the *linear* contribution to the propagator (14) for the *viscous* theory, with the *one-loop* correction (12) of the *perfect-fluid* propagator. This can be achieved with 1% accuracy, and we find

$$\beta_s + \beta_\nu = \frac{27}{10} k_m^2 \sigma^2_{dm}(0). \tag{15}$$

It must be emphasized that there are no free parameters in our approach, as the effective couplings are uniquely determined. Moreover, the deep UV region, which is out of the reach of perturbation theory, gives a negligible contribution to the integral of eq. (7), consistently with the expectation of decoupling of virialized structures. The dominant contribution comes from the region slightly above k_m, for which perturbation theory is expected to give reliable results.

By matching heuristically viscous fluid dynamics to the one-loop contribution (14), effective couplings were explored in [6]. This procedure can be obtained as a limiting case from the renormalization-group formulation of Ref. [8]. The latter approach also allows one to break the degeneracy between β_ν and β_s [8]. The numerical results that we present below depend only on the sum $\beta_\nu + \beta_s$ to a very good approximation.

5 Density and velocity power spectra

The procedure that we outlined in the previous section for an EdS background can be generalized to the ΛCDM case with an appropriate change of variables [6]. We present here results for the density, velocity and mixed power spectra for a large-scale viscous theory with parameters determined

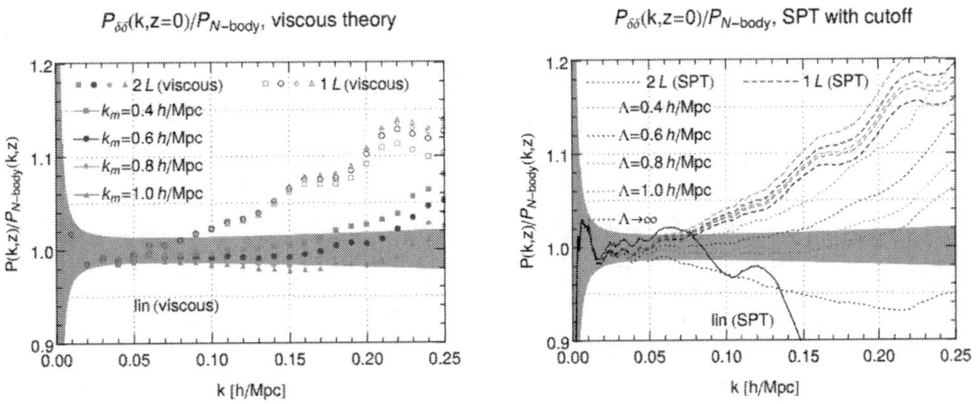

Figure 1. Density power spectrum at redshift $z = 0$, normalized to the N-body power spectrum [14]. The left panel shows the results obtained in the viscous description, for various values of the scale k_m. The coloured lines show the linear spectra, the open and filled symbols the one- and two-loop spectra. For comparison, the right panel shows the corresponding one- and two-loop results in standard perturbation theory. Here Λ is a sharp cutoff in k-space. The grey shaded region shows the estimated uncertainty of the N-body simulation. Figure taken from Ref. [6].

through eq. (15). The non-linear spectra for the low-energy theory can be computed through perturbation theory, with the crucial modification that (i) all momentum integrals possess a UV cut off k_m, and (ii) the Euler equation includes pressure and viscosity contributions. The presence of the cutoff prevents the inclusion of spurious effects from the deep UV, so that fast convergence is expected. It must be emphasized that the precise value of the scale k_m should not influence the final results for physical quantities, such as the power spectra: the explicit dependence of the effective pressure and viscosity on k_m compensates the k_m-dependence introduced by the cutoff. In some sense, the scale k_m may be viewed as a technical device that guarantees the good convergence of our scheme. Nevertheless, because of the perturbative truncation we expect a residual dependence that is a measure of the remaining theoretical uncertainty.

In order to test the accuracy of our results we compare them to the power spectrum extracted from N-body simulations [14]. In the left panel of fig. 1 we show the density power spectrum at redshift $z = 0$, obtained from solving the viscous fluid equations and averaging over an initially Gaussian random distribution of the density field. The different symbols show our results obtained for the four values $k_m = 0.4, 0.6, 0.8, 1.0\,h/\text{Mpc}$. In addition, open symbols are the one-loop results, and the filled symbols show the power spectrum up to two-loops. The loop expansion is perturbative in the density contrast. However, the effect of the time-dependent pressure and viscosity on the propagation of density and velocity perturbations is taken into account non-perturbatively. We normalized all curves to the result of the N-body simulation. The grey-shaded area corresponds to an estimate of the error of the simulation. The agreement between N-body data and the two-loop results is at the percent level for $k \lesssim 0.2\,h/\text{Mpc}$. In addition, the variation of the results when changing the scale k_m is mild, at the $\pm 2\%$ level for $k \lesssim 0.2\,h/\text{Mpc}$. Our results can be compared to the ones obtained in standard perturbation theory (SPT) [1], depicted in the right panel of 1. For SPT we show the one- and two-loop contributions obtained when imposing a cutoff Λ in k-space. The cutoff dependence in SPT is sizable at two loops, and amounts to $\pm 5\%$ ($\pm 10\%$) for $0.4\,h/\text{Mpc} < \Lambda < 1\,h/\text{Mpc}$ ($0.4\,h/\text{Mpc} < \Lambda \lesssim 5\,h/\text{Mpc}$). It is apparent that, as compared to SPT, the viscous description reduces the uncertainty generated by the treatment of UV modes significantly (left panel in fig. 1). The remaining uncertainty within

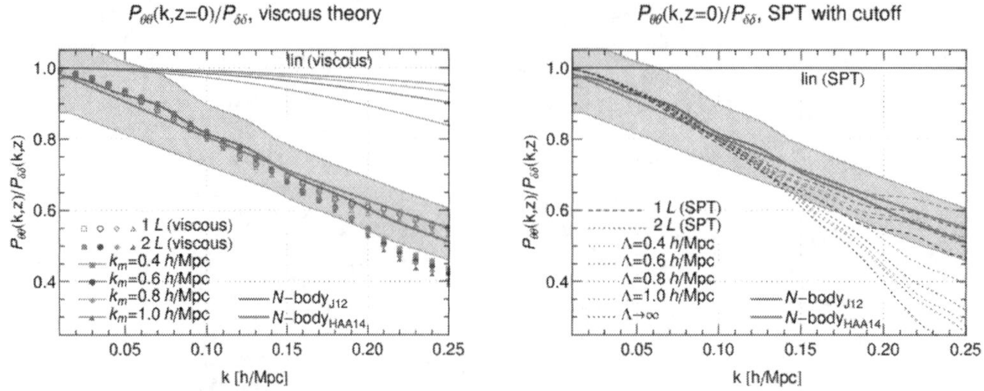

Figure 2. Power spectrum of the velocity-divergence, normalized to the density power spectrum, for the viscous theory (left) and standard perturbation theory (right). The various lines show the dependence on k_m and on Λ, respectively, as in fig. 1. The red lines are N-body results ((J12) [16], (HAA14) [17]), and the shaded region is the quoted uncertainty of the velocity power spectra extracted from N-body data.

the viscous framework is consistent with the expected impact of effects not captured by the fluid dynamical description employed here [15].

We also consider the power spectrum for the velocity divergence and the cross power spectrum. The velocity spectrum at $z = 0$ is shown in fig. 2, where we again compare the viscous description (left panel) with the result obtained in SPT (right panel) for various values of k_m and Λ, respectively. All curves are normalized to the density power spectra corresponding to the same approximation. Fig. 3 shows a similar comparison for the cross power spectrum. The extraction of power spectra for the velocity divergence from N-body simulations is less robust than for the density, because of complications arising when extracting a continuous velocity field from a discrete set of particles. We show the results obtained from two different methods [16, 17] by the two thick red lines in figs. 2 and 3. The results are sensitive to certain parameters of the simulation, especially the resolution, and we show the uncertainty quoted in refs. [16, 17] as shaded regions. Within this relatively large uncertainty, the results obtained in the viscous description (left panels in figs. 2 and 3) at two loops are again in agreement up to about $k \lesssim 0.2\,h/\text{Mpc}$. In addition, the variation when changing k_m is much smaller than the corresponding cutoff dependence in SPT (right panels in figs. 2 and 3). The residual k_m-dependence yields a quantitative estimate of the theoretical uncertainty. The difference of the perturbative prediction and the N-body results, as well as the theoretical uncertainty, increase quickly with k. Given that the predictions within the scheme discussed here do not require to fit any free parameters, the level of agreement between the two approaches for $k \lesssim 0.2\,h/\text{Mpc}$ is remarkable.

6 Conclusions

We conclude that the description of dark matter as a viscous fluid discussed here yields a robust framework for predicting power spectra for $k \lesssim 0.2\,h/\text{Mpc}$ at $z = 0$ without the need to adjust any free parameters. The framework can be constructed and extended through a formal analysis based on the renormalization group. We refer the reader to ref. [8] for details.

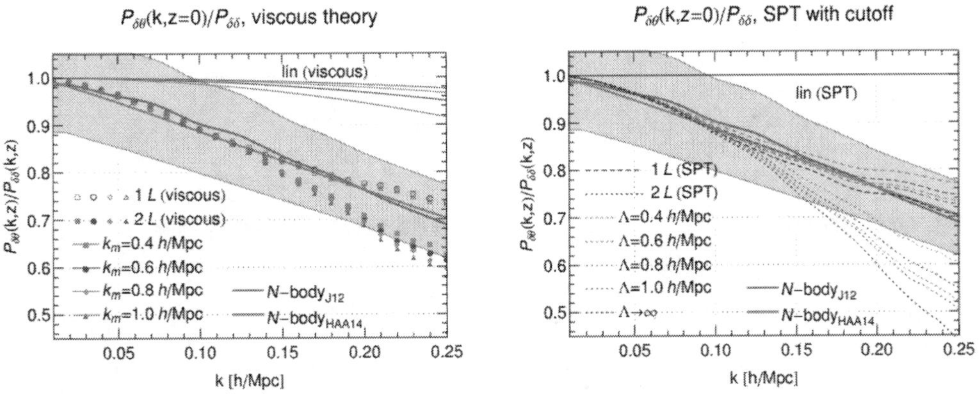

Figure 3. Cross density-velocity power spectrum, corresponding to the same approximations as in fig. 2.

References

[1] F. Bernardeau, S. Colombi, E. Gaztanaga and R. Scoccimarro, Phys. Rept. **367** (2002) 1 [astro-ph/0112551].

[2] D. Baumann, A. Nicolis, L. Senatore and M. Zaldarriaga, JCAP **1207** (2012) 051 [arXiv:1004.2488 [astro-ph.CO]].

[3] J. J. M. Carrasco, M. P. Hertzberg and L. Senatore, JHEP **1209** (2012) 082 [arXiv:1206.2926 [astro-ph.CO]].

[4] M. Pietroni, G. Mangano, N. Saviano and M. Viel, JCAP **1201** (2012) 019 [arXiv:1108.5203 [astro-ph.CO]].

[5] P.J.E. Peebles, *The Large-scale Structure of the Universe*, Princeton University Press (1980).

[6] D. Blas, S. Floerchinger, M. Garny, N. Tetradis and U. A. Wiedemann, JCAP **1511** (2015) 049 [arXiv:1507.06665 [astro-ph.CO]].

[7] S. Matarrese and M. Pietroni, JCAP **0706** (2007) 026 [arXiv:astro-ph/0703563].

[8] S. Floerchinger, M. Garny, N. Tetradis and U. A. Wiedemann, arXiv:1607.03453 [astro-ph.CO].

[9] T. Nishimichi, F. Bernardeau and A. Taruya, arXiv:1411.2970 [astro-ph.CO].

[10] M. Garny, T. Konstandin, R. A. Porto and L. Sagunski, JCAP **1511** (2015) no.11, 032 doi:10.1088/1475-7516/2015/11/032 [arXiv:1508.06306 [astro-ph.CO]].

[11] D. B. Thomas, M. Kopp and C. Skordis, arXiv:1601.05097 [astro-ph.CO].

[12] M. Kunz, S. Nesseris and I. Sawicki, Phys. Rev. D **94** (2016) no.2, 023510 [arXiv:1604.05701 [astro-ph.CO]].

[13] M. Kopp, C. Skordis and D. B. Thomas, arXiv:1605.00649 [astro-ph.CO].

[14] J. Kim, C. Park, G. Rossi, S. M. Lee and J. R. Gott III, *J. Korean Astron. Soc.* **44** (2011) 217 [arXiv:1112.1754 [astro-ph.CO]].

[15] S. Pueblas and R. Scoccimarro, Phys. Rev. D **80** (2009) 043504 [arXiv:0809.4606 [astro-ph]].

[16] E. Jennings, Mon. Not. Roy. Astron. Soc. **427** (2012) L25 [arXiv:1207.1439 [astro-ph.CO]].

[17] O. Hahn, R. E. Angulo and T. Abel, Mon. Not. Roy. Astron. Soc. **454** (2015) 3920 [arXiv:1404.2280 [astro-ph.CO]].

To the problem of the secondary CMB anisotropy separation

Oleg Verkhodanov[1,⋆], *Dmitrij* Solovyov[2,⋆⋆], *Elena* Majorova[1,⋆⋆⋆], and *Margarita* Khabibullina[1,⋆⋆⋆⋆]

[1] *Special Astrophysical Observatory, Russian Academy of Sciences, Nizhnii Arkhyz, 369167 Russia*
[2] *St. Petersburg Branch of the Special Astrophysical Observatory of the Russian Academy of Sciences, St.Petersburg, Russia 196140*

Abstract. We study contribution to the secondary anisotropy maps of cosmic microwave background (CMB) radiation which difficult to account for faint sources. Two effects are investigated. They are the Sunyaev–Zeldovich effect connected with the inverse Compton scattering of CMB photons on hot electrons of cluster of galaxies, and contamination of the background by weak extragalctic sources. First, we study fields of the Planck CMB maps around radio sources of the RATAN-600 catalog. We see weak microwave sources which make an additional contribution to the secondary anisotropy on angular small scales ($< 7'$). An algorithm for selecting candidate objects with the Sunyaev–Zeldovich effect was proposed, based on the use of data on the radio spectral indices and the signal in cosmic-microwave background maps. Second, applying the stacking method, we examine the areas of the CMB maps, constructed according to the Planck Space Observatory data in the neighborhood of different populations of radio sources and giant elliptical galaxies. The samples of objects include giant radio galaxies (GRG), radio sources, selected by the radio spectral index and redshift, as well as the gamma-ray bursts, used as a secondary comparative sample. The signal from this objects exists on CMB maps and its difference in the neighborhood of GRGs from the other types of objects was discovered.

1 Introduction

Publishing of the microwave maps restored from the millimeter and submillimeter observations of the NASA WMAP[1] (Wilkinson Microwave Anisotropy Probe) [1] and ESA Planck[2] [2] opened a new era in the research of the extended foreground emission and the extragalactic sources. The key role in last five years was played by the Planck Space Observatory and the full sky maps which were constructed based on the results of its work at nine frequencies of the microwave range: 30, 44, 70, 100, 143, 217, 353, 545, and 847 GHz. The maps of the Planck satellite of the European Space Agency were opened to the scientific community in 2013 (and the second release in 2015) and at once became an

⋆e-mail: vo@sao.ru
⋆⋆e-mail: drukhi@ya.ru
⋆⋆⋆e-mail: len@sao.ru
⋆⋆⋆⋆e-mail: rita@sao.ru

[1] `http://lambda.gsfc.nasa.gov`
[2] `http://www.cosmos.esa.int/web/planck`

effective means for solving many problems of galactic and extragalactic astronomy. Moreover, the catalogs of the discovered galactic and extragalactic sources of the millimeter and submillimeter radiation were presented in the first data release. The measurements of radiation fluxes were conducted using the High Frequency Instrument (HFI) with frequency bands of 100 GHz and higher, and the Low Frequency Instrument (LFI) with frequency bands lower than 100 GHz. The resolution of the maps at frequencies higher than 100 GHz amounted to about 5′. Although the Planck mission is secondary (by the launch date) relative to the other space mission, WMAP, it has better observational characteristics, which allow scientists to use it for search and investigation of weak radio sources in themicrowave range. Among the most important astrophysical products of the Planck mission, we can mention (1) table of cosmological parameters; (2) cosmic microwave background (CMB) radiation maps; (3) channel data maps (temperature anisotropy and polarization); (4) foregrounds maps; (5) secondary anisotropy maps; and (6) catalogs of millimeter and submillimeter sources. Among the advantageous features of Planck in comparison with the WMAP, let us note the higher resolution (three times higher), higher response (ten times higher), and nine frequency bands, which allowed the radiation components to be separated. The Planck Legacy Archive[3] (PLA) contains both the maps of radiation components and the lists of the detected objects. PLA contains the data base of sets of maps and we mention some of them below. First, four CMB maps of cosmic microwave background (CMB) radiation restored by 4 component separation methods, respectively. They are (a) internal linear combination in spherical wavelets domain — needlets (NILC), (b) template fitting using data at different frequencies to estimate by the expectation maximazation algorithm (Spectral Estimation Via Expectation Maximazation — SEVEM), (c) the method realized in the sperical harmonic domain of amplitude matching of the CMB spectral parameters and background components (Spectral Matching Independent Component Analysis — SMICA), and (d) the Bayesian separation of components in the pixel space by matching a parametric model (Commander).

The second group of PLA contains sky maps at observational bands at 30, 44, 70, 100, 143, 217, 353,545, 857 GHz, where channels from 30 to 353 GHz include the Q,U-polarization data too. The 3d group of maps are sky maps of diffent foregrounds, e.g. CO emission maps, dust, galactic low frequency emission (synchrotron plus free-free), zodiac light maps, etc. Also, there are maps of gravitational lensing, y-comptonization maps, and separately, masks used for restored different CMB temperature and polarization maps.

2 Secondary anisotropy maps

There exist several basic effects which acted on the CMB fluctuations and can be discussed in terms of the secondary anistropy. Among them are the intergated Sachs-Wolfe effect [3], the Sunyaev-Zeldovich effect [4], reionization, gravitational lensing of CMB, contamination of microwave background with extragalactic radio sources. The Planck observatory produced some secondary anisotropy maps for the astronomical community use. In Figures 1 and 2 are shown the Wiener-filtered lensing potential distribution on the sky [5] and the reconstructed the Planck all-sky y-map in Compton parameter units [6], respectively.

In Fig. 3, we demonstate positions on the sky of microwave sources from Planck calalogs at different frequencies. The used Second Planck Catalogue of Compact Sources [7] is the catalogue of sources detected from the full duration of Planck operations. It consists of compact Galactic and extragalactic objects detected over the entire sky. The total number of sources in the catalogue ranges from 1560 at 30 GHz up to 48 181 sources at 857 GHz.

[3]`http://pla.esac.esa.int/pla/`

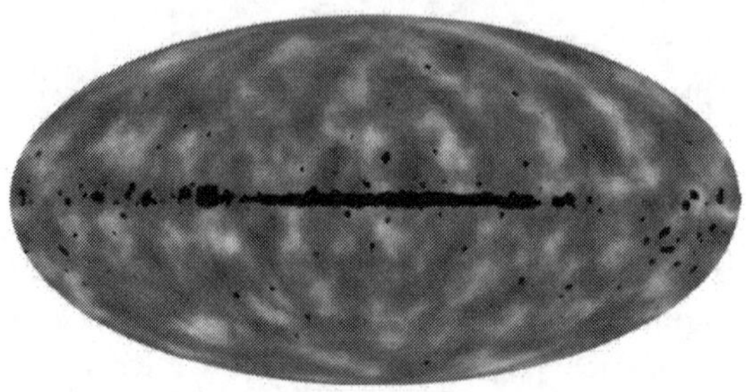

Figure 1. The Wiener-filtered lensing potential estimate with minimal masking (using the NILC component separated map). From [5].

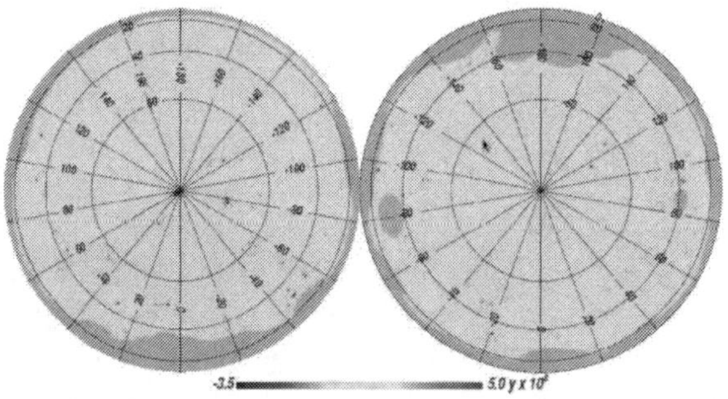

Figure 2. The orthographic projection of the reconstructed Planck all-sky y-map in Compton parameter units. Positive sources in the map correspond to clusters and super-clusters of galaxies with strong tSZ emission. The Coma and Virgo clusters are clearly visible near the north Galactic pole. The region of strongest contamination from Galactic foreground emission in the Galactic plane has been partially masked. From [6].

Figure 3. Positions on the sky of microwave sources from Planck calalogs at different frequencies.

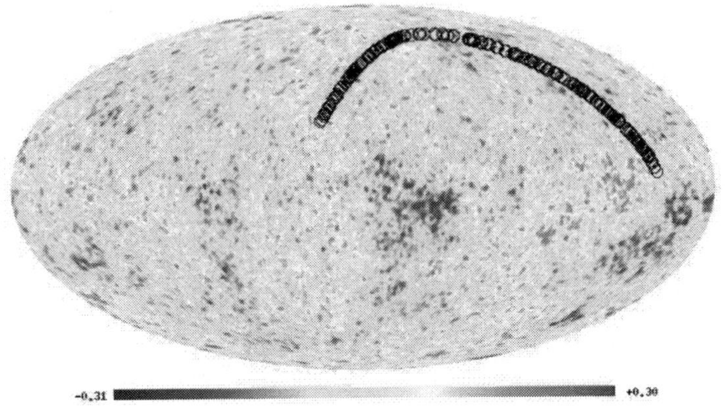

Figure 4. Positions of RCR radio sources on a CMB map (the range of coordinates is 7^h < RA < 17^h, Dec~5°).

3 RCR-sources on the Planck maps

One of the main problems of the sensitive radio surveys data analysis is the presence of faint sources at the noise level (S/N $\lesssim$ 3) which make additional contribution to the angular power spectrum at small angular scales (high multipoles ℓ). This situation is also observed in Planck CMB maps when we consider a signal at $\ell > 2000$. Another problem arose for the Planck data is the missed clusters of galaxies which are observed with the Sunayev–Zeldovich effect at multifrequency maps. Number of expected SZ-sources contributed to the secondary anistropy was larger than 20000 (expected number of clusters of galaxies), but the presented catalog contains 1653 objects [8].

We use the RCR (RATAN Cold Refined) catalog [9, 10] to look for any response from RC-source direction in millimeter and submillimeter wave length ranges. The RC catalog was obtained during the Cold experiment [9] conducted on the RATAN-600 radio telescope in the 1980s at 3.94 GHz. The horizontal and vertical resolutions 60″ and, 15′, respectively. To improve the coordinates and flux densities of the RC catalog sources, seven additional cycles of observations were carried out in the strip of sky corresponding to the Cold survey at several frequencies in 1987–1999, and the data obtained in 1980–1981 were reprocessed for right ascensions 7^h < RA < 17^h and declination Dec~5°(Fig. 4). This yielded a list of 551 objects identified with objects in the NVSS catalog with a positional accuracy of ~0.1″. The limiting sensitivity in the RCR catalog was ~10 mJy.

Using properties of the CMB signal in the fields of RCR-sources, we prepared some diagrams to explore contamination of the background [11]. Figure 5 shows a plot of radio spectral index α versus CMB response T_p corresponding to the thermodynamical temperature on the Planck CMB SMICA map. Note that the $\alpha - T_p$ diagram is interesting in that we can distinguish groups of radio sources with correlated parameters that are basically determined by the 'spectral index — redshift' and 'flux density (here temperature) — redshift' relations. Before constructing the diagram, we removed the low-order harmonics ($\ell < 20$), which mainly include the statistically anisotropic signal from the CMB map (Fig. 6). In principle, it is possible to eliminate harmonics with $\ell \leq 220$, before the first CMB peak, but it is important to remove precisely data with $\ell \leq 20$ after which the map becomes Gaussian; for example, features such as the Cold Spot [12] disappear. We can distinguish four regions in the diagram on Fig. 5 (A, B, C, D), which correspond to the positions of peaks on histograms of the parameters of the objects: regions A, B, and C correspond to the position of the main peak in the histogram of spectral indices (Fig. 7), and region D corresponds to the second peak in the distribution of the spectral indices for the flat-spectrum sources. The histogram of the radio spectral indices has a standard two-peaked distribution determined by the number of objects with steep and flat spectra, and

 DOI: 10.1051/epjconf/201612503019

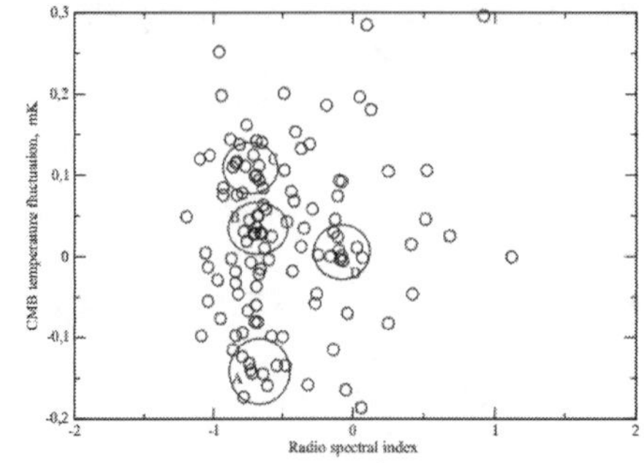

Figure 5. Plot of radio spectral index vs. CMB signal for all the RCR catalog sources. The CMB signals are given in thermodynamical units. The circles mark four groups of objects (A, B, C, D) corresponding to peaks in the histograms and discussed in the text.

Figure 6. CMB SMICA map for the 2015 Planck data release used in this study for our analysis of the distribution of the CMB fluctuation signal. The low-order multipoles ($\ell \leq 20$, i.e., the signal on angular scales larger than 8°, which mainly includes the statistically anisotropic background) were first removed

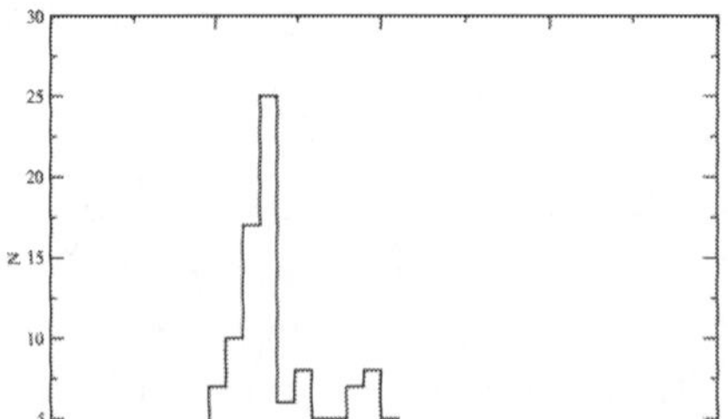

Figure 7. Distribution of spectral indices of all RCR catalog sources at 1.4 GHz.

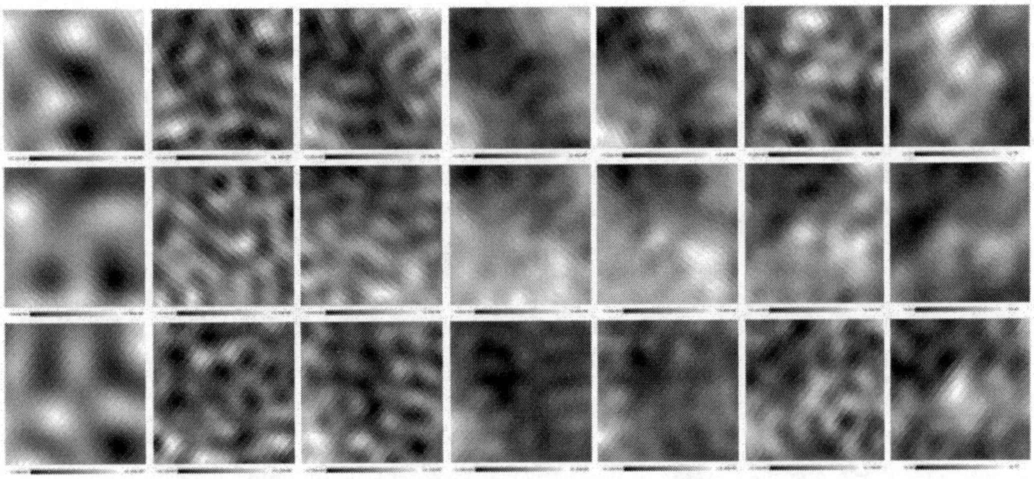

Figure 8. Left to right: CMB maps for the RCR sources falling in group A (Fig. 5) at 44, 70, 100, 143, 217, 353, and 545 GHz, respectively, according to the 2015 Planck data. For the S–Z effect at the Planck frequencies, a dip at $\nu < 217$ GHz and a positive peak at $\nu > 217$ GHz are expected. From top to bottom: the radio sources 092445+050344, 112437+045618, and 151054+045410. The size of the images is $0.5° \times 0.5°$.

described. The main values of the signal in the CMB map are concentrated near zero, but from the positive side: the median CMB signal calculated using the data for all the studied sample sources is +0.02527 mK.

Field A in Fig. 5, which corresponds to radio sources with negative responses exceeding the expected 1σ level and having extreme slopes in their radio spectra, $\alpha = 0.6 - 0.8$. The relatively low negative level of the background on the CMB map could be due to an additional contribution to the inhomogeneity of the Sunyaev-Zeldovich effect, enhanced during the separation of the components and reconstruction of the CMB signal. This effect is also suggested by the existence of a radio source in this region, which is a sign of an accumulation of material (ultimately a cluster of galaxies). Figure 8 presents the maps of three RCR sources from field A at various frequencies as example candidates for objects displaying the Sunyaev-Zeldovich effect.

4 Response from radio galaxies on the Planck microwave maps

We examined the response of an average object, obtained by stacking the areas sized $1° \times 1°$ around the sources from different populations according to the microwave maps of the Planck mission [13]. The main population of objects studied comprises 89 giant radio galaxies with the angular sizes of more than $4'$, the linear sizes of which (larger than 1 Mpc) are confirmed in the literature. For a comparison of GRG properties with the properties of other populations we have studied the subsamples of different classes of objects with the selection by the spectral indices, redshifts, optical morphology (the gE and cD-type galaxies), as well as the gamma-ray bursts. Radio galaxies and gian elliptical galaxies (gE) include following sub-samples: 1) giant radio galaxies (89 objects, > 1 Mpc, $\theta > 4'$); 2) distant radio galaxies (2400) [14]; 3) giant elliplical galaxies (gE and cD) from clusters of galaxies (25) [15]; 4) WENSS radio sources in the range of right ascensions: $0^h \leq \mathrm{RA} < 2^h$; 5) objects from BATSE [16] and BeppoSAX [17] gamma-ray burst catalogs. Applying the stacking method, we examine the areas of the CMB maps, constructed according to the Planck data in the neighborhood of this populations of galaxies. The samples of objects include giant We have studied the topological properties of the

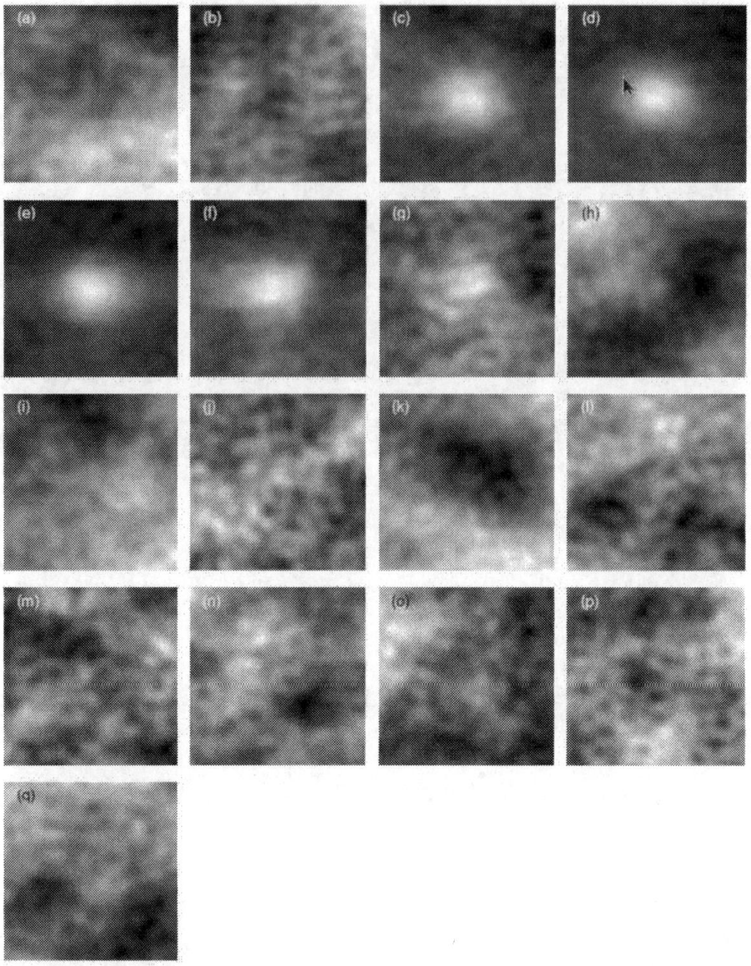

Figure 9. The stacking zones sized 1° × 1° built from the CMB SMICA Planck 2.02 map data. From left to right and from top to bottom are presented the images of the average maps for the object populations (the number of sample objects is marked in the brackets): (a) GRGs with the angular sizes of $\theta > 4'$ and linear sizes of > 1 Mpc (89), (b) the gE and CD galaxy types (25), (c) the radio sources (RS) from the WENSS survey with the radio spectral indices $\alpha < -1.1$ (224), (d) WENSS RS with $-1.1 \leq \alpha < -0.75$ (661), (e) WENSS RS with $-0.75 \leq \alpha < -0.5$ (497), (f) WENSS RS with $-0.5 \leq \alpha < 0$ (238), (g) WENSS RS with $\alpha \geq 0$ (19), (h) distant RG with the redshifts $0.3 \leq z < 0.7$ (1797), (i) HZRG with $0.7 \leq z < 1.0$ (205), (j) HZRG with $1.0 \leq z < 1.5$ (149), (k) HZRG with $1.5 \leq z < 2.0$ (103), (k) HZRG with $2.0 \leq z < 2.5$ (77), (l) HZRG with $z \geq 2.5$ (81), (n) the BATSE events with the duration of $t < 2$s (495), (o) BATSE, $t > 2$s (1540), (p) BeppoSAX, $t < 2$s (87), (q) BeppoSAX, $t > 2$ s (694). The original CMB data are given in the scale of thermodynamic temperatures.

CMB signal in the neighborhood of the average object of the population, namely, we searched for the presence of the maxima and minima in the average area (Fig. 9). The difference of the signal in the neighborhood of GRGs from the other types of objects was discovered.

5 Summary

It is possible to note a few points:

- There is a positive signal for objects of different radio galaxy popultaion;
- It is possible to increase the number of SZ-objects for several times using searching fields around radio sources;
- There is a positive response in the zone of radio sources which number should be tens time larger than Planck source list;
- All objects are detected within $7'$ zones, thus they do not affect the $C(\ell)$ power spectrum at low and intermediate ($\ell < 2000$) multipoles.

For an average GRG in the microwave range, the area around which is built by stacking, it was found that 1) there is a positive signal in the millimeter range; 2) the spectrum in the millimeter range is flat; 3) the signal in the submillimeter range is absent; 4) the topological properties in the average CMB fluctuations map possess special features: a weak insignificant maximum ($< 1\sigma$) is located in the minimum zone; 5) a significant SZ–effect, related with the scattering of CMB photons by the electrons of the GRG jet plasma [18] is not observed. A comparison with other populations of objects has revealed that given the average microwave properties, the distant radio galaxies with redshift ranges $0.3 \leq z < 0.7$.

Acknowledgments. We thank the European Space Agency for their open access to the observations and data in the Planck Legacy Archive. We have used the GLESP package [19] for the analysis of extended emission on a sphere.

References

[1] C.L. Bennett, D. Larson, J.L. Weiland, et al., *Astrophys. J. Suppl.* **208**, 20 (2013).
[2] Planck Collaboration, Astrophys. Astrophys. accepted; arXiv:1502.01582.
[3] R. K. Sachs and A. M. Wolfe, ApJ **147**, 73 (1967).
[4] Ya. B. Zeldovich and R. A. Sunyaev, Astrophys. Sp. Sci. **4**, 301 (1969).
[5] Planck Collaboration, Astron. Astrophys. accepted; arXiv:1502.01591.
[6] Planck Collaboration, Astrophys. Astrophys. accepted; arXiv:1502.01596.
[7] Planck Collaboration Astron. Astrophys. accepted; arXiv:1507.02058.
[8] Planck Collaboration Astron. Astrophys. accepted; arXiv:1502.01598
[9] Yu. N. Parijskij, N. N. Bursov, N. M. Lipovka, et al., Astron. Astrophys. Supp. **87**, 1 (1991).
[10] O. P. Zhelenkova, N. S. Soboleva, E. K. Majorova, and A. V. Temirova, Astrophys. Bull. 68, 26 (2013).
[11] O.V.Verkhodanov, E.K.Maiorova, O.P.Zhelenkova, et al., Astronomy Reports **60**, 630 (2016).
[12] P. D. Naselsky, P. R. Christensen, P. Coles, et al., Astrophys. Bull. **65**, 101 (2010).
[13] O.V. Verkhodanov, D.I. Solovyov, O.S. Ulakhovich, and M.L. Khabibullina. Astrophys. Bull. **71**, 139 (2016).
[14] M. L. Khabibullina and O. V.Verkhodanov, Astrophys. Bull. **64**, 123 (2009).
[15] S. A. Stanford, P. R. Eisenhardt, M. Dickinson, et al., Astrophys. J. Suppl. **142**, 153 (2002).
[16] W. S. Paciesas, C. A. Meegan, G. N. Pendleton, et al. Astrophys. J. Suppl. **122**, 465 (1999), astro-ph/9903205.
[17] D. Riccia, F. Fioreb, and P. Giommia, Nuclear Physics B - Proc. Suppl. **69**, 618 (1999).
[18] S. Colafrancesco, P. Marchegiani, Astron. Astrophys. **535**, A108 (2011).
[19] A. G. Doroshkevich, P. D. Naselsky, O. V. Verkhodanov, et al., Int. J. Mod. Phys. 14, 275 (2003).

EPJ Web of Conferences **125**, 03020 (2016) DOI: 10.1051/epjconf/201612503020
QUARKS-2016

Inflation with light Weyl ghost

Anna Tokareva[1,★]

[1] *INR RAS, 60-letiya Oktyabrya prospekt 7a, Moscow 117312, Russia*

Abstract. Inflationary perturbations are considered in a renormalizable but non-unitary theory of gravity with the additional Weyl term. We obtained that ghost degrees of freedom do not spoil the inflation and the scalar perturbation amplitude at the linear level even in a case of the ghost with mass smaller than Hubble parameter at inflation. The ghost impact to the observables is also estimated to be negligible for the range of masses allowed by the experiment. The non-linear level of the theory and its possible application are also discussed.

1 Introduction

General relativity, providing a good explanation for all known experimental results, fails all attempts to quantize it consistently. The Einstein-Hilbert action leads to the non-renormalizable theory which means the loss of its predictive force at Planck energies. Despite the low energy physics is described by the effective field theory, there are examples showing that some processes below the Planck scale (such as inflation and reheating) may be sensitive to the UV completion of the theory [1, 2].

It is well known that there is a way to construct the renormalizable theory of gravity by the price of unitarity. The quadratic in curvature action for gravity was studied by Stelle in Refs. [3, 4]. This action,

$$S = \frac{M_P^2}{2}\int d^4x\,\sqrt{-g}\left(R - \frac{1}{2m^2}\,C_{\mu\nu\rho\sigma}\,C^{\mu\nu\rho\sigma} + \frac{1}{6m_0^2}R^2\right) \tag{1}$$

when linearised on the flat space background, provides usual massless graviton, scalar with mass m_0 and spin-2 ghost degree of freedom with mass m. Here $C_{\mu\nu\rho\sigma}$ stands for the Weyl tensor and M_P is the reduced Planck mass. The adimentional variant of this action was recently discussed in [5].

Despite the interpretation of non-unitary theories is a complicated problem such theories might be useful in the sense of an effective description. In particular, the non-interacting ghost is not dangerous because it may be consistently quantized with the negative norm and positive energy (see for example Ref. [6] for the discussion). In this paper, we discuss the limit in which the Weyl ghost becomes interacting very weakly and hereafter we impose the negative norm prescription for the ghost states. Since it has only derivative coupling to matter we expect that this limit corresponds to the light ghost. In the next section we estimate the typical time of ghost amplification and find out the allowed range for the ghost mass to be $m \lesssim 10$ MeV providing the lifetime of ghost instability to be much larger than the Universe age.

★e-mail: tokareva@ms2.inr.ac.ru

In order to check the validity of the effective description in terms of non-unitary theory we calculate the evolution of the scalar perturbations in a theory with light Weyl ghost starting from vacuum initial conditions (this case is not yet studied in literature). We obtained the deviations from the standard cosmology to be negligible for the discussing range of masses. The ghost impact to the tensor perturbations has been found in a paper [7]. It looks interesting that the amplitude of tensor perturbations (which is observable in cosmology) in a case of light ghost is damped by a factor $m/\sqrt{2}H_0$, where H_0 stands for the Hubble parameter at the inflationary stage.

The paper is organised in the following way. In the Section 2 we show that the limit of light ghost really corresponds to low rates of ghost creation and decay. In the Section 3 we calculate the scalar perturbations on the inflationary stage in a gravity with the Weyl term for the case of the light ghost. In the Section 4 we follow the evolution of scalar perturbations up to the late Universe and obtain the ghost impact on CMB. In the last section, we come to conclusions.

2 The discussion about the light ghost

In a paper [5] the naturalness problem for the Higgs mass in a case of the heavy ghost was pointed out. This problem may be avoided when the ghost is lighter than 10^{10} GeV which is less than the usual Hubble parameter at inflation. Preserving the Higgs mass naturalness motivates the study of the inflationary perturbations in a case of the ghost being lighter than the Hubble parameter. We must require for the first that the instability connected with the ghost decay (which is actually the ghost number amplification because of the negative width) evolves very slowly, corresponding to the ghost decay width much larger than the Universe lifetime. With negative norm prescription the ghost decay width may be estimated as [8, 9]

$$\Gamma \sim -\frac{m^3}{M_P^2} \tag{2}$$

In order to have a ghost lifetime to be much larger than the age of Universe we must have $m \ll 10$ MeV.

Another process with the negative probability which must be negligible in the considering limit of the light ghost is the ghost production in a hot plasma. Although there is a problem of interpretation with the negative cross sections we take as the estimation of the ghost effects values calculated in a usual way. We approximate the cross section as

$$\sigma \sim -\frac{e^2}{M_P^2}. \tag{3}$$

Here e stands for the electron charge. We obtain the number of produced ghosts which corresponds to their impact on the energy density of Universe being equal to

$$\Omega_{gh} = 0.02\,\frac{m}{10\text{ MeV}}\,\frac{T_{max}}{10^{13}\text{ GeV}}. \tag{4}$$

Here T_{max} is the maximal temperature that has ever been in the Universe. One can see again that for the ghost lighter than 10 MeV and natural values of the reheating temperature a number of ghosts turn out to be negligible.

Ghosts with too low masses cannot be considered because of experimental constraints for the gravity at short distances. The limit coming from recent measurements [10] is $m > 10^{-2}$ eV which leaves a wide range (9 orders of magnitude) for the weakly coupled ghost we discuss in this paper.

3 Scalar perturbations at the inflationary stage

3.1 Common equations

We consider the action

$$S = \frac{M_P^2}{2}\int d^4x\,\sqrt{-g}R - \frac{1}{2}\int d^4x\,\sqrt{-g}\left(\partial_\mu\phi\,\partial^\mu\phi + 2V(\phi)\right) - \frac{M_P^2}{4m^2}\int d^4x\,\sqrt{-g}\,C_{\mu\nu\rho\sigma}\,C^{\mu\nu\rho\sigma}, \quad (5)$$

where g is the determinant of the metric $g_{\mu\nu}$, R is the scalar curvature and $C_{\mu\nu\rho\sigma}$ is the Weyl tensor.[1]

This action describes the usual Einstein gravity corrected by the Weyl curvature term. The scalar ϕ with the proper potential may be appropriate for inflation. If we consider the most general quadratic in curvature corrections to the Einstein gravity action than the R^2 term may be rewritten in the Einstein frame through the additional scalaron field ϕ [11]. So the action (5) with a special form of $V(\phi)$ may describe the general second order corrected gravity which with the specially tuned scalaron mass may provide the phenomenologically viable Starobinsky inflation [12].

We start to work in the newtonian gauge [13] where the metric and inflaton perturbations take the form:

$$ds^2 = a(\eta)^2\left\{-(1+2\Psi)\,d\eta^2 + 2\bar{\Psi}_i\,d\eta\,dx^i + [(1+2\Phi)\,\delta_{ij} + \bar{h}_{ij}]\,dx^i\,dx^j\right\}, \quad \delta\phi = \chi. \quad (6)$$

Here $\bar{h}_{ij}$ correspond to the tensor modes, $\bar{\Psi}_i$ – to the vector modes and the scalar sector is described by Φ, Ψ and χ.

The expansion of the action (5) at quadratic order in the perturbations is presented in Ref. [14]:

$$S^{(S)} = S^{(S)}_{\mathrm{E}} - \frac{M_P^2}{3m^2}\int d^4x\,[\triangle(\Psi-\Phi)]^2, \quad (7)$$

where

$$S^{(S)}_{\mathrm{E}} = \frac{M_P^2}{2}\int d^4x\,a^2\Big[-6\Phi'^2 + 12\mathcal{H}\,\Psi\,\Phi' + 2\partial_i\Phi\,(2\partial^i\Psi + \partial^i\Phi) - 2(\mathcal{H}' + 2\mathcal{H}^2)\,\Psi^2$$
$$+\frac{1}{M_P^2}\left(\chi'^2 - \partial_i\chi\,\partial^i\chi - a^2\,V_{,\phi\phi}\,\chi^2 - 6\phi'\,\Phi'\,\chi - 2\phi'\,\chi'\,\Psi - 2a^2\,V_{,\phi}\,\Psi\,\chi\right)\Big]. \quad (8)$$

Here the primes mean the derivatives with respect to conformal time η, and $\mathcal{H} = a'/a$ is the conformal Hubble parameter. The action describes two coupled dynamical fields: a ghost and an inflaton while Ψ is to be integrated out.

3.2 Initial conditions and solutions for the de Sitter space

In a limit of de Sitter space the action may be diagonalized. Here we use $a(\eta) = -1/(H_0\eta)$ where H_0 is the Hubble parameter at inflation. The action for the inflaton perturbation takes the usual form:

$$S_\chi = \frac{1}{2}\int d^4x\frac{1}{H_0^2\eta^2}(\chi'^2 - (\partial_i\chi)^2), \quad (9)$$

[1] Units: $M_P = 1/\sqrt{8\pi G}$, $c = 1$. Conventions: $(-+++)$; $R^\mu{}_{\nu\rho\sigma} = \partial_\rho\Gamma^\mu_{\nu\sigma} - \partial_\sigma\Gamma^\mu_{\nu\rho} + \cdots$; $R_{\nu\sigma} = R^\mu{}_{\nu\mu\sigma}$; $R = g^{\mu\nu}R_{\mu\nu}$; $G_{\mu\nu} = R_{\mu\nu} - \frac{1}{2}g_{\mu\nu}R$. $C_{\mu\nu\rho\sigma} = R_{\mu\nu\rho\sigma} - \frac{1}{2}(g_{\mu\rho}G_{\nu\sigma} - g_{\mu\sigma}G_{\nu\rho} - g_{\nu\rho}G_{\mu\sigma} + g_{\nu\sigma}G_{\mu\rho}) - \frac{R}{3}(g_{\mu\rho}g_{\nu\sigma} - g_{\mu\sigma}g_{\nu\rho})$. Greek indices run from 0 to 3; latin indices run from 1 to 3.

After integrating out Ψ one obtain the action for the ghost perturbation:

$$S_\Phi = \int d\eta d^3q \frac{3M_P^2}{H_0^2}\left[-\frac{\beta q^4\eta^2}{9+\beta q^4\eta^4}\Phi'(q)\Phi'(-q) + \frac{3q^4\left(\beta^2 q^6\eta^6 + \beta q^4\eta^4(1-4\beta) - 3\beta q^2\eta^2 + 9\right)}{(9+\beta q^4\eta^4)^2}\Phi(q)\Phi(-q)\right] \tag{10}$$

Here $\beta = H_0^2/m^2$. The vacuum initial conditions for the inflaton and ghost perturbations are to be set at $\eta \to -\infty$ in the subhorizon limit. The action for the inflaton behaves as in the usual gravity while the action for the ghost becomes

$$S_\Phi = \int d^4x \frac{3M_P^2}{H_0^2\eta^2}\left[-\Phi'^2 + (\partial_i\Phi)^2\right]. \tag{11}$$

After canonical normalization one can obtain that the vacuum initial conditions lead to the field perturbations at $\eta \to -\infty$:

$$\Phi = \frac{H_0 q\eta}{2\pi\sqrt{6}M_P}e^{-iq\eta}, \quad \chi = \frac{H_0 q\eta}{2\pi}e^{-iq\eta}. \tag{12}$$

In the opposite limit $\eta \to 0$ one obtain an equation

$$\Phi'' + \frac{2}{\eta}\Phi' + \frac{1}{\beta\eta^2}\Phi = 0, \tag{13}$$

while the equation for χ is standard [13]. The solution for ghost consists of the modes: $\Phi \propto \eta^{-1}$ and $\Phi \propto \eta^{-1/\beta} \approx const.$ We see that there is a growing as a scale factor mode in the set of solutions for the ghost field. The initial conditions lead for this mode to be normalised as (up to a factor of order unity)

$$\Phi = \frac{H_0}{2\pi\sqrt{6}M_P}\frac{aH_0}{q}. \tag{14}$$

Later we will see that this instability is not physical, it is connected with the choice of gauge.

3.3 Slow roll inflation

The real inflationary stage lead to small deviations from the exact de Sitter space so the equations on Φ and χ following from the action (7) becomes coupled. In the superhorizon limit and within the slow roll approximation the equation for χ is[2]

$$\Psi = \frac{3M_P^2 H}{V}\dot{\Phi}, \quad \ddot{\chi} + 3H\dot{\chi} = -\dot{\phi}(3\dot{\Phi} - \dot{\Psi}) - 2V_{,\phi}\Psi. \tag{15}$$

Since Φ has a growing mode (14) it induce the growing solution for χ through the upper equation. In the slow roll approximation $V = 3H^2M_P^2$, $V_{,\phi} = -3H\dot{\phi}$ so for χ one obtain

$$\ddot{\chi} + 3H\dot{\chi} = 4H\dot{\phi}\frac{H_0^2}{2\pi\sqrt{6}M_P}\frac{a}{q}, \tag{16}$$

which provides the solution (up to the falling sub-dominate modes):

$$\chi = \frac{H_0}{2\pi} + \frac{\dot{\phi}}{H}\frac{H_0^2}{2\pi\sqrt{6}M_P}\frac{a}{q}. \tag{17}$$

In order to show that the growing modes are artefacts of the choice of gauge in the next section we calculate the metric perturbations in the comoving gauge.

[2] Hereafter we work in the physical time t, $H = \dot{a}/a$ is the usual Hubble parameter

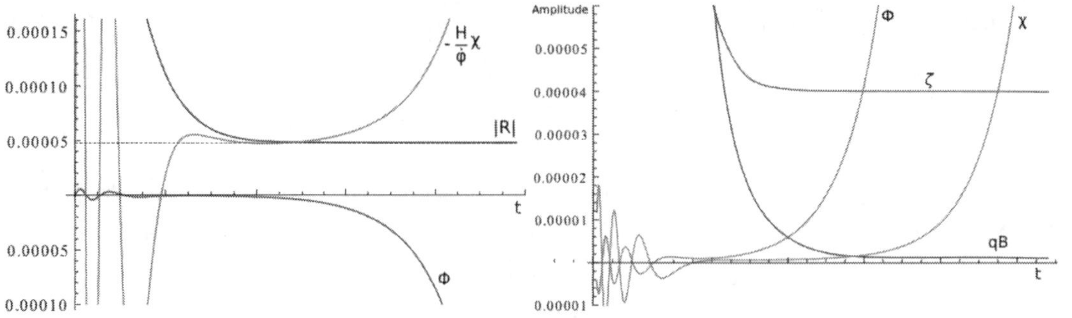

Figure 1. Left plot shows the cancellation between the growing modes of Φ and χ when the metric perturbation in a comoving gauge ζ is calculated. Right plot illustrates the time evolution for the discussed perturbation amplitudes in Newtonian and comoving gauges.

3.4 The perturbations in the comoving gauge

The comoving gauge is defined by the condition $\delta\phi = 0$. The metric takes the form

$$g_{00} = a^2(1 + 2A),\ g_{0i} = 2a^2\partial_i B,\ g_{ij} = a^2(\delta_{ij}(1 + 2\zeta) + \partial_i\partial_j E) \tag{18}$$

Here only scalar part of perturbations is considered. The perturbation E may be set to be zero by the special choice for one of the two gauge transformation parameters. The other parameter fixes the condition $\delta\phi = 0$ [13]. The parameter A is falling approximately as $1/a$ in the superhorizon limit so we concentrate on the perturbations ζ and B connected with variables from Newtonian gauge as

$$\zeta = \Phi - \frac{H}{\dot\phi}\chi,\ \ B = \frac{\chi}{a\dot\phi} \tag{19}$$

Substituting our solutions (14) and (17) we obtain

$$\zeta = -\frac{H_0^2}{2\pi\dot\phi} = \frac{H_0}{2\pi\sqrt{2\epsilon}M_P},\ \ B = \frac{H_0}{2\pi\sqrt{6}qM_P} = -\sqrt{\frac{\epsilon}{3}}\frac{\zeta}{q}. \tag{20}$$

Here $\epsilon = M_P^2(V_{,\phi}/V)^2/2$ is the first slow roll parameter. We see that growing modes in Φ and χ cancel each other providing the value of ζ to be the same as in the usual Einstein gravity defining the amplitude of curvature perturbation that is known to be $\zeta = 10^{-5}$. The difference is in the component B: in usual gravity it is falling as $1/a$ while in the considered model it is constant. The quantity which actually stands in metric, $\partial_i B = n_i\sqrt{\epsilon/3}\zeta \ll 1$ (where n_i ia a unit vector) is much smaller than unity proving that the linear regime is consistent in the comoving gauge being broken in the Newtonian gauge.

All the discussed solutions are plotted in a Figure 1. In order to obtain the possible phenomenological consequences of non-vanishing perturbation B we need to follow its evolution after inflation up to the late Universe. We will do it in the next section.

4 Evolution of cosmological perturbations after inflation

4.1 Equations

The comoving gauge is convenient for understanding of the superhorizon evolution due to the good behaviour of the metric components. But it is much more easy to solve the equations on the variables defined in the Newtonian gauge. The combination $W = \Psi - \Phi$ has a very important property: it is possible to write all Einstein equations as one fourth order equation on W. After solving it one can easily obtain all the perturbations in the comoving gauge using formulas given in [14]. For the inflationary stage (or stage dominated by the scalar field) the master equation has been derived in [14]:

$$\frac{1}{m^2}\left(W^{(4)} - \frac{\ddot{H}}{\dot{H}}W^{(3)}\right) + C_2\,\ddot{W} + C_1\,\dot{W} + C_0\,W = 0\,, \tag{21}$$

where

$$\begin{cases} C_2 = 1 + \dfrac{1}{m^2}\left(\dfrac{2q^2}{a^2} - H^2\right), \\ C_1 = H - \dfrac{\ddot{H}}{\dot{H}} + \dfrac{1}{m^2}\left[\dfrac{H^2\ddot{H}}{\dot{H}} - 3H\dot{H} - \dfrac{q^2}{a^2}\left(\dfrac{\ddot{H}}{\dot{H}} + 4H\right)\right], \\ C_0 = 2\dot{H} - \dfrac{H\ddot{H}}{\dot{H}} + \dfrac{q^2}{a^2} + \dfrac{q^2}{m^2a^2}\left(\dfrac{H\ddot{H}}{\dot{H}} - \dfrac{4}{3}\dot{H} + 2H^2 + \dfrac{q^2}{a^2}\right). \end{cases} \tag{22}$$

This equation was derived for the stage dominated by the scalar field. To proceed with the hot stage we need to obtain a similar equation from the set of Einstein equations. At the stages dominated by the fluid (matter or radiation) they take the form:

$$\frac{q^2}{a^2}\Phi + 3H(\dot{\Phi} - H\Psi) - \frac{1}{3m^2}\frac{q^4}{a^4}W = \frac{\delta\rho}{2M_p^2}, \tag{23}$$

$$-\ddot{\Phi} - 3H\dot{\Phi} + H\dot{\Psi} + \Psi(2\dot{H} + 3H^2) - \frac{q^2}{3a^2}(\Psi + \Phi) - \frac{1}{9m^2}\frac{q^4}{a^4}W = \frac{\delta p}{2M_p^2}, \tag{24}$$

$$-(\Phi + \Psi) = \frac{1}{m^2}(\ddot{W} + H\dot{W} + \frac{q^2}{3a^2}W). \tag{25}$$

The first equation is (00) component of Einstein equations and the other two are the trace and traceless part of (ij) component. The $(0i)$ component may be considered as the definition of velocity perturbation. For adiabatic perturbations the equation of state connects δp and $\delta\rho$ as

$$\delta p = c_s^2\delta\rho, \;\; c_s^2 = -1 - \frac{\ddot{H}}{3H\dot{H}} \tag{26}$$

which allows us to obtain the closed equation on W like in the case of inflation driven by scalar field.

$$\frac{1}{m^2}\left(W^{(4)} - \frac{\ddot{H}}{\dot{H}}W^{(3)}\right) + A_2\,\ddot{W} + A_1\,\dot{W} + A_0\,W = 0\,, \tag{27}$$

where

$$\begin{cases} A_2 = 1 + \dfrac{1}{m^2}\left(\dfrac{q^2}{3a^2}\dfrac{\ddot{H}}{H\dot{H}} - H^2\right), \\ A_1 = H - \dfrac{\ddot{H}}{\dot{H}} + \dfrac{1}{m^2}\left[\dfrac{H^2\ddot{H}}{\dot{H}} - 3H\dot{H} - \dfrac{q^2}{a^2}\left(\dfrac{2\ddot{H}}{3\dot{H}} + 2H\right)\right], \\ A_0 = 2\dot{H} - \dfrac{H\ddot{H}}{\dot{H}} - \dfrac{q^2}{a^2}\left(\dfrac{\ddot{H}}{3H\dot{H}}\right) + \dfrac{q^2}{m^2a^2}\left(\dfrac{H\ddot{H}}{\dot{H}} - 4\dot{H} + 2H^2\right) - \dfrac{q^4}{m^2a^4}\left(1 + \dfrac{\ddot{H}}{3H\dot{H}}\right). \end{cases} \tag{28}$$

One can see that equation (27) differs from (21) only by terms proportional to q^2 so in the superhorizon regime both equations coincide. All modes that we can observe are in the superhorizon regime at the moment of transition from inflation to the matter stage so we do not care about the difference in equations for the subhorizon modes.

4.2 Solutions in the superhorizon regime

Now we consider the common solution of equation (21) (or (27) since they coincide in the superhorizon regime). Until $H \gg m$ holds this equation may be analytically solved. The four solutions are:

$$W_c, \quad \int a dt, \quad \int \frac{dt}{a}, \quad \int a dt_1 \int^{t_1} \frac{H}{a^2} dt_2 \int^{t_2} \frac{a\dot{H}}{H^2} dt_3 \tag{29}$$

The first mode W_c correspond to $\zeta = const$ and this solution provide the main impact to ζ in the limits $q/a \ll H$ and $H \gg m$. In the Section 3.4 it was shown that at inflation ζ behaves exactly as in the usual gravity. The exact conservation of ζ in the considering model was proved in [14]. W_c is connected to ζ as:

$$W_c = -2\epsilon\zeta \ - \text{inflation}, \quad W_c = -\frac{2\zeta}{p+1} \ - \text{stage } a \propto t^p. \tag{30}$$

The second mode growing as $\int a dt$ provides the main constant impact to the comoving perturbation B which is absent in Einstein gravity. One can see that B is conserved until $H \gg m$ holds independently of the stage of the Universe expansion.

The last solution at inflationary stage grows as t while at matter stage it behaves as t^2. This solution may be combined with W_c in such a way that it would give the only subdominant (and decreasing) impact to ζ and B.

When the Hubble parameter becomes smaller then the ghost mass, $H \ll m$, the ghost modes start oscillate even in the superhorizon regime. The solutions for W in that case were obtained in [14] for the stage $a \propto t^p$:

$$W_c, \quad t^{-p-1}, \quad t^{p/2} e^{\pm imt}. \tag{31}$$

The first two modes correspond to the Einstein modes and the growing oscillations are special solutions for Weyl gravity. As it was noted in [14] the growing modes do not give an impact to ζ. Again only mode W_c contribute to ζ like in the usual gravity. But the ghost modes provide the main impact to the comoving perturbation B. Since constant modes can be sewed exactly as in the Einstein gravity [mukhanov] we can sew remaining ghost modes through their contribution to B. The oscillating mode contribute to B as $t^{-p/2} e^{\pm imt}$. Connecting the solution for B at the transition moment $t = 1/m$ one can estimate for $t > 1/m$

$$B = \frac{\zeta}{q} \sqrt{\frac{\epsilon}{3}} (mt)^{-p/2} \left(\alpha_+ e^{imt} + \alpha_- e^{-imt} \right), \tag{32}$$

where $\alpha_\pm$ are expected to be constants of order unity.

4.3 Coming back to the horizon

In order to connect the superhorizon evolution of modes with the late subhorizon regime one can observe that the oscillating ghost modes resolve the equation on W in both cases: $q/a < H$ and $q/a < H$. The only needed condition is $m > H$ which holds at late stages for all potentially observable modes. So the ghost modes come into the horizon remaining unchanged from which follows that the subhorizon evolution for constant mode is the same as in the usual gravity.

Since the presence of ghost breaks the Newtonian gauge we must find all observables using the comoving gauge. The difference from the usual gravity is only in the non-vanishing perturbation B (32). In order to understand if this component provides any observable effect on the temperature anisotropy of CMB one need to consider its impact to the (integrated) Sachs–Wolfe effect. One may observe that the quantity qB in the corresponding equations stands together with ζ. So we can

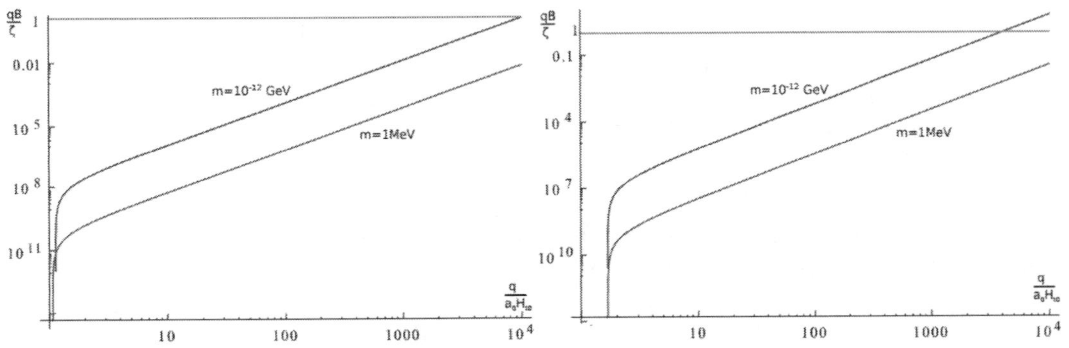

Figure 2. The left plot shows the value qB/ζ characterising the ghost contribution to the CMB anisotropy related to the present Universe. The right plot is for the moment of recombination. Here q/a_0 is the present physical momentum and H_{t0} is the present Hubble parameter. Directly observable modes correspond to $q/a_0H_{t0} < 1000$.

distinguish this case only if they are of the same order: $qB \sim \zeta$. The plots on Figure 2 shows the results related to the moment of recombination (right plot) and to the present moment (left plot). One can see that the deviations from the usual gravity may be expected only for short waves that we can not directly observe because they are in a strongly non-linear regime.

5 Conclusions

Summarising, we exploited the idea that extremely weak interacting ghost may be not so dangerous if we are considering the only effective description of the Universe. The instability arising from the small interaction between matter and ghost may be moved to the time much larger than the Universe age. This means that we can describe early Universe in terms of renormalizable and weakly interacting theory. We obtained the corrections to the observables, connected with scalar perturbations, due to extra ghost degrees of freedom to be negligible for the allowed ghost mass range.

In the considering theory, there are some processes with small but negative probabilities. The interpretation of such processes is an outstanding problem which we try to hide but actually it is impossible. Another problem for the non-unitary theory is that it remains unclear how to connect the quantum theory with bounded and positive energy, reached due to the negative norm prescription, to the classical field theory in which there is no way to avoid unbounded energy. This is an independent problem which is left beyond the scope of our consideration in this paper.

The author thank M. Ivanov, D. Gorbunov, A.A. Starobinsky, V.A. Rubakov, S.Sibiryakov for valuable discussions and critics. The work has been supported by Russian Science Foundation grant 14-12-01430.

References

[1] R. Kallosh, A. D. Linde, D. A. Linde and L. Susskind, Phys. Rev. D **52**, 912 (1995) [hep-th/9502069].
[2] F. Bezrukov, A. Magnin, M. Shaposhnikov and S. Sibiryakov, JHEP **1101**, 016 (2011) [arXiv:1008.5157 [hep-ph]].
[3] K. S. Stelle, Gen. Rel. Grav. **9**, 353 (1978).
[4] K. S. Stelle, Phys. Rev. D **16**, 953 (1977).

[5] A. Salvio and A. Strumia, JHEP **1406**, 080 (2014) [arXiv:1403.4226 [hep-ph]].
[6] S. W. Hawking and T. Hertog, Phys. Rev. D **65**, 103515 (2002) doi:10.1103/PhysRevD.65.103515 [hep-th/0107088].
[7] T. Clunan and M. Sasaki, Class. Quant. Grav. **27**, 165014 (2010) [arXiv:0907.3868 [hep-th]].
[8] A.A. Starobinsky, Ya.B. Zeldovich, JETP Lett. **26**, 252 (1977)
[9] A. Vilenkin, Phys. Rev. D **32**, 2511 (1985).
[10] A. A. Geraci, S. J. Smullin, D. M. Weld, J. Chiaverini and A. Kapitulnik, Phys. Rev. D **78**, 022002 (2008) [arXiv:0802.2350 [hep-ex]].
[11] G. Magnano, M. Ferraris and M. Francaviglia, Gen. Rel. Grav. **19**, 465 (1987). doi:10.1007/BF00760651
[12] A. A. Starobinsky, Phys. Lett. B **91**, 99 (1980).
[13] V. F. Mukhanov, H. A. Feldman and R. H. Brandenberger, Phys. Rept. **215**, 203 (1992).
[14] N. Deruelle, M. Sasaki, Y. Sendouda and A. Youssef, JCAP **1103**, 040 (2011) [arXiv:1012.5202 [gr-qc]].

Horndeski scalar-tensor black hole geodesics

Darya Tretyakova[1,2,⋆], *Dmitry* Melkoserov[1], and *Timur* Adyev[1]

[1] *Institute for Natural Sciences, Ural Federal University, Lenin av. 51, Yekaterinburg, 620083, Russia*
[2] *Sternberg Astronomical Institute, Lomonosov Moscow State University,*

Abstract. We examine massive particles and null geodesics for the scalar-tensor black hole in the Horndeski-Galileon framework. Our analysis shows that first kind relativistic orbits, corresponding to circular and elliptic orbits, are absent for the black hole solution with the static scalar field. This is a highly pathological behavior contradicting to the black hole accretion and Solar System observations.

1 Introduction

Scalar-tensor gravity is a widely accepted alternative to the General Relativity for the role of gravity theory. The most general classical effective scalar-tensor theory was proposed by Horndeski [1], and the same result was aslo obtained by studying Galileons [2, 3]. The Galileon model is a ghost free scalar effective field theory containing higher derivative terms that are protected by the Galileon symmetry while Horndeski was interested in a most general type of scalar-tensor theory yielding second order field equations. The resulting framework is a healthy scalar-tensor theory, which was until recently believed to be the most general scalar-tensor theory for second order field equations. A newly discovered class that goes beyond Horndeski contains no dangerous instabilities, since higher order terms cancel out due to hidden constraints [4]. However Horndeski framework remains interesting and fruitfull, working very well in cosmological applications. The action for the Horndeski/Galileon model in the non-trivial kinetic coupling sector reads

$$S = \int dx^4 \sqrt{-g}\left(\zeta R - \eta\,(\partial\phi)^2 + \beta G^{\mu\nu}\partial_\mu\phi\partial_\nu\phi - 2\Lambda\right) \tag{1}$$

here $G^{\mu\nu}$ is the Einstein tensor, ϕ is the scalar field, constants $\zeta > 0, \eta$ and β are model parameters. Paper [5] shows that purely kinetic coupled gravity is inconsistent with the constraint from the gravitational Cherenkov radiation for any theoretically allowed parameter β, however the action considered in [5] has $\eta = 1$ and $\Lambda = 0$. Introducing Λ in the action (1) (dividing the dark energy into different-behaving components) is a way to relax this inconsistency. The presence of a Vainshtein screening mechanism on the other hand allows the scalar field to couple to matter without mediating unacceptably large fifth forces in the solar system [6, 7]. Henceforth it would be interesting to pursue the study of the solutions of (1) at the astrophysical scale.

Any modification of GR must be consistent with constraints astrophysical scales and at the Solar System level, which are very stringent. The purpose of this paper is thus to threat the local spherically

⋆e-mail: daria.tretiakova@urfu.ru

symmetric solutions of (1) as astronomical objects and see if this picture is in agreement withe the observed one. To do so we study the test particle motion around the compact object described by the geodesic equations.

This paper is organized as follows. In section 2 we briefly summarize the properties of the black hole-like solutions for the action (1). Section 3 contains solution's parameter estimates found in the literature as well as some new ones, obtained in this paper. We proceed with analyzing geodesic equations in section 4. Conclusions are given thereafter.

2 Horndeski/Galileon black hole

For the spherically-symmetric ansatz with the a mild time dependence of the scalar field

$$ds^2 = h(r)dt^2 - \frac{dr^2}{f(r)} - r^2 d\Omega^2 \tag{2}$$

$$\phi(r) = qt + \psi(r), \tag{3}$$

black hole solutions were found in the series of papers [8]-[12]. These solutions possess very similar properties, so we choose to consider the metric from [9], the only one having peculiar terms

$$f(r) = \frac{(1+\gamma^2 r^2)h(r)}{(1+3Ar^2+B/(1+\gamma^2 r^2))}, \tag{4}$$

$$h(r) = 1 - \frac{\mu}{r} + Ar^2 + B\frac{\arctan(r\gamma)}{r\gamma}, \tag{5}$$

where μ plays the role of the mass term. We also used the following notations[1]

$$A = \frac{\gamma^2}{3}\frac{2\zeta\eta - \lambda}{2\zeta\eta + \lambda}, \qquad B = \frac{\lambda^2}{4\zeta^2\eta^2 - \lambda^2}, \tag{6}$$

$$\lambda = \zeta\eta + \beta\Lambda, \qquad \gamma = \sqrt{\eta/\beta}. \tag{7}$$

(8)

By fixing all the parameters and rising γ we approach the corresponding adS solution for $h(r)$ (which also shifts with γ). By lowering γ Schwarzshild-like behavior is recovered. The solution in general interpolates in-between. However the Schwarzshild horizon radius μ is never achieved with $\gamma \to \infty$, since $r_h = 3/4\mu$ in this case and the gap remains.

3 Parameter estimates

In this section we briefly review the constraints on the parameters of the non-minimal derivative coupling sector of Horndeski theory, determined elsewhere. First of all, thermodynamical considerations suggest $\beta > 0$ [8].

The authors of [13] considered slowly rotating neutron stars for the metric (16)-(15) by using several tabulated and realistic equations of state. The existence of a static configuration requires

$$q^2\beta < 4\zeta/3 \tag{9}$$

for $\beta > 0$, in order to have the pressure maximum at the center of the neutron star. The model reproduces the maximal neutron mass observed and provides a viable compact objects astrophysics.

[1]The units are $[\gamma] = m^{-1}$.

Within the constraint, it was shown that there exists neutron stars configurations with the redshift value compatible with typical measurements.

Minamitsuji [14] investigated the stability of BH solutions under massless scalar perturbations in the nonminimal derivative coupling subclass. The quasinormal modes can be computed, and no unstable modes were found. Considering the same BH solutions, Cisterna et al. [15] found that these black holes are stable under odd-parity gravitational perturbations as well. Kobayashi et al. [16] have shown that the static scalar field solution in question is stable in some range of the Λ vs. $\zeta\gamma$ parameters.

The solution (16)-(15) possesses the Schwarzschild - anti de Sitter structure if we suggest that Λ does not exceed the observed value $\Lambda \approx 10^{-52}m^{-2}$. For the ΛCDM expansion history, there essentially always exists a family of models which interpolate between the cosmological constant as the only source of acceleration and models which are fully dynamical, so the conjecture above makes sense. Given this value of Λ we can further neglect it with respect to the other terms and set $\Lambda \approx 0$. Then

$$\lambda \approx \zeta\eta, \qquad A \approx \frac{\gamma^2}{9}, \qquad B \approx \frac{1}{3}. \tag{10}$$

With this simplifications we have now only one additional parameter γ left for the fine tuning. The correction term

$$\Delta(r) = B\frac{arctan(r\gamma)}{r\gamma}, \tag{11}$$

now gives $\Delta \to B$ near $r = 0$ and $\Delta \to 0$ when $r \to \infty$.

We can also consider weak-field observations. One of the well-studied gravitational effects is the frequency shift for the satellite on the Earth orbit

$$\frac{\delta\nu}{\nu} = 1 - \sqrt{\frac{h(R+h)}{h(R)}} \approx \frac{V(R+h) - V(R)}{c^2}, \tag{12}$$

up to the first order in the weak field approximation with h being the satellite orbit height and R - the Earth surface radius, V - the corresponding gravitational potential. For the metric (4)-(5) there should be an additional shift, related to the deviation of h(r) from the Schwarzshild solution

$$\begin{aligned} 2\frac{\delta\nu}{\nu} &\approx \delta_{Schw} + \delta_1 + \delta_2 = \\ &\left(\frac{\mu}{R} - \frac{\mu}{R+h}\right) + A((R+h)^2 - R^2) + (\Delta(R+h) - \Delta(R)). \end{aligned} \tag{13}$$

Modern frequency measurements are in agreement with GR, hence we can make bounds using the frequency measurement accuracy 10^{-14} achieved in the GP–A redshift experiment[2] [17]. Numerical estimates show that $\delta_1 + \delta_2$ does not exceed the accuracy of the relative frequency measurement when

$$\gamma < 10^{-14}. \tag{14}$$

This estimate allows us to set $A \approx 0$ (and hence $\gamma^2 \approx 0$) when considering accretion and null geodesics for the Sgr A[3] black hole ore any smaller one since the conditions $Ar^2 << \mu/r$ and $Ar^2 << \Delta$ are well

[2] $h = 15 \times 10^3 km, M_\oplus = 5,972 \times 10^{24} kg, R_\oplus = 6371 km,$
$G_0 = 6.67384 \times 10^{-11} m^3 kg^{-1} c{-2}$

[3] $M_{SgrA} = 4 \times 10^6 M_\odot$

satisfied within the corresponding $100r_{Schw}$ distance. All the above results in the metric

$$h(r) = 1 - \frac{\mu}{r} + \frac{1}{3}\frac{\arctan(r\gamma)}{r\gamma}, \tag{15}$$

$$f(r) = \frac{3}{4}h(r). \tag{16}$$

(17)

4 Geodesic equations

Geodesic equation for the radial coordinate reads

$$\left(\frac{dr}{d\varphi}\right)^2 = f(r)P(r), \tag{18}$$

$$P(r) = \left[\frac{E^2 - jh(r)}{h(r)}\right]\frac{r^4}{L^2} - r^2, \tag{19}$$

where $j = 0$ for massless particles and $j = 1$ for the massive ones. We will further use the inverse radius $u = r^{-1}$. Working with geodesics we are beyond the black hole horizon and hence $h(u) \neq 0$ (and hence $f(u) \neq 0$). Therefore we can write (18) for the circular orbits as

$$P(u) = 0, \tag{20}$$

$$P'(u) = 0. \tag{21}$$

4.1 Null geodesics

An observational parameter, which will be important in the near future, is the photon sphere radius of the black hole, which relates directly to it's shadow size. Due to the smallness of γ the following expansion will be valid not too far from the black hole:

$$\arctan(r\gamma) \approx r\gamma - \frac{(r\gamma)^3}{3}, \tag{22}$$

$$\Delta(r) \approx \frac{1}{3} - Ar^2 \tag{23}$$

$$h(r) \approx \frac{4}{3} - \frac{\mu}{r}. \tag{24}$$

(25)

Writing the null geodesic equation in terms of the inverse radius $u = r^{-1}$ we thus obtain

$$P(u) = \frac{E^2}{h(u)}\frac{u^{-4}}{L^2} - u^{-2} \tag{26}$$

$$\left(\frac{du}{d\varphi}\right)^2 = \frac{3}{4}\frac{E^2}{L^2} - u^2 + \frac{3}{4}\mu u^3 = R(u) \tag{27}$$

We are seeking for the last unstable photon orbit, so we are interested in the critical case when two positive real roots of $R(u)$ coincide, which gives in our case

$$u_1 = u_2 = \frac{8}{9\mu}, \quad u_3 = -\frac{4}{9\mu}. \tag{28}$$

The solution of the geodesic equation takes then the following form:

$$u = -\frac{4}{9\mu} + \frac{4}{3\mu}\left[tanh\left(\frac{\varphi - \varphi_0}{2}\right)\right]^2. \tag{29}$$

Considering the limit $\varphi \to \infty$ we can find the radius of photon sphere as

$$r_s = \frac{9\mu}{8} \tag{30}$$

whereas $r_s = 3\mu/2$ for the Schwarzshild metric. The observational characteristics of the metric in question differ sufficiently and require therefore further exploration, so we now proceed with time-like geodesics.

4.2 Massive particles

For the bound orbits we have $E^2 < 1$. Substituting (15) into (20)-(21) we obtain

$$\mu u^3 - u^2 + \frac{j\mu}{L^2}u - \frac{j - E^2}{L^2} - \Delta\left(u^2 + \frac{j}{L^2}\right) = 0, \tag{31}$$

$$3\mu u^2 - 2u + \frac{j\mu}{L^2} - \Delta'\left(u^2 + \frac{j}{L^2}\right) - 2u\Delta = 0, \tag{32}$$

where the prime denotes the derivative with respect to the radial coordinate. One can simplify Δ' if we recall that $\gamma^2/u^2 \equiv \gamma^2 r^2 \sim Ar^2$, and neglect this term as we did before. Now extract Δ from (31) and substitute into (32):

$$F(u) = (1 - 3\varepsilon)\mu u^3 - (1 - \frac{5}{3}\varepsilon)u^2 + (1 - \varepsilon)\frac{\mu}{L^2}u - \frac{(1 + \varepsilon/3) - E^2}{L^2}, \tag{33}$$

$$\varepsilon = 1 - \frac{2}{3 + \frac{1}{u^2L^2}}, \tag{34}$$

where $F(u) = 0$. Note that $\varepsilon(u, L)$ is bound as $1/3 < \varepsilon < 1$ while u is running between 0 and ∞: This forces the coefficient before u^3 in $F(u)$ to be negative and determines the behavior of the expression. We can also rewrite $F(u)$ as

$$F(u) = \frac{1}{3}\frac{-4L^4u^4 + 9E^2L^2u^2 - 8L^2u^2 + 3E^2 - 4}{(3L^2u^2 + 1)L^2} \tag{35}$$

which is a direct consequence of $B = 1/3$ since higher orders in u cancel out in the equation. This results in the following solution for the circular geodesics:

$$u = \pm\frac{1}{4L}\sqrt{18E^2 - 16 \pm 2\sqrt{81E^4 - 96E^2}}. \tag{36}$$

We see that for $E^2 < 1$ there are no real positive roots, so circular orbits for massive particles are excluded. This may seem an artifact of the approximation, however one can verify numerically for the general metric (5), that unlike the Schwarzshild case conditions (20) and (21) can never be satisfied simultaneously.

Circular orbits represent a narrow subclass of relativistic bound orbits. However, as they represent a subclass of the elliptic orbits, the latter should be excluded as well (the analysis does not involve the eccentricity at all) and so all the first class orbits, twisted between two radii are eliminated. This conclusion can be also obtained by analysing the general behavior of P(u). This leaves us only with the trajectories, ending up in the black hole singularity (ε).

5 Discussion and conclusions

Departing from GR we are left aside of the Birkhoff's theorem. In this situation we are forced to choose between multiple spherically symmetric vacua and therefore some hints on how to determine the realistic solution would come in handy. A configuration used by the authors of [9] is rather interesting. By introducing a mild linear dependence in the time coordinate for the scalar field one evades the scalar field being singular for it's derivative on the horizon [18] and makes the field equations to bifurcate the no-hair theorem at the same time. Despite this dependence the shift symmetry is keeping the field equations time-independent and consistent with the static ansatz. This permits asymptotically flat (or de-Sitter) solutions and crucially gives regular scalar tensor black holes.

The metric (4)-(5) is static, but has $q = 0$ for the the scalar field which is not regular at the horizon. Our analysis of this metric revealed the absence of stable orbits which do not end up in the singularity. This is a highly pathological behavior contradicting to the black hole accretion observations [19, 20]. Furthermore, the Schwarzschild solution in GR also describes the exterior of any spherically symmetric body in the weak field limit (hence the Solar System), and so must do it's analog in extended gravity theory. However, with in-falling orbits only, the metric (4)-(5) fails to reproduce the solar system motion. This is the situation that to our knowledge did not appear before in extended gravity solutions: usually we can always suppress the deviations from the GR by the fine tuning. Another interesting example, demonstrating that studying local spherically symmetric solutions can provide useful information on cosmological models is given in [21]. The paper implements some PPN-based bounds on the model parameters due to the black hole-like metric for the framework, initially constructed for cosmological purposes. The parameter region left does not satisfy the purposes of the model, rendering it far less attractive. Hence geodesic analysis remains a useful tool to explore extended gravity solutions, even though the observations of the test body motion in the vicinity ob black holes are not sensitive enough to distinguish extended gravity from the GR. Black hole geodesics consideration can give us the direction towards astrophysically viable black holes in Horndeski-Galileon framework.

Acknowledgements

This work was supported by Russian Foundation for Basic Research via grant RFBR №16-02-00682.

References

[1] G.W. Horndeski, Int. J. Theor. Phys. **10**, 363 (1974)
[2] C. Deffayet, S. Deser, G. Esposito-Farese, Phys. Rev. **D80**, 064015 (2009), 0906.1967
[3] C. Deffayet, G. Esposito-Farese, A. Vikman, Phys. Rev. **D79**, 084003 (2009), 0901.1314
[4] C. Lin, S. Mukohyama, R. Namba, R. Saitou, JCAP **1410**, 071 (2014), 1408.0670
[5] R. Kimura, K. Yamamoto, JCAP **1207**, 050 (2012), 1112.4284
[6] R. Kase, S. Tsujikawa, JCAP **1308**, 054 (2013), 1306.6401
[7] E. Babichev, C. Deffayet, G. Esposito-Farese, Phys. Rev. Lett. **107**, 251102 (2011), 1107.1569
[8] M. Rinaldi, Phys. Rev. **D86**, 084048 (2012), 1208.0103
[9] E. Babichev, C. Charmousis, JHEP **1408**, 106 (2014), 1312.3204
[10] C. Charmousis, D. Iosifidis, J. Phys. Conf. Ser. **600**, 012003 (2015), 1501.05167
[11] E. Babichev, C. Charmousis, M. Hassaine, JCAP **1505**, 031 (2015), 1503.02545
[12] M. Minamitsuji, Phys. Rev. **D89**, 064017 (2014), 1312.3759

[13] A. Cisterna, T. Delsate, L. Ducobu, M. Rinaldi, Phys. Rev. **D93**, 084046 (2016), 1602.06939
[14] M. Minamitsuji, Gen. Rel. Grav. **46**, 1785 (2014), 1407.4901
[15] A. Cisterna, M. Cruz, T. Delsate, J. Saavedra, Phys. Rev. **D92**, 104018 (2015), 1508.06413
[16] T. Kobayashi, H. Motohashi, T. Suyama, Phys. Rev. **D89**, 084042 (2014), 1402.6740
[17] R.F.C. Vessot, M.W. Levine, E.M. Mattison, E.L. Blomberg, T.E. Hoffman, G.U. Nystrom, B.F. Farrel, R. Decher, P.B. Eby, C.R. Baugher et al., Phys. Rev. Lett. **45**, 2081 (1980)
[18] C. Charmousis, T. Kolyvaris, E. Papantonopoulos, M. Tsoukalas, JHEP **07**, 085 (2014), 1404.1024
[19] J.A. Muñoz, E. Mediavilla, C.S. Kochanek, E.E. Falco, A.M. Mosquera, The Astrophysical Journal **742**, 67 (2011), 1107.5932
[20] R. Fender, T. Belloni, Science **337**, 540 (2012), 1208.1138
[21] D.A. Tretyakova, B.N. Latosh, S.O. Alexeyev, Class. Quant. Grav. **32**, 185002 (2015), 1504.06723

The very best limits on cosmological magnetic fields

Federico R Urban[1,*]

[1] *National Institute of Chemical Physics and Biophysics, Rävala 10*
10143 Tallinn, Estonia

Abstract. I introduce you to the very best limits on cosmological magnetic fields. The numbers: magnetic fields coherent across a Jeans' length (around 2.4 Mpc today), are bound to be weaker than about 1 nG (2σ statement), whereas for a field as wide as the entire observable Universe the number is approximately 0.5 nG. These limits are obtained from rotation measures data, and as such they do not depend on the epoch at which the fields were generated.

1 Why we care

The Universe is swamped with all sorts of magnetic fields (MF): planetary MFs, galactic MFs, cluster MFs, possibly void MFs, etc. [1, 2]. Cosmological MFs (fields which imbue the entire Universe), on the other hand, remain elusive, but there is a whole world of physicists from different disciplines out there, ultra-high energy cosmic rays physics, structure formation, early and very early Universe theory, physics beyond the Standard Model, radio-astronomy, who would die to know more about them [3–5].

Cosmological MFs manifest themselves by rotating the plane of polarisation of electromagnetic waves propagating from far away sources to the Earth. Now, since we know that: (1) Faraday rotation measures (RM) of distant objects do not show any evolution with redshift z [6], and (2) cosmological MFs (cosmoMF) generate RMs as we move away from here [7, 8], then (3) we can limit the strength of these cosmoMFs by comparing simulated and observed RMs distributions.

This contribution is based on my work with my collaborators Maxim Pshirkov and Peter Tinyakov [9].

2 The new method

Observations.

The only formula we need to know is that of Faraday rotation: the plane of polarisation of a linearly polarised electromagnetic wave which moves through a magnetised plasma rotates by an angle φ proportional to the square of the wavelength λ: $\varphi = \mathrm{RM}\,\lambda^2$, where

$$\mathrm{RM} = 812 \int_D^0 \frac{n_e(z) B_\parallel(z)}{(1+z)^2} \left| \frac{\mathrm{d}l(z)}{\mathrm{d}z} \right| \mathrm{d}z \,. \qquad (1)$$

[*] e-mail: federico.urban@kbfi.ee

Here n_e is the density of free electrons measured in cm^{-3}, $B_{\parallel}$ is the component of the MF (in μG) parallel to the line of sight $l(z)$, and D is the distance to the source in kpc; here and everywhere the RM is measured in rad/m^2.

What do we know about those RMs? First, the data: the largest set of RM of extragalactic sources to date was compiled in [10] from the NRAO VLA Sky Survey (NVSS) data [11]. The total number of observed sources was 37,543, of which 4002 have known redshifts [12]. These RMs consist of several contributions: we can split the observed RM as $\mathrm{RM_{obs}} = \mathrm{RM_{gal}} + \mathrm{RRM}$, where the first term comes from the regular MF of the Milky Way, and the second term stands for "residual RM": RM instrinsic to the source, measurement errors, turbulent galactic MF, and cosmoMF (xRRM) [13]. We massaged the data a bit to get the best grip on the redshift dependence, see [6] for all the tricks: after the massage we end up with a total of 2593 sources (our data points).

Simulations.

Like we said, if there is a cosmoMF the xRRM systematically grows with redshift due to its accumulation along the line of sight. To compute this we need n_e, which we take from the observed Lyman-α forest distribution of neutral hydrogen absorption lines [7, 14, 15]. Next, since the conductivity of the Universe is extremely large, and since diffusion of the MF is inefficient at scales much larger than 1 AU, we can take a MF frozen into the plasma [16, 17]. And then we run a simple code which spits out xRRM for a MF with strength B and coherence length l_c, the latter ranging from a tenth of the Jeans length λ_J to the Hubble size $1/H_0$.

These simulations can generate only the contribution from the cosmoMF, but in real life there is a mickle of other stuff we do not (want to) know, so here is the stratagem: we exploit the data at low redshifts to obtain information about these unsung, redshift-independent, heroes, since there the cosmoMF is subdominant there, and use it together with our simulated xRRM at high redshift. In practice the algorithm is: (i) split the Universe in two redshift bands; (ii) make a first RM batch by randomly picking RRMs from the low-z set; (iii) generate a second RM batch by simulating 100 xRRMs for each of the data sources of the high-z set; (iv) blindly pluck one value from each of the batches and incoherently (that is, each with its own sign) add them a bunch of times to generate the final theoretical |RRM| distribution; (v) compare with actual data by means of a Kolmogorov-Smirnov test, which yields a p-value for the statistical (in)compatibility of the two distributions.

3 The result

See the self-explanatory Fig. 1.

4 How we fare

Currently, the strongest upper limits on the strength of present-day cosmoMFs come from microwave background observations [18] and read $B \lesssim 2.8$ nG for a coherence length $l_c = 1$ Mpc — these limits apply only to primordial MFs, i.e., fields generated in the very early Universe. The limits coming RM data are looser [7, 19]: λ_J-scale fields are bound by $B \lesssim 6$ nG. Our limits on the other hand read 1 nG (this is a 2σ statement) for $l_c = \lambda_J$, whereas $B \lesssim 0.5$ nG for the bloated field case, $l_c = 1/H_0$.

Conclusion: all in all, we win 😎

Acknowledgements.

This work was supported by IISN project No. 4.4502.13 and Belgian Science Policy under IAP VII/37; my participation to QUARKS2016 was supported by ERC grant PUT808.

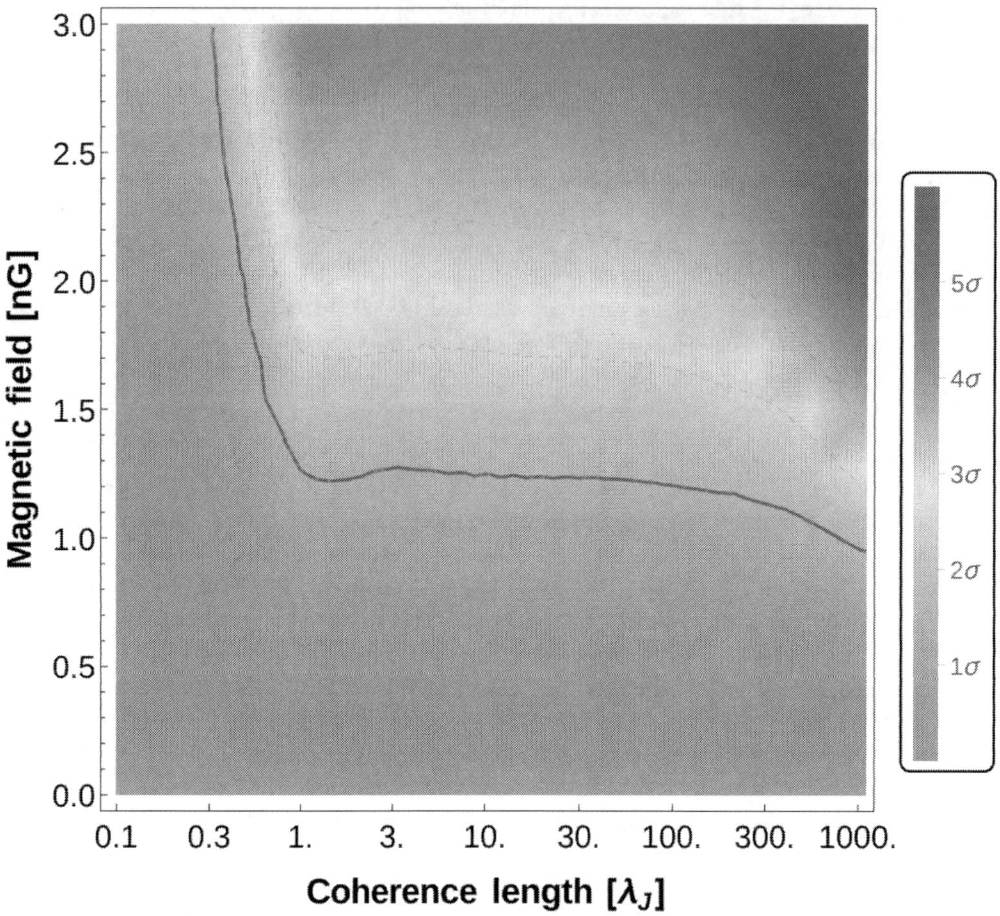

Figure 1. Confidence interval contours obtained from the p-values of the Kolmogorov-Smirnov tests as a function of the MF reference strength (y-axis) and coherence length (x-axis).

References

[1] J. P. Vallée, New Astron. Rev. **48** (2004) no.10, 763. doi:10.1016/j.newar.2004.03.017.

[2] D. Ryu, D. R. G. Schleicher, R. A. Treumann, C. G. Tsagas and L. M. Widrow, Space Sci. Rev. **166** (2012) 1 doi:10.1007/s11214-011-9839-z [arXiv:1109.4055 [astro-ph.CO]].

[3] O. Deligny, A. Letessier-Selvon and E. Parizot, Astropart. Phys. **21** (2004) 609 doi:10.1016/j.astropartphys.2004.04.012 [astro-ph/0303624].

[4] L. M. Widrow, D. Ryu, D. R. G. Schleicher, K. Subramanian, C. G. Tsagas and R. A. Treumann, Space Sci. Rev. **166** (2012) 37 doi:10.1007/s11214-011-9833-5 [arXiv:1109.4052 [astro-ph.CO]].

[5] R. Durrer and A. Neronov, Astron. Astrophys. Rev. **21** (2013) 62 doi:10.1007/s00159-013-0062-7 [arXiv:1303.7121 [astro-ph.CO]].

[6] M. S. Pshirkov, P. G. Tinyakov and F. R. Urban, Mon. Not. Roy. Astron. Soc. **452** (2015) 2851 doi:10.1093/mnras/stv1273 [arXiv:1407.3909 [astro-ph.GA]].

[7] P. Blasi, S. Burles and A. V. Olinto, Astrophys. J. **514** (1999) L79 doi:10.1086/311958 [astro-ph/9812487].
[8] T. Akahori and D. Ryu, Astrophys. J. **738** (2011) 134 doi:10.1088/0004-637X/738/2/134 [arXiv:1107.0142 [astro-ph.CO]].
[9] M. S. Pshirkov, P. G. Tinyakov and F. R. Urban, arXiv:1504.06546 [astro-ph.CO].
[10] A. R. Taylor, J. M. Stil and C. Sunstrum, Astrophys. J. **702** (2009) 1230 doi:10.1088/0004-637X/702/2/1230.
[11] J. J. Condon, W. D. Cotton, E. W. Greisen, Q. F. Yin, R. A. Perley, G. B. Taylor and J. J. Broderick, Astron. J. **115** (1998) 1693. doi:10.1086/300337.
[12] A. M. Hammond, T. Robishaw and B. M. Gaensler, arXiv:1209.1438 [astro-ph.CO].
[13] D. H. F. M. Schnitzeler, Mon. Not. Roy. Astron. Soc. **409** (2010) L97 doi:10.1111/j.1745-3933.2010.00957.x [arXiv:1011.0737 [astro-ph.GA]].
[14] P. Coles and B. Jones, Mon. Not. Roy. Astron. Soc. **248** (1991) 1.
[15] H. Bi and A. F. Davidsen, Astrophys. J. **479** (1997) 523 doi:10.1086/303908 [astro-ph/9611062].
[16] L. Hollenstein, R. K. Jain and F. R. Urban, JCAP **1301** (2013) 013 doi:10.1088/1475-7516/2013/01/013 [arXiv:1208.6547 [astro-ph.CO]].
[17] P. G. Tinyakov and F. R. Urban, arXiv:1309.2270 [astro-ph.CO].
[18] P. A. R. Ade *et al.* [Planck Collaboration], arXiv:1502.01594 [astro-ph.CO].
[19] P. P. Kronberg, M. L. Bernet, F. Miniati, S. J. Lilly, M. B. Short and D. M. Higdon, Astrophys. J. **676** (2008) 7079 doi:10.1086/527281 [arXiv:0712.0435 [astro-ph]].

High energy particle collisions near black holes

O. B. Zaslavskii ⋆

[1] *Department of Physics and Technology, Kharkov V.N. Karazin National University, 4 Svoboda Square, Kharkov 61022, Ukraine*

Abstract. If two geodesic particles collide near a rotating black hole, their energy in the centre of mass frame $E_{c.m.}$ can become unbound under certain conditions (the so-called BSW effect). The special role is played here by so-called critical geodesics when one of particles has fine-tuned energy and angular momentum. The nature of geodesics reveals itself also in fate of the debris after collisions. One of particles moving to a remote observer is necessarily near-critical. We discuss, when such a collision can give rise not only unboud $E_{c.m.}$ but also unbound Killing energy E (so-called super-Penrose process).

Investigation of high energy collisions of particles near black holes comes back to works [1] - [3]. In recent years, an interest to this issue revived after the observation made by Bañados, Silk and West (the BSW effect, after the names of its authors) that particle collision near the Kerr black hole can lead, under certain additional conditions, to the unbounded growth of the energy in their centre of mass $E_{c.m.}$ [4]. Later on, in a large series of works, this observation was generalized and extended to other objects and scenarios. The energy that appears in the BSW effect is relevant for an observer who is present just near the point of collision in the vicinity of the black hole horizon. Meanwhile, what is especially physically important is the Killing energy E of debris after such a collision measured by an observer at infinity. Strong redshift "eats" significant part of $E_{c.m.}$, so it was not quite clear in advance, to what extent the energy E may be high. If E exceeds the initial energy of particles, we are faced with the energy extraction from a black hole. This is a so-called collisional Penrose process (the Penrose process that occurs due to particle collisions). Thus there are two related but different issues: (i) investigation of the effect of unbounded $E_{c.m.}$, (ii) study of properties of E and the question about the maximum possible efficiency of the collisional Penrose process.

Consider the generic axially symmetric metric. It can be written as

$$ds^2 = -N^2dt^2 + g_{\phi\phi}(d\phi - \omega dt)^2 + dl^2 + g_{zz}dz^2. \quad (1)$$

Here, the metric coefficients do not depend on t and ϕ. On the horizon $N = 0$. Alternatively, one can use coordinates θ and r, similar to Boyer–Lindquist ones for the Kerr metric, instead of l and z. In (1) we assume that the metric coefficients are even functions of z, so the equatorial plane $\theta = \frac{\pi}{2}$ ($z = 0$) is a symmetry one.

In the space-time under discussion there are two conserved quantities $u_0 \equiv -E$ and $u_\phi \equiv L$ where $u^\mu = \frac{dx^\mu}{d\tau}$ is the four-velocity of a test particle, τ is the proper time and $x^\mu = (t, \phi, l, z)$ are coordinates..The aforementioned conserved quantities have the physical meaning of the energy per unit mass (or frequency for a lightlike particle) and azimuthal component of the angular momentum,

⋆e-mail: zaslav@ukr.net

respectively. It follows from the symmetry reasonings that there exist geodesics in such a background which lie entirely in the plane $\theta = \frac{\pi}{2}$. We consider a pair of particles labeled by the subscript $i = 1, 2$ and having the equal rest masses $m_1 = m_2 = m$. We also assume that both particles are approaching the horizon. Then, using the first integrals for such geodesics one can calculate the energy in the centre of mass frame $E_{c.m.} = \sqrt{2}m\sqrt{1 - u_{\mu(1)}u^{\mu(2)}}$[4]. After simple manipulations, one obtains

$$\frac{E_{c.m.}^2}{2m^2} = \frac{X_1X_2 - Z_1Z_2}{N^2} + 1 - Y = \frac{L_1L_2}{g_{\phi\phi}}, \tag{2}$$

where $X = X_i \equiv E_i - \omega L_i$,

$$Z_i = \sqrt{(E_i - \omega L_i)^2 - N^2 b_i}, \; b_i = 1 + \frac{L_i^2}{g_{\phi\phi}}. \tag{3}$$

In the horizon limit $N \to 0$ one obtains

$$\left(\frac{E_{c.m.}^2}{2m^2}\right)_H = 1 + \frac{b_{1(H)}(L_{2(H)} - L_2)}{2(L_{1H} - L_1)} + \frac{b_{2(H)}(L_{(1)H} - L_1)}{2(L_{2(H)} - L_2)} - \frac{L_1L_2}{(g_{\phi\phi})_H}, \; L_{i(H)} \equiv \frac{E_i}{\omega_H}. \tag{4}$$

By a very meaning of derivation, it is supposed in (4) that $L_1 \neq L_{2(H)}$, $L_2 \neq L_{2(H)}$. If $L_1 = L_{1(H)}(1 - \varepsilon)$, $\varepsilon \ll 1$, $L_2 \neq L_{2(H)}$,

$$\left(\frac{E_{c.m.}^2}{2m^2}\right)_H \approx \frac{b_{1(H)}(L_{2(H)} - L_2)}{2L_{1(H)}\varepsilon} \tag{5}$$

becomes unbounded. This is just the BSW effect. What is important is that the effect is universal and takes place near generic rotating axially symmetric black holes [6]. Moreover, this happens near not only extremal black holes (as assumed usually) but near nonextremal one as well [7].

One can suggest a quite generic explanations based on geometric properties of the particles' trajectories near the horizon [8]. Let us consider the collision of two particles near the future horizon of a black hole. We assume that both particles move towards the horizon. Let us introduce in the point P under consideration and its vicinity the tetrad with lightlike vectors l^μ, N^μ and spacelike vectors a^μ, b^μ orthogonal to them. Here, the vectors l^μ, N^μ are normalized, say, as $l^\mu N_\mu = -1$.

Then,

$$g_{\alpha\beta} = -l_\alpha N_\beta - l_\beta N_\alpha + \sigma_{\alpha\beta} \tag{6}$$

where $\sigma_{\alpha\beta} = a_\alpha b_\beta + a_\beta b_\alpha$, $l^\alpha \sigma_{\alpha\beta} = N^\alpha \sigma_{\alpha\beta} = 0$. We assume that it is the vector l^μ that becomes the generator of the future horizon. In general, we can use the decomposition of the four-velocity u^μ in the form

$$u_i^\mu = \frac{l^\mu}{2\alpha_i} + \beta_i N^\mu + s_i^\mu, \; s_i^\mu = A_i a^\mu + B_i b^\mu \tag{7}$$

where $i = 1, 2$ labels the particles and α_i, β_i, A_i and B_i are coefficients. The time-like vector u^μ is normalized as usual, $(uu) = -1$, hereafter the symbol (...) denotes the scalar product. Then, it follows from (7) that

$$\beta_i = -(u_i l), \tag{8}$$

$$\alpha_i = -\frac{1}{2}(u_i N)^{-1}. \tag{9}$$

As vectors u^μ, l^μ, N^μ are assumed to be future-directed, $\alpha_i > 0$, $\beta_i > 0$ (motion "forward in time"). The normalization condition entails

$$s_i^\mu s_{i\mu} = \frac{\beta_i}{\alpha_i} - 1. \tag{10}$$

The case $\beta_i = \alpha_i$, $s_i^\mu = 0$ corresponds to pure radial motion (see below).

Then,

$$-(u_1 u_2) = \frac{1}{2}(\frac{\beta_1}{\alpha_2} + \frac{\beta_2}{\alpha_1}) - (s_1 s_2). \tag{11}$$

The energy in the centre of mass frame [4] is equal to $E_{c.m.}^2 = m_1^2 + m_2^2 - 2m_1 m_2 (u_1 u_2)$ (m_i are rest masses of particles), so

$$E_{c.m.}^2 = m_1^2 + m_2^2 + m_1 m_2 [\frac{\beta_1}{\alpha_2} + \frac{\beta_2}{\alpha_1} - 2(s_1 s_2)]. \tag{12}$$

The frame of the centre of mass falls down with both particles [4], so the possible effect of unbound acceleration is not direct manifestation of the blueshift. In general, as it is seen from (12), $E_{c.m.}^2$ remains finite even in the vicinity of the horizon for any nonzero α_1, α_2. However, if, say, $\alpha_1 \to 0$ that corresponds to a so-called critical particle, $E_{c.m.}^2 \to \infty$.

We can also suggest simple kinematic explanation. We consider the geodesic motion of massive particles in the equatorial plane $\theta = \frac{\pi}{2}$. Then, the relation holds [9]

$$E - \omega L = \frac{N}{\sqrt{1 - V^2}}, \tag{13}$$

where V is the velocity measured by a local zero angular momentum observer (ZAMO) [5]. One can observe that for generic E, L in the horizon limit $N \to 0$ the velocity $V \to 1$. But for $E = \omega_H L$ (ω_H is the horizon value), $V < 1$. Thus we have collision between one "slow" and one "rapid" particles. Their relative velocity $w \to 1$, the Lorentz factor of relative motion $\gamma = \frac{1}{\sqrt{1-w^2}} \to \infty$, the energy $E_{c.m.} \to \infty$.

Now, we turn to the question, what can see an observer at infinity. If E exceeds the initial energy of particles, we are faced with the energy extraction from a black hole. This is a so-called collisional Penrose process (the Penrose process that occurs due to particle collisions). Strong redshift "eats" significant part of $E_{c.m.}$, so it was not quite clear in advance, to what extent the energy E may be high. It turned out that energy extraction from the Kerr black hole is possible but it was find to be relatively modest [10], [11], [12]. More general situation not restricted to the Kerr metric was considered in [13] where quite general upper bounds were derived that depend on the details of the metric. But, also, indefinitely large E turned out to be impossible.

The aforementioned results on the energy extraction were obtained for the standard BSW scenario: both colliding particles move towards a black hole, one of which is "critical" with fine-tuned parameters (energy and angular momentum), the second particle is "usual" (not fine-tuned). However, numeric findings in [16] showed that if the critical particle moves away from a black hole, the efficiency of the process significantly increases and amounts to 13,92 for the Kerr metric. In what follows, we call this Schnittman process (scenario). Later, numerically [17] and analytically [18] it was found that if colliding particles move in opposite directions and both of them are usual, a formally infinite efficiency becomes possible (so-called super-Penrose process). The problem is, however, that a usual particle with a finite energy cannot move away from a black hole although for the critical particle this is possible ([10], [13], [14]). Therefore, one could think that an outgoing usual particle could be created in some precedent collision. However, more careful treatment showed that the kind of particles under discussion cannot appear as a result of preceding collisions with finite masses and angular momenta. For a divergent mass of an initial particle this is possible but this reduces the physical value of the process [19], [20]. One is led to conclusion that starting from initial conditions in which usual outgoing particles near the black hole horizon are absent one cannot obtain them by means of additional collisions. As a result, a super-Penrose process near black holes is impossible.

Account for more involved scenarios in which particles intermediate between the critical and usual ones participate, only confirmed this conclusion [14]. (There is another option when collision occurs near a white black hole [15] but we do not discuss here these rather exotic objects.)

Quite recently, the estimates of the efficiency of the energy extraction from the Kerr black hole found in [16] numerically were derived analytically for the Kerr metric [21], [22]. General approach was developed in [23]. Maximum efficiency of extraction in the Kerr metric requires heavy particles that fall into a black hole [22] but for more general metrics this is not necessarily so [23].

What is especially important is that extraction can become infinite if instead of a black hole collision happens in the background without the horizon but on the threshold of its formation. This was shown in a general form in [24] and analyzed in detail in [25].

It turns out that collisions between ingoing and outgoing particles near generic axially symmetric blakc holes lead to some new qualitative possibilities absent for the Kerr one. In particular, the maximum efficiency of the energy extraction becomes possible even without heavy produced particles that are necessary in the Kerr case [22].

The conservation of the energy and angular momentum gives us

$$E_1 + E_2 = E_3 + E_4,\ L_1 + L_2 = L_3 + L_4. \tag{14}$$

Let a particle move towards a centre, bounce back and collide with another ingoing particle. Examination of (14) in the background (1), where $N > 0$ but is small at some r_0 shows that there are scenarios in which formally $E_3, L_3 \to \infty$ and $E_4, L_4 \to -\infty$. Thus extraction becomes unbounded. One reservation is in order. All particles are considered in the test approximation, so self-gravitation is neglected. There exists another mechanism of the energy extraction related to the electrically charged particles [26], [27] but we did not discuss it here.

How to take into account the effect of self-gravitaiton for particle collision under discussion remains unresolved very interesting problem.

Acknowledgments. I thank organizers for support that made my participation in the seminar possible.

References

[1] T. Piran, J. Katz, and J. Shaham, Astrophys. J. **196**, L107 (1975).
[2] T. Piran and J. Shaham, Astrophys. J. **214**, 268 (1977).
[3] T. Piran and J. Shanam, Phys. Rev. D **16**, 1615 (1977).
[4] M. Bañados, J. Silk and S.M. West, Phys. Rev. Lett. **103** (2009) 111102.
[5] J. M. Bardeen, W. H. Press, and S. A. Teukolsky, Astrophys. J. **178**, 347 (1972).
[6] O.B. Zaslavskii, Phys. Rev. D **82** (2010) 083004.
[7] A.A. Grib and Yu.V. Pavlov, Pis'ma v ZhETF 92, **147** (2010) [JETP Letters **92**, 125 (2010)].
[8] O. B. Zaslavskii, Class. Quantum Grav. **28**, 105010(2011).
[9] O. B. Zaslavskii, Phys. Rev. **D 84**, 024007 (2011)
[10] T. Harada, H. Nemoto and U. Miyamoto, Phys. Rev. D **86** (2012) 024027.
[11] M. Bejger, T. Piran, M. Abramowicz, and F. Håkanson, Phys. Rev. Lett. **109** (2012) 121101.
[12] T. Harada, K. Ogasawara, U. Miyamoto, [arXiv:1606.08107].
[13] O. Zaslavskii, Phys. Rev. D **86** (2012) 084030.
[14] O. B. Zaslavskii, Phys. Rev. D **93**, 024056 (2016).
[15] A. A. Grib and Yu. V. Pavlov, Gravitation and Cosmology **21**, 13 (2015).
[16] J. D. Schnittman, Phys.Rev.Lett. **113**, 261102 (2014).
[17] E. Berti, R. Brito and V. Cardoso, Phys. Rev. Lett. **114**, 251103 (2015).

[18] O. Zaslavskii, Mod. Phys. Lett. A **30** (2015) 1550076.
[19] E. Leiderschneider and T. Piran, arXiv:1501.01984.
[20] O. B. Zaslavskii, Europhys. Lett. **111** 50004 (2015).
[21] E. Leiderschneider, T. Piran, Phys. Rev. D **93**, 043015 (2016), [arXiv:1510.06764].
[22] K. Ogasawara, T. Harada, U. Miyamoto, Phys. Rev. D **93**, 044054 (2016).
[23] O. B. Zaslavskii, [arXiv:1607.00651].
[24] O. B. Zaslavskii, Phys. Rev. D **88**, 044030 (2013).
[25] M. Patil, T. Harada, Ken-ichi Nakao, P. S. Joshi, M. Kimura, Phys. Rev. **D** 93, 104015 (2016).
[26] O. B. Zaslavskii, Phys. Rev. D **86**, 124039 (2012).
[27] H. Nemoto, U. Miyamoto, T. Harada, and T. Kokubu, Phys. Rev. D **87**, 127502 (2013).

EPJ Web of Conferences **125**, 03024 (2016) DOI: 10.1051/epjconf/201612503024
QUARKS-2016

THE SPATIAL STRUCTURE OF THE WMAP/PLANCK HAZE

Yana Zhezher[1,2,★]

[1] *Institute for Nuclear Research, Russian Academy of Sciences, 117312, Moscow, Russia*
[2] *Faculty of Physics, Lomonosov Moscow State University, 119991, Moscow, Russia*

Abstract. It was recently proposed that the two phenomena, WMAP/Planck haze and Fermi bubbles may have a common origin. In the present paper we analyze the spatial structure of the haze using the Planck 2015 data. It is found that the spatial dimensions and locations of WMAP/Planck haze and Fermi bubbles are compatible within experimental uncertainties for symmetric approximation case, and WMAP/Planck haze appear to be more flattened in galactic latitude direction for the asymmetric approximation case. WMAP/Planck haze possesses north-south asymmetry, it is more intensive in the northern hemisphere. No substructures similar to Fermi bubbles cocoon are identified in Planck data.

1 Introduction

The Wilkinson Microwave Anisotropy Probe spacecraft (or WMAP) was launched in 2001. The main objective of the mission was precise measurement the of cosmic microwave background anisotropy. The product of the spacecraft are so-called full-sky maps, which may also be used for the galactic matter emission study. The mechanisms of galactic emission are known, and described theoretically, so a full-sky map can be built for every emission mechanism. One may expect that if all relevant mechanisms are included, the full-sky map in the specific wave band is equal to the sum of all galactic emission components, CMB radiation and radiation of the bright sources.

A component separation procedure was conducted in 2003 by D. Finkbeiner [1]. As a result, an excess of emission in the Galactic center region was found, which was called WMAP haze (after the confirmation by the Planck mission - WMAP/Planck haze). The radiation is assumed to be radially symmetric, extending to more than 20°.

The Fermi satellite with Large Area Telescope (LAT) apparatus was launched in 2008 in order to study full sky in gamma-ray band. A similar analysis may be conducted using the gamma-ray sky maps produced using Fermi-LAT event data. It was made in 2010 by M. Su et al. [2], and an excess of radiation was found. It was later called the Fermi bubbles – a symmetric radiation extending to 50° above and below the Galactic Plane.

For now, there is no single explanation for the WMAP/Planck haze phenomenon. One class of theories assumes that several emission phenomena observed in the Galactic Center region have the same origin: WMAP/Planck haze, Fermi bubbles, soft X-ray emission and 2.3 GHz radio-emission. This scenario has a characteristic property, which is a coincidence of spatial structures of all emission maps. The purpose of this paper is the examination of this hypothesis.

★e-mail: zhezher.yana@physics.msu.ru

2 Method

Template fitting is a basic minimising scheme which was used in original WMAP analysis.

The initial map may be represented as:

$$d_i = a_j P_{ji} + x_i, \tag{1}$$

where a_j is an amplitude vector, P_{ji} is template matrix and x_i is an unknown component.

White noise makes a main contribution to noise maps of full-sky maps. It is statistically uncorrelated and it's covariation matrix is diagonal:

$$\langle d_i d_j \rangle = N_i \delta_{ij}, \tag{2}$$

where N_i is a vector with length N_{pix}.

The minimum of chi-squared criterion is obtained:

$$a_{\alpha'} = \left(P_{\alpha' i} N_i^{-1} P_{\alpha i}\right)^{-1} \left(P_{\alpha i} N_i^{-1} d_i\right). \tag{3}$$

Three interstellar medium emission mechanisms were considered:

- free-free emission – emission due to interaction between electrons and protons at the regions of partially ionised hydrogen (H_{II}). H_α emission (Lyman line transition of ionized hydrogen) is used as a template for this mechanism [3].
- synchrotron emission – an emission of electromagnetic waves by charged particles which propagate in a magnetic field with relativistic speed [4], [5].
- dust emission – emission of particles, including among their components silicon and carbon. An important dust property which allows to detect it is an ability to absorb and scatter light, it is called extinction [6].

One should notice, that the template for an unknown component should be also included in the template matrix P_{ji}, otherwise other components may compensate the part of studied emission. A traditional template for the WMAP/Planck haze is 2-D Gaussian ellipse with characteristic lengths $\sigma_l = 15^\circ$ and $\sigma_b = 20^\circ$.

Bright objects and the Galactic Plane are covered with a mask which covers 30.8 % of the sky.

3 Results

Planck 2015 Low Frequency Instrument (LFI) full-sky maps were used as an initial data [7]. LFI creates maps on three different frequencies: 30, 44 and 70 GHz. The original WMAP analysis showed that the WMAP/Planck haze is the most intensive in at the frequencies 20-50 GHz, so 30 and 44 GHz separated emission were used for the extended emission analysis.

According to extracted data,new component of extended emission is more intensive in northern hemisphere. But northern hemisphere is believed to be contaminated by unresolved sources..

The latest Fermi collaboration research [8] have shown the existence of substructures in the Fermi bubbles. The Planck data investigation doesn't indicate the existence of such substructures in the WMAP/Planck haze.

The size of the WMAP/Planck haze are approximated by two ellipses symmetric below and above the Galactic Plane. The derived values for symmetric case are $\sigma_l = \sigma_b = b_0 = 17.9^\circ$ in a good

DOI: 10.1051/epjconf/201612503024

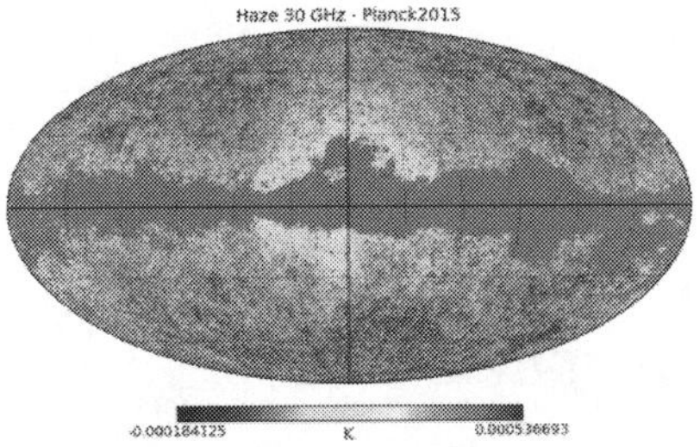

Figure 1. Extracted WMAP/Planck haze emission on 30 GHz.

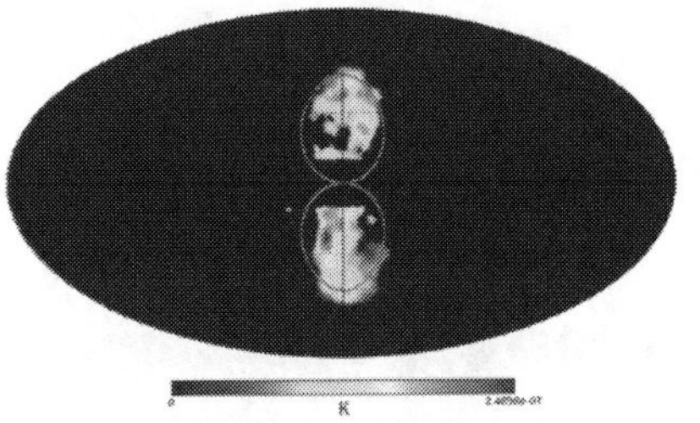

Figure 2. Comparison of the WMAP/Planck haze and Fermi bubbles sizes.

agreement with the Fermi bubbles sizes. For asymmetric case size estimations are: $\sigma_l = 55.6^\circ \pm 0.6^\circ, \sigma_b = b_0 = 15.88^\circ \pm 0.18^\circ$.

The preliminary results for spectrum are derived. Taking the initial spectrum shape as $T_\nu \propto \nu^{\beta h}$, estimation for β is $\beta = -2.86$, which is less than Planck results [9]: $\beta = -2.55 \pm 0.5$, and closer to the synchrotron scenario.

4 Conclusion

The indication on the similarity of spatial sizes of the WMAP/Planck haze and the Fermi bubbles is obtained. For the future work, it is proposed compare WMAP/Planck haze to other types of extended emission observed in the Galactic Center region and test other theoretical scenarios, which are dark matter and pulsar origin of WMAP/Planck haze.

Acknowledgments

The work is supported by the Russian Science Foundation grant 14-22-00161.

References

[1] D.P. Finkbeiner, ApJ **614** (2004) 186-193, astro-ph/0311547.
[2] M. Su, T. R. Slatyer, D. P. Finkbeiner, Astrophys.J. **724** 1044-1082 (2010), astro-ph/1005.5480.
[3] D. P. Finkbeiner, Astrophys.J.Suppl **146** 407 (2003), astro-ph/0301558.
[4] C. G. T. Haslam et al., Astronomy and Astrophysics **100** 2 209-219 (1981).
[5] C. G. T. Haslam et al., Astronomy and Astrophysics Supplement Series **47** 1, 2, 4-51, 53-142 (1982).
[6] D. P. Finkbeiner, M. Davis, D. J. Schegel, Astrophys.J. **524** 867-886 (1999), astro-ph/9905128.
[7] J. A. Tauber et al., Astronomy&Astrophysics **520** A1 (2010).
[8] Fermi-LAT Collaboration, Astrophys.J. **793** 1 34 (2014), astro-ph/1407.7905.
[9] P. A. R. Ade et al., astro-ph/1208.5483v1.

The mass-gap in Quantum Chromodynamics

S.A. Larin[1,3,⋆]

[1] *Institute for Nuclear Research of the Russian Academy of Sciences, 60th October Anniversary Prospect 7a, Moscow 117312, Russia*

Abstract. It is shown that the fundamental Lagrangian of Quantum Chromodynamis should be modified by the adding gluon masses to produce the mass-gap in accordance with the Källen-Lehmann spectral representation. On mass-shell renormalizability and unitarity of the resulting theory is demonstrated.

1 Introduction

The obtaining the mass-gap in Quantum Chromodynamics (QCD) is the well known long standing task. QCD was quite quickly established as the theory of strong interactions after the discovery [1] of asymptotic freedom since physicists got the small parameter, the strong coupling constant at high energies, and it became possible to perform perturbative calculations. The gauge bosons of the theory, the gluons, are considered to be massless to have gauge invariance and correspondingly renormalizability. The introduction of the non-zero Lagrangian gluon masses into the fundamental QCD Lagrangian was considered to be forbidden because of the violating either renormalizability or untarity of the corresponding theory. Giving masses to gauge bosons via the Higgs mechanism [2] is also not allowed since colored Higgs particles are rejected by experiments.

Quite recently it was found that the non-abelian Yang-Mills theory [3] with masses of the Proca type is in fact on mass-shell renormalizable [4]. Unitarity of the theory of the Proca type is obvious since it contains only physical degrees of freedom (no ghosts).

In the present paper it is shown that it is impossible to obtain the necessary mass-gap in QCD with zero Lagrangian gluon masses. The fundamental QCD Lagrangian should be modified by the adding non-zero Lagrangian gluon masses to get the theory of the Proca type to produce the mass-gap in accordance with the Källen-Lehmann spectral representation [5]. On mass-shell renormalizability of the resulting theory is discussed.

2 The main part

2.1 Definitions and notations

The QCD Lagrangian is

$$L_{QCD} = -\frac{1}{4}F^a_{\mu\nu}F^{a\,\mu\nu} + i\overline{\psi}_f\gamma^\mu(\partial_\mu - igA^a_\mu T^a)\psi_f - m_f\overline{\psi}_f\psi_f \quad (1)$$

⋆e-mail: larin@inr.ac.ru

$$-\frac{1}{\xi}(\partial^{\mu}A^{a}_{\mu})^{2}+\partial^{\mu}\overline{c}^{a}(\partial_{\mu}c^{a}-gf^{abc}c^{b}A^{c}_{\mu})$$

plus counterterms. The notations are standard. The summation over the flavour index $f = u, ..., t$ is assumed. ξ is the gauge parameter of the usually used general covariant gauge chosen because of its convenience for perturbative calculations. $m_f = m_f(\mu)$ is the renormalized (current) quark mass, $g = g(\mu)$ is the renormalized strong coupling constant, $g^2/(16\pi^2) \equiv a_s$, where μ is the renormalization point.

It is known [6] that the covariant gauge chosen in the Lagrangian (1) does not fix the gauge ambiguity uniquely and is valid strictly speaking only within perturbation theory. Hence the theory described by the Lagrangian (1) with the covariant gauge condition (chosen because of its convenience for perturbative calculations) should not be seen as the complete theory of strong interaction, but it can be considered as the theory which reproduces the perturbative part of the complete theory [7]. The theories, see e.g. [8], which do not reproduce conventional perturbative QCD expansions are not considered here.

2.2 Obtaining the mass-gap

To demonstrate the necessity of the introducing the Lagrangian gluon masses for the obtaining the mass-gap let us consider the vacuum polarization function $\Pi(q^2)$

$$(-q^2 g_{\mu\nu}+q_\mu q_\nu)\Pi(q^2)=i\int \mathrm{d}x e^{iqx}\langle 0|\; T\left(j_\mu(x)j_\nu(0)\right)\;|0\rangle. \qquad (2)$$

where $j_\mu = \sum_f e_f \overline{\psi}_f \gamma_\mu \psi_f$ is the quark electromagnetic current, e_f being the electromagnetic quark charge.

According to the general principles of local quantum field theory the function $\Pi(q^2)$ satisfies the fundamental Källen-Lehmann [5] spectral representation

$$\Pi(q^2)=\frac{1}{12\pi^2}\int_{4m_\pi^2}^{\infty} ds\,\frac{R(s)}{s-q^2-i0}+C, \qquad (3)$$

where the ratio $R(s) = \sigma_{tot}(e^+e^- \to hadrons)/\sigma(e^+e^- \to \mu^+\mu^-)$ is the normalized total cross-section of electron-positron annihilation into hadrons, m_π is a pion mass. The representation is written with one subtraction and C is the subtraction constant (the number of necessary subtractions is not important for our considerations).

We would like to stress here that this spectral reprsentation is more fundamental then just a dispersion relation. It can not be improved or modified. To write a dispersion relation for a function one should know analytic propertirs of a function. The Källen-Lehmann representation itself determines the analytic properties of $\Pi(q^2)$ which should be an analytic function in the complex q^2-plane with the cut starting from the first physical threshold, i.e. as it is dictated by experiments from the two-pion threshold $q^2 = 4m_\pi^2$. One gets for the discontinuity of $\Pi(q^2)$ on the cut

$$\Delta\Pi(q^2)\equiv\Pi(q^2+i0)-\Pi(q^2-i0)=\begin{cases} i\,R(q^2)/(6\pi) & at\;\; s>4m_\pi^2\\ 0 & at\;\; s<4m_\pi^2.\end{cases} \qquad (4)$$

On the other hand one obtains within Perturbative QCD for the discontinuity

$$\Delta\Pi(q^2)_{pQCD}=\theta(q^2)\,\rho_{gl}(q^2)+\theta(q^2-4M_u^2)\,\rho_q(q^2). \qquad (5)$$

The gluon spectral density $\rho_{gl}(q^2)$ contributes for $q^2 > 0$ as it is shown by the theta-function $\theta(q^2)$. This is the famous zero threshold. It arises from those absorptive parts of Feynman diagrams of $\Pi(q^2)$ which are produced by purely gluonic cuts of the diagrams (i.e. by the Cutcosky cuts which cross only gluon propagators of diagrams). As it is well known such diagrams appear for the first time at the four-loop level in the order a_s^3 (corresponding cuts cross 3 gluon propagators).

The quark spectral density

$$\rho_q(q^2) = 3\sum_f \theta(q^2 - 4M_f^2)e_f^2(1 + 2M_f^2/q^2)\sqrt{1 - 4M_f^2/q^2} + O(a_s) \tag{6}$$

arises from the quark cuts of the diagrams (i.e. from the cuts which cross two or more quark propagators of the diagrams). It contributes for $q^2 > 4M_u^2$ where M_u is the perturbative pole mass of the lightest u-quark, defined as the perturbative pole of the quark propagator. The perturbative quark pole mass $M_f = m_f(\mu) + O(a_s)$ naturally appears as the pole of the quark propagator after summation of perturbative loop corrections to the propagator. It is a renormalization group invariant quantity, i.e. independent on the renormalization point μ and on the subtraction scheme. In this sense it behaves as a physical quantity and that is why it is natural to use a perturbative pole quark mass to parametrize the theory.

We do not discuss here the important by themselves questions of convergence or divergence of corresponding perturbative QCD series at low or high energies. We will just accept the constructive approach that our conventional perturbation theory is adequate to the exact solution of the theory, i.e. it correctly reproduces the perturbative expansion of the exact solution. We assume here that the exact solution is in principle obtainable if we know enough mathematics.

Thus one gets within QCD that $\Delta\Pi(q^2)$ is non-zero in the energy region $0 < q^2 < 4m_\pi^2$ since the perturbative contribution $\Delta\Pi(q^2)_{pQCD}$ is non-zero in this interval. Here we would like to stress that one should get in QCD an exact zero below the two-pion threshold as it is dictated by experiments. There are also non-perturbative contributions to $\Delta\Pi(q^2)$, i.e. contributions of the type e^{-1/a_s} which are invisible in the perturbative expansion at the point $a_s = 0_+$ (that is $e^{-1/a_s} = 0 \cdot a_s + 0 \cdot a_s^2 + ...$).

But we note that the non-perturbative contributions can not exactly cancel the perturbative ones in the continuous interval $0 < q^2 < 4m_\pi^2$ because of the different analytical dependence on the coupling constant a_s. To get the condition $\Delta\Pi(q^2) = 0$ at $0 < q^2 < 4m_\pi^2$ in agreement with experiments we should shift perturbative gluon and quark thresholds above the two-pion threshould $q^2 = 4m_\pi^2$. That is why one should introduce the non-zero gluon masses into the fundamental QCD Lagrangian.

One could object that peturbative contributions in the exact solution for $\Delta\Pi(q^2)$ could be moved above the two-pion threshould with the help of the suitably arranged θ-function. E.g. the gluon perturbative contribution could be moved above the two-pion threshould in the following way

$$\theta\left(q^2 - M^2 f(a_s)\right)\, \rho_{gl}(q^2), \tag{7}$$

where $f(a_s)$ is some function which is zero at $a_s = 0$ and M^2 is a dimensional parameter such that $M^2 f(a_s)$ is a renormalization group invariant quantity which produces the necessary mass-gap below the two-pion threshold.

But perturbative expansion of the contribution of eq.(7) would contain terms with $\delta(q^2)$ and its derivatives arising from the differentiating the θ-function and we do not see such terms in real perturbative expansion. Thus the contributions of the type of eq.(7) are excluded in the exact solution of the theory.

The standard naive objection here is that nobody trusts perturbation theory ar such low energies, since the corresponding perturbative series is heavily divergent below the two-pion threshold. But only the principal existence of the pertubative series with well defined coefficients below the two-pion threshold is of importance here independently on the question of its convergence: the non-zero perturbative expansion means the non-zero exact function generating this expansion.

Thus one gets the restrictions on the perturbative pole masses of gluons and quarks

$$M_{gl} > \frac{2}{\sqrt{3}} m_\pi, \tag{8}$$

$$M_u > m_\pi.$$

The restriction on M_u seems to be quite strong for the lightest u-quark but it is not excluded from the first principles.

2.3 Constructing QCD with gluon masses

To construct QCD with massive gluons we will follow the approach of [4]. Presently this is the only known way to get (on mass-shell) renormalizable theory of massive gluons without color scalars (color scalars are rejected by experiments). Within this approach one starts from a renormalizable theory with scalar fields using the Englert-Brout-Higgs mechanism of spontaneous symmetry breaking [2] and after transition to the unitary gauge removes remaining massive scalar fields. Thus we add to the massless QCD Lagrangian (1) the scalar part to begin with the following general Lagrangian

$$L_{QCD+SCALARS} = -\frac{1}{4} F^a_{\mu\nu} F^{a\,\mu\nu} + i\bar{\psi}_f \gamma^\mu D_\mu \psi_f - m_f \bar{\psi}_f \psi_f + \tag{9}$$

$$\left(D_\mu \Phi\right)^+ D^\mu \Phi + \left(D_\mu \Sigma\right)^+ D^\mu \Sigma - \lambda_1 \left(\Phi^+ \Phi - v_1^2\right)^2 - \lambda_2 \left(\Sigma^+ \Sigma - v_2^2\right)^2$$

$$-\lambda_3 \left(\Phi^+ \Phi + \Sigma^+ \Sigma - v_1^2 - v_2^2\right)^2 - \lambda_4 \left(\Phi^+ \Sigma\right)\left(\Sigma^+ \Phi\right) + L_{gf} + L_{gc}$$

plus counterterms, where we introduced two triplets $\Phi(x)$ and $\Sigma(x)$ of complex scalar fields in the fundamental representation of the $SU(3)$ color group to get all gluon massive. L_{gf} is the gauge fixing part of the Lagrangian in some chosen gauge and L_{gc} is the corresponding gauge compensating part with the Faddeev-Popov ghost fields.

We can choose the following shifts of scalar fields by the quantities v_1 and v_2 to generate masses of all eight gluons

$$\Phi(x) = \begin{pmatrix} \phi_1(x) + i\phi_2(x) + v_1 \\ \phi_3(x) + i\phi_4(x) \\ \phi_5(x) + i\phi_6(x) \end{pmatrix}, \quad \Sigma(x) = \begin{pmatrix} \sigma_1(x) + i\sigma_2(x) \\ \sigma_3(x) + i\sigma_4(x) + v_2 \\ \sigma_5(x) + i\sigma_6(x) \end{pmatrix}. \tag{10}$$

Choosing for simplicity $v_1 = v_2 \equiv v$ one obtains the following massive terms for gluons in the Lagrangian

$$L_M = m_{gl}^2 \Big[(A^1)^2 + (A^2)^2 + (A^3)^2 + \frac{1}{2}(A^4)^2 + \tag{11}$$

$$\frac{1}{2}(A^5)^2 + \frac{1}{2}(A^6)^2 + \frac{1}{2}(A^7)^2 + \frac{1}{3}(A^8)^2 \Big],$$

where $m_{gl}^2 \equiv g^2 v^2$ is the gluon mass parameter of the theory.

Now one can make transition to the unitary gauge. All ghost fields as usual disappear from the Lagrangian. Following the approach of [4] one can remove in the unitary gauge all Higgs fields from the Lagrangian preserving on mass-shell renormalizability of the theory.

To give the derivation of this statement let us consuder as an example the simplified case, the generalization to the above QCD+SCALARS case will be straightforward. Let us consider the known model given by the initial $SU(2)$-invariant Lagrangian of interaction of vector bosons and scalar fields possessing the spontaneously broken symmetry

$$L = -\frac{1}{4}F^a_{\mu\nu}F^a_{\mu\nu} + \left(D_\mu\Phi\right)^+ D_\mu\Phi - \lambda\left(\Phi^+\Phi - v^2\right)^2 \tag{12}$$

with the doublet of scalar fields $\Phi(x)$ in the fundamental representation of the group.

Here $D_\mu\Phi = \left(\partial_\mu - ig\frac{\tau^a}{2}W^a_\mu\right)\Phi$ is the covariant derivative, τ^a are the Pauli matrices, $\lambda > 0$, $v^2 > 0$. (This model can be considered as the Standard Model of electroweak interactions without $U(1)$-interaction and fermions. The derivation given below can be applied also to the complete Standard Model, the γ_5- matrix being treated within dimensional regularization according to the technique of [9].)

To obtain the complete Lagrangian one makes the standard shift of the scalar field fixes the gauge and adds ultraviolet counterterms. Let us consider two gauges: the widely used R_ξ-gauge [10], [11] with an arbitrary parameter ξ and the unitary gauge.

The theory in the R_ξ-gauge describes three physical massive vector bosons with the mass $m = gv/\sqrt{2}$, and the physical Higgs field χ with the mass $M = 2\lambda v$. Here are also Goldstone ghosts ϕ^a and Faddeev-Popov ghosts c^a with masses ξm^2. The structure of the counterterms (consistent with gauge invariance and Slavnov-Taylor identities [12, 13] to ensure unitarity) is well known, see e.g. [7].

The propagators in the unitary gauge defined by the gauge condition $\phi^a = 0$ are obtained from those of the R_ξ-gauge by taking the limit $\xi \to \infty$. The theory in the unitary gauge is renormalizable only on mass-shell, i.e. Green functions are divergent at $\epsilon \to 0$ (ϵ being the parameter of dimensional regularization) but the S-matrix elements are finite.

To consider renormalization for our purpose it is convenient to use the Bogoliubov-Parasiuk-Hepp subtraction scheme [14]. As it is well known in this scheme a counterterm of e.g. a primitively divergent Feynman diagram is the truncated Taylor expansion of the diagram itself at some fixed values of external momenta. Hence counterterms of mass dependent diagrams are also mass dependent. Needless to say that subtractions should respect Slavnov-Taylor identities.

Let us consider S-matrix elements in the R_ξ-gauge without external Higgs bosons (i.e. with external W-bosons only in this simplified model). We will analyze the dependence of diagrams on the Higgs mass M by using for convinience the expansion in large M (after renormalization but before the removing regularization). The algorithm for the large mass expansion of Feynman diagrams is given e.g. in [15] (where it is quite reasonably checked in calculations of the 4-loop diagrams for the Z-boson decay into hadrons). It can be rigorously derived with the technique of [16].

We separate all diagrams into physical ones which are not nullified in the limit $\xi \to \infty$ and unphysical ones which are nullified. In this limit the propagator of the W-boson reduces to the known unitary form

$$\lim_{\xi\to\infty} < T(W^a_\mu W^b_\nu) >= -i\delta^{ab}\lim_{\xi\to\infty}\left(\frac{g_{\mu\nu} - k_\mu k_\nu/m^2}{k^2 - m^2} + \frac{k_\mu k_\nu/m^2}{k^2 - \xi m^2}\right) =$$

$$-i\delta^{ab}\frac{g_{\mu\nu} - k_\mu k_\nu/m^2}{k^2 - m^2}$$

The propagators of the Goldstone bosons ϕ^a and ghosts c^a vanish in this limit and correspondingly all diagrams which contain these propagators are also nullified.

Thus in our notations the physical diagrams are the diagrams which do not contain Goldstone bosons propagators or ghosts propagators and the unphysical diagrams are the diagrams which contain such propagators.

Within the large-M expansion the physical diagrams with χ-propagators contain either terms with integer negative powers of M^2: $\frac{1}{M^{2n}}$, $n = 1,2,3,...$ or terms with non-integer powers of M^2 (non-integer powers contain ϵ): $\frac{1}{M^{2(k+l\epsilon)}}$, k – *integer*, l – *positive integer*. This is because each vertex with the large factor M^2 has three or four attached χ-propagators due to the structure of the Higgs boson self-coupling.

In contrast, unphysical diagrams can have polynomial in M terms due to the four-ϕ vertex with the large factor M^2. But they are ξ-dependent (they are nullified in the limit $\xi \to \infty$) and this polynomial terms cancel in S-matrix elements.

In the renormalizable R_ξ-gauge one can present ultraviolet renormalization in a standard form of the Bogoliubov-Parasiuk R-operation for individual diagrams. This ensures that after renormalization the M-dependent terms are finite at $\epsilon \to 0$ separately from M-independent terms. Thus if one removes all M-dependent terms one is left with a finite expression.

On the Lagrangian level it means in the unitary gauge that one removes from L_U all terms containing the field χ and also all M-dependent terms in the counterterms. This should be done in the unitary gauge because in the R_ξ-gauge some diagrams containing propagators of Higgs particle can give contributions not depending on the Higgs mass M. In contrast in the unitary gauge all diagrams containing Higgs propagators give only contributions depending on M so there is one to one correspondence between M-dependent diagrams and terms in the Lagrangian containing the Higgs field χ

The resulting theory is on mass-shell finite. This is the massive Yang-Mills theory of the Proca type

$$L_{YM} = -\frac{1}{4}(\partial_\mu W^a_\nu - \partial_\nu W^a_\mu + \frac{z_1}{z_2} g f^{abc} W^b_\mu W^c_\nu)^2 + m^2 W^a_\mu W^a_\mu$$

plus counterterms.

Thus the Higgs mechanism can be considered as an efficient mathematical tool to observe on mass-shell renormalizability of the massive Yang-Mills theory of the Proca type which is far from to be obvious directly.

Let us now return to our Lagrangian (9) with two scalar triplets which after spontaneous symmetry breaking has four Higgs particles. Following the above approach we can remove in the unitary gauge all four Higgs fields from the Lagrangian preserving on mass-shell renormalizability of the theory. The Lagrangian of the resulting QCD with massive gluons is

$$L_{massive\ QCD} = L_M - \frac{1}{4} F^a_{\mu\nu} F^a_{\mu\nu} + i\bar{\psi}_f \gamma_\mu D_\mu \psi_f - m_f \bar{\psi}_f \psi_f \tag{13}$$

plus counterterms, where the part L_M with gluon mass terms is given in eq.(11).

We would like to note that on mass-shell renormalizability does not mean that one should consider quarks and gluons as free external particles which would contradict to confinement. It means only that in the $SU(3) \times SU(2) \times U(1)$ theory the S-matrix elements with the physical external particles are finite.

One can calculate the one-loop β-function in this theory [4] to obtain that asymptotic freedom remains valid in the considered theory with massive gluons.

3 Conclusions

Thus we have demonstrated that it is impossible to produce the mass-gap in accordance with the fundamental Källen-Lehmann spectral representation within QCD with zero Lagrangian gluon masses. Correspondingly one should introduce the non-zero gluon masses into the fundamental QCD Lagrangian to generate the necessary mass-gap.

We have also demonstrated that it is possible to introduce the gluon masses into the QCD Lagrangian preserving on mass-shell renormalizability and unitarity of the theory.

Acknowledgments

The author is grateful to the collaborators of the Theory division of INR for helpful discussions.

References

[1] D.J. Gross, F. Wilczek, Phys. Rev. Lett. 30 (1973) 1343;
H.D. Politzer, Phys. Rev. Lett. 30 (1973) 1346.
G. 't Hooft, report at the Marseille Conference on Yang-Mills Fields, 1972.
[2] P.W. Higgs, Phys. Lett. 12 (1964) 132.
F.Englert and R. Brout, Phys. Rev. Lett. 13 (1964) 321.
[3] C.N. Yang and R.L. Mills, Phys. Rev. 96 (1954) 191.
[4] S.A. Larin, e-Print: hep-ph/0503198; Phys.Part.Nucl. 44 (2013) 386; e-Print: arXiv:1304.8107 [hep-ph]; PoS BaldinISHEPP XXII (2015) 018.
[5] G. Källen, Helv. Phys. Acta 25 (1952) 417; H. Lehmann, Nuovo Cim. 11 (1954) 342.
[6] V.N. Gribov, Nucl. Phys. B139 (1978) 1.
[7] L.D. Faddeev and A.A. Slavnov, Gauge fields. Introduction to quantum theory, Front. Phys. 83 (1990) 1.
[8] N. Vandersickel and D. Zwanziger, Phys. Rept. 520 (2012) 175.
[9] S.A. Larin, Phys. Lett. B 303 (1993) 113; e-Print: hep-ph/9302240.
[10] G. 't Hooft, Nucl. Phys. B 35 (1971) 167.
[11] K. Fujikawa, B.W. Lee and A.I. Sanda, Phys. Rev. D6 (1972) 2923.
[12] A.A. Slavnov, Theor. Math. Phys. 10 (1972) 99.
[13] J.C. Taylor, Nucl. Phys. B 33 (1971) 436.
[14] N.N. Bogoliubov and O.S. Parasiuk, Acta Math. 97 (1957) 227.
K. Hepp, Comm. Math. Phys. 2 (1966) 301.
[15] S.A. Larin, T. van Ritbergen and J.A.M. Vermaseren, Nucl. Phys. B 438 (1995) 278.
[16] S.A. Larin, Phys. Lett. B 469 (1999) 220.

Physics of the charmonium-like state $X(3872)$

Nikolay Achasov[1,⋆]

[1] *Sobolev Institute for Mathematics, 630090 Novosibirsk, Russia*

Abstract. We construct spectra of decays of the resonance $X(3872)$ with good analytical and unitary properties which allows to define the branching ratio of the $X(3872) \to D^{*0}\bar{D}^0 + c.c.$ decay studying only one more decay, for example, the $X(3872) \to \pi^+\pi^- J/\psi(1S)$ decay, and show that our spectra are effective means of selection of models for the resonance $X(3872)$.

Then we discuss the scenario where the $X(3872)$ resonance is the $c\bar{c} = \chi_{c1}(2P)$ charmonium which "sits on" the $D^{*0}\bar{D}^0$ threshold.

We explain the shift of the mass of the $X(3872)$ resonance with respect to the prediction of a potential model for the mass of the $\chi_{c1}(2P)$ charmonium by the contribution of the virtual $D^*\bar{D} + c.c.$ intermediate states into the self energy of the $X(3872)$ resonance. This allows us to estimate the coupling constant of the $X(7872)$ resonance with the $D^{*0}\bar{D}^0$ channel, the branching ratio of the $X(3872) \to D^{*0}\bar{D}^0 + c.c.$ decay, and the branching ratio of the $X(3872)$ decay into all non-$D^{*0}\bar{D}^0 + c.c.$ states. We predict a significant number of unknown decays of $X(3872)$ via two gluons: $X(3872) \to$ *gluon gluon* $\to$ *hadrons*.

1 Introduction

The $X(3872)$ resonance became the first in discovery of the resonant structures XYZ ($X(3872)$, $Y(4260)$, $Z_b^+(10610)$, $Z_b^+(10650)$, $Z_c^+(3900)$), the interpretations of which as hadron states assumes existence in them at least pair of heavy and pair of light quarks in this or that form.

Thousands of articles on this subject already were published in spite of the fact that many properties of new resonant structures are not defined yet and not all possible mechanisms of dynamic generation of these structures are studied, in particular, the role of the anomalous Landau thresholds is not studied.

Anyway, this spectroscopy took the central place in physics of hadrons.

Below we give reasons that $X(3872)$, $I^G(J^{PC}) = 0^+(1^{++})$, is the $\chi_{c1}(2P)$ charmonium and suggest a physically clear program of experimental researches for verification of our assumption.

2 How to learn the branching ratio $X(3872) \to D^{*0}\bar{D}^0 + c.c.$ [1]

The mass spectrum $\pi^+\pi^- J/\psi(1S)$ looks as the ideal Breit-Wigner one in the $X(3872) \to \pi^+\pi^- J/\psi(1S)$ decay, see Figure 1(a).

⋆e-mail: achasov@math.nsc.ru

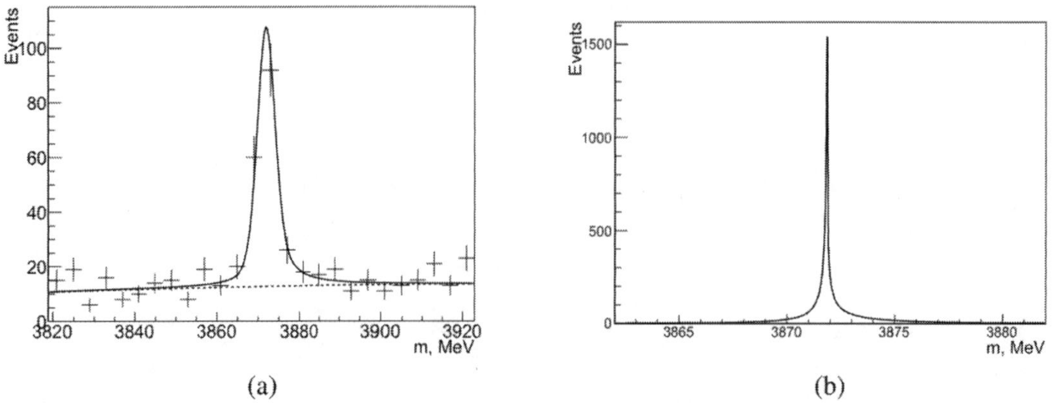

Figure 1. (a) The Belle data [2] on the invariant $\pi^+\pi^- J/\psi(1S)$ mass (m) distribution. The solid line is our theoretical one with taking into account the Belle energy resolution. The dotted line is second-order polynomial for the incoherent background. (b) Our undressed theoretical line.

The mass spectrum $\pi^+\pi^-\pi^0 J/\psi(1S)$ in the $X(3872) \to \pi^+\pi^-\pi^0 J/\psi(1S)$ decay looks in a similar way [3, 4].

The mass spectrum $D^{*0}\bar{D}^0 + c.c.$ in the $X(3872) \to D^{*0}\bar{D}^0 + c.c.$ decay [5] looks as the typical resonance threshold enhancement, see Figure 2.

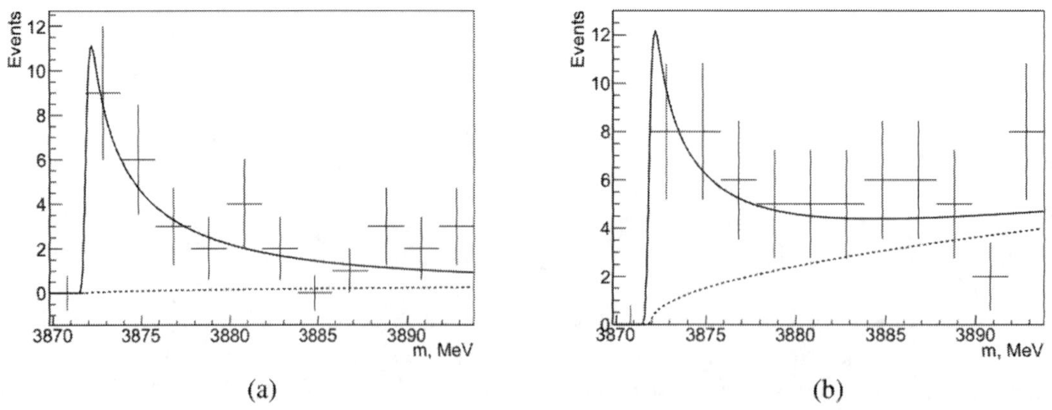

Figure 2. The Belle data [5] on the invariant $D^{*0}\bar{D}^0 + c.c.$ mass (m) distribution. The solid line is our theoretical one with taking into account the Belle energy resolution. The dotted line is a square root function for the incoherent background. (a) $D^{*0} \to D^0\pi^0$. (b) $D^{*0} \to D^0\gamma$.

If structures in the above channels are manifestation of the same resonance, it is possible to define the branching ratio $BR(X(3872) \to D^{*0}\bar{D}^0 + c.c.)$ treating data of the two above decay channels only.

We believe that the $X(3872)$ is the axial vector, 1^{++} [6, 7]. In this case the S wave dominates in the $X(3872) \to D^{*0}\bar{D}^0 + c.c.$ decay and hence is described by the effective Lagrangian

$$L_{XD^{*0}D^0}(x) = g_A X^\mu \Big(D^0_\mu(x)\bar{D}^0(x) + \bar{D}^0_\mu(x) D^0(x) \Big). \tag{1}$$

The width of the $X \to D^{*0}\bar{D}^0 + c.c.$ decay

$$\Gamma(X \to D^{*0}\bar{D}^0 + c.c.,\, m) = \frac{g_A^2}{8\pi}\frac{\rho(m)}{m}\left(1 + \frac{\mathbf{k}^2}{3m_{D^{*0}}^2}\right), \tag{2}$$

where $\mathbf{k}$ is momenta of D^{*0} (or $\bar{D}^0$) in the $D^{*0}\,\bar{D}^0$ center mass system, m is the invariant mass of the $D^{*0}\,\bar{D}^0$ pair,

$$\rho(m) = \frac{2|\mathbf{k}|}{m} = \frac{\sqrt{(m^2 - m_+^2)(m^2 - m_-^2)}}{m^2}, \quad m_\pm = m_{D^{*0}} \pm m_{D^0}. \tag{3}$$

The second term in the right side of Eq. (2) is very small in our energy region and can be neglected. This gives us the opportunity to construct the mass spectra for the $X(3872)$ decays with the good analytical and unitary properties as in the scalar meson case [8, 9].

The mass spectrum in the $D^{*0}\bar{D}^0 + c.c.$ channel

$$\frac{dBR(X \to D^{*0}\bar{D}^0 + c.c.,\, m)}{dm} = 4\frac{1}{\pi}\frac{m^2\Gamma(X \to D^{*0}\bar{D}^0,\, m)}{|D_X(m)|^2}. \tag{4}$$

The branching ratio of $X(3872) \to D^{*0}\bar{D}^0 + c.c.$

$$BR(X \to D^{*0}\bar{D}^0 + c.c.) = 4\frac{1}{\pi}\int_{m_+}^{\infty}\frac{m^2\Gamma(X \to D^{*0}\bar{D}^0,\, m)}{|D_X(m)|^2}dm. \tag{5}$$

In others $\{i\}$ (non-$D^{*0}\bar{D}^0$) channels the $X(3872)$ state is seen as a narrow resonance that is why we write the mass spectrum in the i channel in the form

$$\frac{dBR(X \to i,\, m)}{dm} = 2\frac{1}{\pi}\frac{m_X^2\Gamma_i}{|D_X(m)|^2}, \tag{6}$$

where Γ_i is the width of the $X(3872) \to i$ decay.

The branching ratio of $X(3872) \to i$

$$BR(X \to i) = 2\frac{1}{\pi}\int_{m_0}^{\infty}\frac{m_X^2\Gamma_i}{|D_X(m)|^2}dm, \tag{7}$$

where m_0 is the threshold of the i channel.

The inverse propagator $D_X(m)$

$$D_X(m) = m_X^2 - m^2 + Re(\Pi_X(m_X^2)) - \Pi_X(m^2) - \imath m_X\Gamma, \tag{8}$$

where $\Gamma = \Sigma\Gamma_i < 1.2$ MeV is the total width of the $X(3872)$ decay into all non-($D^{*0}\bar{D}^0 + c.c.$) channels.

$$\Pi_X(s) = \frac{g_A^2}{8\pi^2}\left(I^{D^{*0}\bar{D}^0}(s) + I^{D^{*+}D^-}(s)\right), \quad m^2 = s. \tag{9}$$

When $m_+ = m_{D^*} + m_D \le m$,

$$I^{D^*\bar{D}}(m) = \left\{\frac{(m^2 - m_+^2)}{m^2}\frac{m_-}{m_+}\ln\frac{m_{D^*}}{m_D} + \rho(m)\left[\imath\pi + \ln\frac{\sqrt{m^2 - m_-^2} - \sqrt{m^2 - m_+^2}}{\sqrt{m^2 - m_-^2} + \sqrt{m^2 - m_+^2}}\right]\right\}, \tag{10}$$

where $m_- = m_{D^*} - m_D$.

When $m_- \le m \le m_+$,

$$I^{D^*\bar{D}}(m) = \left\{ \frac{(m^2 - m_+^2)}{m^2} \frac{m_-}{m_+} \ln \frac{m_{D^*}}{m_D} - 2|\rho(m)| \arctan \frac{\sqrt{m^2 - m_-^2}}{\sqrt{m_+^2 - m^2}} \right\}, \tag{11}$$

where $|\rho(m)| = \sqrt{(m_+^2 - m^2)(m^2 - m_-^2)}/m^2$.

When $m \le m_-$ and $m^2 \le 0$,

$$I_X^{D^*\bar{D}}(m) = \left\{ \frac{(m^2 - m_+^2)}{m^2} \frac{m_-}{m_+} \ln \frac{m_{D^*}}{m_D} - \rho(m) \ln \frac{\sqrt{m_+^2 - m^2} - \sqrt{m_-^2 - m^2}}{\sqrt{m_+^2 - m^2} + \sqrt{m_-^2 - m^2}} \right\}. \tag{12}$$

Our branching ratios satisfy the unitarity

$$1 = BR(X \to D^{*0}\bar{D}^0 + c.c.) + BR(X \to D^{*+}\bar{D}^- + c.c.) + \sum_i BR(X \to i)\,. \tag{13}$$

Fitting the Belle data [2], we take into account the Belle results [2]: $m_X = 3871.84\,\text{MeV} = m_{D^{*0}} + m_{D^0} = m_+$ and $\Gamma_{X(3872)} < 1.2$ MeV 90%CL, that corresponds to $\Gamma < 1.2$ MeV, which controls the width of the $X(3872)$ signal in the $\pi^+\pi^- J/\psi(1S)$ channel and in every non-$(D^{*0}\bar{D}^0 + c.c.)$ channel. The results of our fit are in the Table 1.

Table 1. BR_{seen} is a branching ratio for $m \le 3891.84$ MeV, Γ in MeV, g_A in GeV.

Γ	$1.2_{-0.4}$	mode	$D^{*0}\bar{D}^0 + c.c.$	$D^{*+}D^- + c.c.$	$Others$
$g_A^2/8\pi$	1.4^{+5}_{-1}	BR	$0.6^{+0.02}_{-0.1}$	$0.31^{+0.13}_{-0.16}$	$0.1^{+0.3}_{-0.1}$
χ^2/Ndf	45/42	BR_{seen}	$0.3^{+0.1}_{-0.2}$	$0.03^{+0.004}_{-0.02}$	$0.09^{+0.3}_{-0.1}$

Our approach can serve as the guide in selection of theoretical models for the $X(3872)$ resonance. Indeed, if 3871.68 MeV $< M_X <$ 3871.95 MeV and $\Gamma_{X(3872)} = \Gamma < 1.2$ MeV then for $g_A^2/4\pi < 0.2$ GeV^2 $BR(X \to D^{*0}\bar{D}^0 + c.c.; m \le 3891.84\,\text{MeV}) < 0.3$. That is, unknown decays of $X(3872)$ into non-$D^{*0}\bar{D}^0$ states are considerable or dominant.

For example, in Ref. [10] the authors considered $m_X = 3871.68$ MeV, $\Gamma = 1.2$ MeV and $g_{XDD^*} = g_A\sqrt{2} = 2.5$ GeV, that is, $g_A^2/8\pi = 0.1$ GeV^2. In this case $BR(X \to D^{*0}\bar{D}^0 + c.c.) \approx 0.2$, that is, unknown decays $X(3872)$ into non-$(D^{*0}\bar{D}^0 + c.c.)$ states are dominant. For details see Table 2.

Table 2. Branching ratios for the model from Ref. [10]. Γ in MeV, m_X in MeV, g_A in GeV.

m_X	3871.68	mode	$X \to D^{*0}\bar{D}^0 + c.c.$	$X \to D^{*+}D^- + c.c.$	$X \to Others$
Γ	1.2	BR	0.176	0.045	0.779
$g_A^2/8\pi$	0.1	BR_{seen}	0.14	0.011	0.761

3 $X(3872)$, $I^G(J^{PC}) = 0^+(1^{++})$, as the $\chi_{c1}(2P)$ charmonium [11]

Contrary to almost standard opinion that the $X(3872)$ resonance is the $D^{*0}\bar{D}^0 + c.c.$ molecule or the $qc\bar{q}\bar{c}$ four-quark state, we discuss the scenario where the $X(3872)$ resonance is the $c\bar{c} = \chi_{c1}(2P)$ charmonium which "sits on" the $D^{*0}\bar{D}^0$ threshold.

The two dramatic discoveries have generated a stream of the $D^{*0}\bar{D}^0 + D^0\bar{D}^{*0}$ molecular interpretations of the $X(3872)$ resonance.

The mass of the $X(3872)$ resonance is 50 MeV lower than predictions of the most lucky naive potential models for the mass of the $\chi_{c1}(2P)$ resonance,

$$m_X - m_{\chi_{c1}(2P)} = -\Delta \approx -50\,\text{MeV}, \tag{14}$$

and the relation between the branching ratios

$$BR(X \to \pi^+\pi^-\pi^0 J/\psi(1S)) \sim BR(X \to \pi^+\pi^- J/\psi(1S)), \tag{15}$$

that is interpreted as a strong violation of isotopic symmetry.

But the bounding energy is small, $\epsilon_B < (1 \div 3)$ MeV. That is, the radius of the molecule is large, $r_{X(3872)} > (3 \div 5)$ fm $= (3 \div 5) \cdot 10^{-13}$ cm. As for the charmonium, its radius is less one fermi, $r_{\chi_{c1}(2P)} \approx 0.5$ fm $= 0.5 \cdot 10^{-13}$ cm. That is, the molecule volume is $100 \div 1000$ times as large as the charmonium volume, $V_{X(3872)}/V_{\chi_{c1}(2P)} > 100 \div 1000$.

The enthusiasts of the molecular scenario do not discuss this question with rare exception. Despite the fact that this means a probability of production of a giant molecule in hard processes, at small distances, is suppressed in comparison with a probability of production of heavy a charmonium by a factor $\sim V_{\chi_{c1}(2P)}/V_{X(3872)}$.

How to explain sufficiently abundant inclusive production of the rather extended molecule X(3872) in a hard process $pp \to X(3872) + anything$ with rapidity in the range 2,5 - 4,5 and transverse momentum in the range 5-20 GeV [12] ? Really,

$$\sigma(pp \to X(3872) + anything)BR(X(3872) \to \pi^+\pi^- J/\psi) = 5.4\,\text{nb} \tag{16}$$

and

$$\sigma(pp \to \psi(2S) + anything)BR(\psi(2S) \to \pi^+\pi^- J/\psi) = 38\,\text{nb}. \tag{17}$$

But, according to Ref.[7]

$$BR(\psi(2S) \to \pi^+\pi^- J/\psi) = 0.34 \tag{18}$$

while

$$0.023 < BR(X(3872) \to \pi^+\pi^- J/\psi) < 0.066 \tag{19}$$

according to Ref. [14]. So,

$$0.74 < \frac{\sigma(pp \to X(3872) + anything)}{\sigma(pp \to \psi(2S) + anything)} < 2.1. \tag{20}$$

The extended molecule is produced in hard process as intensively as the compact charmonium. It's a miracle.

By the way, Belle Collaboration supports this conclusion [13]. Indeed,

$$BR(B^0 \to X(3872)K^+\pi^-) \times BR(X(3872) \to J/\psi\pi^+\pi^-) = (7.9 \pm 1.3 \pm 0.4) \times 10^{-6} \tag{21}$$

and

$$BR(B^0 \to \psi(2S)K^+\pi^-) = (5.79 \pm 1.4) \times 10^{-4}. \tag{22}$$

So,

$$0.2 < \frac{BR(B^0 \to X(3872)K^+\pi^-)}{BR(B^0 \to \psi(2S)K^+\pi^-)} < 0.6. \tag{23}$$

As for the problem of the mass shift, Eq. (14), the contribution of the D^-D^{*+} and $\bar{D}^0D^{*0}$ loops, see Figure 3, into the self energy of the $X(3872)$ resonance, $\Pi_X(s)$, solves it easily. Let us calculate $I^{D^*\bar{D}}(s)$ in Eq. (9) with help of a cut-off Λ.

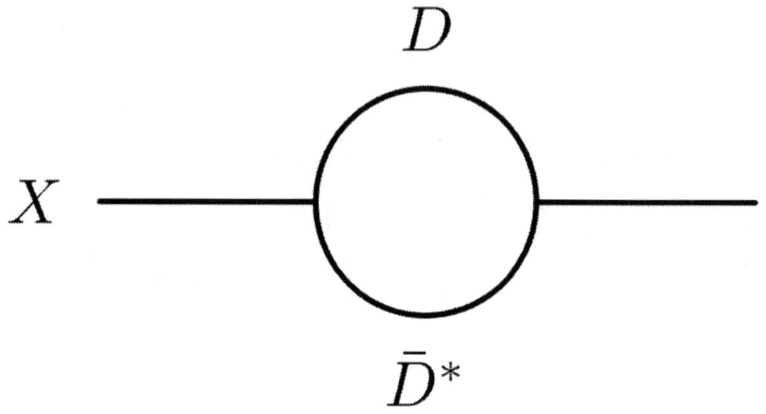

Figure 3. The contribution of the $\bar{D}^0D^{*0}$ and D^-D^{*+} loops into the self energy of the $X(3872)$ resonance.

$$I^{D^*\bar{D}}(s) = \int\limits_{m_+^2}^{\Lambda^2} \frac{\sqrt{(s'-m_+^2)(s'-m_-^2)}}{s'(s'-s)} ds' \approx 2\ln\frac{2\Lambda}{m_+} - 2\sqrt{\frac{m_+^2-s}{s}}\arctan\sqrt{\frac{s}{m_+^2-s}}, \tag{24}$$

where $s < m_+^2$, $\Lambda^2 \gg m_+^2$. The inverse propagator of the X(3872) resonance

$$D_X(s) = m^2_{\chi_{c1}(2P)} - s - \Pi_X(s) - \imath m_X\Gamma. \tag{25}$$

The renormalization of mass

$$m^2_{\chi_{c1}(2P)} - m_X^2 - \Pi_X(m_X^2) = 0 \tag{26}$$

results in

$$\Delta\,(2m_X + \Delta) = \Pi_X(m_X^2) \approx \left(g_A^2/8\pi^2\right) 4\ln(2\Lambda/m_+). \tag{27}$$

If $\Delta = m_{\chi_{c1}(2P)} - m_X \approx 50$ MeV, then $g_A^2/8\pi \approx 0.2$ GeV^2 for $\Lambda = 10$ GeV and $BR(X \to D^0\bar{D}^{*0} + \bar{D}^0D^{*0}) \approx 0.3$. [1]

Thus, we expect that a number of unknown mainly two-gluon decays of $X(3872)$ into non-$D^{*0}\bar{D}^0+$ *c.c.* states are considerable [2]. The discovery of these decays would be the strong (if not decisive) confirmation of our scenario.

[1]The assumption of the determining role of the $D^*\bar{D}+c.c.$ channels in the shift of the mass of the $\chi_{c1}(2P)$ meson is based on the following reasoning. Let us imagine that D and D^* mesons are light, for example, as the K and K^* mesons. Then the width of $X(3872)$ meson is equal 50 MeV for $g_A^2/8\pi = 0.2\,GeV^2$ that much more than the width of its decay into all non-($D^{*0}\bar{D}^0$+c.c.) channels, $\Gamma < 1.2$ MeV. That is, in our case the coupling of the $X(3872)$ meson with the $D^*\bar{D} + c.c.$ channels is rather strong.

[2]Note that in the $\chi_{c1}(1P)$ case the width of the two-gluon decays equals 0.56 MeV [7] that agrees with $\Gamma < 1.2$ MeV satisfactory.

As for $BR(X \to \omega J/\psi) \sim BR(X \to \rho J/\psi)$, Eq. (15), this could be a result of dynamics. In our scenario the $\omega J/\psi$ state is produced via the three gluons, see Figure 4.

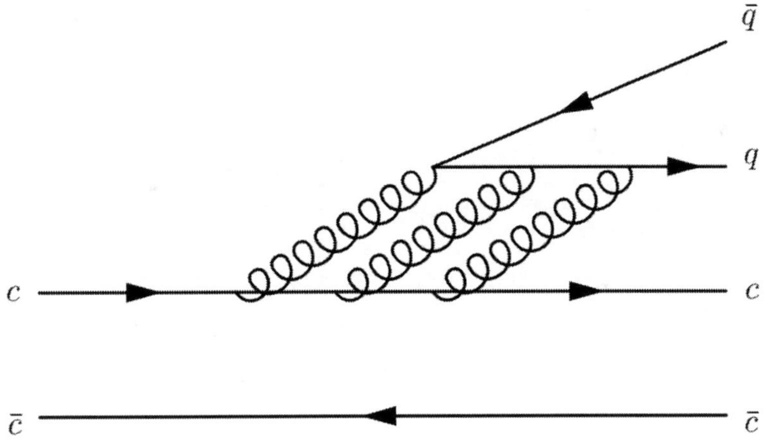

Figure 4. The three-gluon production of the ω and ρ mesons (the ρ meson via the contribution $\sim m_u - m_d$). All possible permutations of gluons are assumed.

As for the $\rho J/\psi$ state, it is produced both via the one photon, see Figure 5, and via the three gluons (via the contribution $\sim m_u - m_d$), see Figure 4.

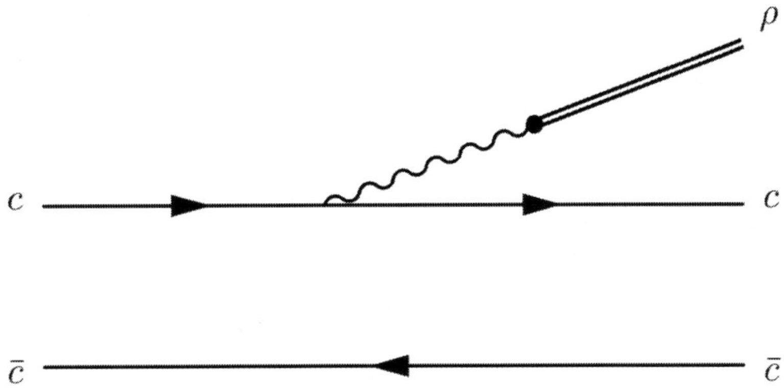

Figure 5. The one-photon production of the ρ meson. All possible permutations of photon are assumed.

Close to our scenario is an example of the $J/\psi \to \rho\eta'$ and $J/\psi \to \omega\eta'$ decays. According to Ref. [7]

$$BR(J/\psi \to \rho\eta') = (1.05 \pm 0.18) \cdot 10^{-4} \text{ and } BR(J/\psi \to \omega\eta') = (1.82 \pm 0.21) \cdot 10^{-4}. \quad (28)$$

Note that in the $X(3872)$ case the ω meson is produced on its tail ($m_X - m_{J/\psi} = 775$ MeV), while the ρ meson is produced on a half.

It is well known that the physics of charmonium ($c\bar{c}$) and bottomonium ($b\bar{b}$) is similar. Let us compare the already known features of X(3872) with the ones of $\Upsilon_{b1}(2P)$.

Recently, the LHCb Collaboration published a landmark result [15]

$$\frac{BR(X \to \gamma\psi(2S))}{BR(X \to \gamma J/\psi)} = C_X \left(\frac{\omega_{\psi(2S)}}{\omega_{J/\psi}}\right)^3 = 2.46 \pm 0.7\,, \tag{29}$$

where $\omega_{\psi(2S)}$ and $\omega_{J/\psi}$ are the energies of the photons in the $X \to \gamma\psi(2S)$ and $BR(X \to \gamma J/\psi)$ decays, respectively.

On the other hand, it is known [7] that

$$\frac{BR(\chi_{b1}(2P) \to \gamma\Upsilon(2S))}{BR(\chi_{b1}(2P) \to \gamma\Upsilon(1S))} = C_{\chi_{b1}(2P)} \left(\frac{\omega_{\Upsilon(2S)}}{\omega_{\Upsilon(1S)}}\right)^3 = 2.16 \pm 0.28\,, \tag{30}$$

where $\omega_{\Upsilon(2S)}$ and $\omega_{\Upsilon(1S)}$ are the energies of the photons in the $\chi_{b1}(2P) \to \gamma\Upsilon(2S)$ and $\chi_{b1}(2P) \to \gamma\Upsilon(1S)$ decays, respectively.

Consequently,

$$C_X = 136.78 \pm 38.89 \tag{31}$$

and

$$C_{\chi_{b1}(2P)} = 80 \pm 10.37 \tag{32}$$

as all most lucky versions of the potential model predict for the quarkonia, $C_{\chi_{c1}(2P)} \gg 1$ and $C_{\chi_{b1}(2P)} \gg 1$.

According to Ref. [7]

$$BR(\chi_{b1}(2P) \to \omega\Upsilon(1S)) = \left(1.63\pm^{0.4}_{0.34}\right)\%\,. \tag{33}$$

If the one-photon mechanism dominates in the $X(3872) \to \rho J/\psi$ decay, see Figure 5, then one should expect

$$BR(\chi_{b1}(2P) \to \rho\Upsilon(1S)) \sim (e_b/e_c)^2 \cdot 1.6\,\% = (1/4) \cdot 1.6\,\% = 0.4\%\,, \tag{34}$$

where e_c and e_b are the charges of the c and b quarks, respectively.

If the three-gluon mechanism (its part $\sim m_u - m_d$) dominates in the $X(3872) \to \rho J/\psi$ decay, see Figure 4, then one should expect

$$BR(\chi_{b1}(2P) \to \rho\Upsilon(1S)) \sim 1.6\%\,. \tag{35}$$

4 Conclusion

We believe that discovery of a significant number of unknown decays of $X(3872)$ into non-$D^{*0}\bar{D}^0$+$c.c.$ states via two gluons and discovery of the $\chi_{b1}(2P) \to \rho\Upsilon(1S)$ decay could decide destiny of X(3872).

Once more, we discuss the scenario where the $\chi_{c1}(2P)$ charmonium sits on the $D^{*0}\bar{D}^0$ threshold but not a mixing of the giant $D^*\bar{D}$ molecule and the compact $\chi_{c1}(2P)$ charmonium. Note that the mixing of such states requests the special justification. That is, it is necessary to show that the transition of the giant molecule into the compact charmonium is considerable at insignificant overlapping of their wave functions. Such a transition $\sim \sqrt{V_{\chi_{c1}(2P)}/V_{X(3872)}}$ and a branching ratio of a decay via such a transition $\sim V_{\chi_{c1}(2P)}/V_{X(3872)}$.

Acknowledgments

This work was supported in part by RFBR, Grant No. 16-02-00065, and by Presidium of the Russian Academy of Sciences, Project No. 0314-2015-0011.

References

[1] N. N. Achasov and E. V. Rogozina, Pis'ma v ZhETF **100**, 252 (2014) [JETP Letters **100**, 227 (2014)]
[2] S. K. Choi, et al. (Belle Callaboration), Phys. Rev. D **84**, 052004 (2011)
[3] K. Abe, et al. (Belle Callaboration), arXiv:hep-ex/0505037 (2005)
[4] P. del Amo Sanchez, et al. (BABAR Callaboration), Phys. Rev. D **82**, 011101 (R) (2010)
[5] T. Aushev, et al. (Belle Callaboration), Phys. Rev. D **81**, 031103 (2010)
[6] R. Aaij, et al. (LHCb Callaboration), Phys. Rev. Lett. **110**, 222001 (2013)
[7] K. A. Olive, et al. (Particle Data Group), Chin. Phys. C **38**, 090001 (2014)
[8] N. N. Achasov, S. A. Devyanin and G.N. Shestakov, Yad. Fiz. **32**, 1098 (1980) [Sov. J. Nucl. Phys. **32**, 566 (1980)]
[9] N. N. Achasov and A. V. Kiselev, Phys. Rev. D **70**, 11901(R) (2004)
[10] L. Maiani, et al., Phys. Rev. D **87** 111102(R) (2013)
[11] N. N. Achasov and E. V. Rogozina, Mod. Phys. Lett. A **30** 1550181 (2015)
[12] R. Aaij, et al. (LHCb Collaboration), Eur. Phys. J. C **72** 1972 (2012)
[13] A. Bala, et al. (Belle Callaboration), Phys. Rev. D **91** 051101(R) (2015)
[14] C.-Z. Yuan (Belle Collaboration), *Proceedings of the XXIX PHYSICS IN COLLISION* Kobe, Japan, arXiv: 0910.3138 [hep-ex] (2009)
[15] R. Aaij, et al. (LHCb Collaboration), Nucl. Phys. B **886**, 665 (2014)

EPJ Web of Conferences **125**, 04003 (2016) DOI: 10.1051/epjconf/201612504003
QUARKS-2016

Transition formfactors $\gamma^* \to \gamma f_2(1270)$ and $\gamma^* \to \gamma a_2(1320)$ in the e^+e^- collisions

N. N. Achasov[1,⋆] and A. V. Kiselev[1,2,⋆⋆]

[1] *Sobolev Institute for Mathematics, 630090 Novosibirsk, Russia*
[2] *Novosibirsk State University, 630090 Novosibirsk, Russia*

Abstract. Transition form factors $\gamma^* \to \gamma f_2(1270)$ and $\gamma^* \to \gamma a_2(1320)$ and corresponding $e^+e^- \to \gamma^* \to f_2\gamma$ and $e^+e^- \to \gamma^* \to a_2\gamma$ cross sections are considered up to high energies. It is shown that the QCD asymptotics of the amplitudes of the $e^+e^- \to \gamma^* \to f_2\gamma$ and $e^+e^- \to \gamma^* \to a_2\gamma$ reactions can be reached only by taking into account a compensation of contributions of $\rho(770)$, $\omega(782)$ with contributions of their radial excitations. Recent BABAR measurement of the $e^+e^- \to f_2\gamma \to \pi^+\pi^-\gamma$ cross section at 10.58 GeV gives hope for detailed investigation of the $\gamma^*(s) \to f_2\gamma$ and $\gamma^*(s) \to a_2\gamma$ transition form factors at high energy region. Recent Belle data on the $\gamma^*(Q^2)\gamma \to f_2$ transition up to $Q^2 = 30$ GeV^2 are taken into account.

It is accepted that $f_2(1270)$ and $a_2(1320)$ tensor mesons are two-quark states, $a_2 = (u\bar{u} - d\bar{d})/\sqrt{2}$ and $f_2 = (u\bar{u} + d\bar{d})/\sqrt{2}$.

The $\gamma^* \to f_2(1270)\gamma$ and $\gamma^* \to a_2(1320)\gamma$ transition form factors manifest themselves in $e^+e^- \to \gamma^* \to T\gamma$ (time-like region of γ^*) and $\gamma^*(Q^2)\gamma \to T$ (space-like region) reactions, where $T = f_2(1270)$ or $a_2(1320)$. These reactions are connected with each other, so progress in investigation of one reactions can shed light on the other reactions.

Recently BABAR presented a measurement $\sigma(e^+e^- \to f_2\gamma \to \pi^+\pi^-\gamma)$ at hight-energy point [1], $\sigma(e^+e^- \to f_2\gamma \to \pi^+\pi^-\gamma, 10.58\text{ GeV}) = 37^{+24}_{-18}$, and Belle measured the $\gamma^*(Q^2)\gamma \to f_2$ transition form factor in the space-like region for Q^2 up to 30 GeV^2 [2].

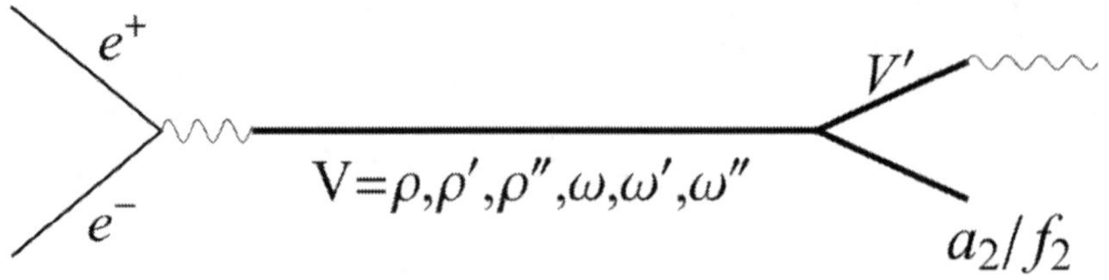

Figure 1. Diagram of $e^+e^- \to a_2(f_2)\gamma$ process.

⋆e-mail: achasov@math.nsc.ru
⋆⋆e-mail: kiselev@math.nsc.ru

The diagram of the $e^+e^- \to \gamma^* \to T\gamma$ process is shown in Fig. 1. V and V' are appropriate vector mesons, ρ, ω and their radial excitations.

It is known that in $\gamma\gamma \to f_2 \to \pi\pi$ reaction tensor mesons are produced mainly by the photons with the opposite helicity states. This is naturally described by the Lagrangian [3]

$$L = g_{T\gamma\gamma} T_{\mu\nu} F_{\mu\sigma} F_{\nu\sigma} \tag{1}$$

where $F_{\mu\nu} = \partial_\mu A_\nu - \partial_\nu A_\mu$, $T_{\mu\nu}$ is a tensor meson T field.

In the frame of GVDM it is assumed that the effective Lagrangian of the reaction $f_2 \to VV$ is

$$L = g_{f_2 VV} T_{\mu\nu} F^V_{\mu\sigma} F^V_{\nu\sigma} \tag{2}$$

where $F^V_{\mu\sigma} = \partial_\mu V_\sigma - \partial_\sigma V_\mu$ and $V = \rho_1, \rho_2, \rho_3, \omega_1, \omega_2, \omega_3$, where $\rho_1 = \rho$, $\rho_2 = \rho'$, $\rho_2 = \rho''$ and $\omega_1 = \omega$, $\omega_2 = \omega'$, $\omega_3 = \omega''$.

Assuming the GVDM mechanism $e^+e^- \to \gamma^* \to (\rho + \rho' + \rho'' + \omega + \omega' + \omega'') \to f_2(\rho + \rho' + \rho'' + \omega + \omega' + \omega'') \to f_2\gamma$ one obtains [4]

$$\sigma_{e^+e^-\to f_2\gamma}(s) = \frac{4\pi^2}{9}\alpha^3\Big(1 - \frac{m^2_{f_2}}{s}\Big)^3\Big(\frac{s^2}{m^4_{f_2}} + 3\frac{s}{m^2_{f_2}} + 6\Big)\times$$

$$\left|\frac{m^2_\rho g_{f_2\rho\rho}}{f^2_\rho D_\rho(s)} + \frac{m^2_{\rho'} g_{f_2\rho'\rho'}}{f^2_{\rho'} D_{\rho'}(s)} + \frac{m^2_{\rho''} g_{f_2\rho''\rho''}}{f^2_{\rho''} D_{\rho''}(s)} + \frac{m^2_\omega g_{f_2\omega\omega}}{f^2_\omega D_\omega(s)} + \frac{m^2_{\omega'} g_{f_2\omega'\omega'}}{f^2_{\omega'} D_{\omega'}(s)} + \frac{m^2_{\omega''} g_{f_2\omega''\omega''}}{f^2_{\omega''} D_{\omega''}(s)}\right|^2. \tag{3}$$

Here $D_V(s)$ are inverse propagators

$$D_V(s) = m^2_V - s - i\sqrt{s}\Gamma_V(s), \tag{4}$$

the forms of widths $\Gamma_V(s)$ are described below. The constants f_V are related to e^+e^- widths as usual:

$$\Gamma(V \to e^+e^-) = \frac{4\pi\alpha^2}{3f^2_V} m_V. \tag{5}$$

In $q\bar{q}$ model $g_{f_2\rho_i\rho_i} = g_{f_2\omega_i\omega_i}$, $m_{\omega_i} = m_{\rho_i}$, $f_{\omega_i} = 3f_{\rho_i}$, this gives

$$\Gamma_{f_2\to\gamma\gamma} = \Big(\frac{10}{9}\Big)^2 \frac{\pi\alpha^2}{5}\left|\frac{g_{f_2\rho\rho}}{f^2_\rho} + \frac{g_{f_2\rho'\rho'}}{f^2_{\rho'}} + \frac{g_{f_2\rho''\rho''}}{f^2_{\rho''}}\right|^2 m^3_{f_2} = 3.03 \pm 0.35 \text{ keV [5]}, \tag{6}$$

$$\sigma_{e^+e^-\to f_2\gamma}(s) = \frac{20\pi}{9}\alpha\frac{\Gamma_{f_2\to\gamma\gamma}}{m^3_{f_2}}|F_{\gamma^*\to f_2\gamma}(s)|^2\Big(1 - \frac{m^2_{f_2}}{s}\Big)^3\Big(\frac{s^2}{m^4_{f_2}} + 3\frac{s}{m^2_{f_2}} + 6\Big). \tag{7}$$

Here $F_{\gamma^*\to f_2\gamma}(s)$ is a transition form factor. Since ω_i contributions are suppressed, we neglect the difference between Γ_{ω_i} and Γ_{ρ_i} in $D_{\omega_i}(s)$ to simplify the formula.

Let's denote $a_{f_2} = g_{f_2\rho'\rho'} f^2_\rho / g_{f_2\rho\rho} f^2_{\rho'}$ and $b_{f_2} = g_{f_2\rho''\rho''} f^2_\rho / g_{f_2\rho\rho} f^2_{\rho''}$. One has

$$F_{\gamma^*\to f_2\gamma}(s) = \frac{1}{1 + a_{f_2} + b_{f_2}}\Big(\frac{m^2_\rho}{D_\rho(s)} + a_{f_2}\frac{m^2_{\rho'}}{D_{\rho'}(s)} + b_{f_2}\frac{m^2_{\rho''}}{D_{\rho''}(s)}\Big). \tag{8}$$

The QCD asymptotics of $\sigma(e^+e^- \to \textit{light hadrons})$ at $s \to \infty$ is equal to $\sum_f Q^2_f 4\pi\alpha^2/s$ to within logarithms (Q_f are charges of u, d, and s quarks). Due to asymptotic freedom at $s \to \infty$ imaginary part of $D_V(s)$ vanishes

$$Im(D_V(s)) = -\sqrt{s}\Gamma_V(s) \to 0; \quad \frac{1}{D_V(s)} \to \frac{1}{m^2_V - s}. \tag{9}$$

It means that $F_{\gamma^*\to f_2\gamma}(s) \sim 1/s^2$, i.e.

$$m_\rho^2 + a_{f_2} m_{\rho'}^2 + b_{f_2} m_{\rho''}^2 = 0 \tag{10}$$

to suppress asymptotics $\sim 1/s$ in Eq. 8.

Note the analogous condition was obtained in Refs. [6, 7] after taking into account the QCD based asymptotics of the $\gamma^*(Q^2)\gamma \to a_2$ amplitude.

The Belle Collaboration presented 3 data sets for each helicity difference $\lambda \equiv |\lambda_{\gamma^*} - \lambda_\gamma| = 0, 1, 2$. These data were described in Ref. [7] with $b_{f_2} = 0$. To take ρ'' into account it is enough to modify $\gamma^*(Q^2)\gamma \to f_2$ transition form factor (Eq. (10) of that paper):

$$F_T(Q) \to F_T(Q) = \frac{1}{1 + a_{f_2} + b_{f_2}} \left(\frac{1}{1 + \frac{Q^2}{m_\rho^2}} + \frac{a_{f_2}}{1 + \frac{Q^2}{m_{\rho'}^2}} + \frac{b_{f_2}}{1 + \frac{Q^2}{m_{\rho''}^2}} \right). \tag{11}$$

This gives two minimums of the χ^2 function, constructed for the Belle data: $b_{f_2} = 0.082 \pm 0.025$ and $b_{f_2} = 0.740 \pm 0.011$, see Fig. 2.

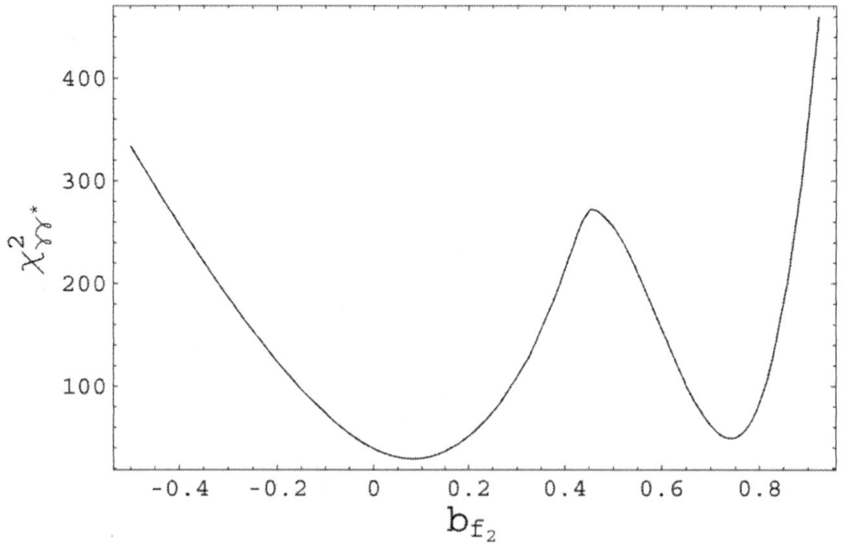

Figure 2. χ^2 function for Belle data on $\gamma^*\gamma \to f_2$ transition form factor.

Let's consider three cases:
1) $(a_{f_2}, b_{f_2}) = (-0.393, 0.082)$ - best minimum of the χ^2 function ($\chi^2/n.d.f. = 29.3/28$);

2) $(a_{f_2}, b_{f_2}) = (-1.3, 0.74)$ - second minimum of the χ^2 function ($\chi^2/n.d.f. = 49.1/28$);

3): $(a_{f_2}, b_{f_2}) = (-0.28, 0)$ - no ρ'' contribution ($\chi^2/n.d.f. = 39.1/29$).

Recently the $e^+e^- \to f_2\gamma \to \pi^+\pi^-\gamma$ cross section was measured at BABAR at 10.58 GeV: $\sigma(\pi^+\pi^-\gamma)_{10.58} \equiv \sigma(e^+e^- \to f_2\gamma \to \pi^+\pi^-\gamma,\ 10.58\text{ GeV}) = 37^{+24}_{-18}$ fb.

Taking into account $Br(f_2 \to \pi^+\pi^-) \approx 56.5\%$ and $D_V(s) \to m_V^2 - s$ at $s \to \infty$, we find $\sigma(\pi^+\pi^-\gamma)_{10.58} = 1.0$ fb for $b_{f_2} = 0.082$, $\sigma(\pi^+\pi^-\gamma)_{10.58} = 3.8$ fb for $b_{f_2} = 0.74$ and $\sigma(\pi^+\pi^-\gamma)_{10.58} = 1.5$

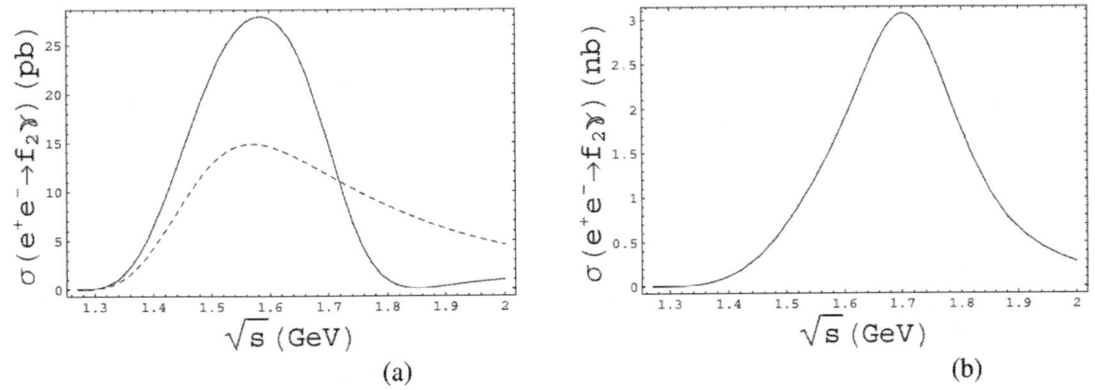

Figure 3. The $e^+e^- \to f_2\gamma$ cross section up to 2 GeV. (a) Solid line is for $(a_{f_2}, b_{f_2}) = (-0.393, 0.082)$, dashed line is for $(a_{f_2}, b_{f_2}) = (-0.28, 0)$ (ρ'' is absent). (b) The plot for $(a_{f_2}, b_{f_2}) = (-1.3, 0.74)$.

fb for $b_{f_2} = 0$. The elaboration of the pioneer measurement of $\sigma(\pi^+\pi^-\gamma)_{10.58}$ is important for the $\gamma^* \to \gamma f_2/a_2$ transition form factor investigations.

The $e^+e^- \to f_2\gamma$ cross section Eq. (7) for all three cases under consideration is shown in Figs. 3 and 4. In vector propagators Eq. (4) we use constant widths taken from Ref. [5] below 2 GeV and zero widths above 5 GeV. Dashed lines show continuation of these curves to the intermediate region.

The treatment of the $e^+e^- \to \gamma^* \to a_2\gamma$ cross section is similar:

$$\sigma_{e^+e^-\to a_2\gamma}(s) = \frac{\pi^2}{9}\alpha^3\Big(1 - \frac{m_{a_2}^2}{s}\Big)^3\Big(\frac{s^2}{m_{a_2}^4} + 3\frac{s}{m_{a_2}^2} + 6\Big)\times$$
$$\Big|\frac{m_\rho^2 g_{a_2\rho\omega}}{f_\rho f_\omega D_\rho(s)} + \frac{m_\omega^2 g_{a_2\rho\omega}}{f_\rho f_\omega D_\omega(s)} + \frac{m_{\rho'}^2 g_{a_2\rho'\omega'}}{f_{\rho'} f_{\omega'} D_{\rho'}(s)} + \frac{m_{\omega'}^2 g_{a_2\rho'\omega'}}{f_{\rho'} f_{\omega'} D_{\omega'}(s)} +$$
$$\frac{m_{\rho''}^2 g_{a_2\rho''\omega''}}{f_{\rho''} f_{\omega''} D_{\rho''}(s)} + \frac{m_{\omega''}^2 g_{a_2\rho''\omega''}}{f_{\rho''} f_{\omega''} D_{\omega''}(s)}\Big|^2, \quad (12)$$

$$\Gamma_{a_2\to\gamma\gamma} = \frac{4\pi\alpha^2}{5}\Big|\frac{g_{a_2\rho\omega}}{f_\rho f_\omega} + \frac{g_{a_2\rho'\omega'}}{f_{\rho'} f_{\omega'}} + \frac{g_{a_2\rho''\omega''}}{f_{\rho''} f_{\omega''}}\Big|^2 m_{a_2}^3 = 1.00 \pm 0.06 \text{ keV [5]}\ . \quad (13)$$

Denoting $a_{a_2} = g_{a_2\rho'\omega'} f_\rho f_\omega / g_{a_2\rho\omega} f_{\rho'} f_{\omega'}$ and $b_{a_2} = g_{a_2\rho''\omega''} f_\rho f_\omega / g_{a_2\rho\omega} f_{\rho''} f_{\omega''}$ one gets

$$\sigma_{e^+e^-\to a_2\gamma}(s) = \frac{20\pi}{9}\alpha\frac{\Gamma_{a_2\to\gamma\gamma}}{m_{a_2}^3}\big|F_{\gamma^*\to a_2\gamma}(s)\big|^2\Big(1 - \frac{m_{a_2}^2}{s}\Big)^3\Big(\frac{s^2}{m_{a_2}^4} + 3\frac{s}{m_{a_2}^2} + 6\Big) \quad (14)$$

with transition form factor

$$F_{\gamma^*\to a_2\gamma}(s) = \frac{1}{2(1 + a_{a_2} + b_{a_2})}\times$$
$$\Big(\frac{m_\rho^2}{D_\rho(s)} + \frac{m_\rho^2}{D_\omega(s)} + a_{a_2}\frac{m_{\rho'}^2}{D_{\rho'}(s)} + a_{a_2}\frac{m_{\rho'}^2}{D_{\omega'}(s)} + b_{a_2}\frac{m_{\rho''}^2}{D_{\rho''}(s)} + b_{a_2}\frac{m_{\rho''}^2}{D_{\omega''}(s)}\Big) \quad (15)$$

and the asymptotic condition $m_\rho^2 + a_{a_2}m_{\rho'}^2 + b_{a_2}m_{\rho''}^2 = 0$.

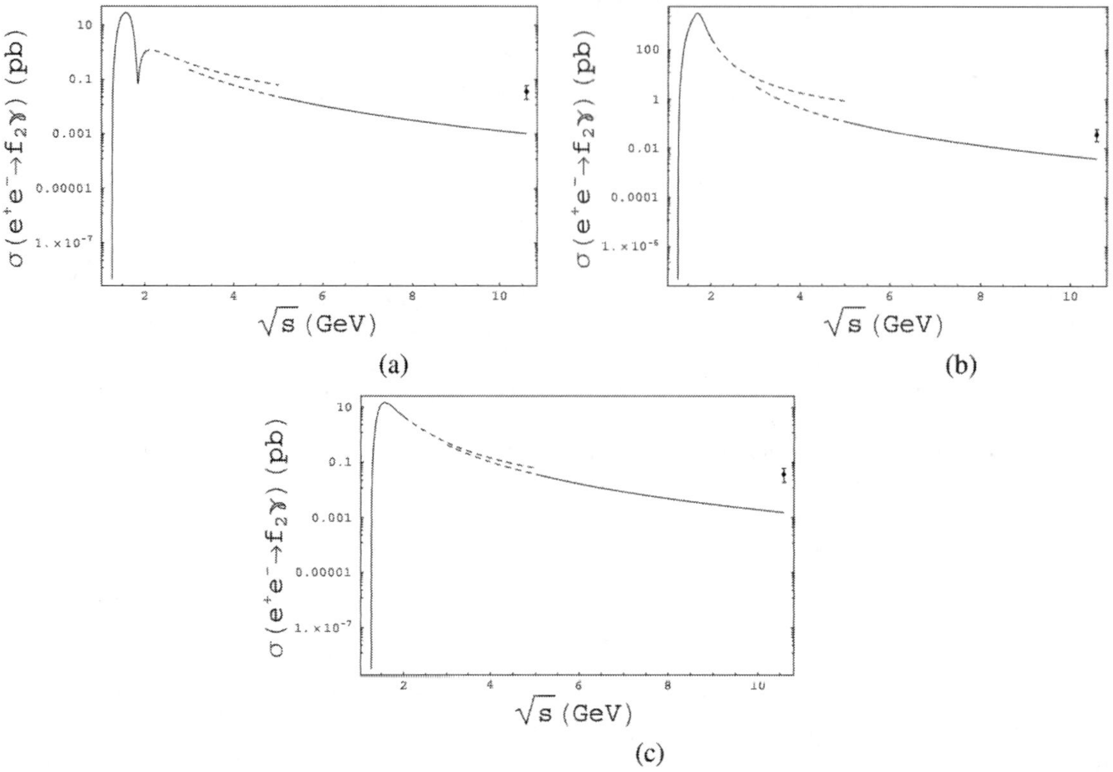

Figure 4. The $e^+e^- \to f_2\gamma$ cross section up to 10.58 GeV. In each logarithmic plot solid line below 2 GeV is drawn with constant vector widths taken from Ref. [5], solid line above 5 GeV is drawn with zero vector widths, dashed lines are their continuations. The experimental BaBar point [1] is shown also. (a) $(a_{f_2}, b_{f_2}) = (-0.393, 0.082)$, (b) $(a_{f_2}, b_{f_2}) = (-1.3, 0.74)$, (c) $(a_{f_2}, b_{f_2}) = (-0.28, 0)$ (ρ'' is absent).

In the naive $q\bar{q}$ model for $i = 1, 2, 3$ $g_{a_2\rho_i\omega_i} = g_{f_2\rho_i\rho_i}$. It means that in our approach $a_{a_2} = a_{f_2}$, $b_{a_2} = b_{f_2}$.

The resulting cross-section for the same sets of (a_{a_2}, b_{a_2}) as for $e^+e^- \to f_2(1270)\gamma$ reaction is shown in Fig. 5.

It is natural that our $q\bar{q}$ consideration agrees with QCD requirement for high s

$$\frac{\sigma(e^+e^- \to f_2\gamma, s)}{\sigma(e^+e^- \to a_2\gamma, s)} = \frac{25}{9} \tag{16}$$

and the well-known relation being in perfect agreement with the data

$$\frac{\Gamma(f_2 \to \gamma\gamma)}{\Gamma(a_2 \to \gamma\gamma)} = \frac{25}{9}\frac{m_{f_2}^3}{m_{a_2}^3} \approx \frac{25}{9}\,. \tag{17}$$

The above-mentioned BABAR measurement $\sigma(e^+e^- \to f_2\gamma \to \pi^+\pi^-\gamma,\ 10.58\ \text{GeV}) = 37^{+24}_{-18}$ fb gives

$$\sigma(e^+e^- \to a_2\gamma,\ 10.58\ \text{GeV}) = 24^{+15}_{-11}\ \text{fb}\,. \tag{18}$$

Brief summary and conclusions:

 DOI: 10.1051/epjconf/201612504003

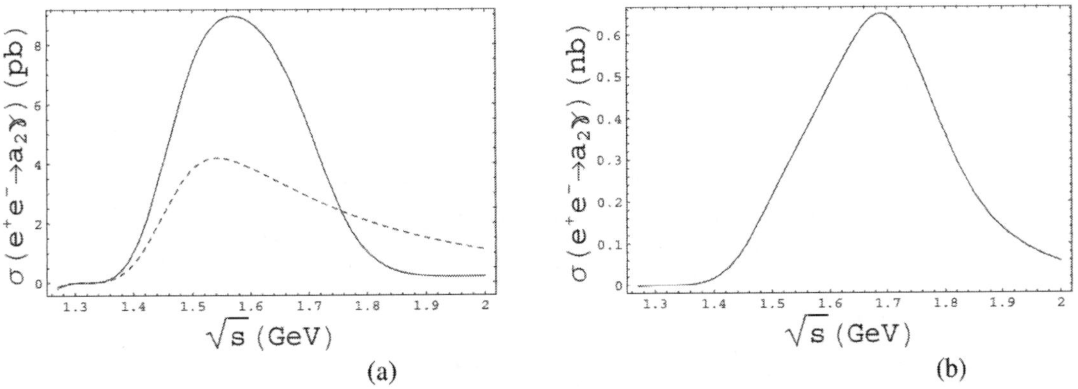

Figure 5. The $e^+e^- \to a_2\gamma$ cross section. (a) Solid line is for $(a_{a_2}, b_{a_2}) = (-0.393, 0.082)$, dashed line is for $(a_{a_2}, b_{a_2}) = (-0.28, 0)$. (b) The plot for $(a_{a_2}, b_{a_2}) = (-1.3, 0.74)$.

1. It is shown that the QCD asymptotics of the amplitudes of the reactions $e^+e^- \to \gamma^* \to f_2\gamma$ and $e^+e^- \to \gamma^* \to a_2\gamma$ can be reached only by taking into account a compensation of contributions of $\rho(770)$, $\omega(782)$ with contributions of their radial excitations.

2. The ρ and ω excitation contribution to the $e^+e^- \to f_2\gamma$ and $e^+e^- \to a_2\gamma$ cross sections is essential and allows the experimental investigation of $e^+e^- \to \eta\pi^0\gamma$ and $e^+e^- \to \pi^0\pi^0\gamma$ processes, for example, at VEPP-2000 collider. Note that the best channel to study the a_2 production is the $e^+e^- \to a_2\gamma \to \rho\pi\gamma$ because $Br(a_2 \to \rho\pi) = 70\%$ and the background is expected to be small.

3. At the BABAR energy (10.58 GeV) our estimation of the $e^+e^- \to f_2\gamma \to \pi^+\pi^-\gamma$ cross section, based on the Belle data about the $\gamma * (Q^2)\gamma \to f_2 \to \pi^+\pi^-$ cross section, is much less than the average BABAR value, but the experimental error is half of the average value, so no conclusions concerning this disagreement could be done now. Anyway, the pioneer BABAR measurement gives hope that the detailed investigation of the $\gamma^*(s) \to f_2\gamma$ and $\gamma^*(s) \to a_2\gamma$ transition form factors at high energy region is possible.

This work was supported in part by RFBR, Grant No. 16-02-00065, and the Presidium of RAS project No. 0314-2015-0011.

References

[1] J. P. Lees et al (BABAR Collaboration), Phys. Rev. D92 (2015) 072015.
[2] M. Masuda *et al.* (Belle Collaboration), Phys. Rev. D93 (2016) 032003.
[3] N.N. Achasov and V.A. Karnakov, Z. Phys. C30 (1986) 141.
[4] N.N. Achasov, A.I. Goncharenko, A.V. Kiselev, and E.V. Rogozina, Phys. Rev. D88, (2013) 114001; Erratum-ibid. D89 (2014) 059906.
[5] K.A. Olive *et al.* (Particle Data Group), Chin. Phys. C38 (2014) 090001.
[6] N.N. Achasov, A.V. Kiselev, and G.N. Shestakov, Phys. At. Nucl. 79 (2016), 397.
[7] N.N. Achasov, A.V. Kiselev and G.N. Shestakov, Pis'ma Zh. Eksp. Teor. Fiz. 102 (2015) 655 [JETP Lett. **102** (2015) 571], arXiv:1509.09150.

The no-wall holographic model for vector quarkonia

Sergey Afonin[1,*] and *Ilya* Pusenkov[1,**]

[1] *Saint Petersburg State University, 7/9 Universitetskaya nab., St.Petersburg, 199034, Russia*

Abstract. We use the no-wall holographic approach (a relative of the soft-wall one) to construct a universal description of vector mesons with arbitrary quark masses. The proposed model predicts a specific dependence of the parameters of radial Regge trajectories on the quark masses in a reasonable agreement with the meson phenomenology and some theoretical expectations.

The bottom-up holographic approach to QCD introduced in Ref. [1] has many interesting applications in the phenomenology of the strong interactions. Usually this approach is applied to the spectroscopy of light hadrons and the related physics [2]. The heavy-quark sector is addressed relatively seldom (see, e.g., [3, 4] and references therein). In this sector, only the unflavored vector mesons have a rich established radial spectrum [5] that can be compared with theoretical predictions. Our aim is a description of this spectrum within the no-wall holographic approach [6].

The action of the simplest holographic no-wall model is defined by

$$S = \int d^4x\, dz\, \sqrt{g} \left\{ \sum_i \left(|DX_i|^2 - m_i^2 |X_i|^2 \right) - \frac{1}{4g_5^2} F_{MN}^2 \right\}, \tag{1}$$

where

$$g = |\det g_{MN}|, \qquad F_{MN} = \partial_M V_N - \partial_N V_M, \qquad D_M X_i = \partial_M X_i - i V_M X_i, \tag{2}$$

and $M, N = 0, 1, 2, 3, 4$. The action (1) is defined in the AdS_5 space with the metrics

$$g_{MN} dx^M dx^N = \frac{R^2}{z^2} (\eta_{\mu\nu} dx^\mu dx^\nu - dz^2). \tag{3}$$

Here $\eta_{\mu\nu} = \text{diag}(1, -1, -1, -1)$ and R denotes the AdS_5 radius. Below we set $R = 1$ for simplicity. The holographic coordinate $z \geq 0$ has the interpretation of inverse energy scale. The UV boundary $z = 0$ represents the 4D Minkovski space. The 5D vector field $V_M(x, z)$ is dual to the 4D conserved vector current $j_\mu = \bar{\psi}_q \gamma_\mu \psi_q$ for any quark flavor q. The gauge invariance of the action (1) allows to choose a convenient axial gauge $V_z = 0$. On the UV boundary, the scalar fields X_i are identified with sources of various QCD operators with canonical dimension i. The corresponding 5D masses are [7]

$$m_i^2 = i(i - 4). \tag{4}$$

*e-mail: s.afonin@spbu.ru
**e-mail: i.pusenkov@spbu.ru

By assumption, the fields X_i acquire z-dependent vacuum expectation values $\langle X_i \rangle$ which represent the x-independent solutions of the equation of motion,

$$\partial_z\left(\frac{1}{z^3}\partial_z X_i\right) = \frac{m_i^2}{z^5}X_i, \tag{5}$$

with the UV boundary condition

$$\langle X_i \rangle|_{z=0} = 0. \tag{6}$$

The vector physical modes are defined by the plane-wave ansatz $V_\mu(x,z) = e^{ipx}v(z)\varepsilon_\mu$. The mass spectrum of these modes is given by the eigenvalues $M_n^2 = p_n^2$ of the corresponding equation of motion,

$$\partial_z\left(\frac{1}{z}\partial_z v_n\right) + \frac{M_n^2}{z}v_n = \frac{2g_5^2}{z^3}v_n \sum_i \langle X_i \rangle^2. \tag{7}$$

The change of variables $v_n = \sqrt{z}\,\psi_n$ brings this equation into the Schrödinger form

$$-\partial_z^2\psi_n + \left(\frac{3}{4z^2} + 2g_5^2 f(z)\right)\psi_n = M_n^2\psi_n, \tag{8}$$

where

$$f(z) = \frac{1}{z^2}\sum_i \langle X_i \rangle^2 \tag{9}$$

determines the analogue of potential in the Schrödinger equation.

Let us consider the dimension-two operator, $i = 2$, and neglect all others. The solution of Eq. (5) satisfying (6) reads

$$\langle X_2 \rangle = C_2^{(1)} z^2 + C_2^{(2)} z^2 \ln z. \tag{10}$$

Choosing $C_2^{(2)} = 0$, the equation (8) coincide with the equation on the mass spectrum in the standard holographic soft-wall model [8].

The potential in Eq. (8) is written near the UV boundary, where, by assumption, the holographic correspondence allows to use the QCD inputs. In order to calculate the mass spectrum we must continue the function $f(z)$ to the infrared domain, $z \to \infty$. The linear radial Regge spectrum observed in the unflavored light and heavy quarkonia [9] can be obtained only if

$$f(z)|_{z\to\infty} = a^2 z^2. \tag{11}$$

The UV asymptotics of the 5D field dual to the dim2 operator provides the correct IR asymptotics automatically.

The dependence of the linear mass spectrum on the quark mass appears from the v.e.v. $\langle X_3 \rangle$ as the field X_3 is dual to the quark bilinear operator $\bar{q}q$. The quark mass emerges from the following AdS/CFT prescription: The solution of classical e.o.m. for a scalar field Φ corresponding to an operator O of canonical dimension i near the 4D boundary $z \to 0$ has the form [10]

$$\Phi(x,z) \to z^{4-i}\left[\Phi_0(x) + O(z^2)\right] + z^i\left[\frac{\langle O(x) \rangle}{2i-4} + O(z^2)\right], \tag{12}$$

where $\Phi_0(x)$ acts as a source for $O(x)$ with the v.e.v. $\langle O(x) \rangle$. In the real QCD, the quark mass m acts as the source for the operator $\bar{q}q$.

For the canonical dimension $i = 3$, the solution of Eq. (5) satisfying (6) is $\langle X_3 \rangle = C_3^{(1)} z + C_3^{(2)} z^3$. According to the prescription (12), this solution can be rewritten in terms of the physical quantities,

$$\langle X_3 \rangle = \xi m z + \frac{\sigma}{2\xi} z^3, \tag{13}$$

where σ stays for the quark condensate $\langle \bar{q} q \rangle$ and the normalization factor ξ is [11],

$$\xi^2 = \frac{N_c}{4\pi^2}. \tag{14}$$

The solution (13) allows to extract explicitly the m-dependent terms in (9),

$$f(z) \to \xi^2 m^2 + m\sigma z^2 + \tilde{f}(z), \tag{15}$$

where the contribution $\frac{\sigma^2}{4\xi^2} z^6$ is absorbed into the new sum $\tilde{f}(z)$. The Regge form of the spectrum appears in the case of the asymptotics

$$\tilde{f}(z)\Big|_{z\to\infty} = a^2 z^2 + \delta, \tag{16}$$

which will be considered as our choice of $\tilde{f}(z)$.

We arrive then at the following equation on the mass spectrum,

$$-\partial_z^2 \psi_n + \left[\frac{3}{4z^2} + 2g_5^2(\sigma m + a^2) z^2 + 2g_5^2 \xi^2 m^2 + 2g_5^2 \delta \right] \psi_n = M_n^2 \psi_n. \tag{17}$$

The spectrum of Eq. (17) is

$$M_n^2 = 4\sqrt{2}\, g_5 \sqrt{\sigma m + a^2}(n+1) + 2g_5^2 \xi^2 m^2 + 2g_5^2 \delta. \tag{18}$$

This expression demonstrates the parametric dependence of the linear spectrum on the quark mass m. The spectrum (18) can be written in a more compact form,

$$M_n^2 = \sqrt{\alpha + \beta m}(n+1) + \gamma m^2 + \delta. \tag{19}$$

The value of constant γ follows from the relations (14) and the normalization factor for the 5D massless vector fields, $g_5^2 = \frac{12\pi^2}{N_c}$ [1],

$$\gamma = 2g_5^2 \xi^2 = 6. \tag{20}$$

The ensuing phenomenological fits are elaborated in Ref. [3]. It can be also shown [3] that the behavior of the obtained slope in the chiral limit, $m \to 0$, matches a prediction of the dual Veneziano-like amplitudes with the chiral symmetry breaking [12].

The spectrum (19) predicts that the binding energy grows linearly with the quark mass in the heavy quark limit, $m \to \infty$, due to the relation (20) showing that $\gamma > 4$. This behavior seems to agree with the experimental data [5]. For instance, one has for the ground states $M_\psi - 2m_c \approx 0.56$ GeV and $M_\Upsilon - 2m_b \approx 1.1$ GeV if $m_{c,b}$ are taken at 2 GeV (taking m_b at 5 GeV leads to $M_\Upsilon - 2m_b \approx 2.5$ GeV). Such a behavior admits a simple interpretation: When a non-relativistic quark of mass m is created and moves with the velocity v, the binding energy must compensate its kinetic energy $\frac{mv^2}{2}$.

It should be emphasized that the mass relation of the kind (18) emerges within a whole class of holographic models and we have presented a particular model from this class. An example of another model of this sort is constructed in Ref. [3].

The proposed approach can be applied to various physical problems including the hadron spectroscopy, finite-temperature and density effects, hadron formfactors *etc.*

Acknowledgments. The work was supported by the Saint Petersburg State University research grant 11.38.189.2014 and by the RFBR grant 16-02-00348-a.

References

[1] J. Erlich, E. Katz, D. T. Son and M. A. Stephanov, Phys. Rev. Lett. **95**, 261602 (2005); L. Da Rold and A. Pomarol, Nucl. Phys. B **721**, 79 (2005).
[2] J. Erlich, Contemp. Phys. **56**, 159 (2015) [arXiv:1407.5002 [hep-ph]].
[3] S. S. Afonin and I. V. Pusenkov, Phys. Lett. B **726**, 283 (2013).
[4] T. Gutsche, V. E. Lyubovitskij, I. Schmidt and A. Vega, Phys. Rev. D **87**, 056001 (2013).
[5] K. A. Olive *et al.* (Particle Data Group), Chin. Phys. C **38**, 090001 (2014).
[6] S. S. Afonin, Int. J. Mod. Phys. A **26**, 3615 (2011); Phys. Lett. B **678**, 477 (2009).
[7] E. Witten, Adv. Theor. Math. Phys. **2**, 253 (1998); S. S. Gubser, I. R. Klebanov and A. M. Polyakov, Phys. Lett. B **428**, 105 (1998).
[8] A. Karch, E. Katz, D. T. Son and M. A. Stephanov, Phys. Rev. D **74**, 015005 (2006).
[9] S. S. Afonin and I. V. Pusenkov, Phys. Rev. D **90**, 094020 (2014); Mod. Phys. Lett. A **29**, 1450193 (2014).
[10] I. R. Klebanov and E. Witten, Nucl. Phys. B **556**, 89 (1999).
[11] A. Cherman, T. D. Cohen and E. S. Werbos, Phys. Rev. C **79**, 045203 (2009).
[12] M. Ademollo, G. Veneziano and S. Weinberg, Phys. Rev. Lett. **22**, 83 (1969).

Missed radiative corrections in muon g-2 and proton charge radius measurements

Andrej Arbuzov[1,3,⋆] and *Tatiana* Kopylova[2]

[1]*Bogoliubov Laboratory for Theoretical Physics, JINR, Dubna 141980, Russia*
[2]*Dubna State University, Dubna 141982, Russia*

Abstract. QED radiative corrections to the muon anomalous magnetic moment and elastic electron-proton scattering are discussed. It is shown that a collective effect due to mutual interaction of muons within experimental conditions might provide a contribution to the observed muon magnetic moment. This effect is parameterized by an effective mean shift of muons off their mass shells. Higher order corrections to elastic electron-proton at low energies are systematically treated within the leading and next-to-leading logarithmic approximation. The corrections are relevant for the modern experiments on proton form factor and charge radius definition.

1 Introduction

This report is devoted to discussion of effects due to higher order QED radiative corrections (RC) in observables which are being studied in modern low-energy experiments. A very high precision is achieved some experiments, that makes a challenge for construction of adequately accurate theoretical predictions. In particular various effects of radiative corrections have to be re-considered. By *missed* RC we mean effects which were either not taken into account or treated in a wrong way. The report is based on two papers [1] and [2].

2 Off-mass-shell muon anomalous magnetic moment

We have [3] the following difference between the experimental data and theoretical predictions for the anomalous magnetic moment of muon:

$$\Delta a_\mu^{\text{exp-SM}} = 288(63)(49)\cdot 10^{-11} \tag{1}$$

It is almost twice as large as the pure weak contribution $\Delta a_\mu^{\text{weak}} = 153.6(1.0)\cdot 10^{-11}$. An explanation of the deviation by a contribution due to some effects beyond the SM, requires introduction of a new energy scale being rather close to the EW one, while ongoing searches (in particular at LHC) more and more disfavor finding new physics at such a scale. Attempts to explain the difference by some effect of strong interactions result in continuous efforts in calculations of the corresponding non-perturbative contributions and trying to fix them using relations to experimentally observables quantities, see *e.g.*

⋆e-mail: arbuzov@theor.jinr.ru

review [4]. Presently, there remains some valuable uncertainty in QCD effects, but the possibility to explain in this way the difference (1) is very unlikely. Another possibility could be an error in the experimental analysis. It will be verified by the forthcoming experiments Muon $g-2$ [5] at Fermilab (USA) and J-PARC $g-2$ [6] at KEK (Japan).

In order to resolve the puzzle, we have to look for the whole spectrum of possibilities. Here we discuss the scenario that some additional (but standard) interactions of muons within given experimental conditions contribute to the observed gyromagnetic ratio. Obviously interactions of a muon with his neighbors in the beam can lead to a (small) effective shift of its mass shell. Such collective effects of electromagnetic interactions between charged particles become large for intensive beams [7]. In the $g-2$ experiment the beam was not dense but the effect we are looking for is small. Remind also a weakened conservation of energy and momentum required to describe oscillations of atmospheric neutrinos originated from muon (and pion) decays.

The calculations of the one-loop fermion from factor with one external fermion line being off-shell was presented in the book [8]. Here we will consider only the off-diagonal situation (one fermion off-shell and the other one on-shell).

We choose the following kinematics:

$$k^2 = 0, \qquad p^2 = -m^2, \qquad (p+k)^2 + m^2 = 2pk = \kappa m^2, \tag{2}$$

where k is the photon 4-momentum; p and $p+k$ are the electron momenta. One of them is off shell. Note that metric $(-,+,+,+)$ is used. Dimensionless parameter κ describes the degree of the off-shellness, $\kappa < 0$. We will consider small values $|\kappa| \ll 1$.

At the one-loop level, the fermion-photon vertex receives contributions from scalar, vector and tensor Feynman integrals J_0, J_σ, $J_{\sigma\tau}$. The scalar and the tensor integrals are infrared and ultraviolet divergent, respectively. But as it is known from the standard calculations of $(g_f - 2)$, just the vector integral contributes to this quantity. For the off-mass-shell case it reads [8]

$$\frac{m^2}{i\pi^2} J_\sigma = \left(J_0 - \frac{\ln|\kappa|}{\kappa - 1}\right) p_\sigma + \left(2J_0 - \frac{\kappa - 2}{\kappa - 1}\ln|\kappa| - 2\right)\frac{k_\sigma}{\kappa},$$

$$J_0 = \frac{1}{\kappa}\Big[\mathrm{Li}_2(1) - \mathrm{Li}_2(1-\kappa)\Big], \qquad \mathrm{Li}_2(x) \equiv -\int\limits_0^x \frac{\ln(1-y)}{y} dy. \tag{3}$$

In the limit $\kappa \to 0$ and $k \to 0$, the vector integral leads to the well know one-loop result $\Delta a_f^{(1)} = \alpha/(2\pi)$ received first by J. Schwinger [9]. Expanding in κ gives the first correction due to the off-shellness:

$$\Delta a_f^{(1,\kappa)} = \frac{\alpha}{2\pi}\Big[1 + \delta a_f^{(\kappa)}\Big], \qquad \delta a_f^{(\kappa)} = \left(\frac{1}{4} + \frac{\ln|\kappa|}{2}\right)\kappa + O(\kappa^2). \tag{4}$$

Assuming that the difference (1) is due to the off-shellness effect, we get the equation to define the value of κ:

$$\Delta a_\mu^{\text{exp-SM}} \approx 3 \cdot 10^{-9} = \frac{\alpha}{2\pi}\delta a_\mu^{(\kappa)}. \tag{5}$$

Numerical solution gives $\kappa \approx -3.5 \cdot 10^{-7}$. Note that for small values of κ the shift has the proper sign and the solution exists. Such a value corresponds to off-shellness of a muon of the order $m|\kappa| \sim 35$ eV. One can see that such an off-shellness is very small compared to the muon mass. But to get it due to an external electric field one requires a rather dense muon beam which is not the case of the former Brookhaven $g-2$ experiment. The new experiments at Fermilab and J-PARC will have quite different experimental condition. This will allow to test indirectly the potential collective effect. It is well known that collective effects are important in experiments with high-intensity beams. We claim such effects can also contribute to observables which are measured at an extremely high precision.

 DOI: 10.1051/epjconf/201612504005

3 Higher order radiative corrections to elastic electron-proton scattering

Here we will discuss the treatment of QED radiative corrections in the Mainz Microtron (MAMI) experiment [10]. Elastic electron-proton scattering was measured there with point-to-point errors of the order of a few permille. That allowed to extract information on electric and magnetic form factors of the proton with high precision. Extrapolation of the electric form factor to the zero momentum transfer provided the value of the proton charge radius. The latter appeared being inconsistent with the value extracted from the muonic hydrogen spectrum [11]. This fact stimulates verification of all relevant elements of the studies. In particular, the effects of radiative corrections have to be treated with care.

We consider the process

$$e(p_1) + p(P_1) \longrightarrow e(p_2) + p(P_2) + (n\gamma,\ e^+e^-). \tag{6}$$

The initial electron energy $E_1 = p_1^0 \equiv E$ is of the order 1 GeV, $E \gg m_e$. The momentum transfer squared $Q^2 = -(p_2 - p_1)^2$ lies the range $0.003 < Q^2 < 1$ GeV^2. The condition $Q^2 \gg m_e^2$ holds for the whole range.

One-loop QED corrections to this process are well known. They are naturally separated into the following parts: i) real and virtual corrections to the electron line, ii) real and virtual corrections to the proton line, iii) interference of amplitudes of the first two types, iv) the effect due to vacuum polarization. The corresponding analytic results were reproduced in [10]. Among one-loop corrections, there is still an open discussion about the proper treatment of double photon exchange contributions, see *e.g.* papers [12, 13] and references therein. We agree with the importance of this point, but it goes beyond the scope of our present study.

To estimate the numerical effect of radiative corrections one has to take into account concrete experimental conditions. Of course, to get the final answer one should include the corrections into the whole program of the data analysis. But our task here will be just to present analytic results with simple estimates of their impact. So we will simplify the set-up (still following the main features of the experiment):

— we assume that the measurement is based on the detection of the final electron energy and momentum,

— the electron is detected "bare", *i.e.* without possible accompanying photons,

— there is just a simple cut on the lost energy: $p_1^0 - p_2^0 \geq \Delta E$ where Δ is a dimensionless parameter, $\Delta \ll 1$ and $\Delta E \gg m_e$.

The typical magnitude of the $O(\alpha)$ corrections to the differential cross section is defined by three major factors:

$$\delta^{(1)} = \frac{d\sigma^{(1)}}{d\sigma^{(0)}} \sim \frac{\alpha}{2\pi} \cdot \ln\left(\frac{Q^2}{m_e^2}\right) \cdot \ln\Delta. \tag{7}$$

The enhancement by the so called large logarithm $L \equiv \ln\left(Q^2/m_e^2\right)$ and by the logarithm of the cut-off parameter make the size of the one-loop correction to be of the order of a few percent. Since the experimental uncertainties are well below this order, the one-loop corrections were taken into account in the data analysis [10].

Below we (re-)consider the following higher order QED effects:

1) higher order effects in vacuum polarization;

2) cut-off dependence of the photonic corrections;

3) light pair corrections in the leading logarithmic approximation;

4) complete next-to-leading $O\left(\alpha^2 L^1\right)$ corrections to the lepton line.

Running of the QED coupling constant can be naturally represented as

$$\alpha(Q^2) = \frac{\alpha(0)}{1 - \Pi(Q^2)}, \qquad \Pi(Q^2) = \Pi_e(Q^2) + \Pi_\mu(Q^2) + \Pi_{\rm hadr}(Q^2) + \ldots \tag{8}$$

where $\alpha(0) \equiv \alpha \approx 1/137.036$. The magnitude of $\Pi(Q^2)$ for the range of momentum transfer under consideration is about 0.01. The bulk of the vacuum polarization effect comes from the one-loop e^+e^- pair insertion into the photon propagator,

$$\Pi_e(Q^2) = \frac{\alpha(0)}{\pi}\left(\frac{1}{3}L - \frac{5}{9}\right) + \left(\frac{\alpha(0)}{\pi}\right)^2\left(\frac{1}{4}L + \zeta(3) - \frac{5}{24}\right) + O(\alpha^3). \tag{9}$$

One can note that the $O\left(\alpha^2\right)$ contribution is of the next-to-leading order, since it contains only the first power of the large logarithm L. So it makes only a $\sim 10^{-5}$ effect well below the precision tag. The resummation of the vacuum polarization effect gives

$$\delta\sigma_{\rm vac.pol.} = \sigma^{(0)}\left(\frac{\alpha(Q^2)}{\alpha(0)}\right)^2 = \frac{\sigma^{(0)}}{|1 - \Pi(Q^2)|^2}. \tag{10}$$

Polarization of vacuum by virtual $\mu^+\mu^-$ pairs is not as large as by the e^+e^- ones. But in the bulk of the kinematical domain the suppression is only logarithmic. So,

$$\Pi_\mu(Q^2) = \frac{\alpha}{\pi}\left[\frac{\upsilon_\mu}{2}\left(1 - \frac{\upsilon_\mu^2}{3}\right)\ln\frac{\upsilon_\mu + 1}{\upsilon_\mu - 1} + \frac{\upsilon_\mu^2}{3} - \frac{8}{9}\right] + O\left(\alpha^2\right), \quad \upsilon_\mu \equiv \sqrt{1 + \frac{4m_\mu^2}{Q^2}} \tag{11}$$

has to be taken into account at least in the first order in α. For $Q^2 = 1$ GeV it reaches $2 \cdot 10^{-3}$.

Instead of the resummed geometrical series of Eq. 10, the A1 collaboration in Ref. [10] used exponentiation of the effect of the vacuum polarization by leptons, which is close numerically for the given Q^2 range.

Fig. 1 shows different contributions to the vacuum polarization correction $\delta_{\rm vac.pol.} = \delta\sigma_{\rm vac.pol.}/\sigma^{(0)}$. This figure was obtained with the help of the Fortran package `alphaQED` by F. Jegerlehner [24]. One can see that vacuum polarization by muons and hadrons contributes by up to one percent. That is a very large effect for the given precision tag. Moreover, the momentum dependence of the total vacuum polarization correction is different from the pure electron one.

As concerning the hadronic contribution to vacuum polarization, it can be either treated as a part of radiative corrections or as a part of the proton form factor. To our mind, the former treatment has two advantages. First, this contribution is always there as for point-like as well as for non-point-like particles. Second, in higher order corrections it is not factorized out as can be seen already in Eq. (10). From the first glance the hadronic contribution should not affect the value of the proton charge radius since it is defined at the zero momentum transfer, where this effect is vanishing. Nevertheless, the effect has a pronounced Q^2 dependence in the explored domain and it certainly affects the extrapolation to the zero momentum transfer point. For this reason we recommend to treat the hadronic vacuum polarization as a part of radiative corrections along this the corresponding leptonic contributions.

The Yennie-Frautschi-Suura theorem [14] proves that emission of each soft photon can be treated as an independent process. As the result, multiple emission of soft photons can be resummed into an exponent. By construction in this case, the maximal energy of each photon is limited independently. But in the given experimental set-up, we have a cut-off on the total lost energy. The corresponding effect was considered *e.g.* in Ref. [15]. For double soft photon emission in gives the following shift:

$$e^{\delta_{\rm soft}} \longrightarrow e^{\delta_{\rm soft}} - \left(\frac{\alpha}{\pi}\right)^2\frac{\pi^2}{3}(L - 1)^2. \tag{12}$$

At $Q^2 = 1$ GeV^2 this leads to a visible relative shift of the cross section of about $-3.5 \cdot 10^{-3}$. The same concerns emission of three and more soft photons, where the numerical effect of the proper cut on the lost energy is also not negligible. The proper exponentiation of radiative corrections in the leading logarithmic approximation is based on the exact solution of the renormalization group equation, see [16]. But for the practical application under consideration it is sufficient to compute effect order by order [17].

The contribution of e^+e^- pairs can be easily estimated with the help of the leading logarithmic approximation (LLA) in QED [16–19]:

$$\delta_{\text{pair}}^{LLA} = \frac{2}{3}\left(\frac{\alpha}{2\pi}L\right)^2 P_\Delta^{(0)} + \frac{4}{3}\left(\frac{\alpha}{2\pi}L\right)^3 \left\{\left(P^{(0)} \otimes P^{(0)}\right)_\Delta + \frac{2}{9}P_\Delta^{(0)}\right\} + O\left(\alpha^2 L, \alpha^4 L^4\right) \tag{13}$$

where the so-called Δ-parts of splitting functions (see *e.g.* Refs. [17, 18]) read

$$P_\Delta^{(0)} = 2\ln\Delta + \frac{3}{2}, \qquad \left(P^{(0)} \otimes P^{(0)}\right)_\Delta = \left(P_\Delta^{(0)}\right)^2 - \frac{\pi^2}{3}. \tag{14}$$

Note that in the third order in α we have an effect due to simultaneous (either virtual or soft) radiation of a pair and a photon. To have a better control on the precision level, we can include also the next-to-leading pair corrections in the order $O\left(\alpha^2 L\right)$ where some enhancement due to the experimental cut-off takes place. The corresponding effect will be estimated below.

In order to control the precision of theoretical estimates we can compute the complete set of next-to-leading order (NLO) corrections to the given process by means of the renormalization group approach. The NLO QED structure functions were first introduced in [20]. The corresponding fragmentation functions were used in Refs. [21, 22] to evaluate NLO corrections to the muon decay spectrum. Here we can follow the approach developed in Ref. [23], where NLO QED corrections were computed in a similar set-up for Bhabha scattering.

The relevant photonic and e^+e^- pair contributions to the NLO electron structure (str) and fragmentation (frg) functions have the form[1]

$$\begin{aligned} D_{ee}^{\text{str,frg}}(z) &= \delta(1-z) + \frac{\alpha}{2\pi}\left(LP^{(0)}(z) + d_1(z)\right) + \left(\frac{\alpha}{2\pi}\right)^2\left(\frac{1}{2}L^2 P^{(0)} \otimes P^{(0)}(z) + \frac{1}{3}L^2 P^{(0)}(z)\right. \\ &\left. + LP^{(0)} \otimes d_1(z) + LP_{ee}^{(1,\gamma)\text{str,frg}}(z) + LP_{ee}^{(1,\text{pair})\text{str,frg}}(z)\right) + O\left(\alpha^2 L^0, \alpha^3\right). \end{aligned} \tag{15}$$

Explicit expressions for splitting functions $P_{ee}^{(n)}$ and d_1 can be found in [23]. The master formula for NLO photonic corrections to elastic electron-proton scattering reads

$$d\sigma = \int_{\bar{z}}^{1} dz D_{ee}^{\text{str}}(z)\left(d\sigma^{(0)}(z) + d\bar{\sigma}^{(1)}(z) + O\left(\alpha^2 L^0\right)\right)\int_{\bar{y}}^{1}\frac{dy}{Y}D_{ee}^{\text{frg}}\left(\frac{y}{Y}\right), \tag{16}$$

where $d\bar{\sigma}^{(1)}$ is the $O(\alpha)$ correction to the *ep* scattering with a "massless electron", calculated using the $\overline{\text{MS}}$ scheme to subtract the lepton mass singularities. The energy fraction of the incoming parton is z, and Y is the the energy fraction of the outgoing (observed) electron. As concerning the factorization scale, it is natural to choose it to be equal to the momentum transfer: $L \equiv \ln(Q^2/m_e^2)$.

Here we are interested in the contributions due to virtual and soft photons, so both integrals have the same lower limit being equal to $1-\Delta$. First we can perform convolution of the structure and fragmentation functions entering Eq. (16) with each other $D_{ee}^{\text{str}} \otimes D_{ee}^{\text{frg}}(z)$. If $z = 1-\Delta$ and $\Delta \ll 1$, the result of the convolution gives the probability density to find such a situation where one looses in total due to photon emission ΔE_{beam} from the total energy of the process under consideration.

[1] We dropped the singlet channel contributions which are suppressed in the given experimental set-up.

 DOI: 10.1051/epjconf/201612504005

Convolution of the function found above with the Born part of the kernel cross section gives us the corresponding cross section (with the upper limit on the lost energy):

$$d\sigma^{\rm NLO} = d\sigma^{(0)}(1)\Big\{1 + 2\frac{\alpha}{2\pi}\Big[LP^{(0)}_\Delta + (d_1)_\Delta\Big] + 2\Big(\frac{\alpha}{2\pi}\Big)^2\Big[L^2\left(P^{(0)}\otimes P^{(0)}\right)_\Delta + \frac{1}{3}L^2P^{(0)}_\Delta$$
$$+2L(P^{(0)}\otimes d_1)_\Delta + L(P^{(1,\gamma)}_{ee})_\Delta + L(P^{(1,\mathrm{pair})}_{ee})_\Delta\Big]\Big\} + d\overline{\sigma}^{(1)}(1)\,2\frac{\alpha}{2\pi}LP^{(0)}_\Delta + O\left(\alpha^3L^3\right), \qquad (17)$$

$$(d_1)_\Delta = -2\ln^2\Delta - 2\ln\Delta + 2,$$
$$(P^{(0)}\otimes d_1)_\Delta = -4\ln^3\Delta - 7\ln^2\Delta + \ln\Delta\,(1 + 8\zeta(2)) + 3 - 8\zeta(3) + 4\zeta(2),$$
$$(P^{(1,\gamma)}_{ee})_\Delta = \frac{3}{8} - 3\zeta(2) + 6\zeta(3), \qquad (P^{(1,\mathrm{pair})}_{ee})_\Delta = -\frac{20}{9}\ln\Delta - \frac{1}{6} - \frac{4}{3}\zeta(2). \qquad (18)$$

The values of the Riemann zeta function are $\zeta(2) = \pi^2/6$ and $\zeta(3) \approx 1.202$. The Δ-parts of the structure and fragmentation splitting functions $(P^{(1,\gamma(\mathrm{pair}))\mathrm{str}}_{ee})_\Delta$ and $(P^{(1,\gamma(\mathrm{pair}))\mathrm{frg}}_{ee})_\Delta$ coincide, so the notation is simplified. By construction in the $\overline{\mathrm{MS}}$ scheme, the complete first order correction is reproduced since

$$d\overline{\sigma}^{(1)}(1) \;=\; d\sigma^{(1)}(1) - 2d\sigma^{(0)}(1)\frac{\alpha}{2\pi}\Big[LP^{(0)}_\Delta + (d_1)_\Delta\Big]. \qquad (19)$$

The factor 2 before the subtracted term on the right hand side reflects the presence of mass singularities in both the initial and final state corrections.

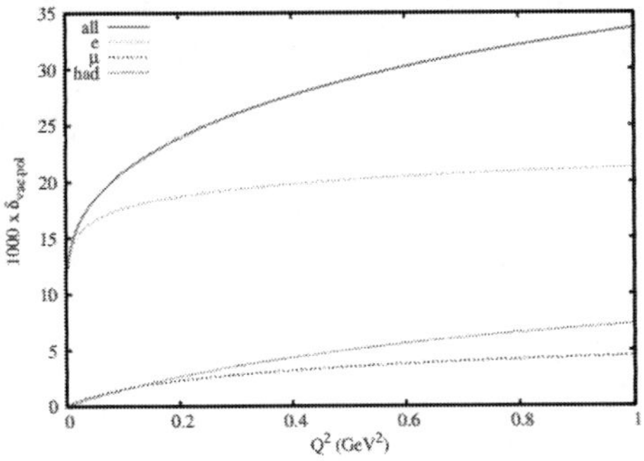

Figure 1. Vacuum polarization corrections due to electrons (e), muons (μ), hadrons (had), and the combined effect (all).

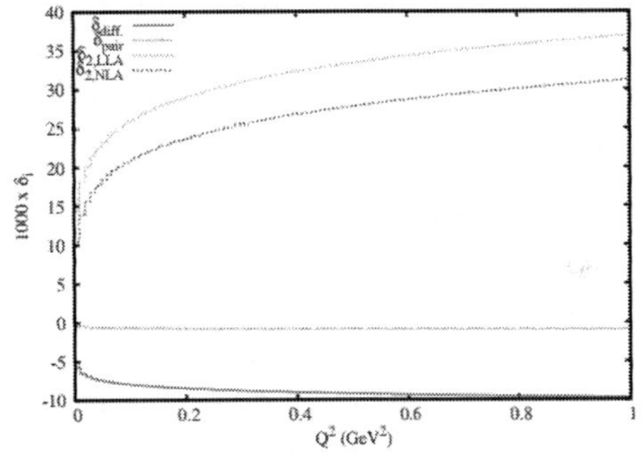

Figure 2. Relative higher order QED corrections to electron line in ep scattering cross section *vs* momentum transferred squared.

Relative QED corrections to the electron line

$$\delta_i = \frac{d\sigma^{(i)}}{d\sigma^{(0)}} \qquad (20)$$

are presented in Fig. 2. Index i runs over:

a) "2,LLA", *i.e.* pure photonic $O\left(\alpha^2L^2\right)$ corrections from Eq. (17),

b) "2,NLA", *i.e.* the sum of pure photonic $O\left(\alpha^2L^2\right)$ and $O\left(\alpha^2L^1\right)$ corrections from Eq. (17),

c) "pair", *i.e.* the leading log pair corrections from Eq. (13) supplemented by subleading pair corrections extracted from Eq. (17),

d) "diff.", *i.e.* the shift from the exponentiated one-loop result:

$$\delta_{\rm diff.} \;=\; \frac{d\sigma^{\rm NLO}}{d\sigma^{(0)}(1)} + \delta^{(3)}_{\rm LLA} + \delta^{(3)}_{\rm LLA,pair} + \delta^{(4)}_{\rm LLA} - \exp\{\delta^{(1)}\}. \qquad (21)$$

We claim that the advanced treatment of RC to elastics electron-proton scattering, which is presented above, should be applied in the analysis of experimental data in future experiments. An adequate treatment of all other relevant effects (double photon exchange, radiative corrections to the proton line, details of the experimental set-up, *etc.*) is also required.

4 Conclusions

Two types of effects due to radiative corrections are discussed. We have shown that these effects might be relevant for modern high-precision experiments.

We are grateful to S. Karshenboim for fruitful discussions.

References

[1] A.B. Arbuzov and T.V. Kopylova, Phys. Part. Nucl. Lett. **11**, 339 (2014).
[2] A.B. Arbuzov and T.V. Kopylova, Eur. Phys. J. C **75**, 603 (2015).
[3] K.A. Olive *et al.*, Particle Data Group Collaboration, Chin. Phys. C **38**, 090001 (2014).
[4] S. Actis *et al.*, Working Group on Radiative Corrections and Monte Carlo Generators for Low Energies Collaboration, Eur. Phys. J. C **66**, 585 (2010).
[5] B.L. Roberts, Chin. Phys. C **34**, 741 (2010).
[6] T. Mibe, J-PARC g-2 Collaboration, Nucl. Phys. Proc. Suppl. **218**, 242 (2011).
[7] I.F. Ginzburg, G.L. Kotkin, S.I. Polityko and V.G. Serbo, Phys. Lett. B **286**, 392 (1992).
[8] A.I. Akhiezer and V.B. Berestetskii, *Elements of Quantum Electrodynamics* (Oldbourne Press, London, 1962).
[9] J. S. Schwinger, Phys. Rev. **73**, 416 (1948).
[10] J. C. Bernauer *et al.* [A1 Collaboration], Phys. Rev. C **90**, 015206 (2014).
[11] A. Antognini *et al.*, Science **339**, 417 (2013).
[12] J. Arrington, P. G. Blunden and W. Melnitchouk, Prog. Part. Nucl. Phys. **66**, 782 (2011).
[13] G. Lee, J.R. Arrington and R.J. Hill, Phys. Rev. D **92**, 013013 (2015).
[14] D.R. Yennie, S.C. Frautschi and H. Suura, Annals Phys. **13**, 379 (1961).
[15] A.B. Arbuzov, E.A. Kuraev and B.G. Shaikhatdenov, Mod. Phys. Lett. A **13**, 2305 (1998).
[16] E.A. Kuraev and V.S. Fadin, Sov. J. Nucl. Phys. **41**, 466 (1985).
[17] A.B. Arbuzov, Phys. Lett. B **470**, 252 (1999).
[18] M. Skrzypek, Acta Phys. Polon. B **23**, 135 (1992).
[19] A. B. Arbuzov, JHEP **0107**, 043 (2001).
[20] F.A. Berends, W.L. van Neerven and G.J. H. Burgers, Nucl. Phys. B **297** 429 (1988), *Erratum-ibid.* B **304**, 921 (1988).
[21] A. Arbuzov and K. Melnikov, Phys. Rev. D **66**, 093003 (2002).
[22] A. Arbuzov, JHEP **0303**, 063 (2003).
[23] A.B. Arbuzov and E.S. Scherbakova, JETP Lett. **83**, 427 (2006).
[24] F. Jegerlehner, Nuovo Cim. C **034S1**, 31 (2011).

EPJ Web of Conferences **125**, 04006 (2016) DOI: 10.1051/epjconf/201612504006
QUARKS-2016

Linear radial Regge trajectories for mesons with any quark flavor

Sergey Afonin[1,*] and *Ilya* Pusenkov[1,**]

[1] *Saint Petersburg State University, 7/9 Universitetskaya nab., St.Petersburg, 199034, Russia*

Abstract. In the Regge phenomenology, the radial spectrum of light mesons is given by a linear relation $M_n^2 = a(n+b)$, where a is a universal slope, the dimensionless intercept b depends on quantum numbers, and n enumerates the excited states in radial recurrences. The usual extensions of this relation to heavy quarkonia in the framework of hadron string models typically lead to strong nonlinearities which seem to be at variance with the available experimental data. Introducing a radially static string picture of mesons, we put forward a linear generalization $(M_n - m_1 - m_2)^2 = a(n+b)$, where $m_{1,2}$ are quark masses. The vector channel contains enough experimental states to check this new relation and a good agreement is observed. It is shown that this generalization leads to a simple estimate of current quark masses from the radial spectra.

The Regge phenomenology grew out of the dual amplitudes [1] and gave rise to various hadron string models. This approach continues to play an important role in the study of hadron spectroscopy. The dual amplitudes and some related string approaches predict the following behavior of meson masses,

$$M_n^2 = a(J+n+c), \qquad J, n = 0, 1, 2, \ldots, \tag{1}$$

where J is the spin, n enumerates the radially excited states, a represents a universal slope and c is a constant. The relation (1) reproduces the classical Regge behavior $M^2 \sim J$ observed in the light baryons and mesons. Also this relation predicts the equidistant daughter Regge trajectories. The number n enumerates the states on these "radial" Regge trajectories. The available experimental data on light non-strange mesons seem to confirm the linear Regge behavior (1) [2–5].

Various attempts to generalise the relation (1) to the sector of heavy quarks resulted in the emergence of strong non-linearities with respect to J and n. Let us look, however, at the experimental data. A relatively rich set of data on radial recurrences in this sector exists only for the unflavored vector heavy quarkonia. Using the relevant data from the Particle Data [6] which are displayed in Table 1 (see Ref. [7] for detailed discussions on the data choice) we plot the masses squared of known ω, φ, ψ, and Υ-mesons as a function of consecutive number n in Figs. 1 – 4. Aside from the ground state, the masses approximately lie on the linear radial trajectories

$$M_n^2 = a(n+b), \tag{2}$$

where we re-denoted the vector intercept $b = 1 + c$. The slope and intercept in (2) depend strongly on the quark flavor, see Table 2.

*e-mail: s.afonin@spbu.ru
**e-mail: i.pusenkov@spbu.ru

Table 1. The masses of known ω, ϕ, ψ and Υ mesons (in MeV) which are used in our analysis. The experimental error is not displayed if it is less than 1 MeV.

n	0	1	2	3	4
M_ω	783	1425 ± 25	1670 ± 30	1960 ± 25	2205 ± 30
M_ϕ	1020	1680 ± 20	—	2175 ± 15	—
M_ψ	3097	3686	4039 ± 1	4421 ± 4	—
M_Υ	9460	10023	10355	10579 ± 1	10865 ± 8

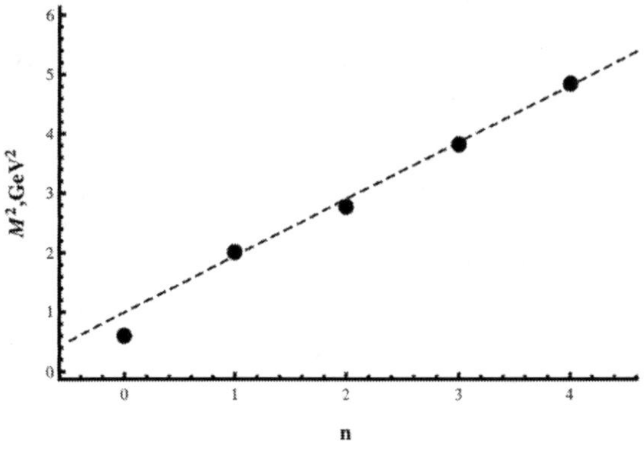

Figure 1. The spectrum of ω-mesons. The experimental points (for this and subsequent figures) are taken from Table 1.

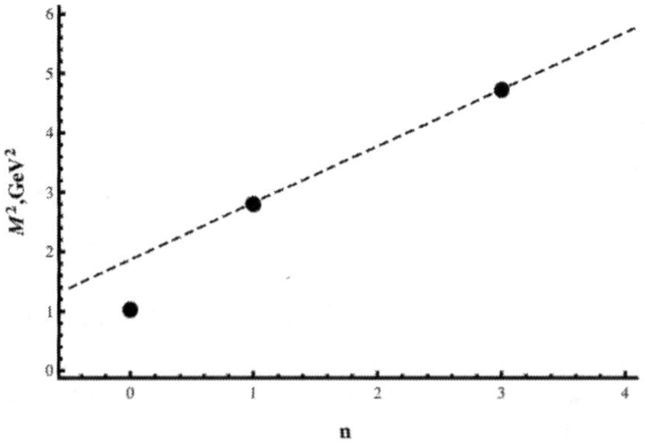

Figure 2. The spectrum of ϕ-mesons.

The universal linearity seen in Figs. 1 – 4 suggests that some universal gluodynamics lies behind the observed behavior. We propose a simple string scheme explaining this universality. The central idea is that at the conditions when a quark-antiquark pair form a resonance in the QCD vacuum, the pair can be viewed as a radially static system. The binding is provided by the exchange of some massless particle (the pion or gluon in concrete realizations of the scheme). And one must quantize the motion of this particle (not the radial motion of quarks as in the standard hadron string approaches!). The total energy (mass) of the system is

$$M = m_1 + m_2 + p + \sigma r. \tag{3}$$

DOI: 10.1051/epjconf/201612504006

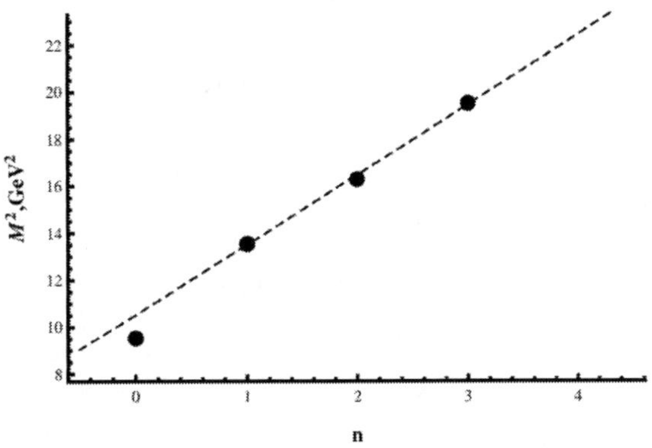

Figure 3. The spectrum of ψ-mesons.

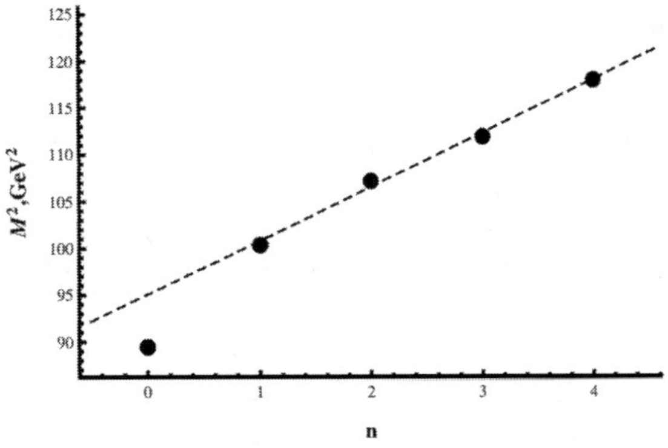

Figure 4. The spectrum of Υ-mesons.

Table 2. The radial Regge trajectories (2) (in GeV^2) for the data from Table 1 (see text).

M_n^2	Fit (a)	Fit (b)
M_ω^2	$1.03(n + 0.74)$	$0.95(n + 1.04)$
M_ϕ^2	$1.19(n + 1.07)$	$0.95(n + 1.96)$
M_ψ^2	$3.26(n + 3.03)$	$2.98(n + 3.53)$
M_Υ^2	$6.86(n + 11.37)$	$5.75(n + 16.54)$

Here m_1 and m_2 are the masses of quark and antiquark separated by the distance r, p is the momentum of exchanged particle, and σ represents the standard string tension. Let us apply the semiclassical quantization to the momentum p

$$\int_0^l p\,dr = \pi(n + b), \qquad n = 0, 1, 2, \ldots. \tag{4}$$

Here l is the maximal quark separation and the constant b depends on the boundary conditions. Substituting p from (3) to (4) and making use of the definition $\sigma = \frac{M}{l}$ we obtain the linear radial trajectory

$$(M_n - m_1 - m_2)^2 = 2\pi\sigma(n + b). \tag{5}$$

Table 3. The quark masses (in GeV), the slope a (in GeV^2) and the dimensionless intercept parameter b in the relation (6).

	Fit I	Fit II
$m_{u,d}$	0	0.36
m_s	0.13	0.49
m_c	1.17	1.55
m_b	4.33	4.69
a	1.10	0.49
b	0.57	0.00

In our unflavored case $m_1 = m_2 \equiv m$ and the relation (5) can be simplified to

$$(M_n - 2m)^2 = a(n + b), \tag{6}$$

which is our generalization of the linear spectrum (2) to non-zero quark masses. In this relation, the universal gluodynamics (the slope a) and dependence on quantum numbers (the dimensionless intercept b) are clearly separated from the contribution of quark masses. For this reason, the parameters a and b in the relation (6) should be flavor-independent.

Let us test our expectations. We will consider two cases — with the light quark mass set to zero (Fit I) and with all quark masses unfixed (Fit II). The results of interpolation of the data in Table 1 by the ansatz (6) are given in Table 3. The ensuing two variants for the spectrum are depicted in Fig. 5 and Fig. 6. Taking into account a simplicity of the relation (6), the agreement is remarkable.

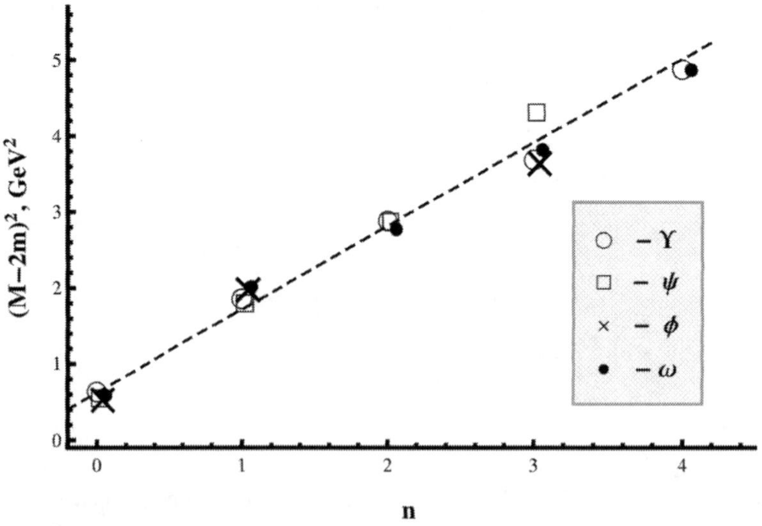

Figure 5. The spectrum (6) for $m_{u,d}$ fixed (Fit I).

The results in Table 3 demonstrate that the radial meson trajectories are able to "measure" the current quark masses with surprisingly good accuracy. If we set $m_{u,d} = 0$, the current masses of other quarks turn out to be very close to their phenomenological values [6]. If we keep all quark masses unfixed, they acquire an additional contribution about 360 MeV. This contribution may be interpreted as an averaged value of constituent quark mass emerging due to the chiral symmetry breaking in QCD. A more detailed analysis (see Ref. [7]) shows that the Fit I works better and the interpretation with

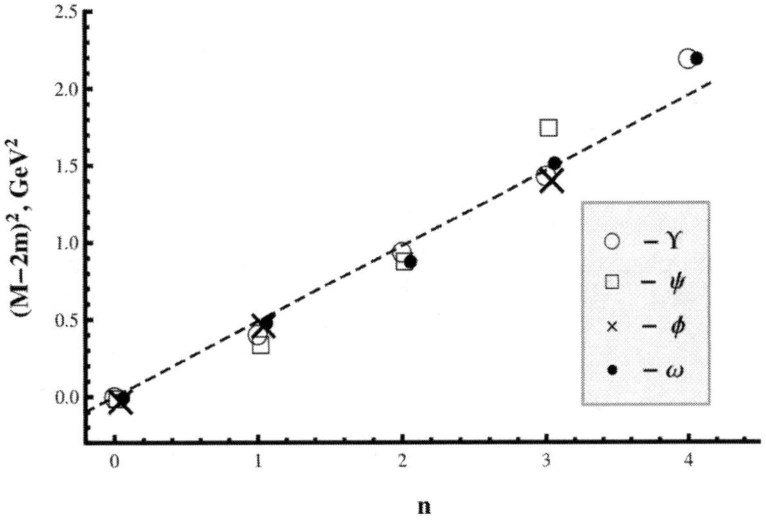

Figure 6. The spectrum (6) for $m_{u,d}$ unfixed (Fit II).

the current quark masses looks preferable. Many related discussions and further fits can be found in Ref. [7]. An independent fit of the relation (6) was also performed by the authors of Ref. [8].

In summary, we proposed a new generalization of linear radial Regge trajectories to the case of massive quarks. Within this generalization, the form of contribution to the meson masses due to confinement is universal for any quarkonia. Our generalization is well consistent with the experimental data on the unflavored vector mesons. Although our considerations were simple and did not take into account various effects which may cause some mass shifts, the quality of final fits is comparable with typical results of semirelativistic potential models, hadron strings and other technically nontrivial approaches.

A natural question emerging after our analysis is whether the relation (6) can be extended to other types of mesons? Unfortunately, the available experimental data are too scarce for making any definite conclusion. A possible extensions of (6) is

$$(M_n - m_1 - m_2)^2 = a(n + x + b), \tag{7}$$

where we may have $x = \beta L$ or $x = \beta J$ (here L and J mean the orbital momentum of valent quarks and total spin). The constant β should be fixed from the phenomenology. An intriguing possibility $\beta = 1$ would lead to a large degeneracy observed in the light non-strange mesons [3–5]. The fact that the slope $a = 2\pi\sigma$ in (5) coincides with the slope of rotating open string may give a theoretical explanation for this degeneracy.

Acknowledgments. The work was supported by the Saint Petersburg State University research grant 11.38.189.2014 and by the RFBR grant 16-02-00348-a.

References

[1] P.D.B. Collins, Phys. Rept. C **1**, 103 (1971).

[2] A. V. Anisovich, V. V. Anisovich and A. V. Sarantsev, Phys. Rev. D **62**, 051502(R) (2000); D. V. Bugg, Phys. Rept. **397**, 257 (2004).

[3] E. Klempt and A. Zaitsev, Phys. Rep. **454**, 1 (2007).
[4] S. S. Afonin, Phys. Lett. B **639**, 258 (2006); Eur. Phys. J. A **29**, 327 (2006).
[5] S. S. Afonin, Phys. Rev. C **76**, 015202 (2007); Mod. Phys. Lett. A **22**, 1359 (2007); M. Shifman and A. Vainshtein, Phys. Rev. D **77**, 034002 (2008).
[6] K. A. Olive *et al.* (Particle Data Group), Chin. Phys. C **38**, 090001 (2014).
[7] S. S. Afonin and I. V. Pusenkov, Phys. Rev. D **90**, 094020 (2014); Mod. Phys. Lett. A **29**, 1450193 (2014).
[8] P. Masjuan, E. R. Arriola and W. Broniowski, EPJ Web Conf. **73**, 04021 (2014).

Holographic Wilson loops in anisotropic quark-gluon plasma.

Dmitry Ageev[1,⋆]

[1] *Steklov Mathematical Institute, Russian Academy of Sciences, Gubkina str. 8, 119991, Moscow, Russia*

Abstract. The nonequilibrium properties of the anisotropic quark-gluon plasma are condidered from the holographic viewpoint. Lifshitz-like solution is considered as a holographic dual of anisotropic QGP. The black brane formation in such background is considered as a thermalization in dual theory. As a probe of thermalization we consider rectangular spatial Wilson loops with different orientation.

1 Introduction and holographic setup

Gauge/gravity duality is a powerful tool for strong coupling systems equilibration study. The quark-gluon plasma (QGP) produced in heavy-ion collisions (HIC) [1] and created after a very short time after the collision (about $0.2 \div 1 fm/c$) is one of such systems.

The holographic duality provides suitable background for description of this period of evolution [2, 3]. The anisotropy and anisotropy effects in the holographic description of QGP is also convenient too [4, 5].

We choose the bottom-up approach for the general AdS/CFT correspondence in background depending on some parameters. This means fitting parameters in background (assuming the dependence on the minimal number of such parameters) in such a way, that the hologprahic calculations fit the experimental data.

Holographic description [6] of the HIC modeled by shock wall collisions in the Lifshitz-like background [10, 11], parametrized by the critical exponent ν is based on background

$$ds^2 = \frac{-dt^2 + dx^2}{z^2} + \frac{dy_1^2 + dy_2^2}{z^{2/\nu}} + \frac{dz^2}{z^2}. \quad (1)$$

This background perfectly matches experimental data [6] for holographic multiplicity calculation. The experimental data is reproduced from holography for the parameter $\nu = 4$. This is the case of special interest for us. Vaidya solutions in these Lifshitz-like spacetimes have been found in [8]. The thermalization time of the 2-point correlators and the thermalization of the holographic entanglement entropy have been estimated in [7, 8].

In this proceedings, based on [9], we investigate numerically spatial Wilson loops behaviour in the Lifshitz-like background. An infalling shell of matter deforms Lifshitz-like spacetime and this corresponds to the formation of the black hole in this background. Wilson loops under consideration

⋆e-mail: ageev@mi.ras.ru

are rectangles with two infinitely long sides. Lifshitz-like spacetime depends on some parameter ν, controlling the spatial anisotropy. We study (pseudo)potentials calculated from holographic Wilson loops and their evolution in asymptotically Lifshitz-like backgrounds. We calculate the dependence of the equilibration time on spatial orientations of Wilson loops. Also we investigate the dependence on the anisotropy parameter ν.

A non-zero temperature generalization of the Lifshitz-like spacetime (1) is a black brane background with the metric [8]

$$ds^2 = \frac{-f(z)dt^2 + dx^2}{z^2} + \frac{dy_1^2 + dy_2^2}{z^{2/\nu}} + \frac{dz^2}{z^2 f(z)}, \tag{2}$$

where the function f is defined by

$$f = 1 - mz^{2+2/\nu}. \tag{3}$$

The metric of the black brane in the 5d AdS spacetime corresponds to $\nu = 1$.

The temperature of the black brane solution (2) is

$$T = \frac{1}{\pi}\frac{(\nu+1)}{2\nu} m^{\frac{\nu}{2\nu+2}}. \tag{4}$$

A straightforward generalization of the stationary black brane metric is the Lifshitz-Vaidya solution:

$$ds^2 = -\frac{f(v,z)dv^2 - 2dvdz + dx^2}{z^2} + \frac{dy_1^2 + dy_2^2}{z^{2/\nu}}, \tag{5}$$

with the function f taken in the following form

$$f(z,v) = 1 - \frac{M}{2}\left(1 + \tanh\frac{v}{\alpha}\right) z^{\frac{2}{\nu}+2}, \tag{6}$$

where $\alpha = 0.2$ for all calculations in this paper.

In this work we focus on rectangular spatial Wilson loops with an infinite extent. The basic formula for the computation of the expectation value of the Wilson loop operator from holography [12, 13] (specified by the contour C) reads as:

$$W[C] = \langle \mathrm{Tr}_F\, e^{i\oint_C dx_\mu A_\mu} \rangle = e^{-S_{string}[C]}, \tag{7}$$

where S is the action of the string with boundary conditions defined by C. The Nambu-Goto action [1] can be represented as

$$S_{string} = \frac{1}{2\pi\alpha'} \int d\sigma^1 d\sigma^2 \sqrt{-\det(h_{\alpha\beta})}, \tag{8}$$

with the induced metric of the world-sheet $h_{\alpha\beta}$ given by

$$h_{\alpha\beta} = g_{MN}\partial_\alpha X^M \partial_\beta X^N, \tag{9}$$

where $\alpha,\beta = 1,2$. In (9) g_{MN} is the background metric, $M,N = 1,\dots,5$, $X^M = X^M(\sigma^1,\sigma^2)$ specify the string worldsheet and σ^1,σ^2 parametrize the worldsheet.

[1]We keep α' in formulas. For all intermediate calculation we set $\alpha' = 1/2\pi$ whereas we take its appropriate value in final results to fit observable data.

Using a spatial rectangular Wilson loop with a finite extent X one can define the so called pseudopotential $\mathcal{V}(X)$ for large Y

$$W(X,Y) = \langle \mathrm{Tr} e^{i\oint_{X\times Y} dx_\mu A_\mu} \rangle = e^{-\mathcal{V}(X)Y}, \tag{10}$$

So, it is straightforward to write down the expression which relates the pseudopotential and the bulk string action:

$$\mathcal{V}(X) = \frac{S_{string}}{Y}. \tag{11}$$

2 Non-equilibrium Wilson loops

In [9] we have considered rectangular Wilson loops with different orientations to capture the anisotropy effect. Here we consider one of the orientation cases in greater details. Cases of orientation considered in [9] are

- a rectangular loop in the xy_1 (or xy_2) plane with a short side of the length ℓ

$$x \in [0, \ell < L_x], \quad y_1 \in [0, L_{y_1}]; \tag{12}$$

- a rectangular loop in the xy_1 plane with a long side of the length L_x along the longitudinal x direction,

$$x \in [0, L_x], \quad y_1 \in [0, \ell < L_{y_1}]; \tag{13}$$

- a rectangular loop in the transversal y_1y_2 plane with a long side of the length L_{y_2} along the other transversal direction y_2, namely

$$y_1 \in [0, \ell < L_{y_1}], \quad y_1 \in [0, L_{y_2}]. \tag{14}$$

2.1 Rectangular strip infinite along the y_1-direction

We consider only this case of orientation in details. For investigation of another orientations see [9].

The orientation of interest is the rectangular Wilson loop on the xy_1-plane, case (12). We choose the following worldsheet parametrization

$$\sigma^1 = x, \quad \sigma^2 = y_1, \tag{15}$$

assuming $v = v(x)$, $z = z(x)$ and $z(\pm\ell/2) = 0$.

The Nambu-Goto action in this case can be written as

$$\frac{2\pi\alpha' S_{x,y_{1(\infty)}}}{L_{y_1}} = \int \frac{dx}{z^{1+1/\nu}} \sqrt{1 - fv'^2 - 2v'z'}. \tag{16}$$

The equations of motion of dynamical system with the action (16) are

$$\begin{aligned} v'' &= \frac{1}{2}\frac{\partial f}{\partial z}v'^2 + \frac{(\nu+1)}{\nu z}(1 - fv'^2 - 2v'z'), \\ z'' &= -\frac{\nu+1}{\nu}\frac{f}{z} + \frac{\nu+1}{\nu}\frac{f^2 v'^2}{z} - \frac{1}{2}\frac{\partial f}{\partial v}v'^2 \\ &- \frac{1}{2}fv'^2\frac{\partial f}{\partial z} - v'z'\frac{\partial f}{\partial z} + 2\frac{(\nu+1)}{\nu z}fv'z', \end{aligned} \tag{17}$$

and the boundary conditions of interest are

$$z(\pm\ell) = 0, \quad v(\pm\ell) = t, \tag{18}$$

where ℓ is the length of the Wilson loop along the x-direction.

To solve these equations numerically we set the initial value problem instead in form

$$z(0) = z_* \tag{19}$$
$$v(0) = v_* \tag{20}$$
$$z'(0) = v'(0) = 0. \tag{21}$$

and then define the values of z_* and v_* that satisfy boundary conditions (18).

Let us briefly consider the structure of the boundary condition/initial value parameters space. In Fig.1 we plot the constant level lines of $z(\ell) = 0$ and ℓ beign fixed in parametric space z_* and v_* on the left plot, and the same for $v(\ell)$ on the right plot. We see, that generally for each ℓ and t one can find the unique values pair fixing the values z_* and v_* so that boundary conditions (18) are satysfied. White curve on each plot is the kind of "attractor" for constant level lines. In the parametric space the parameter values lying upper than this curve corresponds to the geodisics refracting from the shell and going to the singularity. So one should neglect these parameters values as they does not correspond to right boundary conditions.

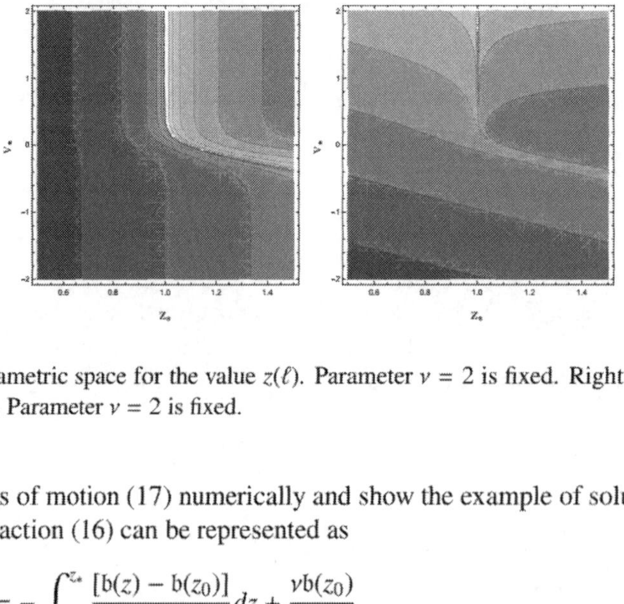

Figure 1. Left plot: parametric space for the value $z(\ell)$. Parameter $\nu = 2$ is fixed. Right plot: parametric space for the value $v(\ell)$, time t. Parameter $\nu = 2$ is fixed.

We solve equations of motion (17) numerically and show the example of solution in Fig.2.

The renormalized action (16) can be represented as

$$\frac{2\pi\alpha' S_{x,y_{1(\infty)},ren}}{L_{y_1}} = -\int_{z_0}^{z_*} \frac{[\mathfrak{b}(z) - \mathfrak{b}(z_0)]}{z^{1+1/\nu}} dz + \frac{\nu \mathfrak{b}(z_0)}{z_*^{1/\nu}},$$

where we have introduced the quantity $\mathfrak{b}$ defined by

$$\mathfrak{b}(z) = \frac{1}{z'}\left(\frac{z_*}{z}\right)^{1+1/\nu}. \tag{22}$$

and the pseudopotential is expressed as :

$$\mathcal{V}_{x,y_{1(\infty)}} = \frac{S_{x,y_{1(\infty)},ren}}{L_{y_1}}. \tag{23}$$

 DOI: 10.1051/epjconf/201612504007

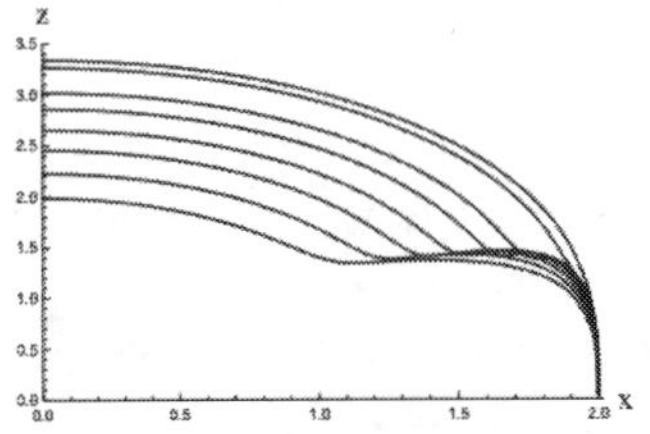

Figure 2. Profile of the string $z(x)$, $z(2) = 0$ at different moments of the boundary time for $\nu = 4$. In (6) we take $M = 1$.

It is also instructive to define and study the following quantity

$$\delta\mathcal{V}_1(x,t) = \mathcal{V}_{x,y_{1(\infty)}}(x,t) - \mathcal{V}_{x,y_{1(\infty)}}(x,t_f), \tag{24}$$

expressing the deviation of $\mathcal{V}$ from thermal equilibrium.

Evaluating (24) on numerical solutions of (17) we can see the dynamics of pseudopotential equilibration (see Fig.3 and Fig.4)

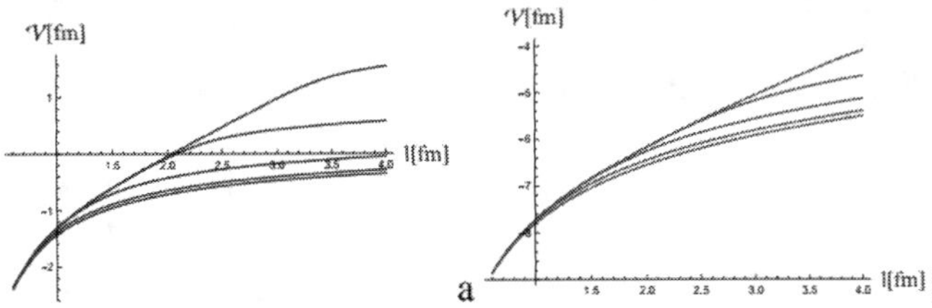

Figure 3. The pseudopotential $\mathcal{V}_{x,y_{1(\infty)}}$ as a function of ℓ at fixed values of t for $\nu = 1$ and $\nu = 4$. ((a) and (b), respectively). Different curves correspond to time $t = 0.1, 0.5, 0.9, 1.4, 2$ (from down to top, respectively). In (6) we take $M = 1$.

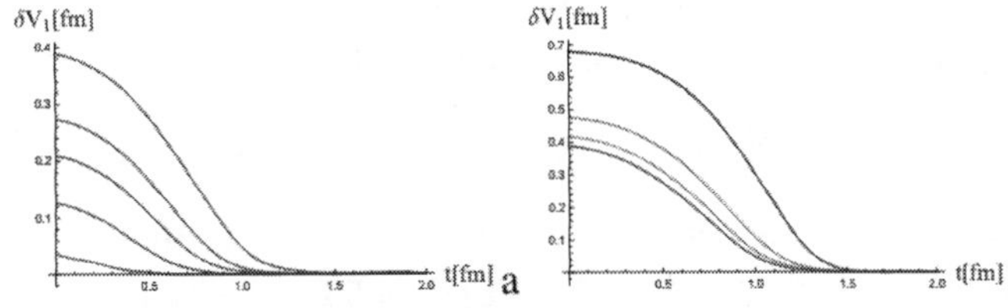

Figure 4. a. The time dependence of the quantity $-\delta\mathcal{V}_1(x,t)$, given by (24), for different values of the length ℓ, $\nu = 4$. b. The quantity $-\delta\mathcal{V}_1(x,t)$ as a function of t at $\ell = 2$ for $\nu = 1, 2, 3, 4$ (from top to down, respectively). In (6) we take $M = 1$.

3 Conclusion

The equilibration process of Wilson loop in the anisotropic quark-gluon plasma using the gauge/gravity duality is analysed. Lifshitz-like background is taken on the gravity side and the critical exponent ν controls the anisotropy. The expectation values of Wilson loops and their dynamic is considered for certain orientation of contour. Numerical issues, concerning the solution of equation of motion for the Nambu-Goto string are discussed.

The pseudopotential out of equilibrium has the universal behaviour: it tends to achievement of the saturation corresponding to large values of the boundary time. The dynamical exponent influences the thermalization process of Wilson loops: this behavior described above strengthen as the value of ν increases.

Acknowledgments

This work is supported by the Russian Science Foundation (project 14-50-00005).

References

[1] J. Casalderrey-Solana, H. Liu, D. Mateos, K. Rajagopal, U. A. Wiedemann, Gauge/String Duality, Hot QCD and Heavy Ion Collisions, arXiv:1101.0618

[2] I. Ya. Aref'eva, Holographic approach to quark-gluon plasma in heavy ion collisions, *Phys. Usp.* **57** (2014) 527.

[3] O. DeWolfe, S. S. Gubser, C. Rosen and D. Teaney, Heavy ions and string theory, *Prog. Part. Nucl. Phys.* **75**, 86 (2014).

[4] M. Strickland, Thermalization and isotropization in heavy-ion collisions, *Pramana* **84**, (2015) 671.

[5] D. Giataganas, Observables in Strongly Coupled Anisotropic Theories, *PoS Corfu* **2012**, 122 (2013).

[6] I. Ya. Aref'eva and A. A. Golubtsova, Shock waves in Lifshitz-like spacetimes, *JHEP* **1504** (2015) (011).

[7] I. Ya. Aref'eva, Formation time of quark-gluon plasma in heavy-ion collisions in the holographic shock wave model, *Theor. Math. Phys.*, 184:3 (2015), 1239-1255, arXiv:1503.02185.

[8] I. Y. Aref'eva, A. A. Golubtsova and E. Gourgoulhon, Analytic black branes in Lifshitz-like backgrounds and thermalization, arXiv: 1601.06046

[9] D. S. Ageev, I. Y. Aref'eva, A. A. Golubtsova and E. Gourgoulhon, "Holographic Wilson loops in Lifshitz-like backgrounds," arXiv:1606.03995.

[10] M. Taylor, Non-relativistic holography, arXiv:0812.0530.

[11] T. Azeyanagi, W. Li and T. Takayanagi, On string theory duals of Lifshitz-like fixed points, *JHEP* **06** (2009) 084.

[12] J. M. Maldacena, Wilson loops in large N field theories, *Phys.Rev.Lett.* **80** (1998) 4859-4862.

[13] S. J. Rey, S. Theisen and J. T. Yee, Wilson-Polyakov loop at finite temperature in large N gauge theory and anti-de Sitter supergravity, *Nucl. Phys.* **B 527**, 171 (1998) .

EPJ Web of Conferences **125**, 04008 (2016) DOI: 10.1051/epjconf/201612504008
QUARKS-2016

On the four-loop strong coupling beta-function in the SM

Alexander Bednyakov[1,⋆] and *Andrey* Pikelner[2,⋆⋆]

[1] *Joint Institute for Nuclear Research, 141980 Dubna, Russia*
[2] *II. Institut für Theoretische Physik, Universität Hamburg, Luruper Chaussee 149, 22761 Hamburg, Germany*

Abstract. In the talk the leading four-loop contribution to the beta-function of the strong coupling in the SM is discussed. Some details of calculation techniques are provided. Special attention is paid to the ambiguity due to utilized γ_5 treatment and a particular prescription with anticommuting γ_5 is advocated. As a by-product of our computation the four-loop beta-function in QCD with "gluino" is also obtained.

The Standard Model (SM) of fundamental interactions being renormalizable can, in principle, by used to make predictions at scales far above the Z-boson mass $Q^2 \gg M_Z$. At such scales it is convenient to use "running", or scale-dependent, couplings $a(Q)$, which are obtained from a set of measurable quantities $\{O\}$ by means of the following two-step procedure:

$$\begin{array}{ccccc} \underline{\text{PDG [1] 20XX}} & & \underline{\text{Fixed } \mu_0} & & \text{Evolve from } \mu_0 \\ \{O\} = M_b, M_W, M_Z, & \longrightarrow & g_i(\mu_0), y_i(\mu_0), \lambda(\mu_0) & \longrightarrow & \text{to scale } \mu \\ M_H, M_t, G_F & & \text{in } \overline{\text{MS}} \text{ scheme} & & \end{array}$$

The first step is called *matching* and boils down to the extraction/fitting of the model parameters $a(\mu_0 \simeq M_Z)$ at the electroweak scale (in what follows, we employ $\overline{\text{MS}}$-scheme). The second step — "running" — allows one to utilize renormalization-group equations (RGEs) to re-summ potentially large logarithms $\log \mu^2/\mu_0^2$ contributing to finite-order relations between $a(\mu_0)$ and $a(\mu)$.

One of the most important applications of such a procedure is the vacuum stability analysis of the SM (see, e.g., [2, 3] and references therein). It turns out that for large values of Higgs field ϕ the effective potential can be approximated as

$$V_{\text{eff}}(\phi \gg v) \simeq \frac{\lambda(\mu = \phi)}{4}\phi^4, \tag{1}$$

where the scale dependence of self-coupling $\lambda(\mu)$ is governed by the following (one-loop) RGEs

$$(4\pi)^2 \frac{d\lambda}{d\ln\mu^2} = 12\lambda + 6y_t^2\lambda - 3y_t^4 + \dots, \qquad (4\pi)^2 \frac{dy_t}{d\ln\mu^2} = \frac{9}{4}y_t^3 - 4g_s{}^2 y_t + \dots, \tag{2}$$

in which the "de-stabilizing" contribution due to top-quark Yukawa coupling y_t is emphasized. The importance of the strong coupling g_s can be deduced from RGE for y_t - strong interactions tend to decrease the latter with μ.

⋆e-mail: bednya@theor.jinr.ru
⋆⋆e-mail: pikelner@theor.jinr.ru, on leave from BLTP, JINR

At present, the state-of-the-art analysis utilizes full two-loop matching [4] together with three-loop evolution via RGEs [5–7]. In this talk, we discuss one little step towards the full four-loop analysis — calculation of leading N^{3}LO corrections to β_{a_s}. The latter is defined here as (h counts powers of couplings)

$$\frac{d\ a_s}{d\ \log\mu^2} = \beta_{a_s}a_s = -a_s\sum_{i=0}^{3}\beta_i h^{i+2}. \tag{3}$$

For convenience, we introduce a set of SM parameters (with ξ being a gauge-fixing parameter)

$$(16\pi^2)a = \left\{g_s^2, y_t^2, \lambda, (16\pi^2)\xi\right\}. \tag{4}$$

Since we are interested in the leading corrections to β_3 (3), the electroweak gauge interactions are neglected together with Yukawa interactions of all SM fermions but the top-quark.

For completeness, let us mention here that the *matching* procedure for the strong coupling constant is different than that mentioned earlier. One usually considers five-flavor ($n_f = 5$) QCD as an effective theory obtained from a more fundamental one (e.g., QCD with "active" top quark) and find the relations of the form:

$$a_s^{(5)}(\mu) = a_s(\mu)\zeta_{a_s}(\mu, M),$$

where M corresponds to the mass of a heavy field. The ("threshold") corrections to the so-called decoupling constant ζ_{a_s} are known in pure QCD up to four loops [8–10], while two-loop electroweak contribution is considered in Ref. [11].

Before going to the result, let us discuss some technicalities and important issues encountered in our calculation. To simplify our life we made use of the background-field gauge (BFG) [12, 13]. The advantage of BFG lies in the QED-like relation between the gauge coupling renormalization constant Z_{a_s} and that of the background gluon field $Z_{\hat{G}}$:

$$Z_{a_s} = 1/Z_{\hat{G}}, \qquad Z_\xi = Z_{\tilde{G}}. \tag{5}$$

Obviously, this allows one to obtain the final result solely from massless propagator-type integrals. In (5), we also indicate the relation between the renormalization constants of quantum gluon field $\tilde{G}$ and gauge-fixing parameter. It is worth mentioning that, since in $\overline{\text{MS}}$-scheme beta-functions do not depend on masses, one can avoid any special infra-red rearrangement (IRR) [14] tricks.

For diagram generation we employ the package `DIANA` [15], which internally uses `QGRAF` [16]. The color [17] and Dirac algebra are carried out by means of `FORM`. All the generated two-point functions are mapped onto three auxiliary topologies, each containing 11 propagators and 3 irreducible numerators. The corresponding diagrams are evaluated by means of the C++ version of the `FIRE` package [18], which performs integration-by-parts (IBP) [19] reduction based on the reduction rules prepared by the `LiteRed`[20] package. The IBP reduction leads to a small set of master integrals. The expressions for the latter are known in analytical form up to the finite parts [21].

Let us also note that as an independent cross-check of our setup, we prepared a simple QCD-like model with additional fermions in the adjoint representation of SU(3) color group ("gluino"). We calculated four-loop correction $\Delta\beta_3 \equiv \beta_3(n_f, n_g) - \beta_3(n_f)$ to the beta-function of the strong coupling

$$\begin{aligned}
\Delta\beta_3 &= n_g\left[\frac{d_A^{abcd}d_A^{abcd}}{N_A}\left(\frac{256}{9}-\frac{832}{3}\zeta_3\right) - C_A^4\left(\frac{68507}{243}-\frac{52}{9}\zeta_3\right)\right] \\
&+n_f n_g\left[C_A^2 C_F T_F\left(\frac{23480}{243}-\frac{352}{9}\zeta_3\right) + C_A C_F^2 T_F\left(-\frac{152}{27}-\frac{64}{9}\zeta_3\right) + \right. \\
&\left. C_A^3 T_F\left(\frac{30998}{243}+\frac{128}{3}\zeta_3\right) + \frac{d_F^{abcd}d_A^{abcd}}{N_A}\left(-\frac{704}{9}+\frac{512}{3}\zeta_3\right)\right] \\
&+n_g^2\left[C_A^4\left(\frac{26555}{486}-\frac{8}{9}\zeta_3\right) + \frac{d_A^{abcd}d_A^{abcd}}{N_A}\left(-\frac{176}{9}+\frac{128}{3}\zeta_3\right)\right] \\
&+n_g^2 n_f\left[C_A^3 T_F\frac{934}{243} + C_A^2 C_F T_F\frac{308}{243}\right] + C_A^4 n_g^3\frac{23}{27} \\
&+n_f^2 n_g\left[C_A^2 T_F^2\frac{1252}{243} + C_A C_F T_F^2\frac{1232}{243}\right]
\end{aligned} \tag{6}$$

in terms of the SU(3) casimirs and $n_f(n_g)$ corresponding to the number of quarks(gluino). The beta-function for such a model at four loops was predicted by A.F. Pikelner [22] along the lines of Ref. [23] and can be used, e.g, in the derivation of $\{\beta\}$-expansions [24]. We found perfect agreement and, thus, both confirmed the prediction and verified our computer setup.

Let us now discuss an important obstacle – the ambiguities in the dimensionally regularized expressions due to γ_5. It is known that there is a clash between anticommutativity $\{\gamma_\mu, \gamma_5\} = 0$ and strictly four-dimensional relation

$$\mathrm{tr}\,(\gamma^\mu\gamma^\nu\gamma^\rho\gamma^\sigma\gamma_5) = -4i\epsilon^{\mu\nu\rho\sigma} \tag{7}$$

in $D \neq 4$ (see, e.g.,[25]). A self-consistent BMHV-algebra [26, 27] breaks D-dimensional Lorentz invariance and requires too much effort when applied to multi-loop problems involving chiral fermions. External axial currents in QCD can be conveniently treated within the prescription due to Larin [28]. Another approach [29] is based on anticommuting γ_5 but promote every fermionic trace "tr" to a non-cyclic linear functional, which depends on the choice of utilized *reading point/prescription*, i.e., the position, at which we start(end) reading the trace.

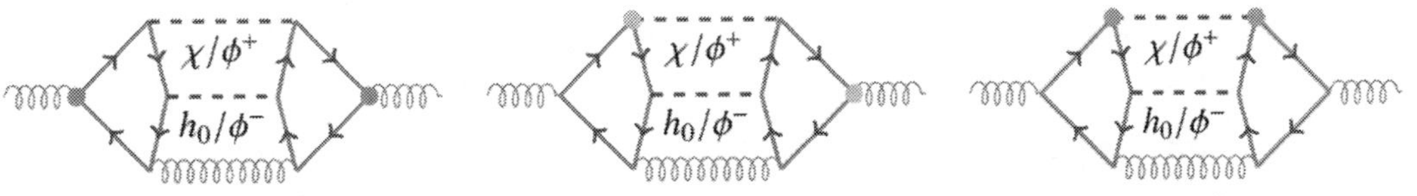

Figure 1. A typical diagram giving rise to a non-trivial contribution due to traces involving odd number of γ_5. Three non-equivalent reading prescriptions are indicated by dots. In our problem it does not matter, whether we start or end the traces at the indicated points. All internal "cut" points turn out to be equivalent.

Since the relevant diagrams (48 non-planar and 24 planar graphs, see, e.g., Fig. 1) involve only single poles in the regularization parameter $\epsilon \equiv (4 - D)/2$, we expected that there should be no ambiguity in β_3. We made a (incorrect) assumption that it is safe to read a trace from any position and use anticommuting γ_5, Eq. (7) and the contraction[1]

$$\epsilon^{\mu\nu\rho\sigma}\epsilon_{\alpha\beta\gamma\delta} = -\mathcal{T}^{[\mu\nu\rho\sigma]}_{[\alpha\beta\gamma\delta]}, \qquad \mathcal{T}^{\mu\nu\rho\sigma}_{\alpha\beta\gamma\delta} = \delta^{\mu}_{\alpha}\delta^{\nu}_{\beta}\delta^{\rho}_{\gamma}\delta^{\sigma}_{\delta}$$

to get a unique result. However, similar calculation was carried out by M. Zoller [30] and an agreement was found only in the "naive" part, in which contributions due to traces with odd number of γ_5 are neglected. The discrepancy triggered further investigation of the issue and it was found that, indeed, the results for the diagrams giving rise to non-trivial γ_5 contribution do depend on the choice of "cut" points, at which one breaks a closed Dirac trace.

The result for the $1/\epsilon$ part of the diagrams can be casted into

$$\frac{a_s^2 a_t^2 T_F^2}{\epsilon}(X_1 + X_2\zeta_3)\cdot R$$

and for non-planar ones we have $X_1 = -1/18$, $X_2 = 1/6$, while in the planar case $X_1 = 1/6$, $X_2 = 0$.

The coefficient R depends on the "cut" points and it turns out that there are three non-equivalent cases, indicated by dots in Fig.1. If both traces are cut at external gluon vertices, one has $R = 1$. If only one external vertex is chosen as a "cut" point, $R = 2$. Finally, for both traces terminated at internal vertices we have $R = 3$.

A natural question arises whether it is possible to single out a unique prescription. In our original paper [31] we advocate the choice $R = 3$. The main argument comes from the calculation of finite, $O(\epsilon^0)$, parts of the diagrams. It is known that IRR procedure (e.g.,of Ref. [30]), usually utilized to find RGEs in $\overline{\text{MS}}$, is only aimed to calculate the pole part of a diagram and does not guarantee that the $O(\epsilon^0)$ terms remain the same after its application. Since we effectively do not do any IRR tricks, we can safely calculate the finite parts and check, whether it is transverse in D-dimensions or not[2].

It turns out that the case with $R = 3$ leads to transverse gluon self-energy, while the case $R = 2$ gives rise to a correction to the longitudinal part, thus, explicitly breaking gauge invariance. In spite of the fact that the prescription $R = 2$ also produce zero upon multiplication by the product of external momenta $q_\mu q_\nu$, we exclude it by simple symmetry argument (we do not want to give preference to either external vertex).

At the end of the day we obtain the following gauge-parameter independent expression [31]:

$$\begin{aligned}
\beta_3 &= \beta_3^{\rm QCD}(n_f = 2n_G) + a_s^3 a_t\left[T_F C_F^2(6 - 144\zeta_3) + T_F C_A C_F\left(\frac{523}{9} - 72\zeta_3\right) + \frac{1970}{9}T_F C_A^2\right.\\
&- \left.\frac{1288}{9}T_F^2 C_F n_G - \frac{872}{9}T_F^2 C_A n_G\right] + a_s a_t^3 T_F\left(\frac{423}{2} + 12\zeta_3\right) + 60 a_s a_t^2 a_\lambda T_F - 72 a_s a_t a_\lambda^2 T_F\\
&- a_s^2 a_t^2\left[T_F^2\left(48 - 96\zeta_3 + \underbrace{R}_{3}\cdot\left[\frac{16}{3} + 32\zeta_3\right]\right) + T_F C_F(117 - 144\zeta_3) + 222 T_F C_A\right], \qquad (8)
\end{aligned}$$

where n_G corresponds to the number of SM families.

It is interesting to compare the relative sizes of different four-loop terms (8) and recent five-loop pure QCD contribution to β_4 [32]. From Fig. 2 one can see that a_s^4 ammounts for about 94% of $\beta_3 + \beta_4$ both at the top-mass and Planck scales. The mixed $a_s^3 a_t$ and $a_s^2 a_t^2$ terms have opposite signs and partially compensate each other. The contributions due to five loops [32] and that from γ_5 are also of different signs and are both less than a percent.

[1]Non-trivial contributions due to γ_5 can only appear when even number of such traces are present.

[2]There seems to be no problem with gauge-invariance in the pole part.

DOI: 10.1051/epjconf/201612504008

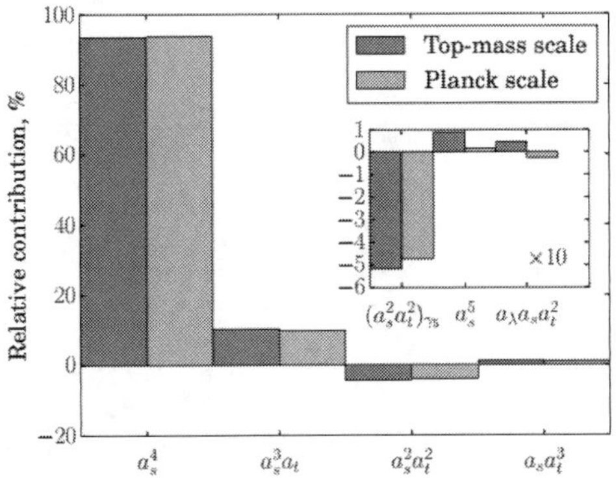

Figure 2. Relative size of the calculated four-loop contributions and the pure QCD five-loop term $O(a_s^5)$ with respect to the sum $\beta_3 + \beta_4$. Non-trivial part due to γ_5 is indicated.

To summarize, we calculated different four-loop corrections to beta-functions for α_s both in the SM and in hypothetical QCD with "gluino". The γ_5 ambiguities were studied and a reading prescription for "odd" fermion traces, consistent with gauge symmetry, was singled out. In our future studies, we plan to extend the result for β_3 to the full SM case and compute leading electroweak threshold corrections at three loops.

Acknowledgments

The authors thank the Organizing committee of the QUARKS-2016 seminar for warm hospitality. This work is supported in part by RFBR grant 14-02-00494-a and the Heisenberg-Landau programme.

References

[1] K. Olive et al. (Particle Data Group), Chin.Phys. **C38**, 090001 (2014)
[2] G. Degrassi, S. Di Vita, J. Elias-Miro, J.R. Espinosa, G.F. Giudice et al., JHEP **1208**, 098 (2012), 1205.6497
[3] A.V. Bednyakov, B.A. Kniehl, A.F. Pikelner, O.L. Veretin, Phys. Rev. Lett. **115**, 201802 (2015), 1507.08833
[4] B.A. Kniehl, A.F. Pikelner, O.L. Veretin, Nucl.Phys. **B896**, 19 (2015), 1503.02138
[5] L.N. Mihaila, J. Salomon, M. Steinhauser, Phys.Rev.Lett. **108**, 151602 (2012), 1201.5868
[6] A. Bednyakov, A. Pikelner, V. Velizhanin, Phys.Lett. **B722**, 336 (2013), 1212.6829
[7] K. Chetyrkin, M. Zoller, JHEP **1304**, 091 (2013), 1303.2890
[8] K.G. Chetyrkin, J.H. Kuhn, C. Sturm, Nucl. Phys. **B744**, 121 (2006), hep-ph/0512060
[9] Y. Schroder, M. Steinhauser, JHEP **0601**, 051 (2006), hep-ph/0512058
[10] B.A. Kniehl, A.V. Kotikov, A.I. Onishchenko, O.L. Veretin, Phys. Rev. Lett. **97**, 042001 (2006), hep-ph/0607202
[11] A.V. Bednyakov, Phys. Lett. **B741**, 262 (2015), 1410.7603
[12] L. Abbott, Nucl.Phys. **B185**, 189 (1981)
[13] A. Denner, G. Weiglein, S. Dittmaier, Nucl.Phys. **B440**, 95 (1995), hep-ph/9410338

[14] A. Vladimirov, Theor.Math.Phys. **43**, 417 (1980)
[15] M. Tentyukov, J. Fleischer, Comput.Phys.Commun. **132**, 124 (2000), hep-ph/9904258
[16] P. Nogueira, J.Comput.Phys. **105**, 279 (1993)
[17] T. van Ritbergen, A. Schellekens, J. Vermaseren, Int.J.Mod.Phys. **A14**, 41 (1999), hep-ph/9802376
[18] A.V. Smirnov, Comput. Phys. Commun. **189**, 182 (2014), 1408.2372
[19] K. Chetyrkin, F. Tkachov, Nucl.Phys. **B192**, 159 (1981)
[20] R. Lee (2012), 1212.2685
[21] P.A. Baikov, K.G. Chetyrkin, Nucl. Phys. **B837**, 186 (2010), 1004.1153
[22] A. Pikelner, paper in preparation.
[23] L. Clavelli, P.W. Coulter, L.R. Surguladze, Phys. Rev. **D55**, 4268 (1997), hep-ph/9611355
[24] A.L. Kataev, S.V. Mikhailov, Phys. Rev. **D91**, 014007 (2015), 1408.0122
[25] F. Jegerlehner, Eur.Phys.J. **C18**, 673 (2001), hep-th/0005255
[26] G. 't Hooft, M. Veltman, Nucl.Phys. **B44**, 189 (1972)
[27] P. Breitenlohner, D. Maison, Commun. Math. Phys. **52**, 11 (1977)
[28] S. Larin, Phys.Lett. **B303**, 113 (1993), hep-ph/9302240
[29] J.G. Korner, D. Kreimer, K. Schilcher, Z. Phys. **C54**, 503 (1992)
[30] M.F. Zoller, JHEP **02**, 095 (2016), 1508.03624
[31] A.V. Bednyakov, A.F. Pikelner (2015), 1508.02680
[32] P.A. Baikov, K.G. Chetyrkin, J.H. Kühn (2016), 1606.08659

Pomeron interaction in the perturbative QCD

J.Bartels[1] and M.A.Braun[2]

[1] *Hamburg University, Germany,*
[2] *S.Petersburg State University, Russia*

Abstract. Formation of the triple pomeron vertices in processes P→PP→PPP and PP→P→PP is studied in the BFKL-Bartels approach. It is demonstarted that successive splittings of the pomeron indeed generate the standard triple pomeron vertex in agreement with the dipole approach. In contrast, in the process PP→P→PP the improper part of the vertex is found missing. Significance of this result is discussed in the framework of the theory of interacting pomerons.

1 Motivation

In the perturbative QCD strong interactions are mediated by exchange of QCD pomerons ("BFKL pomerons",P), which are bound states of pairs of reggeized gluons ("reggeons") [1]. The latter are extension of the standard gluons to complex angular momenta. Pomerons can interact either by splitting of a pomeron in two (P→PP) or by fusing of a pair of pomerons into one (PP→P). For the splitting the corresponding triple pomeron vertex was constructed both in the BFKL approach [2, 3] and dipole picture [4]. It is illustrated in Fig. 1, in which the pomerons coupled to the projectile and targets are shown as ladders of interacting reggeons. In the transverse coordinate space the three participating

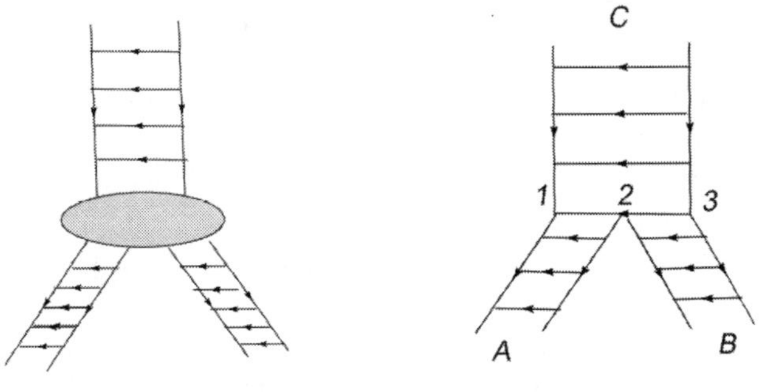

Figure 1. The triple pomeron vertex

pomerons are connected as shown in the right part of Fig. 1. The connecting line passing through points r_1, r_2 and r_3 corresponds to factor [5]

$$\Gamma(r_1, r_2, r_3) = -\frac{g^4 N_c}{4\pi^3} \frac{r_{13}^2 \nabla_1^2 \nabla_3^2}{r_{12}^2 r_{23}^2}, \tag{1}$$

where $r_{12} = r_1 - r_2$ and so on.

The triple pomeron vertex has played an important role in applications of both the BFKL and dipole approaches to observables. It lies at the origin of the structure function of the heavy nucleus, as given by the Balitski-Kovchegov equation [6, 7], which sums pomeron fan diagrams with this vertex (Fig. 2, left). It is also responsible for the pomeron self-mass (Fig. 2, center), which was calculated

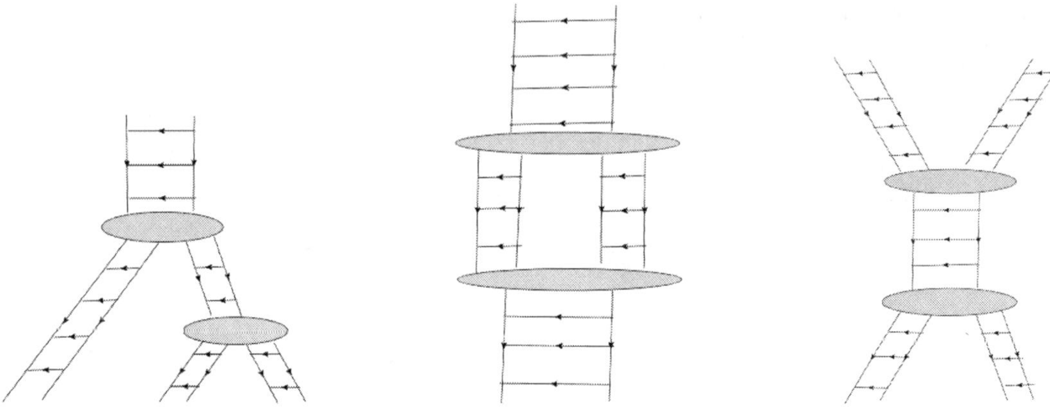

Figure 2. Pomeron fans, self-mass, and transition PP→P→PP

in [8–11].

The splitting and fusing of pomerons by vertex Γ was formalized in the effective pomeron action introduced in [12] in which pomerons are described as bilocal fields $\phi(r_1, r_2)$ with the interaction Lagrangian

$$L_I = \frac{2\alpha_s^2 N_c}{\pi} \int \frac{d^2 r_1 d^2 r_2 d^2 r_3}{r_{12}^2 r_{23}^2 r_{31}^2} \phi(r_1, r_2)\phi(r_2, r_3) r_{13}^4 \nabla_1^2 \nabla_3^2 \phi^\dagger(r_1, r_3) + h.c.\,. \tag{2}$$

As folows from this action in the second order, apart from the self-mass contribution of Fig.2,center, transition PP→P→PP, Fig. 2,right, takes place.

However in the BFKL approach the triple pomeron vertex was derived only for the case of a single splitting of a pomeron in two. Neither two nor more consecutive splittings corresponding to Fig. 2. left and right, have been studied and it was not demonstrated that these processes are governed by two and more vertices Γ. For fan diagrams Fig. 2, left, there exists an alternative derivation in the dipole picture, which, translated into the BFKL language, indicated that indeed in consecutive splittings exactly vertices Γ appear [13, 14]. However in view of the rather indirect correspondence between the dipole and BFKL approaches it would be desirable to derive this result within the latter approach.

More complicated is the situation with the transition PP→P→PP. In this case the dipole approach cannot be used and the derivation of Fig. 2, right, is a problem.

The point is that within the BFKL approach Γ is obtained after summation of transitions of two, three and four reggeons from the colourless short range projectile ($q\bar{q}$ loop) to the two final pomerons.

Suppressing evolution in rapidity, which does not influence the triple pomeron vertex, the latter is obtained from the sum of diagrams shown in Fig. 3.

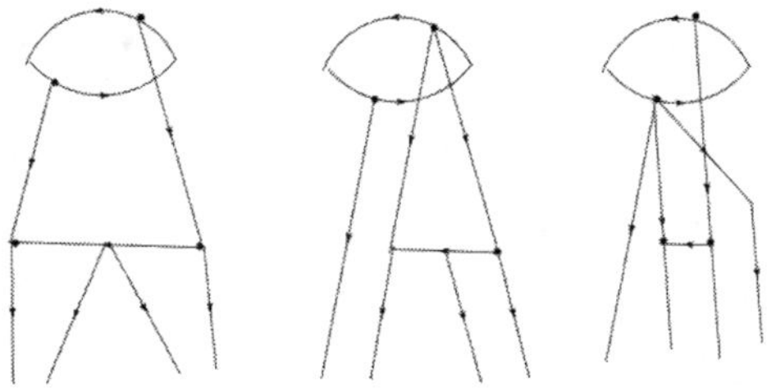

Figure 3. The triple pomeron vertex. Points refer to the coordinate space

Of the three diagrams which sum into Γ, transitions P→PP→PPP (Fig. 2,left) and PP→P→PP (Fig. 2,right) contain only the first.

We demonstrate, in the first place, that additional diagrams for P→PP→PPP provide the necessary contributions which transform the kernel $K_{2\to 4}$ for the second spliting into Γ in accordance with the results obtained in the dipole approach. On the other hand, in the transition PP→P→PP additional diagrams with three and four intermediate reggeons seem to give only a part of the contributions to convert one of the simple kernels $K_{2\to 4}$ into Γ. The rest of the necessary contributions remains missing. In case no new contributions are found, in the transition PP→P→PP the triple pomeron vertex degenerates into the splitting kernel $K_{2\to 4}$, which is in fact one half of the proper part of the standard Γ.

2 Three-pomeron vertex in the reggeon approach

The basis of the derivation of the three-pomeron vertex in the reggeon approach is the so-called bootstrap property of reggeon interaction. It consists in fusing of three and four reggeons in the adjoint colour repersentation into a single reggeon when they are coupled in a special manner to the colourless projectile. This special coupling is provided by a $q\bar{q}$ loop. In this case the amplitudes for coupling of three and four reggeons to the loop, D_{30} and D_{40} are expressed via amplitude D_{20} with only two reggeons coupled to the loop:

$$D_{30}^{a_1a_2a_3}(1,2,3) = -\frac{1}{2}f^{a_1a_2a_3}g\Big(D_{20}(2,31) - D_{20}(1,23) - D_{20}(3,12)\Big), \tag{3}$$

$$D_{40}^{a_1a_2a_3a_4}(1,2,3,4) = -d^{a_1a_2a_3a_4}D_{40}^{1234}(1,2,3,4) - d^{a_2a_1a_3a_4}D_{40}^{2134}(1,2,3,4), \tag{4}$$

where

$$D_{40}^{1234} = g^2\Big(D_{20}(1,234) + D_{20}(123,4) - D_{20}(14,23)\Big), \tag{5}$$

$$D_{40}^{2134} = g^2\Big(D_{20}(2,134) + D_{20}(124,3) - D_{20}(12,34) - D_{20}(13,24)\Big) \tag{6}$$

for two states of 4 reggeons, which differ in their order on the surface of a cylinder (indicated by the upper index). Here 12 means $q_1 + q_2$, a_i are colour indices and $d^{a_1a_2a_3a_4} = \mathrm{Tr}\,(t_1^a t_2^a t_3^a t_4^a + t_4^a t_3^a t_2^a t_1^a)$. The bootstrap ensures that evolution does not change these relations, that is the full amplitudes D_3 and D_4 for transition from the loop to 3 and 4 reggeons with all BFKL interactions and transitions from 2 to 3 and from 2 to 4 reggeons is given by the same formulas Eqs. (3), (5) and (6) with D_{20} substituted by the BFKL pomeron D_2 ('reggeization') As a result transitions to 2 final pomerons all turn out to start also from the pomeron state and so can be expressed via the triple pomeron vertex. In fact the equation for this amplitude has a structure of a Shroedinger equation in rapidity and transverse momenta with certain inhomogeneous terms, which apart from the direct interaction with the $q\bar{q}$ loop, represent transitions into the two-pomeron state from two, three and four evolved reggeons coming from the loop. The three-pomeron vertex can be found from these latter inhomogeneous terms

$$D^{(0)}_{2\to 2P} + D^{(0)}_{3\to 2P} + D^{(0)}_{4\to 2P} = \Gamma D_{20} + R^{(6)}_{2P}, \tag{7}$$

where $R^{(6)}_{2P}$ is the reggeized D_{40} in order g^6.

It is essential that this result does not use the explicit form of function $D_{20}(1,2)$ and is valid if one substitutes D_{20} with arbitrary function $F_2(1,2)$ which vanishes when q_1 or q_2 vanishes. It is only important that the impact factors F_3 and F_4 for 3 and 4 reggeons are related to F_2 in the same way as D_{30} and D_{40} to D_{20}.

3 Pomeron fan diagrams

In the fan diagrams (Fig. 2,left) in the first splitting all contributions from Fig. 3 take part, so that the standard vertex Γ is formed. However already in the second splitting there seems to enter only the first contribution. However in fact one finds two additional diagrams in which instead of the first splitting into 4 reggeons there appear splittings into 5 and 6 reggeons, shown in Fig. 4.

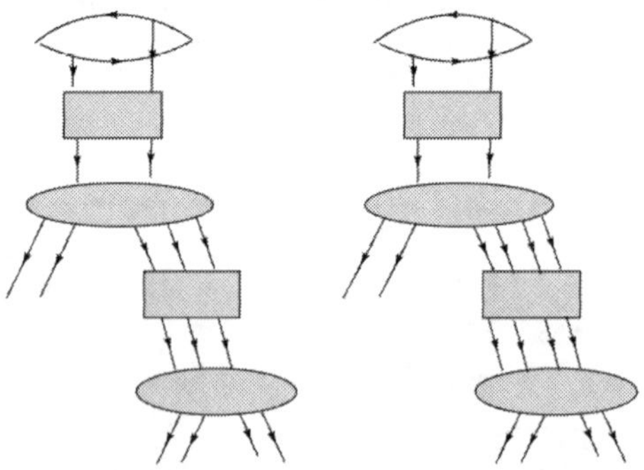

Figure 4. Additional pomeron fan diagrams to be added to Fig.2, left

As we shall demonstrate they convert vertex $K_{2\to 3}$ to the standard vertex Γ.

4 reggeons after the first splitting

After the first splitting we have the general 4-reggeon aplitude D_4. Its part corresponding to the final state of two pomeron states made of reggeons 1,2 and 3,4 is given by $D_4^{a_1a_2a_3a_4} = \delta_{a_1a_2}\delta_{a_3a_4}D_4(1,2,3,4)$. Amplitude $D_4(1,2,3,4)$ can be expresed via function

$$G(1,2,3) = K_{3\leftarrow 2}(1,2,3)D_2 + D_2(1,23)\big(\omega(2)-\omega(23)\big) + D_2(12,3)\big(\omega(2)-\omega(12)\big), \tag{8}$$

where $K_{2\to 3}$ describes transition from 2 to 3 reggeons and ω is the reggeon trajectory.

$$D_4(1,2,3,4) = \frac{1}{2}g^2\big\{G(1,23,4) + G(2,13,4) + G(1,24,3) + G(2,14,3)\big\}.$$

Coupling to the pomeron $P(1,2)$ we 0btain $g^2\big(G(1,23,4)+G(1,24,3)\big)\otimes P(1,2)$. Here $\otimes$ means integration over the pair of intermediate momenta. Introducing function $f(3,4)=G(1,23,4)\otimes g^2P(1,2)$ we find that before the second splitting the final 4-gluon amplitude is coupled to a symmetric function $F_2(3,4)=f(3,4)+f(4,3)$. It takes the role of the loop D_{20} for the second splitting.

5 reggeons after the first splitting

Here before the second splitting we find amplitude D_5 [17] for transition of the initial pomeron into five outgoing reggeons, two of which 1 and 2 are to transform into the outgoing pomeron and the remaining three, 3,4 and 5, serve as ingoing for the second splitting. D_5 contains 10 terms, all are expressed via D_4. Only three terms contribute to our colour structure. The first of them can be written in terms of fuction G as follows

$$(1)=\delta_{a_1a_2}f^{a_3a_4a_5}\frac{1}{2}g^2N_c^2\big\{G(1,234,5)+G(2,134,5)+G(1,25,34)+G(2,15,34)\big\}. \tag{9}$$

Coupling 1 and 2 to the outgoing pomeron we get $(1)\otimes P(1,2)=f^{a_3a_4a_5}F(34,5)$, where F was defined above.

The remaining two terms give results corresponding to interchanges $(4\leftrightarrow 5)$ and $(3\leftrightarrow 4)$ and in this way will add terms corresponding to the full interaction with the $q\bar{q}$ loop: $(2)\otimes P(1,2)=-f^{a_3a_4a_5}(F(35,4)$ and $(3)\otimes P(1,2)=f^{a_3a_4a_5}F(45,3)$.

In the sum we get the impact factor for transition to 3 reggeons 3,4 and 5

$$F_3=\frac{1}{2}gf^{a_3a_4a_5}\big((1)+(2)+(3)\big)=\frac{1}{2}gf^{a_3a_4a_5}\big(F(34,5)+F(45,3)-F(35,4)\big) \tag{10}$$

exactly folowing the form (3).

6 gluons after the first splitting

The corresponding amplitude D_6 satisfies the Shroedinger equation with 6 inhomogeneous terms [17]. Only two of them give a non-zero contribution in our colour configuration: $T_1=\sum d^{a_1a_2a_3a_4}\delta_{a_5a_6}I(1,2,3,4|5,6)$ and $T_2=\sum d^{a_2a_1a_3a_4}\delta_{a_5a_6}J(1,2,3,4|5,6)$. The sums go over all partitions of reggeons into two groups with 4 reggeons in one of them and 2 in the other.

We start with T_1. The relevant term from it is $T_1=\delta_{a_1a_2}d^{a_3a_4a_5a_6}I(3,4,5,6,1,2)$. Here according to [17]

$$I(3,4,5,6|1,2)=-g^2\big(D_4(3,456,1,2)+D_4(345,6,1,2)-D_4(36,45,1,2)\big). \tag{11}$$

Via function G.

$$D_4(3,456,1,2)=\frac{1}{2}g^2\big[G(3,456,2)+G(456,13,2)+G(3,456,1)+G(456,32,1)\big], \tag{12}$$

$$D_4(345,6,1,2)=\frac{1}{2}g^2\big[G(345,16,2)+G(6,345,2)+G(345,26,1)+G(6,345,1)\big], \tag{13}$$

$$D_4(36,45,1,2)=\frac{1}{2}g^2\big[G(36,451,2)+G(45,361,2)+G(36,452,1)+G(45,362,1)\big]. \tag{14}$$

Convolution with the pomeron $P(1,2)$ gives

$$T_3\otimes P(1,2)=-g^2d^{a_3a_4a_5a_6}\big(F_2(456,3)+F_2(6,345)-F_2(36,45)\big). \tag{15}$$

Comparing with (5) we observe that it is the correct structure for D_{40}. for the order of gluons 1234.

Now T_2. The relevant term from it is $T_2=\delta_{a_1a_2}d^{a_4a_3a_5a_6}J(4.3,5,6,1,2)$ where from [17]

$$J(4,3,5,6|2,)=-g^2\big(D_4(456,3,2,1)+D_4(346,5,2,1)-D_4(34,56,2,1)-D_4(45,36,2,1)\big). \tag{16}$$

In terms of G

$$D_4(456,3,2,1) = \frac{1}{2}g^2\Big[G(456,32,1) + G(42,356,1) + G(456,31,2) + G(41,356,2)\Big],$$

$$D_4(346,5,2,1) = \frac{1}{2}g^2\Big[G(346,52,1) + G(52,346,1) + G(346,51,2) + G(51,346,2)\Big],$$

$$D_4(34,56,2,1) = \frac{1}{2}g^2\Big[G(34,562,1) + G(56,342,1) + G(34,561,2) + G(56,341,2)\Big],$$

$$D_4(45,36,2,1) = \frac{1}{2}g^2\Big[G(45,362,1) + G(36,452,1) + G(45,361,2) + G(36,451,2)\Big].$$

Convoluting with pomeron $P(1,2)$ we find

$$T_4 \otimes P(1,2) = -g^2 d^{a_4a_3a_5a_6}\Big[F_2(4,356) + F_2(5,345) - F_2(35,46) - F_2(34,56)\Big]. \tag{17}$$

This has the same structure as the term D_{40} for the order or reggeons 2134., Eq, (6).

So collecting our results we see that the initial states for the second evolution into 4 reggeons 3456 are exactly the same as for the evolution into 4 reggeons in Fig. 3 with the only difference that loops D_{20} are substituted by functions F_2. T As a result due to Eq. (7) the second splitting in Fig. 2,left, will be accomplished by the same vertex Γ as the first splitting, in accordance with the results from the dipole picture.

4 Transitions PP→P→PP

In this section we discuss the possibility to obtain the structure shown in Fig. 2,right, in which two initial pomerons go into the intermediate pomeron and then into two final pomerons via the vertex Γ. As stressed in the introduction, vertex Γ is local in rapidity, has a fixed order g^8 and does not depend on y. So, to locate it in diagrams, we can neglect evolution in rapidity and study diagrams of order g^8 apart from the g-dependence of impact factors. This means that our aim reduces to study if we can find diagrams of order g^8 which, added to the contribution of the diagram in Fig. 5,A, transform it into the diagram in Fig. 5,B.

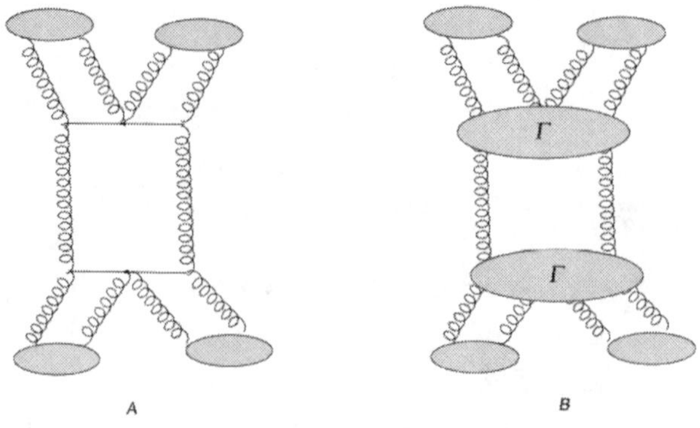

Figure 5. PP→P→PP with two $K_{2\to4}$ (A) and with two Γ (B) in the lowest order

All typical diagrams in this order are illustrated in Fig. 6. Apart from the trivial term with two

DOI: 10.1051/epjconf/201612504009

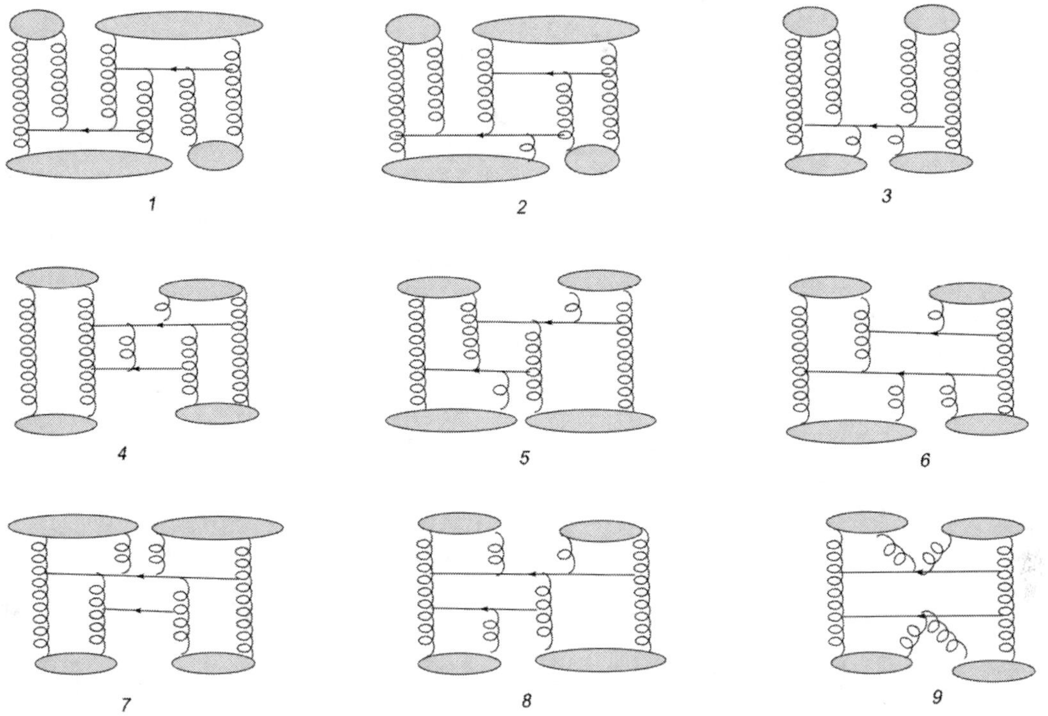

Figure 6. Diagrams of order g^8

kernels $K_{2\to4}$ posible candidates for the necessary additional contributions are reduced to diagrams 7, 8 and similar ones with the targets and projectile exchanged (3 and 6). Note that after convolution with, say, the initial pomerons these terms are to depend on only one variable.

Diagram in Fig.6,8

The diagram in Fig. 6,8 contains the kernel $K_{3\to4}$ [18]. It is composed of three parts shown graphically in Fig. 7. . After convolution of $K_{3\to4}$ with the upper impact factors we find one term in

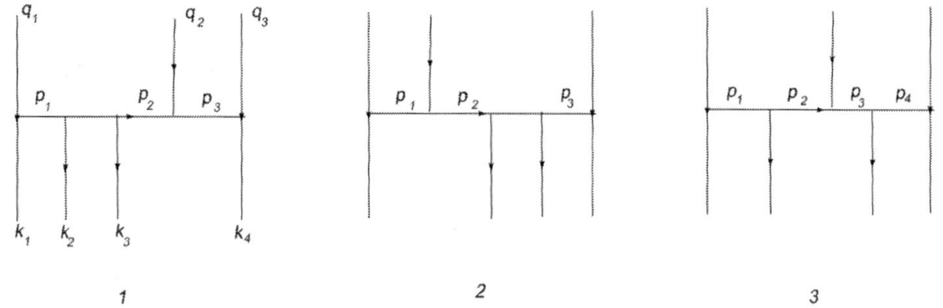

Figure 7. Kernel $K_{3\to4}$

part 1 and another one in part 2 which give the necessary structure $g\big(F_2(q_1, q_{23}) + F_2(q_3, q_{12})\big)/2$. So contribitions from 3 intermediate reggeons give two of the necessary contributions to F_3 but the third term $-gF_2(q_2, q_{13})/2$ is missing.

Diagram in Fig.6,7

The diagram in Fig. 6,7 contains kernel $K_{4\to 4}$ It is composed of six parts shown graphically in Fig. 8. . Four of them 1,2,3 and 6 can be expressed via the already known 3 parts of $K_{3\to 4}$. parts 4

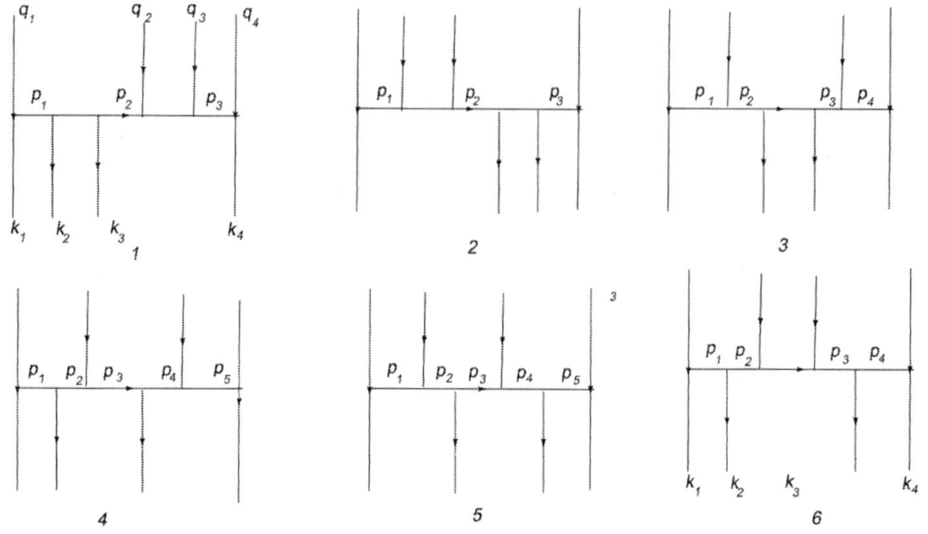

Figure 8. Kernel $K_{4\to 4}$

and 5 are new. From parts one and two we find contributions $-g^2 d^{a_1 a_2 a_3 a_4}\big(F_2(q_1, q_{234}) + F_2(q_4, q_{123})\big)$.

From part 3 we find a term with the correct momentum structure $C_3 F_2(q_{12}, q_{34})$. but with the wrong color factor $C_3 = (d^{a_1 a_2 a_3 a_4} - d^{a_2 a_1 a_3 a_4})/2$ which does not correspond to (6) and violates the bootstrap.

No more terms depending on only one variable appear.

So we conclude that diagram Fig. 6,7 indeed gives two of the terms in (4). However the rest 5 terms are missing.

5 Conclusion

We have found that in successive splittings of the pomeron (fan diagrams) new diagrams with production of extra reggeons and higher splitting kernels $K_{2\to n}$ with n =5 and 6 provide the contributions which are needed to form the three-pomeron vertex at each splitting. This result agrees with conclusions in the framework of the dipole model, as expressed in the Balitski-Kovchegov equation.

In contrast, we have not been able to discover contributions needed to form two three-pomeron vertices in the transition PP→P→PP. A part of these contribution necessary to form one of the vertices was found from diagrams generated by kernels $K_{3\to 4}$ and $K_{4\to 4}$ but the rest is missing. So even the possibility to form one of the vertices Γ is questionable. If the necessary contributions are indeed absent then the vertex connecting the two initial and two final pomerons with the intermediate one reduces to just the simple kernel $K_{2\to 2}$, that is to one half of the proper part of Γ.

References

[1] V.S.Fadin, E.A.Kuraev, L.N.Lipatov, Phys. lett. **B 60** (1975) 50; Ya.Ya. Balitsky, L.N.Lipatov, Sov. J. Nucl. Phys., **28** (1978) 822
[2] J.Bartels, Z.Phys. **C 60** (1993) 471
[3] J.Bartels and M.Wuesthoff, Z.Phys. **C 66** (1995) 157
[4] A.H.Mueller, Nucl. Phys. **B 415** (1994) 373; **B 437** (1995) 107; A.H.Mueller and B.Patel, Nucl. Phys. **B 425** (1994) 471
[5] M.A. Braun and G.P.Vacca, Eur. Phys. J. C 6 (1997) 147
[6] I.Balitski, Nucl. Phys. **B 463** (1996) 99
[7] Yu.V. Kovchegov, Phys.Rev **D 60** (1999) 034008
[8] J.Bartels, M.Ryskin and G.P.Vacca, Eur. Phys. J. **C 27** (2003) 101
[9] R. Peshanski. Phys. Lett. **B 409** (1997) 491
[10] M.A.Braun, Eur. Phys. J. **C 63** (2009) 287
[11] M.A.Braun and A.N. Tarasov, arXiv: 1304.3014
[12] M.A.Braun, Eur. Phys. J. **48** (2006) 511
[13] M.A.Braun, Eur.Phys.J. **c 16** (2000) 337
[14] J.Bartels, L.N.Lipatov, G.P.Vacca, Nucl. Phys. **B 706** (2005) 391
[15] J.Bartels, Nucl. Phys. **B 175** (1980) 365; J. Kwiecinski, M.Praszaloviicz, Phys. Lett. **B 94** (1980) 413
[16] M.A.Braun, Eur. Phys. J. **C 6** (1999) 321
[17] J.Bartels and C.Ewerz, JHEP **9909** (1999) 026
[18] J.Bartels, Nucl. Phys. **B 175** (1980) 365

Hydro and Jets in Relativistic Heavy-Ion Collisions

L.V. Bravina[1,3,4,a], B.H. Brusheim Johansson[1], J. Crkovská[5,6], G.Kh. Eyyubova[2,7], V.L. Korotkikh[2], I.P. Lokhtin[2], L.V. Malinina[2,7], E.N. Nazarova[2], S.V. Petrushanko[2], A.M. Snigirev[2], and E.E. Zabrodin[1,2,3,4]

[1]*Department of Physics, University of Oslo, PB 1048 Blindern, N-0316 Oslo, Norway*
[2]*Skobeltsyn Institute of Nuclear Physics, Moscow State University, RU-119991 Moscow, Russia*
[3]*Frankfurt Institute for Advanced Studies, Ruth-Moufang-Str. 1, D-60438 Frankfurt a.M., Germany*
[4]*National Research Nuclear University "MEPhI" (Moscow Engineering Physics Institute), RU-115409 Moscow, Russia*
[5]*Czech Technical University in Prague, FNSPE, CR-11519 Prague, Czech Republic*
[6]*Institut de Physique Nucléaire, CNRS-IN2P3, Univ. Paris-Sud, Université Paris-Saclay, F-91406 Orsay Cedex, France*
[7]*Joint Institute for Nuclear Researches, Dubna, Russia*

Abstract. We apply HYDJET++ model, which contains the treatment of both soft and hard processes, to study the heavy-ion collisions at LHC energies. The interplay of parametrised hydrodynamics and jets describes many features of the development of particle anisotropic flow including the break-up of mass hierarchy of elliptic and triangular flow, the falloff of the flow at certain transverse momentum and violation of the number-of-constituent-quark (NCQ) scaling at LHC energies compared to the lower ones. Other signals, such as long-range dihadron correlations (ridge) and event-by-event (EbyE) fluctuations of the flow are also discussed. Model calculations demonstrate a good agreement with the available experimental data.

1 Introduction

The main difficulties emerging in analysis of data on relativistic heavy-ion collisions (HIC's) are (i) absence of the unified theory for the description of these collisions, (ii) that perturbative quantum chromodynamics (QCD) does not apply for "soft" interactions with small transverse momentum transfer, which play an essential role in physics of HIC's, and (iii) that the final state of HIC's is very complicated to extract easily the signatures of the new state of matter, the quark-gluon plasma (QGP). Therefore, models are indispensable for evaluating the current experiments and for planing the new ones. These models can be subdivided into macroscopic, i.e. thermal [1–3] and hydrodynamic models [4–6], and microscopic Monte Carlo models [7–10], incorporating partonic and hadronic degrees of freedom in a consistent fashion. Note that for better description of data, hydrodynamic models nowadays are employing the microscopic transport models as afterburner.

[a]e-mail: larissa.bravina@fys.uio.no

In contrast to microscopic transport models, macroscopic models usually deal with soft processes with particle transverse momentum up to $p_T \approx 1.5 - 2$ GeV/c only, and do not include jets. HYDJET++ model [11] was, probably, the first attempt to combine a hydro-like state and hard processes within a single event generator. This idea appears to be very promising. After the discussion of basic principles of the model in Sec. 2, the most interesting results obtained by the HYDJET++ are listed in Sec. 3. Finally, conclusions are drawn in Sec. 4.

2 HYDrodynamics with JETs (HYDJET++)

In HYDJET++ the soft state is represented by a hydro-like system [12, 13]. It is coupled to a hard multiparton state, where the hard partons are suffering the collisional and gluon radiation losses [14] when propagating the dense and hot medium. For the sake of simplicity, both states are treated independently. In the soft sector sector HYDJET++ employs the popular Bjorken or Hubble expansion scenarios for the overlapping zone of heavy-ion collisions. At certain temperature $T = T^{ch}$ and chemical potentials μ_i, where $i = B, S, Q$ represents baryon charge, strangeness and electric charge, respectively, the thermalised fireball breaks into individual hadrons in a sudden chemical freeze-out process. The particle number density $\rho_i^{eq}(T, \mu_i)$ is given by the well-known series expansion [4]

$$\rho_i^{eq}(T,\mu_i) = \frac{g_i}{2\pi^2} m_i^2 T \sum_{k=1}^{\infty} \frac{(\mp)^{k+1}}{k} \exp\left(\frac{k\mu_i}{T}\right) K_2\left(\frac{km_i}{T}\right) , \qquad (1)$$

where g_i is the spin degeneracy factor, m_i is the particle mass, and K_2 is the modified Bessel function of a second kind. To calculate the particle yields one has to know the fireball volume. It is proportional to the mean number of wounded nucleons at a given impact parameter b, see [11–13] for details. After the chemical freeze-out, the system of hadrons continues to expand until the thermal contact between the particles is lost. This is the stage of thermal freeze-out. The only final-state interactions taken into account in the model are two- and three-body decays of resonances. HYDJET++ benefits from the extensive table of particles which contains more than 360 meson and baryon states including the charmed ones. The mass distribution of the resonances is given by a non-relativistic Breit-Wigner formula

$$P(m)dm \propto \frac{1}{(m - m_0)^2 + \Delta m^2/4} dm , \qquad (2)$$

where m_0 and Δm are the resonance nominal mass and width, respectively.

The hard multiparton stage emerges in the model as a superposition of individual hard nucleon-nucleon (NN) collisions. For each collision, the PYQUEN routine [14] generates the initial parton spectra (by means of the tune Pro-Q20 of PYTHIA [15]) and the production vertexes. The parton rescattering stage is accompanied by collisional energy loss and radiation of gluons. In high-momentum transfer limit, the collisional loss reads

$$\frac{dE^{col}}{dl} = \frac{1}{4T\lambda\sigma} \int_{\mu_D^2}^{t_{\max}} dt \frac{d\sigma}{dt} t , \qquad (3)$$

with the dominant contribution to the differential scattering cross section

$$\frac{d\sigma}{dt} \cong C \frac{2\pi\alpha_s^2(t)}{t^2} \frac{E^2}{E^2 - m_p^2} . \qquad (4)$$

Here $\lambda = 1/(\sigma\rho)$ is the mean free path of a parton in a medium with density ρ and integral cross section of parton-medium interaction σ. In the last equation E and m_p are energy and mass of the parton, α_s is the QCD running coupling constant, and $C = 4/9, 1, 9/4$ for qq, gq and gg interaction, respectively. Then, the energy spectrum of coherent medium-induced gluon radiation and the corresponding dominant part of radiative energy loss of massless parton in the framework of BDMS formalism [16, 17] become

$$\frac{dE^{rad}}{dl} = \frac{2\alpha_s(\mu_D^2)C_R}{\pi L}\int\limits_{\omega_{\min}}^{E} d\omega\left[1 - y + \frac{y^2}{2}\right]\ln\left|\cos(\omega_1\tau_1)\right|, \tag{5}$$

$$\omega_1 = \sqrt{i\left(1 - y + \frac{C_R}{3}y^2\right)\bar{\kappa}\ln\frac{16}{\bar{\kappa}}} \quad \text{with} \quad \bar{\kappa} = \frac{\mu_D^2\lambda_g}{\omega(1-y)}, \tag{6}$$

containing $\tau_1 = L/(2\lambda_g)$, λ_g is the gluon mean free path, and $y = \omega/E$ is the parton energy fraction carried away by the gluon. Partons produced in initial hard scatterings with the transverse momentum transfer lower than a certain limit p_T^{min} are excluded from the hard component. Their hadronization products are added to the thermalised component of particle spectrum.

Let us discuss generation of the anisotropic flow in non-central heavy ion collisions in the model. Recall, that the flow harmonics are the Fourier coefficients of series expansion of the invariant cross section in azimuthal plane [18]

$$E\frac{d^3N}{d^3p} = \frac{1}{\pi}\frac{d^2N}{dp_T^2dy}\left\{1 + 2\sum_{n=1}^{\infty} v_n\cos\left[n(\phi - \Psi_n)\right]\right\}, \tag{7}$$

$$v_n = \langle\cos\left[(n(\phi - \Psi_n)\right]\rangle. \tag{8}$$

Here y is the particle rapidity, ϕ is the azimuthal angle between the particle transverse momentum p_T and the participant plane, and Ψ_n is the azimuth of the participant plane of n-th order. The flow harmonic coefficients are called directed, v_1, elliptic, v_2, triangular, v_3, quadrangular, v_4, pentagonal, v_5, hexagonal, v_6, flow and so forth. The transverse radius of the fireball for non-central heavy ion collision depends on impact parameter b, azimuthal angle ϕ and spatial eccentricity $\epsilon(b) = (R_y^2 - R_x^2)/(R_y^2 + R_x^2)$ as [13]

$$R_{ell}(b,\phi) = R_{fr.-out}(b)\sqrt{\frac{1-\epsilon^2(b)}{1+\epsilon(b)\cos 2\phi}}. \tag{9}$$

The parameter $R_{fr.-out}(b)$ is linked to the freeze-out radius of the fireball in a central collision, R_0, as $R_{fr.-out}(b) = R_0\sqrt{1-\epsilon(b)}$. The momentum anisotropy arises from the pressure gradients, which are stronger in the direction of short axis of the ellipsoid. Another free parameter, $\delta(b)$, connects the flow azimuthal angle ϕ_{fl} with the azimuthal angle ϕ. Namely,

$$\delta(b) = \frac{1-C^2}{1+C^2}, \quad C = \frac{\tan\phi_{fl}}{\tan\phi}. \tag{10}$$

Triangular flow needs further modification of the transverse radius [20], which is controlled by new free parameter $\epsilon_3(b)$

$$R_{triang}(b,\phi) = R_{ell}(b,\phi)\{1 + \epsilon_3(b)\cos\left[3(\phi - \Psi_3)\right]\}. \tag{11}$$

According to experimental data, elliptic and triangular flow are not correlated in the model. Further details of HYDJET++ can be found in [11–13].

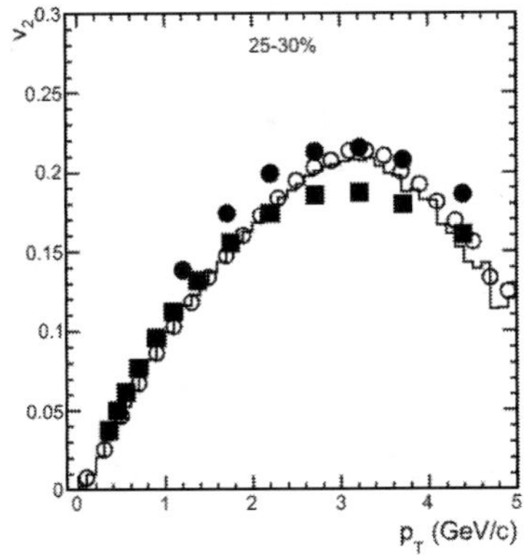

Figure 1. The $v_2(p_T)$ distributions in the HYDJET++ model for 25-30% centrality collisions (histogram) compared to event-plane flow restoration method (open circles) and CMS data (full symbols). See text for details.

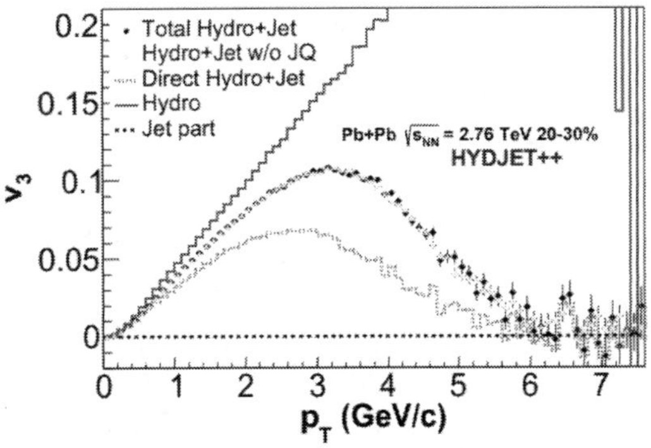

Figure 2. The $v_3(p_T)$ distributions for charged hadrons from (i) all processes (solid circles), (ii) soft+hard processes w/o jet quenching (open triangles), (iii) directly frozen hadrons (dash-dotted line), (iv) soft processes (solid histogram), (v) jets (dashed line) in Pb+Pb at 2.76 TeV with centrality 20-30%.

3 Interplay of soft processes and jets

For our further analysis Pb+Pb collisions at $\sqrt{s}$ = 2.76 TeV are chosen. Figure 1 displays elliptic flow of charged hadrons produced in collisions with centrality 25-30%. Experimental data by the CMS Collaboration [19] obtained by the 2-cumulant and Lee-Yang zeroes (LYZ) methods are plotted also. To estimate possible distortion of the flow spectra caused by the reconstruction procedure, the generated particle spectra were used as an input for the flow determination. The event-plane (EP) method was employed. The restored spectrum is very close to the original one. Model calculations agree well with the data. The good agreement with the experiment is obtained for the triangular flow and higher flow harmonics as well, see [20].

Note that falloff of the differential flow $v_2(p_T)$ at $p_T \geq 3$ GeV/c is a common feature of all flow harmonics. Since we apply ideal parameterised hydrodynamics which has no such peculiarity, the most plausible explanation of the drop of $v_n(p_T)$ is the jet influence. One can see this in Fig. 2, which shows the partial contributions of different subprocesses to the triangular flow $v_3(p_T)$ in collisions with centrality 20-30%. In addition to the total v_3, this figure presents also the flow in case of absence of jet-medium interaction, flow of particles produced merely in soft processes, flow of particles without the resonance feed-down, and flow of hadrons decoupled from jets. The latter is essentially zero, whereas its hydrodynamic counterpart increases with rising p_T. After certain transverse momentum, however, the total spectrum of hadrons is dominated by particles produced in hard processes which can carry very weak flow because of the jet quenching [21]. Since these particles prevail in the hadron spectrum, the total flow drops.

The same mechanism explains also violation of the meson-baryon hierarchy in p_T-differential $v_n(p_T)$, seen in Fig. 3. This figure depicts the triangular flow of most abundant charged hadrons, i.e. pions, $\pi^{\pm}$, kaons, $K^{\pm}$, and (anti)protons, $p+\bar{p}$. At transverse momenta below 2 GeV/c, triangular flow of mesons is stronger than the flow of $p+\bar{p}$. Moreover, in ideal hydrodynamics $v_3(p_T)$ of different hadron species do not cross each other. But then jets come into play. For a given hadron species the crossing point of hard and soft parts of its p_T-spectrum depends on the hadron mass. The heavier the hadron, the larger the transverse momentum at which the jet-produced particles become more

DOI: 10.1051/epjconf/201612504010

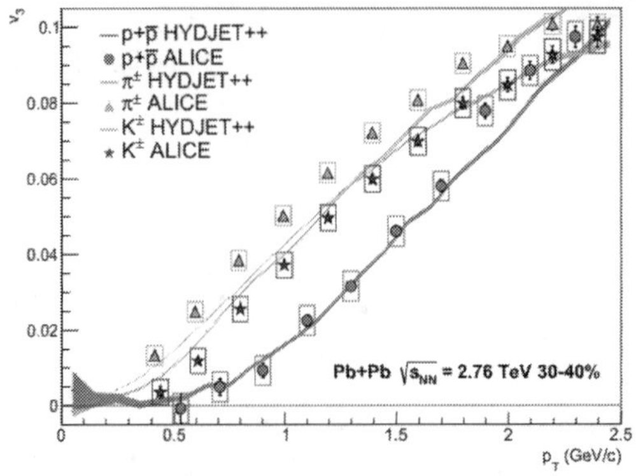

Figure 3. p_T-differential triangular flow of charged pions, kaons, and protons plus antiprotons in Pb+Pb collisions at 2.76 TeV with centrality 20-30%. solid curves indicate HYDJET++ calculations, symbols show ALICE data from [22].

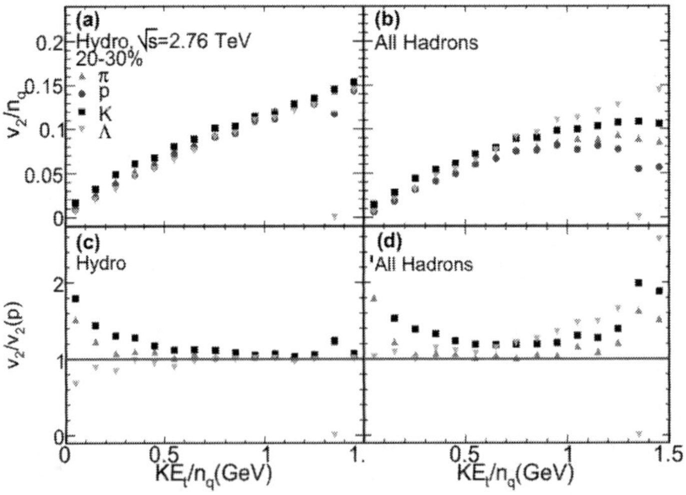

Figure 4. Upper row: The KE_{T}/n_q dependence of elliptic flow for (a) hadrons produced in soft processes only, and (b) all hadrons calculated in HYDJET++ model for Pb+Pb collisions at 2.76 TeV with centrality 20-30%. Bottom row: The distributions from the upper row are normalised to the elliptic flow of $p + \bar{p}$.

abundant. Therefore, the p_T-differential flow of mesons experiences the saturation and subsequent drop at smaller p_T compared to heavier baryons. This effect is called violation of the mass ordering of the flow.

Also, increasing number of hard processes with the rise of collision energy of heavy ions explains the worsening of the number-of-constituent-quark scaling of elliptic flow at LHC. The effect was first observed in Au+Au collisions at RHIC [23, 24]. If the elliptic flow v_2 and the transverse kinetic energy $KE_{\mathrm{T}} = m_{\mathrm{T}} - m_0$ are divided by a number of constituent quarks n_q, 3 for baryons and 2 for mesons, then the distributions $v_2/n_q(KE_{\mathrm{T}}/n_q)$ for all hadrons coincide with good accuracy up to $KE_{\mathrm{T}}/n_q \approx 1$ GeV. This circumstance was taken as a strong evidence of the predominant production of

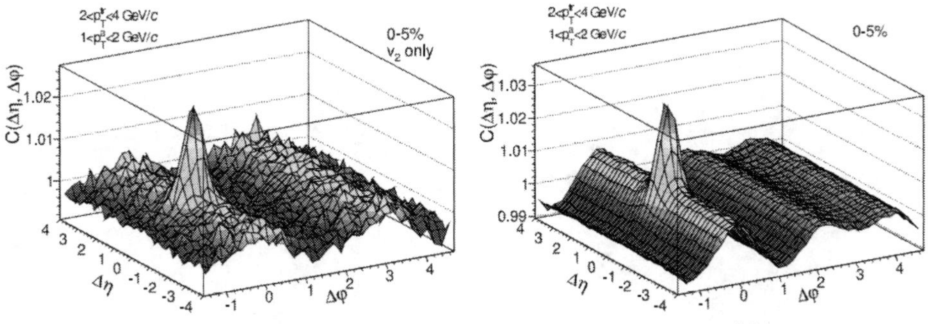

(a) (b)

Figure 5. 2D correlation function in HYDJET++ in Pb+Pb collisions at $\sqrt{s_{\rm NN}} = 2.76$ TeV for $2 < p_{\rm T}^{\rm tr} < 4$ GeV/c and $1 < p_{\rm T}^{\rm a} < 2$ GeV/c for centrality 0-5% with (a) only elliptic flow and (b) with both elliptic and triangular flow.

elliptic flow from the quark coalescence at the partonic stage, most presumably, quark-gluon plasma (QGP). The violation of NCQ scaling was predicted by HYDJET++ [25, 26] before its experimental observation by the ALICE Collaboration [27].

The reduced functions $v_2(KE_T/n_q)/n_q$ of $\pi^\pm, K^\pm, p$ and Λ calculated in HYDJET++ for 20-30% central Pb+Pb collisions at 2.76 TeV are shown in Fig. 4. Left upper panel displays the results for the soft sector. As one can see, all distributions are indeed on the top of each other up to 1.5 GeV. To see the degree of the scaling fulfillment explicitly all particle flows in the bottom row are normalized to the flow of protons, $v_2^h/n_q : v_2^p/n_q$. Hadrons from jets make the scaling performance significantly worse, as seen im right panels of Fig. 4. The NCQ scaling at LHC holds only approximately within the range $0.4 \le KE_T/n_q \le 0.8$ GeV despite the fact that in pure hydrodynamic sector with final state interactions the scaling fulfillment is very good. The detailed study of consequences of the interplay between the soft processes and jets on the development of elliptic and triangular flow and higher order harmonics within the HYDJET++ can be found in [20, 25, 26, 28–31].

Finally, we would like to discuss the issue of correlations and fluctuations presented in the model. Among the sources of particle correlations in HYDJET++ are decays of resonances, jets, femtoscopic correlations, correlations imposed by energy-momentum conservation and so forth. However, as was shown in [32], the long-range correlations arise in the model merely because of the collective flow. The di-hadron correlation function $C(\Delta\eta, \Delta\varphi)$, where $\Delta\eta = \eta^{tr} - \eta^a$ and $\Delta\varphi = \varphi^{tr} - \varphi^a$ is presented in Fig. 5(a,b) for the calculations of Pb+Pb collisions with centrality 0-5%. - The indices "tr" and "a" indicate the so-called "trigger" and "associated" particle, respectively. - Two cases are considered: (a) without and (b) with the triangular flow in addition to the elliptic one. We see that the long-range azimuthal correlations appear both at near-side ($\Delta\varphi \approx 0$) and at away-side ($\Delta\varphi \approx \pi$) in both figures. But the experimentally observed [33, 34] double-hump structure at the away-side emerges in the model only in the presence of elliptic and triangular flow simultaneously.

The event-by-event (EbyE) distributions of anisotropic flow harmonics in Pb+Pb collisions at LHC were studied by ATLAS Collaboration in [35] by means of the *unfolding* procedure. This method suppresses strongly the non-flow effects caused, e.g. by decays of resonances, finite event multiplicity or jet fragmentation. To get rid of such effects, unfolding was implemented in our study of the EbyE flow fluctuations within the HYDJET++ in [36]. Results are shown in Fig. 6, where the model calculations of probability density distributions of elliptic and triangular EbyE flow are compared with the ATLAS data. Agreement is good. Initial $P(V_n)$ distributions are broader than the unfolded ones; unfolding

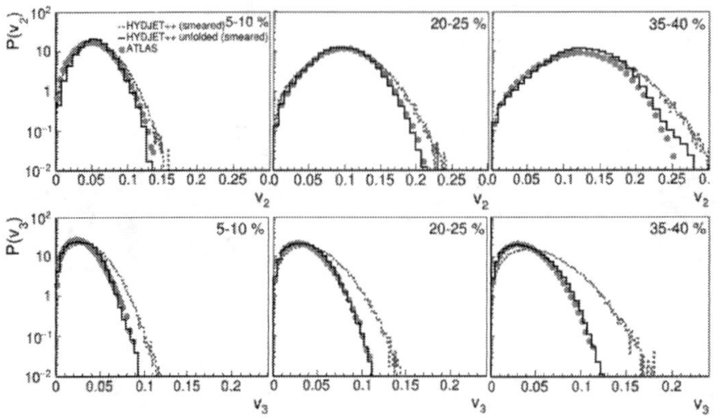

Figure 6. The probability density distributions of elliptic flow V_2 (upper row) and triangular flow V_3 (bottom row) in three centrality intervals: 5 – 10% (left), 20 – 25% (middle) and 35 – 40% (right). Dashed and solid histograms present the results for simulated HYDJET++ events before and after the unfolding procedure, respectively. The full circles are the ATLAS data from [35].

procedure makes it narrower. Obtained results strongly support the idea of dynamical origin of the EbyE flow fluctuations in the model.

4 Conclusions

We show the importance of hard processes for the correct treatment of anisotropic flow harmonics in heavy-ion collisions at ultrarelativistic energies within the ideal hydrodynamic model. Jets account for (i) falloff of the flow harmonics in the range of intermediate transverse momenta, (ii) changing of the mass ordering of the hadron elliptic and triangular flow, and (iii) the violation of the number-of-constituent-quark scaling at energies of LHC and higher.

The interplay between v_2 and v_3 is able to describe both qualitatively and quantitatively the di-hadron long-range correlations, i.e. ridge at near-side and double-hump structure at the away-side. The unfolding procedure was employed to study event-by-event flow fluctuations in the model and compare the obtained results with the experimental data. The comparison revealed dynamical origin of the flow fluctuations, originating primarily from the correlation between the momenta and coordinates of final particles and the velocities of hadronic fluid elements.

Acknowledgments

This work was supported in parts by the Grant INGO II LG15001 of the Ministry of Education, Youth and Sports of the Czech Republic and the Grant Agency of Czech Republic No.13-20841S, and the Grant of the President of Russian Federation for Scientific Schools Supporting No. 7989.2016.2. L.B. acknowledges financial support of the Alexander von Humboldt Foundation.

References

[1] E. Fermi, Prog. Theor.Phys. **5**, 570 (1950)

[2] G. Torrieri *et al.*, Comput. Phys. Commun. **167**, 229 (2005)
[3] A. Kisiel, T. Taluc, W. Broniowski, and W. Florkowski, Comput. Phys. Commun. **174**, 669 (2006)
[4] L. D. Landau, Izv. Akad. Nauk Ser. Fiz. **17**, 51 (1953);
S. Z. Belenkij and L. D. Landau, Nuovo Cim. Suppl. **3S10**, 15 (1956)
[5] H. Song, S. A. Bass, and U. Heinz, Phys. Rev. C **83**, 024912 (2011)
[6] B. Schenke, S. Jeon, and C. Gale, Phys. Rev. C **82**, 014903 (2010)
[7] A. Capella, U. Sukhatme, C.-I. Tan, and J. Tran Thanh Van, Phys. Rep. **236**, 225 (1994)
[8] S. A. Bass *et al.*, Prog. Part. Nucl. Phys. **41**, 255 (1998);
M. Bleicher *et al.*, J. Phys. G **25**, 1859 (1999)
[9] N. S. Amelin and L. V. Bravina, Sov. J. Nucl. Phys. **51**, 133 (1990);
E. E. Zabrodin, C. Fuchs, L. V. Bravina, and Amand Faessler, Phys. Lett. B **508**, 184 (2001);
J. Bleibel, L. V. Bravina, and E. E. Zabrodin, Phys. Rev. D **93**, 114012 (2016)
[10] Z.-W. Lin, *et al.*, Phys. Rev. C **72**, 064901 (2005)
[11] I. P. Lokhtin *et al.*, Comput. Phys. Commun. **180**, 779 (2009)
[12] N. S. Amelin *et al.*, Phys. Rev. C **74**, 064901 (2006)
[13] N. S. Amelin *et al.*, Phys. Rev. C **77**, 014903 (2008)
[14] I. P. Lokhtin and A. M. Snigirev, Eur. Phys. J. C **46**, 211 (2006)
[15] T. Sjostrand, S. Mrenna, and P. Skands, JHEP **0605**, 026 (2006)
[16] R. Baier, Yu. L. Dokshitzer, A. H. Mueller, and D. Schiff, Phys. Rev. C **60**, 064902 (1999)
[17] R. Baier, Yu. L. Dokshitzer, A. H. Mueller, and D. Schiff, Phys. Rev. C **64**, 057902 (2001)
[18] S. A. Voloshin and Y. Zhang, Z. Phys. C **70**, 665 (1996)
[19] S. Chatrchyan *et al.* (CMS Collaboration), Phys. Rev. C **87**, 014902 (2013)
[20] L. V. Bravina *et al.*, Eur. Phys. J. C **74**, 2807 (2014)
[21] X.-N. Wang and M. Gyulassy, Phys. Rev. Lett. **68**, 1480 (1992)
[22] J. Adam *et al.*, (ALICE Collaboration), arXiv:1606.06057 [nucl-ex]
[23] J. Adams *et al.* (STAR Collaboration), Phys. Rev. Lett. **92**, 052302 (2004)
[24] S. S. Adler *et al.* (PHENIX Collaboration), Phys. Rev. Lett. **91**, 182301 (2003)
[25] G. Eyyubova *et al.*, Phys. Rev. C **80**, 064907 (2009)
[26] E. Zabrodin *et al.*, J. Phys. G. **37**, 094060 (2010)
[27] F. Noferini *et al.* (ALICE Collaboration), Nucl. Phys. A **904-905**, 438c (2013)
[28] I. P. Lokhtin *et al.*, Eur. Phys. J. C **72**, 2045 (2012)
[29] L. Bravina, B. H. Brusheim Johansson, G. Eyyubova, and E. Zabrodin, Phys. Rev. C **87**, 034901 (2013)
[30] L. V. Bravina *et al.*, Phys. Rev. C **89**, 024909 (2014)
[31] J. Crkovská *et al.*, arxiv:1606.03250 [hep-ph]
[32] G. Eyyubova *et al.*, Phys. Rev. C **91**, 064907 (2015)
[33] K. Aamodt *et al.* (ALICE Collaboration), Phys. Rev. Lett. **107**, 032301 (2011)
[34] S. Chatrchyan *et al.* (CMS Collaboration), Eur. Phys. J. **C72**, 2012 (2012)
[35] G. Aad *et al.* (ATLAS Collaboration), JHEP **11**, 183 (2013)
[36] L. V. Bravina *et al.*, Eur. Phys. J. C **75**, 588 (2015)

EPJ Web of Conferences **125**, 04011 (2016) DOI: 10.1051/epjconf/201612504011
QUARKS-2016

Azimuthal flows in hadron collisions from quark-gluon string repulsion

Igor Altsybeev[1,⋆] and *Grigory* Feofilov[1,⋆⋆]

[1] *Saint-Petersburg State University*

Abstract. Color flux tubes (quark-gluon strings), formed at early stages of hadron-hadron collisions, may overlap in case of sufficiently high densities and interact, producing long-range azimuthal correlations. In the hypothesis of repulsive interaction, each string may acquire, before the hadronization, the additional transverse boost, which is an efficient sum of all accounted string-string interactions. This modifies transverse momenta to the particles formed in string decay, leading to modification of event-wise observables, like azimuthal asymmetry of two-particle correlations, over a wide range of rapidity.
In this report we discuss results of Monte Carlo model with string repulsion, where efficient string-string interaction radius is introduced. We show that the effect of string repulsion can be the main dynamic origin of the elliptic flow and of the higher harmonics, which are reflected in the complicated structures observed in two-particle long-range correlation topology in nucleus-nucleus collisions at RHIC and at LHC.

1 Introduction

The very first proposals for studies of azimuthal and long-range correlations were formulated in [1]. It was emphasized that the data on the azimuthal asymmetry in soft multiparticle production may contain very non-trivial information. In [2] it was shown that anisotropies in transverse-momentum distributions provide an unambiguous signature of transverse collective flow in ultrarelativistic nucleus-nucleus collisions.

The azimuthal anisotropy in two-particle correlation functions observed at RHIC and LHC is characterized by the Fourier flow coefficients [3, 4]:

$$v_n = \langle \cos[n(\varphi - \Psi_n)] \rangle. \quad (1)$$

Here φ is the azimuthal angle of the particle, Ψ_n is the angle of the initial state spatial plane of symmetry, and n is the order of the harmonic. The planes of symmetry Ψ_n are not known experimentally and the anisotropic flow coefficients are estimated from measured correlations between the observed particles. The second Fourier coefficient v_2 is called elliptic flow and has been studied in details in recent years in number of experiments. The first results on the elliptic flow of charged particles at midrapidity in Au+Au collisions at 130 GeV were reported by the STAR Collaboration at RHIC [5]. It was assumed that the observed azimuthal anisotropy appears at very early time and is caused by

⋆e-mail: Igor.Altsybeev@cern.ch
⋆⋆e-mail: feofilov@hiex.phys.spbu.ru

the pressure gradient and anisotropic expansion of the matter formed in the collisions [2–4, 6]. The anisotropy effects were confirmed later in Pb+Pb collisions at the LHC. For example, large values of elliptic flow at the LHC were observed by the ALICE Collaboration [7]. In addition to the elliptic flow, direct (v_1), triangular (v_3) and higher flow coefficients are also obtained [4]. It was established that the initial state spatial anisotropy is converted, due to some collective behaviour, to the momentum anisotropy.

In the experimental studies of flows, two main methods are usually applied. In the first method, flow coefficients are extracted from a two-particle angular correlations function, usually studied in terms of difference between two particles in azimuthal angle $\Delta\varphi$ and pseudorapidity $\Delta\eta$. Correlations can be studied either between 'trigger' and associated particles (which are taken from different p_T ranges or by some PID selection criteria) or just by taking all pairs of particles in an event (untriggered correlations). Another method is based on multi-particle cumulants using moments of the flow vectors [8], which allows to eliminate a contribution from correlations that are not related to the initial geometry (so-called non-flow).

A new set of striking phenomena was obtained in the two-particle correlation studies: the so-called "ridge" was first observed in Au+Au collisions at $\sqrt{s_{NN}} = 200$ GeV at RHIC [9–11]. Another puzzling experimental effect in the two-particle angular correlations was obtained later in the first measurement of the triangular v_3 and higher order harmonics of charged particle flow in Pb-Pb collisions at $\sqrt{s} = 2.76$ TeV [12, 13]: a double-peak structure in the away-side azimuthal region was discovered in the most central events. It was shown in [12] that this structure can be naturally described by the anisotropic flow Fourier coefficients, in particular, by adding the higher harmonics. Detailed studies of p_T, centrality and rapidity dependences of elliptic and higher order flow harmonics were performed by CMS [14], ATLAS [15] and ALICE [12, 16] at the LHC. Recent review on novel phenomena in particle correlations in relativistic heavy-ion collisions can be found in [17].

The azimuthal anisotropy in two-particle correlations in Au+Au and Pb+Pb collisions is generally related to manifestations of high-order flows and is interpreted as the result of a hydrodynamic expansion of the medium (of a nearly perfect liquid) formed in the collision. We have to mention, that contrary to this interpretation, the collective flow effects in relativistic ion collisions could be also explained without the hydrodynamic assumptions in so-called particle free streaming approach [18]. The last one was motivated by the fact that the effects, similar to the collective flow and associated with hydrodynamics, were not expected for elementary pp or p+Pb collisions. Remarkable experimental observations were done by CMS in very high multiplicity proton-proton collisions at the LHC: a positive long-range correlation was reported between two particles produced at similar azimuthal angles, spanning a large range in rapidity [19]. Recent CMS results on multi-particle correlations in pp collisions show that this structure has collective nature [20]. The strong effects were also found in p+Pb collisions [21]. Also, unexpectedly for light colliding systems, the existence of so-called double-ridge structures was found in d+Au collisions at RHIC [22, 23] and in p+Pb collisions at the LHC [24, 25]. These experimental findings produced a number of challenges to the existing theoretical approaches.

The characteristic feature of these azimuthal asymmetries of multi-particle angular correlations is that they are extending over a long range in rapidity. Causality then requires that these long-range correlations, if they exist, originate from the earliest times of the collision [26]. Therefore, we would like to mention here just briefly those models that are related intrinsically to the initial stages of the QGP formation and to the observed long-range rapidity correlation phenomena (one may see also the reviews [27] and [28]).

In the model of color glass condensate (CGC) the boost invariant Glasma flux tubes are considered [26] which produce particles isotropically with equal probability along the length of the flux tube. Interactions among those particles produces pressure leading to collective radial flow. The average transverse velocity is obtained here from the blast wave fits to the RHIC data obtained by the PHENIX collaboration. Thus the common transverse expansion experienced by particles in a flux tube collimates the particles forming a ridge like structure extended in rapidity.

The string-like approach considers the quark-gluon strings, formed in the result of color exchange between partons of colliding hadrons, as particle emitting sources. These sources are stretched longitudinally between the beam remnants and at certain density in the transverse plane they may start to interact, forming clusters and producing some collective effects. One may mention here the string-fusion and percolation phenomena extensively studied in nucleus-nucleus collisions in [29, 30]. In [29] the string percolation phenomenology is compared to the CGC results on effective string or glasma flux tube intrinsic correlations, including the ridge phenomena and long-range forward-backward azimuthal correlations. Color string percolation model and its similarities with the CGC were discussed in [30].

In the theoretical analysis of long-range correlations there are two major factors that should be taken into account. In the string-like approach both are intrinsically present: (i) formation of the particle production sources extended in rapidity and (ii) certain initial anisotropy of configuration of these sources (color-flux tubes or strings) in the transverse plane. It is the last factor that might be responsible for the azimuthal anisotropy effects of transport of particles through the given non-uniform medium. Among these approaches are the recent studies [31–33] of the azimuthal anisotropy of two-particle correlation functions, observed in hadron collisions, that are based on the assumption of quark-gluon string formation at the very early stages of hadron collisions and further interaction of particles produced by these sources with the strong colour fields inside string clusters. It is shown by the direct Monte Carlo simulations [33] that the elliptic flow can be successfully described in the color string picture with fusion and percolation, provided the anisotropy of particle emission from the fused string is taken into account. Two possible sources of this anisotropy are considered, propagation of the string in the transverse plane and quenching of produced particles in the strong colour field of the string [33].

Another motivation to understand the dynamics of QGP formation in relativistic heavy-ion collisions can be found in the dual holographic approach (one may see review in [34]). These issues of holography and AdS/CFT duality are being intensively studied at present. The system of QCD strings is considered to be attractively interacting and collapsing, and it is viewed as a QCD analog to AdS/CFT black hole formation [35–37].

In string model approach, following general assumptions from the Regge phenomenology, it is considered that quark-gluon strings (color flux tubes) are formed between the partons of the colliding nuclei at the early stage of hadron-hadron collision. The number of these strings depends on the number of participating nucleons (N_{part}) and on the collision energy. In case of sufficiently high density these strings may overlap and interact by attracting or repulsing,

Possible interaction of color strings (color flux tubes) in the form of repulsion or attraction and formation of flow harmonics in hadron collisions was first considered more then 3 decades ago in [1]. The exact type of string-string interaction is not known. The sign of string-string interactions depends on the direction of color fluxes [1]. This problem of the string-string interaction has not yet been systematically addressed till recently [35, 38, 39]. The magnitude of this interaction in string tension units was found to be small $\sim 10^{-1} - 10^{-2}$ [39]. It is the collective effect of the large number of strings that is responsible for significant compression of the system in the transverse plane, leading in case of attractive interaction to implosion of such high-density configurations of strings [40].

Recently a new interest to interactions bewteen color flux tubes appeared, that was motivated both by the experimental findings of long-rage correlation phenomena in hadron collisions and by the progress in fitting the data in the framework of the various string models. They include Pomeron exchange approach, Lund model and MC event generators based on this model, the collectivity effects like color reconnection and string fusion and percolation phenomena, etc. One may conclude that the field of initial stages of hadron collisions, in particular, in string-like approach, is being extensively explored.

In the present study we continue our investigations [41–44] of the role of string density effects in the formation of azimuthal asymmetry of two-parrticle correlation functions in the case of string-string repulsion. The paper has the following structure. After the Introduction we describe in the Section 2 the approach and the main assumptions that are implemented in our MC model [42]. Two-particle correlation functions at different centralities and harmonic decomposition of the azimuthal correlation function, obtained in the MC model, are presented in Section 3. Results on the p_T spectra and mass-splitting of the elliptic flow coefficients are presented in the Section 4. Discussion and conclusions are given in the last two sections.

2 MC model with repulsive strings

In our study we consider a simplified approach to string-string interaction mechanism for the case of repulsion. Basing on this hypothesis, a Monte Carlo (MC) model [42] was developed. In this MC model, final-state particles from A+A interactions are generated in each event in the following stages:
Stage 1. Simulation of the nuclei.
Initial positions of the nucleons in nuclei are generated in accordance to Woods-Saxon distribution. Nuclei of Pb^{208} are considered, the Woods-Saxon radius is 6.62 fm, diffusion = 0.546 fm. Inside each nucleon partons are distributed in transverse (xy) plane with 2D-Gauss law with σ_{xy} = 0.4 fm. Number of partons inside each nucleon is generated by Poisson law, where the mean number depends on a collision energy and is a model parameter. Two such a nuclei with some impact parameter are generated in each collision.
Stage 2. Simulation of strings configuration.
Interaction between colliding hadrons is implemented at the partonic level: partons from colliding nuclei can interact forming a qurk-gluon string, if the distance between partons in xy plane is less then parton interaction distance d = 0.4 fm (another model parameter). Additionally, there is a 3% probability for a string to be a result of "hard scattering" of partons to emulate of jet-like structures. Other 97% of the strings are considered to be "soft" and long in rapidity, occupying rapidity range $y \in (-4, 4)$.
Stage 3. Repulsion of the strings.
At the next step of the system evolution, "soft" strings interact with each other. In the current MC model, interaction manifests itself as a repulsion. The repulsion mechanism is adopted from [1]. In our MC model, an efficient string-string interaction radius R_{int} is introduced in the transverse plane. We consider this variable R_{int} as a free parameter (it is different from the string radius r_0 which is often considered to be of about r_0 = 0.2 fm). R_{int} determines "interaction zone" around each string, these zones may overlap, and, due to string-string mutual repulsion, in case of high string density each individual string acquires significant transverse boost β (see [1] and [42] for detailed mechanism). The mean value of the string boost is adjusted to be $\overline{\beta} \approx 0.65$ in central Pb-Pb collisions.
Stage 4. Event final state: hadronization of the strings.
Each string breaks in several places into quark-antiquark pairs with the same exponentially distributed p_T and in opposite azimuthal directions. Quarks from neighbour string pieces then combine and form a meson (mostly pions and ρ). *In the laboratory frame, all the particles originated from one string are boosted by the factor* $\overline{\beta}_{\text{string}}$. Due to the numerous string-string interactions taken into account, the initial configuration of quark-gluon strings could be transferred into the particles azimuthal flows.

The modified version of the MC model will be used in Section 4 to consider spectra and mass ordering for the particles with different masses.

3 Some features of MC model with string repulsion

It was shown earlier [43] that the MC model with string repulsion is capable to reproduce the characteristic features of two-particle correlation functions obtained in peripheral, semi-central and central collisions of heavy ions for charged particles, if rather large string-string interaction radius $R_{int} = 2$ fm is used (see Figure 1).

The pads in the Fig. 1 illustrate the onset of collectivity with the increase of string density when passing from peripheral to central A-A collisions. In peripheral events, a structure along $\Delta\varphi$ is visible, which is formed due to the ρ^0 decays into pions. One may see also, that the evolution of two-particle angular correlations with centrality and the relevant transition from low-density to the high density in Au-Au collisions at $\sqrt{s_{NN}} = 200$ GeV is correctly reproduced. (Note that STAR variables $\Delta\rho/\sqrt{\rho_{\text{ref}}}$ [22] are used here for two-particle correlations.)

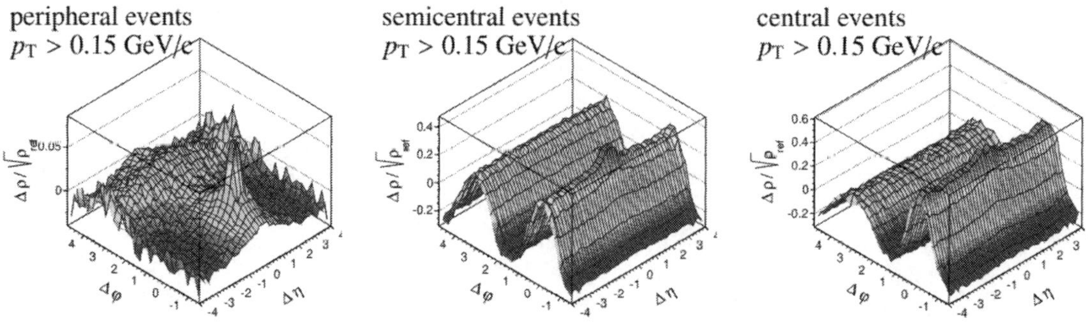

Figure 1. MC model [42] results for two-particle correlation functions obtained in the for peripheral (left), semi-central (middle) and central (right) events (from [43]). Particles with $p_T > 0.15$ GeV/c are analyzed. String-string interaction radius is 2 fm.

It was also obtained in [44] that two-particle azimuthal correlations, observed in very central Pb-Pb collisions at LHC [12], can be obtained in the MC model with repulsive strings, along with the harmonic decomposition of the correlation function (see Figure 2). This decomposition reveals significant values of v_2, v_3 and v_4 coefficients.

4 Transverse momentum spectra and mass ordering

Another important experimental observation is a so-called mass ordering of flow harmonics, when $v_n(p_T)$ points are "blue-shifted" to the right for hadrons with higher masses (see, for example, results for elliptic flow in Pb-Pb collisions in [45]). It is natural that the MC model should be able to describe spectra and flow coefficients for different kind of particles. In the present work we check a dependence v_2 on p_T for pions, kaons and protons in order to see if the mass ordering is preserved.

For this study, string fragmentation in its rest frame was changed: pions, kaons and protons were generated as products of string decay (instead of just pions and ρ-mesons), and their transverse momentum spectra were parametrized by Tsallis formula tuned with data for pp collisions at $\sqrt{s} = 7$ TeV [46]. Obtained spectra are shown in Fig.3 before application of string boosts (dotted lines), and after

 DOI: 10.1051/epjconf/201612504011

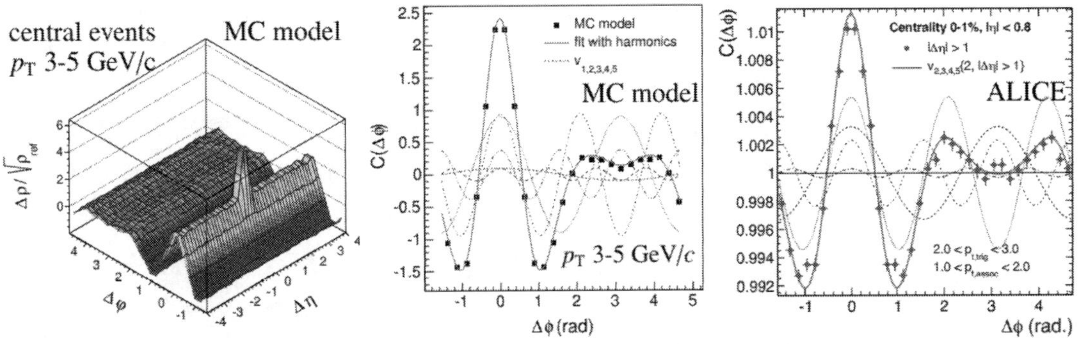

(a) (b) (c)

Figure 2. (a) MC model results for two-particle correlation function for very central nucleus-nucleus collision events for charged particles with p_T in [3, 5] GeV/c [42, 44]. String interaction radius is 2 fm. (b) Double-peak around $\Delta\varphi = \pi$ in harmonics decomposition of two-particle angular correlations in MC model at the most central Pb-Pb events ($b = 0$ fm) for charged particles with $p_T \in [3, 5]$ GeV/c. The solid red line shows the sum of the anisotropic flow Fourier coefficients $v_1 ... v_5$ (dashed lines). (c) Experimental results [12] for harmonics decomposition of two-particle angular correlations in central Pb-Pb collisions at the LHC. Note: the different definitions for two-particle correlation function are used at (b) and (c).

boosting (solid lines). Modifications of spectra for different types of particles are found to be in line with the observed hardening of $\langle p_T \rangle$ observed for hadrons for more central Pb-Pb events [47].

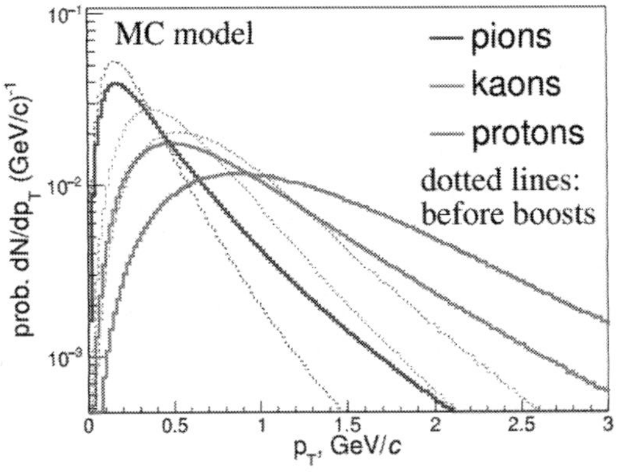

Figure 3. The p_T-spectra for pions, kaons and protons in MC model before (dotted lines) and after string boosts (solid lines), in Pb-Pb collisions with centrality 10-20%.

To check the mass ordering, the $v_2(p_T)$ was calculated with the Scalar Product (SP) method, similarly to what was done with LHC Pb+Pb data in [45]. Hadrons of one of the species (π, K or p) in $|\eta| < 1$ were chosen as particles-of-interest, while reference charged particles were taken from $-3 < \eta < -1$ and $1 < \eta < 3$, providing a pseudorapidity gap of $|\Delta\eta| > 1.0$. This choice of rapidity ranges emulates experimental conditions of the $v_2(p_T)$ analysis of identified hadrons in ALICE experiment [45]. The MC model results are shown in Fig.4 (b) and compared to data (a). It can be seen that the mass ordering observed in the experiment is qualitatively reproduced by the MC model in the region of interest (for p_T values below 3 GeV/c).

Is should be mentioned, that $v_2(p_T)$ dependences for different particle species are quite sensitive to the shape of the p_T spectra, and parameterization of the spectra by Tsallis distribution tuned out to be good for reproduction of qualitative behaviour of the $v_2(p_T)$. Another important point is that the effective string interaction radius R_{int}, which is the main parameter of the MC model, is still needed to be quite large – of about 2 fm.

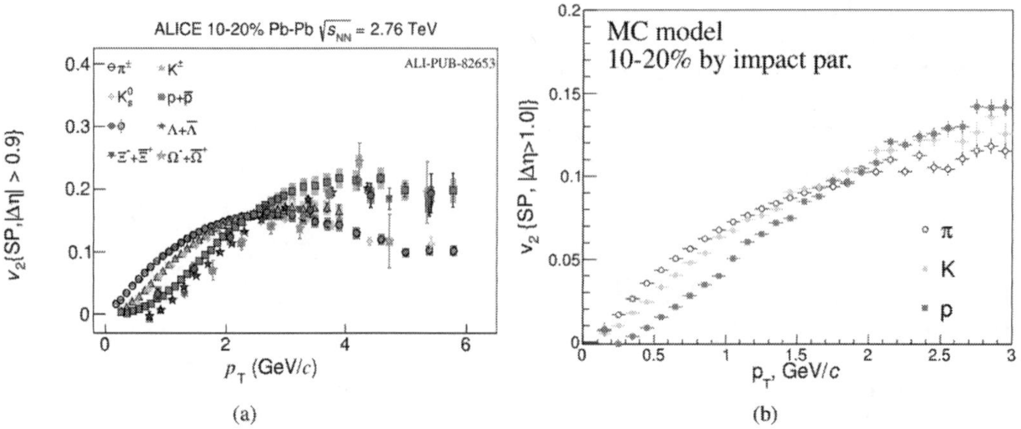

Figure 4. The p_T-differential v_2 for pions, kaons and protons in centrality class 10-20% of Pb-Pb collisions: (a) – ALICE data at $\sqrt{s_{NN}}$ = 2.76 TeV [45], (b) – MC model calculations. The string interaction radius R_{int} = 2 fm.

5 Discussion

Results of application of the MC model with the repulsive string-string interaction show that quite a *large effective string-string interaction radius* of about 2 fm is needed in order to describe qualitatively (and in some cases almost quantitatively) azimuthal and rapidity topology of two-particle correlations picture, including the harmonic decomposition, centrality dependence of flow coefficients, mass ordering, etc. Possible physical mechanisms behind it could be the following:
(i) existence of rather weak (but long-range in transverse plane) interactions between color flux tubes (strings) mediated by exchange of some kind of meson, or
(ii) necessity to have a "dynamical" phase for a string when it moves within "string medium", before decaying into hadrons. It means that the string should experience some pressure from other strings (thus effectively increasing the interaction distances).

The current version of the MC model is based on simple kinematic assumptions and just a few parameters. There is still an open question: is it possible to describe pp and p-Pb effects with minor changes in model assumptions and parameter values? This check is be necessary to do if one wants to establish some uniform mechanism of flow generation.

6 Conclusions

High density string medium could be formed in hadronic collisions. It is shown in the Monte Carlo model that collective effects could appear due to some, rather weak, repulsive type of interaction between color flux tubes (strings). The particle-emitting sources (strings) are boosted by the combined repulsion by neighbours, thus the initial space anisotropy is converted into the anisotropic azimuthal distributions of charged particles.

Similar results could occur at sufficiently high string density p-Pb and pp collisions. More detailed quantitative analysis including mass ordering of flows in p-Pb and pp collisions has to follow.

Acknowledgements. This work is supported by the Saint-Petersburg State University research grant 11.38.242.2015. The authors are grateful to V.Vechernin and V.Kovalenko for useful discussions.

References

[1] V.A. Abramovsky, O.V. Kanchely, JETP letters **31**, (1980) 566.
[2] J.-Y. Ollitrault, PRD 46, 229 (1992), PRD 48, 1132 (1993).
[3] S. Voloshin and Y. Zhang, Z. Phys. C 70, (1996) 665.
[4] A. M. Poskanzer and S. A. Voloshin, Phys. Rev. C 58, (1998) 1671.
[5] K.H. Ackermann et al. (STAR Collaboration), Phys. Rev. Lett. 86 (2001) 402.
[6] Heinz Sorge, Phys. Rev. Lett. 78, (1997) 2309.
[7] ALICE Collaboration, Phys. Rev. Lett. 105, (2010) 252302.
[8] A. Bilandzic, R. Snellings and S. Voloshin, Phys.Rev. C83 (2011) 044913.
[9] F. Wang, STAR Collaboration, Talk at Quark Matter 2004, J. Phys. G 30 (2004) S1299–S1304.
[10] J. Adams, et al. (STAR Collaboration), Phys.Rev.Lett. 95 (2005) 152301.
[11] J. Adams, et al. (STAR Collaboration), Phys. Rev. C 73 (2006) 064907.
[12] ALICE Collaboration, Physics Letters B, Volume 708, Issues 3–5, (2012) 249–264.
[13] ATLAS Collaboration, Phys.Rev. C86 (2012) 014907.
[14] CMS Collaboration, PHYSICAL REVIEW C 87 (2013) 014902.
[15] ATLAS, Phys.Lett. B707 (2012) 330-348.
[16] ALICE Collaboration, arXiv:1605.02035, CERN-EP-2016-115 (2016).
[17] F. Wang, Progress in Particle and Nuclear Physics, 74 (2014) 35–54.
[18] P. Romatschkea, Eur. Phys. J. C (2015) 75:429.
[19] V. Khachatryan et al. [CMS Collaboration], JHEP 1009 (2010) 091.
[20] V. Khachatryan et al. [CMS Collaboration], arXiv:1606.06198 (2016).
[21] S. Chatrchyan et al. [CMS Collaboration], Phys. Lett. B 718, 795 (2013); Phys. Lett. B 724, 213 (2013).
[22] G.Agakishev et al. (STAR Collaboration), Phys.Rev.C 86, (2012) 064902.
[23] A. Adare et al. (PHENIX Collaboration), Phys. Rev. Lett. 111, (2013) 212301.
[24] B. Abelev et al. [ALICE Collaboration], Phys. Lett. B 719, 29 (2013); B. Abelev et al. [ALICE Collaboration, arXiv:1406.2474 [nucl-ex].
[25] G. Aad et al. [ATLAS Collaboration], Phys. Rev. Lett. 110, 182302 (2013); Phys. Lett. B 725, 60 (2013); The ATLAS collaboration, ATLAS-CONF-2014-021.
[26] A. Dumitru, F. Gelis, L. McLerran and R. Venugopalan, Nucl. Phys. A 810, (2008) 91.
[27] Wei Li, Mod. Phys. Lett. A27 (2012) 1230018.
[28] James D. Bjorken, Stanley J. Brodsky, and Alfred Scharff Goldhaber, arXiv:1308.1435 (2013).
[29] J. Dias de Deus and C. Pajares, Phys.Lett B**695** (2011) 211-213.
[30] C. Pajares, Nucl. Phys, A 854 (2011) 125-130, arXiv:1007.3610.
[31] M.A.Braun and C.Pajares, Eur.Phys.J.C 71 (2011) 1558.
[32] M.A. Braun, C. Pajares, V.V. Vechernin, Nucl.Phys.A 906 (2013) 14.
[33] M.A. Braun, C. Pajares, V.V. Vechernin, Eur.Phys.J.A 51 (2015) 44.
[34] I. Arefeva, Phys. Usp. 57, (2014) 527.

[35] E. Shuryak, arXiv:1412.8393 (2014).
[36] E. Shuryak and I. Zahed, arXiv:1311.0836 (2014).
[37] I. Iatrakis, A. Ramamurti and E. Shuryak, PHYSICAL REVIEW D 92, (2015) 014011.
[38] T. Kalaydzhyan, E. Shuryak, arXiv:1404.1888 (2014).
[39] T. Kalaydzhyan and E. Shuryak, arXiv:1402.7363 [hep-ph].
[40] T. Kalaydzhyan and E. Shuryak, arXiv:1503.05213 (2015).
[41] O.Kochebina, G.Feofilov, arXiv:1012.0173 (2010).
[42] I. Altsybeev, AIP Conf. Proc. 1701 (2016) 100002, arXiv:1502.03608.
[43] G.Feofilov, I.Altsybeev, O.Kochebina, AIP Conf. Proc. 1701 (2016) 060011.
[44] G.Feofilov, I.Altsybeev, O.Kochebina,PoS (Baldin ISHEPP XXII) 067 (2014).
[45] ALICE collaboration, JHEP 06 (2015) 190.
[46] ALICE collaboration, EPJC 75, (2015) 226.
[47] ALICE collaboration, Phys. Rev. C 88, (2013) 044910.

Saturation effects in low-x DIS structure functions and related hadronic total cross sections

Francesco Giovanni Celiberto[1,2,⋆], *Laszlo* Jenkovszky[3,⋆⋆], and *Volodymyr* Myronenko[4,⋆⋆⋆]

[1] *Dipartimento di Fisica, Università della Calabria, Arcavacata di Rende, 87036 Cosenza, Italy*
[2] *Istituto Nazionale di Fisica Nucleare, Gruppo Collegato di Cosenza, Arcavacata di Rende, 87036 Cosenza, Italy*
[3] *Bogolyubov Institute for Theoretical Physics (BITP), Ukrainian National Academy of Sciences 14-b, Metrologicheskaya str., Kiev, 03680, Ukraine*
[4] *Deutsches Elektronen-Synchrotron DESY, Hamburg, Germany*

Abstract.
High-energy nucleon total cross sections are related to low-x DIS structure functions by using the additive quark model.

In the additive quark model, the hadron-hadron total cross section can be written as a product of the cross sections of the constituents, σ_{qq} [1, 2], *e.g.*

$$\sigma(s)^t_{pp} = \sigma_{qq}[n_V + n_S(s)]^2, \tag{1}$$

where n_V is the number of valence quarks and $n_S(s)$ is that of sea quarks, their number increasing with energy.

It was suggested in Refs. [1, 2] that the increasing number of sea quarks is related to the Bjorken scaling-violating contribution to the deep inelastic lepton-hadron structure function (DIS SF), namely to the momentum fraction of the relevant quarks given by the integral over the DIS structure function $F_2(x, Q^2)$. In Ref. [1] a simple model for the DIS structure function, known at those times, was used, resulting in the following expression for the total cross section, compatible with the data

$$\sigma(s)^t_{pp} = \sigma_{qq} n_V^2 (1 + 0.0.16\, \ell n(s/Q_0^2)), \tag{2}$$

where σ_{qq} is a free parameter, Q_0^2 was fitted to the DIS data, and $n_V = 3$.

In Ref. [2] the DIS SF was related to hadronic cross sections by means of finite-energy sum rules in Q^2.

The number of quarks in a reaction can be calculated from the SF by means of sum rules. see e.g. [3, 4].

In Ref. [5] following ansatz for the small-x singlet part (labelled by the upper index $S, 0$) of the proton structure function, interpolating between the soft (VMD, Pomeron) and hard (GLAP evolution) regimes was proposed:

⋆e-mail: francescogiovanni.celiberto@fis.unical.it
⋆⋆e-mail: jenk@bitp.kiev.ua
⋆⋆⋆e-mail: volodymyr.myronenko@desy.de

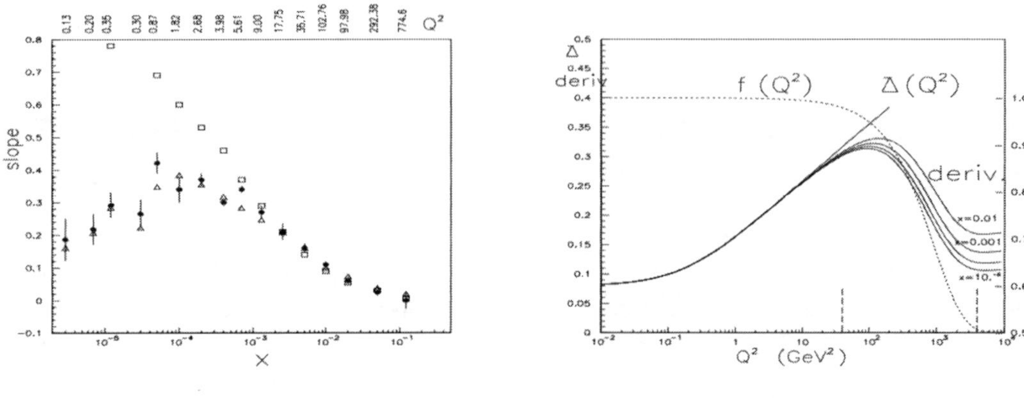

Figure 1. Slope of the structure function $F_2(x, Q^2)$.

$$F_2^{(S,0)}(x, Q^2) = A\left(\frac{Q^2}{Q^2+a}\right)^{1+\widetilde{\Delta}(Q^2)} e^{\Delta(x,Q^2)}, \tag{3.1}$$

with the "effective power"

$$\widetilde{\Delta}(Q^2) = \epsilon + \gamma_1 \ell n\left(1 + \gamma_2 \ell n\left[1 + \frac{Q^2}{Q_0^2}\right]\right), \tag{3.2}$$

and

$$\Delta(x, Q^2) = \left(\widetilde{\Delta}(Q^2)\ell n\frac{x_0}{x}\right)^{f(Q^2)}, \tag{3.3}$$

where

$$f(Q^2) = \frac{1}{2}\left(1 + e^{-Q^2/Q_1^2}\right). \tag{3.4}$$

At small and moderate values of Q^2, the exponent $\widetilde{\Delta}(Q^2)$ (3.2) may be interpreted as a Q^2-dependent "effective Pomeron intercept", as shown in Fig. 1.

The function $f(Q^2)$ has been introduced in order to provide for the transition from the Regge behaviour, where $f(Q^2) = 1$, to the asymptotic solution of the GLAP evolution equation, where $f(Q^2) = 1/2$.

In Ref. [5] the above singlet SF was appended by a non-singlet part, important at large values of x. The parameters were fitted to the DIS data in a wide range of x and Q^2.The values of the fitted parameters are: $A = 0.1623$, $a = 0.2916$ GeV^2, $\gamma_2 = 0.01936$, $Q_0^2 = 0.1887$ GeV^2, $Q_1^2 = 916.1$ GeV^2; $x_0 = 1$, $\epsilon = 0.08$, $\gamma_1 + 2.4$ were fixed (by QCD-related arguments). The resulting fits and more details can be found in Ref. [5].

The proton-proton total cross section is cast by integrating Eqs. (3) between $x = 0$ and $x = 1$. At high energies, only the singlet part of the SF, Eqs. (3) (the "Pomeron") is relevant. Integration can be

DOI: 10.1051/epjconf/201612504012

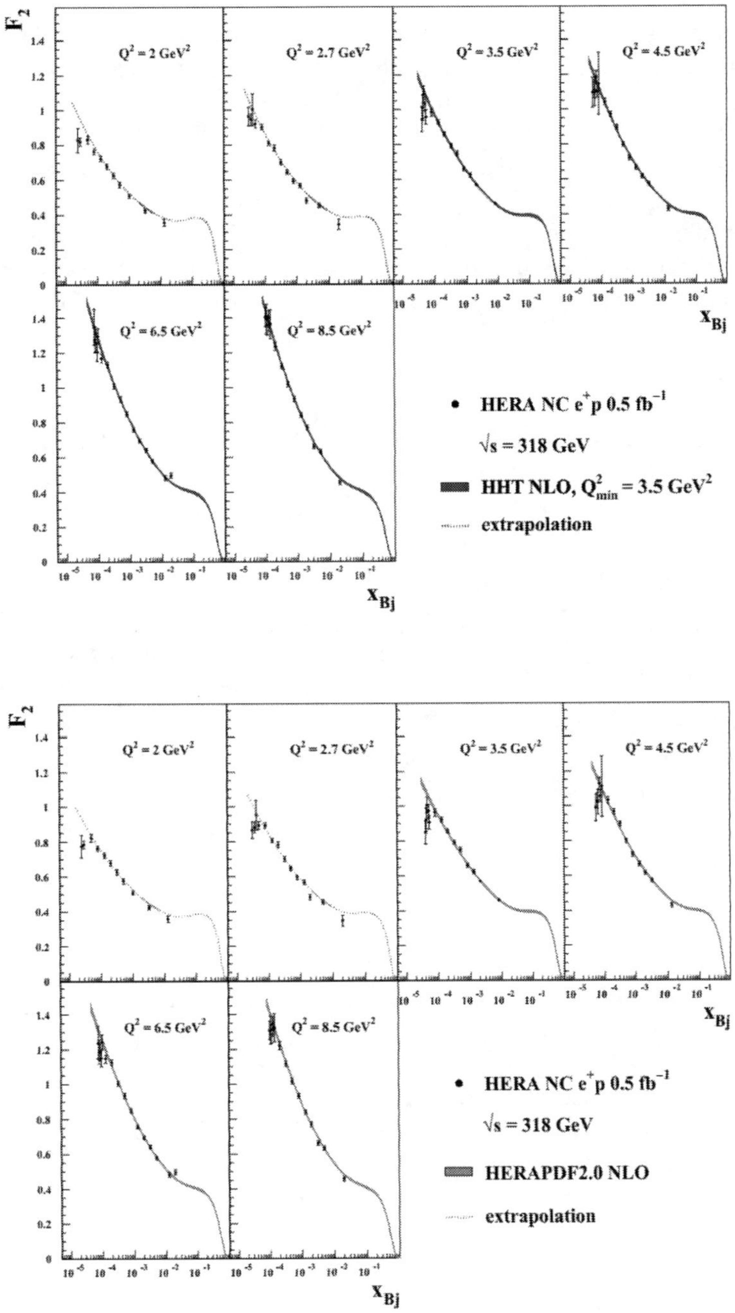

Figure 2. HHT (top) and HERAPDF2.0 (bottom) NLO predictions of the F_2 structure function at $Q^2_{\mathrm{min}} = 3.5\,\mathrm{GeV}^2$, compared to extracted values. For more details, see Ref. [7].

performed numerically. The result is in reasonable agreement with the data on *pp* total cross sections, including those from the LHC.

The operator-product expansion beyond leading twist has diagrams in which two, three or four gluons may be exchanged in the *t*-channel such that these gluons may be viewed as recombining. This recombination could lead to gluon saturation. The colour-dipole framework also inspired a phenomenological model of saturation by Golec–Biernat–Wüsthoff (GBW), in which the onset of saturation is characterised as the transition from a "soft" to a "hard" scattering regime. This occurs along a "critical line" in the x_{Bj}, Q^2 plane.

Recently new results on low-x DIS parton distrubutions (PDF and SF) have appeared [6, 7]; they show an intriguing change of regime towards smallest values of x (Fig. 2) - possible saturation effect?

We intend to investigate its impact on the asymptotic behaviour of the total cross sections.

Acknowledgement

L.J. thanks the Organizers of the Quarks2016 conference for their warm hospitality and support.

References

[1] L.L. Jenkovszky and B.V. Struminsky: *Violation of scale invariance in deep-inelastic lepton-hadron scattering and rise of hadronic cross-sections*, ITP-77-37E Preprint, Kiev, 1977.

[2] L.L. Jenkovszky and B.V. Struminsky: Yadernaya Fizika (English translation: Sov. J. Nucl. Phys.), **38**(1983), 1568; see also: http://inspirehep.net/record/1456006/files/$mpi_a t_l hc_2$015.pdf; http://indico.ictp.it/event/a14280/ , p.171.

[3] F.E. Close, *An introduction to quarks and partons*, Academic Press Inc. (London) Ltd., 1979.

[4] R.G. Roberts, *The structure of the proton*, Cambridge University Press, 1990.

[5] P. Desgrolard, Laszlo L. Jenkovszky, F. Paccanoni, *Interpolating between soft and hard dynamics in deep inelastic scattering*. LYCEN-9808, Eur. Phys. J. C7 (1999) 263-270; hep-ph/9803286.

[6] H. Abramowicz *et al.* [H1 and ZEUS Collaborations], *Combination of measurements of inclusive deep inelastic $e^{\pm}p$ scattering cross sections and QCD analysis of HERA data*, Eur. Phys. J. C **75** (2015) no.12, 580 doi:10.1140/epjc/s10052-015-3710-4 [arXiv:1506.06042 [hep-ex]].

[7] I. Abt, A. M. Cooper-Sarkar, B. Foster, V. Myronenko, K. Wichmann and M. Wing, *Study of HERA ep Data at Low Q^2 and Low x_{B_j} and the Need for Higher-Twist Corrections to Standard pQCD Evolution*, arXiv:1604.02299 [hep-ph].

Sound Attenuation in Quark Matter Due to Pairing Fluctuations

Boris Kerbikov[1,2,3,★]

[1]*Alikhanov Institute for Theoretical and Experimental Physics*
[2]*Lebedev Physical Institute*
[3]*Moscow Institute of Physics and Technology*

Abstract. The sound wave in dense quark matter is subject to strong absorption due to di-quark field fluctuations above T_c. The result is another facet of Mandelshtam-Leontovich slow relaxation time theory.

During the last decade the investigation of quark matter at finite temperature and density became a compelling topic in QCD. Drawn in the (T, μ) plane with μ being the quark chemical potential the QCD phase diagram embodies several domains with quite different properties. Hopefully, we understand more or less the physics of quark-gluon matter in the high-temperature and low-density region. On the experimental side this is because most of the data on heavy-ion collisions obtained at RHIC and LHC correspond to high T and zero, or very small, μ. On the theoretical side the zero μ and high T domain is accessible to Monte-Carlo simulations.

Our focus in the present talk is on the opposite regime of high-density and moderate temperature. Such conditions may be realized in neutron stars and in future experiments at FAIR and NICA. The experimental data on this region are extremely scarce and theoretical understanding is on shaky ground. A powerful way to investigate the nature of a certain substance is to study its response to an external perturbation. This may be, e.g., magnetic field, temperature or pressure gradient, etc. Here we investigate the sound wave absorption in dense moderate temperature quark matter in the pre-critical region above T_c. According to the present understanding attractive interaction between quarks in the color antitriplet state leads at high density to the formation of the color superconducting phase. This phese is formed approximately in the region $\mu \gtrsim 400\,MeV$ and $T < T_c \sim 40\,MeV$. There are certain subtleties concerning the above numbers which are not crucial for our problem. There are two principal points we want to briefly elucidate in this short talk before the final results are presented. These two interrelated questions are:

i) what is the difference between BCS and color superconductivity,

ii)what regime one has to attribute to the above domain of the QCD phase diagram: strong-coupling, weak-coupling, or BCS-BEC crossover.

Both points were overwhelmingly discussed in literature including a few works of the present author [1–5]. We consider the 2SC color superconducting phase when u- and d- quarks participate in pairing, while the density is not high enough to involve the s-quark in pairing. The very short answer to the question (i) is the following. Nonzero gap Δ is what BCS and 2SC have in common. In BCS two other important physical parameters are: the Debye frequency $\omega_D \sim 10^{-2}\,eV$ which measures the

★e-mail: borisk@itep.ru

electron-phonon interaction and he Fermi energy $E_F \sim 2\,eV$. The BCS gap is $\Delta \sim T_c \sim$ few K. The parameters of 2SC are: the ultra-violet momentum cuttoff $\Lambda \sim 0.8\,GeV$ and the chemical potential $\mu \sim 0.4GeV$. The 2SC gap is $\Delta \sim 0.1\,GeV$. The hierarchy of the inherent scales in the two regimes is very different

$$BCS \quad \Delta : \omega_D : E_F \simeq 1 : 10^2 : 10^4, \tag{1}$$

$$2SC \quad \Delta : \Lambda : \mu \simeq 1 : 8 : 4. \tag{2}$$

The fact that the BCS hierarchy is badly broken in 2SC reflects itself in the value of a dimensionless parameter $n^{1/3}\xi \sim k_F\xi$, where n is the number density and ξ is the characteristic length of pair correlation. In the BCS scenario $k_F\xi \gtrsim 10^3$ v.s. $k_F\xi \gtrsim 2$ in $2SC$ [1–5]. In BCS pairs are large compared to their separarion, in 2SC they are small compared to their separation (such molecular-like objects are sometimes called Shafroth pairs). The continuous evolution of this parameter as a function of density and of the interaction strength reflects the transition from BCS regime to Bose-Einstein condensation (BEC). The idea of such transformation is due to F. Dyson, the problem was theoretically investigated in [6] for electron pairs and following [6] in [2–4] for quarks. Starting from the early 2000th dramatic theoretical and experimental progress has been done in the study of the BCS-BEC crossover in ultracold fermionic atoms [7].

Explaining the BCS-BEC crossover we partly elucidated the problem (ii) from the list above. A remark on terminology is needed. The terms strong- and weak-coupling in superconductivity (and in many-body physics) and in QCD have a different meaning. One can find, of course, some anology. Chiral symmetry breaking in QCD is a kind of a counterpart of the formation of fermion bilinear condensate in BCS. Roughly speaking, weak coupling in superconductivity means that the interaction between particles (electrons) is concentrated within a thin layer of momentum space around the Fermi surface and $\omega_D \ll E_F$ (see (1)). Integration in the vicinity of the Fermi surface is performed (in relativistic case) using the variable ξ defined as

$$\xi = \sqrt{\mathbf{p}^2 + m^2} - \mu, \tag{3}$$

$$\int \frac{d\mathbf{p}}{(2\pi)^3} \simeq \int d\xi \rho(\xi) \simeq \int d\xi \left[\rho(\mu) + \left(\frac{\partial\rho}{\partial\xi}\right)_\mu \xi\right], \tag{4}$$

where $\rho(\mu) = \frac{p_0\mu}{2\pi^2}$, p_0 is the Fermi momentum. The first term in (4) gives rise to Cooper logarithm. The second term takes into account the energy dependence of the density of states near the Fermi surface. Only this term in (4) gives nonzero contribution to the sound absorption [8, 9]. In a great number of works the NJL (Nambu and Jona-Lasinio, Vaks and Larkin) model has been used to study color superconductivity. Strong-coupling in this approach is tantamaunt to a large value of the di-quark coupling constant. In absence of a fundamental approach and impossibility to perform lattice calculations at nonzero density the NJL model may serve for the orientation purposes with a hope to confront its predictions with future FAIR and NICA data. As a candidate for a strong-coupling model the NJL suffers from a lack of renormalizability and confinement (the latter drawback is proposed to cure adding the Polyakov loop).

On the other hand, the strong-coupling Migdal-Eliashberg theory of superconductivity was developed already more than half-century ago [10, 11] and has been successfully applied in many problems including HTSC [12].

Migdal-Eliashberg strong-coupling theory leads to the enhancement of pairing fluctuations and to the broadening of the transition region [13]. Implementation of Migdal theorem [11] to color superconductors will be discussed elsewhere. The fluctuation contribution is characterized by Ginzburg-Levanyuk number Gi which for color superconductor becomes very large

$$Gi \simeq \frac{\delta T}{T_c} \simeq \left(\frac{T_c}{\mu}\right)^4 \simeq 10^{-4}. \tag{5}$$

This is a huge number compared to the BCS $Gi \simeq 10^{-12} - 10^{-14}$. An alternative estimate $Gi \sim (k_F\xi)^{-4} \sim 10^{-2} - 10^{-3}$ [4] leads to even larger value.

Precursor pair fluctuations above T_c give the dominant contribution to the quark matter transport coefficients. The leading diagram is the Aslamazov-Larkin (AL) one [8], which includes two propagators of the slow pair collective mode singular at T_c. Within the ordinary theory of super conductivity this so-called fluctuation propagator (FP) was derived in [8]. For color superconductor it was evaluated using Dyson equation and Matsubara formalism in [14] and from the time-dependent Landau-Ginzburg equation with stohastic Langevin forces in [9]. The FP reads

$$L(\mathbf{q},\omega) = -\frac{1}{\nu}\frac{1}{\varepsilon + \frac{\pi}{8T_c}(-i\omega + D\mathbf{q}^2)}. \tag{6}$$

Here $\nu = \rho(\mu) = \frac{p_0\mu}{2\pi^2}$ (see (4)), $\varepsilon = \frac{T - T_c}{T_c}$, D is the diffusion coefficient. At small ω and q the quantity $L(\mathbf{q},\omega)$ can be arbitrary large close to T_c. The AL diagram for the sound absorption is shown in Fig.1 The two way lines correspond to the FP-s, the solid lines are the quarks Matsubara propagators,

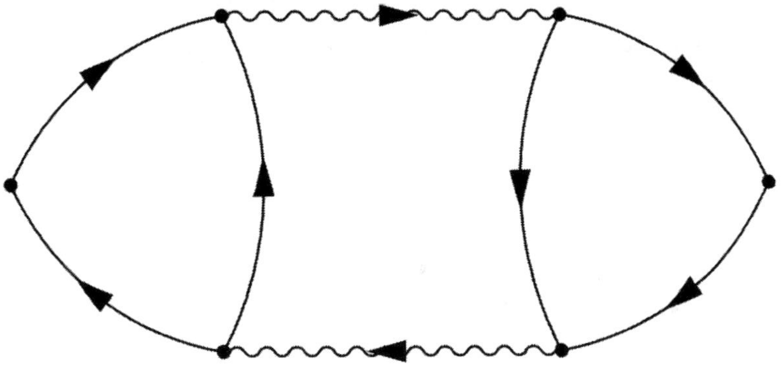

Figure 1. Feynman diagram for the AL polarization operator for the sound absorption.

the in- and out- vertices are equal to the constants g of phonon-quark interaction. The sound attenuation coefficient is equal to the imaginary part of the retarded polarization operator corresponding to the diagram in Fig.1. In this short presentation we leave out the detailed calculation which may be found in [8] for BCS and in [9] for the quark matter. In the fluctuation region above T_c the dominant contribution and the sharp temperature dependence come from the two FP-s. The final result for the imaginary part of the polarization operator reads [9]

$$\begin{aligned}\mathrm{Im}\,\Pi = &-\omega g^2 \frac{m^2}{2^5 p_0^4 \varkappa^3}(v_0^2+1)^2 \times \\ &\times \ln^2\frac{\Lambda}{2\pi T_c}\left(\frac{T_c}{T-T_c}\right)^{3/2}.\end{aligned} \tag{7}$$

Here m is the quark mass, p_0 and v_0 are the Fermi momentum and velocity Λ is the ultra-violet cutoff (2), $\varkappa^2 = \frac{\pi}{8T_c}D$, where D is the diffusion coefficient. As we remarked after Eq.(4), the nonzero contribution to $\mathrm{Im}\,\Pi$ comes not from crust of the Fermi surface (the first term in (4)), but from the energy dependent density of states given by the second term in (4). Important to note that the evaluation of the AL diagram in [8] and [9] making use of (4) are in complete agreement with the result for this polarization operator obtained in the Eliashberg strong-coupling theory [13].

The physics behind the strong energy dissipation of the sound wave in the precritical region is simple and very general. This is another manifestation of Mandelshtam-Leontovich slow relaxation time theory [15–17]. Propagation of the sound wave locally changes the critical temperature and the slow fluctuation pairing can not keep up with this process.

The author is supported by a grant from the Russian Science Foundation number 16-12-10414. The author express his gratitude for discussions and remarks to A. Varlamov, Yu. A. Simonov, M. Lukashov, M. Andreichikov, L. McLerran, P. Petreczky.

References

[1] B. O. Kerbikov, *Color Superconducting State of Quarks* [arXiv: hep-ph/0110197].
[2] B. O. Kerbikov, Phys. Atom. Nucl. **65**, 1918 (2002) [arXiv: hep-ph/0204209].
[3] B. O. Kerbikov, Phys. Atom. Nucl. **68**, 890 (2005) [arXiv: hep-ph/0407292].
[4] B. O. Kerbikov, Surveys in High Energy Physics **20**, 47 (2006) [arXiv: hep-ph/0510302].
[5] B. O. Kerbikov and E. V. Luschevskaya, Phys. Atom. Nucl. **71**, 364 (2008) [arXiv: hep-ph/0607304].
[6] P. Nozieres and S. Schmitt-Rink, J. Low Temp. Phys. **59**, 195 (1985).
[7] M. Greiner, C. A. Regal and D. S. Jin, Nature **426**, 537 (2003);
S. Jochim et al., Science **302**, 2101 (2003).
[8] L. G. Aslamazov and A. I. Larkin, Sov. Phys. Solid State **10**, 875 (1968); Phys. Lett. **26A**, 238 (1968).
[9] B. O. Kerbikov and M. S. Lukashov, *Fluctuation sound absorption in quark matter*, accepted for publication in Mod.Phys.Lett.A [arXiv: 1607.00125 [hep-ph]].
[10] G. M. Eliasberg, Sov. Phys. JETP **14**, 886 (1962).
[11] A. B. Migdal, Sov. Phys. JETP **7**, 996 (1958).
[12] A. I. Larkin and A. A. Varlamov, *Theory of Fluctuations in Superconductors* (OUP, New York, 2005) 432 pp.
[13] B. N. Narozhny, Sov. Phys. JETP **77**, 301 (1993); Phys. Rev. **B49**, 6375 (1994).
[14] B. O. Kerbikov and M. A. Andreichikov, Phys. Rev. **D91**, 074010 (2015).
[15] L. L. Mandelshtam and M. A. Leontovich, Zh. Eksp. Teor. Fiz. **7**, 438 (1937).
[16] L. D. Landau and E. M. Lifshitz, *Fluid Mechanics* (2nd ed., Pergamon Press, Oxford, 1987) 539 pp.
[17] Z. Tisza, Phys. Rev. **61**, 531 (1942).

EPJ Web of Conferences **125**, 04014 (2016) DOI: 10.1051/epjconf/201612504014
QUARKS-2016

Poincaré-invariant constituent quark model for light mesons: capabilities and constraints

A.F. Krutov[1,⋆], *R.G.* Polezhaev[1,⋆⋆], and *V.E.* Troitsky[2,⋆⋆⋆]

[1] *Samara University, 443086 Samara, Russia*
[2] *D.V. Skobeltsyn Institute of Nuclear Physics, Moscow State University, Moscow 119991, Russia*

Abstract. We present a brief survey of some results on electroweak properties of composite systems that are obtained in the frameworks of our version of the instant form of relativistic quantum mechanics (RQM). Our approach describes well the π- and the ρ-mesons in wide range of momentum transfers Q^2. At large Q^2 the obtained pion form factor asymptotics coincides with that of QCD predictions. The method permits to perform analytic continuation of pion form factor to complex plane of momentum transfers that is in accordance with predictions of quantum field theory.

1 Introduction

The purpose of this paper is to present a version of relativistic composite quark model developed by the authors (see, e.g.,[1]) for the study of composite systems of light quarks.The investigation of electroweak properties of light mesons is an important part of the study of the transition region where the perturbative QCD behavior starts to be valid. In this connection such particles are in the focus of experiments on up-to-day accelerators. The most important results here are the following: measurement of the pion form factor at large momentum transfers in JLab (see, e.g., [2, 3]), the obtaining of the ρ-meson decay constant from the reaction $\tau \to \rho\,\nu_\tau$ [4, 5], the magnetic moment measurement of the radiation transition $\rho \to \pi\gamma^*$. The interesting result obtained by BABAR collaboration concerning the deviation of the behavior of the transition form factor $F_{\pi\to\gamma\gamma^*}(Q^2)$ for large momentum transfer from that predicted by perturbative QCD [6] remains unexplained.

The theoretical description of systems containing light quarks, requires the account of relativistic effects even at low energies. The method in the relativistic theory of composite systems which will be used here is based on the direct realization of the Poincaré algebra on the set of dynamic observable systems and dates back to the P.Dirac's paper [7]. This approach is called the relativistic quantum mechanics with fixed number of particles (RQM) (see, e.g.,review [1] and references therein). From the point of view of the principles underlying it, the RQM occupies the intermediate position between the local quantum field theory and nonrelativistic quantum mechanical models.

It should be noted that the field theory and the RQM are formulated as fundamentally different structures and the establishment of a connection between them is a complicated problem that has

⋆e-mail: krutov@samsu.ru
⋆⋆e-mail: polezaev@list.ru
⋆⋆⋆e-mail: troitsky@theory.sinp.msu.ru

not been solved yet. Unlike the field theory, in the Poincaré-invariant quantum mechanics, the finite number of degrees of freedom are separated initially, that represents some model. The covariance of description in the RQM is provided by the construction on the Hilbert state space of the composite system with the finite number of degrees of freedom of the unique unitary transformation of the inhomogeneous group $SL(2,C)$, which is the universal covering of the Poincaré group. The RQM can be realized using different methods and in different forms of dynamics (instant form, point form, light-front form). In this survey basic attention was paid to the description of the electroweak structure of composite systems in the framework of the instant form (IF) of RQM.

One of the main problem in construction of the relativistic constituents models is known to be the problem of construction of operators of transition currents. It should be noted that the approach to describing the electroweak structure of two-particle composite systems presented in this survey has the following characteristic features: the matrix element of the electroweak current of the composite system automatically satisfies the relativistic covariance conditions; the matrix element of the electromagnetic current satisfies the conservation law; the impulse approximation (IA) is formulated in a relativistically invariant way and in the case of the electromagnetic current and with account of the conservation law, the so called modified impulse approximation (MIA) is formulated.

2 Instant form of RQM

Relativistic invariance means that on the Hilbert state space of the system the unitary representation of the Poincaré group (or, more precisely, of the inhomogeneous group $SL(2,C)$ which is the universal covering of the Poincaré group) is realized [8].

Considering infinitesimal transformations and introducing the generators of translations $\hat{P}^{\mu}$ and space-time rotations $\hat{M}^{\mu\nu}$, we arrive at the Poincaré algebra in the common way. The construction of the representation of the Poincaré group in the Hilbert space is reduced to finding the generators $\hat{P}^{\mu}$, $\hat{M}^{\mu\nu}$ in terms of dynamic variables of the system. In the case of the system of noninteracting particles generators in the Poincaré algebra have the clear physical meaning, $\hat{P}^0 \equiv \hat{H}$ is the operator of total energy,$\hat{\vec{P}} = (\hat{P}^1, \hat{P}^2, \hat{P}^3)$ is the operator of total 3- momentum, $\hat{\vec{J}} = (\hat{M}^{23}, \hat{M}^{31}, \hat{M}^{12})$ is the operator of total angular momentum, and $\hat{\vec{N}} = (\hat{M}^{01}, \hat{M}^{02}, \hat{M}^{03})$ are the Lorentz boost generators. However, inclusion of the interaction between particles in this approach involves some problems; the essence of these problems can be illustrated by considering first the quantum nonrelativistic theory and its invariance group, the Galilean group. After the known transition to the central extension of the Galilean group to the covering group $SU(2)$ [9] we obtain the 11-parametric group with the set of generators: $\hat{H}$, $\hat{\vec{P}}$, $\hat{\vec{J}}$, $\hat{\vec{K}}$, $\hat{M}$, where $\hat{\vec{K}}$ are the generators of the Galilean boosts and $\hat{M}$ is the mass operator. The other generators coincide with the corresponding operators of the Poincaré group. The following nonzero commutation relations are contained in the Galilean algebra:

$$[\hat{J}_i, \hat{J}_j] = i\epsilon_{ijk}\,\hat{J}_k\,, \quad [\hat{J}_i, \hat{K}_j] = i\epsilon_{ijk}\,\hat{K}_k\,, \quad [\hat{J}_i, \hat{P}_j] = i\epsilon_{ijk}\,\hat{P}_k\,,$$

$$[\hat{K}_i, \hat{H}] = -i\,\hat{P}_i\,, \quad [\hat{K}_i, \hat{P}_k] = -i\delta_{ik}\,\hat{M}\,. \tag{1}$$

In nonrelativistic quantum mechanics the operator of interaction is added to the operator of total energy, $\hat{H} \quad \to \hat{H} + \hat{V}$. In order to preserve the Galilean invariance of the theory, i.e., to preserve algebra (1) under such re-definition of the operator of total energy, the following conditions should be imposed on the operator of interaction:

$$[\hat{\vec{P}}, \hat{V}] = [\hat{\vec{J}}, \hat{V}] = [\vec{\nabla}_P, \hat{V}] = [\hat{M}, \hat{V}] = 0\,. \tag{2}$$

Since the generator $\hat{H}$ is absent in the right-hand sides of relations (1), it is not necessary to include the interaction to other generators of the group in order to preserve the Galilean algebra.

The matter is different in the case of the Poincaré group. Let us consider one of the generators of the Poincaré algebra (see, e.g., [1]):

$$[\hat{P}^j \hat{N}^k] = i\,\delta^{jk}\,\hat{H}\,. \tag{3}$$

At inclusion of the interaction to the operator of total energy described above, the right-hand side of (3) depends on the interaction; therefore, either both generators in the left-hand side of (3), or one of them should depend on the interaction. Thus, in order to preserve commutation relations of the Poincaré algebra it is necessary to make other generators in this algebra dependent on the interaction. The generators of the algebra are separated into two types in this case: generators independent of interaction which form the so called kinematical subgroup and generators depending on interaction, Hamiltonians. The separation of the generators into kinematical generators and Hamiltonians is not unambiguous. Different methods for the separation of the kinematical subgroup result in different forms of dynamics. Further, the instant form of dynamics will be used in which the kinematical subgroup is comprised of the generators of the group of rotations and shifts of the Euclidean space, $\hat{\vec{J}}$, $\hat{\vec{P}}$, the other generators are the Hamiltonians, i.e., depend on the interaction, $\hat{P}^0$, $\hat{\vec{N}}$.

One of the technical methods for inclusion of interaction to the Poincaré algebra allowing to preserve commutation relations is the additive inclusion of interaction to the mass operator, the so called Bakamajian-Thomas procedure [10] (see also [1]):

$$\hat{M}_0 \to \hat{M}_I = \hat{M}_0 + \hat{V}\,. \tag{4}$$

Here, $\hat{M}_0$ is the operator of the invariant mass of the system without interaction, $\hat{M}_I$ is the mass operator of the system with interaction. In the instant form of dynamics the operator of interaction should satisfy the first three conditions in Eqs. (2). The first three equalities (2) provide the satisfaction of algebraic relations in the Poincaré algebre in the system with an interaction.

The wave function of the system of interacting particles in the RQM is determined as the eigen-function of the total commuting set of operators. In the instant form of dynamics this set consists of the operators

$$\hat{M}_I^2\ (\text{or } \hat{M}_I)\,, \quad \hat{J}^2\,, \quad \hat{J}_3\,, \quad \hat{\vec{P}}\,. \tag{5}$$

$\hat{J}^2$ is the operator of the total angular momentum squared. In the instant form of dynamics the operators $\hat{J}^2$, $\hat{J}_3$, $\hat{\vec{P}}$ coincide with the corresponding operators of the system without interaction. Thus, only the operator $\hat{M}_I^2$ ($\hat{M}_I$) depends on the interaction in (5).

It is possible to introduce a basis in which the motion of the center of mass of two particles is separated and three of four operators of set (5) are diagonal:

$$|\,\vec{P},\ \sqrt{s},\ J,\ l,\ S,\ m_J\,\rangle\,, \tag{6}$$

where $P_\mu = (p_1 + p_2)_\mu$, $P_\mu^2 = s$, $\sqrt{s}$ is the invariant mass of the system of two particles, l is the orbital momentum in the center-of-mass system, $\vec{S}^{\,2} = (\vec{S}_1 + \vec{S}_2)^2 = S(S+1)$, S is the total spin in the center-of-mass system, J is the total angular momentum, m_J is the projection of the total angular momentum.

Basis (6) diagonalizes the operators $\hat{J}^2$, $\hat{J}_3$, $\hat{\vec{P}}$ in (5). Thus, the problem of calculation of the wave function of the system is reduced to the diagonalization of the operator $\hat{M}_I^2$ (or $\hat{M}_I$).

3 Construction of the electroweak current matrix elements

Let us describe now a general method of construction (parameterization) of local operator matrix elements briefly (see e.g., [1, 11] for details).

The main idea of the parametrization can be formulated as follows. Using the variables entering the state vectors which define the matrix elements one has to construct two types of objects.

1. A set of linearly independent matrices which are Lorentz scalars (scalars or pseudoscalars). This set describes transition matrix elements non-diagonal in spin projections in the initial and finite states, as well as the properties defined by the discrete space–time transformations.

2. A set of linearly independent objects with the same tensor dimension as the operator under consideration (for example, four–vector, or four–tensor of some rank). This set describes the matrix element behavior under the action of Lorentz group transformations.

The operator matrix element is written as a sum of all possible objects of the first type multiplied by all possible objects of the second type. The coefficients in this representation as a sum are just the reduced matrix elements – form factors.

To demonstrate this let us consider the parameterization of the matrix elements taken between the states of a free particle of mass M in different simple cases. Let us normalize the state vectors as follows:

$$\langle \vec{p}\,, m\,|\,\vec{p}\,'\,, m'\rangle = 2p_0\,\delta(\vec{p} - \vec{p}\,')\,\delta_{mm'}\,, \quad p_0 = \sqrt{M^2 + \vec{p}^{\,2}}\,, \tag{7}$$

here $\vec{p}, \vec{p}\,'$ are three–momenta of particle, m, m' are spin projections.

Let us consider now the 4-vector operator of electromagnetic current $j_\mu(0)$. To parameterize the matrix element one needs a set of quantities of the appropriate tensor dimension. Using the variables entering the particle state vectors one can construct one pseudovector of the covariant spin operator $\Gamma^\mu(p')$ (see, e.g., [1]):

$$\Gamma_0(p) = (\vec{p}\vec{j})\,, \quad \vec{\Gamma}(p) = M\,\vec{j} + \frac{\vec{p}(\vec{p}\vec{j})}{p_0 + M} \tag{8}$$

and three independent vectors:

$$K_\mu = (p - p')_\mu = q_\mu\,, \quad K'_\mu = (p + p')_\mu\,, \quad R_\mu = \epsilon_{\mu\nu\lambda\rho}\,p^\nu\,p'^{\,\lambda}\,\Gamma^\rho(p')\,. \tag{9}$$

Here $\epsilon_{\mu\nu\lambda\rho}$ is a completely anti-symmetric pseudo-tensor in four dimensional space-time with $\epsilon_{0123} = -1$.

The set in question of linearly independent matrices in spin projections of the initial and the final states giving the set of independent Lorentz scalars is presented by $2j + 1$ quantities:

$$D^j(p,\, p')\,(p_\mu\Gamma^\mu(p'))^n\,, \quad n = 0, 1, \ldots, 2j\,. \tag{10}$$

The operator matrix element contains the matrix elements of the listed quantities multiplied by Wigner's rotation matrix $D^j(p,\, p')$ from the left. Each of such products is to be multiplied by the sum of linearly independent scalars (10):

$$\langle\,\vec{p},\, M,\, j,\, m\,|\,j_\mu(0)|\,\vec{p}\,',\, M,\, j,\, m'\,\rangle = \sum_{m''}\langle\, m|D^j(p,\, p')|m''\rangle\langle\, m''|\,F_1\,K'_\mu + F_2\,\Gamma_\mu(p') +$$

$$F_3\,R_\mu + F_4\,K_\mu|m'\rangle\,, \tag{11}$$

where

$$F_i = \sum_{n=0}^{2j} f_{in}(Q^2)(ip_\mu\Gamma^\mu(p'))^n\,. \tag{12}$$

Let us impose some additional conditions on the operator: self-adjointness, orthogonality of the vectors in the parametrization (11), parity conservation and the conservation condition: $q_\mu\, j_\mu(0) = 0$.

So, finally we have the parametrization of the matrix element which has following form for particle with spin 1/2:

$$\langle\, \vec{p},\, M,\, 1/2,\, m\,|\, j_\mu(0)\,|\, \vec{p}\,',\, M,\, 1/2,\, m'\,\rangle =$$

$$= \sum_{m''} \langle\, m|D^{1/2}(p,\, p')|m''\rangle\langle\, m''\,|\, f_{10}(Q^2)\, K'_\mu + i f_{30}(Q^2)\, R_\mu|m'\rangle\;, \tag{13}$$

The form factors $f_{10}(Q^2)$ and $f_{30}(Q^2)$ are the electric and the magnetic form factors of the particle, respectively. These form factors are connected with Sachs form factors $G_E(Q^2)$ and $G_M(Q^2)$.

The developed procedure can be applied to the construction of the electromagnetic current two-quark system. The following integral representation for the pion form factor in the MIA (see, e.g., [1, 12, 13]) holds:

$$F_\pi(Q^2) = \int d\sqrt{s}\, d\sqrt{s'}\, \varphi(k)\, g_0(s, Q^2, s')\, \varphi(k')\;. \tag{14}$$

Here $\varphi(k)$ is pion wave function in the sense of RQM, $g_0(s, Q^2, s')$ is the free two-particle form factor describing the electromagnetic properties of the two noninteracting quarks without interaction with the quantum numbers of the pion.

The wave function in (14) has the following structure:

$$\varphi(k) = \sqrt[4]{s}\, u(k)k\;, \quad s = 4(k^2 + M^2)\;. \tag{15}$$

M is mass of constituent quarks. Below for the function $u(k)$ we use some phenomenological wave functions.

The function $g_0(s, Q^2, s')$ is written in terms of the quark electromagnetic form factors in the form

$$g_0(s, Q^2, s') = A_1(s, Q^2, s')\left[G_E^q(Q^2) + G_E^{\bar{q}}(Q^2)\right] + A_2(s, Q^2, s')\left[G_M^q(Q^2) + G_M^{\bar{q}}(Q^2)\right]\;. \tag{16}$$

Functions A_1 , A_2 may be obtained explicitly by the methods of relativistic kinematics (see, e.g.,[12]).

Expressions similar to (14) take a place for charge $G_C(Q^2)$, quadrupole $G_Q(Q^2)$ and magnetic $G_M(Q^2)$ form factors of the ρ–meson. The explicit form of corresponding free two-particle form factors with the quantum numbers of the ρ-meson is cumbersome; it can be found in Ref. [14].

The method of the parameterization of the current matrix elements has been generalized in the paper [15] to the case of the nondiagonal in the total angular momentum matrix elements. In particular the expression for the lepton decay constant of the ρ-meson was obtained in the 4-fermion approximation:

$$f_\rho = \frac{\sqrt{3}}{\sqrt{2}\pi}\int_0^\infty dk\, k^2\, u(k)\frac{(\sqrt{k^2+M^2}+M)}{(k^2+M^2)^{3/4}}\left(1+\frac{k^2}{3(\sqrt{k^2+M^2}+M)^2}\right)\;. \tag{17}$$

4 Parameters of the model

For the calculation of the π- and ρ-meson characteristics basing on the relations (14)–(17) we use the following model wave functions (see, e.g., [16]).

1. The Gaussian or harmonic oscillator wave function

$$u(k) = N_{HO}\exp\left(-k^2/2\, b^2_{\pi,\rho}\right). \tag{18}$$

2. The power-law wave function:

$$u(k) = N_{PL}\,(k^2/b^2_{\pi,\rho} + 1)^{-n}\,, \quad n = 2, 3\,. \tag{19}$$

In eqs. (18), (19) $b_{\pi,\rho}$ are parameters of wave functions for pion and ρ-meson, respectively.

The electromagnetic form factors of constituent quarks in (16) are taken in the form [1, 17]:

$$G^q_E(Q^2) = e_q\, f_q(Q^2)\,, \quad G^q_M(Q^2) = (e_q + \kappa_q)\, f_q(Q^2)\,, \tag{20}$$

where e_q is the quark charge and κ_q is the quark anomalous magnetic moment,

$$f_q(Q^2) = \frac{1}{1 + \ln(1 + \langle r^2_q\rangle Q^2/6)}\,, \tag{21}$$

$\langle r^2_q\rangle$ is the MSR of the constituent quark. Values of all parameters used in these expressions are taken from the π-meson calculation, see, e.g., Ref. [18].

So, the following parameters enter our calculations:

1) the parameters that describe the constituent quarks *per se* (the quark mass M, the anomalous magnetic moments of the quarks κ_q, that enter our formulae through the sum $s_q = \kappa_u + \kappa_{\bar{d}}$, and the quark mean square radius (MSR) $\langle r^2_q\rangle$);

2) the parameters $b_{\pi,\rho}$ that enter the quark wave functions (18), (19) and is determined by the quark interaction potential.

In the paper [12] on pion we have shown that in our approach all the parameters of the first group are the functions of the quark mass M and are defined by its value. In particular, for the quark MSR we can use the relation (see, also [19]):

$$\langle r^2_q\rangle \simeq 0.3/M^2\,. \tag{22}$$

To calculate electroweak properties of the ρ meson, we use the same values of quark parameters from the first group as that we have used for the pion [12]. So, the wave function parameters $b_{\pi,\rho}$ are the only free parameters in our calculations.

5 Asymptotics of the pion form factor at high momentum transfers

It is worth to consider the pion form factor asymptotic behavior (14) at $Q^2 \to \infty$ especially. In our paper [17] it was shown that in our approach, the pion form-factor asymptotics at

$$Q^2 \to \infty\,, \quad M(Q^2) \to 0 \tag{23}$$

does not depend on the choice of a wave function but is defined by the relativistic kinematics only. We consider the fact that the asymptotics obtained in our nonperturbative approach does coincide with that predicted by QCD as a very significant one. Our approach occurs to be consistent with the asymptotic freedom, and this feature surely distinguishes it from other nonperturbative approaches.

Let us note that it is obvious that at very high momentum transfers the quark mass decreases as it goes to zero at the infinity. Our approach permits to take into account the dependence $M(Q^2)$ beginning from the range where this becomes necessary to correspond to experimental data. It is possible that this will take place at the values of Q^2 lower than 6 GeV^2. So, it is obtained in Ref. [17] that

$$F_\pi(Q^2) \sim Q^{-2}\,. \tag{24}$$

DOI: 10.1051/epjconf/201612504014

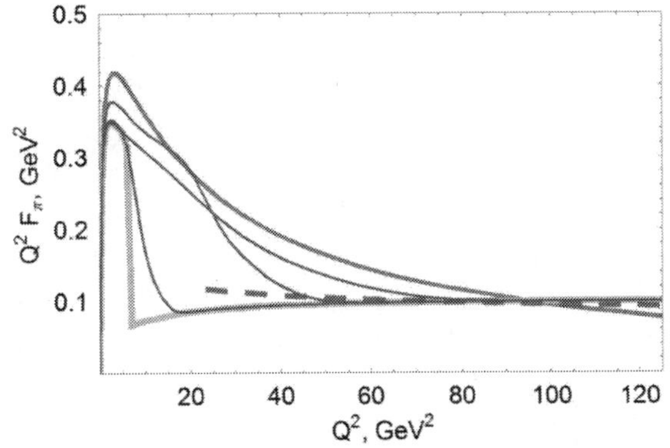

Figure 1. Examples of the allowed solutions for $F_\pi(Q^2)$ (thin lines) demonstrating how the QCD asymptotics, Eq.(25) (dashed line) settles down. The thick gray line bounds from below the range of all solutions allowed by the experimental constraints. The thick full (red) line represents the solution with M =const, Refs. [12, 26].

The asymptotics of the pion electromagnetic form factor F_π at momenta transfer $Q^2 \to \infty$ has been determined [20–22], in the QCD frameworks, as

$$Q^2 F_\pi(Q^2) \to 8\pi\alpha_s^{1\text{-loop}}(Q^2) f_\pi^2, \tag{25}$$

where $\alpha_s^{1\text{-loop}}(Q^2) = 4\pi / \left(\beta_0 \log\left(Q^2/\Lambda^2_{\text{QCD}}\right)\right)$ is the one-loop running strong coupling constant, $\beta_0 = 11 - 2N_f/3$ is the first beta-function coefficient, N_f is the number of active quark flavours and $f_\pi \approx 130$ MeV is the pion decay constant. It is important to note that this asymptotical behaviour, consistent with the quark counting rules [23, 24], includes the one-loop coupling only and is to be modified whenever the one-loop approximation fails, but not by means of a simple replacing of α_s with its more precise value. Involved QCD calculations have been performed to obtain corrections to Eq. (25), see e.g. [25]. The QCD does not predict the value of Q^2 at which this asymptotics should be reached.

The results of the calculation of F_π from Ref. [18] are presented in Fig. 1. As can be seen the our asymptotic coincides with QCD prediction at different scenarios for the zero limit of constituent mass (23).

6 The pion form factor in the region of the JLab experiments

The results of the calculation of the charge pion form factor using the wave functions (18), (19) and the value of constituent-quark mass $M = 0.22$ GeV (this parameter has been fixed as early as in 1998 [12] from the data at $Q^2 \leq 0.26$ (GeV)^2 [27]) are shown in Fig. 2. Let us note that our RQM describes well the experimental data for the pion form factor including the recent points [3]. The upper of our curves corresponds to the model (18), the lower - to the models (19) with $n = 3$.

Let us emphasize that the parameters used in our calculations were obtained from the fitting to the experimental data up to $Q^2 \simeq 0.26$ GeV^2 [27]. At that time the data for higher Q^2 was not correlated in different experiments and had significant uncertainties. The later data for pion form factor in JLab experiments up to Q^2 =2.45 GeV^2 were obtained with rather good accuracy. All experimental points obtained in JLab up to now agree very well with our prediction of 1998.

So, our way of fixing the model parameters constrains effectively the behavior of wave functions both at small and at large relative momenta. The structure of our relativistic integral representation

DOI: 10.1051/epjconf/201612504014

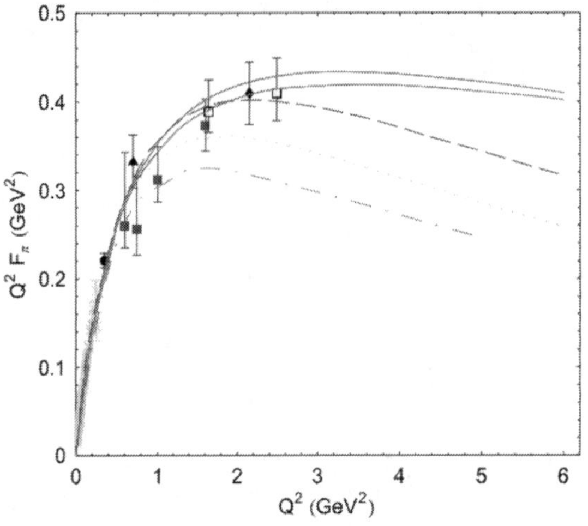

Figure 2. Our predictions for the pion form factor given in 1998 [12] (full red lines; upper line: wave functions (18), lower line: wave functions (19) with $n = 3$) compared with data and with some other models. Experimental data are from Refs. [2, 3, 27–29]. Other theoretical curves are: those by C.-W. Hwang [30] (blue dashed); Cardarelli *et al.* [19] and (precisely coinciding with it) instant-form predictions [31] (magenta dash-dotted); Ref. [32] (green dotted).

(14) is so, that the form–factor behavior in the region of small momentum transfers is determined by the wave function at small relative momenta, and the behavior of the form factor in the region of high momentum transfer — by the wave function at large relative momenta. The constraints for the wave functions provide the limitations for the form factor, and this is seen in the results of the calculation.

7 Lepton decay constant and MSR of the ρ-meson

Let us describe the procedure of calculation of the ρ- meson MSR in detail, starting from the quark parameter, $M = 0.22$ GeV, used in a successful calculation of the pion parameters [12]. As it has been demonstrated in Ref. [12], the actual choice of the wave-function form does not affect the result provided the quark parameters are fixed. In what follows, we illustrate the procedure with the wave function (19) with $n = 3$.

In the lower panel of Fig. 3, the ρ-meson lepton decay constant f_ρ as a function of the only free parameter of the model, b_ρ, is presented. The interval on the vertical axis representing the experimental values of f_ρ, that is $f_\rho^{\exp} = (152 \pm 8)$ MeV [4, 5], is shown. It corresponds to the interval of the values of b_ρ which give, through our calculation, the correct experimental values of the decay constant. This interval, $b_\rho = (0.385 \pm 0.019)$ GeV, is shown on the horizontal axis of Fig. 3.

The calculated MSR of the ρ meson is presented in the upper panel of Fig. 3. The interval of admissible values of b_ρ gives now the corresponding interval of MSR predicted in the present study, $\langle r_\rho^2 \rangle = (0.56 \pm 0.04)$ fm^2.

Table 1 presents a comparison of our results with the results of calculations of electroweak properties of the ρ meson in other approaches. The values of $\langle r_\rho^2 \rangle$, while not measured directly, are important for testing various conjectures about strongly interacting systems. One of the interesting related prediction was introduced as a consequence of the so-called Wu–Yang hypothesis [38] (see also

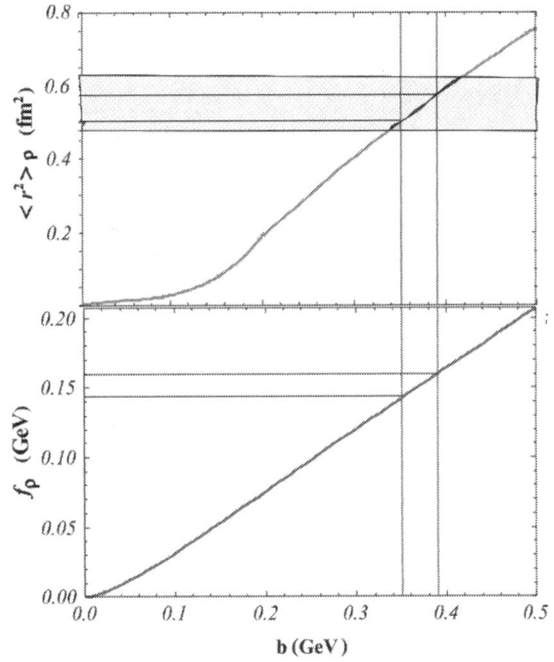

Figure 3. The decay constant f_ρ and the ρ-meson MSR, $\langle r_\rho^2 \rangle$, as functions of the only free parameter of the model, b_ρ. The experimental data fix the value of f_ρ, as shown in the lower panel. This fixes the value of b_ρ, which determines in turn the value of $\langle r_\rho^2 \rangle$, as shown in the upper panel. The values of quark parameters are taken from Ref. [12]; the wave function (19) with $n = 3$ is used.

Table 1. The lepton decay constant f_ρ and MSR of the ρ meson calculated within different approaches.

Model	f_ρ, MeV	$\langle r_\rho^2 \rangle$, fm^2
This work	152±8 (fixed)	0.56±0.04
[4]	154	0.268
[33]	146	0.54
[34]	130	0.312
[35]	—	0.67
[36]	—	0.67
[37]	134	0.296

Ref. [39]), though it is remarkable by itself. Namely, one may define the radius of a hadron either in terms of the electroweak interaction (the mean square charge radius, $\langle r_{\rm ch}^2 \rangle$, calculated for the ρ meson in this paper) or in terms of the strong interaction (this radius, $\langle r_{\rm st}^2 \rangle$, is defined by the slope of the cross section of hadron–proton scattering). The conjecture [39], which may be derived from, though not necessary implies, the hypothesis of Ref. [38], is the equality of the two radii,

$$\langle r_{\rm st}^2 \rangle = \langle r_{\rm ch}^2 \rangle \,. \tag{26}$$

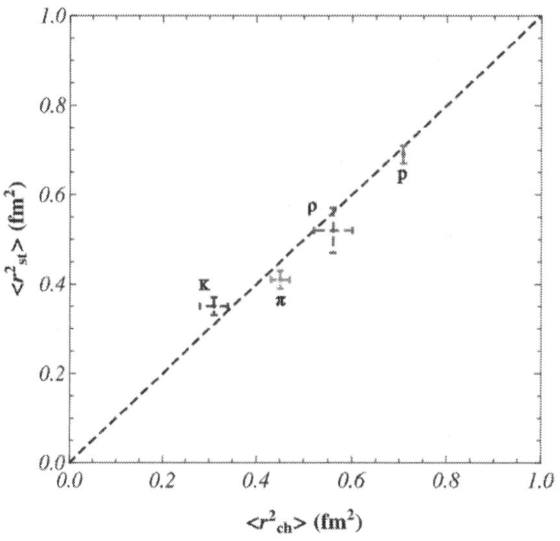

Figure 4. Relation between the strong-interaction hadronic radius $\langle r^2_{\rm st}\rangle$ and the charge radius $\langle r^2_{\rm ch}\rangle$ for light hadrons.

This remarkable equality has been verified experimentally with a great degree of accuracy for the proton, π- and K- mesons.

Even more demonstrative is Fig. 4, analogous to a figure from the paper [39], but presenting more recent data. We can see that the value of the ρ-meson charge radius obtained in this paper fits perfectly the conjecture (26).

8 Analytic continuation of the pion form factor to the complex plain of momentum transfers

As mentioned above, one unsolved theoretical problem of RQM is its relations to the fundamental QFT principles. This paper partially addresses this issue. We compare the corollaries of the QFT principles with the model-independent corollaries of the RQM axioms. In Ref. [40] we compared the predictions for analytic properties of the pion form factor in the complex plane of transferred momenta $t = -Q^2$ that follow from QFT general principles with those obtained by the analytic continuation of the form factor integral representation (14) in the spacelike domain derived in the framework of the instant form of RQM [13]. It follows from the QFT microcausality condition that the pion form factor is an analytic function in the complex plane of the parameter t with a cut running from $4m_\pi^2$ to ∞, where m_π is the pion mass (see, e.g., [41] and the references therein). The obtained in our approach analytic properties of the form factor differ from those obtained in the QFT approach only by the position of the branch point on the real axis: it is $4M^2$ in our case. The main experimentally observed qualitative feature of the form factor behavior in the complex plane is the presence of two resonances, which correspond to ρ and ω mesons and are located on the positive part of the real axis (in the timelike domain). We can demonstrate that wave functions in the momentum representation without poles in the momentum complex plane do not yield the resonance behavior of the pion form factor for timelike transferred momenta. In particular, this is the case for a wave function of the Gaussian type (18) widely used in composite models, and this result holds independently of the model parameter values. We therefore encounter the problem of finding a wave function that produces resonances and establishing the relation between the locus of the resonance in the pion form factor and the locus of the wave function poles. In solving this problem, the main idea is to introduce wave functions with poles such that poles of form factor (14) appear on the nonphysical sheet.

So, the wave functions $u(k)$ (15) resulting in the resonance behavior of the pion form factor for timelike transferred momenta must have a pole at the point k_s:

$$k_s^2 = \frac{M}{4}(z_s - 2M)\,,\ z_s = z_s'' + i\sqrt{t_r - 4M^2}\,, \tag{27}$$

where t_r is resonance location, $z_s'' < 0$.

Let the function $u(k)$ (15) be even and satisfies condition of reality ($\mathrm{Im}\,u(k) = 0$) at real k also. The simplest rational function satisfying all the above conditions is

$$u_M(k) = N\left[(k^2 - k_s{}^2)(k^2 - (k_s{}^*)^2)\right]^{-n}\,. \tag{28}$$

The obtained analytic continuation with wave functions of form (28) therefore contains the following parameters: t_r and z'' from (27), the constituent mass M, and the exponent n of the wave function. We fix these data by fitting experimental data for the parameters of the ρ resonance in the pion form factor.

To describe the experimental data pertaining to the pion form factor, we use the fitting by the celebrated Breit-Wigner formula (see [42]):

$$F_{\pi BW}(t) = \frac{4\kappa^2}{4\kappa^2 - t - 2i\kappa\Gamma}\,, \tag{29}$$

where $\kappa = 0.375$ GeV, $\Gamma = 0.1$ GeV. (we take the parameters from [42]).

We numerically tuned the parameters of our model such that the absolute value of the form factor best agrees with the Breit-Wigner approximation of the experimental data, which gives

$$t_r = 0.57\,\mathrm{GeV}^2\,,\quad z_s'' = -0.01\,,\quad M = 0.32\,\mathrm{GeV}\,,\quad n = 0.86\,. \tag{30}$$

We note that the parameter t_r is the resonance mass, and its value is very close to the value $0.5625\,\mathrm{GeV}^2$ provided by formula (29); the value of the constituent quark mass is typical for different formulations of the composite quark model in which constituent masses are in the interval $(0.2 - 0.33)$ GeV (see, e.g., [12, 19]). The parameter z_s'' determining the position of the form factor singularity on the nonphysical sheet and the wave function parameter n are purely model parameters determined by the tuning procedure.

The results of numerically integrating expression (14) with wave function (28) are shown in Fig. 5. We can see that our analytic continuation of pion form factor (14) provides a good qualitative description of the form factor behavior in the timelike domain already for the simplest wave function choice given by (28).

We can demonstrate the correctness of our constructed analytic continuation as follows. We verify how well the constructed form factor satisfies the known dispersion relation in the spacelike domain

$$F_\pi(t) = \frac{1}{\pi}\int\limits_{4M^2}^{\infty}\frac{Im\,[F_\pi(t')]}{t' - t}dt'\,, \tag{31}$$

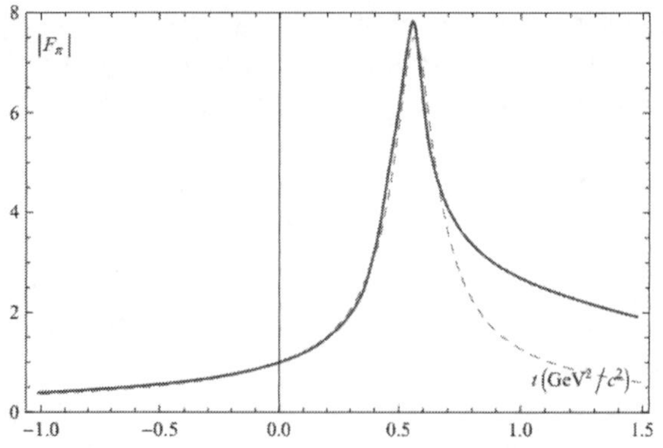

Figure 5. The function $|F_\pi(t)|$ from: the solid curve is expression (14), and the dashed curve is the Breit-Wigner approximation.

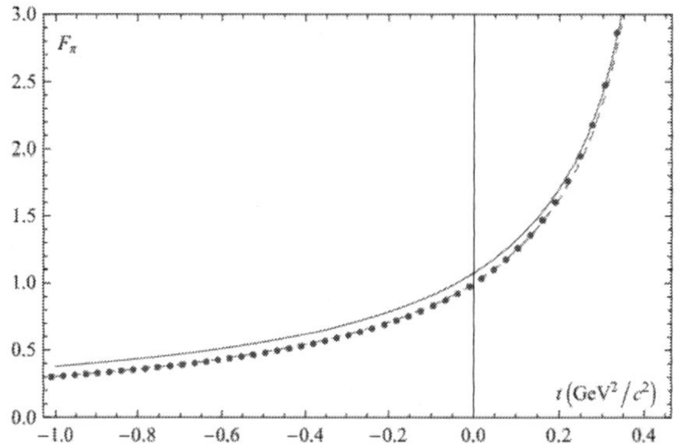

Figure 6. Comparing the calculated pion form factor with the known dispersion relations: points correspond to a straightforward calculation with formula (14), the solid line is obtained from dispersion relation (31), and the dashed line is obtained from dispersion relation (32). The results of calculating by formula (14) practically coincide with those obtained using dispersion relation (32).

and also the dispersion relation with one subtraction (with the weakened condition of decreasing at infinity)

$$F_\pi(t) = 1 + \frac{t}{\pi} \int\limits_{4M^2}^{\infty} \frac{Im\,[F_\pi(t')]\,dt'}{t'(t'-t)}\,. \qquad (32)$$

In Fig. 6, we compare our result with formulas (31) and (32). We use parameter values (30). It can be seen that the constructed pion form factor in the complex plane satisfies the dispersion relations with good accuracy.

The problem of a more detailed description of the character of the resonance behavior of the pion form factor in the framework of our approach is related to the problem of choosing a wave function of interacting quarks that is more realistic than (28).

9 Conclusions

The approach developed in IF RQM has following main features: predictivity: parameters of model are fixed from experimental data at small momentum transfers. The experimental data for the pion form factor at large momentum transfers obtained later are described without tuning of parameters; robustness: the behavior of the pion and ρ-meson electromagnetic form factors does not depend on the choice of wave functions and are determined by the mass of the constituent quarks; the approach gives the asymptotics that agrees with QCD asymptotic behavior at large momentum transfers; the approach gives the self-consistent description of the electroweak properties of the pion and ρ-meson; the approach gives the right analytical properties of the pion form factor in the complex plane of momentum transfers.

One of the authors (VT) thanks S.Troitsky for interesting discussions. Authors (AK and RP) thank the Organizing committee for his kind invitation to the XIX International Seminar on High Energy Physics "QUARKS-2016" and hospitality. This work was supported in part (AK and RP) by the Ministry of Education and Science of the Russian Federation (grant No. 1394, state task).

References

[1] A.F. Krutov and V.E. Troitsky, Phys. Part. Nucl. **40**, 136 (2009).
[2] H.P. Blok *et.al.* The Jefferson Lab F_π Collaboration, Phys.Rev.C. **78**, 045202 (2008).
[3] G.M. Hubert *et.al.* The Jefferson Lab F_π Collaboration, Phys.Rev.C. **78**, 045203 (2008).
[4] C.S. Mello, A.N. da Silva, J.P.B.C. de Melo, T. Frederico, Few-Body Syst,**56**, 509 (2015).
[5] K.A. Olive *et.al.* (Particle Data Group), Chin. Phys. C **38**, 090001 (2014).
[6] B. Aubert *et.al.* (Babar Collaboration), Phys.Rev.D **80**, 052002 (2009).
[7] P.A.M. Dirac, Rev.Mod.Phys. **21**, 392 (1949).
[8] Yu.V. Novozhilov, *Introduction to Elementary Particle Theory* (Nauka, Moskow, 1972, Pergamon Press, Oxford, 1975).
[9] V.I. Fushchich and A.G. Nikitin, *Symmetry of Equations of Quantum Mechanics* (Nauka, Moskow, 1990).
[10] B. Bakamjian, L.H. Thomas, Phys.Rev. **92**, 1300 (1953).
[11] A.F.Krutov, V.E.Troitsky, Theor. Math. Phys. **143**, 704 (2005).
[12] A.F. Krutov and V.E. Troitsky, Eur. Phys. J. C **20**, 71 (2001); [hep-ph/9811318].
[13] A.F. Krutov, V.E. Troitsky, Phys.Rev.C, **65**, 045501 (2002); [hep-ph/0204053].
[14] A.F. Krutov, and V.E. Troitsky, arXiv: hep-ph/0210046v1 (Appendix).
[15] A.F. Krutov, R.G. Polezhaev, and V.E. Troitsky, Phys.Rev. D **93**, 036007 (2016).
[16] F. Coester and W.N. Polyzou, Phys. Rev. C **71**, 028202 (2005).
[17] A.F. Krutov and V.E. Troitsky, Theor. Math. Phys. **116**, 907 (1998).
[18] S.V. Troitsky and V.E. Troitsky, Phys. Rev. D **88**, 093005 (2013).
[19] F. Cardarelli *et al.*, Phys. Lett. B **332**, 1 (1994); F. Cardarelli, I.L. Grach, I.M. Narodetskii, E. Pace, G. Salmé, and S. Simula, Phys. Rev.D **53**, 6682 (1996).
[20] G. R. Farrar and D. R. Jackson, Phys. Rev. Lett. **43**, 246 (1979).
[21] A. V. Efremov and A. V. Radyushkin, Phys. Lett. B **94**, 245 (1980).
[22] G. P. Lepage and S. J. Brodsky, Phys. Lett. B **87**, 359 (1979).
[23] V. A. Matveev, R. M. Muradyan and A. N. Tavkhelidze, Lett. Nuovo Cim. **5**, 907 (1972).
[24] S. J. Brodsky and G. R. Farrar, Phys. Rev. Lett. **31**, 1153 (1973).

[25] A. P. Bakulev, A. V. Pimikov and N. G. Stefanis, Mod. Phys. Lett. A **24**, 2848 (2009). [arXiv:0910.3077 [hep-ph]].
[26] A. F. Krutov, V. E. Troitsky and N. A. Tsirova, Phys. Rev. C **80** 055210 (2009) [arXiv:0910.3604 [nucl-th]].
[27] S.R. Amendolia *et al.*, Phys. Lett.B **146**, 116 (1984);
S.R. Amendolia *et al.*, Nucl. Phys. B **277**, 168 (1986).
[28] H. Ackermann *et al.*, Nucl. Phys. B **137**, 294 (1978).
[29] P. Brauel *et al.*, Z. Phys. C **3**, 101 (1979).
[30] C.-W. Hwang, Phys. Rev. D **64**, 034011 (2001).
[31] Jun He, B. Juliá-Díaz, and Yu-bing Dong, Phys.Lett.B **602**, 212 (2004).
[32] H.-M. Choi and C.-R. Ji, Phys. Rev. D **59**, 074015 (1999).
[33] M.S. Bhagwat and P. Maris, Phys. Rev. C **77**, 025203 (2008).
[34] H.L.L. Roberts, A. Bashir, L.X. Gutierrez-Guerrero, C.D. Roberts and D.J. Wilson, Phys. Rev. C **83**, 065206 (2011).
[35] M. E. Carrillo-Serrano, W. Bentz, I. C. Cloët, and A. W. Thomas, Phys. Rev. C **92**, 015212 (2015).
[36] B. Owen, W. Kamel, D. Leinweber, B. Menadue, S. Mahbub, Phys. Rev. D **91**, 074503 (2015).
[37] B.L.G. Bakker, H.M. Choi and C.R. Ji, Phys. Rev. D **65**, 116001 (2002).
[38] T.T. Wu and C.N. Yang, Phys. Rev. B **137**, 708 (1965).
[39] B. Povh, J. Hüfner, Phys. Lett. B **245**, 653 (1990).
[40] A.F. Krutov, M.A. Nefedov and V.E. Troitsky, Theor. Math. Phys. **174**, 331 (2013).
[41] M. Belička, S. Dubnička, A.Z. Dubničková, A. Liptaj, Phys. Rev. C **83**, 028201 (2011).
[42] S.J. Brodsky, G.F. de Teramond, Phys. Rev. D **77**, 056007 (2008).

Colour Information Loss due to Decoherence in Vacuum

Viatcheslav Kuvshinov[1,⋆] and *Eugene* Bagashov[1]

[1] *Joint Institute for Power and Nuclear Research - Sosny of National Academy of Sciences of Belarus, Minsk, P/O box 119*

Abstract. The model of stochastic vacuum of quantum chromodynamics might be used for an alternative treatment of subhadronic processes. The possibility of considering the stochastic vacuum as an environment for colour charges is discussed. It is shown that the interaction of colour particles with the stochastic vacuum leads to the decoherence process and the loss of information on the initial quantum state of the particle, which gives a new perspective regarding the confinement of quarks phenomenon.
The mentioned effect is demonstrated in cases of a single particle and a superposition of colour charges. Generalization on the multiparticle case is proposed.
Quantum characteristics (purity and von Neumann entropy) are used to analyse the process of interaction. On the base of von Neumann entropy the measure of quantum information is proposed.

1 Introduction

Interactions of some quantum system with the environment can be effectively represented by additional stochastic terms in the Hamiltonian of the system. The density matrix of the system is then obtained by averaging with respect to degrees of freedom of the environment. Interactions with the environment result in decoherence and loss of quantum superpositions. Information on the initial state of the quantum system is lost after sufficiently large time.

Quantum decoherence is the loss of coherence or ordering of the phase angles between the components of a system in a quantum superposition. Decoherence occurs when a system interacts with its environment in a thermodynamically irreversible way. This process can be viewed as the loss of information on a system. Dissipation is a decohering process by which quantum states are changed due to entanglement with an external thermal bath.

2 Stochastic QCD Vacuum

The model of stochastic vacuum of quantum chromodynamics (QCD) is one of the popular phenomenological models which explains quark confinement (Wilson loop decreasing), string tensions and field congurations around static colour charges.

In this model only the second field correlators are important and the others are negligible [1] (Gauss domination). It has been confirmed by lattice calculation [2]. The most important evidence for this is Casimir scaling.

⋆e-mail: kuvshinov2003@gmail.com

The stochastic vacuum approach is based on the assumption that one can calculate vacuum expectation values of gauge invariant quantities as expectation values with respect to some well-behaved stochastic gauge field. It is known that such vacuum provides confining properties, giving rise to QCD strings with constant tension at large distances [1, 3, 4].

2.1 QCD Vacuum as Environment for Colour Particles

We consider QCD stochastic vacuum as the environment for colour quantum particles. To get the density matrices of those particles we average over the external QCD stochastic vacuum implementations.

Thus, instead of considering nonperturbative dynamics of Yang-Mills fields one introduces external environment (in our case represented by QCD stochastic vacuum) and average over its implementations. As a consequence, we obtain decoherence, loss of quantum superpositions and information, and confinement of colour states.

White objects can be obtained as colourless mixtures of states described by the diagonal density matrix as a result of evolution in the QCD stochastic vacuum.

3 Decoherence of Colour States

Consider propagation of heavy spinless colour particle along some fixed path γ. The amplitude of such process is obtained by parallel transport [5–7]

$$\partial_\mu|\phi\rangle = i\hat{A}_\mu|\phi\rangle, \tag{1}$$

$$|\phi(\gamma)\rangle = \mathcal{P}\exp\left(i\int\limits_\gamma dx^\mu \hat{A}_\mu\right)|\phi_{in}\rangle, \tag{2}$$

where $\mathcal{P}$ is the path ordering operator and $\hat{A}_\mu$ is the gauge field vector. In order to consider mixed states we introduce the colour density matrix, taking into account both colour particle and QCD stochastic vacuum (environment):

$$\hat{\rho}(loop,\gamma\bar{\gamma}) =< |\phi(\gamma)\rangle\langle\phi(\gamma)| >, \tag{3}$$

Here we average over all implementations of stochastic gauge field (environment degrees of freedom) - and decoherence arises due to interaction with environment.

In the model of QCD stochastic vacuum only expectation values of path ordered exponents over closed paths are defined (in order to keep the gauge invariance). Closed path corresponds to a process in which the particle-antiparticle pair is created, propagates and finally annihilates.

With the help of (1), (2), (3) we can obtain the expression for density matrix [5, 7]:

$$\hat{\rho}(loop,\gamma\bar{\gamma}) = N_c^{-1} + (|\phi_{in}\rangle\langle\phi_{in}| - N_c^{-1})W_{adj}(\gamma\bar{\gamma}). \tag{4}$$

Here N_c is the number of colours.

Colour density matrix in colour neutral stochastic vacuum can be decomposed into the pieces that transform under trivial and adjoint representations [5, 7]

$$\hat{\rho}_1 = N_c^{-1}\hat{I} + \rho_1^a\hat{T}_a, \tag{5}$$

and Wilson loop in fundamental representation is [1]

$$W_{fund}(loop,\gamma\bar{\gamma}) = \langle \mathrm{Tr}\,\mathcal{P}\exp\left(\int_{loop,\gamma\bar{\gamma}} i\hat{A}_\mu dx^\mu\right)\rangle. \tag{6}$$

In confinement region Wilson loop decays exponentially with the area spanned by loop, so for the rectangular loop spanning the time interval T and distance R we get:

$$\hat{\rho}(loop, \gamma\bar{\gamma}) = \hat{I}N_c^{-1} + (\hat{\rho}_{in} - \hat{I}N_c^{-1})\exp(-\sigma_{adj}RT), \tag{7}$$

$$\hat{\rho}(\gamma : RT \to \infty) = N_c^{-1}\hat{I}, \tag{8}$$

where $\hat{I}$ is the unit operator, $\sigma_{adj} = \sigma_{fund}G_{adj}G_{fund}^{-1}$ is the string tension in the adjoint representation, G_{adj}, G_{fund} are the eigenvalues of quadratic Casimir operators. Under Gaussian dominance string tension is

$$\sigma_{fund} = \frac{g^2}{2}l_{corr}^2F^2, \tag{9}$$

g is coupling constant, l_{corr} correlation length in the QCD stochastic vacuum, F - average of the second cumulant of curvature tensor [2, 4].

As follows from (8) all colour states are mixed with equal probabilities and all information on initial color state is lost.

3.1 Decoherence Characteristics

The decoherence rate of transition from pure colour states to white mixture state can be estimated on the base of purity [8]:

$$P = \mathrm{Tr}\,\hat{\rho}^2. \tag{10}$$

This characteristic represents the closeness of a quantum state to a pure one. In our case,

$$P = N_c^{-1} + (1 - N_c^{-1})\exp\left(-2\sigma_{fund}G_{adj}G_{fund}^{-1}RT\right). \tag{11}$$

When RT tends to 0, $P \to 1$, which corresponds to a pure state. When the composition RT tends to infinity, the purity tends to N_c^{-1}, which corresponds to the white mixture [7].

The rate of purity decrease is

$$T_{dec}^{-1} = -2\sigma_{fund}G_{adj}G_{fund}^{-1}R. \tag{12}$$

Left side of the equation is the inverse characteristic time of decoherence proportional to QCD string tension and distance R. It can be inferred from (4) and (11) that the larger is the distance between particle and antiparticle, the quicker the initial state tends to white mixture as a result of interaction with the QCD stochastic vacuum. Thus white states can be obtained as a result of decoherence process. The information of quark colour states is lost in hadrons due to interactions between quarks and confining non-Abelian gauge fields.

Von Neumann entropy can be used as a measure of the loss of information:

$$S = -\mathrm{Tr}\,(\hat{\rho}\ln\hat{\rho}). \tag{13}$$

Initial density matrix gives us $S = 0$ and we will have $S = \ln N_c$ for large RT.

In order to obtain a more suitable characteristic to work with, we might want to define the information on the basis of the given entropy parameter. As it can be seen, the entropy ranges from 0 (for a pure, unperturbed quantum state, which in our approach corresponds to the maximum amount of quantum information) to $\ln N_c$. So it would be quite natural to define the information measure as

$$I = \frac{\ln N_c - S}{\ln N_c} = 1 - \frac{S}{\ln N_c}. \tag{14}$$

So the overall range of this measure is [0, 1]: in case of zero entropy it is equal to 1 and in case of maximum entropy it is equal to 0. The latter case corresponds to the asymptotically big values of RT (Wilson loop area).

Thus, as we might see, during the interaction with QCD vacuum the entropy increases and the information is being lost.

3.2 Interaction of Colour Superposition with QCD Vacuum

When the initial (pure) colour state is a superposition of colour states, the process of the interaction with vacuum can be expressed with the following equations [9]:

$$|\phi_{in}\rangle = \alpha|1\rangle + \beta|2\rangle + \gamma|3\rangle, \tag{15}$$

$$|\alpha|^2 + |\beta|^2 + |\gamma|^2 = 1. \tag{16}$$

The corresponding density matrix is

$$\hat{\rho}_{in} = |\phi_{in}\rangle\langle\phi_{in}|, \tag{17}$$

$$\hat{\rho}_{in} = \begin{pmatrix} |\alpha|^2 & \alpha\beta^* & \alpha\gamma^* \\ \alpha^*\beta & |\beta|^2 & \beta\gamma^* \\ \alpha^*\gamma & \beta^*\gamma & |\gamma|^2 \end{pmatrix}. \tag{18}$$

After the integration and averaging we arrive at the same result as in (7) and thus in the confinement region ($RT \to \infty$) the same diagonalization of the density matrix would be observed (see 8). Thus density matrix during the interaction with QCD vacuum would transform in the following way:

$$\begin{pmatrix} |\alpha|^2 & \alpha\beta^* & \alpha\gamma^* \\ \alpha^*\beta & |\beta|^2 & \beta\gamma^* \\ \alpha^*\gamma & \beta^*\gamma & |\gamma|^2 \end{pmatrix} \to \begin{pmatrix} N_c^{-1} & 0 & 0 \\ 0 & N_c^{-1} & 0 \\ 0 & 0 & N_c^{-1} \end{pmatrix}. \tag{19}$$

(In this particular case $N_c = 3$). Purity changes from 1 (as the density matrix ρ_{in} is idempotent) to N_c^{-1}, signaling the departure of our system from the pure state and turning into fully mixed state. And the information measure introduced in Sec. 3.1 changes from 1 to 0, which corresponds to the information loss.

Thus the interaction of an arbitrary colour superposition with the QCD stochastic vacuum at large distances leads to an emergence of a mixed state with equal probabilities for different colours (without non-diagonal terms in the corresponding density matrix).

4 Multiparticle States and QCD Vacuum

Density matrix of a system is dependent on the quantity of its degrees of freedom, which is dependent on the number of particles in the system. Accordingly, the density matrix dimesions for multiparticle states would be increased from (N_c)x(N_c) to $(N_c^{N_p})$x$(N_c^{N_p})$, where N_p is the number of particles. It is necessary, therefore, to make the corresponding changes in all the formulas that were considered for a single-particle colour superposition. So the expression (7) in the multiparticle case would look like

$$\hat{\rho} = N_c^{-N_p}\hat{I} + (\hat{\rho}_{in} - N_c^{-N_p}\hat{I}) \exp(-\sigma_{adj}RT). \tag{20}$$

And the expression (8) can be written as

$$\hat{\rho}(\gamma : RT \to \infty) = N_c^{-N_p}\hat{I}. \tag{21}$$

Table 1. Purity and entropy of different states of multiparticle system before the interaction with vacuum

State:	pure separable	mixed separable	pure entangled	mixed entangled
P (purity)	1	$\frac{1}{N_c^{N_p}} \leq P < 1$	1	$\frac{1}{N_c^{N_p}} < P < 1$
S (entropy)	0	$0 < S \leq N_p \ln N_c$	0	$0 < S < N_p \ln N_c$

Table 2. Purity and entropy of different states of multiparticle system after the interaction with vacuum

State:	pure separable	mixed separable	pure entangled	mixed entangled
P (purity)	$\frac{1}{N_c^{N_p}}$	$\frac{1}{N_c^{N_p}}$	$\frac{1}{N_c^{N_p}}$	$\frac{1}{N_c^{N_p}}$
S (entropy)	$N_p \ln N_c$	$N_p \ln N_c$	$N_p \ln N_c$	$N_p \ln N_c$

The purity (10) would change to:

$$\mathrm{P} = N_c^{-N_p} + (1 - N_c^{-N_p}) W_{adj}^2(\gamma), \tag{22}$$

with maximum von Neumann entropy (13) being:

$$S = N_p \ln N_c. \tag{23}$$

The proposed information measure (14) might then be generalized as

$$I = 1 - \frac{S}{N_p \ln N_c}. \tag{24}$$

Here is the list of possible types of states, in which the system may be found (excluding some specific types e.g. as the squeezed states):

- By purity: 1) pure; 2) mixed.
- By separability: 1) separable; 2) non-separable (entangled).

Table 1 represents the quantum characteristics for the four types of states (before the interaction with vacuum). The strictness of the inequalities in the last column is due to the fact that a fully mixed state is always separable.

The same quantities after the interaction are shown in table 2.

Thus we can see that the final state is defined by a number of particles (and the number of possible colours for any individual particle) in the system. But in any case the entropy during the interaction with QCD vacuum rises to its maximum value and the information is being lost in this process. In the limit $W(\gamma) \to 0$ we get fully mixed state of a form (21).

5 Conclusions

Vacuum of quantum chromodynamics can be considered as environment (in the sense of quantum optics) for colour particles. Density matrix, purity and von Neumann entropy for colour particles depend on Wilson loop averaged through QCD vacuum degrees of freedom.

In the case of of stochastic (not coherent) QCD vacuum (only correlators of the second order are important) in confinement region (Wilson loop decays exponentially) we have decoherence of pure colour states into a mixed white states, while purity decays exponentially (decay rate = string tension).

For multiparticles (pure separable, mixed separable and nonsepaparable (entangled) states) when $RT \to \infty$ we obtain diagonalization of density matrix, decreasing of purity and increasing of von Neumann entropy.

References

[1] Y. A. Simonov. The Confinement. Uspekhi Fizicheskih Nauk. **4** (1996) (in Russian).

[2] A. D. Giacomo, H. Dosch, V. I. Shevchenko, and Y. A. Simonov, Physics Reports **372**, no. 4, 319-368 (2002).

[3] J. Ambjørn, P. Olesen. On the Formation of a Color Magnetic Quantum Liquid in QCD. Nuclear Physics B. **170**, no. 1, 60-78 (1980).

[4] D. S. Kuzmenko, Y. A. Simonov, et al. Vacuum, Confinement and QCD Strings in the Vacuum Correlator Method. Uspekhi Fizicheskih Nauk. **174**, no. 1 (2004) (in Russian).

[5] V. Kuvshinov, A. Kuzmin. *Gauge Fields and Theory of Deterministic Chaos.* (Belorussian Science, Minsk, 2006, in Russian).

[6] V. Kuvshinov, P. Buividovich. White Mixed States in QCD Stochastic Vacuum. Nonlinear Phenomena in Complex Systems. **8**, no.3, 313-316 (2005).

[7] V. Kuvshinov, P. Buividovich. Decoherence of Quark Colour States in QCD Vacuum. Acta Physica Polonica B. (Proceedings Supplement), **1**, no. 3, 579-582 (2008).

[8] F. Haake. *Quantum Signatures of Chaos.* (Springer-Verlag, Berlin, 1991).

[9] V. I. Kuvshinov, E. G. Bagashov. Evolution of colour superposition in the stochastic QCD vacuum. Nonlinear Phenom. Complex Syst., **16**, No. 3, 242-246 (2013).

Proton structure, its spin and tensor gluons.

Spyros Konitopoulos[1,*] and *George* Savvidy[1,**]

[1] *Demokritos National Research Center, Ag. Paraskevi, Athens, Greece*

Abstract. We assume that inside the hadrons there are additional partons - tensorgluons - which carry the same charges as the gluons, but have larger spin. That assumption leads to a possible contribution of polarised tensorgluon density to the spin of the nucleon. The nonzero density of tensorgluons can be created inside a nucleon by radiation of tensorgluons by gluons. Tensorgluons can carry a part of nucleon momentum together with gluons. Because tensorgluons have a larger spin, they can influence the spin structure of the nucleon. This contribution appears in the next to leading order, compared to the gluons and can provide a substantial screening effect due to the larger spin of tensorgluons.

1 Introduction

We assume that inside the hadrons there are additional partons - tensorgluons - which carry the same charges as the gluons, but have larger spin [1, 2]. That assumption leads to a possible contribution of polarised tensorgluon density to the spin of the nucleon. The nonzero density of tensorgluons can be created inside a nucleon by radiation of tensorgluons by gluons. The tensorgluons can carry a part of nucleon momentum together with gluons. Because tensorgluons have larger a spin s (s=2,3,...), they can influence the spin structure of the nucleon. Our aim is to analyse a possible contribution of polarised tensorgluon density ΔT_s into the spin of the nucleon, if they are present in the nucleon.

The spin structure of the nucleon remains the essential problem of nonperturbative QCD and hadronic physics. One of its most significant manifestations is the so-called spin crisis, or spin puzzle, related to the surprisingly small fraction of proton polarisation carried by the quarks [3–10]. The contribution from the quarks spin is found out to be small, approximately 25% of the total proton spin, and it is expected that the rest should come from the spin of gluons, quark sea polarisation and the orbital angular momentum of quarks and gluons. This problem attracted the attention to the peculiarities of the underlying QCD description of the nucleon spin, and, in particular, to the role of the gluons [17–22, 26]. One of the most interesting suggestions to come out of the spin crisis was that gluon polarisation $\triangle G$ may contribute significantly to the nucleon spin [20–23]. The amount of gluon polarisation from the experiment is approximately 20% [6, 11, 12, 15, 16] and it is not by itself sufficient to resolve the problem. It remains to identify the rest (50%) of the nucleon spin. The experimental indications of the small value of the polarisation carried by the quarks and gluons point to a possible contribution of the orbital angular momentum of the constituent patrons or *to a possible contribution coming from the polarisation of additional kind of partons in the nucleon.*

*e-mail: spykoni@inp.demokritos.gr
**e-mail: savvidy@inp.demokritos.gr

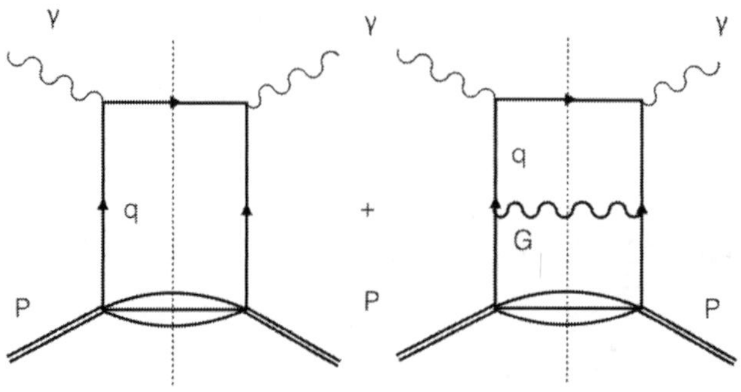

Figure 1. Scattering of a photon on a proton quark q in the leading and next to leading order through the gluon exchange.

Following the suggestion [1, 2] that there are additional patrons in the nucleon, one can propose that the nucleon's spin is accounted by the quark spin $\frac{1}{2}\Delta\Sigma$, gluon spin $1\Delta G$, as well as by the tensorgluon spin $s\Delta T_s$ and the total orbital angular momentum L_z. Physically the quantities $\Delta\Sigma, \Delta G, \Delta T_s$ represent the differences between the quark number densities, gluons and tensorgluons with spin parallel to the nucleon spin and those with spin anti-parallel (20). Because the spin of the nucleon is $\frac{1}{2}\hbar$, one should have the following spin sum rule of helicity weighted distributions:

$$\frac{1}{2}\Delta\Sigma + \Delta G + \sum_s s\,\Delta T_s + L_z = \frac{1}{2}. \tag{1}$$

In this equation the helicity of the gluon is equal to one, of the tensorgluons is $s = 2, 3, ..$ and the summation is over all "active" tensorgluons in the nucleon. The various components of the nucleon spin in the above equation can be measured experimentally [6, 8–10, 12, 14–16]. Our aim is to consider a possible contribution of tensorgluons of spin s, the $s\Delta T_s$ contribution.

Considering the influence of polarised tensorgluons density ΔT_s to the singlet part of the first moment g_1 in (6) we found an additional contribution from tensorgluons of spin s:

$$\begin{aligned} I_0 &= \frac{1}{9}\,(\Delta\Sigma - n_f\frac{\alpha(Q^2)}{2\pi}\Delta G)\left(1 - \frac{\alpha(Q^2)}{2\pi}\frac{3(12s^2-1) - 8n_f}{3(12s^2-1) - 2n_f}\right) + \\ &+ n_f\frac{\alpha(Q^2)}{2\pi}\Delta T_s\,\frac{\sum_{k=1}^{2s+1}\frac{1}{k}}{3(12s^2-1) - 2n_f}\,. \end{aligned} \tag{2}$$

The present paper is organized as follows. In section 2 the basic formulae for polarised electroproduction are recalled and short review of the experimental data is presented. In section 3 we are presenting the evolution equations that describe the Q^2 dependence of polarised parton densities including polarised tensorgluons. In section 4 the solution of the evolution equations are considered in leading and next to leading order. In Appendix we present the polarised splitting functions including tensorgluons and the corresponding anomalous dimensions [1, 2].

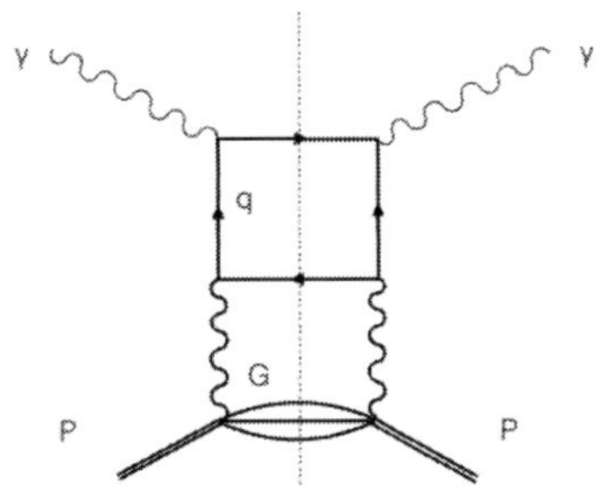

Figure 2. The typical diagram contributing to the scattering of a photon on a proton gluon G through the quark exchange.

2 *The Polarised Electroproduction*

The polarised electroproduction is described by two structure functions G_1 and G_2 in addition to the unpolarised structure functions F_1 and F_2. It is convenient to use the scaling functions $g_1(x, Q^2) = M(p \cdot q)G_1$ and $g_2(x, Q^2) = (p \cdot q)^2 G_2/M$, where p and M are the momentum and mass of the nucleon, q is the momentum of the virtual photon with $Q^2 = -q^2$ and $x = Q^2/2(p \cdot q)$. These spin-dependent structure functions g_i can be extracted from measurements where longitudinally polarised leptons are scattered from a target that is polarised either longitudinally or transversely relative to the electron beam [10]. For longitudinal beam and target polarisation, the difference between the cross sections for spins aligned and antialigned is dominated at high energy by the g_1 structure function. The g_2 structure function can be determined with additional measurement of cross sections for a nucleon polarised in a direction transverse to the beam polarisation. The moments of the spin structure functions are defined as

$$\int_0^1 dx\, x^{n-1}\, g_i(x, Q^2), \quad i = 1, 2. \tag{3}$$

In analogy with the unpolarised F_1 structure function

$$F_1 = \frac{1}{2} \sum_i e_i^2 [q^i(x, Q^2) + \bar{q}^i(x, Q^2)] \tag{4}$$

the structure function g_1 can be expressed at leading order of perturbation theory in terms of differences between quark distributions with spins aligned q_+^i and antialigned q_-^i relative to that of the nucleon $\Delta q^i = q_+^i - q_-^i$. Thus in zero order of the perturbative QCD, i.e. in the limit of the free quark-parton model the spin-dependent structure function g_1 can be expressed as [20, 22]

$$g_1 = \frac{1}{2} \sum_i e_i^2 [\Delta q^i(x, Q^2) + \Delta \bar{q}^i(x, Q^2)]. \tag{5}$$

At large Q^2 the moments (3) are given by the OPE series with the expansion coefficients related to nucleon matrix elements of the twist-2 operators and Wilson coefficient functions in the form of QCD series in strong coupling constant $\alpha = g^2/4\pi$. The lowest ($n = 1$) moments of the spin-dependent

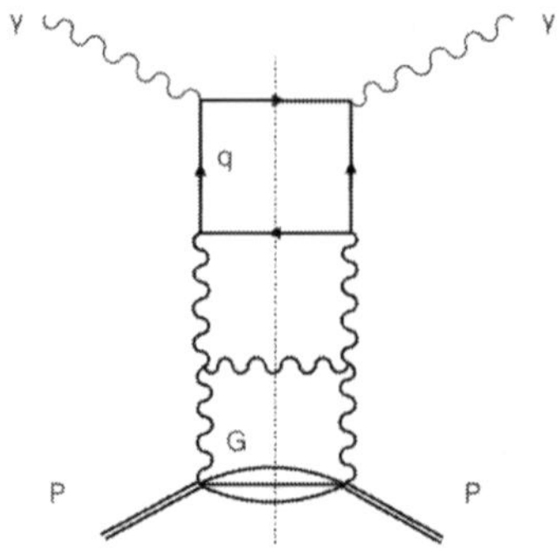

Figure 3. Example of the diagram demonstrating the scattering of a photon on a proton gluon G in the next to leading order through the quark and gluon exchange.

nucleon structure functions g_1 defined in (3) can be represented in the following form [17–20, 22, 30, 32]:

$$\Gamma_1^p = \int_0^1 dx\, g_1^p(x, Q^2) = I_3 + I_8 + I_0$$
$$\Gamma_1^n = \int_0^1 dx\, g_1^n(x, Q^2) = -I_3 + I_8 + I_0, \qquad (6)$$

where

$$I_3 = \frac{1}{12}\, a_3\, (1 - \frac{\alpha(Q^2)}{2\pi} + ...)$$
$$I_8 = \frac{1}{36}\, a_8\, (1 - \frac{\alpha(Q^2)}{2\pi} + ...)$$
$$I_0 = \frac{1}{9}\, a_0\, (1 - \frac{\alpha(Q^2)}{2\pi}\frac{33 - 8n_f}{33 - 2n_f} + ...). \qquad (7)$$

The matrix elements of the non-singlet axial currents $J^k_{5\mu}$ and the singlet axial current $J_{5\mu}$ are defined as follows[1]:

$$< p, s|\bar{\psi}\gamma^\mu\gamma_5\frac{\lambda^k}{2}\psi|p, s > = M s^\mu\, a_k, \quad k = 1, 2, ..., 8$$
$$< p, s|\bar{\psi}\gamma^\mu\gamma_5\psi|p, s > = 2M s^\mu\, a_0. \qquad (8)$$

The λ^k are generators of the flavour group and $\psi = (u, d, s, ...)$ is a vector in flavour space. The a_k, a_0 are Lorentz invariant matrix elements and reflect the unknown, nonperturbative aspect of the QCD dynamics. The nucleon states $< p, s|$ are labeled by the momentum p_μ and the covariant spin vector s_μ $(s \cdot p = 0,\ s^2 = -1)$. There are perturbative corrections to the coefficient functions which are included in expression (7).

To determine the flavour non-singlet components a_3 and a_8 on the right hand side of the equations (6), (7) one should turn to the beta-decay data [9, 10, 30, 31]. The values of a_3 and a_8 extracted from

[1]The quark distributions are defined in terms of SU(3)-flavour structure [9, 10] : *NS hypercharge* $a_8 = \Delta u + \Delta d - 2\Delta s$, *NS isovector* $a_3 = \Delta u - \Delta d$, *Singlet* $a_0 = \Delta u + \Delta d + \Delta s$.

hyperon beta-decays under the assumption of SU(3) flavour symmetry are[2] [9, 10, 30, 31, 33]:

$$\begin{aligned} a_8 &= \Delta u + \Delta d - 2\Delta s = 3F - D \quad (0.58 \pm 0.03) \\ a_3 &= \Delta u - \Delta d = F + D \quad (1.270 \pm 0.003) \end{aligned} \tag{9}$$

The experimental value of the quarks spin density $\Delta\Sigma = a_0$ can be obtained through the measurements of the first moment integrals of g_1 (6) $\Gamma_1^p = 0.135 \pm 0.011$, (7) $\Gamma_1^n = -0.028 \pm 0.006$, of the strong coupling constant $\alpha_s = 0.28 \pm 0.02$ all at $< Q^2 >= 5GeV^2$ [30, 34] and the knowledge of a_3 and a_8 in (9), thus

$$a_0 = \Delta\Sigma = \Delta u + \Delta d + \Delta s = 0.33 \pm 0.03. \tag{10}$$

This value is unexpectedly small, compared with the naive expectation of $\Delta\Sigma_{SU(6)} \simeq 1$ and of the Ellis-Jaffe sum rule $\Delta\Sigma_{EJ} \simeq 0.65$ [13]. By now it is experimentally well established that indeed the matrix element a_0 of the flavour-singlet axial-vector current is small and only of the order of $0.2 - 0.3$. The question raised was the following: why the quark spin content of the nucleon is so small?

One of the possible answers considered in the literature consists in the realisation that the singlet component a_0 receives an additional contribution from the gluon polarisation ΔG [20–23, 35]

$$\Delta G = \int_0^1 dx[G_+(x, Q^2) - G_-(x, Q^2)], \tag{11}$$

which is the amount of spin carried by polarised gluons in the polarised nucleon.

In general, the value of the matrix elements (8) depend on the renormalisation scale μ^2. Only if the operator is conserved, one can show that the matrix elements are μ^2 independent [36]. Because the non-singlet currents $J_{5\mu}^{3,8}$ are exactly conserved in the massless limit, the anomalous dimension of the non-singlet axial current vanishes at all orders. Thus a_3 and a_8 in (7) are renormalisation group invariant (μ^2 independent), they have a real physics meaning and are protected from substantial QCD radiative corrections [36]. Perturbative corrections to the coefficient functions are presented in (7) [17–23].

The singlet axial current $J_{5\mu} = \bar{\psi}\gamma_\mu\gamma_5\psi$ in (8) is not conserved [27, 28]:

$$\partial^\mu J_{5\mu} = n_f \frac{\alpha}{2\pi} trG \wedge G = n_f \frac{\alpha}{2\pi} \partial^\mu K_\mu. \tag{12}$$

Its expectation value $a_0(\mu^2)$ depends on the renormalisation point μ^2 and it is therefore not a physical quantity and receives anomalous contribution [17–23]. Therefore the interpretation of the quark spin-density $a_0 = \Delta\Sigma$ in (7), (8) and (10) as the matrix element of the flavour singlet axial charge was reconsidered [20–23]. The modified singlet axial current

$$J_{5\mu} - n_f \frac{\alpha}{2\pi} K_\mu \tag{13}$$

is conserved, as follows from (12), and its matrix element defines the invariant spin-density, which includes the polarisation of the gluons ΔG in the nucleon

$$< p, s|J_{5\mu} - n_f \frac{\alpha}{2\pi} K_\mu|p, s >= 2Ms^\mu \hat{a}_0, \tag{14}$$

where now [22]

$$\hat{a}_0 = \Delta\Sigma - n_f \frac{\alpha}{2\pi}\Delta G = \Delta u + \Delta d + \Delta s - n_f \frac{\alpha}{2\pi}\Delta G. \tag{15}$$

[2]The invariant matrix elements F and D measured in beta-decay are: $F = 0.459 \pm 0.008$ and $D = 0.798 \pm 0.008$. [9, 10].

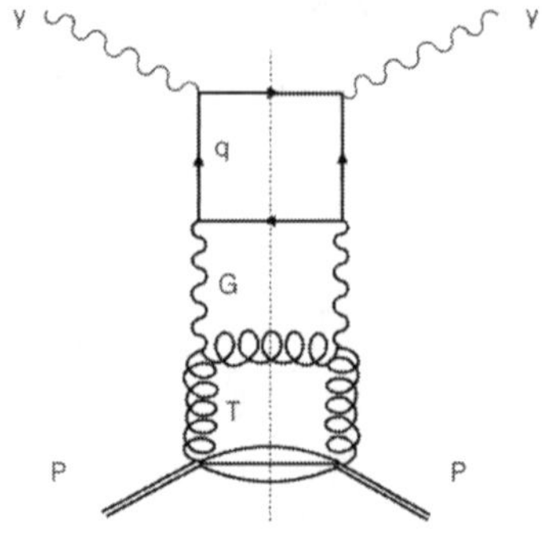

Figure 4. Example of the diagram demonstrating the scattering of a photon on a tensorgluon T, which appears in the next to leading order through the quark and gluon exchange.

Thus the matrix elements of the flavor singlet axial vector current are determined not solely by the first moment of the quark distributions, but also by the first moment of a spin-dependent gluon distribution ΔG. This leads to possible explanations for the small value of $\hat{a}_0$ extracted from polarised deep inelastic experiments (10) that have been suggested in the literature [20, 22], which includes a spin screening (15) from positive gluon polarisation ΔG. From the evolution equations it follows that the gluon polarisation grow as

$$\Delta G(Q^2) \sim \frac{const}{\alpha(Q^2)}$$

at large Q^2. This corresponds to a scaling contribution $\alpha(Q^2)\Delta G(Q^2) = const$ to the first moment $I_{ep,en}$ (6) of g_1 (see formulas (41) and (42)). This suggestion sparked a program to measure ΔG which finally provides the first evidence for nonzero gluon polarisation in the proton [6, 12, 15].

The polarised gluon distribution function ΔG was obtained from a QCD fit to the data on the spin-dependent structure function g_1. The QCD fits for ΔG were found in [6, 12, 15], one with a positive and one with a negative first moment ΔG, which describe the data equally well. The absolute value is of the order of $|\Delta G| \simeq 0.2 - 0.3$ for $Q^2 = 3GeV^2$ and the uncertainty from the fit is of the order of 0.1. *This amount of gluon polarisation is not, by itself, sufficient to resolve the problem of the small value of* $\hat{a}_0$. However that gluon polarisation $\Delta G \approx 0.2 - 0.3$ would still make a significant contribution to the nucleon polarisation [6, 12, 15, 16].

Considering the influence of polarised tensorgluons density ΔT_s to the spin of the nucleon we found that $I_0 = I_0^{'} + I_0^{''}$, the singlet component of the first moments of g_1 in (6) receives additional contribution $I_0^{''}$ from tensorgluons (see Fig. 4):

$$I_0^{'} = \frac{1}{9}\,(\Delta\Sigma - n_f \frac{\alpha(Q^2)}{2\pi}\Delta G)\left(1 - \frac{\alpha(Q^2)}{2\pi}\frac{3(12s^2-1)-8n_f}{3(12s^2-1)-2n_f} + ...\right), \qquad (16)$$

$$I_0^{''} = n_f \frac{\alpha(Q^2)}{2\pi}\,\Delta T_s\,\frac{\sum_{k=1}^{2s+1}\frac{1}{k}}{3(12s^2-1)-2n_f} + ... \qquad (17)$$

In the absence of tensorgluons (s=1) the $I_0^{'}$ in (16) reduces to the standard QCD result (7) and $I_0^{''} = 0$ ($\Delta T = 0$). In the presence of the tensorgluons of spin s the result depends on the spin factor $(12s^2 - 1)$, which appears in the Callan-Simanzik beta function coefficient [1, 2], as well as on the contribution of the tensorgluon density ΔT_s, which scales similar to the quark density $\Delta\Sigma$ (see formulas (36) and (37)). In general, one should sum there over all "active" tensorgluon spins. It is difficult to assist the effect of the tensorgluons because it is not known if ΔT_s is positive or negative. Note that if ΔT_s is negative, i.e. if tensorgluons give negative contribution to the proton helicity, then the sign is correct for additional compensation of $\Delta\Sigma$ to occur.

In the next sections we shall derive evolution equations which include the tensorgluon polarisation and their contribution to the singlet part of the structure functions.

3 *Polarisation of Tensorgluons in Nucleon*

Our aim is to analyse a possible contribution of polarised tensorgluons ΔT_s into the spin of the nucleon if they are present in a nucleon. For that one should generalise the evolution equations for quark, gluons and tensorgluons derived in [1, 2] into the case of polarised tensorgluons. The generalisation of the DGLAP evolution equations which account for polarisation of tensorgluons can be written in the following form:

$$
\begin{aligned}
\frac{dq^i_+(x,t)}{dt} &= \frac{\alpha(t)}{2\pi}\int_x^1 \frac{dy}{y}\Big[q^i_+(y,t)P_{q_+q_+}(\frac{x}{y}) + q^i_-(y,t)P_{q_+q_-}(\frac{x}{y}) + \\
&\quad +G_+(y,t)P_{q_+G_+}(\frac{x}{y}) + G_-(y,t)P_{q_+G_-}(\frac{x}{y})\Big] \\
\frac{dq^i_-(x,t)}{dt} &= \frac{\alpha(t)}{2\pi}\int_x^1 \frac{dy}{y}\Big[q^i_+(y,t)P_{q_-q_+}(\frac{x}{y}) + q^i_-(y,t)P_{q_-q_-}(\frac{x}{y}) + \\
&\quad +G_+(y,t)P_{q_-G_+}(\frac{x}{y}) + G_-(y,t)P_{q_-G_-}(\frac{x}{y})\Big] \\
\frac{dG_+(x,t)}{dt} &= \frac{\alpha(t)}{2\pi}\int_x^1 \frac{dy}{y}\Big[\sum_{i=1}^{2f} q^i_+(y,t)P_{G_+q_+}(\frac{x}{y}) + \sum_{i=1}^{2f} q^i_-(y,t)P_{G_+q_-}(\frac{x}{y}) + \\
&\quad +G_+(y,t)P_{G_+G_+}(\frac{x}{y}) + G_-(y,t)P_{G_+G_-}(\frac{x}{y}) + \\
&\quad +T_+(y,t)P_{G_+T_+}(\frac{x}{y}) + T_-(y,t)P_{G_+T_-}(\frac{x}{y})\Big] \qquad (18)\\
\frac{dG_-(x,t)}{dt} &= \frac{\alpha(t)}{2\pi}\int_x^1 \frac{dy}{y}\Big[\sum_{i=1}^{2f} q^i_+(y,t)P_{G_-q_+}(\frac{x}{y}) + \sum_{i=1}^{2f} q^i_-(y,t)P_{G_-q_-}(\frac{x}{y}) + \\
&\quad +G_+(y,t)P_{G_-G_+}(\frac{x}{y}) + G_-(y,t)P_{G_-G_-}(\frac{x}{y}) + \\
&\quad +T_+(y,t)P_{G_-T_+}(\frac{x}{y}) + T_-(y,t)P_{G_-T_-}(\frac{x}{y})\Big] \\
\frac{dT_+(x,t)}{dt} &= \frac{\alpha(t)}{2\pi}\int_x^1 \frac{dy}{y}\Big[G_+(y,t)P_{T_+G_+}(\frac{x}{y}) + G_-(y,t)P_{T_+G_-}(\frac{x}{y}) + \\
&\quad +T_+(y,t)P_{T_+T_+}(\frac{x}{y}) + T_-(y,t)P_{T_+T_-}(\frac{x}{y})\Big] \\
\frac{dT_-(x,t)}{dt} &= \frac{\alpha(t)}{2\pi}\int_x^1 \frac{dy}{y}\Big[G_+(y,t)P_{T_-G_+}(\frac{x}{y}) + G_-(y,t)P_{T_-G_-}(\frac{x}{y}) + \\
&\quad +T_+(y,t)P_{T_-T_+}(\frac{x}{y}) + T_-(y,t)P_{T_-T_-}(\frac{x}{y}),
\end{aligned}
$$

where $t = \ln(Q^2/Q_0^2)$. This set of equations can be simplified by observing that parity conservation in QCD implies the relations

$$P_{B_+A_\pm}(z) = P_{B_-A_\mp}(z) \tag{19}$$

for any A and B and that the sums and the differences

$$\begin{aligned} q^i &= q^i_+ + q^i_-, \quad G = G_+ + G_-, \quad T = T_+ + T_-, \\ \Delta q^i &= q^i_+ - q^i_-, \quad \Delta G = G_+ - G_-, \quad \Delta T = T_+ - T_- \end{aligned} \tag{20}$$

evolve separately. Physically the quantities $\Delta q^i, \Delta G, \Delta T$ represent the differences between the number densities of quarks, gluons and tensorgluons with spin parallel to the nucleon spin and those with spin anti-parallel. It is also convenient to define the sum and the difference of splitting functions:

$$P_{BA}(z) = P_{B_+A_+}(z) + P_{B_-A_+}(z), \qquad \Delta P_{BA}(z) = P_{B_+A_+}(z) - P_{B_-A_+}(z). \tag{21}$$

We derive then from (18) the master equation for differences:

$$\begin{aligned} \tfrac{d}{dt}\Delta q^i(x,t) &= \tfrac{\alpha(t)}{2\pi}\int_x^1 \tfrac{dy}{y}\Big[\Delta q^i(y,t)\Delta P_{qq}(\tfrac{x}{y}) + \Delta G(y,t)\Delta P_{qG}(\tfrac{x}{y})\Big] \\ \tfrac{d}{dt}\Delta G(x,t) &= \tfrac{\alpha(t)}{2\pi}\int_x^1 \tfrac{dy}{y}\Big[\textstyle\sum_{i=1}^{2n_f}\Delta q^i(y,t)\Delta P_{Gq}(\tfrac{x}{y}) + \Delta G(y,t)\Delta P_{GG}(\tfrac{x}{y}) + \Delta T(y,t)\Delta P_{GT}(\tfrac{x}{y})\Big] \\ \tfrac{d}{dt}\Delta T(x,t) &= \tfrac{\alpha(t)}{2\pi}\int_x^1 \tfrac{dy}{ry}\Big[\Delta G(y,t)\Delta P_{TG}(\tfrac{x}{y}) + \Delta T(y,t)\Delta P_{TT}(\tfrac{x}{y})\Big]. \end{aligned} \tag{22}$$

In the case of non-singlet (NS) quark densities the above equation reduces to the one of the standard QCD:

$$\frac{d}{dt}\Delta q^{NS}(x,t) = \frac{\alpha(t)}{2\pi}\int_x^1 \frac{dy}{y}\Delta q^{NS}(y,t)\Delta P_{qq}(\frac{x}{y}), \tag{23}$$

where $q^{NS}(y,t) = q(x,t) - \bar{q}(x,t)$. The splitting function $P_{q_-q_+} = 0$ and therefore $\Delta P_{qq} = P_{q_+q_+} = P_{qq}$. It follows then that the moments of Δq^{NS} evolve in Q^2 with the same logarithmic exponent as q^{NS} and therefore as the non-singlet parts of F_1 and g_1.

The integral

$$\Delta q^i = \int_0^1 dx[\Delta q^i(x,t) + \Delta\bar{q}^i(x,t)] \tag{24}$$

represents the fraction of the proton's spin carried by the quarks and the anti-quarks of the flavour q^i. The summation over all quark flavour contributions gives the total *fraction of the nucleon spin carried by light quark and antiquarks* (below the contribution of the heavy quarks has been neglected):

$$\Delta\Sigma = \int_0^1 dx\sum_i[\Delta q^i(x,t) + \Delta\bar{q}^i(x,t)] = \Delta u + \Delta d + \Delta s. \tag{25}$$

As we already reviewed in the introduction, the conservation of the SU(3) singlet axial current, valid for massless quarks, is broken in QCD by the Adler-Bell-Jackiw anomaly [27, 28]. While the anomalous dimension of the non-singlet axial current vanishes at all orders [17, 18], the singlet axial current anomalous dimension is different from zero. Due to this non-conservation in the singlet sector there is a gluon contribution at all values of Q^2 in the singlet part of the integral of g_1. This gluon term ΔG in (15)

$$\hat{a}_0 = \Delta\Sigma - n_f\frac{\alpha}{2\pi}\Delta G \tag{26}$$

Figure 5. We assume that inside the hadrons there are additional partons - tensorgluons - which carry the same charges as the gluons, but have larger spin [1, 2]. The nucleon's spin is accounted then by the quark spin $\frac{1}{2}\Delta\Sigma$, gluon spin $1\Delta G$, as well as by the tensorgluon spin $s\Delta T_s$ and the total orbital angular momentum L_z (28).

can in principle be large and we are interested to consider a possible contribution of the tensorgluons polarisation as well. With this aim we shall introduce in addition to the gluon polarisation the corresponding tensorgluon helicity content in the polarised nucleon:

$$\Delta G = \int_0^1 dx[G_+(x,Q^2) - G_-(x,Q^2)], \quad \Delta T = \int_0^1 dx[T_+(x,Q^2) - T_-(x,Q^2)], \tag{27}$$

with $G_\pm, T_\pm$ being the gluon and tensorgluons densities with helicities ±1 and $\pm s$ respectively in a nucleon with helicity $+\frac{1}{2}$. ΔG represents the amount of helicity carried by the gluons in the nucleon and ΔT_s, the amount carried by the tensorgluons.

The nucleon's spin is accounted by the quarks' spin as well as by gluon and tensorgluon spins ΔG, ΔT_s and the total orbital angular momentum L_z. Because the spin of the nucleon is $1/2$ one should have

$$\frac{1}{2}\Delta\Sigma + \Delta G + \sum_s s\,\Delta T_s + L_z = \frac{1}{2}. \tag{28}$$

In this equation the helicity of the gluon is equal to one and of tensorgluon is $s = 2, 3, ..$, and the summation is over all "active" tensor bosons in the nucleon. The various components of the nucleon spin in the above equation can be measured experimentally. The gluon sector ΔG can be measured in deep inelastic scattering experiments in which gluon-gluon or gluon-quark scattering dominates. One of the most interesting suggestions to come out of the spin crisis is that ΔG may contribute significantly to the nucleon. Our aim is to consider a possible contribution of the tensorgluon spin $s\Delta T_s$.

4 *Evolution Equation in Leading and Next to Leading Order*

At leading order the QCD evolution of $\Delta\Sigma$, ΔG and ΔT can written from (22) as

$$\frac{d}{dt}\begin{pmatrix} \Delta\Sigma \\ \Delta G \\ \Delta T \end{pmatrix} = \frac{\alpha(t)}{2\pi}\begin{pmatrix} \Delta\gamma^{(1)}_{qq} & 2n_f\Delta\gamma^{(1)}_{qG} & 0 \\ \Delta\gamma^{(1)}_{Gq} & \Delta\gamma^{(1)}_{GG} & \Delta\gamma^{(1)}_{GT} \\ 0 & \Delta\gamma^{(1)}_{TG} & \Delta\gamma^{(1)}_{TT} \end{pmatrix}\begin{pmatrix} \Delta\Sigma \\ \Delta G \\ \Delta T \end{pmatrix}, \tag{29}$$

where [2] (see Appendix A for the values of the corresponding anomalous dimensions)

$$\Delta\gamma_1 = \begin{pmatrix} 0 & 0 & 0 \\ \frac{3}{2}C_2(R) & C_2(G)\frac{\sum_s(12s^2-1)}{6} - \frac{2}{3}n_f T(R) & C_2(G)\sum_{k=1}^{2s+1}\frac{1}{k} \\ 0 & 0 & 0 \end{pmatrix}, \tag{30}$$

and $C_2(G) = N, C_2(R) = \frac{N^2-1}{2N}, T(R) = \frac{1}{2}$ for the SU(N) groups. Let us represent the $\Delta\gamma_1$ as

$$\Delta\gamma_1 = \begin{pmatrix} 0 & 0 & 0 \\ C_1 & b & C_3 \\ 0 & 0 & 0 \end{pmatrix},$$

where [2]

$$C_1 = \frac{3}{2}C_2(R), \; b = C_2(G)\frac{\sum_s(12s^2-1)}{6} - \frac{2}{3}n_f T(R), \; C_3 = C_2(G)\sum_{k=1}^{2s+1}\frac{1}{k} \tag{31}$$

and

$$\frac{d\alpha(t)}{dt} = -\frac{b}{2\pi}\alpha^2(t). \tag{32}$$

The eigenvalues of the matrix $\Delta\gamma_1$ are $\lambda_1 = b, \; \lambda_{2,3} = 0$ and the corresponding eigenvectors

$$S = \begin{pmatrix} -\frac{C_3}{C_1} & -\frac{b}{C_1} & 0 \\ 0 & 1 & 1 \\ 1 & 0 & 0 \end{pmatrix}, \quad S^{-1} = \begin{pmatrix} 0 & 0 & 1 \\ -\frac{C_1}{b} & 0 & -\frac{C_3}{b} \\ \frac{C_1}{b} & 1 & \frac{C_3}{b} \end{pmatrix}$$

define the transformation $\Delta\gamma_1 = S\Lambda S^{-1}$, where $\Lambda = diag(0,0,b)$, which allows to represent the evolution equation (29) in the following form:

$$\frac{d}{dt}\begin{pmatrix} \Delta\Sigma \\ \Delta G \\ \Delta T \end{pmatrix} = \frac{\alpha(t)}{2\pi}S\Lambda S^{-1}\begin{pmatrix} \Delta\Sigma \\ \Delta G \\ \Delta T \end{pmatrix}. \tag{33}$$

Multiplying this equation from the left by S^{-1} we shall get

$$\frac{d}{dt}\left[S^{-1}\begin{pmatrix} \Delta\Sigma \\ \Delta G \\ \Delta T \end{pmatrix}\right] = \frac{\alpha(t)}{2\pi}\Lambda\left[S^{-1}\begin{pmatrix} \Delta\Sigma \\ \Delta G \\ \Delta T \end{pmatrix}\right],$$

and it is useful to introduce the new coordinates

$$\begin{pmatrix} y_1(t) \\ y_2(t) \\ y_3(t) \end{pmatrix} = S^{-1}\begin{pmatrix} \Delta\Sigma \\ \Delta G \\ \Delta T \end{pmatrix} = \begin{pmatrix} \Delta T \\ -\frac{1}{b}(C_1\Sigma + C_3\Delta T) \\ \frac{C_1}{b}\Sigma + \Delta G + \frac{C_3}{b}\Delta T \end{pmatrix}. \tag{34}$$

The evolution equation takes a diagonal form:

$$\frac{d}{dt}\begin{pmatrix} y_1(t) \\ y_2(t) \\ y_3(t) \end{pmatrix} = \frac{\alpha(t)}{2\pi}\begin{pmatrix} 0 & 0 & 0 \\ 0 & 0 & 0 \\ 0 & 0 & b \end{pmatrix}\begin{pmatrix} y_1(t) \\ y_2(t) \\ y_3(t) \end{pmatrix}. \tag{35}$$

The first equation gives $\frac{dy_1(t)}{dt} = \frac{d\Delta T(t)}{dt} = 0$, and we conclude that the helicity density of tensorgluons does not evolve with Q^2:

$$\Delta T(t) = \Delta T. \tag{36}$$

From this and from the second equation $dy_2(t)/dt = 0$ it follows that the quark helicity distribution also does not evolve with Q^2 ($d\Delta\Sigma(t)/dt = 0$):

$$\Delta\Sigma(t) = \Delta\Sigma. \tag{37}$$

From the last equation $\frac{dy_3(t)}{dt} = \frac{\alpha(t)}{2\pi}\, b\, y_3(t)$ it follows then that

$$\frac{d\,\alpha(t)\, y_3(t)}{dt} = \frac{d\alpha(t)}{dt}\, y_3(t) + \alpha(t)\frac{dy_3(t)}{dt} = -\frac{b}{2\pi}\alpha^2(t)\, y_3(t) + \alpha(t)\frac{\alpha(t)}{2\pi}\, b\, y_3(t) = 0, \tag{38}$$

where we use (32). Thus

$$\alpha(t)\, y_3(t) = Const \tag{39}$$

and therefore from the last equation in (34)

$$\frac{\alpha(t)}{2\pi}\Delta G(t) = Const - \frac{\alpha(t)}{2\pi b}[C_3\Delta T + C_1\Delta\Sigma] \rightarrow Const. \tag{40}$$

as $t \rightarrow \infty$ and $\alpha(t) \rightarrow 0$. Here we use the fact that $\Delta\Sigma$ and ΔT are t-independent (36), (37). In the absence of tensorgluons the equation reduces to the Altarelli-Ross equation:

$$\frac{\alpha(t)}{2\pi}\Delta G(t) = Const - \frac{\alpha(t)}{2\pi b}C_1\Delta\Sigma \rightarrow Const. \tag{41}$$

Therefore it is natural to introduce the quantity [22]

$$\Delta\Gamma = \frac{\alpha(t)}{2\pi}\Delta G(t) \tag{42}$$

and consider its behaviour in the next to leading order. One should notice that the coefficient in front of the tensorgluon density C_3 in (40) grows with spin as $\ln s$, therefore its contribution can still be numerically large for higher spins. *The main conclusion which follows from the above consideration is that the densities $\Delta\Sigma$ and ΔT are Q^2 independent and that the gluon density grows with Q^2 as $\Delta G \approx 1/\alpha(Q^2)$.*

Expanding the splitting functions and anomalous dimensions into perturbative series

$$\Delta P_{ij} = \Delta P^{(1)}_{ij} + \frac{a_s(t)}{2\pi}\Delta P^{(2)}_{ij} + \left(\frac{a_s(t)}{2\pi}\right)^2\Delta P^{(3)}_{ij} + \dots \tag{43}$$

one can get from (22):

$$\frac{d}{dt}\begin{pmatrix} \Delta\Sigma(t) \\ \Delta\Gamma(t) \\ \Delta T(t) \end{pmatrix} = \left(\frac{a_s(t)}{2\pi}\right)^2 \begin{pmatrix} \Delta\gamma^{(2)}_{qq} & 2n_f\Delta\gamma^{(3)}_{qG} & 0 \\ \Delta\gamma^{(1)}_{Gq} & \Delta\gamma^{(2)}_{GG} & \Delta\gamma^{(1)}_{GT} \\ 0 & \Delta\gamma^{(3)}_{TG} & \Delta\gamma^{(2)}_{TT} \end{pmatrix} \begin{pmatrix} \Delta\Sigma(t) \\ \Delta\Gamma(t) \\ \Delta T(t) \end{pmatrix}. \tag{44}$$

Here we have taken into account that there are no direct interactions between tensorgluons and quarks $\gamma^{(2)}_{Tq} = \gamma^{(2)}_{qT} = 0$ [1, 2]. The Q^2 evolution of the singlet component of the first moment (6) is given by the short-distance operator product expansion [17–19, 22]:

$$I_0 = \frac{1}{9}c(a_s(t))E(a_s, a_s(t))O(a_s), \tag{45}$$

where we take into account the presence of the tensorgluons:

$$c(a_s(t)) = \left(1 + c_\Sigma a_s(t),\ c_\Gamma + c'_\Gamma a_s(t),\ c_T + c'_T a_s(t)\right), \quad O(a_s) = \begin{pmatrix} \Delta\Sigma(t) \\ \Delta\Gamma(t) \\ \Delta T(t) \end{pmatrix} \tag{46}$$

and

$$E(a_s, a_s(t)) = 1 + \frac{a_s - a_s(t)}{4\pi b}\Delta\gamma\,, \tag{47}$$

γ is the matrix of anomalous dimensions in equation (44). Substituting the matrix of anomalous dimensions (47) , (44) and the vectors (46) into the (45) we shall get

$$\begin{aligned} I_0 &= \frac{1}{9}\Big\{\Big[1 + \frac{a_s - a_s(t)}{4\pi b}\big(\Delta\gamma^{(2)}_{qq} + c_\Gamma\Delta\gamma^{(1)}_{Gq} - bc_\Sigma\big) + c_\Sigma a_s\Big]\Delta\Sigma + \\ &+\Big[1 + \frac{a_s - a_s(t)}{4\pi b}\big(\Delta\gamma^{(2)}_{GG} + \frac{1}{c_\Gamma}\Delta\gamma^{(3)}_{qG} + \frac{c_T}{c_\Gamma}\Delta\gamma^{(3)}_{TG} - b\frac{c'_\Gamma}{c_\Gamma}\big) + \frac{c'_\Gamma}{c_\Gamma}a_s\Big]c_\Gamma\Delta\Gamma + \\ &+\Big[c_T + \frac{a_s - a_s(t)}{4\pi b}\big(c_T\Delta\gamma^{(2)}_{TT} + c_\Gamma\Delta\gamma^{(1)}_{GT} - bc'_T\big) + c'_T a_s\Big]\Delta T\Big\}. \end{aligned} \tag{48}$$

One can see the appearance of the tensorgluon contribution ΔT into the first moment of the singlet polarised structure function. In this expression the anomalous dimensions associated with the quark and gluon interactions were computed in [17–19, 22]. These values are $\Delta\gamma^{(2)}_{qq} = 0$ and

$$\begin{aligned} c_\Gamma &= -n_f,\ \ c_\Sigma = -\frac{3}{4\pi}C_2(R),\ \ c'_\Gamma = c_\Gamma c_\Sigma \\ \Delta\gamma^{(1)}_{Gq} &= \frac{3}{2}C_2(R),\ \ \Delta\gamma^{(3)}_{qG} = c^2_\Gamma\Delta\gamma^{(1)}_{Gq},\ \ \Delta\gamma^{(2)}_{GG} = 0. \end{aligned} \tag{49}$$

Because there are no interactions between photon and tensorgluons through the quark exchanges, the coefficients c_T and c'_T vanish. Hence, keeping terms proportional to $\alpha(t)$ and using the above values for the anomalous dimensions, the first moment of the singlet part of the proton structure function (48) takes the following form:

$$\begin{aligned} I'_0 &= \frac{1}{9}\Big[1 - \frac{a_s(t)}{4\pi b}(c_\Gamma\Delta\gamma^{(1)}_{Gq} - bc_\Sigma)\Big](\Delta\Sigma - n_f\Delta\Gamma) \\ I''_0 &= -\frac{1}{9}\Big[\frac{a_s(t)}{4\pi b}c_\Gamma\Delta\gamma^{(1)}_{GT}\Big]\Delta T, \end{aligned}$$

where $I_0 = I'_0 + I''_0$. The coefficient of the beta function b of the running coupling constant is given in (31). Given that $C_2(G) = 3$, $C_2(R) = \frac{4}{3}$ and $T(R) = \frac{1}{2}$ we finally get:

$$I'_0 = \frac{1}{9}\,(\Delta\Sigma - n_f\frac{\alpha_s(t)}{2\pi}\Delta G)\left(1 - \frac{a_s(t)}{2\pi}\frac{3\sum(12s^2-1) - 8n_f}{3\sum(12s^2-1) - 2n_f} + ...\right) \tag{50}$$

$$I''_0 = n_f\sum_s\frac{a_s(t)}{2\pi}\Delta T_s\,\frac{\sum_{k=1}^{2s+1}\frac{1}{k}}{3\sum(12s^2-1) - 2n_f} + ... \tag{51}$$

Comparing this result with the standard QCD case one can see that the assumed presence of the tensorgluons inside the proton modifies the singlet part of the polarised proton structure function in two ways. On the one hand, there is a contribution $3\sum(12s^2 - 1)$ in the numerator and the denominator to the radiative correction I'_0 which effectively lowers its value. On the other hand, there is an additional contribution I''_0 of the tensorgluons ΔT_s. The overall sign of this contribution depends on the values of the spin s of the tensorgluon and of its density ΔT_s. Note that if ΔT_s is negative, i.e. if tensorgluons give negative contribution to the proton helicity, then the sign is correct for the additional compensation of $\Delta\Sigma$ to occur.

Acknowledgements

On of us (G.S.) would like to thank the organisers of the International Seminar on High Energy Physics "Quarks-2016" for invitation and kind hospitality.

Appendix A

Since $P_{B_+A_\pm}(z) = P_{B_-A_\mp}(z)$, we only need to evaluate $P_{B_+A_+}(z)$, $P_{B_-A_+}(z)$. Then

$$P_{BA}(z) = P_{B_+A_+}(z) + P_{B_-A_+}(z), \quad \Delta P_{BA}(z) = P_{B_+A_+}(z) - P_{B_-A_+}(z).$$

Hence for the $\triangle P_{qq}$ [24]:

$$\begin{aligned} P_{q_-q_+}(z) &= 0 \\ \triangle P_{qq} = P_{q_+q_+}(z) = P_{qq} = &= C_2(R)\Big[\frac{1+z^2}{(1-z)_+} + \frac{3}{2}\delta(z-1)\Big], \end{aligned} \tag{52}$$

for the $\triangle P_{Gq}(z)$

$$\begin{aligned} P_{G_+q_+}(z) = C_2(R)\frac{1}{z}, &\qquad P_{G_-q_+}(z) = C_2(R)\frac{(1-z)^2}{z} \\ P_{Gq}(z) = P_{G_+q_+}(z) + P_{G_-q_+}(z) &= C_2(R)\frac{1+(1-z)^2}{z} \\ \triangle P_{Gq}(z) = P_{G_+q_+}(z) - P_{G_-q_+}(z) &= C_2(R)\frac{1-(1-z)^2}{z}, \end{aligned} \tag{53}$$

for the $\triangle P_{qG}(z)$

$$\begin{aligned} P_{q_+G_+}(z) = \frac{z^2}{2}, &\qquad P_{q_-G_+}(z) = \frac{(z-1)^2}{2} \\ P_{qG}(z) = P_{q_+G_+}(z) + P_{q_-G_+}(z) &= \frac{z^2+(z-1)^2}{2} \\ \triangle P_{qG}(z) = P_{q_+G_+}(z) - P_{q_-G_+}(z) &= \frac{z^2-(z-1)^2}{2} \end{aligned} \tag{54}$$

as well as for the $\triangle P_{GG}(z)$

$$\begin{aligned} P_{G_+G_+}(z) &= C_2(G)\Big[(1+z^4)\Big(\frac{1}{z} + \frac{1}{(1-z)_+}\Big) + \Big(\frac{\sum_s(12s^2-1)}{6} - \frac{2}{3}\frac{T(R)}{C_2(G)}\Big)\delta(z-1)\Big] \\ P_{G_-G_+}(z) &= C_2(G)\frac{(1-z)^3}{z} \\ \triangle P_{GG}(z) &= C_2(G)\Big[(1+z^4)\Big(\frac{1}{z} + \frac{1}{(1-z)_+}\Big) - \frac{(1-z)^3}{z} + \Big(\frac{\sum_s(12s^2-1)}{6} - \frac{2}{3}\frac{T(R)}{C_2(G)}\Big)\delta(z-1)\Big]. \end{aligned}$$

The behaviour at small $x \to 0$ of the unpolarised and polarised splitting functions is different

$$\lim_{x\to 0} P_{GG}(z) = C_2(G)\Big(\frac{1}{z} + \frac{1}{z}\Big), \qquad \lim_{x\to 0} \triangle P_{GG}(z) = C_2(G)\Big(\frac{1}{z} - \frac{1}{z}\Big) = 0. \tag{55}$$

The polarised splitting function $\triangle P_{GG}(z)$ has moderate behaviour at small x's. The polarised splitting functions which are including tensorgluons were found in [1, 2]. For the non-diagonal functions they are:

$$
\begin{aligned}
P_{T_+G_+}(z) &= C_2(G)\frac{z^{2s+1}}{(1-z)^{2s-1}}\\
P_{T_-G_+}(z) &= C_2(G)\frac{(1-z)^{2s+1}}{z^{2s-1}}\\
\triangle P_{TG}(z) &= C_2(G)\Big[\frac{z^{2s+1}}{(1-z)^{2s-1}} - \frac{(1-z)^{2s+1}}{z^{2s-1}}\Big],
\end{aligned} \tag{56}
$$

and at small x both functions are singular

$$\lim_{x\to 0} P_{TG}(z) = C_2(G)\frac{1}{z^{2s-1}}, \qquad \lim_{x\to 0} \triangle P_{TG}(z) = -C_2(G)\frac{1}{z^{2s-1}}. \tag{57}$$

Comparing this behaviour with the gluon splitting functions (55) one can see that here the polarised splitting function has enhanced singular behaviour in the small x region. Finally for the P_{TG} function we have

$$
\begin{aligned}
P_{G_+T_+}(z) &= C_2(G)\frac{1}{z(1-z)^{2s-1}}\\
P_{G_-T_+}(z) &= C_2(G)\frac{(1-z)^{2s+1}}{z}\\
\triangle P_{GT}(z) &= C_2(G)\Big[\frac{1}{z(1-z)^{2s-1}} - \frac{(1-z)^{2s+1}}{z}\Big].
\end{aligned} \tag{58}
$$

In the small x region we have behaviour which is similar to gluon (55)

$$\lim_{x\to 0} P_{GT}(z) = C_2(G)\frac{2}{z}, \qquad \lim_{x\to 0} \triangle P_{GT}(z) = 0. \tag{59}$$

The tensor-tensor splitting function is helicity conserving:

$$P_{T_-T_+}(z) = 0 \tag{60}$$

$$\triangle P_{TT}(z) = P_{T_+T_+}(z) = C_2(G)\Big[\frac{z^{2s+1}}{(1-z)_+} + \frac{1}{(1-z)z_+^{2s-1}} + \sum_{j=1}^{2s+1}\frac{1}{j}\delta(z-1)\Big]$$

and is singular in the small x region both functions are singular

$$\lim_{x\to 0} P_{TT}(z) = C_2(G)\frac{1}{z^{2s-1}}, \qquad \lim_{x\to 0} \triangle P_{TT}(z) = C_2(G)\frac{1}{z^{2s-1}}. \tag{61}$$

The matrix of anomalous dimensions is [24]:

$$\int_0^1 dz z^{n-1}\begin{pmatrix} \triangle P_{qq}(z) & 2n_f\triangle P_{qG}(z) & 0\\ \triangle P_{Gq}(z) & \triangle P_{GG}(z) & P_{GT}(z)\\ 0 & \triangle P_{TG}(z) & \triangle P_{TT}(z)\end{pmatrix} = \begin{pmatrix} \triangle\gamma_{qq}(n) & 2n_f\triangle\gamma_{qG}(n) & 0\\ \triangle\gamma_{Gq}(n) & \triangle\gamma_{GG}(n) & \triangle\gamma_{GT}(n)\\ 0 & \triangle\gamma_{TG}(n) & \triangle\gamma_{TT}(n)\end{pmatrix},$$

where

$$\begin{aligned}
\int_0^1 dz \Delta P_{qq}(z) &= 0, \qquad n=1, \\
\int_0^1 dz \Delta P_{qG}(z) &= 0, \\
\int_0^1 dz \Delta P_{Gq}(z) &= \frac{3}{2} C_2(R), \\
\int_0^1 dz \Delta P_{GG}(z) &= \frac{\sum_s (12s^2-1)}{6} C_2(G) - \frac{2}{3} T(R),
\end{aligned}$$

and for the tensorgluons [1, 2]

$$\begin{aligned}
\int_0^1 dz \Delta P_{GT}(z) &= C_2(G) \sum_{k=1}^{2s+1} \frac{1}{k}, \qquad n=1, \\
\int_0^1 dz \Delta P_{TG}(z) &= 0, \\
\int_0^1 dz \Delta P_{TT}(z) &= 0.
\end{aligned} \tag{62}$$

References

[1] G. Savvidy, *Asymptotic freedom of non-Abelian tensor gauge fields,* Phys. Lett. B **732** (2014) 150.

[2] G. Savvidy, *Proton structure and tensor gluons,* J. Phys. A **47** (2014) 35, 355401 [arXiv:1310.0856 [hep-th]].

[3] J. Ashman *et al.* [European Muon Collaboration], *An Investigation of the Spin Structure of the Proton in Deep Inelastic Scattering of Polarized Muons on Polarized Protons,* Nucl. Phys. B **328** (1989) 1.

[4] P. L. Anthony *et al.* [E155 Collaboration], *Measurements of the Q^2 dependence of the proton and neutron spin structure functions g_1^p and g_1^n,* Phys. Lett. B **493** (2000) 19.

[5] M. G. Alekseev *et al.* [COMPASS Collaboration.], *The Spin-dependent Structure Function of the Proton g_1^p and a Test of the Bjorken Sum Rule,* Phys. Lett. B **690** (2010) 466.

[6] C. Adolph *et al.* [COMPASS Collaboration], *Leading and Next-to-Leading Order Gluon Polarization in the Nucleon and Longitudinal Double Spin Asymmetries from Open Charm Muoproduction,* Phys. Rev. D **87** (2013) 5, 052018 [arXiv:1211.6849 [hep-ex]].

[7] A. Airapetian *et al.* [HERMES Collaboration], *Precise determination of the spin structure function g(1) of the proton, deuteron and neutron,* Phys. Rev. D **75** (2007) 012007.

[8] D. de Florian, R. Sassot, M. Stratmann and W. Vogelsang, *Extraction of Spin-Dependent Parton Densities and Their Uncertainties,* Phys. Rev. D **80** (2009) 034030.

[9] M. Anselmino, A. Efremov and E. Leader, *The Theory and phenomenology of polarised deep inelastic scattering,* Phys. Rept. **261** (1995) 1 [Erratum-ibid. **281** (1997) 399] .

[10] C. A. Aidala, S. D. Bass, D. Hasch and G. K. Mallot, *The Spin Structure of the Nucleon,* Rev. Mod. Phys. **85** (2013) 655 [arXiv:1209.2803 [hep-ph]].

[11] P. Jimenez-Delgado, W. Melnitchouk and J. F. Owens, *Parton momentum and helicity distributions in the nucleon,* J. Phys. G **40** (2013) 093102 [arXiv:1306.6515 [hep-ph]].

[12] G. K. Mallot, *Measurements of Delta G,* AIP Conf. Proc. **915** (2007) 325 [hep-ex/0612055].

[13] J. R. Ellis and R. L. Jaffe, *A Sum Rule for Deep Inelastic Electroproduction from Polarized Protons,* Phys. Rev. D **9** (1974) 1444 [Erratum-ibid. D **10** (1974) 1669].

[14] J. R. Ellis and M. Karliner, *Direct estimate of the gluon polarization in the nucleon,* Mod. Phys. Lett. A **21** (2006) 721 [hep-ph/0501115].

[15] E. R. Nocera *et al.* [NNPDF Collaboration], *A first unbiased global determination of polarized PDFs and their uncertainties,* Nucl. Phys. B **887** (2014) 276 [arXiv:1406.5539].

[16] R. D. Ball *et al.* [The NNPDF Collaboration], *Unbiased determination of polarized parton distributions and their uncertainties,* Nucl. Phys. B **874** (2013) 36 [arXiv:1303.7236].

[17] J. Kodaira, S. Matsuda, T. Muta, K. Sasaki and T. Uematsu, *QCD Effects in Polarized Electroproduction,* Phys. Rev. D **20** (1979) 627.

[18] J. Kodaira, S. Matsuda, K. Sasaki and T. Uematsu, *QCD Higher Order Effects in Spin Dependent Deep Inelastic Electroproduction,* Nucl. Phys. B **159** (1979) 99.

[19] J. Kodaira, *QCD Higher Order Effects in Polarized Electroproduction: Flavor Singlet Coefficient Functions,* Nucl. Phys. B **165** (1980) 129.

[20] R. L. Jaffe, *The Axial Anomaly and the Sum Rules for Spin Dependent Electroproduction,* Phys. Lett. B **193** (1987) 101.

[21] A. V. Efremov and O. V. Teryaev, *Spin Structure of the Nucleons and Triangle Anomaly,* JINR Report E2-88-287, In *Prague 1988, Proceedings, Hadron Interactions Theory and Phenomenology* pp.302-312.

[22] G. Altarelli and G. G. Ross, *The Anomalous Gluon Contribution to Polarized Leptoproduction,* Phys. Lett. B **212** (1988) 391.

[23] R. D. Carlitz, J. C. Collins and A. H. Mueller, *The Role of the Axial Anomaly in Measuring Spin Dependent Parton Distributions,* Phys. Lett. B **214** (1988) 229.

[24] G. Altarelli and G. Parisi, *Asymptotic Freedom in Parton Language,* Nucl. Phys. B **126** (1977) 298.

[25] G. Altarelli and B. Lampe, *The Light Cone Expansion and the Parton Model in Polarized Muon Production,* Z. Phys. C **47** (1990) 315.

[26] Y. L. Dokshitzer, V. A. Khoze, A. H. Mueller and S. I. Troian, *Basics of perturbative QCD,* Gif-sur-Yvette, France: Ed. Frontieres (1991) pp 65-88.

[27] S. L. Adler, *Axial vector vertex in spinor electrodynamics,* Phys. Rev. **177** (1969) 2426.

[28] J. S. Bell and R. Jackiw, *A PCAC puzzle: π_0 to $\gamma\gamma$ in the sigma model,* Nuovo Cim. A **60** (1969) 47.

[29] A. V. Efremov, J. Soffer and O. V. Teryaev, *Spin Structure of Nucleon and the Axial Anomaly,* Nucl. Phys. B **346** (1990) 97.

[30] F. E. Close and R. G. Roberts, *Consistent analysis of the spin content of the nucleon,* Phys. Lett. B **316** (1993) 165 [hep-ph/9306289].

[31] F. E. Close, *On the Transformation Between Current and Constituent Quarks and Consequences for Polarized Electroproduction Structure Functions,* Nucl. Phys. B **80** (1974) 269.

[32] S. A. Larin, T. van Ritbergen and J. A. M. Vermaseren, *The Alpha-s**3 approximation of quantum chromodynamics to the Ellis-Jaffe sum rule,* Phys. Lett. B **404** (1997) 153.

[33] S. D. Bass and A. W. Thomas, *The Nucleon's octet axial-charge g(A)**(8) with chiral corrections,* Phys. Lett. B **684** (2010) 216 [arXiv:0912.1765 [hep-ph]].

[34] V. Y. Alexakhin *et al.* [COMPASS Collaboration], *The Deuteron Spin-dependent Structure Function g_1^d and its First Moment,* Phys. Lett. B **647** (2007) 8 [hep-ex/0609038].

[35] B. Lampe, *A Short review and some new results on polarization,* Nucl. Phys. Proc. Suppl. **54A** (1997) 134 [hep-ph/9608400].
[36] D. J. Gross and F. Wilczek, *Asymptotically Free Gauge Theories. 2.*, Phys. Rev. D **9** (1974) 980.

EPJ Web of Conferences **125**, 04017 (2016) DOI: 10.1051/epjconf/201612504017
QUARKS-2016

Photon splitting in a strongly magnetized, charge-asymmetric plasma

Chistyakov M.V.[1,⋆], *Rumyantsev* D.A.[1,⋆⋆], and *Shlenev* D.M.[1,2,⋆⋆⋆]

[1] *Division of Theoretical Physics, Department of Physics, Yaroslavl State P. G. Demidov University, Sovietskaya 14, 150000 Yaroslavl, Russian Federation*
[2] *Division of Physics, A. F. Mozhaiskiy Space Military Academy, Yaroslavl Branch, Moskovskiy Prosp. 28, 150001 Yaroslavl, Russian Federation*

Abstract. The process of the photon splitting, $\gamma \to \gamma\gamma$, is investigated in the presence of strongly magnetized charge-asymmetric cold plasma. The dispersion properties of photons and the new polarization selection rules are obtained in such plasma. The absorption rate of the leading photon splitting channel are calculated with taking account of the photon dispersion and wave function renormalization. In addition, a comparison of the photon splitting and the Compton scattering processes is performed.

1 Introduction

The process of photon splitting into two photons is a prominent example of the external active medium influence on the reactions with elementary particles. Though it is forbidden in vacuum by charge conjugation symmetry of QED known as Furry's theorem, it becomes allowed in the presence of an external electromagnetic field and/or plasma. In spite of the rather long history of the investigation [1–5] (for a review see e.g. [6–8]) the process of photon splitting still attracts a great interest stimulated by its possible applications.

It is remarkable, that such an exotic process may lead to observable physical manifestations in astrophysical objects. Its principal effect is to degrade photon energies and thereby to soften the gamma-ray spectra from neutron stars. In particular, it is supposed, that the photon splitting could explain a peculiarities in the gamma-ray spectra of some radio pulsars [9].

There is one more interesting application of the process under consideration concerning the radio quiescence of soft gamma repeaters (SGRs) and anomalous X-ray pulsars (AXPs). Because the photon splitting has no threshold, high-energy photons propagating at very small angles to the magnetic field in the neutron star magnetospheres may split before reaching the threshold of the pair production. Therefore, this process could change the production efficiency of electron-positron pairs, required for a detectable radio emission [10–13].

⋆e-mail: mch@uniyar.ac.ru
⋆⋆e-mail: rda@uniyar.ac.ru
⋆⋆⋆e-mail: allen_caleb@rambler.ru

The process discussed here also plays an important role in the model of the SGR burst [14, 15]. Namely, a formation is assumed of a "magnetically trapped fireball" in the near magnetosphere of a magnetar with hot e^+e^--photon plasma in a local thermodynamic equilibrium. Notice that in all of these astrophysical models the photon splitting process in the presence of strong magnetic field is considered in the environment which is not necessarily approximated by vacuum. Instead, the presence of considerably dense and hot electron-positron plasma is possible. This plasma could significantly change the photon splitting rate.

Note, that the scale of the magnetic field of the above objects may reach the critical value $B_e = m^2/e \simeq 4.41 \times 10^{13}$ G [1] or even exceed it, $B \gg B_e$ [16].

There are two ways in which a magnetized plasma influences the process under consideration. On the one hand, it modifies the photon dispersion properties. On the other hand, it can change the process amplitude. The first effect has been studied previously in Refs. [2, 17]. It was shown (see [2]) that in the case of the cold weakly magnetized plasma the process kinematics and the polarization selection rules remain the same as in the magnetized vacuum if the electron density is not too large ($n_e \lesssim 10^{19}\text{cm}^{-3}$). In Ref. [17], the probability of the photon splitting was calculated by including the plasma effects on the photon dispersion but using the process amplitude obtained in the weak magnetic field without plasma. It was found that the plasma influence in such an approach was negligibly small, except the very narrow region of the plasma and magnetic field parameter space.

The modification of the photon splitting amplitude in the presence of a magnetic field and plasma was considered in Refs. [18, 19] on the basis of the Euler-Heisenberg effective Lagrangian with taking account of the thermal corrections in the one-loop and two-loop approximation correspondingly. It was noted that in the low temperature limit the process $\gamma \to \gamma\gamma$ could compete with the other absorption reactions such as the Compton scattering process.

Another approach was applied in Ref. [20]. Using the expansion of the electron propagator over the magnetic field strength, the amplitude and the absorption coefficient of the photon splitting were calculated in the high energy limit. The main conclusion of the paper was that the plasma effect was negligible. However, the estimations of the photon splitting absorption coefficients in a magnetic field presented in the paper in the high energy limit were incorrect because these expressions were applicable only in the low energy approximations. Note that in Refs. [18–20] the effects of the photon dispersion in magnetized plasma were not taken into account. However, the situation may change significantly in the limit of a strong magnetic field and plasma.

The self-consistent investigation of the photon splitting/merging and the Compton scattering including modifications of both dispersion relations and the process amplitude in magnetized plasma is provided in the recent paper [8]. The case of charge-symmetric plasma was studied in Ref. [8], however, a very interesting case of charge-asymmetric plasma with non zero chemical potential was not investigated in that paper. From the point of of possible astrophysical applications, such situation is, in our opinion, more realistic, and it is realized, for example, in the surface layer of a neutron star, where the process $\gamma \to \gamma\gamma$ can play an important role in forming an emission spectrum in this region (see, for example, Ref. [21]).

2 Photon dispersion in the magnetized medium

We begin to consider the processes $\gamma \to \gamma\gamma$ and $e\gamma \to e\gamma$ with investigation of the photon dispersion properties. The propagation of an electromagnetic radiation in any active medium is conveniently described in terms of normal modes (eigenmodes). In turn, the polarization and dispersion properties of the normal modes are defined by the eigenvectors and eigenvalues of the polarization operator $\mathcal{P}_{\alpha\beta}$,

[1] We use the natural units $c = \hbar = k = 1$, m is the electron mass, $e > 0$ is the elementary charge.

respectively. In this case, it is convenient to decompose the tensor $\mathcal{P}_{\alpha\beta}$ in terms of the basis of 4-vectors [22] constructed from the electromagnetic field tensor reduced to a dimensionless form, and the 4-momentum of the photon q_α:

$$b_\mu^{(1)} = (\varphi q)_\mu, \qquad b_\mu^{(2)} = (\tilde{\varphi} q)_\mu, \tag{1}$$
$$b_\mu^{(3)} = q^2 (\Lambda q)_\mu - q_\mu q_\perp^2, \qquad b_\mu^{(4)} = q_\mu .$$

Wherein $(b^{(1)}b^{*(1)}) = -q_\perp^2$, $(b^{(2)}b^{*(2)}) = -q_\parallel^2$, $(b^{(3)}b^{*(3)}) = -q^2 q_\parallel^2 q_\perp^2$, $(b^{(4)}b^{*(4)}) = q^2$.

On the other hand, the presence of plasma can change the set of eigenvectors of the polarization operator compared in the case of a pure magnetic field. Then, in the plasma rest frame, we can represent the tensor $\mathcal{P}_{\alpha\beta}$ as an expansion over new eigenvectors $r_\alpha^{(\lambda)}$ with eigenvalues $\varkappa^{(\lambda)}$ [23]:

$$\mathcal{P}_{\alpha\beta} = \sum_{\lambda=1}^{3} \varkappa^{(\lambda)} \frac{r_\alpha^{(\lambda)} (r_\beta^{(\lambda)})^*}{(r^{(\lambda)})^2}, \tag{2}$$

$$r_\beta^{(\lambda)} = \sum_{i=1}^{3} A_i^{(\lambda)} b_\beta^{(i)}, \tag{3}$$

where $A_i^{(\lambda)}$ are some complex coefficients.

Hereafter we use the following notations: the four-vectors with the indices $\perp$ and $\parallel$ belong to the Euclidean $\{1, 2\}$ subspace and the Minkowski $\{0, 3\}$ subspace correspondingly. Then for arbitrary 4-vectors A_μ, B_μ one has

$$A_\perp^\mu = (0, A_1, A_2, 0), \quad A_\parallel^\mu = (A_0, 0, 0, A_3),$$
$$(AB)_\perp = (A\Lambda B) = A_1 B_1 + A_2 B_2,$$
$$(AB)_\parallel = (A\widetilde{\Lambda} B) = A_0 B_0 - A_3 B_3,$$

where the matrices $\Lambda_{\mu\nu} = (\varphi\varphi)_{\mu\nu}$, $\widetilde{\Lambda}_{\mu\nu} = (\tilde{\varphi}\tilde{\varphi})_{\mu\nu}$ are constructed with the dimensionless tensor of the external magnetic field, $\varphi_{\mu\nu} = F_{\mu\nu}/B$, and the dual tensor, $\tilde{\varphi}_{\mu\nu} = \frac{1}{2}\varepsilon_{\mu\nu\rho\sigma}\varphi^{\rho\sigma}$. The matrices $\Lambda_{\mu\nu}$ and $\widetilde{\Lambda}_{\mu\nu}$ are connected by the relation $\widetilde{\Lambda}^{\mu\nu} - \Lambda^{\mu\nu} = g^{\mu\nu} = \mathrm{diag}(1, -1, -1, -1)$, and play the roles of the metric tensors in the perpendicular ($\perp$) and the parallel ($\parallel$) subspaces respectively.

In the case of strongly magnetized cold plasma with the temperature T and the chemical potential μ, in the limits $T \ll \mu - m$ and $eB \gg m^2, \mu^2, T^2$, we obtain the following coefficients $A_i^{(\lambda)}$ in the decomposition (3) with an accuracy of $O(1/(eB)^2)$ [22, 23]:

$$A_1^{(1,3)} = \mp \sqrt{q_\perp^4 + (6\Delta N\, \omega)^2 \frac{q^2}{q_\parallel^2}} - q_\perp^2, \tag{4}$$

$$A_2^{(1,3)} = \mathrm{i}\, \frac{\Delta N\, k_z\, q_\perp^2}{2eB\, \mathcal{D}(q_\parallel)\, q_\parallel^2} \left[\pm \sqrt{q_\perp^4 + (6\Delta N\, \omega)^2 \frac{q^2}{q_\parallel^2}} + q_\perp^2 \right],$$

$$A_3^{(1,3)} = -\mathrm{i}\, \frac{6\Delta N\, \omega}{q_\parallel^2}, \quad A_1^{(2)} = -\mathrm{i}\, \frac{\Delta N\, k_z}{2eB\, \mathcal{D}(q_\parallel)}, \quad A_2^{(2)} = 1, \quad A_3^{(2)} = 0,$$

where

$$\mathcal{D}(q_\parallel) = -\mathcal{J}(q_\parallel) - H\left(\frac{q_\parallel^2}{4m^2}\right), \tag{5}$$

DOI: 10.1051/epjconf/201612504017

$$\mathcal{J}(q_\parallel) = -\frac{1}{2\sqrt{z(1-z)}}\left(\arctan\left[\frac{v_F - v_\phi + zv_F(v_\phi^2 - 1)}{(v_\phi^2 - 1)\sqrt{z(1-z)}}\right] + \right. \tag{6}$$

$$\left. + \arctan\left[\frac{v_F + v_\phi + zv_F(v_\phi^2 - 1)}{(v_\phi^2 - 1)\sqrt{z(1-z)}}\right]\right), \quad 0 \leqslant z \leqslant 1,$$

$$\mathcal{J}(q_\parallel) = -\frac{1}{4\sqrt{z(z-1)}}\left(\ln\left[\frac{v_F - v_\phi + (v_\phi^2 - 1)(zv_F - \sqrt{z(z-1)})}{v_F - v_\phi + (v_\phi^2 - 1)(zv_F + \sqrt{z(z-1)})}\right] + \right. \tag{7}$$

$$\left. + \ln\left[\frac{v_F + v_\phi + (v_\phi^2 - 1)(zv_F - \sqrt{z(z-1)})}{v_F + v_\phi + (v_\phi^2 - 1)(zv_F + \sqrt{z(z-1)})}\right]\right) - \frac{i\pi\,\Theta(v_F|v_\phi| - 1)}{2\sqrt{z(z-1)}}, \qquad z > 1,$$

$$z = \frac{q_\parallel^2}{4m^2}, \quad v_F = \frac{\sqrt{\mu^2 - m^2}}{\mu}, \quad v_\phi = \frac{\omega}{k_z}, \quad \Delta N = 2p_F = 2\sqrt{\mu^2 - m^2}, \tag{8}$$

$\Theta(x)$ is the theta-function,

$$H(z) = \frac{1}{\sqrt{z(1-z)}}\arctan\sqrt{\frac{z}{1-z}} - 1, \quad 0 \leqslant z \leqslant 1, \tag{9}$$

$$H(z) = -\frac{1}{2\sqrt{z(z-1)}}\ln\frac{\sqrt{z} + \sqrt{z-1}}{\sqrt{z} - \sqrt{z-1}} - 1 + \frac{i\pi}{2\sqrt{z(z-1)}}, \quad z > 1.$$

As one can see from the decomposition of the eigenvectors (3) over the basis (1) with the coefficients (4), in the case $\mu \neq 0$ and in a strong magnetic field, the photons have an elliptical polarization. However, in the cold, almost degenerate, moderately relativistic plasma, $\Delta N/(2m) \simeq v_F \ll 1$, and from Eqs. (4) we obtain that the physical polarization vectors of the photons are just the same as in the charge-symmetric magnetized plasma [8]:

$$\varepsilon_\alpha^{(1)}(q) = \frac{(q\varphi)_\alpha}{\sqrt{q_\perp^2}}, \qquad \varepsilon_\alpha^{(2)}(q) = \frac{(q\tilde{\varphi})_\alpha}{\sqrt{q_\parallel^2}}. \tag{10}$$

In this approximation, the corresponding eigenvalues have the form

$$\varkappa^{(1)} = -\frac{\alpha}{3\pi}q_\perp^2 + O\left(\frac{1}{(eB)^2}\right), \tag{11}$$

$$\varkappa^{(2)} = \frac{2\alpha}{\pi}eB\,\mathcal{D}(q_\parallel) + O\left(\frac{1}{eB}\right), \tag{12}$$

$$\varkappa^{(3)} = O\left(\frac{1}{(eB)^2}\right). \tag{13}$$

DOI: 10.1051/epjconf/201612504017

In Figs. 1 and 2, the dispersion law of the mode 2 photon in both strong magnetic field and cold magnetized plasma is depicted as the solution of the equation

$$q^2 - \varkappa^{(2)} = 0 \tag{14}$$

for various values of the chemical potential, propagation angles and photon momenta.

Moreover, as can be seen from Fig. 1, in degenerate plasma there arises the shift of the e^+e^- pair-creation threshold which is defined in a pure magnetic field by the relation $q_\parallel^2 = 4m^2$. One can see that in the region $|v_\phi| > 1/v_F$ ($|k_z| < 2p_F$) the last terms in Eqs. (10) and (7) cancel each other and the only contribution to the imaginary part of $\varkappa^{(2)}(q)$ comes from the logarithm function in Eq. (7). It is the fact that leads to the shift of the pair creation threshold from $q_\parallel^2 = 4m^2$ to

$$q_\parallel^2 = 2\left(\mu^2 - p_F|k_z| + \mu\,\sqrt{(p_F - |k_z|)^2 + m^2}\right). \tag{15}$$

This result is in agreement with a simple kinematical analysis of the process $\gamma_2 \to e^+e^-$ in degenerate plasma. Indeed, using the energy and momentum conservation laws with obviuos conditions $E \geqslant \mu$ and $|p_z| \geqslant p_F$ for the electron we come to the result (15).

It follows from Eq. (6) and (9) that the eigenvalue of the polarization operator $\varkappa^{(2)}$ becomes large near the electron-positron pair production threshold (see Eq. (15) and Figs. 1 and 2). This means that the renormalization of the wave function for a photon of this polarization should be taken into account:

$$\varepsilon_\alpha^{(2)}(q) \to \varepsilon_\alpha^{(2)}(q)\,\sqrt{Z_2}, \quad Z_2^{-1} = 1 - \frac{\partial \varkappa^{(2)}(q)}{\partial \omega^2}. \tag{16}$$

It is natural to expect that the presence of plasma can change the selection rules for the polarizations suggested in Ref. [2]. As one can see from Eqs. (11)-(13), only the eigenvalue $\varkappa^{(2)}(q)$ is modified in charge-asymmetric cold plasma. It means that the dispersion law of the mode 1 photon is the same as in the magnetized vacuum, where its 4-momentum is space-like and the deviation from the vacuum dispersion law, $q^2 = 0$, is negligibly small. On the other hand, the dispersion properties of the mode 2 photon essentially differ from the ones in magnetized vacuum. In this case the dispersion law in the form of the relation between $q_\parallel^2$ and $q_\perp^2$ additionally depends on the angle between the magnetic field direction and the photon momentum $q_\parallel^2 = q_\parallel^2(q_\perp^2, \theta)$.

One can see from Figs. 1 and 2, that in the presence of magnetized plasma there exists the kinematical region, where $q^2 > 0$ contrary to the case of a pure magnetic field. It is connected with the appearance of the plasma frequency in the presence of the real electrons which can be defined from the equation

$$\omega_{pl}^2 - \varkappa^{(2)}(\omega_{pl}, \mathbf{k} \to 0) = 0. \tag{17}$$

These facts lead to new polarization selection rules: in the region $q^2 > 0$, a new photon splitting channel $\gamma_2 \to \gamma_1\gamma_1$ forbidden in the magnetic field without plasma, is possible, while the splitting channels $\gamma_1 \to \gamma_2\gamma_2$ and $\gamma_1 \to \gamma_1\gamma_2$ allowed in the pure magnetic field, are forbidden. In the region $q^2 < 0$, polarization selection rules are the same ones as in the magnetized vacuum [2].

It should be noted that, in the limit of cold plasma, $T \ll \mu - m$, the photons with energies near $\omega \sim T \ll m$, will participate in the splitting. Thus, the "new" splitting channel, $\gamma_2 \to \gamma_1\gamma_1$, will be dominating in our conditions (see also Sec. 4).

The partial amplitude for the channel $\gamma_2 \to \gamma_1\gamma_1$ could be obtained from the results of the papers [7, 8] by the substitution of the corresponding polarization vectors (10), and it can be represented in the following form:

$$\mathcal{M}_{2\to 11} = i4\pi \left(\frac{\alpha}{\pi}\right)^{3/2} \frac{(q'\varphi q'')(q'\tilde{\varphi} q'')}{[q_\parallel^2 q_\perp'^2 q_\perp''^2]^{1/2}} \mathcal{D}(q_\parallel), \tag{18}$$

where $\mathcal{D}(q_\parallel)$ is given by the formula (5).

3 The probability of the photon splitting

To analyze the efficiency of the process under consideration and to compare it with other competitive reactions we calculate the photon absorption rate due to the photon splitting which can be defined in the following way:

$$W_{\lambda\to\lambda'\lambda''} = \frac{g_{\lambda'\lambda''}}{32\pi^2\omega_\lambda} \int \left|\mathcal{M}_{\lambda\to\lambda'\lambda''}\right|^2 Z_\lambda Z_{\lambda'} Z_{\lambda''} (1 + f_{\omega'})(1 + f_{\omega''}) \times \tag{19}$$

$$\times \delta(\omega_\lambda(\mathbf{k}) - \omega_{\lambda'}(\mathbf{k} - \mathbf{k}'') - \omega_{\lambda''}(\mathbf{k}'')) \frac{d^3k''}{\omega_{\lambda'}\omega_{\lambda''}},$$

where $f_\omega = [e^{\omega/T} - 1]^{-1}$ is the photons distribution function, the factor $g_{\lambda'\lambda''} = 1 - (1/2)\,\delta_{\lambda'\lambda''}$ is inserted to account for the possible identity of the final photons.

As it was found in Sec. 2, in the presence of strongly magnetized plasma the "new" photon splitting channel $\gamma_2 \to \gamma_1\gamma_1$ forbidden in pure magnetic field, becomes open. According to Eqs. (18) and (19), the absorption rate of the channel $\gamma_2 \to \gamma_1\gamma_1$ can be presented as

$$W_{2\to 11} = \frac{\alpha^3}{8\pi^2} Z_2\, \mathcal{D}^2(q_\parallel) \frac{q_\perp^2}{\omega} \mathcal{F}\left(\sqrt{\frac{q_\parallel^2}{q_\perp^2}}\right) \Theta(q^2), \tag{20}$$

where

$$\mathcal{F}(z) = \frac{4z^3}{\pi} \int_{1-z}^{1+z} dx \int_0^{y_0} dy \frac{y^2\sqrt{y_0^2 - y^2}}{(x^2 + z^2y^2)[(2-x)^2 + z^2y^2]} \times$$

$$\{(1 + f_{\omega_+})(1 + f_{\omega-\omega_+}) + (\omega_+ \to \omega_-)\},$$

$$\times \omega_\pm = \frac{1}{2z^2}\left[\omega(z^2 - 1 + x) \pm k_z z^2 \sqrt{y_0^2 - y^2}\right],$$

$$y_0 = \frac{\sqrt{z^2-1}}{z^2}\sqrt{z^2 - (1-x)^2}, \quad z = \sqrt{\frac{q_\parallel^2}{q_\perp^2}}.$$

If one can neglect the effect of the final photon stimulation emission ($f_{\omega'} = f_{\omega''} = 0$), the expression for the function $\mathcal{F}(z)$ is considerably simplified:

$$\mathcal{F}(z) = 2\ln z - 1 + z^{-2}.$$

In Fig. 3, the absorption rate of the channel $\gamma_2 \to \gamma_1\gamma_1$ as the function of the initial photon energy is depicted for the case of the photon propagation across the magnetic field direction, for the chemical potential of 1 MeV and 0.75 MeV. One can see that the probability of the "new" channel, $\gamma_2 \to \gamma_1\gamma_1$, decreases rapidly with the chemical potential decreasing. This is due to an increase of the kinematically allowed region ($q^2 > 0$) for the channel under consideration leading to the growth of the process phase space volume with the chemical potential increasing.

The dependence of the $\gamma_2 \to \gamma_1\gamma_1$ splitting rate on the angle between the initial photon momentum and the magnetic field direction, for chemical potential of 0.75 MeV is depicted in Fig. 4.

It is interesting to note that the qualitative behavior of the probability in the case of a cold, almost degenerate plasma is similar to the case of hot charge-symmetric plasma considered in Ref. [8]. This result is likely due to the fact, that in the cases $T \ll \mu - m$ and $\mu = 0$ and in the region of $q^2 > 0$, the photon dispersion law is qualitatively the same, whereas the quantitative result is determined, basically, by the value of the phase volume of the reaction.

4 Photon scattering

Compton scattering is another type of the photon reaction, competing with the photon splitting. The process of the photon scattering in a strongly magnetized plasma, in degenerate plasma ($T \ll \mu - m$), when the initial photon propagates across the magnetic field, for different chemical potential values and with taking account of the photon dispersion and the wave function renormalization was investigated recently (see Ref. [24]). Here we generalize our results to the case of an arbitrary angle between the initial photon momenta and the magnetic field direction.

In the degenerate plasma ($T \ll \mu - m$), the absorption rates have the following form [2]:

$$W_{1\to 1} \simeq \frac{\alpha^2}{4\pi}\frac{B_e}{B}\frac{\omega^3}{m^2 v_F^2}\left\{(1 - v_F n_z)^2 \int\limits_{t_1}^{t_2} dt\, t\, \mathcal{F}\left(\frac{\omega}{T}, t\right) + (n_z \to -n_z)\right\}, \tag{21}$$

$$W_{2\to 1} \simeq \frac{\alpha^2}{4\pi}\frac{B_e}{B}\frac{q_\perp^2}{q_\parallel^2}\frac{\omega^3}{m^2 v_F^2} Z_2\, \Theta(\omega - \omega_{pl})\left\{(v_F - n_z)^2 \int\limits_{t_1}^{t_2} dt\, t\, \mathcal{F}\left(\frac{\omega}{T}, t\right) + (n_z \to -n_z)\right\}, \tag{22}$$

$$W_{1\to 2} \simeq \frac{\alpha^2}{4\pi}\frac{B_e}{B}\frac{\omega^3}{m^2}\left\{\Theta\left(\frac{\mu}{m}(1 - n_z v_F)\,\omega - \omega_{pl}\right) \int\limits_{t_1^*}^{t_2^*} dt\, t\, \mathcal{F}\left(\frac{\omega}{T}, t\right) \times \right. \tag{23}$$
$$\left. \times\left(1 + \frac{m^2}{p_F^2} t - \frac{n_z}{v_F}\right)^2 \left(1 - \frac{\varkappa^{(2)}(q')}{q_\parallel'^2}\right)\Bigg|_{q_\parallel'^2 = \omega^2[(1-t)^2 - (n_z - t/v_F)^2],\, k_z' = \omega(n_z - t/v_F)} + (n_z \to -n_z)\right\},$$

[2]The details of calculations of the photon absorption rate for the channel $e\gamma_2 \to e\gamma_2$ can be found in Ref. [23].

$$W_{2\to 2} \simeq \frac{\alpha^2}{2\pi}\frac{B}{B_e}\left(\frac{m}{\mu}\right)^6 \frac{q_\parallel^2}{\omega v_F^4} Z_2 \left\{ \frac{\Theta\left(\frac{\mu}{m}(1-n_z v_F)\,\omega - \omega_{pl}\right)}{(1-n_z v_F)^4} \int\limits_{t_1^*}^{t_2^*} dt\, t\, \mathcal{F}\left(\frac{\omega}{T}, t\right) \times \right. \tag{24}$$

$$\times \left[v_F^2(1-t)^2 - (n_z v_F - t)^2\right] \left[\left(1 - \frac{q_\perp^2}{4m^2}\frac{B_e}{B} - \frac{{q'_\parallel}^2 - \varkappa^{(2)}(q')}{4m^2}\frac{B_e}{B}\right)^2 + \right.$$

$$+ \frac{\mu^2\omega^2}{2m^4}\left(\frac{B_e}{B}\right)^2 \frac{q_\perp^2}{{q'_\parallel}^2}\left(1 - \frac{\varkappa^{(2)}(q')}{{q'_\parallel}^2}\right)\frac{(1-n_z v_F)^2}{(1-n_z^2)^4} \times$$

$$\left. \left. \times \left(1 - n_z^2 - t\left(1 - \frac{n_z}{v_F}\right) - \frac{\mu^2}{m^2}(1-n_z v_F)^2\right)^2 \right] \right|_{{q'_\parallel}^2 = \omega^2[(1-t)^2 - (n_z - t/v_F)^2],\, k'_z = \omega(n_z - t/v_F)} + (n_z \to -n_z) \right\},$$

where

$$\mathcal{F}(x,t) = \left[1 - \exp(-xt)\right]^{-1}\left[1 - \exp(-x(1-t))\right]^{-1},$$

$$t_{2,1} = \pm\frac{v_F}{1 \pm v_F}(1 \pm n_z), \quad t_{2,1}^* = -\frac{v_F}{1 - v_F^2}\left(v_F - n_z \mp \sqrt{(1 - v_F n_z)^2 - \frac{\omega_{pl}^2}{\omega^2}(1 - v_F^2)}\right),$$

$$n_z = v_\phi^{-1} = \frac{\sqrt{q_\perp^2}}{\omega}\cot\theta.$$

One can see from Eqs. (21)-(25), that the absorption coefficients have significantly different dependences on the magnetic field strength, the energy and the chemical potential for the 1-mode and 2-mode photons. On the one hand, the absorption rates $W_{1\to 1}$, $W_{2\to 1}$ and $W_{1\to 2}$ are inversely proportional to the magnitude of the magnetic field and strongly depend on the initial photon frequency to give a ground to expect the efficient diffusion of the 1-mode photon, as it takes place in hot charge-symmetric plasma case [8]. On the other hand, one can see that the absorption rate $W_{2\to 2}$ is directly proportional to the magnetic field value and strongly depends on the chemical potential in contrast to the other channels. It is interesting to note that in the emission region with smaller μ [3] it is seen that the large value of $W_{2\to 2}$ and its weak dependence on the energy could lead to the trapping of the 2-mode. Therefore, such photons quickly come into the equilibrium with medium. Thus, the mode-1 photons are interesting for the not-blackbody spectrum formation.

Previous investigations of the radiation transfer problem in strongly magnetized plasma have shown that along with the Compton scattering process, the photon splitting $\gamma \to \gamma\gamma$ could play a significant role as a mechanism of the photon production [8, 14]. As it was noted earlier (see also Sec. 3), in our conditions the major channels of the mode 2 photon conversion will be the splitting $\gamma_2 \to \gamma_1\gamma_1$ and the scattering $e\gamma_2 \to e\gamma_1$. To compare these processes, we have estimated the probability of the photon splitting and the Compton scattering rates near $\omega \sim \omega_{pl} \simeq \sqrt{(2\alpha e B/\pi)} v_F$ and for nonrelativistic cold plasma ($v_F \ll 1$) in the following way

$$W_{2\to 11} \simeq \frac{\alpha^3 v_F^2}{8\pi^2}\,\omega_{pl}\left(1 - \frac{\omega_{pl}^2}{\omega^2}\right)\sin^2\theta \times \tag{25}$$

$$\times\left\{-\ln\left[\left(1 - \frac{\omega_{pl}^2}{\omega^2}\right)\sin^2\theta\right] - \cos^2\theta - \frac{\omega_{pl}^2}{\omega^2}\sin^2\theta\right\}\frac{\Theta(\omega - \omega_{pl})}{(1 - \mathrm{e}^{-\omega_{pl}/(2T)})^2},$$

[3] This situation is realized, for example, in the surface layer of a neutron star with the density $\rho \sim 10^6$ g/cm^3.

$$W_{2\to 1} \simeq \frac{2\alpha^3 v_F^2}{\pi^2} T \left(1 - \frac{\omega_{pl}^2}{\omega^2}\right) \sin^2\theta \, \frac{\Theta(\omega - \omega_{pl})}{1 - e^{-\omega_{pl}/T}} . \tag{26}$$

It follows from Eqs. (25) and (26), that $W_{2\to 11} \sim W_{2\to 1}$ in some region of the plasma parameters. Thus, we can see now that in the presence of cold plasma the process of the photon splitting could not be the only intensive source of the photon production but also an effective absorption mechanism. Therefore, the both processes of the photon splitting and the photon scattering are important for the X-ray spectra formation.

5 Conclusion

The obtained results show that the plasma influence modifies the polarization selection rules in comparison with a pure magnetic field. In particular, the "new" splitting channel $\gamma_2 \to \gamma_1\gamma_1$ which is forbidden without plasma, is allowed in the plasma presence.

On the other hand, the presence of plasma suppresses the probabilities of the channels $\gamma_1 \to \gamma_1\gamma_2$ and $\gamma_1 \to \gamma_2\gamma_2$ in comparison with a pure magnetic field.

The comparison of the photon splitting process and the Compton scattering shows that the influences of these reactions on the 1-mode radiation transfer are competitive in the cold plasma ($T \ll \mu - m$). As a result, it could lead to the modification of the mechanism of the emission spectra formation of neutron stars.

Acknowledgements
We are grateful to A. V. Kuznetsov for useful discussions and valuable remarks. D. R. expresses his deep gratitude to the organizers of the Seminar "Quarks-2016" for warm hospitality.

The study was supported in part by the Russian Foundation for Basic Research (Project No. 14-02-00233-a). The work of D. R. was performed with the support by the Project No. 984 within the base part of the State Assignment for the Yaroslavl University Scientific Research.

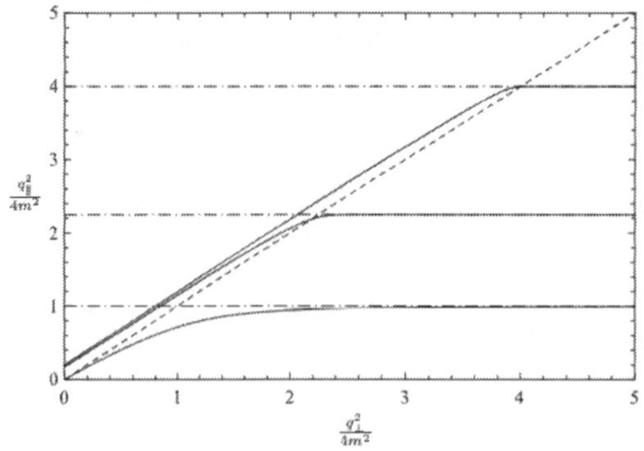

Figure 1. The photon dispersion in a strong magnetic field ($B/B_e = 200$) and degenerate ($T = 0$) plasma vs. chemical potential $\mu = 1$ MeV (upper curve), $\mu = 0.75$ MeV (middle curve) and $\mu = 0.5$ MeV (lowest curve). The dotted line corresponds to the vacuum dispersion law, $q^2 = 0$. The angle between the photon momentum and the magnetic field direction is $\pi/2$.

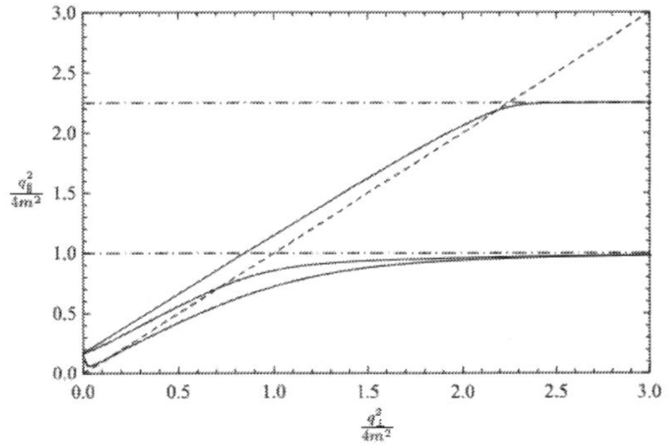

Figure 2. The photon dispersion in a strong magnetic field ($B/B_e = 200$) and degenerate plasma at $\mu = 0.75$ MeV vs. the angle between the photon momentum and the magnetic field direction $\theta = \pi/2$ (upper curve), $\theta = \pi/4$ (middle curve) and $\theta = \pi/12$ (lowest curve). The dotted line corresponds to the vacuum dispersion law, $q^2 = 0$.

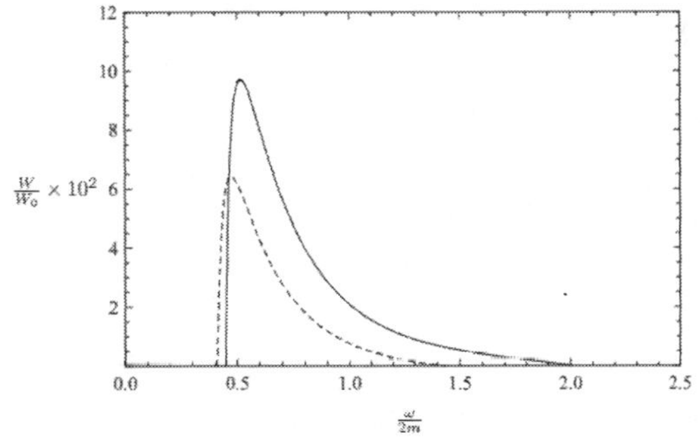

Figure 3. The photon absorption rate due to the channel $\gamma_2 \to \gamma_1\gamma_1$ in a strong magnetic field ($B/B_e = 200$) and cold plasma ($T = 50$ keV) at chemical potential 1 MeV (solid line) and 0.75 Mev (dashed line). Here, $\theta = \pi/2$, $W_0 = (\alpha/\pi)^3 m \simeq 3.25 \cdot 10^2 \text{cm}^{-1}$.

References

[1] Z. Bialynicka-Birula, L. I. Bialynicka-Birula, Phys. Rev. **D 2**, 2341 (1970).
[2] S. L. Adler, Ann. Phys. (N.Y.) **67**, 599 (1971).
[3] V. O. Papanian, V. I. Ritus, Sov. Phys. JETP **34**, 1195 (1972).
[4] M. V. Chistyakov, A. V. Kuznetsov, N. V. Mikheev, Phys.Lett. **B434**, 67 (1998).
[5] J. I. Weise, Phys. Rev. **D 69**, 105017 (2004).
[6] V. O. Papanian, V. I. Ritus: Three-Photon Interaction in an Intense Field. In: *Issues in Intense-Field Quantum Electrodynamics*, ed. by V.L. Ginzburg (Nova Science Publishers, New York, 1989) pp. 153-179.
[7] D. A. Rumyantsev, M. V. Chistyakov, *JETP* **101**, 635 (2005).
[8] D. A. Rumyantsev, M. V. Chistyakov, N. S. Stus' Phys. Rev. **D 86**, 043007 (2012).
[9] A. K. Harding, M. G. Baring, P. L. Gonthier, Astrophys.J. **476**, 246 (1997).
[10] M. G. Baring, Astrophys. J. **440**, L69 (1995).
[11] M. G. Baring, A. K. Harding, Astrophys. J. Lett. **507**, L55 (1998).
[12] V. M. Malofeev, O. I. Malov, D. A. Teplykh, et al., Astron. Rep., **49**, 242 (2005).

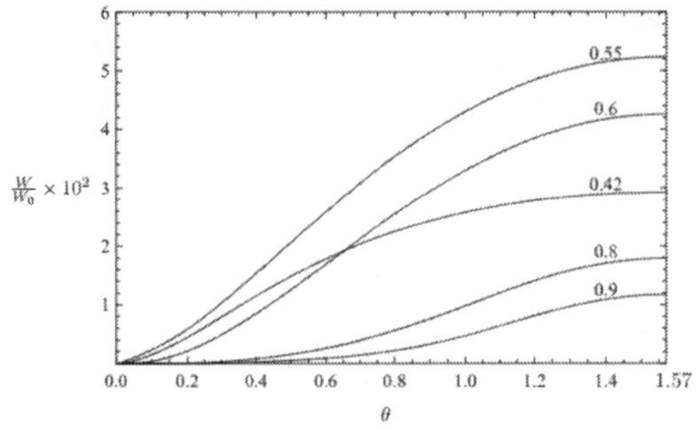

Figure 4. The dependence of the absorption rate of the channel $\gamma_2 \to \gamma_1\gamma_1$ on the angle between the initial photon momentum and the magnetic field direction at different initial photon energies ($B/B_e = 200$, $T = 50$keV, $\mu = 0.75$MeV). The numbers above the curves correspond to the values of the ratio $\omega/2m$.

[13] Y. N. Istomin, D. N. Sobyanin, Astronomy Letters, **33**, 660 (2007).
[14] C. Thompson, R. C. Duncan, Mon. Not. Roy. Astron. Soc. **275**, 255 (1995).
[15] C. Thompson, R. C. Duncan, Astrophys. J. **561**, 980 (2001).
[16] R. C. Duncan, C. Thompson, Astrophys. J. **392**, L9 (1992).
[17] T. Bulik, Acta Astronomica. **48**, 695 (1998).
[18] P. Elmfors, B. Skagerstam, Phys. Lett. **B427**, 197 (1998).
[19] H. Gies, Phys. Rev. **D61**, 085021 (2000).
[20] J. M. Martinez Resco, M. A. Valle Basagoiti, Phys. Rev. **D64**, 016006 (2001).
[21] V. Suleimanov, J. Poutanen, K. Werner, Astronomy & Astrophysics. **545**, A120 (2012).
[22] A. E. Shabad, Tr. Fiz. Inst. Akad. Nauk SSSR **192**, 5 (1988).
[23] N. V. Mikheev, D. A. Rumyantsev, M. V. Chistyakov, Zh. Eksp. Teor. Fiz. **146**, 289 (2014) [J. Exp. Theor. Phys. **119**, 251 (2014)].
[24] D. A. Rumyantsev, M. V. Chistyakov, Int. J. Mod. Phys. **A 24**, 3995 (2009).

EPJ Web of Conferences **125**, 04018 (2016) DOI: 10.1051/epjconf/201612504018
QUARKS-2016

Electromagnetic properties of neutrinos: three new phenomena in neutrino spin oscillations

Alexander Studenikin[1,2,⋆]

[1] *Department of Theoretical Physics, Faculty of Physics, Lomonosov Moscow State University, Moscow 119991, Russia*
[2] *Joint Institute for Nuclear Research, Dubna 141980, Moscow Region, Russia*

Abstract. In studies of neutrino electromagnetic properties we discuss three very interesting aspects related to neutrino spin oscillations. First we consider neutrino mixing and oscillations in the mass and flavour bases under the influence of a constant magnetic field with nonzero transversal and longitudinal components. Then we discuss the effect of neutrino spin oscillations induced by electroweak interactions of neutrino with moving matter in case there is matter transversal current or polarization. In the final part of the paper we discuss recently developed approach to description of neutrino spin and spin-flavour oscillations in a constant magnetic field that is based on the use of the exact neutrino stationary states in the magnetic field.

1 Introduction

It is well known that massive neutrinos have nontrivial electromagnetic properties, and at least the magnetic moment is not zero [1]. Thus, neutrinos do participate also in the electromagnetic interaction (see [2] for a review). The best terrestrial laboratory upper bound on neutrino magnetic moments is obtained by the GEMMA reactor neutrino experiment [3]. The best astrophysical upper bound was derived from considering stars cooling [4]. The neutrino magnetic moment procession in the transversal magnetic field $\mathbf{B}_\perp$ was first considered in [5], then spin-flavor precession in vacuum was discussed in [6], the importance of the matter effect was emphasized in [7]. The effect of resonant amplification of neutrino spin oscillations in $\mathbf{B}_\perp$ in the presence of matter was proposed in [8, 9], the impact of the longitudinal magnetic field $\mathbf{B}_{||}$ was discussed in [10].

Here below we discuss three very interesting aspects related to the neutrino spin and spin-flavour oscillations:

1) we consider in details [11] neutrino mixing and oscillations in arbitrary constant magnetic field that have $\mathbf{B}_\perp$ and $\mathbf{B}_{||}$ nonzero components in mass and flavour bases ,

2) we show that neutrino spin and spin-flavour oscillations can be induced not only by the neutrino interaction with a magnetic field but also by neutrino interactions with matter in the case when there is a transversal matter current or matter polarization (see [12] and references therein),

3) we develop a new (and more precise than the usual one) approach to description of neutrino spin and spin-flavor oscillations in the presence of an arbitrary magnetic field; our approach [13] is

⋆e-mail: studenik@srd.sinp.msu.ru

based on the use of the stationary states in the magnetic field for classification of neutrino spin states, contrary to the customary approach when the neutrino helicity states are used for this purpose.

2 Neutrino spin oscillations in mass and flavour bases

2.1 Oscillations in mass basis

We start [11] with two neutrino physical states ν_1 and ν_2 having masses m_1 and m_2 and introduce neutrino electromagnetic interaction via magnetic moment matrix $\mu_{\alpha\beta}$, $\alpha, \beta = 1, 2$:

$$H_{EM} = \frac{1}{2}\mu_{\alpha\beta}\bar{\nu}_\beta \sigma_{\mu\nu}\nu_\alpha F^{\mu\nu} + h.c. \,, \tag{1}$$

where $F^{\mu\nu}$ is the electromagnetic field tensor and $\sigma_{\mu\nu} = i/2(\gamma_\mu\gamma_\nu - \gamma_\nu\gamma_\mu)$. In a uniform magnetic field the Hamiltonian (1) becomes

$$H_{EM} = -\mu_{\alpha\alpha'}\bar{\nu}_{\alpha'}\boldsymbol{\Sigma}\mathbf{B}\nu_\alpha + h.c., \quad \Sigma_i = \begin{pmatrix} \sigma_i & 0 \\ 0 & \sigma_i \end{pmatrix}, \tag{2}$$

where σ_i are the Pauli matrices.

In the neutrino oscillation framework, one is interested in evolution of chiral neutrino components within the common neutrino beam state. Since in the ultrarelativistic limit the latter are approximated by free neutrino states with definite helicity $s = \pm 1$ the 4-component basis $(\nu_{1,s=1}, \nu_{1,s=-1}, \nu_{2,s=1}, \nu_{2,s=-1})$ is adopted to describe neutrino beam. With the standard column vector notation, $\nu_m \equiv (\nu_{1,s=1}, \nu_{1,s=-1}, \nu_{2,s=1}, \nu_{2,s=-1})^T$ the neutrino evolution equation relevant to electromagnetic interaction has the Schrödinger-like form

$$i\frac{d}{dt}\nu_m(t) = H_{eff}\nu_m(t). \tag{3}$$

The effective Hamiltonian consists of the vacuum and interaction parts

$$H_{eff} = H_{vac} + H_B \tag{4}$$

where the interaction part is composed with matrix elements of the field interaction Hamiltonian taken over the helicity neutrino states: $H_B = \langle \nu_{\alpha,s}|H_{EM}|\nu_{\alpha',s'}\rangle$.

Let us calculate the effective interaction Hamiltonian under assumption that neutrino moves along the z-axis. From the magnetic field interaction Hamiltonian (1) we have:

$$H^B_{\alpha,s;\alpha',s'} = \langle \nu_{\alpha,s}|H_{EM}|\nu_{\alpha',s'}\rangle = -\frac{\mu_{\alpha,\alpha'}}{2}\int d^3x \nu^\dagger_\alpha \gamma_0 \begin{pmatrix} \boldsymbol{\Sigma}\mathbf{B} & 0 \\ 0 & \boldsymbol{\Sigma}\mathbf{B} \end{pmatrix} \nu_{\alpha'}. \tag{5}$$

In the spinor representation the free neutrino states are given by

$$\nu_{\alpha,s} = C_\alpha \sqrt{\frac{E_\alpha + m_\alpha}{2E_\alpha}} \begin{pmatrix} u_s \\ \frac{\boldsymbol{\Sigma}\mathbf{p}_\alpha}{E_\alpha + m_\alpha} u_s \end{pmatrix} e^{i\mathbf{p}_\alpha x}, \tag{6}$$

where $\mathbf{p}_\alpha$ is the neutrino ν_α momentum. The two-component spinors u_s define neutrino helicity states, and are given by

$$u_{s=1} = \begin{pmatrix} 1 \\ 0 \end{pmatrix}, \quad u_{s=-1} = \begin{pmatrix} 0 \\ 1 \end{pmatrix}. \tag{7}$$

Recall that in the ultrarelativistic limit these are correspondent to the right-handed ν_R and left-handed ν_L chiral neutrinos, respectively.

Substituting (6) into the effective Hamiltonian formula (5) we get

$$H^B_{\alpha,s;\alpha',s'} = -\frac{1}{2}\mu_{\alpha\alpha'}C_\alpha C_{\alpha'} \int d^3x B \left(u_s^\dagger \quad \frac{\Sigma p_\alpha}{E_\alpha+m_\alpha}u_s^\dagger\right)\begin{pmatrix}\Sigma & 0\\ 0 & -\Sigma\end{pmatrix}\begin{pmatrix}u_{s'}\\ \frac{\Sigma p_{\alpha'}}{E_{\alpha'}+m_{\alpha'}}u_{s'}\end{pmatrix}$$

$$\times \frac{\sqrt{(E_\alpha+m_\alpha)(E_{\alpha'}+m_{\alpha'})}}{2\sqrt{E_\alpha E_{\alpha'}}}\exp(i\Delta px). \quad (8)$$

Decomposing the magnetic field vector into longitudinal and transversal with respect to neutrino motion components $\mathbf{B} = \mathbf{B}_\parallel + \mathbf{B}_\perp$ it is possible to show that

$$\mathbf{B}\left(u_s^\dagger \quad \frac{\Sigma p_\alpha}{E_\alpha+m_\alpha}u_s^\dagger\right)\begin{pmatrix}\Sigma & 0\\ 0 & -\Sigma\end{pmatrix}\begin{pmatrix}u_{s'}\\ \frac{\Sigma p_{\alpha'}}{E_{\alpha'}+m_{\alpha'}}u_{s'}\end{pmatrix} =$$

$$u_s^\dagger\left(\Sigma B_\parallel\left(1 - \frac{p_\alpha p_{\alpha'}}{(E_\alpha+m_\alpha)(E_{\alpha'}+m_{\alpha'})}\right) + \Sigma B_\perp\left(1 + \frac{p_\alpha p_{\alpha'}}{(E_\alpha+m_\alpha)(E_{\alpha'}+m_{\alpha'})}\right)\right)u_{s'}. \quad (9)$$

Let us apply the ultrarelativistic condition $\frac{m_\alpha}{E_\alpha} \ll 1$ to the part of the integrand in (8):

$$\left(1 \mp \frac{p_\alpha p_{\alpha'}}{(E_\alpha+m_\alpha)(E_{\alpha'}+m_{\alpha'})}\right)\frac{\sqrt{(E_\alpha+m_\alpha)(E_{\alpha'}+m_{\alpha'})}}{2\sqrt{E_\alpha E_{\alpha'}}} \approx \begin{cases}\gamma^{-1}_{\alpha\alpha'}\\ 1\end{cases} \quad (10)$$

where $\gamma_{\alpha\alpha'} = \frac{1}{2}\left(\frac{m_\alpha}{E_\alpha} + \frac{m_{\alpha'}}{E_{\alpha'}}\right)$ is the transition gamma-factor.

Introducing an angle β between $\mathbf{B}$ and $\mathbf{p}_\alpha$ vectors and assuming that $\mathbf{B}_\perp$ is aligned along the x-axis we further obtain:

$$u^\dagger_{s=1}\Sigma\mathbf{B}u_{s=1} = B\cos\beta, \quad u^\dagger_{s=1}\Sigma\mathbf{B}u_{s=-1} = B\sin\beta, \quad (11)$$

$$u^\dagger_{s=-1}\Sigma\mathbf{B}u_{s=1} = B\sin\beta, \quad u^\dagger_{s=-1}\Sigma\mathbf{B}u_{s=-1} = -B\cos\beta. \quad (12)$$

As it was expected, in neutrino transitions without change of helicity only the $B_\parallel = B\cos\beta$ component of the magnetic field contribute to the effective potential, whereas in transitions with change of the neutrino helicity the transversal component $B_\perp = B\sin\beta$ matters.

Performing remaining some simple algebra one can readily write out the H_B matrix. For the effective Hamiltonian with the diagonal vacuum part H_{vac} we get

$$i\frac{d}{dt}\begin{pmatrix}\nu_{1,s=1}\\ \nu_{1,s=-1}\\ \nu_{2,s=1}\\ \nu_{2,s=-1}\end{pmatrix} = \begin{pmatrix} E_1 + \mu_{11}\frac{B_\parallel}{\gamma_{11}} & \mu_{11}B_\perp & \mu_{12}\frac{B_\parallel}{\gamma_{12}} & \mu_{12}B_\perp\\ \mu_{11}B_\perp & E_1 - \mu_{11}\frac{B_\parallel}{\gamma_{11}} & \mu_{12}B_\perp & -\mu_{12}\frac{B_\parallel}{\gamma_{12}}\\ \mu_{12}\frac{B_\parallel}{\gamma_{12}} & \mu_{12}B_\perp & E_2 + \mu_{22}\frac{B_\parallel}{\gamma_{22}} & \mu_{22}B_\perp\\ \mu_{12}B_\perp & -\mu_{12}\frac{B_\parallel}{\gamma_{12}} & \mu_{22}B_\perp & E_2 - \mu_{22}\frac{B_\parallel}{\gamma_{22}}\end{pmatrix}\begin{pmatrix}\nu_{1,s=1}\\ \nu_{1,s=-1}\\ \nu_{2,s=1}\\ \nu_{2,s=-1}\end{pmatrix}. \quad (13)$$

This equation governs all possible oscillations of the four neutrino mass states determined by the masses m_1 and m_2 and helicities $s = 1$ and $s = -1$ in the presence of a magnetic field. Thus, it follows that: 1) the change of helicity is due to the magnetic (or transition) moment interaction with $\mathbf{B}_\perp$, 2) the longitudinal field $\mathbf{B}_\parallel$, coupled to the magnetic moment, shifts the neutrino energy, 3) an additional mixing between neutrino states with different masses is induced by the magnetic moment interaction with $\mathbf{B}_\parallel$.

2.2 Oscillations in flavour basis

Once having physics in the mass basis in hands, our next step is to bring it to observational terms [11]. This means that we must elaborate a generalization of the mixing matrix for transitions between neutrino vector written in two four-component bases ν_m and $\nu_f = (\nu_e^R, \nu_e^L, \nu_\mu^R, \nu_\mu^L)^T$ so that

$$\nu_f = U\nu_m. \tag{14}$$

This procedure appears to be not quite direct since we should hold the condition that the polarization of the fields must preserve under transformation of the bases elements. That is why we put (still keeping in mind that chiral components are almost helicity ones):

$$\nu_e^{R,L} = \nu_{1,s=\pm1}\cos\theta + \nu_{2,s=\pm1}\sin\theta, \quad \nu_\mu^{R,L} = -\nu_{1,s=\pm1}\sin\theta + \nu_{2,s=\pm1}\cos\theta. \tag{15}$$

Then, using Eqs. (14) and (15), it is easy to obtain that

$$U = \begin{pmatrix} \cos\theta & 0 & \sin\theta & 0 \\ 0 & \cos\theta & 0 & \sin\theta \\ -\sin\theta & 0 & \cos\theta & 0 \\ 0 & -\sin\theta & 0 & \cos\theta \end{pmatrix}. \tag{16}$$

Given the transition matrix (16), derivation of the evolution equation in the flavor basis is straightforward:

$$i\frac{d}{dt}\nu_f = UHU^\dagger \nu_f, \tag{17}$$

so that the effective interaction Hamiltonian $\tilde{H}_B^f \equiv UH_BU^\dagger$ has the following structure,

$$\tilde{H}_B^f = \begin{pmatrix} (\frac{\mu}{\gamma})_{ee}B_\parallel & \mu_{ee}B_\perp & (\frac{\mu}{\gamma})_{e\mu}B_\parallel & \mu_{e\mu}B_\perp \\ \mu_{ee}B_\perp & -(\frac{\mu}{\gamma})_{ee}B_\parallel & \mu_{e\mu}B_\perp & -(\frac{\mu}{\gamma})_{e\mu}B_\parallel \\ (\frac{\mu}{\gamma})_{e\mu}B_\parallel & \mu_{e\mu}B_\perp & (\frac{\mu}{\gamma})_{\mu\mu}B_\parallel & \mu_{\mu\mu}B_\perp \\ \mu_{e\mu}B_\perp & -(\frac{\mu}{\gamma})_{e\mu}B_\parallel & \mu_{\mu\mu}B_\perp & -(\frac{\mu}{\gamma})_{\mu\mu}B_\parallel \end{pmatrix}. \tag{18}$$

Here we introduce the following formal notations intended to manifest an analogy with the standard spin and spin-flavor oscillation formalism,

$$\begin{aligned} \Big(\frac{\mu}{\gamma}\Big)_{ee} &= \frac{\mu_{11}}{\gamma_{11}}\cos^2\theta + \frac{\mu_{22}}{\gamma_{22}}\sin^2\theta + \frac{\mu_{12}}{\gamma_{12}}\sin 2\theta \\ \Big(\frac{\mu}{\gamma}\Big)_{e\mu} &= \frac{\mu_{12}}{\gamma_{12}}\cos 2\theta + \frac{1}{2}\Big(\frac{\mu_{22}}{\gamma_{22}} - \frac{\mu_{11}}{\gamma_{11}}\Big)\sin 2\theta \\ \Big(\frac{\mu}{\gamma}\Big)_{\mu\mu} &= \frac{\mu_{11}}{\gamma_{11}}\cos^2\theta + \frac{\mu_{22}}{\gamma_{22}}\sin^2\theta - \frac{\mu_{12}}{\gamma_{12}}\sin 2\theta \end{aligned} \tag{19}$$

$$\begin{aligned} \mu_{ee} &= \mu_{11}\cos^2\theta + \mu_{22}\sin^2\theta + \mu_{12}\sin 2\theta \\ \mu_{e\mu} &= \mu_{12}\cos 2\theta + \frac{1}{2}(\mu_{22} - \mu_{11})\sin 2\theta \\ \mu_{\mu\mu} &= \mu_{11}\cos^2\theta + \mu_{22}\sin^2\theta - \mu_{12}\sin 2\theta \end{aligned} \tag{20}$$

It should be noted that equations (20) follows from the general expression that settles relations between two neutrino bases [2].

It is interesting to consider a particular case when only the longitudinal magnetic field $\mathbf{B}_{\parallel}$ is present in an environment ($\mathbf{B}_{\perp} = 0$). Then the Hamiltonian (18) is reduced to the following one,

$$\tilde{H}_B^f = \begin{pmatrix} (\frac{\mu}{\gamma})_{ee}B_{\parallel} & 0 & (\frac{\mu}{\gamma})_{e\mu}B_{\parallel} & 0 \\ 0 & -(\frac{\mu}{\gamma})_{ee}B_{\parallel} & 0 & -(\frac{\mu}{\gamma})_{e\mu}B_{\parallel} \\ (\frac{\mu}{\gamma})_{e\mu}B_{\parallel} & 0 & (\frac{\mu}{\gamma})_{\mu\mu}B_{\parallel} & 0 \\ 0 & -(\frac{\mu}{\gamma})_{e\mu}B_{\parallel} & 0 & -(\frac{\mu}{\gamma})_{\mu\mu}B_{\parallel} \end{pmatrix}. \quad (21)$$

Obviously, the neutrino states with different flavor and same chirality decouple and form subsystems independently mixed by the magnetic field. For example, one would have two states (ν_e^L, ν_μ^L) mixed in accordance with the equation

$$i\frac{d}{dt}\begin{pmatrix}\nu_e^L \\ \nu_\mu^L\end{pmatrix} = \begin{pmatrix} -\frac{\Delta m^2}{4E}\cos 2\theta - (\frac{\mu}{\gamma})_{ee}B_{\parallel} & \frac{\Delta m^2}{4E}\sin 2\theta - (\frac{\mu}{\gamma})_{e\mu}B_{\parallel} \\ \frac{\Delta m^2}{4E}\sin 2\theta - (\frac{\mu}{\gamma})_{e\mu}B_{\parallel} & \frac{\Delta m^2}{4E}\cos 2\theta - (\frac{\mu}{\gamma})_{\mu\mu}B_{\parallel} \end{pmatrix}\begin{pmatrix}\nu_e^L \\ \nu_\mu^L\end{pmatrix}. \quad (22)$$

From this it follows that the neutrino magnetic moment interactions with the longitudinal magnetic field can generate the neutrino flavour mixing (an additional mixing to the usual effect due to neutrino mixing angle θ) without changing neutrino chirality. For the flavour neutrino oscillation probability in the adiabtic case we get

$$P_{\nu_e^L \to \nu_\mu^L} = \frac{\left(\frac{\Delta m^2}{2E}\sin 2\theta - 2(\frac{\mu}{\gamma})_{e\mu}B_{\parallel}\right)^2}{\left(\frac{\Delta m^2}{2E}\sin 2\theta - 2(\frac{\mu}{\gamma})_{e\mu}B_{\parallel}\right)^2 + \left(\frac{\Delta m^2}{2E}\cos 2\theta + 2\frac{\mu_{12}}{\gamma_{12}}B_{\parallel}\sin 2\theta\right)^2}\sin^2\left(\frac{1}{2}\sqrt{D}x\right), \quad (23)$$

where $D = \left(\frac{\Delta m^2}{2E}\sin 2\theta - 2(\frac{\mu}{\gamma})_{e\mu}B_{\parallel}\right)^2 + \left(\frac{\Delta m^2}{2E}\cos 2\theta + 2\frac{\mu_{12}}{\gamma_{12}}B_{\parallel}\sin 2\theta\right)^2$. It follows that $B_{\parallel}$ not only generates flavour neutrino mixing but also can produce the resonance amplification of the corresponding oscillations.

3 Neutrino spin precession and oscillations due to matter transversal motion

Consider, as an example, an electron neutrino spin procession in the case when neutrinos with the Standard Model interaction are propagating through moving and polarized matter composed of electrons (electron gas) in the presence of an electromagnetic field given by the electromagnetic-field tensor $F_{\mu\nu} = (\mathbf{E}, \mathbf{B})$. As discussed in [14, 15] (see also [16–18]), the the generalized Bargmann-Michel-Telegdi equation describes the evolution of the three-dimensional neutrino spin vector $\vec{S}$,

$$\frac{d\mathbf{S}}{dt} = \frac{2}{\gamma}\left[\mathbf{S} \times (\mathbf{B}_0 + \mathbf{M}_0)\right], \quad (24)$$

where the magnetic field $\mathbf{B_0}$ in the neutrino rest frame is determined by the transversal and longitudinal (with respect to the neutrino motion) magnetic and electric field components in the laboratory frame,

$$\mathbf{B}_0 = \gamma\left(\mathbf{B}_{\perp} + \frac{1}{\gamma}\mathbf{B}_{\parallel} + \sqrt{1-\gamma^{-2}}\left[\mathbf{E}_{\perp} \times \frac{\beta}{\beta}\right]\right), \quad (25)$$

$\gamma = (1 - \beta^2)^{-\frac{1}{2}}$, β is the neutrino velocity.

The matter term $\mathbf{M}_0$ in Eq. (24) is also composed of the transversal $\mathbf{M}_{0_\parallel}$ and longitudinal $\mathbf{M}_{0_\perp}$ parts,

$$\mathbf{M}_0 = \mathbf{M}_{0_\parallel} + \mathbf{M}_{0_\perp}, \tag{26}$$

$$\mathbf{M}_{0_\parallel} = \gamma\beta\frac{n_0}{\sqrt{1-v_e^2}}\left\{\rho_e^{(1)}\left(1-\frac{\mathbf{v}_e\beta}{1-\gamma^{-2}}\right)\right\} - \frac{\rho_e^{(2)}}{1-\gamma^{-2}}\left\{\zeta_e\beta\sqrt{1-v_e^2}+\left(\zeta_e\mathbf{v}_e\frac{(\beta\mathbf{v}_e)}{1+\sqrt{1-v_e^2}}\right)\right\}, \tag{27}$$

$$\mathbf{M}_{0_\perp} = -\frac{n_0}{\sqrt{1-v_e^2}}\left\{\mathbf{v}_{e_\perp}\left(\rho_e^{(1)}+\rho_e^{(2)}\frac{(\zeta_e\mathbf{v}_e)}{1+\sqrt{1-v_e^2}}\right)+\zeta_{e_\perp}\rho_e^{(2)}\sqrt{1-v_e^2}\right\}. \tag{28}$$

Here $n_0 = n_e\sqrt{1-v_e^2}$ is the invariant number density of matter given in the reference frame for which the total speed of matter is zero. The vectors $\mathbf{v}_e$, and ζ_e $(0 \le |\zeta_e|^2 \le 1)$ denote, respectively, the speed of the reference frame in which the mean momentum of matter (electrons) is zero, and the mean value of the polarization vector of the background electrons in the above mentioned reference frame. The coefficients $\rho_e^{(1,2)}$ calculated within the extended Standard Model supplied with $SU(2)$-singlet right-handed neutrino ν_R are respectively, $\rho_e^{(1)} = \frac{\tilde{G}_F}{2\sqrt{2}\mu}$, $\rho_e^{(2)} = -\frac{G_F}{2\sqrt{2}\mu}$, where $\tilde{G}_F = G_F(1+4\sin^2\theta_W)$. For neutrino evolution between two neutrino states $\nu_e^L \Leftrightarrow \nu_e^R$ in presence of the magnetic field and moving matter we get the following equation

$$i\frac{d}{dt}\begin{pmatrix}\nu_e^L\\ \nu_e^R\end{pmatrix} = \begin{pmatrix}\frac{1}{\gamma}\left|\mathbf{M}_{0\parallel}+\mathbf{B}_{0\parallel}\right| & \left|\mathbf{B}_\perp+\frac{1}{\gamma}\mathbf{M}_{0\perp}\right| \\ \left|\mathbf{B}_\perp+\frac{1}{\gamma}\mathbf{M}_{0\perp}\right| & -\frac{1}{\gamma}\left|\mathbf{M}_{0\parallel}+\mathbf{B}_{0\parallel}\right|\end{pmatrix}\begin{pmatrix}\nu_e^L\\ \nu_e^R\end{pmatrix}. \tag{29}$$

Thus, the probability of the neutrino spin oscillations in the adiabatic approximation is given by (see [14, 15])

$$P_{\nu_L\to\nu_R}(x) = \sin^2 2\theta_{\text{eff}}\sin^2\frac{\pi x}{L_{\text{eff}}},\quad \sin^2 2\theta_{\text{eff}} = \frac{E_{\text{eff}}^2}{E_{\text{eff}}^2+\Delta_{\text{eff}}^2},\quad L_{\text{eff}} = \frac{2\pi}{\sqrt{E_{\text{eff}}^2+\Delta_{\text{eff}}^2}}, \tag{30}$$

where $E_{\text{eff}} = \left|\mathbf{B}_\perp + \frac{1}{\gamma}\mathbf{M}_{0\perp}\right|$, $\Delta_{\text{eff}} = \frac{1}{\gamma}\left|\mathbf{M}_{0\parallel}+\mathbf{B}_{0\parallel}\right|$. It follows that even without presence of an electromagnetic field, $\mathbf{B}_\perp = \mathbf{B}_{0\parallel} = 0$, neutrino spin oscillations can be induced in the presence of matter when the transverse matter term $\mathbf{M}_{0\perp}$ is not zero. This possibility is realized in the case when the transverse component of the background matter velocity or its transverse polarization is not zero. It is obvious that for neutrinos with nonzero transition magnetic moments a similar effect of spin-flavour oscillations exists under the same background conditions. A possibility of neutrino spin procession and oscillations induced by the transversal matter current or polarization was first discussed in [14, 15]. The existence of this effect has been recently confirmed in [19–21] where neutrinos propagation in anisotropic media is studied.

4 Neutrino spin oscillations and stationary spin states in magnetic field

We develop a new approach [13], that is more precise than the usual one, to description of neutrino spin and spin-flavor oscillations in the presence of an arbitrary magnetic field. The derived probability of neutrino oscillations does not coincide with the usual one and the difference might have important phenomenological consequences. Within this customary approach the helicity operator is used for classification of a neutrino spin states in a magnetic field. However, the helicity operator does not commute with the neutrino evolution Hamiltonian in a magnetic field. This case resembles situation of the flavour neutrino oscillations in the nonadiabatic case when the neutrino mass states are not stationary. The proposed alternative approach to neutrino spin oscillations is based on the exact solutions of the corresponding Dirac equation for a massive neutrino wave function in the presence of a magnetic field that stipulates the description of the neutrino spin states with the corresponding spin operator that commutes with the neutrino dynamic Hamiltonian in the magnetic field.

Here we again consider a simple model with two generations of flavour neutrinos ν_e and ν_μ that are the orthogonal superpositions of mass states ν_1 and ν_2 $\nu_f = \sum_i U_{fi}\nu_i$, where U_{fi} are elements the mixing matrix given by (15) and $f = e, \mu$, $i = 1, 2$. We start with consideration of a massive ν_i with the magnetic moment μ_i that propagates along $\mathbf{n_z}$ direction in presence of constant homogeneous arbitrary orientated magnetic field $\mathbf{B} = (B_\perp, 0, B_\parallel)$. The neutrino wave function in the momentum representation is given by a plane wave solution of the modified Dirac equation

$$(\gamma p - m_i - \mu_i \Sigma \mathbf{B})\nu_i(p) = 0. \tag{31}$$

For the neutrino energy spectrum we obtain

$$E_i^{\pm} = \sqrt{m_i^2 + p^2 + \mu_i{}^2\mathbf{B}^2 \pm 2\mu_i \sqrt{m_i^2\mathbf{B}^2 + p^2 B_\perp^2}}, \tag{32}$$

where "±" denotes two different eigenvalues of the Hamiltonian $H_i = \gamma_0(m_i + \gamma\mathbf{p} + \mu_i\Sigma\mathbf{B})$, which describes dynamics of the neutrino system under consideration.

We define different neutrino spin states in the mass basis as eigenstates of the spin operator

$$S_i = \frac{1}{|\mathbf{B}|}(\Sigma\mathbf{B} - \frac{i}{m_i}\gamma_0\gamma_5\,[\Sigma \times \mathbf{p}]\,\mathbf{B}), \tag{33}$$

which commutes with the Hamiltonian H_i. Hence, we specify the neutrino spin states as the stationary states for the Hamiltonian, contrary to the case when the helicity operator is used.

Consider the mass state ν_i as a superposition of neutrinos ν_i^+ and ν_i^- in a definite spin state, $\nu_i = c_i^+\nu_i^+ + c_i^-\nu_i^-$. The complex coefficients denote two different eigenstates of the spin operator S_i and $|c_i^+|^2 + |c_i^-|^2 = 1$. Thus, the neutrino mass states evolve following to

$$\nu_i(x) = \left[c_i^+ e^{-iE_i^+ t}\xi_i^+ + c_i^- e^{-iE_i^- t}\xi_i^-\right] e^{ipx}, \tag{34}$$

where the neutrino initial state at $t = 0$ is given by $\nu_i^{\pm}(t = 0) = \xi_i^{\pm} e^{ipx}$. In the following calculations the term e^{ipx} is neglected because it is irrelevant for the neutrino oscillation probability.

Next we assume that the initial neutrino state $\nu(t = 0)$ is a pure electron state which is defined as the superposition of the mass states,

$$\nu(t = 0) = [c_1^+\xi_1^+ + c_1^-\xi_1^-]\cos\theta + [c_2^+\xi_2^+ + c_2^-\xi_2^-]\sin\theta. \tag{35}$$

Using (34) we see that this state depends on time as

$$\nu(t) = \left[c_1^+ e^{-iE_1^+ t}\xi_1^+ + c_1^- e^{-iE_1^- t}\xi_1^-\right]\cos\theta + \left[c_2^+ e^{-iE_2^+ t}\xi_2^+ + c_2^- e^{-iE_2^- t}\xi_2^-\right]\sin\theta. \tag{36}$$

Therefore, the probability to observe the muon neutrino state ν_μ at time t is given by $P_{\nu_e \to \nu_\mu}(t) = \left|\langle \nu_\mu | \nu(t) \rangle\right|^2$, where $\nu_\mu = -\left[c_1^+\xi_1^+ + c_1^-\xi_1^-\right]\sin\theta + \left[c_2^+\xi_2^+ + c_2^-\xi_2^-\right]\cos\theta$. It is clear that $\xi_i^{s\dagger}\xi_i^{s'} =$

$\delta^{ss'}$ and $\xi_2^{s\dagger}\xi_1^{s'} = \xi_1^{s\dagger}\xi_2^{s'} = 0$, where $s, s' = \pm$. Using (35) and (36) we get for the oscillation probability

$$P_{\nu_e\to\nu_\mu}(t) = \Big\{ -|c_2^+|^2|c_2^-|^2 \sin^2\frac{E_2^+ - E_2^-}{2}t + |c_2^+|^2|c_1^+|^2 \sin^2\frac{E_2^+ - E_1^+}{2}t+ \\ + |c_2^+|^2|c_1^-|^2 \sin^2\frac{E_2^+ - E_1^-}{2}t + |c_2^-|^2|c_1^+|^2 \sin^2\frac{E_2^- - E_1^+}{2}t+ \\ +|c_2^-|^2|c_1^-|^2 \sin^2\frac{E_2^- - E_1^-}{2}t - |c_1^+|^2|c_1^-|^2 \sin^2\frac{E_1^+ - E_1^-}{2}t\Big\} \sin^2 2\theta. \quad (37)$$

It is usually assumed that the initial state of relativistic neutrino is a negative-helicity state, which means that

$$\frac{\mathbf{\Sigma p}}{|\mathbf{p}|}\xi_i = -\xi_i. \quad (38)$$

Next we consider the left-handed spinors because only the left-handed fermions participate in the production and detection processes and we suppose that each of the mass states of the initial electron neutrino are left-handed. In our case the helicity operator is equal to $\mathbf{\Sigma p}/|\mathbf{p}| = \sigma_3$, therefore the initial neutrino state is given by $\psi_L = (0,1,0,0)^T$. Let us write the initial neutrino state ψ_L as a superposition of the eigenvectors of the spin operator S_i. From (33) we get

$$S_i = \begin{pmatrix} \cos\phi & \sin\phi & 0 & -\frac{p}{m_i}\sin\phi \\ \sin\phi & -\cos\phi & \frac{p}{m_i}\sin\phi & 0 \\ 0 & \frac{p}{m_i}\sin\phi & \cos\phi & \sin\phi \\ -\frac{p}{m_i}\sin\phi & 0 & \sin\phi & -\cos\phi \end{pmatrix}, \quad (39)$$

where ϕ is the angle between $\mathbf{B}$ and $\mathbf{p}$. It is obvious that $S_i^2 = \left(1 + \frac{p^2}{m_i^2}\sin^2\phi\right)\hat{I}_{4\times4}$. In order to define the spin projector operators we introduce the normalized spin operator following to

$$\tilde{S}_i = \sqrt{\frac{1}{1 + \frac{p^2}{m_i^2}\sin^2\phi}}S_i \equiv N_i S_i, \quad \tilde{S}_i^2 = 1, \quad N_i = \sqrt{\frac{1}{1 + \frac{p^2}{m_i^2}\sin^2\phi}}. \quad (40)$$

The spin projector operators are $P_i^\pm = \frac{1\pm\tilde{S}_i}{2}$, and we use them to split the initial neutrino state ψ_L in two neutrino states with definite spin quantum numbers

$$\psi_i^+ = P_i^+\psi_L = \frac{1}{2}\begin{pmatrix} N_i\sin\phi \\ 1 - N_i\cos\phi \\ \frac{p}{m_i}N_i\sin\phi \\ 0 \end{pmatrix}, \quad \psi_i^- = P_i^-\psi_L = \frac{1}{2}\begin{pmatrix} -N_i\sin\phi \\ 1 + N_i\cos\phi \\ -\frac{p}{m_i}N_i\sin\phi \\ 0 \end{pmatrix}. \quad (41)$$

Note that $\psi_i = \psi_i^+ + \psi_i^- = c_i^+\eta_i^+ + c_i^-\eta_i^-$, where $\eta_i^\pm$ is a basis in the spin operator S_i eigenspace. From the condition $\psi_i^{\dagger\pm}\psi_L^\pm = |c_i^\pm|^2$ we get that $|c_i^+|^2 = \frac{1-N_i\cos\phi}{2}$, $|c_i^-|^2 = \frac{1+N_i\cos\phi}{2}$.

Now we can insert the obtained expressions for $|c_i^\pm|^2$ in (37). In the forthcoming evaluation of the probability $P_{\nu_e\to\nu_\mu}(t)$ we consider the case when the magnetic field $\mathbf{B}$ is nearly a transversal one and $B_\perp \gg B_\parallel$, therefore $\sin\phi \approx 1$, $\cos\phi \approx 0$. Then we get (it is also supposed that $\frac{p^2}{m_i^2} \gg 1$)

$$N_i \approx 1 - \frac{\frac{p^2}{2m_i^2}\sin^2\phi}{1 + \frac{p^2}{m_i^2}\sin^2\phi} \approx \frac{m_i^2}{p\sin^2\phi}. \quad (42)$$

Thus, for typical combinations of the coefficients $|c_i^{\mp}|^2$ of eq.(37) in the linear approximation over $\cos\phi \ll 1$ we get

$$|c_i^{+}|^2|c_i^{-}|^2 = \frac{1 - N_i^2\cos^2\phi}{4} \approx \frac{1}{4}, \tag{43}$$

$$|c_2^{s}|^2|c_1^{s}|^2 \approx \frac{1}{4}(1 - s(N_1 + N_2)\cos\phi) \approx \frac{1}{4}\left(1 - s\frac{m_1^2 + m_2^2}{p^2}\cos\phi\right), \tag{44}$$

$$|c_2^{+}|^2|c_1^{-}|^2 \approx \frac{1}{4}(1 - (N_2 - N_1)\cos\phi) \approx \frac{1}{4}\left(1 - \frac{m_2^2 - m_1^2}{p^2}\cos\phi\right), \tag{45}$$

$$|c_2^{-}|^2|c_1^{+}|^2 \approx \frac{1}{4}(1 - (N_1 - N_2)\cos\phi) \approx \frac{1}{4}\left(1 - \frac{m_1^2 - m_2^2}{p^2}\cos\phi\right). \tag{46}$$

Using these expressions finally from (37) we get

$$\begin{aligned} P_{\nu_{eL}\to\nu_\mu}(t) = \frac{1}{4}\sin^2 2\theta\Big\{ &-\sin^2\frac{E_2^+ - E_2^-}{2}t + \sin^2\frac{E_2^+ - E_1^+}{2}t + \sin^2\frac{E_2^+ - E_1^-}{2}t + \\ &+\sin^2\frac{E_2^- - E_1^+}{2}t + \sin^2\frac{E_2^- - E_1^-}{2}t - \sin^2\frac{E_1^+ - E_1^-}{2}t\Big\} + \\ &+\frac{m_1^2 + m_2^2}{p^2}\sin^2 2\theta\cos\phi\left\{\sin^2\frac{E_2^- - E_1^-}{2}t - \sin^2\frac{E_2^+ - E_1^+}{2}t\right\} + \\ &+\frac{m_2^2 - m_1^2}{p^2}\sin^2 2\theta\cos\phi\left\{\sin^2\frac{E_2^- - E_1^+}{2}t - \sin^2\frac{E_2^+ - E_1^-}{2}t\right\}. \end{aligned} \tag{47}$$

Note that two last terms here are suppressed by the presence of $\cos\phi \ll 1$. If we also account for a rather general condition $2\mu B_\perp \ll p$ then for the neutrino energies we get

$$E_i^{\pm} \approx \sqrt{m_i^2 + p^2 \pm 2\mu_i p B_\perp\left(1 + \frac{m_i^2}{2p^2}\right)} \approx p\sqrt{1 + \frac{m_i^2}{p^2} \pm \frac{2\mu_i B_\perp}{p}} \approx p + \frac{m_i^2}{2p} \pm \mu_i B_\perp. \tag{48}$$

Finally, for the neutrino oscillations probability in the flavour basis we get (here $\Delta m^2 \equiv m_2^2 - m_1^2$)

$$\begin{aligned} P_{\nu_{eL}\to\nu_\mu}(t) \approx \sin^2 2\theta\sin^2\frac{\Delta m^2}{4p}t + \frac{1}{2}\left(\sin^2\frac{\mu_2 - \mu_1}{2}B_\perp t + \sin^2\frac{\mu_2 + \mu_1}{2}B_\perp t\right)\sin^2 2\theta\cos\frac{\Delta m^2}{2p}t \\ -\frac{1}{4}\sin^2 2\theta\left(\sin^2\mu_1 B_\perp t + \sin^2\mu_2 B_\perp t\right). \end{aligned} \tag{49}$$

It should be emphasized that, as it follows from the above derivations, the obtained probability $P_{\nu_{eL}\to\nu_\mu}(t)$ accounts for the transitions from the initial left-handed electron neutrino to the final muon neutrino that can be in both left- and right-handed states. Note that in the mass basis the transition $\nu_{1L} \to \nu_{2R}$ is not possible when $\mu_{12} = 0$ (see (29)).

Consider the difference $\Delta P(t) = P_{\nu_{eL}\to\nu_\mu} - P_{\nu_{eL}\to\nu_{\mu L}}$ of the probability $P_{\nu_{eL}\to\nu_\mu}(t)$ given by (50) and the usual result $P_{\nu_{eL}\to\nu_{\mu L}}(t) = \sin^2 2\theta\sin^2\frac{\Delta m^2}{4E_\nu}t$ for $\nu_{eL} \to \nu_{\mu L}$. Obviously, this difference is just a probability for the transition $P_{\nu_{eL}\to\nu_{\mu R}}$, and we get (here $\Delta\mu_\pm = \mu_2 \pm \mu_1$)

$$\begin{aligned} P_{\nu_{eL}\to\nu_{\mu R}} = \frac{1}{2}\left(\sin^2\frac{\Delta\mu_-}{2}B_\perp t + \sin^2\frac{\Delta\mu_+}{2}B_\perp t\right)\sin^2 2\theta\cos\frac{\Delta m^2}{2p}t \\ -\frac{1}{4}\sin^2 2\theta\left(\sin^2\mu_1 B_\perp t + \sin^2\mu_2 B_\perp t\right). \end{aligned} \tag{50}$$

Note that in the usual approach (see, for instance, in [2]) transitions $\nu_{eL} \to \nu_{\mu R}$ in $B_{\perp}$ are not possible if the transition magnetic moments are zero. However, within the developed approach it is shown that the previous statement is not correct.

The author is thankful to the organizers of the 19th International Seminar on High Energy Physics QUARKS-2016 for the invitation to attend this very interesting event. This work is supported by the Russian Basic Research Foundation grants No. 14-22-03043, 15-52-53112 and 16-02-01023.

References

[1] K.Fujikawa and R.Shrock, Phys. Rev. Lett. **45**, 963 (1980)

[2] C.Giunti and A.Studenikin, Rev. Mod. Phys. **87**, 531 (2015)

[3] A.Beda, V.Brudanin, V.Egorov, D.Medvedev et al, Adv. High Energy Phys. **2012**, 350150 (2012)

[4] G.Raffelt, Phys. Rev. Lett. **64**, 2856 (1990)

[5] A.Cisneros, Astrophys. Space Sci. **10**, 87 (1971)

[6] J.Schechter and J.W.F.Valle Phys. Rev. D **24**. 1883 (1981)

[7] L. Okun, M.Voloshin and M.Vysotsky, Sov. Phys. JETP **64**, 446 (1986)

[8] E. Akhmedov, Phys. Lett. B **213**, 64 (1988)

[9] C.-S.Lim and W. Marciano, Phys. Rev. D **37**, 1368 (1988)

[10] E.Akhmedov and M.Khlopov, Mod. Phys. Lett.A bf 3, 451 (1988)

[11] R. Fabbricatore, A. Grigoriev and A. Studenikin, J. Phys. Conf. Ser. **718**, 062058 (2016) [arXiv:1604.01245 [hep-ph]]

[12] A. Studenikin, J. Phys. Conf. Ser. **718** , 062076 (2016) [arXiv:1603.00337 [hep-ph]]

[13] A. Dmitriev, R. Fabbricatore and A. Studenikin, PoS CORFU **2014**, 050 (2015) [arXiv:1506.05311 [hep-ph]]

[14] A. Studenikin, Phys. Atom. Nucl. **67**, 993 (2004)

[15] A.Studenikin, *Preprint* hep-ph/0407010

[16] A.Egorov, A.Lobanov and A.Studenikin, Phys. Lett. B **491**, 137 (2000)

[17] A.Lobanov and A.Studenikin, Phys. Lett. B **515**, 94 (2001)

[18] M.Dvornikov and A.Studenikin, JHEP **09**, 016 (2002)

[19] A.Kartavtsev, G.Raffelt and H.Vogel, Phys. Rev. D **91**, 125020 (2015)

[20] C.Volpe, Int. J. Mod. Phys. E **24**, 1541009 (2015)

[21] V.Cirigliano, G.Fuller and A.Vlasenko, Phys. Lett. B **747**, 27 (2015)

EPJ Web of Conferences **125**, 04019 (2016)
QUARKS-2016
DOI: 10.1051/epjconf/201612504019

Squeezed and Entangled Gluon States in QCD Jet

Valery Shaparau[1,*] and *V.I.* Kuvshinov[1,**]

[1]*Joint Institute for Power and Nuclear Research - Sosny*
National Academy of Science of Belarus
PO box 119, 220109 Minsk, BELARUS

Abstract. Theoretical justification for the occurrence of multimode squeezed and entangled colour states in QCD is given. We show that gluon entangled states which are closely related with corresponding squeezed states can appear by the four-gluon self-interaction. Correlations for the collinear gluons are revealed two groups of the colour correlations which is significant at consider of the quark-antiquark pair productions.

1 Introduction

Many experiments at $e^+e^-, p\bar{p}, ep$ colliders are devoted to hadronic jet physics, since detailed studies of jets are important for better understanding and testing both perturbative and non-perturbative QCD and also for finding manifestations of new physics. Although the nature of jets is of a universal character, e^+e^-- annihilation stands out among hard processes, since jet events admit a straightforward and clear-cut separation in this process. In the reaction $e^+e^- \to hadron$ four evolution phases are recognized by various time and space scales (Fig.1). These are (I) the production of a quark-antiquark pair: $e^+e^- \to q\bar{q}$; (II) the emissions of gluons and quarks from primary partons — perturbative evolution of the quark-gluon cascade; (III) the non-perturbative evolution and the hadronization of quarks and gluons; (IV) the decays of unstable particles.

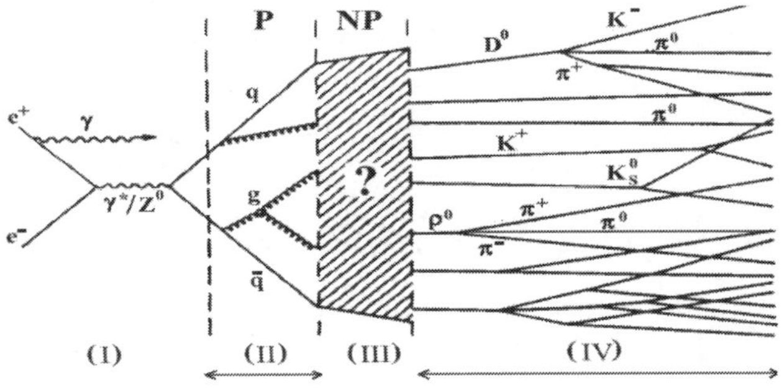

Figure 1. Jet evolution

[*]e-mail: shaparau@sosny.bas-net.by
[**]e-mail: v.kuvshinov@sosny.bas-net.by

The second phase of e^+e^-- annihilation has been well understood and sufficiently accurate predictions for it have been obtained within the perturbative QCD (PQCD) [1]. But predictions of the PQCD are limited by small effective coupling $\alpha(Q^2) < 1$ and third phase is usually taken into account either through a constant factor which relates partonic features with hadronic ones (within local parton-hadron duality) or through the application of various phenomenological models of hadronization. As a consequence, theoretical predictions both for intrajet and for interjet characteristics remain unsatisfactory. For example, the width of the multiplicity distribution (MD) according to the predictions of PQCD is larger than the experimental one. The discrepancies between theoretical calculations and experimental data, for example the width of MD, suggest that the non-perturbative evolution of the quark-gluon cascade plays important role. New gluon states, generated at the non-perturbative stage, contribute to various features of jets. In particular, such a contribution to MD can be in the form of the sub-Poissonian distribution [2, 3].

It is known that such property is inherent for the squeezed states (SS), which are well studied in quantum optics (QO) [4]-[6]. Squeezed states posses uncommon properties: they display a specific behaviour of the factorial and cumulant moments [7] and can have both sub-Poissonian and super-Poissonian statistics corresponding to antibunching and bunching of photons.

Therefore we believe that the non-perturbative stage of gluon evolution can be one of sources of the gluon SS in QCD by analogy with nonlinear medium for photon SS in QO. Gluon MD in the range of the small transverse momenta (thin ring of jet) is Poissonian [8]. Quark-gluon MD in the whole jet at the end of the perturbative cascade can be represented as a combination of Poissonian distributions each of which corresponds to a coherent state. Studying a further evolution of gluon states at the non-perturbative stage of jet evolution we obtain new gluon states. These states are formed as a result of non-perturbative self-interaction of the gluons expressed by nonlinearities of Hamiltonian. Using the Local parton hadron duality it is easy to show that in this case behaviour of hadron multiplicity distribution in jet events is differentiated from the negative binomial one that is confirmed by experiments for pp, $p\overline{p}$-collisions [9]-[11].

At finite squeezed parameter r a continuous variables entangled state is known from quantum optics as a two-mode squeezed state [4, 14]

$$|\mathrm{f}\rangle = \hat{S}_{12}(r)|0\rangle_1|0\rangle_2 = \frac{1}{\cosh r}\sum_{n=0}^{\infty}(\tanh r)^n\,|n\rangle_1|n\rangle_2, \tag{1}$$

where $\hat{S}_{12}(r) = \exp\{r(\hat{a}_1^+\hat{a}_2^+ - \hat{a}_1 a_2)\}$ is operator of two-mode squeezing. It is not difficult to demonstrate that the state vector $|\mathrm{f}\rangle$ describes the entangled state. Each of these entangled states has a uncommon property: a measurement over one particle have an instantaneous effect on the other, possibly located at a large distance.

The dimensionless coefficient

$$y = \left[\frac{|\overline{\hat{a}_1\hat{a}_2^+}|^2 + |\overline{\hat{a}_1\hat{a}_2}|^2}{2(\overline{\hat{a}_1^+\hat{a}_1} + 1/2)(\overline{\hat{a}_2^+\hat{a}_2} + 1/2)}\right]^{1/2} \tag{2}$$

is the measure of entanglement for two-mode states [15], $0 \leq y < 1$ (entanglement is not observe when $y = 0$). Here $\overline{\hat{a}_i\hat{a}_j^+} = \langle\hat{a}_i\hat{a}_j^+\rangle - \langle\hat{a}_i\rangle\langle\hat{a}_j^+\rangle$, $\hat{a}_i, \hat{a}_j^+$ are the annihilation and creation operators correspondingly. Averaging the annihilation and creation operators in the expression (2) over the vector $|\mathrm{f}\rangle$ (1) at small squeeze factor we have

$$y \simeq \sqrt{2}r. \tag{3}$$

Two-mode gluon states with two different colours can lead to $q\bar{q}$-entangled states. Interaction of the quark entangled states with stochastic vacuum (quantum measurement) has a remarkable property, namely, as soon as some measurement projects one quark onto a state with definite colour, the other quark also immediately obtains opposite colour that leads to coupling of quark-antiquark pair, string tension inside $q\bar{q}$-pare and free propagation of colourless hadrons. Therefore the investigation of the gluon entangled states connected with the corresponding squeezed ones is issue of the day.

2 Multimode squeezed states of the gluons

By analogy with QO [6] a multimode squeezing condition for gluons with different colors $i_1, \ldots, i_p$ is written as

$$\hat{N}\left\langle \Delta(\hat{X}_{\lambda}^{i_1,\ldots,i_p})_{\frac{1}{2}})^2 \right\rangle < 0, \tag{4}$$

where $\Delta(\hat{X}_{\lambda}^{i_1,\ldots,i_p}) = \hat{X}_{\lambda}^{i_1,\ldots,i_p} - \langle \hat{X}_{\lambda}^{i_1,\ldots,i_p} \rangle$, $\hat{N}$ is a normal ordering operator, the phase-sensitive Hermitian operators $(\hat{X}_{\lambda}^{i_1,\ldots,i_p})_1 = \frac{1}{2\sqrt{p}}\sum_{j=1}^{p}[\hat{b}_{\lambda}^{i_j} + (\hat{b}_{\lambda}^{i_j})^+]$ and $(\hat{X}_{\lambda}^{i_1,\ldots,i_p})_2 = \frac{1}{2i\sqrt{p}}\sum_{j=1}^{p}[\hat{b}_{\lambda}^{i_j} - (\hat{b}_{\lambda}^{i_j})^+]$ are linear combination of the annihilating (creating) operators $\hat{b}_{\lambda}^{i_j}$ $(\hat{b}_{\lambda}^{i_j+})$, $i_1, \ldots, i_p = \overline{1,8}$ are gluon color charges, λ is a polarization index. Averaging in (4) is performed over final state vector

$$|\mathrm{f}\rangle \simeq |\mathrm{in}\rangle - i\,t\left(\hat{H}_{\mathrm{I}}^{(3)}(0) + \hat{H}_{\mathrm{I}}^{(4)}(0)\right)|\mathrm{in}\rangle \tag{5}$$

which describes gluon system later small time t. Operators $\hat{H}_{\mathrm{I}}^{(3)}(t)$ and $\hat{H}_{\mathrm{I}}^{(4)}(t)$ describing the three- and four-gluon selfinteractions include combinations of three and four annihilating and creating operators [16]:

$$\begin{aligned}
\hat{H}_{\mathrm{I}}^{(3)}(t) = {} & i\frac{g}{(2\pi)^{3/2}}f_{abc}\sum_{\lambda_1,\lambda_2,\lambda_3}\int d\tilde{k}_1 d\tilde{k}_2 d\tilde{k}_3 \times \Big\{\vec{k}_1\vec{\varepsilon}_{\lambda_2}(k_2)\varepsilon^{\nu}_{\lambda_1}(k_1)\varepsilon^{\lambda_3}_{\nu}(k_3)\delta(\vec{k}_1+\vec{k}_2+\vec{k}_3)\\
& \times[\hat{b}^a_{\lambda_1}(k_1)\hat{b}^b_{\lambda_2}(k_2)\hat{b}^c_{\lambda_3}(k_3)\,e^{-2i(k_{01}+k_{02}+k_{03})t} - \hat{b}^{a+}_{\lambda_1}(k_1)\hat{b}^{b+}_{\lambda_2}(k_2)\hat{b}^{c+}_{\lambda_3}(k_3)\,e^{2i(k_{01}+k_{02}+k_{03})t}]\\
& + \hat{b}^{a+}_{\lambda_1}(k_1)\hat{b}^b_{\lambda_2}(k_2)\hat{b}^c_{\lambda_3}(k_3)\,e^{2i(k_{01}-k_{02}-k_{03})t}\,\delta(\vec{k}_1-\vec{k}_2-\vec{k}_3)\\
& \times\Big[\vec{k}_3\vec{\varepsilon}_{\lambda_1}(k_1)\varepsilon^{\nu}_{\lambda_2}(k_2)\varepsilon^{\lambda_3}_{\nu}(k_3) + \vec{k}_2\vec{\varepsilon}_{\lambda_3}(k_3)\varepsilon^{\nu}_{\lambda_1}(k_1)\varepsilon^{\lambda_2}_{\nu}(k_2) - \vec{k}_1\vec{\varepsilon}_{\lambda_2}(k_2)\varepsilon^{\nu}_{\lambda_1}(k_1)\varepsilon^{\lambda_3}_{\nu}(k_3)\Big]\\
& + \hat{b}^{a+}_{\lambda_1}(k_1)\hat{b}^{b+}_{\lambda_2}(k_2)\hat{b}^c_{\lambda_3}(k_3)\,e^{2i(k_{01}+k_{02}-k_{03})t}\,\delta(\vec{k}_1+\vec{k}_2-\vec{k}_3)\\
& \times\Big[\vec{k}_1\vec{\varepsilon}_{\lambda_3}(k_3)\varepsilon^{\nu}_{\lambda_1}(k_1)\varepsilon^{\lambda_2}_{\nu}(k_2) + \vec{k}_3\vec{\varepsilon}_{\lambda_1}(k_1)\varepsilon^{\nu}_{\lambda_2}(k_2)\varepsilon^{\lambda_3}_{\nu}(k_3) - \vec{k}_1\vec{\varepsilon}_{\lambda_2}(k_2)\varepsilon^{\nu}_{\lambda_1}(k_1)\varepsilon^{\lambda_3}_{\nu}(k_3)\Big]\Big\},
\end{aligned} \tag{6}$$

$$\begin{aligned}
\hat{H}_{\mathrm{I}}^{(4)}(t) = {} & \frac{g^2}{4(2\pi)^3}\sum_{\lambda_1,\lambda_2,\lambda_3,\lambda_4}\int d\tilde{k}_1 d\tilde{k}_2 d\tilde{k}_3 d\tilde{k}_4\Big\{\varepsilon^{\mu}_{\lambda_1}(k_1)\varepsilon^{\lambda_3}_{\mu}(k_3)\varepsilon^{\nu}_{\lambda_2}(k_2)\varepsilon^{\lambda_4}_{\nu}(k_4)f_{abc}f_{ade}\\
& \times\Big[\delta(\vec{k}_1+\vec{k}_2+\vec{k}_3+\vec{k}_4)\Big(\hat{b}^b_{\lambda_1}(k_1)\hat{b}^c_{\lambda_2}(k_2)\hat{b}^d_{\lambda_3}(k_3)\hat{b}^e_{\lambda_4}(k_4)\,e^{-2i(k_{01}+k_{02}+k_{03}+k_{04})t}\\
& + \hat{b}^{b+}_{\lambda_1}(k_1)\hat{b}^{c+}_{\lambda_2}(k_2)\hat{b}^{d+}_{\lambda_3}(k_3)\hat{b}^{e+}_{\lambda_4}(k_4)\,e^{2i(k_{01}+k_{02}+k_{03}+k_{04})t}\Big)\\
& + 4\hat{b}^{b+}_{\lambda_1}(k_1)\hat{b}^c_{\lambda_2}(k_2)\hat{b}^d_{\lambda_3}(k_3)\hat{b}^e_{\lambda_4}(k_4)\,e^{2i(k_{01}-k_{02}-k_{03}-k_{04})t}\,\delta(\vec{k}_1-\vec{k}_2-\vec{k}_3-\vec{k}_4)\\
& + 4\hat{b}^{b+}_{\lambda_1}(k_1)\hat{b}^{c+}_{\lambda_2}(k_2)\hat{b}^{d+}_{\lambda_3}(k_3)\hat{b}^e_{\lambda_4}(k_4)\,e^{2i(k_{01}+k_{02}+k_{03}-k_{04})t}\,\delta(\vec{k}_1+\vec{k}_2+\vec{k}_3-\vec{k}_4)\Big]\\
& + 2\hat{b}^{b+}_{\lambda_1}(k_1)\hat{b}^{c+}_{\lambda_2}(k_2)\hat{b}^d_{\lambda_3}(k_3)\hat{b}^e_{\lambda_4}(k_4)\,e^{2i(k_{01}+k_{02}-k_{03}-k_{04})t}\,\delta(\vec{k}_1+\vec{k}_2-\vec{k}_3-\vec{k}_4)\\
& \times\Big[\varepsilon^{\mu}_{\lambda_1}(k_1)\varepsilon^{\lambda_3}_{\mu}(k_3)\varepsilon^{\nu}_{\lambda_2}(k_2)\varepsilon^{\lambda_4}_{\nu}(k_4)(f_{abc}f_{ade} + f_{abe}f_{adc})\\
& + \varepsilon^{\mu}_{\lambda_1}(k_1)\varepsilon^{\lambda_2}_{\mu}(k_2)\varepsilon^{\nu}_{\lambda_3}(k_3)\varepsilon^{\lambda_4}_{\nu}(k_4)f_{abd}f_{ace}\Big]\Big\}.
\end{aligned} \tag{7}$$

Here g is a self-interaction constant, $d\tilde{k} = \dfrac{d^3k}{\sqrt{2k_0}}$, k_0 is a gluon energy, $\varepsilon^{\mu}_{\lambda}$ is a polarization vector, f_{ahb} are structure constants of $SU_c(3)$ group.

Initial state vector $|\text{in}\rangle$ describe gluon system at end of perturbative stage [8] and is product of the coherent states of the gluons with different colours and polarization indexes

$$|\text{in}\rangle \equiv |\alpha\rangle = \prod_{\lambda=1}^{3}\prod_{b=1}^{8}|\alpha^b_\lambda\rangle. \tag{8}$$

By analogy with QO gluon coherent state vector $|\alpha^b_\lambda\rangle$ is the eigenvector of the corresponding annihilation operator $\hat{b}^b_\lambda$ with the eigenvalue $\alpha^b_{\lambda_1}$ which can be written in terms of the gluon coherent field amplitude $|\alpha^b_{\lambda_1}|$ and phase γ^b_λ of the given gluon field $\alpha^b_{\lambda_1} = |\alpha^b_{\lambda_1}|e^{i\gamma^b_{\lambda_1}}$. In each gluon coherent state $|\alpha^b_\lambda\rangle$ the gluon number with fixed colour b and polarization λ is arbitrary (the average multiplicity of given gluon is equal to square of the gluon coherent field amplitude $\langle n^b_\lambda\rangle = |\alpha^b_\lambda|^2$) and phase of considering state γ^b_λ is fixed.

Averaging the annihilation and creation operators $\hat{b}^{i_j}_\lambda, \hat{b}^{i_j+}_\lambda$ in (4) over the evolved vector $|\text{f}\rangle$ which is defined according to (5) and taking into account of chosen initial state vector we write the multimode squeezing condition in the form

$$\left\langle \hat{N}\left(\Delta(\hat{X}^{i_1,\ldots,i_p}_\lambda)_{\frac{1}{2}}\right)^2\right\rangle = \mp\frac{it}{4p}\sum_{j,k=1}^{p}\left\{\langle\alpha|[\hat{b}^{i_k}_\lambda,[\hat{b}^{i_j}_\lambda,\hat{H}_I(0)]]|\alpha\rangle - \langle\alpha|[[\hat{H}_I(0),\hat{b}^{i_j+}_\lambda],\hat{b}^{i_k+}_\lambda]|\alpha\rangle\right\} < 0, \tag{9}$$

where $\hat{H}_I(0) = \hat{H}^{(3)}_I(0) + \hat{H}^{(4)}_I(0)$.

It can be shown that only the four-gluon self-interaction can yield the multimode squeezing effect since

$$[[\hat{H}^{(3)}_I(0),\hat{b}^{i_j+}_\lambda],\hat{b}^{i_k+}_\lambda] = 0, \quad [\hat{b}^{i_k}_\lambda,[\hat{b}^{i_j}_\lambda,\hat{H}^{(3)}_I(0)]] = 0. \tag{10}$$

Indeed, the multimode squeezing condition can be written in the explicit form as

$$\begin{aligned}
\left\langle \hat{N}\left(\Delta(\hat{X}^{i_1,\ldots,i_p}_\lambda)_{\frac{1}{2}}\right)^2\right\rangle &= \pm\frac{it}{8\,p\,k_0}\frac{g^2}{(2\pi)^3}\sum_{j,k=1}^{p}\sum_{\lambda_1,\lambda_2=1}^{3}\int\prod_{i=1}^{2}\frac{d^3k_i}{\sqrt{2k_{0i}}}\\
&\times\left[\left(\varepsilon^{\lambda_1}_\mu(k_1)\,\varepsilon^{\mu}_{\lambda_2}(k_2)\right)\left(\varepsilon^{\lambda}_\nu(k)\,\varepsilon^{\nu}_{\lambda}(k)\right) - \left(\varepsilon^{\lambda_1}_\mu(k_1)\varepsilon^{\mu}_{\lambda}(k)\right)\left(\varepsilon^{\lambda_2}_\nu(k_2)\varepsilon^{\nu}_{\lambda}(k)\right)\right]\\
&\times\Big\{ f_{ai_jb}f_{ai_kc}\left[\langle\alpha|\hat{b}^{b+}_{\lambda_1}(k_1)\hat{b}^{c+}_{\lambda_2}(k_2)|\alpha\rangle - \langle\alpha|\hat{b}^{b}_{\lambda_1}(k_1)\hat{b}^{c}_{\lambda_2}(k_2)|\alpha\rangle\right]\\
&\times\left[\delta(2\vec{k}-\vec{k}_1-\vec{k}_2) - \delta(2\vec{k}+\vec{k}_1+\vec{k}_2)\right] + \left(f_{ai_jb}f_{ai_kc} + f_{ai_kb}f_{ai_jc}\right)\\
&\times\langle\alpha|\hat{b}^{b+}_{\lambda_1}(k_1)\hat{b}^{c}_{\lambda_2}(k_2)|\alpha\rangle\left[\delta(2\vec{k}-\vec{k}_1+\vec{k}_2) - \delta(2\vec{k}+\vec{k}_1-\vec{k}_2)\right]\Big\} < 0.
\end{aligned} \tag{11}$$

In particular for the collinear gluons we have corresponding squeezing condition

$$\left\langle \hat{N}\left(\Delta(X^{i_1,\ldots,i_p}_\lambda)_{\frac{1}{2}}\right)^2\right\rangle = \pm\frac{g^2\,t}{8\,p\,(2\pi)^3\,k_0^2}\sum_{j,k=1}^{p}\sum_{\lambda_1\neq\lambda} f_{ai_kb}f_{ai_jc}|\alpha^b_{\lambda_1}||\alpha^c_{\lambda_1}|\sin(\gamma^b_{\lambda_1}+\gamma^c_{\lambda_1}) < 0, \tag{12}$$

Here $|\alpha^b_{\lambda_1}|$ and $\gamma^b_{\lambda_1}$ are an amplitude and a phase of the initial gluon coherent field, f_{ahb} is a structure constant of the color group $SU_c(3)$. The multimode squeezing condition (12) is fulfilled for any cases apart from if all initial gluon coherent fields are real or imaginary. Obviously, the larger are both the amplitudes of the initial gluon coherent fields with different colour and polarization indexes and coupling constant, the larger is multimode squeezing effect.

By analogy with two-mode photon state (1) we can make corresponding gluon states with squeezed operator

$$\hat{S}_{12}(r) = \exp\left\{r\,(\hat{b}^h_\lambda \hat{b}^g_\lambda - \hat{b}^{h+}_\lambda \hat{b}^{g+}_\lambda)\right\}. \tag{13}$$

In case of the evolved gluon system during small time t we have

$$\left\langle \hat{N}\left((\widehat{\Delta X}^{h,g}_\lambda)_{\frac{1}{2}}\right)^2 \right\rangle = \frac{r(r \mp 1)}{2}. \tag{14}$$

Two-mode squeezing condition is

$$-\frac{1}{8} < \left\langle \hat{N}\left((\widehat{\Delta X}^{h,g}_\lambda)_1\right)^2 \right\rangle < 0 \text{ or } 0 < \left\langle \hat{N}\left((\widehat{\Delta X}^{h,g}_\lambda)_2\right)^2 \right\rangle < 1. \tag{15}$$

In particular for the collinear gluons we have corresponding two-mode squeezing condition

$$-\frac{1}{8} < \frac{g^2\, t}{16\,(2\pi)^3\, k_0^2}(f_{ahb}f_{ahc} + f_{agb}f_{agc} + f_{ahb}f_{agc} + f_{agb}f_{ahc}) \sum_{\lambda_1 \neq \lambda} |\alpha^b_{\lambda_1}||\alpha^c_{\lambda_1}| \sin(\gamma^b_{\lambda_1} + \gamma^c_{\lambda_1}) < 0. \tag{16}$$

Thus non-perturbative four-gluon selfinteraction is source of the multimode squeezing effect.

3 Gluon entangled states

One of entangled condition is

$$0 < y < 1, \tag{17}$$

where entangled measure is defined by dimentionless value y (2) which in our case is

$$y = \left[\frac{|\overline{\hat{b}^h_\lambda \hat{b}^{g+}_\lambda}|^2 + |\overline{\hat{b}^h_\lambda \hat{b}^g_\lambda}|^2}{2(\overline{\hat{b}^{h+}_\lambda \hat{b}^h_\lambda} + 1/2)(\overline{\hat{b}^{g+}_\lambda \hat{b}^g_\lambda} + 1/2)}\right]^{1/2} \tag{18}$$

Entangled condition is

$$-\frac{1}{8} < \left\langle \hat{N}\left((\widehat{\Delta X}^{h,g}_\lambda)_1\right)^2 \right\rangle < \frac{1-\sqrt{2}}{4} \text{ or } 0 < \left\langle \hat{N}\left((\widehat{\Delta X}^{h,g}_\lambda)_2\right)^2 \right\rangle < \frac{1+\sqrt{2}}{4}, \tag{19}$$

Obviosly the entangled condition (19) imposes greater restrictions than the squeezing condition (15).

Indeed, corresponding entangled condition for collinear gluons is

$$-\frac{1}{8} < \frac{g^2\, t}{16\,(2\pi)^3\, k_0^2}(f_{ahb}f_{ahc} + f_{agb}f_{agc} + f_{ahb}f_{agc} + f_{agb}f_{ahc}) \sum_{\lambda_1 \neq \lambda} |\alpha^b_{\lambda_1}||\alpha^c_{\lambda_1}| \sin(\gamma^b_{\lambda_1} + \gamma^c_{\lambda_1}) < \frac{1-\sqrt{2}}{4}. \tag{20}$$

Obviously squeezed gluon states are simultaneously entangled if the amplitudes of the initial gluon coherent fields are small enough.

Thus by analogy with quantum optics as a result of four-gluon self-interaction we obtain two-mode squeezed gluon states which are also entangled.

4 Colour correlations of the collinear gluons

Second-order colour correlation function is defined as

$$C^{h,g}_{\lambda(2)}(k) = \rho^{h,g}_{\lambda(2)}(k) - \rho^{h}_{\lambda(1)}(k)\rho^{g}_{\lambda(1)}(k) = it\Big[\alpha^{g}_{\lambda}\alpha^{h}_{\lambda}\langle\alpha|[[\hat{H}_I(0),\hat{b}^{h+}_{\lambda}],\hat{b}^{g+}_{\lambda}]|\alpha\rangle - \alpha^{g*}_{\lambda}\alpha^{h*}_{\lambda}\langle\alpha|[\hat{b}^{g}_{\lambda},[\hat{b}^{h}_{\lambda},\hat{H}_I(0)]]\alpha\rangle\Big] \tag{21}$$

Corresponding function for the collinear gluons is

$$C^{h,g}_{\lambda(2)}(k) = -\frac{g^2 t}{16\pi^3 k_0^2}\sum_{\substack{\lambda_1=1\\ \lambda_1\neq\lambda}}^{3} f_{hab}f_{gac}|\alpha^{h}_{\lambda}||\alpha^{g}_{\lambda}||\alpha^{b}_{\lambda_1}||\alpha^{c}_{\lambda_1}|\sin(\gamma^{h}_{\lambda}+\gamma^{g}_{\lambda}-\gamma^{b}_{\lambda_1}-\gamma^{c}_{\lambda_1}) \tag{22}$$

It is obviously this function is sin-like function of the difference between the phases of the investigated and other colour phases of the coherent gluon states. Indeed, the colour correlation is positive at $0 < (\gamma^{b}_{\lambda_1}+\gamma^{c}_{\lambda_1})-(\gamma^{h}_{\lambda}+\gamma^{g}_{\lambda}) < \pi$ (the gluon bunching) and one is negative at $0 < (\gamma^{h}_{\lambda}+\gamma^{g}_{\lambda})-(\gamma^{b}_{\lambda_1}+\gamma^{c}_{\lambda_1}) < \pi$ (the gluon antibunching). Moreover correlation of colours h and g is defined by the structure constants f_{hab}, f_{gac} of $SU_c(3)$ colour group.

The study of color correlations for collinear gluons revealed two groups of colour correlations:

1) First group: $h = 1,2,3 \quad g = 1,2,3; \quad h = 4, \ g = 5; \quad h = 6, g = 7$ is determined only by gluons with different polarizations. For example at $h = 1$ or 2 and $g = 2$ or 1 we have

$$C^{1,2}_{\lambda(2)}(k) = C^{2,1}_{\lambda(2)}(k) = \frac{g^2 t}{16\pi^3 k_0^2}\sum_{\substack{\lambda_1=1\\ \lambda_1\neq\lambda}}^{3} |\alpha^{1}_{\lambda}||\alpha^{2}_{\lambda}||\alpha^{1}_{\lambda_1}||\alpha^{2}_{\lambda_1}|\sin(\gamma^{1}_{\lambda}+\gamma^{2}_{\lambda}-\gamma^{1}_{\lambda_1}-\gamma^{2}_{\lambda_1}); \tag{23}$$

2) Second group: $h = 1,2,3$ and $g = 4,5,6,7; \quad h,g = 4,5,6,7,8$ is defined in addition by gluons with other colours. For example at $h = 1$ or 4 and $g = 4$ or 1 we have

$$\begin{aligned} C^{1,4}_{\lambda(2)}(k) = \frac{g^2 t}{(4\pi)^3 k_0^2}\sum_{\substack{\lambda_1=1\\ \lambda_1\neq\lambda}}^{3} |\alpha^{1}_{\lambda}||\alpha^{4}_{\lambda}|\Big\{|\alpha^{1}_{\lambda_1}||\alpha^{4}_{\lambda_1}|\sin(\gamma^{1}_{\lambda}+\gamma^{4}_{\lambda}-\gamma^{1}_{\lambda_1}-\gamma^{4}_{\lambda_1}) \\ - 3|\alpha^{2}_{\lambda_1}||\alpha^{5}_{\lambda_1}|\sin(\gamma^{1}_{\lambda}+\gamma^{4}_{\lambda}-\gamma^{2}_{\lambda_1}-\gamma^{5}_{\lambda_1}) + 3|\alpha^{3}_{\lambda_1}||\alpha^{6}_{\lambda_1}|\sin(\gamma^{1}_{\lambda}+\gamma^{4}_{\lambda}-\gamma^{3}_{\lambda_1}-\gamma^{6}_{\lambda_1}) \\ + \sqrt{3}|\alpha^{6}_{\lambda_1}||\alpha^{8}_{\lambda_1}|\sin(\gamma^{1}_{\lambda}+\gamma^{4}_{\lambda}-\gamma^{6}_{\lambda_1}-\gamma^{8}_{\lambda_1})\Big\} \end{aligned} \tag{24}$$

5 Conclusion

Investigating of the gluon fluctuations we have proved theoretically the possibility of existence of the multimode gluon squeezed states. The emergence of such remarkable states becomes possible owing to the four-gluon self-interaction. The three-gluon self-interaction does not lead to squeezing effect.

We have shown that QCD jet non-perturbative evolution leads both to squeezing and entanglement of gluons. It should be noted that the greater are both the amplitudes of the initial gluon coherent fields with different colour and polarization indexes and coupling constant, the greater is multimode squeezing effect of the colour gluons. We have demonstrated that entanglement condition of the gluon states with two colours imposes greater restrictions than the squeezing condition and is also defined by the the amplitudes and phases of initial coherent gluon fields.

Because two-mode gluon states with two different colours can lead to $q\bar{q}$-entangled states role of colour correlations could be very significant for explanation of the confinement phenomenon. The study of color correlations for collinear gluons revealed two groups of colour correlations: first group of the correlations ($h = 1, 2, 3$ and $g = 1, 2, 3$; $h = 4$, $g = 5$; $h = 6$, $g = 7$) depends only on distinguished of the gluon polarization, in the second group ($h = 1, 2, 3$ and $g = 4, 5, 6, 7$; $h, g = 4, 5, 6, 7, 8$) correlation behaviour is defined in addition by gluons with other colours.

References

[1] Yu.L. Dokshitzer, V.A. Khoze, A.H. Mueller, S. I. Troyan, *Basics of Perturbative QCD* (Frontières, France, 1991)
[2] V.I. Kuvshinov, Acta Phys. Pol. B **10**, 19 (1979)
[3] E.S. Kokoulina, V.I. Kuvshinov, Acta Phys. Pol. B **13**, 553 (1982)
[4] D.F. Walls, G.J. Milburn, *Quantum Optics* (Springer-Verlag, NY., USA, 1995)
[5] M.O. Scully, M.S. Zubairy, *Quantum Optics* (Cambridge University Press, 1997)
[6] S.Ya. Kilin, *Quantum Optics* (Minsk, 1990). (in Russian)
[7] I.M. Dremin, Phys. Lett. A **193**, 209 (1994)
[8] S. Lupia, W. Ochs and J. Wosiek, Nucl. Phys. B **540**, 405 (1999)
[9] G.J. Alner et al. (UA5 coll.), Phys. Rep. **154**, 247 (1987)
[10] P. Abreu et al. (DELPHI coll.), Z. Phys. C **50**, 185 (1991)
[11] P.D. Acton et al. (OPAL coll.), Z. Phys. C **53**, 539 (1992)
[12] J.S. Bell, Physics **1**, 195 (1964)
[13] A. Einstein, B. Podolsky and N. Rosen , Phys. Rev. **45**, 777 (1935)
[14] E.A. De Wolf , Progress in Optics. **42**, 1 (2001)
[15] V.V. Dodonov, A.S.M. de Castro, S.S. Mizrahi, Phys. Lett. A **296**, 73 (2002)
[16] V.I. Kuvshinov, V.A. Shaparau, Ukrainian Journal of Physics **48**, 1255 (2003)

EPJ Web of Conferences **125**, 04020 (2016) DOI: 10.1051/epjconf/201612504020
QUARKS-2016

On the equivalence of GPD representations

Dieter Müller[1,⋆] and *Kirill* Semenov-Tian-Shansky[2,⋆⋆]

[1] *Theoretical Physics Division, Rudjer Bošković Institute, HR-10002 Zagreb, Croatia*
[2] *Petersburg Nuclear Physics Institute, Gatchina, 188300, St.Petersburg, Russia*

Abstract. Phenomenological representations of generalized parton distributions (GPDs) implementing the non-trivial field theoretical requirements are employed in the present day strategies for extracting of hadron structure information encoded in GPDs from the observables of hard exclusive reactions. Showing out the equivalence of various GPD representations can help to get more insight into GPD properties and allow to build up flexible GPD models capable of satisfactory description of the whole set of available experimental data. Below we review the mathematical aspects of establishing equivalence between the the double partial wave expansion of GPDs in the conformal partial waves and in the t-channel SO(3) partial waves and the double distribution representation of GPDs.

1 Introduction

One of the the prominent goals of the present day hadron physics is the extraction of generalized parton distributions (GPDs) (see Refs. [1–3] for reviews). These non-perturbative quantities provide a detailed description of hadronic structure in terms of QCD quark and gluon degrees of freedom. GPDs can be accessed in hard exclusive reactions such as the deeply virtual Compton scattering (DVCS) (see Ref. [4] for a recent review) and hard meson electroproduction off hadrons (HMP) (see Ref. [5]).

The realistic strategy for extracting GPDs from the data relies on employing of phenomenologically motivated GPD representations and fitting procedures for the whole set of observable quantities (see *e.g.* Ref. [6] and references therein). The non-trivial requirements (forward limit, polynomiality and support properties, hermiticity, positivity constraints, evolution properties, *etc.*) following from the fundamental properties of the underlying quantum field theory, provide a clue to build up a consistent phenomenological representation for GPDs.

All GPD representations present the same field theoretical object, and therefore, as the basic theoretical requirements are satisfied, should be equivalent. Moreover, comparing the manifestation of GPD properties within different representations could provide additional insight of GPDs and their physical interpretation. However, sometimes explicitly working out the mapping of a GPD within one representation to that in a different representation could be mathematically cumbersome.

Below we present an overview of the current status of equivalence studies between the double distribution representation of GPDs and the representations based on the expansion of GPDs in conformal partial waves.

⋆e-mail: dieter.muller@irb.hr
⋆⋆e-mail: cyrstsh@gmail.com

2 Double distribution representation of GPDs and the conformal partial wave expansion

2.1 Double distribution representation

The double distribution (DD) representation of GPDs introduced in [7, 8] was historically one of the first parametrizations for GPDs suitable for phenomenological applications. It arises directly from the diagrammatical considerations and inherits most of the basic field theoretic requirements by presenting a GPD as a one-dimensional section of a two-varibale double distribution. It provides an elegant way to implement both the polynomiality, the support properties and the forward limit constraint. The DD representation is known to be not uniquely defined. However, by means of a 'gauge' transformation [9] it can be put into the conventional form known as the "DD+D" 'gauge' with the double distribution part supplemented by the so-called D-term addendum [10]. For the case of a quark GPD H with the support $x \in [-\eta; 1]$ it reads

$$H^q(x \geq -\eta, \eta, t) = \int_0^1 dy \int_{-1+y}^{1-y} dz\, \delta(x - y - \eta z) f(y, z, t) + D(x/|\eta|, t), \tag{1}$$

where x, the skewness parameter η and the momentum transfer squared t stand for the usual GPD variables.

As first pointed out in [9], the relation between GPD and the corresponding double distribution is a particular case of the Radon transform. This transformation can be inverted in order to recover DDs from GPDs by the standard means of the Radon tomography [9, 11]:

$$f(y, z) = \frac{-1}{2\pi^2} \int_{-\infty}^{\infty} d\eta \, \mathrm{PV} \int_{-\infty}^{\infty} \frac{dx}{x} H_{\mathrm{DD}}(x + y + z\eta, \eta), \tag{2}$$

where $H_{\mathrm{DD}} = H - D$ stands for the DD part of (1) and 'PV' denotes the principal value regularization prescription. Note that the inversion (2) requires the knowledge of GPD for $|\eta| > 1$ which is given by the cross channel counterpart of GPD, the so-called generalized distribution amplitude (GDA):

$$\begin{aligned} &H(x, \eta, t) \\ &= \theta(1 - |\eta|) H_{\mathrm{GPD}}(x, \eta, t) + \theta(\eta - 1) \frac{1}{\eta} H_{\mathrm{GDA}}\left(\frac{x}{\eta}, \frac{1}{\eta}, t\right) + \theta(-\eta - 1) \frac{1}{\eta} H_{\mathrm{GDA}}\left(\frac{x}{\eta}, \frac{-1}{\eta}, t\right). \end{aligned} \tag{3}$$

The example of the application of the inverse Radon transform to recover the double distributions from GPDs for the toy model case of photon GPDs [12] is presented in Ref. [13]. However, the mathematical subtleties encountered while applying the inversion (2) for the realistic GPD models deserve further investigation.

2.2 Conformal partial wave expansion of GPDs and the dual parametrization

The alternative strategy for building up a GPD representation resides on the expansion of GPDs over the conformal partial wave (PW) basis, which ensures the diagonalization of the leading order (LO) evolution operator. The conformal PW expansion of a quark GPD H^q is formally given by

$$H^q(x \geq -\eta, \eta, t) = \sum_{n=0}^{\infty} (-1)^n p_n(x, \eta) H_n(\eta, t); \quad H_n(\eta, t) = \int_{-1}^{1} dx \eta^n c_n(x/\eta) H(x, \eta, t), \tag{4}$$

where we adopt the notations of Ref. [14]. The conformal moments $H_n(\eta, t)$ are formed with respect to the Gegenbauer polynomials $c_n(x) = N_n C_n^{\frac{3}{2}}(x) \equiv \frac{\Gamma(3/2)\Gamma(n+1)}{2^n \Gamma(n+\frac{3}{2})} C_n^{\frac{3}{2}}(x)$ and the conformal PWs p_n also are the suitably normalized Gegenbauer polynomials dotted by the weight $(1 - x^2/\eta^2)$ and the support θ-function. The expansion (4) is to be understood as an ill-defined sum of generalized functions and require proper resummation in order to be defined rigourously.

Two different receipts were proposed in the literature to handle the conformal PW expansion (4). The first one [14, 15] is based on the techniques of the Mellin-Barnes integral. The resummation of the series is performed by trading the sum in (4) for the Mellin-Barnes integral and the analytic continuation of PWs and conformal moments to the complex values of the conformal spin. The second way consists in the use of the so-called Shuvaev-Noritzsch transform [16, 17]. In this case one introduces the forward-like functions of an auxiliary momentum-fraction-type variable whose Mellin moments generate the conformal moments of GPDs. GPD can then be expressed as convolution of the forward-like functions with certain integral kernels.

In order to achieve further factorization of the GPD variable dependencies it turns out extremely instructive to further expand the conformal moments over the basis of the t-channel SO(3) rotation group partial waves.

In the context of the Shuvaev transform techniques the resulting GPD representation is known as the dual parametrization of GPDs [18–20]. For the case of a spinless target hadron the cross channel SO(3) rotation group expansion goes over the usual Legendre polynomials $P_l(\cos\theta_t) \simeq P_l\left(\frac{1}{\eta}\right)$ (up to the power suppressed and small target mass corrections):

$$N_n^{-1}\frac{(n+1)(n+2)}{2n+3}H_n(\eta,t) = \eta^{n+1}\sum_{l=0}^{n+1} B_{nl}(t)P_l\left(\frac{1}{\eta}\right). \tag{5}$$

For the charge even combination of GPDs $H^{(+)}(x,\eta,t) = \text{sign}(x)[H^q(|x|,\eta,t) + H^{\bar{q}}(|x|,\eta,t)]$ we then come to the following double PW expansion

$$H^{(+)}(x,\eta,t) = 2\sum_{\substack{n=1\\ \text{odd}}}^{\infty}\sum_{\substack{l=0\\ \text{even}}}^{n+1} B_{nl}(t)\,\theta(|\eta|-|x|)\left(1-\frac{x^2}{\eta^2}\right)C_n^{3/2}\left(\frac{x}{\eta}\right)P_l\left(\frac{1}{\eta}\right). \tag{6}$$

The Mellin transform of the set of the forward-like functions $Q_{2\nu}(y,t)$ of an auxiliary variable y generates the generalized form factors B_{nl}: $B_{nl}(t) = \int_0^1 dy\, y^n Q_{n+1-l}(y,t)$. A GPD can then be represented as a series of integral transformations of the forward-like functions. The charge even combination (6) reads:

$$H^{(+)}(x,\eta,t) = \sum_{\nu=0}^{\infty}\int_0^1 dy\,[K_{2\nu}(x,\eta|y) - K_{2\nu}(-x,\eta|y)]\,y^{2\nu}Q_{2\nu}(y,t)\,. \tag{7}$$

The integral kernels $K_{2\nu}(x,\eta|y)$ and $K_{2\nu}(-x,\eta|y)$ are defined non-vanishing for $-\eta \le x \le 1$ and $-1 \le x \le \eta$ respectively. The closed analytic expressions for these kernels were originally calculated by means of Shuvaev's dispersion technique [18] and can be expressed in terms of the elliptic integrals.

The double partial wave expansion was also implemented within the Mellin-Barnes integral approach. In the literature it is usually referred as the SO(3) partial wave expansion [21].

The cross-channel SO(3)-PW expansion of the corresponding conformal moments (4) reads

$$H_n(\eta,t) = \sum_{\nu=0}^{\frac{n+1}{2},\,(n/2)} \eta^{2\nu}H_{n,n+1-2\nu}(t)\hat{d}^{n+1-2\nu}(\eta), \quad \text{for odd (even) } n, \tag{8}$$

where the SO(3)-PWs in the spinless target hadron case are expressed by the reduced Wigner d_{00}^l-functions $\hat{d}_{00}^l(\eta) = \frac{\Gamma(\frac{1}{2})\Gamma(1+J)}{2^J\Gamma(\frac{1}{2}+J)}\eta^l P_l\left(\frac{1}{\eta}\right)$.

The final expression for the conformal and cross-channel SO(3) partial wave expansion of the quark part of the GPD H within the Mellin-Barnes integral approach reads:

$$H(x \geq -\eta, \eta, t) = \sum_{\nu=0}^{\infty} \frac{1}{2i} \int_{c+2\nu-i\infty}^{c+2\nu+i\infty} dj \, \frac{p_j(x,\eta)}{\sin(\pi[j+1])} H_{j,j+1-2\nu}(t)\, \eta^{2\nu} \hat{d}_{00}^{j+1-2\nu}(\eta) - \sum_{\nu=1}^{\infty} \eta^{2\nu} p_{2\nu-1}(x,\eta) H_{2\nu-1,0}(t) \,. \quad (9)$$

The detailed discussion on the analytic continuation of both the conformal PWs p_n and the double partial waves (dPWs) $H_{n,n+1-2\nu}$ to the complex values $n = j$ of the conformal spin can be found in Ref. [14]. The value of the intercept c of the Mellin Barnes integral contour is specified by the relative position of the rightmost pole of the conformal partial wave and that of the corresponding dPW.

3 Establishing equivalence of the GPD representations

The complete equivalence of the existing GPD representations was understood already long ago. However, this still has not been demonstrated in the strict mathematical sense for all variety of GPD representations. Below we would like to summarize the recent progress in this field.

For clearness we show the relation between the DD representation of GPDs and the GPD representations based on the conformal PW expansion in the form of the "commutative diagram" depicted on Fig. 1. We start with the consideration of the two "dialects" of the conformal PW and the cross channel SO(3) PW expansion: the dual parametrization and the Mellin-Barnes integral technique, specify the dictionary and present the transition rules. Then we comment on the established relation with the DD representation of GPDs.

3.1 Dual parametrization in two equivalent forms

The complete equivalence of the dual parametrization of GPDs and the Mellin-Barnes techniques for the SO(3) conformal partial wave expansion of GPDs was explicitly established in Ref. [22]. The dPWAs $H_{nl}(t)$ (8) of the Mellin-Barnes transform approach can be put in correspondence with the generalized form factors $B_{nl}(t)$ given by the Mellin moments of the forward-like functions introduced in the context of the dual parametrization (6):

$$H_{n,n+1-2\nu}(t) = \frac{\Gamma(3+n)\Gamma\left(\frac{3}{2}+n-2\nu\right)}{2^{2\nu}\Gamma\left(\frac{5}{2}+n\right)\Gamma(2+n-2\nu)} B_{n,n+1-2\nu}(t). \quad (10)$$

The inversion of the corresponding Mellin transform allows to reconstruct the forward-like functions from the dPWAs,

$$y^{2\nu} Q_{2\nu}(y,t) = \frac{1}{2\pi i} \int_{c-i\infty}^{c+i\infty} dj \, y^{-j-1} \frac{2^{2\nu}\Gamma(5/2+j+2\nu)\Gamma(2+j)}{\Gamma(3+j+2\nu)\Gamma(3/2+j)} H_{j+2\nu,j+1}(t) \,. \quad (11)$$

One also can establish the following Mellin-Barnes representation for the dual parametrization convolution kernels occurring in (7) $K_{2\nu}(x,\eta|y)$:

$$K_{2\nu}(x,\eta|y) = K_{2\nu}^{J\neq 0}(x,\eta|y) - \eta^{2\nu} p_{2\nu-1}(x,\eta) \frac{\Gamma\left(\frac{1}{2}\right)\Gamma(2+2\nu)}{2^{2\nu}\Gamma\left(\frac{3}{2}+2\nu\right)} \frac{1}{y} \,; \quad (12a)$$

$$K_{2\nu}^{J\neq 0}(x,\eta|y) = \frac{1}{2i} \int_{c-i\infty}^{c+i\infty} dj \, \eta^{2\nu} \frac{p_{j+2\nu}(x,\eta)}{\sin(\pi[1+j])} \frac{\Gamma(3+j+2\nu)\Gamma\left(\frac{3}{2}+j\right)}{2^{2\nu}\Gamma\left(\frac{5}{2}+j+2\nu\right)\Gamma(2+j)} y^j \hat{d}_{00}^{j+1}(\eta) \,. \quad (12b)$$

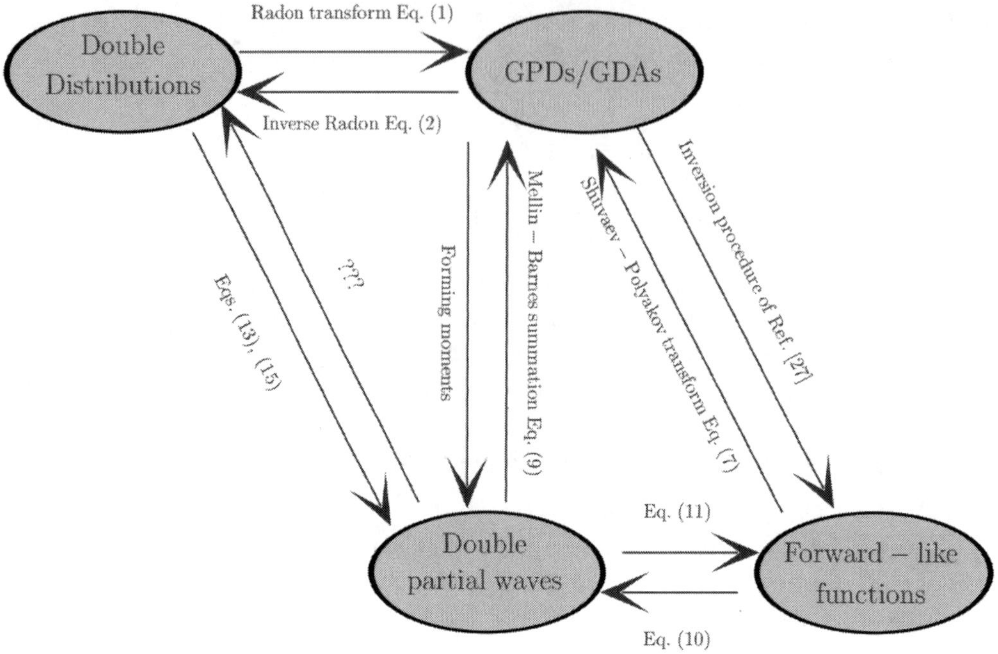

Figure 1. Relations between the double distribution parametrization of GPDs and the two equivalent forms of the double partial wave expansion of GPDs: SO(3) partial wave expansion within the Mellin-Barnes integral approach and the dual parametrization of GPDs. See comments in the text.

Both in (11) and in (12b) the contour intercept c is chosen as explained in Sec. 2.2. It is straightforward to check that the Mellin-Barnes representation for kernels (12b) leads to the familiar expression for the dual parametrization convolution kernels [18, 19] in terms of elliptic integrals.

As the example of application of the reparametrization procedure (11) the successful Kumerički-Müller (KM) model [23], providing global fit to the DVCS world data set for unpolarized proton target [24], was rewritten in terms of the dual parametrization technique.

The established equivalence also contributed much into the clarification of the $J = 0$ fixed pole manifestation issue in DVCS (see discussion in [25]). It was shown that the $J = 0$ fixed pole universality hypothesis of Ref. [26] is equivalent to the conjecture that the D-term form factor is given by the inverse moment sum rule for the Compton form factor. Unfortunately, there exists no theoretical proof for the $J = 0$ fixed pole universality conjecture. Therefore, any supplementary D-term added to a GPD results in an additional $J = 0$ fixed pole contribution and implies the violation of the $J = 0$ fixed pole universality hypothesis.

3.2 Mapping DD representation to the conformal PW expansion

Now we turn to the link between the conformal PW expansion of GPDs and the the DD representation. The first example of the reparametrization procedure for this case was proposed in Ref. [27]. This procedure is based on the Taylor expansion of GPDs in the vicinity of $\eta = 0$, allowing to map any particular GPD to the forward-like function as it appears in the dual parametrization. An example of application was given in Ref. [19], where several first forward-like functions reexpressing the Radyshkin double distribution Ansatz within the dual parametrization approach were computed.

A much more general form of reparametrization procedure linking DD representation to the double partial wave expansion was constructed in Ref. [22]. It consists in the explicit analytic continuation of the corresponding double partial wave amplitudes to the complex values of conformal spin.

Below for simplicity we quote the result for the GPD within the DD representation in the DD+D 'gauge' (1). The set of dPWAs corresponding to a DD representation with a given DD $f(y,z)$ can be presented in a form

$$H^{(+)}_{j+2\nu,j+1}=\int_0^1 dy \int_{-1+y}^{1-y} dz\, y^j f(y,z)\left[1+j+\frac{y\vec{\partial}}{\partial y}\right]\mathcal{K}^{-1}_{j+2\nu,j+1}(y,z)+\delta_{j+1,0}\,D_{2\nu-1,0}(t)\,, \tag{13}$$

where the integral kernel turns to be a polynomial in y and z. It can be expressed through action of the appropriate differential operator on the Gegenbauer polynomials.

$$\begin{aligned}\mathcal{K}^{-1}_{j+2\nu,j+1}(y,z) &= \frac{(1+2\nu)_j(2+j)_\nu}{\Gamma(2+j)\left(\frac{3}{2}+j+\nu\right)_\nu} \\ &\times \int_{-1}^1 d\omega\, \frac{(2+j)_{2+j}\left(1-\omega^2\right)^{1+j}}{2^{3+2j}\,\Gamma(2+j)}\; {}_2F_1\!\left(\begin{matrix}-2\nu, 3+2j+2\nu\\ 2+j\end{matrix}\middle|\frac{1-\omega y-z}{2}\right).\end{aligned} \tag{14}$$

The closed form of the kernel $\mathcal{K}^{-1}_{j+2\nu,j+1}(y,z)$ can be worked out in terms of the rather cumbersome Appell hypergeometric function F_4, which finally reduces to the finite double sum

$$\begin{aligned}\mathcal{K}^{-1}_{j+2\nu,j+1}(y,z) &= \frac{(-1)^\nu 2^{-2\nu}(2+j)_{2\nu-1}}{\Gamma(\nu+1)\left(\frac{3}{2}+j+\nu\right)_\nu} F_4(-\nu,\frac{3}{2}+j+\nu,\frac{5}{2}+j,\frac{1}{2},y^2,z^2) \\ &= \frac{(2+j)_{2\nu-1}}{\left(\frac{3}{2}+j+\nu\right)_\nu}\sum_{m=0}^{\nu}\sum_{n=0}^{\nu-m}\frac{(-1)^{\nu-m-n}\,y^{2m}z^{2n}}{m!n!(\nu-m-n)!\,2^{2\nu}}\frac{\left(\frac{3}{2}+j+\nu\right)_{m+n}}{\left(\frac{5}{2}+j\right)_m\left(\frac{1}{2}\right)_n},\end{aligned} \tag{15}$$

where $(k)_l \equiv \frac{\Gamma(k+l)}{\Gamma(l)}$ stand for the Pochhammer symbol.

The transformation (13), (14) provides the analytic continuation of the GPD moments in j for the regular part of the GPD. The $j=-1$ Kronecker delta contribution induced by the D-term part can be calculated from a $\delta(y)$-proportional addenda to DD by the convolution with the kernel $z\frac{\partial}{\partial z}K^{-1}_{2\nu-1,0}(y=0,z)$, which explicitly reads as

$$D_{2\nu-1,0}(t) = \frac{\Gamma\left(\frac{1}{2}\right)\Gamma(2+2\nu)}{2^{2\nu}\Gamma\left(\frac{1}{2}+2\nu\right)}\int_{-1}^1 dz\, D(z,t)\; {}_2F_1\!\left(\begin{matrix}1-2\nu, 2+2\nu\\ 2\end{matrix}\middle|\frac{1-z}{2}\right). \tag{16}$$

4 Conclusions and outlook

Thus all but one equivalence relations depicted on the "commutative diagram" Fig. 1 have been constructed explicitly. The dual parametrization approach is found to be completely equivalent to the Mellin-Barnes type integral based techniques for GPDs. Also the general transformation allowing to recast the DD representation in the form of double partial wave expansion is proposed.

The only so far missing equivalence relation of Fig. 1 is the inverse relation connecting DPWAs of the double partial wave expansion (8) and the double distributions. It can be designed as the proper combination of the Mellin-Barnes summation procedure and the inverse Radon transform (2). The corresponding studies are underway and the result will be published elsewhere [28].

In particular, it would be extremely instructive to map the successful KM GPD model into the double distribution representation. This can lead to lifting off the unjustified "ad hoc" assumptions in DD modeling and hence to building of more flexible GPD models capable of description of the existing and oncoming DVCS and HMP data.

Acknowledgments

The work of K.S. has been supported by by the Russian Science Foundation (Grant No. 14-22-00281).

References

[1] M. Diehl, Phys. Rept. **388**, 41 (2003).
[2] A. V. Belitsky and A. V. Radyushkin, Phys. Rept. **418**, 1 (2005).
[3] D. Mueller, Few Body Syst. **55**, 317 (2014)
[4] N. d'Hose, S. Niccolai and A. Rostomyan, Eur. Phys. J. A **52**, no. 6, 151 (2016).
[5] L. Favart, M. Guidal, T. Horn and P. Kroll, Eur. Phys. J. A **52**, no. 6, 158 (2016).
[6] K. Kumericki, S. Liuti and H. Moutarde, Eur. Phys. J. A **52**, no. 6, 157 (2016).
[7] D. Müller, D. Robaschik, B. Geyer, F.-M. Dittes, and J. Hořejši, Fortschr. Phys. **42**, 101 (1994).
[8] A. V. Radyushkin, Phys. Rev. **D56**, 5524 (1997).
[9] O. V. Teryaev, Phys. Lett. B **510**, 125 (2001).
[10] M. V. Polyakov and C. Weiss, Phys. Rev. **D60**, 114017 (1999).
[11] A. V. Radyushkin, Phys. Lett. B **449**, 81 (1999).
[12] S. Friot, B. Pire and L. Szymanowski, Phys. Lett. B **645**, 153 (2007).
[13] I. R. Gabdrakhmanov and O. V. Teryaev, Phys. Lett. B **716**, 417 (2012).
[14] D. Müller and A. Schäfer, Nucl. Phys. **B739**, 1 (2006).
[15] M. Kirch, A. Manashov, and A. Schäfer, Phys. Rev. **D72**, 114006 (2005).
[16] A. G. Shuvaev, Phys. Rev. **D60**, 116005 (1999).
[17] J. D. Noritzsch, Phys. Rev. **D62**, 054015 (2000).
[18] M. V. Polyakov and A. G. Shuvaev, *"On 'dual' parametrizations of generalized parton distributions"*, hep-ph/0207153, (2002).
[19] M. V. Polyakov and K. M. Semenov-Tian-Shansky, Eur.Phys.J. **A40**, 181 (2009).
[20] K. M. Semenov-Tian-Shansky, Eur. Phys. J. A **45**, 217 (2010).
[21] K. Kumerički, D. Müller, K. Passek-Kumerički, and A. Schäfer, Phys. Lett. B **648**, 186 (2007).
[22] D. Müller, M. V. Polyakov, and K. M. Semenov-Tian-Shansky, JHEP **1503**, 052 (2015).
[23] K. Kumerički and D. Müller, Nucl. Phys. **B841**, 1 (2010).
[24] E.-C. Aschenauer, S. Fazio, K. Kumerički and D. Müller, JHEP **1309**, 093 (2013).
[25] D. Müller and K. M. Semenov-Tian-Shansky, Phys. Rev. D **92**, no. 7, 074025 (2015).
[26] S. J. Brodsky, F. J. Llanes-Estrada, and A. P. Szczepaniak, Phys.Rev. **D79**, 033012 (2009).
[27] K. M. Semenov-Tian-Shansky, Eur. Phys. J. **A36**, 303 (2008).
[28] D. Müller and K. M. Semenov-Tian-Shansky, *in preparation*.

EPJ Web of Conferences **125**, 04021 (2016)
QUARKS-2016
DOI: 10.1051/epjconf/201612504021

Investigation of the influence of nuclear matter on hard neutrino nucleus interaction using the HARDPING Monte Carlo Event Generator

Ya.A. Berdnikov[1,2], *A.Ya.* Berdnikov[1], *V.T.* Kim[1,2], *A.E.* Ivanov[1], *D.P.* Suetin[1,2,⋆], and *K.D.* Tiangov[1]

[1] *Peter the Great Saint-Petersburg Polytechnic University, St. Petersburg, 195251, Russia*
[2] *The Petersburg Nuclear Physics Institute, National Research Center "Kurchatov Institute", Gatchina, Leningrad district, 188300, Russia*

Abstract. Hadron production in neutrino-nucleus interactions is implemented in Monte Carlo event generator HARDPING (HARD Probe INteraction Generator). Such effects as formation length, energy loss and multiple rescattering for produced hadrons and their constituents are taken into account in HARDPING. Available data from WA/59 and SCAT collaborations on hadron production in neutrino-nucleus collisions is described by HARDPING with a reasonable agreement.

Hadronization of quarks and gluons is one of the most intriguing processes of nonperturbative QCD. Use of nuclear targets may allow to study important features of space-time picture of hadronization, like hadron formation length and energy loss, see, e.g, for a review [1- 7] and references therein.

The understanding of quark propagation in nuclear medium is crucial to interpret nucleus-nucleus, proton-nucleus and lepton-nucleus interactions. To simplify interpretation of observable effects first of all let's consider hadron production in lepton scattering off nuclei.

It is well known from theoretical and experimental studies of lepton-nucleus collisions at high energies that hadrons are not produced at the point of collision but only after certain "formation" length [1]. In the Lund string fragmentation model, the production of hadrons is described as a two-stage process. On the first perturbative stage pre-hadron is formed at the end of the string. On the next nonperturbative stage hadron is formed. Prior to formation of pre-hadron, the knocked-out quark propagates through nuclear matter with a very small cross section. It takes some time to form a pre-hadron at the perturbative stage (formation time, t_p or formation length, l_p). When the pre-hadron is formed, it interacts with nuclear matter with pre-hadron cross section, which is different from hadronic cross section. Also it takes an extra time to form the final hadron from pre-hadron state. Therefore, formation length consists of two parts l_p and l_h, which correspond to pre-hadron and hadron stages accordingly. There are two approaches to calculate such formation length. The first one is based on a simplified description of nonperturbative stage [1], while the second one [3] is based on Lund string model, which neglects the energy loss effects during perturbative stage. Present work takes into account both effects: energy loss in perturbative stage and Lund string model hadronization in nonperturbative stage. In the approach presented in this article effect of energy loss was built into HARDPING [8-10] using PYTHIA event generator implementation of parton shower for the perturbative stage and Lund string model for the nonperturbative one.

⋆e-mail: suetindaniil@gmail.com

 DOI: 10.1051/epjconf/201612504021

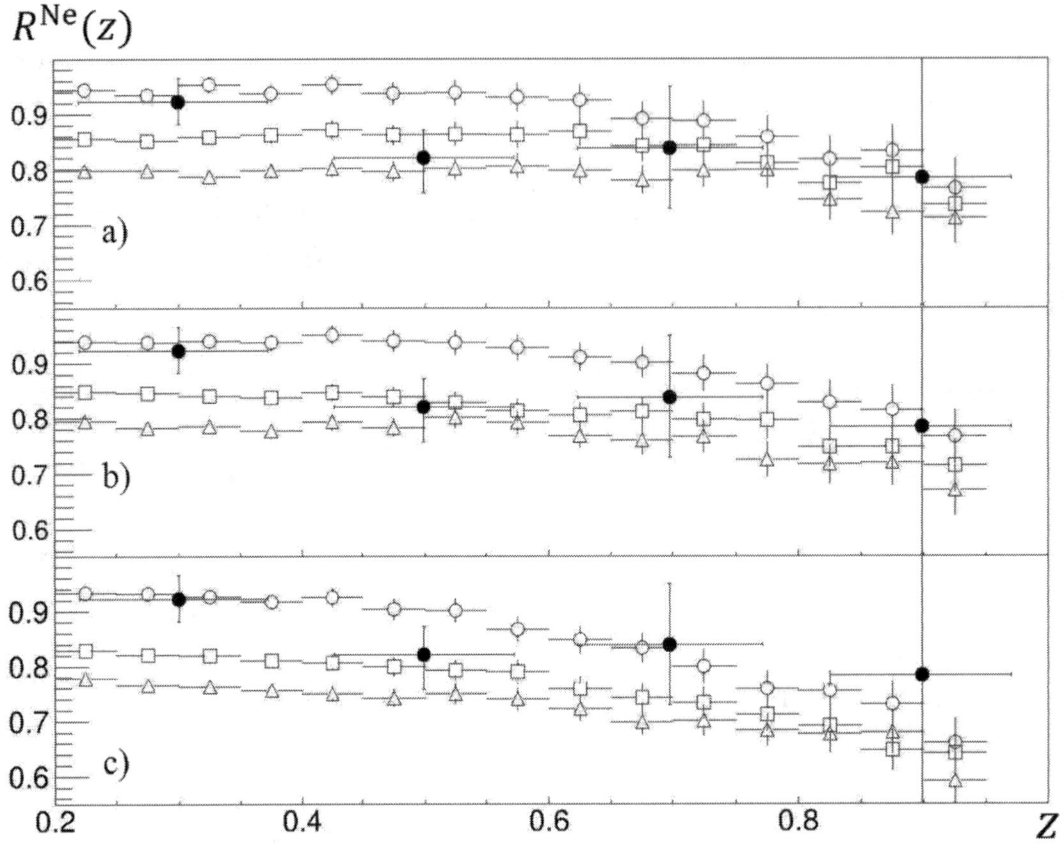

Figure 1. Ratio of hadron yield on neon (Ne) target to analogous yield on proton (p) target as function of z for neutrino interactions (2) in kinematic region $4 < \nu < 8$ GeV. Black circles represent WA/59 collaboration data [11], blank circles, squares and triangles represent HARDPING simulation data with string tension coefficient κ equal to 1.5, 2.5 and 3.5 GeV/fm accordingly. Figure shows results of simulation with pre-hadron nucleon cross-section equal to a) 5 mb; b) 10 mb; c) 20mb.

During the perturbative stage, which corresponds to formation length l_p, a pre-hadron state is formed. It can interact with intranuclear nucleons with inelastic pre-hadronic cross-section. At the end of the nonperturbative stage the observed hadron is formed. Produced pre-hadrons and hadrons can undergo soft collisions with intranuclear nucleons (with small momentum transfer: $|\, t \,| \sim 1\ GeV^2$). Therefore one has to take into account their soft multiple rescattering and energy loses in nuclear matter [8-10]. The goal of present work is to study these effects for neutrino-nuclei collisions using HARDPING.

The experimental data from WA/59 collaboration on hadron production in neutrino-nuclei collisions are presented in terms of nuclear modification factor R_{Ne}, which represents ratio of hadron yield on neon (Ne) target to analogous yield on proton (p) target, as function of virtual boson energy ($\nu = E_\mu - E_\nu$) or virtual boson energy fraction taken away by hadron ($z = E_h/\nu$):

$$R^{\mathrm{Ne}}(\nu) = \frac{1}{N_{\mathrm{Ne}}}\frac{dN^{ch}_{\mathrm{Ne}}(\nu)}{d\nu} \Big/ \frac{1}{N_{\mathrm{p}}}\frac{dN^{ch}_{\mathrm{p}}(\nu)}{d\nu} \qquad (1)$$

DOI: 10.1051/epjconf/201612504021

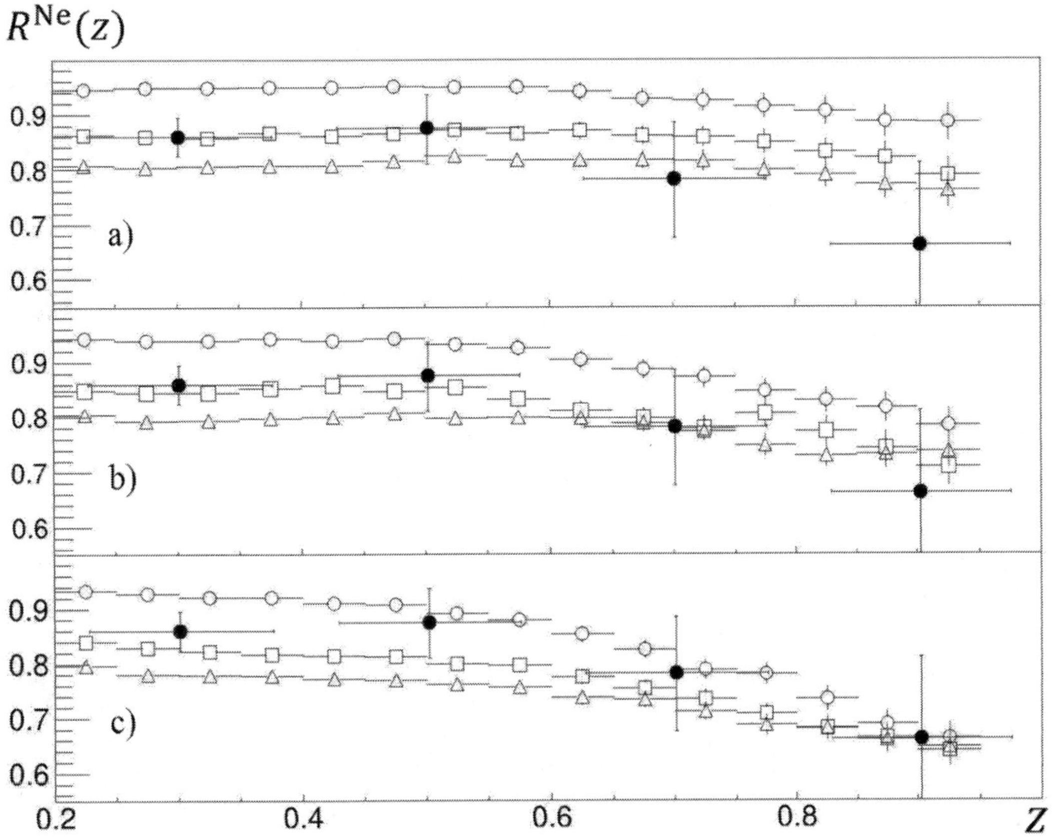

Figure 2. Ratio of hadron yield on neon (Ne) target to analogous yield on proton (p) target as function of z for anti-neutrino interactions (2) in the kinematic region $4 < \nu < 8$ GeV. Black circles represent WA/59 collaboration data [11], blank circles, squares and triangles represent HARDPING simulation data with string tension coefficient κ equal to 1.5, 2.5 and 3.5 GeV/fm accordingly. Figure shows results of simulation with pre-hadron nucleon cross-section equal to a) 5 mb; b) 10 mb; c) 20mb.

$$R^{\rm Ne}(z) = \frac{1}{N_{\rm Ne}} \frac{dN^{ch}_{\rm Ne}(z)}{dz} \Big/ \frac{1}{N_{\rm p}} \frac{dN^{ch}_{\rm p}(z)}{dz} \tag{2}$$

where $N_{\rm Ne}$ and $N_{\rm p}$ are numbers of neutrino charged current interactions on Ne and p targets accordingly; $N^{ch}_{\rm Ne}$ and $N^{ch}_{\rm p}$ are yields of hadron on Ne and p targets accordingly.

Neutrino and anti-neutrino beams in WA/59 originate from 400 GeV protons impinging on a beryllium target. Simulations of neutrino-nuclei collisions obtained by HARDPING were compared with WA/59 data [11]. The results are shown in figs. 1, 2, 3.

Experimental results from SCAT collaboration on hadron production in neutrino-nuclei collisions are presented in terms of nuclear modification factor R^{ch}, which represents ratio of hadron yield on nuclear (A = 28) target to analogous yield on deuterium (D) target, as function of virtual boson energy ($\nu = E_\mu - E_\nu$) or virtual boson energy fraction taken away by hadron ($z = E_h/\nu$):

$$R^{ch}(\nu) = \frac{1}{N_{\rm A}} \frac{dN^{ch}_{\rm A}(\nu)}{d\nu} \Big/ \frac{1}{N_{\rm D}} \frac{dN^{ch}_{\rm D}(\nu)}{d\nu} \tag{3}$$

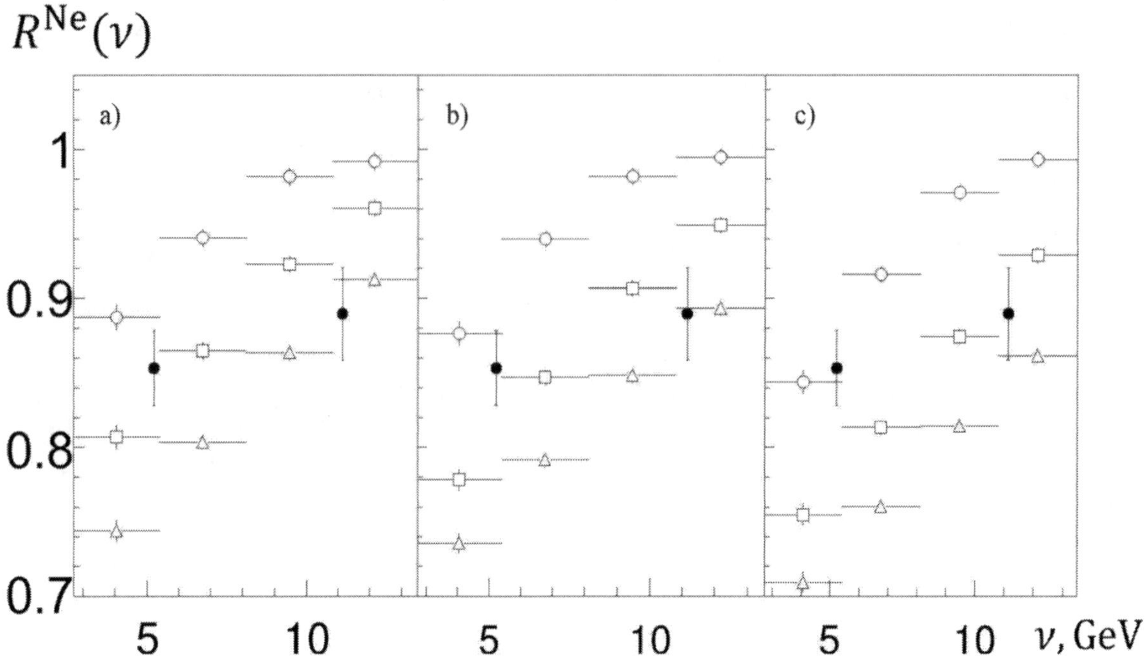

Figure 3. Ratio of hadron yield on neon (Ne) target to analogous yield on proton (p) target as function of ν for neutrino and anti-neutrino interactions (1) in the kinematic region $z > 0.2$. Black circles represent WA/59 collaboration data [11], blank circles, squares and triangles represent HARDPING simulation data with string tension coefficient κ equal to 1.5, 2.5 and 3.5 GeV/fm accordingly. Figure shows results of simulation with pre-hadron nucleon cross-section equal to a) 5 mb; b) 10 mb; c) 20mb.

$$R^{ch}(z) = \frac{1}{N_{\mathrm{A}}} \frac{dN_{\mathrm{A}}^{ch}(z)}{dz} \Big/ \frac{1}{N_{\mathrm{D}}} \frac{dN_{\mathrm{D}}^{ch}(z)}{dz} \tag{4}$$

where N_{A} and N_{D} are numbers of neutrino charged current interactions on A and D targets accordingly, N_{A}^{ch} and N_{D}^{ch} are yields of hadron on A and D targets accordingly. Neutrino and anti-neutrino beams in SCAT originate from 70 GeV protons. Simulations of neutrino-nuclei collisions obtained by HARDPING were compared with SCAT data [12]. The results are shown in Figs. 4 and 5.

Performed simulations show a reasonable agreement of MC HARDPING with experimental data from WA/59 and SCAT [11, 12]. This allows to fix values of such model parameters as pre-hadron nucleon cross-section $\sigma = 10$ mb and string tension $\kappa = 2.5$ GeV/fm, with which the best agreement with the two experiments was achieved. In conclusion HARDPING Monte Carlo event generator describes:

- Experimental data from WA/59 collaboration [11] on hadron production in neutrino- and anti-neutrino nuclei scattering at average incident particle energy of 54 and 40 GeV accordingly;
- Experimental data from SCAT collaboration [12] on hadron production in neutrino-nuclei scattering at average neutrino energy of 10 GeV.

References

[1] B.Z Kopeliovich, J Nemchik, E. Predazzi, and A. Hayashigaki, Nucl. Phys. A **740**, 211 (2004)

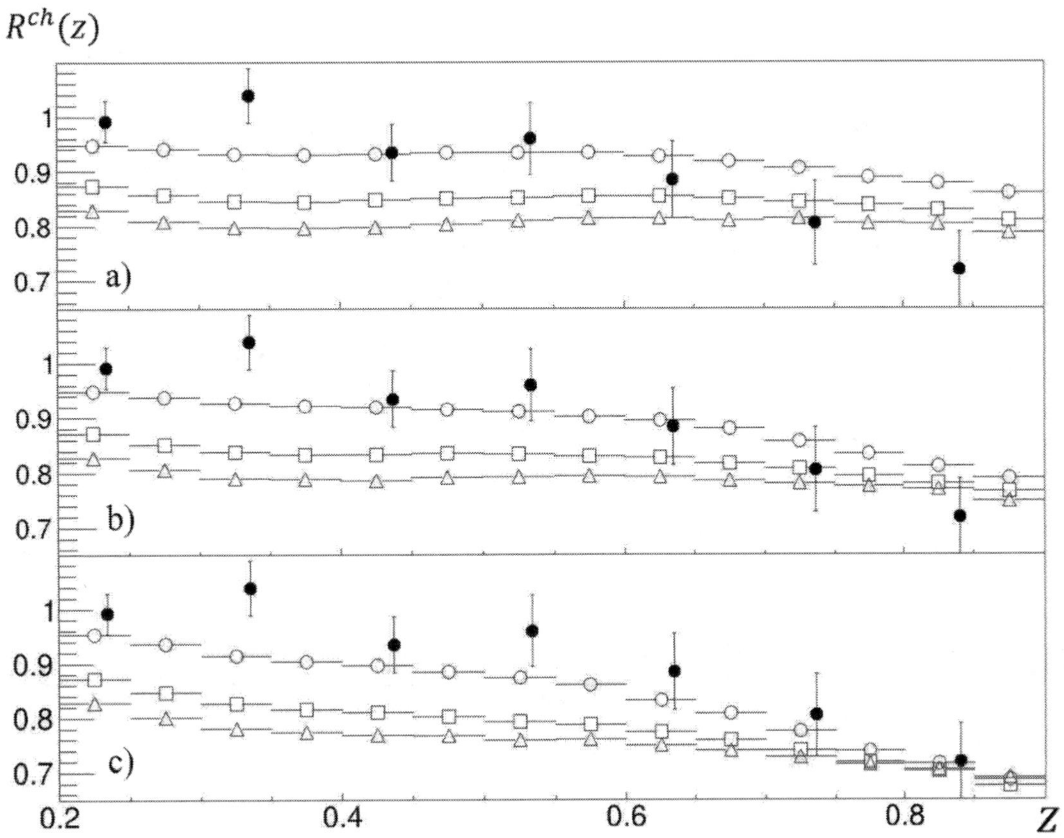

Figure 4. Ratio of hadron yield on nuclear (A = 28) target to analogous yield on deuterium target as function of z for neutrino interactions (4). Black circles represent SCAT collaboration data [12], blank circles, squares and triangles represent HARDPING simulation data with string tension coefficient κ equal to 1.5, 2.5 and 3.5 GeV/fm accordingly. Figure shows results of simulation with pre-hadron nucleon cross-section equal to a) 5 mb; b) 10 mb; c) 20mb.

[2] S. Domdey, D. Grunewald, B.Z. Kopeliovich, Nucl. Phys. A **825**, 200-211 (2009)

[3] A. Accardi, V. Muccifora, H.J. Pirner, Nucl. Phys. A **720**, 131-157 (2003)

[4] A. Accardi, D. Grunewald, V. Muccifora, H.J. Pirner, Nucl. Phys. A **761**, 67-96 (2005)

[5] R. Baier, D. Schiff and B. G. Zakharov, Ann. Rev. Nucl. Part. Sci. **50**, 37 (2000)

[6] R. Baier, Y. L. Dokshitzer, A. H. Mueller et al., Nucl. Phys. B **484**, 265 (1997)

[7] B. G. Zakharov, JETP Lett. **65**, 615 (1997)

[8] Ya.A. Berdnikov, A.E. Ivanov, V.T. Kim, V.A. Murzin D.P Suetin, Nucl. Phys. B. Proc. Suppl. **245**, 267-270 (2013)

[9] Ya.A. Berdnikov, A.E. Ivanov, V.T. Kim, D.P Suetin, Bulletin of the Russian Academy of Sciences: Physics **79(7)**, 925-929 (2015)

[10] Ya.A. Berdnikov, A.E. Ivanov, V.T. Kim, D.P Suetin, AIP Conference Proceedings **1701**, 080005 (2016)

[11] W. Burkot, , T. Coghen, J. Czyżewski, Zeitschrift für Physik **C70(1)**, 47-53 (1996)

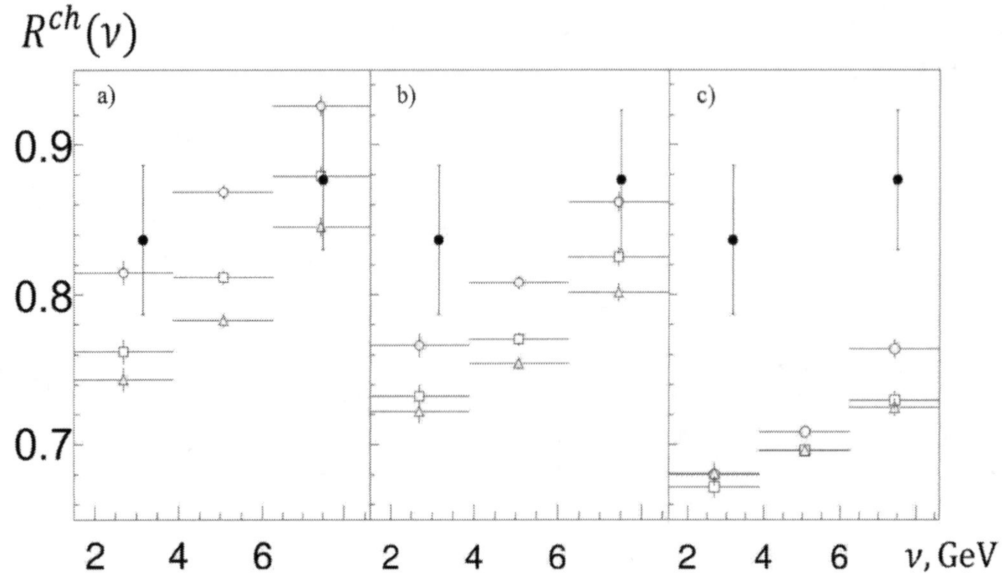

Figure 5. Ratio of hadron yield on nuclear (A = 28) target to analogous yield on deuterium target as function of ν for neutrino and anti-neutrino interactions (3) in the kinematic region $z > 0.2$. Black circles represent SCAT collaboration data [12], blank circles, squares and triangles represent HARDPING simulation data with string tension coefficient κ equal to 1.5, 2.5 and 3.5 GeV/fm accordingly. Figure shows results of simulation with pre-hadron nucleon cross-section equal to a) 5 mb; b) 10 mb; c) 20mb.

[12] N. M. Agababyan, V. V. Ammosov, M. R. Atayan, Physics of Atomic Nuclei **66**, 310-318 (2003)

 DOI: 10.1051/epjconf/201612504022

Long-range rapidity correlations between mean transverse momenta in the model with string fusion

Vladimir Vechernin[1,*]

[1] *Saint Petersburg State University*

Abstract. The long-range correlation between mean-event transverse momenta, being robust against the volume fluctuations and the details of the centrality determination, enables to obtain the signatures of string fusion at the initial stage of hadronic interaction in relativistic heavy ion collisions. The dependence of the correlation strength between mean-event transverse momenta on the collision centrality and initial energy is analyzed in a simple model with quark-gluon string fusion on the transverse lattice. It is shown that above RHIC energy the dependence reveals the decline of the correlation coefficient for most central collisions, reflecting the attenuation of color field fluctuations due to the string fusion at large string density. It is also found that contrary to the correlation between transverse momenta of single particles the strength of the correlation between mean-event transverse momenta of particles in two separated rapidity intervals is not decreasing with the total number of produced strings, remaining significant even in the case of Pb-Pb collisions, in which the total number of strings can reach several thousand.

1 Introduction

The study of the correlations between observables in two separated rapidity windows (the so called long-range forward-backward correlations) has been proposed [1] as a signature of the string fusion and percolation phenomenon [2–5], which is one of the collectivity effects in ultrarelativistic heavy ion collisions. Later it was realized [6–11] that the investigations of the forward-backward correlations between intensive observables, such e.g. as mean-event transverse momenta, enable to obtain more clear signal about the initial stage of hadronic interaction, including the process of string fusion, compared to usual forward-backward multiplicity correlations.

In the present work, as an example, we consider the correlation between mean-event transverse momenta of charged particles in separated rapidity intervals [8, 10, 11]:

$$p_{tF} = \frac{1}{n_F}\sum_{i=1}^{n_F} |\mathbf{p}^i_{tF}| \qquad\qquad p_{tB} = \frac{1}{n_B}\sum_{i=1}^{n_B} |\mathbf{p}^i_{tB}| \tag{1}$$

The calculations are fulfilled in a simple model with string fusion by introducing a lattice in transverse plane [7, 12]. Previously, in the framework of the model we found the analytical asymptotics of the correlation coefficient, $b_{p_tp_t}$, between mean transverse momenta in separated rapidity intervals at large

*e-mail: v.vechernin@spbu.ru

string density and fulfilled, using MC simulations, the calculation of the correlation coefficient as a function of string density [10, 11]. In present paper we discuss the impact of the obtained results on the $b_{p_t p_t}$ correlation coefficient centrality dependence at RHIC and LHC energies.

2 Forward-Backward Rapidity Correlations

Usually the forward-backward (FB) rapidity correlation are studied as the correlation between two observables, F and B, measured event-by-event in some rapidity intervals, $\delta\eta_F$ and $\delta\eta_B$, separated by some rapidity gap, η_{gap}. Numerically it can be described by the correlation function:

$$\langle B\rangle_F = f(F)\,, \tag{2}$$

giving the dependence of the mean value of the variable B on the variable F. Usually this dependence is approximated by a linear function

$$\langle B\rangle_F = a + b_{FB}F\,, \tag{3}$$

the so-called linear regression. In this case the strength of the correlation can be characterized by one number - the correlation coefficient b_{FB}, for which one can find the following expression:

$$b_{FB} = \frac{\langle FB\rangle - \langle F\rangle\langle B\rangle}{\langle F^2\rangle - \langle F\rangle^2} = \frac{cov(F,B)}{D_F}\,, \tag{4}$$

To reduce the influence of different widths of the rapidity windows, $\delta\eta_F$ and $\delta\eta_B$, on the correlation coefficient the analysis sometimes performs using the relative variables, $F/\langle F\rangle$ and $B/\langle B\rangle$. Clear that the correlation coefficient in this variables, b^{rel}_{FB}, is simply connected with the b_{FB}: $b^{rel}_{FB} = b_{FB}\langle F\rangle/\langle B\rangle$. In our analysis we will use only symmetric observation windows, $\delta\eta_F = \delta\eta_B$, in this case $b^{rel}_{FB} = b_{FB}$ for a symmetric collision. Nevertheless note, that this does not eliminates the dependence of the correlation coefficient on the rapidity width of observation window (see e.g. [13]) what is important when one compares the results of different experiments.

Due to the locality of strong interaction in rapidity the FB correlation naturally divides into two parts: the short- and the long-range ones. The Short-Range Correlation (SRC) takes place only between particles produced from a same source (string). It goes to zero for large enough rapidity gaps, $\eta_{gap} \gtrsim 2$. The origin of the Long-Range Correlations (LRC) at large η_{gap} is event-by-event variance in the number and properties of particle sources (cut pomerons [14], strings [13] and so on), formed in high energy hadronic interaction. In present paper we will consider only these LRC, as we are interested in the study of changing of source properties (the string properties due to their fusion).

Traditionally in the FB correlation analysis the multiplicities of charged particles with rapidities belonging to these FB windows were used as the observables, $F = n_F$ and $B = n_B$. Unfortunately the n-n LRC is dominated mainly by the fluctuation in the number of initial sources (strings) - the so-called "volume" fluctuation, originating, for example, due to the fluctuation of the collision impact parameter. The influence of the other interesting effects, connected with fluctuation of source properties (e.g. due to possible string fusion processes), on this type of correlation is rather small [15]. The reason is the extensive nature of these observables.

More clear signal about a variation of sources properties we can expect for the LRC correlations involving intensive observables, e.g. the mean p_t, (1). The simple "volume" fluctuations do not lead to the correlation between the intensive observables and we need the fluctuation in "quality" of sources for its explanation. So we have to study the p_t-p_t LR correlation with $F = p_{tF}$ and $B = p_{tB}$, (1), as FB observables, if we aim to discover the possible signatures of string fusion phenomenon.

3 The model

The discrete version of string fusion model [4, 5] with a lattice in transverse plane was introduced in [12] and then was successfully exploited for a description of various phenomena in ultra relativistic nuclear collisions [7–11, 16–18]. In this approach one splits the impact parameter plane into M cells with the area equal to the transverse area of single string, σ_{str}, and supposes the fusion of all strings with the centers in a given cell. Then each string configuration is characterized by the set of integers:

$$C_\eta = \{\eta_1, ..., \eta_M\} \ , \tag{5}$$

where η_i is a number of initial strings fused in a given i-th cell . This leads to the splitting of the transverse area into domains with different, fluctuating values of color field within them. What is similar to the attempts to take into account the density variation in transverse plane in models based on the BFKL evolution [19] and on the CGC approach [20].

In present calculations we supposed that the η_i in each cell fluctuates around the same mean value $\overline{\eta}$ with a scaled variance ω_η. Then in accordance with the string fusion prescription [4, 5] the mean number of charged particles in forward and backward observation windows, produced from the fragmentation of fused strings in the i-th cell, is given by the expression

$$\overline{n}_i^F = \mu_F \sqrt{\eta_i} \ , \tag{6}$$

and the same for $\overline{n}_i^B$, where μ_F is a mean number of particles produced from decay of a single string in the forward window. Because we study only LRC, we also suppose the independent fragmentation of the i-th fused string in forward and backward windows. So the corresponding numbers of charge particles n_i^F and n_i^B fluctuates independently event-by-event around the above mean values with some scaled variance ω_μ.

Finally, in accordance with the string fusion ideas [4, 5] we suppose that the transverse momentum distribution of particles, produced from the η_i fused strings, has enlarged mean transverse momentum:

$$\overline{p}(\eta_i) = p_0 \sqrt[4]{\eta_i} \ , \tag{7}$$

where p_0 is a mean transverse momentum for particles produced from a decay of a single string. For popular transverse momentum distributions $\varphi(p)$ with one dimensional parameter $\tilde{p}$ (see table 1), due to dimensional reasons, we have $\tilde{p} \sim \sqrt[4]{\eta_i}$. We have also the following connection between the variance and the mean transverse momentum for particles produced from a decay of the fused string in the i-th cell:

$$d_{p_i}(\eta_i) = \gamma \overline{p}^2(\eta_i) \ , \tag{8}$$

where the dimensionless parameter γ depends only on the shape of the distribution (see table 1).

So each event in the model is characterized by the following configuration

$$C = \left\{C_\eta, C_n^F, C_n^B, C_p^F, C_p^B\right\} \ , \tag{9}$$

where

$$C_n^F = \left\{n_1^F, ..., n_M^F\right\} \ , \quad C_n^B = \left\{n_1^B, ..., n_M^B\right\} \ , \tag{10}$$

$$C_p^F = \left\{p_1^{1F}, ..., p_1^{n_1^F F}; ...; p_M^{1F}, ..., p_M^{n_M^F F}\right\} \tag{11}$$

and the same for the C_p^B. Here n_i^F and n_i^B are the numbers of charged particles, produced in forward and backward windows from the fragmentation of fused string in i-th cell, and p_i^{jF} and p_i^{jB} are their

Table 1. The values of the dimensionless parameter γ, see formula (8), for the popular transverse momentum distributions $\varphi(p)$, normalized as $\int \varphi(p)\, d^2p = 1$ $(m > 4)$.

$\varphi(p)$	γ
$\sim \exp(-p^2/\bar{p}^2)$	$\frac{4-\pi}{\pi}$
$\sim \exp(-p/\bar{p})$	$\frac{1}{2}$
$\sim 1/(1+p/\bar{p})^m$	$\frac{m-1}{2(m-4)}$

transverse momenta. Then, for the physical observables, defined in Sect. 1, we have

$$p_{tF} = \frac{1}{n_F}\sum_{i=1}^{M}\sum_{j=1}^{n_i^F} p_i^{jF}\,, \qquad n_F = \sum_{i=1}^{M} n_i^F \tag{12}$$

and the same for the p_{tB} and n_B. To calculate the correlation function (2) and the correlation coefficient (4) we have to average over these configurations with a proper weights (see [11, 21] for details).

4 The asymptotic of the correlation coefficient

In the framework of the model one can find the explicit analytical asymptotic of the LR correlation coefficient between mean-event transverse momenta [11] at large uniform mean string density $\bar{\eta}/\sigma_{str}$ with $\bar{\eta} \gg 1$:

$$b_{p_tp_t} = \frac{\omega_\eta \mu_F}{\omega_\eta \mu_F + 16\gamma\sqrt{\bar{\eta}}}\,. \tag{13}$$

The dimensionless parameter $\bar{\eta}$ characterizes the mean degree of overlap of strings. In the derivation of the asymptotic the existence of two small parameters: $1/M$ and $1/\bar{\eta}$ in the model was used. One can estimate the typical value of M, as ratio of the total transverse interaction area S to the area of a single string $\sigma_{str} = \pi r_{str}^2$:

$$M \sim S/\sigma_{str} \sim R_A^2/r_{str}^2\,.$$

For Pb-Pb collisions $R_A = 7 \div 8$ fm, $r_{str} = 0.2 \div 0.3$ fm, so in this case we have $M \sim 10^3$. The value of $\bar{\eta} \sim 10$ at LHC and $\bar{\eta} \sim 3$ at RHIC energies [10, 22] (as we'll see below the asymptotic (13) is a good approximation only at $\bar{\eta} \gtrsim 7 \div 8$).

The most interesting feature of the asymptotic is the decrease of the correlation coefficient $b_{p_tp_t}$ with growth of the string density. The physical reason of such behavior is the attenuation of color field fluctuations due to the string fusion processes at large string density.

Another important property of the asymptotic is the M-scaling. The expression (13) for the $b_{p_tp_t}$ depends only on the string density $\bar{\eta}$ and does not depend on the total number of cells, M, and hence on the total number of strings. Note that this nontrivial property takes place only for the correlation between *mean* event transverse momenta, defined above, (1), in contrast with the behavior of the p_t-p_t correlation coefficient, $\beta_{p_tp_t}$, between transverse momenta of two single particles, defined as

$$\beta_{p_tp_t} \equiv \frac{I_2(p_{t1}, p_{t2})}{I(p_{t1})I(p_{t2})} - 1\,,$$

DOI: 10.1051/epjconf/201612504022

where $I(p_{t1})$ and $I_2(p_{t1}, p_{t2})$ are single and double inclusive cross sections. In [7] it was shown that this correlation coefficient decreases with the total number of sources (strings), and hence with M:

$$\beta_{p_t p_t} \sim 1/N_{sources} \sim 1/M$$

So, its value becomes very small, e.g. for Pb-Pb interactions at LHC energy, in which the number the strings is of order of few thousand [10, 22]. The reason of the difference between these two types of p_t-p_t correlation, characterized by $b_{p_t p_t}$ and $\beta_{p_t p_t}$, is that the mean p_{tF} and p_{tB}, being intensive variables, stronger correlates with the value of string tension and better reflects the specific event transverse energy in observation rapidity windows $\delta\eta_F$ and $\delta\eta_B$, than the transverse momenta of single particles.

We see also that the asymptotic (13) depends only on the one combination of all model parameters, $\frac{\omega_\eta \mu_F}{\gamma\sqrt{\bar{\eta}}}$, what we'll refer on as $\frac{\omega_\eta \mu_F}{\gamma\sqrt{\bar{\eta}}}$ - scaling. Note also, that in contrast with the correlations between multiplicities n_F and n_B and between mean transverse momentum p_{tB} and multiplicity n_F, considered in [12], there is no dependence of the $b_{p_t p_t}$ asymptotic on the parameter ω_μ. Recall that ω_μ characterizes the fluctuations in the number of particles produced from a string.

Instead of the ω_μ we have the dependence of the $b_{p_t p_t}$ asymptotic on the dimensionless parameter γ, which characterizes the transverse momentum distribution from one initial string (see table 1). Since for a distribution with one dimensional parameter it does not depends on a string fusion, then in this condition it can be evaluated from data, as $\gamma = \langle\langle p_t^2\rangle\rangle/\langle\langle p_t\rangle\rangle^2 - 1$, where the $\langle\langle ...\rangle\rangle$ means averaging over tracks from all events.

5 Results of the MC simulations

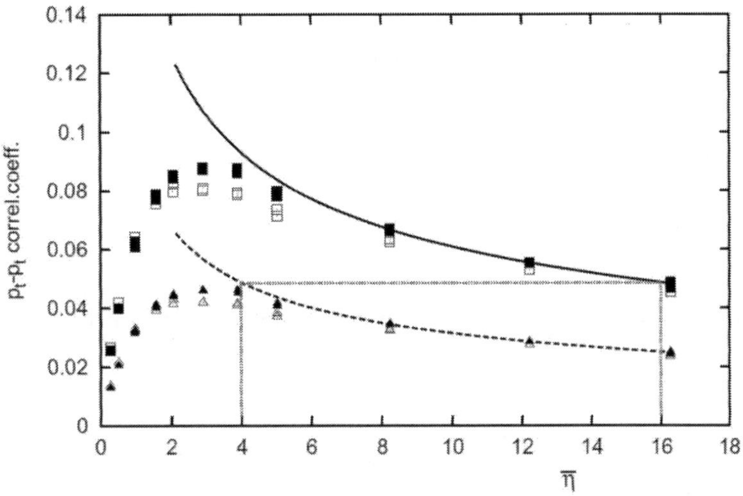

Figure 1. The $b_{p_t p_t}$ correlation coefficient (points), obtained by MC simulations in the model with string fusion on a transverse lattice with $M = 45$ cells for $\mu_F = 1$ (△) and $\mu_F = 2$ (□), as a function of mean string density $\bar{\eta}$. The filled points are calculated in the version with preliminary averaging of string density within lattice clusters (see [10, 23] for details). The curves - corresponding asymptotics calculated by the formula (13) at the parameter values: $\omega_\eta = \omega_\mu = 1$ and $\gamma = 0.6$. The dotted lines illustrate the $\mu_F/\sqrt{\bar{\eta}}$-scaling for asymptotics (see text).

The MC simulations in the framework of this simple model with string fusion on a transverse lattice enable to calculate the value of the correlation coefficient $b_{p_t p_t}$, (4), for an arbitrary value and

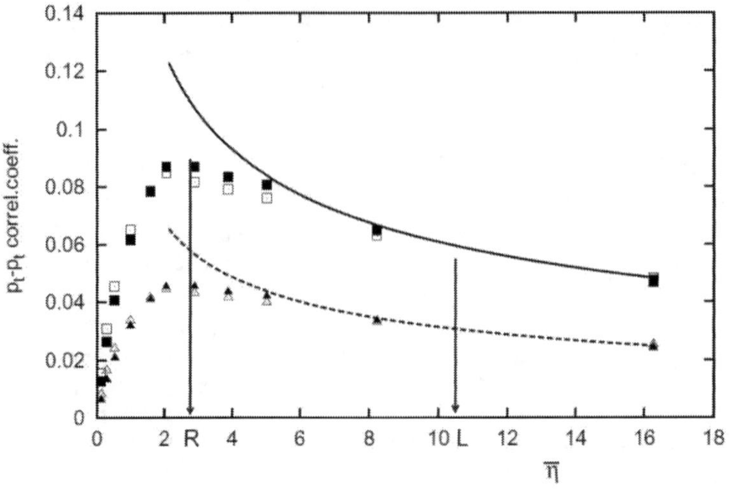

Figure 2. The same as in figure 1, but for a lattice with $M = 450$ cells. The arrows show the string densities, R and L, available correspondingly at RHIC and LHC [10, 22].

distribution of mean string density [10]. We restrict to the uniform string distribution over lattice with the mean density $\overline{\eta}$ and the poissonian fluctuations around this value ($\omega_{\eta} = 1$) in each cell. The results for the $b_{p_tp_t}$ correlation coefficient calculated on the base the MC simulation for $\mu_F = 1$ and $\mu_F = 2$ in the cases $M = 45$ and $M = 450$ are presented in figures 1 and 2 together with the corresponding asymptotics calculated by the formula (13).

First of all we see in these figures that the asymptotics well approximate the correlation coefficient $b_{p_tp_t}$ at values of $\overline{\eta}$ greater than $7 \div 8$. Comparing figures 1 and 2 we see also a good fulfillment of the M-scaling at any value of string density $\overline{\eta}$, and not only in the asymptotic region. (More thorough analysis [12] shows that the M-scaling is broken only in narrow region $\overline{\eta} \lesssim 1/M$, when the total number of strings is small, $\overline{N} = M\overline{\eta} \sim 1$, so that the condition of formation at least one string in any inelastic collision starts to influence the results.)

In contrast to the M-scaling the $\mu_F/\sqrt{\overline{\eta}}$-scaling is valid only in the asymptotic region $\overline{\eta} \gg 1$. One can see this from the dotted lines in figure 1. The lines illustrate that in this region the increase of the μ_F from 1 to 2 can be compensated by fourfold increase of the $\overline{\eta}$.

6 Energy and centrality dependencies

The general behavior of the p_t-p_t correlation coefficient, $b_{p_tp_t}$, as a function of heavy ion collision centrality for RHIC and LHC energies is presented in figure 3. Qualitatively, it can be understood based on the dependency, shown in figure 2, as follows.

The maximum string density, achieved for 0-10% of the most central Au-Au collisions at energy 200 GeV/nucleon in RHIC, corresponds to the value $\overline{\eta}$=2.88±0.09 and the one for 0-5% of the most central Pb-Pb collisions at energy 2.76 TeV/nucleon in LHC reaches $\overline{\eta}$=10.56±1.05 [22]. So, depending on the collision centrality, it varies in the intervals (0, R) for RHIC and (0, L) for LHC, shown in figure 2. In this figure we see that the increase of the string density $\overline{\eta}$ with centrality from 0 to R leads to the monotonic growth of the correlation coefficient, which at RHIC energies reaches the highest value in most central collisions, the left panel in figure 3.

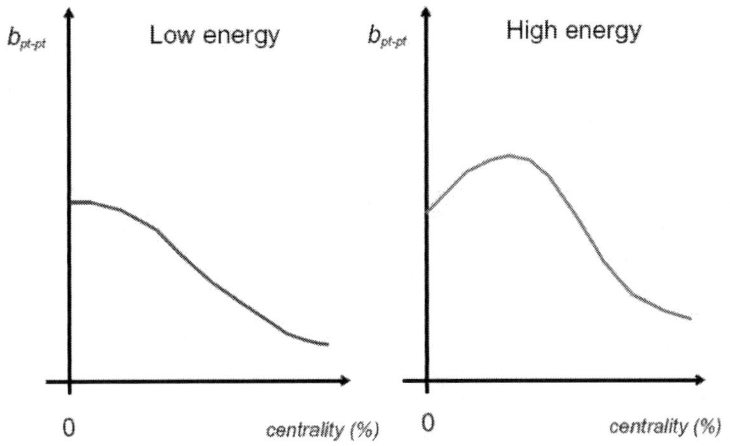

Figure 3. The general form of the dependence of $b_{p_tp_t}$ on the collision centrality at RHIC (left panel) and LHC (right panel), following from the existence of the $b_{p_tp_t}$ maximum at $\overline{\eta} = 3 \div 4$ in figures 1 and 2 (see text).

At LHC energies, depending on the centrality, the string density $\overline{\eta}$ varies in the interval (0, L). In figure 2 we see that in this case the p_t-p_t correlation coefficient, $b_{p_tp_t}$, reaches the highest value at some intermediate value of centrality, which corresponds to the string density $\overline{\eta} \simeq 3$. The further increase of $\overline{\eta}$ with centrality from 3 to L leads to the decrease of the correlation coefficient, in contrast with the situation at RHIC energies. So at LHC energies the p_t-p_t correlation coefficient reaches the highest value in semicentral collisions at some intermediate value of centrality, see the right panel in figure 3,

This decrease of the correlation coefficient for the most central collisions at LHC energies corresponds to the decrease of the asymptotic (13) with the growth of $\overline{\eta}$, discussed in Sect. 4. It is clear from the figure 2 that the RHIC data, corresponding to the $\overline{\eta}$ region (0, R), can not be described by the asymptotic formula (13), which starts to work at $\overline{\eta} \gtrsim 7 \div 8$. This asymptotic, decreasing with string density, behavior is reached only for central heavy ion collisions at LHC energies. The model calculations show that the physical reason of this decrease is the attenuation of color field fluctuations due to the string fusion at large string density.

So the results show that the non-monotonic behaviour of $b_{p_tp_t}$ with centrality is achieved in heavy ion collisions at LHC, while at RHIC the string density is not enough to provide a decline of the correlation coefficient for most central collisions (see figure 3). This qualitative conclusion is confirmed by the results [24], obtained in more realistic dipole-based Monte Carlo string fusion model [25, 26].

7 Summary and conclusions

The p_t-p_t correlations between mean-event transverse momenta are studied in a simple model with quark-gluon string fusion on the transverse lattice. The dependence of the correlation coefficient, $b_{p_tp_t}$, characterizing the correlation strength, on the collision centrality is analyzed for different initial energies. It is shown that at LHC energy the dependence reveals the decrease of the correlation coefficient for most central collisions, reflecting the attenuation of color field fluctuations due to the string fusion at large string density, while at RHIC energy the string density is not yet enough to provide a decline of the correlation coefficient for most central collisions.

It is also found that contrary to the correlation between transverse momenta of single particles the strength of the correlation between mean-event transverse momenta of particles in two separated rapidity intervals is not decreasing with the total number of produced strings, remaining significant even in the case of Pb-Pb collisions, in which the total number of strings can reach several thousand.

So the long-range correlation between mean-event transverse momenta, being robust against the volume fluctuations and the details of the centrality determination, enables to obtain the signatures of string fusion at the initial stage of hadronic interaction in relativistic heavy ion collisions at LHC energy.

Acknowledgements

The work was supported by the Saint-Petersburg State University grant 11.38.242.2015 and by the RFBR grant 15-02-02097-a.

References

[1] N.S. Amelin et al., Phys. Rev. Lett. **73**, 2813 (1994)
[2] T.S. Biro, H.B. Nielsen, J. Knoll, Nucl. Phys. B **245**, 449 (1984)
[3] A. Bialas, W. Czyz, Nucl. Phys. B **267**, 242 (1986)
[4] M.A. Braun, C. Pajares, Phys. Lett. B **287**, 154 (1992)
[5] M.A. Braun, C. Pajares, Nucl. Phys. B **390**, 542 (1993)
[6] M.A. Braun, C. Pajares, Phys. Rev. Lett. **85**, 4864 (2000)
[7] M.A. Braun, R.S. Kolevatov, C. Pajares, V.V. Vechernin, Eur. Phys. J. C **32**, 535 (2004).
[8] ALICE collaboration et al., J. Phys. G **32**, 1295 (2006), [Sect. 6.5.15]
[9] V.V. Vechernin, R.S. Kolevatov, Phys. Atom. Nucl. **70**, 1797 (2007)
[10] V.V. Vechernin, R.S. Kolevatov, Phys. Atom. Nucl. **70**, 1809 (2007)
[11] V.V. Vechernin, Theor. Math. Phys., **184**, 1271 (2015)
[12] V.V. Vechernin, R.S. Kolevatov, Vestn. Peterb. Univ., Ser.4: Fiz. Khim., No.4, 11 (2004), arXiv: hep-ph/0305136
[13] V. Vechernin, Nucl. Phys. A **939**, 21 (2015)
[14] A. Capella, A. Krzywicki, Phys. Rev. D **18**, 4120 (1978)
[15] M.A. Braun, C. Pajares, Eur. Phys. J. C **16**, 349 (2000)
[16] M.A. Braun, C. Pajares, Eur. Phys. J. C **71**, 1558 (2011)
[17] M.A. Braun, C. Pajares, V.V. Vechernin, Nucl. Phys. A **906**, 14 (2013)
[18] M.A. Braun, C. Pajares, V.V. Vechernin, Eur. Phys. J. A **51**, 44 (2015)
[19] E. Levin, A.H. Rezaeian, Phys.Rev. D **84**, 034031 (2011)
[20] A.Kovner., M. Lublinsky, Phys.Rev. D **83**, 034017 (2011)
[21] V.V. Vechernin, Theor. Math. Phys. (2016) [in press]
[22] J. Dias de Deuset al., Eur. Phys. J. C **72**, 2123 (2012)
[23] V. Vechernin, I. Lakomov, Proceedings of Science **Baldin-ISHEPP-XXI**, 072 (2012)
[24] V. Kovalenko, V. Vechernin, EPJ Web of Conferences **66**, 04015 (2014)
[25] V.N. Kovalenko, Phys. Atom. Nucl. **76**, 1189 (2013)
[26] V. Kovalenko, V. Vechernin, Proceedings of Science **Baldin-ISHEPP-XXI**, 077 (2012)

Universal effective coupling constant ratios of 3D scalar ϕ^4 field theory and pseudo-ϵ expansion

A. I. Sokolov[1,⋆], *M. A.* Nikitina[1,2], and *A.* Kudlis[1,2,⋆⋆]

[1]*Department of Quantum Mechanics, Saint Petersburg State University, Ulyanovskaya 1, Petergof, Saint Petersburg 198504, Russia*
[2]*ITMO University, Kronverkskii ave 49, Saint Petersburg 197101, Russia*

Abstract. The ratios $R_{2k} = g_{2k}/g_4^{k-1}$ of renormalized coupling constants g_{2k} entering the small-field equation of state approach universal values R_{2k}^* at criticality. They are calculated for the three-dimensional $\lambda\phi^4$ field theory within the pseudo-ϵ expansion approach. Pseudo-ϵ expansions for R_6^*, R_8^* and R_{10}^* are derived in the five-loop approximation, numerical estimates are obtained with a help of the Padé–Borel–Leroy resummation technique. Its use gives $R_6^* = 1.6488$, the number which perfectly agrees with the most recent lattice result $R_6^* = 1.649$. For the octic coupling the pseudo-ϵ expansion is less favorable numerically. Nevertheless the Padé–Borel–Leroy resummation leads to the estimate $R_8^* = 0.890$ close to the values $R_8^* = 0.871$, $R_8^* = 0.857$ extracted from the lattice and field-theoretical calculations. The pseudo-ϵ expansion for R_{10}^* turns out to have big and rapidly increasing coefficients. This makes correspondent estimates strongly dependent on the Borel–Leroy shift parameter b and prevents proper evaluation of R_{10}^*.

1 Introduction

The critical behavior of the systems undergoing continuous phase transitions is characterized by a set of universal parameters including, apart from critical exponents, renormalized effective coupling constants g_{2k} and the ratios $R_{2k} = g_{2k}/g_4^{k-1}$. These ratios enter the scaling equation of state via the small-magnetization expansion of the free energy:

$$F(z,m) - F(0,m) = \frac{m^3}{g_4}\left(\frac{z^2}{2} + z^4 + R_6 z^6 + R_8 z^8 + R_{10} z^{10} \ldots\right), \tag{1}$$

where $z = M\sqrt{g_4/m^{1+\eta}}$ is dimensionless magnetization, g_4 is the renormalized quartic coupling constant, $m \sim (T - T_c)^\nu$ being proportional to the inverse correlation length. The ratios R_{2k} along with the renormalized quartic coupling constant g_4 determine also the nonlinear susceptibilities of the system.

The scaling equation of state of the Ising model which is described by the three-dimensional (3D) Euclidean scalar field theory with $\lambda\phi^4$ self-interaction was intensively studied theoretically. In particular, higher-order renormalized coupling constants g_6, g_8,... and the ratios R_{2k} were evaluated

⋆e-mail: ais2002@mail.ru
⋆⋆e-mail: andrewkudlis@gmail.com

by a number of analytical and numerical methods [1–27]. Within the renormalization-group approach the numerical estimates for R_6, R_8 and R_{10} were obtained from the five-loop perturbative expansions of the higher-order renormalized couplings calculated in physical dimensions [10]. Corresponding 3D RG expansions for the universal ratios are as follows

$$R_6 = \frac{9}{\pi} g_4 \Big(1 - \frac{3}{\pi} g_4 + 1.38996295 g_4^2 - 2.50173246 g_4^3 + 5.275903\, g_4^4\Big), \tag{2}$$

$$R_8 = -\frac{81}{2\pi} g_4 \Big(1 - \frac{65}{6\pi} g_4 + 7.77500131 g_4^2 - 18.5837685 g_4^3 + 48.16781 g_4^4\Big), \tag{3}$$

$$R_{10} = \frac{243}{\pi} g_4 \Big(1 - \frac{20}{\pi} g_4 + 23.1841758 g_4^2 - 74.2747105 g_4^3 + 238.6138 g_4^4\Big). \tag{4}$$

These series are seen to have big and rapidly growing coefficients what makes them, along with a big value of the Wilson fixed point coordinate $g_4^* \approx 0.99$, rather inconvenient for getting numerical estimates.

The main aim of this work is to improve the situation converting the RG expansions (2), (3), (4) into alternative series looking more friendly from the numerical point of view. To do this we address the pseudo-ϵ expansion approach which proved to be very efficient numerically when used to evaluate the critical exponents and other universal parameters of various 3D and 2D systems [27–40].

2 Pseudo-ϵ expansions and resummation

The Hamiltonian of the Euclidean field theory describing the critical behavior of the 3D Ising-like systems reads:

$$H = \frac{1}{2} \int d^3x \left(m_0^2 \varphi^2 + \nabla\varphi^2 + \frac{\lambda}{12}\varphi^4 \right), \tag{5}$$

where φ is a real scalar field, bare mass squared m_0^2 being proportional to $T - T_c^{(0)}$, and $T_c^{(0)}$ – mean field transition temperature. The β function and critical exponents of the model (5) has been calculated [41] in the six-loop approximation within the massive theory with the propagator, quartic vertex and φ^2 insertion normalized in a conventional way at zero external momenta. Later, R. Guida and J.Zinn-Justin have found the RG expansions for the higher-order effective coupling constants of this model up to the five-loop terms [10].

To derive the pseudo-ϵ expansions for the universal coupling constant ratios one needs the pseudo-ϵ expansion for the value of renormalized quartic coupling g_4 at criticality, i. e. for the Wilson fixed point location g_4^*. Such an expansion may be found by an insertion of the formal small parameter τ into the first (linear) term of the perturbative series for β function and by solving the equation for the fixed point coordinate $\beta(g_4, \tau) = 0$ iteratively in τ. The pseudo-ϵ expansion for g_4^* thus obtained was reported in [27]. Substituting it into RG series (2), (3), (4) and reexpanding corresponding expressions in powers of τ we arrive to

$$R_6^* = 2\tau(1 - 0.244170096\tau + 0.120059430\tau^2 - 0.1075143\tau^3 + 0.1289821\tau^4). \tag{6}$$

$$R_8^* = -9\tau(1 - 1.98491084\tau + 1.76113570\tau^2 - 1.9665851\tau^3 + 2.741546\tau^4). \tag{7}$$

$$R_{10}^* = 54\tau(1 - 4.02194787\tau + 7.55009811\tau^2 - 11.784685\tau^3 + 20.05363\tau^4). \tag{8}$$

Since these expansions originate from divergent RG series and their coefficients do not monotonically decrease, proper numerical estimates may be extracted from them by means of some resummation procedure. Below we apply the Padé–Borel–Leroy resummation technique which proved to be rather efficient when used to evaluate critical exponents on the base of corresponding pseudo-ϵ expansions [28, 35, 38–40]. This technique employs the Borel–Leroy transformation of the original series

$$f(x) = \sum_{i=0}^{\infty} c_i x^i = \int_0^{\infty} e^{-t} t^b F(xt) dt, \qquad F(y) = \sum_{i=0}^{\infty} \frac{c_i}{(i+b)!} y^i , \tag{9}$$

with subsequent analytical continuation of the Borel–Leroy transform $F(y)$ with a help of Padé approximants. The shift parameter b is commonly used for optimization of the resummation procedure.

3 Numerical results

3.1 Sextic universal ratio

It is easy to see that the series (6) should demonstrate good approximating properties because of the smallness of its coefficients. So, one can expect that use of Padé–Borel–Leroy resummation technique will lead to high-precision numerical results in this case. To get most comprehensive information about the numerical value R_6^* and its accuracy we employ all the non-trivial Padé approximants except for the approximant [1/4] which turned out to be spoilt by a positive axis pole for the relevant values of b. The Padé–Borel–Leroy estimates of R_6^* given by approximants [4/1], [3/2], and [2/3] under various b are listed in Table 1. As is well known, diagonal and near-diagonal Padé approximants possess the

Table 1. Padé–Borel–Leroy estimates of R_6^* under various values of the shift parameter b.

b	0	1	2	3	4
[4/1]	1.6485	1.6494	1.6500	1.6505	1.6509
[3/2]	1.64786	1.64822	1.64847	1.64867	1.6488294
[2/3]	1.64819	1.64832	1.64851	1.64868	1.6488294
b	5	7	10	15	20
[4/1]	1.6512	1.6516	1.6521	1.6526	1.65287
[3/2]	1.648958	1.64915	1.64936	1.64957	1.64970
[2/3]	1.648961	1.64917	1.64940	1.64964	1.64980

best approximating properties. That is why the results obtained with a help of approximants [3/2] and [2/3] are expected to be the most reliable ones. Indeed, for various values of b the numbers given by mentioned near-diagonal approximants are very close. Moreover, under $b = 4$ these numbers coincide: as seen from figure 1 curves $R_6^*(b)_{[3/2]}$ and $R_6^*(b)_{[2/3]}$ touch (do not cross) each other at this point. Then it is natural to take the value $R_6^* = 1.6488$ corresponding to the touch point as a final pseudo-ϵ expansion estimate for the sextic coupling constant ratio at criticality.

What may be referred to as an accuracy or, more precisely, an apparent accuracy of the numerical value thus obtained? It looks reasonable to accept as an error bar a range of variation of R_6^* values given by the most stable – [2/3] – Padé–Borel–Leroy approximant under b varying within the whole interval $[0, \infty)$. For b running from 0 to infinity $R_6^*(b)_{[2/3]}$ grows from 1.6482 to 1.6502. Hence, we adopt

$$R_6^* = 1.6488 \pm 0.0014. \tag{10}$$

This number is consistent with the five-loop 3D RG estimate $R_6^* = 1.644 \pm 0.006$ [10] and is in a brilliant agreement with the value $R_6^* = 1.649 \pm 0.002$ extracted from advanced lattice calculations [25].

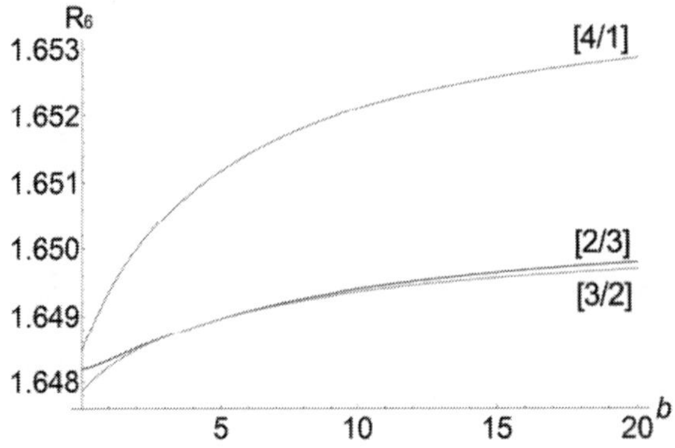

Figure 1. The values of R_6^* as functions of the parameter b obtained on the base of various Padé approximants.

3.2 Octic universal ratio

In the case of the eighth-order coupling we have pseudo-ϵ expansion with obviously less favorable structure. As seen from Eq. (7), the coefficients of the series in brackets are not small and grow what complicates obtaining of numerical estimates under the physical value of the expansion parameter $\tau = 1$. To overcome this difficulty we use the Padé–Borel–Leroy resummation technique addressing as before the most reliable approximants. Since approximant [3/2] suffers from the dangerous (positive-axis) pole, approximants [4/1] and [2/3] are taken as working ones. The numerical values of R_8^* given by these approximants are presented in Table 2. One can see that the estimates given by working approximants practically coincide under $b = 0$. The number corresponding to this value of the shift parameter is equal to 0.89. It may be referred to as a final pseudo-ϵ expansion estimate for R_8^*.

Table 2. Padé–Borel–Leroy estimates of R_8^* as functions of the shift parameter b.

b	0	1	2	3	4
[4/1]	0.89064	0.81273	0.75638	0.71374	0.68036
[2/3]	0.89163	0.82032	0.74242	0.67275	0.61365
b	5	7	10	15	20
[4/1]	0.65352	0.61304	0.57230	0.53131	0.50661
[2/3]	0.56408	0.48711	0.40829	0.32883	0.28125

In contrast to the previous case, Padé–Borel–Leroy estimates for R_8^* are sensitive to the shift parameter. Figure 2 clearly demonstrates this fact. In principle, it does not look surprising since there is a big difference between the lowest-order ($R_8^* = -9$, see Eq. (7)) and final estimates. Since the higher-order perturbative contributions turn out to decrease the modulo of R_8^* by an order of magnitude and change its sign, the final estimate arises as a small difference of big numbers and its accuracy can not be high. Despite of this, our final result $R_8^* = 0.89$ differs only slightly from to the values $R_8^* = 0.871$, $R_8^* = 0.857$ given by the lattice [25] and field-theoretical [10] calculations.

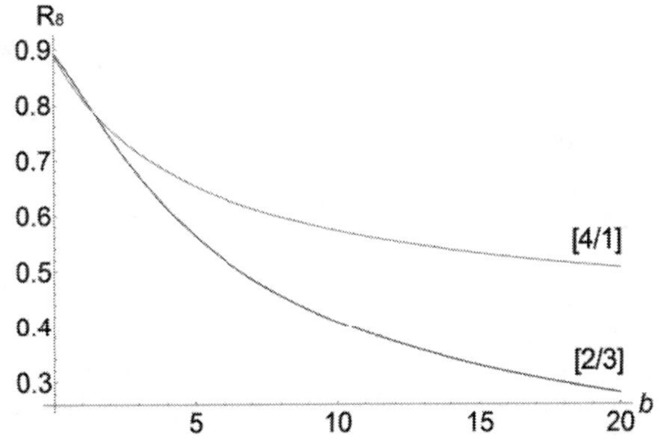

Figure 2. The numerical values of R_8^* given by two working approximants as functions of parameter b.

3.3 Universal ratio of tenth order

For the ratio R_{10} the situation is really dramatic. As can be seen from (8) the coefficients of corresponding pseudo-ϵ expansion rapidly grow up. On the other hand, the rate of their growth is much smaller than that of the original RG series (4). That is why addressing the Padé–Borel–Leroy resummation procedure in this case a priori does not look meaningless. We processed the series (8) with a help of the technique mentioned using all non-trivial Padé approximants – [4/1], [3/2], [2/3] and [1/4]. Obtained estimates as functions of the shift parameter b are listed in Table 3. One can see from

Table 3. Padé–Borel–Leroy estimates of R_{10}^* as functions of the shift parameter b.

b	0	1	2	3	4
[4/1]	−9.2984	−6.3268	−4.1835	−2.5655	−1.3010
[3/2]	−1.6574	−1.7327	−1.4839	−1.0656	−0.5518
[2/3]	−10.1211	−9.0282	−8.4022	−7.9887	−7.6891
[1/4]	4.4362	3.0367	2.3928	2.0355	1.8126
b	5	7	10	15	20
[4/1]	−0.2858	1.2428	2.7778	4.3188	5.2460
[3/2]	0.0180	1.2432	3.14965	6.2645	9.1808
[2/3]	−7.4582	−7.1195	−6.7817	−6.4372	−6.2248
[1/4]	1.6621	1.4739	1.3222	1.1993	1.1369

this Table that the numbers given by various Padé approximants are strongly scattered. Moreover, all the Padé–Borel–Leroy estimates are very sensitive to the parameter b. In such a situation the way to determine the optimal value of b we used above could not lead to proper results.

For instance, although under $b = 7$ the numbers given by approximants [4/1] and [3/2] almost coincide and are close to that given by approximant [1/4], corresponding value $R_{10}^* = 1.24$ is in a marked disagreement with the alternative field-theoretical estimates: $R_{10}^* = -1.1 \pm 0.1$ [10], $R_{10}^* = -2.1 \pm 1.3$ [10], $R_{10}^* = -1.8 \pm 1.4$ [16], $R_{10}^* = -2.3 \pm 1.6$ [13], $R_{10}^* = -1.1 \pm 0.1$ [13], with the results of lattice calculations: $R_{10}^* = -0.75 \pm 0.38$ [12], $R_{10}^* = -1.2 \pm 0.4$ [18, 23], $R_{10}^* = -1.39 \pm 0.4$ [25], and with the "exact RG" prediction $R_{10}^* = -1.7 \pm 0.4$ [11]. So, the obvious improvement of the original RG series for R_{10}^* the pseudo-ϵ expansion technique provides turns out to be insufficient to make τ-series suitable for getting numerical estimates, at least within the Padé–Borel–Leroy resummation ideology.

At the same time, it is worthy to note that if one takes some other criterium for fixing the optimal value of b, a meaningful estimate for R_{10}^* can be found. Namely, as seen from Figure 3, the near-diagonal approximant [3/2] as function of b has a local minimum forming the "domain of stability" (plateau) around $b = 1$. The value $R_{10}^* = -1.73$ corresponding to this minimum is in agreement with many of the field-theoretical and lattice estimates mentioned above. Of course, this fact can not be considered as a serious argument in favor of using Padé–Borel–Leroy resummed pseudo-ϵ expansion (8) for evaluation of R_{10}^*.

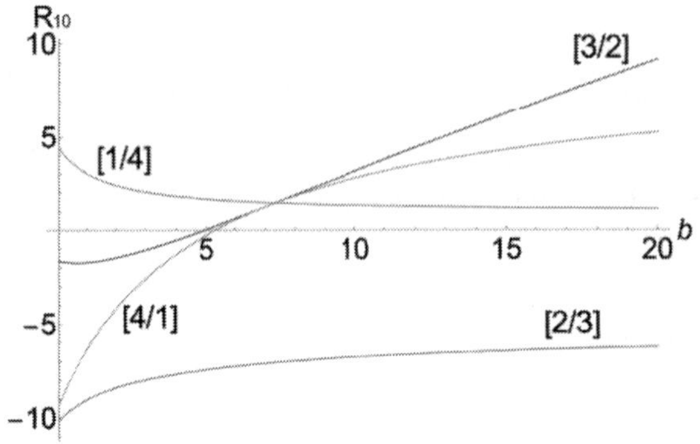

Figure 3. The dependence of the numerical values of R_{10}^* given by different approximants on parameter b.

4 Conclusion

To summarize, we have calculated the five-loop pseudo-ϵ expansions of the universal ratios R_6^*, R_8^*, R_{10}^* for 3D Euclidean scalar φ^4 field theory. To find numerical estimates of these quantities the Padé–Borel–Leroy resummation technique has been applied. The pseudo-ϵ expansion for R_6^* has proven to be very convenient for getting accurate numerical estimates. For the octic coupling the situation is much less favorable but it is still possible to arrive to proper numerical results. The coefficients of the pseudo-ϵ expansion for R_{10}^* turned out to be rapidly increasing what leads to strong scattering of corresponding numerical estimates and makes the Padé-Borel-Leroy resummation technique practically useless in this case. We plan to use alternative, more advanced resummation procedures for evaluation of the quantities of interest in near future.

Acknowledgments

We gratefully acknowledge the support of Saint Petersburg State University via Grant 11.38.636.2013, of the Russian Foundation for Basic Research under Project No. 15-02-04687, and of the Dynasty foundation (M. A. N.).

References

[1] C. Bagnuls, C. Bervillier, Phys. Rev. B **32**, 7209 (1985).
[2] C. Bagnuls, C. Bervillier, D. I. Meiron, B. G. Nickel, Phys. Rev. B **35**, 3585 (1987).
[3] C. M. Bender, S. Boettcher, Phys. Rev. D **48**, 4919 (1993).
[4] N. Tetradis, C. Wetterich, Nucl. Phys. B **422**, 541 (1994).
[5] C. M. Bender, S. Boettcher, Phys. Rev. D **51**, 1875 (1995).
[6] T. Reisz, Phys. Lett. B **360**, 77 (1995).
[7] A. I. Sokolov, Fiz. Tverd. Tela **38**, 640 (1996) [Phys. Solid State **38**, 354 (1996)].
[8] S.-Y. Zinn, S.-N. Lai, M. E. Fisher, Phys. Rev. E **54**, 1176 (1996).
[9] A. I. Sokolov, E. V. Orlov, V. A. Ul'kov, Phys. Lett. A **227**, 255 (1997).
[10] R. Guida, J. Zinn-Justin, Nucl. Phys. B **489**, 626 (1997).
[11] T. R. Morris, Nucl. Phys. B **495**, 477 (1997).
[12] P. Butera, M. Comi, Phys. Rev. E **55**, 6391 (1997).
[13] R. Guida, J. Zinn-Justin, J. Phys. A **31**, 8103 (1998).
[14] A. I. Sokolov, E. V. Orlov, Phys. Rev. B **58**, 2395 (1998).
[15] P. Butera, M. Comi, Phys. Rev. B **58**, 11552 (1998).
[16] A. Pelissetto, E. Vicari, Nucl. Phys. B **522**, 605 (1998).
[17] A. I. Sokolov, Fiz. Tverd. Tela **40**, 1284 (1998) [Phys. Solid State **40**, 1169 (1998)].
[18] M. Campostrini, A. Pelissetto, P. Rossi, E. Vicari, Phys. Rev. E **60**, 3526 (1999).
[19] A. I. Sokolov, E. V. Orlov, V. A. Ul'kov, S. S. Kashtanov, Phys. Rev. E **60**, 1344 (1999).
[20] A. Pelissetto, E. Vicari, Nucl. Phys. B **575**, 579 (2000).
[21] J. Zinn-Justin, Phys. Reports **344**, 159 (2001).
[22] M. Campostrini, M. Hasenbusch, A. Pelissetto, E. Vicari, J. Phys. A **34**, 2923 (2001).
[23] M. Campostrini, A. Pelissetto, P. Rossi, E. Vicari, Phys. Rev. E **65**, 066127 (2002).
[24] A. Pelissetto, E. Vicari, Phys. Reports **368**, 549 (2002).
[25] P. Butera, M. Pernici, Phys. Rev. B **83**, 054433 (2011).
[26] P. Butera, M. Pernici, Phys. Rev. B **86**, 011139 (2012).
[27] A. I. Sokolov, M. A. Nikitina, Phys. Rev. E **89**, 052127 (2014).
[28] J. C. Le Guillou, J. Zinn-Justin, Phys. Rev. B **21** 3976 (1980).
[29] R. Folk, Yu. Holovatch, T. Yavorskii, Phys. Rev. B **62**, 12195 (2000).
[30] P. Calabrese, P. Parruccini, J. High Energy Phys. **05**, 018 (2004).
[31] P. Calabrese, E. V. Orlov, D. V. Pakhnin, A. I. Sokolov, Phys. Rev. B **70**, 094425 (2004).
[32] Yu. Holovatch, D. Ivaneiko, B. Delamotte, J. Phys. A **37**, 3569 (2004).
[33] M. Dudka, Yu. Holovatch, T. Yavorskii, J. Phys. A **37**, 10727 (2004).
[34] P. Calabrese, P. Parruccini, Phys. Rev. B **71**, 064416 (2005).
[35] A. I. Sokolov, Fiz. Tverd. Tela **47**, 2056 (2005) [Phys. Sol. State **47**, 2144 (2005)].
[36] A. I. Sokolov, Teor. Mat. Fiz. **176**, 140 (2013) [Theor. Math. Phys. **176**, 948 (2013)].
[37] M. A. Nikitina, A. I. Sokolov, Phys. Rev. E **89**, 042146 (2014).
[38] A. I. Sokolov, M. A. Nikitina, Phys. Rev. E **90**, 012102 (2014).
[39] M. A. Nikitina, A. I. Sokolov, Teor. Mat. Fiz. **186**, 230 (2016) [Theor. Math. Phys. **186**, 192 (2016)].
[40] A. I. Sokolov, M. A. Nikitina, Physica A **444**, 177 (2016).
[41] G. A. Baker, B. G. Nickel, D. I. Meiron, Phys. Rev. B **17**, 1365 (1978).

Sigma-models with complex homogeneous target spaces

Dmitri Bykov[1,2,*]

[1] *Max-Planck-Institut für Gravitationsphysik, Albert-Einstein-Institut, Am Mühlenberg 1, D-14476 Potsdam-Golm, Germany*
[2] *Steklov Mathematical Institute of Russ. Acad. Sci., Gubkina str. 8, 119991 Moscow, Russia*

Abstract. We review the recently proposed class of σ-models with complex homogeneous target spaces, whose equations of motion admit zero-curvature representations.

1 The models

The action of a σ-model describing maps X from a 2D worldsheet $\mathscr{C}$ to a target space $\mathscr{M}$ with metric h is given by

$$S = \frac{1}{2} \int\limits_{\mathscr{C}} d^2z \, h_{ij}(X) \, \partial_\mu X^i \, \partial_\mu X^j \tag{1}$$

Its critical points $X(z, \bar{z})$ are called *harmonic maps*. We will be interested in the case when the target space $\mathscr{M}$ is homogeneous: $\mathscr{M} = G/H$, G compact and semi-simple. We will use the following standard decomposition of the Lie algebra $\mathfrak{g}$ of G:

$$\mathfrak{g} = \mathfrak{h} \oplus \mathfrak{m}, \tag{2}$$

where $\mathfrak{m} \perp \mathfrak{h}$ with respect to the Killing metric on $\mathfrak{g}$. For a reductive homogeneous space one has the following relations:

$$[\mathfrak{h}, \mathfrak{h}] \subset \mathfrak{h} \quad \Rightarrow \quad \mathfrak{h} \text{ is a subalgebra of } \mathfrak{g}$$
$$[\mathfrak{h}, \mathfrak{m}] \subset \mathfrak{m} \quad \Rightarrow \quad \mathfrak{m} \text{ is a representation of } \mathfrak{h}$$

A homogeneous space G/H is called *symmetric* if

$$[\mathfrak{m}, \mathfrak{m}] \subset \mathfrak{h} \tag{3}$$

Equivalently, there exists a $\mathbb{Z}_2$-grading on $\mathfrak{g}$, i.e. a Lie algebra homomorphism σ of $\mathfrak{g}$, such that $\sigma(a) = a$ for $a \in \mathfrak{h}$ and $\sigma(b) = -b$ for $b \in \mathfrak{m}$.

The action of a σ-model with homogeneous target space G/H is globally invariant under the Lie group G. Therefore, there exists a conserved Noether current $K^\mu \in \mathfrak{g}$:

$$\partial_\mu K^\mu = 0 \tag{4}$$

[*]e-mail: dmitri.bykov@aei.mpg.de,dbykov@mi.ras.ru

Since the group G acts transitively on its quotient space G/H, the equations of motion are in fact *equivalent* to the conservation of the current.

It was observed by Pohlmeyer [1] that in the case when the target space is *symmetric*, the current K is, moreover, flat (when viewed as a one-form and with proper normalization):

$$dK - K \wedge K = 0 \tag{5}$$

To get an idea, why this can be the case, recall that the Maurer-Cartan equation has the solution

$$K = -g^{-1}dg, \qquad g \in G \tag{6}$$

What is the relation between g and a point in the configuration space $[\tilde{g}] \in G/H$? The answer is given by Cartan's embedding $G/H \hookrightarrow G$:

$$g = \widehat{\sigma}(\tilde{g})\tilde{g}^{-1} \tag{7}$$

$\widehat{\sigma}$ is a Lie group homomorphism induced by the Lie algebra involution σ.

Another observation of Pohlmeyer was that the two conditions

$$\begin{aligned} &d * K = 0 \qquad \text{(Conservation)} \\ &dK - K \wedge K = 0 \qquad \text{(Flatness)} \end{aligned} \tag{8}$$

may be rewritten as an equation of flatness of a connection

$$A_u = \frac{1+u}{2}K_z dz + \frac{1+u^{-1}}{2}K_{\bar{z}} d\bar{z}, \tag{9}$$

where we have decomposed the current $K = K_z dz + K_{\bar{z}} d\bar{z}$. We have

$$dA_u - A_u \wedge A_u = 0 \tag{10}$$

This leads to an associated linear system (Lax pair)

$$(d - A_u)\Psi = 0 \tag{11}$$

The existence of a linear system described above is often a sufficient condition for the classical integrability of the model. The linear system was used in [2] to solve the equations of motion for the principal chiral model (target space G), with worldsheet $\mathbf{CP}^1$. A more rigorous approach was developed in [3]. Solutions of the e.o.m. for σ-models with symmetric target spaces may be obtained by restricting the solutions of the principal chiral model. These constructions could not be directly generalized to the case of homogeneous, but not symmetric target spaces (in particular, because there is no Cartan involution).

We will consider a different class of models, with target spaces $\mathcal{M}$ of the following type:

- $\mathcal{M} = G/H$ is a homogeneous space; for simplicity we take G compact and semi-simple

$$\mathfrak{g} = \mathfrak{h} \oplus \mathfrak{m}, \quad [\mathfrak{h}, \mathfrak{h}] \subset \mathfrak{h}, \quad [\mathfrak{h}, \mathfrak{m}] \subset \mathfrak{m}$$

- $\mathcal{M}$ has an integrable G-invariant complex structure $\mathscr{I}$

$$\mathfrak{m} = \mathfrak{m}_+ + \mathfrak{m}_-, \qquad [\mathfrak{h}, \mathfrak{m}_\pm] \subset \mathfrak{m}_\pm, \qquad [\mathfrak{m}_\pm, \mathfrak{m}_\pm] \subset \mathfrak{m}_\pm$$

Here $\mathfrak{m}_\pm$ may be thought of as the holomorphic/anti-holomorphic tangent spaces to $\mathcal{M}$, i.e. $\mathscr{I} \circ \mathfrak{m}_\pm = \pm i\, \mathfrak{m}_\pm$.

- The Killing metric h is Hermitian (i.e. of type $(1,1)$) w.r.t. $\mathscr{I}$

$$h(\mathfrak{m}_\pm, \mathfrak{m}_\pm) = 0$$

Complex homogeneous spaces were classified in [4] a long time ago. They are toric bundles over flag manifolds. Consider for simplicity the case of $G = SU(N)$. Then the relevant manifolds are of the form

$$\mathcal{M} = \frac{SU(N)}{S(U(n_1)\times \ldots \times U(n_m))}, \qquad \sum_{i=1}^{m} n_i \leq N,$$

One has to choose the integers n_i in such a way that $\mathcal{M}$ is even-dimensional. If $\sum\limits_{i=1}^{m} n_i = N$, this is the manifold of partial flags in $\mathbf{C}^N$. Otherwise it is a $U(1)^{2s}$-bundle over a flag manifold, where $2s = N - \sum\limits_{i=1}^{m} n_i$.

Given a homogeneous space of the type just described, one can introduce the action of the model [5]:

$$\mathcal{S} = \int\limits_{\mathscr{C}} d^2z\, \|\partial X\|^2 + \int\limits_{\mathscr{C}} X^*\omega =$$
$$= \int\limits_{\mathscr{C}} d^2z \left(h_{ij}\partial_\mu X^i \partial_\mu X^j + \epsilon_{\mu\nu}\omega_{ij}\partial_\mu X^i \partial_\nu X^j\right),$$

where $\omega = h \circ \mathscr{I}$ is the Kähler form. Note, however, that, in general, the metric h is <u>not Kähler</u>, hence the form ω is not closed: $d\omega \neq 0$. Therefore the second term in the action contributes to the e.o.m.!

Let K be the Noether current constructed using the above action. As we already discussed, the e.o.m. are equivalent to its conservation:

$$d * K = 0$$

The key observation is that, for the models considered, it is also flat:

$$dK - K \wedge K = 0$$

These two equations mean, in essence, that the described models are sub-models of the principal chiral model (PCM). In particular, the solutions of these models are a subset of solutions of the PCM. The Lax pair representation can be constructed in parallel with the Pohlmeyer procedure.

Symmetric spaces of the group $SU(N)$ are the Grassmannians

$$\mathbb{G}_{n|N} := \frac{SU(N)}{S(U(n)\times U(N-n))}$$

In this case the canonical one-parametric family of flat connections has the form

$$\widetilde{A}_\lambda = \frac{1-\lambda}{2}\widetilde{K}_z dz + \frac{1-\lambda^{-1}}{2}\widetilde{K}_{\bar{z}} d\bar{z},$$

where $\widetilde{K}$ is the canonical Noether current, i.e. the one constructed using the standard action

$$\mathcal{S} = \frac{1}{2}\int\limits_{\mathscr{C}} d^2z\, h_{ij}(X)\, \partial_\mu X^i\, \partial_\mu X^j \tag{12}$$

The models, which we described above, feature an additional term in their action: $\int_{\mathscr{C}} X^*\omega$, the integral of the Kähler form. Therefore the Noether current K defined using this action will be different from $\widetilde{K}$, the difference being a 'topological' current:

$$K = \widetilde{K} + *dM$$

Nevertheless both K and $\widetilde{K}$ are flat. The one-parametric family of connections constructed using the current K has the form

$$A_u = \frac{1+u}{2} K_z dz + \frac{1+u^{-1}}{2} K_{\bar{z}} d\bar{z},$$

A natural question arises: How are $\widetilde{A}_\lambda$ and A_u related? The answer is: $\widetilde{A}_\lambda$ and A_u are related by a gauge transformation Ω:

$$\widetilde{A}_\lambda = \Omega A_u \Omega^{-1} - \Omega d\Omega^{-1}$$

Ω can be written out explicitly (g is the 'dynamical' group element):

$$\Omega = g\Lambda g^{-1}, \quad \text{where} \quad \Lambda = \mathrm{diag}(\underbrace{\lambda^{-1/2}, \ldots, \lambda^{-1/2}}_{n}, \underbrace{\lambda^{1/2}, \ldots, \lambda^{1/2}}_{N-n})$$

Rather important is the nontrivial relation between the spectral parameters:

$$\lambda = u^{1/2}$$

This relation may be confirmed by analyzing the limiting behavior of the holonomies of the connection as $u \to 0$ (such analysis can be borrowed from [6]).

2 An example. The flag manifold.

As an example let us consider one of the simplest homogeneous, but non-symmetric spaces – the flag manifold

$$\mathcal{F}_3 = \frac{U(3)}{U(1)^3} \tag{13}$$

It is the space of ordered triples of lines through the origin in $\mathbb{C}^3$, and can be parametrized by three orthonormal vectors u_i, $i = 1, 2, 3$, $\bar{u}_i \circ u_j = \delta_{ij}$, modulo phase rotations: $u_k \sim e^{i\alpha_k} u_k$.

To formulate the model, we need to pick a particular complex structure on $\mathcal{F}_3$. The (co)tangent space to $\mathcal{F}_3$ is spanned at each point by the one-forms

$$J_{ij} := u_i \circ d\bar{u}_j, \quad i \neq j \tag{14}$$

One can pick any three non-mutually conjugate one-forms and *define* the action of the complex structure operator $\mathscr{I}$ on them:

$$\mathscr{I} \circ J_{12} = \pm i J_{12}, \quad \mathscr{I} \circ J_{23} = \pm i J_{23}, \quad \mathscr{I} \circ J_{31} = \pm i J_{31} \tag{15}$$

Altogether there are $2^3 = 8$ possible choices, so that there are 8 invariant almost complex structures. However, only 6 of them are *integrable*.

Pick the integrable complex structure $\mathscr{I}$, in which J_{12}, J_{13}, J_{23} are holomorphic one-forms. Then the action can be written as [7]

$$S = \int d^2z \left(|(J_{12})_{\bar{z}}|^2 + |(J_{13})_{\bar{z}}|^2 + |(J_{23})_{\bar{z}}|^2 \right) \tag{16}$$

The e.o.m. are:

$$\mathscr{D}_z(J_{12})_{\bar{z}} = 0, \qquad \mathscr{D}_z(J_{31})_{\bar{z}} = 0, \qquad \mathscr{D}_z(J_{23})_{\bar{z}} = 0 \qquad \text{and c.c. ones} \tag{17}$$

From the action (16) it is clear that the holomorphic curves defined by $(J_{12})_{\bar{z}} = (J_{13})_{\bar{z}} = (J_{23})_{\bar{z}} = 0$ minimize the action, hence are solutions of the e.o.m. From (17) it follows that $(J_{12})_{\bar{z}} = (J_{31})_{\bar{z}} = (J_{23})_{\bar{z}} = 0$ is a solution as well. This defines a curve, holomorphic in a different, non-integrable almost complex structure I. We have seen that the curves, holomorphic in at least two different almost

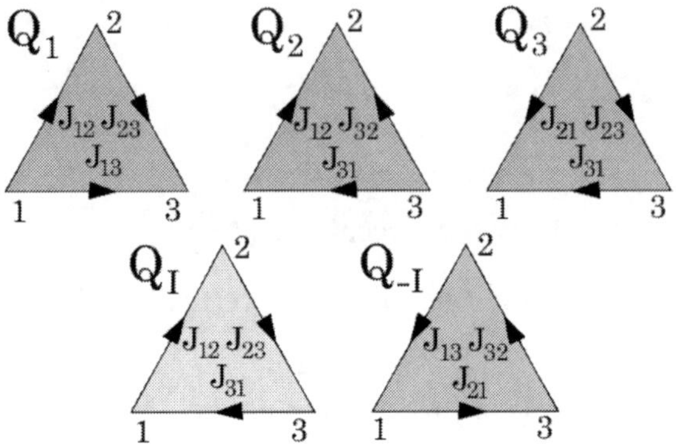

Figure 1. The triangles indicate the complex structures, whose associated holomorphic curves are solutions of the σ-model. The three top triangles correspond to integrable complex structures, whereas the two lower ones correspond to the non-integrable ones.

complex structures, satisfy the e.o.m. As we discussed, there are 8 almost complex structures on the flag manifold. Are there any other holomorphic curves that still solve the e.o.m.? The answer is positive. The relevant complex structures are shown in Fig. 1.

We have already discussed why the Q_I-holomorphic curves and Q_1-holomorphic curves satisfy the e.o.m. To see why the Q_2- and Q_3-holomorphic curves satisfy the e.o.m., one should note that the differences between the respective Kähler forms are closed forms, i.e. for example $\omega_1 - \omega_2 = \Omega_{top}$ with $d\Omega_{top} = 0$. Therefore the two actions $\mathcal{S}_1$ and $\mathcal{S}_2$ differ by a topological term:

$$\mathcal{S}_1 - \mathcal{S}_2 = \int_{\mathscr{C}} \Omega_{top} \tag{18}$$

This leads to an interesting bound on the instanton numbers of the holomorphic curves. To see this, note that the flag manifold may be embedded as

$$i : \mathcal{F}_3 \hookrightarrow \mathbf{CP}^2 \times \mathbf{CP}^2 \times \mathbf{CP}^2 \tag{19}$$

The second cohomology $H^2(\mathcal{F}_3, \mathbb{R}) = \mathbb{R}^2$ can be described via the pull-backs of the Fubini-Study forms of the $\mathbf{CP}^2$'s, and the corresponding instanton numbers are $n_i = \int_{\mathscr{C}} i^*(\Omega_{FS}^{(i)}),\ i = 1, 2, 3.$

These are subject to the condition

$$n_1 + n_2 + n_3 = 0. \tag{20}$$

The bounds on the topological numbers n_i for the holomorphic curves, which follow from the non-negativity of the actions $\mathcal{S}_i$, are shown in Fig. 2.

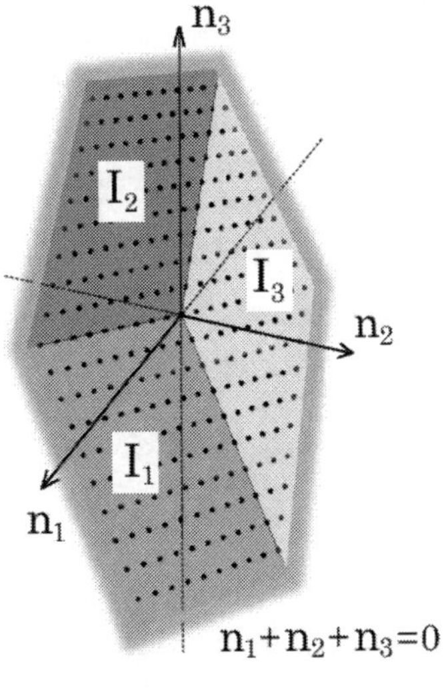

Figure 2. Admissible instanton numbers for various complex structures on $\frac{U(3)}{U(1)^3}$ are shown by black dots.

In [8] we were able to construct all solutions to the e.o.m. of the flag manifold σ-model for the case when the worldsheet is a sphere, $\mathscr{C} = \mathbf{CP}^1$. In that case all solutions to the e.o.m. are parametrized by the following data:

- One of the projections $\pi_i : \mathcal{F}_3 \to (\mathbf{CP}^2)_i, \quad i = 1, 2, 3$
- A harmonic map $v_{har} : \mathbf{CP}^1 \to (\mathbf{CP}^2)_i$ to the base of the projection
- A holomorphic map $w_{hol} : \mathbf{CP}^1 \to \mathbf{CP}^1$ to the fiber of the projection, .

For every triple (i, v_{har}, w_{hol}) there exists a solution of the e.o.m., and all solutions are obtained in this way [8]. The crucial point is that the harmonic maps to the base manifold $\mathbf{CP}^2$ are known explicitly [9] (and the holomorphic maps $\mathbf{CP}^1 \to \mathbf{CP}^1$ are just rational functions).

3 Discussion

It has been known for a long time that the e.o.m. of σ-models with symmetric target spaces admit zero-curvature representations – this fact is the cornerstone of integrability of these models. In our recent works we have considered modified σ-models with complex homogeneous target spaces, not necessarily symmetric ones, for which there exist Lax pairs as well [5]. This class of models was reviewed above in section 1. We have also considered in detail a concrete example of such model, namely the case when the target space is the flag manifold $\frac{U(3)}{U(1)^3}$ [7]. When the worldsheet is a sphere $\mathbf{CP}^1$, all solutions of the e.o.m. have been constructed [8]. These results were summarized in section 2. The crucial test of integrability of the proposed models would lie in the construction of classical solutions for the case when the worldsheet is a torus $S^1 \times S^1$ (as was done in [6] for $\mathcal{M} = SU(2)$). Other directions for future research include supersymmetrization of the models and the clarification of the role of Lie algebra gradings (such as $\mathbb{Z}_2$ for symmetric spaces, or $\mathbb{Z}_4$ for semi-symmetric spaces).

Acknowledgements. I would like to thank the organizers of the conference 'Quarks 2016' for the inspiring and productive atmosphere. I am indebted to Prof. A.A.Slavnov and to my parents for support and encouragement. My work was supported in part by the grant RFBR 14-01-00695-a.

References

[1] K. Pohlmeyer, Communications in Mathematical Physics **46**, 207 (1976)
[2] V.E. Zakharov, A.V. Mikhailov, Sov. Phys. JETP **47**, 1017 (1978), [Zh. Eksp. Teor. Fiz.74,1953(1978)]
[3] K. Uhlenbeck, Journal of Differential Geometry **30**, 1 (1989)
[4] H.C. Wang, Amer. J. Math. **76**, 1 (1954)
[5] D. Bykov, Phys. Lett. B **760C**, 341 (2016), 1605.01093
[6] N.J. Hitchin, Journal of Differential Geometry **31**, 627 (1990)
[7] D. Bykov, Nucl. Phys. **B894**, 254 (2015), 1412.3746
[8] D. Bykov, Nucl. Phys. **B902**, 292 (2016), 1506.08156
[9] A.M. Din, W.J. Zakrzewski, Nucl. Phys. **B174**, 397 (1980)

Relaxation of the chiral imbalance in dense matter of a neutron star

Maxim Dvornikov[1,2,3,⋆]

[1] *Pushkov Institute of Terrestrial Magnetism, Ionosphere and Radiowave Propagation (IZMIRAN), Kaluzhskoe HWY 4, 142190 Troitsk, Moscow, Russia*
[2] *Physics Faculty, National Research Tomsk State University, 36 Lenin Avenue, 634050 Tomsk, Russia*
[3] *University of Hamburg, II Institute for Theoretical Physics, 149 Luruper Chaussee, D-22761 Hamburg, Germany*

Abstract. Using the quantum field theory methods, we calculate the helicity flip of an electron scattering off protons in dense matter of a neutron star. The influence of the electroweak interaction between electrons and background nucleons on the helicity flip is examined. We also derive the kinetic equation for the chiral imbalance. The derived kinetic equation is compared with the results obtained by other authors.

1 Introduction

Some neutron stars (NS) can possess extremely strong magnetic fields $B \gtrsim 10^{15}$ G. These NSs are called magnetars [1]. Despite long observational history of magnetars and numerous theoretical models for the generation of their magnetic fields, nowadays there is no commonly accepted mechanism explaining the origin of magnetic fields in these compact stars.

In Refs. [2–4] we developed the new model for the generation of magnetic fields in magnetars. In this model a large-scale magnetic field in NS can be amplified to the strenth predicted in magnetars owing to the magnetic field instability driven by the electron-nucleon (*eN*) electroweak interaction. Using the proposed model we could account for some properties of magnetars, e.g., ages of these compact stars. Despite the plausibility of the model in Refs. [2–4], some of its ingredients should be substantiated by more detailed calculations based on reliable methods of the quantum field theory (QFT). This work is devoted to the further development of the proposed description of the magnetic fields generation in magnetars. In particular, we will be interested in the evolution of the chemical potentials of the electron gas in NS in the presence of background nucleons.

2 Electron-proton collisions in dense plasma

In NS, the helicity of a massive electron can be changed in electron-proton (*ep*) and electron-electron electromagnetic scatterings as well as in the interaction of an electron with the anomalous magnetic moment of a neutron. As found in Ref. [5], the rate of the former reaction in dense matter of NS is higher than that of the later ones. Therefore, in our analysis, we shall account for only *ep* collisions.

⋆e-mail: maxdvo@izmiran.ru

2.1 Helicity flip rate in ep collisions

The matrix element for the ep collision, due to the electromagnetic interaction, has the form,

$$\mathcal{M} = \frac{\mathrm{i}e^2}{(k_1 - k_2)^2} \bar{e}(p_2)\gamma^\mu e(p_1) \cdot \bar{p}(k_2)\gamma_\mu p(k_1), \tag{1}$$

where $e > 0$ is the absolute value of the electron charge, $\gamma^\mu = \left(\gamma^0, \boldsymbol{\gamma}\right)$ are the Dirac matrices, $p^\mu_{1,2} = (E_{1,2}, \mathbf{p}_{1,2})$ and $k^\mu_{1,2} = (\mathcal{E}_{1,2}, \mathbf{k}_{1,2})$ are the four momenta of electrons and protons. The momenta of incoming particles are marked with the label 1 and that of the outgoing particles with the label 2.

We shall consider a process when a proton is in the unpolarized states before and after the collision whereas an electron changes its polarization in the scattering. The square of the matrix element in Eq. (1) reads

$$|\mathcal{M}|^2 = \frac{e^4}{(k_1 - k_2)^4} \mathrm{tr}\left[\rho_p(k_1)\gamma^\mu \rho_p(k_2)\gamma^\nu\right] \cdot \mathrm{tr}\left[\rho_e(p_1)\gamma_\mu \rho_e(p_2)\gamma_\nu\right], \tag{2}$$

where the spin density matrices of protons $\rho_p(k_{1,2})$ and electrons $\rho_e(p_{1,2})$ are [6, pp. 106–111]

$$\rho_p(k_{1,2}) = \frac{1}{2}\left(\gamma_\mu k^\mu_{1,2} + M\right), \quad \rho_e(p_{1,2}) = \frac{1}{2}\left(\gamma_\mu p^\mu_{1,2} + m\right)\left(1 + \gamma^5 \gamma_\mu a^\mu_{1,2}\right), \tag{3}$$

where M and m are the masses of a proton and an electron, and $\gamma^5 = \mathrm{i}\gamma^0\gamma^1\gamma^2\gamma^3$. The polarization vector of an electron a^μ in Eq. (3) has the form,

$$a^\mu = \left(\frac{(\boldsymbol{\zeta} \cdot \mathbf{p})}{m}, \boldsymbol{\zeta} + \frac{\mathbf{p}\,(\boldsymbol{\zeta} \cdot \mathbf{p})}{m(E + m)}\right). \tag{4}$$

where $\boldsymbol{\zeta}$ is the invariant three vector of the particle polarization, which describes the polarization in the particle rest frame. We shall suppose that electrons are in pure spin states with $\boldsymbol{\zeta}^2 = 1$.

Let us fix the polarizations of an electron. If we study the $R \to L$ transition, i.e. we take that $\boldsymbol{\zeta}_1 = \mathbf{n}_1$ and $\boldsymbol{\zeta}_2 = -\mathbf{n}_2$, one gets that $a^\mu_{1,2} = (\pm p_{1,2}, \pm E_{1,2}\mathbf{n}_{1,2})/m$, where $\mathbf{n}_{1,2} = \mathbf{p}_{1,2}/p_{1,2}$ are the unit vectors towards the electron momenta and $p_{1,2} = |\mathbf{p}_{1,2}|$. Using the approximation of the elastic ep scattering, i.e. assuming that $E_1 = E_2$, and keeping only the leading term in the electron mass m in the computation of the traces of the Dirac matrices in Eq. (2), one gets the matrix element squared in the form,

$$|\mathcal{M}|^2 = 2e^4 m^2 \left[1 - (\mathbf{n}_1 \cdot \mathbf{n}_2)\right] \frac{\left[\mathcal{E}_1\mathcal{E}_2 + M^2 + (\mathbf{k}_1 \cdot \mathbf{k}_2)\right]}{\left[(\mathcal{E}_1 - \mathcal{E}_2)^2 - (\mathbf{k}_1 - \mathbf{k}_2)^2\right]^2}, \tag{5}$$

which will be used in the following calculations.

The total probability of the ep collision in the ep plasma has the form [6, pp. 247–252],

$$W = \frac{V}{8(2\pi)^8} \int \frac{\mathrm{d}^3p_1\mathrm{d}^3p_2\mathrm{d}^3k_1\mathrm{d}^3k_2}{E_1E_2\mathcal{E}_1\mathcal{E}_2} \delta^4\left(p_1 + k_1 - p_2 - k_2\right)|\mathcal{M}|^2$$
$$\times f_e(E_1 - \mu_\mathrm{R})\left[1 - f_e(E_2 - \mu_\mathrm{L})\right] f_p(\mathcal{E}_1 - \mu_p)\left[1 - f_p(\mathcal{E}_2 - \mu_p)\right], \tag{6}$$

where we summed over the polarizations of the outgoing proton. Here $f_{e,p}(E) = \left[\exp(\beta E) + 1\right]^{-1}$ are the Fermi-Dirac distributions of electrons and protons, μ_p is the chemical potential of protons, and V is the normalization volume. In Eq. (6) we assume that incoming and outgoing electrons have different chemical potentials: μ_R and μ_L respectively. Protons and electrons are taken to be in the thermal equilibrium having the same temperature T.

Assuming that electrons are highly degenerate and ultrarelativistic, we can calculate the integrals over the electron momenta,

$$I_e = \int \frac{d^3p_1 d^3p_2}{2E_1 2E_2} \frac{(p_1 \cdot p_2)}{E_1 E_2} \delta^4(p_1 - p_2 - q)\theta(\mu_R - E_1)\theta(E_2 - \mu_L), \tag{7}$$

where $q^\mu = k_2^\mu - k_1^\mu$ and and $(p_1 \cdot p_2) = E_1 E_2 [1 - (\mathbf{n}_1 \cdot \mathbf{n}_2)]$. Using the identity

$$\int \frac{d^3p}{2E} = \int d^4p\theta(p_0)\delta(p^2), \tag{8}$$

and integrating over p_2 we can obtain that

$$I_e = -\int \frac{d^3p}{2E} \frac{Eq_0 - (\mathbf{p} \cdot \mathbf{q})}{E(E - q_0)} \delta\left(2Eq_0 - 2(\mathbf{p} \cdot \mathbf{q}) - q^2\right)\theta(\mu_R - E)\theta(E - q_0 - \mu_L). \tag{9}$$

Here we recall that we use the approximation of the elastic scattering, i.e. $E_1 = E_2$. Hence we should set $q_0 = \mathcal{E}_2 - \mathcal{E}_1 = 0$ in Eq. (9). Then we represent $d^3p = 2\pi d\cos\theta E^2 dE$ and integrate over $\cos\theta$ using the remaining delta function

$$\delta\left(2E|\mathbf{q}|\cos\theta - \mathbf{q}^2\right) = \frac{1}{2E|\mathbf{q}|}\delta\left(\cos\theta - \frac{|\mathbf{q}|}{2E}\right). \tag{10}$$

Finally, we obtain

$$I_e = \frac{\pi|\mathbf{q}|}{4} \int_0^\infty \frac{dE}{E^2}\theta(\mu_R - E)\theta(E - \mu_L) \approx \frac{\pi|\mathbf{q}|}{4}\frac{\mu_R - \mu_L}{\mu_e^2}, \tag{11}$$

where we take that $\mu_e \approx \mu_R \approx \mu_L$ is the mean chemical potential of the electron gas. Moreover we assume that $\mu_R > \mu_L$ for the total probability to be positive.

Using Eqs. (5) and (11), one gets that the total probability in Eq. (6) takes the form,

$$W = \frac{V\pi e^4 m^2}{4(2\pi)^8}\frac{\mu_R - \mu_L}{\mu_e^2} \int \frac{d^3k_1 d^3k_2}{\mathcal{E}_1\mathcal{E}_2} \frac{\mathcal{E}_1\mathcal{E}_2 + M^2 + (\mathbf{k}_1 \cdot \mathbf{k}_2)}{\left[(\mathbf{k}_1 - \mathbf{k}_2)^2 + \omega_p^2\right]^{3/2}}$$
$$\times f_p(\mathcal{E}_1 - \mu_p)\left[1 - f_p(\mathcal{E}_2 - \mu_p)\right], \tag{12}$$

Here, in the denominator, we replace $(\mathbf{k}_1 - \mathbf{k}_2)^2 \to (\mathbf{k}_1 - \mathbf{k}_2)^2 + \omega_p^2$, where ω_p is the plasma frequency in the ep plasma, to avoid the infrared divergencies. Now one can compute the integrals over the proton momenta,

$$I_p = \int \frac{d^3k_1 d^3k_2}{\mathcal{E}_1\mathcal{E}_2} \frac{\mathcal{E}_1\mathcal{E}_2 + M^2 + (\mathbf{k}_1 \cdot \mathbf{k}_2)}{\left[(\mathbf{k}_1 - \mathbf{k}_2)^2 + \omega_p^2\right]^{3/2}} f_p(\mathcal{E}_1 - \mu_p)\left[1 - f_p(\mathcal{E}_2 - \mu_p)\right]$$
$$\approx 32\sqrt{2\tilde{\mu}_p}M^{3/2}\pi^2 TJ, \tag{13}$$

where

$$J = \int_0^1 dx \frac{x^2\sqrt{1 - x^2}}{(x^2 + L^{-2})^{3/2}} = \frac{1}{\sqrt{1 + L^2}} K\left(\frac{L}{\sqrt{1 + L^2}}\right)\left[L + \frac{2}{L}\right] - \frac{2}{L}E\left(\frac{L}{\sqrt{1 + L^2}}\right)\sqrt{1 + L^2}. \tag{14}$$

Here $L^2 = 8M\tilde{\mu}_p/\omega_p^2$, $\boldsymbol{K}(z)$ and $\boldsymbol{E}(z)$ are the complete elliptic integrals, and $\tilde{\mu}_p = \mu_p - M$ is the nonrelativistic part of the protons chemical potential. To derive Eq. (13) we assume that protons are nonrelativistic and the scattering is elastic. Moreover, the proton gas is supposed to have a small nonzero temperature.

Assuming that in a degenerate plasma one has $\omega_p^2 = 4\alpha_{\text{em}}\mu_e^2/3\pi$ (see, e.g., Ref. [7]), where $\alpha_{\text{em}} = e^2/4\pi$ is the fine structure constant, and $\tilde{\mu}_p \approx \mu_e^2/2M$ due to the electroneutrality of the NS matter, we get that $L^2 = 3\pi/\alpha_{\text{em}} \approx 1291 \gg 1$. Decomposing J in Eq. (14) in this limit, one obtains

$$J = \ln 4L - 2 + O\left(L^{-1}\right) = \frac{1}{2}\left[\ln\left(\frac{48\pi}{\alpha_{\text{em}}}\right) - 4\right] > 0. \tag{15}$$

Finally, using Eqs. (12)-(15), we get the total probability for the helicity flip in an ep collision,

$$W = W_0\left(\mu_{\text{R}} - \mu_{\text{L}}\right)\theta\left(\mu_{\text{R}} - \mu_{\text{L}}\right), \quad W_0 = \frac{Ve^4}{32\pi^5}\frac{m^2M}{\mu_e}T\left[\ln(16L^2) - 4\right]. \tag{16}$$

Note that the total probability is always positive. That is why we introduce the step function $\theta(\mu_{\text{R}} - \mu_{\text{L}})$ in Eq. (16). In the next section we shall compare Eq. (16) with the analogous expression accounting for the electroweak interaction between electrons and neutrons.

2.2 Helicity flip rate of electrons, electroweakly interacting with nuclear matter

If we take into account the electroweak interaction between electrons and nucleons, the matrix element for the ep scattering has the same form as in Eq. (1). However, instead of the electron spinors in vacuum, we should use the exact solutions of the Dirac equation for an electron, interacting with a background matter, found in Appendix A.

We shall start with the analysis of $R \to L$ transitions. According to Eq. (1) it is necessary to compute the following quantity:

$$J^\mu = (J_0, \boldsymbol{J}) = \bar{u}_-(p_2)\gamma^\mu u_+(p_1). \tag{17}$$

Using Eqs. (31) and (33) one gets

$$\begin{aligned} J_0 &= -\frac{mP_0\left[p_1 + p_2 + E_+(p_1) + E_-(p_2) - 2\bar{V}\right]}{2\sqrt{E_{0+}(p_1)E_{0-}(p_2)}\left[E_-(p_2) + p_2 - V_{\text{R}}\right]\left[E_+(p_1) + p_1 - V_{\text{L}}\right]}, \\ \boldsymbol{J} &= -\frac{m\mathbf{P}\left[p_1 - p_2 + E_+(p_1) - E_-(p_2) - 2V_5\right]}{2\sqrt{E_{0+}(p_1)E_{0-}(p_2)}\left[E_-(p_2) + p_2 - V_{\text{R}}\right]\left[E_+(p_1) + p_1 - V_{\text{L}}\right]}, \end{aligned} \tag{18}$$

where

$$P_0 = w_-^\dagger(\mathbf{p}_2)w_+(\mathbf{p}_1), \quad \mathbf{P} = w_-^\dagger(\mathbf{p}_2)\boldsymbol{\sigma}w_+(\mathbf{p}_1). \tag{19}$$

Here $w_\pm$ are the two component spinors corresponding to different helicities, which are defined in Ref. [6, p. 86], $\boldsymbol{\sigma}$ are the Pauli matrices, $V_{\text{L,R}}$ are the effective potentials of the interaction of left and right chiral projections of an electron field with the background matter, defined in Eq. (29), $\bar{V} = (V_{\text{L}} + V_{\text{R}})/2$, and $V_5 = (V_{\text{L}} - V_{\text{R}})/2$. To obtain Eq. (18) we use the Dirac matrices in the chiral representation [8, pp. 691–696].

For ultrarelativistic electrons one obtains from Eq. (30) that $E_\pm(p_{1,2}) = p_{1,2} + V_{\text{R,L}}$. As in Sec. 2.1, here we also study the elastic scattering, in which colliding particles do not transfer to exited states, i.e. $E_+(p_1) = E_-(p_2)$ and $\mathcal{E}_1 = \mathcal{E}_2$. The equality of energies of incoming and outgoing ultrarelativistic

DOI: 10.1051/epjconf/201612505003

electrons is equivalent to $p_1 - p_2 = 2V_5$. Thus, one obtains that $J = 0$. The computation of P_0 in Eq. (19) gives $|P_0|^2 = [1 - (\mathbf{n}_1 \cdot \mathbf{n}_2)]/2$. Thus, using Eq. (18), one gets that the square of matrix element in Eq. (1) has the form,

$$|\mathcal{M}|^2 = e^4 m^2 \frac{(p_1 + p_2)^2 [1 - (\mathbf{n}_1 \cdot \mathbf{n}_2)]}{8(p_1 - V_5)^2 (p_2 + V_5)^2} \frac{\left[\mathcal{E}_1 \mathcal{E}_2 + M^2 + (\mathbf{k}_1 \cdot \mathbf{k}_2)\right]}{\left[(\mathcal{E}_1 - \mathcal{E}_2)^2 - (\mathbf{k}_1 - \mathbf{k}_2)^2\right]^2}, \tag{20}$$

where we keep only the leading term in the electron mass.

The total probability of the process has the form,

$$W = \frac{V}{2(2\pi)^8} \int \frac{\mathrm{d}^3 p_1 \mathrm{d}^3 p_2 \mathrm{d}^3 k_1 \mathrm{d}^3 k_2}{\mathcal{E}_1 \mathcal{E}_2} \delta^4 (p_1 + k_1 - p_2 - k_2) |\mathcal{M}|^2 \\ \times f_e(E_1 - \mu_\mathrm{R}) [1 - f_e(E_2 - \mu_\mathrm{L})] f_p(\mathcal{E}_1 - \mu_p) \left[1 - f_p(\mathcal{E}_2 - \mu_p)\right]. \tag{21}$$

Note that Eq. (21) is slightly different from Eq. (6). It can be accounted for by the different way to normalize electron basis spinors; cf. Eq. (32).

As in Sec. 2.1, first, in Eq. (21) we compute the integrals over the electron momenta,

$$I_e = \int \frac{\mathrm{d}^3 p_1 \mathrm{d}^3 p_2 (p_1 + p_2)^2 [1 - (\mathbf{n}_1 \cdot \mathbf{n}_2)]}{16(p_1 - V_5)^2 (p_2 + V_5)^2} \\ \times \delta^4 (p_1 - p_2 - q) \theta(\mu_\mathrm{R} - p_1 - V_\mathrm{R}) \theta(p_2 + V_\mathrm{L} - \mu_\mathrm{L}), \tag{22}$$

where we suppose that incoming and outgoing electrons are ultrarelativistic, highly degenerate, and have different chemical potentials. We remind that we study $R \to L$ transitions. After the integration over $\mathbf{p}_2$ with help of the delta function, one obtains that

$$I_e = \int \frac{\mathrm{d}^3 p (p + |\mathbf{p} - \mathbf{q}|)^2}{16(p - V_5)^2 (|\mathbf{p} - \mathbf{q}| + V_5)^2} \left[1 - \frac{(\mathbf{p} \cdot \mathbf{p} - \mathbf{q})}{p|\mathbf{p} - \mathbf{q}|}\right] \\ \times \delta(p - 2V_5 - |\mathbf{p} - \mathbf{q}|) \theta(\mu_\mathrm{R} - p - V_\mathrm{R}) \theta(|\mathbf{p} - \mathbf{q}| + V_\mathrm{L} - \mu_\mathrm{L}). \tag{23}$$

Then we represent $\mathrm{d}^3 p = 2\pi \mathrm{d}\cos\theta p^2 \mathrm{d}p$ and integrate over $\cos\theta$ using the remaining delta function

$$\delta(p - 2V_5 - |\mathbf{p} - \mathbf{q}|) = \frac{p - 2V_5}{p|\mathbf{q}|} \delta\left(\cos\theta - \frac{|\mathbf{q}|^2 + 4pV_5 - V_5^2}{2p|\mathbf{q}|}\right). \tag{24}$$

Finally we get the following expression for I_e:

$$I_e = \pi \frac{|\mathbf{q}|^2 - V_5^2}{4|\mathbf{q}|} \int_{\mu_\mathrm{L} - V_\mathrm{R}}^{\mu_\mathrm{R} - V_\mathrm{R}} \frac{\mathrm{d}p}{(p - V_5)^2} = \pi \frac{|\mathbf{q}|^2 - V_5^2}{4|\mathbf{q}|} \frac{(\mu_\mathrm{R} - \mu_\mathrm{L})}{\left(\mu_\mathrm{R} - \bar{V}\right)\left(\mu_\mathrm{L} - \bar{V}\right)} \\ \approx \frac{\pi|\mathbf{q}|}{4} \frac{(\mu_\mathrm{R} - \mu_\mathrm{L})}{\mu_e^2}, \tag{25}$$

where we suppose that $\mu_\mathrm{R,L} \gg \bar{V}$, $\mu_\mathrm{R} \approx \mu_\mathrm{L} \approx \mu_e$, and $|\mathbf{q}| \gg V_5$. The most important consequence of Eq. (25) consists in the fact that the eN interaction does not contribute to the difference of the chemical potentials in the numerator.

Eventually, on the basis of Eqs. (21) and (25), one gets the expression for the total probability which coincides with that in Eq. (12). Performing the integration over the protons momenta as in

Sec. 2.1, we obtain that the total probability for the helicity flip in an ep collision has the same form as in Eq. (16). If we study $L \to R$ transitions, making analogous calculations as in the $R \to L$ case, we derive the expression for W, which also coincides with that in Eq. (16). For the sake of brevity we omit these computations.

At the end of this section we shall analyze the approximations made to derive the total probability in the presence of the electroweak eN interaction. First, we assume that ep collisions are elastic. As found in Ref. [9, pp. 205–208], the inelastic contributions to the relativistic collision integral in case of the scattering due to the long range Coulomb interaction are suppressed compared to elastic ones. This situation takes place in our work.

The differential probability of the helicity flip (the analogue of the differential cross section) is $\mathrm{d}W \sim \mathrm{d}\chi/\chi$, where χ is the scattering angle. The maximal probability corresponds to $\chi \sim 0$. That is why we introduce $\omega_p \neq 0$ in Eq. (12) to regularize this infrared divergence. One can estimate χ_{min} using Eq. (15) as $\chi_{\mathrm{min}} \sim \alpha_{\mathrm{em}}/48\pi = 4.8 \times 10^{-5}$. Note that the Coulomb cut-off parameter $L_C = \ln(\chi_{\mathrm{min}}^{-1}) \sim 10$ is typical for relativistic plasmas. The energy of an electron can change by $\Delta E = |E_1 - E_2| \sim \chi^2 \langle E \rangle$ in a collision, where $\langle E \rangle \sim 10^2\,\mathrm{MeV}$ is the mean electron energy. If we define χ_5 corresponding to $\Delta E \sim 2V_5$, i.e. when the inelastic effects become comparable with the electroweak interaction contribution, it reads, $\chi_5 = 3.2 \times 10^{-4}$. Thus, one gets that $\chi_5 \gg \chi_{\mathrm{min}}$, i.e. there will be many particles with a big contribution to $\mathrm{d}W$, for which the energy change in a collision is less than $2V_5$. Hence the elasticity is a good approximation even in the presence of the eN interaction. Moreover, the approximation of the elastic ep scattering was also used in Ref. [10], where Γ_f was computed in the situation when protons are nondegenerate. If inelastic effects are taken into account, it can result in some dependence of W_0 in Eq. (16) on $V_{\mathrm{L,R}}$. However, V_5 still will not contribute to the factor $\mu_{\mathrm{R}} - \mu_{\mathrm{L}}$ in Eq. (16).

Second, the proton energies in Eq. (22) are taken as in vacuum: $\mathcal{E}_{1,2} = \sqrt{k_{1,2}^2 + M^2}$, whereas for electrons we exactly account for the contribution of the electroweak interaction to the electron wave function in Eqs. (17) and (18) and and to the electron energies in Eq. (23). However, protons like electrons can electroweakly interact with background neutrons. To justify our choice of the proton energies, we define the analogues of $V_{\mathrm{L,R}}$ for protons and denote them as $V_{\mathrm{L,R}}^{(p)}$. Using Eq. (30), we can estimate the contribution of the electroweak pN interaction to the proton energies as $|\Delta(\mathcal{E}_{1,2})_{\mathrm{EW}}| \sim \bar{V}_p \mp k_{1,2} V_5^{(p)}/M$, where $\bar{V}_p = [V_{\mathrm{L}}^{(p)} + V_{\mathrm{R}}^{(p)}]/2$ and $V_5^{(p)} = [V_{\mathrm{L}}^{(p)} - V_{\mathrm{R}}^{(p)}]/2$. Now, in the energy conservation delta function in Eq. (22), one has $|\Delta(\mathcal{E}_1 - \mathcal{E}_2)_{\mathrm{EW}}| \lesssim p_{\mathrm{F_p}} V_5^{(p)}/M \sim 0.1 V_5$ since $p_{\mathrm{F_p}} \sim 10^2\,\mathrm{MeV}$ is the Fermi momentum for protons in NS, $M \sim 1\,\mathrm{GeV}$, and $V_5^{(p)} \sim V_5$. Analogously for electrons we have $|\Delta(E_1 - E_2)_{\mathrm{EW}}| = 2V_5$. Thus the contribution of the electroweak pN interaction to the conservation of energy is negligible compared to that of the eN interaction: $|\Delta(\mathcal{E}_1 - \mathcal{E}_2)_{\mathrm{EW}}| \ll |\Delta(E_1 - E_2)_{\mathrm{EW}}|$.

2.3 Kinetics of the chiral imbalance

Basing on Eq. (16), one gets the kinetic equations for the total numbers of right and left electrons $N_{\mathrm{R,L}}$ as

$$
\begin{aligned}
\frac{\mathrm{d}N_{\mathrm{R}}}{\mathrm{d}t} &= -W(R \to L) + W(L \to R) = -W_0\,(\mu_{\mathrm{R}} - \mu_{\mathrm{L}}), \\
\frac{\mathrm{d}N_{\mathrm{L}}}{\mathrm{d}t} &= -W(L \to R) + W(R \to L) = -W_0\,(\mu_{\mathrm{L}} - \mu_{\mathrm{R}}),
\end{aligned}
\tag{26}
$$

where $\mu_{\mathrm{R,L}}$ are the chemical potentials of right and left electrons. Let us introduce the number densities as $n_{\mathrm{R,L}} = N_{\mathrm{R,L}}/V$. Using the standard relation between $n_{\mathrm{R,L}}$ and $\mu_{\mathrm{R,L}}$ [2], we get that $\mathrm{d}\,(n_{\mathrm{R}} - n_{\mathrm{L}})\,/\mathrm{d}t \approx$

$2\dot{\mu}_5\mu_e^2/\pi^2$, where $\mu_5 = (\mu_R - \mu_L)/2$ is the chiral imbalance. Finally, we get the kinetic equation for μ_5,

$$\frac{d\mu_5}{dt} = -\Gamma_f\mu_5, \quad \Gamma_f = \frac{\alpha_{em}^2}{\pi}\left[\ln\left(\frac{48\pi}{\alpha_{em}}\right) - 4\right]\left(\frac{m}{\mu_e}\right)^2\left(\frac{M}{\mu_e}\right)T, \tag{27}$$

where we use Eqs. (16) and (26).

One can see in Eq. (27) that the helicity flipping term contains μ_5 rather than $\mu_5 + V_5$ as recently suggested in Ref. [11]. Note that this our result is based on the explicit QFT calculation of the total probability of the ep scattering, where we exactly account for the eN electroweak interaction.

We also mention that, in Eq. (27), we corrected the value of Γ_f used in Refs. [2–4]. The reason for the discrepancy of Γ_f consists in the fact that in Ref. [2] we relied on the results of Ref. [5], where the scattering of unpolarized electrons off protons was studied. However, in our case it is essential to have the fixed opposite polarizations of incoming and outgoing electrons. Hence, the matrix elements used in the present work and in Ref. [5] are different. This fact, explains, e.g., that Γ_f in Eq. (27) is linear in T whereas that used in Refs. [2–4] is proportional to T^2.

We also mention that Γ_f was recently calculated in Ref. [10]. The value of Γ_f obtained in Ref. [10] is independent of T since it was assumed that protons are nondegenerate. This assumption is valid when the early stages of the NS evolution are considered. In Refs. [2–4], we studied the magnetic field generation in a thermally relaxed NS at $t \gtrsim 10^2$ yr after the onset of the supernova collapse. At this time, the proton component of the NS matter should be taken as degenerate. Note that $\Gamma_f \sim \alpha_{em}^2$ in Eq. (27) as in Ref. [10].

3 Discussion

The detailed computation of the helicity flip rate in ep collisions was the last missing ingredient of the new model, initially proposed in Ref. [2], for the generation of magnetic fields in magnetars driven by the parity violating eN electroweak interaction. Note that, in Refs. [2, 3], we estimated the helicity flip rate Γ_f basing on the classical plasma physics. However, the particle spin is a purely quantum object and its evolution should be treated appropriately. In the present work we used the QFT methods to compute the helicity flip rate. It can explain the discrepancy of our results from those in Refs. [2, 3].

The second important result obtained in the present work was the analysis of the influence of the electroweak interaction of colliding electrons with background nucleons on the helicity flip process. Using the approximation of the elastic scattering and assuming that electrons are ultrarelativistic, we have found that the kinetic equation for the chiral imbalance should have the same form as in Refs. [2, 3], see Eq. (27), contrary to the claim in Ref. [11]. It should be also noted that the computation of the anomalous current along the magnetic field made in Refs. [2, 3] and the calculation of the electron helicity flip in the present work imply the same definition of the chemical potentials $\mu_{L,R}$; see, e.g., the distribution functions of left and right electrons in Eq. (21)-(23) and in Eq. (7) in Ref. [2]. Thus the results of the present work are consistent with Refs. [2–4].

I am thankful to the organizers of Quarks 2016 for the inivitation, to the Tomsk State University Competitiveness Improvement Program, as well as to RFBR (research project No. 15-02-00293) and DAAD (grant No. 91610946) for partial support.

A Solution of the Dirac equation for an electron, electroweakly interacting with nuclear matter

Let us consider the electroneutral matter in NS consisting of neutrons, protons, and electrons. This matter is supposed to be at rest and unpolarized. The Dirac equation for a test electron, described by the bispinor wave function ψ, electroweakly interacting with neutrons and protons, has the form,

$$\left[i\gamma^\mu\partial_\mu - m - \gamma^0 (V_L P_L + V_R P_R)\right]\psi = 0, \tag{28}$$

where

$$V_L = \frac{G_F}{\sqrt{2}}\left[n_n - n_p(1-4\xi)\right](1-2\xi), \quad V_R = -\frac{G_F}{\sqrt{2}}\left[n_n - n_p(1-4\xi)\right]2\xi, \tag{29}$$

are the effective potentials of the interaction of left and right chiral projections, $n_{n,p}$ are the constant and uniform densities of neutrons and protons, $\xi = \sin^2\theta_W \approx 0.23$ is the Weinberg parameter, and $P_{L,R} = (1 \mp \gamma^5)/2$ are the chiral projectors.

We shall look for the solution of Eq. (28) in the form, $\psi = e^{-iEt+i\mathbf{pr}}u/\sqrt{V}$, where u is the basis spinor. The energy levels of an electron have the form,

$$E = \bar{V} + E_0, \quad \bar{V} = \frac{V_L + V_R}{2}, \quad E_0^2 = (p - sV_5)^2 + m^2, \quad V_5 = \frac{V_L - V_R}{2}, \tag{30}$$

where $s = \pm 1$. If electrons are ultrarelativistic, one gets from Eq. (30) that $E_\pm = p + V_{R,L}$. Note that the energy levels in Eq. (30) coincide with those found in Ref. [12].

The basis spinors $u_\pm$ for different helicities can be also obtained in the explicit form,

$$u_+ = N_+\begin{pmatrix} w_+ \\ -\frac{m}{E_+ + p - V_L} w_+ \end{pmatrix}, \quad u_- = N_-\begin{pmatrix} -\frac{m}{E_- + p - V_R} w_- \\ w_- \end{pmatrix}, \tag{31}$$

where $w_\pm$ are the basis two component spinors which are given in Ref. [6, p. 86]. The spinors $w_\pm$ are the eigenvectors of the helicity operator: $(\boldsymbol{\sigma}\cdot\mathbf{p})\,w_\pm = \pm|\mathbf{p}|w_\pm$. Eq. (31) implies the chiral representation for the Dirac matrices [8, pp. 691–696].

If we normalize the electron wave function as

$$\int d^3x \psi^\dagger \psi = 1, \tag{32}$$

we can find the normalization constants $N_\pm$ in Eq. (31) in the form,

$$N_\pm = \sqrt{\frac{E_\pm + p - V_{L,R}}{2E_{0\pm}}}, \tag{33}$$

where $E_\pm$ and $E_{0\pm}$ are given in Eq. (30).

References

[1] R. Turolla, S. Zane, and A. L. Watts, Rep. Prog. Phys. **78**, 116901 (2015).
[2] M. Dvornikov and V. B. Semikoz, Phys. Rev. D **91**, 061301 (2015).
[3] M. Dvornikov and V. B. Semikoz, J. Cosmol. Astropart. Phys. 05 (2015) 032.
[4] M. Dvornikov and V. B. Semikoz, Phys. Rev. D **92**, 083007 (2015).
[5] D. C. Kelly, Astrophys. J. **179**, 599 (1973).
[6] V. B. Berestetskii, E. M. Lifschitz, and L. P. Pitaevskii, *Quantum Electrodynamics*, 2nd ed. (Pergamon, Oxford, 1982).
[7] E. Braaten and D. Segel, Phys. Rev. D **48**, 1478 (1993).
[8] C. Itzykson and J.-B. Zuber, *Quantum field theory* (McGraw-Hill, New York, 1980).
[9] E. M. Lifschitz and L. P. Pitaevskii, *Physical Kinetics* (Pergamon, Oxford, 1981).
[10] D. Grabowska, D. B. Kaplan, and S. Reddy, Phys. Rev. D **92**, 085035 (2015).
[11] G. Sigl and N. Leite, J. Cosmol. Astropart. Phys. 01 (2016) 025.
[12] A. Grigoriev, et al., Grav. Cosmol. **14**, 248 (2008).

EPJ Web of Conferences **125**, 05004 (2016) DOI: 10.1051/epjconf/201612505004
QUARKS-2016

Signatures of chaos and non-integrability in two-dimensional gravity with dynamical boundary

Maxim Fitkevich[1,2,⋆]

[1] *Institute for Nuclear Research of the Russian Academy of Sciences, 60th October Anniversary Prospect, 7a, 117312 Moscow, Russia*
[2] *Moscow Institute of Physics and Technology, Institutskii per. 9, Dolgoprudny 141700, Moscow Region, Russia*

Abstract. We propose a model of two-dimensional dilaton gravity with a boundary. In the bulk our model coincides with the classically integrable CGHS model; the dynamical boundary cuts off the CGHS strong-coupling region. As a result, classical dynamics in our model reminds that in the spherically-symmetric gravity: wave packets of matter fields either reflect from the boundary or form black holes. We find large integrable sector of multisoliton solutions in this model. At the same time, we argue that the model is globally non-integrable because solutions at the verge of black hole formation display chaotic properties.

1 Introduction

About half a century ago Hawking discovered [1] that black holes thermally evaporate exhibiting an apparent loss of quantum coherence. This so-called information paradox contradicts to the principles of quantum theory where a pure initial quantum state $|\psi_{in}\rangle$ cannot evolve into a one. To resolve this puzzle, theorists try to keep track of quantum information within the black hole interior. The current understanding of the information paradox relies on the AdS/CFT argument [2] which relates black hole formation to a thermalization of a quantum state in a dual CFT. The latter process, albeit unitary, dynamically hides quantum coherence in the an approximately thermal final state. This suggests that investigation of dynamical properties of black hole formation may be helpful for resolving the information paradox.

Recently, it was argued [3],[4] that particle collisions near the black hole horizon display chaotic behaviour: their final states are very sensitive to the initial conditions. This dynamical chaos may be significant for the apparent loss of quantum coherence during black hole formation and evaporation.

In this note based on the forthcoming publication [5] we propose a toy model of dilaton gravity which has several desired properties. On one hand, the model possesses a large sector of exact solutions which describe either gravitational scattering or collapse of matter. On the other hand, the solutions in our model are extremely sensitive to the initial conditions at the verge of black hole formation implying that the model is globally non-integrable.

⋆e-mail: fitkevich@phystech.edu

2 The model

The basis for construction of our model is a two-dimensional CGHS gravity [6]. In the CGHS model the metric $g_{\mu\nu}$ is coupled to a dilaton field ϕ and a massless free scalar field f representing matter, the action is[1]

$$S_{CGHS} = \int d^2x\,\sqrt{-g}\left[e^{-2\phi}\left(R + 4(\nabla\phi)^2 + 4\lambda^2\right) - \frac{1}{2}(\nabla f)^2\right] . \tag{1}$$

Here R is the Ricci scalar, λ^2 is the cosmological constant. The model (1) is exactly solvable at the classical level. The metric and the dilaton can be explicitly obtained from constraints, which leaves the field f as the only dynamical degree of freedom.

Unfortunately, the CGHS model itself cannot be used as a consistent model at the semiclassical level. In Minkowski vacuum

$$g_{\mu\nu} = \eta_{\mu\nu}\,, \qquad \phi = -\lambda x^1\,, \tag{2}$$

the dilaton field tends to infinity at $x^1 \to -\infty$ implying infinitely large gravitational coupling constant e^{ϕ} in this region, see Fig. 1a and cf. Eq (1). This makes the model strongly coupled at large negative x^1 and restricts applicability of the semiclassical calculations.

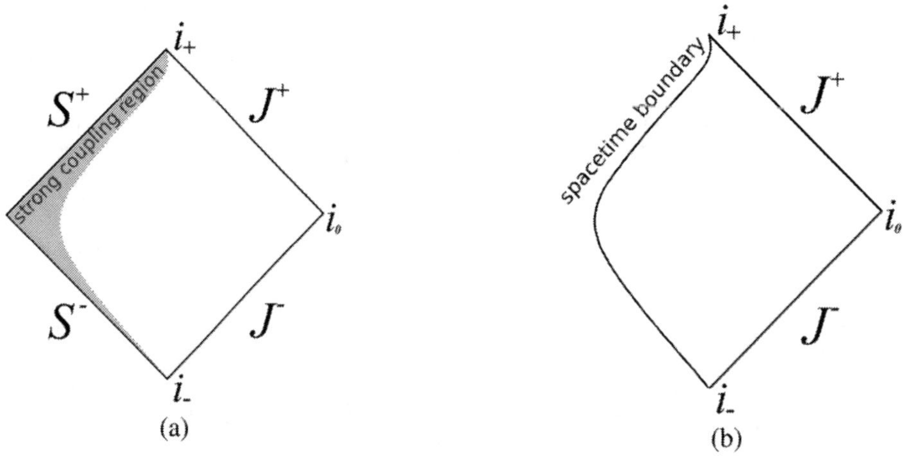

Figure 1. Penrose diagrams for the Minkowski spacetime in the (a) CGHS model and (b) model with a boundary.

In order to avoid this difficulty we add a dynamical boundary at $\phi = \phi_0$ cutting off the strong coupling region at large ϕ. Namely, we restrict integration in Eq. (1) to the manifold $\mathcal{M}$ where $\phi < \phi_0$ and add a Gibbons-Hawking boundary term,

$$S = S_{CGHS}\Big|_{\mathcal{M}} + 2\oint_{\partial\mathcal{M}} d\sigma\,\sqrt{|h|}e^{-2\phi}\,(K + 2\lambda)\,, \tag{3}$$

where $K = g^{\mu\nu}\nabla_\mu n_\nu$ is the extrinsic curvature of the boundary, n^ν is the outer normal. A new parameter e^{ϕ_0} can be regarded as a gravitational coupling constant in our model. Previously a similar model with dynamical mirror has been proposed in [8]. Unlike the authors of [8] we analyse integrable properties of this model at the classical level rather than trying to quantize it.

[1]The first term in Eq. (1) can be obtained by spherical reduction of the d-dimensional Einstein gravity in the limit $d \to +\infty$ [7].

3 Field equations

The starting point in simplifying the field equations is to find a general solution in the bulk. We work in conformal coordinates (u, v), where the interval has the form $ds^2 = -e^{2\rho}dv\,du$. Besides, we choose the frame with $\rho = \phi$. Field equation for $f(u, v)$ in these coordinates can be explicitly solved,

$$\Box f \equiv -4e^{-2\rho}\partial_v\partial_u f = 0 \qquad \Rightarrow \qquad f(v, u) = f_{out}(u) + f_{in}(v) . \tag{4}$$

Expressions for the other fields follow from the constraints,

$$e^{-2\rho} \equiv e^{-2\phi} = -\lambda^2 vu + g(v) + h(u) , \tag{5}$$

where

$$g(v) = \frac{1}{2}\int_0^v dv' \int_{v'}^{\infty} dv''\, f_v^2(v'') , \qquad h(u) = -\frac{1}{2}\int_{-\infty}^{u} du' \int_{-\infty}^{u'} du''\, f_u^2(u'') . \tag{6}$$

Equations (4)–(6) give the general solution in the bulk.

Now, we impose conditions at dynamical boundary $\phi = \phi_0$. Varying Eq. (3) with respect to the boundary values of $g_{\mu\nu}$ and f, we obtain,

$$n^\mu \nabla_\mu f\Big|_{\phi=\phi_0} = 0 , \qquad n^\mu \nabla_\mu \phi\Big|_{\phi=\phi_0} = \lambda , \tag{7}$$

The first of these equations implies that the energy flux through the boundary equals zero. The second condition guarantees that the flat spacetime (2) is a solution. Integrating the first of Eqs. (7) we obtain a diffeomorphism between in- and out-going wave packets,

$$f_{out}(U(v)) = f_{in}(v) , \tag{8}$$

where $u = U(v)$ is the trajectory of the boundary in the (u, v) frame: $\phi(U(v), v) = \phi_0$.

Using Eqs. (5), one derives an equation for $U(v)$ from the conditions (7),

$$U_v = \frac{e^{2\phi_0}}{\lambda^2}(g_v - \lambda^2 U)^2 , \tag{9}$$

where $\partial_v g$ can be derived from Eq. (6). Introducing auxiliary function $\psi(v)$, satisfying

$$e^{-2\phi_0}\frac{\psi_v}{\psi} = \partial_v g - \lambda^2 U , \tag{10}$$

we further simplify this equation,

$$\psi_{vv}(v) = \mathcal{T}(v)\psi(v) , \tag{11}$$

where $\mathcal{T} = e^{2\phi_0}g_{vv}$ depends on the incoming wave packet. At $\mathcal{T} = 0$ one obtains the Minkowski solution

$$\psi(v) = \lambda v \qquad \text{and} \qquad U(v) = -\frac{e^{-2\phi_0}}{\lambda^2 v} , \tag{12}$$

which is equivalent to Eq. (2) after proper change of coordinates

$$-\lambda u = e^{-\lambda\bar{u}} , \qquad \lambda v = e^{\lambda\bar{v}} , \qquad \text{where} \qquad (\bar{v}, \bar{u}) = (x^0 + x^1, x^0 - x^1) . \tag{13}$$

In what follows we use $(\bar{v}, \bar{u})$ as asymptotically flat coordinates.

4 Integrable sector

Exact solutions in our model are obtained in the following way. Suppose $\psi(v)$ in Eq. (11) is analytic everywhere except for N isolated singularities v_i, where $\psi \propto (v - v_i)^{-s_i}$. In the vicinity of v_i Eq. (11) gives,

$$\mathcal{T} = \frac{\psi_{vv}}{\psi} \sim \frac{s_i(s_i + 1)}{(v - v_i)^2} . \tag{14}$$

Thus, function $\mathcal{T}(v)$ has only the poles of rank 2, and we write,

$$\mathcal{T}(v) = \sum_{i=1}^{N} \left(\frac{s_i(s_i + 1)}{(v - v_i)^2} + \frac{\mathcal{T}_i}{v - v_i} \right) , \qquad \mathcal{T} \leq 0 \quad \text{at} \quad v > 0 . \tag{15}$$

In what follows we consider for simplicity only the singularities of rank $s_i = 1$. Substituting Laurent series,

$$\psi \sim \frac{\psi^i_{-1}}{v - v_i} + \psi^i_0 + \psi^i_1(v - v_i) + \psi^i_2(v - v_i)^2 + \dots , \tag{16}$$

$$\mathcal{T} \sim \frac{2}{(v - v_i)^2} + \frac{\mathcal{T}_i}{v - v_i} + \mathcal{T}^i_0 + \mathcal{T}^i_1(v - v_i) + \dots \qquad \text{at} \quad v \to v_i \tag{17}$$

into Eq. (11), one obtains an infinite chain of equations for the Laurent coefficients,

$$2 = 2 , \tag{18a}$$

$$\psi_0 = -\frac{1}{2}\mathcal{T}_i\psi_{-1} , \tag{18b}$$

$$\psi_1 = \left(-\frac{1}{2}\mathcal{T}^i_0 + \frac{1}{4}\mathcal{T}_i^2 \right)\psi_{-1} , \tag{18c}$$

$$\mathcal{T}^i_1\psi_{-1} + \mathcal{T}_i\psi_1 + \mathcal{T}^i_0\psi_0 = 0 , \tag{18d}$$

$$\psi_2 = \dots . \tag{18e}$$

We note that two equations in the set (18) are special since ψ_{-1} is not present in Eq. (18a), as well as ψ_2 in Eq. (18d). Thus, ψ_{-1} and ψ_2 are free parameters of the solution, while Eqs. (18a) and (18d) are constraints on the function $\mathcal{T}(v)$. Using Eqs. (18b), (18c), we rewrite Eq. (18d) in the form,

$$\mathcal{T}^i_1 = \mathcal{T}^i_0\mathcal{T}_i - \frac{1}{4}\mathcal{T}^3_i . \tag{19}$$

The other constraints are obtained from a requirement for solutions to be localized, i.e. $\mathcal{T}(v) = o(v^{-2})$ at $v \to +\infty$. This gives two conditions,

$$\sum_{i=1}^{N} \mathcal{T}_i = 0 , \qquad \sum_{i=1}^{N} (2 + v_i\mathcal{T}_i) = 0 . \tag{20}$$

Solving system of algebraic equations (19), (20) one obtains coefficients $\mathcal{T}_i$. i.e. the exact solution for $\mathcal{T}(v)$ in the form (15).

Now, consider the solution with two singularities, $s_1 = s_2 = 1$. Reality of $\mathcal{T}(v)$ gives $v_1 = v_2^* = v_0 + iT$. Solving the conditions (20), one obtains,

$$\mathcal{T} = -\frac{8T^2}{[(v - v_0)^2 + T^2]^2} . \tag{21}$$

Then Eqs. (18) give the Laurent coefficients of $\psi(v)$, and we find,

$$\psi = 1 + \frac{T^2 - 3v_0^2}{8v_0T^2}v - \frac{\frac{4}{3}T^2}{(v-v_0)^2 + T^2}\left[1 + \frac{T^2 - 3v_0^2}{8v_0T^2}(3v - 2v_0)\right]. \quad (22)$$

It is straightforward to implement the procedure of this Section for finding exact solutions with N singularities of different ranks.

5 Reflecting and collapsing solutions

We restrict our analysis to the localized wave packets $\mathcal{T} \propto f_v^2(v)$. In this case $\psi(v) \sim Cv$ at $v \to +\infty$, see Eq. (11). We will see that the coefficient C determines regimes of behaviour and plays a role of a critical parameter; in the case (22) we obtain $C = \frac{T^2 - 3v_0^2}{8v_0T^2}$. Solution (22) at different values of C is shown in Fig. 2.

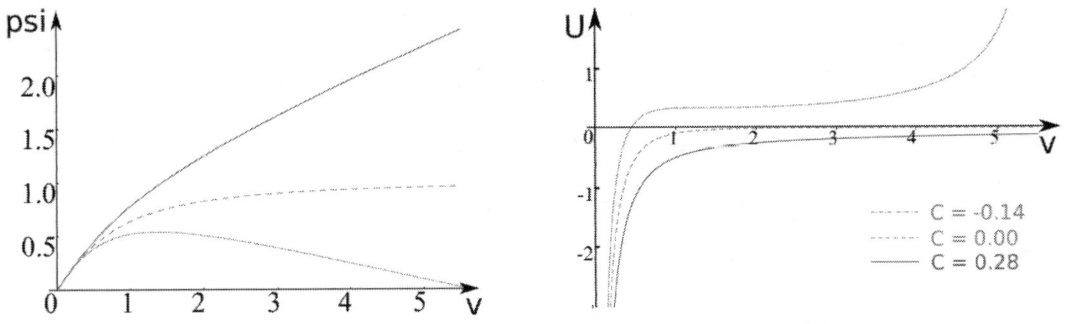

(a) (b)

Figure 2. Plots of (a) $\psi(v)$ and (b) $U(v)$ for the exact solution (22) at $T = 1$ and different values of the parameter $C = C(v_0)$.

At $C > 0$ one shows that $U(v) \sim -\frac{e^{-2\phi_0}}{\lambda^2 v} + o(v^{-1})$ at $v \to +\infty$, see Eq. (10). This coincides asymptotically with the Minkowski solution (12); thus $C > 0$ corresponds to total reflection of the matter from the boundary.

Solutions with $C < 0$ describe collapse. Indeed, in this case the function $\psi(v)$ has a maximum $\psi_v = 0$ which corresponds to the singularity $dU/dv = 0$ where the transformation $u = U(v)$ cannot be inverted. This is the singularity point of the outgoing wave packet $f_{out}(u)$ in Eq. (8). Appearance of this singularity signals that the black hole is formed in the final state of the process.

In the critical case $C = 0$ the boundary becomes asymptotically null $U(v) = o(v^{-1})$ at $v \to +\infty$. One can show that energy difference between the incident and reflected wave packets in this case is equal to $M_{cr} = 2\lambda e^{-2\phi_0}$ and corresponding bulk solution at i^+ describes CGHS black hole with mass M_{cr}. The critical solutions with $C = 0$ separate the families of the reflecting and collapsing solutions. One can show that solutions at small negative C form a black hole with minimal mass M_{cr}. On the other hand, the solutions with small positive C describe total reflection.

6 Nonintegrability in the critical regime

The CGHS model in the bulk is exactly solvable and has an infinite number of conserved quantities in the in- and out-sectors. In fact, any functionals of f_{in} and f_{out}, e.g.

$$I_k^{in} = \int dv' \, (\partial_v f_{in}(v'))^k \,, \qquad I_k^{out} = \int du' \, (\partial_u f_{out}(u'))^k \,,$$

can be regarded as such conserved quantities. The question is whether these quantities can be smoothly joined at the boundary to represent the integrals in our model? This would require the map $\{I_i^{in}\} \mapsto \{I_j^{out}\}$ in Eq. (8) to be smooth. We are going to argue, however, that this is not the case: the map $\{I_i^{in}\} \mapsto \{I_j^{out}\}$ is essentially singular in the vicinity of the near-critical solutions.

To understand the singular behaviour in the critical regime, consider the trajectory of the boundary in asymptotic coordinates (13),

$$\bar{u}(v) = \frac{2\phi_0}{\lambda} - \frac{1}{\lambda} \ln\left(\frac{\psi_v(v)}{\lambda\psi(v)} - \frac{2g_v(v)}{M_{cr}} \right) , \qquad \lambda v = e^{\lambda\bar{v}} \,, \tag{23}$$

where Eq. (10) was used. The typical trajectory in the near–critical regime is shown in Fig. 3a. It has a large "plateau" part $v_1 \leq v \leq v_2$ which size grows to infinity as $C \to 0$. Besides, one can argue that the trajectory $\bar{u}(\bar{v})$ within the "plateau" becomes null in the limit $C \to 0$. Thus, the part of the initial wave packet $f_{in}(v)$ which gets reflected from the "plateau" gets blue shifted and shrinks into a narrow pulse in the outgoing packet. The latter part of the outgoing wave packet shrinks to a delta-peak at $C \to 0$; in what follows we call it "thunderpop".

Let us describe reflection of wave packets from the boundary in the limit $C \to 0$. At large v we expect,

$$\psi(v) = 1 + Cv + \psi^{(1)} + \psi^{(2)} + \dots \,, \tag{24}$$

where $\psi^{(2)} \ll \psi^{(1)} \ll 1$ at $v \to +\infty$. Besides, we assume that Cv and $\psi^{(1)}$ are of the same order. Solving Eq. (11) perturbatively and recalling that $g_{vv} \propto v^{-2}$ is small, we obtain,

$$\psi^{(1)}(v) = e^{2\phi_0} \int_v^\infty dv' \int_{v'}^\infty dv'' g_{vv}(v'') = e^{2\phi_0}(g(v) - g(\infty)) \,, \tag{25}$$

$$\psi_v^{(2)}(v) = -C\lambda e^{2\phi_0} (g(v) - g(\infty) - vg_v(v)) + \\ +e^{4\phi_0} \left((g(v) - g(\infty))g_v(v) + \int_v^\infty dv' (g_v(v'))^2 \right) . \tag{26}$$

Next, we calculate the outgoing wave packet using Eq. (8),

$$f_{\bar{u}}^{out}(\bar{u}(v)) = \lambda^3 e^{2\phi_0} f_v(v) \left(\frac{\psi(v)}{\psi_v(v)} \right)^2 U(v) \,. \tag{27}$$

Substituting the perturbative solution (25), (26), we find,

$$f_{\bar{u}}^{out}(\bar{u}(v)) \approx \frac{Cf_v(v)}{\left(C + e^{2\phi_0} \frac{g_v(v)}{\lambda}\right)^2} \,, \tag{28}$$

$$\bar{u}(v) \approx \frac{2\phi_0}{\lambda} - \frac{1}{\lambda} \ln C + \text{corrections} \,. \tag{29}$$

Thus the boundary is indeed almost null, $\bar{u} \approx$ const, in the large "plateau" region $Cv \sim \psi^{(1)}$.

To illustrate instability of the critical solution, we slightly change the initial wave packet by modulating its tail,

$$\mathcal{T} = \mathcal{T}_{sol} \cos^2 (\omega \ln (\lambda v)) \;, \qquad \text{at} \quad v > v_1 \;, \tag{30}$$

where $v_1 \gg v_0$ and $\mathcal{T}_{sol}$ is given by Eq. (21). In the limit $\omega \to 0$ the original solution is restored. Let us check whether the integrals I_k^{out} are smooth functions of ω and C parametrizing the solutions.

For simplicity, we consider the third integral

$$I_3^{out} = \int_0^{+\infty} dv \frac{C^2 f_v^3(v)}{\left(C + e^{2\phi_0} \frac{g_v(v)}{\lambda}\right)^4} \;. \tag{31}$$

Substituting the asymptotics of (30) we obtain a combination of power-law and periodic functions in the integrand of Eq. (31). Rescaling the integration variable $v \mapsto ve^{2\pi n/\omega}$, one proves an identity

$$I_3(\omega, e^{6\pi n/\omega} C) = e^{-2\pi n/\omega} I_3(\omega, C) \;. \tag{32}$$

This functional equation has a solution

$$I_3(\omega, \ln C) = \frac{J(\omega \ln C)}{C^{\frac{1}{3}}} \;, \tag{33}$$

where $J(x)$ is periodic $J(x) = J(x + 6\pi n)$.

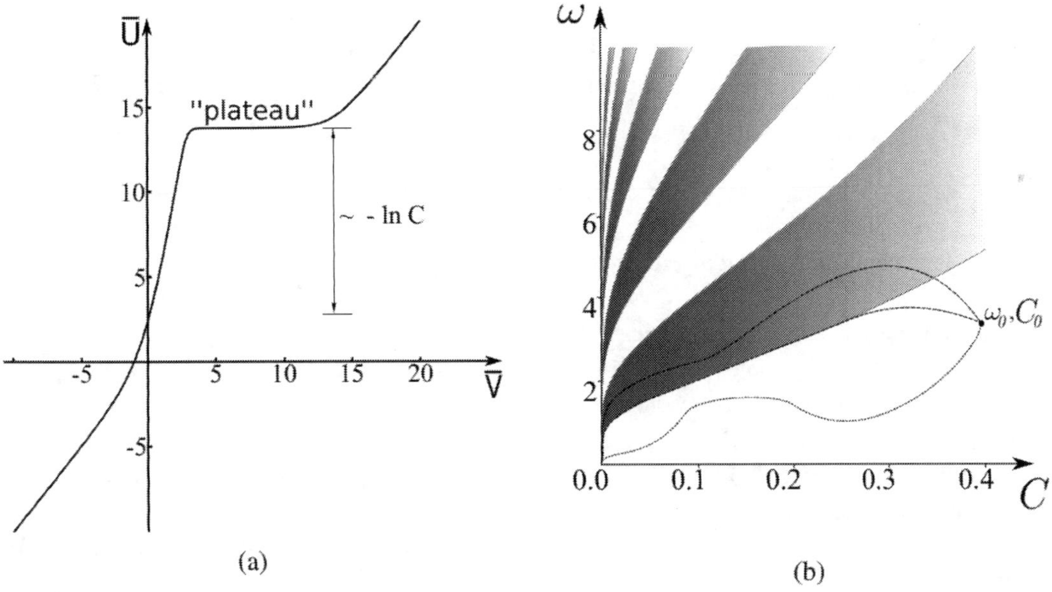

Figure 3. (a) Apparent boundary trajectory for the solution (22) at $T = 1$ and $C = 10^{-6}$. (b) Function I_3 on the plane (C, ω). White and blue regions correspond to $I_3 > 0$ and $I_3 < 0$, respectively. At the boundaries of these regions $I_3 = 0$.

Given representation (33), one immediately concludes that the limit of $I_3(\omega, C)$ at $\omega \to 0$ and $C \to 0$ essentially depends on the method of evaluating: one can make it equal to any real number by choosing the limiting path $\omega(C)$, see Fig. 3b. In other words, I_3 has an essential singularity at $\omega \to 0$, $C \to 0$ and the internal structure of the thunderpop is sensitive to small variations of f_{in}.

To summarize, the structure of the final pulse is unstable under the small variations of initial data. Despite our intuition the final pulse could have a non-trivial peak structure, which depends on values of ω and C. If we choose $\omega \simeq \frac{6\pi n}{\ln C}$, the pulse contains about n peaks. In the above we argued that action-angle can not be introduced for our boundary model because I_3^{out} is an essentially singular function of the initial parameters ω and C.

7 Conclusions

We considered a toy model of dilaton gravity with a dynamical boundary. As opposed to the original CGHS model, gravitational interactions in our model are weak in the Minkowski vacuum. The classical dynamics of this model in described by the single differential equation with an infinite number of exact solutions. The latter solutions describe either reflection of matter from the boundary or collapse. An important feature of the model is an existence of a minimal black hole mass M_{cr}. We studied the class of critical solutions separating regimes of total reflection and black hole formation.

We argued that despite an infinite number of exact solutions, our model is globally non-integrable. Namely, the natural conserved integrals in the outgoing and ingoing sectors cannot be smoothly glued at the boundary: the procedure fails in the vicinity of critical solution where the map of the integrals becomes singular.

Our two-dimensional model with dynamical boundary might be a useful tool for future researches of the semiclassical gravity.

Acknowledgments

We thank Sergey Sibiryakov for useful discussion. The work is supported by the RFBR grant 14-02-31768.

References

[1] S. W. Hawking, "Breakdown of Predictability in Gravitational Collapse," Phys. Rev. D **14** (1976) 2460.

[2] J. M. Maldacena, "Eternal black holes in anti-de Sitter," JHEP **0304** (2003) 021.

[3] G. 't Hooft, "The black hole interpretation of string theory," Nucl. Phys. B **335** (1990) 138.

[4] J. Polchinski, "Chaos in the black hole S-matrix," arXiv:1505.08108 [hep-th].

[5] M. Fitkevich and D. Levkov, *in preparation* (2016).

[6] C. G. Callan, Jr., S. B. Giddings, J. A. Harvey and A. Strominger, "Evanescent black holes," Phys. Rev. D **45** (1992) R1005.

[7] D. Grumiller, W. Kummer and D. V. Vassilevich, "Dilaton gravity in two-dimensions," Phys. Rept. **369** (2002) 327.

[8] T. D. Chung and H. L. Verlinde, "Dynamical moving mirrors and black holes," Nucl. Phys. B **418** (1994) 305.

Vacuum polarization in the model of Dirac fermions with anomalous magnetic moment interacting with background axial-vector condensate and magnetic field

A. F. Bubnov[1,⋆], *N. V.* Gubina[1], *V. Ch.* Zhukovsky[1,⋆⋆]

[1] *Department of Theoretical Physics, Faculty of Physics, Moscow State University, 119991 Moscow, Russia*

Abstract. In this paper, we consider vacuum polarization effects in the model of charged fermions with anomalous magnetic moment and axial-vector interaction term in a uniform magnetic field. Nontrivial orrections to the effective Lagrangian from the anomalous moment and axial-vector term are calculated with account for various configurations of parameters of the model.

1 Introduction

As is known a charged particle described by the Dirac equation should possess an intrinsic moment, i.e., spin, and also a kinematic Dirac magnetic moment whose value is equal to the Bohr magneton $\mu_0 = \frac{e\hbar}{2mc}$. In the lowest order with respect to the fine-structure constant $\alpha = \frac{e^2}{\hbar c}$ (Schwinger [1]), the total value of the electron magnetic moment with account fot the vacuum contribution becomes

$$\mu_{Sch} = \mu_0 + \Delta\mu_{Sch}, \quad \Delta\mu_{Sch} = \frac{\alpha}{2\pi} \cdot \mu_0, \tag{1}$$

where $\Delta\mu_{Sch}$ is the Schwinger vacuum magnetic moment of the electron, which was shown [2] to depend on the electron energy and the strength of the external magnetic field. In the general case, a Dirac particle can obtain an anomalous magnetic moment (AMM) $\Delta\mu$ not necessarily equal to its vacuum value $\Delta\mu_{Sch}$ ([3]). Then the total magnetic moment of a Dirac particle is equal to

$$\mu_{tot} = \mu_0 \ + \ \Delta\mu, \ \ \Delta\mu \simeq \kappa\mu_0. \tag{2}$$

Here κ, a 4-invariant quantity, is the value of the anomalous moment, which is assumed to be the given constant of the model (in the case of the vacuum magnetic moment of the electron $\kappa = \frac{\alpha}{2\pi}$). The interaction of the anomalous magnetic moment with an external electromagnetic field $F_{\alpha\beta}$ can be described phenomenologically by adding the Pauli–Schwinger term $\Delta\mu\frac{1}{2}\sigma^{\alpha\beta}F_{\alpha\beta}$ [4] (see also [5]) to the Dirac Lagrangian. Along with the AMM, another additional parameter can be taken into account in the general description of fermion. This new term characterizes the violation of Lorentz invariance in the system. The Standard Model does not include a mechanism that allows the violation of Lorentz and CPT symmetries. These violations, however, can be present in extended models related to theories

⋆e-mail: bfandrey@mail.ru
⋆⋆e-mail: zhukovsk@phys.msu.ru

including terms of interaction with axial-vector condensate discussed in connection with possible manifestations of a fermion induced torsion field [6]. or some more fundamental theories valid in higher dimensions. Thus, for example, such terms can occur in gravitation theory and cosmology (see [7],[8]), where the violation of symmetries is connected with the spatial anisotropy that is caused by some vector field with a nonzero vacuum average, or, as follows from [9], in the string theory, in which it is assumed that our world is situated on a brane that exists in a universe of a larger number of dimensions. The resulting theory can be effectively described in the framework of the Standard Model Extension (SME)[10, 11].

In this paper, we calculate a one-loop effective Lagrangian of electromagnetic field in the framework of the SME, generalized by introducing the Pauli–Schwinger AMM term in the Dirac Lagrangian for a fermion in a constant and uniform magnetic field.

We introduce an additional CPT-odd term to the Dirac equation of the form $\bar{\psi}\gamma^5\gamma^\mu b_\mu\psi$, where b^μ is the constant axial four-vector [12–14]). The Dirac equation of this form and the energy spectrum for a fermion were considered earlier [15]. Modern experiments give the following estimates (see, e.g. [16]): $|b_0| \lesssim 10^{-14}$GeV, $|\vec{b}| \lesssim 10^{-31}$GeV.

This calculation of the effective Lagrangian is motivated by the study of the so called chiral magnetic effect [17] with the vacuum current induced by an external magnetic field in the presence of the chiral chemical potential, which is similar to the above mentioned Lorentz-violating term.

One should also note recent studies of interaction of massless fermions with the background field, when the chirality of massless fermions makes it possible to observe an energy current along the external magnetic field, or the so-called axial magnetic effect, which helps to explain the low energy description of Weyl semi-metals conductive properties [18], [19]. The axial magnetic effect in QCD is also examined with respect to phase structure of quark-gluon plasma and the effect of deconfinment [20]. Note that the problem of the description of the dynamics of fermions that possess an anomalous moment has attracted considerable attention recently (see, e.g. [3], [21], [22], [23]. Background interaction of the form of the Pauli–Schwinger term can also occur in the framework of the SME and in parity-violating interactions of cosmic fields with atoms, molecules, and nuclei [3][1].

2 Energy spectrum of electron

Consider a charged fermion in a constant uniform magnetic field in the presence of an additional background axial-vector interaction of the form $\bar{\psi}\gamma^5\gamma^\mu b_\mu\psi$ with the constant axial vector b^μ in the Lagrangian of the model.

In this paper, we consider two variants: the axial fourvector b^μ has the form $(b^0, \mathbf{0})$ or $(0, \mathbf{b})$ (only the time or space components are nonzero, an evident generalization with both components nonvanishing is also presented). In the latter case we assume that the three-dimensional vector $\mathbf{b}$ is aligned along (or opposite to) the magnetic-field strength vector $\mathbf{H}$.

In the former case, we also take the condensate to include the fermion AMM $\kappa\mu_0$ (2) (which can occur for cosmic fields that violate the interaction parity [3]).

The total Lagrangian of a negatively charged Dirac fermion that interacts with the above condensates has the following form:

$$\mathcal{L} = \bar{\psi}\left(i\gamma^\mu D_\mu - m\right)\psi - \bar{\psi}\gamma^5\gamma^\mu b_\mu\psi + \bar{\psi}\left(\frac{\Delta\mu}{2}\sigma^{\alpha\beta}F_{\alpha\beta}\right)\psi, \quad (3)$$

[1]It should also be noted that the influence of CPT-odd action on the dipole moment of bound electrons that results in the appearance of an anomalous moment of atomic orbitals and specific asymmetry of the angular distribution of radiation for a hydrogen-like atom was considered in the framework of the SME [24].

where the first term is the standard Dirac Lagrangian, the second part corresponds to the interaction with the axial-vector background that violates the Lorentz invariance of the theory and, as can be easily verified, CPT parity, and the last term is the part that phenomenologically describes the interaction of the AMM $\Delta\mu$ of the Dirac particle with the electromagnetic field.

The stationary solutions of the equation, following from the total Lagrangian (3)

$$\left(i\gamma^{\mu}D_{\mu} - m + \frac{\Delta\mu}{2}\sigma^{\mu\nu}F_{\mu\nu} - \gamma^{5}\gamma^{\alpha}b_{\alpha}\right)\psi = 0, \tag{4}$$

where $\sigma^{\mu\nu} = \frac{i}{2}(\gamma^{\mu}\gamma^{\nu} - \gamma^{\nu}\gamma^{\mu})$, and $F_{\mu\nu}$ is a constant and uniform external field, will be found in the case, when an external constant uniform magnetic field with the strength $\mathbf{H}$, is directed along the z axis: $\mathbf{H} = H\mathbf{e}_z$, $H > 0$, and the electric field is absent. Consider the following possible particular cases (in what follows, $n = 0, 1, 2, \ldots$ is the Landau principal quantum number, $\zeta = \pm 1$ is the fermion spin projection on the magnetic field direction, $\epsilon = \pm$ is the sign of the energy).

a) AMM $\Delta\mu = 0$, **axial condensate** $b_0 \neq 0$, $\vec{b} = 0$ ([15], [17], [25], [26], [27] [2])

$$E^2 = m^2 + (\text{sgn}(p)\sqrt{p^2 + 2eHn} + \zeta b_0)^2. \tag{5}$$

b) AMM $\Delta\mu = 0$, **axial condensate** $b_0 = 0$, $\vec{b} \neq 0, \vec{b} = (0, 0, b)$, $p \equiv p_z$ ([15], [28])

$$E^2 = 2eHn + (\sqrt{m^2 + p^2} + \zeta b)^2. \tag{6}$$

c) AMM $\Delta\mu \neq 0$, **axial condensate** $b_0 = 0$, $\vec{b} = 0$ ([15], [28], [29])

$$E^2 = p^2 + (\sqrt{2eHn + m^2} + \zeta\Delta\mu H)^2. \tag{7}$$

d) AMM $\Delta\mu \neq 0$, **axial condensate**, $b_0 \neq 0$, $\vec{b} = 0$ ([15])

$$E^2 = m^2 + p^2 + 2eHn + (\Delta\mu H)^2 + b_0^2 \pm 2\sqrt{[m(\Delta\mu H) + b_0 p]^2 + 2eHn\left[(\Delta\mu H)^2 + b_0^2\right]}. \tag{8}$$

3 The effective Lagrangian

The effective Lagrangian can be calculated basing upon the proper time (fifth parameter) method [30]

$$\mathcal{L}_{\text{eff}} = \frac{1}{4\sqrt{\pi}}\frac{eH}{(2\pi)^2}\int_{-\infty}^{\infty} dp \sum_{n,\zeta,\epsilon}\int_{\frac{1}{\Lambda^2}}^{\infty}\frac{ds}{s^{3/2}}\,\mathrm{e}^{-sE^2}, \tag{9}$$

where $1/\Lambda^2$ is the cutoff parameter. In this formula summation over the quantum numbers gives

$$\sum_{n,\zeta,\epsilon}\mathrm{e}^{-2s(eH)n} = 2\left[1 + 2\sum_{n=1}^{\infty}\mathrm{e}^{-2s(eH)n}\right] = 2\coth(eHs). \tag{10}$$

[2]Formally speaking, in this paper, the axial vector b_0 is considered in place of the chiral chemical potential μ_5, which was used in [17], [25] for description of the quark spectrum in the case of asymmetry with respect to the number of left and right quarks due to axial anomaly.

Here, the first term corresponds to the quantum numbers of the ground (nondegenerate) state $n = 0,\ \zeta = +1$ (for $e > 0$) and $n = 0,\ \zeta = -1$ (for $e < 0$). As a result, with $b_\mu = 0$ and $\Delta\mu = 0$, after regularization and renormalization one obtains the well-known Heisenberg–Euler effective Lagrangian (see, e.g., [1, 5])

$$\mathcal{L}_{\rm eff} = -\frac{1}{8\pi^2}\int_0^\infty \frac{ds}{s}\, e^{-sm^2}\left[(eH)(eE)\cot(esE)\coth(esH) - \frac{1}{s^2} + \frac{1}{3}((eE)^2 - (eH)^2)\right].$$

In what follows we shall use the above formulated particular solutions for the Dirac energy eigenvalues (5), (6), (7) and (8).

a) Consider the eigenvalue solution $\Delta\mu = 0,\ \vec{b} = 0,\ b_0 \neq 0$

In the approximation of small parameter $b_0 \ll 1$ in the expansion in powers of b_0 the linear term will be absent due to the oddness of the expression in ζ for $n \neq 0$, and at $n = 0$ due to oddness in the integration variable p that in this case enters the energy in combination $(p + \zeta b)^2$ (at fixed ζ).Then expansion in b_0 of expression (9) after summation over all quantum numbers (10) gives

$$\Delta\mathcal{L}^b = b_0^2\frac{(eH)^2}{(2\pi)^2}\int_{\frac{1}{\Lambda^2}}^\infty ds \frac{e^{-sm^2}}{\sinh^2(eHs)}. \tag{11}$$

Integrating by parts and performing regularization we arrive at the expression for the b_0^2-contribution to the effective Lagrangian

$$\Delta\mathcal{L}^b = \frac{b_0^2}{4\pi^2}\left[-m^2 eH\int_0^\infty ds\, e^{-sm^2}\left(\coth(eHs) - \frac{1}{eHs}\right) + \Lambda^2 - m^2\ln\frac{\Lambda^2}{m^2}\right]. \tag{12}$$

Note that the terms with the cut-off Λ do not actually depend on the magnetic field, and thus can be eliminated by subtraction, yielding the final result

$$\Delta\mathcal{L}^b = -b_0^2\frac{(eH)m^2}{4\pi^2}\int_0^\infty ds\, e^{-sm^2}\left(\coth(eHs) - \frac{1}{eHs}\right). \tag{13}$$

In the weak field limit, $eH \ll m^2$ after integration one obtains

$$\Delta\mathcal{L} = \frac{1}{12\pi^2}b_0^2(eH)^2\int_0^\infty dx\, e^{-x}\frac{x}{m^2} = \frac{b^2(eH)^2}{12\pi^2 m^2}. \tag{14}$$

In the strong field limit, $eH \gg m^2$, using the substitution $x = sm^2$ and integrating approximately

$$\int_0^\infty dx\, e^{-x}(\coth(eHx/m^2) - \frac{1}{eHx/m^2}) \approx 1,$$

one gets the result

$$\Delta\mathcal{L} \approx \frac{b_0^2 eH}{4\pi^2}. \tag{15}$$

A more accurate approximation is obtained if the results are represented via Euler ψ- function (see [27]).

$$\Delta\mathcal{L} = -\frac{b_0^2}{4\pi^2}m^2\left(\psi(\frac{m^2}{2H}) - \ln(\frac{m^2}{2H}) + \frac{H}{m^2}\right). \tag{16}$$

b) Consider the eigenvalue solution $\Delta\mu = 0,\ \vec{b} = (0,0,b),\ b \neq 0,\ b_0 = 0$

In the approximation of small b the expansion is considered up to b^2 only (the linear term is absent since the expression is odd with respect to ζ).

The final renormalized expression takes the form

$$\Delta\mathcal{L}^b = b^2 \frac{(eH)m^2}{4\pi^2} \int_0^\infty ds\, e^{-sm^2} \left(\coth(eHs) - \frac{1}{eHs} \right). \tag{17}$$

The terms that do not dependent on the field H were taken away by renormalization.

This result can be easily rewritten for a more general option with an axial condensate given in the form

$$b = (b_0, 0, 0, b_3), \text{where both } b_0 \neq 0,\ b_3 \neq 0.$$

Then, by introducing a vector $\beta^\mu = \frac{1}{2}\varepsilon^{\mu\nu\alpha\beta} b_\nu F_{\alpha\beta}$, with nonzero components $\beta^0 = \varepsilon^{0312} b_3 F_{12} = b_3 H,\ \ \beta^3 = \varepsilon^{3012} b_0 F_{12} = -b_0 H,$ we deal with the same intermediate expressions as in the previous situations, and the unified result can then be represented in the invariant form

$$\Delta\mathcal{L}^b = -\beta^2 \frac{e^2 m^2}{4\pi^2} \int_0^\infty ds\, e^{-sm^2} \left(\frac{\coth(eHs)}{eH} - \frac{1}{(eH)^2 s} \right), \tag{18}$$

here $\beta^2 = \beta^\mu \beta_\mu$ and H is an invariant quantity equal to the strength of a magnetic field in the system, where $\vec{H} \| Oz$.

c) Consider the eigenvalue solution $\Delta\mu \neq 0,\ \vec{b} = 0,\ b_0 = 0$

Assuming $\Delta\mu H$ to be small, consider the expansion up to $(\Delta\mu H)^2$ in (9) for $\Delta\mathcal{L}$. In what follows we shall use the notation $\Delta\mu H = h$. The linear in h contribution for $n \neq 0$ is vanishing, since the expression for $\Delta\mathcal{L}$ is odd with respect to $\zeta = \pm 1$. However, contribution of the ground state ($n = 0$) is nonzero, since the sign is fixed: $\zeta = +1$ (for positively charged femion) or $\zeta = -1$ (for negatively charged femion).

Let us first consider the terms that are proportional to h^2.

After regularization and renormalization by addition of the appropriate counter terms, the final total expression can be found in the following form: [3]

$$\Delta\mathcal{L}^{hq} = -\frac{h^2 eH}{2(2\pi)^2} \int_0^\infty ds \frac{e^{-sm^2}}{s} \left(\coth(eHs) - \frac{1}{eHs} \right) + \frac{h^2}{2(2\pi)^2} \left(\Lambda^2 + m^2 \log \frac{\Lambda^2}{m^2} \right). \tag{19}$$

Consider the contribution from the linear term

$$\Delta\mathcal{L}^{hl} = h \frac{eHm}{(2\pi)^2} \int_{\frac{1}{\Lambda^2}}^\infty \frac{ds}{s} e^{-sm^2} = h \frac{eHm}{(2\pi)^2} \log \frac{\Lambda^2}{m^2}. \tag{20}$$

Recall that $\Delta\mu \simeq \kappa\mu_0$, then one may rewrite the counterterms for $\Delta\mathcal{L}^{hq}$ in the form

$$\frac{h^2}{2(2\pi)^2} \left(\Lambda^2 + m^2 \log \frac{\Lambda^2}{m^2} \right) = \left(\frac{\kappa}{4(2\pi)} \right)^2 (eH)^2 \left(\log \frac{\Lambda^2}{m^2} + \frac{\Lambda^2}{m^2} \right). \tag{21}$$

At the same time, we have for the linear term

[3] This result was obtained earlier in [31], where only terms qudratic in $\Delta\mu$ were observed

DOI: 10.1051/epjconf/201612505005

$$\Delta\mathcal{L}^{hl} = h\frac{eHm}{(2\pi)^2}\int_{\frac{1}{\Lambda^2}}^{\infty}\frac{ds}{s}\,\mathrm{e}^{-sm^2} = h\frac{eHm}{(2\pi)^2}\log\frac{\Lambda^2}{m^2} = \frac{\kappa}{2(2\pi)^2}(eH)^2\log\frac{\Lambda^2}{m^2}. \tag{22}$$

Thus, the divergences in the effective Lagrangian $\Delta\mathcal{L}^h = \Delta\mathcal{L}^{hq} + \Delta\mathcal{L}^{hl}$ could be eliminated by making renormalization for the factor κ in the anomalous magnetic moment expression, yielding the final result

$$\Delta\mathcal{L}^h = -\frac{h^2 eH}{2(2\pi)^2}\int_0^{\infty} ds\frac{\mathrm{e}^{-sm^2}}{s}\left(\coth(eHs) - \frac{1}{eHs}\right). \tag{23}$$

d) Consider the eigenvalue solution $h \neq 0,\ \vec{b} = 0,\ b_0 \neq 0$

In this more general case we shall follow [15] and introduce, for convenience, a mixing angle Θ, and we go over to the effective mass $\widetilde{m}$ and momentum $\widetilde{p}$:

$$\Theta = \arctan\frac{b_0}{h},\quad \begin{pmatrix}\widetilde{m}\\ \widetilde{p}\end{pmatrix} = \begin{pmatrix}\cos\Theta & \sin\Theta\\ -\sin\Theta & \cos\Theta\end{pmatrix}\begin{pmatrix}m\\ p\end{pmatrix},\quad \begin{pmatrix}m\\ p\end{pmatrix} = \begin{pmatrix}\cos\Theta & -\sin\Theta\\ \sin\Theta & \cos\Theta\end{pmatrix}\begin{pmatrix}\widetilde{m}\\ \widetilde{p}\end{pmatrix}.$$

The energy spectrum (8) can be written in these new effective variables as follows:

$$E^2 = (\sqrt{\widetilde{m}^2 + 2eHn} + \widetilde{\mu}H\zeta)^2 + \widetilde{p}^2. \tag{24}$$

We shall name the quantity $\widetilde{\mu}$ as an effective anomalous magnetic moment. Taking the anomaly of the ground state into account, we consider the cases $n \neq 0$ and $n = 0$ separately. In the last case, the spectrum of the ground state is rewritten in terms of the effective variables, as follows:

$$E^2 = (-\widetilde{m} + \widetilde{\mu}H)^2 + \widetilde{p}^2. \tag{25}$$

Expanding expression (9) in small parameter $\widetilde{\mu}H$ and calculating the integral over momentum, the contributions to the effective Lagrangian linear with respect to $\widetilde{\mu}H$ can be written as

$$\Delta\mathcal{L}^{\widetilde{\mu}H,l} = \frac{(eH)(\widetilde{\mu}H)m}{(2\pi)^2}\cos\Theta\int_{\frac{1}{\Lambda^2}}^{\infty}\frac{ds}{s}\,\mathrm{e}^{-sm^2}. \tag{26}$$

This linear in $\widetilde{\mu}H$ contribution has the same divergence as in case **c**) Eq. (20).

After summing the terms of the $\Delta\mathcal{L}$ expansion quadratic in $\widetilde{\mu}H$ including quadratic contributions both from AMM $\Delta\mu$ and the parameter violating Lorentz invariance b, the final renormalized expression for the effective Lagrangian has the form

$$\Delta\mathcal{L}^q = -(\widetilde{\mu}H)^2\frac{eH}{(2\pi)^2}\int_0^{+\infty} ds\,\mathrm{e}^{-m^2 s}\left(m^2\sin^2\Theta + \frac{1}{2s}\cos^2\Theta\right)\left(\coth(eHs) - \frac{1}{eHs}\right), \tag{27}$$

where the renormalization is performed similar to case **c**).

Note that the final expression is correctly transformed into particular cases **c**) ($\mu \neq 0$ and $b = 0$) or **a**) ($b_0 \neq 0$ and $\mu = 0$):

$\underline{\Theta = 0}$ leads to the variant $b = 0$, and $\widetilde{\mu}H = h$

$$\Delta\mathcal{L} = \Delta\mathcal{L}^h = -\frac{1}{2}h^2\frac{eH}{(2\pi)^2}\int_0^{+\infty}\frac{ds}{s}\,\mathrm{e}^{-m^2 s}\left(\coth(eHs) - \frac{1}{eHs}\right). \tag{28}$$

$\underline{\Theta = \frac{\pi}{2}}$ leads to the variant $\mu = 0$, and $\widetilde{\mu}H = b_0$

$$\Delta\mathcal{L} = \Delta\mathcal{L}^b = -b_0^2\frac{eH}{(2\pi)^2}m^2\int_0^{+\infty} ds\,\mathrm{e}^{-m^2 s}\left(\coth(eHs) - \frac{1}{eHs}\right). \tag{29}$$

4 Conclusions

In this report, we calculated the first nonvanishing contributions to the effective Lagrangian, taking into account the interaction of a fermion with an axial vector and an anomalous magnetic moment in an external magnetic field. These parameters may be considered as some sort of condensate values that follow from higher dimensional fundamental interactions. The results of this short report will be extended and generalized in our forthcoming publications to other possible effects of the nontrivial structure of the QFT vacuum and its manifestations, such as the chiral magnetic effects, and others.

References

[1] *Schwinger J.* // Proc. Nat. Acad. Sci. USA, **40**, 132 (1954).
[2] *Ternov I.M., Bagrov V.R., Dorofeyev O.F., Khalilov V.R., Rodionov V.N.* // J. Phys. A: Math. Gen. 1978 **11.** P. 739
[3] *Roberts B. M., Stadnik Y. V., Dzuba V. A., Flambaum V., Leefer N., and Budker D.* // arXiv:1409.2564v2
[4] *Pauli W.* // Rev. Mod. Phys, 1941, **13**, P. 203
[5] *Ternov I.M., Zhukovsky V.Ch., Borisov A.V.* // Quantum processes in a strong external field (in Russian): (Moscow University 1989)
[6] *Shapiro I. L.* // Phys. Rept. 2002 **357**, P. 113
[7] *Ahmadi F., Jalalzadeh S., Sepangi H. R.* // Class. Quantum Grav. 2006 **23.** P. 4069
[8] *Bertolami O., Carvalho C.* // Phys. Rev. D 2006 **74.** P. 084020
[9] *Kostelecky V.A., Samuel S.* // Phys. Rev. D. 1989 **39.** P. 683
[10] *Colladay D., Kostelecky V.A.* // Phys. Rev. D. 1998 **58.** P. 116002
[11] *Jackiw R., Kostelecky V.A.* // Phys. Rev. Lett. 1999 **82.** P. 3572
[12] *Zhukovsky V.Ch., Lobanov A.E., Murchikova E.M.* // Phys. Rev. D. 2006 **73.** P. 065016
[13] *Lobanov A. E., Murchikova E. M.* // Univ. Phys. (2008) 63: 91
[14] *Ebert D., Zhukovsky V. Ch., and Razumovsky A. S.* // Phys. Rev. D 2004 **70**, P. 025003
[15] *Frolov I.E., Zhukovsky V.Ch.* // J. Phys. A. 2007 **40.** P. 10625
[16] *Roberts M., Stadnik Y. V.* // Phys. Rev. D. 2014 **90.** P. 096005
[17] *Fukushima K., Kharzeev D.E., Warringa H.J.* // Phys. Rev. D 2008 **78.** P. 074033
[18] *Chernodub M. N. et.al.* // hep-th 1311.0878v2
[19] *Chernodub M. N., Zubkov M.* // hep-th 1508.03114v1
[20] *Braguta V. et.al.* // EPJ Web of Conferences 95, 03002 (2015)
[21] *Bezerra de Mello E. R.* // JHEP **0406**, 016 (2004).
[22] *Kadyshevsky V. G. and Rodionov V. N.* // Theor. and Math. Phys. 2003 **136** 3 P. 1346
[23] *Gitman D. M. and Saa A. V.* // Class. Quant. Grav. **10**, 1447 (1993).
[24] *Kharlanov O. G., and Zhukovsky V. Ch* // J. of Math.Phys. 2007 **48.9** P. 092302
[25] *Frolov I. E., Zhukovsky V. Ch., Klimenko K. G.* // Phys. Rev. D 2010 **82**, P. 076002
[26] *Sitenko Y.A., Rulik K.Y.* // Eur. Phys. J. 2003 **C28.** P. 405
[27] *Bubnov A. F., Zhukovsky V. CH.* // Moscow Univ. Phys. (2010) 65: 105
[28] *Ternov I.M., Bagrov V.G., Zhukovsky V.Ch.* // Moscow Univ. Bull., 1966. **21**. P. 21.
[29] *Rodionov V. N.*, J. of Exp. and Theor. Phys. 2004. **98**. P. 395
[30] *Fock V. A.* // Sow. Phys. 1937 **12**. P. 404.
[31] *O'Connell R. F.* // Phys. Rev. 1968 **176.** n.5. P. 1433.

EPJ Web of Conferences **125**, 05006 (2016) DOI: 10.1051/epjconf/201612505006
QUARKS-2016

Renormalization group analysis of a turbulent compressible fluid near $d = 4$: Crossover between local and non-local scaling regimes.

N. V. Antonov[1,a], N. M. Gulitskiy[1,b], M. M. Kostenko[1,c], and T. Lučivjanský[2,d]

[1]*Department of Theoretical Physics, Faculty of Physics, Saint-Petersburg State University, 7/9 Universitetskaya nab., St. Petersburg, 199034 Russia*
[2]*Faculty of Sciences, P.J. Šafárik University, Moyzesova 16, 040 01 Košice, Slovakia, and Fakultät für Physik, Universität Duisburg-Essen, D-47048 Duisburg, Germany*

Abstract. We study scaling properties of the model of fully developed turbulence for a compressible fluid, based on the stochastic Navier-Stokes equation, by means of the field theoretic renormalization group (RG). The scaling properties in this approach are related to fixed points of the RG equation. Here we study a possible existence of other scaling regimes and an opportunity of a crossover between them. This may take place in some other space dimensions, particularly at $d = 4$. A new regime may there arise and then by continuity moves into $d = 3$. Our calculations have shown that there really exists an additional fixed point, that may govern scaling behaviour.

1 Introduction

A majority of works on fully developed turbulence is concerned with an incompressible fluid. The renormalization group approach to such problems has been successful in verifying Kolmogorov scaling and provides an efficient tool for a calculation of universal quantities. However, a similar treatment has been only scarcely applied to compressible fluids. In this paper we present an application of the field theoretic renormalization group (RG) onto the scaling regimes of a compressible fluid, whose behavior is governed by a proper generalization of stochastic Navier Stokes equation [1]. Similar models of compressible fluid were considered in [2–4]. In [2] the phenomenological corrections to the Kolmogorov spectrum were verified in the framework of the skeleton equatios for consistency, while the model, considered in [3], appears to be in fact unrenormalizible. All these papers shows us a necessity of the further investigations of compressibility.

Following [5], we employ double expansion scheme. Here the formal expansion parameters are y, which describes the scaling behavior of a random force, and $\varepsilon = 4 - d$, i.e., a deviation from the dimension of space $d = 4$.

[a]e-mail: n.antonov@spbu.ru
[b]e-mail: n.gulitskiy@spbu.ru
[c]e-mail: kontramot@mail.ru
[d]e-mail: tomas.lucivjansky@uni-due.de

2 Description of the model

The Navier-Stokes equation for a compressible fluid can be written in the following form:

$$\rho[\partial_t \boldsymbol{v} + (\boldsymbol{v}\cdot\boldsymbol{\nabla})\boldsymbol{v}] = \nu_0[\boldsymbol{\nabla}^2\boldsymbol{v} - \boldsymbol{\nabla}(\boldsymbol{\nabla}\cdot\boldsymbol{v})] + \mu_0\boldsymbol{\nabla}(\boldsymbol{\nabla}\cdot\boldsymbol{v}) - \boldsymbol{\nabla}p + \boldsymbol{f}, \tag{1}$$

where ρ is the fluid density, $\boldsymbol{v}$ is the velocity field, ∂_t is a time derivative $\partial/\partial t$, $\boldsymbol{\nabla}^2$ is the Laplace operator, ν_0 and μ_0 are molecular viscosity coefficients, p is pressure field, and $\boldsymbol{f}$ is an external field per unit mass. The model must be augmented by two additional equations, namely a continuity equation and an equation of state between deviations δp and $\delta\rho$ from the equilibrium values. They read

$$\partial_t\rho + \boldsymbol{\nabla}\cdot(\rho\boldsymbol{v}) = 0; \tag{2a}$$

$$\delta p = c_0^2\delta\rho. \tag{2b}$$

In order to obtain the renormalizable field theoretic model expression (1) is divided by ρ, and fluctuations in viscous terms are neglected [6]. Further, by using the continuity equation and the equation of state (2), the problem can be recasted in terms of two coupled equations:

$$(\partial_t + \boldsymbol{v}\cdot\boldsymbol{\nabla})\boldsymbol{v} = \nu_0[\boldsymbol{\nabla}^2\boldsymbol{v} - \boldsymbol{\nabla}(\boldsymbol{\nabla}\cdot\boldsymbol{v})] + \mu_0\boldsymbol{\nabla}(\boldsymbol{\nabla}\cdot\boldsymbol{v}) - \boldsymbol{\nabla}\phi + \boldsymbol{f}; \tag{3a}$$

$$(\partial_t + \boldsymbol{v}\cdot\boldsymbol{\nabla})\phi = c_0^2\ln(\rho/\overline{\rho}). \tag{3b}$$

Here ϕ is related to the density fluctuations via the relation $\phi = c_0^2\ln(\rho/\overline{\rho})$. Parameter c_0 is an adiabatic speed of sound, $\overline{\rho}$ denotes the mean value of ρ.

The turbulence is modeled by an external force – it is assumed to be a random variable, which should mimic the input of the energy into the system from the outer large scale L. Its precise form is believed to be unimportant and is usually considered to be a random Gaussian variable with zero mean and correlator

$$\langle f_i(t,\boldsymbol{x})f_j(t',\boldsymbol{x}')\rangle = \frac{\delta(t-t')}{(2\pi)^d}\int_{k>m} \mathrm{d}^d\boldsymbol{k}\, D_{ij}(\boldsymbol{k})\mathrm{e}^{i\boldsymbol{k}\cdot(\boldsymbol{x}-\boldsymbol{x}')}, \quad \text{where} \tag{4a}$$

$$D_{ij}(\boldsymbol{k}) = g_{10}\nu_0^3 k^{4-d-y}\Big\{P_{ij}(\boldsymbol{k}) + \alpha Q_{ij}(\boldsymbol{k})\Big\}. \tag{4b}$$

Here d is the space dimension, $P_{ij}(\boldsymbol{k}) = \delta_{ij} - k_ik_j/k^2$ and $Q_{ij}(\boldsymbol{k}) = k_ik_j/k^2$ are the transverse and longitudinal projectors, $k = |\boldsymbol{k}|$, a parameter $m = L^{-1}$ provides an infrared (IR) cutoff, amplitude α is a free parameter ($\alpha \neq 0$ corresponds to a compressible fluid, while $\alpha = 0$ provides us an incompressible fluid case), an exponent y plays a role of a formally small expansion parameter, and g_{10} is a coupling constant; Dirac delta function ensures Galilean invariance [7].

3 Field theoretic formulation of the model

According to the general theorem [8, 9], the stochastic problem is equivalent to the field theoretic model with a doubled set of fields $\tilde{\psi}, \psi$ and de Dominicis-Janssen action functional, written in a compact form as

$$\mathcal{S}(\varphi) = \frac{v_i' D_{ik}^f v_k'}{2} + v_i'\Big\{-\partial_t v_i - v_j\partial_j v_i + \nu_0[\delta_{ik}\partial^2 - \partial_i\partial_k]v_k + u_0\nu_0\partial_i\partial_k v_k - \partial_i\phi\Big\}$$
$$+ \phi'[-\partial_t\phi + v_j\partial_j\phi + v_0\nu_0\partial^2\phi - c_o^2(\partial_i v_i)]. \tag{5}$$

Here we have employed a condensed notation, in which integrals over the spatial variable **x** and the time variable t, as well as summation over repeated indices, are implicitly assumed. The action (5) is amenable to the standard methods of the quantum field theory, such as the Feynman diagrammatic technique and the renormalization group procedure.

In a standard approach, if we apply quantum field methods to the stochastic differential equations, the space dimension d plays a passive role and an actual perturbative parameter is y; for more details see the monographs [7, 8]. Our approach closely follows the analysis of the incompressible Navier-Stokes equation near space dimension $d = 2$ (see [5, 10–12]). In this case three additional divergences appear in the Green's function $v'v'$. They can be absorbed by a suitable local counterterm $v_i'\nabla^2 v_i'$, and a regular expansion in both y and $\varepsilon' = d-2$ was constructed. Up to now the present model (5) has been investigated at the fixed space dimension $d = 3$, for which the action (5) contains all terms that can be generated during the renormalization procedure [1, 13–15]. However, using the dimensional analysis it can be shown that at $d = 4$ there appears an additional divergence, also in the Green's function $v'v'$. Therefore, to keep the model renormalizable at $d = 4$ the kernel function in (4) has to be generalized to the following form:

$$D_{ij}(\boldsymbol{k}) \to g_{10}\nu_0^3 k^{4-d-y}\Big\{P_{ij}(\boldsymbol{k}) + \alpha Q_{ij}(\boldsymbol{k})\Big\} + g_{20}\nu_0^3\delta_{ij}, \tag{6}$$

where the new term on the right hand side absorbs divergent contributions from $v'v'$. In contrast to [5] no momentum dependence is needed.

4 Feynman diagrammatic technique

The perturbation theory of the model can be expressed in the standard Feynman diagrammatic expansion [8, 16]. Bare propagators are read off from the inverse matrix of the Gaussian (free) part of the action functional, while the nonlinear part of the differential equation defines the interaction vertices. Their graphical representation is depicted in Fig. 1. Explicit expressions of propagators in frequency-momentum representation can be found, e.g., in [1], and they are right for actual calculations.

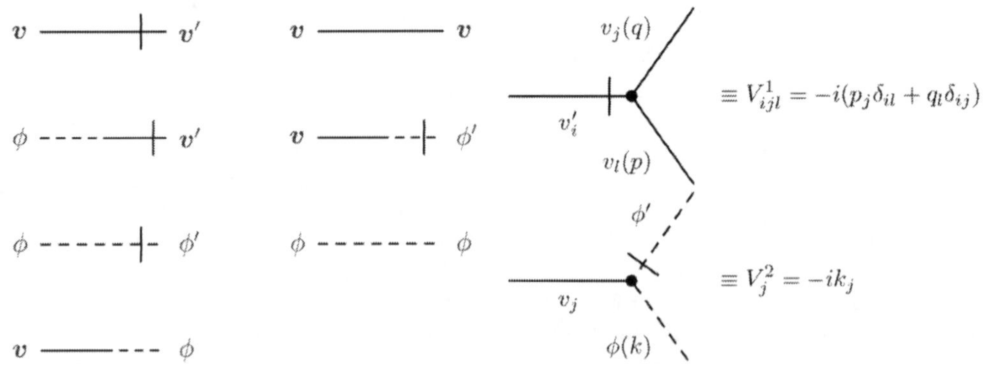

Figure 1. Graphical representation of the bare propagators and interaction vertices in the model (5)

DOI: 10.1051/epjconf/201612505006

The ultraviolet (UV) renormalizability is very efficiently revealed by an analysis of the 1-irreducible Green's functions. Corresponding generating functional can be written in the form

$$\Gamma(\varphi) = \mathcal{S}(\varphi) + \widetilde{\Gamma}(\varphi), \tag{7}$$

where for the functional arguments we have used the same symbols $\varphi = \{\boldsymbol{v}, \boldsymbol{v}', \phi, \phi'\}$ as for the corresponding random fields [8]; $\mathcal{S}(\varphi)$ is the action functional (5) and $\widetilde{\Gamma}(\varphi)$ is the sum of all the 1-irreducible diagrams with loops [8]. As it has been shown in [1] and discussed in [14], the model (5) – (6) is invariant with respect to the Galilean symmetry, which results to the UV finitness of the two Green's functions: $v_i\partial_t v_i$ and $v_i'(v_j\partial_j)v_i$. We have carried on the perturbative analysis in the one-loop order, consequently the expressions for the 1-irreducible Green's functions, which requires UV renormalization, can be formally written in the following way:

$$\Gamma_{v'v} = i\omega - (\delta_{ij}p^2 - p_ip_j)Z_1\nu - p_ip_jZ_2u\nu + \quad , \tag{8}$$

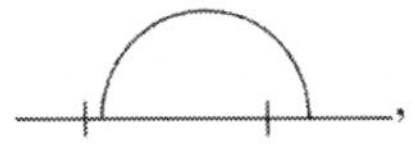

$$\Gamma_{\phi\phi'} = i\omega - p^2Z_3v\nu + \quad , \tag{9}$$

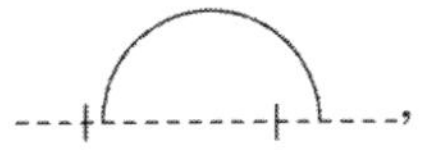

$$\Gamma_{v'\phi} = -iZ_4p_i + \quad , \tag{10}$$

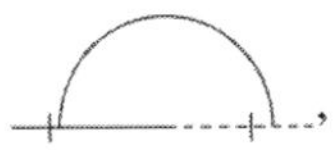

$$\Gamma_{\phi'v} = -iZ_5p_ic^2 + \quad + \quad + \quad , \tag{11}$$

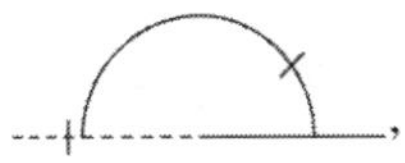

$$\Gamma_{v'v'} = g_1\nu^3p^{4-d-y}Z_6\Big\{P_{ij}(\boldsymbol{p}) + \alpha Q_{ij}(\boldsymbol{p})\Big\} + g_2\nu^3\delta_{ij}Z_7 + \frac{1}{2} \quad , \tag{12}$$

where **p** always represents a corresponding external momentum. A factor $1/2$ in front of the diagram in (12) denotes a symmetry coefficient of the given graph. Collecting all the mentioned facts and taking into account that non-local terms should not be renormalized, it is straightforward to show that the theory is UV renormalizable. From the direct comparison of the relations between renormalized parameters it follows that

$$\begin{array}{llll} Z_\nu = Z_1, & Z_{g_1} = Z_1^{-3}, & Z_u = Z_2Z_1^{-1}, & Z_\phi = Z_4, \\ Z_{\phi'} = Z_4^{-1}, & Z_v = Z_3Z_1^{-1}, & Z_c = (Z_4Z_5)^{1/2}, & Z_{g_2} = Z_6Z_1^{-3}. \end{array} \tag{13}$$

Employing dimensional regularization within minimal subtraction scheme (MS) [17] the renormalization constants can be calculated and the UV divergences manifests themselves in pole terms in y and $\varepsilon = 4-d$. In higher loops pole terms in form of general linear combination in $ay + b\varepsilon$ may appear.

5 UV renormalization of the model and fixed points

The large scale behavior with respect to spatial and time scales is governed by the IR attractive stable fixed point $g^* \equiv \{g_1^*, g_2^*, u^*, v^*\}$. Here and henceforth the asterisk refers to a coordinate of the fixed point (FP). Their coordinates are determined from the relations [8, 16]

$$\beta_{g_1}(g^*) = \beta_{g_2}(g^*) = \beta_u(g^*) = \beta_v(g^*) = 0, \tag{14}$$

where $\beta_x = \widetilde{D}_\mu x$ for any variable x, and differential operator $\widetilde{D}_\mu$ denotes operation $\mu\partial_\mu$ at fixed bare parameters $\{g_{10}, g_{20}, u_0, v_0, \nu_0, \alpha_0, c_0\}$; μ is the "reference mass" (additional free parameter of the renormalized theory) in the MS renormalization scheme. The eigenvalues of the matrix of first derivatives

$\Omega_{ij} \equiv \partial\beta_i/\partial g_j$, where $i, j \in \{g_1, g_2, u, v\}$, determine whether the given FP is IR stable or not. Points with all positive eigenvalues are candidates for macroscopic regimes and in principle can be observed experimentally. An explicit forms of the β-functions are

$$\beta_{g_1} = g_1(-y - \gamma_{g_1}), \quad \beta_{g_2} = g_2(-\varepsilon - \gamma_{g_2}), \quad \beta_u = -u\gamma_u, \quad \beta_v = -v\gamma_v, \tag{15}$$

where $\gamma_x = \widetilde{D}_\mu \ln Z_x$ are the anomalous dimensions [8]. A direct analysis of the system of equations (14) reveals the existence of three IR stable fixed points: FPI, FPII and FPIII. FPI is the free (Gaussian) fixed point, for which all interactions are irrelevant and no scaling and universality is expected. Its coordinates are

$$g_1^* = 0, \quad g_2^* = 0, \quad \text{whereas} \quad u^* \quad \text{and} \quad v^* \quad \text{are undetermined.} \tag{16}$$

The corresponding eigenvalues of the matrix Ω are

$$\lambda_1 = 0, \quad \lambda_2 = 0, \quad \lambda_3 = -\varepsilon, \quad \lambda_4 = -y. \tag{17}$$

Though trivial, this point is necessary for the correct use of perturbative renormalization group.

Further, there is a local fixed point FPII, for which the charge g_2 attains a non-zero value, and corresponding coordinates are

$$g_1^* = 0, \quad g_2^* = \frac{8\varepsilon}{3}, \quad u^* = 1, \quad v^* = 1. \tag{18}$$

The eigenvalues of the matrix Ω are

$$\lambda_1 = \frac{7\varepsilon}{18}, \quad \lambda_2 = \frac{5\varepsilon}{6}, \quad \lambda_3 = \varepsilon, \quad \lambda_4 = \frac{3\varepsilon - 2y}{2}. \tag{19}$$

For the last fixed point, FPIII, both non-local and local parts of the random force are relevant:

$$g_1^* = \frac{16y(2y - 3\varepsilon)}{9(y(2+\alpha) - 3\varepsilon)}, \quad g_2^* = \frac{16\alpha y^2}{9(y(2+\alpha) - 3\varepsilon)}, \quad u^* = 1, \quad v^* = 1; \tag{20}$$

the required eigenvalues are

$$\lambda_1 = \frac{y[2y(10\alpha + 11) - 3\varepsilon(3\alpha + 11)]}{54[y(2+\alpha) - 3\varepsilon]}, \quad \lambda_2 = \frac{y[2y(2\alpha + 3) - \varepsilon(\alpha + 9)]}{6[y(\alpha + 2) - 3\varepsilon]}, \quad \lambda_{3,4} = \frac{A \pm \sqrt{B}}{C}, \tag{21}$$

where A, B and C are given by the following expressions:

$$A = -27\varepsilon^3 + 9(9+\alpha)\varepsilon^2 y - 9(8 + 3\alpha)\varepsilon y^2 + 2y^3(\alpha^2 + 7\alpha + 10); \tag{22}$$

$$B = [-3\varepsilon + (2+\alpha)y]^2[81\varepsilon^4 - 54\varepsilon^3 y - 9(3 + 20\alpha)\varepsilon^2 y^2 + 12(1 + 17\alpha + 3\alpha^2)\varepsilon y^3 - 4(-1 + 14\alpha + 5\alpha^2)y^4]; \tag{23}$$

$$C = 6[-3\varepsilon + (2+\alpha)y]^2. \tag{24}$$

From the physical interpretation of the kernel function (4) it follows that the charges g_1^* and g_2^* can not attain negative values. Using this fact together with an explicit form of the eigenvalues $\lambda_1 \ldots \lambda_4$ it can be shown, that the point FPIII is stable for $y > 0$ and $y > 3\varepsilon/2$. Note, that the crossover between two nontrivial points happens along the line $y = 3\varepsilon/2$, which is in accordance with [18].

6 Conclusion

In this paper the compressible extension of the stochastic Navier Stokes equation has been studied using the field theoretic approach. Crucial points of the Feynman diagrammatic technique and perturbative renormalization group have been discussed. One loop approximation provides that, depending of the exponent y and deviation from the dimension of $\boldsymbol{x}$ space $\varepsilon = 4 - d$, the model possesses three stable fixed points in the IR region (i.e., three possible scaling regimes) – trivial (Gaussian, FPI), local (FPII) and nonlocal (FPIII).

This shows us, that the simple analysis around $d = 3$, which indicates existence of only one nontrivial fixed point [1], is incomplete in this case.

Acknowledgements

The authors are indebted to M. Yu. Nalimov, L. Ts. Adzhemyan, M. Hnatič, J. Honkonen, and V. Škultéty for discussions.

The work was supported by the Saint Petersburg State University within the research grant 11.38.185.2014 and by the Russian Foundation for Basic Research within the Project 16-32-00086. N. M. G. and M. M. K. were also supported by the Dmitry Zimin's "Dynasty" foundation.

References

[1] N. V. Antonov, M. Yu. Nalimov and A. A. Udalov, Theor. Math. Phys. **110**, 305 (1997)

[2] V. S. L'vov and A. V. Mikhailov, Preprint No. 54, Inst. Avtomat. Electron., Novosibirsk (1977)

[3] I. Staroselsky, V. Yakhot, S. Kida, and S. A. Orszag, Phys. Rev. Lett., **65**, 171 (1990)

[4] S. S. Moiseev, A. V. Tur, and V. V. Yanovskii, Sov. Phys. JETP **44**, 556 (1976)

[5] J. Honkonen and M. Yu. Nalimov, Z. Phys. B **99**, 297 (1996)

[6] D. Yu. Volchenckov and M. Yu. Nalimov, Theor. Math. Phys. **106**, 307 (1996)

[7] L. Ts. Adzhemyan, N. V. Antonov, A. N. Vasil'ev: *The Field Theoretic Renormalization Group in Fully Developed Turbulence* (Gordon & Breach, London, 1999)

[8] A. N. Vasil'ev, *The Field Theoretic Renormalization Group in Critical Behavior Theory and Stochastic Dynamics* (Boca Raton, Chapman Hall/CRC, 2004)

[9] U. Täuber, *Critical Dynamics: A Field Theory Approach to Equilibrium and Non-Equilibrium Scaling Behavior* (Cambridge University Press, New York, 2014)

[10] L. Ts. Adzhemyan, J. Honkonen, M. V. Kompaniets, A. N. Vasil'ev, Phys. Rev. E **68**, 055302(R) (2003)

[11] L. Ts. Adzhemyan, J. Honkonen, M. V. Kompaniets, A. N. Vasil'ev, Phys. Rev. E **71**, 036305 (2005)

[12] L. Ts. Adzhemyan, M. Hnatich and J. Honkonen, Eur. Phys. J B **73**, 275 (2010)

[13] L. Ts. Adzhemyan, M. Yu. Nalimov and M. M. Stepanova, Theor. Math. Phys. **104**, 971 (1995)

[14] N. V. Antonov and M. M. Kostenko, Phys. Rev. E **90**, 063016 (2014)

[15] N. V. Antonov and M. M. Kostenko, Phys. Rev. E **92**, 053013 (2015)

[16] J. Zinn-Justin, *Quantum Field Theory and Critical Phenomena* (4th edition, Oxford University Press, Oxford, 2002)

[17] N. V. Antonov, N. M. Gulitskiy, M. M. Kostenko and T. Lučivjanský, in preparation

[18] N. V. Antonov, Phys. Rev. Lett. **92**, 161101 (2004)

Shape invariance and SUSY separation of variables

M.V. Ioffe[1a], E.V. Kolevatova[1b], and V. Vereshagin[1c]

[1] *Saint-Petersburg State University, 198504 St.-Petersburg, Russia*

Abstract. The main ingredients of conventional Supersymmetrical Quantum Mechanics (SUSY QM) are presented. The generalization with supercharges of second order in derivatives - Second Order SUSY - is formulated, and the property of shape invariance is defined. The generalization to two-dimensional coordinate space, after using just these two elements of the modern SUSY QM approach, provides the opportunity to solve analytically some two-dimensional problems. Two different procedures of supersymmetrical separation of variables are formulated. They are illustrated by two-dimensional generalization of the Morse model.

1 Introduction

The idea of Supersymmetry (SUSY) appeared in the framework of Elementary Particle Theory and Quantum Field Theory in the papers [1]. The main motivation was to extend the Poincare algebra by nontrivial unification of space-time and internal symmetries, i.e. to overcome the limitation of "no-go theorem" [2]. New symmetry was named "super" because of its several exclusive physical properties: unification of internal symmetries with space-time symmetries; new kind of algebra which includes both commutators and anticommutators; inclusion of bosons and fermions into the same multiplets.

One-dimensional SUSY QM was proposed by E.Witten [3] as a toy model of (0+1) Quantum Field Theory which was useful to study the problems of spontaneous breaking of SUSY. The corresponding Lagrangian is:

$$L = \dot{x}^2(t) - \left[(W'(x(t))^2 - W''(x(t))\right] - 2W''(x(t))\psi^+\psi^-,$$

where x is a coordinate, $\psi^\pm$ are fermionic variables, and $W(x)$ - superpotential defining the interaction. In the framework of this model, the criterion of spontaneous breaking was formulated in terms of the so called Witten index.

Very soon, it became clear that SUSY QM provides the really new approach to many problems of Quantum Mechanics itself [4], [5]. Now, it is one of the most effective approaches for analytical methods in Quantum Mechanics. The new notion of shape invariance [6] formulated in SUSY QM has to be noticed here since it allows to solve problems in a pure algebraic way. A new impetus was given to the development of multidimensional $d > 1$ Quantum Mechanics [5]. In particular, before this time the most known and most effective regular approach for study of $d = 2$ systems was the separation of variables. It reduces the multidimensional (multiparticle) problem to a few problems of

[a]E-mail: m.ioffe@spbu.ru
[b]E-mail: e.v.kolev@yandex.ru
[c]E-mail: vvv@av2467.spb.edu

lower dimensionality of space. SUSY QM provides the new way for study of 2-dim models which are not amenable to standard separation of variables: several two-dimensional models not amenable to conventional separation of variables were shown to be exactly solvable [7].

2 SUSY Quantum Mechanics of E.Witten

The simplest superalgebra includes three elements - superHamiltonian $\widehat{H}$ and mutually conjugate supercharges $\widehat{Q}^{+}$, $\widehat{Q}^{-} = (\widehat{Q}^{+})^{\dagger}$ - with the following superalgebra:

$$[\widehat{H}, \widehat{Q}^{+}] = [\widehat{H}, \widehat{Q}^{-}] = 0; \{\widehat{Q}^{+}, \widehat{Q}^{-}\} = \widehat{H}; \{\widehat{Q}^{\pm}, \widehat{Q}^{\pm}\} = 0$$

This algebra admit the simplest realization by 2×2 matrix differential operators in one variable:

$$\widehat{H} = \begin{pmatrix} h^{(0)}(x) & 0 \\ 0 & h^{(1)}(x) \end{pmatrix}; \widehat{Q}^{+} = \begin{pmatrix} 0 & 0 \\ q^{-} & 0 \end{pmatrix};$$
$$\widehat{Q}^{-} = \begin{pmatrix} 0 & q^{+} \\ 0 & 0 \end{pmatrix}$$

where

$$h^{(i)} = -\partial^2 + V^{(i)}(x), \quad q^{\mp} \equiv \pm\partial + \partial W(x); \quad \partial \equiv d/dx.$$

The nilpotency of $\widehat{Q}^{\pm}$ is ensured by their matrix structure. Anticommutator of $\widehat{Q}^{\pm}$ gives $\widehat{H}$ if $h^{(0)}$, $h^{(1)}$ are factorized:

$$h^{(0)} = q^{+}q^{-}; \quad h^{(1)} = q^{-}q^{+}.$$

This factorization ensures also the commutation of $\widehat{H}$ with $\widehat{Q}^{\pm}$, which in components takes the form of the intertwining relations:

$$h^{(0)}q^{+} = q^{+}h^{(1)}; \quad q^{-}h^{(0)} = h^{(0)}q^{-},$$

which are the most important relations of SUSY QM approach. In components, the above relations of SUSY algebra reproduce the well known Factorization Method of Schrödinger [8].

One has almost coinciding spectra of $h^{(0)}$ and $h^{(1)}$ or, equivalently, double degeneracy of the energy spectrum of $\widehat{H}$:

$$h^{(0)}\Psi_n^{(0)} = E_n\Psi_n^{(0)}; \ \Psi_n^{(1)} = q^{-}\Psi_n^{(0)}; \ n = 0, 1, ...$$

$$h^{(1)}\Psi_n^{(1)} = q^{-}q^{+}q^{-}\Psi_n^{(0)} = E_nq^{-}\Psi_n^{(0)} = E_n\Psi_n^{(1)}$$

$$\Psi_n^{(0)}(x) = q^{+}\Psi_n^{(1)}(x); \ n = 0, 1, ...$$

Although SUSY QM explores the typical terminology of supersymmetrical models in Elementary Particle Physics: supersymmetry, superpartners, boson-fermion's multiplets, supertransformations etc., the physical meaning of many terms may be very different.

3 Shape invariance and exact solvability

The notion of shape invariance appeared [6] for the first time in d=1 SUSY QM. This property provides very elegant approach to solve a variety of Schrödinger equations in a pure algebraic way. Let $h^{(0)}$, $h^{(1)}$ depend not only on x, but also on some parameter a. Let both Hamiltonians participate in the intertwining relations given above. Let the dependence of Hamiltonians on parameter a obeys the property of "shape invariance":

$$h^{(0)}(x;a) = h^{(1)}(x;\tilde{a}) + R(a); \quad \tilde{a} = \tilde{a}(a),$$

where the shift $R(a)$ does not depend on coordinate x. If the ground state of $h^{(1)}(x;a)$ is fixed as $E_0^{(1)}(a) = 0$, and if its zero mode is known:

$$q^+(x;a)\Psi_0^{(1)}(x;a) = 0,$$

the entire spectral problem for $h^{(1)}(x;a)$ is solved algebraically due to intertwining relations. In particular, the first excited state is:

$$h^{(1)}(x;a)(q^-(x;a)\Psi_0^{(1)}(x;\tilde{a})) = R(a)(q^-(x;a)\Psi_0^{(1)}(x;\tilde{a}))$$

Analogously, the higher excited states can be built by the action of the next q^-:

$$h^{(1)}(x;a)\Big(q^-(x;a)q^-(x;\tilde{a})\Psi_0^{(1)}(x;\tilde{\tilde{a}})\Big) =$$

$$= (R(a) + R(\tilde{a}))\Big(q^-(x;a)q^-(x;\tilde{a})\Psi_0^{(1)}(x;\tilde{\tilde{a}})\Big).$$

As a remark, all well known exactly solvable potentials are shape-invariant. And vice a versa, under some natural conditions on the superpotentials, shape invariant potentials give just a list of known exactly solvable models [9]. Nevertheless, the intertwining relations are the source of new solvable potentials - superpartners of the old ones.

A generalization of shape invariance in $d = 1$ is based on the polynomial SUSY QM [10] - the special deformation of Witten's SUSY algebra. Briefly speaking, if one'll choose intertwining operators (components of supercharge) to be of second order:

$$q^+ \equiv \partial^2 + f(x)\partial + b(x),$$

the SUSY algebra takes the same Witten's form, but the anticommutator of supercharges becomes now second order polynomial in H

$$\{\widehat{Q}^+, \widehat{Q}^-\} = \widehat{H}^2 + \alpha\widehat{H} + \beta$$

It is important that not all such $q^{\pm}$ are factorizable, and therefore some new solutions of intertwining relations can be built. The shape invariance property with these supercharges also leads to a new class of potentials:

$$V^{(0,1)}(x) = \frac{2c(2a \mp 1)}{3x^{2/3}} - \frac{5}{36x^2} + c^2x^{2/3}.$$

The Schrödinger equation with this potential was solved analytically, it turned out [11] to be an implicitly solvable model.

4 Two-dimensional generalizations of SUSY QM

The direct multidimensional generalization was built in a series of papers [12] by another choice of realization for the elements $\widehat{H}$, $\widehat{Q}^+$, $\widehat{Q}^-$ of SUSY algebra. It was made in two ways: by handmade extension of 1-dim formulas, and by means of the superfield approach of Quantum Field Theory. The main idea of the latter was to consider the model of scalar real multicomponent superfield:

$$\Phi_k(t,\theta,\bar{\theta}) = x_k(t) + i(\bar{\theta}\psi_k(t) + \theta\bar{\psi}_k(t)) + \frac{1}{2}\bar{\theta}\theta F_k(t),$$

where the number of components $k = 1, 2, ..., d$. It was shown that in this case the superHamiltonian $\widehat{H}$ includes a chain of matrix components of the Schrödinger form but with matrix potential. The intertwining relations between matrix Hamiltonians lead to interrelations between spectra of neighbouring components of the Superhamiltonian.

Matrix Hamiltonians occur in different branches of Physics: particles with spin, coupled channels systems etc. The interesting results were obtained, such as the spectrum of Pauli operator for spin $1/2$ particle in a class of nonhomogeneous external electromagnetic field [13] and the relations between amplitudes and phases of $NN-$ and $N\bar{N}-$scattering in triplet and singlet states [14].

Is it possible to avoid matrix Hamiltonians from the $d-$dimensional SUSY QM ? Any attempt to intertwine two scalar Hamiltonians by means of first order operators $q^{\pm}$ leads to potentials with standard separation of variables. The polynomial SUSY QM with second order supercharges turned out to be fruitful in construction of 2-dim SUSY QM without matrix Hamiltonians at all.

Concerning solvability of two-dimensional quantum models, the only regular way is the standard separation of variables:

$$h(\vec{x}) = h_1(x_1) + h_2(x_2); \quad \Psi(\vec{x}) = \sum c_{ik}\psi_{1i}(x_1)\cdot\psi_{2k}(x_2).$$

The complete classification of such models was given by L.P.Eisenhart [15]: separation is possible in the following coordinates - Cartesian, polar, elliptic and parabolic. The general form of potentials amenable to separation is known explicitly up to arbitrary functions of one variable.

It is very interesting to expand the list of solvable two-dimensional models beyond the class of models with separation of variables. Two-dimensional polynomial SUSY Quantum Mechanics [16] is useful to find new solvable systems. The main idea is to intertwine scalar Hamiltonians $h^{(0)}(\vec{x})$, $h^{(1)}(\vec{x})$ by second order operators:

$$h^{(0)}q^+ = q^+h^{(1)}; \quad q^-h^{(0)} = h^{(1)}q^-$$

Here, intertwined Hamiltonians have still the Schrödinger form:

$$h^{(0,1)} = -\partial_i^2 + V^{(0,1)}(\vec{x}),$$

but intertwining operators have the most general second order form now:

$$q^+ = g_{ik}(\vec{x})\partial_i\partial_k + C_i(\vec{x})\partial_i + B(\vec{x}); \quad q^- \equiv (q^+)^{\dagger}.$$

The intertwining relations are equivalent to a complicate system of nonlinear second order differential equations for functions $g_{ik}(\vec{x})$, $C_i(\vec{x})$, $B(\vec{x})$, and potentials $V^{(0,1)}(\vec{x})$. Finally, with some ansatzes it was reduced to the functional-differential equation, which does not admit any regular procedure of solution.

Nevertheless, some particular solutions were found for Lorentz form of metric in supercharges $g_{ik}(\vec{x}) = diag(1,-1)$, when the system is simplified essentially. Wide list of such pairs of potentials was found in a series of papers [16], [7].

5 SUSY QM and new separation of variables: $d = 2$

The polynomial two-dimensional SUSY QM allowed to build two new procedures [17], [18], [7], [5] for solution of the Schrödinger equation which can be useful when the standard separation of variables does not work:
1) Partial solvability, or quasi-exact solvability (QES) in another terminology;
2) Complete solvablity;
Both methods were called as supersymmetrical separations of variables.

The first variant is realized when both Hamiltonians $h^{(0,1)}(\vec{x})$ DO NOT allow separation of variables, but the supercharge $q^{+}(\vec{x})$ DOES allow. In this case, the general scheme is the following. Let's suppose that we know zero modes of q^{+} :

$$q^{+}\Omega_n(\vec{x}) = 0; \quad n = 0, 1, ..., N; \quad q^{+}\vec{\Omega}(\vec{x}) = 0.$$

The important property of intertwining relations $h^{(0)}q^{+} = q^{+}h^{(1)}$ is that the space of zero modes is closed under the action of $h^{(1)}$:

$$h^{(1)}\vec{\Omega}(\vec{x}) = \hat{C}\vec{\Omega}(\vec{x}),$$

where $\hat{C}$ is the constant matrix. If this matrix can be diagonalized, i.e.

$$\hat{B}\hat{C} = \hat{\Lambda}\hat{B}; \quad \hat{\Lambda} = diag(\lambda_0, \lambda_1, ..., \lambda_N),$$

we obtain the set of eigenvalues of $h^{(1)}$ algebraically:

$$h^{(1)}(\hat{B}\vec{\Omega}(\vec{x})) = \hat{\Lambda}(\hat{B}\vec{\Omega}(\vec{x})).$$

To realize this scheme, we have to find zero modes $\Omega_n(\vec{x})$ and to find constant matrix B diagonalizing $\hat{C}$.

For the Lorentz metric in supercharge, the special similarity (not unitary) transformation allows to separate variables in operator q^{+}. By this reason the following procedure is called as SUSY-separation of variables. Similarly to the standard separation, the next task is to solve two one-dimensional problems. Fortunately, for several problems this is possible. The problem of diagonalization of matrix $\hat{C}$ was solved as well by some specific procedure [17]. It is clear in advance that there is no reasons to expect exact solvability of the problem by this procedure. This is quasi-exact solvability (QES) when only a part of spectrum and corresponding wave functions can be built.

The applicability of the full scheme to the concrete models depends on analytical solvability of two one-dimensional problems. A good example - generalized 2-dim Morse potential:

$$\begin{aligned} V &= \alpha^2 a(2a \mp 1)\sinh^{-2}\left(\frac{\alpha x_-}{2}\right) + \\ &+ A\left[e^{-2\alpha x_1} - 2e^{-\alpha x_1} + e^{-2\alpha x_2} - 2e^{-\alpha x_2}\right]. \end{aligned}$$

The model is called as two-dimensional Morse due to its form:

$$V(\vec{x}) = V_{Morse}(x_1) + V_{Morse}(x_2) + v(x_1, x_2),$$

where the first two terms have the well known one-dimensional Morse form. In this case, separation of variables in q^{+} leads to a pair of one-dimensional Morse problems, which are exactly solvable. The matrix $\hat{C}$ was found explicitly - it turned out to be triagonal. Therefore, the energy eigenvalues coincide with its elements. Construction of the wave functions is a much more difficult task, but it was solved as well [17]. Each wave function constructed by SUSY-separation produces a set of additional wave functions due to shape invariance property of the model:

$$h^{(0)}(\vec{x};a) = h^{(1)}(\vec{x};\tilde{a}) + \mathcal{R}(a);\ \tilde{a} = a - 1/2;$$
$$\mathcal{R}(a) = \alpha^2(4a-1).$$

The same two-dimensional Morse potential is a good illustration of the second procedure of SUSY-separation of variables [18], [7], [5]. Let's suppose that, due to the specific choice of parameters of the problem, one of the intertwined Hamiltonians DOES allow the standard separation of variables, but its superpartner still DOES NOT. In our case, $h^{(1)}$ allows separation if $a = -1/2$, when the mixed term in potential vanishes. Due to intertwining, wave functions $\Psi^{(0)}$ of $h^{(0)}$ are obtained as usual:

$$\Psi^{(0)} = q^{+}\Psi^{(1)}.$$

Because of nontrivial action of supercharges, which are the second order differential operators, some additional wave functions could exist, in general. But this possibility is not realized in our case. The whole spectrum of $h^{(0)}$ and corresponding normalized wave functions were found for two-dimensional Morse potential. Due to shape invariance, again the result can be extended to all negative half-integer values of a.

Both procedures of of SUSY-separation of variables described above work also for two-dimensional Pöschl-Teller [19] potential and for two-dimensional Scarf II [20] potential.

Acknowledgements

The work of E.V.K. was partially supported by the Saint Petersburg State University grant N 11.38.223.2015. The work of V.V.V. was partially supported by the Saint Petersburg State University grant N 11.38.189.2014.

References

[1] Y.A. Gol'fand, E.P. Likhtman, JETP Letters **13** 323 (1971); J.-L. Gervais, B. Sakita, Phys. Rev. D **4** 2291 (1971); P. Ramond, Phys. Rev. D **3** 2415 (1971); A. Neveu, J. Schwarz, Nucl. Phys. B **31** 86 (1971); D.V. Volkov, V.P. Akulov, Phys. Lett. B **46** 109 (1973); J. Wess, B. Zumino, Nucl. Phys. B **70** 39 (1974);

[2] S. Coleman, J. Mandula, Phys. Rev. **159** 1251 (1967).

[3] E. Witten, Nucl. Phys. B, **188** 513 (1981).

[4] F. Cooper, A. Khare, U. Sukhatme, Phys. Rep., **25** 268 (1995); B.K. Bagchi, *Supersymmetry in Quantum and Classical Mechanics* (Chapman, Boca Raton, 2001); D.J. Fernandez C, AIP Conf. Proc., **1287** 3 (2010).

[5] A.A. Andrianov, M.V. Ioffe, J. Phys. A **45** 503001 (2012).

[6] L.E. Gendenshtein, JETP Lett. **38** 356 (1983).

[7] M.V. Ioffe, J. Phys. A **37** 10363 (2004); M.V. Ioffe, SIGMA **6** 075 (2010).

[8] L. Infeld, T.E. Hull, Rev. Mod. Phys. **23** 21 (1951).

[9] J. Bougie, A. Gangopadhyaya, J.V. Mallow, Phys. Rev. Lett. **105** 210402 (2010); J. Bougie, A. Gangopadhyaya, J.V. Mallow, J. Phys. A **44** 275307 (2011).

[10] A.A. Andrianov, M.V. Ioffe, V.P. Spiridonov, Phys. Lett. A **174** 273 (1993); A.A. Andrianov, F. Cannata, J.-P. Dedonder, M.V. Ioffe, Int. J. Mod. Phys. A **10** 2683 (1995).

[11] F. Cannata, M.V. Ioffe, E.V. Kolevatova, D.N. Nishnianidze, Ann. Phys. **356** 438 (2015).

[12] A.A. Andrianov, N.V. Borisov, M.V. Ioffe, JETP Lett. **39** 93 (1984); A.A. Andrianov, N.V. Borisov, M.V. Ioffe, Phys. Lett. A **105** 19 (1984); A.A. Andrianov, N.V. Borisov, M.V. Ioffe, Theor. Math. Phys. **61** 1078 (1984); A.A. Andrianov, N.V. Borisov, M.I. Eides, M.V. Ioffe, Phys. Lett. A **109** 143 (1985); A.A. Andrianov, N.V. Borisov, M.I. Eides, M.V. Ioffe, Theor. Math. Phys. **61** 965 (1984).

[13] A.A. Andrianov, M.V. Ioffe, Phys. Lett. B **205** 507 (1988); M.V. Ioffe, A.I. Neelov, J. Phys. A **36** 2493 (2003); F. Cannata, M.V. Ioffe, A.I. Neelov, D.N. Nishnianidze, J. Phys. A **37** 10339 (2004).

[14] A.A. Andrianov, N.V. Borisov, M.V. Ioffe, Phys. Lett. B **181** 141 (1986); A.A. Andrianov, N.V. Borisov, M.V. Ioffe, Theor. Math. Phys. **72** 748 (1986).

[15] L.P. Eisenhart, Phys. Rev. **74** 87 (1948); L.P. Eisenhart, Ann. Math. **35** 284 (1934).

[16] A.A. Andrianov, M.V. Ioffe, D.N. Nishnianidze, *Phys. Lett. A* **201** (1995) 103; A.A. Andrianov, M.V. Ioffe, D.N. Nishnianidze, Polynomial supersymmetry and dynamical symmatries in Quantum Mechanics Theor. Math. Phys. **104** 1129 (1995);
A.A. Andrianov, M.V. Ioffe, D.N. Nishnianidze, *Higher order SUSY in Quantum Mechanics and integrability of two-dimensional Hamiltonians* preprint solv-int/9605007 (1996); A.A. Andrianov, M.V. Ioffe, D.N. Nishnianidze, J. Phys. A **32** 4641 (1999).

[17] F. Cannata, M.V. Ioffe, D.N. Nishnianidze, J. Phys. A **35** 1389 (2002).

[18] M.V. Ioffe, D.N. Nishnianidze, Phys. Rev. A **76** 052114 (2007);

[19] M.V. Ioffe, P.A. Valinevich, J. Phys. A **38** 2497 (2005); M.V. Ioffe, D.N. Nishnianidze, P.A. Valinevich, J. Phys. A **43** 485303 (2010).

[20] M.V. Ioffe, E.V. Krupitskaya, D.N. Nishnianidze, Ann. Phys. **327** 764 (2012); M.V. Ioffe, E.V. Krupitskaya, D.N. Nishnianidze, EPL **98** 10013 (2012).

EPJ Web of Conferences **125**, 05008 (2016) DOI: 10.1051/epjconf/201612505008
QUARKS-2016

Harmonic superspaces for $\mathcal{N} = (1,1), 6D$ SYM theory

Evgeny Ivanov ★

[1] *Bogoliubov Laboratory of Theoretical Physics, JINR, 141980, Dubna, Moscow Region, Russia*

Abstract. This is a short account of the off-shell $\mathcal{N} = (1,0)$ and on-shell $\mathcal{N} = (1,1), 6D$ harmonic superspace formalism and its applications for the analysis of higher-dimension invariants in $\mathcal{N} = (1,1)$ SYM theory.

1 Motivations and contents

For last years, there is a permanent interest in the maximally extended (with 16 supercharges) supersymmetric gauge theories in diverse dimensions (see, e.g., [1]),

$$\mathcal{N} = 4, 4D \implies \mathcal{N} = (1,1), 6D \implies \mathcal{N} = (1,0), 10D.$$

The famous $\mathcal{N} = 4, 4D$ SYM theory was the first example of an UV *finite* theory. Perhaps, it is also *completely integrable* [2]. The $\mathcal{N} = (1,1), 6D$ SYM is not renormalizable by formal counting (the coupling constant is dimensionful) but it is also expected to possess unique properties. In particular, it respects the so called "dual conformal symmetry", like its $4D$ counterpart [3]. It provides the effective theory descriptions of some particular low energy sectors of string theory, such as D5-brane dynamics. The full effective action of D5-brane, generalizing the $\mathcal{N} = (1,1)$ SYM action, was conjectured to be of non-abelian Born-Infeld type [4, 5]. The $\mathcal{N} = (1,1)$ SYM is anomaly free, as distinct from $\mathcal{N} = (1,0)$ SYM.

The $\mathcal{N} = (1,1)$ and $\mathcal{N} = (1,0)$ SYM theories can be viewed as a laboratory for studying $\mathcal{N} = 8$ supergravity and its some lower $\mathcal{N}$ analogs, which are also non-renormalizable.

The recent perturbative calculations in $\mathcal{N} = (1,1)$ SYM show a lot of unexpected cancelations of the UV divergencies. The theory is UV finite up to 2 loops, while at 3 loops only a single-trace (planar) counterterm of canonical dim 10 is required. The allowed double-trace (non-planar) counterterms do not appear [6] - [8]. Various arguments to explain this were put forward [9] - [12], though the complete understanding is still lacking. Some new non-renormalization theorems could be expected in this connection.

The maximal off-shell supersymmetry one can gain in $6D$ is $\mathcal{N} = (1,0)$ supersymmetry. The natural off-shell formulation of $\mathcal{N} = (1,0)$ SYM theory is achieved in harmonic $\mathcal{N} = (1,0), 6D$ superspace [13, 14] as a generalization of the harmonic $\mathcal{N} = 2, 4D$ one [15, 16]. This harmonic $6D$ formalism was further developed in [17] - [20] and [21]. The $\mathcal{N} = (1,1)$ SYM theory in the harmonic formulation is a hybrid of two $\mathcal{N} = (1,0)$ theories, [$\mathcal{N} = (1,1)$ SYM] = [$\mathcal{N} = (1,0)$ SYM] + [$6D$ hypermultiplets], with the second hidden on-shell $\mathcal{N} = (0,1)$ supersymmetry. How to construct higher-dimension $\mathcal{N} = (1,1)$ invariants in the $\mathcal{N} = (1,0)$ superfield approach?

★e-mail: eivanov@theor.jinr.ru

One way is the "brute-force" method. One starts with the appropriate dimension $\mathcal{N} = (1,0)$ SYM invariant and then completes it to $\mathcal{N} = (1,1)$ invariant by adding the proper hypermultiplet terms. This approach is very complicated technically.

The things are simplified due to the fact that for finding superfield counterterms it is enough to stay on the mass shell. In a recent paper [21] a new approach to constructing higher-dimension $\mathcal{N} = (1,1)$ invariants was developed. It uses the concept of the *on-shell* $\mathcal{N} = (1,1)$ harmonic superspace with the double set of the harmonic variables $u_i^{\pm}, u_A^{\hat{\pm}}, i = 1,2; A = 1,2$ [22]. The novel point of the construction in [21] is solving the $\mathcal{N} = (1,1)$ SYM constraints [23, 24] through $\mathcal{N} = (1,0)$ superfields. The $d = 8$ and $d = 10$ invariants were built in a simple way and an essential difference between the single- and double-trace $d = 10$ invariants was found. The present contribution is a brief account of the 6D harmonic methods, with the focus on their recent uses in [21].

2 6D superspaces and superfields

2.1 6D superspaces

- The standard $\mathcal{N} = (1,0)$, 6D superspace is parametrized by the coordinates:

$$z = (x^M, \theta_i^a), \quad M = 0,\ldots,5, \; a = 1,\ldots,4, \; i = 1,2, \tag{2.1}$$

- The harmonic $\mathcal{N} = (1,0)$ superspace is obtained by adding $SU(2)$ harmonics to (2.1):

$$Z := (z,u) = (x^M, \theta_i^a, u^{\pm i}), \quad u_i^- = (u_i^+)^*, u^{+i}u_i^- = 1, \; u^{\pm i} \in SU(2)_R/U(1). \tag{2.2}$$

- The *analytic* $\mathcal{N} = (1,0)$ superspace is an invariant subspace of (2.2):

$$\zeta := (x^M_{(\mathrm{an})}, \theta^{+a}, u^{\pm i}) \subset Z, \quad x^M_{(\mathrm{an})} = x^M + \frac{i}{2}\theta_k^a \gamma^M_{ab} \theta_l^b u^{+k} u^{-l}, \quad \theta^{\pm a} = \theta_i^a u^{\pm i}. \tag{2.3}$$

The differential operators in the analytic basis $Z_A := (x^M_{(\mathrm{an})}, \theta^{+a}, u^{\pm i}, \theta^{-a})$ are defined as

$$\begin{aligned}
&D_a^+ = \partial_{-a}, \; D_a^- = -\partial_{+a} - 2i\theta^{-b}\partial_{ab}, \; D^0 = u^{+i}\frac{\partial}{\partial u^{+i}} - u^{-i}\frac{\partial}{\partial u^{-i}} + \theta^{+a}\partial_{+a} - \theta^{-a}\partial_{-a} \\
&D^{++} = \partial^{++} + i\theta^{+a}\theta^{+b}\partial_{ab} + \theta^{+a}\partial_{-a}, \; D^{--} = \partial^{--} + i\theta^{-a}\theta^{-b}\partial_{ab} + \theta^{-a}\partial_{+a}, \\
&\partial_{\pm a}\theta^{\pm b} = \delta_a^b, \quad \partial^{++} = u^{+i}\frac{\partial}{\partial u^{-i}}, \quad \partial^{--} = u^{-i}\frac{\partial}{\partial u^{+i}}.
\end{aligned} \tag{2.4}$$

2.2 Basic superfields

The basic object of $\mathcal{N} = (1,0)$ SYM theory is the analytic gauge connection $V^{++}(\zeta)$

$$\nabla^{++} = D^{++} + V^{++}, \quad \delta V^{++} = -\nabla^{++}\Lambda, \quad \Lambda = \Lambda(\zeta). \tag{2.5}$$

The second harmonic (non-analytic) connection $V^{--}(Z)$,

$$\nabla^{--} = D^{--} + V^{--}, \quad \delta V^{--} = -\nabla^{--}\Lambda,$$

is related to V^{++} by the harmonic flatness condition

$$\begin{aligned}
&[\nabla^{++}, \nabla^{--}] = D^0 \;\Leftrightarrow\; D^{++}V^{--} - D^{--}V^{++} + [V^{++}, V^{--}] = 0 \\
&\Rightarrow V^{--} = V^{--}(V^{++}, u^{\pm}).
\end{aligned} \tag{2.6}$$

The off-shell content of $\mathcal{N} = (1,0)$ SYM theory is revealed in the Wess-Zumino gauge:

$$V^{++} = \theta^{+a}\theta^{+b}A_{ab} + 2(\theta^+)^3_a\lambda^{ai}u^-_i - 3(\theta^+)^4\mathcal{D}^{ik}u^-_iu^-_k\,. \tag{2.7}$$

Here A_{ab} is the gauge field, λ^{ai} is the gaugino and $\mathcal{D}^{ik} = \mathcal{D}^{ki}$ are the auxiliary fields.

The $\mathcal{N} = (1,0)$ SYM covariant derivatives are given by the expressions

$$\begin{aligned}
&\nabla^-_a = [\nabla^{--}, D^+_a] = D^-_a + \mathcal{A}^-_a, \qquad \nabla_{ab} = \frac{1}{2i}[D^+_a, \nabla^-_b] = \partial_{ab} + \mathcal{A}_{ab}\,,\\
&\mathcal{A}^-_a(V) = -D^+_aV^{--}, \quad \mathcal{A}_{ab}(V) = \frac{i}{2}D^+_aD^+_bV^{--},\\
&[\nabla^{++}, \nabla^-_a] = D^+_a\,, \quad [\nabla^{++}, D^+_a] = [\nabla^{--}, \nabla^-_a] = [\nabla^{\pm\pm}, \nabla_{ab}] = 0\,.
\end{aligned} \tag{2.8}$$

The covariant superfield strengths are defined as

$$\begin{aligned}
&[D^+_a, \nabla_{bc}] = \frac{i}{2}\varepsilon_{abcd}W^{+d}, \qquad [\nabla^-_a, \nabla_{bc}] = \frac{i}{2}\varepsilon_{abcd}W^{-d}\,,\\
&W^{+a} = -\frac{1}{6}\varepsilon^{abcd}D^+_bD^+_cD^+_dV^{--}, \quad W^{-a} := \nabla^{--}W^{+a}\,,\\
&\nabla^{++}W^{+a} = \nabla^{--}W^{-a} = 0, \quad \nabla^{++}W^{-a} = W^{+a}\,,\\
&D^+_bW^{+a} = \delta^a_bF^{++}\,, \quad F^{++} = \frac{1}{4}D^+_aW^{+a} = (D^+)^4V^{--}\,,\\
&\nabla^{++}F^{++} = 0\,, \quad D^+_aF^{++} = 0\,.
\end{aligned} \tag{2.9}$$

The hypermultiplet is accommodated by the analytic superfield $q^{+A}(\zeta)\,, (A = 1,2)$:

$$q^{+A}(\zeta) = q^{iA}(x)u^+_i - \theta^{+a}\psi^A_a(x) + \text{ An infinite tail of auxiliary fields}\,. \tag{2.10}$$

2.3 $\mathcal{N} = (1,0)$ superfield actions

The $\mathcal{N} = (1,0)$ SYM action was constructed by Boris Zupnik [14]:

$$\begin{aligned}
&S^{SYM} = \frac{1}{f^2}\sum_{n=1}^{\infty}\frac{(-1)^{n+1}}{n}\mathrm{Tr}\int d^6x\,d^8\theta\,du_1\ldots du_n\frac{V^{++}(z,u_1)\ldots V^{++}(z,u_n)}{(u^+_1u^+_2)\ldots(u^+_nu^+_1)}\,,\\
&\delta S^{SYM} = 0 \Rightarrow F^{++} = 0\,.
\end{aligned} \tag{2.11}$$

Here, $(u^+_1u^+_2)^{-1}, \ldots (u^+_nu^+_1)^{-1}$ are harmonic distributions [16].

The hypermultiplet action, with q^{+A} in adjoint of the gauge group, is written as

$$\begin{aligned}
&S^q = -\frac{1}{2f^2}\mathrm{Tr}\int d\zeta^{-4}q^{+A}\nabla^{++}q^+_A\,, \quad \nabla^{++}q^+_A = D^{++}q^+_A + [V^{++}, q^+_A]\,,\\
&\delta S^q = 0 \Rightarrow \nabla^{++}q^{+A} = 0\,.
\end{aligned} \tag{2.12}$$

The $\mathcal{N} = (1,0)$ superfield form of the $\mathcal{N} = (1,1)$ SYM action ia a sum of the two superfield actions given above:

$$\begin{aligned}
&S^{(V+q)} = S^{SYM} + S^q = \frac{1}{f^2}\left(\int dZ\mathcal{L}^{\mathrm{SYM}} - \frac{1}{2}\mathrm{Tr}\int d\zeta^{-4}q^{+A}\nabla^{++}q^+_A\right),\\
&\delta S^{(V+q)} = 0 \Rightarrow F^{++} + \frac{1}{2}[q^{+A}, q^+_A] = 0\,, \quad \nabla^{++}q^{+A} = 0\,.
\end{aligned} \tag{2.13}$$

It is invariant under the second hidden $\mathcal{N} = (0,1)$ supersymmetry:

$$\delta V^{++} = \epsilon^{+A}q^+_A\,, \quad \delta q^{+A} = -(D^+)^4(\epsilon^-_AV^{--})\,, \quad \epsilon^{\pm}_A = \epsilon_{aA}\theta^{\pm a}\,. \tag{2.14}$$

These transformations have the correct closure only on shell.

3 Higher-dimensional invariants

3.1 Dimension $d = 6$

In the pure $\mathcal{N} = (1,0)$ SYM the $d = 6$ invariant is defined uniquely [17]:

$$S^{(6)}_{SYM} = \frac{1}{2g^2}\mathrm{Tr}\int d\zeta^{-4}du\,(F^{++})^2 \sim \mathrm{Tr}\int d^6x[(\nabla^M F_{ML})^2 + \ldots]. \tag{3.1}$$

It vanishes on shell, when $F^{++} = 0$. Using the results of [18], its $\mathcal{N} = (1,1)$ completion is defined up to a real parameter

$$\mathcal{L}^{d=6} = \frac{1}{2g^2}\mathrm{Tr}\int du d\zeta^{-4}\left(F^{++} + \frac{1}{2}[q^{+A}, q^+_A]\right)\left(F^{++} + 2\beta[q^{+A}, q^+_A]\right). \tag{3.2}$$

But it vanishes on the full $\mathcal{N} = (1,1)$ SYM mass shell! This proves the *one-loop finiteness* of $\mathcal{N} = (1,1)$ SYM theory.

3.2 Dimension $d = 8$

All superfield operators of the canonical dimension $d = 8$ in the $\mathcal{N} = (1,0)$ SYM theory vanish on shell, in accord with ref. [24]. Can adding the hypermultiplet terms change something? We have found that there exist no $\mathcal{N} = (1,0)$ off-shell invariants of the dimension $d = 8$ which would respect the on-shell $\mathcal{N} = (1,1)$ invariance.

Surprisingly, the $d = 8$ superfield expression which is non-vanishing on shell and respects the on-shell $\mathcal{N} = (1,1)$ supersymmetry can be constructed by *giving up* the requirement of *off-shell* $\mathcal{N} = (1,0)$ supersymmetry.

An example of such an invariant in $\mathcal{N} = (1,0)$ SYM theory is very simple

$$\tilde{S}^{(8)}_1 \sim \mathrm{Tr}\int d\zeta^{-4}\,\varepsilon_{abcd}W^{+a}W^{+b}W^{+c}W^{+d}. \tag{3.3}$$

Indeed, $D^+_a W^{+b} = \delta^b_a F^{++}$, which vanishes on shell, with $F^{++} = 0$. Thus, W^{+a} is on-shell analytic, and so the above action respects $\mathcal{N} = (1,0)$ supersymmetry on shell. Also, a double-trace on-shell invariant exists.

These invariants admit $\mathcal{N} = (1,1)$ completions. For (3.3) such a completion reads

$$\begin{aligned}\mathcal{L}^{+4}_{(1,1)} &= \mathrm{Tr}_{(S)}\Big\{\frac{1}{4}\varepsilon_{abcd}W^{+a}W^{+b}W^{+c}W^{+d} + 3iq^{+A}\nabla_{ab}q^+_A W^{+a}W^{+b}\\ &- q^{+A}\nabla_{ab}q^+_A\, q^{+B}\nabla^{ab}q^+_B - W^{+a}[D^+_a q^-_A, q^+_B]q^{+A}q^{+B}\\ &- \frac{1}{2}[q^{+C}, q^+_C][q^-_A, q^+_B]q^{+A}q^{+B}\Big\}.\end{aligned} \tag{3.4}$$

Here, $\mathrm{Tr}_{(S)}$ stands for the *symmetrized* trace. This Lagrangian is analytic, $D^+_a\mathcal{L}^{+4}_{(1,1)} = 0$, on the full mass shell $F^{++} + \frac{1}{2}[q^{+A}, q^+_A] = 0$, $\nabla^{++}q^{+A} = 0$, and so it is on-shell $\mathcal{N} = (1,1)$ supersymmetric.

Though the nontrivial on-shell $d = 8$ invariants exist, the perturbative expansion for the amplitudes in the $\mathcal{N} = (1,1)$ SYM theory does not involve divergences at the two-loop level. The reason is that these $d = 8$ invariants do *not* possess the full off-shell $\mathcal{N} = (1,0)$ supersymmetry which the physically relevant counterterms should obey.

4 $\mathcal{N} = (1,1)$ on-shell harmonic superspace

Despite the fact that the $d = 8$ terms mentioned above cannot appear as counterterms in $\mathcal{N} = (1,1)$ SYM theory, they can come out, e.g., as quantum corrections to the effective Wilsonian action. For the pure $\mathcal{N} = (1,0)$ SYM theory this was recently observed in [20]. It was desirable to work out some systematic way of constructing such higher-order $\mathcal{N} = (1,1)$ invariants. This becomes possible within the on-shell harmonic $\mathcal{N} = (1,1)$ superspace.

As the first step, extend the $\mathcal{N} = (1,0)$ superspace to the $\mathcal{N} = (1,1)$ one,

$$z = (x^{ab}, \theta^a_i) \;\Rightarrow\; \hat{z} = (x^{ab}, \theta^a_i, \hat{\theta}^A_a)\,. \tag{4.1}$$

Then we define the covariant spinor derivatives,

$$\nabla^i_a = \frac{\partial}{\partial\theta^a_i} - i\theta^{bi}\partial_{ab} + \mathcal{A}^i_a\,, \quad \hat{\nabla}^{aA} = \frac{\partial}{\partial\hat{\theta}_{Aa}} - i\hat{\theta}^A_b\partial^{ab} + \hat{\mathcal{A}}^{aA}\,. \tag{4.2}$$

The constraints of the $\mathcal{N} = (1,1)$ SYM theory can now be written as follows [23, 24]:

$$\begin{aligned} &\{\nabla^{(i}_a, \nabla^{j)}_b\} = \{\hat{\nabla}^{a(A}, \hat{\nabla}^{bB)}\} = 0\,, \quad \{\nabla^i_a, \hat{\nabla}^{bA}\} = \delta^b_a\phi^{iA} \\ &\Rightarrow \quad \nabla^{(i}_a\phi^{j)A} = \hat{\nabla}^{a(A}\phi^{B)i} = 0 \quad \text{(By Bianchis)}\,. \end{aligned} \tag{4.3}$$

As the next step, we introduce the $\mathcal{N} = (1,1)$ harmonic superspace [22],

$$Z = (x^{ab}, \theta^a_i, u^\pm_k) \;\Rightarrow\; \hat{Z} = (x^{ab}, \theta^a_i, \hat{\theta}^A_b, u^\pm_k, u^{\hat{\pm}}_A)\,, \tag{4.4}$$

pass to the analytic basis and choose the "hatted" spinor derivatives short, $\nabla^{\hat{+}a} = D^{\hat{+}a} = \frac{\partial}{\partial\theta^{\hat{-}}_a}$. The set of constraints (4.3) is rewritten as

$$\begin{aligned} &\{\nabla^+_a, \nabla^+_b\} = 0\,, \quad \{D^{\hat{+}a}, D^{\hat{+}b}\} = 0\,, \quad \{\nabla^+_a, D^{\hat{+}b}\} = \delta^b_a\phi^{+\hat{+}}\,, \\ &[\nabla^{\hat{+}\hat{+}}, \nabla^+_a] = 0\,,\ [\tilde{\nabla}^{++}, \nabla^+_a] = 0\,,\ [\nabla^{\hat{+}\hat{+}}, D^{a\hat{+}}] = 0\,,\ [\tilde{\nabla}^{++}, D^{a\hat{+}}] = 0\,, \\ &[\tilde{\nabla}^{++}, \nabla^{\hat{+}\hat{+}}] = 0\,, \end{aligned} \tag{4.5}$$

$$\begin{aligned} &\nabla^+_a = D^+_a + \mathcal{A}^+_a(\hat{Z})\,,\ \tilde{\nabla}^{++} = D^{++} + \tilde{V}^{++}(\hat{\zeta})\,, \quad \nabla^{\hat{+}\hat{+}} = D^{\hat{+}\hat{+}} + V^{\hat{+}\hat{+}}(\hat{\zeta})\,, \\ &\hat{\zeta} = (x^{ab}_{\text{an}}, \theta^{\pm a}, \theta^{\hat{+}}_c, u^\pm_i, u^{\hat{\pm}}_A)\,. \end{aligned} \tag{4.6}$$

5 Solving $\mathcal{N} = (1,1)$ SYM constraints through $\mathcal{N} = (1,0)$ superfields

The starting point of our analysis in [21] was the WZ gauge for the extra connection $V^{\hat{+}\hat{+}}(\hat{\zeta})$

$$V^{\hat{+}\hat{+}} = i\theta^{\hat{+}}_a\theta^{\hat{+}}_b\hat{\mathcal{A}}^{ab} + \varepsilon^{abcd}\theta^{\hat{+}}_a\theta^{\hat{+}}_b\theta^{\hat{+}}_c\varphi^A_d u^{\hat{-}}_A + \varepsilon^{abcd}\theta^{\hat{+}}_a\theta^{\hat{+}}_b\theta^{\hat{+}}_c\theta^{\hat{+}}_d\mathcal{D}^{AB}u^{\hat{-}}_A u^{\hat{-}}_B\,, \tag{5.1}$$

where $\hat{\mathcal{A}}^{ab}, \varphi^A_d$ and $\mathcal{D}^{(AB)}$ are some $\mathcal{N} = (1,0)$ harmonic superfields, still arbitrary.

Then the above constraints are reduced to some sets of harmonic equations. We have solved them and, as the eventual result, obtained that the first harmonic connection V^{++} coincides with the previous $\mathcal{N} = (1,0)$ one, $V^{++} = V^{++}(\zeta)$, while the dependence of all other geometric $\mathcal{N} = (1,1)$ objects on the "hatted" variables is strictly fixed

$$\begin{aligned} V^{\hat{+}\hat{+}} &= i\theta^{\hat{+}}_a\theta^{\hat{+}}_b\mathcal{A}^{ab} - \frac{1}{3}\epsilon^{abcd}\theta^{\hat{+}}_a\theta^{\hat{+}}_b\theta^{\hat{+}}_c D^+_d q^{-\hat{-}} + \frac{1}{8}\epsilon^{abcd}\theta^{\hat{+}}_a\theta^{\hat{+}}_b\theta^{\hat{+}}_c\theta^{\hat{+}}_d[q^{+\hat{-}}, q^{-\hat{-}}] \\ \phi^{+\hat{+}} &= q^{+\hat{+}} - \theta^{\hat{+}}_a W^{+a} - i\theta^{\hat{+}}_a\theta^{\hat{+}}_b\nabla^{ab}q^{+\hat{-}} + \frac{1}{6}\varepsilon^{abcd}\theta^{\hat{+}}_a\theta^{\hat{+}}_b\theta^{\hat{+}}_c[D^+_d q^{-\hat{-}}, q^{+\hat{-}}] \\ &+ \frac{1}{24}\varepsilon^{abcd}\theta^{\hat{+}}_a\theta^{\hat{+}}_b\theta^{\hat{+}}_c\theta^{\hat{+}}_d[q^{+\hat{-}}, [q^{+\hat{-}}, q^{-\hat{-}}]]\,. \end{aligned} \tag{5.2}$$

Here, $q^{+\hat{\pm}} = q^{+A}(\zeta)u_A^{\hat{\pm}}$, $q^{-\hat{\pm}} = q^{-A}(\zeta)u_A^{\hat{\pm}}$ and $W^{+a}, q^{\pm A}$ are just the $\mathcal{N} = (1,0)$ superfields explored previously. In the course of solving the constraints, there naturally appear the superfield equations of motion

$$\nabla^{++}q^{+A} = 0\,, \quad F^{++} = \frac{1}{4}D^+_a W^{+a} = -\frac{1}{2}[q^{+A}, q^+_A]\,. \tag{5.3}$$

Also, the structure of the spinor covariant derivatives is fully fixed

$$\begin{aligned} \nabla^+_a &= D^+_a - \theta^{\hat{+}}_a q^{+\hat{-}} + \theta^{\hat{-}}_a \phi^{+\hat{+}}\,, \\ \nabla^-_a &= D^-_a - D^+_a V^{--} - \theta^{\hat{+}}_a q^{-\hat{-}} + \theta^{\hat{-}}_a \phi^{-\hat{+}}\,, \quad \phi^{-\hat{+}} = \nabla^{--}\phi^{+\hat{+}}\,. \end{aligned} \tag{5.4}$$

The advantage of using the constrained $\mathcal{N} = (1,1)$ strengths $\phi^{\pm\hat{+}}$ for constructing invariants is the simple transformation rules of $\phi^{\pm\hat{+}}$ under the hidden $\mathcal{N} = (0,1)$ supersymmetry

$$\delta\phi^{\pm\hat{+}} = -\epsilon^{\hat{+}}_a \frac{\partial}{\partial\theta^{\hat{+}}_a}\phi^{\pm\hat{+}} - 2i\epsilon^{\hat{-}}_a \theta^{\hat{+}}_b \partial^{ab}\phi^{\pm\hat{+}} - [\Lambda^{(comp)}, \phi^{\pm\hat{+}}]\,, \tag{5.5}$$

where $\Lambda^{(comp)}$ is some composite gauge parameter which does not contribute under trace.

6 Invariants in $\mathcal{N} = (1,1)$ superspace

The single-trace $\underline{d = 8}$ invariant (3.4) admits a simple rewriting in $\mathcal{N} = (1,1)$ superspace

$$S_{(1,1)} = \int d\zeta^{-4}\mathcal{L}^{+4}_{(1,1)}\,, \; \mathcal{L}^{+4}_{(1,1)} = -\mathrm{Tr}\,\frac{1}{4}\int d\hat{\zeta}^{-4}d\hat{u}\,(\phi^{+\hat{+}})^4, \quad d\hat{\zeta}^{-4} \sim (D^{\hat{-}})^4 \tag{6.1}$$

$$\delta\mathcal{L}^{+4}_{(1,1)} = -2i\partial^{ab}\mathrm{Tr}\int d\hat{\zeta}^{-4}d\hat{u}\left[\epsilon^{\hat{-}}_a\theta^{\hat{+}}_b\frac{1}{4}(\phi^{+\hat{+}})^4\right].$$

Analogously, the double-trace $\underline{d = 8}$ invariant is given by

$$\hat{\mathcal{L}}^{+4}_{(1,1)} = -\frac{1}{4}\int d\hat{\zeta}^{-4}d\hat{u}\,\mathrm{Tr}\,(\phi^{+\hat{+}})^2\,\mathrm{Tr}\,(\phi^{+\hat{+}})^2. \tag{6.2}$$

Now it is easy to construct the single- and double-trace $\underline{d = 10}$ invariants

$$\begin{aligned} S^{(10)}_1 &= \mathrm{Tr}\int dZ d\hat{\zeta}^{-4}d\hat{u}\,(\phi^{+\hat{+}})^2(\phi^{-\hat{+}})^2, \quad \phi^{-\hat{+}} = \nabla^{--}\phi^{+\hat{+}}\,, \\ S^{(10)}_2 &= -\int dZ d\hat{\zeta}^{-4}d\hat{u}\,\mathrm{Tr}\big(\phi^{+\hat{+}}\phi^{-\hat{+}}\big)\,\mathrm{Tr}\big(\phi^{+\hat{+}}\phi^{-\hat{+}}\big)\,. \end{aligned} \tag{6.3}$$

It is notable that the single-trace $\underline{d = 10}$ invariant admits a representation as an integral over the *full* $\mathcal{N} = (1,1)$ superspace

$$S^{(10)}_1 \sim \mathrm{Tr}\int dZ d\hat{Z} d\hat{u}\,\phi^{+\hat{+}}\phi^{-\hat{-}}\,, \quad \phi^{-\hat{-}} = \nabla^{\hat{-}\hat{-}}\phi^{-\hat{+}}\,, \tag{6.4}$$

with $d\hat{Z} \sim (D^{\hat{-}})^4(D^{\hat{+}})^4$. On the other hand, the double-trace $\underline{d = 10}$ invariant *cannot* be written as the full integral and so looks as being UV *protected*.

This could explain why in the perturbative calculations of the amplitudes in the $\mathcal{N} = (1,1)$ SYM single-trace 3-loop divergence is seen, while no double-trace structures at the same order were observed [6], [7], [8]. However, this does not seem to be like the standard non-renormalization theorems because the quantum calculation of $\mathcal{N} = (1,0)$ supergraphs should give invariants in the *off-shell* $\mathcal{N} = (1,0)$ superspace, not in the *on-shell* $\mathcal{N} = (1,1)$ superspace. So the above property seems not enough to explain the absence of the double-trace divergences and some additional piece of reasoning is needed.

7 Summary and outlook

Following refs. [14], [17], [18] and [21], the off-shell $\mathcal{N} = (1, 0)$ and on-shell $\mathcal{N} = (1, 1)$ harmonic superfield approaches were reviewed and argued to be efficient for constructing higher-dimensional invariants in the $\mathcal{N} = (1, 0)$ and $\mathcal{N} = (1, 1)$ SYM theories. The $\mathcal{N} = (1, 1)$ SYM constraints were solved in terms of harmonic $\mathcal{N} = (1, 0)$ superfields. This allowed to explicitly construct the full set of the dimensions $\underline{d = 8}$ and $\underline{d = 10}$ superfield invariants possessing $\mathcal{N} = (1, 1)$ on-shell supersymmetry.

All possible $\underline{d = 6}$ $\mathcal{N} = (1, 1)$ invariants were demonstrated to be on-shell vanishing, thereby proving the UV finiteness of $\mathcal{N} = (1, 1)$ SYM at one loop.

The off-shell $\underline{d = 8}$ invariants which would be non-vanishing on shell, are absent. Assuming that the $\mathcal{N} = (1, 0)$ supergraphs yield integrals over the full $\mathcal{N} = (1, 0)$ harmonic superspace, this means the absence of two-loop counterterms as well.

Two $\underline{d = 10}$ invariants were constructed as integrals over the whole $\mathcal{N} = (1, 0)$ harmonic superspace. The single-trace invariant can be rewritten as an integral over the $\mathcal{N} = (1, 1)$ superspace, while the double-trace one cannot. This property combined with an additional reasoning (e.g., based on the algebraic renormalization approach [25]) could explain why the double-trace invariant is UV protected.

<u>*Some further lines of development*</u>:

(a). To construct the $\underline{d \geq 12}$ invariants in the $\mathcal{N} = (1, 1)$ SYM theory using the on-shell $\mathcal{N} = (1, 1)$ harmonic superspace techniques.
(b). To apply the same method for constructing the Born-Infeld action with the manifest off-shell $\mathcal{N} = (1, 0)$ and hidden on-shell $\mathcal{N} = (0, 1)$ supersymmetries.
(c). To develop an analogous on-shell harmonic $\mathcal{N} = 4, 4D$ superspace approach to the $\mathcal{N} = 4, 4D$ SYM theory in the $\mathcal{N} = 2$ superfield formulation (by solving the $\mathcal{N} = 4$ SYM constraints in terms of $\mathcal{N} = 2$ superfields) and apply it to the problem of constructing the $\mathcal{N} = 4$ SYM effective action.
(d). Applications in supergravity? The absence of the double-trace divergent structures in the 3-loop amplitude in $\mathcal{N} = (1, 1)$ SYM theory is similar to the absence of 3-loop and 4-loop divergences for the four-graviton amplitudes in $\mathcal{N} = 4, 4D$ and $\mathcal{N} = 5, 4D$ supergravities [26], [27], [28], [29]. All these UV divergence cancelations could find a common explanation within the harmonic superspace approach [1].

Acknowledgements

I thank the Organizers of the International Seminar QUARKS-2016 for the very kind hospitality in Pushkin. I am grateful to my co-authors Guillaume Bossard, Andrei Smilga and Boris Zupnik. A partial support from the RFBR grant No 15-02-06670 and a grant of Heisenberg-Landau program is acknowledged.

References

[1] N. Seiberg, *Notes on theories with 16 supercharges*, Nucl. Phys. Proc. Suppl. **67** (1998) 158, hep-th/9705117.
[2] N. Beisert, *On Yangian Symmetry in Planar $\mathcal{N} = 4$ SYM*, In *Trieste 2010, Gribov-80 Memorial Volume*, pp. 413-438, arXiv:1004.5423 [hep-th].

[1]For a recent relevant discussion see [30].

[3] T. Dennen, Yu-tin Huang, *Dual Conformal Properties of Six-Dimensional Maximal Super Yang-Mills Amplitudes*, JHEP **1101** (2011) 140, arXiv:1010.5874 [hep-th].

[4] A.A. Tseytlin, *On non-abelian generalization of Born-Infeld action in string theory*, Nucl. Phys. B **501** (1997) 41-52, arXiv:hep-th/9701125.

[5] J.M. Drummond, P.J. Heslop, P.S. Howe, S.F. Kerstan, *Integral invariants in $\mathcal{N} = 4$ SYM and the effective action for coincident D-branes*, JHEP **0308** (2003) 016, arXiv:hep-th/0305202.

[6] Z. Bern, L.J. Dixon, V.A. Smirnov, *Iteration of planar amplitudes in maximally supersymmetric Yang-Mills theory at three loops and beyond*, Phys. Rev. D **72** (2005) 085001, arXiv:hep-th/0505205.

[7] Z. Bern, J.J.M. Carrasco, L.J. Dixon, H. Johansson, R. Roiban, *The Complete Four-Loop Four-Point Amplitude in N=4 Super-Yang-Mills Theory*, Phys. Rev. D **82** (2010) 125040, arXiv:1008.3327 [hep-th.

[8] Z. Bern, J.J.M. Carrasco, L.J. Dixon, H. Johansson, R. Roiban, *Simplifying Multiloop Integrands and Ultraviolet Divergences of Gauge Theory and Gravity Amplitudes*, Phys. Rev. D **85** (2012) 105014, arXiv:1201.5366 [hep-th].

[9] G. Bossard, P.S. Howe, U. Lindström, K.S. Stelle, L. Wulff, *Integral invariants in maximally supersymmetric Yang-Mills theories*, JHEP **1105** (2011) 021, arXiv:1012.3142 [hep-th].

[10] N. Berkovits, M.B. Green, J.G. Russo, P. Vanhove, *Non-renormalization conditions for four-gluon scattering in supersymmetric string and field theory*, JHEP **0911** (2009) 063, arXiv:0908.1923 [hep-th].

[11] J. Bjornsson, M.B. Green, *5 loops in 24/5 dimensions*, JHEP **1008** (2010) 132, arXiv:1004.2692 [hep-th].

[12] J. Bjornsson, *Multi-loop amplitudes in maximally supersymmetric pure spinor field theory*, JHEP **1101** (2011) 002, arXiv:1009.5906 [hep-th].

[13] P.S. Howe, K.S. Stelle, P.C. West, *$\mathcal{N} = 1, d = 6$ harmonic superspace*, Class. Quant. Grav. **2** (1985) 815.

[14] B.M. Zupnik, *Six-dimensional Supergauge Theories in the Harmonic Superspace*, Sov. J. Nucl. Phys. **44** (1986) 512 [Yad.Fiz. **44** (1986) 794-802];
B.M. Zupnik, *The Action of the Supersymmetric $\mathcal{N} = 2$ Gauge Theory in Harmonic Superspace*, Phys. Lett. B **183** (1987) 175-176.

[15] A.S. Galperin, E.A. Ivanov, S. Kalitzin, V.I. Ogievetsky, E.S. Sokatchev, *Unconstrained $\mathcal{N} = 2$ matter, Yang-Mills and supergravity theories in harmonic superspace*, Class. Quant. Grav. **1** (1984) 469-498.

[16] A.S. Galperin, E.A. Ivanov, V.I. Ogievetsky, E.S. Sokatchev, *Harmonic Superspace*, Cambridge University Press, 2001, 306 p.

[17] E.A. Ivanov, A.V. Smilga, B.M. Zupnik, *Renormalizable supersymmetric gauge theory in six dimensions*, Nucl. Phys. B **726** (2005) 131-148, arXiv:hep-th/0505082.

[18] E.A. Ivanov, A.V. Smilga, *Conformal properties of hypermultiplet actions in six dimensions*, Phys. Lett. B **637** (2006) 374-381, arXiv:hep-th/0510273.

[19] I.L. Buchbinder, N.G. Pletnev, *Construction of 6D supersymmetric field models in $\mathcal{N} = (1,0)$ harmonic superspace*, Nucl.Phys. B **892** (2015) 21-48, arXiv:1411.1848 [hep-th].

[20] I.L. Buchbinder, N.G. Pletnev, *Leading low-energy effective action in the 6D hypermultiplet theory on a vector/tensor background*, Phys. Lett. B **744** (2015) 125-130, arXiv:1502.03257 [hep-th].

[21] G. Bossard, E. Ivanov, A. Smilga, *Ultraviolet behavior of 6D supersymmetric Yang-Mills theories and harmonic superspace*, JHEP **1512** (2015) 085, arXiv:1509.08027 [hep-th].

[22] G. Bossard, P.S. Howe, K.S. Stelle, *The Ultra-violet question in maximally supersymmetric field theories*, Gen. Rel. Grav. **41** (2009) 919, arXiv:0901.4661 [hep-th].

[23] P.S. Howe, G. Sierra, P.K. Townsend, *Supersymmetry in six dimensions*, Nucl. Phys. B **221** (1983) 331-348.

[24] P.S. Howe, K.S. Stelle, *Ultraviolet Divergences in Higher Dimensional Supersymmetric Yang-Mills Theories*, Phys. Lett. B **137** (1984) 175-180.

[25] O. Piguet, S.P. Sorella, *Algebraic renormalization: Perturbative renormalization, symmetries and anomalies*, Lect. Notes Phys. M **28** (1995) 1.

[26] Z. Bern, S. Davies, T. Dennen, Yu-tin Huang, *Absence of three-loop four-point divergences in $\mathcal{N} = 4$ supergravity*, Phys. Rev. Lett. **108** (2012) 201301, arXiv:1202.3423 [hep-th].

[27] P. Tourkine, P. Vanhove, *An R^4 non-renormalization theorem in $\mathcal{N} = 4$ supergravity*, Class. Quant. Grav. **29** (2012) 115006, arXiv:1202.3692 [hep-th].

[28] Z. Bern, S. Davies, T. Dennen, Yu-tin Huang, *Ultraviolet cancellations in half-maximal supergravity as a consequence of the double-copy structure*, Phys. Rev. D **86** (2012) 105014, arXiv:1209.2472 [hep-th].

[29] Z. Bern, S. Davies, T. Dennen, *Enhanced ultraviolet cancellations in $\mathcal{N} = 5$ supergravity at four loops*, Phys. Rev. D **90** (2014)105011, arXiv:1409.3089 [hep-th].

[30] Andrei Smilga, *Ultraviolet divergences in non-renormalizale supersymmetric theories*, arXiv:1603.06811 [hep-th].

Emergent Lorentz invariance in fermion sector

Ivan Kharuk[1,2,⋆]

[1] *Institute for Nuclear Research of the Russian Academy of Sciences, 60th October Anniversary Prospect, 7a, 117312 Moscow, Russia*
[2] *Moscow Institute of Physics and Technology, Institutskii per. 9, Dolgoprudny 141700, Moscow Region, Russia*

Abstract. By using holographic description of strongly interacting field theories we show that under common assumptions Lorentz invariance emerges as an effective low–energy symmetry of the theory, despite fundamental theory at hight energies being Lorentz–violating. We consider fermions sector and show that the notion of chirality also automatically arises in the infrared.

1 Introduction

The idea that Lorentz–invariance (LI) can be not a fundamental symmetry of nature but accidental or emergent symmetry of low–energy physics has been subject of studies for a while. In condensed matter physics such scenarios are rather common [1–3]: if one takes non–relativistic system and examines its renormalization group (RG) flow towards infrared (IR), then, typically, the effective Lagrangian would be LI, up to some suppressed corrections. This observation provides motivation for studying analogous models in particle physics, thus allowing to abandon LI at high energies. The last idea is interesting in context of models of modified gravity, particularly in case of Hórava – Lifshitz gravity [4], as such assumptions allow to improve ultraviolet (UV) behaviour of gravity and may help to explain some phenomena in cosmology.

The pioneering work in that direction in particle physics were made by Nielsen, Chadha and Ninomiya [5, 6], who studied quantum electrodynamics and Yang–Mills theory with Lorentz–violating (LV) coupling of dimension four and showed that they vanish in the IR limit of RG flow. However, their model fails to satisfy the experimental constrains, coming from precision test of LI [7, 8] – as in any weakly–coupled system, the marginal LV operators have logarithmic dependence on RG scale, that is not enough to stay within existing bounds on LV corrections. Indeed, if such corrections would be of order one at some energy scale Λ_* (smaller or equal to Plank scale), then they would be only mildly suppressed at experimentally accessible energies.

There are two ways to overcome that problem. The first one is to impose additional symmetries on the model, that would forbid marginal LV couplings. In this case, the resulting theory in the IR would automatically be LI, as would contain only irrelevant LV operators. The other way is to turn to strong coupling regime, that may change the dependence of marginal coupling on RG scale from logarithmic to power–law one, that allows to avoid contradictions with experimental constrains. The second option

⋆e-mail: ivan.kharuk@phystech.edu

is the one we are interested in, and we will show, by using gauge–gravity duality to describe strongly coupled system, that LV corrections do become power–law suppressed. What is more, we also show that the notion of chirality emerges together with LI, thus solving the puzzle how Weyl fermions – one of the irreducible representations of Lorentz group – can appear in non–relativistic system.

The scenario we will be considering is the following. At energy higher than Λ_* the theory is governed by some fundamental description, that does not include LI as a symmetry. Then, at energies of order Λ_*, the theory becomes strongly coupled and passes through strong RG evolution down to energy scale Λ_{IR}, at which it enters a confining phase. This implies that the Standard Model particles are composite ones. This property should become observable at energies of order Λ_{IR}, so that it should be bigger than experimentally accessible energies, as no such signal has been observed yet. We will be considering fermion fields, see also [9] for a more complete discussion on the that model for boson fields.

The paper is organised as follows. In the first section we consider a non–relativistic fermion coupled to a relativistic strongly interacting sector. In the second section we consider a more general set–up, describing RG flow from UV theory with anisotropic Lifshitz scaling down to LI IR fixed point. Finally, we conclude and summarize Sec. (4).

2 Chiral LI fermions from non–relativistic systems

2.1 Holography for fermions

The holographic construction for fermions is presented in [10, 11], here we will use some of its results and adopt the model for our purposes. From here and in what follows capital Latin letters from the beginning of the alphabet are used as indices in local Lorentz frame, from the middle – for (d+1)–dimensional indices in AdS. Greek letters are used for flat directions, t and x^i, where small Latin letters are used for purely spatial directions. In Poincare patch AdS metric has the form

$$g_{MN} = \left(\frac{l}{u}\right)^2 \eta_{\mu\nu}, \quad i = 1, \ldots, d-1, \tag{1}$$

where $\eta_{\mu\nu} = (-, +, ..., +)$ is flat mostly plus metric and l is AdS radius. For simplicity, we will assume (d+1) to be even, thought all of the results can be generalized to odd dimensionality of space–time.

In order to make possible introduction of fundamental LV fermion and to introduce confinement into the theory, we set two branes in the AdS, one in the UV and one in the IR, located at ϵ and L respectively. To proceed one needs to solve for Dirac equation, coming from the standard action

$$S = -\int d^{d+1}x\,\sqrt{-g}\; i(\bar{\psi} \not{D} \psi - M\bar{\psi}\psi) + S_\delta\,, \tag{2}$$

where all quantities are written in local Lorentz frame and S_δ,

$$S_\delta = -i \int_{u=\epsilon} d^d x\, (\epsilon/l)^{-\frac{d-1}{2}}\; \bar{\psi}_- \psi_+\,, \tag{3}$$

is introduced to cancel the term, coming from integration by parts after varying (2) with respect to ψ. The important step step to be done is to decompose ψ into ψ_+ and ψ_-, that are eigenvectors of Γ^u, the analogy of γ^5 in four dimensions:

$$\psi = \psi_+ + \psi_-, \quad \Gamma^u \psi_\pm = \pm\psi_\pm. \tag{4}$$

That allows to solve the Dirac equation,

$$\psi_-(\vec{p},u) = (pu)^{(d+1)/2}\left(\chi_1(\vec{p})J_{Ml+1/2}(pu) + \chi_2(\vec{p})Y_{Ml+1/2}(pu)\right)\ , \tag{5a}$$

$$\psi_+(\vec{p},u) = (pu)^{(d+1)/2}\frac{ip_\mu\Gamma^\mu}{p}\left(\chi_1(\vec{p})J_{Ml-1/2}(pu) + \chi_2(\vec{p})Y_{Ml-1/2}(pu)\right)\ , \tag{5b}$$

where $p = \sqrt{-p_\mu p^\mu}$. Depending on the value of Ml, one or the other of the components is leading in the $u \to 0$ limit. We will focus on the case $Ml \geq -\frac{1}{2}$, when ψ_- acts as a source for dual theory, the opposite choice is equivalent to that one up to some trivial substitutions. In this case on the CFT side we have dual fermion operator of dimension

$$\dim O_\psi = Ml + \frac{d}{2}, \tag{6}$$

that can be read–out from the behaviour of ψ_- near UV boundary. Note also that in the regime $-\frac{1}{2} \leq Ml \leq \frac{1}{2}$ the dual CFT theory admits two alternative quantizations.

There are two possible boundary conditions (BC) on the IR brane, that we well denote as $\mathcal{B}_+$ and $\mathcal{B}_-$:

$$\mathcal{B}_+ : \qquad \psi_-|_L = 0, \qquad \psi_+|_L - \text{arbitrary}, \tag{7a}$$

$$\mathcal{B}_- : \qquad \psi_+|_L = 0, \qquad \psi_-|_L - \text{arbitrary}. \tag{7b}$$

The spectrum of the theory depends on which BC are imposed. Direct calculations show that there are no massless modes in $\mathcal{B}_+$ case, while in $\mathcal{B}_-$ there is one of the form

$$\psi_+ = \chi_0\left(\frac{u}{l}\right)^{Ml+\frac{d}{2}}, \qquad \psi_- = 0. \tag{8}$$

As it was explained in the introduction, in this set–up we would like to introduce fundamental massless LV fermion. Of course, this changes the resulting spectrum – in $\mathcal{B}_+$ case there would be a gapless mode, coming from LV fermion, while in the $\mathcal{B}_-$ case, as it can be directly checked, there would be no massless excitations. As our primary goal is to study emergence of chiral fermions, and as they can emerge only from the gapless mode, in this section we will consider $\mathcal{B}_+$ boundary conditions.

2.2 Emergence of LI chiral fermions

Lets now introduce LV fermion. In AdS this is done by adding LV action for ψ_- on the UV brane. For simplicity, we take its action to be

$$S_{UV} = -b\int d^dx\, i(\bar{\psi}_-\gamma^0\partial_0\psi_- + v\bar{\psi}_-\gamma^i\partial_i\psi_-)\ , \tag{9}$$

where parameter $v \neq 1$ explicitly breaks LI and b is some parameter (of dimension length), introduced to make action dimensionless.

From the CFT point of view it is equivalent to promoting the source χ to a dynamical field at UV and introducing the interaction between the initial CFT and χ,

$$S = S_{CFT} + S_\chi + \Lambda_{int}\int d^dx\bar{\chi}O_\psi, \tag{10}$$

where S_χ is LV action for χ,

$$S_\chi = -b \int d^d x\, i(\bar{\chi}\gamma^0\partial_0\chi + v\bar{\chi}\gamma^i\partial_i\chi)\,, \tag{11}$$

that directly translates to (9).

The dimension of elementary fermion field χ is $\frac{d-1}{2}$, and the dimension of O_ψ is given by (6). That allows to find the dimension of the coupling constant Λ_{int}, that turns out to be relevant only in the range

$$-\frac{1}{2} \le ML \le \frac{1}{2}. \tag{12}$$

If $Ml \ge \frac{1}{2}$, the coupling is irrelevant, so that one expects those two sectors to decouple in the IR and, hence, there is no reason for LI to emerge at low–energy limit. Direct calculations show that this is indeed the case in that regime. Thus, further we will focus only on the regime (12).

The boundary condition on the UV brane then reads

$$\left[b(\gamma^0\partial_0 + v\gamma^i\partial_i)\psi_- + \psi_+\right]\Big|_{u=l} = 0\,. \tag{13}$$

Together with IR boundary condition that leads the solution [11]

$$\psi_- = \chi_1 f_-(u)\,, \quad \psi_+ = \frac{ip_\mu\gamma^\mu}{p}\chi_1 f_+(u)\,, \tag{14}$$

where

$$f_\pm(u) = (pu)^{(d+1)/2}\left[J_{Ml\mp1/2}(pu) - \frac{J_{Ml-1/2}(pL)}{Y_{Ml-1/2}(pL)}Y_{Ml\mp1/2}(pu)\right], \tag{15}$$

where J_α and Y_α are first and second kind Bessel functions respectively.

Our main interest is to study the low–energy behaviour of that theory. Assuming $pl \ll pL \ll 1$ and expanding the solution in Tylor series, we get the dispersion relation

$$\omega = \pm|\mathbf{k}|\left[1 + (v-1)(1-2Ml)\frac{b}{l}\left(\frac{l}{L}\right)^{1-2Ml}\right], \tag{16}$$

where $\omega = -p_0$, $k_i = p_i$ and $\mathbf{k} = \sqrt{k_i k^i}$. It is LI up to corrections, that are power–law suppressed, as it was advertised in the introduction. If the ration $\frac{l}{L}$ is small enough, than this model passes all experimental constrains. What is more, more careful analysis also shows that it has exactly $2^{\frac{d}{2}-2}$ components (others being suppressed by the same factor), that coincides with the structure and number of components of Weyl fermion.

This also have a very clear interpretation on the CFT side. As it was mentioned earlier, in regime (12) dual theory admits two alternative quantizations, one with dimension of dual operator given by (6), and one with

$$\dim O'_\psi = -Ml + \frac{d}{2}. \tag{17}$$

The LV corrections to dispersion relation are reproduced by deforming the initial CFT by

$$\delta S_{CFT} \sim \int d^d x \bar{O}'_\psi\gamma^0\partial_0 O'_\psi. \tag{18}$$

Thus, the above model also describes RG flow between two alternative quantizations of bulk theory, that helps to restore LI in IR.

3 Fermions in Lifshitz flow

3.1 Geometry of Lifshitz flow

The other possibility to consider is the case when fundamental theory at high energies has anisotropic isometries, that is of interest, for example, in context of Hórava gravity. This is more general set–up for studying non–relativistic field theories as there is no grantee a priori that time and space are similar in any sense in some fundamental theory. By definition, Lifshitz space, introduced in [12, 13], has anisotropic isometries

$$t \mapsto \lambda^z t, \quad x_i \mapsto \lambda x_i, \quad u \mapsto \lambda u, \tag{19}$$

By extrapolation of the logic of AdS/CFT duality, such spaces are claimed to be dual to "anisotropic CFT" and are expected to catch the strong coupling dynamics of such theories.

In order to make contact with phenomenology, one should construct a domain–wall geometry, that would interpolate between Lifshitz space at $u \to 0$ and AdS at $u \to \infty$. Such models were constructed and studied in [14], and here we will provide a very brief overview of its properties.

The required geometry can be obtained as a solution of Einstein equations with negative cosmological constant and a vector field with mass M_V. Provided that inequality

$$d - 1 \leq (M_V l)^2 \leq \frac{d(d-1)^2}{3d-4} \tag{20}$$

is satisfied, LI fixed point in the infrared acts as attractor for RG flows, thus providing required geometry. From the dual viewpoint, this is equivalent to adding an operator of dimension

$$\dim \mathcal{O}^{\mu}_{V} = d + \alpha_V, \quad \alpha_V = -\frac{d}{2} + \sqrt{\left(\frac{d}{2} - 1\right)^2 + (M_V l)^2}\,. \tag{21}$$

that is irrelevant as long as $\alpha_V < 0$, that can be checked to coincide with (20). In this set–up we are interested in the spectrum of pure CFT, without any sources, so that, in contrast to previous section, we would have no UV brane and impose $\mathcal{B}_-$ boundary conditions on IR brane.

3.2 Emergence of LI fermions in Lifshitz flows

We would like to study the behaviour of fermions on such domain–wall background. The action of the model we are considering is

$$S = -\int d^{d+1}x \sqrt{-g}\, i\bar{\psi}(\not{D} - \xi \not{D}^{(V)} - M)\psi, \tag{22}$$

where $\not{D}$ is a Dirac operator and $\not{D}^{(V)}$ introduces interaction between fermion and vector field,

$$\not{D}^{(V)} \equiv \frac{(M_V l)^2}{4} V_N V_M e^M_A \Gamma^A D^N\,. \tag{23}$$

The coupling constant ξ can depend on the species of the fermion (if there were several of them, as it is required in realistic models), thus allowing to trace possible dependence of LV corrections on that coupling.

Avoiding technical details, the proof that there is a gapless chiral LI fermion in the IR goes as follows. First, one calculates the explicit view of Dirac operator and coupling $D^{(V)}$. Then the field ψ

should be again split into ψ_+ and ψ_-, the eigenvalues of Γ^u, as in previous section. At this stage it is possible to set $\omega = k_i = 0$ and show that there is a solution of a form

$$\psi_+ = \psi_+^{(0)}, \quad \psi_- = 0, \tag{24}$$

where the exact expression for $\psi_+^{(0)}$ is unimportant for our purposes. The normalization integral for that mode is saturated in AdS region, that is, in the IR. But this means that we can consider the above solution as zero–th order approximation to the exact solution in some vicinity of $\omega = k_i = 0$ and use perturbation series technique to calculate higher corrections. Implementing that idea in first order perturbations yields dispersion relation

$$\omega^{(1)} = |\mathbf{k}|\left[a_1(1+2Ml)\left(\frac{l_*}{L}\right)^{2\alpha_V} + a_2(1+2Ml)\left(\frac{l_*}{L}\right)^{1+2Ml}\right], \tag{25}$$

with some coefficients a_1 and a_2 being functions of background metric and coupling constant ξ. Clearly, as long as $\alpha_V > 0$ and $Ml > -\frac{1}{2}$, LV correction are suppressed, so that LI do emerges in that regime, that matches nicely with the results, obtained in previous section. What is more, it can be shown that the corresponding mode is chiral.

Again, this results have clear interpretation on the CFT side. The resulting RG flow describes the evolution of CFT, deformed by

$$\delta S_{CFT} = \int d^dx(c_1\, l_*^{\alpha_V} O_V^0 + c_2\, l_*^{1+2Ml} \bar{O}_\psi \gamma^0 \partial_0 O_\psi)\,, \tag{26}$$

where $c_{1,2}$ are some dimensionless constants. They generate exactly the same corrections that appear in (25) [1].

4 Conclusions

Considering LV models in particle physics, we demonstrated that turning to strong coupling regime allows to overcome the problems, present in a weakly coupled scenarios. Namely, LV corrections become power–law suppressed, that allows to avoid contradictions with experimental data, provided that the hierarchy between Λ_* and Λ_{IR} is large enough. This means that the scenario of [9] is also applicable in fermion sector. What is more, the emergence of LI in this case is followed by the emergence of the notion of chirality, that is crucial for realistic theories, as Weyl fermion play central role in the Standard Model.

We observed that the exponent of suppression is bounded from above by 2. Whether it is a fundamental constrain or not is left as an opened question. It is important to study this question as if there are theories with exponents grater than 2 it would be much harder to close them experimentally. Also, it would be interesting to include gauge fields into consideration, but this is left for future work, as such models would require more complicated bulk sector.

To summarize, our calculations show that fundamental theory can be not just LV, but also anisotropic one, still having a well–defined IR limit, with LI as an effective low–energy symmetry, having no contradictions with experimental bounds on LV effects.

[1]Only double insertion of operator O_V^0 modifies the dispersion relation due to presence of bulk symmetry $V_M \mapsto -V_M$ for a bulk vector field.

References

[1] O. Vafek, Z. Tesanovic, M. Franz, Phys. Rev. Lett. **89**, 157003 (2002) [cond-mat/0203047].
[2] D. J. Lee, I. F. Herbut, Phys. Rev. B **66**, 094512 (2002) [cond-mat/0201088].
[3] I. F. Herbut, V. Juricic, B. Roy, Phys. Rev. B **79**, 085116 (2009) [arXiv:0811.0610 [cond-mat.str-el]].
[4] P. Horava, Phys. Rev. D **79**, 084008 (2009) [arXiv:0901.3775 [hep-th]].
[5] H. B. Nielsen, M. Ninomiya, Nucl. Phys. B **141**, 153 (1978).
[6] S. Chadha, H. B. Nielsen, Nucl. Phys. B **217**, 125 (1983).
[7] D. Mattingly, Living Rev. Rel. **8**, 5 (2005) [gr-qc/0502097].
[8] V. A. Kostelecky, N. Russell, Rev. Mod. Phys. **83**, 11 (2011) [arXiv:0801.0287 [hep-ph]].
[9] G. Bednik, O. Pujolàs and S. Sibiryakov, JHEP **1311**, 064 (2013) [arXiv:1305.0011 [hep-th]].
[10] M. Henningson and K. Sfetsos, Phys. Lett. B **431**, 63 (1998) [hep-th/9803251].
[11] R. Contino, A. Pomarol, JHEP **0411**, 058 (2004) [arXiv:hep-th/0406257].
[12] S. Kachru, X. Liu and M. Mulligan, Phys. Rev. D **78**, 106005 (2008) [arXiv:0808.1725[hep-th]].
[13] M. Taylor, arXiv:0812.0530 [hep-th].
[14] H. Braviner, R. Gregory and S. F. Ross, Class. Quant. Grav. **28**, 225028 (2011) [arXiv:1108.3067 [hep-th]].

EPJ Web of Conferences **125**, 05010 (2016) DOI: 10.1051/epjconf/201612505010
QUARKS-2016

Holographic dictionary and defects in the bulk

Mikhail Khramtsov[1,*]

[1] *Steklov Mathematical Institute, Russian Academy of Sciences, Gubkina str. 8, 119991, Moscow, Russia*

Abstract. We study the holographic dual of the AdS_3 spacetime with a conical defect. We calculate the boundary two-point correlator using the holographic Gubser-Klebanov-Polyakov/Witten dictionary for a scalar field in the bulk. We consider the general case, when the conical defect breaks conformal symmetry at the boundary. The results are compared with previous studies based on the geodesic approximation. They are in good agreement for short correlators, and main discrepancy comes in the region of long correlations. It is shown that in the case when the spacetime is the $AdS_3/\mathbb{Z}_N$ orbifold, both methods give the same result which also produces the result expected from the orbifold CFT.

1 Introduction and summary

One of the key features of the AdS_3/CFT_2-correspondence [1, 2] is the fact that the $3D$ gravity is much simpler than its higher-dimensional counterparts, primarily because of its topological nature. There are no propagating gravitational degrees of freedom, and thus solutions of Einstein equations at least asymptotically look like topological quotients of the form AdS'_3/Γ, where Γ is a representation of a topological identification in the AdS isometry group, and AdS'_3 is the part of AdS_3 on which Γ which acts discretely [3]. The dynamics of the bulk objects described by these solutions is almost trivial in a sense that it can only be innately realized in the action of the identification Γ, or it can be induced by acting with isometries (such e. g. as Lorentzian symmetry) on the basic static configurations. Yet this underwhelming at first sight picture is still highly useful in the holographic approach for study of problems such as the thermalization problem [4–9], entanglement physics [10–13], chaos in QFT [14]. Moreover, there are also many powerful analytic method suitable for studying the 2D CFT on the boundary. Recent developments in techniques based on the extensive utilization of the Virasoro symmetry allow to study some properties of 3D gravity as quantum field theory [15, 16].

One of the simplest non-vacuum solutions of the three-dimensional gravity, which we consider in the present paper, is a point particle. The point particles [17–20] produce conical singularities in the spacetime. In the context of the asymptotically AdS solution [18], the point particle can be represented as a wedge cut out from the spacetime with an endpoint running along the worldline of the particle. Such spacetime is still a locally AdS spacetime, however it has a non-trivial topology. The latter leads to the fact that two given points in the bulk (or at the boundary) can be connected, generally speaking, by several geodesics. This has a crucial impact on the holographic dual of the AdS_3 with a point

*e-mail: khramtsov@mi.ras.ru

particle, as was shown in the recent work [21–25] based on the geodesic approximation [26]. The main object of study therein were two-point correlation functions and the entanglement entropy in the boundary dual to the AdS_3-deficit spacetime in the framework of geodesic approximation.

The goal of the present investigation is to study two-point boundary correlators in the framework of the holographic GKPW prescription [27, 28] on AdS_3 with a conical defect, and clarify the applicability of the geodesic approximation in the AdS-deficit spacetime by comparison of GKPW correlators to the geodesic correlators obtained in the earlier work. We find that in a certain special case, namely when the space is an orbifold $AdS_3/\mathbb{Z}_N$ both GKPW and geodesic correlators are given by a finite sum over images which precisely matches the expected result from the orbifold CFT. In the general case we illustrate that the discontinuities in the geodesic result correspond to the non-conformal regime. In this regime the most substantial discrepancy of two computations comes from the region of long-range correlations. Similar picture is observed in the time dependence of the correlators.

2 Scalar field in the AdS-deficit spacetime

2.1 The background

In $3D$ gravity the geometry with a conical singularity at the origin arises as a solution of the Einstein equations with a point-like source, which was obtained by Deser, Jackiw and t'Hooft originally in the flat space [17] and generalized to the case of constant curvature in [18]. In the AdS_3 case, when the negative cosmological constant is added, for a static source the solution exists only when the source is located at the origin of AdS spatial slice. A static particle in the AdS_3 spacetime is described by AdS_3 metric in global coordinates which can be written as follows (we set AdS radius to 1):

$$ds^2 = \frac{1}{\cos^2\rho}\left(-dt^2 + d\rho^2 + \sin^2\rho d\vartheta^2\right), \tag{1}$$

where we have the holographic coordinate $\rho \in [0, \frac{\pi}{2})$, and $\vartheta \in [0, 2\pi A)$ is the angular coordinate. The size of the angle deficit is related to the mass of the particle μ by the relation

$$A = 1 - 4G\mu, \tag{2}$$

where G is the Newtonian constant. Then the angle deficit equals to $\gamma = 2\pi(1 - A) = 8\pi G\mu$. When $A = 1$, the spacetime is the pure AdS, and the case of $A = 0$ is the BTZ black hole threshold.

2.2 The GKPW prescription for boundary correlators in AdS-deficit space

To calculate correlation functions in the boundary dual of the AdS-deficit spacetime, we use the Gubser-Klebanov-Polyakov/Witten (GKPW) holographic prescription [27, 28]. We use the version of the prescription for the Lorentzian signature proposed by Skenderis and van Rees [29]. We require that the partition functions of bulk and boundary theories are equal in the semiclassical limit (after the renormalization):

$$\left\langle e^{i\int dt d\vartheta\, \varphi_0 O}\right\rangle_{\mathrm{CFT}} = e^{iS_{on-shell}[\phi]}|_{\phi|_{\mathrm{bd}}=\varphi_0}\,; \tag{3}$$

where in the right hand side we have the classical action of the field evaluated on the solution of the Dirichlet boundary value problem with the boundary value of the field given by φ_0. The correlation functions in the boundary theory are then obtained by varying the right hand side with respect to φ_0. We are interested in the two-point Lorentzian correlators in the boundary theory, and therefore we need to make sure that those functional derivatives of the generating functional give a specific two-point correlator with correct $i\epsilon$-prescription. To be more concrete, let us specify the contour C. Suppose it has three segments:

$$C_- \; : \; t = -T + i\kappa \, ; \; \kappa \in [0, \infty) \, , \quad C_0 \; : \; t \in [-T, T] \, , \quad C_+ \; : \; t = T - i\kappa \, .$$

The CFT path integral then equals to

$$\mathcal{Z}_{\text{boundary}}[\phi_0] = \int \mathcal{D}\phi_- \mathcal{D}\phi_+ \, \langle 0|\phi_-, -T\rangle \, \langle \phi_-, -T| e^{i \int_C d^2 x \varphi_0 O} |\phi_+, T\rangle \, \langle \phi_+, T|0\rangle \; , \tag{4}$$

where the source ϕ_0 satisfies the condition $\varphi_0(\pm T) = 0$.

Wave function insertions can be interpreted as Euclidean path integrals with no sources present and with specified boundary conditions. It is natural then to construct a write a holographic dual to this quantity as $\mathcal{Z}_{\text{bulk}} = e^{-S_- + iS_0 - S_+}$, where

$$S_0 = \frac{1}{2} \int_{-T}^{T} dt \int d^{d-1}x \sqrt{-g} \left(-(\partial\phi)^2 - m^2 \phi^2 \right) |_{\phi = \Phi_0} \, , \tag{5}$$

$$S_- = \frac{1}{2} \int_{-\infty}^{0} d\tau \int d^{d-1}x \sqrt{g} \left((\partial\phi)^2 + m^2 \phi^2 \right) |_{\phi = \Phi_-} \, , \tag{6}$$

$$S_+ = \frac{1}{2} \int_{0}^{\infty} d\xi \int d^{d-1}x \sqrt{-g} \left((\partial\phi)^2 + m^2 \phi^2 \right) |_{\phi = \Phi_+} \, . \tag{7}$$

The action S_0 is that of a scalar field theory on a AdS-deficit spacetime with Lorentzian signature, computed on a solution Φ_0 of the field equation, which satisfies the Dirichlet boundary conditions with $\phi(\text{boundary}) = \varphi_0$. The solutions of the equations of motion $\Phi_\pm$ for the Euclidean actions $S_\pm$ are not supposed to generate any boundary sources, so they are required to simply decay when approaching to the boundary of the Euclidean AdS-deficit bulk spacetime. In addition, all solutions Φ are required to be regular at the origin of the spacetime.

The discussed on-shell actions have by definition vanishing bulk terms, but they also have generally non-trivial surface terms [29]. Those can be of two kinds: AdS boundary terms and corner terms from integration along the time segments. While the terms of the first kind are exactly what we need to study boundary theory, the terms of the second kind should vanish, since we do not have a consistent analytic continuation into the complex time plane of the usual relation (3) otherwise. The cancellation of the corner terms happens is we impose the matching conditions on the solutions Φ:

$$\Phi_-|_{\tau=0} = \Phi_0|_{t=-T} \, , \qquad \Phi_0|_{t=T} = \Phi_+|_{\xi=0} \, ; \tag{8}$$

$$i\partial_t \Phi_0|_{t=-T} + \partial_\tau \Phi_-|_{\tau=0} = 0 \, , \qquad i\partial_t \Phi_0|_{t=T} + \partial_\tau \Phi_+|_{\xi=0} = 0 \, . \tag{9}$$

These conditions basically mean C^1-smoothness of Φ as a solution of the EOM on the entire contour. The next step is to find the solutions Φ themselves.

To construct the bulk solution, we define the bulk-boundary propagator as a function $R_{\omega,l}(\rho)$, through which the Lorentzian solution Φ_0 can be expressed in following way:

$$\Phi_0(\rho, t, \vartheta) = \frac{1}{(2\pi)^2} \sum_{l \in \mathbb{Z}} \int d\omega \, e^{-i\omega t + i \frac{l}{A} \vartheta} \varphi_0(\omega, l) R_{\omega,l}(\rho) \, , \tag{10}$$

where $\varphi_0(\omega, l)$ is Fourier image of $\varphi_0(t, \vartheta) = \Phi_0(t, \rho, \vartheta)|_{\rho=\pi/2}$. We also introduce the notations $h_\pm = \frac{1}{2}(1 \pm \sqrt{1 + m^2})$, $\nu = h_+ - h_-$.

For the boundary condition to be satisfied, we normalize R in such way that $R(\rho = \frac{\pi}{2} - \epsilon) = \epsilon^{2h_-} + \ldots$. It is clear that in our case R is defined by the fundamental mode solution which is regular at the origin. The non-normalizable (growing) part of the solution has the unit as a coefficient in front of the leading power of ρ, providing the saturation of the boundary condition, and also has analytic behaviour in the frequency space in all (growing) orders, since all poles from gamma-functions compensate each other.

However, the normalizable (decaying) piece is not analytic in the complex frequency plane. The leading coefficient can be read off from the representation of the mode solution as a sum of normalizable and non-normalizable solutions[1]. It equals to $\alpha(\omega, l)\beta(\omega, l)$, where

$$\alpha(\omega, l) := \frac{1}{\nu!(\nu-1)!} \frac{\Gamma\left(h_+ + \frac{1}{2}\left(\frac{|l|}{A} + \omega\right)\right)\Gamma\left(h_+ + \frac{1}{2}\left(\frac{|l|}{A} - \omega\right)\right)}{\Gamma\left(h_- + \frac{1}{2}\left(\frac{|l|}{A} + \omega\right)\right)\Gamma\left(h_- + \frac{1}{2}\left(\frac{|l|}{A} - \omega\right)\right)}, \tag{11}$$

$$\beta(\omega, l) := -\left(\psi\left(h_+ + \frac{1}{2}\left(\frac{|l|}{A} + \omega\right)\right) + \psi\left(h_+ + \frac{1}{2}\left(\frac{|l|}{A} - \omega\right)\right)\right) + \dots; \tag{12}$$

Here ψ is the digamma function - the main source of singularities. Those digamma functions which are independent of ω are denoted as $\dots$ and will be ignored in future since they do not contribute to singularities. The poles of β are located at

$$\omega^{\pm}_{nl} = \pm(2h_+ + 2n + \frac{|l|}{A}), \qquad n \in \mathbb{Z}_+; \tag{13}$$

In order for (10) to be well-defined, the contour of integration around the frequency poles needs to be specified. We rewrite (10) as

$$\Phi_0(\rho, t, \vartheta) = \sum_{l\in\mathbb{Z}} \int_C \frac{d\omega}{(2\pi)^2} \mathrm{e}^{-i\omega t + i\frac{l}{A}\vartheta} \varphi_0(\omega, l) R_{\omega,l}(\rho) + \sum_{\pm}\sum_{l\in\mathbb{Z}}\sum_{n=0}^{\infty} c^{\pm}_{nl} \mathrm{e}^{-i\omega^{\pm}_{nl} t + i\frac{l}{A}\vartheta} \mathrm{Res}(R_{\omega,l}(\rho))|_{\omega=\omega_{nl}}, \tag{14}$$

We have here integration over some reference contour C, and arbitrary coefficients $c^{\pm}_{nl}$ deform the contour by adding or removing extra poles to the sum over residues defined by C. One can show, that if one takes C to be of Feynman form, than from the condition $\varphi_0(\pm T, \vartheta) = 0$, and matching conditions (8,9) it follows that all $c^{\pm}_{nl}$ are zero. The shape of the time contour C through matching conditions for the bulk field defines the shape of the contour C in the frequency space, resulting in any desirable behaviour of correlators.

Now we can use the bulk-boundary propagator to calculate the bulk on-shell action and compute the correlation functions of a scalar operator O of conformal dimension $\Delta = 2h_+ = 1 + \nu$. Varying over φ_0 twice, we obtain the Feynman propagator:

$$\begin{aligned}\langle T O_\Delta(t_1, \vartheta_1) O_\Delta(t_2, \vartheta_2)\rangle = {} & \frac{2(\Delta-1)\theta(t_1-t_2)}{\pi} \sum_{l\in\mathbb{Z}}\sum_{n=0}^{\infty} \alpha(\omega_{nl}, |l|) \mathrm{e}^{-i\omega^{+}_{nl}(t_1-t_2) + i\frac{l}{A}(\vartheta_1-\vartheta_2)} + \\ & + \frac{2(\Delta-1)\theta(t_2-t_1)}{\pi} \sum_{l\in\mathbb{Z}}\sum_{n=0}^{\infty} \alpha(\omega_{nl}, |l|) \mathrm{e}^{-i\omega^{-}_{nl}(t_1-t_2) + i\frac{l}{A}(\vartheta_1-\vartheta_2)}.\end{aligned} \tag{15}$$

It is easy to extract Wightman correlators:

$$\langle O_{1+\nu}(t, \vartheta) O_{1+\nu}(0, 0)\rangle = \frac{2}{\pi(\nu-1)!^2} \sum_{l\in\mathbb{Z}}\sum_{n=0}^{\infty} \frac{(n+\nu)!}{n!} \frac{\left(n + \frac{|l|}{A} + \nu\right)!}{\left(n + \frac{|l|}{A}\right)!} \times \mathrm{e}^{-i(1+\nu+2n)t - i\frac{|l|}{A}t + i\frac{l}{A}\vartheta}; \tag{16}$$

[1]See [30] for thorough discussion of (non)-normalizable mode solutions in the pure AdS case

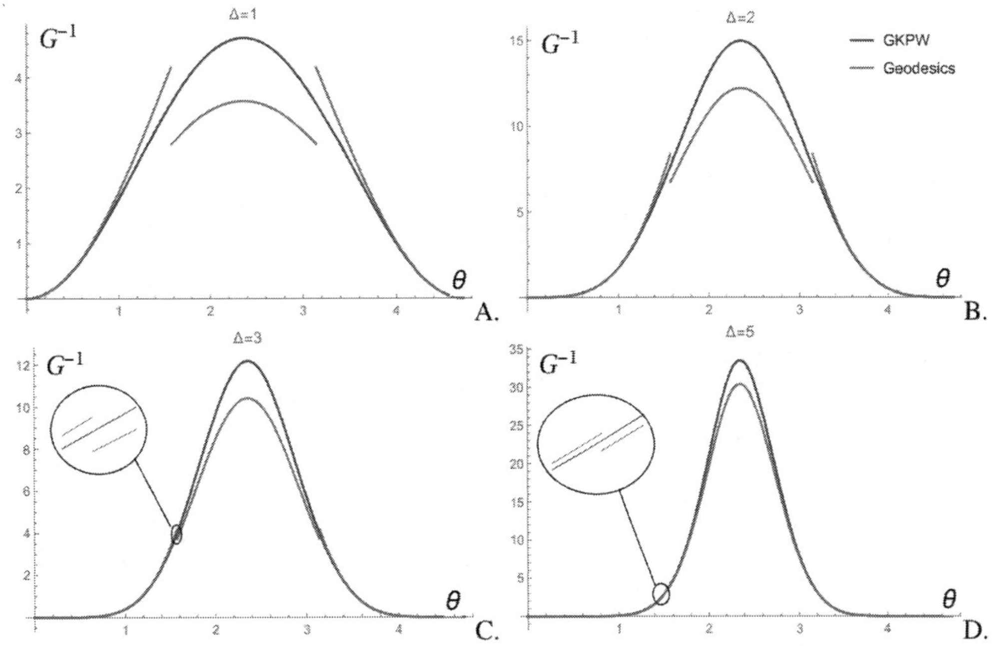

Figure 1: Inverse equal time correlators obtained via GKPW prescription and the geodesic image method for $A = \frac{3}{4}$ for different conformal weights. Contributions of discontinuities in the geodesic result (represented by the brown line) diminish as Δ increases.

where we have used (11) to write the result more explicitly and omitted the ϵ-prescription. In case of $\nu = 0$ the summation can be performed explicitly:

$$\langle O_1(t,\vartheta)O_1(0,0)\rangle = \frac{1}{\pi}\frac{\sin\frac{t}{A}}{\sin t}\frac{1}{\cos\frac{t}{A}-\cos\frac{\theta}{A}}. \tag{17}$$

Then the result for arbitrary integer $\nu = \Delta - 1$ by differentiating the sum and formally written as

$$\langle O_{1+\nu}(t,\vartheta)O_{1+\nu}(0,0)\rangle = \frac{\nu}{2^\nu(\nu-1)!\pi}(-1)^\nu\frac{\partial^\nu}{\partial(\cos t)^\nu}\left(\frac{\sin\frac{t}{A}}{\sin t}\frac{1}{\cos\frac{t}{A}-\cos\frac{\theta}{A}}\right). \tag{18}$$

3 Comparison with geodesic approximation

The geodesic prescription for correlation functions [26] dual to particles (both static and moving) in AdS_3 has been constructed in [21–25]. It takes into account geodesics which wind around conical defects functions by including the contributions of image geodesics generated by the identification isometry. As a result, one gets an image method for the two-point correlation function. However, by construction, the number of geodesic images is always finite, unlike in case of image method for a Green's function problem. Also, the correlator given by the geodesic image method sum has discontinuities in its dependence on the angular coordinate. This is because the number of possible windings of a geodesic between two given points in general case depends on their position. As we see, this fact contributes to the deviation from the GKPW correlator, see [32] for more details. In this section we compare the GKPW correlators given by (17,18) with correlators given by the geodesic approximation.

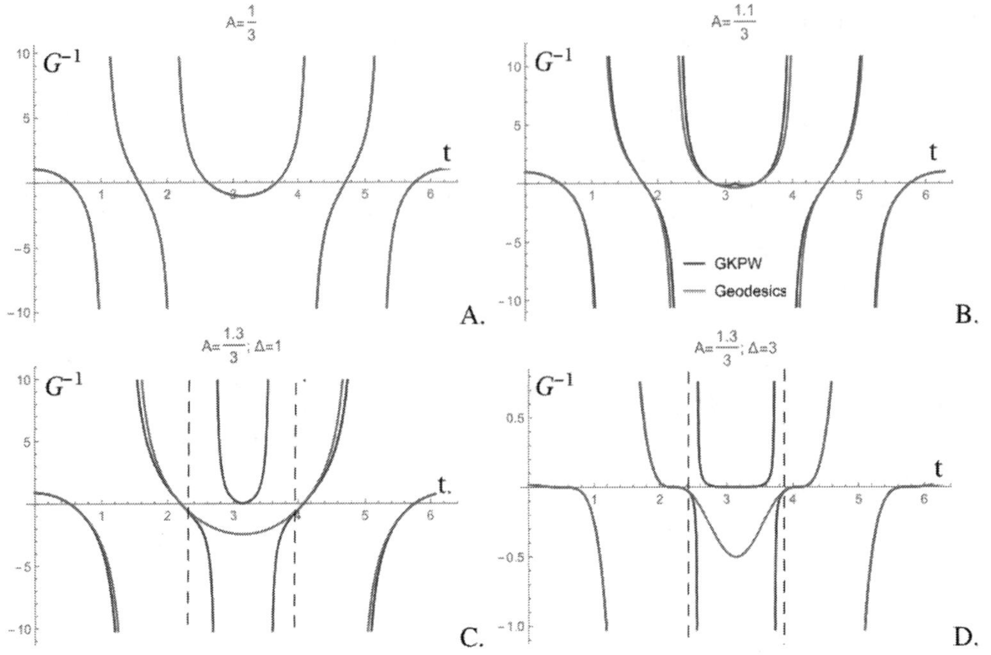

Figure 2: Time dependence of the inverse correlators obtained by GKPW and geodesic prescriptions. Plots A-C show the increase of discrepancy between two prescriptions in case of $\Delta = 1$ when the deficit parameter is close to the orbifold value $A = \frac{1}{3}$. Plot D shows the discrepancy for $\Delta = 3$. The value of angular variable is fixed $\vartheta = \frac{\pi}{6}$.

3.1 Special case: $AdS_3/\mathbb{Z}_N$-orbifold

First, we consider a special case when $A = 1/N$, where $N = 1, 2, 3, \ldots$. In this case the spacetime is a conical orbifold $AdS_3/\mathbb{Z}_N$. It is important to note that the boundary of this spacetime preserves asymptotic Virasoro symmetry [13, 31]. In our case, one can show that in this case a sum over images emerges naturally:

$$\frac{2}{(\nu-1)!^2}\sum_{l\in\mathbb{Z}}\sum_{n=0}^{\infty}\frac{(n+\nu)!}{n!}\frac{(n+N|l|+\nu)!}{(n+N|l|)!}e^{-i(1+\nu+2n)t-i|l|Nt+ilN\vartheta} = \frac{\nu^2}{2^\nu N}\sum_{k=0}^{N-1}\left(\frac{1}{\cos t-\cos\left(\vartheta+2\pi\frac{k}{N}\right)}\right)^{\nu+1} \tag{19}$$

and thus we get for arbitrary integer $\Delta > 1$ ($\Delta = 1$ case yields a different constant prefactor):

$$\langle O_\Delta(t,\vartheta)O_\Delta(0,0)\rangle = \frac{2(\Delta-1)^2}{\pi N}\sum_{k=0}^{N-1}\left(\frac{1}{2(\cos t-\cos\left(\vartheta+2\pi\frac{k}{N}\right))}\right)^{\Delta}. \tag{20}$$

The geodesic approximation [23–25] gives the same result up to a constant prefactor. Moreover, this formula coincides with the two-point correlator in the orbifold CFT obtained by usual image method [13, 32]. Thus, in the orbifold case we have a full agreement of GKPW, geodesic and CFT computations.

3.2 General case: temporal and angular dependences

In the general case the angle deficit breaks the asymptotic Virasoro symmetry at the boundary. This effect is sharply manifest in correlators given by the geodesic image method [32]. The geodesic two-point correlator acquires discontinuities in its angular dependence, which is reflected in the fig. 1. We see that there is a discrepancy between two holographic prescriptions, and GKPW prescriptions has the largest difference from the geodesic prescription in the region of long-range correlations, where the longest geodesics contribute. This difference, however, diminishes for large Δ.

The correlators calculated via GKPW and geodesic prescriptions also start exhibit slight differences in their time dependence if one moves the deficit parameter from the orbifold value. As illustrated in fig. 2A-C, once one deforms the orbifold, the GKPW inverse correlator obtains an additional zero. For general non-orbifold angle deficits geodesic and GKPW correlators have different number of poles. This effect does not depend on the value of angle and, in general, on the number of geodesic images: we see in Fig.2C that the number of images has decreased, and GKPW expression reflects this as well, but still keeps its extra zero. The comparison of plots C and D in Fig.2 shows that while increasing the conformal dimension does not insure that the analytic structure of correlators in the region between the dashed lines is not completely in agreement for increasing Δ, geodesic correlator still approaches the GKPW expression quantitatively.

Acknowledgements

The paper is based on the talk given at the 19th International Seminar on High Energy Physics "Quarks 2016" at Pushkin, Russia on 29 May - 4 June, 2016. I am very grateful to the organizers of the Quarks conference for the opportunity to give a talk and for inspiring atmosphere. I am also thankful to Dmitrii Ageev, Irina Aref'eva and Maria Tikhanovskaya for many useful discussions. The work is supported by the Russian Science Foundation (project 14-50-00005, Steklov Mathematical Institute).

References

[1] J. M. Maldacena, Adv. Theor. Math. Phys. **2**, 231-252 (1998), [hep-th/9711200].
[2] O. Aharony, S. S. Gubser, J. M. Maldacena, H. Ooguri and Y. Oz, Phys. Rept. **323**, 183 (2000) [hep-th/9905111].
[3] A. Maloney and E. Witten, JHEP **1002**, 029 (2010) [arXiv:0712.0155 [hep-th]].
[4] V. Balasubramanian *et al.*, Phys. Rev. D **84**, 026010 (2011) [arXiv:1103.2683 [hep-th]].
[5] E. Caceres and A. Kundu, JHEP **1209**, 055 (2012) [arXiv:1205.2354 [hep-th]].
[6] V. Balasubramanian *et al.*, JHEP **1304**, 069 (2013) [arXiv:1212.6066 [hep-th]].
[7] I. Aref'eva, A. Bagrov and A. S. Koshelev, JHEP **1307**, 170 (2013) [arXiv:1305.3267 [hep-th]].
[8] I. Ya. Aref'eva, Phys. Usp. **57**, 527 (2014).
[9] I. Y. Aref'eva, TMF, 182 (2015), 3, arXiv:1503.02185 [hep-th].
[10] S. Ryu and T. Takayanagi, Phys. Rev. Lett. **96**, 181602 (2006) [hep-th/0603001].
[11] V. E. Hubeny, M. Rangamani and T. Takayanagi, JHEP **0707**, 062 (2007) [arXiv:0705.0016 [hep-th]].
[12] T. Albash and C. V. Johnson, New J. Phys. **13**, 045017 (2011) [arXiv:1008.3027 [hep-th]].
[13] V. Balasubramanian, B. D. Chowdhury, B. Czech and J. de Boer, JHEP **1501**, 048 (2015) [arXiv:1406.5859 [hep-th]].
[14] J. Maldacena, S. H. Shenker and D. Stanford, "A bound on chaos," arXiv:1503.01409 [hep-th].

[15] A. L. Fitzpatrick, J. Kaplan and M. T. Walters, JHEP **1408** (2014) 145 [arXiv:1403.6829 [hep-th]].
[16] K. B. Alkalaev and V. A. Belavin, arXiv:1510.06685 [hep-th].
[17] S. Deser, R. Jackiw, and G. 't Hooft, Ann. Phys. **152** (1984) 220
[18] S. Deser and R. Jackiw, Ann. Phys. **153** (1984) 405.
[19] S. Deser, R. Jackiw, Commun.Math.Phys. 118 (1988) 495
[20] G. 't Hooft, Class. Quant. Grav. **13** (1996) 1023.
[21] I. Y. Arefeva and A. A. Bagrov, Theor. Math. Phys. **182**, 1 (2015)
[22] I. Arefeva, A. Bagrov, P. Saterskog and K. Schalm, arXiv:1508.04440 [hep-th].
[23] D. S. Ageev, I. Y. Aref'eva and M. D. Tikhanovskaya, arXiv:1512.03362 [hep-th].
[24] D. S. Ageev and I. Y. Aref'eva, arXiv:1512.03363 [hep-th].
[25] I. Y. Aref'eva, M. A. Khramtsov and M. D. Tikhanovskaya, arXiv:1604.08905 [hep-th].
[26] V. Balasubramanian and S. F. Ross, Phys. Rev. D **61**, 044007 (2000) [hep-th/9906226].
[27] S. S. Gubser, I. R. Klebanov, A. M. Polyakov, Phys. Lett. **B428**, 105-114 (1998), [hep-th/9802109].
[28] E. Witten, Adv. Theor. Math. Phys. **2**, 253-291 (1998), [hep-th/9802150].
[29] K. Skenderis and B. C. van Rees, JHEP **0905**, 085 (2009) [arXiv:0812.2909 [hep-th]].
[30] V. Balasubramanian, P. Kraus and A. E. Lawrence, Phys. Rev. D **59**, 046003 (1999) [hep-th/9805171].
[31] J. de Boer, M. M. Sheikh-Jabbari and J. Simon, Class. Quant. Grav. **28**, 175012 (2011) [arXiv:1011.1897 [hep-th]].
[32] I. Y. Aref'eva and M. A. Khramtsov, JHEP **1604**, 121 (2016) [arXiv:1601.02008 [hep-th]].

Vortex dynamics in nonrelativistic version of Abelian Higgs model: Effects of the medium on the vortex motion

Arkadii Kozhevnikov[1,2,⋆]

[1] *Laboratory of Theoretical Physics, S. L. Sobolev Institute for Mathematics*
[2] *Novosibirsk State University*

Abstract. The closed vortex dynamics is considered in the nonrelativistic version of the Abelian Higgs Model. The effect of the exchange of excitations propagating in the medium on the vortex string motion is taken into account. The obtained are the effective action and the equation of motion both including the exchange of the propagating excitations between the distant segments of the vortex and the possibility of its interaction with the static fermion asymmetric background. They are applied to the derivation of the time dependence of the basic geometrical contour characteristics.

1 Introduction

The string-like (or vortex) solutions of the field-theoretical models are widely discussed in application to numerous physical systems, from the really observed in condensed matter physics (quantized vortices in He^4 [1], Abrikosov lines in type II superconductors [2], vortices in Bose - Einstein condensates [3] to hypothetical cosmic strings [4]. Of particular interest are the situations when the dynamics of the vortex contours can be deduced from the field equations of the underlying field theory. The dynamics of cosmic strings was studied in the framework of the Nambu-Goto action [5] which was shown to result from the relativistic version of the Abelian Higgs model (AHM) [6]. Transverse motions of the Nambu-Goto string are typically with the velocity of light [5]. The solution in the form of the static cosmic string was obtained from the static version of AHM [7]. Another limiting case is the nonrelativistic one, which can be derived from the nonrelativistic Abelian Higgs model (NRAHM). The static limit of this model coincides with the Ginzburg-Landau theory [8]. Another interesting aspect of the time-dependent gauge vortex solutions is that they can elucidate the interrelations among such characteristics of the closed contours as the helicity of gauge field forming the vortex, the writhe and torsion (twist) numbers and their possible dependence on time [9–11].

The aim of the present talk is to consider the dynamical evolution of the closed gauge vortex string in NRAHM, including the possibility of its interaction with the static fermion asymmetric background. The corresponding equation of motion is obtained, side by side with expressions for the time derivatives of the curvature and torsion of the gauge vortex contour [12–14].

⋆e-mail: kozhev@math.nsc.ru

2 Nonrelativistic Abelian Higgs Model with the gauge vortex

Nonrelativistic Abelian Higgs model incorporating gauge vortices is given by the Lagrangian density

$$\begin{aligned}\mathcal{L} &= \frac{1}{8\pi}(\boldsymbol{E}^2 - \boldsymbol{H}^2) - \frac{g}{2}(|\psi|^2 - n_0)^2 + \frac{1}{2}[\psi^*(i\hbar\partial_t - q\varphi + qa_0)\psi + \text{c.c.}] - \\ &\frac{1}{2m}\left|\left(-i\hbar\boldsymbol{\nabla} - \frac{q}{c}\boldsymbol{A} + \frac{q}{c}\boldsymbol{a}\right)\psi\right|^2 - \rho_0\varphi. \end{aligned} \tag{1}$$

The source of the vortex with the contour $\boldsymbol{X}(\sigma, t)$ is the singular phase of the Higgs field χ_s, $a_\mu = -\frac{\hbar c}{q}\partial_\mu \chi_s$:

$$[\boldsymbol{\nabla}\times\boldsymbol{\nabla}]\chi_s = 2\pi \oint d\sigma \boldsymbol{X}'\delta^{(3)}(\boldsymbol{x} - \boldsymbol{X}(\sigma, t)). \tag{2}$$

The solution of Eq. (2) is

$$\begin{aligned} a_0 &= \frac{\hbar}{2q}\oint \frac{([\dot{\boldsymbol{X}}\times\boldsymbol{X}'], \boldsymbol{x}-\boldsymbol{X})}{|\boldsymbol{x}-\boldsymbol{X}|^3}d\sigma, \\ \boldsymbol{a} &= \frac{\hbar c}{2q}\oint \frac{[\boldsymbol{X}'\times(\boldsymbol{x}-\boldsymbol{X})]}{|\boldsymbol{x}-\boldsymbol{X}|^3}d\sigma. \end{aligned} \tag{3}$$

In what follows, all fields are represented as the sum of the background fields and the fluctuations. As usual, the background fields are found from the condition that the terms in the action which are linear in the fluctuations, should vanish while the terms quadratic in the latter will give the quantum corrections to the effective action. The expressions for the background fields are [12]:

$$\begin{aligned} \boldsymbol{A}(\boldsymbol{x}, t) &= \int \frac{d^3k}{(2\pi)^3}\frac{i\Phi_0/\lambda_L^2}{\boldsymbol{k}^2(\boldsymbol{k}^2 + 1/\lambda_L^2)} \times \oint d\sigma[\boldsymbol{k}\times\boldsymbol{X}']e^{i(\boldsymbol{k},\boldsymbol{x}-\boldsymbol{X})}, \\ \boldsymbol{H}(\boldsymbol{x}, t) &= \int \frac{d^3k}{(2\pi)^3}\frac{\Phi_0/\lambda_L^2}{\boldsymbol{k}^2 + 1/\lambda_L^2} \times \oint d\sigma \boldsymbol{X}' e^{i(\boldsymbol{k},\boldsymbol{x}-\boldsymbol{X})}, \\ \boldsymbol{E}(\boldsymbol{x}, t) &= -\int \frac{d^3k}{(2\pi)^3}\frac{\Phi_0/(c\lambda_L^2)}{\boldsymbol{k}^2 + 1/\lambda_L^2}\oint d\sigma e^{-i(\boldsymbol{k},\boldsymbol{x}-\boldsymbol{X})}\left\{[\dot{\boldsymbol{X}}\times\boldsymbol{X}'] - \frac{\boldsymbol{k}(\boldsymbol{k}[\dot{\boldsymbol{X}}\times\boldsymbol{X}'])}{\boldsymbol{k}^2}\right\}, \end{aligned} \tag{4}$$

where $\Phi_0 = \frac{2\pi\hbar c}{q}$, $\lambda_L = \sqrt{\frac{mc^2}{4\pi n_0 q^2}}$, and the constrain $\rho_0 + qn_0 = 0$ provides the net charge neutrality. The modulus of the Higgs field is assumed to be uniform except the vortex lines where it tends to zero.

When evaluating the action of the background fields,

$$S_{\text{bg}} = \int d^4x\left[\frac{1}{8\pi}(\boldsymbol{E}^2 - \boldsymbol{H}^2) - \frac{1}{8\pi\lambda_L^2}(\boldsymbol{A} - \boldsymbol{a})^2 + n_0 q a_0\right],$$

it is useful to represent it in the form $S_{\text{bg}} = S^{(0)}_{\text{bg}} + \Delta S_{\text{bg}}$, where

$$\begin{aligned} S^{(0)}_{\text{bg}} &= \frac{\Phi_0^2}{8\pi}\int dt \frac{d^3k}{(2\pi)^3}\left(\frac{1/\lambda_L^2}{\mathbf{k}^2 + 1/\lambda_L^2}\right)^2 \oint d\sigma_1 d\sigma_2 e^{-i\boldsymbol{k}\boldsymbol{X}_{12}} \times \\ &\left\{[\dot{\boldsymbol{X}}_1\times\boldsymbol{X}'_1]\cdot[\dot{\boldsymbol{X}}_2\times\boldsymbol{X}'_2]/c^2 - (\boldsymbol{X}'_1\boldsymbol{X}'_2)(1 + \lambda_L^2\boldsymbol{k}^2)\right\} + n_0 q\int d^4x a_0 = \\ &\frac{\Phi_0^2}{32\pi^2\lambda_L^2}\int dt d\sigma_1 d\sigma_2 e^{-|\boldsymbol{X}_{21}|/\lambda_L}\left\{\frac{[\dot{\boldsymbol{X}}_1\times\boldsymbol{X}'_1][\dot{\boldsymbol{X}}_2\times\boldsymbol{X}'_2]}{2c^2\lambda_L} - \frac{(\boldsymbol{X}'_1\boldsymbol{X}'_2)}{|\boldsymbol{X}_{21}|}\right\} + n_0 q\int d^4x a_0 \end{aligned} \tag{5}$$

is the infrared finite contribution while

$$\Delta S_{\rm bg} = -\frac{\Phi_0^2}{8\pi c^2}\int dt \int \frac{d^3k}{(2\pi)^3 \boldsymbol{k}^2}\left(\frac{1/\lambda_L^2}{\mathbf{k}^2+1/\lambda_L^2}\right)^2 \times$$
$$\oint d\sigma_1 d\sigma_2 e^{-ik X_{12}}(\boldsymbol{k}[\dot{\boldsymbol{X}}_1 \times \boldsymbol{X}_1'])(\boldsymbol{k}[\dot{\boldsymbol{X}}_2 \times \boldsymbol{X}_2']) \qquad (6)$$

represents the infrared divergent contribution. Hereafter the notation are: $\boldsymbol{X}_{1,2} \equiv \boldsymbol{X}(\sigma_{1,2})$, dot and prime stand for the derivatives over time and the corresponding contour parameter $\sigma_{1,2}$, respectively.

Using $\psi = n^{1/2}e^{i\chi}$, the fields are represented in terms of the background fields and fluctuations: $\varphi \to \varphi + \delta\varphi$, $\boldsymbol{A} \to \boldsymbol{A} + \delta\boldsymbol{A}$, $n \to n_0 + \delta n$, $\chi \to \delta\chi$, where δn and $\delta\chi$ are the Higgs mode and Goldstone mode, respectively. The action of fluctuations is

$$S_{\rm f}^{(0)} = \int d^4x \left\{\frac{1}{2} f^T M f - \frac{1}{8\pi}\delta\boldsymbol{A}\left(-\frac{1}{c^2}\partial_t^2 + \boldsymbol{\nabla}^2 - \frac{1}{\lambda_L^2}\right)\delta\boldsymbol{A} + q a_0 \delta n\right\} + ..., \qquad (7)$$

where $f^T = (\delta n, \delta\chi, \delta\varphi)$, and .... represents the derivatives $\partial^3 X/\partial t^3$ or higher which can be neglected in nonrelativistic case [12]. The matrix M looks like

$$M = \begin{pmatrix} \frac{\hbar^2}{4mn_0}\boldsymbol{\nabla}^2 - g & -\hbar\partial_t & -q \\ \hbar\partial_t & \frac{\hbar^2 n_0}{m}\boldsymbol{\nabla}^2 & 0 \\ -q & 0 & -\frac{1}{4\pi}\boldsymbol{\nabla}^2 \end{pmatrix},$$

which permits one to obtain the propagator matrix in the four-momentum representation:

$$M^{-1} = \frac{1}{\left(\frac{\epsilon_B^2}{\hbar^2} + \frac{c^2}{\lambda_L^2} - \omega^2 - i0\right)} \begin{pmatrix} -\frac{n_0}{m}\boldsymbol{k}^2 & -\frac{i\omega}{\hbar} & -\frac{4\pi n_0 q}{m} \\ \frac{i\omega}{\hbar} & -\frac{m}{\hbar^2 \boldsymbol{k}^2 n_0}\left(\frac{\epsilon_B^2}{\hbar^2} + \frac{c^2}{\lambda_L^2}\right) & \frac{4\pi i\omega q}{\hbar \boldsymbol{k}^2} \\ -\frac{4\pi n_0 q}{m} & -\frac{4\pi i\omega q}{\hbar \boldsymbol{k}^2} & \frac{4\pi}{\boldsymbol{k}^2}\left(\frac{\epsilon_B^2}{\hbar^2} - \omega^2\right) \end{pmatrix},$$

where $\epsilon_B^2 \equiv \epsilon_B^2(\boldsymbol{k}) = \left(\frac{\hbar^2 \boldsymbol{k}^2}{2m}\right)^2 + \hbar^2 c_s^2 \boldsymbol{k}^2$ is the square of the Bogolyubov spectrum [15], $c_s = \sqrt{\frac{n_0 g}{m}}$ is the sound velocity. The freely propagating excitations obey the equations of motion:

$$0 = \hbar\partial_t\delta\chi + g\delta n + q\delta\varphi - \frac{\hbar^2}{4mn_0}\boldsymbol{\nabla}^2\delta n,$$
$$0 = \partial_t\delta n + \frac{\hbar n_0}{m}\boldsymbol{\nabla}^2\delta\chi.$$

Upon excluding $\delta\chi$ one obtains

$$\left(-\partial_t^2 + c_s^2\boldsymbol{\nabla}^2 - \frac{c^2}{\lambda_L^2} - \frac{\hbar^2}{4m^2}\boldsymbol{\nabla}^4\right)\delta n = 0$$

resulting in the above dispersion law. The static limit with $q = 0$ gives $\boldsymbol{\nabla}^2\delta n - \xi^{-2}\delta n = 0$, where $\xi = \hbar/2mc_s$ is the correlation length which, by the order of magnitude, is the distance where the Higgs field goes to zero near the string core. Formally, ξ is the ultraviolet cutoff reflecting the ignorance of the Higgs field profile at short distances.

The dominant quantum correction to the action due to exchange of excitations propagating in the medium is

$$\Delta S_{\rm f} = \frac{1}{2}\int \frac{d^4k}{(2\pi)^4}\left(\frac{\epsilon_B^2}{\hbar^2} + \frac{c^2}{\lambda_L^2} - \omega^2 - i0\right)^{-1} \frac{n_0 q^2}{m}|a_{0k}|^2 \boldsymbol{k}^2. \qquad (8)$$

DOI: 10.1051/epjconf/201612505011

It is treated by using the derivative expansion for $I = \int_{-\infty}^{\infty} \frac{d\omega}{2\pi} \cdot \frac{|f_k|^2}{\omega^2 - \Omega_k^2 + i0}$:

$$I = \int_{-\infty}^{\infty} dt d\tau \frac{d\omega}{2\pi} \frac{f_k(t) f_k^*(t+\tau) e^{i\omega\tau}}{\omega^2 - \Omega_k^2 + i0} = \int dt f_k(t) \sum_{l=0}^{\infty} \frac{(-i)^l}{l!} \frac{d^l f_k^*}{dt^l} \left[\frac{\partial^l}{\partial\omega^l} \frac{1}{\omega^2 - \Omega_k^2} \right]_{\omega=0} =$$
$$\int dt \left[-\frac{|f_k(t)|^2}{\Omega_k^2} - \frac{1}{2\Omega_k^4} \left(f_k \frac{\partial^2 f_k^*}{\partial t^2} + \text{c.c.} \right) + \cdots \right].$$

Then the leading correction to the effective action arising from the exchange of fluctuations can be represented in the form

$$\Delta S_f = \frac{\Phi_0^2}{8\pi\lambda_L^2} \int dt d\sigma_1 d\sigma_2 \frac{d^3k}{(2\pi)^3} \frac{e^{ik\cdot X_{21}}}{\frac{\hbar^2 k^4}{4m^2} + c_s^2 k^2 + \frac{c^2}{\lambda_L^2}} \frac{(\boldsymbol{k}[\dot{\boldsymbol{X}}_1 \times \boldsymbol{X}_1'])(\boldsymbol{k}[\dot{\boldsymbol{X}}_2 \times \boldsymbol{X}_2'])}{\boldsymbol{k}^2}.$$

It has the finite part contributing to the effective action, and the infrared divergent piece cancelling analogous divergence in the background action Eq. (6). In fact, the sum of the infrared divergent contribution of the background action and that of due to the exchange of the Higgs mode represented in the form

$$\Delta S_{\text{eff}} = \frac{\Phi_0^2}{8\pi c^2} \int dt d\sigma_2 d\sigma_2 [\dot{\boldsymbol{X}}_1 \times \boldsymbol{X}_1']_i [\dot{\boldsymbol{X}}_2 \times \boldsymbol{X}_2']_j \nabla_{21i} \nabla_{21j} I(\sigma_1, \sigma_2),$$
$$I(\sigma_1, \sigma_2) = \frac{\partial}{\partial\lambda_L^2} \left[\lambda_L^2 \int \frac{d^3k}{(2\pi)^3} \left(\frac{1}{\boldsymbol{k}^2} - \frac{1}{\boldsymbol{k}^2 + 1/\lambda_L^2} \right) e^{ikX_{21}} \right] -$$
$$\int \frac{d^3k}{(2\pi)^3} e^{ikX_{21}} \left[\frac{1}{\boldsymbol{k}^2} - \frac{\boldsymbol{k}^2 + 1/\xi^2}{\boldsymbol{k}^4 + \boldsymbol{k}^2/\xi^2 + 1/\xi^2\lambda_s^2} \right],$$

demonstrates explicitly the cancellation of the infrared divergences. Here, the new length parameter arises: $\lambda_s = \lambda_L \cdot \frac{c_s}{c} \ll \lambda_L$. In what follows, the assumption $\xi \ll \lambda_s \ll \lambda_L$ is adopted.

All expressions in the vortex action are of the type $\int dt \oint d\sigma_1 d\sigma_2 F(\sigma_1, \sigma_2) \exp^{-|X_{21}|/\lambda_L}$, i.e. non-local. The expansion $\boldsymbol{X}_2 = \boldsymbol{X}_1 + z\boldsymbol{X}_1' + \frac{z^2}{2}\boldsymbol{X}_1'' + \frac{z^3}{6}\boldsymbol{X}_1''' + \cdots$ in $z = \sigma_2 - \sigma_1$ admits the local limit. The local form of the action is

$$S_{\text{eff}} \approx \left(\frac{\Phi_0}{4\pi\lambda_L} \right)^2 \int dt d\sigma \left\{ \frac{1}{2c_s^2} [\dot{\boldsymbol{X}} \times \boldsymbol{X}']^2 \ln \frac{\lambda_s}{\xi} - \boldsymbol{X}'^2 \ln \frac{\lambda_L}{\xi} \right\} + n_0 q \int d^4x a_0.$$

The leading kinetic term $[\dot{\boldsymbol{X}} \times \boldsymbol{X}']^2/c_s^2$ is due to the exchange of Higgs field excitations. It is strongly enhanced as compared with the bare one. The similar kinetic term in the background action is suppressed as c_s^2/c^2 and does not contain the logarithmic enhancement factor $\ln \lambda_s/\xi$.

3 Equation of the vortex motion

When obtaining the equations of motion the local limit should be taken after making the variations. They are fulfilled by using the expressions $\delta \int d^4x a_0 = -\frac{\Phi_0}{c} \int dt d\sigma (\delta\boldsymbol{X}[\dot{\boldsymbol{X}} \times \boldsymbol{X}'])$,

$$\delta I \equiv \delta \oint d\sigma_1 d\sigma_2 (\boldsymbol{X}_1' \boldsymbol{X}_2') \frac{e^{-|X_{21}|/\lambda_L}}{|\boldsymbol{X}_{21}|} = 2 \oint d\sigma_1 d\sigma_2 (\delta\boldsymbol{X}_1 \cdot [\boldsymbol{X}_1' \times [\boldsymbol{X}_{21} \times \boldsymbol{X}_2']]) \frac{1 + |\boldsymbol{X}_{21}|/\lambda_L}{|\boldsymbol{X}_{21}|^3} \times$$
$$e^{-|X_{21}|/\lambda_L} = 2 \oint d\sigma \int_{-\infty}^{\infty} dz \frac{e^{-|z|/\lambda_L}}{2|z|} \delta\boldsymbol{X} \cdot [\boldsymbol{X}' \times [\boldsymbol{X}' \times \boldsymbol{X}'']] = -2 \ln \frac{\lambda_L}{\xi} \oint d\sigma \delta\boldsymbol{X} \cdot \boldsymbol{X}'',$$

and analogous (cumbersome) expressions for the variation of the terms containing the time derivatives. The variation of the effective action is

$$\begin{aligned}\delta S_{\text{eff}} &= -\left(\frac{\Phi_0}{4\pi\lambda_L}\right)^2 \int dtd\sigma\delta \boldsymbol{X}\cdot\left\{\frac{1}{c_s^2}\left[\boldsymbol{X}'\times\frac{\partial}{\partial t}[\dot{\boldsymbol{X}}\times\boldsymbol{X}']\right]\ln\frac{\lambda_s}{\xi}-\boldsymbol{X}''\ln\frac{\lambda_L}{\xi}\right\}-\\ &\frac{n_0 q\Phi_0}{c}\int dtd\sigma(\delta\boldsymbol{X}[\dot{\boldsymbol{X}}\times\boldsymbol{X}']),\end{aligned}$$

resulting in the equation of the vortex motion

$$[\dot{\boldsymbol{X}}\times\boldsymbol{X}'] = \frac{\hbar}{2m}\left(\boldsymbol{X}''\ln\frac{\lambda_L}{\xi}+\frac{1}{c_s^2}\left[\frac{\partial}{\partial t}[\dot{\boldsymbol{X}}\times\boldsymbol{X}']\times\boldsymbol{X}'\right]\ln\frac{\lambda_s}{\xi}\right). \tag{9}$$

When the exchange of excitation is turning off, $c_s\to\infty$, the solution of Eq. (9) is the ring vortex moving in the direction perpendicular to its plane with velocity [12]

$$\dot{\boldsymbol{X}} = \boldsymbol{b}\frac{\hbar}{2m}|\boldsymbol{X}''|\ln\frac{\lambda_L}{\xi}. \tag{10}$$

One can include the static background of chiral fermions (with asymmetry characterized by the chemical potential μ_f) coupled to the gauge field A_μ. This results in the term $-\mu_f N_{\text{CS}}$ [16–20], where

$$N_{\text{CS}} = \frac{1}{\Phi_0^2}\int d^3x(\boldsymbol{A},[\boldsymbol{\nabla}\times\boldsymbol{A}]) = \frac{h_A}{\Phi_0^2}$$

is the Abelian Chern-Simons number related with helicity

$$\begin{aligned}h_A &= i\Phi_0^2\oint d\sigma_1 d\sigma_2\int\frac{d^3k}{(2\pi)^3\boldsymbol{k}^2}\left(\frac{1/\lambda_L^2}{\boldsymbol{k}^2+1/\lambda_L^2}\right)^2\times\\ &\left(\boldsymbol{k},\left[\frac{\partial\boldsymbol{X}(\sigma_1)}{\partial\sigma_1}\times\frac{\partial\boldsymbol{X}(\sigma_2)}{\partial\sigma_2}\right]\right)\exp\left[-i(\boldsymbol{k},\boldsymbol{X}(\sigma_1)-\boldsymbol{X}(\sigma_2)\right].\end{aligned}$$

The variation of the Chern-Simons induced contribution to the equation of the vortex motion is deduced from the variation of helicity:

$$\begin{aligned}\delta h_A &= 2\Phi_0^2\int\frac{d^3k}{(2\pi)^3}\left(\frac{1/\lambda_L^2}{\boldsymbol{k}^2+1/\lambda_L^2}\right)^2\oint d\sigma_1 d\sigma_2(\delta\boldsymbol{X}_1,[\boldsymbol{X}_1'\times\boldsymbol{X}_2'])e^{-i(\boldsymbol{k},\boldsymbol{X}_{12})} =\\ &\frac{\Phi_0^2}{4\pi\lambda_L^3}\oint d\sigma_1 d\sigma_2(\delta\boldsymbol{X}_1,[\boldsymbol{X}_1'\times\boldsymbol{X}_2'])e^{-|\boldsymbol{X}_{12}|/\lambda_L}\\ &= \frac{\Phi_0^2}{8\pi\lambda_L^3}\oint d\sigma([\boldsymbol{X}'\times\boldsymbol{X}'''],\delta\boldsymbol{X})\int_{-\infty}^{\infty}z^2e^{-|z|/\lambda_L}dz = \frac{\Phi_0^2}{2\pi}\oint d\sigma([\boldsymbol{X}'\times\boldsymbol{X}'''],\delta\boldsymbol{X}).\end{aligned}$$

Taking all the above considerations into account one obtains the equation of motion which, in the gauge $(\dot{\boldsymbol{X}},\boldsymbol{X}')=0$, has the following form:

$$\frac{1}{c_0^2}\ddot{\boldsymbol{X}}-\boldsymbol{X}''+\frac{1}{\gamma}[\dot{\boldsymbol{X}}\times\boldsymbol{X}']+\frac{\tilde{\mu}}{\gamma}[\boldsymbol{X}'\times\boldsymbol{X}''']=0, \tag{11}$$

DOI: 10.1051/epjconf/201612505011

where

$$\begin{aligned}
\gamma &= \frac{\hbar}{2m}\ln\frac{\lambda_L}{\xi},\\
T_0 &= \frac{\hbar}{2mc_s^2}\ln\frac{\lambda_s}{\xi},\\
\tilde{\mu} &= \frac{\mu_f}{4\pi^2\hbar n_0},\\
c_0 &= \left(\frac{\gamma}{T_0}\right)^{1/2} = c_s\left(\frac{\ln\lambda_L/\xi}{\ln\lambda_s/\xi}\right)^{1/2}.
\end{aligned} \tag{12}$$

It looks like the wave equation added with the two additional terms. The term $\propto [\dot{\boldsymbol{X}}\times\boldsymbol{X}']$ is the analog of the Magnus force acting on the moving vortex due to the nonzero circulation of the supercurrent. The term $\propto [\boldsymbol{X}'\times\boldsymbol{X}''']$ is induced by the anomalous current $\boldsymbol{j}_{\mathrm{an}} \propto \mu_f \boldsymbol{H}$ widely discussed in connection with chiral magnetic effect [21]. This current is subjected to the action of the Lorentz force $\propto [\boldsymbol{j}_{\mathrm{an}}\times\boldsymbol{H}]$:

$$f_L \propto \mu_f \int_{-\infty}^{\infty} dz[\boldsymbol{X}'(\sigma)\times\boldsymbol{X}'(\sigma+z)]e^{-|z|/\lambda_L} \propto \mu_f[\boldsymbol{X}'\times\boldsymbol{X}'''].$$

The static solution is helix with the constant curvature κ =const and the torsion related with the amount of fermion asymmetry $\tau = \gamma/\tilde{\mu}$ [13]. See the next section for the definition of these geometric quantities.

4 Time derivatives of the basic contour characteristics

The basic contour vectors of normal $\boldsymbol{n}$, bi-normal $\boldsymbol{b}$, and tangent $\boldsymbol{X}'$ constitute the right triple, $\boldsymbol{X}' = [\boldsymbol{n}\times\boldsymbol{b}]$ and obey the Frenet-Serret equations:

$$\frac{\partial}{\partial\sigma}\begin{pmatrix}\boldsymbol{X}'\\ \boldsymbol{n}\\ \boldsymbol{b}\end{pmatrix} = \begin{pmatrix}0 & \kappa & 0\\ -\kappa & 0 & \tau\\ 0 & -\tau & 0\end{pmatrix}\begin{pmatrix}\boldsymbol{X}'\\ \boldsymbol{n}\\ \boldsymbol{b}\end{pmatrix}$$

They are of unit length. Here, κ and τ are the curvature and the torsion, respectively. One can write the equations determining the time derivatives of the triple $(\boldsymbol{n}, \boldsymbol{b}, \boldsymbol{X}')$ using the fact that they are the unit orthogonal vectors:

$$\frac{\partial}{\partial t}\begin{pmatrix}\boldsymbol{X}'\\ \boldsymbol{n}\\ \boldsymbol{b}\end{pmatrix} = \begin{pmatrix}0 & a_{Xn} & a_{Xb}\\ a_{nX} & 0 & a_{nb}\\ a_{bX} & a_{bn} & 0\end{pmatrix}\begin{pmatrix}\boldsymbol{X}'\\ \boldsymbol{n}\\ \boldsymbol{b}\end{pmatrix} \equiv \hat{A}\begin{pmatrix}\boldsymbol{X}'\\ \boldsymbol{n}\\ \boldsymbol{b}\end{pmatrix} \tag{13}$$

Using the relation the commutativity of derivatives $\partial_t\partial_\sigma - \partial_\sigma\partial_t = 0$ one can further restrict the matrix $\hat{A}$:

$$\hat{A} = \begin{pmatrix}0 & a_{Xn} & a_{Xb}\\ -a_{Xn} & 0 & a_{nb}\\ -a_{Xb} & -a_{nb} & 0\end{pmatrix},$$

and obtain the expressions

$$\begin{aligned}
a_{nb} &= (\tau a_{Xn} + a'_{Xb})/\kappa,\\
\dot{\kappa} &= a'_{Xn} - \tau a_{Xb},\\
\dot{\tau} &= \kappa a_{Xb} + a'_{nb}.
\end{aligned}$$

Introducing the vector $\boldsymbol{W} = [\dot{\boldsymbol{X}} \times \boldsymbol{X}'] = W_n \boldsymbol{n} + W_b \boldsymbol{b}$, $\dot{\boldsymbol{X}} = W_n \boldsymbol{b} - W_b \boldsymbol{n}$ one gets $\frac{\partial X'}{\partial t} = \boldsymbol{b}(W_n' - \tau W_b) - \boldsymbol{n}(W_b' + \tau W_n)$ which permits one to identify

$$\begin{aligned} a_{Xn} &= -W_b' - \tau W_n, \\ a_{Xb} &= W_n' - \tau W_b. \end{aligned}$$

Using the above relations the time derivatives of the curvature and torsion can be obtained:

$$\begin{aligned} \dot{\kappa} &= -(W_b' + \tau W_n)' - \tau(W_n' - \tau W_b), \\ \dot{\tau} &= \kappa(W_n' - \tau W_b) + \left\{\left[(W_n' - \tau W_b)' - \tau(W_b' + \tau W_n)\right]/\kappa\right\}'. \end{aligned} \tag{14}$$

The system of equations is closed by adding the equations of motion

$$\begin{aligned} W_n &= \kappa(\gamma - \tilde{\mu}\tau) + T_0\left\{\dot{W}_b + \left[(W_n' - \tau W_b)' - \tau(W_b' + \tau W_n)\right] \times \right. \\ &\quad \left. W_n/\kappa\right\}, \\ W_b &= \tilde{\mu}\kappa' - T_0\left\{\dot{W}_n - \left[(W_n' - \tau W_b)' - \tau(W_b' + \tau W_n)\right] W_b/\kappa\right\}. \end{aligned} \tag{15}$$

The interesting particular case is one with $T_0 = 0$ and $\tilde{\mu} = 0$ corresponding to the neglect of the exchange of excitations and the fermion asymmetry. Then, after re-scaling $\kappa \to \gamma^{-1/2}\kappa$, $\tau \to \gamma^{-1/2}\tau$, $\sigma \to \gamma^{1/2}\sigma$ one gets

$$\begin{aligned} \dot{\kappa} &= -(2\kappa'\tau + \kappa\tau'), \\ \dot{\tau} &= \left[\kappa\kappa' + \left(\frac{\kappa''}{\kappa} - \tau^2\right)'\right]. \end{aligned}$$

These are the Da-Rios-Betchov equations [22] first proposed in hydrodynamics of ideal fluid in 1906.

The equations (14) and (15) are strongly nonlinear, hence one needs some approximations to get consequences from them. We choose the solution exact in μ_f and of the first order in T_0. The derivation is straightforward, but the expressions are still cumbersome. Instead, one can find such useful interesting quantity as the time derivative of the total torsion:

$$\oint \dot{\tau} d\sigma = \frac{\hbar\mu_f}{64\pi^2 m^2 c_s^2 n_0} \times \ln\frac{\lambda_L}{\xi} \times \ln\frac{\lambda_s}{\xi} \oint \kappa^4 \tau' d\sigma. \tag{16}$$

The total torsion is related with the twist:

$$\mathrm{Tw} = \frac{1}{2\pi}\oint \tau d\sigma = \frac{1}{2\pi}\oint (\boldsymbol{X}', [\boldsymbol{n}' \times \boldsymbol{n}]) d\sigma.$$

On the other hand, the time derivative of helicity is

$$\begin{aligned} \dot{h}_A &= \frac{\Phi_0^2}{2\pi}\oint d\sigma([\dot{\boldsymbol{X}} \times \boldsymbol{X}'], \boldsymbol{X}''') = \frac{\Phi_0^2}{2\pi}\oint d\sigma(\boldsymbol{W}, \kappa'\boldsymbol{n} + \kappa\tau\boldsymbol{b}) = \\ &\frac{\Phi_0^2}{2\pi}\oint d\sigma(\kappa' W_n + \kappa\tau W_b) = -\frac{\Phi_0^2}{2\pi}\oint \dot{\tau} d\sigma \equiv -\Phi_0^2 \frac{d\mathrm{Tw}}{dt}. \end{aligned}$$

The writhe number and its variation are, respectively,

$$\mathrm{Wr} = \frac{1}{4\pi}\oint d\sigma_1 \oint d\sigma_2 \frac{(\boldsymbol{X}_{12}, [\boldsymbol{X}_1' \times \boldsymbol{X}_2'])}{|\boldsymbol{X}_{12}|^3} = -i\int \frac{d^3k}{(2\pi)^3}\oint d\sigma_1 d\sigma_2 \frac{(\boldsymbol{k}, [\boldsymbol{X}_2' \times \boldsymbol{X}_1'])}{\boldsymbol{k}^2} \times e^{i\boldsymbol{k}\cdot\boldsymbol{X}_{21}}$$

and

$$\begin{aligned}\delta \mathrm{Wr} &= -i\int \frac{d^3k}{(2\pi)^3}\frac{\boldsymbol{k}}{\boldsymbol{k}^2}\oint d\sigma_1 d\sigma_2 \left\{[\delta X_2' \times X_1'] + [X_2' \times \delta X_1'] + i[X_2' \times X_1'](\boldsymbol{k}\cdot X_{21})\right\} e^{i\boldsymbol{k}\cdot X_{21}} = \\ &\int \frac{d^3k}{(2\pi)^3}\oint d\sigma_1 d\sigma_2 \,(\delta X_{21}\cdot [X_2' \times X_1'])\, e^{i\boldsymbol{k}\cdot X_{21}} = \\ &\oint d\sigma_1 d\sigma_2 \,(\delta X_{21}\cdot [X_2' \times X_1'])\,\delta^{(3)}(X_{21}). \end{aligned} \tag{17}$$

The ambiguity $0\times\infty$ in Eq. (17) is regularized with the help of expression

$$\delta^{(3)}(X) = \lim_{\xi\to 0}\frac{e^{-X^2/2\xi^2}}{(2\pi)^{3/2}\xi^3},$$

so that the regularized variation of the writhe becomes

$$\begin{aligned}\delta \mathrm{Wr}_{\mathrm{reg}} &= \lim_{\xi\to 0}\frac{1}{(2\pi)^{3/2}\xi^3}\oint d\sigma_1 d\sigma_2 e^{-X_{21}^2/2\xi^2}(\delta X_{21}, [X_2' \times X_1']) = \\ &\lim_{\xi\to 0}\frac{1}{(2\pi)^{3/2}2\xi^3}\oint d\sigma \int_{-\infty}^{\infty} dz z^2 e^{-z^2/2\xi^2} \times (\delta X, [X' \times X''']) = \\ &\frac{1}{2\pi}\oint d\sigma(\delta X, [X' \times X''']) = \frac{\delta h_A}{\Phi_0^2}.\end{aligned}$$

The time derivative of writhe,

$$\frac{d\mathrm{Wr}}{dt} = \frac{\dot{h}_A}{\Phi_0^2},$$

permits one to establish the "conservation law":

$$\frac{d}{dt}(\mathrm{Wr} + \mathrm{Tw}) = 0$$

The sum Wr+Tw is known from topology as the topological invariant called self-linking [23, 24]. It is obtained here as the consequence of the dynamical equations govern the gauge vortex motion.

5 Discussion and conclusion

We have shown that the vortex string motion admits the non-relativistic regime if one takes into account the effects of the medium resulting in the exchange of the excitations between the different segments of the string. However, the characteristic velocity of the transverse motion c_0 in the last line of Eq. (12) exceeds the velocity of sound. This may result in the Vavilov-Cherenkov damping. We leave this problem for a separate work. The kinetic term becomes strongly modified due to the exchange of fluctuations. The ring vortex motion due to the term $[\dot{X} \times X']$ in the medium consisting of the charged scalar field condensate is possible. The interaction with the fermion asymmetric background results in additional term $[X' \times X''']$ in equation of motion because of the anomalous current $\boldsymbol{j}_{\mathrm{an}} \propto \mu_f \boldsymbol{B}$. The anomaly is the quantum effect hence the appearance of the Planck constant in the equation for such macroscopic quantity as the time derivative of the total twist, Eq. (16). Interesting enough, but the topological conservation of the sum Tw+Wr is shown here to result from the dynamics of the field system.

As for the application to physics, one can hardly hope to observe the closed or curved vortices in type II superconductors immersed into the static external magnetic fields. However, their production could be quite possible in the fast temperature quench [25]. See, for example, Ref. [26], where such a mechanism was considered in the framework of the 2D Abelian Higgs Model. Also, numerous simulations of the phase transitions in the early Universe point to a considerable portion of the closed cosmic strings [4]. The type of the model considered in the present work is one where the charged particles interacting with the gauge field, condense uniformly except for the vortex core. The total neutrality is provided by some static background of the opposite charge. The cosmic strings of such type are not excluded in the models of the dark (hidden) sector [27, 28]. However, the relation of the present work parameters q, n_0, and g with ones from the models of the hidden sector, goes beyond the scope of this communication.

References

[1] R. J. Donnelly, *Quantized Vortices in Helium II* (Cambridge University Press, 1991).
[2] A. A. Abrikosov, ZhETF **32**, 1442 (1957).
[3] K. W. Madison, F. Chavy, W. Wohlleben, and J. Dalibard, Phys. Rev. Lett. **84**, 806 (2000).
[4] A. Vilenkin and E. P. S. Shellard, *Cosmic Strings and other Topological Defects*(Cambridge University Press, 1994).
[5] J. Scherk, Rev. Mod. Phys. **47**, 123 (1975).
[6] D. Förster, Nucl. Phys. **B81**, (1974) 84.
[7] H. B. Nielsen and P. Olesen, Nucl. Phys. B61, 45 (1973).
[8] V. L. Ginzburg and L. D. Landau, ZhETF **20**, 1064 (1950).
[9] H. K. Moffatt, J. Fluid Mech. **35**, 117 (1969).
[10] M. A. Berger and G. B. Field, J. Fluid Mech. **147**, 133 (1984).
[11] H. K. Moffatt and R. L. Ricca, Proc. R. Soc. London A**439**, 411 (1992).
[12] A. A. Kozhevnikov, Int. Journal of Mod. Phys. B**24**, 605 (2010).
[13] A. A. Kozhevnikov, Phys. Lett. B**461**, 256 (1999).
[14] A. A. Kozhevnikov, Phys. Lett. B**750**, 122 (2015).
[15] N. N. Bogolyubov, J. Phys. (USSR) **11**, 23 (1947).
[16] A. N. Redlich and L. C. R. Wijewardhana, Phys. Rev. Lett. **54**, 970 (1985).
[17] V. A. Rubakov and A. N. Tavkhelidze, Theor. Math. Phys. **65**, 250 (1985); Phys. Lett. B**165**, 109 (1985).
[18] M. Joyce and M. E. Shaposhnikov, Phys. Rev. Lett. **79**, 1193 (1997).
[19] M. Laine, JHEP 0510:056 (2005).
[20] V. B. Semikoz and J. W. F. Valle, JCAP **11**, 048 (2011).
[21] D. E. Kharzeev, Progress in Particle and Nuclear Physics **75**, 133 (2014).
[22] L. S. Da Rios, Rend. Circ. Mat. Palermo **22**, 117 (1906).
[23] F. B. Fuller, Proc. Natl. Acad. Sci. USA **68**, 815, (1971).
[24] W. F. Pohl, J. Math. Mech. **17**, 975, (1968).
[25] W. H. Zurek, Phys. Rep. **276**, 177 (1996).
[26] G. J. Stephens, Luís M. A. Bettencourt, and W. H. Zurek, Phys. Rev. Lett. **88**, 137004 (2002).
[27] T. Vachaspati, Phys. Rev. D**80**, 063502 (2009).
[28] J. M. Hyde, A. J. Long, and T. Vachaspati, Phys. Rev. D**89**, 065031 (2014).

Non-Hermitian quantum theory with maximal mass

Rodionov Vasilij N.[1,⋆] and *Kravtsova* Galina A.[2,⋆⋆]

[1] *Plekhanov Russian University of Economics*
[2] *Moscow State University*

Abstract. Non-Hermitian fermion Hamiltonian with γ_5 mass extension is studied. The connection of this $\mathcal{PT}$-symmetrical pseudo-Hermitian model with the geometrical theory with maximal mass is established. The interaction of modified Dirac particles with the external magnetic field is considered. In particular, modified neutrinos is under studying. The results of neutrinos mass values measurements are explained.

1 Introduction

It is well known fact that the most important recent experimental data are the following:

1). approximately only 5 % of the mass density of the Universe is consistent with the mass of known particles;

2). neutrinos have a nonzero mass.

This new knowledge put serious questions for investigators. The Standard Model can not present the cardinal decision of these problems. Here we suggest to find this decision with the help of so-called theory with the maximal mass. We suppose that this theory will can become something like "Standard Model in the curved p-space". But this is a far future, now we study the simple variant of this model.

It is possible to say that the theory with the maximal mass consists of two parts: Geometrical Theory and Algebraic one. We should consider shortly both of them here.

2 Geometrical Theory with Maximal Mass

The history of theory with maximal mass began in 1965, when Markov suggested a hypothesis that mass spectrum of elementary particles should be cut off at the Planck mass $M_{Planck} \approx 10^{19} GeV$ [1]. He named the particle with the maximal mass *maximon*.

The next step was made by Kadyshevsky in 1978. He proposed "geometrical approach" for building of the theory [2]. It means that Minkowski planar 4-dimensional p-space is replaced by the curved momentum 4-space which is the hyperboloid surface in a 5-dimensional p-space,

$$p_0^2 - p_1^2 - p_2^2 - p_3^2 + p_5^2 = M^2, \tag{1}$$

⋆e-mail: rodyvn@mail.ru
⋆⋆e-mail: gakr@chtc.ru

where M is the curvature radius of the 5-dimensional hyperboloid. It is called the space of anti-de Sitter. Note, that in the approximation

$$|p_0|,\ |\vec{p}| \ll M, p_5 \cong M \tag{2}$$

anti-de Sitter geometry is transformed into Minkowski geometry in the 4-dimensional pseudo-Euclidean p-space (the so called "planar limit").

The postulate that the objects with masses m larger than M cannot be considered as elementary particles, since no local fields correspond to them, is accepted:

$$m \leq M. \tag{3}$$

In this theory it becomes the fundamental principle together with the ordinary postulates of quantum theory. The parameter M is named *maximal mass* here. It corresponds to the mass of maximon.

Note, that for a free particle, $p_0^2 - \vec{p}^2 = m^2$, condition (3) is automatically satisfied on surface (1).

The development of this theory ([3]–[14]) leads to the following facts. One of the main features of the Geometrical Theory fermion sector is that the Dirac operator becomes modified [3],

$$\mathcal{D}(p, p_5) \equiv p_n\gamma^n - (p_5 - M)\gamma^5 + 2M sin\frac{\mu}{2}, \tag{4}$$

where $\cos\mu = \sqrt{1 - \frac{m^2}{M^2}}$. Easy to see, that in "planar limit" $M \to \infty$ (see also (2)) modified $D(p, p_5)$ becomes ordinary Dirac operator D. Modified equations of motion corresponding to operator (4) in the Hamiltonian form are the following,

$$\left(p_0 - \hat{\alpha}\mathbf{p} - \hat{\beta}m_1 - \hat{\beta}\gamma^5 m_2\right)\Psi(x, t, x_5) = 0. \tag{5}$$

Here the matrices $\hat{\beta} = \gamma_0$, $\gamma^i = \hat{\beta}\hat{\alpha}^i$, and we have notation

$$m_1 = 2M \sin\mu/2, \qquad m_2 = 2M \sin^2\mu/2, \tag{6}$$

and $p_\mu = i\partial_\mu$.

The another specific feature of Geometrical Theory is the existence of so-called the *exotic Dirac operator* in it [3]. It has the following form,

$$D_{exotic}(p, M) = p_\nu\gamma^\nu + (p_5 + M)\gamma^5 - 2M\cos(\mu/2), \tag{7}$$

and the corresponding equation of motion

$$\left(p_0 - \hat{\alpha}\mathbf{p} - \hat{\beta}m_3 - \hat{\beta}\gamma^5 m_4\right)\Psi^{ex}(x, t, x_5) = 0, \tag{8}$$

where $m_3 = 2M\cos\mu/2, m_4 = 2M\cos^2\mu/2$. In planar limit exotic operator and exotic equation don't come to ordinary Dirac operator and Dirac equation of motion. So one can propose that they associated with some fermion fields that has no analogues in Standard Model. They were named as *exotic fermions*. It is proposed that they put a contribution to the *Dark Matter* in the Universe.

In the quantum mechanical approximation, the Hamiltonians corresponding to modified equations of motion for ordinary and exotic fermion fields can be represented in the following form:

$$\hat{H} = \vec{\alpha}\vec{p} + \hat{\beta}(m_1 + m_2\gamma_5), \tag{9}$$

$$\hat{H}_{exotic} = \vec{\hat{\alpha}}\vec{p} + \hat{\beta}(m_3 + m_4\gamma_5). \tag{10}$$

Easy to see that these expressions turn out to be non-Hermitian due to the γ_5-mass terms. In summary, we can conclude that the Geometric Theory leads to the emergence of non-Hermitian Lagrangian in the fermion sector. It seems to be crucial point of the theory, but the solution of the problem was found.

3 Algebraic Theory with Maximal Mass

At the end of 90th and in the early 2000-ies a group of mathematicians found a solution to this problem. They proved that if the model isn't Hermitian but is PT-symmetrical [15], [16] or pseudo-Hermitian $\left(\eta_0 H \eta_0^{-1} = H^\dagger, \quad \eta_0 \text{ is linear Hermitian operator}\right)$ [17], so the average of Hamiltonian can become real in new scalar product, which is modified due to special operator C. The formula for this operator was found. This method appears to be very powerful, so such models become very popular (see, for example, [18]–[37]).

So it is possible to apply this technique for investigation of non-Hermitian fermionic model which arises in Geometrical Theory with Maximal Mass.

Let's try to study our Hamiltonian (9) at the algebraic theory point of view. [1] We'll call it the Algebraic Theory with the physical mass (mass of the particle) $m^2 = m_1^2 - m_2^2$. One can easily see that these Hamiltonians are not Hermitian, but PT -symmetrical and pseudo-Hermitian with

$$\eta_0 = e^{\alpha\gamma_5}, \quad \text{where } \alpha \text{ is defined as} \quad m_1 = m\,\mathrm{ch}\,\alpha; \qquad m_2 = m\,\mathrm{sh}\,\alpha \tag{11}$$

(see [38], [39], [40]). It is so-called non-disturbed PT-symmetry in the theory when m is real, t.e. $m_1^2 \geq m_2^2$.

Very important feature of Algebraic Theory is that there exists a parameter of maximal mass in it. Indeed, due to simple mathematical manipulations one can find ([39], [40]), that

$$m \leq \frac{{m_1}^2}{2m_2} \equiv m_{max}. \tag{12}$$

This formal estimation becomes the physical one due to the postulate which we propose: *m_{max} must be equal to M in Geometrical Theory.* So these two theories become equivalent (at least, on this simple level). This fact allows to Algebraic Theory to find a physical sense, and to Geometrical Theory to solve the problem of non-Hermitian Hamiltonian in fermion sector. But for understanding this equality it is necessary to put the postulate $m_{max} = M$ in the theory for all particles [39], [40]. In this case M have to be maximal mass for all the elementary particles mass spectrum.

Due to parametrization (11) the parameters m_1, m_2 can be re-expressed through another two ones: m and m_{max},

$$m_1^{\mp} = \sqrt{2}m_{max}\sqrt{1 \mp \sqrt{1 - m^2/m_{max}^2}}; \tag{13}$$

$$m_2^{\mp} = m_{max}\left(1 \mp \sqrt{1 - m^2/m_{max}^2}\right). \tag{14}$$

So if we know the physical mass of a particle and the mass of maximon, we can calculate the parameters m_1, m_2 for Hamiltonian (9). But obviously we have two values of the parameters m_1, m_2 for

[1] See also [27]

each value of m. It means that in Algebraic Theory exotic particles also arise, such as in Geometrical one [40]. The high sign in formulas (13), (14) corresponds to ordinary particles, and the down sign to exotic fermions. This fact is the indirect confirmation of the equality Geometrical (in fermion sector) and Algebraic Theories. Even more, if in the formulas (13), (14) we use M instead of m_{max}, we get the correspondence (equality) of the parameters in Algebraic and Geometric Theories,

$$m_1^- \leftrightarrow m_1, \quad m_1^+ \leftrightarrow m_3, \quad m_2^- \leftrightarrow m_2, \quad m_2^+ \leftrightarrow m_4. \tag{15}$$

As we note above, the exotic particles are the candidates for Dark Matter in the Universe. Of course, maximon have to put contribution to Dark Matter too.

Easy to see also that modified Dirac Hamiltonian $\hat{H}$ in algebraic theory with maximal mass and ordinary Dirac Hamiltonian $\hat{H}_0$ are connected by the following transformations,

$$\hat{H}_0 = \eta \hat{H} \eta^{-1}, \qquad \hat{H}_0 = \eta^{-1} \hat{H}^\dagger \eta, \qquad \eta = e^{\alpha\gamma_5/2} \tag{16}$$

(see [38], [39], [41]). Note that η isn't unitary, so this modified Dirac model and ordinary Dirac model are not equivalent. [2]

Using a common algebraic theory [17], one can find C-operator for the definition of modified scalar product,

$$C = \eta_0{}^{-1}\mathcal{P} = e^{-\alpha\gamma_5}\gamma_0. \tag{17}$$

It is possible to demonstrate that the new scalar product is positive defined and the average values of the Hamiltonian (9), (10) are real [40].

Also the eigenvalues and eigenvectors for free modified Dirac Hamiltonian were found [40], [43], [44]. The law of the current density continuity was obtained [45].

All these problems were solved for the algebraic model with maximal mass, where fermionic fields interact with the external magnetic field. Namely, let's study modified fermion without spin in the homogeneous magnetic field $\vec{\mathcal{H}} = (0, 0, \mathcal{H})$. The electromagnetic field potentials in gauge [46] can be chosen in the following form: $A_0 = 0,\ A_1 = 0,\ A_2 = \mathcal{H}x_1,\ A_3 = 0$. Considering the equation

$$H\widetilde{\psi} = E\widetilde{\psi}, \tag{18}$$

where

$$H = (\vec{\alpha}\vec{\mathcal{P}}) + \beta m_1 + \beta\gamma_5 m_2, \tag{19}$$

and $\mathcal{P}_\mu = i\partial_\mu - eA_\mu$; $e = -|e|$, one can find ([45], [47], [48], [49], see also review [50])

$$E = \pm\sqrt{m_1{}^2 - m_2{}^2 + 2\gamma n + p_3{}^2}, \tag{20}$$

where $n = 0, 1, 2...$ and $\gamma = |e|\mathcal{H}$. Taking into account that $m^2 = m_1{}^2 - m_2{}^2$, we can see that this result coincides with the eigenvalues of the Hermitian Hamiltonian describing Landau relativistic levels (see, e.g., [46]).

The situation is cardinally changed when the spin of the fermion is taken into account. We consider the Hamiltonian [47], [48], [51]

$$H_{\Delta\mu} = \vec{\alpha}\vec{\mathcal{P}} + \beta(m_1 + \gamma_5 m_2) + \Delta\mu\beta(\vec{\sigma}\vec{\mathcal{H}}), \tag{21}$$

where $\Delta\mu = (\mu - \mu_0) = \mu_0(g - 2)/2$. Here, μ is the magnetic moment of the fermion, g is the gyromagnetic factor of the fermion, $\mu_0 = |e|/2m$ is the Bohr magneton, and and $\sigma^{\mu\nu} = i/2(\gamma^\mu\gamma^\nu - \gamma^\nu\gamma^\mu)$.

[2]See, for comparison, [42].

Thus, the phenomenological constant $\Delta\mu$ introduced by Pauli is a part of the equation and can be interpreted from the point of view of the quantum field theory. Taking into account the quantum electrodynamic contribution into the anomalous magnetic moment of the electron to e^2, we have $\Delta\mu = \frac{\alpha}{2\pi}\mu_0$, where $\alpha = e^2 = 1/137$ is the fine structure constant, and it is still assumed that the external field potential satisfies the free Maxwell equation. As it was shown in paper [45], [48], with the help of the modified Dirac-Pauli equation we can find the exact values of the energy spectrum for the Hamiltonian (21),

$$E(\zeta, p_3, 2\gamma n, \mathcal{H}) = \pm\sqrt{p_3{}^2 - m_2{}^2 + \left[\sqrt{m_1{}^2 + 2\gamma n} + \zeta\Delta\mu\mathcal{H}\right]^2}, \tag{22}$$

where $\zeta = \pm 1$ characterizes the fermion spin projection onto the magnetic field direction.

For the case of neutral particles with anomalous magnetic moment it is necessary to replace the value of the quantized transverse momentum of a charged particle in a magnetic field by the regular value, $2\gamma n \to p_1{}^2 + p_2{}^2 = p_\perp{}^2$. As a result, we obtain [47], [51]

$$E(\zeta, p_3, p_\perp, \mathcal{H}) = \pm\sqrt{p_3{}^2 - m_2{}^2 + \left[\sqrt{m_1{}^2 + p_\perp{}^2} + \zeta\Delta\mu\mathcal{H}\right]^2}. \tag{23}$$

Namely, the estimations were found for the neutrinos ([40], [50]), [52].

Formulas (22), (23) show that taking into account of the spin with the magnetic field interaction allows us to separate the variables m_1 and m_2 ($m_1^{\mp}$ and $m_2^{\mp}$), and therefore, separate ordinary and exotic particles. This means that for the observation of exotic particles it is not necessary ultra-high energy. It is necessary the presence of the anomalous magnetic moment of the particles and it's interaction with the magnetic field. It's very important to emphasize that the experiment, in which the exotic fermions may be observable, proves the existence of Maximal mass.

4 Algebraic Theory and Experiment

The neutrinos mass observation may become experiment mentioned above [53]. Just the existence of the exotic particles in the theory allows to explain the experiments for the neutrino mass observations. Really, using (23) one could write for neutrino mass square in the magnetic field:

$$\tilde{m}_\nu^2(\mathcal{H}) = \tilde{m}_\nu^2 \pm 4\Delta\mu\mathcal{H}M. \tag{24}$$

We want draw attention to the fact that we deal with experiment measurements of neutrinos mass with taking into account influence of the magnetic fields. The result of experiment depends on the spin direction relatively magnetic field direction, and the exotic neutrinos make the principal contribution.

According to scheme of the the Troitsk experimental installation for measurement of mass the electron antineutrino gas-tritium is injected into a long tube, where a strong magnetic field (up to H = 0.8T = 8000Gauss) is existed [54], [55]. The results of Troitsky experiment at 95% confidence (C.L.) are the following [56]

$$m_\nu^2 = -0.67 \pm 1.89_{stat} \pm 1.68_{syst}\, eV^2. \tag{25}$$

If we suggest that the beam of neutrinos partially consist of pseudo-Hermitian components exotic neutrinos ($\tilde{m}_\nu^2$) that we can obtain the following evaluation $\tilde{m}_\nu^2(\mathcal{H}) = \tilde{m}_\nu^2 - 4\Delta\mu\mathcal{H}M$. If we consider estimation, which follows from the ordinary experimental dates $\tilde{m}_\nu = m_\nu = 1eV$ and suppose $\tilde{m}_\nu^2(\mathcal{H}) = -1eV^2$ so we can evaluate that $4\Delta\mu\mathcal{H}M \cong 2eV^2$ and moreover, estimate $M \cong 2\cdot 10^{14} - 10^{15}\, GeV$.

5 Conclusions

In summary, we can conclude that the Algebraic Theory makes possible to describe the experimental data of the mass squared neutrinos observations and to estimate the value of maximal mass M in the low energies region.

So the experiments for observations of mass squared neutrinos may prove the existence of exotic particles and confirm the restriction of elementary particles mass spectrum by the maximal mass M.

References

[1] M.A.Markov, Prog. Theor Phys. Suppl., Commemoration Issue for the Thirtieth Anniversary of Meson Theory and Dr. H. Yukawa, p. 85 (1965); Sov. Phys. JETP, **24**, p. 584 (1967).

[2] V.G.Kadyshevsky, Nucl. Phys., **B141**, p 477 (1978); in Proceedings of International Integrative Conference on Group theory and Mathematical Physics, Austin, Texas, 1978; Fermilab-Pub. 78/70-THY, Sept. 1978; Phys. Elem. Chast. Atom. Yadra, **11**, p.5 (1980).

[3] V.G.Kadyshevsky, M.D.Mateev, V.N.Rodionov, A.S.Sorin, *Towards a maximal mass model.* CERN TH/2007-150; hep-ph/0708.4205.

[4] V.G.Kadyshevsky, M.D.Mateev, V.N.Rodionov, A.S.Sorin, Doklady Physics **51**, p.287 (2006); hep-ph/0512332.

[5] V.G.Kadyshevsky, Phys. Part. Nucl., **29**, p.227 (1998).

[6] V.G.Kadyshevsky, M.D.Mateev, Phys. Lett., **B106**, p.139 (1981).

[7] V.G.Kadyshevsky, M.D.Mateev, Nuovo Cimento, **A87**, p.324 (1985).

[8] M.V.Chizhov, A.D.Donkov, V.G.Kadyshevsky, M.D.Mateev, Nuovo Cimento, **A87**, p.350 (1985).

[9] M.V.Chizhov, A.D.Donkov, V.G.Kadyshevsky, M.D.Mateev, Nuovo Cimento, **A87**, p.373, (1985).

[10] V.G.Kadyshevsky, Phys. Part. Nucl., **29**, p.227 (1998).

[11] V.G.Kadyshevsky, D.V.Fursaev, JINR Rapid Commun., N 6, p.5 (1992).

[12] R.M.Ibadov, V.G.Kadyshevsky, Preprint JINR-P2-86-835 (1986).

[13] V.G.Kadyshevsky, D.V.Fursaev, JINR-P2-87-913 (1987); Sov.Phys.Dokl.**34**, p. 534 (1989).

[14] V.G.Kadyshevsky, V.N.Rodionov, Phys. Part. Nucl., **36**, no.1, S34-S37 (2005).

[15] C.M.Bender and S.Boettcher, Phys. Rev. Lett., **80**, p.5243 (1998).

[16] C.M.Bender, S.Boettcher and P.N. Meisinger, J. Math. Phys., **40**, p.2210 (1999).

[17] A.Mostafazadeh, Int. J. Geom. Meth. Mod. Phys., **7**, p.1191-1306 (2010); arXiv: 0810.5643 (2008).

[18] A.Khare, B.P.Mandal, Phys. Lett.A, **272**, p.53 (2000).

[19] M.Znojil, G.Levai, Mod. Phys. Lett.A, **16**, p.2273 (2001).

[20] A.Mostafazadeh, J. Phys.A **38** p.6657 (2005), Erratum-ibid. **38**, p.8185 (2005).

[21] C.M.Bender, D.C.Brody, J.Chen, H.F.Jones, K.A.Milton, and M.C.Ogilvie, Phys. Rev.D, **74**, p.025016 (2006). and see refs therein.

[22] C.M.Bender, K.Besseghir, H.F.Jones, and X. Yin, J.Phys.A, **42**, p.355301,(2009); arXiv: 0906.1291 (2009).

[23] A.Khare, B.P.Mandal, Spl. Issue of Pramana J. Phys., **73**, p.387 (2009).

[24] P.Dorey, C.Dunning, and R.Tateo, J. Phys A: Math. Theor., **34** p.5679 (2001).

[25] C.M.Bender, D.C.Brody, and H.F.Jones, Phys. Rev.D **70** p.025001 (2004)(2004); Erratum-ibid. **71** p.049901 (2005).

[26] C.M.Bender, S.F.Brandt, J.Chen, and Q.Wang, Phys. Rev.D, **71**, p.065010 (2005).

[27] C.M.Bender, H.F.Jones and R.J.Rivers, Phys. Lett.B, **625** p.333 (2005).

[28] C.M.Bender, S.Boettcher, Phys. Rev. Lett., **89**, p.270401-1 (2002); Erratum-ibid. **92**, p.119902 (2004).

[29] C.M.Bender, J.Brod, A.Refig, and M.Reuter, J. Phys. A: Math. Gen., **37**. p.10139, (2004); quant-ph/0402026 (2004).

[30] A.Mostafazadeh, J. Math Phys., **43** p.205 (2002); **43** p.2814 (2002); **43** p.3944 (2002).

[31] A.Mostafazadeh, A. Batal, J. Phys A: Math. Theor., **37**, p.11645 (2004).

[32] A.Mostafazadeh, J. Phys A: Math. Theor., **36**, p.7081 (2003).

[33] M.Mostafazadeh, Class. Q. Grav. **20**, p.155 (2003).

[34] M.Mostafazadeh, Ann. Phys., **309**, p.1 (2004).

[35] M.Mostafazadeh, F.Zamani, Ann. Phys., **321**, p.2183; p.2210 (2004).

[36] M.Mostafazadeh, Int. J. Mod. Phys. A, **21**, no.12 p.2553 (2006).

[37] F.Zamani, M.Mostafazadeh, J. Math. Phys., **50**, p.052302 (2009).

[38] V.N.Rodionov. *PT-symmetric pseudo-Hermitian relativistic quantum mechanics with maximal mass.* arXiv:hep-th/1207.5463 (2012).

[39] V.N.Rodionov, G.A.Kravtsova, Mosc. Phys. Bull., **69**, no.3, p.223 (2014).

[40] V.N.Rodionov, G.A.Kravtsova, Theor. Math. Phys., **182**, p.100 (2016).

[41] V.N.Rodionov. *Non-Hermitian PT-symmetric quantum mechanics of relativistic particles with the restriction of mass.* arXiv:1303.7053 (2013).

[42] I.P.Volobuev, V.G.Kadyshevsky, M.D.Mateev, and M.R.Mir-Kasimov, Theor. Math. Phys., **40**, p.363 (1979).

[43] V.N.Rodionov. *On limitation of mass spectrum in non-Hermitian PT-symmetric models with the γ_5-dependent mass term.* arXiv:1309.0231 (2013).

[44] V.N.Rodionov. *Non-Hermitian PT-symmetric relativistic Quantum mechanics with a maximal mass in an external magnetic field.* arXiv:1404.0503 (2014);

[45] V.N.Rodionov. *Exact Solutions for Non-Hermitian Dirac-Pauli Equation in an intensive magnetic field.* arXiv:1406.0383 (2014);

[46] I.M.Ternov, V.R.Khalilov, and V.N.Rodionov. textitInteraction of charged particles with intensive electromagnetic field. (Moscow State Univ. Press., Moscow, 1982).

[47] V.N.Rodionov, Phys. Scr., **90**, p.045302 (2015).

[48] V.N.Rodionov. Int. J. Theor.Phys., **54**, p.3907 (2015).

[49] V.N.Rodionov. *Non-Hermitian PT-Symmetric Relativistic Quantum Theory in an Intensive Magnetic Field.* In: *Non-Hermitin Hamiltonians in Quantum Physica. Selected Contributions from the 15th International Conference on Non-Hermitian Hamiltonians in Quantum Physics, Palermo, Italy, 18-25 May 2015. Eds.: F.Bagarello, R.Passante, C.Trapani.* (Springer Proc. Phys., **184**, Springer, Switzerland, 2016) p.357.

[50] V.N.Rodionov, G.A.Kravtsova, Phys. Part. Nucl., **47**, no.2, p.135 (2016); arXiv:1603.03063 (2016).

[51] V.N.Rodionov. *Non-Hermitian PT-Symmetric Dirac-Pauli Hamiltonians with Real Energy Eigenvalues in the Magnetic Field.* arXiv:1409.5412 (2014).

[52] V.N.Rodionov. *Non-Hermitian PT-symmetric relativistic quantum theory in an intensive magnetic field.* arXiv:1603.07443 (2016).

[53] V.N.Rodionov. *Towards the detecting of pseudo-Hermitian anomalies for negative square masses neutrinos in intensive magnetic fields.* arXiv:1603.08425 (2016).

[54] V.M.Lobashev, Nucl. Phys. A, **719**, p.C153 (2003).

DOI: 10.1051/epjconf/201612505012

[55] V.N.Aseev et al., Phys. Rev. D, **84**, p.112003 (2011).
[56] V.N.Aseev, A.I.Belesev et al., Phys. At. Nucl., **75**, p.464 (2012).

EPJ Web of Conferences **125**, 05013 (2016) DOI: 10.1051/epjconf/201612505013
QUARKS-2016

Applicability of the Wigner functional approach to evolution of quantum fields

Andrey Leonidov[1,2,3,⋆] and *Anna* Radovskaya[1,⋆⋆]

[1] *P.N. Lebedev Physics Institute of the Russian Academy of Sciences, Moscow, Russia*
[2] *Moscow Engineering Physics Institute, Moscow, Russia*
[3] *Moscow Institute of Physics and Technology, Moscow, Russia*

Abstract. Evolution of highly excited quantum field is considered in the framework of Keldysh formalism . It is demonstrated that leading order (LO) term of semiclassical approximation appears as well-known Classical Statistical Approximation (CSA). In simple case of spatially homogeneous scalar field analytical expressions for leading and next-to-leading (NLO) order are presented. It is shown that the range of applicability of CSA strongly depends on the properties of the initial state of the system.

1 Introduction

The problem of description of the evolution of highly excited quantum has recently attracted considerable attention of theorists working on problems in various domains of theoretical physics. The relevant examples include, in particular, studies of initial stages of ultrarelativistic heavy ion collisions [1, 8], heating of the Universe after inflation [9, 11], particle production in ultrastrong fields [12, 15], thermalization in ultracold quantum gases [16, 17]. All the above-mentioned systems share the basic common feature - their initial state can not be described as the vacuum one. This means that the standard S-matrix formalism is not applicable here. In fact, for highly excited fields it is necessary to give a correct specification of the initial state and be able to describe its subsequent temporal evolution. The Keldysh technique [18, 20] provides all the necessary technical tools for solving such problems.

In the previous work [21] we demonstrated that the Keldysh formalism applied to description of evolution of excited fields results in its Leading Order in so-called Classical Statistical Approximation (CSA). CSA is based on the fact that summation of leading quantum corrections can be cast in the form of integration over initial conditions for classical trajectories with the weight given by the Wigner functional. It is interesting to note that similar approach was used in atomic and chemical physics [22],[23].

In this paper we analyze the applicability of the Wigner functional approach to evolution of highly excited quantum fields by calculatingf the Next-to-Leading corrections to several observables. We show that this applicability does directly depend on the properties of the initial state of the system and find out range of the validity of the Wigner functional approach considering homogeneous scalar field model as an example.

⋆e-mail: leonidov@lpi.ru
⋆⋆e-mail: raan@lpi.ru

2 General formalism

In this section we will shortly remind general formalism for calculation of observables within the Keldysh technique suitable to our problem. Out of equilibrium an expectation value of observable $F(\hat{\varphi})$ at the moment t_1 can be calculated as a trace with density matrix as

$$\langle F(\hat{\varphi})\rangle_{t_1} = tr(F(\hat{\varphi})\hat{\rho}(t_1)) = \int \mathfrak{D}\xi(\vec{x})\, F(\xi)\langle\xi|\hat{U}(t_1,t_0)\hat{\rho}(t_0)\hat{U}(t_0,t_1)|\xi\rangle \tag{1}$$

where evolution of the density matrix $\hat{\rho}(t)$ is governed by the evolution operator $\hat{U}(t,t_0)$

$$\hat{\rho}(t) = \hat{U}(t,t_0)\hat{\rho}(t_0)\hat{U}(t_0,t), \tag{2}$$

and $|\xi\rangle$ is an eigenstate of the field operator $\hat{\varphi}(\vec{x})|\xi\rangle = \xi(\vec{x})|\xi\rangle$ and $\int \mathfrak{D}\xi(\vec{x})$ is a path integral over all possible 3-d functions originating from unity operator $\hat{1} = \int \mathfrak{D}\xi(\vec{x})\,|\xi\rangle\langle\xi|$.

The matrix elements of the evolution operator can be represented as the path integrals over 4-d functions $\mathcal{D}\eta(t,\vec{x})$

$$\langle\xi|\hat{U}(t_1,t_0)|\xi_1\rangle = \int\limits_{\eta_F(t_0,\vec{x})=\xi_1(\vec{x})}^{\eta_F(t_1,\vec{x})=\xi(\vec{x})} \mathcal{D}\eta_F(t,\vec{x})e^{iS[\eta_F]},$$

$$\langle\xi_2|\hat{U}(t_0,t_1)|\xi\rangle = \int\limits_{\eta_B(t_0,\vec{x})=\xi_2(\vec{x})}^{\eta_B(t_1,\vec{x})=\xi(\vec{x})} \mathcal{D}\eta_B(t,\vec{x})e^{-iS[\eta_B]}$$

Here $\eta_F(t_1,\vec{x})$ and $\eta_F(t_1,\vec{x})$ are the field that lies on the forward (η_F) and backward (η_B) sides of Keldysh contour (see [21] for details). Combining all together we obtain

$$\langle F(\hat{\varphi})\rangle_{t_1} = \int \mathfrak{D}\xi \int \mathfrak{D}\xi_1 \int \mathfrak{D}\xi_2\, \langle\xi_1|\hat{\rho}(t_0)|\xi_2\rangle \times \tag{3}$$

$$F(\xi) \int\limits_{\eta_F(t_0,\vec{x})=\xi_1(\vec{x})}^{\eta_F(t_1,\vec{x})=\xi(\vec{x})} \mathcal{D}\eta_F(t,\vec{x}) \int\limits_{\eta_B(t_0,\vec{x})=\xi_2(\vec{x})}^{\eta_B(t_1,\vec{x})=\xi(\vec{x})} \mathcal{D}\eta_B(t,\vec{x})\, e^{iS[\eta_F]-iS[\eta_B]}.$$

The next step is to perform technical trick and extend Keldesh contour to infinity. After that we can introduce new field as

$$\phi_c = \frac{\eta_F+\eta_B}{2}, \qquad \phi_q = \eta_F - \eta_B \tag{4}$$

and define the Keldysh action $S_K[\eta_F,\eta_B] = S[\eta_F] - S[\eta_B]$ in order to obtain general expression

$$\langle F(\hat{\varphi})\rangle_{t_1} = \int \mathfrak{D}\chi_1 \int \mathfrak{D}\xi_1 \int \mathfrak{D}\xi_2\, \langle\xi_1|\hat{\rho}(t_0)|\xi_2\rangle \times \tag{5}$$

$$\int\limits_{\phi_c(t_0,\vec{x})=\frac{\xi_1(\vec{x})+\xi_2(\vec{x})}{2}}^{\phi_c(\infty,\vec{x})=\chi_1(\vec{x})} \mathcal{D}\phi_c \int\limits_{\phi_q(t_0,\vec{x})=\xi_1(\vec{x})-\xi_2(\vec{x})}^{\phi_q(\infty,\vec{x})=0} \mathcal{D}\phi_q\, F(\phi_c(t_1))\, e^{iS_K[\phi_c,\phi_q]}.$$

Eq.(5) represents our main tool for calculation of observables however we need to specify Lagrangian in order to calculate certain results.

3 Scalar φ^4 theory

The simplest model which has necessary nontrivial dynamical behavior is a scalar φ^4 model with quartic interaction term.

$$\mathcal{L} = \frac{1}{2}\partial_\mu\varphi\partial^\mu\varphi - \frac{g^2}{4}\varphi^4 + J\varphi, \qquad S = \int d^4x\,\mathcal{L}. \tag{6}$$

Here $J(t,\vec{x})$ is an auxiliary source which we keep in order to perform semiclassical decomposition.

For the Lagrangian (6) the Keldysh actions reads

$$S_K[\phi_c,\phi_q] = \int d^3x\,\dot{\phi}_c(t_0,\vec{x})(\xi_1(\vec{x}) - \xi_2(\vec{x})) - \int d^3x \int_{t_0}^{\infty} dt\left(\phi_q A[\phi_c] - \frac{g^2}{4}\phi_c\phi_q^3\right),$$
$$A[\phi_c] = [\ddot{\phi}_c + g^2\phi_c^3 - J]. \tag{7}$$

Here term $\phi_q(t_0,\vec{x}) = \xi_1(\vec{x}) - \xi_2(\vec{x})$ appears from integration by parts. Note that $A[\phi_c] = 0$ corresponds to projecting onto the classical equation of motion for the Lagrangian Eq.(6). It is important that the source J connected to both fields on upper and lower sides of Keldysh contour in similar way so we get term $J(\eta_F - \eta_B) = J\phi_q$.

It is clear that path integrals of (5) with this action can not be calculated explicitly. The systematic procedure we employ is expansion in ϕ_q in (7) around its saddle-point value. This expansion is, in fact, a semiclassical one. This can be seen by restoring $\hbar$ in the action and replacing $\phi_q \to \hbar\phi_q$ so the only remaining dependence on $\hbar$ is in ϕ_q^3 which is proportional to $\hbar^2$ and

$$e^{-i\frac{g^2\hbar^2}{4}\int_{t_0}^{\infty}dt\int d^3x\,\phi_c\phi_q^3} = \underbrace{1}_{LO}\underbrace{-\frac{ig^2\hbar^2}{4}\int_{t_0}^{\infty}dt\int d^3x\,\phi_c\phi_q^3}_{NLO} + \ldots. \tag{8}$$

This expansion shows that the LO contribution contains quantum fluctuation up to one loop order.

3.1 Leading Order

The Leading Order (LO) contribution to observables corresponds to the first term in decomposition (8). In this case we are able to perform functional integration over field ϕ_q resulting with the functional delta-function which project ϕ_c onto the solution of classical equation of motion of the Lagrangian (6). As the next step we insert the "initial velocity" as $1 = \int \mathfrak{D}p(\vec{x})\delta(\tilde{p}(\vec{x}) - \dot{\phi}_c(t_0,\vec{x}))$ in order to integrate over ϕ_c (see [21] for details). After substitution $\alpha = \frac{\xi_1+\xi_2}{2}$, $\beta = \xi_1 - \xi_2$ we end up with

$$\langle F(\hat{\varphi})\rangle_{t_1} = \int \mathfrak{D}\alpha(\vec{x})\mathfrak{D}p(\vec{x})f_W[\alpha(\vec{x}),p(\vec{x})),t_0]F(\phi_{cl}(t_1,\vec{x})), \tag{9}$$

where the Wigner functional is commonly defined as

$$f_W[\alpha(\vec{x}),p(\vec{x}),t_0] = \int \mathfrak{D}\beta(\vec{x})\left\langle\alpha + \frac{\beta}{2}\right|\hat{\rho}(t_0)\left|\alpha - \frac{\beta}{2}\right\rangle \times \exp\left(i\int d^3x\,p(\vec{x})\beta(\vec{x})\right). \tag{10}$$

Here ϕ_{cl} is the solution of classical equation of motion

$$\partial_\mu \partial^\mu \phi_{cl} + g^2 \phi_{cl}^3 = 0 \tag{11}$$

with initial conditions given by

$$\phi_{cl}(t_0, \vec{x}) = \alpha(\vec{x}), \quad \dot{\phi}_{cl}(t_0, \vec{x}) = p(\vec{x}) \tag{12}$$

and at zero axillary source $J(t, \vec{x})$.

Expression (9) was earlier derived by other methods in [1, 4], see also [19],[24] . The general recipe for calculation at the LO level is the following: in order to obtain expectation value of observable at a given time moment t_1 one need to find solution of classical equation of motion as a function of the field and its time derivative at the initial time and average over initial conditions with weight of Wigner functional. Note that choice of Wigner functional (or density matrix at initial time) lies outside the method itself and depends on problem that should be solved. For certain choice of Wigner functional the expression (9) represent the Classical Statistical Approximation widely used in the context of ultrarelativistic heavy ion collisions.

Let us introduce new notation for averaging over initial conditions

$$\langle O \rangle_{i.c.} = \int \mathfrak{D}\alpha(\vec{x}) \mathfrak{D}p(\vec{x}) f_W[\alpha(\vec{x}), p(\vec{x})), t_0]\, O. \tag{13}$$

Then we can rewrite (9) simply as

$$\langle F(\hat{\varphi}) \rangle_{t_1}^{LO} = \langle F(\phi_{cl}(t_1, \vec{x})) \rangle_{i.c.}. \tag{14}$$

3.2 Next-to-Leading Order

In order to decide if expansion over ϕ_q is valid we should consider Next-to-Leading Order (NLO) term in (8). With this ϕ_q^3 part the path integration over ϕ_q can not be done as easy as at LO level. However one can note that each ϕ_q can be replaced by functional derivative over source J due to $\phi_q J$ term in the Keldysh action (7)

$$\frac{\delta}{\delta J(t, \vec{x})} e^{iS_K[\phi_c, \phi_q]} = -i\phi_q(t, \vec{x}) e^{iS_K[\phi_c, \phi_q]}. \tag{15}$$

After that we can perform functional integration over ϕ_q and ϕ_c in order to obtain answer for expectation value of the observable up to NLO level

$$\langle F(\hat{\varphi}) \rangle_{t_1}^{LO+NLO} = \left\langle F(\phi_{cl}(t_1, \vec{x})) + \frac{g^2}{4} \int_{t_0}^{t_1} dt_2 \int d^3x_2 \phi_{cl}(t_2, \vec{x}_2) \frac{\delta^3 F(\phi_{cl}(t_1, \vec{x}))}{\delta J^3(t_2, \vec{x}_2)} \bigg|_{J=0} \right\rangle_{i.c.}.$$

The expression above shows that there is no necessity in any new information for evaluation of NLO correction. We should find classical trajectory as a function of initial conditions, perform three variations over auxiliary source, integrate over intermediate time and average with the Wigner functional. These calculations seem to be rather tedious however there is a technical trick.

Let us define k-th variation of the classical solution over source J as

$$\frac{\delta^k \phi_{cl}(t_1, \vec{x}_1)}{\delta J^k(t_2, \vec{x}_2)} = \Phi_k(t_1, \vec{x}_1; t_2, \vec{x}_2). \tag{16}$$

Then

$$\frac{\delta^3 F(\phi_{cl}(t_1,\vec{x}_1))}{\delta J^3} = \frac{\partial F}{\partial \phi_{cl}}\Phi_3(t_1,\vec{x}_1;t_2,\vec{x}_2) + 3\frac{\partial^2 F}{\partial \phi_{cl}^2}\Phi_1(t_1,\vec{x}_1;t_2,\vec{x}_2)\Phi_2(t_1,\vec{x}_1;t_2,\vec{x}_2) + \frac{\partial^3 F}{\partial \phi_{cl}^3}\Phi_1^3(t_1,\vec{x}_1;t_2,\vec{x}_2). \quad (17)$$

Variations $\Phi_k(t_1,\vec{x}_1;t_2,\vec{x}_2)$ can be found by variation of the classical EoM

$$\frac{\delta^3}{\delta J^3(t_2,\vec{x}_2)}\Big(\partial_\mu\partial^\mu\phi_{cl}(t_1,\vec{x}_1) + g^2\phi_{cl}^3(t_1,\vec{x}_1) = J(t_1,\vec{x}_1)\Big),$$

that gives

$$\begin{aligned} L_{t_1}\Phi_1(t_1,\vec{x}_1;t_2,\vec{x}_2) &= \delta(t_1-t_2)\delta^{(3)}(\vec{x}_1-\vec{x}_2) \\ L_{t_1}\Phi_2(t_1,\vec{x}_1;t_2,\vec{x}_2) &= -6g^2\phi_{cl}(t_1,\vec{x}_1)\Phi_1^2(t_1,\vec{x}_1;t_2,\vec{x}_2) \\ L_{t_1}\Phi_3(t_1,\vec{x}_1;t_2,\vec{x}_2) &= -6g^2\Phi_1^3(t_1,\vec{x}_1;t_2,\vec{x}_2) - 18g^2\phi_{cl}(t_1,\vec{x}_1)\Phi_1(t_1,\vec{x}_1;t_2,\vec{x}_2)\Phi_2(t_1,\vec{x}_1;t_2,\vec{x}_2) \\ L_{t_1} &= \partial_{t_1}^2 - \partial_{\vec{x}_1}^2 + 3g^2\phi_{cl}^2(t_1,\vec{x}_1). \end{aligned} \quad (18)$$

With help of differential equations (18) we are able to calculate NLO correction to any observable without knowledge of exact dependence of the classical solution ϕ_{cl} on auxiliary source $J(t,\vec{x})$. Although the recipe for NLO calculation is fulfilled we can not use it directly with Lagrangian (6). Analytical solution of equation of motion even for zero source can not be found. In the next section we consider simplification of described model that allow

4 Spatially homogenous analytical solution

Let us consider φ^4 toy model . We suppose that all spatial gradients small enough to be neglected. Although this assumption takes us away from implementation to real physical systems it let us perform a lot of work analytically and reveal general features that can be extended to full theory.

In spatially homogeneous case $\partial_i\varphi(t,\mathbf{x}) = 0$, than

$$S = V_3\int dt\left(\frac{1}{2}\dot{\varphi}^2 - \frac{g^2}{4}\varphi^4 + J\varphi\right), \quad (19)$$
$$V_3 = \int d^3x.$$

The equation of motion is

$$\ddot{\varphi} + g^2\varphi^3 = J. \quad (20)$$

We can find classical solution of EoM (20) for J=0 in terms of the Jacobi elliptical function cn with module $k^2 = \frac{1}{2}$ as

$$\phi_{cl}(t) = \phi_m cn\left(\frac{1}{2}, g\phi_m t + C\right). \quad (21)$$

The period of this function is $T_{cl} = \frac{4}{g\phi_m}K(1/2)$, where K(1/2) is the complete elliptic integral of the first kind. Here ϕ_m and C are the amplitude and the phase of the solution.

First of all we calculate the field expectation value $\langle\hat{\varphi}\rangle_{t_1}$ as a simplest example. In our spatially homogeneous case the eq.(3.2) has the form of

$$\langle\hat{\varphi}\rangle_{t_1}^{LO+NLO} = \left\langle \phi_{cl}(t_1) + \frac{g^2 V_3}{4}\int\limits_{t_0}^{t_1} dt_2 \phi_{cl}(t_2)\Phi_3(t_1,t_2)\right\rangle_{i.c.}, \tag{22}$$

where ϕ_{cl} is the Jacobi elliptical function (21) and V_3 is the full volume of the system (19).

Spatially homogeneous case of variations eq.(18) depends only on two times. Since we know analytical solution for ϕ_{cl} and the first variation $\Phi_1(t_1,t_2)$ (see [21]) it is clear that all three variations are periodic functions (with period of the classical solution (21)) multiplied by some power of t_1. It is easy to show that k-th variation grows as t_1^k term at most.

The limits of integration over t_2 in NLO term of eq.(22) show that this term is subleading at small times t_1 and we can consider behavior of NLO term only at $t_1 >> t_0$. At this limit we can neglect phase C in eq.(21) and introduce new dimensionless variable $z = g\phi_m t$. It help us obtain dimensionless variations $f_n(z_1,z_2)$ as

$$\phi_{cl}(t_1) = \phi_m f_0(z_1), \qquad \Phi_k(t_1,t_2) = g^{-k}\phi_m^{1-2k} f_k(z_1,z_2), \qquad k = 1,2,3. \tag{23}$$

Dimensionless case of eq.(22) has the form

$$\langle\hat{\varphi}(t)\rangle_{t_1}^{LO+NLO} = \left\langle \phi_m f_0(z_1) + \frac{1}{4g^2V_3^2\phi_m^5}\int\limits_{z_0}^{z_1} dz_2 f_0(z_2) f_3(z_1,z_2)\right\rangle_{i.c.}. \tag{24}$$

Because of periodicity of $f_k(z_1,z_2)$ the integral over z_2 behave at late times ($z_1 >> z_0$) as

$$\int\limits_{z_0}^{z_1} dz_2 f_0(z_2) f_3(z_1,z_2) \approx z_1^3 \psi(z_1), \tag{25}$$

where $\psi(z_1)$ is a periodic function with same period T_{cl}. Finally, we have

$$\langle\hat{\varphi}(t)\rangle_{t_1}^{LO+NLO} \approx \left\langle \phi_m \left[f_0(g\phi_m t_1) + \frac{g t_1^3}{4V_3^2\phi_m^3}\psi(g\phi_m t_1)\right]\right\rangle_{i.c.}. \tag{26}$$

It seems that NLO term of eq.(26) is dominated at large times t_1 before averaging on initial conditions. Therefore the choice of the Wigner function play a crucial role and dynamics of the system strongly depends on the initial state. This feature is rather general for far-from-equilibrium systems.

As it is shown in [21] in case of gaussian choice for the Wigner function

$$f_W(\alpha,p) = \frac{1}{\pi\alpha_0 p_0} e^{-\frac{(\alpha-A)^2}{\alpha_0^2}} e^{-\frac{p^2}{p_0^2}},$$
$$\langle\quad\rangle_{i.c.} \equiv \int\frac{dp}{2\pi}\int d\alpha f_W(t_0,\alpha,p), \tag{27}$$

it is possible to perform the averaging over initial conditions analytically. Let us make a change of variables $(\alpha,p) \to (\phi_{max}, C)$ (see (21)):

$$\int\frac{dp}{2\pi}\int d\alpha \to \int |J|\, d\phi_{max}\, dC.$$
$$|J(\phi_{max})| = g\phi_{max}^2. \tag{28}$$

Analytical integration over ϕ_{max} and C is possible in the saddle point approximation, where

$$f_W(\phi_{max}, C, 0) \approx \frac{1}{\alpha_0 p_0 \pi} e^{-\frac{(\phi_{max}-A)^2}{\alpha_0^2} - \frac{C^2 A^4 g^2}{p_0^2}}. \tag{29}$$

Let us introduce Fourier transform as

$$cn\left(\frac{1}{2}; g\phi_{max}t + C\right) = \sum_{k=-\infty}^{\infty} u_k e^{\frac{2\pi i k}{T}(g\phi_{max}t + C)}, \tag{30}$$

$$u_m = \frac{1}{T}\int_0^T cn\left(\frac{1}{2}; t\right) e^{-imt\frac{2\pi}{T}} dt,$$

$$\psi_m = \frac{1}{T}\int_0^T \psi(t) e^{-imt\frac{2\pi}{T}}, \tag{31}$$

where T is the period of the Jacobi elliptical function $cn\left(\frac{1}{2}; t\right)$. Performing these calculations for the field expectation value we obtain

$$\langle\hat{\varphi}(t)\rangle_{t_1}^{LO+NLO} \approx 2A \sum_{k=0}^{\infty} \left(u_k + \frac{g t_1^3}{4V_3^2 A^3}\psi_k\right) e^{-\frac{\pi^2 p_0^2}{g^2 A^4 T^2}k^2} e^{-\frac{\alpha_0^2 \pi^2 g^2}{T^2}k^2 t_1^2} cos\left(\frac{2Ag\pi k}{T} t_1\right). \tag{32}$$

The common time behavior of both LO and NLO terms is oscillation with the amplitude that fall in time according to the cosine and the exponent of the eq.(32). However NLO term additionally grows as t_1^3. The asymptotic behavior of both terms are identical because of term $\exp(-t_1^2)$, but early time evolution (usually most interesting part of the problem) fully depend on initial state.

Parameter A in the Wigner function (27) denotes the measure of excitation of the initial field configurations. It means that starting points of all classical trajectories $\phi_{cl}(t_0) \equiv \alpha$ are grouped around some value A. In order to describe highly-excited fields at the initial state (that is what we want for heavy ion collisions for example) this parameter should be large $A >> 1$. As one can see from (32) it means that NLO term is suppressed.

Similar analysis can be performed for the $T^{\mu\nu}$ components

$$T^{\mu\nu} = \partial^\mu\varphi\partial^\nu\varphi - g^{\mu\nu}\left(\frac{1}{2}\partial_\sigma\varphi\partial^\sigma\varphi - \frac{g^2}{4}\varphi^4\right). \tag{33}$$

At the classical level energy and pressure are:

$$\varepsilon_{cl} = \frac{1}{2}\dot{\phi}_{cl}^2 + \frac{g^2}{4}\phi_{cl}^4, \; p_{cl} = \frac{1}{2}\dot{\phi}_{cl}^2 - \frac{g^2}{4}\phi_{cl}^4. \tag{34}$$

At the LO level we have

$$\varepsilon_{LO}(t_1) = \left\langle\frac{1}{2}\dot{\phi}_{cl}^2(t_1) + \frac{g^2}{4}\phi_{cl}^4(t_1)\right\rangle_{i.c.}, \tag{35}$$

$$p_{LO}(t_1) = \left\langle\frac{1}{2}\dot{\phi}_{cl}^2(t_1) - \frac{g^2}{4}\phi_{cl}^4(t_1)\right\rangle_{i.c.}. \tag{36}$$

These result had been obtained by [1]. However, one need to note that the NLO answers of [1] correspond to LO observables in our work.

At the NLO level $\varepsilon_{NLO} = 0$. It is conserved, but at the initial moment $\varepsilon_{NLO}(t_0) = 0$ due to integration over t_2 in (3.2). For p_{NLO} we have

$$\langle \hat{p}(t)\rangle_{t_1}^{LO+NLO} = \left\langle g^2\phi_m^4 \left[\pi_{LO}(g\phi_m t_1) + \frac{g t_1^3}{4V_3^2\phi_m^3}\pi_{NLO}(g\phi_m t_1)\right]\right\rangle_{i.c.}, \tag{37}$$

where

$$\pi_{LO}(z_1) = \frac{1}{2}(f_0'(z_1))^2 - \frac{1}{4}f_0^4(z_1) \tag{38}$$

and

$$\int_{z_0}^{z_1} dz_2 f_0(z_1)\Big[f_0'(z_1)f_3'(z_1,z_2) + 3f_1'(z_1,z_2)f_2'(z_1,z_2) - f_0^3(z_1)f_3(z_1,z_2) - 9f_0^2(z_1)f_1(z_1,z_2)f_2(z_1,z_2) - 6f_0(z_1)f_1^3(z_1,z_2)\Big] \approx z_1^3\pi_{NLO}(z_1).$$

Again π_{LO}, π_{NLO} are periodic functions with period T_{cl}.

One can note the dimensionless combination

$$c_1(g\phi_m t_1) + \frac{g t_1^3}{4V_3^2\phi_m^3}c_2(g\phi_m t_1), \tag{39}$$

which in both examples eq.(26) and eq.(37) is similar and corresponds to LO + NLO terms. As $\phi_m = \phi_m(\alpha, p)$ depends on initial conditions we can see that integration with the Wigner function which gives small amplitude ϕ_m (i.e. has small height and wide tails) results in case where NLO term became significant.

5 Conclusions

Let us summarize the main results presented in the talk.

- Systematic formalism for calculating corrections to the semiclassical approximation for correlators of highly excited quantum field in the framework of Keldysh technique is presented. It is shown that the results obtained in the leading semiclassical approximation reproduce those within the Classical Statistical Aproximation (CSA) thus establishing equivalence between the two approaches. An important feature of the CSA approximation is that it is generically not a small-coupling one.
- Generic expressions for leading corrections to the CSA approximation for the scalar field case were derived.
- Analytical expressions for the mean field, energy and pressure of the homogeneous scalar field were derived. The critical role of the character of initial conditions for applicability of the CSA approximation was discussed.

AR was supported by the Russian Science Foundation under grant No. 14-02-31748

References

[1] K. Dusling, T. Epelbaum, F. Gelis, R. Venugopalan, Nucl.Phys. A**850**, 69-109 (2011)
[2] T. Epelbaum, F. Gelis, Nucl.Phys. A**872**, 210-244 (2011)
[3] K. Dusling, T. Epelbaum, F. Gelis, R. Venugopalan, Phys.Rev. D**86**, 085040 (2012)
[4] K. Fukushima, F. Gelis, L. McLerran, Nucl.Phys. A**786**, 107-130 (2007)
[5] J. Berges, K. Boguslavski, S. Schlichting, et al., J. High Energ. Phys. **1405**, 54 (2014)
[6] A. Kurkela and Y. Zhu, Phys. Rev. Lett. **115**, 182301 (2015)
[7] L. Keegan, A. Kurkela, A. Mazeliauskas and D. Teaney, arXiv:1605.04287 [hep-ph]
[8] A. Kurkela, arXiv:1601.03283 [hep-ph]
[9] D. T. Son, hep-ph/9601377.
[10] S. Y. Khlebnikov and I. I. Tkachev, Phys. Rev. Lett. **77**, 219 (1996)
[11] D. Boyanovsky, Phys. Rev. D **92**, 023527 (2015)
[12] E. T. Akhmedov, H. Godazgar and F. K. Popov, Phys. Rev. D **93**, 024029 (2016)
[13] E. T. Akhmedov and F. K. Popov, JHEP **1509**, 085 (2015)
[14] E. T. Akhmedov, N. Astrakhantsev and F. K. Popov, JHEP **1409**, 071 (2014)
[15] F. Gelis and N. Tanji, Prog. Part. Nucl. Phys. **87**, 1 (2016)
[16] J. Berges and T. Gasenzer, Phys. Rev. A **76**, 033604 (2007)
[17] Kean Loon Lee, Nick P. Proukakis, arXiv:1607.06939 [cond-mat.quant-gas]
[18] L.V. Keldysh, Zh.Eksp.Teor.Fiz. 47 (1964) 1515-1527; Sov.Phys.JETP 20 (1965) 1018
[19] J. Berges, AIP Conf.Proc. **739**,3-62 (2005)
[20] J. Berges, e-Print: arXiv:1503.02907
[21] A. V. Leonidov and A. A. Radovskaya, JETP Lett. **101**, 215 (2015)
[22] H. Lee, M.O. Scully, Journal of Chemical Physics **73**, 2238 (1980)
[23] L.Bonnet, Journal of Chemical Physics **139**, 114108 (2013)
[24] S. Jeon, Annals Phys. **340** (2014) 119-170

EPJ Web of Conferences **125**, 05014 (2016) DOI: 10.1051/epjconf/201612505014
QUARKS-2016

Some aspects of $\mathcal{N} = 1$ SYM renormalization

Konstantin Stepanyantz[1,★]

[1]*Moscow State University, Physical Faculty, Department of Theoretical Physics, 119991, Moscow, Russia*

Abstract. Using the BRST invariant version of the higher covariant derivative regularization, we demonstrate that in $\mathcal{N} = 1$ supersymmetric gauge theories the three-point vertices with two ghost legs and a single leg of the quantum gauge superfield are finite in all orders. This theorem is proved by the help of the Slavnov–Taylor identities and the supergraph technique. Its correctness is verified by explicit one-loop calculation. Using finiteness of the considered vertices we express the NSVZ relation in terms of the anomalous dimensions of the gauge superfield, of the Faddeev–Popov ghosts, and of the matter superfields.

1 Introduction

The ultraviolet behaviour of supersymmetric theories is better than in the non-supersymmetric case due to non-renormalization theorems. For $\mathcal{N} = 1$ supersymmetric theories the statement that the superpotential has no divergent quantum corrections [1] is usually called the non-renormalization theorem. However, it is also possible to consider the relation between the β-function and the anomalous dimension of the matter superfields as a non-renormalization theorem. This relation is usually called the exact NSVZ β-function [2–5], because for the pure $\mathcal{N} = 1$ SYM theory it gives the exact expression for the β-function in the form of the geometric series.

In this paper, following Ref. [6], we argue that in $\mathcal{N} = 1$ SYM theories the three-point ghost-gauge vertices are finite in all orders. This statement can be also considered as a non-renormalization theorem.

The non-renormalization theorem originate from large symmetry of the theory. That is why deriving them it is necessary to quantize the theory in such a way that most symmetries of the theory remain unbroken. In particular, it is desirable to use the invariant regularization. However, the dimensional regularization breaks supersymmetry [7], while the dimensional reduction [8] is not self-consistent [9] and due to this can also lead to supersymmetry breaking [10]. The consistent regularization applicable for supersymmetric theories is the higher covariant derivative regularization [11, 12]. It can be formulated in explicitly $\mathcal{N} = 1$ supersymmetric way [13, 14] and reveals some interesting features of quantum corrections [15–18]. In particular, it allows deriving the NSVZ β-function at least in the Abelian case by direct summation of superdiagrams [19, 20]. That is why in proving the above mentioned non-renormalization theorem we will use the higher covariant derivative regularization.

★e-mail: stepan@m9com.ru

2 BRST invariant version of the higher covariant derivative regularization for $\mathcal{N} = 1$ supersymmetric gauge theories

We consider the $\mathcal{N} = 1$ SYM theory, which is described by the action

$$S = \frac{1}{2e_0^2}\mathrm{Re}\,\mathrm{tr}\int d^4x\, d^2\theta\, W^a W_a + \frac{1}{4}\int d^4x\, d^4\theta\, \phi^{*i}(e^{2V})_i{}^j\phi_j$$
$$+\Big\{\int d^4x\, d^2\theta\Big(\frac{1}{4}m_0^{ij}\phi_i\phi_j + \frac{1}{6}\lambda_0^{ijk}\phi_i\phi_j\phi_k\Big) + \text{c.c.}\Big\}, \qquad (1)$$

assuming that the theory is invariant under the gauge transformations $\phi \to e^A\phi$; $e^{2V} \to e^{-A^+}e^{2V}e^{-A}$, where $A = ie_0 A^B T^B$ is an arbitrary chiral superfield. We will also use the background field method. The quantum-background splitting in the supersymmetric case is made by the substitution $e^{2V} \to e^{\mathbf{\Omega}^+}e^{2V}e^{\mathbf{\Omega}}$. Then, the background gauge superfield $\mathbf{V}$ is defined as $e^{2\mathbf{V}} = e^{\mathbf{\Omega}^+}e^{\mathbf{\Omega}}$.

We will regularize this theory by higher covariant derivatives. For this purpose it is necessary to add a term with higher degrees of the covariant derivatives to the classical action. Then divergences remain only in the one-loop approximation. These remaining divergences are regularized by the Pauli–Villars determinants [21]. We choose the higher derivative term

$$S_\Lambda = \frac{1}{2e_0^2}\mathrm{Re}\,\mathrm{tr}\int d^4x\, d^2\theta\, e^{\Omega}e^{\mathbf{\Omega}}W^a e^{-\mathbf{\Omega}}e^{-\Omega}\Big[R\Big(-\frac{\bar\nabla^2\nabla^2}{16\Lambda^2}\Big) - 1\Big]_{Adj} e^{\Omega}e^{\mathbf{\Omega}}W_a e^{-\mathbf{\Omega}}e^{-\Omega}$$
$$+\frac{1}{4}\int d^4x\, d^4\theta\, \phi^+ e^{\mathbf{\Omega}^+}e^{\Omega^+}\Big[F\Big(-\frac{\bar\nabla^2\nabla^2}{16\Lambda^2}\Big) - 1\Big]e^{\Omega}e^{\mathbf{\Omega}}\phi, \qquad (2)$$

and the gauge fixing term

$$S_{\text{gf}} = \frac{1}{e_0^2}\mathrm{tr}\int d^4x\, d^4\theta\Big(16\xi_0 f^+\Big[e^{\mathbf{\Omega}^+}K^{-1}\Big(-\frac{\bar{\boldsymbol\nabla}^2\boldsymbol\nabla^2}{16\Lambda^2}\Big)e^{\mathbf{\Omega}}\Big]_{Adj}f + e^{\mathbf{\Omega}}fe^{-\mathbf{\Omega}}\boldsymbol\nabla^2 V + e^{-\mathbf{\Omega}^+}f^+e^{\mathbf{\Omega}^+}\bar{\boldsymbol\nabla}^2 V\Big), \qquad (3)$$

where the regulators R, F, and K rapidly grow at infinity. The corresponding action for the Faddeev–Popov is

$$S_{\text{FP}} = \frac{1}{e_0^2}\mathrm{tr}\int d^4x\, d^4\theta\,\Big(e^{\mathbf{\Omega}}\bar{c}e^{-\mathbf{\Omega}} + e^{-\mathbf{\Omega}^+}\bar{c}^+e^{\mathbf{\Omega}^+}\Big)\Big\{\Big(\frac{V}{1-e^{2V}}\Big)_{Adj}\Big(e^{-\mathbf{\Omega}^+}c^+e^{\mathbf{\Omega}^+}\Big) + \Big(\frac{V}{1-e^{-2V}}\Big)_{Adj}\Big(e^{\mathbf{\Omega}}ce^{-\mathbf{\Omega}}\Big)\Big\}. \qquad (4)$$

(Also it is necessary to add the Nielsen–Kallsodh ghosts.) The total action of the gauge fixed theory is invariant under the BRST transformations [22, 23]

$$\delta V = -\varepsilon\Big\{\Big(\frac{V}{1-e^{2V}}\Big)_{Adj}\Big(e^{-\mathbf{\Omega}^+}c^+e^{\mathbf{\Omega}^+}\Big) + \Big(\frac{V}{1-e^{-2V}}\Big)_{Adj}\Big(e^{\mathbf{\Omega}}ce^{-\mathbf{\Omega}}\Big)\Big\}; \qquad \delta\phi = \varepsilon c\phi; \qquad \delta f = 0;$$
$$\delta\bar{c} = \varepsilon\bar{D}^2(e^{-2V}f^+e^{2V}); \qquad \delta\bar{c}^+ = \varepsilon D^2(e^{2V}fe^{-2V}); \qquad \delta c = \varepsilon c^2; \qquad \delta c^+ = \varepsilon(c^+)^2, \qquad (5)$$

where ε is an anticommuting scalar parameter.

We will defined the renormalization constants by the following equations

$$\frac{1}{\alpha_0} = \frac{Z_\alpha}{\alpha}; \qquad \frac{1}{\xi_0} = \frac{Z_\xi}{\xi}; \qquad \boldsymbol{V} = \boldsymbol{V}_R; \qquad V = Z_V Z_\alpha^{-1/2} V_R; \qquad \phi_i = (\sqrt{Z_\phi})_i{}^j (\phi_R)_j;$$
$$\bar{c}c = Z_c Z_\alpha^{-1} \bar{c}_R c_R; \qquad m^{ij} = m_0^{mn} (Z_m)_m{}^i (Z_m)_n{}^j; \qquad \lambda^{ijk} = \lambda_0^{mnp} (Z_\lambda)_m{}^i (Z_\lambda)_n{}^j (Z_\lambda)_p{}^k. \tag{6}$$

The renormalized superfields are denoted by the subscript R; α, λ, m, and ξ are also the renormalized parameters. One can also impose the conditions $(Z_m)_i{}^j = (Z_\lambda)_i{}^j = (\sqrt{Z_\phi})_i{}^j$; $Z_\xi = Z_V^{-2}$ to these renormalization constants.

3 Non-renormalization of the three-point vertices with two ghost legs and a single leg of the quantum gauge superfield

We will prove that the three-point vertices with 2 ghost legs and 1 leg of the quantum gauge superfield are finite in all orders. There are 4 such vertices: $\bar{c}\,Vc$, $\bar{c}^+ Vc$, $\bar{c}\,Vc^+$, and $\bar{c}^+ Vc^+$. They have the same renomalization constant $Z_\alpha^{-1/2} Z_c Z_V$. Therefore, the above statement can be written as

$$\frac{d}{d\ln\Lambda}(Z_\alpha^{-1/2} Z_c Z_V) = 0. \tag{7}$$

(In the one-loop approximation this has first been noted in [24].) Thus, there is a subtraction scheme in which $Z_\alpha^{-1/2} Z_c Z_V = 1$. It is important to note that Z_c is divergent. (This is demonstrated by an explicit calculation below.) Therefore, the Green functions of the type $\bar{c}\,V^n c$ are divergent for $n \neq 1$.

The finiteness of the considered vertices will be proved using the Slavnov–Taylor identities [25, 26]. The Slavnov–Taylor identity can be obtained by making the substitution coinciding with the BRST transformations in the generating functional and has the form

$$0 = \int d^4x\, d^4\theta_x \frac{\delta\Gamma}{\delta V_x^A}\left\langle \delta V_x^A \right\rangle + \int d^4x\, d^2\theta_x \Big(\left\langle \delta \bar{c}_x^A \right\rangle \frac{\delta\Gamma}{\delta \bar{c}_x^A} + \left\langle \delta c_x^A \right\rangle \frac{\delta\Gamma}{\delta c_x^A} + \langle \delta\phi_i \rangle \frac{\delta\Gamma}{\delta\phi_i} \Big)$$
$$+ \int d^4x\, d^2\bar{\theta}_x \Big(\left\langle \delta \bar{c}_x^{*A} \right\rangle \frac{\delta\Gamma}{\delta \bar{c}_x^{*A}} + \left\langle \delta c_x^{*A} \right\rangle \frac{\delta\Gamma}{\delta c_x^{*A}} + \left\langle \delta\phi^{*i} \right\rangle \frac{\delta\Gamma}{\delta\phi^{*i}} \Big), \tag{8}$$

where we did not omit the ε-dependence. We will also use one more identity that is obtained by making the substitution $\bar{c} \to \bar{c} + a$, where a is an arbitrary chiral superfield:

$$\varepsilon \frac{\delta\Gamma}{\delta \bar{c}_x^A} = \frac{1}{4}\bar{D}^2 \left\langle \delta V_x^A \right\rangle; \qquad \varepsilon \frac{\delta\Gamma}{\delta \bar{c}_x^{*A}} = \frac{1}{4} D^2 \left\langle \delta V_x^A \right\rangle, \tag{9}$$

where the background gauge superfield is set to 0, for simplicity.

To derive the Slavnov–Taylor identity for the ghost-gauge three-point functions, we differentiate the Slavnov–Taylor identity (8) with respect to $\bar{c}_y^{*B}$, c_z^C, and c_w^D using the equations

$$\frac{\delta^2\Gamma}{\delta \bar{c}_y^{*B} \delta c_x^A} = -\frac{D_y^2 \bar{D}_x^2}{16} G_c \delta_{xy}^8 \delta_{AB}; \qquad \frac{\delta}{\delta c_x^A}\left\langle \delta V_y^B \right\rangle = -\varepsilon \cdot \frac{1}{4} G_c \bar{D}^2 \delta_{xy}^8 \delta_{AB}. \tag{10}$$

Then we obtain the identity relating the three-point Green functions

$$\varepsilon G_c(\partial_w^2/\Lambda^2) \bar{D}_w^2 \frac{\delta^3\Gamma}{\delta \bar{c}_y^{*B} \delta V_w^D \delta c_z^C} - \varepsilon G_c(\partial_z^2/\Lambda^2) \bar{D}_z^2 \frac{\delta^3\Gamma}{\delta \bar{c}_y^{*B} \delta V_z^C \delta c_w^D} + \frac{1}{2} G_c \left(\partial_y^2/\Lambda^2\right) D_y^2 \frac{\delta^2}{\delta c_z^C \delta c_w^D} \left\langle \delta c_y^B \right\rangle = 0. \tag{11}$$

To simplify this identity, we will use explicit expressions for the Green functions which can be found from dimensional and chirality considerations:

$$\frac{\delta^3\Gamma}{\delta\bar{c}_x^{*A}\delta V_y^B\delta c_z^C} = -\frac{ie_0}{16}f^{ABC}\int\frac{d^4p}{(2\pi)^4}\frac{d^4q}{(2\pi)^4}\Big(f(p,q)\partial^2\Pi_{1/2} - F_\mu(p,q)(\gamma^\mu)_{\dot{a}}{}^b\bar{D}^{\dot{a}}D_b + F(p,q)\Big)_y$$
$$\times\Big(D_x^2\delta_{xy}^8(q+p)\,\bar{D}_z^2\delta_{yz}^8(q)\Big), \tag{12}$$

where $\partial^2\Pi_{1/2} \equiv -D^a\bar{D}^2D_a/8$ is the supersymmetric transversal projection operator and

$$\delta_{xy}^8(p) \equiv \delta^4(\theta_x - \theta_y)e^{ip_\alpha(x^\alpha - y^\alpha)}. \tag{13}$$

To calculate the ghost correlator entering in the Slavnov–Taylor identity, we introduce the chiral source $\mathcal{J}$ and add the term

$$-\frac{e_0}{2}\int d^4x\,d^2\theta\,f^{ABC}\mathcal{J}^Ac^Bc^C + \text{c.c.} \tag{14}$$

to the action. Then from the dimensional and chirality considerations we find

$$\frac{\delta^2}{\delta c_z^C\delta c_w^D}\left\langle\delta c_y^B\right\rangle = -i\varepsilon\frac{\delta^3\Gamma}{\delta c_z^C\delta c_w^D\delta\mathcal{J}_y^B} = -\frac{ie_0\varepsilon}{4}f^{BCD}\int\frac{d^4p}{(2\pi)^4}\frac{d^4q}{(2\pi)^4}H(p,q)\bar{D}_z^2\delta_{zy}^8(q+p)\bar{D}_w^2\delta_{yw}^8(q), \tag{15}$$

where $[H(p,q)] = 1$ and, by construction, $H(p,q) = H(p,-q-p)$.

Substituting the explicit expressions for the Green function (12) and the ghost correlator (15) into the Slavnov–Taylor identity we can rewrite it in the form

$$G_c(q)F(q,p) + G_c(p)F(p,q) = 2G_c(q+p)H(-q-p,q), \tag{16}$$

where we use the compact notation $G_c(-q^2/\Lambda^2) \to G_c(q)$. The identity (16) will be used below for proving finiteness of the three-point ghost-gauge vertices.

However, let us first prove that the function $H(p,q)$ is finite. H is contributed by diagrams in which one leg corresponds to the chiral source $\mathcal{J}$ and two other legs correspond to the chiral ghost superfields c. They contain the expression

$$\int d^4y\,d^2\theta_y\,\mathcal{J}_y^A\cdot\frac{\bar{D}_y^2D_y^2}{4\partial^2}\delta_{y1}^8\cdot\frac{\bar{D}_y^2D_y^2}{4\partial^2}\delta_{y2}^8 = -2\int d^4y\,d^4\theta_y\,\mathcal{J}_y^A\cdot\frac{D_y^2}{4\partial^2}\delta_{y1}^8\cdot\frac{\bar{D}_y^2D_y^2}{4\partial^2}\delta_{y2}^8. \tag{17}$$

Therefore, the considered contribution can be presented as an integral over the whole superspace, which includes the integration

$$\int d^4\theta = -\frac{1}{2}\int d^2\theta\bar{D}^2 + \text{total derivatives in the coordinate space.} \tag{18}$$

Therefore, two left supersymmetric derivatives should act to the chiral external lines. Consequently, a non-trivial result can be obtained only if two right supersymmetric derivatives also act to the external lines. Thus, the result should be proportional to the second degree of the external momenta and is finite in the UV region. Thus, the function $H(p,q)$ is UV finite.

Then we proceed to proving the non-renormalization theorem for the three-point ghost-gauge vertices. For this purpose the Slavnov–Taylor identity is multiplied by the renormalization constant

Z_c (such that $(G_c)_R = Z_c G$ is finite). Then we differentiate the result with respect to $\ln\Lambda$ and take the limit $\Lambda \to \infty$. Then, due to the finiteness of the functions $(G_c)_R$ and H

$$\Big((G_c)_R(q)\frac{d}{d\ln\Lambda}F(q,p) + (G_c)_R(p)\frac{d}{d\ln\Lambda}F(p,q)\Big)\Big|_{\Lambda\to\infty} = 0. \tag{19}$$

Setting $p = -q$ in this identity we obtain $dF(-q,q)/d\ln\Lambda\Big|_{\Lambda\to\infty} = 0$. This implies that the corresponding renormalization constant is finite,

$$\frac{d}{d\ln\Lambda}(Z_\alpha^{-1/2}Z_c Z_V) = 0. \tag{20}$$

Consequently, the function $F(p,q)$ is finite. Therefore, all three-point gauge-ghost vertices are also finite, because they also have the renormalization constant $Z_\alpha^{-1/2}Z_c Z_V$.

4 One-loop calculation

The two-point ghost Green function in the one-loop approximation (in the Euclidean space after the Wick rotation) is

$$G_c(p) = 1 + e_0^2 C_2 \int \frac{d^4k}{(2\pi)^4}\Big(\frac{\xi_0}{K_k} - \frac{1}{R_k}\Big)\Big(-\frac{1}{6k^4} + \frac{1}{2k^2(k+p)^2} - \frac{p^2}{2k^4(k+p)^2}\Big) + O(e_0^4, e_0^2\lambda_0^2), \tag{21}$$

where $R_k \equiv R(k^2/\Lambda)$ and $K_k \equiv K(k^2/\Lambda^2)$. Evidently, this function is divergent in the ultraviolet region (for infinite Λ).

The three-point ghost-gauge Green functions in the one-loop approximation have been calculated in [6]. Here we present only results for the functions F:

$$F(p,q) = 1 + \frac{e_0^2 C_2}{4}\int \frac{d^4k}{(2\pi)^4}\Big\{ -\frac{(q+p)^2}{R_k k^2(k+p)^2(k-q)^2} - \frac{\xi_0\, p^2}{K_k k^2(k+q)^2(k+q+p)^2}$$
$$+\frac{\xi_0\, q^2}{K_k k^2(k+p)^2(k+q+p)^2} + \Big(\frac{\xi_0}{K_k} - \frac{1}{R_k}\Big)\Big(-\frac{2(q+p)^2}{k^4(k+q+p)^2} + \frac{2}{k^2(k+q+p)^2}$$
$$-\frac{1}{k^2(k+q)^2} - \frac{1}{k^2(k+p)^2}\Big)\Big\} + O(\alpha_0^2, \alpha_0\lambda_0^2); \tag{22}$$

(23)

We see that this expression is finite in the ultraviolet region. This fact is in agreement with the non-renormalization theorem discussed above. Expressions for the functions F_μ and f are much larger and can found in [6]. They are also finite, but in the one-loop approximation this result is trivial and follows from simple dimensional considerations. It was also verified in [6] that the other Green functions of the considered structure are also finite.

The one-loop expressions for the function H has the form

$$H(p,q) = 1 - \frac{e_0^2 C_2}{4}\int \frac{d^4k}{(2\pi)^4}\Big\{\frac{p^2}{R_k k^2(k+q)^2(k+q+p)^2} + \frac{(q+p)^2}{k^4(k+q+p)^2}\Big(\frac{\xi_0}{K_k} - \frac{1}{R_k}\Big)$$
$$+\frac{q^2}{k^4(k+q)^2}\Big(\frac{\xi_0}{K_k} - \frac{1}{R_k}\Big)\Big\} + O(e_0^4, e_0^2\lambda_0^2). \tag{24}$$

It is finite in the ultraviolet region and proportional to the second degree of the external momenta. (This fact was very important for proving finiteness of the three-point ghost-gauge vertices.)

Finally, we note that correctness of the one-loop calculations has been confirmed by verifying the Slavnov–Taylor identities, and in particular, Eq. (16).

5 NSVZ β-function and non-renormalization of the ghost-gauge vertices

In $\mathcal{N} = 1$ supersymmetric theories the β-function is related with anomalous dimension of the matter superfields by the equation [2, 4, 5]

$$\beta(\alpha) = -\frac{\alpha^2\Big(3C_2 - T(R) + C(R)_i{}^j\gamma_j{}^i(\alpha)/r\Big)}{2\pi(1 - C_2\alpha/2\pi)}, \tag{25}$$

which has originally been found from various general arguments, such as instantons, anomalies, etc. For $\mathcal{N} = 1$ SQED, regularized by higher derivatives, the NSVZ relation [27, 28] was derived by explicit summation of supergraphs [19, 20]. To generalize this result to the case of the dimensional reduction is a complicated and so far unsolved problem [29–32].

Qualitatively, the NSVZ β-function in the Abelian case appears as follows [16]. Let us consider a supergraph without external lines. By attaching two external lines of the background gauge superfield we obtain a contribution to the β-function. From the other hand, cutting the matter line in original supergraph we obtain a contribution to the anomalous dimension of the matter superfields. It has been shown in [19, 20] that two these contributions coincide up to a multiplicative factor.

Let us equivalently rewrite the NSVZ β-function for the non-Abelian case in a different form, which admits similar qualitative interpretation. This can be done using the non-renormalization of the three-point ghost-gauge vertices proved above. First, we present the NSVZ relation (for the renormalization group functions defined in terms of the bare coupling constant) in the form

$$\frac{\beta(\alpha_0, \lambda_0)}{\alpha_0^2} = -\frac{3C_2 - T(R) + C(R)_i{}^j(\gamma_\phi)_j{}^i(\alpha_0, \lambda_0)/r}{2\pi} + \frac{C_2}{2\pi}\cdot\frac{\beta(\alpha_0, \lambda_0)}{\alpha_0}. \tag{26}$$

The β-function in the right hand side can be expressed in terms of the renormalization constant Z_α,

$$\beta(\alpha_0, \lambda_0) = \frac{d\alpha_0(\alpha, \lambda, \Lambda/\mu)}{d\ln\Lambda}\Big|_{\alpha,\lambda=\text{const}} = -\alpha_0\frac{d\ln Z_\alpha}{d\ln\Lambda}\Big|_{\alpha,\lambda=\text{const}}. \tag{27}$$

Then, using the equation $d(Z_\alpha^{1/2}Z_V Z_c)/d\ln\Lambda = 0$ we obtain

$$\beta(\alpha_0, \lambda_0) = -2\alpha_0\frac{d\ln(Z_c Z_V)}{d\ln\Lambda}\Big|_{\alpha,\lambda=\text{const}} = 2\alpha_0\Big(\gamma_c(\alpha_0, \lambda_0) + \gamma_V(\alpha_0, \lambda_0)\Big), \tag{28}$$

where γ_c and γ_V are anomalous dimensions of the Faddeev–Popov ghosts and of the quantum gauge superfield (defined in terms of the bare coupling constants), respectively. Substituting this expression into the right hand side of the NSVZ relation (26) we obtain

$$\frac{\beta(\alpha_0, \lambda_0)}{\alpha_0^2} = -\frac{1}{2\pi}\Big(3C_2 - T(R) - 2C_2\gamma_c(\alpha_0, \lambda_0) - 2C_2\gamma_V(\alpha_0, \lambda_0) + C(R)_i{}^j(\gamma_\phi)_j{}^i(\alpha_0, \lambda_0)/r\Big).$$

From this form of the NSVZ β-function we see that the matter and ghost contributions to the right hand side look very similar.

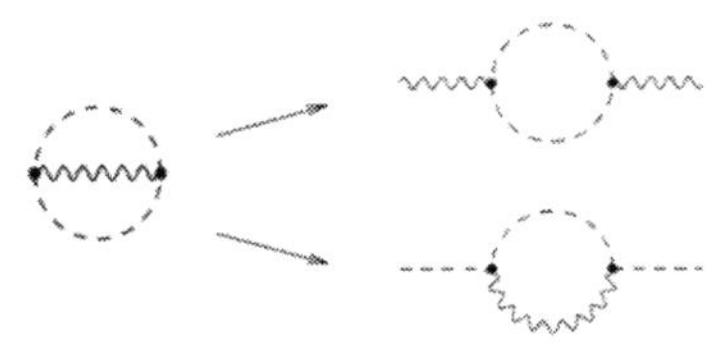

Figure 1. Graphical interpretation of the new form of the NSVZ relation in the non-Abelian case

Let us assume that, similarly to the Abelian case, the NSVZ relation is valid for the RG functions defined in terms of the bare coupling constants in the case of using the higher covariant derivative regularization. The RG functions defined in terms of the renormalized coupling constant are scheme dependent and satisfy the NSVZ relation only in a certain subtraction scheme. In the Abelian case the NSVZ scheme with the higher derivative regularization has been constructed in [33–35]. One can similarly try to construct the NSVZ scheme in the non-Abelian case. Repeating argumentation of [33–35] it is possible to verify that in the non-Abelian case the RG functions defined in terms of the bare coupling constant coincide with the ones defined in terms of the renormalized coupling constants if the boundary conditions

$$Z_\alpha(\alpha,\lambda,x_0)=1;\qquad (Z_\phi)_i{}^j(\alpha,\lambda,x_0)=\delta_i^j;\qquad Z_c(\alpha,\lambda,x_0)=1, \tag{29}$$

where x_0 is a fixed value of $\ln\Lambda/\mu$, are imposed on the renormalization constants. (For example, it is possible and convenient to choose $x_0=0$.) We also assume that the renormalization constants satisfy the equation

$$Z_V=Z_\alpha^{1/2}Z_c^{-1}, \tag{30}$$

If our assumption (that the RG functions defined in terms of the bare coupling constants satisfy the NSVZ relation in the case of using the higher covariant derivative regularization) is true, then the conditions (29) and (30) give the NSVZ scheme with the higher covariant derivative regularization.

6 Conclusion

For $\mathcal{N}=1$ SYM theories the three-point vertices with two ghost legs and one leg of the quantum gauge superfield are finite. This has been proved using the Slavnov–Taylor identities in all orders and has been verified by an explicit one-loop calculation. Due to the non-renormalization of the triple ghost-gauge vertices the renormalization constants can be chosen so that $Z_\alpha^{-1/2}Z_VZ_c=1$. The corresponding anomalous dimensions can also be related in all orders.

The NSVZ β-function can be rewritten in terms of the anomalous dimensions of the quantum gauge field, of the Faddeev–Popov ghosts, and of the matter superfields. Then the ghost contribution has a structure similar to the matter superfield contribution. The resulting expression for the NSVZ β-function has a very simple qualitative interpretation.

Using the above results it is possible to suggest a simple prescription giving the NSVZ scheme in the non-Abelian case, if the higher covariant derivative method is used for regularization.

Acknowledgments

This work is supported by the RFBR grant No. 14-01-00695.

References

[1] M. T. Grisaru, W. Siegel and M. Rocek, Nucl. Phys. B **159** 429 (1979).
[2] V. A. Novikov, M. A. Shifman, A. I. Vainshtein and V. I. Zakharov, Nucl. Phys. B **229** 381 (1983).
[3] D. R. T. Jones, Phys. Lett. B **123** 45 (1983).
[4] V. A. Novikov, M. A. Shifman, A. I. Vainshtein and V. I. Zakharov, Phys. Lett. B **166** 329 (1986); Sov. J. Nucl. Phys. **43** 294 (1986).
[5] M. A. Shifman and A. I. Vainshtein, Nucl. Phys. B **277** 456 (1986); Sov. Phys. JETP **64** 428 (1986).
[6] K. V. Stepanyantz, Nucl. Phys. B **909** 316 (2016).
[7] R. Delbourgo and V. B. Prasad, J. Phys. G **1** 377 (1975).
[8] W. Siegel, Phys. Lett. B **84** 193 (1979).
[9] W. Siegel, Phys. Lett. B **94** 37 (1980).
[10] L. V. Avdeev, Phys. Lett. B **117** 317 (1982).
[11] A. A. Slavnov, Nucl. Phys. B **31** 301 (1971).
[12] A. A. Slavnov, Theor.Math.Phys. **13** 1064 (1972).
[13] V. K. Krivoshchekov, Theor. Math. Phys. **36** 745 (1978).
[14] P. C. West, Nucl. Phys. B **268** 113 (1986).
[15] A. A. Soloshenko and K. V. Stepanyantz, Theor. Math. Phys. **140** 1264 (2004).
[16] A. V. Smilga and A. Vainshtein, Nucl. Phys. B **704** 445 (2005).
[17] A. B. Pimenov, E. S. Shevtsova and K. V. Stepanyantz, Phys. Lett. B **686** 293 (2010).
[18] K. V. Stepanyantz, arXiv:1108.1491 [hep-th].
[19] K. V. Stepanyantz, Nucl. Phys. B **852** 71 (2011).
[20] K. V. Stepanyantz, JHEP **1408** 096 (2014).
[21] A. A. Slavnov, Theor. Math. Phys. **33** 977 (1977).
[22] C. Becchi, A. Rouet and R. Stora, Commun. Math. Phys. **42** 127 (1975).
[23] I. V. Tyutin, Lebedev Institute preprint No. 39 (1975), arXiv:0812.0580 [hep-th].
[24] S. S. Aleshin, A. E. Kazantsev, M. B. Skoptsov and K. V. Stepanyantz, JHEP **1605** 014 (2016).
[25] J. C. Taylor, Nucl. Phys. B **33** 436 (1971).
[26] A. A. Slavnov, Theor. Math. Phys. **10** 99 (1972).
[27] A. I. Vainshtein, V. I. Zakharov and M. A. Shifman, JETP Lett. **42** 224 (1985).
[28] M. A. Shifman, A. I. Vainshtein and V. I. Zakharov, Phys. Lett. B **166** 334 (1986).
[29] I. Jack, D. R. T. Jones and C. G. North, Phys. Lett. B **386** 138 (1996).
[30] I. Jack, D. R. T. Jones and C. G. North, Nucl. Phys. B **486** 479 (1997).
[31] I. Jack, D. R. T. Jones and A. Pickering, Phys. Lett. B **435** 61 (1998).
[32] S. S. Aleshin, A. L. Kataev and K. V. Stepanyantz, Pisma Zh. Eksp. Teor. Fiz. **130** 83 (2016).
[33] A. L. Kataev and K. V. Stepanyantz, Nucl. Phys. B **875** 459 (2013).
[34] A. L. Kataev and K. V. Stepanyantz, Phys. Lett. B **730** 184 (2014).
[35] A. L. Kataev and K. V. Stepanyantz, Theor. Math. Phys. **181** 1531 (2014).

Fermions in 5D brane world models

Mikhail Smolyakov[1,*]

[1] *Skobeltsyn Institute of Nuclear Physics, Lomonosov Moscow State University, 119991, Moscow, Russia*

Abstract. In the present manuscript the fermion fields in the background of 5D brane world models with compact extra dimension are examined. It is shown that the only case that allows one to perform the Kaluza-Klein decomposition in a mathematically consistent way without unnatural fine-tunings and possible pathologies, is the one which does not admit localization of the zero mode. The report is based on the results presented in [1].

1 Setup and equations of motion

Let us take a five-dimensional space-time with the compact extra dimension forming the orbifold S^1/Z_2 with the coordinate $-L \leq z \leq L$ and the points $-z$ and z identified. Let us consider the following form of the background metric, which is standard in five-dimensional brane world models:

$$ds^2 = e^{2\sigma(z)}\eta_{\mu\nu}dx^\mu dx^\nu - dz^2, \tag{1}$$

where $\sigma(-z) = \sigma(z)$. We do not specify the explicit form of the solution for $\sigma(z)$.

As a model-independent example, we consider a theory with the action of the general form

$$S = \int d^4xdz\,\sqrt{g}\Big(E^M_N i\bar{\Psi}_1\Gamma^N\nabla_M\Psi_1 + E^M_N i\bar{\Psi}_2\Gamma^N\nabla_M\Psi_2 \\ -F(z)\left(\bar{\Psi}_1\Psi_1 - \bar{\Psi}_2\Psi_2\right) - G(z)\left(\bar{\Psi}_2\Psi_1 + \bar{\Psi}_1\Psi_2\right)\Big), \tag{2}$$

where $M, N = 0, 1, 2, 3, 5$, $\Gamma^\mu = \gamma^\mu$, $\Gamma^5 = i\gamma^5$, ∇_M is the covariant derivative containing the spin connection, E^M_N is the vielbein. The fields are also supposed to satisfy the orbifold symmetry conditions

$$\Psi_1(x,-z) = \gamma^5\Psi_1(x,z), \qquad \Psi_2(x,-z) = -\gamma^5\Psi_2(x,z), \tag{3}$$

the functions $F(z)$ and $G(z)$ are such that $F(-z) = -F(z)$ and $G(-z) = G(z)$. Due to the absence of chirality in five-dimensional space-time, we have to take two five-dimensional spinor fields Ψ_1 and Ψ_2 [2–4]. Here the function $F(z)$ is responsible for the localization of the zero Kaluza-Klein mode, whereas the function $G(z) \not\equiv 0$ provides its nonzero mass — for $G(z) \equiv 0$ there always exist the modes with zero four-dimensional mass [5].

*e-mail: smolyakov@theory.sinp.msu.ru

For the case of metric (1), action (2) can be rewritten in the form (see, for example, [4, 6] for the explicit form of the vielbein and spin connections)

$$S = \int d^4xdze^{4\sigma}\Big(e^{-\sigma}i\bar{\Psi}_1\gamma^\mu\partial_\mu\Psi_1 - \bar{\Psi}_1\gamma^5\left(\partial_5 + 2\sigma'\right)\Psi_1 + e^{-\sigma}i\bar{\Psi}_2\gamma^\mu\partial_\mu\Psi_2$$
$$-\bar{\Psi}_2\gamma^5\left(\partial_5 + 2\sigma'\right)\Psi_2 - F(z)\left(\bar{\Psi}_1\Psi_1 - \bar{\Psi}_2\Psi_2\right) - G(z)\left(\bar{\Psi}_2\Psi_1 + \bar{\Psi}_1\Psi_2\right)\Big).$$

The equations of motion for the fields Ψ_1 and Ψ_2 take the form

$$e^{-\sigma}i\gamma^\mu\partial_\mu\Psi_1 - \gamma^5\left(\partial_5 + 2\sigma'\right)\Psi_1 - F\Psi_1 - G\Psi_2 = 0, \tag{4}$$
$$e^{-\sigma}i\gamma^\mu\partial_\mu\Psi_2 - \gamma^5\left(\partial_5 + 2\sigma'\right)\Psi_2 + F\Psi_2 - G\Psi_1 = 0. \tag{5}$$

From (4) and (5) it is not difficult to obtain the second-order differential equations for the components of the fields Ψ_1 and Ψ_2:

$$-e^{-2\sigma}\Box\Psi_1 + \Psi_1'' + 5\sigma'\Psi_1' + (6\sigma'^2 + 2\sigma'')\Psi_1 - (G^2 + F^2 - e^{-\sigma}(e^\sigma F)'\gamma^5)\Psi_1$$
$$+e^{-\sigma}(e^\sigma G)'\gamma^5\Psi_2 = 0, \tag{6}$$
$$-e^{-2\sigma}\Box\Psi_2 + \Psi_2'' + 5\sigma'\Psi_2' + (6\sigma'^2 + 2\sigma'')\Psi_2 - (G^2 + F^2 + e^{-\sigma}(e^\sigma F)'\gamma^5)\Psi_2$$
$$+e^{-\sigma}(e^\sigma G)'\gamma^5\Psi_1 = 0. \tag{7}$$

We see that equations (6), (7) remain coupled in the general case, whereas we expect that in a consistent theory the corresponding equations should be five-dimensional second-order differential equations, which contain the derivatives in the four-dimensional coordinates only in the form $\Box = \eta^{\mu\nu}\partial_\mu\partial_\nu$ and provide a correct mass spectrum for the Kaluza-Klein modes. Equations for each component of the fields Ψ_1 and Ψ_2 separately can be obtained only for the special choice

$$G(z) \equiv Me^{-\sigma(z)}, \tag{8}$$

where M is a constant. In this case equations (6), (7), (4) and (5) give for the lowest mode

$$\Psi_1 = C_f \exp\left[-\int_0^z F(y)dy - 2\sigma(z)\right]\psi_L(x), \quad i\gamma^\mu\partial_\mu\psi_L - M\psi_R = 0, \quad \gamma^5\psi_L = \psi_L, \tag{9}$$

$$\Psi_2 = C_f \exp\left[-\int_0^z F(y)dy - 2\sigma(z)\right]\psi_R(x), \quad i\gamma^\mu\partial_\mu\psi_R - M\psi_L = 0, \quad \gamma^5\psi_R = -\psi_R, \tag{10}$$

where C_f is a normalization constant. The fields ψ_L and ψ_R are localized in the vicinity of the same point in the extra dimension because they have the same wave function; taken together they make up a four-dimensional Dirac fermion with mass M. Depending on the form of the function $F(z)$, the zero mode can be localized at any point of the orbifold (see, for example, [3, 5, 7]).

In order to decouple equations (6) and (7) for $(e^\sigma G)' \not\equiv 0$, first let us consider the left-handed parts of the spinor fields $\Psi_1^L = \gamma^5\Psi_1^L$, $\Psi_2^L = \gamma^5\Psi_2^L$. Equations (6) and (7) for Ψ_1^L and Ψ_2^L can be rewritten in the operator form as

$$\begin{pmatrix}\hat{L} & 0\\ 0 & \hat{L}\end{pmatrix}\begin{pmatrix}\Psi_1^L\\ \Psi_2^L\end{pmatrix} + \hat{\Lambda}_L\begin{pmatrix}\Psi_1^L\\ \Psi_2^L\end{pmatrix} = 0, \tag{11}$$

where the operator $\hat{L}$ and the matrix $\hat{\Lambda}_L$ are defined as

$$\hat{L} = -e^{-2\sigma}\Box + \partial_5^2 + 5\sigma'\partial_5 + 6\sigma'^2 + 2\sigma'' - G^2 - F^2, \qquad \hat{\Lambda}_L = \begin{pmatrix}e^{-\sigma}(e^\sigma F)' & e^{-\sigma}(e^\sigma G)'\\ e^{-\sigma}(e^\sigma G)' & -e^{-\sigma}(e^\sigma F)'\end{pmatrix}. \tag{12}$$

The form of equation (11) implies that the decoupling of the equations of motion for the components of the fermion fields is equivalent to the diagonalization of the matrix $\hat{\Lambda}_L$ (and the corresponding matrix $\hat{\Lambda}_R$). This can be done for [1]

$$F(z) \equiv 0, \tag{13}$$

$$G(z) \equiv \gamma\, \mathrm{sign}(z) F(z) + M e^{-\sigma}, \quad \text{any} \quad F(z); \tag{14}$$

$$F(z) \equiv \gamma\, \mathrm{sign}(z) e^{-\sigma}, \quad \text{any} \quad G(z); \tag{15}$$

$$G(z) \equiv K_1\, \delta(z) + K_2\, \delta(z-L) + M e^{-\sigma}, \quad \text{any} \quad F(z); \tag{16}$$

where γ, M, K_1 and K_2 are constants.

To examine in more detail the general case, in which the functions $F(z)$ and $G(z)$ do not satisfy the conditions (8) or (13)–(16), let us again, for simplicity, consider the left parts Ψ_1^L, Ψ_2^L of the spinor fields and represent equations (6) and (7) as

$$\hat{L}\Psi_1^L + e^{-\sigma}(e^{\sigma}F)'\Psi_1^L + e^{-\sigma}(e^{\sigma}G)'\Psi_2^L = 0, \tag{17}$$

$$\hat{L}\Psi_2^L - e^{-\sigma}(e^{\sigma}F)'\Psi_2^L + e^{-\sigma}(e^{\sigma}G)'\Psi_1^L = 0, \tag{18}$$

where the operator $\hat{L}$ is defined in (12). If the appropriate diagonalization of the matrix $\hat{\Lambda}_L$ in (12) is impossible, then the only obvious way is to obtain separate equations for the fields Ψ_1^L, Ψ_2^L. For example, equation for Ψ_1^L takes the form

$$\left(\hat{L} - e^{-\sigma}(e^{\sigma}F)'\right)\frac{\hat{L} + e^{-\sigma}(e^{\sigma}F)'}{e^{-\sigma}(e^{\sigma}G)'}\Psi_1^L - e^{-\sigma}(e^{\sigma}G)'\Psi_1^L = 0. \tag{19}$$

Analogous equations can be obtained for Ψ_2^L, Ψ_1^R, Ψ_2^R. Thus, we get fourth-order differential equations of motion (recall that the operator $\hat{L}$ contains second derivatives). In the general case such equations describe more degrees of freedom than the second-order differential equations and, in principle, may contain serious pathologies like tachyons or even ghosts [8] (see the detailed discussion of this issue in [1]). The appearance of the fourth-order equations of motion for the general form of $F(z)$ and $G(z)$ is a nonperturbative effect, so it is hard to believe that the perturbation analysis, which is often used to examine fermion sector in brane world models, can adequately describe the theory in the general case. Indeed, the perturbation theory can provide solutions for some of the physical degrees of freedom of the theory, which can be obtained perturbatively. However, the rest of the possible physical degrees of freedom may appear to be lost when one uses perturbation theory in such nontrivial cases. Thus, the general case with $(e^{\sigma}G)' \not\equiv 0$ and $F(z) \not\equiv 0$, naively leading to equations of form (19), should be carefully and thoroughly examined before considering its phenomenological consequences.

Equation (19) is still a fourth-order differential equation even for $F(z) \equiv 0$, whereas it was noted above that the system of equations (6) and (7) can be decoupled for $F(z) \equiv 0$. This would-be contradiction can be easily resolved — in fact, the corresponding fourth-order differential equations can be reduced to second-order differential equations [1].

2 Kaluza-Klein decomposition

Below the Kaluza-Klein decompositions will be performed for the following three cases:

1. $G(z) \equiv M e^{-\sigma(z)}$; any $F(z)$.
2. $F(z) \equiv 0$; any $G(z)$.
3. $G(z) \equiv K\, \delta(z-L)$; any $F(z)$.

The fine-tuned cases (14) and (15) appear to be more complicated and deserve an additional study.

2.1 $G(z) \equiv Me^{-\sigma(z)}$, any $F(z)$

It is convenient to represent the five-dimensional fields as [1]

$$\Psi_1 = \sum_{n=0}^{\infty}\left(f_n(z)\psi_L^n(x) + \sqrt{\frac{m_n+M}{m_n-M}}\tilde{f}_n(z)\hat{\psi}_R^n(x)\right), \quad \Psi_2 = \sum_{n=0}^{\infty}\left(f_n(z)\psi_R^n(x) - \sqrt{\frac{m_n+M}{m_n-M}}\tilde{f}_n(z)\hat{\psi}_L^n(x)\right), \tag{20}$$

where m_n is the Kaluza-Klein mass which will be defined later. According to the orbifold symmetry conditions (3) for the fields Ψ_1 and Ψ_2, the functions $f_n(z)$ and $\tilde{f}_n(z)$ satisfy the symmetry conditions $f_n(-z) = f_n(z), \tilde{f}_n(-z) = -\tilde{f}_n(z)$. Substituting the decomposition into equations (6), (7), we get the following equations for the wave functions $f_n(z)$, $\tilde{f}_n(z)$:

$$e^{-2\sigma}(m_n^2 - M^2)f_n + f_n'' + 5\sigma' f_n' + (6\sigma'^2 + 2\sigma'')f_n - (F^2 - \sigma' F - F')f_n = 0, \tag{21}$$

$$e^{-2\sigma}(m_n^2 - M^2)\tilde{f}_n + \tilde{f}_n'' + 5\sigma' \tilde{f}_n' + (6\sigma'^2 + 2\sigma'')\tilde{f}_n - (F^2 + \sigma' F + F')\tilde{f}_n = 0. \tag{22}$$

Since all the coefficients in (21) and (22) are all even in z and real, we can always find real solutions to these equations satisfying the symmetry conditions presented above. It is easy to show that whenever equations (21), (22) hold, the following system of equations also holds (see [1]):

$$f_n' + (2\sigma' + F)f_n = (m_n + M)e^{-\sigma}\tilde{f}_n, \qquad \tilde{f}_n' + (2\sigma' - F)\tilde{f}_n = -(m_n - M)e^{-\sigma}f_n. \tag{23}$$

With the help of (21) and (22) it is possible to show that the orthonormality conditions

$$\int_{-L}^{L} e^{3\sigma} f_n f_k dz = \delta_{nk}, \qquad \int_{-L}^{L} e^{3\sigma}\tilde{f}_n\tilde{f}_k dz = \frac{m_n - M}{m_n + M}\delta_{nk} \tag{24}$$

can be imposed, whereas $\int_{-L}^{L} e^{3\sigma} f_n\tilde{f}_k dz = 0$ for all n and k because of the symmetry conditions.

Substituting decomposition (20) into action (4), taking into account (23) and integrating over the coordinate of the extra dimension, we get the effective four-dimensional action

$$S_{eff} = \int d^4x\Bigg(i\bar{\psi}\gamma^\mu\partial_\mu\psi - M\bar{\psi}\psi + \sum_{n=1}^{\infty}\Big(i\bar{\psi}_n\gamma^\mu\partial_\mu\psi_n + i\bar{\hat{\psi}}_n\gamma^\mu\partial_\mu\hat{\psi}_n$$
$$-M(\bar{\psi}_n\psi_n - \bar{\hat{\psi}}_n\hat{\psi}_n) - \sqrt{m_n^2 - M^2}\left(\bar{\psi}_n\hat{\psi}_n + \bar{\hat{\psi}}_n\psi_n\right)\Big)\Bigg), \tag{25}$$

where $\psi = \psi_L^0 + \psi_R^0$ (see also equations (9) and (10)) and $\psi_n = \psi_L^n + \psi_R^n$, $\hat{\psi}_n = \hat{\psi}_L^n + \hat{\psi}_R^n$ for $n \geq 1$. To diagonalize the mass matrix, we will use the rotation

$$\psi_n(x) = \psi_{1,n}(x)\cos(\theta_n) + \psi_{2,n}(x)\sin(\theta_n), \qquad \hat{\psi}_n(x) = \psi_{1,n}(x)\sin(\theta_n) - \psi_{2,n}(x)\cos(\theta_n) \tag{26}$$

for $n \geq 1$, where $\tan(2\theta_n) = \frac{\sqrt{m_n^2 - M^2}}{M}$; supplemented by the subsequent redefinition $\psi_{2,n} \to \gamma^5\psi_{2,n}$. Finally, we get

$$S_{eff} = \int d^4x\left(i\bar{\psi}\gamma^\mu\partial_\mu\psi - M\bar{\psi}\psi + \sum_{n=1}^{\infty}\sum_{i=1}^{2}\left(i\bar{\psi}_{i,n}\gamma^\mu\partial_\mu\psi_{i,n} - m_n\bar{\psi}_{i,n}\psi_{i,n}\right)\right). \tag{27}$$

DOI: 10.1051/epjconf/201612505015

2.2 $F(z) \equiv 0$, any $G(z)$

From the very beginning it is convenient to use the combinations $\Psi_A = \frac{1}{\sqrt{2}}(\Psi_1 + \Psi_2)$, $\Psi_B = \frac{1}{\sqrt{2}}(\Psi_1 - \Psi_2)$. With these notations equations of motion, following from (6), (7) with $F(z) \equiv 0$, can be rewritten as

$$-e^{-2\sigma}\Box\Psi_{A,B} + \Psi''_{A,B} + 5\sigma'\Psi'_{A,B} + (6\sigma'^2 + 2\sigma'')\Psi_{A,B} - G^2\Psi_{A,B} \pm (G' + \sigma' G)\gamma^5\Psi_{A,B} = 0. \quad (28)$$

Solutions to these equations, corresponding to a four-dimensional mass m, can be represented as

$$\Psi_A = f(z)\psi_L^A(x) + f(-z)\psi_R^A(x), \qquad \Psi_B = f(z)\psi_R^B(x) + f(-z)\psi_L^B(x), \quad (29)$$

where the equation for the function $f(z)$, corresponding to the four-dimensional mass m, takes the form

$$e^{-2\sigma}m^2 f(z) + f''(z) + 5\sigma' f'(z) + (6\sigma'^2 + 2\sigma'')f(z) - (G^2 - \sigma' G - G')f(z) = 0. \quad (30)$$

According to the general theory [9], the solutions to equation (30) make up an orthonormal set of eigenfunctions $f_n(z)$, the lowest eigenvalue m_0 being simple. One can also show that $m_n^2 \geq 0$.

The Kaluza-Klein decomposition for the fields Ψ_1 and Ψ_2 take the form [1]

$$\Psi_1 = \frac{1}{\sqrt{2}}\left(f_{+,0}(z)\psi_L(x) - f_{-,0}(z)\psi_R(x) + \sum_{n=1}^{\infty}\sum_{k=1}^{2}\left(f_{+,n,k}(z)\psi_L^{n,k}(x) - f_{-,n,k}(z)\psi_R^{n,k}(x)\right)\right), \quad (31)$$

$$\Psi_2 = \frac{1}{\sqrt{2}}\left(f_{+,0}(z)\psi_R(x) + f_{-,0}(z)\psi_L(x) + \sum_{n=1}^{\infty}\sum_{k=1}^{2}\left(f_{+,n,k}(z)\psi_R^{n,k}(x) + f_{-,n,k}(z)\psi_L^{n,k}(x)\right)\right), \quad (32)$$

where $f_{+,0}(z) = f_0(z) + f_0(-z)$, $f_{-,0}(z) = f_0(z) - f_0(-z)$, $f_{+,n,k}(z) = f_{n,k}(z) + f_{n,k}(-z)$, $f_{-,n,k}(z) = f_{n,k}(z) - f_{n,k}(-z)$; whereas the corresponding first-order equations, which will be necessary for obtaining the effective four-dimensional action, take the form [1]

$$f_0'(z) + 2\sigma'(z)f_0(z) + G(z)f_0(z) = m_0 e^{-\sigma(z)}f_0(-z), \quad (33)$$

$$f_{n,1}'(z) + 2\sigma'(z)f_{n,1}(z) + G(z)f_{n,1}(z) = m_{n,1}e^{-\sigma(z)}f_{n,1}(-z), \quad (34)$$

$$f_{n,2}'(z) + 2\sigma'(z)f_{n,2}(z) + G(z)f_{n,2}(z) = -m_{n,2}e^{-\sigma(z)}f_{n,2}(-z). \quad (35)$$

Here $m_0 \geq 0$, $m_{n,k} > 0$. Since solutions to equation (33), (34) and (35) satisfy equation (30), the following orthonormality conditions can be imposed:

$$\int_{-L}^{L} e^{3\sigma}f_{n,k}(z)f_{j,l}(z)dz = \int_{-L}^{L} e^{3\sigma}f_{n,k}(-z)f_{j,l}(-z)dz = \frac{1}{2}\delta_{n,j}\delta_{k,l}, \qquad \int_{-L}^{L} e^{3\sigma}f_0^2(z)dz = \frac{1}{2} \quad (36)$$

$$\int_{-L}^{L} e^{3\sigma}f_0(z)f_{n,k}(z)dz = \int_{-L}^{L} e^{3\sigma}f_0(-z)f_{n,k}(-z)dz = 0, \qquad n \geq 1. \quad (37)$$

Substituting the Kaluza-Klein decomposition (31) and (32) into (4), using equations (33), (34), (35), the orthonormality conditions (36)–(37) when integrating over the coordinate of the extra dimension z, defining $\psi = \psi_L + \psi_R$, $\psi_{n,1} = \psi_L^{n,1} + \psi_R^{n,1}$, $\psi_{n,2} = \psi_L^{n,2} + \psi_R^{n,2}$ and then using the redefinition $\psi_{n,2} \to \gamma^5\psi_{n,2}$, we arrive at

$$S_{eff} = \int d^4x\left(i\bar{\psi}\gamma^\mu\partial_\mu\psi - m_0\bar{\psi}\psi + \sum_{n=1}^{\infty}\sum_{k=1}^{2}\left(i\bar{\psi}_{n,k}\gamma^\mu\partial_\mu\psi_{n,k} - m_{n,k}\bar{\psi}_{n,k}\psi_{n,k}\right)\right). \quad (38)$$

One can expect that in the most cases all the eigenvalues $m_{n,k}$, $n \geq 1$ are simple — it was noted in [10] that in general the double eigenvalues are not common, whereas the Fourier case is not typical. However, the double eigenvalues are still possible. For example, the case $G(z) \equiv Me^{-\sigma}$, $F(z) \equiv 0$, for which the Kaluza-Klein decomposition can be performed in two different ways presented in the previous subsection and in this subsection, leads to double eigenvalues. Of course, both ways lead to the same four-dimensional effective theory and can be connected explicitly [1].

2.3 $G(z) \equiv K\,\delta(z - L)$, any $F(z)$

Now we consider the third case with $G(z) \equiv K\,\delta(z - L)$, where $K > 0$ is a constant. It describes the Higgs field located exactly on the brane. Without loss of generality we take this brane to be the one at $z = L$.

It is clear that everywhere except the point $z = L$ the following five-dimensional equations hold:

$$-e^{-2\sigma}\Box\Psi_1 + \Psi_1'' + 5\sigma'\Psi_1' + (6\sigma'^2 + 2\sigma'')\Psi_1 - (F^2 - e^{-\sigma}(e^{\sigma}F)'\gamma^5)\Psi_1 = 0, \tag{39}$$

$$-e^{-2\sigma}\Box\Psi_2 + \Psi_2'' + 5\sigma'\Psi_2' + (6\sigma'^2 + 2\sigma'')\Psi_2 - (F^2 + e^{-\sigma}(e^{\sigma}F)'\gamma^5)\Psi_2 = 0. \tag{40}$$

These equations are not coupled and suggest the decomposition (for simplicity, we keep only a single mode)

$$\Psi_1 = f(z)\psi_L(x) + \tilde{f}(z)\psi_R(x), \qquad \Psi_2 = f(z)\hat{\psi}_R(x) - \tilde{f}(z)\hat{\psi}_L(x). \tag{41}$$

According to the symmetry conditions and equations (39), (40), the functions $f(-z) = f(z)$ and $\tilde{f}(-z) = -\tilde{f}(z)$ are supposed to satisfy the equations

$$e^{-2\sigma}m^2 f + f'' + 5\sigma' f' + (6\sigma'^2 + 2\sigma'')f - (F^2 - \sigma' F - F')f = 0, \tag{42}$$

$$e^{-2\sigma}m^2\tilde{f} + \tilde{f}'' + 5\sigma'\tilde{f}' + (6\sigma'^2 + 2\sigma'')\tilde{f} - (F^2 + \sigma' F + F')\tilde{f} = 0, \tag{43}$$

where $\Box\psi_{L,R} + m^2\psi_{L,R} = 0$ and $\Box\hat{\psi}_{L,R} + m^2\hat{\psi}_{L,R} = 0$. Note that here m is not an eigenvalue, but just a parameter, and the functions $f(z)$ and $\tilde{f}(z)$ are not eigenfunctions — equations (42) and (43) do not hold at $z = L$. So $f(z)$ and $\tilde{f}(z)$ are just solutions to equations (42) and (43) with some parameter m which is not defined yet.

Analogously to the previous cases, it is possible to show that whenever equations (42), (43) hold, the following system of equations also holds:

$$f'(z) + (2\sigma' + F)f(z) = me^{-\sigma}\tilde{f}(z), \qquad \tilde{f}'(z) + (2\sigma' - F)\tilde{f}(z) = -me^{-\sigma}f(z), \tag{44}$$

again everywhere except the point $z = L$. Substituting (41) into equations (4), (5) with $G(z) \equiv K\delta(z-L)$ and using (44), we get everywhere except $z = L$

$$i\gamma^\mu\partial_\mu\psi_L - m\psi_R = 0, \qquad i\gamma^\mu\partial_\mu\psi_R - m\psi_L = 0, \qquad i\gamma^\mu\partial_\mu\hat{\psi}_L - m\hat{\psi}_R = 0, \qquad i\gamma^\mu\partial_\mu\hat{\psi}_R - m\hat{\psi}_L = 0, \tag{45}$$

which are the standard four-dimensional Dirac equations. At the point $z = L$ equations (4) and (5) give

$$\hat{\psi}_R(x) = -\lim_{\epsilon\to+0}\frac{2\tilde{f}(L-\epsilon)}{Kf(L)}\psi_R(x), \qquad \hat{\psi}_L(x) = -\lim_{\epsilon\to+0}\frac{Kf(L)}{2\tilde{f}(L-\epsilon)}\psi_L(x) \tag{46}$$

for any x. In derivation of (46), we have used the regularization $\tilde{f}(z)\delta(z-L) \sim \mathrm{sign}(z-L)\delta(z-L) = 0$ (in order to support the existence of the delta-function in the five-dimensional action, the function $\tilde{f}(z)$ should be discontinuous at $z = L$). Finally, using (46) we can find that equations (45) are consistent if the conditions

$$\lim_{\epsilon\to+0}\frac{Kf(L)}{2\tilde{f}(L-\epsilon)} = -\beta_i, \qquad i = 1,2, \qquad \beta_1 = 1, \qquad \beta_2 = -1 \tag{47}$$

hold. This condition defines the allowed values of m, i.e., the mass spectrum of the theory. It can be shown explicitly that the values $m^2 < 0$ are impossible in the model at hand, see [1] for details.

Let us label the solutions to equation (47) as $m_{n,i}$ and suppose that all the masses are different. The complete decomposition of the five-dimensional fields has the form [1]

$$\Psi_1 = f_0(z)\psi_L(x) + \tilde{f}_0(z)\psi_R(x) + \sum_{n=1}^{\infty}\sum_{i=1}^{2}\left(f_{n,i}(z)\psi_L^{n,i}(x) + \tilde{f}_{n,i}(z)\psi_R^{n,i}(x)\right), \quad (48)$$

$$\Psi_2 = f_0(z)\psi_R(x) - \tilde{f}_0(z)\psi_L(x) + \sum_{n=1}^{\infty}\sum_{i=1}^{2}\beta_i\left(f_{n,i}(z)\psi_R^{n,i}(x) - \tilde{f}_{n,i}(z)\psi_L^{n,i}(x)\right), \quad (49)$$

where we have used (46) and (47). Here the modes with $i = 1$ are defined by (47) with $\beta_1 = 1$, whereas the modes with $i = 2$ are defined by (47) with $\beta_2 = -1$.

With (47), it is possible to choose the following nonstandard orthonormality conditions (see [1] for the proof):

$$\int_{-L}^{L} e^{3\sigma}\left(f_{n,i}(z)f_{k,j}(z) + \beta_i\beta_j\tilde{f}_{n,i}(z)\tilde{f}_{k,j}(z)\right)dz = \int_{-L}^{L} e^{3\sigma}\left(\beta_i\beta_j f_{n,i}(z)f_{k,j}(z) + \tilde{f}_{n,i}(z)\tilde{f}_{k,j}(z)\right)dz = \delta_{nk}\delta_{ij}.$$

The last useful step is to find the equations instead of equations (44), which are valid in the whole extra dimension. Indeed, equation for the mass spectrum (47) relates the values of $f_n(L)$ and $\tilde{f}_n(L)$, so we can use it to supply (44) by extra terms in order to obtain the systems of equations [1]

$$f_0' + (2\sigma' + F)f_0 = m_0e^{-\sigma}\tilde{f}_0, \qquad \tilde{f}_0' + (2\sigma' - F)\tilde{f}_0 - \frac{K}{\beta_0}f_0\,\delta(z-L) = -m_0e^{-\sigma}f_0 \quad (50)$$

$$f_{n,i}' + (2\sigma' + F)f_{n,i} = m_{n,i}e^{-\sigma}\tilde{f}_{n,i}, \qquad \tilde{f}_{n,i}' + (2\sigma' - F)\tilde{f}_{n,i} - \frac{K}{\beta_i}f_{n,i}\,\delta(z-L) = -m_{n,i}e^{-\sigma}f_{n,i}, \quad (51)$$

which are valid for any z. Substituting (48), (49) into five-dimensional action (4), using the orthonormality conditions and equations (50)–(51) and defining $\psi = \psi_L^0 + \psi_R^0$, $\psi_{n,i} = \psi_L^{n,i} + \psi_R^{n,i}$, we get

$$S_{eff} = \int d^4x\left(i\bar{\psi}\gamma^\mu\partial_\mu\psi - m_0\bar{\psi}\psi + \sum_{n=1}^{\infty}\sum_{i=1}^{2}\left(i\bar{\psi}_{n,i}\gamma^\mu\partial_\mu\psi_{n,i} - m_{n,i}\bar{\psi}_{n,i}\psi_{n,i}\right)\right). \quad (52)$$

3 Discussion

Let us briefly discuss the results presented above.

1. As was shown in Section 1, the general case may contain rather pathological behavior in the form of fourth-order differential equations of motion for the components of the five-dimensional spinor fields. At the moment it is unclear how to perform the Kaluza-Klein decomposition in this case — in all the cases discussed in Section 2 the independent second-order equations of motion for the components of the five-dimensional fields (or of their linear combinations) provided the complete set of possible physical degrees of freedom of the four-dimensional effective theory, whereas the corresponding first-order differential equations allow one to get rid of the terms with the functions $F(z)$, $G(z)$ and with the derivative in the coordinate of the extra dimension in (4) and to obtain the four-dimensional effective action in a consistent way. It is not clear how to perform an analogous analysis starting from the fourth-order equations like equation (19).

2. The case $F(z) \not\equiv 0$, $G(z) \equiv Me^{-\sigma}$ does not contain any pathological behavior and the consistent Kaluza-Klein decomposition can be performed in this case. However, this case may contain some serious drawbacks, at least for $\sigma(z) \not\equiv 0$. Indeed, the existence of the functions $\sigma(z) \not\equiv 0$, $F(z) \not\equiv 0$ and $G(z) \equiv Me^{-\sigma}$, whatever the origin of these functions is, imply that the extra dimension is not uniform in z. Thus, one can expect that the backreaction of the bulk fields on the background metric (or possible quantum corrections) are also nonuniform in z, which may violate the fine-tuned relation between the vacuum profile of the Higgs field and the form of the background metric. Even if the violation is very small, the effect is nonperturbative, so it will lead to the problems discussed in the previous item.

   Analogous reasonings can be applied to the fine-tuned cases (14) and (15).

3. The case $G(z) \equiv K\,\delta(z-L)$, $F(z) \not\equiv 0$ also admits a consistent Kaluza-Klein decomposition. However, it leads to the discontinuous wave functions of the modes. This discontinuity imply that such a wave function is just an approximation for a continuous, but very rapidly varying in the vicinity of $z = L$ wave function, whereas the delta-function-like $G(z)$ is an approximation of a very narrow and peaked, but continuous profile of the Higgs field. However, any modification of the delta-function (such as, for example, $\delta(z-L) \to$ Gaussian function) in this case leads to impossibility to diagonalize the matrix $\hat{\Lambda}_L$ in (12) and, consequently, to fourth-order equations of motion for the components of the five-dimensional fermion fields.

4. Contrary to the previous cases, the choice $F(z) \equiv 0$ looks the most natural and safe from this point of view — indeed, the Kaluza-Klein decomposition can be performed in a mathematically consistent way for any $G(z)$ without the necessity for any fine-tuning. However, this case is not interesting from phenomenological point of view.

Acknowledgements

The work was supported by grant 14-12-00363 of the Russian Science Foundation.

References

[1] M.N. Smolyakov, JHEP **1606**, 165 (2016)
[2] C. Macesanu, Int. J. Mod. Phys. A **21**, 2259 (2006)
[3] S.L. Dubovsky, V.A. Rubakov and P.G. Tinyakov, Phys. Rev. D **62**, 105011 (2000)
[4] R. Casadio and A. Gruppuso, Phys. Rev. D **64**, 025020 (2001)
[5] V.A. Rubakov and M.E. Shaposhnikov, Phys. Lett. B **125**, 136 (1983)
[6] S. Chang, J. Hisano, H. Nakano, N. Okada and M. Yamaguchi, Phys. Rev. D **62**, 084025 (2000)
[7] A.A. Andrianov, V.A. Andrianov, P. Giacconi and R. Soldati, JHEP **0307**, 063 (2003)
[8] R.P. Woodard, Lect. Notes Phys. **720**, 403 (2007)
[9] E.A. Coddington and N. Levinson, *Theory of Ordinary Differential Equations* (McGraw-Hill, New York, 1955)
[10] A. Zettl, *Sturm-Liouville Theory*, in *Mathematical Surveys and Monographs*, v. 121 (American Mathematical Society, 2005)

The Feynman-Vernon Influence Functional Approach in QED

Alexander Biryukov[1,⋆] and *Mark* Shleenkov[1,⋆⋆]

[1] *Samara University*

Abstract. In the path integral approach we describe evolution of interacting electromagnetic and fermionic fields by the use of density matrix formalism. The equation for density matrix and transitions probability for fermionic field is obtained as average of electromagnetic field influence functional. We obtain a formula for electromagnetic field influence functional calculating for its various initial and final state. We derive electromagnetic field influence functional when its initial and final states are vacuum. We present Lagrangian for relativistic fermionic field under influence of electromagnetic field vacuum.

1 Introduction

At the present time the development of non-perturbative methods of non-linear dynamics description is actual for quantum systems interacting with strong electromagnetic field [1–5]. One of these methods is Feynman-Vernon influence functional approach in the path integral formalism [6]. This method development was presented in many papers [7–9]. In the Feynman-Vernon influence functional approach the main point is its calculation for the model of investigated system. In this paper we calculate Feynman-Vernon influence functional of quantum electromagnetic field vacuum on fermionic field.

2 The Feynman-Vernon Influence Functional of quantum electromagnetic field on fermionic field

The QED Lagrangian for a spin-1/2 field interacting with the electromagnetic field is given by

$$\mathcal{L}_{full}(x) = \overline{\psi}(x)(\imath\gamma^{\mu}\partial_{\mu} - m)\psi(x) - \frac{1}{4}F_{\mu\nu}(x)F^{\mu\nu}(x) - j_{\mu}(x)A^{\mu}(x) \qquad (1)$$

where γ^{μ} are Dirac matrices, $\psi(x)$ is a bispinor field of spin-1/2 particles (e.g. electron–positron field), $A_{\mu}(x)$ is electromagnetic four-potential;

$$F_{\mu\nu}(x) = \partial_{\mu}A_{\mu}(x) - \partial_{\nu}A_{\mu}(x) \qquad (2)$$

is the electromagnetic field tensor,

$$j_{\mu}(x) = e\overline{\psi}(x)\gamma_{\mu}\psi(x) \qquad (3)$$

⋆e-mail: biryukov_1@mail.ru
⋆⋆e-mail: shleenkov@list.ru

is four-current. We use natural units system $c = \hbar = 1$ in formulas (1)–(3) and further.

By the use of the second quantization formalism we present fields as field operators $\hat{\psi}(\mathbf{x}, t)$, $\hat{\bar{\psi}}(\mathbf{x}, t)$, $\hat{j}_\mu(\mathbf{x}, t)$, $\hat{A}^\mu(\mathbf{x}, t)$ where $\mathbf{x}$ and t describe coordinate of 3D space and time moment. Here we introduce the creation and annihilation operators:

$$\hat{A}^\mu(\mathbf{x}, t) = \sum_{\mathbf{k},\lambda=\pm 1} \frac{1}{\sqrt{2V\omega_{\mathbf{k}}^{(b)}}} \left(\varepsilon_\lambda^\mu \hat{a}_{\mathbf{k}\lambda} e^{ikx} + \varepsilon_\lambda^{\mu *} \hat{a}_{\mathbf{k}\lambda}^\dagger e^{-ikx}\right) \tag{4}$$

$$\hat{\psi}(\mathbf{x}, t) = \sum_{\mathbf{p},\sigma=\pm\frac{1}{2}} \frac{1}{\sqrt{2V\omega_{\mathbf{p}}^{(f)}}} \left(\hat{b}_{\mathbf{p}\sigma} u_\sigma(p) e^{ipx} + \hat{c}_{\mathbf{p}\sigma}^\dagger u_\sigma(-p) e^{-ipx}\right), \tag{5}$$

$$\hat{\psi}(\mathbf{x}, t) = \sum_{\mathbf{p},\sigma=\pm\frac{1}{2}} \frac{1}{\sqrt{2V\omega_{\mathbf{p}}^{(f)}}} \left(\hat{b}_{\mathbf{p}\sigma}^\dagger u_\sigma(p) e^{-ipx} + \hat{c}_{\mathbf{p}\sigma} u_\sigma(-p) e^{ipx}\right), \tag{6}$$

$$\hat{j}_\mu^+(\mathbf{k}, t) = \int \hat{j}_\mu(\mathbf{x}, t) e^{i\mathbf{k}\mathbf{x}} d\mathbf{x}, \qquad \hat{j}_\mu^-(\mathbf{k}, t) = \int \hat{j}_\mu(\mathbf{x}, t) e^{-i\mathbf{k}\mathbf{x}} d\mathbf{x} \tag{7}$$

$$\hat{j}_\mu(\mathbf{x}) = \hat{j}_\mu(\mathbf{x}, \hat{b}_{\mathbf{p}\sigma}, \hat{b}_{\mathbf{p}\sigma}^\dagger, \hat{c}_{\mathbf{p}\sigma}, \hat{c}_{\mathbf{p}\sigma}^\dagger) \tag{8}$$

where $\hat{a}_{\mathbf{k}\lambda}^\dagger$ and $\hat{a}_{\mathbf{k}\lambda}$ are creation and annihilation operators of photon with wave vector $\mathbf{k}$ and polarization (the eigenvalue of the z-component of the photon spin) $\lambda = \pm 1$, ε_λ^μ is unitary 4-vector of polarization and $\varepsilon_\lambda^{\mu *} \varepsilon_{\lambda\mu} = -1$; $\hat{b}_{\mathbf{p}\sigma}^\dagger$ and $\hat{b}_{\mathbf{p}\sigma}$ are creation and annihilation operators of electron with wave vector $\mathbf{p}$ and polarization $\sigma = \pm\frac{1}{2}$; $\hat{c}_{\mathbf{p}\sigma}^\dagger$ and $\hat{c}_{\mathbf{p}\sigma}$ are creation and annihilation operators of positron with wave vector $\mathbf{p}$ and polarization $\sigma = \pm\frac{1}{2}$, V is volume of space.

For evolution of the interacting systems we construct Hamiltonian of full system $\hat{H}_{full}$ on the base of operators $\hat{a}_{\mathbf{k}\lambda}^\dagger$, $\hat{a}_{\mathbf{k}\lambda}$, $\hat{b}_{\mathbf{p}\sigma}^\dagger$, $\hat{b}_{\mathbf{p}\sigma}$, $\hat{c}_{\mathbf{p}\sigma}^\dagger$, $\hat{c}_{\mathbf{p}\sigma}$.

$$\hat{H}_{full} = \hat{H}_{Dirac} + \hat{H}_{field} + \hat{H}_{int} \tag{9}$$

where

$$\hat{H}_{Dirac} = \sum_{\mathbf{p},\sigma} \omega_{\mathbf{p}}^{(f)} \left(\hat{b}_{\mathbf{p}\sigma}^\dagger \hat{b}_{\mathbf{p}\sigma} + \hat{c}_{\mathbf{p}\sigma}^\dagger \hat{c}_{\mathbf{p}\sigma}\right) \tag{10}$$

— Hamiltonian of Dirac field;

$$\hat{H}_{field} = \sum_{\mathbf{k},\lambda} \omega_{\mathbf{k}}^{(b)} \hat{a}_{\mathbf{k}\lambda}^\dagger \hat{a}_{\mathbf{k}\lambda} \tag{11}$$

— Hamiltonian of electromagnetic field;

$$\hat{H}_{int} = \sum_{\mathbf{k},\lambda} \frac{1}{\sqrt{2\omega_{\mathbf{k}}^{(b)} V}} \left(\varepsilon_\lambda^\mu \hat{j}_\mu^+(\mathbf{k}, t) \hat{a}_{\mathbf{k}\lambda} + \varepsilon_\lambda^{*\mu} \hat{j}_\mu^-(\mathbf{k}, t) \hat{a}_{\mathbf{k}\lambda}^\dagger\right) \tag{12}$$

— interaction part of Hamiltonian.

The evolution equation of statistical operator $\hat{\rho}$ is the following

$$\hat{\rho}(t_f) = \hat{U}(t_f, t_{in}) \hat{\rho}(t_{in}) \hat{U}^\dagger(t_f, t_{in}) \tag{13}$$

where $\hat{\rho}(t_{in})$ is statistical operator, describing initial state at moment t_{in} and $\hat{U}(t_f, t_{in})$ is operator of evolution from initial time moment t_{in} to final time moment t_f

$$\hat{U}(t_f, t_{in}) = \hat{T} \exp[-\frac{\imath}{\hbar} \int_{t_{in}}^{t_f} \hat{H}_{full}(\tau) d\tau]. \tag{14}$$

For density matrix calculation we need to formulate equation (13) in functional representation. We introduce coherent representation $|\alpha_{\mathbf{k}\lambda}\rangle$ for electromagnetic field by the following way

$$\hat{a}_{\mathbf{k}\lambda}|a_{\mathbf{k}\lambda}\rangle = a_{\mathbf{k}\lambda}|a_{\mathbf{k}\lambda}\rangle, \qquad \langle a_{\mathbf{k}\lambda}|\hat{a}^{\dagger}_{\mathbf{k}\lambda} = \langle a_{\mathbf{k}\lambda}|a^{*}_{\mathbf{k}\lambda}, \tag{15}$$

$$\langle a'_{\mathbf{k}'\lambda'}|a_{\mathbf{k}\lambda}\rangle = \delta_{\mathbf{k}'\mathbf{k}}\delta_{\lambda'\lambda} \exp\left\{-\frac{1}{2}\left(a'^{*}_{\mathbf{k}'\lambda'}a'_{\mathbf{k}\lambda} + a^{*}_{\mathbf{k}\lambda}a_{\mathbf{k}\lambda} - a'^{*}_{\mathbf{k}'\lambda'}a_{\mathbf{k}\lambda} - a^{*}_{\mathbf{k}\lambda}a_{\mathbf{k}'\lambda'}\right)\right\}. \tag{16}$$

$$\int |a_{\mathbf{k}\lambda}\rangle\langle a_{\mathbf{k}\lambda}| da^{*}_{\mathbf{k}\lambda} da_{\mathbf{k}\lambda} = \hat{1}. \tag{17}$$

where $\alpha_{\mathbf{k}\lambda}$ — complex value, which describe one mode of quantum electromagnetic field with wave vector $\mathbf{k}$ and polarization λ.

We introduce grassman representation $|b_{\mathbf{p},\sigma}\rangle$ and $|c_{\mathbf{p},\sigma}\rangle$ for Dirac field by the following way

$$\hat{b}_{\mathbf{p},\sigma}|b_{\mathbf{p},\sigma}\rangle = b_{\mathbf{p},\sigma}|b_{\mathbf{p},\sigma}\rangle, \qquad \langle b_{\mathbf{p},\sigma}|\hat{b}^{\dagger}_{\mathbf{p},\sigma} = \langle b_{\mathbf{p},\sigma}|b^{*}_{\mathbf{p},\sigma}, \qquad \int |b_{\mathbf{p}\sigma}\rangle\langle b_{\mathbf{p}\sigma}| db^{*}_{\mathbf{p}\sigma} db_{\mathbf{p}\sigma} = \hat{1}; \tag{18}$$

$$\hat{c}_{\mathbf{p},\sigma}|c_{\mathbf{p},\sigma}\rangle = c_{\mathbf{p},\sigma}|c_{\mathbf{p},\sigma}\rangle, \qquad \langle c_{\mathbf{p},\sigma}|\hat{c}^{\dagger}_{\mathbf{p},\sigma} = \langle c_{\mathbf{p},\sigma}|c^{*}_{\mathbf{p},\sigma}, \qquad \int |c_{\mathbf{p}\sigma}\rangle\langle c_{\mathbf{p}\sigma}| dc^{*}_{\mathbf{p}\sigma} dc_{\mathbf{p}\sigma} = \hat{1}; \tag{19}$$

$$\langle b'_{\mathbf{p}'\sigma'}|b_{\mathbf{p}\sigma}\rangle = \delta_{\mathbf{p}'\mathbf{p}}\delta_{\sigma'\sigma} \exp\left\{-\frac{1}{2}\left(b'^{*}_{\mathbf{p}'\sigma'}b'_{\mathbf{p}'\sigma'} + b^{*}_{\mathbf{p}\sigma}b_{\mathbf{p}\sigma} - b'^{*}_{\mathbf{p}'\sigma'}b_{\mathbf{p}\sigma} - b^{*}_{\mathbf{p}\sigma}b'_{\mathbf{p}'\sigma'}\right)\right\}. \tag{20}$$

$$\langle c'_{\mathbf{p}'\sigma'}|c_{\mathbf{p}\sigma}\rangle = \delta_{\mathbf{p}'\mathbf{p}}\delta_{\sigma'\sigma} \exp\left\{-\frac{1}{2}\left(c'^{*}_{\mathbf{p}'\sigma'}c'_{\mathbf{p}'\sigma'} + c^{*}_{\mathbf{p}\sigma}c_{\mathbf{p}\sigma} - c'^{*}_{\mathbf{p}'\sigma'}c_{\mathbf{p}\sigma} - c^{*}_{\mathbf{p}\sigma}c'_{\mathbf{p}'\sigma'}\right)\right\}. \tag{21}$$

where $b_{\mathbf{p},\sigma}$ and $c_{\mathbf{p},\sigma}$ — grassman variable.

In holomorphic representation $|a_{\mathbf{k}\lambda}, b_{\mathbf{p}\sigma}, c_{\mathbf{p}\sigma}\rangle = |a_{\mathbf{k}\lambda}\rangle \otimes |b_{\mathbf{p}\sigma}\rangle \otimes |c_{\mathbf{p}\sigma}\rangle$ we present the evolution equation in the following

$$\begin{aligned}\rho(a^{*}_{f}, b^{*}_{f}, c^{*}_{f}, a'_{f}, b'_{f}, c'_{f}; t_f) = \int da'^{*}_{in} da'_{in} db'^{*}_{in} db'_{in} dc'^{*}_{in} dc'_{in} da^{*}_{in} da_{in} db^{*}_{in} db_{in} dc^{*}_{in} dc_{in} \times \\ \times \langle a_f, b_f, c_f|\hat{U}(t_f, t_{in})|a_{in}, b_{in}, c_{in}\rangle \rho(a^{*}_{in}, b^{*}_{in}, c^{*}_{in}, a'_{in}, b'_{in}, c'_{in}; t_{in}) \times \\ \times \langle a'_{in}, b'_{in}, c'_{in}|\hat{U}^{\dagger}(t_f, t_{in})|a'_f, b'_f, c'_f\rangle\end{aligned} \tag{22}$$

where the density matrix:

$$\rho(a^{*}_{f}, b^{*}_{f}, c^{*}_{f}, a'_{f}, b'_{f}, c'_{f}; t_f) = \langle a_f, b_f, c_f|\hat{\rho}(t_f)|a'_f, b'_f, c'_f\rangle \tag{23}$$

For simplification we relabel variables $\alpha_{\mathbf{k}\lambda}$ and $\theta_{\mathbf{p}\sigma}$ to α and θ.

Be the use of amplitude transition properties [10], we present the kernel of evolution operator $\langle a_f, b_f, c_f|\hat{U}(t_f, t_{in})|a_{in}, b_{in}, c_{in}\rangle$ as a path integral:

$$\langle a_f, b_f, c_f|\hat{U}(t_f, t_{in})|a_{in}, b_{in}, c_{in}\rangle = \int \mathfrak{D}a^*(\tau)\mathfrak{D}a(\tau)\mathfrak{D}b^*(\tau)\mathfrak{D}b(\tau)\mathfrak{D}c^*(\tau)\mathfrak{D}c(\tau)\times \\ \times\exp\left\{\imath S_{full}\left[a^*(\tau), a(\tau), b^*(\tau), b(\tau), c^*(\tau), c(\tau)\right]\right\}, \tag{24}$$

where full system action is the following

$$S_{full}\left[a^*(\tau), a(\tau), b^*(\tau), b(\tau), c^*(\tau), c(\tau)\right] = \\ = S_f\left[b^*(\tau), b(\tau), c^*(\tau), c(\tau)\right] + S_b\left[a^*(\tau), a(\tau)\right] + S_{int}\left[a^*(\tau), a(\tau), b^*(\tau), b(\tau), c^*(\tau), c(\tau)\right], \tag{25}$$

where

$$S_f\left[b^*(\tau), b(\tau), c^*(\tau), c(\tau)\right] = \int_{t_{in}}^{t_f}\left(\frac{\dot{b}^*(\tau)b(\tau) - b^*(\tau)\dot{b}(\tau)}{2\imath} + \frac{\dot{c}^*(\tau)c(\tau) - c^*(\tau)\dot{c}(\tau)}{2\imath} - \right. \\ \left. - \omega^{(f)}\left(b^*(\tau)b(\tau) + c^*(\tau)c(\tau)\right)\right)d\tau \tag{26}$$

is the action S_f of fermionic field,

$$S_b\left[a^*(\tau), a(\tau)\right] = \int_{t_{in}}^{t_f}\left(\frac{\dot{a}^*(\tau)a(\tau) - a^*(\tau)\dot{a}(\tau)}{2\imath} - \omega^{(b)}a^*(\tau)a(\tau)\right)d\tau \tag{27}$$

is the action S_b of bosonic field,

$$S_{int}\left[a^*(\tau), a(\tau), b^*(\tau), b(\tau), c^*(\tau), c(\tau)\right] = \tag{28}$$

$$= \int_{t_{in}}^{t_f}\left[\varepsilon_\mu\tilde{j}^{\mu+}\left[b^*(\tau), b(\tau), c^*(\tau), c(\tau)\right]a(\tau) + \varepsilon^*_\mu\tilde{j}^{\mu-}\left[b^*(\tau), b(\tau), c^*(\tau), c(\tau)\right]a^*(\tau)\right]d\tau; \tag{29}$$

is the action S_{int} of interaction part.

Using (24)–(29) we present equation (22) for density matrix evolution in the following form:

$$\rho(a^*_f, b^*_f, c^*_f, a'_f, b'_f, c'_f; t_f) = \int da'^*_{in}da'_{in}db'^*_{in}db'_{in}dc'^*_{in}dc'_{in}da^*_{in}da_{in}db^*_{in}db_{in}dc^*_{in}dc_{in}\times \\ \times\rho(a^*_{in}, b^*_{in}, c^*_{in}, a'_{in}, b'_{in}, c'_{in}; t_{in})\times \\ \times\mathfrak{D}a^*(\tau)\mathfrak{D}a(\tau)\mathfrak{D}b^*(\tau)\mathfrak{D}b(\tau)\mathfrak{D}c^*(\tau)\mathfrak{D}c(\tau)\mathfrak{D}a'^*(\tau)\mathfrak{D}a'(\tau)\mathfrak{D}b'(\tau)\mathfrak{D}b'(\tau)\mathfrak{D}c'(\tau)\mathfrak{D}c'(\tau)\times \\ \times\exp\left\{\imath\left(S_{full}\left[a^*(\tau), a(\tau), b(\tau), b(\tau)\right] - S_{full}\left[a'^*(\tau), a'(\tau), b'(\tau), b'(\tau)\right]\right)\right\}. \tag{30}$$

We can choose at initial moment t_{in}

$$\rho(a^*_{in}, b^*_{in}, c^*_{in}, a'_{in}, b'_{in}, c'_{in}; t_{in}) = \rho_f(b^*_{in}, c^*_{in}, b'_{in}, c'_{in}; t_{in})\times\rho_b(a^*_{in}, a'_{in}; t_{in}). \tag{31}$$

This is true for many models.

We consider the quantum transition from pure fermionic field state $|n\rangle$ at initial time moment t_{in}

$$\rho_{syst}(b^*_{in}, c^*_{in}, b'_{in}, c'_{in}; t_{in}) = \psi_n(b^*_{in}, c^*_{in})\psi^*_n(b'_{in}, c'_{in}) \tag{32}$$

$$\psi^*_n(b_{in}, c_{in}) = \langle n|b_{in}, c_{in}\rangle \qquad \psi_n(b^*_{in}, c^*_{in}) = \langle b_{in}, c_{in}|n\rangle \tag{33}$$

and electromagnetic field from coherent state $|\zeta\rangle$ at t_{in}

$$\rho(a^*_{in}, a'_{in}; t_{in}) = \zeta(a^*_{in})\zeta^*(a'_{in}) \tag{34}$$

$$\zeta^*(a'_{in}) = \langle\zeta|a'_{in}\rangle \qquad \zeta(a^*_{in}) = \langle a_{in}|\zeta\rangle \tag{35}$$

to pure fermionic field state $|m\rangle$ at time moment t_f:

$$\psi^*_m(b'_f, c'_f) = \langle m|b'_f, c'_f\rangle \qquad \psi_m(b^*_f, c^*_f) = \langle b_f, c_f|m\rangle \tag{36}$$

and to coherent state $|\chi\rangle$ at time moment t_f

$$\rho(a_f, a'_f) = \chi(a_f)\chi^*(a'_f) \tag{37}$$

$$\chi^*(a'_f) = \langle\chi|a'_f\rangle \qquad \chi(a^*_f) = \langle a_f|\chi\rangle \tag{38}$$

The transition probability $P(m, \chi, t_f; n, \zeta, t_{in})$ is determined from (30) by the following formula

$$P(m, \chi, t_f; n, \zeta, t_{in}) = \int\int da'^*_f da'_f db'^*_f db'_f dc'^*_f dc'_f da^*_f da_f db^*_f db_f dc^*_f dc_f \times$$
$$\times \psi^*_m(b_f, c_f)\chi^*(a_f)\rho(a^*_f, b^*_f, c^*_f; a'_f, b'_f, c'_f; t)\psi_m(b'^*_f, c'^*_f)\chi(a'^*_f) \tag{39}$$

By the use of (30)-(38) we present the transition probability in the path integral representation

$$P(m, \chi, t_f; n, \zeta, t_{in}) = \int da^*_f da_f da'^*_f da'_f db^*_f db_f db'^*_f db'_f dc^*_f dc_f dc'^*_f dc'_f \times$$
$$\times \mathcal{D}a'^*(\tau)\mathcal{D}a'(\tau)\mathcal{D}a^*(\tau)\mathcal{D}a(\tau)\mathcal{D}b'^*(\tau)\mathcal{D}b'(\tau)\mathcal{D}b^*(\tau)\mathcal{D}b(\tau)\mathcal{D}c'^*(\tau)\mathcal{D}c'(\tau)\mathcal{D}c^*(\tau)\mathcal{D}c(\tau)\times$$
$$\times \psi^*_m(b_f, c_f)\chi^*(a_f)\exp\left[\frac{i}{\hbar}S_{full}[a^*(\tau), a(\tau), b^*(\tau), b(\tau), c^*(\tau), c(\tau)]\right]\psi_n(b^*_{in}, c^*_{in})\zeta(a^*_{in})\times$$
$$\times \psi^*_n(b'_{in}, c'_{in})\zeta^*(a'_{in})\exp\left[-\frac{i}{\hbar}S'_{full}[a'^*(\tau), a'(\tau), b'^*(\tau), b'(\tau), c'^*(\tau), c'(\tau)]\right]\psi_m(b'^*_f, c'^*_f)\chi(a'^*_f). \tag{40}$$

Eq. (40) for quantum transition probability we present in the following form

$$P(m, \chi; t|n, \zeta; 0) = \int \mathcal{D}b'^*(\tau)\mathcal{D}b'(\tau)\mathcal{D}b^*(\tau)\mathcal{D}b(\tau)\mathcal{D}c'^*(\tau)\mathcal{D}c'(\tau)\mathcal{D}c^*(\tau)\mathcal{D}c(\tau)\times$$
$$\times da^*_f da_f da'^*_f da'_f db^*_f db_f db'^*_f db'_f dc^*_f dc_f dc'^*_f dc'_f \psi^*_m(b_f, c_f)\psi_m(b'^*_f, c'^*_f)\times$$
$$\times \exp\left\{\imath\left(S_f[b^*(\tau), b(\tau), c^*(\tau), c(\tau)] - S_f[b'^*(\tau), b'(\tau), c'^*(\tau), c'(\tau)]\right)\right\}\times$$
$$\times F[\chi_f|b'^*(\tau), b'(\tau), b^*(\tau), b(\tau), c'^*(\tau), c'(\tau), c^*(\tau), c(\tau)|\zeta_{in}]\psi_n(b^*_{in}, c^*_{in})\psi^*_n(b'_{in}, c'_{in}) \tag{41}$$

where

$$F[\chi_f|b'^*(\tau), b'(\tau), b^*(\tau), b(\tau), c'^*(\tau), c'(\tau), c^*(\tau), c(\tau)|\zeta_{in}] = \int da^*_{in}da_{in}da'^*_{in}da'_{in}\times$$
$$\times \chi^*(a_f)U_{infl}\left[a^*_f, b^*_f, c^*_f, t_f, b^*(\tau), b(\tau), c^*(\tau), c(\tau), a_{in}, b_{in}, c_{in}, t_{in}\right]\times$$
$$\times \zeta(a^*_{in})\zeta^*(a'_{in})U^*_{infl}\left[a'^*_{in}, b'^*_{in}, c'^*_{in}, t_{in}, b'^*(\tau), b'(\tau), c'^*(\tau), c'(\tau), a_f, b_f, c_f, t_f\right]\chi(a'^*_f) \tag{42}$$

is influence functional of electromagnetic field on fermionic subsystems.

$$U_{infl}\left[a_f^*, b_f^*, c_f^*, t_f, b^*(\tau), b(\tau), c^*(\tau), c(\tau), a_{in}, b_{in}, c_{in}, t_{in}\right] =$$
$$= \int \mathfrak{D}a^*(\tau)\mathfrak{D}a(\tau) \exp\left\{\imath S_{infl}[a^*(\tau), a(\tau), b^*(\tau), b(\tau), c^*(\tau), c(\tau)]\right\} \quad (43)$$

is electromagnetic field transition amplitude from initial state $|a_{in}\rangle$ to final state $|a_f\rangle$.

$$S_{infl}[a^*(\tau), a(\tau), b^*(\tau), b(\tau), c^*(\tau), c(\tau)] =$$
$$= S_b\left[a^*(\tau), a(\tau)\right] + S_{int}\left[a^*(\tau), a(\tau), b^*(\tau), b(\tau), c^*(\tau), c(\tau)\right]. \quad (44)$$

is action of fermionic field under electromagnetic field influence. It describe electromagnetic field influence on fermionic field.

The influence functional (42) present the influence of one mode of electromagnetic field with wave vector $\mathbf{k}$ and polarization λ. We note it by index $F_{\mathbf{k},\lambda}[\chi_f|b'^*(\tau), b'(\tau), b^*(\tau), b(\tau), c'^*(\tau), c'(\tau), c^*(\tau), c(\tau)|\zeta_{in}]$.

The influence functional of all electromagnetic field modes calculate as production of one mode influence functionals:

$$F[\chi_f|b'^*(\tau), b'(\tau), b^*(\tau), b(\tau), c'^*(\tau), c'(\tau), c^*(\tau), c(\tau)|\zeta_{in}] =$$
$$= \prod_{\mathbf{k}\lambda} F_{\mathbf{k},\lambda}[\chi_f|b'^*(\tau), b'(\tau), b^*(\tau), b(\tau), c'^*(\tau), c'(\tau), c^*(\tau), c(\tau)|\zeta_{in}]. \quad (45)$$

For calculating influence functional by eq. (42) we need to calculate amplitude $U_{infl}\left[a_f^*, b_f^*, c_f^*, t_f, b^*(\tau), b(\tau), c^*(\tau), c(\tau), a_{in}, b_{in}, c_{in}, t_{in}\right]$ and to specify $\chi(a)$, $\zeta(a)$.

3 Electromagnetic field influence functional calculation

In this section we calculate functional $U_{infl}\left[a_f^*, b_f, t_f, b^*(\tau)b(\tau), c^*(\tau)c(\tau), a_{in}, b_{in}, t_{in}\right]$. For this we present equation (43) as finite-multiple integral [9].

$$U_{infl}\left[a_f^*, b_f, t_f, b^*(\tau)b(\tau), c^*(\tau)c(\tau), a_{in}, b_{in}, t_{in}\right] =$$
$$= \lim_{n\longrightarrow\infty} \int_C \dots \int_C da_n^* da_n \cdots da_1^* da_1 exp\left[\imath(\Delta S_n^{(infl)} + \dots + \Delta S_1^{(infl)} + \Delta S_0^{(infl)})\right], \quad (46)$$

where

$$\Delta S_k^{(infl)} = -\frac{1}{2\imath}(|a_{k+1}|^2 + |a_k|^2) + (1 - \imath\Delta\tau\omega^{(b)})a_{k+1}^* a_k + \Delta\tau e f_k^{\mu+} \varepsilon_\mu^* a_{k+1}^* + \Delta\tau e f_k^{\mu-} \varepsilon_\mu a_k, \quad (47)$$

time interval t is spliting on equal intervals $\Delta\tau$. So $n\Delta\tau = t_f$ and $a_{n+1}^* = a_f^*$, $a_0 = a_{in}$.

We integrate through intermediate variables $a_1, \dots, a_n$ and obtain the following result in continual limit $\Delta t \to 0$, $n \to \infty$:

$$U_{infl}\left[a_f^*, b_f, t_f, b^*(\tau)b(\tau), c^*(\tau)c(\tau)a_{in}, b_{in}, t_{in}\right] =$$
$$= \exp\left[-\frac{1}{2}(|a_f|^2 + |a_{in}|^2) + A a_f^* a_{in} + B a_f^* + C a_{in} + S_{infvac}\right]. \quad (48)$$

where function $A = \exp\left[-i\omega^{(b)}t\right]$ and functional B, C, S_{inf}

$$B[b^*(\tau), b(\tau), c^*(\tau), c(\tau)] = -ie \int_{t_in}^{t_f} j^{\mu+}(b^*(\tau), b(\tau), c^*(\tau), c(\tau))\varepsilon^*_\mu e^{-i\omega^{(b)}(t-\tau)}d\tau, \tag{49}$$

$$C[b^*(\tau), b(\tau)c^*(\tau), c(\tau)] = -ie \int_{t_in}^{t_f} j^{\mu-}(b^*(\tau), b(\tau), c^*(\tau), c(\tau))\varepsilon_\mu e^{-i\omega^{(b)}\tau}d\tau, \tag{50}$$

$$S_{infvac}[b^*(\tau), b(\tau), c^*(\tau), c(\tau)] =$$

$$= -\int_{t_{in}}^{t_f}\int_{t_in}^{\tau} j^{\mu+}(b^*(\tau), b(\tau), c^*(\tau), c(\tau))\varepsilon^*_\mu j^{\nu-}(b^*(\tau), b(\tau, c^*(\tau), c(\tau))\varepsilon_\nu e^{-i\omega^{(b)}(\tau-\tau')}d\tau d\tau'. \tag{51}$$

Now we calculate the influence functional of electromagnetic vacuum. For this we use equation (42) and specify initial and final state of electromagnetic field as vacuum:

$$\zeta_v(a_{in}) = \exp[-\frac{|a_{in}|^2}{2}], \qquad \chi^*_v(a_f) = \exp[-\frac{|a_f|^2}{2}]. \tag{52}$$

By the use of (42) and (48),(52) we calculate the influence functional of one mode **k** and λ electromagnetic field on one mode **p** and σ fermionic field:

$$F_{vac\mathbf{k},\lambda} = \exp[S_{infvac\mathbf{k}\lambda} + S'_{infvac\mathbf{k}\lambda}], \tag{53}$$

where functionals $S_{infvac\mathbf{k}\lambda}, S'_{infvac\mathbf{k}\lambda}$ is determined by (51).

We define influence functional of miltimode electromagnetic vacuum using (45), (51) and (53).

$$F_{vacall}[b'^*(\tau), b'(\tau), b^*(\tau), b(\tau), c'^*(\tau), c'(\tau), c^*(\tau), c(\tau)] = \exp[S_{infvac} + S'_{infvac}], \tag{54}$$

where

$$S_{infvac} = -\sum_{\mathbf{k},\lambda}\frac{1}{2\omega_\mathbf{k}V}\int_{t_{in}}^{t_f}\int_{t_{in}}^{\tau}\left(\varepsilon^\mu_\lambda j^+_\mu(\mathbf{k},\tau)\varepsilon^{*\nu}_\lambda j^-_\nu(\mathbf{k},\tau')e^{\iota\omega_\mathbf{k}(\tau-\tau')}d\tau d\tau'\right), \tag{55}$$

functional S'_{infvac} has the same structure as S_{infvac} with variables $b'^*(\tau), b'(\tau), c'^*(\tau), c'(\tau)$ i.e. j'^+_μ, j'^-_ν.

4 The fermionic field evolution under electromagnetic vacuum influence

We describe the fermionic field evolution by the use of probability transition (41). Using equation (54) for influence functional of multimode electromagnetic vacuum, we obtain equation for probability transition

$$P(m; t_f|n; t_{in}) = K^*(m; t_f|n; t_{in})K(m; t_f|n; t_{in}), \tag{56}$$

where

$$K(m; t_f|n; t_{in}) = \int \mathcal{D}b^*(\tau)\mathcal{D}b(\tau)\mathcal{D}c^*(\tau)\mathcal{D}c(\tau)db^*_f db_f dc^*_f dc_f \times$$

$$\times\psi^*_m(b_f, c_f)\exp\left\{\iota\left(S_{fullvac}[b^*(\tau), b(\tau), c^*(\tau), c(\tau)]\right)\right\}\psi_n(b^*_{in}, c^*_{in}), \tag{57}$$

$K^*(m; t_f|n; t_{in})$ is conjugated transition amplitude,

$$S_{full}\left[b^*(\tau), b(\tau), c^*(\tau), c(\tau)\right] = S_f\left[b^*(\tau), b(\tau), c^*(\tau), c(\tau)\right] + S_{infvac}\left[b^*(\tau), b(\tau), c^*(\tau), c(\tau)\right], \quad (58)$$

In equations (56)–(58) we going from grassman variables $b^*(\tau), b(\tau), c^*(\tau), c(\tau)$ to functions of spinor field $\psi(x), \overline{\psi}(x)$.

In equation (55) we convert summation over discrete $\mathbf{k}$ to integral over their continuum values and Fourier transform $j^+_\mu(\mathbf{k}, \tau)$, $j^-_\nu(\mathbf{k}, \tau')$ to function $j_\mu(\mathbf{x}, \tau)$, $j_\nu(\mathbf{x}', \tau')$ by the use of equations (7) and (8). So, we have

$$S_{infvac}[\psi(x), \psi(x)] = \frac{\imath}{2\pi}\int_{t_{in}}^{t_f}\int_{t_{in}}^{\tau}\int\int j_\mu(\mathbf{x}, \tau)D^{\mu\nu}(\mathbf{x}-\mathbf{x}', \tau-\tau')j_\nu(\mathbf{x}', \tau')d\mathbf{x}d\mathbf{x}'d\tau d\tau', \quad (59)$$

where

$$D^{\mu\nu}(\mathbf{x}-\mathbf{x}', \tau-\tau') = \frac{1}{(2\pi)^3}\int\frac{2\pi\imath d\mathbf{k}}{\omega_\mathbf{k}}\left(\sum_\lambda \varepsilon^\mu_\lambda \varepsilon^{*\nu}_\lambda\right)e^{-\imath\mathbf{k}(\mathbf{x}-\mathbf{x}')}e^{\imath\omega(\tau-\tau')} \quad (60)$$

— is the photon propagator [11].

In equation (59) we transform iterated integral to double integral. Then we consider the limit $t_f \to \infty$, $t_{in} \to -\infty$. Now we obtain the action $S_{infvac}[\psi(x), \psi(x)]$:

$$S_{infvac}[\psi(x), \psi(x)] = \frac{\imath}{2\pi}\int\int j_\mu(x)D^{\mu\nu}(x-x')j_\nu(x')d^4x'd^4x. \quad (61)$$

We transform equation (58) in the same form. So the effective action (58) for action of fermionic field under electromagnetic field vacuum influence is presenting in the following

$$S_{fullvac}\left[\psi(x), \psi(x)\right] = \int \mathcal{L}_{fullvac}(\gamma^\mu\partial_\mu\psi(x), \psi(x))dx, \quad (62)$$

where

$$\mathcal{L}_{fullvac} = \psi(x)(\imath\gamma^\mu\partial_\mu - m)\psi(x) + \frac{\imath}{2\pi}j_\mu(x)\int D^{\mu\nu}(x-x')j_\nu(x')dx', \quad (63)$$

is density of effective Lagrangian,

$$\mathcal{L}_{infvac} = \frac{\imath}{2\pi}j_\mu(x)\int D^{\mu\nu}(x-x')j_\nu(x')dx', \quad (64)$$

is density of Lagrangian, which describe interaction of spinor field and vacuum of electromagnetic field.

The transition amplitude (57) is equivalent to S-matrix. In addition it contain contribution of electromagnetic vacuum influence on fermionic field dynamics:

$$K(m|n) = \int \mathcal{D}\overline{\psi}(\tau)\mathcal{D}\psi(\tau)\overline{\psi}_m(x)\exp\left\{\imath\left(S_{fullvac}[\overline{\psi}(\tau), \psi(\tau)]\right)\right\}\psi_n(x), \quad (65)$$

where action $S_{fullvac}[\overline{\psi}(\tau), \psi(\tau)]$ is determined by equation (62).

5 Summary

We consider the evolution of fermionic field interacting with electromagnetic field in path integral approach by the use of electromagnetic field influence functional inroduction. We obtain formula for influence functional calculation in specific model of electromagnetic field, interaction with fermionic fields and initial and final states. We calculate the electromagnetic vacuum influence functional. This lead to addition term in Lagrangian of spinor field and modify fermionic field dynamics. The developed method allow to describe interacting fermionic system without perturbation expansions.

The work was supported by a grant of the Russian Ministry of Science and Education No. 1349.

References

[1] Sweatlock L.A. Highly confined electromagnetic fields in arrays of strongly coupled Ag nanoparticles / L.A. Sweatlock, S.A. Maier, H.A. Atwater, J.J. Penninkhof, A. Polman // Phys.Rev. B – 2005. – 71. – 235408.

[2] Wellers Ch. Resonant IR multi-photon dissociation spectroscopy of a trapped and sympathetically cooled biomolecular ion species / Ch. Wellers, A. Borodin, S. Vasilyev, D. Offenberg, S. Schiller // Phys. Chem. Chem. Phys. – 2011. – 13 – 18799-18809.

[3] Richter M. Extreme Ultraviolet Laser Excites Atomic Giant Resonance / M. Richter, M.Ya. Amusia, S.V. Bobashev, T. Feigl, P. N. Juranic, M. Martins, A. A. Sorokin, K. Tiedtke // Phys. Rev. Let. – 2009. – 102 – 163002.

[4] Farrell J.P. Strong Field Ionization to Multiple Electronic States in Water / J. P. Farrell1, S. Petretti, J. Förster, B. K. McFarland, L. S. Spector, Y. V. Vanne, P. Decleva, P. H. Bucksbaum, A. Saenz, M. Gühr // Phys. Rev. Let. – 2011. – 107 – 083001.

[5] Kupersztych J. Anomalous Multiphoton Photoelectric Effect in Ultrashort Time Scales / M. Raynaud, J. Kupersztych // Phys. Rev. Let. – 2005. – 95 – 147401.

[6] Richard P. Feynman, A. R. Hibbs. Quantum Mechanics and Path Integrals / McGraw-Hill Companies, 1965, 365 pages.

[7] Albeverio S. A rigorous approach to the Feynman-Vernon influence functional and its applications I / S. Albeverio, L. Cattaneo, S. Mazzucchi, L. Persio // J. Math. Phys. – 2007 – v. 48 – 10 – pp. 102109-102109-22

[8] A. Biryukov, M. Shleenkov. The Influence Functional approach to the quantum systems dynamics. PoS(QFTHEP 2013)076.

[9] A. A. Biryukov, M. A. Shleenkov The Influence Functional of quantum electromagnetic field Theoretical Physics, 12, p. 28–49, (2011)

[10] L. D. Fadeev, A. A. Slavnov Gauge Fields: An Introduction To Quantum Theory Second Edition (Frontiers in Physics). Westview Press, 1993., 236 p.

[11] V.B. Berestetskii Quantum Electrodynamics / V.B. Berestetskii, E.M. Lifshitz, L.P. Pitaevskii // FML – 1980. – 704 p.

Exotic branes in Double Field Theory

Edvard Musaev[1,2,*]

[1] *Max-Planck-Institut für Gravitationsphysik (Albert-Einstein-Institut), Am Mühlenberg 1, DE-14476 Potsdam, Germany*
[2] *Kazan Federal University, Institute of Physics, General Relativity Department, Kremlevskaya 16a, 420111, Kazan, Russia*

Abstract. The non-geometric Q- and R-monopole are shown to be a particular case of the DFT-monopole solution. The notion of magnetic charge for the solutions is defined and shown to be equal to the magnetic charge of the NS5-brane solution. This is a talk presented by the author at the conference QUARKS'16 in St.-Petersburg.

1 T-duality and Double Field Theory

It is known that string theory respects a hidden symmetry, called T-duality, on the quantum perturbative level. The most intuitive way of understanding the action of this symmetry is consider swapping of momentum and winding modes of a closed string on a d-torus $\mathbb{T}^d$. In this case the T-duality symmetry group is $O(d,d)$. It has long been conjectured that these symmetries must carry some geometric meaning which would allow to define them on non-toroidal backgrounds.

The recently constructed $O(d,d)$-covariant gravity-like theory called Double Field Theory exploits the old ideas by Tseytlin [1] and Siegel [2] that left and right movers on the string worldsheet should be considered independently and generalizes the construction of generalized geometry by Hitchin, Cavalcanti and Gualtiery (see [3] for review and further links). The pioneering works in this direction were [4, 5] by Hull, Hohm and Zwiebach who presented the background independent duality covariant formulation of the action of DFT in terms of the so-called generalized metric $\mathcal{H}_{MN}$. Soon after the same was done for the U-duality groups in [6–8] for the internal sector only, and in [9–12] for the full theory including p-forms. Supersymmetric extension and inclusion of fermions has been presented in [13, 14] for DFT and in [15, 16] for EFT.

The idea behind Double Field Theory is that one doubles the space-time by adding dual coordinates $\tilde{x}_\mu$ corresponding to winding modes of closed strings. The resulting space-time (better to say 'space' to avoid issues with imaginary time) is parametrized by $(d+d)$ coordinates $\mathbb{X}^M = \{x^\mu, \tilde{x}_\mu\}$. The fields of the NS-NS sector, i.e. the metric, the B-field, and the dilaton, combine into the generalized metric $\mathcal{H}_{MN}$ that transforms covariantly under $O(d,d)$ transformations O and depends on all the coordinates

$$\mathcal{H}_{MN} = \begin{bmatrix} g - Bg^{-1}B & g^{-1}B \\ Bg^{-1} & g^{-1} \end{bmatrix}, \quad \mathcal{H} \to O^T \mathcal{H} O. \tag{1}$$

[*] e-mail: edvard.musaev@aei.mpg.de

By an $O(d) \times O(d)$ rotation this can equivalently be parametrized by the metric and a bivector field β, that is useful for defining non-geometric fluxes

$$\mathcal{H}_{MN} = \begin{bmatrix} g & \beta g^{-1} \\ g^{-1}\beta & g^{-1} - \beta g^{-1}\beta \end{bmatrix}. \tag{2}$$

Locally T-duality transformations are represented by generalized Lie derivative that combines these with conventional diffeomorphisms and gauge transformations of the B-field, and is defined by its action on a generalized vector V^M along a generalized vector U^M

$$\mathcal{L}_U V^M = U^N \partial_N V^M - V^N \partial_N U^M + \eta^{MN} \eta_{KL} \partial_N U^K V^L, \tag{3}$$

where η_{MN} is the (skew-diagonal) invariant tensor of $O(d, d)$. Algebra of these transformations generalizes that of the Dorfmann brackets and closes on the so-called C-bracket, that generalizes the Courant bracket, upon a special constraint

$$\begin{aligned} [\mathcal{L}_{V_1}, \mathcal{L}_{V_2}] &= \mathcal{L}_{[V_1,V_2]_C} + F_0, \\ Jac(\mathcal{L}_{V_1}, \mathcal{L}_{V_2}, \mathcal{L}_{V_3}) &= F_1. \end{aligned} \tag{4}$$

Here the additional terms F_0 and F_1 represent breaking of the closure and Jacobi conditions, and vanish upon setting

$$\eta^{MN} \partial_M \Phi \partial_N \Psi = 0 \tag{5}$$

for any fields Φ, Ψ. This constraint called the section condition effectively reduces the number of coordinates to d and returns the dynamics back to the conventional supergravity.

Given this equivalence there has been a lot of discussion in the community whether DFT is just a more convenient rewriting of supergravity or it may actually give new results. The results discussed below suggest that DFT (and similarly EFT) is indeed able in principle to capture dynamics of non-geometric backgrounds. One of the ways of doing that is to consider backgrounds with fields satisfying the section condition, but which depend explicitly on dual (winding) coordinates. These have no description in the framework of the conventional supergravity as it does not know anything about the dual coordinates. Hence, these are purely DFT backgrounds, some particular examples of which will be shown to carry non-geometric flux. In what follows we briefly describe the results of [17] sending those interested in the technical details to the original paper.

2 Exotic branes in string theory

Probably the most famous example of non-geometric fluxes is given by Q- and R-fluxes, which are known to correspond to non-commutative and non-associative behavior of strings respectively. These appear as nodes of the T-duality orbit starting from the geometric flux $H = dB$. Hence, one writes

$$H_{xyz} \xrightarrow{T_x} \tau^x{}_{yz} \xrightarrow{T_y} Q^{xy}{}_z \xrightarrow{T_z} R^{xyz}, \tag{6}$$

where $\tau^x{}_{yz}$ is the so-called geometric flux, related to curvature of the background. Existence of this orbit was conjecture in [18] where a simple toy model of twisted torus has been considered. Here we would like to proceed with realistic backgrounds starting with NS5-brane with non-zero H-flux. This is written in terms of the harmonic function $H = H(x^1, \ldots, x^d)$ that is a solution of the (flat) Laplace equation with a Dirac-delta-like source term in the RHS

$$\left(\partial_1^2 + \cdots + \partial_d^2\right) H(x^1, \ldots, x^d) \sim \delta(x^1, \ldots, x^d). \tag{7}$$

In what follows we will not take into account the source term and focus only on the vacuum solution, however an extensive discussion of the issue of sources for geometric and non-geometric solutions can be found in [19]. The harmonic function

$$H(x^1,\dots,x^d) = 1 + \frac{h_0}{\left(x_1^2+\dots+x_d^2\right)^{\frac{d-2}{2}}}, \tag{8}$$

where for $d = 2$ the power function should be replaced by the log and hence is not well defined. This is the well known issue of the non-geometric 5_2^2-brane, carrying Q-flux (see [20] for discussion).

Table 1. *Under T-dualities an NS5-brane stretched in directions marked by × turns into a Kaluza-Klein monopole and a 5_2^2-brane. Dotted circles denote special cycles along which the T-duality acts, these are compactified.*

	1	2	3	4	5	6	7	8	9
NS5	·	·	·	·	×	×	×	×	×
KKM	·	·	·	⊙	×	×	×	×	×
5_2^2	·	·	⊙	⊙	×	×	×	×	×

To act by a T-duality along an isometry direction of the NS5-brane solution one should first smear the brane along, say, $x^4 =: z$, that is equivalent to compactification of this direction. I.e. the background with smeared harmonic function is a solution of the theory compatified on a circle [21]

$$\frac{h_0}{\left(x_1^2+\dots+x_d^2\right)^{\frac{d-2}{2}}} \Longrightarrow \frac{2\pi R_d h_0}{\left(x_1^2+\dots+x_{d-1}^2\right)^{\frac{d-3}{2}}} \tag{9}$$

and the resulting background is that of the Kaluza-Klein monopole

$$\begin{aligned} ds^2 &= ds^2_{056789} + H\left(d\rho^2+\rho^2 d\theta^2 + (dx^3)^2\right) + H^{-1}(dx^4+\omega)^2, \\ B^{(2)} &= 0, \quad \rho^2 = (x^1)^2+(x^2)^2. \end{aligned} \tag{10}$$

This solution is characterized by the geometric flux $\tau_{mn}{}^k$ defined as a structure constant of the algebra of right-invariant vector fields (vielbeins) $[e_m, e_n] = \tau_{mn}{}^k e_k$.

Smearing along another direction, say x^3, and performing T-duality T_{x^3} one arrives to a non-geometric background usually referred to as 5_2^2-brane [20]

$$\begin{aligned} ds^2 &= H(d\rho^2+\rho^2 d\theta^2) + \frac{H}{H^2+\sigma^2\theta^2} ds^2_{34} + ds^2_{056789}, \\ B^{(2)} &= \frac{\sigma\theta}{H^2+\sigma^2\theta^2} dx^3\wedge dx^4, \quad e^{-2\phi} = \frac{H}{H^2+\sigma^2\theta^2}. \end{aligned} \tag{11}$$

This solution was constructed in [20] as a realistic example of the toy background of [18], which was conjectured to carry a portion of Q-flux. The most interesting property of the solution is the non-trivial monodromy around the brane $\theta \to \theta + 2\pi$ that mixes the metric and the B-field acting as a T-duality transformation on the torus (x^3, x^4). Namely, the size of the special 2-torus does not come back to itself

$$\begin{aligned} \theta = 0: &\quad G_{33} = G_{44} = H^{-1}, \\ \theta = 2\pi: &\quad G_{33} = G_{44} = \frac{H}{H^2+(2\pi\sigma)^2}. \end{aligned} \tag{12}$$

In the DFT notation the monodromy takes the form of an $O(2,2)$ rotation acting as

$$\mathcal{H}(\theta' = \theta + 2\pi) = O^{tr}\mathcal{H}(\theta)O, \quad O = \begin{bmatrix} \mathbf{1}_2 & 0 \\ \beta(\theta') & \mathbf{1}_2 \end{bmatrix}, \tag{13}$$

where $\beta(\theta') = \beta(\theta')^{34}\partial_3 \wedge \partial_4$ is a bivector.

Hence, the background is defined globally only up to a T-duality transformation and in principle could be described by the T-fold construction. However, formally performing T-duality along the remaining coordinate we arrive to a background, which carry a portion of R-flux and which cannot be described by supergravity even locally. To deal with such kind of backgrounds and to define T-duality along a non-isometry direction one has to use the formalism of Double Field Theory.

3 Exotic objects in DFT

3.1 DFT monopole

Dynamics of DFT is encoded by the following effective potential

$$V = e^{-2d}\left(\frac{1}{8}\mathcal{H}^{MN}\partial_M\mathcal{H}^{KL}\partial_N\mathcal{H}_{KL} - \frac{1}{2}\mathcal{H}^{KL}\partial_L\mathcal{H}^{MN}\partial_N\mathcal{H}_{KM} - 2\partial_M d\partial_N\mathcal{H}^{MN} + 4\mathcal{H}^{MN}\partial_M d\partial_N d\right), \tag{14}$$

which drops back to the bosonic Lagrangian density of supergravity upon the section condition. In [22] a solution of DFT equations of motion called DFT-monopole has been found. To write the solution it is convenient to understand the generalized metric as a Kaluza-Klein decomposition and to write the following DFT interval

$$ds^2_{DFT} = (g_{\mu\nu} - B_\mu{}^\rho B_{\rho\nu})dx^\mu dx^\nu + 2B_\mu{}^\nu dx^\mu d\tilde{x}_\nu + g^{\mu\nu}d\tilde{x}_\mu d\tilde{x}_\nu. \tag{15}$$

One should note that this is just a notation and the above quantity in principle does not have to indeed measure distance in the extended space.

With that in hands one writes the solution in the Taub-NUT form as follows

$$\begin{aligned} ds^2_{DFT} &= H(1 + H^{-2}A^2)dz^2 + H^{-1}d\tilde{z}^2 + 2H^{-1}A_i(dy^i d\tilde{z} - \delta^{ij}d\tilde{y}_j dz) \\ &+ H(\delta_{ij} + H^{-2}A_iA_j)dy^i dy^j + H^{-1}\delta^{ij}d\tilde{y}_i d\tilde{y}_j \\ &+ \eta_{rs}dx^r dx^s + \eta^{rs}d\tilde{x}_r d\tilde{x}_s, \end{aligned} \tag{16}$$

where the functions H, A_i and the invariant dilation are given by $(i, j, k, l = 1, 2, 3)$

$$\begin{aligned} H(y) &= 1 + \frac{h}{\sqrt{\delta_{ij}y^i y^j}}, \\ 2\partial_{[i}A_{j]} &= \epsilon_{ijk}\partial_k H, \\ e^{-2d} &= He^{-2\varphi_0}. \end{aligned} \tag{17}$$

The coordinates (x^r) with r, s running from 5 to 9 and 0, labeling the remaining directions; in addition we keep the notation $z = x^4$.

The solution depends on 20 parameters $(z, y^i, x^r, \tilde{z}, \tilde{y}_i, \tilde{x}_r)$, which in principle can be identified with either physical or dual coordinates of DFT. This is equivalent to choosing a T-duality frame or performing a T-duality transformation. The important difference here is, that these coordinates do not have to be compact.

The most intriguing feature of the DFT-monopole solution is that it combines the three backgrounds of Type II supergravity, H-, KK-monopoles and the 5^2_2-brane, and the locally non-geometric background of R-brane upon a choice of a solution of the section condition. While the behaviour of string on the background of 5^2_2-brane is known to be non-commutative, the R-brane background was conjectured to show signs of non-associativity. Indeed, commutator and Jacobiator of string coordinates on backgrounds with Q- and/or R-fluxes can be written as (see e.g. [23])

$$\begin{aligned}[x^m, x^n] &= Q^{mn}{}_k x^k,\\ [x^m, x^n, x^k] &= R^{mnk}.\end{aligned} \tag{18}$$

Since a solution of the section condition is determined by a choice of physical and dual coordinates, one may write the following general rule for embedding of the backgrounds with H, τ, Q and R fluxes into the DFT-monopole solution

Table 2. *Rules for identification of the parameters $(z, y^i, \tilde{z}, \tilde{y}_i)$ with the physical coordinates (x^z, x^1, x^2, x^3). The remaining parameters are identified with the dual coordinates.*

$(x^z, x^1, x^2, x^3) = (z, y^1, y^2, y^3)$,	NS5-brane (H-monopole),
$(x^z, x^1, x^2, x^3) = (\tilde{z}, y^1, y^2, y^3)$,	KK-monopole,
$(x^z, x^1, x^2, x^3) = (\tilde{z}, y^1, y^2, \tilde{y}^3)$,	Q-monopole,
$(x^z, x^1, x^2, x^3) = (\tilde{z}, \tilde{y}^1, \tilde{y}^2, y^3)$,	R-monopole.

Definition of fluxes in DFT in similar to that of the geometric flux $\tau_{mn}{}^k$ in supergravity, i.e. these are structure constants of the algebra of generalized vielbeins $E_M{}^A$

$$\begin{aligned}\mathcal{H}_{MN} &= E_M{}^{\bar{A}} E_M{}^{\bar{B}} \mathcal{H}_{\bar{A}\bar{B}},\\ [E_{\bar{A}}, E_{\bar{B}}]_C &= \mathcal{F}_{\bar{A}\bar{B}}{}^{\bar{C}} E_{\bar{C}}.\end{aligned} \tag{19}$$

Components of the generalized flux $\mathcal{F}_{\bar{A}\bar{B}}{}^{\bar{C}} = \{\mathcal{F}_{\bar{a}\bar{b}\bar{c}}, \mathcal{F}_{\bar{a}\bar{b}}{}^{\bar{c}}, \mathcal{F}_{\bar{a}}{}^{\bar{b}\bar{c}}, \mathcal{F}^{\bar{a}\bar{b}\bar{c}}\}$ are identified with the fluxes of the T-duality orbit in question. In addition in DFT generalized flux with one index $\mathcal{F}_A$ is defined

$$\begin{aligned}\mathcal{F}^{\bar{A}}{}_{\bar{B}\bar{C}} &= 2E^{\bar{A}}_M E^N_{[\bar{B}} \partial_N E^M_{\bar{C}]} - E^{\bar{A}}_M \eta^{MN} \eta_{KL} \partial_N E^K_{[\bar{B}} E^L_{\bar{C}]},\\ \mathcal{F}_{\bar{A}} &= \partial_M E^M_{\bar{A}} + 2E^M_{\bar{A}} \partial_M d.\end{aligned} \tag{20}$$

Using this we have for the backgrounds of H- and KK-monopoles

$$\begin{aligned}&\text{H-monopole:} && \mathcal{F}_{\bar{z}\bar{a}\bar{b}} = 2e^z{}_{\bar{z}} e^k{}_{\bar{a}} e^l{}_{\bar{b}} \partial_{[k} A_{l]}, \quad \mathcal{F}^{\bar{a}}{}_{\bar{b}\bar{c}} = -\delta^{\bar{a}}{}_{[\bar{b}} \mathfrak{f}_{\bar{c}]}, \quad \mathcal{F}_{\bar{a}} = \frac{3}{2} H^{-1} \partial_{\bar{a}} H,\\ &\text{KK-monopole:} && \mathcal{F}^{\bar{a}}{}_{\bar{b}\bar{c}} = 2e^{\bar{a}}{}_z e^k{}_{\bar{b}} e^l{}_{\bar{c}} \partial_{[k} A_{l]} - \frac{1}{3} \delta^{\bar{a}}{}_{[\bar{b}} \mathfrak{f}_{\bar{c}]}, \quad \mathcal{F}_{\bar{a}} = -\frac{3}{2} H^{-1} \partial_{\bar{a}} H,\end{aligned} \tag{21}$$

with $(\bar{a}, \bar{b}, \bar{c})$ running from 1 to 3.

Note that for the above cases the harmonic function $H = H(y^1, y^2, y^3)$ depends only on physical coordinates and hence the story does not differ much from the conventional supergravity. In contrast, for the non-geometric backgrounds of Q- and R-monopoles H becomes a function of dual coordinates, that is an expected behavior given contributions from worldsheet instantons [24–26]. From the technical side this is crucial for reproducing non-geometric fluxes, whose definition includes derivatives with respect to dual coordinates.

3.2 Q- and R-monopoles

As follows from the Table 2 choosing the physical coordinates as $(x^z, x^1, x^2, x^3) = (\tilde{z}, y^1, y^2, \tilde{y}^3)$ we obtain the background of Q-monopole

$$
\begin{aligned}
ds^2 &= \eta_{rs}dx^r dx^s + H^{-1}\Big[(d\tilde{z} + A_\theta d\theta)^2 + d\tilde{y}_3^2\Big] + H\Big(d\rho^2 + \rho^2 d\theta^2\Big),\\
A_\theta &= h\left(1 - \frac{y^3}{\sqrt{\rho^2 + (y^3)^2}}\right), \quad e^{-2(\varphi-\varphi_0)} = H^{-1}, \quad B^{(2)} = 0\\
\rho^2 &= (y^1)^2 + (y^2)^2,
\end{aligned}
\tag{22}
$$

where the polar coordinates (ρ, θ) in the physical plane (x^1, x^2) have been adopted. Interestingly, the solution is purely metric, i.e. it does not have non-trivial 2-form or bivector fields similarly to the KK-monopole. However, it still carries non-geometric Q-flux, thas is justified by calculating components of the generalized flux

$$
\begin{aligned}
&Q_{\bar{\theta}}{}^{\bar{3}\bar{z}} = -H^{-\frac{3}{2}}\partial_\rho H, \quad \mathcal{F}_{\bar{\alpha}}{}^{\bar{\beta}\bar{3}} = -\frac{1}{2}\delta^{\bar{\beta}}{}_{\bar{\alpha}}H^{-\frac{3}{2}}\tilde{\partial}^3 H\\
\text{Q-monopole:} \quad &\mathcal{F}^{\bar{z}}{}_{\bar{\rho}\bar{z}} = \mathcal{F}^{\bar{3}}{}_{\bar{\rho}\bar{3}} = -\mathcal{F}^{\bar{\theta}}{}_{\bar{1}\bar{2}} = \frac{1}{3}\mathcal{F}_{\bar{\rho}} = \frac{1}{2}H^{-\frac{3}{2}}\partial_\rho H\\
&\mathcal{F}_{\bar{z}}{}^{\bar{z}\bar{3}} = \frac{1}{3}\mathcal{F}^{\bar{3}} = \frac{1}{2}\mathcal{F}^{\bar{z}}{}_{\bar{1}\bar{2}} = \frac{1}{2}H^{-\frac{3}{2}}\tilde{\partial}^3 H,
\end{aligned}
\tag{23}
$$

where one notes that $\tilde{\partial}^3 = \partial/\partial_{\tilde{x}_3}$ is the derivative along a winding mode coordinate, the indices $\{\alpha, \beta\}$ run through $\{1, 2\}$ and the barred indices are flat indices.

We see, that the background indeed carries a portion of Q-flux and does not carry non-trivial components of other fluxes. Note that the harmonic function has always the same form (17), and for this background it becomes a function of the dual coordinate $\tilde{x}_3 := y^3$.

For the background of R-monopole the harmonic function becomes a function of two dual coordinates $\tilde{x}_1 := y^1$, $\tilde{x}_2 := y^2$, and the background fields become

$$
\begin{aligned}
ds^2 &= \eta_{rs}dx^r dx^s + H^{-1}\Big(d\tilde{z}^2 + d\rho^2 + \rho^2 d\theta^2\Big) + H(dy^1)^2,\\
\beta^{\theta z} &= A_\theta, \quad e^{-2(\varphi-\varphi_0)} = 1.\\
A_\theta &= h\left(1 - \frac{y^1}{\sqrt{\tilde{\rho}^2 + (y^1)^2}}\right), \quad \rho^2 = (\tilde{y}_2)^2 + (\tilde{y}_3)^2 = (x^2)^2 + (x^3)^2.
\end{aligned}
\tag{24}
$$

It is important to note the different meaning of the coordinates ρ and $\tilde{\rho}$, with the latter defined as

$$
\begin{aligned}
\tilde{x}_2 &= y^2 = \tilde{\rho}\cos\tilde{\theta},\\
\tilde{x}_3 &= y^3 = \tilde{\rho}\sin\tilde{\theta},\\
x^1 &= y^1.
\end{aligned}
\tag{25}
$$

With that notations in hands components of the generalized flux can be written in the following form

$$
\begin{aligned}
&R^{\bar{z}\bar{\alpha}\bar{\beta}} = \epsilon^{\bar{\alpha}\bar{\beta}}H^{-\frac{3}{2}}\partial_1 H, \qquad \mathcal{F}_{\bar{z}}{}^{\bar{z}\bar{\alpha}} = \mathcal{F}_{\bar{1}}{}^{\bar{1}\bar{\alpha}} = -\frac{1}{2}H^{-1}\tilde{\partial}^{\bar{\alpha}}H,\\
\text{R-monopole:} \quad &\mathcal{F}_{\bar{a}}{}^{\bar{b}\bar{c}} = \delta_1{}^{[\bar{b}}\delta_{\bar{a}}{}^{\bar{c}]}H^{-\frac{3}{2}}\partial_1 H, \quad \mathcal{F}_{\bar{\alpha}}{}^{\bar{\beta}\bar{\gamma}} = \frac{1}{2}\epsilon^{\bar{\beta}\bar{\gamma}}\epsilon_{\bar{\alpha}\bar{\delta}}H^{-1}\tilde{\partial}^{\bar{\delta}}H, \qquad \mathcal{F}_{\bar{1}}{}^{\bar{z}\bar{\alpha}} = \epsilon^{\bar{\alpha}\bar{\beta}}H^{-1}\tilde{\partial}^{\bar{\beta}}H\\
&\mathcal{F}^{\bar{\alpha}} = \frac{3}{2}H^{-1}\tilde{\partial}^{\bar{\alpha}}H \qquad \mathcal{F}_{\bar{1}} = \frac{3}{2}H^{-\frac{3}{2}}\partial_1 H,
\end{aligned}
\tag{26}
$$

where α,β now take the values 2 and 3. One notices that the R-flux itself is written in terms of derivative of the harmonic function along the physical coordinate, that is the consequence of using the β-frame of DFT. In addition one finds components of the generalized flux, which as in the previous case, are proportional to derivatives of the harmonic function with respect to dual coordinates.

In principle, one is able to obtain the above results in the B-frame as well, however, we found these calculations too lengthy.

4 Conclusions

This letter briefly describes the result of [17] where the solutions of Double Field Theory characterised by non-geometric Q- and R-fluxes have been constructed as particular cases of the DFT-monopole of Berman and Rudolph [27]. By making use of the notion of generalized flux it is explicitly shown, that these solutions indeed carry a portion of Q-flux or R-flux, which satisfy generalized Bianchi identities of [28].

The most interesting feature of these solutions is that the fields depend on dual coordinates. Although, this is an expected behaviour due to worldsheet instanton corrections, one should no forget, that DFT is a low energy field theory and does not have to know about stringy effects. The described effect however gives a hint, that it actually may know about strings more, than conventional supergravity. Performing smearing procedure along these dual coordinates one obtains the known 5^2_2-brane solution of [20] from Q-monopole and a co-dimension-1 solution from R-monopole. For both of these the harmonic function becomes not well defined at spatial infinity and hence the solutions are pathological.

In contrast in DFT the solutions behave well at infinity, that allows to define and explicitly calculate magnetic charge, that reads

$$4\pi\mu = \int_\Sigma \mathcal{F}_{MNK} d\mathbb{X}^M \wedge d\mathbb{X}^N \wedge d\mathbb{X}^K. \tag{27}$$

Here the integration is performed along a special 3-cycle $\Sigma \equiv \mathbb{S}^2 \times \mathbb{S}^1$ which is defined differently for the backgrounds. However, for all the solutions considered in this paper such defined magnetic charge gives the same result $\mu = Q_{NS5}$, that is just the magnetic charge of the NS5-brane (without smearing). This is expected as the initial definition is $O(d,d)$-invariant.

Table 3. *The 3-cycle $\Sigma \equiv \mathbb{S}^2 \times \mathbb{S}^1$ is a product of a 1-circle and a 2-sphere. Here the bullet • denotes the direction of the 1-circle, while the crosses × denote the directions in which the 2-sphere lives.*

Σ	x^1	x^2	x^3	x^z	$\tilde{x}_1$	$\tilde{x}_2$	$\tilde{x}_3$	$\tilde{x}_z$
H	×	×	×	•				
KK	×	×	×					•
Q	×	×					×	•
R	×					×	×	•

The main idea behind such defined integration surface is (i) that it cannot vary both in a physical and the corresponding dual directions, that follows from the section condition, (ii) the rest is just chosen by the generalized flux itself and one has nothing else to write.

Note that the coordinates x^z or $\tilde{x}_z$ are always compact as the fields does not depend on them. In general one may consider the full solution that depends on (one of) these coordinates and to obtain similar results. Such solution of the DFT equations of motion has already been constructed in [27] and called the localized DFT monopole.

Acknowledgements

The authors' attendance of the conference QUARKS'16 was sponsored by the Alexander von Humboldt foundation. This work was supported by the Alexander von Humboldt Foundation and in part by the Russian Government programme of competitive growth of Kazan Federal University.

References

[1] A.A. Tseytlin, Phys.Lett. **B242**, 163 (1990)
[2] W. Siegel, Phys.Rev. **D48**, 2826 (1993), hep-th/9305073
[3] N. Hitchin, *Lectures on generalized geometry* (2010), arXiv:1008.0973
[4] O. Hohm, C. Hull, B. Zwiebach, JHEP **1007**, 016 (2010), 1003.5027
[5] O. Hohm, C. Hull, B. Zwiebach, JHEP **1008**, 008 (2010), 1006.4823
[6] D.S. Berman, M.J. Perry (2010), 1008.1763
[7] D.S. Berman, H. Godazgar, M.J. Perry, Phys.Lett. **B700**, 65 (2011), 1103.5733
[8] D.S. Berman, H. Godazgar, M. Godazgar, M.J. Perry, JHEP **1201**, 012 (2012), 1110.3930
[9] O. Hohm, H. Samtleben, Phys.Rev. **D89**, 066016 (2014), 1312.0614
[10] O. Hohm, H. Samtleben, Phys.Rev. **D89**, 066017 (2014), 1312.4542
[11] O. Hohm, H. Samtleben, Phys.Rev. **D90**, 066002 (2014), 1406.3348
[12] A. Abzalov, I. Bakhmatov, E.T. Musaev, JHEP **06**, 088 (2015), 1504.01523
[13] I. Jeon, K. Lee, J.H. Park, Phys.Rev. **D85**, 081501 (2012), 1112.0069
[14] M. Cederwall (2016), 1603.04684
[15] H. Godazgar, M. Godazgar, O. Hohm, H. Nicolai, H. Samtleben, JHEP **1409**, 044 (2014), 1406.3235
[16] E. Musaev, H. Samtleben, JHEP **1503**, 027 (2015), 1412.7286
[17] I. Bakhmatov, A. Kleinschmidt, E.T. Musaev (2016), 1607.05450
[18] J. Shelton, W. Taylor, B. Wecht, JHEP **0510**, 085 (2005), hep-th/0508133
[19] D. Andriot, A. Betz, JHEP **07**, 059 (2014), 1402.5972
[20] J. de Boer, M. Shigemori (2012), 1209.6056
[21] T. Ortin, *Gravity and Strings*, Cambridge Monographs on Mathematical Physics (Cambridge University Press, 2015), ISBN 9780521768139, 9780521768139
[22] D.S. Berman, F.J. Rudolph, JHEP **05**, 015 (2015), 1409.6314
[23] D. Lust, JHEP **12**, 084 (2010), 1010.1361
[24] J.A. Harvey, S. Jensen, JHEP **10**, 028 (2005), hep-th/0507204
[25] S. Jensen, JHEP **1107**, 088 (2011), 1106.1174
[26] T. Kimura, S. Sasaki, JHEP **08**, 126 (2013), 1305.4439
[27] D.S. Berman, D.C. Thompson (2013), 1306.2643
[28] D. Geissbuhler, D. Marques, C. Nunez, V. Penas (2013), 1304.1472

EPJ Web of Conferences **125**, 05018 (2016) DOI: 10.1051/epjconf/201612505018
QUARKS-2016

Supergravity with broken Lorentz invariance

Arthur Marakulin[1,*] and *Sergey* Sibiryakov[1,2,3,**]

[1] *Institute for Nuclear Research of the Russian Academy of Sciences, Moscow, Russia*
[2] *CERN Theory division, Geneva, Switzerland;*
[3] *Institut de Physique, EPFL, Lausanne, Switzerland*

Abstract. The supersymmetric extension of the Lorentz violating Einstein-aether theory of gravity is considered. The most general Lagrangian of the linearized Einstein-aether supergravity is constructed using the superfield formalism. The constraints imposed by supersymmetry on the parameters of the theory are obtained.

1 Introduction

Incompatibility of the principles of quantum field theory with general relativity is one of the most important problems in modern theoretical physics. A potential way out of this situation consists in restricting the domain of validity of some basic postulates of general relativity and abandoning them at high energy scales. A promising approach to quantization of gravity based on abandoning the Lorentz invariance has been proposed by Horava [1]. The low-energy limit of the Horava theory, called khrono-metric model, presents a special case of the Einstein-aether gravity [2]. In the latter model violation of the Lorentz invariance is described by the time-like vector field u^m with unit norm ($u^m u_m = -1$) called aether that minimally couples to the Einstein-Hilbert action for gravity,

$$S = S_{GR} - \frac{1}{2}\int d^4x\sqrt{-g}\Big(c_1\,(\nabla_n u_m)^2 + c_2\,(\nabla_m u_m)^2 + c_3\nabla_n u_m \nabla^m u^n - c_4 u^r u^s \nabla_r u_m \nabla_s u^m\Big)\,. \quad (1)$$

A phenomenologically acceptable theory with Lorentz violation in the ultraviolet must incorporate a mechanism that ensures emergence of Lorentz invariance at low energies. One of the possible mechanisms is based on non-relativistic supersymmetry [3]. It has been shown [4] that in the supersymmetric extension of the Einstein-aether theory based on chiral vector super-aether field interactions between the aether and the fields of the Standard Model are suppressed.

In this work we consider the coupling of supersymmetric aether to supergravity. This cannot be realized within the minimal $N = 1$ supergravity which does not allow to define a chiral aether vector superfield ($\bar{\nabla}_{\dot{\alpha}} U^a = 0$): anti-commutator of two supercovariant derivatives with spinor indices acting on a vector superfield does not vanish prohibiting to impose the chirality condition. This problem is resolved in the non-minimal formulation of the $N = 1$ supergravity with Ziegel parameter $n = -1$ [5].

*e-mail: marakulin@physics.msu.ru
**e-mail: sergey.sibiryakov@cern.ch

We use the latter formulation to construct a superfield Lagrangian for the aether field coupled to linearized gravity and prove that it is unique, up to the choice of a single parameter.

2 Superfield formalism

Non-minimal linearized $n = -1$ supergravity is formulated [5] in terms of real superfield V^a and so-called linear compensator Γ ($\bar{D}^2\Gamma = 0$) as a gauge theory with the action

$$S_{SG} = \int d^8z\left(\frac{1}{4}\left((\partial_k V_b)^2 - (\Delta_k V_b)^2\right) + \Delta_k V_k\left(\Gamma + \bar{\Gamma}\right) + i\partial_k V_k\left(\Gamma - \bar{\Gamma}\right) - \Gamma^2 - \bar{\Gamma}^2 - \Gamma\bar{\Gamma}\right), \quad (2)$$

where $\Delta_k V_n = \frac{1}{4}\bar{\sigma}_k^{\dot{\alpha}\alpha}[\bar{D}_{\dot{\alpha}}, D_\alpha]V_n$. The action is invariant under super-gauge transformations

$$\delta V_{\alpha\dot{\alpha}} = \bar{D}_{\dot{\alpha}} L_\alpha - D_\alpha \bar{L}_{\dot{\alpha}} \quad (3)$$

$$\delta\bar{\Gamma} = \frac{1}{4} D^\alpha \bar{D}^2 L_\alpha \quad (4)$$

We construct linearized Einstein-aether supergravity theory in terms the latter superfields and the chiral vector super-aether field which is considered as consisting of the real vacuum and small perturbation: $U^a = W^a + \mathcal{U}^a$. The super-aether field gauge transformation

$$\delta_L \mathcal{U}_a = \frac{1}{4}(\sigma_{ab})^\gamma_\beta D_\gamma \bar{D}^2 L^\beta W^b + \text{h. c.}, \quad (5)$$

is consistent with the chirality condition. A straightforward but tedious calculation yields the most general Lagrangian invariant under gauge transformations with one free parameter C:

$$\begin{aligned} L = &\frac{1}{4}\left((\partial_k V_b)^2 - (\Delta_k V_b)^2\right) + \Delta_k V_k\left(\Gamma + \bar{\Gamma}\right) + i\partial_k V_k\left(\Gamma - \bar{\Gamma}\right) - \Gamma^2 - \bar{\Gamma}^2 - \Gamma\bar{\Gamma} \\ &+ C\Big\{\mathcal{U}_a\bar{\mathcal{U}}^a - W^a W^c\Big[\frac{3}{16}\eta_{ac}\left((\Delta_k V_b)^2 - (\partial_k V_b)^2\right) - \frac{1}{4}\eta_{ac}\left((\Delta_b V_b)^2 - (\partial_b V_b)^2\right) \\ &+ \Delta_a V_c\left(\Gamma + \bar{\Gamma}\right) + i\partial_a V_c\left(\Gamma - \bar{\Gamma}\right) + \frac{1}{8}\eta_{ac}\left(\Gamma^2 + \bar{\Gamma}^2\right) - \frac{1}{4}\eta_{ac}\Gamma\bar{\Gamma}\Big]\Big\} \end{aligned} \quad (6)$$

This parameter is a coupling constant of the super-aether to supergravity.

3 Component expressions

In the real world supersymmetry must be broken. As shown in [4], a soft breaking of supersymmetry gives mass to the imaginary component of the super-aether, whereas its real part remains massless. At low energies, the theory is described by the action (1), where the parameters c_i, $i = 1, 2, 3, 4$, however, are constrained by supersymmetry. To find these constraints, we write the bosonic part of the Lagrangian (6) in component fields. We use the bosonic parts of the gravitational superfield in the Wess-Zumino gauge [5],

$$V^m(x)|_{\text{bos}} = \theta\sigma_a\bar{\theta}e^a_m(x) + \theta^2\bar{\theta}^2 A^m(x), \quad (7)$$

where e_{mn} is the symmetrized vielbein; the linear compensator in terms of the "left" coordinates $\left(y^m = x^m + i\theta\sigma^m\bar{\theta}, \theta^\alpha\right)$

$$\Gamma(y)|_{\text{bos}} = \theta^2 B(y) + \theta\sigma^a\bar{\theta}\left(v_a(y) + i\mu_a(y)\right); \quad (8)$$

and the aether superfield

$$\mathcal{U}_b(y)|_{\text{bos}} = v_b(y) + \theta^2 G_b(y) + \theta\sigma^k\bar{\theta} f_{bk}(y), \tag{9}$$

where w_a is the aether vacuum expectation value, v_a is the first order aether perturbation, and

$$f_{bk} = w_c A_p \text{tr}\left(\sigma_{bc}\sigma_{pk}\right) + i w_c \partial_m e_{np} \text{tr}\left(\sigma_k \bar{\sigma}_p \sigma_{bc} \sigma_{nm}\right) + w_c \left(\nu_p + i\mu_p\right) \text{tr}\left(\bar{\sigma}_{bc}\bar{\sigma}_{kp}\right). \tag{10}$$

Expanding in the Grassmann coordinates we obtain the bosonic part of the Lagrangian in components. Integrating out the auxiliary fields we reduce to the linearized Lagrangian of the Einstein-aether theory

$$\begin{aligned} L_{bos} = \left(1 + \frac{C}{4}\right)\left(\frac{1}{2}(\partial_k e)^2 - \frac{1}{2}(\partial_k e_{mn})^2 + \partial_m e_{mp}\partial_n e_{np} - \partial_m e_{mp}\partial_p e\right) \\ + C\Big(- (\partial_k v_b)^2 + 2w_a \partial_c v_b (\partial_a e_{bc} - \partial_b e_{ac}) \\ - w_a w_c \left(\partial_p e_{ak} - \partial_a e_{pk}\right)\left(\partial_p e_{ck} - \partial_c e_{pk}\right)\Big), \end{aligned} \tag{11}$$

where e_{mn} is the symmetrized vielbein, w_a is the aether vacuum expectation value and v_a is the first order aether perturbation. By comparing with the linearized form of the action (1) one finds that supersymmetry fixes all but one coupling constants in this action to zero: $c_1 = 4C/(4 + C)$, $c_{2,3,4} = 0$.

4 Discussion

In this paper the supersymmetric extension of the Lorentz violating Einstein-aether theory of gravity was considered. The most general Lagrangian of the linearized Einstein-aether supergravity was constructed using the superfield formalism. The constraints imposed by supersymmetry on the theory parameters are obtained.

Modern experimental tests provide tight bounds on the parameters of Lorentz violation. In general, the Einstein-aether model gives rise to non-zero values of the post-Newtonian parameters $\alpha_{1,2}$ that can be expressed in terms of c_i [2]. These are constrained from observations in the Solar System: $|\alpha_1| \lesssim 10^{-4}$, $|\alpha_2| \lesssim 4 \times 10^{-7}$ [6]. In the supersymmetric model $c_1 \neq 0$, $c_{2,3,4} = 0$, so the constraints translate into a bound on c_1: $|c_1| \lesssim 10^{-7}$.

The work is supported by the RFBR grant 14-02-31429.

References

[1] P. Horava, Physical Review D, **79**, 8, 084008 (2009).
[2] T. Jacobson, PoS **QG-PH**, 020 (2007).
[3] S. Groot Nibbelink and M. Pospelov, Phys. Rev. Lett. **94**, 081601 (2005).
[4] O. Pujolas and S. Sibiryakov, JHEP, **1201**, 062 (2012).
[5] I. L. Buchbinder and S. M. Kuzenko, *Ideas and methods of supersymmetry and supergravity, or A walk through superspace.* (Institute of Physics Pub., 1998).
[6] B. Z. Foster, and T. Jacobson. Physical Review D, **73**, 6, 064015, 1-9 (2006).

EPJ Web of Conferences **125**, 05019 (2016) DOI: 10.1051/epjconf/201612505019
QUARKS-2016

Massless Interacting Scalar Fields in de Sitter space

Diana López Nacir[1,⋆], *Francisco D.* Mazzitelli[2,⋆⋆], and *Leonardo G.* Trombetta[2,⋆⋆⋆]

[1] *Theoretical Physics Department, CERN, CH-1211 Genève 23, Switzerland.*
[2] *Centro Atómico Bariloche and Instituto Balseiro, CONICET, Comisión Nacional de Energía Atómica, R8402AGP Bariloche, Argentina.*

Abstract. We present a method to compute the two-point functions for an $O(N)$ scalar field model in de Sitter spacetime, avoiding the well known infrared problems for massless fields. The method is based on an exact treatment of the Euclidean zero modes and a perturbative one of the nonzero modes, and involves a partial resummation of the leading secular terms. This resummation, crucial to obtain a decay of the correlation functions, is implemented along with a double expansion in an effective coupling constant $\sqrt{\lambda}$ and in $1/N$. The results reduce to those known in the leading infrared approximation and coincide with the ones obtained directly in Lorentzian de Sitter spacetime in the large N limit. The new method allows for a systematic calculation of higher order corrections both in $\sqrt{\lambda}$ and in $1/N$.

1 Introduction

The analysis of interacting quantum fields in de Sitter (dS) spacetime is relevant to understand different aspects in cosmology, both in the early universe and in the present period of accelerated expansion. Indeed, quantum effects could induce large corrections during inflation, including non-Gaussianities in the cosmic microwave background. They could also produce nonnegligible contributions to the present cosmological constant, or could lead to instabilities of dS solutions to the semiclassical Einstein equations, in which the mean value of the stress tensor associated to the quantum fields is used as a source. The case of light or massless fields is of particular importance, because it is in this case that perturbative calculations fail due to the infrared (IR) growth of the correlation functions.

The case of a free scalar field already shows some peculiarities that occur in dS spacetime. For massive fields, it is possible to define a dS-invariant vacuum state (the so called Bunch-Davies vacuum). For this state, the Feynman propagator reads

$$G_F^{(m)}(x,x') = \frac{H^2\Gamma(\frac{3}{2}-\nu)\Gamma(\frac{3}{2}+\nu)}{(4\pi)^2}\,{}_2F_1\left(\frac{3}{2}-\nu,\frac{3}{2}+\nu;2;1-\frac{r}{4}-i\epsilon\right), \tag{1}$$

⋆e-mail: diana.laura.lopez.nacir@cern.ch
⋆⋆e-mail: fdmazzi@cab.cnea.gov.ar
⋆⋆⋆e-mail: lgtrombetta@cab.cnea.gov.ar

where $\nu = \sqrt{9/4 - m^2/H^2}$ and $r = [-(\eta - \eta')^2 + |\vec{x} - \vec{x}'|^2]/\eta\eta'$ (in comoving coordinates). In the limit $r \to +\infty$, i.e. at large spatial separations or late times, the massive free propagator decays as

$$G_F^{(m)}(r) \sim r^{-\frac{m^2}{3H^2}}. \quad (2)$$

Expanding the propagator for masses much smaller than the Hubble constant $m^2 \ll H^2$, the result behaves as $G_F^{(0)} \simeq 3H^4/(8\pi^2 m^2)$. There is a divergence for $m \to 0$ (which is not present in Minkowski spacetime), and after subtracting this divergent term, the corresponding massless Feynman propagator

$$\hat{G}_F^{(0)}(r) = G_F^{(0)}(r) - \frac{3H^4}{8\pi^2 m^2} \sim \log(r) \quad (3)$$

shows a logarithmic divergence as $r \to +\infty$. We see that the massless limit is not smooth: the behavior of the propagator is completely different in the IR. Moreover, for strictly massless fields there is no dS-invariant vacuum state [1].

When considering interacting fields (for instance a $\lambda\phi^4$ theory), the usual perturbative calculations involve the above mentioned propagators. It can be shown that a diagram with L loops will be proportional to $(\lambda H^2/m^2)^L$ [2]. Therefore, the usual perturbative calculations break down for light ($m^2 \ll \lambda H^2$) or massless quantum fields.

There have been several attempts to cure the IR problem, all of them introducing some sort of nonperturbative approach. For instance, in the so called stochastic approach [3] (which is based on a classical treatment of the long wavelength modes), nonperturbative results can be obtained at the leading IR order (i.e. at leading order in the coupling constant) by neglecting the interactions of the short wavelength modes. In the context of QFT in Lorentzian de Sitter spacetime, the nonperturbative approaches are based on the mean field (Hartree) approximation [4], the large N expansion [5], the analysis of the Schwinger-Dyson equation using a physical momentum decomposition [6] or the exact renormalization group equation [7]. A well known (nonperturbative) result is that interaction terms in the potential turn on a dynamical mass, and one can obtain finite and de Sitter invariant correlation functions even for minimally coupled massless fields. However, in none of these approaches, there is a clear systematic way of computing corrections beyond the leading IR approximation. The main inconvenients are of technical nature.

An alternative and simpler approach emerges for quantum fields in Euclidean de Sitter space. As this is a compact space, the modes of a quantum field are discrete, and the origin of the IR problems can be traced back to the zero mode [8]: the IR behavior improves if the zero mode is treated exactly while the nonzero modes (UV modes in what follows) are treated perturbatively. Using this idea, it has been shown that in the massless $\lambda\phi^4$ theory one obtains a dynamical squared mass for the field that in the leading IR limit is proportional to $\sqrt{\lambda}H^2$. Indeed, in the Euclidean space, this mass cures the IR problems. Moreover, a systematic perturbative procedure for calculating the n-point functions has been delineated in Ref. [9], where it was shown that, for massless fields, the effective coupling is $\sqrt{\lambda}$ instead of λ. It has been pointed out that this procedure together with an analytical continuation could be used to cure the IR problems also in the Lorentzian de Sitter spacetime, and in particular to obtain n-point functions that respect the de Sitter symmetries. However, so far explicit calculations have been restricted to obtaining corrections to the variance of the zero mode, which has no analog quantity in the Lorentzian de Sitter spacetime. The main concern here is to present an explicit calculation of the inhomogeneous two-point functions of the scalar fields which, after the analytical continuation, lead to two-point functions respecting the de Sitter symmetries.

In a recent work [10], we presented a generalization of the approach of Ref. [9] to the case of $O(N)$ scalar field theory, including a detailed calculation of the corrections to the two-point functions up to second order in the parameter $\sqrt{\lambda}$. We extended the nonperturbative treatment performing a resummation of the leading IR secular terms, that produces the proper decay of the two-point functions at large distances. When this resummation is combined with an expansion both in $\sqrt{\lambda}$ and $1/N$, it is possible to compute systematically the corrections coming from the interactions of both IR and UV sectors. In what follows we will summarize our findings in Ref. [10], emphazising the main concepts and avoiding technical details.

2 $O(N)$ model in Euclidean de Sitter space

We will consider an $O(N)$-symmetric model with action

$$S = \int d^d x \sqrt{g}\left[\frac{1}{2}\phi_a\left(-\Box + m^2\right)\phi_a + \frac{\lambda}{8N}(\phi_a\phi_a)^2\right], \tag{4}$$

where the fields ϕ_a, with $a = 1, .., N$, are the components of an element of the adjoint representation of the $O(N)$ group, and the sum over repeated indices is implied. Euclidean de Sitter space is obtained from Lorentzian de Sitter space in global coordinates by performing an analytical continuation $t \to -i(\tau - \pi/2H)$ and a compactification in imaginary time $\tau = \tau + 2\pi H^{-1}$. The resulting metric is that of a d-sphere of radius H^{-1}

$$ds^2 = H^{-2}\left[d\theta^2 + \sin(\theta)^2 d\Omega^2\right]. \tag{5}$$

where $\theta = H\tau$. Due to the symmetries and compactness of this space, the field can be expanded in d-dimensional spherical harmonics

$$\phi_a(x) = \sum_{\vec{L}} \phi_{\vec{L},a} Y_{\vec{L}}(x). \tag{6}$$

We generalize to the $O(N)$ model the method developed in [8] and [9] for a single field, for which the constant zero modes ϕ_{0a} play a crucial role. We split the fields as $\phi_a(x) = \phi_{0a} + \hat{\phi}_a(x)$ as well as the free propagator

$$G^{(m)}_{ab}(x, x') = G^{(m)}_{0,ab} + \hat{G}^{(m)}_{ab}(x, x'), \tag{7}$$

where now $\hat{G}^{(m)}_{ab}$ has the property of being finite in the IR ($m^2 \to 0$). The interaction part of the action takes the following form:

$$S_{int} = \frac{\lambda V_d}{8N}|\phi_0|^4 + \tilde{S}_{int}[\phi_{0a}, \hat{\phi}_a]. \tag{8}$$

Here $|\phi_0|^2 = \phi_{0a}\phi_{0a}$ and V_d is the total volume of Euclidean de Sitter space in d-dimensions, which is finite thanks to the compactification.

In order to compute the quantum correlation functions of the theory we define the generating functional in the presence of sources J_{0a} and $\hat{J}_a$,

$$Z[J_0, \hat{J}] = \mathcal{N}\int d^N\phi_0 \int \mathcal{D}\hat{\phi}\, e^{-S - \int_x (J_{0a}\phi_{0a} + \hat{J}_a\hat{\phi}_a)} = exp\left(-\tilde{S}_{int}\left[\frac{\delta}{\delta J_0}, \frac{\delta}{\delta \hat{J}}\right]\right) Z_0[J_0]\hat{Z}_f[\hat{J}], \tag{9}$$

where we introduced the shorthand notation $\int_x = \int d^d x \sqrt{g}$. Here $Z_0[J_0]$ is defined as the exact generating functional of the theory with the zero modes alone, and it gives the leading IR contribution. Note that, as the zero modes are constant on the sphere, their kinetic terms vanish, and $Z_0[J_0]$ involves only ordinary integrals, which allows it to be exactly computed in several interesting cases following Ref. [8]. The corrections beyond the leading IR approximation come from $\tilde{S}_{int}$ which encondes the interaction among the zero and UV modes.

The effective potential gives valuable information about how the quantum fluctuations around a background field $\bar{\phi}$ influence its behavior. We are interested in particular in the generation of a dynamical mass due to quantum effects. Up to quadratic order it can be shown that

$$V_{eff}(\bar{\phi}_0) = V_0 + \frac{1}{2}\frac{N}{V_d\langle\phi_0^2\rangle}|\bar{\phi}_0|^2 + O(|\bar{\phi}_0|^4) \equiv V_0 + \frac{1}{2}m_{dyn}^2|\bar{\phi}_0|^2 + O(|\bar{\phi}_0|^4). \tag{10}$$

This is an exact property of the Euclidean theory valid for all N and λ, which shows that the dynamical mass m_{dyn}, defined by the above equation, is related to the inverse of the variance of the zero modes.

A case of great interest is when the fields are massless at tree level, $m = 0$, as it is in this case in which the perturbative treatment becomes ill-defined. The nonperturbative treatment of the zero modes ensures that these modes acquire a dynamical mass, avoiding the IR divergence associated to the free two-point functions in the massless limit. Indeed, the n-point functions of the zero modes can be exactly computed from $Z_0[J_0]$ to be

$$\langle\phi_0^{2p}\rangle_0 = \frac{\int_0^\infty d\phi_0\,\phi_0^{N-1+2p}e^{-\frac{V_d\lambda}{8N}\phi_0^4}}{\int_0^\infty d\phi_0\,\phi_0^{N-1}\,e^{-\frac{V_d\lambda}{8N}\phi_0^4}} = 2^{\frac{3p}{2}}\left(\frac{N}{V_d\lambda}\right)^{\frac{p}{2}}\frac{\Gamma\left[\frac{N+2p}{4}\right]}{\Gamma\left[\frac{N}{4}\right]}, \tag{11}$$

which exhibit no IR divergences. This equation shows a scaling of the form $\phi_0 \sim \lambda^{-1/4}$, making the perturbative expansion of the UV modes to be in powers of $\sqrt{\lambda}$.

At the leading IR order the interaction between the zero and UV modes in Eq. (9) can be neglected,

$$m_{dyn,0}^2 = \sqrt{\frac{N\lambda}{2V_d}}\frac{1}{2}\frac{\Gamma\left[\frac{N}{4}\right]}{\Gamma\left[\frac{N+2}{4}\right]}. \tag{12}$$

For $N = 1$, we recover the result of [8], which is also the one from the stochastic approach [3].

3 Corrections from the UV modes to the two-point functions

We will now show the results of computing the two-point functions of the full scalar fields including up to the second perturbative correction coming from the UV modes. Corrections to the leading order result come from expanding the exponential with $\tilde{S}_{int}$ in Eq. (9).

We split the two-point functions of the total fields ϕ_a into UV and IR parts

$$\langle\phi_a(x)\phi_b(x')\rangle = \langle\phi_{0a}\phi_{0b}\rangle + \langle\hat{\phi}_a(x)\hat{\phi}_b(x')\rangle = \frac{1}{Z[0,0]}\left[\left.\frac{\delta^2 Z[J_0,\hat{J}]}{\delta J_{0a}\delta J_{0b}}\right|_{J_0,\hat{J}=0} + \left.\frac{\delta^2 Z[J_0,\hat{J}]}{\delta\hat{J}_a(x)\delta\hat{J}_b(x')}\right|_{J_0,\hat{J}=0}\right], \tag{13}$$

where the cross-terms vanish by orthogonality. In what follows we compute each part separately.

The expression for the UV part of the propagator can be simplified considerably using that the integrals of free UV propagators in Euclidean space can be expressed in terms of derivatives of a single propagator with respect to its mass:

$$\int\dots\int_{x_2,\dots,x_{k-1}}\hat{G}^{(m)}(x_1,x_2)\dots\hat{G}^{(m)}(x_{k-1},x_k) = \frac{(-1)^k}{(k-2)!}\frac{\partial^{k-2}\hat{G}^{(m)}(x_1,x_k)}{\partial(m^2)^{k-2}}. \tag{14}$$

Using this, one can show that the UV part reads

$$\langle\hat{\phi}_a(x)\hat{\phi}_b(x')\rangle = \delta_{ab} \quad \Bigg\{ \quad \hat{G}^{(0)}(x,x') + \left[\frac{\lambda(N+2)}{2N^2}\langle\phi_0^2\rangle_0 + \frac{\lambda}{2N}(N+2)[\hat{G}^{(0)}]_{ren} - \frac{\lambda^2}{8N^3}(N+2)^2 V_d[\hat{G}^{(m)}]_{ren}\left(\langle\phi_0^4\rangle_0 - \langle\phi_0^2\rangle_0^2\right)\right]\left.\frac{\partial\hat{G}^{(m)}(x,x')}{\partial m^2}\right|_0 + \frac{\lambda^2}{8N^3}(N+8)\langle\phi_0^4\rangle_0 \left.\frac{\partial^2\hat{G}^{(m)}(x,x')}{\partial(m^2)^2}\right|_0\Bigg\}, \tag{15}$$

while the two-point functions for the zero modes (IR part) are

$$\begin{aligned}\langle\phi_{0a}\phi_{0b}\rangle &= \delta_{ab}\Bigg\{\frac{\langle\phi_0^2\rangle_0}{N} + \frac{\lambda}{4N^2}(N+2)\left[\langle\phi_0^2\rangle_0^2 - \langle\phi_0^4\rangle_0\right]V_d[\hat{G}^{(m)}]_{ren} \\ &\quad + \frac{\lambda^2}{32N^3}(N+2)^2\left[\langle\phi_0^6\rangle_0 - 3\langle\phi_0^2\rangle_0\langle\phi_0^4\rangle_0 + 2\langle\phi_0^2\rangle_0^3\right]V_d^2[\hat{G}^{(m)}]_{ren}^2 \\ &\quad - \frac{\lambda^2}{16N^3}(N+8)\left[\langle\phi_0^6\rangle_0 - \langle\phi_0^2\rangle_0\langle\phi_0^4\rangle_0\right]V_d\left(\frac{\partial[\hat{G}^{(m)}]}{\partial m^2}\right)_{fin}\Bigg\} \\ &= \frac{\delta_{ab}}{V_d m_{dyn}^2(IR)},\end{aligned} \tag{16}$$

where [...] indicates the coincidence limit has been taken and the *ren* and *fin* subscripts denote that the quantities have been rendered finite by renormalization. The last equality follows after interpreting the corrections as a modification to the mass $m_{dyn}^2(IR)$ of the zero modes which, as mentioned before, determines the curvature of the effective potential. Eqs. (15) and (16) contain the main corrections to the renormalized UV and IR propagators, for any values of d and N. These are the main results of this section.

In the limit $N \to \infty$ these results are compatible with the two-point functions of the full fields, Eq. (13), being equal to massive free de Sitter propagators with a dynamical mass

$$m_{dyn}^2 = \sqrt{\frac{\lambda}{2V_d}} + \frac{\lambda}{4}[\hat{G}^{(0)}]_{ren} + O(\lambda^{3/2}). \tag{17}$$

Beyond the LO contribution in $1/N$, the two-point functions have a more complicated structure than that of a free field.

4 Resumming the leading IR secular terms to the two-point functions

It is worth to note that the free UV propagators that build up the expressions of the two-point functions of the UV modes, Eq. (15), are massless. After performing the analytical continuation to the Lorentzian spacetime, this leads to an IR enhanced behavior at large distances. Therefore, it is necessary to extend the nonperturbative treatment to resum the leading IR secular terms. In order to achieve this, we need to perform a resummation of diagrams that give mass to the UV propagators present in Eq. (15). It can be shown that it is enough to take into account only a subclass of diagrams: those coming from the interaction term that is quadratic in both ϕ_0 and $\hat{\phi}$,

$$S_{int}^{(2)}[\phi_0,\hat{\phi}] = \frac{\lambda}{4N}\int d^dx\sqrt{g}\,(\delta_{ab}\delta_{cd} + \delta_{ac}\delta_{bd} + \delta_{ad}\delta_{bc})\,\phi_{0a}\phi_{0b}\hat{\phi}_c\hat{\phi}_d. \tag{18}$$

The remaining terms in $\tilde{S}_{int}$ are still treated perturbatively.

We start by rewritting the generating functional Eq. (9) by grouping this term with the other terms quadratic in $\hat{\phi}$, as part of the "free" generating functional of the UV modes,

$$\begin{aligned} Z[J_0, \hat{J}] &= \mathcal{N} e^{-\tilde{\tilde{S}}_{int}\left[\frac{\delta}{\delta J_0}, \frac{\delta}{\delta \hat{J}}\right]} \int d^N\phi_0\, e^{-\left[\frac{\lambda V_d}{8N}|\phi_0|^4 + V_d J_{0a}\phi_{0a}\right]} \\ &\times \int \mathcal{D}\hat{\phi}\, \exp\left(-\frac{1}{2}\iint_{x,y} \hat{\phi}_a \hat{G}^{-1}_{ab}(\phi_0)\hat{\phi}_b + \int_x \hat{J}_a\hat{\phi}_a\right), \end{aligned} \tag{19}$$

where now the "free" UV propagator $\hat{G}_{ab}(\phi_0)$ has a ϕ_0-dependent mass,

$$\hat{G}^{-1}_{ab}(\phi_0)(x, x') = \left[-\Box\delta_{ab} + m^2_{ab}(\phi_0)\right]\frac{\delta^{(d)}(x - x')}{\sqrt{g}}, \tag{20}$$

and where $\tilde{\tilde{S}}_{int} = \tilde{S}_{int} - S^{(2)}_{int}$ has the remaining interaction terms that are treated perturbatively.

In order to compare with the results of the previous section, it is enough to keep terms up to order λ. Therefore, it is necessary to include perturbatively only the first correction coming from the term

$$S^{(4)}_{int}[\hat{\phi}] = \frac{\lambda}{8N}\int_x |\hat{\phi}|^4, \tag{21}$$

and so the generating functional of Eq. (19) reduces to

$$Z^{(1)}[J_0, \hat{J}] = \mathcal{N}\left[1 - \frac{\lambda}{8N}\delta_{cd}\delta_{ef}\int_x \frac{\delta^4}{\delta\hat{J}^4_{cdef}(x)}\right]\left(\hat{Z}_f\left[\hat{J}, m^2\right]\left[\det \hat{G}^{(m)}_{rs}\right]^{1/2}\right)_{m\left(\frac{\delta}{\delta J_0}\right)} Z_0[J_0]. \tag{22}$$

Now we proceed to calculate the connected two-point functions of the UV modes,

$$\langle\hat{\phi}_a(x)\hat{\phi}_b(x')\rangle^{(1)} = \frac{1}{Z^{(1)}[0,0]}\frac{\delta^2 Z^{(1)}[J_0, \hat{J}]}{\delta\hat{J}_a(x)\delta\hat{J}_b(x')}\bigg|_{J_0, \hat{J}=0} = \langle\hat{\phi}_a(x)\hat{\phi}_b(x')\rangle^{(0)} + \Delta\langle\hat{\phi}_a(x)\hat{\phi}_b(x')\rangle, \tag{23}$$

which we split in two contributions, according to the interaction term that we are treating perturbatively. The calculation is technically involved, and we refer the reader to Ref. [10] for details. After a careful diagramatic analysis and proper renormalization we arrive at

$$\begin{aligned} \langle\hat{\phi}_a(x)\hat{\phi}_b(x')\rangle^{(0)} &= \delta_{ab}\Bigg\{\sum_{p=0}^{\infty}\frac{1}{p!}\frac{\partial^p\hat{G}^{(m)}(x,x')}{\partial(m^2)^p}\bigg|_0\left(\frac{\lambda}{2N}\right)^p\left[1 + \frac{(3^p - 1)}{N}\right]\langle\phi_0^{2p}\rangle_0 \\ &- \frac{\lambda^2(N+2)}{4N}V_d[\hat{G}^{(0)}]_{ren}\times \\ &\sum_{p=0}^{\infty}\frac{1}{p!}\frac{\partial^p\hat{G}^{(m)}(x,x')}{\partial(m^2)^p}\bigg|_0\left(\frac{\lambda}{2N}\right)^p\left[1 + \frac{(3^p - 1)}{N}\right]\left[\langle\phi_0^{2(p+1)}\rangle_0 - \langle\phi_0^2\rangle_0\langle\phi_0^{2p}\rangle_0\right]\Bigg\}, \end{aligned} \tag{24}$$

and

$$\begin{aligned} \Delta\langle\hat{\phi}_a(x)\hat{\phi}_b(x')\rangle &= \delta_{ab}\frac{\lambda(N+2)}{2N}[\hat{G}^{(0)}]_{ren} \\ &\times\sum_{p=0}^{\infty}\frac{1}{p!}\frac{\partial^{p+1}\hat{G}^{(m)}(x,x')}{\partial(m^2)^{p+1}}\bigg|_0\left(\frac{\lambda}{2N}\right)^p\left[1 + \frac{(3^p - 1)}{N}\right]\langle\phi_0^{2p}\rangle. \end{aligned} \tag{25}$$

All of the series above can be resummed order by order in $1/N$, for which we need to expand the summands in $1/N$. Doing this, the full result for the connected two-point functions of the UV modes up to order λ and N^{-1}, with a partial resummation of the infinite subset of diagrams, is:

$$\langle\hat{\phi}_a(x)\hat{\phi}_b(x')\rangle^{(1)} = \delta_{ab} \Bigg\{ \hat{G}^{(m)}(x,x') + \frac{\lambda}{4}[\hat{G}^{(0)}]_{ren}\frac{\partial\hat{G}^{(m)}(x,x')}{\partial m^2} \\ +\frac{1}{2N}\Bigg[2\hat{G}^{(\sqrt{3}m)}(x,x') - 2\hat{G}^{(m)}(x,x') \\ -\sqrt{\frac{\lambda}{2V_d}}\frac{\partial\hat{G}^{(m)}(x,x')}{\partial m^2} + \frac{\lambda}{2V_d}\frac{\partial^2\hat{G}^{(m)}(x,x')}{\partial(m^2)^2} \\ +\frac{\lambda}{4}[\hat{G}^{(0)}]_{ren}\left(7\frac{\partial\hat{G}^{(m)}(x,x')}{\partial m^2} - 6\frac{\partial\hat{G}^{(\sqrt{3}m)}(x,x')}{\partial m^2}\right)\Bigg]\Bigg\}_{m_{dyn,0}}, \tag{26}$$

where $m^2_{dyn,0} = \sqrt{\frac{\lambda}{2V_d}}$. Here, all the UV propagators at separated points now have a mass squared of order $\sqrt{\lambda}$. The only instance of a massless UV propagator has its coincidence limit taken, and it is therefore just a finite constant factor with no IR issues. These are our main results. It was verified as a cross-check that this expression reduces to the perturbative one of Eq. (15), upon expanding the latter at NLO in $1/N$. A noteworthy observation is the presence of some propagators whose squared mass is three times that of the others, something which could not have been anticipated from the perturbative result. The large distance behaviour of the two-point functions ultimately depends on the masses of the free propagators that build up the expression, $m^2_{dyn,0}$ and $3m^2_{dyn,0}$, which determine how fast it decays.

One can see that the Lorentzian 2-point functions corresponding to our results Eqs. (16) and (26) coincide with the ones of Ref. [6] when expanded up to the corresponding order, the latter given by

$$\langle\phi_a(x)\phi_b(x')\rangle = \delta_{ab}\left[\left(1-\frac{5}{16N}\right)G^{(m_+)}(x,x') + \frac{5}{16N}G^{(m_-)}(x,x')\right], \tag{27}$$

with masses $m_+^2 = m^2_{dyn,0}\left(1+\frac{1}{4N}\right)$ and $m_-^2 = 5m_+^2$. These are valid up to the NLO in the large N expansion as well, but only at the leading IR order.

5 Conclusions

In this work we considered an interacting $O(N)$ scalar field model in d-dimensional Euclidean de Sitter space, paying particular attention to the IR problems that appear for massless and light fields. We presented an extension of the approach of Refs. [8, 9] to the $O(N)$ model. The zero modes are treated exactly while the corrections due to the interactions with the UV modes are computed perturbatively. The calculation of the two-point functions of the field shows that the exact treatment of the zero modes cures the IR divergences of the usual massless propagator: the two-point functions becomes de Sitter invariant.

Although the massless UV propagator is de Sitter invariant, its Lorentzian counterpart exhibits a growing behavior at large distances, invalidating the perturbative expansion in this limit. This problem can be fixed in the leading order large N limit by resumming the higher order corrections: one can show that the final result corresponds to two-point functions of free fields with a self-consistent mass. However, the NLO contains derivatives of the free propagator of the UV modes. The behavior of the correlation functions in the IR limit can be improved by performing a resummation of a class of diagrams that give mass to the UV propagator. Higher order corrections can be systematically computed in a perturbative expansion in powers of both $\sqrt{\lambda}$ and $1/N$. We presented explicit results up to second order in $\sqrt{\lambda}$ and NLO in $1/N$.

Our results reduce to the ones obtained by other nonperturbative approaches at leading IR order in $\sqrt{\lambda}$, and coincides with that of the Hartree approximation in the large N limit up to the second order in $\sqrt{\lambda}$. Beyond the leading IR order, a consistent treatment of the UV sector becomes necessary. The use of the Euclidean path integral (which is simpler than its *in-in* counterpart) together with the double perturbative expansion (in $\sqrt{\lambda}$ and $1/N$) performed in our calculations, allowed us to further include the contribution of the UV modes. Moreover, in this framework, the precision of the calculation can be systematically improved by computing higher order corrections.

We are presently working on generalizing these methods to the case of negative square mass to study the restoration of symmetry due to the strong infrared effects in de Sitter space.

Acknowledgements

The work of FDM and LGT has been supported by CONICET and ANPCyT. LGT was also partially supported by the ICTP/IAEA Sandwich Training Educational Programme and UBA. DNL was supported by ICTP during the initial stage of this work. FDM would like to thank ICTP for hospitality during part of this work. DLN would like to thank the organizers of QUARKS 2016.

References

[1] B. Allen, Phys. Rev. D **32** 3136 (1985); E. T. Akhmedov, Int. J. Mod. Phys. D **23**, 1430001 (2014) [arXiv:1309.2557 [hep-th]].

[2] C. P. Burgess, R. Holman, L. Leblond and S. Shandera, JCAP **1010**, 017 (2010) [arXiv:1005.3551 [hep-th]].

[3] A. A. Starobinsky, Lect. Notes in Physics **246**, 107 (1986); A. A. Starobinsky and J. Yokoyama, Phys. Rev. D **50**, 6357 (1994) [astro-ph/9407016].

[4] T. Arai, Phys. Rev. D **86**, 104064 (2012) [arXiv:1204.0476 [hep-th]]; D. L. López Nacir, F. D. Mazzitelli, and L. G. Trombetta, Phys. Rev. D **89**, 024006 (2014) [arXiv:1309.0864 [hep-th]]; ibidem, Phys. Rev. D **89**, 084013 (2014) [arXiv:1401.6094 [hep-th]].

[5] F. D. Mazzitelli and J. P. Paz, Phys. Rev. D **39**, 2234 (1989); A. Riotto and M. S. Sloth, JCAP **0804**, 030 (2008) [arXiv:0801.1845 [hep-ph]]; J. Serreau, Phys. Rev. Lett. **107**, 191103 (2011) [arXiv:1105.4539 [hep-th]]; T. Arai, Phys. Rev. D **88**, 064029 (2013) [arXiv:1304.5631 [hep-th]]; J. Serreau and R. Parentani, Phys. Rev. D **87**, 085012 (2013) [arXiv:1302.3262 [hep-th]].

[6] F. Gautier and J. Serreau, Phys. Lett. B **727**, 541 (2013) [arXiv:1305.5705 [hep-th]]; ibidem, Phys. Rev. D **92**, 105035 (2015) [arXiv:1509.05546 [hep-th]].

[7] J. Serreau, Phys. Lett. B **730**, 271 (2013) [arXiv:1306.3846 [hep-th]]; M. Guilleux and J. Serreau, Phys. Rev. D **92**, 084010 (2015) [arXiv:1506.06183 [hep-th]].

[8] A. Rajaraman, Phys. Rev. D **82**, 123522 (2010) [arXiv:1008.1271].

[9] M. Beneke and P. Moch, Phys. Rev. D **87**, 064018 (2012) [arXiv:1212.3058].

[10] D. L. López Nacir, F. D. Mazzitelli, and L. G. Trombetta, submitted to JHEP [arXiv: 1606.03481]

EPJ Web of Conferences **125**, 05020 (2016) DOI: 10.1051/epjconf/201612505020
QUARKS-2016

Multiloop calculations with differential equations in ϵ-form.

Roman N. Lee*
630090, Budker Institute of Nuclear Physics, Novosibirsk, Russia

Abstract. We review the method of reduction of differential system for multiloop integrals to ϵ-form.

1 Introduction

New methods of the multiloop perturbative calculations are extremely important in the context of searches for New Physics. On of the most elaborated approaches to the multiloop calculations is the differential equations technique, which was first introduced in Refs. [1–5]. Recent observation made by Henn in Ref. [6] makes this approach yet more powerful. According to this observation, in many cases the differential system for the master integrals can be reduced to the form where the dependence on ϵ of the right-hand side is restricted to a single factor ϵ. This form (we call it *ϵ-form* for brevity) allows one to express the coefficients of ϵ expansion series in terms of Goncharov's polylogarithms. Moreover, the coefficients possess a remarkable property of homogeneous transcendental weight. The advantages of the ϵ-form are even more impressive for the case of several variables. This form gives one a possibility to rewrite differential systems with respect to all variable in one unified 'dlog' form. Since Henn's paper, the differential equations in ϵ-form have been used in many papers, pioneered by Refs.[7, 8].

Finding an appropriate change of functions is not easy. Some partial recipes have been formulated in Refs. [6, 9, 10]. In particular, Ref. [9] assumes linear in ϵ form (but with nonzero ϵ^0 terms) of the right-hand side of the differential system. We stress that, within the approach based on the ϵ-form, the most of the problem's complexity is moved towards achieving this form. Therefore, a general agorithm of finding an appropriate change of functions is very crucial. In Ref. [11] an algorithm of finding the ϵ-form based on rational transformations of the differential system has been presented. In this contribution we briefly review this algorithm.

2 Reduction to ϵ-form

Suppose that we have a differential system depending on the parameter ϵ

$$\partial_x \mathbf{J}(x) = \mathbb{M}(\epsilon, x)\, \mathbf{J}(x), \tag{1}$$

where the elements of the matrix $\mathbb{M}(\epsilon, x)$ are rational functions of x and ϵ. The systems appearing in the multiloop calculations appear to have only regular singular points, i.e., the the general solution is bounded by $|x - x_0|^\alpha$ in the vicinity of any singularity x_0.

*e-mail: r.n.lee@inp.nsk.su

The algorithm of Ref. [11] is aimed to find the rational transformation (when it exists)

$$\mathbf{J}(x) \to \mathbb{T}(\epsilon, x)\widetilde{\mathbf{J}}(x)$$

which reduces the system (1) to ϵ-form

$$\partial_x \widetilde{\mathbf{J}}(x) = \epsilon \sum_k \frac{\mathbb{M}_k}{x - x_k} \widetilde{\mathbf{J}}(x), \tag{2}$$

The algorithm includes three major steps. The first step is to reduce the differential system to the Fuchsian form, i.e., to a form when the elements of $\mathbb{M}$ have only simple poles with respect to x. The second step is to normalize the eigenvalues of the matrix residues, i.e., to secure that real parts of all the eigenvalues belong to the interval $[-1/2, 1/2)$. For the systems reducible to ϵ-form this means that all eigenvalues are made proportional to ϵ. Finally, a constant transformation is searched for in order to factor out ϵ, i.e., to reduce the system to ϵ-form.

2.1 Reducing to Fuchsian form and normalizating eigenvalues.

Two first steps of the reduction are not specific for the systems depending on parameter and are very natural for generic differential systems. When it concerns the *local* reduction, the corresponding algorithms are known for a long time. In particular, the possibility to reduce the system to Fuchsian form in the vicinity of a given regular singular point was considered in Ref. [12]. The specific algorithm is given by Barkatou&Pflügel [13, 14]. In fact, this algorithm allows one the reduce the system to Fuchsian form in all *finite* regular singular points. However, the behavior of the system at infinity is not controlled and may become essentially worse[1].

The Barkatou&Pfluegel algorithm is based on the transformation generated by the matrix

$$\mathbb{T}(x) = \overline{\mathbb{P}} + \frac{\mathbb{P}}{x - x_0}, \tag{3}$$

where $\mathbb{P}$ is some properly chosen projector. This transformation is singular in one finite point, $x = x_0$, and at infinity. One can also generalize this transformation so that it has singularities in any two given points:

$$\mathbb{T}(x) = \overline{\mathbb{P}} + \frac{x - x_1}{x - x_0}\mathbb{P}. \tag{4}$$

In what follows we will call this transformation *a balance* between x_0 and x_1.

Suppose that the expansion of $\mathbb{M}$ at $x = x_0$ has the form

$$\mathbb{M}(x) = \frac{\mathbb{A}_0}{(x - x_0)^{p+1}} + \frac{\mathbb{A}_1}{(x - x_0)^p} + \ldots \tag{5}$$

with $p > 0$. Then, the decisive step of Barkatou&Pfluegel algorithm can be re-worded as the construction of the projector $\mathbb{P}$ with some prescribed properties. The advance made in Ref. [11] is based on the observation that this step, as well as the algorithm for the normalization of the eigenvalues, imposes requirements only on the image of the projector $\mathbb{P}$, but not on its kernel. Then the idea of the reduction is to apply some constraint also on $\ker \mathbb{P}$ in such a way that the transformation (4) does not increase the pole order at $x = x_1$. If the expansion at $x = x_1$ has the form $\mathbb{M}(x) = \frac{\mathbb{B}_0}{(x-x_1)^{q+1}} + \ldots$, then the constraint is simply that $\ker \mathbb{P}$ has to be an invariant space of $\mathbb{B}_0$.

[1]We remark that it is not always possible to reduce the system to global Fuchsian form even though all its singular points may be regular. This is known due to the Bolibrukh's negative solution [15] of the 21st Hilbert problem. Still, the algorithm of the reduction for the systems where global Fuchsian form is possible is very desirable.

2.2 Constant transformation

Let us now assume that it appeared to be possible to secure that the system is Fuchsian and normalized at all singular points. I.e., we have

$$\partial_x \mathbf{J} = \mathbb{M}(\epsilon, x)\mathbf{J} = \sum_k \frac{\mathbb{M}_k(\epsilon)}{x - x_k}\mathbf{J}, \tag{6}$$

and the eigenvalues of all matrices $\mathbb{M}_k$ are proportional to ϵ. We need to find an x-independent transformation matrix which simultaneously transforms all matrices $\mathbb{M}_k(\epsilon)$ to the form $\epsilon\mathbb{S}_k$, where $\mathbb{S}_k$ are constant matrices. Let $\mathbb{T}(\epsilon)$ be such a matrix. Then we write

$$\mathbb{T}^{-1}(\epsilon)\frac{\mathbb{M}_k(\epsilon)}{\epsilon}\mathbb{T}(\epsilon) = \mathbb{S}_k = \mathbb{T}^{-1}(\mu)\frac{\mathbb{M}_k(\mu)}{\mu}\mathbb{T}(\mu). \tag{7}$$

Multiplying this equation by $\mathbb{T}(\epsilon)$ from the left and by $\mathbb{T}^{-1}(\mu)$ from the right, we obtain a linear system

$$\begin{aligned}
\frac{\mathbb{M}_1(\epsilon)}{\epsilon}\mathbb{T}(\epsilon,\mu) &= \mathbb{T}(\epsilon,\mu)\frac{\mathbb{M}_1(\mu)}{\mu},\\
&\vdots\\
\frac{\mathbb{M}_m(\epsilon)}{\epsilon}\mathbb{T}(\epsilon,\mu) &= \mathbb{T}(\epsilon,\mu)\frac{\mathbb{M}_m(\mu)}{\mu}
\end{aligned} \tag{8}$$

for the elements of the matrix $\mathbb{T}(\epsilon,\mu) = \mathbb{T}(\epsilon)\mathbb{T}^{-1}(\mu)$. If the general solution of this system (found routinely) determines an invertible matrix, the transformation we are looking for can be chosen as $\mathbb{T}(\epsilon) = \mathbb{T}(\epsilon,\mu_0)$, where μ_0 is some arbitrarily chosen number, provided $\mathbb{T}(\epsilon,\mu)$ is nonsingular at $\mu = \mu_0$.

One might alternatively use the approach based on 'sampling'. Since we are looking for constant transformation $\mathbb{T}(\epsilon)$ which satisfies

$$\mathbb{T}^{-1}(\epsilon)\mathbb{M}(\epsilon, x)\mathbb{T}(\epsilon) = \epsilon\mathbb{M}(x), \tag{9}$$

where $\mathbb{M}(x)$ is the matrix independent of ϵ, we might generate equations for sufficiently many sampling points $a_1, \dots a_m$:

$$\begin{aligned}
\frac{\mathbb{M}(\epsilon, a_1)}{\epsilon}\mathbb{T}(\epsilon,\mu) &= \mathbb{T}(\epsilon,\mu)\frac{\mathbb{M}(\mu, a_1)}{\mu},\\
&\vdots\\
\frac{\mathbb{M}(\epsilon, a_m)}{\epsilon}\mathbb{T}(\epsilon,\mu) &= \mathbb{T}(\epsilon,\mu)\frac{\mathbb{M}(\mu, a_m)}{\mu}
\end{aligned} \tag{10}$$

and use them instead of equations (8). This approach has an advantage due to the possibility to use it in the to the case of differential systems over several variables.

2.3 What may go wrong?

- In some, very special, situations it might be not possible to construct the projector with a given kernel due to the condition which should satisfy kernel and image of the projection operator: $\ker\mathbb{P}\cap \operatorname{Im}\mathbb{P} = \{0\}$. One might try to achieve global Fuchsian form by introducing the apparent singularities, but then the second step is likely to be not possible due same kind of obstruction. Such obstructions naturally correspond to the systems with regular singularities which can not be reduced to global Fuchsian form (corresponding to the negative solution of 21st Hilbert problem).

- The normalization might not result in the proportionality of the eigenvalues to ϵ. This situation might be detected before making the second step. Namely, one has to check that all eigenvalues of the matrix residues have the form $n + k\epsilon$, where n is integer. If this is not the case, the ϵ-form might only be achievable after the variable change. Then, in terms of the original variable, the resulting transformation and system typically involve algebraic functions (square roots).
- Third step may result in degenerate matrix $\mathbb{T}(\epsilon, \mu)$ for all μ (typically $\mathbb{T}(\epsilon, \mu) = 0$). If this is so, the ϵ-form of the system is not possible. However, to the best of our knowledge, there is no examples for which this step is not possible to accomplish for any irreducible diagonal block of the matrix $\mathbb{M}$. As to the reduction of the off-diagonal elements, such an obstacle is possible but not very crucial, see next Section.

3 Using block-triangular form

The size n of the matrices $\mathbb{M}(\epsilon, x)$ appearing in the differential equations for master integrals may be quite large ($\sim$ several tens). This may constitute computational complications for the transformations that we need. Fortunately, the very process of the derivation of the differential equations, the IBP reduction, shows that the matrix $\mathbb{M}(\epsilon, x)$ has a block-triangular form with diagonal blocks corresponding to the master integrals of a given sector.

The idea of using this block-triangular form is to apply first the algorithm of the above section to each diagonal block. Suppose that this application results in ϵ-form of all diagonal blocks. Then one may certainly achieve the global reduction to Fuchsian form and normalization of the eigenvalues for the whole matrix, see Section 7 of Ref. [11] for details.

The third step might be not possible to accomplish for the whole matrix. But even if this is the case, one still has almost all benefits of ϵ-form. Indeed, for the column-vector $\mathbf{J}_1$ of integrals of a certain sector, one then has the system

$$\partial_x \mathbf{J}_1 = \epsilon \mathbb{A}(x) \mathbf{J}_1 + \mathbf{g}(x, \epsilon) , \tag{11}$$

where the inhomogeneous term $\mathbf{g}(x, \epsilon)$ is a linear combination of simpler master integrals, the coefficients being the rational functions of x and ϵ. We assume that these simpler integrals have been already calculated at this point in a sense that we know their expansion in ϵ at fixed x up to a sufficiently high order as well as their asymptotic behavior at fixed ϵ and $x \to x_k$. Then we also know the expansion and the asymptotic behavior of $\mathbf{g}(x, \epsilon)$, irrespectively on whether the coefficients of the linear combination are proportional to ϵ. This is the only information that we need to proceed with construction of the solution for $\mathbf{J}_1$.

4 Conclusion

We have reviewed the algorithm of the reduction of differential system to ϵ-form formulated in Ref.[11]. There is already one published application of this algorithm, see Ref. [16], with several more applications coming in the nearest future.

The algorithm formally applies to the differential systems with respect to one variable. However, it appears that it is also extremely helpful for the multivariate case, although it requires more heuristic work.

Acknoledgments

I am grateful to A.A. Pomeransky for valuable discussions. This work is supported by RFBR grant No. 15-02-07893.

References

[1] A.V. Kotikov, Phys. Lett. **B254**, 158 (1991)
[2] A.V. Kotikov, Phys. Lett. **B259**, 314 (1991)
[3] A.V. Kotikov, Phys. Lett. **B267**, 123 (1991)
[4] E. Remiddi, Nuovo Cim. **A110**, 1435 (1997), hep-th/9711188
[5] T. Gehrmann, E. Remiddi, Nucl. Phys. B **580**, 485 (2000), hep-ph/9912329
[6] J.M. Henn, Phys.Rev.Lett. **110**, 251601 (2013), 1304.1806
[7] J.M. Henn, V.A. Smirnov, JHEP **1311**, 041 (2013), 1307.4083
[8] J.M. Henn, A.V. Smirnov, V.A. Smirnov, JHEP **1307**, 128 (2013), 1306.2799
[9] M. Argeri, S. Di Vita, P. Mastrolia, E. Mirabella, J. Schlenk et al., JHEP **1403**, 082 (2014), 1401.2979
[10] T. Gehrmann, A. von Manteuffel, L. Tancredi, E. Weihs, JHEP **1406**, 032 (2014), 1404.4853
[11] R.N. Lee, J. High Energy Phys. **1504**, 108 (2015), 1411.0911
[12] J. Moser, Mathematische Zeitschrift **72**, 379 (1960)
[13] M.A. Barkatou, E. Pflügel, *Computing super-irreducible forms of systems of linear differential equations via Moser-reduction: a new approach*, in *Proceedings of the 2007 international symposium on Symbolic and algebraic computation* (ACM, 2007), pp. 1–8
[14] M.A. Barkatou, E. Pflügel, Journal of Symbolic Computation **44**, 1017 (2009)
[15] A.A. Bolibrukh, Proc. Steklov Inst. Math. **206**, 145 (1995)
[16] R.N. Lee, K.T. Mingulov, Phys. Lett. B **757**, 207 (2016), 1602.02463

EPJ Web of Conferences **125**, 05021 (2016) DOI: 10.1051/epjconf/201612505021
QUARKS-2016

Critical behavior of $U(n)$-χ^4-model with antisymmetric tensor order parameter coupled with magnetic field

N.V. Antonov[1,⋆], *M.V.* Kompaniets[1,⋆⋆], and *N.M.* Lebedev[1,⋆⋆⋆]

[1] *St. Petersburg State University, Uljanovskaja 1, St. Petersburg, Petrodvorez, 198504 Russia*

Abstract. The critical behavior of $U(n)$-χ^4-model with antisymmetric tensor order parameter at charged regime is studied by means of the field theoretic renormalization group (RG) at the leading order of ε-expansion (one-loop approximation). It is shown that RG equations have no infrared (IR) attractive charged fixed points. It is also shown that anomalous dimension of the order parameter in charged regime appears to be gauge dependent.

1 Introduction

Investigation of critical behavior of Fermi particle systems is permanently in focus of attention due to the problem of superconductive phase transition. The nature and properties of this phase transition was an open question for decades. There are two main approaches to the investigation of this problem. First one is based on microscopic description of quantum gas, and allows to perturbatively reveal phenomenon of superconductivity. Second one can be applied to the description of a small neighborhood of transition point and is based on Ginzburg–Landau effective Lagrangian studied by means of field theoretic renormalization group [1].

Because mean field theory can not be constructed in terms of Grassmann variables, the problem is to trace the connection between this two approaches. That connection has been traced in the work [5], where the authors have shown that critical modes of the simple microscopic model of fermions with n possible spin projections and density-density interaction can be expressed in terms of antisymmetric tensor fields, whose means appears to be an order parameter. It was shown that in the vicinity of phase transition point the behavior of such system can be described by ϕ^4-like theory, with two independent interaction terms.

In the special case of $n = 2$ and dimension $d = 3$ that model coincide with the well known Ginzburg–Landau model [2],[3] and give same predictions. Also for $n = 2$ and $d = 4$ the model is formally close to the Higgs model [4].

For the case of fermions with higher spins ($n > 2$) resulting model was studied by the means of field theoretic renormalization group up to five-loops accuracy [5]-[7]. It was shown that RG equations of the model have no IR attractive fixed points, and corresponding RG flows (solutions of

⋆e-mail: antmur@rambler.ru
⋆⋆e-mail: m.kompaniets@spbu.ru
⋆⋆⋆e-mail: nikita.m.lebedev@gmail.com

the RG equations for invariant couplings) always passes out from stability region of the model. Such situation is usually interpreted as a first order phase transition.

On the other hand it is expected that charged order parameter of superconductive phase transition should be also coupled with a magnetic field. Critical behavior of such model with n-component vector order parameter was studied in [8], [9], where it has been shown that charged fixed points at one-loop level exist only for sufficiently large n.

In this work we apply field theoretic RG to the $U(n)$-χ^4-model with antisymmetric tensor order parameter coupled with a magnetic field. In the sections below we will show that charged fixed points exists only for $n > 19$. At one loop-level they apper to be a saddle points, and lead to gauge dependent values of the critical exponent η.

2 The model

We study a model of a complex antisymmetric tensor field interacting with a magnetic field in d−dimensional Euclidean $\mathbf{x}$ space. Action functional of the model has the form:

$$S(\chi, \mathbf{A}) = tr((\nabla + ie_0\mathbf{A}_0)\chi_0^+(\nabla - ie_0\mathbf{A}_0)\chi_0) + \tau_0 tr(\chi_0^+\chi_0) + \frac{g_{10}}{4}(tr(\chi_0\chi_0^+))^2 + \frac{g_{20}}{4}tr(\chi_0\chi_0^+\chi_0\chi_0^+)+$$

$$+\frac{1}{2}(\nabla \times \mathbf{A}_0)^2 + \frac{1}{2\alpha_0}(\nabla\mathbf{A}_0)^2. \quad (1)$$

Here χ_{ik} and χ^{+ik} is an antisymmetric matrix fields (so that $\chi_{ik} = -\chi_{ki}$ and $\chi^{+ik} = -\chi^{+ki}$, $i, k = 1, \ldots, n$), $\mathbf{A}_0$ is a magnetic field, e_0 is effective charge, g_{10}, g_{20} are the coupling constants, τ_0 is a deviation of temperature from the critical value and α_0 is a gauge fixing parameter. Also here (and in analogous formulas below) integration over the d−dimensional $\mathbf{x}$ space is implied.

Note that field χ is twice covariant ($\chi \to \mathcal{U}\mathcal{U}\chi$), while field χ^+ is twice contravariant ($\chi^+ \to \chi^+\mathcal{U}^+\mathcal{U}^+$) with respect to transformations of the group $\mathcal{U} \in U(n)$.

Stability of the model require interaction part of the action to be positively defined. This requirement imposes following restriction on the coupling constants:

$$2g_{10} + g_{20} > 0, \quad ng_{10} + g_{20} > 0, \quad e^2 > 0. \quad (2)$$

For $n = 2$ and $n = 3$ model is equivalent to $U(1)$ and $U(3)$ invariant ϕ^4 models with a scalar and vector order parameter, respectively. Since analogues model of a vector order parameter has been studied at one-loop level in [9] we have a way to check our results by comparing calculations in those particular cases.

3 UV renormalization

In frameworks of dimensional regularization ultraviolet (UV) divergences in Green functions have the form of poles in $2\varepsilon = 4 - d$ (the model is logarithmic for $d = 4$). Their elimination requires renormalization of the model. Corresponding renormalized action can be obtained by multiplicative renormalization of fields and parameters of the model:

$$\tau_0 = \tau Z_\tau; \quad \chi_0 = \chi Z_\chi; \quad g_{0i} = g_i\mu^{2\varepsilon}Z_{g_i}; \quad \mathbf{A}_0 = \mathbf{A}Z_A; \quad e_0 = e\mu^\varepsilon Z_e; \quad \alpha_0 = \alpha Z_\alpha. \quad (3)$$

Here τ, g_i, e, α are renormalized analogues of bare parameters; χ, $\mathbf{A}$ are renormalized fields and μ is a renormalization mass (arbitrary scale parameter). Also from the Ward identities follow relations:

$$Z_e = Z_a^{-1}; \; Z_\alpha = Z_A^2. \quad (4)$$

Then corresponding renormalized action has the form:

$$S_R(\chi,\mathbf{A}) = tr((\nabla + ie\mu^{\varepsilon}\mathbf{A})\chi^+(\nabla - ie\mu^{\varepsilon}\mathbf{A})\chi)Z_\chi^2 + \tau tr(\chi^+\chi)Z_\tau Z_\chi^2 + \frac{1}{2}(\nabla\times\mathbf{A})^2 Z_A^2 + \frac{1}{2\alpha}(\nabla\mathbf{A})^2 +$$

$$+\frac{g_1\mu^{2\varepsilon}}{4}(tr(\chi\chi^+))^2 Z_\chi^4 Z_{g_1} + \frac{g_2\mu^{2\varepsilon}}{4} tr(\chi\chi^+\chi\chi^+)Z_\chi^4 Z_{g_2} \tag{5}$$

We use minimal subtraction (MS) scheme where all renormalization constants have the form:

$$Z_i = 1 + \sum_{p=1}^{\infty} A_{ip}(g_{1,2}, e)\,\varepsilon^{-p}. \tag{6}$$

They are calculated from the 1-irreducible Green functions $\langle AA\rangle$, $\langle\chi^+\chi\rangle$, $\langle\chi^+\chi\chi^+\chi\rangle$. One-loop calculations gives:

$$Z_A^2 = 1 - \frac{e^2 n(n-1)}{6\varepsilon};\quad Z_\chi^2 = 1 - \frac{e^2(3-\alpha)}{\varepsilon}; \tag{7}$$

$$Z_\chi^2 Z_{g_1} = 1 + \frac{n^2-n+8}{4}g_1 + (n-1)g_2 + \frac{3}{4}\frac{g_2^2}{g_1} - \frac{2e^2\alpha}{\varepsilon} + \frac{12e^4}{g_1\varepsilon}; \tag{8}$$

$$Z_\chi^2 Z_{g_2} = 1 + 3g_1 + \frac{2n-5}{4}g_2 - \frac{2e^2\alpha}{\varepsilon}. \tag{9}$$

Here and below we pass to rescaled couplings: $g_{1,2} = g_{1,2}/16\pi^2$, $e^2 = e^2/16\pi^2$.

4 RG functions

The RG functions of the model are defined by standard relations [1]:

$$\gamma_i \equiv \widetilde{\mathcal{D}}_\mu \ln Z_i;\quad \beta_i \equiv \widetilde{\mathcal{D}}_\mu g_i;\quad \beta_e \equiv \widetilde{\mathcal{D}}_\mu e, \tag{10}$$

where $\mathcal{D}_x \equiv x\partial_x$ for any variable x, and $\widetilde{\mathcal{D}}_x$ is the operation $\mathcal{D}_x$ at fixed bare parameters. In frameworks of MS scheme the RG functions can be expressed by the following simple expressions:

$$\gamma_i = -\left(2\mathcal{D}_{g_1} + 2\mathcal{D}_{g_2} + \mathcal{D}_e\right)A_{i1}(g_{1,2}, e); \tag{11}$$

$$\beta_{g_i} = g_i[-2\varepsilon - \gamma_{g_i}],\ i = 1,2;\quad \beta_e = e[-\varepsilon - \gamma_e]. \tag{12}$$

From equations (11,12), explicit expressions (7-9) and Ward identities (4) we obtain:

$$\gamma_A = -\gamma_e = \frac{1}{2}\gamma_\alpha = \frac{e^2 n(n-1)}{6}; \tag{13}$$

$$\gamma_\chi = -e^2(3-\alpha); \tag{14}$$

$$\beta_{g_1} = -2\varepsilon g_1 + \frac{n^2-n+8}{2}g_1^2 + 2(n-1)g_1 g_2 + \frac{3}{2}g_2^2 - 12e^2 g_1 + 24e^4; \tag{15}$$

$$\beta_{g_2} = -2\varepsilon g_2 + 6g_1 g_2 + \frac{2n-5}{2}g_2^2 - 12e^2 g_2; \tag{16}$$

$$\beta_e = -e\varepsilon + \frac{e^3 n(n-1)}{6}. \tag{17}$$

DOI: 10.1051/epjconf/201612505021

5 Fixed points

As it follows from RG equation [1], possible asymptotic regimes of the model are determined by asymptotic behavior of the system of equations for invariant coupling constants:

$$\mathcal{D}_s\bar{g}(s,g) = \beta(\bar{g}), \quad \bar{g}(1,g) = g. \tag{18}$$

Here $s = p/\mu$ is a non-dimensionalized momentum, $g = \{g_i, e\}$ is a full set of couplings and $\bar{g}(1,g)$ are the corresponding invariant variables. The IR ($s \to 0$) asymptotic of Green functions is determined by fixed points g_* of system (18), which are determined by the requirement:

$$\beta(g_*) = 0. \tag{19}$$

The type of fixed point is determined by eigenvalues of the matrix:

$$\omega = \partial\beta/\partial g|_{g=g_*}. \tag{20}$$

For IR attractive fixed point real part of all its eigenvalues is positive.

Analysis of expressions (15-17) reveals two sets of fixed point corresponding to trivial ($e_* = 0$) and nontrivial values of e_*. Both sets contain four fixed points.

First one coincide with the set of fixed points of analogues model in absence of a magnetic field. It has been found and described in details in [5]. This set consist of the following fixed points. The trivial point:

$$g_1^* = 0; \quad g_2^* = 0, \tag{21}$$

which is always IR repulsive. The point:

$$g_1^* = \frac{4\varepsilon}{8 - n + n^2}; \quad g_2^* = 0, \tag{22}$$

which is IR attractive for $n = 2$, and is a saddle point for any $n > 2$. And two points with both nontrivial coordinates:

$$g_1^* = \frac{2\left(77 + 8n - 4n^2 \pm \sqrt{-(-5+2n)^2\,(-49 - 16n + 8n^2)}\right)\epsilon}{392 + 151n - 19n^2 - 24n^3 + 4n^4};$$

$$g_2^* = \frac{4\left(-70 + 103n + 5n^2 - 24n^3 + 4n^4 \mp 6\sqrt{-(-5+2n)^2\,(-49 - 16n + 8n^2)}\right)\epsilon}{(-5+2n)\,(392 + 151n - 19n^2 - 24n^3 + 4n^4)}, \tag{23}$$

but they are real only for $n = 2$. In this case they are either IR unstable or lie outside of physical region of parameters, which means that they can't be reached by RG flows.

Another set of fixed points is the original result, that we present by this article. It corresponds to the nontrivial value of e_* equal to:

$$e_*^2 = \frac{6\varepsilon}{n(n-1)}, \tag{24}$$

and consist of fixed points of two types. Two points with coordinates:

$$g_1^* = \frac{2\varepsilon\left(-4n^6 + 16n^5 - 87n^4 + 286n^3 + 2561n^2 - 2772n\right)}{(n-1)^2n^2\,(4n^4 - 24n^3 - 19n^2 + 151n + 392)} \pm$$

$$\pm\frac{\varepsilon\sqrt{-n^2\,(2n^2 - 7n + 5)^2\,(8n^6 - 32n^5 + 2295n^4 - 12014n^3 - 265n^2 + 48024n + 105840)}}{(n-1)^2n^2\,(4n^4 - 24n^3 - 19n^2 + 151n + 392)};$$

$$g_2^* = \frac{4\varepsilon\left(4n^8 - 32n^7 + 201n^6 - 939n^5 + 773n^4 + 3771n^3 - 6298n^2 + 2520n\right)}{(n-1)^2n^2(2n-5)\left(4n^4 - 24n^3 - 19n^2 + 151n + 392\right)} \mp$$

$$\mp \frac{6\sqrt{-n^2\left(2n^2 - 7n + 5\right)^2\left(8n^6 - 32n^5 + 2295n^4 - 12014n^3 - 265n^2 + 48024n + 105840\right)}}{(n-1)^2n^2\left(4n^4 - 24n^3 - 19n^2 + 151n + 392\right)}. \tag{25}$$

For any $n > 1$ they have nontrivial imaginary part and therefore can not be reached by RG flows. Last two points have coordinates:

$$g_1^* = \frac{2\left(n^4 - 2n^3 + 37n^2 - 36n \pm \sqrt{(n-1)^2n^2\left(n^4 - 2n^3 - 359n^2 + 360n - 2160\right)}\right)\epsilon}{(n-1)^2n^2\left(n^2 - n + 8\right)};$$

$$g_2^* = 0; \tag{26}$$

They are real for $n > 19$, but appear to be IR unstable (saddle) points with eigenvalues:

$$\omega_1 = -\frac{2\left(n^6 - 3n^5 + 41n^4 - 77n^3 + 110n^2 - 72n \mp 6\sqrt{(n-1)^2n^2\left(n^4 - 2n^3 - 359n^2 + 360n - 2160\right)}\right)\varepsilon}{(n-1)^2n^2\left(n^2 - n + 8\right)}; \tag{27}$$

$$\omega_2 = \pm\frac{2\sqrt{(n-1)^2n^2\left(n^4 - 2n^3 - 359n^2 + 360n - 2160\right)}\varepsilon}{(n-1)^2n^2}; \tag{28}$$

$$\omega_e = 2\varepsilon. \tag{29}$$

Eigenvalue ω_1 is always negative, while the sign of eigenvalue ω_2 coincide with the sign in front of square root in (26).

6 Conclusions

We have shown that at one-loop level taking interaction with magnetic field into account lead to the appearance of four additional fixed points with nontrivial coordinate e_*, and two of them have real coordinates for $n > 19$. Nevertheless both of them appear saddle points. It means that there is only one possible scenario of RG flows behavior: they will pass outside of stability region given by (2). Such situation is usually interpreted as a first-order phase transition. In turn it means that interaction with magnetic field do not change the qualitative picture of phase transition obtained in [5]-[7].

Another interesting result is that anomalous dimension γ_χ, and critical exponent $\eta = 2\gamma_\chi(g_*)$ as well, appears to be gauge dependent (14). It is not surprising because just the same situation takes place in case of the vector order parameter [9]. However, η become gauge dependent only at charged fixed point ($e_*^2 \neq 0$). At the same time from renormalization of α follows that its RG flow should satisfy equation:

$$\mathcal{D}_s\bar{\alpha} = \bar{\alpha}\gamma_\alpha, \tag{30}$$

and should reach zero at charged fixed point. This means that gauge $\alpha = 0$ is the only gauge invariant with respect to renormalization (see [9] for more detailed explanation).

Acknowledgments

The authors are thankful to G.A. Kalagov and M.Yu. Nalimov for helpful discussions. N.V. Antonov is also thankful to R.Folk for the discussion of [9].

This work was supported by SPbSU grant 11.38.185.2014. N.M. Lebedev also thankful to RFBR project 16-32-00086 mol_a, and to the "Dynasty" Foundation .

References

[1] A.N. Vasil'ev, *Quantum-Field Renormalization Group in the Critical Behavior Theory and in Stochastic Dynamics*, (St. Petersburg Institute for Nuclear Physics, St. Petersburg, 1998).

[2] L. D. Landau, V. L. Ginzburg, JETP, **20**, 1064 (195);
L. D. Landau *Collected papers*, (Pergamon Press, Oxford, 1965) p. 546

[3] N.H.March, W.H.Young, S. Sampanthar, *The Many-Body Problems in Quantum Mechanics*, (Cambridge at the University Press, 1967)

[4] P.W.Higgs, Phys.Lett., **12**, p.132 (1964);
P.W.Higgs, Phys.Rev.Lett. **13**, p.508 (1964);
P.W.Higgs Phys.Rev., **145**, p.1156 (1966)

[5] Nalimov M Yu, Komarova M V and Honkonen J, Teor.Math.Phys. **176**:1, 89–97 (2013)

[6] G. A. Kalagov, M. V. Kompaniets, M. Yu. Nalimov, Theor.Math.Phys. **181**:2, 1448–1458 (2014)

[7] G. A. Kalagov, M. V. Kompaniets, M. Yu. Nalimov, *ArXiv:1505.07360,* (2015)

[8] Halperin B.I., Lubensky T.C., Ma S.K., Phys. Rev. Lett., **32**, 292 (1974)

[9] M.Dudka, R.Folk, G.Moser, Condensed Matter Physics **10**:2, 189–200 (2007)

Bound states in a model of interaction of Dirac field with material plane

Yu. M. Pismak[1,★] and D. Yu. Shukhobodskaia[1]

[1] *Saint-Petersburg State University*

Abstract. In the framework of the Symanzik approach model of the interaction of the Dirac spinor field with the material plane in the 3 + 1-dimensional space is constructed. The model contains eight real parameters characterizing the properties of the material plane. The general solution of the Euler-Lagrange equations of the model and dispersion equations for bound states are analyzed. It is shown that there is a choice of parameters of the model in which the connected states are characterized by dispersion law of a massless particle moving along the material plane with the dimensionless Fermi velocity not exceeding one.

1 Introduction

Recently, the physics of quantum macro-phenomena has attracted the attention of many researchers. For success of the theoretical investigations in this field it is important to build a model in which the most essential features of the mechanism of generating the studied quantum effect would be presented. If such model is constructed in the framework of usual quantum field theory[1], one needs a considerable modification of its methods, since the properties of the vacuum change essentially because of interaction of quantum fields fluctuations with macroscopic objects. In this way quantum macro-effects may appear in dynamics of material bodies which cannot be explained in by means of classical physics.

It was first shown theoretically in 1948 by Casimir [2]. In his model according to calculations carried out by him, the interaction of ideally conducting uncharged plates of planar capacitor with photon vacuum fluctuations generates an attraction force between ones. This phenomenon, called the Casimir effect (CE), is observed experimentally [3–6], and the emperical results obtained for materials with high conductivity are, with a high degree of accuracy, in agreement with theoretical ones.

At distances of 10 to 1000 nm the lows as classical and quantum physics are essential. They form a special nanophysics with its highly unusual phenomena, one of the striking example of which is the CE. Investigation of nanophysics is important not only from the general theoretical point of view, but there is every reason to hope for an effective practical use of their results in the creation of new technologies and devices.

The theoretical phenomena such as CE are studied very intensively in numerous works [6, 7]. However often, these investigations are based on the simplified models which applicability area is very limited. This make it difficult to achieve a holistic understanding specifity of nanophysics and

★e-mail: ypismak@yahoo.com

the internal relationship of its various manifestations. Therefore, they are not meaningless attempts to build and research models assigned for studying a wide range of nanophysical phenomena.

It is natural to assume that this type of models should be constructed on the basis of the most important, universal, well-established physical principles. From all the used currently methods, the closest to that is the Symansik approach [8] for modeling the interaction of quantum field system with classical macro object, which is presented by an additional action functional (defect action), concentrated in the spatial domain where, this inhomogeneity - a macroscopic object - is located. An important assumption is also that the standard requirements for quantum field models (locality, renormalizability, symmetry properties) are fulfilled.

The Symanzik approach has been used in many investigations and, in particular, for modeling the interaction of QED fields with two-dimensional objects [9–14].In this paper we use it for modeling the interaction of spinor Dirac field with material plane and studying the property of localized near one bound states.

2 Statement of problem

Within the Symanzik approach, the general form of the action functional modeling the interaction of the quantum field with material object (defect) can be written as:

$$S(\varphi) = S_V(\varphi) + S_{def}(\varphi).$$

Here, S_V is the action of the basic quantum field model, and S_{def} is the defect action:

$$S_V(\varphi) = \int L(\varphi(x))d^D x,\; S_{def}(\varphi) = \int_\Gamma L_{def}(\varphi(x))d^{D'} x$$

where Γ is a subspace of dimension $D' \leq D$ in a D-dimensional space [8]. In the quantum electrodynamic (QED) the action $S_V(\varphi)$ is the functional of photon vector field $A_\mu(x)$ and Dirac spinor fields $\bar{\psi}(x), \psi(x)$,

$$S_V(\varphi) = S(\bar{\psi}, \psi, A) = -\frac{1}{4}F_{\mu\nu}F^{\mu\nu} + \bar{\psi}(i\hat{\partial} - m + ie\hat{A})\psi$$

with $F_{\mu\nu} = \partial_\mu A_\nu - \partial_\nu A_\mu$. Gauge invariance, locality, and renormalizability as the basic principles of QED impose strong constraints on the possible form of the defect action $S_{def}(\varphi)$. Taking them into account by modeling the interaction of QED fields with a two-dimensional surface without charges and currents, one obtains the following result: $S_{def}(\varphi) = S_{def}(A) + S_{def}(\bar{\psi}, \psi)$. If the surface is defined by equation $\Phi(x) = 0$, $x = (x_0, x_1, x_2, x_3)$. then $S_{def}(A)$ is the Chern-Simons action

$$S_{def}(A) = \frac{a}{2}\int \varepsilon^{\lambda\mu\nu\rho}\partial_\lambda\Phi(x)A_\mu(x)F_{\nu\rho}(x)\delta(\Phi(x))\, d^4x,$$

where $\varepsilon^{\lambda\mu\nu\rho}$ is the totally antisymmetric tensor ($\varepsilon^{0123} = 1$), and the parameter a is a dimensionless coupling constant. This expression is the most general form of a gauge-invariant functional of A_μ concentrated on a defect surface being invariant in respect to reparameterization of one and not having any negative-dimension parameters. The most general form of the defect action $S_{def}(\bar{\psi}, \psi)$ is written as

$$S_{def}(\bar{\psi}, \psi) = \sum_{j=1}^{16}\int \alpha_j\bar{\psi}(x)\Gamma_j\psi(x)\delta(\Phi(x))\, d^4x,$$

where Γ_j are the 16 basis Dirac matrices and α_j are dimensionless coupling constants.

As the material defect we consider the plane $x_3 = 0$. In this case

$$S(\overline{\psi}, \psi) = \int \overline{\psi}(x)(i\hat{\partial} - m + \Omega(x_3))\psi(x)dx, \tag{1}$$

and the matrix $\Omega(x_3) = Q\delta(x_3)$ describes the coupling of Dirac fields with the plane. Because $\Omega(x_3)$ and $\delta(x_3)$ have the dimension of mass, the matrix Q is dimensionless. We suppose that the defect plane is isotropic and homogeneous. It means that $S(\overline{\psi}, \psi)$ must be be invariant under translation in x_0-, x_1-, x_2-directions and rotation around x_3-axe. The most general form of the matrix Q fulfilling this condition is

$$Q = r_1\mathbf{1} + ir_2\gamma_5 + r_3\gamma_3 + r_4\gamma_5\gamma_3 + r_5\gamma_0 + r_6\gamma_5\gamma_0 + ir_7\gamma_0\gamma_3 + ir_8\gamma_1\gamma_2$$

where $\mathbf{1}$ is the unit 4x4 - matrix, $i = \sqrt{-1}$, γ_j, $j = 0, 1, 2, 3$ are the Dirac matrices and $\gamma_5 = i\gamma_0\gamma_1\gamma_2\gamma_3$. For the $\gamma-$matrixes we use the following representation

$$\gamma_0 = \begin{pmatrix} 1 & 0 & 0 & 0 \\ 0 & 1 & 0 & 0 \\ 0 & 0 & -1 & 0 \\ 0 & 0 & 0 & -1 \end{pmatrix}, \quad \gamma_1 = \begin{pmatrix} 0 & 0 & 0 & 1 \\ 0 & 0 & 1 & 0 \\ 0 & -1 & 0 & 0 \\ -1 & 0 & 0 & 0 \end{pmatrix},$$

$$\gamma_2 = \begin{pmatrix} 0 & 0 & 0 & -i \\ 0 & 0 & i & 0 \\ 0 & i & 0 & 0 \\ -i & 0 & 0 & 0 \end{pmatrix}, \quad \gamma_3 = \begin{pmatrix} 0 & 0 & 1 & 0 \\ 0 & 0 & 0 & -1 \\ -1 & 0 & 0 & 0 \\ 0 & 1 & 0 & 0 \end{pmatrix}.$$

Taking the functional derivatives of action (1) with respect to $\overline{\psi}(x)$ and $\psi(x)$ we obtain the Euler-Lagrange equations :

$$(i\hat{\partial} - m + \Omega(x_3))\psi(x) = 0, \; (\partial_\mu\overline{\psi}(x))\gamma^\mu + \overline{\psi}(x)(m - \Omega(x_3)) = 0.$$

It can be easily verified that $\overline{\psi}(x) = \psi^*(x)\gamma_0$, if $\gamma_0\Omega^+(x_3) = \Omega(x_3)\gamma_0$. This condition is fulfilled if parameters r_j, $1 \leq j \leq 8$ are real. We supposed that it is the case.

Let us introduce the following notations

$$\bar{x} = (x_0, x_1, x_2), \; \overline{px} = p_0x_0 - p_1x_1 - p_2x_2.$$

for parts of 4-component vectors and represent $\psi(x) = \psi(\bar{x}, x_3)$ as $\psi(x) = \psi_s(x) + \psi_a(x)$, where

$$\psi_s(x) = \frac{1}{2}(\psi(\bar{x}, x_3) + \psi(\bar{x}, -x_3)), \; \psi_a(x) = \frac{1}{2}(\psi(\bar{x}, x_3) - \psi(\bar{x}, -x_3)).$$

Analysis of the possible singularities of spinor $\psi(x)$ at the plane $x_3 = 0$, made in [14], leads to the following: for $x_3 \neq 0$, the field $\psi(x)$ satisfies the Dirac equation

$$(i\hat{\partial} - m)\psi(x) = 0, \tag{2}$$

and the relation

$$-i\gamma_3\psi_a(\bar{x}) + Q\psi_s(\bar{x}) = 0 \tag{3}$$

holds at $x_3 = 0$.

It is shown in [14] that the general solution of (2) by $x_3 \neq 0$ can be represented as

$$\psi(x) = \frac{1}{(2\pi)^3}\int e^{i\bar{p}\bar{x}}\psi_+(\bar{p}, x_3)d\bar{p}, \text{ for } x > 0, \; \psi(x) = \frac{1}{(2\pi)^3}\int e^{i\bar{p}\bar{x}}\psi_-(\bar{p}, x_3)d\bar{p}, \text{ for } x < 0$$

where

$$\psi_\pm(\bar{p}, x_3) = U(\bar{p}, x_3)\chi_\pm(\bar{p}),\ U(\bar{p}, x_3) = e^{-i\gamma_3(\hat{\bar{p}}+m)x_3} = e^{-i\kappa(\bar{p})x_3}P^+(\bar{p}) + e^{i\kappa(\bar{p})x_3}P^-(\bar{p}).$$

Here, $\chi_\pm(\bar{p})$ are arbitrary spinors depending only on $\bar{p}$, and we used the notations

$$\kappa(\bar{p}) = \sqrt{\bar{p}^2 - m^2},\ P^\pm(\bar{p}) = \frac{1}{2}\left(\mathbf{1} \pm \frac{\gamma_3(\hat{\bar{p}} + m)}{\kappa(\bar{p})}\right).$$

It follows from (3) that $\chi_+(\bar{p})$, $\chi_-(\bar{p})$ satisfy the relation

$$-i\gamma_3(\chi_+ - \chi_-) + Q(\chi_+ + \chi_-) = 0,$$

which implies that

$$\chi_- = S\chi_+,\ S = (i\gamma_3 + Q)^{-1}(i\gamma_3 - Q). \tag{4}$$

The matrices $P^\pm(\bar{p})$ are projectors:

$$P^\pm(\bar{p})P^\pm(\bar{p}) = P^\pm(\bar{p}),\ P^+(\bar{p}) + P^-(\bar{p}) = 1,\ P^+(\bar{p})P^-(\bar{p}) = P^-(\bar{p})P^+(\bar{p}) = 0.$$

Each of them has two two-dimensional eigensubspaces L^+ and L^-, which are represented as linear combinations of two mutually orthogonal eigenvectors $v_1^\pm$ and $v_2^\pm$:

$$L^\pm(a_1^\pm, a_2^\pm) = a_1^\pm v_1^\pm + a_2^\pm v_2^\pm,$$
$$P^\pm(\bar{p})L^\pm(a_1^\pm, a_2^\pm) = L^\pm(a_1^\pm, a_2^\pm),\ P^\pm(\bar{p})L^\mp(a_1^\mp, a_2^\mp) = 0.$$

Here, $a_i^\pm$, $i = 1, 2$, are complex parameters, and $v_i^\pm$, $i = 1, 2$, are spinors, which we choose as

$$v_1^+ = n_1^+\left(-h_1, h_2^-, 1, h_3^-\right),\ v_2^+ = n_2^+\left(h_1, h_2^+, 1, h_3^+\right),$$
$$v_1^- = n_1^-\left(-h_1, h_2^+, 1, -h_3^+\right),\ v_2^- = n_2^-\left(h_1, h_2^-, 1, -h_3^-\right)$$
$$h_1 = \frac{\sqrt{p_0^2 - m^2}}{m + p_0},\ h_2^\pm = \frac{p_0^2 - m^2 \pm \sqrt{p_0^2 - m^2}\kappa(\bar{p})}{(m + p_0)(p_1 - ip_2)},\ h_3^\pm = \frac{\kappa(\bar{p}) \pm \sqrt{p_0^2 - m^2}}{p_1 - ip_2}$$

with normalization constants $n_1^\pm, n_2^\pm$ defined by conditions $\bar{v}_1^\pm v_1^\pm = \bar{v}_2^\pm v_2^\pm = 1$. These spinors are eigenstates of the helicity operators

$$\sigma(\vec{p}) = \frac{i}{2|\vec{p}|}(\vec{p}\,\vec{s}),\ \vec{s} = (\gamma_2\gamma_3, -\gamma_1\gamma_3, \gamma_1\gamma_2).$$

They satisfy the relations

$$\sigma(\vec{p})|_{p_3=\mp\kappa(\bar{p})}v_1^\pm = -\frac{1}{2}v_1^\pm,\ \sigma(\vec{p})|_{p_3=\mp\kappa(\bar{p})}v_2^\pm = \frac{1}{2}v_2^\pm$$

We represent the spinors $\chi_\pm(\overline{p})$ as

$$\chi_\pm(\overline{p}) = b_\pm^1(\overline{p})v_1^-(\overline{p}) + b_\pm^2(\overline{p})v_2^-(\overline{p}) + c_\pm^1(\overline{p})v_1^+(\overline{p}) + c_\pm^2(\overline{p})v_2^+(\overline{p}). \tag{5}$$

If $\kappa > 0$ the fields $\bar{\psi}, \psi$ describe the scattering processes of Dirac particles on the defect plane. For the model with $r_5 = r_6 = r_7 = r_8 = 0$ they are studied in [14]. In this paper we consider the properties of bound states localized in vicinity of $x_3 = 0$.

3 Bound states

The state is localized close to the plane $x_3 = 0$ if $\bar{p}^2 - m^2 < 0$ and $\kappa(\bar{p}) = i|\kappa(\bar{p})|$ is imaginary. In this case the spinors $\chi_{\pm}(\overline{p})$ (5) must fulfill the conditions $b^1_+ = b^2_+ = c^1_- = c^2_- = 0$. Other case the asymptotic of fields $\bar{\psi}(x), \psi(x)$ could be infinite by $x_3 \to \pm\infty$. It follows from equation (4) that

$$P^-(\bar{p})S\chi_+ = 0 \tag{6}$$

It is the system of homogeneous linear equations for coefficients b^1_+, b^2_+. Its solvability condition has the form

$$(p_1^2 + p_2^2)R_3 - (i\kappa R_1 + 2(mr^-_{18} - p_0 r^+_{45}))(i\kappa R_2 + 2(mr^+_{18} + p_0 r^-_{45})) = 0 \tag{7}$$

where we used the notations $r^{\pm}_{jk} = r_j \pm r_k$ and

$$R_1 = 1 + {r^-_{18}}^2 + {r^-_{27}}^2 + {r^+_{36}}^2 - {r^+_{45}}^2,\ R_2 = 1 + {r^+_{18}}^2 + {r^+_{27}}^2 + {r^-_{36}}^2 - {r^-_{45}}^2,\ R_3 = 4(r_6^2 + r_7^2 + r_8^2 - r_4^2).$$

The equation (7) is the dispersion relation for bound state.

The equation (4),(6) enable to express the spinor parameters $b^1_-, b^2_-, c^1_+, c^2_+$ in terms of one of them, which can be fixed by choosing normalization condition. The currents in bound state defined by spinors χ_+,χ_- are presented as

$$J^{\pm}_k(\bar{p}, x_3) = e^{-|\kappa x_3|}\, j^{\pm}_k(\bar{p}),\ j_k(\bar{p}) = \bar{\chi}_{\pm}(\bar{p})\gamma_k\chi_{\pm}(\bar{p})$$

with $j^{\pm}_3 = 0$ and by $p_0 > m$

$$j^+_1 = \frac{4\omega\,(|b^1_-|^2 + |b^2_-|^2)\,\omega\, p_1 - (|b^1_-|^2 - |b^2_-|^2)\,|\kappa|\, p_2}{(p_0 + m)(\omega^2 + |\kappa|^2)},$$
$$j^+_2 = \frac{4\omega\,(|b^1_-|^2 + |b^2_-|^2)\,\omega\, p_2 + (|b^1_-|^2 - |b^2_-|^2)\,|\kappa|\, p_1}{(p_0 + m)(\omega^2 + |\kappa|^2)},$$
$$j^-_1 = \frac{4\omega\,(|c^2_+|^2 + |c^1_-|^2)\,\omega\, p_1 - (|c^2_+|^2 - |c^1_+|^2)\,|\kappa|\, p_2}{(p_0 + m)(\omega^2 + |\kappa|^2)},$$
$$j^-_2 = \frac{4\omega\,(|c^2_+|^2 + |c^1_+|^2)\,\omega\, p_2 + (|c^2_+|^2 - |c^1_+|^2)\,|\kappa|\, p_1}{(p_0 + m)(\omega^2 + |\kappa|^2)},$$

where $\omega = \sqrt{p_0^2 - m^2}$. We see that the currents flow along the plane $x_3 = 0$ and are parallel to the momentum $\vec{p} = (p_1, p_2)$ if $b^1_-| = |b^2_-|, |c^2_+| = |c^1_+|$. The currents above and belove of the plane $x_3 = 0$ are parallel if $b^1_-|/|b^2_-| = |c^2_+|/|c^1_+|$.

If by a special choosing the parameters in the model $R_1 = R_2 = r^+_{18} = r^-_{18} = 0$, then the dispersion relation (7) has the form

$$p_0^2 - v_F^2(p_1^2 + p_2^2) = 0 \tag{8}$$

where V_F is a constant, which is expressed in the term of parameters r_k, $1 \leq k \leq 8$ of the model. The dispersion low (8) describes the movement of massless particle with Fermi-velocity V_f. The motion of such particles explains numerous effects in graphene.

The constants $R_1, R_2, r^+_{18}, r^-_{18}$ are equal to zero by $r_1 = r_8 = r_3 = 0$ and $r^+_{45} = \pm\sqrt{1 + {r^-_{27}}^2 + r_6^2}$, $r^-_{45} = \pm\sqrt{1 + {r^+_{27}}^2 + r_6^2}$. For such choosing of parameters the Fermi-velocity V_F

DOI: 10.1051/epjconf/201612505022

can obtain two different values: $V_F = V_F^+$ or $V_F = V_F^-$,

$$V_F^{\pm} = \sqrt{\frac{1}{2}\left(1 \pm \frac{1 + r_{27}^+ r_{27}^- - r_6^2}{\sqrt{1 + {r_{27}^-}^2 + r_6^2}\sqrt{1 + {r_{27}^+}^2 + r_6^2}}\right)}.$$

The Fermi-velocity V_F^+ (V_F^-) corresponds to the positive (negative) value of $r_{45}^+ r_{45}^-$. By changing the values of parameters r_2, r_6, r_7 one can obtain each value of the Fermi-velocity in the interval (0,1). The current flowing along the plane $x_3 = 0$ is parallel to the momentum $\vec{p}$ only by $r_6 = 0$.

4 Conclusions

In the framework of the Symanzik approach, we built the most general model of spinor field interacting with the homogeneous isotropic material plane. The action of the model includes both the usual spinor Dirac action and additional defect contribution. It contains 8 real dimensionless parameters that characterize the property of material of defect. The Euler-Lagrange equations of the model describe both the scattering of Dirac particles on the plane and the bound states localized in its vicinity. The dispersion relation for the bound states is considered. It was shown that by a special choose of the model parameter it describe the motion of massless particle with Fermi-velocity $0 < V_F < 1$. The currents generated by particles moving along the material plane are parallel to the moment only by special values of the defect action parameters. The presented approach could be used for the theoretical description of the interaction of electrons, positrons and neutrons with two-dimensional materials (graphene, thin films, sputters, sharp boundaries of a solid body). It could be also fruitful for modeling of wide class system characterizing by two-dimensional quantum macro effects (the Hall and plasmon-polariton effects, topological insulators, nanophotonics [15]). Simple modifications of the model allows to take into account the effects of external electromagnet ic fields.

We have constructed a model of the interaction of a spinor field with a material plane in the framework of Symanzik's approach. The action functional of the model includes the standard spinor Dirac action in (3+1) dimensional space-time and an additional contribution from the defect, concentrated in the part of space that it occupies (in considered case it was the plane $x_3 = 0$). The action contains three parameters characterizing the properties of the material of the plane. In the model, we calculated the characteristics of a Dirac particle scattering on the defect plane and also studied the properties of states localized near it.

The model and the obtained results can be used to describe theoretically processes of the interaction of electrons, with two-dimensional materials (graphene, thin films, sputter, sharp boundaries of a solid body). Simple modifications of the model allow taking the effects of the action of external electromagnetic fields into account.

Applying Symanzik's approach to QED enables a construction of a model for a wide class of quantum macroscopic phenomena in two-dimensional materials. These models can prove usefulness in investigating the Hall effect, plasmon-polariton effects, nanophotonics, properties of topological insulators, thin films, sputters, and sharp boundaries [15].

Acknowledgments. This paper was supported in part by RFBR grant 16-02- 00943-a.

References

[1] N. N. Bogoliubov and D. V. Shirkov, *Quantum Fields*, Nauka, Moscow (1993);
C. Itzykson and J.-B. Zuber *Quantum Fields Theory*, McGraw-Hill, New York (1980).

[2] H. B. G Casimir, Proc. K. Ned. Akad. Wet B. **51**, 793-795 (1948).

[3] U. Mohideen and A. Roy, Phys. Rev. Lett. **81**, 4549-4552 (1998); arXiv:physics/9805038v2 (1998);
A. Roy, C.-Y. Lin and U. Mohideen, Phys. Rev. D **60**, 111101(R) (1999); arXiv:quant-ph/9906062v3 (1999).

[4] B. W. Harris, F. Chen and U. Mohideen, Phys. Rev. A **62**, 052109 (2000); arXiv:quant-ph/0005088v1 (2000).

[5] G. Bressi, G. Carugno, R. Onofrio and G. Ruoso, Phys. Rev. Lett. **88**, 041804 (2002); arXiv:quant-ph/ 0203002v1 (2002).

[6] M. Bordag, G. L. Klimchitskaya, U. Mohideen, and V. M. Mostepanenko *Advances in the Casimir Effect*, (Intl. Ser. Monogr. Phys., Vol. 145), Oxford Univ. Press, Oxford (2009); G. L. Klimchitskaya, U. Mohideen and V. M. Mostepanenko, Rev. Mod. Phys. **81**, 1827-1885 (2009); arXiv:0902.4022v1 [cond-mat.other] (2009).

[7] K. A. Milton, J. Phys. A: Math. Gen. **37**, R209-R277 (2004).

[8] K. Symanzik, Nucl.Phys. B **190**, 1-44 (1981). .

[9] V. N. Markov and Yu. M. Pis'mak, J. Phys. A: Math. Gen. **39**, 6525-6532 (2006); arXiv: hep-th/0505218v3 (2005).

[10] V. N. Marachevsky and Yu. M. Pis'mak, Phys. Rev. D **81**, 065005 (2010); arXiv:0907.1985v2 [hep-th] (2009).

[11] D.Yu. Pis'mak and Yu. M. Pis'mak, Theor. Math. Phys. **175**,816-826 (2013); Phys. Part. Nuclei **44**, 450-461 (2013); Theor. Math. Phys. **166**, 1423-1431 (2011).

[12] D.Yu. Pis'mak, Yu. M. Pis'mak and F. J. Wegner, Phys. Rev. E **92**, 013204 (2015); arXiv:1406.1598v1 [hep-th] (2014).

[13] I. V. Fialkovsky, V. N. Markov and Yu. M. Pis'mak, J. Phys. A: Math. Gen. **39**, 6357-6363 (2006); Int. J. Mod. Phys. A **21**, 2601-2616 (2006); arXiv:hep-th/0311236v2 (2003).

[14] D. Yu. Pismak, Yu. M. Pismak, Theor. Math. Phys. **184**, 3, 1329-1341 (2015).

[15] G. Bracco and B. Holst (Eds.), *Surface Science Techniques*, (Springer Ser. Surf. Sci., Vol. 51), Springer, Berlin (2013); M. L. Brongersma and P. G. Kik (Eds.), *Surface Plasmon Nanophotonics*, (Springer Ser. Opt. Sci., Vol. 131), Springer, Berlin (2007); S. A. Maier *Plasmonics: Fundamental and Applications*, Springer, Berlin (2007).

One-particle reducibility in effective scattering theory

V. Vereshagin *

Saint-Petersburg State University, 198504 St.-Petersburg, Russia

Abstract. To construct the reasonable renormalization scheme suitable for the effective theories one needs to resolve the "problem of couplings" because the number of free parameters in a theory should be finite. Otherwise the theory would loose its predictive power. In the case of effective theory already the first step on this way shows the necessity to solve the above-mentioned problem for the 1-loop 2-leg function traditionally called self energy. In contrast to the customary renormalizable models the corresponding Feynman graph demonstrates divergencies that require introducing of an infinite number of prescriptions. In the recent paper [1] it has been shown that the way out of this difficulty requires the revision of the notion of one-particle reducibility. The point is that in *effective scattering* theory one can introduce two different notions: the *graphic* reducibility and the *analytic* one. Below we explain the main ideas of the paper [1] and recall some notions and definitions introduced earlier in [2] and [3].

1 Introduction

First of all, it should be recalled that the field theory is said to be effective if:

1. It is based on the Dyson formula for S-matrix elements (f and i stand for the final and initial states, respectively):

$$S_{fi} = \langle f|T_w exp\left\{i\int L_{int}dx\right\}|i\rangle. \tag{1}$$

   The symbol T_w in this formula denotes Wick's (manifestly covariant) T-product. The non-covariant terms in the interaction Lagrangian and in propagators are to be discarded (see [4], [5]). In the case of effective theories only discussed below this does not lead to any uncertainties.

2. The interaction Lagrangian is defined as the interaction picture operator; it contains *all* the monomials consistent with the given algebraic symmetry.

It is clear that the effective field theory is renormalizable: all necessary counterterms are presented in the Lagrangian by construction. This fact, however, does not make life easier because one needs to formulate an infinite number of renormalization prescriptions (RP's) in order to fix the finite parts of relevant counterterms. This (so-called "problem of couplings") is the main obstacle on the way of constructing the physically meaningful theory of S matrix.

*E-mail: vvv@av2467.spb.edu

As known (see, e.g. [4]), the full set of parameters that appear in the effective field theory can be divided into two independent subsets: the essential parameters (those independent combinations of the Lagrangian coupling constants which the S-matrix elements depend upon) and the redundant ones (those needed to calculate the Green functions off the mass shell). As long as we are only interested in finiteness and physical definiteness of S-matrix elements (amplitudes of the physical processes) there is no necessity in formulating the RP's for redundant parameters. It is this scheme (called in [3] *the effective scattering theory*) which we discuss below.

To make use of this idea one needs to point out the full list of essential parameters. Here two notes are in order. First, as shown in [2] (see also [3]), the sum of S-matrix graphs of the given loop order L depends on the Lth level resultant parameters which are the only numbers to be fixed in the process of step-by-step renormalization. Second, the list of those resultant parameters may be further shortened if certain additional restrictions (say, polynomial boundedness requirements) are imposed. Strictly speaking, only the parameters from that shortened list should be called essential. For this reason the combinations that appear before attracting the additional conditions like the polynomial boundedness are called resultant.

There is an obvious problem that precludes constructing the renormalization scheme completely formulated in terms of S matrix (on shell) parameters: we need to know the two leg function (so-called self energy) at arbitrary value of p^2, not only on the mass shell. In the paper [1] it has been shown that this problem turns out quite solvable if one redefines the notion of one-particle irreducibility (1PI). Below we explain this point in terms of the single scalar effective scattering theory.

2 Preliminary notes and definitions

First we would like to recall some definitions and notions discussed earlier in [2], [3] and [1].

We work in terms of $(4n-10)$ scalar variables needed to describe the n-leg *full effective vertex* $V(p_1,\ldots,p_n)$ with external lines marked by their (off shell!) momenta $p_1,\ldots,p_n$ and depending on $(4n-10)$ scalar variables

$$[\kappa_1,\ldots,\kappa_n;\nu_1,\ldots,\nu_{3n-10}]\,. \tag{2}$$

Here

$$\kappa_k = p_k^2 - m^2 \equiv \kappa(p_k) \tag{3}$$

(below we call the variables of this kind as the shell deviations; note that these are the only variables which the triple effective vertex may depend upon) and

$$\nu_r \equiv \sum_{i,j=1}^{n} s_{ij}^r(p_i\cdot p_j) \qquad (r=1,\ldots,3n-10) \tag{4}$$

stand for the independent scalar products (s_{ij}^r are the relevant numerical coefficients). Since the effective vertex is just a series in scalar variables κ_i $(i=1,\ldots,n)$ and ν_k $(k=1,\ldots,3n-10)$, it can be written as follows

$$V(p_1,\ldots,p_n) \equiv V(\kappa_1,\ldots,\kappa_n;\nu_1,\ldots,\nu_{3n-10})$$
$$= \sum_{i_1,i_n=0}^{\infty} V_{i_1,\ldots,i_n}(\nu_1,\ldots,\nu_{3n-10})\kappa_1{}^{i_1}\ldots\kappa_n{}^{i_n}\,. \tag{5}$$

The item

$$V_{0,\ldots,0}(\nu_1,\ldots,\nu_{3n-10}) \tag{6}$$

is called *minimal vertex* or, the same, *vertex minimal with respect to all its lines*. It contrast to the full effective vertex (5) it does not contain factors $p_i^2 - m^2$.

To better understand the importance of this notion it is instructive to consider the 4-leg tree graph Γ constructed from two full 3-leg effective vertices, say, $V(p_1, p_2, k)$ and $V(p_3, p_4, k)$ connected to one another by the propagator line k:

$$\Gamma(p_1, \ldots, p_4) = V(p_1, p_2, k)\frac{1}{k^2 - m^2}V(p_3, p_4, k)\,. \tag{7}$$

Here $k = p_1 + p_2 = p_3 + p_4$.

Using the relation (5) one obtains:

$$\Gamma(p_1, \ldots, p_4) = \sum_{\substack{i_1,i_2=0\\ j_1,j_2=0}}^{\infty} \kappa^{i_1}(p_1)\kappa^{i_2}(p_2)\kappa^{j_1}(p_3)\kappa^{j_1}(p_4)$$
$$\times\Bigg\{V_{i_1,i_2,0}\frac{1}{\kappa(k)}V_{j_1,j_2,0} +$$
$$\sum_{r=1}^{\infty}\kappa^{r-1}(k)\Big[V_{i_1,i_2,0}V_{j_1,j_2,r} + V_{i_1,i_2,r}V_{j_1,j_2,0}\Big]+$$
$$+\sum_{r_1,r_2=1}^{\infty}\kappa^{r_1+r_2-1}(k)V_{i_1,i_2,r_1}V_{j_1,j_2,r_2}\Bigg\}\,. \tag{8}$$

Here V_{ikr} are the coupling constants of the 3-leg vertex.

It is important to remember that the expressions (8) and (7) are identical because they describe the same graph. However, the interpretation of what may be called a subgraph depends on our definition of the notion of *reducibility*.

Indeed, according to the conventional understanding of one-particle reducibility the graph Γ corresponding to the expression (7) is 1PR: it is constructed from two subgraphs (triple effective vertices) connected with one another by the propagator line.

In contrast, the expression (8) can be considered as the sum of two items:

$$\Gamma = \Gamma_1 + \Gamma_2, \tag{9}$$

where

$$\Gamma_1 = \sum_{\substack{i_1,i_2=0\\ j_1,j_2=0}}^{\infty} \kappa^{i_1}(p_1)\kappa^{i_2}(p_2)\kappa^{j_1}(p_3)\kappa^{j_1}(p_4)\times$$
$$\times\left[V_{i_1,i_2,0}\frac{1}{\kappa(k)}V_{j_1,j_2,0}\right] \tag{10}$$

looks like the customary (traditionally defined) 1PR graph, while Γ_2

$$\Gamma_2 = \sum_{\substack{i_1,i_2=0\\ j_1,j_2=0}}^{\infty} \kappa^{i_1}(p_1)\kappa^{i_2}(p_2)\kappa^{j_1}(p_3)\kappa^{j_1}(p_4)\times$$

$$\times\left\{\sum_{r=1}^{\infty}\kappa^{r-1}(k)\Big[V_{i_1,i_2,0}V_{j_1,j_2,r}+V_{i_1,i_2,r}V_{j_1,j_2,0}\Big]+\right.$$

$$\left.+\sum_{r_1,r_2=1}^{\infty}\kappa^{r_1+r_2-1}(k)V_{i_1,i_2,r_1}V_{j_1,j_2,r_2}\right\} \qquad (11)$$

may be considered as the 1PI 4-leg effective vertex (of the tree level!) with coupling constants constructed as products of 3-leg constants of the Hamiltonian level. The disappearance of the propagator line $1/k$ due to the presence of the "killing" factor k in adjacent vertex (or both vertices) is called *the reduction of the line k*.

Suppose now that we need to know the graph $\Gamma(p_1,\ldots,p_4)$ on the mass shell $p_i^2 = m^2$ or, the same, we are only interested in its contribution with accuracy up to the terms of order $\kappa(p_i)$. This means that we can put

$$\kappa(p_i) = 0, \qquad (i = 1,\ldots,4). \qquad (12)$$

(but $\kappa(k) = (p_1+p_2)^2 - m^2 \neq 0$!). In this case both triple vertices in the graph (10) turn out minimal with respect to all their lines. The same is true with regard to the 4-leg effective vertex (graph (11)). So, the singular (pole) part (10) of the contribution of the initially considered graph (7) can be considered as the *true one-particle reducible subgraph*, and the regular part (11) – as the *induced minimal secondary 4-leg vertex of the tree level* (see [2]).

It is not difficult to generalize the above analysis for more complicated cases. The important observation is that the definition of the graph reducibility is connected with the shape of the analytic recording of the graph contribution.

Let us now consider 1PI 2-leg 1-loop graph S conventionally called (*self energy*) with the external leg momentum p. In the framework of effective theory the relevant triple vertices are described by the following Lagrangian (see [1]):

$$H_3 = \frac{1}{3!}\sum_{s=0}\tilde{D}^{jk;s} : \phi\left(\partial^{[s]}\phi^j\right)\left(\partial_{[s]}\phi^k\right) : , \qquad (13)$$

Here : ... : denotes the normal product, and the shortened notation

$$\partial^{[s]} \stackrel{\text{def}}{=} \partial^{\mu_1}\ldots\partial^{\mu_s}.$$

is used. In momentum space the triple vertex (13) can be written as follows:

$$V(\kappa_1,\kappa_2,\kappa_3) = i(2\pi)^4\delta(k_1+k_2+k_3)$$

$$\sum_{i,j,k=0}^{\infty} D^{ijk}(\kappa_1^i,\kappa_2^j,\kappa_3^k) \qquad (14)$$

where D^{ijk} are certain combinations of the coupling constants $\tilde{D}^{jk;s}$.

The expression for the one-loop self-energy graph in effective theory reads (both external lines are considered incoming):

$$S(p^2) = C(p^2, \Lambda)\, \delta(p+q) + \sum_{\substack{ijk=0 \\ lmn=0}}^{\infty} D^{ijk} D^{lmn} \kappa_k^i \kappa_q^l \times$$
$$\times \int dr\, dt\, \delta(p+r-t)\, \delta(q+t-r) \kappa_t^{j+n-1} \kappa_r^{k+m-1}. \quad (15)$$

Here $C(p^2, \Lambda)$ stands for the counterterm series:

$$C(p^2) = \left[C^{[\log]}(p^2) \cdot \log\Lambda + \sum_{n=0}^{\infty} C^{[n]}(p^2) \Lambda^{2n} \right], \quad (16)$$

Λ is the cutoff parameter, and every $C^{[x]}(p^2)$ $(x = \log, 0, 1, \ldots)$ is a power series in p^2:

$$C^{[x]}(p^2) = \sum_{n=0}^{\infty} c_n^{[x]} p^{2n}.$$

It can be shown that the sum in (15) contains the only nontrivial integral – that with $j+n = k+m = 0$; all the others behave as powers of the cutoff parameter Λ with coefficients that are just power series in p^2. This integral reads:

$$I(p^2) = \delta(p+q) \left[J(p^2) + a_1 \log\Lambda + a_2 \right]. \quad (17)$$

Here

$$J(p^2) \equiv - \int dr \frac{p^2 + 2rp}{(r^2 - m^2)^2 \cdot [(r+p)^2 - m^2]} \quad (18)$$

and a_1, a_2 are just arbitrary constants (they may depend on m^2). All the diverging terms (as well as a_1 and a_2) can be absorbed by the yet unfixed counterterm coefficients $c_n^{[x]}$ $(x = \log, 1, 2, \ldots)$ and the expression (15) can be rewritten as follows

$$S(p^2) = \sum_{il=0}^{\infty} D^{i00} D^{l00} \kappa_p^i \kappa_p^l J(p^2) + \sum_{n=0}^{\infty} c_n (p^2)^n. \quad (19)$$

Here c_n are the new (finite) counterterm coefficients to be fixed with the help of renormalization prescriptions.

Let us now reorder the terms in (19) as follows:

$$S(p^2) = \sum_{i=0}^{\infty} G^i J(p^2) {\kappa_p}^i + \sum_{i=0}^{\infty} \widetilde{d}_i {\kappa_p}^i. \quad (20)$$

Here

$$G^i = \sum_{k=0}^{i} D^{k00} D^{(i-k)00}, \quad (21)$$

and the coefficients $\widetilde{d}_i$ are certain combinations of c_n and various degrees of m^2. These coefficients remain unfixed until one formulates the corresponding RP's.

The problem is that the number of $\widetilde{d_i}$ is infinite. At the same time there are only two physically grounded requirements which can be used as the relevant RP's. First, we need to comply with the requirement

$$S(p^2)\Big|_{p^2=m^2} = 0\,. \tag{22}$$

This is necessary for fixing the pole position of the two-leg Green function. Second, for the correct normalization of the wave function one needs

$$\frac{\partial}{\partial p^2} S(p^2)\Big|_{p^2=m^2} = 0\,. \tag{23}$$

These two requirements only allow us to fix the constants $\widetilde{d_0}$ and $\widetilde{d_1}$:

$$\widetilde{d_0} = -G^0 J(m^2)\,. \tag{24}$$

$$\widetilde{d_1} = G^0 J'(m^2) - G^1 J(m^2)\,; \tag{25}$$

the counterterm coefficients $\widetilde{d_i}$ with $i = 2, 3, \ldots$ remain unfixed though they are certainly nonzero.

As shown in [1], the root of the problem lies in erroneous interpretation of what is the two-leg graph in the framework of effective theory. Let us consider this point in more detail.

It is pertinent to note that both the requirements (22) and (23) are based on the result of formal computation of the full propagator $P(p^2)$. Conventionally this is just a summation of the infinite Dyson's chain constructed from the links (2-leg insertions) connected to one another by simple propagators. Every link is considered as the full 1PI 2-leg function $S(p^2)$. At this point it is tacitly implied that every interim propagator is really presented it the chain. This is not always the case in effective theory just because some of them might be "killed" by the corresponding factors stemming from the adjacent vertices. This leads to a mishmash of orders in the perturbation theory based on the loop counting. Also, as shown above, this prevents one to obtain finite S-matrix elements with the help of finite number of RP's.

Let us analyze the calculation of the Dyson's chain for the (conventionally defined as 1PI) full 2-leg function S (conventional "self energy"). One has:

$$P(p^2) = \pi + \pi S\,\pi + \pi S\,\pi S\,\pi + \ldots = \stackrel{\text{formally}}{=} \frac{\pi}{1 - \pi S} = \frac{\pi}{p^2 - m^2 - S}\,. \tag{26}$$

This result is only valid under the condition

$$\left|\pi(p^2) S(p^2)\right| < 1\,. \tag{27}$$

When this condition is broken one needs to perform the analytic continuation of the right side of (27). In the case in question this condition is certainly broken: this is clearly visible from the expression (20). At the same time, the analytic continuation of the right side looks too difficult (if ever possible) because the pole is not the only singularity.

3 The true self energy function

In the paper [1] it has been suggested an alternative understanding of the meaning of expression (20). Let us rewrite (20) as follows

$$S(\kappa_p) = G^0\overline{J}(\kappa_p) + d_0 + \kappa_p d_1 + \\ + \sum_{j=1}^{\infty} \kappa_p^j G^j \overline{J}(\kappa_p) + \sum_{j=2}^{\infty} \kappa_p^j d_j \,. \tag{28}$$

Here

$$\overline{J}(\kappa_p) \equiv J(p^2) - J(m^2)\,, \tag{29}$$

the integral $J(p^2)$ has been defined in (18), and

$$d_0 \equiv 0\,; \qquad d_1 \equiv -G^0\, J'(m^2)\,; \\ d_j \equiv \widetilde{d}_j + J(m^2). \qquad (j \geqslant 2)\,. \tag{30}$$

Note that the counterterm coefficients d_j $(j = 2, 3, \ldots)$ remain unfixed.

Let us rewrite now the expression (28) as follows:

$$S(\kappa_p) = \Sigma(\kappa_p) + \sum_{j=1}^{\infty} \kappa_p^j R^j(\kappa_p) + \sum_{j=2}^{\infty} \kappa_p^j d_j \,. \tag{31}$$

Here

$$R^j(\kappa_p) \equiv G^j \overline{J}(\kappa_p)\,, \tag{32}$$

and

$$\Sigma(\kappa_p) \equiv G^0 \left(\overline{J}(\kappa_p) - \kappa_p \overline{J}'(\kappa_p)\right). \tag{33}$$

The above-defined function (33) is exactly what has been called in [1] the "true self energy function" (TSE). It meets the conditions of the type (22) and (23). Surely, those conditions only fix the function $\Sigma(\kappa_p)$ itself. They do not allow us to fix completely the form (19) because, first, the second item depends on the non-minimal parameters G^i $(i = 1, 2, \ldots)$ and, second, the last one contains the higher counterterms d_i with $i = 2, 3, \ldots$. Nevertheless, as will be shown below, the form (31) turns out quite sufficient for computing the dressed lines in graphs of the effective scattering theory. The reason is that in this case we are only interested in the renormalization of S-matrix elements; the Green functions may contain infinite terms *not affecting the amplitudes of physical processes*.

So, from this moment we can consider $\Sigma(\kappa_p)$ defined in Eq. (33) as finite correctly normalized function TSF. What about the higher counterterms that are presented in (31)? They are non-minimal. Therefore they can be absorbed by the adjacent vertices of the tree level; the latter ones can be considered as 1-loop n-leg counterterms with $n \geqslant 3$. The problem of fixing such counterterms should be addressed at a later step of the renormalization process.

4 Analytic reducibility

The idea of the future analysis is based on the above-discussed notion of line reduction.

It is pertinent to note that the compliance of the TSE function with the conditions of the type (22) and (23) automatically entails the maintenance of those conditions for the full one-loop 2-leg function

$S(\kappa_p)$. From this it immediately follows that the insertion of $S(\kappa_q)$ in external line of arbitrary S-matrix graph can be neglected: such insertions make no influence on the amplitudes of physical processes.

Let us now consider the effect of insertion of $S(\kappa_r)$ in the internal line of the 1PR tree level graph constructed from two effective vertices $V_1(p_1,\dots,p_m,r)$ and $V_2(q_1,\dots,q_n,r)$ connected to one another by the minimal line r[a]. The analytic expression for this graph reads:

$$\Gamma(p_1,\dots,p_m;q_1,\dots,q_n) =$$
$$= V_1(p_1,\dots,p_m,r)\frac{1}{\kappa_r}S(r^2)\frac{1}{\kappa_r}V_2(q_1,\dots,q_n,r)\,. \quad (34)$$

According to the conventional ("graphical") definition this is the 1PR graph. However, as will become clear below, it is more natural to rely upon the alternative ("analytical") definition of what is reducibility (G1PR and A1PR).

Let us substitute in (34) the explicit expression (31). This gives (with account taken of the relation (21)):

$$\begin{aligned}\Gamma(p,q) =& \\ = \frac{V_1(p,r)}{\kappa_r}&\left[\Sigma(\kappa_r) + \sum_{j=1}^{\infty}\kappa_r^j R^j(\kappa_r) + \sum_{j=2}^{\infty}\kappa_r^j d_j\right]\frac{V_2(q,r)}{\kappa_r} = \\ &= V_1(p,r)\frac{1}{\kappa_r}\Sigma(\kappa_r)\frac{1}{\kappa_r}V_2(q,r)+ \\ &+ \sum_{j=2}^{\infty} d_j\left[V_1(p,r)\kappa_r^{j-2}V_2(q,r)\right] + \\ &+ \sum_{j=2}^{\infty} G^j\overline{J}(\kappa_r)\left[V_1(p,r)\kappa_r^{j-2}V_2(q,r)\right] + \\ &+ G_1\left[V_1(p,r)\frac{\overline{J}}{\kappa_r}V_2(q,r)\right]. \quad (35)\end{aligned}$$

Let us analyze this result term by term.

In the left side of (35) we have the conventionally 1PR (or, the same G1PR) graph for the process $m \to n$ with one-loop dressed internal line. In contrast, the analytic form of the same expression may be interpreted as the finite sum of individual graphs constructed from the minimal secondary effective vertices. Some of those graphs are A1PI while another ones – A1PR. Indeed, the first line in the right side of (35)) (the second line in the Eq. (35)) looks like the A1PR graph with the internal line dressed with the 1-loop TSE insertion. The second item in the right side (see the third line in (35)) looks like the sum of local vertices of the counterterm type. Those counterterm vertices have $m+n$ legs and the loop index $l = 1$ This index was caused by the reduction of loop inner lines. The third item (the fourth line in (35)) is, in fact, nothing but the initial one-loop graph with both lines r reduced. At last, the last line in (35) is nothing but the initial graph with one of the lines r reduced.

Clearly, the expression (35) is still divergent because the above considered A1PI one-loop L-leg graphs ($L = m{+}1, n{+}1, m{+}n$) have not been renormalized; the necessary minimal 1-loop counterterms have not been taken into account. This point will be discussed in forthcoming publications. The main ideas have been discussed in [3].

[a]The non-minimal line r presents no interest because it can be reduced independently of the structure of $S(\kappa_r)$.

At the moment it is important to realize that there is no problem with 1-loop dressing of the lines (both external and internal) of effective scattering theory graphs that occur in the case when we rely upon the notion of analytic reducibility. Besides, the meaning of the term "self energy" becomes clear in this case because the Dyson's sum of full A1PI 2-leg graphs gives for 2-leg Green function the expression

$$G_2(p^2) = \frac{1}{(p^2 - m^2) - \Sigma(p^2)} + Q(\kappa_p) \tag{36}$$

(here $Q(\kappa_p)$ is a power series in κ_p such that $Q(0) = 0$). This expression is quite acceptable from the physical point of view.

We thank M. A. Braun, S. A. Paston and M. I. Vyazovsky for interesting discussions. The work of V.V.V. was partially supported by the Saint Petersburg State University grant N 11.38.189.2014.

References

[1] Vladimir V. Vereshagin, Phys. Rev. D **89**, 125022 (2014).
[2] A. Vereshagin and V. Vereshagin, Phys. Rev. D **69**, 025002 (2004).
[3] K. Semenov-Tian-Shansky, A. Vereshagin, and V. Vereshagin, Phys. Rev. D **73**, 025020 (2006).
[4] S. Weinberg, *The Quantum Theory of Fields*, Vol. 1, (Cambridge University Press, Cambridge, 1996).
[5] N. N. Bogoliubov and D. V. Shirkov, *Introduction to the Theory of Quantized Fields* (Wiley-Interscience, New York, 1980).

The Casimir force for 2d sinusoidal gratings

Valery N. Marachevsky[1,★] and *Alexandra D.* Nelson[1]

[1] *Department of Theoretical Physics, Saint Petersburg State University, Ulianovskaya 1, Petrodvorets, 198504 St. Petersburg, Russia*

Abstract. The Casimir free energy for 2d gratings separated by a vacuum slit is expressed in terms of Rayleigh coefficients, a novel general approach valid for arbitrary 2d surface profiles of gratings is outlined. The normal Casimir force in the system of two identical Si gratings with 2d sinusoidal surface profiles separated by a vacuum slit is computed for several amplitudes of surface profiles, distance dependence of the force is studied. A comparison with results for flat boundaries is performed.

1 Introduction

The Casimir effect is a quantum fluctuation effect in the presence of boundaries. It was theoretically predicted by H.Casimir in 1948 [1] who considered two parallel perfectly conducting plates separated by a vacuum slit and derived the result for the attractive force between the plates.

The theory for two dispersive media which are periodic in one direction, translation invariant in the orthogonal direction and separated by a vacuum slit (i.e. systems and gratings with 1d spatial periodicity) was developed in Refs. [2, 3], the Casimir free energy was expressed in terms of Rayleigh coefficients [4]. Both normal and lateral Casimir forces can be measured in experiments with gratings. The theory was found to be in a very good agreement with experimental measurements of the lateral Casimir force between two sinusoidal 1d Au gratings with different amplitudes [5]. Important difference (up to 66%) in lateral force results based on Rayleigh decompositions and results based on Proximity Force Approximation (PFA) exploiting Lifshitz theory for flat geometries [6] was predicted theoretically and observed in experiments [5]. Details of the method can be found in Refs.[7, 8]. A modal approach to Casimir forces in periodic structures was developed in Ref.[9].

Measurements of the Casimir-Polder potential of Rb atom above a grating made of parallel Au stripes on the sapphire substrate in the presence of additional repulsive potential created by an external laser field were performed during diffraction of Bose-Einstein condensates of Rb atoms [10], results of this experiment were found to be in a good agreement with theoretical results based on Rayleigh decompositions. General analytic properties of Rayleigh coefficients were found in Ref.[11]. Existence of repulsive Casimir-Polder potential for an anisotropic atom in the presence of 20 nm thin 1d Au grating was shown in Ref.[11].

In the current paper the 1d formalism is generalized to systems with 2d periodicity. Periods of gratings coincide in a selected spatial direction though they may differ in orthogonal spatial directions.

★e-mail: maraval@mail.ru

Our formalism is applicable to arbitrary surface profiles of 2d gratings. Gratings are assumed to be homogeneous, their material properties are defined by frequency dependent permittivities.

An outline of the paper is the following. In Sec. 2 we express the Casimir free energy of two gratings with coinciding periods in terms of Rayleigh coefficients and outline a general approach valid for arbitrary 2d surface profiles of gratings. In Sec. 3 we evaluate the Casimir energy and the Casimir normal force between two Si gratings with 2d sinusoidal surface profiles for various amplitudes of surface profiles, distance dependence of the force is studied. A comparison with flat case E.Lifshitz result for the force between two semispaces separated by a vacuum slit [6] is performed.

2 General formalism

Consider two 2d gratings which are periodic in orthogonal spatial directions with periods d_x, d_y and separated by a vacuum slit. In general, materials of the gratings are defined by frequency dependent permittivities. For simplicity the condition $\mu = 1$ for magnetic permeability is assumed in the whole space. In our formalism it is possible to evaluate the Casimir free energy and the Casimir force for arbitrary surface profiles of 2d gratings, periods of the gratings in a selected spatial direction coincide though they may differ in orthogonal directions (Fig. 1).

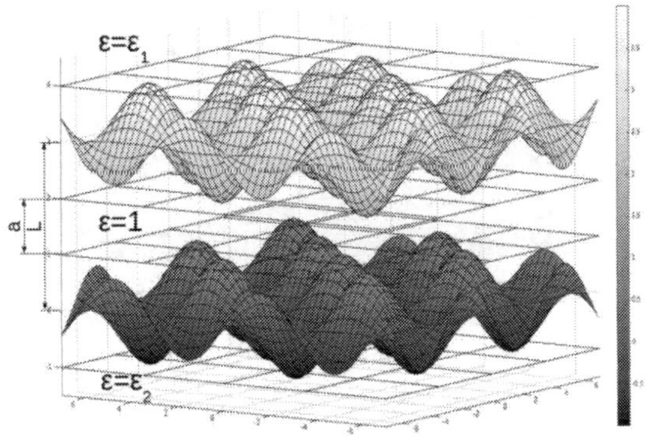

Figure 1. 2d gratings separated by a vacuum slit

The Casimir free energy for 2d gratings separated by a vacuum slit can be written as follows:

$$\mathcal{F} = k_B T \sum_{n=0}^{\infty}{}' \int_{-\pi/d_x}^{\pi/d_x} \frac{dk_x}{2\pi} \int_{-\pi/d_y}^{\pi/d_y} \frac{dk_y}{2\pi} \ln\det\left(I - R_{1down}(i\omega_n, k_x, k_y) R_{2up}(i\omega_n, k_x, k_y)\right), \tag{1}$$

here T is the temperature of the system, $\omega_n = \frac{2\pi k_B T n}{\hbar}$ are Matsubara frequencies; R_{1down}, R_{2up} are matrices of reflection coefficients from the lower and the upper gratings correspondingly, the prime means that the zero Matsubara term is taken with the coefficient $1/2$. To obtain reflection matrices for systems with periodic surface profiles or spatially periodic material properties it is convenient to consider Rayleigh decompositions [4].

Consider diffraction of an electromagnetic plane wave on a single grating (Fig. 2). For 2d periodic geometries the solutions of Maxwell equations can be decomposed in Fourier series on variables x and y. One can write exact solutions of Maxwell equations for $z > z_{max}$ in terms of Rayleigh reflection

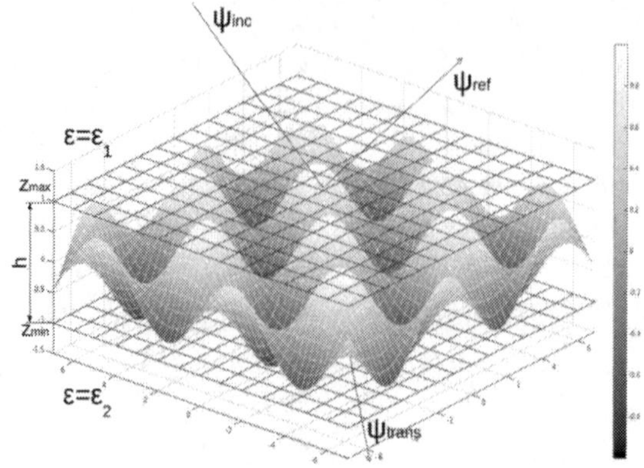

Figure 2. Reflection of the plane wave from the 2d grating

coefficients $R^{(e)}_{np,mq}, R^{(m)}_{np,mq}$ ($e^{-i\omega t}$ factors are omitted):

$$E^{+}_{y,pq} = I^{(e)}_{pq}\, e^{i\alpha_p x} e^{i\beta_q y} e^{-i\gamma_{pq} z} + \sum_{n,m=-\infty}^{+\infty} R^{(e)}_{np,mq} e^{i\alpha_n x} e^{i\beta_m y} e^{i\gamma_{nm} z}$$

$$H^{+}_{y,pq} = I^{(m)}_{pq}\, e^{i\alpha_p x} e^{i\beta_q y} e^{-i\gamma_{pq} z} + \sum_{n,m=-\infty}^{+\infty} R^{(m)}_{np,mq} e^{i\alpha_n x} e^{i\beta_m y} e^{i\gamma_{nm} z}, \qquad (2)$$

and for $z < z_{min}$ in terms of Rayleigh transmission coefficients $T^{(e)}_{np,mq}, T^{(m)}_{np,mq}$:

$$E^{-}_{y,pq} = \sum_{n,m=-\infty}^{+\infty} T^{(e)}_{np,mq} e^{i\alpha_n x} e^{i\beta_m y} e^{-i\gamma_{nm} z}$$

$$H^{-}_{y,pq} = \sum_{n,m=-\infty}^{+\infty} T^{(m)}_{np,mq} e^{i\alpha_n x} e^{i\beta_m y} e^{-i\gamma_{nm} z}, \qquad (3)$$

where

$$\alpha_p = k_x + p\frac{2\pi}{d_x}$$

$$\beta_q = k_y + q\frac{2\pi}{d_y} \qquad (4)$$

$$\gamma_{pq} = \sqrt{\varepsilon(\omega)\frac{\omega^2}{c^2} - \alpha_p^2 - \beta_q^2},$$

d_x, d_y are periods of gratings in orthogonal spatial directions, $k_x \in [-\pi/d_x, \pi/d_x], k_y \in [-\pi/d_y, \pi/d_y], p \in Z, q \in Z$.

Consider the vector of fourier components of longitudinal fields E_x, E_y, H_x, H_y. In numerical calculations it is sufficient to consider fourier components with numbers (n, m) possessing values $n \in (-N, N)$, $m \in (-M, M)$, values of N and M are determined by the accuracy of calculations. Let's denote by $[E_x]$ the vector with $(2N + 1)(2M + 1)$ fourier components of the field E_x, vectors for other components of the electromagnetic field are denoted in analogy.

In the range of values $z_{min} < z < z_{max}$ Maxwell equations can be rewritten in the form of first order differential equations [12]:

$$\partial_z \begin{pmatrix} [E_x] \\ [E_y] \\ [H_x] \\ [H_y] \end{pmatrix} = M \begin{pmatrix} [E_x] \\ [E_y] \\ [H_x] \\ [H_y] \end{pmatrix}, \tag{5}$$

M is a square matrix depending on $i\omega, k_x, k_y$ and grating surface profile in the range $z_{min} < z < z_{max}$. At $z = z_{max}$, $z = z_{min}$ the continuity conditions are imposed on all Fourier components of longitudinal fields E_x, E_y, H_x and H_y. As a result, the system of linear equations on Rayleigh coefficients $R^{(e)}$, $R^{(m)}$, $T^{(e)}$, $T^{(m)}$ can be obtained.

Arranging coefficients $R^{(e)}$, $R^{(m)}$ in the matrix

$$R_{1down} = \begin{pmatrix} R^{(e)}_{ns,ml}(\delta_{ps}\delta_{ql}, 0) & R^{(e)}_{ns,ml}(0, \delta_{ps}\delta_{ql}) \\ R^{(m)}_{ns,ml}(\delta_{ps}\delta_{ql}, 0) & R^{(m)}_{ns,ml}(0, \delta_{ps}\delta_{ql}) \end{pmatrix} \tag{6}$$

we obtain the matrix of reflection coefficients from the lower grating. The arguments of matrix elements in (6) are amplitudes of incident waves $\left(I^{(e)}_{pq}, I^{(m)}_{pq}\right)$.

For evaluation of the Casimir free energy we need matrices of reflection coefficients from the lower and the upper gratings. Suppose one finds the reflection matrix from the lower grating. In a special case when the upper grating is a mirror image of the lower grating it is sufficient to make a change of coordinates $z \to L - z_1, y \to -y_1$ in Rayleigh decompositions for the lower grating (L is a distance between middle planes of corrugation regions of the gratings, Fig.1) to obtain the reflection matrix for the upper grating.

3 Casimir energy and normal force

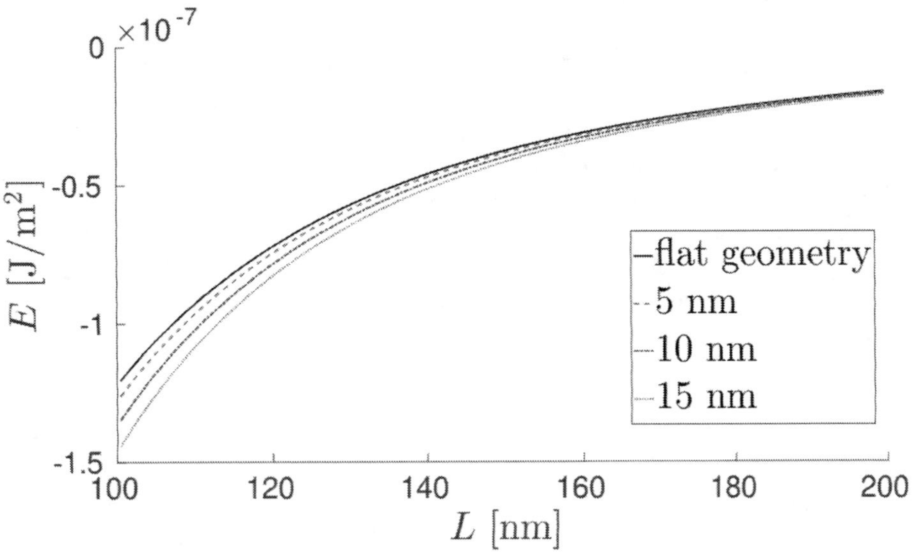

Figure 3. Casimir energy E of two 2d sinusoidal Si gratings with corrugation amplitudes $b = 5, 10, 15$ nm. The Lifshitz energy of two Si semispaces is shown by a black solid line. Grating periods are $d_x = d_y = 100$ nm.

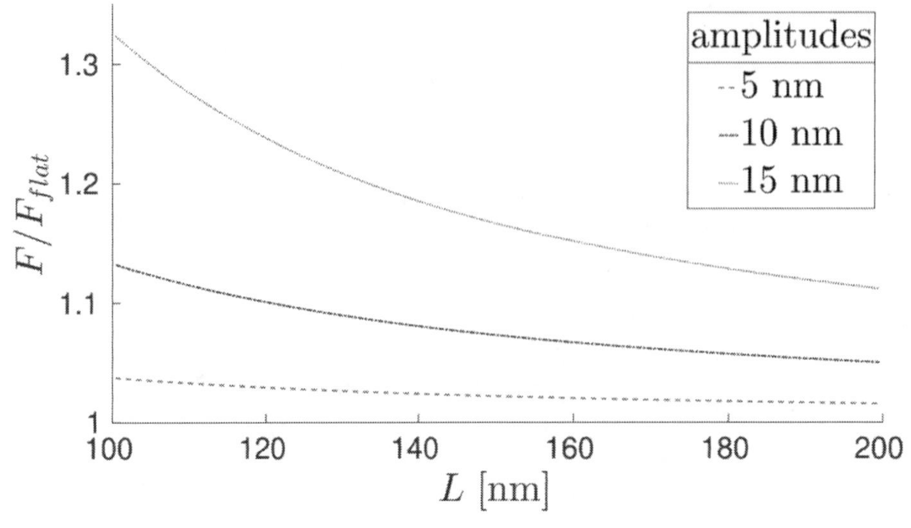

Figure 4. Ratios of the normal force F between two 2d sinusoidal Si gratings to the Lifshitz force F_{flat} between two Si semispaces separated by a distance L.

Consider two homogeneous identical Si gratings with sinusoidal 2d surface profiles separated by a vacuum slit, L is a distance between middle planes of corrugation regions of the gratings (Fig.1). The surface profile of each grating in the corrugation region is defined by a function $f(x, y) = b\cos(2\pi x/d_x)\cos(2\pi y/d_y)$. Note that in this case $z_{max} = -z_{min} = b$. In our calculations we use a frequency dispersion model of Si given by formula (6) in Ref.[13]. We study $T = 0$ case and evaluate Casimir energies E and normal Casimir forces $F = -\partial E/\partial L$ for gratings with periods $d_x = d_y = 100$ nm and amplitudes $b = 5, 10, 15$ nm. Casimir energies of two 2d sinusoidal Si gratings separated by a vacuum slit are evaluated by making use of the $T = 0$ limit of the formula (1), results are shown on Fig.3 where also the Lifshitz energy [6] of two Si semispaces separated by a vacuum slit L is added for comparison. Ratios of the normal Casimir force between two 2d sinusoidal Si gratings separated by a vacuum slit to the Lifshitz force between two Si semispaces separated by a vacuum slit $F_{flat}(L)$ are evaluated in the range $L \in [d_x, 2d_x]$, results are shown on Fig.4.

The normal Casimir force between two sinusoidal gratings is found to be larger than the Lifshitz force $F_{flat}(L)$ between two semispaces (Fig.4). The magnitude of the Casimir force between sinusoidal gratings increases with the increase of the corrugation surface profile amplitude b. This behavior of the Casimir force between gratings is consistent with general expectations: one can construct a sinusoidal grating profile from a dielectric semispace by extracting parts of dielectric bounded by the 2d sinusoidal curve from the semispace in half of the sinusoidal curve period and putting these parts onto the surface of the semispace in another half of the period. This argument explains why the normal force between two sinusoidal gratings should be larger than the force between two semispaces from which these sinusoidal gratings were constructed.

4 Conclusions

In this paper we outline a novel formalism for evaluation of the Casimir free energy in the system of two gratings with 2d periodic surface profiles which are separated by a vacuum slit. The developed

 DOI: 10.1051/epjconf/201612505024

formalism is valid for arbitrary 2d surface profiles of gratings. We apply our formalism to evaluate the normal Casimir force in the system of two identical 2d sinusoidal Si gratings separated by a vacuum slit at various separations between the gratings and for several amplitudes of sinusoidal surface profiles. A comparison with Lifshitz force between two semispaces separated by a vacuum slit is performed.

Acknowledgments

This research was supported in part by Saint Petersburg State University grant 11.38.237.2015. Research was carried out using computational resources provided by Resource Center "Computer Center of SPbU" (http://cc.spbu.ru/en).

References

[1] H. B. G. Casimir, Proc. Kon. Ned. Akad. Wetensch. **51**, 793–795 (1948).

[2] A. Lambrecht and V. N. Marachevsky, Phys.Rev.Lett. **101**, 160403, 1–4 (2008).

[3] A. Lambrecht and V. N. Marachevsky, Int. J. Mod. Phys. A **24**, 1789–1795 (2009).

[4] O. M. Rayleigh, Proc. Roy. Soc. A **79**, 399–416 (1907).

[5] H.-C. Chiu, G. L. Klimchitskaya, V. N. Marachevsky, V. M. Mostepanenko and U. Mohideen, Phys. Rev. B **80**, 121402(R) (2009); Phys. Rev. B **81**, 115417 (2010).

[6] E. M. Lifshitz, ZhETF **29**, 94 (1955); E. M. Lifshitz, Soviet Phys. JETP **2**, 73 (1956).

[7] V. N. Marachevsky, J.Phys.A: Math.Theor. **45**, 374021 (2012).

[8] V. N. Marachevsky, Theor.Math.Phys. **185** (1), 1492-1501 (2015).

[9] P. S. Davids, F. Intravaia, F. S. S. Rosa and D. A. R. Dalvit, Phys.Rev.A **82**, 062111 (2010).

[10] H. Bender, C. Stehle, C. Zimmermann, S. Slama, J. Fiedler, S. Scheel, S. Y. Buhmann and V. N. Marachevsky, Phys. Rev. X **4**, 011029 (2014).

[11] S. Y. Buhmann, V. N. Marachevsky and S. Scheel, Int.J.Mod.Phys.A **31**, 1641029 (2016).

[12] D. Maystre (ed.), *Selected Papers on Diffraction Gratings* (SPIE, Bellingham, 1993); R. Petit (ed.), *Electromagnetic Theory of Gratings* (Springer-Verlag, Berlin, 1980); M. C. Hutley, *Diffraction Gratings* (Academic Press, London, 1982); E. G. Loewen and E. Popov, *Diffraction Gratings and Applications* (Marcel Dekker, New York, 1997).

[13] I. Pirozhenko and A. Lambrecht, Phys. Rev. A **77**, 013811 (2008).

EPJ Web of Conferences **125**, 05025 (2016) DOI: 10.1051/epjconf/201612505025
QUARKS-2016

A note on field redefinitions and higher-spin equations

Massimo Taronna[1,*,**]

[1] *Physique Théorique et Mathématique*
Université Libre de Bruxelles and International Solvay Institutes
ULB-Campus Plaine CP231, 1050 Brussels, Belgium

Abstract. In this note we provide some details on the quasi-local field redefinitions which map interactions extracted from Vasiliev's equations to those obtained via holographic reconstruction. Without loss of generality, we focus on the source to the Fronsdal equations induced by s-0-0 couplings.

1 Introduction

The problem of admissible functional-classes has been of recent interest in the context of higher-spin (HS) theories [1]. In particular, in [2, 3] the quadratic interaction term sourcing Fronsdal's equations was extracted from Vasiliev's equations obtaining a pseudo-local expression of the schematic form:[1]

$$\Box\phi_{\mu(s)} + \ldots = \sum_{l=0}^{\infty} \frac{j_l}{l!l!}\,\Box^l\,(\nabla\ldots\nabla\phi\,\nabla\ldots\nabla\phi)_{\mu(s)}\,, \tag{1}$$

with coefficients $j_l \sim \frac{1}{l^3}$ for $l \to \infty$ for any choice of the spin s. The asymptotic behaviour of the coefficients raised the important question whether the backreaction extracted is or not strongly coupled[2]. Furthermore, a key question whose study was undertaken in [5, 6] was whether it is possible to extract the coefficients of the canonical Metsaev vertices with finitely many derivatives from the above tails. Indeed most of the coefficients at the cubic order are unphysical, since they can be removed by local field redefinitions. In this respect, the full list of Metsaev-like couplings was indeed extracted holographically in [7] and amounts to a finite number of coefficients for any triple of spins, to be contrasted to the above infinite set (see also [8, 9] for the analogous string theory computation and corresponding cubic couplings).[3]

*e-mail: massimo.taronna@ulb.ac.be
**Postdoctoral Researcher of the Fund for Scientific Research-FNRS Belgium.

[1] Notice that in the following we use the schematic notation $\Box^l \sim \ldots \nabla_{\mu(l)}\phi \ldots \nabla^{\mu(l)}\phi$. We give precise formulas for the above contractions in the spinor language in the following section in eq. (38). In this section all formulas are schematic and provide some intuition on their generic structure.

[2] Preliminary questions of this type were raised in [4].

[3] It is important to stress that in a fully non-linear HS theory it is expected that the appropriate field frame which makes HS geometry manifest will entail *all* of the above coefficients. The situation should be similar to the Einstein-Hilbert cubic couplings which are dressed by improvement terms that can be removed by a field redefinition at the cubic order. This is a further key motivation to understand these higher-derivative tails.

Remarkably, the pseudo-local nature of the above currents implies that the only way to relate them to their Metsaev-like counterparts is via a pseudo-local field redefinition of the same schematic form:

$$\phi_{\mu(s)} \longrightarrow \phi'_{\mu(s)} = \phi_{\mu(s)} + \sum_{l=0}^{\infty} \frac{a_l}{l!l!} \Box^l (\nabla \ldots \nabla\phi \, \nabla \ldots \nabla\phi)_{\mu(s)} \,. \tag{2}$$

This result has motivated a renewed interest in the analysis of the admissible functional classes in HS theories. Indeed, an arbitrary pseudo-local redefinition is sufficient to remove all pseudo-local current interactions [2, 10]. A proposal based on the invariance of the holographic Witten-diagrams was put forward in [5, 6], while in [11] (see [12] for further details) it was proposed to study classes of functions in z and y oscillators which are closed under star product multiplication. The aim of this note is to elaborate on these results from various perspectives, and to present the explicit form of the field redefinitions mapping the pseudo-local back-reaction (1) to its canonical (local) form.

In the following we list/summarise some relevant points of this analysis, together with the main results of this note, leaving the details of the derivation to the following sections:

- Defining the canonical s-derivative current made out of two scalars as:

$$J^{\text{can}}_{\mu(s)} = i^s \, \phi \, \overleftrightarrow{\nabla}_{\mu(s)} \phi \,, \tag{3}$$

  the redefinition which allows to bring the pseudo-local backreaction (1) to its canonical form:

$$\underbrace{\Box\phi_{\mu(s)} + \ldots}_{\mathcal{F}_{\mu(s)}} = \alpha_s \, J^{\text{can}}_{\mu(s)} \,, \tag{4}$$

  has the structure (2) with the following choice of coefficients:

$$a_{l-1} = \frac{2l+s+2}{2} \left(\frac{l!}{(l+s+1)!} \right)^2 [p_s(l) + \# \alpha_s] \,. \tag{5}$$

  Above $p_s(l)$ a polynomial of degree $2(s-1)$ in l while we have left the coefficients α_s arbitrary. In this way it is transparent to compare redefinitions that give different answers for the overall coefficient of the canonical current in (4). Moreover, only one of the redefinitions considered above gives the coupling constant which matches the one derived in [7, 13]. Keeping track of field normalisations (see Appendix B), in the type A theory the choice expected from holography is:

$$\alpha_s = 2^{1-s} \, g^2 \, N_0^2 \,, \tag{6}$$

  which has a simple spin-dependence up to a spin-independent factor proportional to the normalisation of the scalar field kinetic term and to the HS coupling constant g.

- The value (6) for the coupling constant can be fixed using Noether procedure to the quartic-order (see e.g. [2] for the 3d computation using admissibility condition) and was reconstructed from Holography in [7, 13]. So far, however, it was not possible to fix all cubic couplings using only the Noether procedure. That this should be possible in principle is suggested by the result of Metsaev in the light-cone gauge [14, 15].[4] Furthermore, the implications of the field redefinition mapping the theory to its canonical form at cubic order should be analysed at the quartic order. Such non-local redefinitions would generate a non-local quartic coupling. To appreciate the issue it might be worth

[4] In covariant language cubic couplings, including highest derivative ones, should be fixed by global part of the HS symmetry from the equation $\delta^{(1)} S^{(3)} \approx 0$.

noting that the above redefinitions can generate quartic couplings which differ from each other by single-trace blocks in the corresponding conformal block expansion (see e.g. [6, 13, 16, 17]). It is also conceivable that, at the quartic order another non-local redefinition will be needed to compensate the cubic redefinition and the additional non-local tails which would arise. The problem of finding a *non-perturbative* redefinition which relates the above tails to standard HS equations is so far open. In this note we restrict the attention to the lowest non-trivial order.

- The leading asymptotic behaviour for $l \to \infty$ of the coefficients in the field redefinition (5) is spin-independent and equal to $\frac{1}{l^3}$. Therefore, all the redefinitions considered above belong to the *same* functional space as the current in eq. (1) itself. This seems to suggest that the "admissible" functional class for field-redefinitions may be smaller than the corresponding "admissible" functional space for interactions.
- Notice that choosing a different coupling constant α_s for the canonical current amounts to a subleading contribution in (5) with spin-dependent behaviour $\sim \frac{1}{l^{2s+1}}$ for the coefficients a_l. Changing the overall coefficient α_s of the current in (4) by ϵ *does not* change the leading asymptotic behaviour $\sim \frac{1}{l^3}$ of the series expansion of the redefinition (5). This implies that the specification of an asymptotic behaviour for the coefficients a_l is *not* sufficient to specify a proper functional class beyond the proposal of [6]. Some further condition on the redefinitions must be introduced in order recover a unique admissible choice for α_s when enlarging the functional space beyond the proposal of [5, 6].

  Notice that the above analysis of the coefficients has a simple interpretation. Given a certain field redefinition with coefficients a_l bound to have a certain asymptotic behaviour, the corresponding improvement can be obtained by a simple action of the covariant adjoint derivative whose effect is to produce some other pseudo-local tail with coefficients $\tilde{a}_l$ which can be expressed linearly in terms of the coefficients a_l, a_{l-1} and a_{l-2} [6]:

$$\tilde{a}_l = A^{(s)}_l\, a_l + B^{(s)}_{l-1}\, a_{l-1} + C^{(s)}_{l-2}\, a_{l-2}\,. \tag{7}$$

  This means that $\tilde{a}_l < a_l$. On the other hand it is *impossible* to obtain coefficients $\tilde{a}_l$ growing for $l \to \infty$ more slowly than the original set of coefficients a_l. The only possibility is to have some fine-tuning so that the coefficients $\tilde{a}_l$ go much faster to zero than the original coefficients. This implies that in order to remove by a field redefinition a backreaction with a given asymptotic behaviour for its coefficients the best one can do is to have a redefinition with the same asymptotic behaviour $\tilde{a}_l \sim a_l$. This simple argument indirectly implies that the redefinition proposed in [11] should also be compatible with the asymptotic behaviour presented in this note. In more detail, a simple test of the functional class proposal of [11, 12] would be to check if the redefinitions (57) for different values of α_s than (6) are indeed not admissible.
- It might be of some interest also to consider a different perspective on the same problem. It is indeed possible to avoid to talk about the subtle issue of field redefinitions and study the limit of the finite derivative truncations of a given back-reaction:

$$J = \sum_{l=0}^{\infty} \frac{a_l}{l!l!}\, \Box^l\, (\nabla \ldots \nabla\phi\, \nabla \ldots \nabla\phi) \equiv \lim_{k\to\infty} \underbrace{\sum_{l=0}^{k} \frac{a_l}{l!l!}\, \Box^l\, (\nabla \ldots \nabla\phi\, \nabla \ldots \nabla\phi)}_{J_k}\,. \tag{8}$$

  In the above procedure each finite-derivative truncation is well-defined and one can extract the canonical-current piece of each truncation unambiguously. Analogously, one can compute for each truncation the corresponding Witten diagram using standard techniques from local field theories and take the limit only afterwards [5, 6].

We declare that the limit exists when the limit is finite (no infinite coefficient is generated) and is *independent* of local redefinitions f_k or g_k performed on each given truncation under the assumption that they converge to admissible redefinitions f_∞ and g_∞ according to[5] [6]. Using a diagrammatic language, the existence of the limit can be summarised by the following commutative diagram:

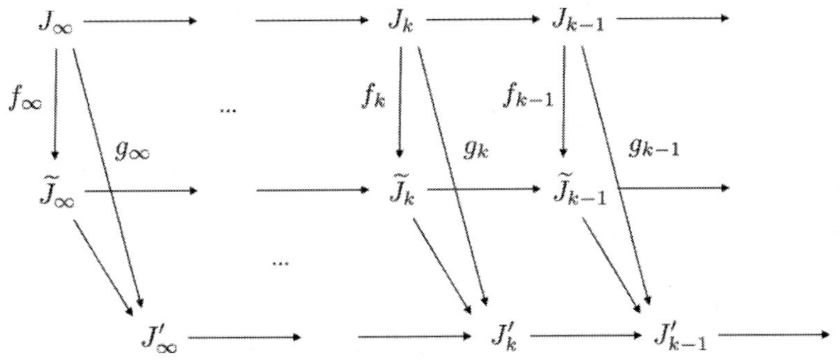

, (10)

where f_∞ and g_∞ belong to the functional class defined in [6] while $\widetilde{J}_k$ and J'_k are different local forms of the truncation which differ by a *local* field redefinition.

If this limit exists we can resum the higher-derivative tail and extract the coefficient of the canonical Metsaev-like coupling. If the limit does not exist, as observed in [5] for the backreaction (1), it might still be possible to define the sum via some analytic continuation. This is a standard situation where one can define the sum of infinite series formally introducing a cut off procedure. An example of this procedure is:

$$\sum_{l=0}^{\infty} 1\,, \tag{11}$$

which can be regularised introducing a cut-off as:

$$\sum_{l=0}^{\infty} e^{-\epsilon(l+\Lambda)} = \frac{e^{\epsilon-\Lambda\epsilon}}{e^{\epsilon}-1} \sim \frac{1}{\epsilon} + \left(\frac{1}{2}-\Lambda\right) + O(\epsilon)\,. \tag{12}$$

As expected, the finite part of the result is cut-off dependent and hence ambiguous. For a back-reaction (8) with a divergent sum one similarly ends up with expressions which can be defined formally by analytic continuation. For each given spin there exist a choice of cut-off which reproduces the result expected by the holographic reconstruction (6). The question then becomes the same which is usually asked about a renormalisable theory. Namely, whether the choice of regularisation which gives results compatible with holographic reconstruction (6) is spin-dependent. If the choice of cut-off is spin *independent*, the choice for spin 2 will fix at the same time the whole backreaction unambiguously. However, if the proper choice of cut-off compatible with the holographic reconstruction is spin-dependent the corresponding analytic continuation is not predictive.

[5] We briefly recall that in order to check whether a redefinition f_∞ belongs to the functional class of [6] one first needs to compute the associated improvement $J^{(f)}$ generated by the field redefinition at this order. The corresponding redefinition is then considered admissible iff the limit of the projections of each local truncation of $J^{(f)}$ on the local canonical coupling is vanishing:

$$\lim_{k\to\infty} J^{(f)}_k = (\lim_{k\to\infty} a_k)\,\mathbf{J} + \lim_{k\to\infty} \Delta_k\,, \qquad \lim_{k\to\infty} a_k = 0\,. \tag{9}$$

Here $\mathbf{J}$ is a fixed, but otherwise arbitrary, (*local*) representative for the non-trivial canonical coupling.

In the following we give the regularised results for the backreaction for $s = 2, 4, 6$ using the results of [5, 6]:

$$\alpha_2(\Lambda) = \frac{1}{36}\left(1 - 6\Lambda^2\right), \tag{13}$$

$$\alpha_4(\Lambda) = \frac{-2100\Lambda^6 + 14280\Lambda^5 - 31290\Lambda^4 + 26600\Lambda^3 - 1680\Lambda^2 + 10080\Lambda + 34567}{1058400}, \tag{14}$$

$$\alpha_6(\Lambda) = \frac{1}{92207808000}\Big(- 291060\Lambda^{10} + 6338640\Lambda^9 - 57387330\Lambda^8 + 280637280\Lambda^7 - 802849740\Lambda^6 + 1351860048\Lambda^5 - 1257850440\Lambda^4 + 525866880\Lambda^3 - 3991680\Lambda^2 + 79833600\Lambda + 415046341\Big). \tag{15}$$

It is now easy to verify that the choice of the cut-off which matches the holographically reconstructed result (6), in the appropriate normalisation, for α_2 is not consistent with α_4 or α_6 with the same normalisation. Therefore, the cut-off Λ must be spin-dependent, to compensate the highly spin-dependent form of the above regularised expressions. This makes the corresponding analytic continuations unpredictive.

This result is not in contradiction with the analysis of field redefinitions presented in this note. The above feature may be a different reincarnation of the fact that all redefinitions removing the higher-derivative tail, regardless the value of α_s, have the same asymptotic behaviour at $l \to \infty$. Similarly, this does not allow to single out a unique value of α_s.

- It might be interesting to compare the complicated redefinition (5) which matches the holographically reconstructed result starting from (1) with other redefinitions which would generate the required coupling constant but from a free theory. One may indeed start with *free* Fronsdal equations and find the non-local redefinition which would generate the appropriate cubic couplings. This redefinition should not be admissible but, remarkably, it has a faster asymptotic behaviour for $l \to \infty$ than (5):

$$a_{l-1} = \alpha_s\, s!(s-1)!\,(2l+s+2)\left(\frac{l!}{(l+s+1)!}\right)^2 . \tag{16}$$

This expression is simpler than (5), and falls off faster as $l \to \infty$: $a_l \sim \frac{1}{l^{2s+1}}$. Notice that the above redefinition (16) allows to generate the holographic backreaction from the free theory with the choice (6). The above redefinition however should not be considered admissible as it does not leave the cubic Witten diagrams invariant.

- In the parity violating case, the backreaction (1) is multiplied by a factor proportional to the parity violating phase θ. Surprisingly (see [3, 18]) this factor is given by $\cos(2\theta)$:

$$J_s(\theta) = \sum_{l=0}^{\infty} \frac{j_l \cos(2\theta)}{l!l!} \Box^l (\nabla \ldots \nabla\phi\, \nabla \ldots \nabla\phi)_s\,, \tag{17}$$

It was then observed in [3] that term by term each element of the pseudo-local series in (1) vanishes identically for $\theta = \frac{\pi}{4}$. The interpretation of this θ-dependence is at the moment unclear as it seems to be in contraddiction with the holographic expectations [19]. In [11] it was proposed that up to an admissible field redefinition one is left with a canonical current and a θ-independent coefficient. Let us assume that the field redefinition reproducing the holographically expected coupling constants is admissible. This means that the difference between the above non-local current (17) and a local canonical current with a fixed non-vanishing coefficient[6] α_s is an admissible improvement:

$$J_s(\theta) - \alpha_s j_s^{\text{can.}} = \Delta_s^{(J)}(\theta) . \tag{18}$$

Since the above must be true for any value of θ, we can now set $\theta = \frac{\pi}{4}$ and use that $J_s(\frac{\pi}{4}) = 0$. This however implies that the canonical current itself is an admissible improvement:

$$\alpha_s j_s^{\text{can.}} = \Delta_s^{(J)}(\tfrac{\pi}{4}) , \tag{19}$$

This further implies that $\Delta J_s(\theta) + x\,\Delta_s^{(J)}(\frac{\pi}{4})$ is an admissible improvement whose associated admissible field redefinition puts *any* coefficient (parameterised by x) in front of the canonical current. This contraddicts our assumption on the admissibility of the above field-redefinition. Furthermore, the above shows that in the $\theta = \frac{\pi}{4}$ case the initial redefinition itself must generate the canonical backreaction *from* the free Fronsdal equations, and therefore has to match (16). Let us also stress that the only assumptions we used is the existence of a field redefinition which is both allowed and capable of changing the θ dependence of the initial pseudo-local current (17). The contraddiction we find seems to imply that no such admissible redefinition may exist.

This note is organised as follows. After a brief review of the main formalism in Section 2, we move to the analysis of the redefinitions and to the study of their structure in Section 3. We end with a short summary and some outlook in Section 5.

2 Pseudo-local currents

In [2, 3, 5, 6, 10] a convenient generating function formalism was developed, first in 3d and then in 4d, to manipulate quasi-local current interactions and corresponding field redefinitions. In this section we recall the basic ingredients of the formalism, and refer to [2, 5] for further details. The main object is the zero-form $C(y, \bar{y}|x)$, which is a formal expansion in the spinorial oscillators y_α and $\bar{y}_{\dot\alpha}$ satisfying the linearised unfolded equations

$$\widetilde{D}C(y, \bar{y}|x) = 0 . \tag{20}$$

Here D and $\widetilde{D}$ are the adjoint and twisted-adjoint covariant derivatives expressed in terms of spinorial oscillators as:[7]

$$D = \nabla - h^{\alpha\dot\alpha}(y_\alpha \bar\partial_{\dot\alpha} - \bar{y}_{\dot\alpha}\partial_\alpha) , \tag{21}$$

$$\widetilde{D} = \nabla + ih^{\alpha\dot\alpha}(y_\alpha \bar{y}_{\dot\alpha} - \partial_\alpha \bar\partial_{\dot\alpha}) . \tag{22}$$

In the unfolded language, the zero-form is the main ingredient upon which one constructs ordinary current interactions. Furthermore, upon solving the twisted adjoint covariant constancy condition one recovers the relations between components of the zero form and derivatives of the linearised Weyl tensors associated to the HS fields:

$$C_{\alpha(k+2s),\dot\alpha(k)}(x) \sim \nabla^k C_{\alpha(2s)}(x) , \qquad C_{\alpha(k),\dot\alpha(k+2s)}(x) \sim \nabla^k C_{\dot\alpha(2s)}(x) , \tag{23}$$

Above, $C_{\alpha(2s)}(x)$ and $C_{\dot\alpha(2s)}(x)$ are the self-dual and anti-self dual part of the linearised HS Weyl tensor for $s > 0$ and $\Phi(x) \equiv C(0, 0|x)$ is the actual scalar field.

[6] Whether or not α_s depends on θ is not important for the following argument. The only assumption is that α_s does not vanish for any value of θ.

[7] $h^{\alpha\dot\alpha}$ is the vierbein, $\nabla = d - \omega^{\alpha\alpha} y_\alpha \partial_\alpha - \varpi^{\dot\alpha\dot\alpha} \bar{y}_{\dot\alpha} \bar\partial_{\dot\alpha}$ is the AdS$_4$ covariant derivative, while $\varpi^{\alpha\alpha}$ and $\varpi^{\dot\alpha\dot\alpha}$ are the (anti-)self-dual components of the spin-connection of AdS$_4$.

DOI: 10.1051/epjconf/201612505025

A current which sources the HS Fronsdal operator is a bilinear functional $J_{\alpha(s)\dot{\alpha}(s)}(C,C)$ of the zero-form C, which is conserved on the equations of motion (20):

$$\nabla^{\beta\dot{\beta}} J_{\beta\alpha(s-1)\dot{\beta}\dot{\alpha}(s)} \approx 0\,. \tag{24}$$

For practical purposes, it is convenient to specify the most general form for a current in terms of a generating function Kernel $J(Y,\xi,\eta)$:

$$\mathbf{J}(C,C) = \int d^4\xi\, d^4\eta\, J(Y,\xi,\eta)\, C(\xi|x)C(\eta|x)\,. \tag{25}$$

Above we have introduced the Fourier transform of the 0-form with respect to the spinorial variables y and $\bar{y}$,

$$C(y,\bar{y}|x) = \int d^2\xi d^2\bar{\xi}\, e^{iy^\alpha\xi_\alpha + i\bar{y}^{\dot{\alpha}}\bar{\xi}_{\dot{\alpha}}}\, C(\xi,\bar{\xi}|x)\,, \tag{26}$$

where for ease of notation we denote both Fourier transforms and original 0-forms by the same letter. In this way, contractions of indices are encoded as simple monomials:

$$(y\xi)^n(y\eta)^m(\bar{y}\bar{\xi})^{\bar{n}}(\bar{y}\bar{\eta})^{\bar{m}}(\xi\eta)^l(\bar{\xi}\bar{\eta})^{\bar{l}} \sim C_{\alpha(n)\nu(l);\dot{\alpha}(\bar{n})\dot{\nu}(\bar{l})}C_{\alpha(m)}{}^{\nu(l)}{}_{\dot{\alpha}(m)}{}^{\dot{\nu}(\bar{l})}\,. \tag{27}$$

In addition to the above way of representing a generic current interaction it is also convenient to introduce generating functions of coefficients via contour integrals, as originally proposed in [10]. Restricting the attention to the canonical current sector, which in four-dimensions is uniquely specified by the absence of trace components, one can then write the most general current kernel as

$$J^{\text{can.}}(Y,\xi,\eta) = \oint_{\tau_i,s,r} g(\alpha_1,\alpha_2,\beta,\gamma)\, e^{i(sy\zeta^- + \tau_1\xi\eta + r\bar{y}\bar{\zeta}^+ + \tau_2\bar{\xi}\bar{\eta})}\,, \tag{28}$$

in terms of a function of 4 complex variables:

$$\alpha_1 = \tau_1^{-1}\,, \qquad \alpha_2 = \tau_2^{-1}\,, \qquad \beta = s^{-1}\,, \qquad \gamma = r^{-1}\,, \tag{29}$$

$$\zeta^\pm = \xi \pm \eta\,, \qquad \bar{\zeta}^\pm = \bar{\xi} \pm \bar{\eta}\,. \tag{30}$$

These parametrise the four contractions of indices relevant to the canonical current sector in 4d. One can easily translate between the contour integral form and the generating function form via:

$$\alpha_2^{m+1}\alpha_1^{n+1}\beta^{s_1+1}\gamma^{s_2+1} \to \frac{(i\bar{\xi}\bar{\eta})^m(i\xi\eta)^n(i\bar{y}\bar{\zeta}^+)^{s_2}(iy\zeta^-)^{s_1}}{m!\,n!\,s_1!\,s_2!}\,. \tag{31}$$

Notice that in this generating function calculus the function $g(\alpha_1,\alpha_2,\beta,\gamma)$ is defined as a formal series, and this should be thought at this level as a regularity condition. This regularity condition will be assumed here since it is equivalent to pseudo-locality of the back-reaction. Notice also that a constant term or terms proportional to $1/\alpha_1$ drop out of the contour integration. In particular, two generating functions will give the same pseudo-local current if they differ by terms of this type. For the details we refer to [2, 10] and use the symbol $\sim$ to indicate equality modulo the above equivalence relation.

To conclude this section we present the corresponding expressions for the explicit Fronsdal currents of the type s-s_1-s_2:

$$\frac{1}{4(s-1)}\underbrace{\left[\Box + 2(s^2-2s-2) + \ldots\right]}_{\mathcal{F}}\phi = J^{\text{Fr.}}\,, \tag{32}$$

as extracted from Vasiliev's equations[8] in [3]:

$$J^{\text{Fr.}} = \frac{i}{2}e^{2i\theta}(\beta\gamma)^{s+1}\sum_{n,m} c^{(s)}_{n,m}\left(\frac{p_1(n) - m\,p_2(n)}{(n+s-1)^2(n+s)^2(n+s+1)}\right)\alpha_1^n\alpha_2^m + h.c.\,, \tag{33}$$

with

$$\begin{aligned} p_1^{(s)}(n) &= \Big[[2(s-1)]n^3 + [2+s(6s-5)]n^2 + \left[s\left(1-4s+6s^2\right)\right]n \\ &\quad + [s(1+s)(2+s(2s-3))]\Big]\,, \end{aligned} \tag{34}$$

$$p_2^{(s)}(n) = \Big[[2(s-1)]n^2 + [2+s(4s-3)]n + s(1+s(2s-1))\Big]\,, \tag{35}$$

and

$$c^{(s)}_{n,m} = \frac{1}{4(s-1)}\frac{(-1)^{m+s}+(-1)^n}{2}\,. \tag{36}$$

Notice that powers of α_1 and α_2 translate into contractions among the 0-form, hence powers of $\alpha_1\alpha_2$ parameterise powers of $\Box$ in the metric-like language. We can then define $\tau = \alpha_1\alpha_2$ parametrising the pseudo-local tail of the given interaction term. In general, when restricting attention to sources to the Fronsdal equation in the spin-s sector, one fixes the dependence on β and γ as in (33) while working with a generating function of the type:

$$J_s = i\cos(2\theta)(\beta\gamma)^{s+1}\,\alpha_1^n\,g(z)\,, \qquad z = \alpha_1\alpha_2\,. \tag{37}$$

The function $g(z)$ parameterises the infinite non-local tail while the dependence on β, γ and a single α_1 fixes the canonical current tensor structure. In the s-0-0 case the dictionary can be given quite explicitly as:

$$J_s \to \cos(2\theta)\sum_{l,k} a_{l,k}\nabla_{\mu(s-k)\nu(l)}\Phi\nabla_{\mu(k)}{}^{\nu(l)}\Phi\,, \tag{38}$$

$$\nabla_{\mu(s-k)\nu(l)}\Phi\nabla_{\mu(k)}{}^{\nu(l)}\Phi \equiv h_\mu^{\alpha\dot\alpha}\dots h_\mu^{\alpha\dot\alpha}C_{\alpha(s-k)\nu(l),\dot\alpha(s-k)\dot\nu(l)}C_{\alpha(k)}{}^{\nu(l)}{}_{,\dot\alpha(s-k)}{}^{\dot\nu(l)}\,, \tag{39}$$

with

$$a_{l,k} = \frac{(-1)^{k+l}s!s!}{l!l!k!k!(s-k)!(s-k)!}\,g_l\,, \qquad g(z) = \sum_{l=0}^{\infty} g_l\,z^{l+1}\,. \tag{40}$$

The above dictionary holds also for redefinitions of the Fronsdal field and allows to easily translate from the generating function language to the standard tensorial language. Notice that the notation $\nabla_{\mu(s-k)\nu(l)}\Phi\nabla_{\mu(k)}{}^{\nu(l)}\Phi$ is defined by eq. (39) and includes symmetrisation and traceless projection. In particular $\nabla_{\mu(s)} \neq \nabla_\mu\dots\nabla_\mu$. We give further details in Appendix A.

In the following we shall restrict our attention to the function $g(z)$ while assuming the above dictionary to go back to standard tensorial notation. It is important to stress that the $\frac{1}{(l!)^2}$ factor in (40) arises via the above contour integrations and does not appear in the function $g(z)$.

[8]The normalisation for the Fronsdal tensor comes from the solution to torsion as described in [3].

3 Pseudo-Local Field Redefinitions

In this section we employ the generating function formalism to study the non-local field redefinitions which relate the back-reaction extracted from Vasiliev's equation in [3] to canonical currents. For simplicity we work in the A-type model, setting $\theta = 0$. The discussion generalises straightforwardly to any choice of θ as this only appears as an overall factor. The effect of a field redefinition of the spin-s field quadratic in the 0-form can be encoded in an arbitrary function $k(z)$ (analogous to $g(z)$ above) via the differential operator[9]

$$\delta J_s = \frac{(\beta\gamma)^{s+1}}{4(s-1)}\Big[z^2(z+1)^2 k_s''(z) + z(z+1)(2s+(3+\bar{n})z+3+\bar{n})k_s'(z) \\ + \left(\bar{n}(z+1)(z+s+1) + s^2(z+1) + 2s + (z+1)^2\right)k_s(z)\Big], \quad (42)$$

where $\bar{n} = n - m$ with n and m giving the power of α_1 and α_2 respectively as in (33). In this way, the problem of finding redefinitions which remove the pseudo-local tail is thus mapped into a ordinary differential equation.

In the following we apply this formalism to the current (33). We first give the explicit field redefinition which maps the pseudo-local current (33) to a canonical current with an arbitrary overall coefficient:

$$\underbrace{\Box\phi + \ldots}_{\mathcal{F}} = \alpha_s\, J^{\text{can}}\,. \quad (43)$$

We then compare redefinitions which give different overall coefficients α_s for the canonical current. For ease of notation and without loss of generality we restrict the attention to the currents with $\bar{n} = 0$.

Spin-2:

Using the generating function formalism reviewed above, the back-reaction (33) for $s = 2$ and $\bar{n} = 0$ can be encoded by (see eq. (37)):

$$g_2(z) = \sum_{l=0}^{\infty} \frac{(-1)^l}{2\,(l+1)^2(l+2)}\, z^l = \frac{z - z\,\text{Li}_2(-z) - (z+1)\log(z+1)}{2z^2}\,. \quad (44)$$

The field redefinition that reduces the above pseudo-local current to its canonical form can be obtained solving the differential equation:

$$z(z+1)\left(z(z+1)k_2''(z) + (3z+7)k_2'(z)\right) + (z+3)^2 k_2(z) = -4\,g_2(z) + \alpha_2 z + \beta_2\,. \quad (45)$$

The constant β_2 parameterises non-trivial pseudo-local redefinitions which result in a vanishing[10] contribution to the backreaction upon performing the contour integration. On the other hand, α_2 is an arbitrary constant in front of the canonical current (which in this case is the stress tensor (43)). In the following we fix the constant $\beta_2 = 2\alpha_2 + 2$ in such a way that the slowest contribution to the coefficients in the $l \to \infty$ limit is set to zero. Furthermore, one should carefully pick the unique

[9]The corresponding redefinition at the level of unfolded equations reads:

$$\delta\omega_s(y,\bar{y}) = h^{\alpha\dot{\alpha}}\partial_\alpha\bar{\partial}_{\dot{\alpha}}k_s(y,\bar{y})\,. \quad (41)$$

[10]We have not considered terms on the right-hand side of the type z^{-m}, since the corresponding redefinitions vanish upon performing the contour integration. This can be checked by solving the differential equation (see [2]).

solution to the above differential equation which is analytic in z, as this corresponds to the standard regularity condition on pseudo-local functionals [5, 10]. The final result has the following series expansion around $z = 0$:

$$k_2(z) = \sum_{l=0}^{\infty} \frac{(4\alpha_2 + l(l+3) + 2)}{(l+1)^2(l+2)(l+3)^2} (-z)^l , \tag{46}$$

where we have left α_2 arbitrary.

Spin-4:

The source to the spin-4 Fronsdal operator extracted from (33) for $\bar{n} = 0$ can be encoded in the following generating-function:

$$g_4(z) = \sum_{n=0}^{\infty} \frac{(3l+11)}{3(l+3)^2(l+4)^2} (-z)^l \sim -\frac{1}{3z^4} \left[(2z+1)\mathrm{Li}_2(-z) + (z+1)\log(z+1)\right] . \tag{47}$$

The redefinition bringing the above back-reaction to its canonical form (43) with an arbitrary coefficient α_4, can be obtained solving the differential equation:

$$z(z+1)\left(z(z+1)k_4''(z) + (3z+11)k_4'(z)\right) + (z^2+18z+25)k_4(z) = -12\, g_4(z) + \alpha_4 z + \beta_4 . \tag{48}$$

Again, β_4 parameterises pseudo-local redefinitions which do not change the back-reaction, and can be chosen to improve the asymptotic behaviour. In this case we choose $\beta_4 = \frac{1}{2}(3\alpha_4 + 4)$. The solution to the above differential equation is then:

$$k_4(z) = \sum_{l=0}^{\infty} \frac{2}{5} \frac{p_4(l) + 120(6\alpha_4 + 5)}{(l+1)^2(l+2)^2(l+3)(l+4)^2(l+5)^2} (-z)^l , \tag{49}$$

with $p_4(l)$ a polynomial of order 5 in the variable l:

$$p_4(l) = l\left(5l^5 + 77l^4 + 470l^3 + 1445l^2 + 2345l + 1898\right) . \tag{50}$$

Spin-6:

The source to the spin-6 Fronsdal operator extracted from (33) for $\bar{n} = 0$ can be encoded in the generating-function

$$g_6(z) = \sum_{l=0}^{\infty} \frac{3(5l+28)}{10(l+5)^2(l+6)^2} (-z)^l \sim -\frac{3}{10\,z^6} \left[(z+1)\log(z+1) + (3z+2)\mathrm{Li}_2(-z)\right] . \tag{51}$$

The redefinition bringing the above back-reaction to its canonical form (43) with an arbitrary coefficient α_6 can be obtained solving the differential equation:

$$z(z+1)\left(z(z+1)k_6''(z) + 3(z+5)k_6'(z)\right) + z(z+38) + 49k_6(z) = -20\, g_6(z) + \alpha_6 z + \beta_6 . \tag{52}$$

Again, β_6 parameterises pseudo-local redefinitions which do not change the back-reaction. In this case we choose $\beta_6 = \frac{1}{3}(4\,\alpha_6 + 6)$. The solution to the above differential equation is then given by:

$$k_6(z) = \sum_{l=0}^{\infty} \frac{p_6(l) + 2 \cdot 6! \cdot 5!\,(\alpha_6 + 1)}{(l+1)^2(l+2)^2(l+3)^2(l+4)(l+5)^2(l+6)^2(l+7)^2}\,(-z)^l\,, \tag{53}$$

with $p_6(l)$ a polynomial of order 10 in the variable l:

$$p_6(l) = l\Big(3l^9 + \frac{323l^8}{3} + 1686l^7 + \frac{105862l^6}{7} + 85755l^5 + 320009l^4 + 792684l^3 + \frac{3844612l^2}{3} + 1288512l + \frac{5072640}{7}\Big). \tag{54}$$

Generic even spin:

In the generic spin-s case, the source quadratic in the 0-form to the Fronsdal operator extracted in [3] can be encoded by the generating-function

$$g_s(z) = \sum_{l=0}^{\infty} \frac{s(2l(s-1) + s(2s-3) + 2)}{8(s-1)(l+s-1)^2(l+s)^2}\,(-z)^n \sim -\frac{c_s}{z^s}\Big[(z+1)\log(z+1) + \tfrac{1}{2}(s(z+1) - 2)\mathrm{Li}_2(-z)\Big]\,, \tag{55}$$

with

$$c_s = \frac{1}{4}\frac{s}{s-1}\,. \tag{56}$$

The corresponding redefinition mapping the above pseudo-local back-reaction to a canonical current (43) with overall coefficient α_s (43) is:

$$k_s(z) = \sum_{l=0}^{\infty} \frac{2l+s+2}{2}\left(\frac{l!}{(l+s+1)!}\right)^2 [p_s(l) + a_s\,\alpha_s]\,(-z)^l\,, \tag{57}$$

with $p_s(l)$ a polynomial of degree $2(s-1)$ in the variable l and a_s a spin-dependent constant. The general form of the polynomial for arbitrary spin is cumbersome, and we do not present it explicitly.

4 Analytic continuation

In this section we study the analytic continuation of the formal series obtained by considering the limit of the lower derivative truncations of the backreaction (1).

The main observation of [5, 6] is that for any truncation of the pseudo-local interaction term (33) it is possible to compute the corresponding projection on the canonical current piece which is parametrised by finitely many structures in correspondence with Metsaev's cubic couplings. Each higher-derivative term gives a contribution to the canonical coupling weighted by some proportionality factor $C^{(s)}_l$ which measures the overlap of the higher-derivative term on the canonical structure. Schematically the projection reads:

$$\mathcal{P}\left(\Box^l\,(\nabla\ldots\nabla\phi\,\nabla\ldots\nabla\phi)_{\mu(s)}\right) = C^{(s)}_l\,(\nabla\ldots\nabla\phi\,\nabla\ldots\nabla\phi)_{\mu(s)}\,, \tag{58}$$

where indeed the right-hand side is proportional to the canonical structure times a certain overall coefficient which was computed in [5, 6].

After the projection the canonical structure factorises and one is left with an overall coefficient which combines together all contributions from each higher-derivative term. Below we give the corresponding coefficient for some low spin examples:

$$s = 2\,, \qquad -\frac{1}{12}\sum_{l=1}^{\infty} l\,, \tag{59}$$

$$s = 4\,, \qquad -\frac{1}{3\cdot 7!}\sum_{l=1}^{\infty} \frac{l(l+1)(l+2)^2(3l+11)(5l(l+4)+3)}{(l+3)(l+4)}\,, \tag{60}$$

$$s = 6\,, \qquad -\frac{3}{5\cdot 11!}\sum_{l=1}^{\infty} \frac{(l+4)!}{(l-1)!}\,\frac{(l+3)(5l+28)(7l(l+6)(3l(l+6)+19)+20)}{(l+5)(l+6)}\,, \tag{61}$$

$$s = 8\,, \qquad -\frac{7}{6\cdot 15!}\sum_{l=1}^{\infty} \frac{(l+6)!}{(l-1)!}\,\frac{(l+4)(7l+53)(l(l+8)(11l(l+8)(13l(l+8)+274)+14631)+420)}{(l+7)(l+8)}\,. \tag{62}$$

These series are divergent and can be regularised by a standard ζ-function regularisation introducing a regulator of the type $e^{-\epsilon(l+\Lambda)}$ which allows to resum them. Dropping the divergent parts and setting ϵ to zero one then obtains the following regularised expressions for the corresponding overall coefficients of the canonical current:[11]

$$\alpha_2(\Lambda) = \frac{1}{36}\left(1 - 6\Lambda^2\right) \tag{63}$$

$$\alpha_4(\Lambda) = \frac{-2100\Lambda^6 + 14280\Lambda^5 - 31290\Lambda^4 + 26600\Lambda^3 - 1680\Lambda^2 + 10080\Lambda + 34567}{1058400} \tag{64}$$

$$\begin{aligned}\alpha_6(\Lambda) = \frac{1}{92207808000}\Big(&-291060\Lambda^{10} + 6338640\Lambda^9 - 57387330\Lambda^8 + 280637280\Lambda^7\\ &- 802849740\Lambda^6 + 1351860048\Lambda^5 - 1257850440\Lambda^4 + 525866880\Lambda^3\\ &- 3991680\Lambda^2 + 79833600\Lambda + 415046341\Big)\end{aligned} \tag{65}$$

$$\begin{aligned}\alpha_8(\Lambda) = \frac{1}{39269461271040000}\Big(&-51531480\Lambda^{14} + 2307024720\Lambda^{13} - 45901996140\Lambda^{12}\\ &+ 535422167280\Lambda^{11} - 4065466421016\Lambda^{10} + 21086692426800\Lambda^9\\ &- 76212086580630\Lambda^8 + 191985197049360\Lambda^7 - 330826659683520\Lambda^6\\ &+ 372883114251648\Lambda^5 - 249836835568320\Lambda^4 + 77398236115200\Lambda^3\\ &- 62270208000\Lambda^2 + 2615348736000\Lambda + 16930453296697\Big)\,.\end{aligned} \tag{66}$$

These results allow in principle to fix the cutoff in order to recover the holographically reconstructed result of [7, 13]. The main observation is however that the associated choice for the cutoff is *spin-dependent*. Furthermore the regularised coupling constant is given by complicated polynomials of order $\Lambda^{2(s-1)}$. The lack of a spin-independent regularisation makes the corresponding analytic continuations unpredictive as a consequence of the strongly coupled nature of the higher-derivative tails. It is therefore not clear how to identify a regularisation which preserves all HS symmetries without

[11] In the formulas for $\alpha_s(\Lambda)$ we take into account the normalisation $\frac{1}{4(s-1)}$ in (32) so to arrive to a source of the type $\mathcal{F}_{\mu(s)} = \alpha_s(\Lambda) J^{\text{can.}}_{\mu(s)}$.

solving the Noether procedure up to the quartic order. On the other hand, the simple form of the local holographically reconstructed couplings [7] predicts a very simple structure for the corresponding local interactions.

5 Conclusions

In this note we have determined explicitly the general form of the redefinition which reduces the back-reaction extracted in [3] from Vasiliev's equations to a canonical current with an arbitrary overall coefficient (43). We have also given some details about the analytic continuation of the formally divergent sums. Our analysis provides a convenient test ground to probe functional class proposals for admissible non-local interactions.

The main conclusion of this note is that the dictionary between Vasiliev's equations and standard HS equations like Fronsdal's equations will not be complete unless a full non-perturbative functional class able to fix the overall coefficient in front of the canonical interaction terms will be provided. Furthermore, it is worth stressing that in this note we have restricted the attention to the cubic order which admits a local completion at least for fixed spins. At the quartic and higher orders we expect the situation to be even more subtle, since non-localities may not be anymore removed by redefinitions, and very restrictive consistency requirements will be put in place by consistency [6, 20].

We conclude this note with a short list of observations and comments:

- The main difficulty of HS theories is the absence of a scale beyond the AdS-radious. This implies that the behaviour of a pseudo-local tail can only be controlled by its coefficients owing to $\Lambda[\nabla, \nabla] \sim 1$. A functional class proposal would then prescribe a given asymptotic behaviour for the coefficients of higher and higher order terms as in the proposal of [6].
- It would be nice to study the limit $\alpha' \to \infty$ at the level of effective cubic string field theory couplings in AdS. Some issues about commutativity of limits may arise in this context when taking the limit $\alpha' \to \infty$ before or after removing the higher-derivative tails. It might well be that if one considers the naive $\alpha' \to \infty$ limit of a non-local string coupling this would indeed be strongly coupled. This would mean that the tensionless limit should be well-defined only in a particular field frame, while singular in others. This kind of situation suggests that allowed redefinitions for any finite value of α' could become not allowed after the limit. If so the infinities observed in this note could be resolved upon taking the limit from string theory using α' as regulator for the strongly coupled series.
- Beyond the particular problem of mapping the tail in (1) to a canonical form, the simplicity of the 4d theory in the spinorial language is expected to manifest in a simple form of all quadratic sources when rewritten in the unfolded form. Such rewriting, and the structures involved beyond the $\star$-product, is at the moment unknown (see however [21, 22]). For this reason an interesting problem would be to unfold the non-linear Fronsdal equations in 4d coming from the cubic couplings extracted holographically in [7]. This result would provide us with the complete list of cubic vertices in the unfolded equations possibly giving us a hint of higher-order completions.

Acknowledgments

I am grateful to N. Boulanger, S. Didenko, D. Francia, A. Sagnotti, E. Sezgin, C. Sleight, Z. Skvortsov, D. Sorokin, P. Sundell and M. Vasiliev for useful discussions. The research of M. Taronna is partially supported by the Fund for Scientific Research-FNRS Belgium, grant FC 6369 and by the Russian Science Foundation grant 14-42-00047 in association with Lebedev Physical Institute. This research was also supported by the Munich Institute for Astro- and Particle Physics (MIAPP) of the DFG cluster of excellence "Origin and Structure of the Universe".

A Canonical Currents Quadratic in the Curvatures

In this appendix we would like to give a few more details on the explicit tensorial form of the currents discussed in this note. Canonical currents are encoded as generating functions by $(\beta\gamma)^{s+1}\alpha_1^n\,\tau$, since any dependence on $\alpha_1\alpha_2$ can be removed by a local field redefinition. Such monomial corresponds to the following generating function kernels:

$$(\beta\gamma)^{s+1}\alpha_1^n\,\tau \to \frac{1}{s!^2 n!}(-i\partial_\xi\cdot\partial_\eta)^n(-iy(\partial_\xi-\partial_\eta))^s(-i\bar{y}(\partial_{\bar{\xi}}+\partial_{\bar{\eta}}))^s$$
$$\sum_{p_1,q_1}\frac{1}{p_1!q_1!}C_{\alpha(p_1)\dot{\alpha}(q_1)}(x)\xi^{\alpha(p_1)}\bar{\xi}^{\dot{\alpha}(q_1)}\sum_{p_2,q_2}\frac{1}{p_2!q_2!}C_{\beta(p_2)\dot{\beta}(q_2)}\eta^{\beta(p_2)}\bar{\eta}^{\dot{\beta}(q_2)}\Big|_{\eta=0,\xi=0}. \quad (67)$$

It is then easy to perform all required differentiations ending up with

$$(\beta\gamma)^{s+1}\alpha_1^n\,\tau \to (-1)^s\,\frac{y^{\alpha(s)}\bar{y}^{\dot{\alpha}(s)}}{s!^2}\sum_{p,q=0}^{s}(-1)^q\binom{s}{p}\binom{s}{q}\frac{(-i)^n}{n!}C_{\alpha(s-p)\beta(n)\dot{\alpha}(s-q)}C_{\alpha(p)}{}^{\beta(n)}{}_{\dot{\alpha}(q)}\,, \quad (68)$$

together with its conjugate piece which can be obtained by replacing α_1 with α_2. Notice that the spin of a zero form $C_{\alpha(n)\dot{\alpha}(m)}$ is given by $\frac{|n-m|}{2}$ so that the first 0-form has spin $s_1=\frac{|n-(p-q)|}{2}$ while the second has spin $s_2=\frac{|n+(p-q)|}{2}$. The number of derivatives is instead $s_1+s_2+\min(s-p+n,s-q)+\min(p+n,q)$. A particularly interesting case is when the canonical current has the maximum number of derivatives $s+s_1+s_2$. For $s_1=s_2$ this coupling is realised with $p=q$ and $s_1=s_2=\frac{n}{2}$. The s-0-0 current is a particular case of the latter for $n=0$. In the s-s_1-s_2 case $n=0$ reproduces Bel-Robinson type currents [23].[12]

B Fixing the 2pt normalisations in Vasiliev's equations holographically

An important subtlety to correctly interpret coefficients of currents at the equations of motion level is to determine the kinetic term normalisation for the Fronsdal fields. Such normalisation indeed play a key role to fix the cubic couplings in [7]. In the following we will recall all these normalisation fixing the notation. Furthermore we will fix the kinetic-term normalisation for all HS fields by matching holographically the sources coming from the 0-0-s cubic couplings.

First of all it is important to fix the convention for spinorial indices as in [3] with

$$h^{\alpha\dot{\alpha}}_{\mu}h^{\mu}_{\beta\dot{\beta}}=\epsilon_{\beta}{}^{\alpha}\epsilon_{\dot{\beta}}{}^{\dot{\alpha}}\,, \qquad h^{\alpha\dot{\alpha}}_{\mu}h^{\nu}_{\alpha\dot{\alpha}}=\delta^{\nu}_{\mu}\,, \quad (69)$$

and in Poincaré coordinates:

$$h^{\alpha\dot{\alpha}}_{\mu}\,dx^{\mu}=\frac{1}{2z}\sigma^{\alpha\dot{\alpha}}_{\mu}\,dx^{\mu}\,, \qquad h^{\mu}_{\alpha\dot{\alpha}}=z\sigma^{\mu}_{\alpha\dot{\alpha}}\,, \qquad g_{\mu\nu}=\frac{1}{2z^2}\eta_{\mu\nu}\,. \quad (70)$$

Here $\epsilon^{\alpha\beta}=-\epsilon^{\beta\alpha}$, $\epsilon^{12}=1$ and $\epsilon=i\sigma_2$ with $\sigma^{\alpha\dot{\alpha}}_i$, $i=1,2,3$ being the Pauli matrices. Notice that in this note we work with the choice $\Lambda=2$. Fronsdal equations then read:

$$\frac{1}{4(s-1)}\Big[\Box+2(s^2-2s-2)+\ldots\Big]\phi_{\alpha(s)\dot{\alpha}(s)}=J_{\alpha(s)\dot{\alpha}(s)}\,. \quad (71)$$

[12] In the $n=0$ case the same generating function can also be written in more conventional form as

$$(\beta\gamma)^{s+1}\,\tau \quad\longrightarrow\quad C(y,\bar{y})C(-y,\bar{y})\,.$$

where we have included the factor $\frac{1}{4(s-1)}$ coming from solving torsion as described in [3]. The mass term is dressed by a factor of 2 coming from our conventions for Λ. The mapping between spinorial and vectorial indices is achieved via the vielbein as:

$$\phi_{\mu(s)} \equiv \phi_{\alpha(s)\dot{\alpha}(s)}(x)\ \underbrace{h^{\alpha\dot{\alpha}}_{\mu} \cdots h^{\alpha\dot{\alpha}}_{\mu}}_{s}\,, \tag{72}$$

which follows from (69).

The holographically reconstructed equations can be extracted from the 0-0-s cubic coupling:

$$\mathcal{L} \sim \sum_s \left[\frac{N_s^2}{2^{s+2}}\,\phi_{\mu(s)}\Box\phi^{\mu(s)} + \ldots\right] - \underbrace{\frac{2^4}{\sqrt{N}}}_{g} \sum_s N_s N_0^2\,\frac{2^{-\frac{s}{2}}}{\Gamma(s)}\,i^s\,\phi^{\mu(s)}\left(\tfrac{1}{2^s}\phi\,\overleftrightarrow{\nabla}_{\mu(s)}\phi + \ldots\right), \tag{73}$$

with an arbitrary normalisation N_s for the Fronsdal kinetic term. They read:[13]

$$(\Box - 4)\phi(x) = g\,N_s\,\frac{2^{2-\frac{s}{2}}\,i^s}{\Gamma(s)}\,\phi^{\mu(s)}\,\nabla_{\mu(s)}\phi = g\,N_s\,\frac{2^{2-\frac{s}{2}}}{\Gamma(s)}\,\phi^{\alpha(s)\dot{\alpha}(s)}\,C_{\alpha(s)\dot{\alpha}(s)}\,, \tag{74}$$

where we have used that on the scalar sector of the zero form:

$$\widetilde{D}C = 0 \qquad \longrightarrow \qquad C_{\alpha(s)\dot{\alpha}(s)}(x) = i^n\,\underbrace{\nabla_{\alpha\dot{\alpha}} \cdots \nabla_{\alpha\dot{\alpha}}}_{s}\,\phi(x)\,. \tag{75}$$

The above equations allow to determine holographically the normalisation of the kinetic term used by Vasiliev's equations simply from the 0-0-s source to the scalar equation which can be extracted from the standard twisted-adjoint structure constants:

$$\widetilde{D}C = \omega \star C - C \star \pi(\omega)\,. \tag{76}$$

Upon translating the above equations to standard Klein-Gordon equations we arrive to:

$$(\Box - 4)\phi(x) = \frac{4}{\Gamma(s)^2}\,\phi^{\alpha(s)\dot{\alpha}(s)}\,C_{\alpha(s)\dot{\alpha}(s)}\,, \tag{77}$$

from which comparing with (74) we can determine the normalisation N_s used by Vasiliev's equations:

$$N_s = \frac{1}{g}\,\frac{2^{\frac{s}{2}}}{\Gamma(s)}\,, \qquad\qquad s > 0\,. \tag{78}$$

In the latter normalisation one then gets the following coupling constants for the source to the Fronsdal equations:

$$\left[\Box + 2(s^2 - 2s - 2) + \ldots\right]\phi_{\mu(s)}(x) = \underbrace{\frac{g\,N_0^2}{N_s}\,\frac{2^{\frac{s}{2}+1}}{\Gamma(s)}}_{=2g^2N_0^2}\,i^s\left(\tfrac{1}{2^s}\phi\,\overleftrightarrow{\nabla}_{\mu(s)}\phi + \ldots\right), \tag{79}$$

which, taking into account the normalisation (3) for the canonical current, determines the value for the coefficient α_s in (4) up to a spin-independent constant:

$$\alpha_s = 2^{1-s} g^2 N_0^2\,. \tag{80}$$

Similar results can be obtained for all other couplings of [7].

[13]Notice that a factor 2^{-s} comes from the non-canonically normalised metric in the contractions between HS field and derivatives. A further factor of 2 comes from the non canonical normalisation for the Laplacian and another factor of 2 comes from the variation with respect to the scalar field.

References

[1] M.A. Vasiliev, Phys. Lett. **B243**, 378 (1990)
[2] P. Kessel, G. Lucena Gómez, E. Skvortsov, M. Taronna, JHEP **11**, 104 (2015), 1505.05887
[3] N. Boulanger, P. Kessel, E.D. Skvortsov, M. Taronna, J. Phys. **A49**, 095402 (2016), 1508.04139
[4] N. Boulanger, S. Leclercq, P. Sundell, JHEP **08**, 056 (2008), 0805.2764
[5] E.D. Skvortsov, M. Taronna, JHEP **11**, 044 (2015), 1508.04764
[6] M. Taronna, *Pseudo-local Theories: A Functional Class Proposal*, in *International Workshop on Higher Spin Gauge Theories Singapore, Singapore, November 4-6, 2015* (2016), 1602.08566, https://inspirehep.net/record/1424813/files/arXiv:1602.08566.pdf
[7] C. Sleight, M. Taronna, Phys. Rev. Lett. **116**, 181602 (2016), 1603.00022
[8] M. Taronna, Master Thesis (2010), 1005.3061
[9] A. Sagnotti, M. Taronna, Nucl. Phys. **B842**, 299 (2011), 1006.5242
[10] S.F. Prokushkin, M.A. Vasiliev, Theor. Math. Phys. **123**, 415 (2000), [Teor. Mat. Fiz.123,3(2000)], hep-th/9907020
[11] M.A. Vasiliev (2016), 1605.02662
[12] M.A. Vasiliev, JHEP **06**, 031 (2015), 1502.02271
[13] X. Bekaert, J. Erdmenger, D. Ponomarev, C. Sleight, JHEP **11**, 149 (2015), 1508.04292
[14] R.R. Metsaev, Mod. Phys. Lett. **A6**, 2411 (1991)
[15] R.R. Metsaev, Mod. Phys. Lett. **A6**, 359 (1991)
[16] X. Bekaert, J. Erdmenger, D. Ponomarev, C. Sleight, JHEP **1503**, 170 (2015), 1412.0016
[17] X. Bekaert, J. Erdmenger, D. Ponomarev, C. Sleight, *Bulk quartic vertices from boundary four-point correlators*, in *International Workshop on Higher Spin Gauge Theories Singapore, Singapore, November 4-6, 2015* (2016), 1602.08570, http://inspirehep.net/record/1424869/files/arXiv:1602.08570.pdf
[18] F. Kristiansson, P. Rajan, JHEP **04**, 009 (2003), hep-th/0303202
[19] J. Maldacena, A. Zhiboedov, Class. Quant. Grav. **30**, 104003 (2013), 1204.3882
[20] M. Taronna, Ph.D. thesis, Pisa, Scuola Normale Superiore (2012), 1209.5755, https://inspirehep.net/record/1188191/files/arXiv:1209.5755.pdf
[21] M.A. Vasiliev, Nucl. Phys. **B324**, 503 (1989)
[22] M.A. Vasiliev, Phys. Lett. **B238**, 305 (1990)
[23] O.A. Gelfond, E.D. Skvortsov, M.A. Vasiliev, Theor. Math. Phys. **154**, 294 (2008), hep-th/0601106

DOI: 10.1051/epjconf/201612505026

Localized quench in 1+1 conformal field theory

Maria Tikhanovskaya[1],*

[1] *Steklov Mathematical Institute, Russian Academy of Sciences, Gubkin str. 8, 119991 Moscow, Russia*

Abstract. We study the boundary dual of AdS_3 spacetime with a point particle. Particles in AdS_3 generate topological defects, which allows to formulate the geodesic image method for boundary correlators. We propose the generalization of the geodesic recipe to arbitrary time intervals in case of the bulk spacetime deformed by a point particle.

1 Introduction

The AdS/CFT - correspondence [1] is the correspondence between $(d + 1)$ dimensional gravity of the AdS spacetime and d - dimensional quantum field theory on the boundary of the space. The useful and important tool in holographic calculations is the geodesic approximation [2]. This approximation allows to get in a simple way the correlation function.

The geodesic approximation directly relates correlation functions of the boundary QFT to geometry of the bulk spacetime. However, the geodesic prescription is valid only either for Euclidean spacetimes, or for spacelike-separated points in the Lorentzian case. Timelike geodesics in asymptotically AdS spacetimes cannot reach the boundary, therefore the timelike region is unavailable to the prescription, unless there is an analytical continuation from the original Euclidean form. Because of this, the information carried by large-time dynamics, or even real-time correlators in general, cannot be obtained from the geodesics approximation in general spacetimes. In [3] a non-trivial Euclidean continuation was constructed for the Vaidya spacetime. Another method that was used in [3, 4] is making use of discontinuous timelike geodesics which go through Poincare horizon.

In our work the prescription for calculation two point correlation function in geodesic approximation for the case when the particle deforms the AdS_3 spacetime is presented. Moving particle in the bulk realizes a quantum quench in the boundary theory. Here the formula for correlation function in geodesic approximation has the form:

$$G_\Delta(a, b) = \sum e^{-\Delta \mathcal{L}_{ren}(a,b)}. \tag{1}$$

We sum over all geodesics that connect a and b points. In the work [5] it was shown that geodesic approximation gives the exact answer (i.e. coincides with the answer from the [6, 7]) for the correlation function if the AdS_3 spacetime is orbifold and does not reproduce exact answer otherwise. We generalize the prescription to timelike-separated points by utilizing a set of auxiliary geodesics.

*e-mail: tikhanovskaya@mi.ras.ru

2 AdS_3 spacetime with particles

The AdS_3 is maximally symmetric solution of Einstein equations with the negative cosmological constant. In the embedding coordinates the AdS spacetime is a hyperboloid. If we parametrize the embedding coordinates by global coordinates (t, χ, ϕ) the metric of AdS_3 spacetime can be written as follows:

$$ds^2 = -\cosh^2\chi\, dt^2 + d\chi^2 + \sinh^2\chi\, d\phi^2, \qquad \phi \in (0, 2\pi) \tag{2}$$

where t is the time coordinate, χ is the radial coordinate and ϕ is the angular coordinate with period 2π. The AdS_3 conformal boundary corresponds to $\chi \to \infty$.

2.1 AdS_3 spacetime with static particle

We briefly discuss the Deser-Jackiw solution of Einstein equations with delta function in the RHS and cosmological constant equals to -1 [8, 9]. It describes a static particle in the center of the AdS_3 spacetime. Solution of these equations has the form:

$$ds^2 = -\cosh^2\chi\, dt^2 + d\chi^2 + \sinh^2\chi\, d\phi^2, \qquad \phi \in (0, 2\pi A). \tag{3}$$

The metric is the same as the metric for empty spacetime but an angle of the space has different ranges. Here A is the parameter associated with the mass parameter:

$$A = 1 - 4G\mu \;\in\; (0, 1) \tag{4}$$

where G is the Newtonian constant and μ is the mass of the particle. In the Fig.1A we show the point particle in the center of the AdS_3 spacetime. This particle cuts out the wedge and faces of the wedge must be glued.

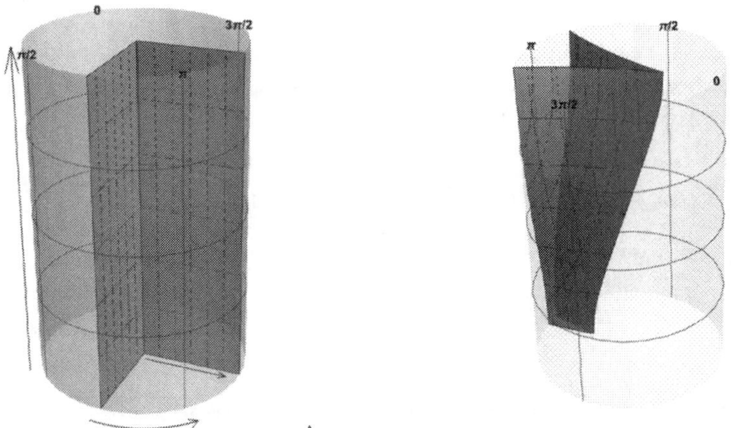

A. B.

Figure 1. A. AdS_3 with static massive particle in the center. Particle cats out the wedge from the spacetime. B. AdS_3 with moving massive particle. Here the mass parameter $\alpha = \pi/8$ and the boost parameter $\xi = 1$.

2.2 AdS_3 spacetime with moving particle

In this section we explain how the moving particle can be obtained in the AdS_3 spacetime. In [10] it was shown that using the group language for description the AdS_3 spacetime with a particle we can boost this particle by Lorentz transformation and get the spacetime with a moving particle in a simple way. The particle moves periodically with the $T = 2\pi$ period. Massive particle does not reach the boundary.

Consider the action of the isometry in the AdS_3 spacetime. The $*$-transformation is the isometry corresponding to the identification of the faces of the wedge. In the case of moving particle the isometry transformation is as follows [10]:

$$\tan t^* = \mathcal{B}_\xi(\alpha)\sec t \tanh\chi \cos\phi + \tan t\left(1 + 2\sinh^2\xi \sin^2\frac{\alpha}{2}\right); \tag{5}$$

$$\tan\phi^* = -\frac{2\tan\phi}{\mathcal{F}_\xi(\alpha)};$$

$$\cosh\chi^* = \cosh\chi\left[\left(\mathcal{B}_\xi(\alpha)\tanh\chi\cos\phi + \sin t\left(1 + 2\sinh^2\xi\sin^2\frac{\alpha}{2}\right)\right)^2 + \cos^2 t\right]^{\frac{1}{2}}. \tag{6}$$

where $\mathcal{B}_\xi$ and $\mathcal{F}_\xi$ are defined as

$$\mathcal{B}_\xi(\alpha) = \sinh\xi\left(\sin\alpha\tan\phi - 2\cosh\xi\sin^2\frac{\alpha}{2}\right), \tag{7}$$

$$\begin{aligned}\mathcal{F}_\xi(\alpha) &= \cosh\xi\,(2\sin\alpha\tan\phi - \cos\alpha + \cos\phi) + \sec\phi\cos(\alpha+\phi)\\ &+ \cos\alpha\cosh 2\xi - 2\sinh^2\xi.\end{aligned} \tag{8}$$

We are interested in the boundary theory so in the formulae (5) we take a limit of radial coordinate (i.e. $\chi \to \infty$) and get the action of isometry on the boundary (b means the boundary):

$$\tan t_b^* = \mathcal{B}_\xi(\alpha)\sec t\cos\phi + \tan t\left(1 + 2\sinh^2\xi\sin^2\frac{\alpha}{2}\right), \tag{9}$$

$$\tan\phi_b^* = -\frac{2\tan\phi}{\mathcal{F}_\xi(\alpha)}.$$

In the Fig.1B we illustrate the AdS_3 spacetime with a moving massive particle. With the expression for the induced boundary identification at hand, one can now formulate the geodesic images prescription for spacelike geodesics, which was explained in [10].

3 Holographic image method and correlation function

Consider the case of the empty AdS_3 spacetime. For the calculation of two point correlation function we use the geodesics approximation [2]. In this approximation one defines the correlator:

$$G_\Delta(\phi_a, t_a; \phi_b, t_b) = e^{-\Delta\mathcal{L}_{ren}(\phi_a,t_a;\phi_b,t_b)}, \tag{10}$$

where a and b are boundary points with coordinates (ϕ_a, t_a) and (ϕ_b, t_b). If the interval between a and b points is spacelike then the $\mathcal{L}_{ren}$ is the renormalized length of the geodesic connecting these points [10]. In case of timelike interval between a and b points we have to use the the prescription from the work [11].

The length of geodesic between two points in the AdS_3 is:

$$\mathcal{L}(a,b) = \ln\left[2(\cos(t_a - t_b) - \cos(\phi_a - \phi_b))\right] \tag{11}$$

then the correlation function (10) has the form:

$$G_{\Delta,AdS}(\phi_a, t_a; \phi_b, t_b) = \left(\frac{1}{2(\cos(t_a - t_b) - \cos(\phi_a - \phi_b))}\right)^{\Delta}. \qquad (12)$$

In case of spacetime deformed by particle we have to change the general formulae for correlation function (12) and use the follows:

$$\begin{aligned} G_{\Delta,l.s.}(\phi_a, t_a; \phi_b, t_b) &= \Phi_\Delta(\phi_a, t_a; \phi_b, t_b) G_{\Delta,\mathrm{AdS}}(\phi_a, t_a; \phi_b, t_b)\Theta_0(\phi_a, t_a; \phi_b, t_b) + \\ &+ \sum_{n=1}^{n_{\max}} \Phi_\Delta(\phi^*_{a,n}, t^*_{a,n}; \phi_b, t_b) G_{\Delta,\mathrm{AdS}}(\phi^*_{a,n}, t^*_{a,n}; \phi_b, t_b)\Theta_0(\phi^*_{a,n}, t^*_{a,n}; \phi_b, t_b) Z_n(\phi^*_{a,n}, t^*_{a,n}; \phi_b, t_b) + \\ &+ \sum_{n=1}^{\bar{n}_{\max}} \Phi_\Delta(\phi^\#_{a,n}, t^\#_{a,n}; \phi_b, t_b) G_{\Delta,\mathrm{AdS}}(\phi^\#_{a,n}, t^\#_{a,n}; \phi_b, t_b)\bar{\Theta}_0(\phi^\#_{a,n}, t^\#_{a,n}; \phi_b, t_b) \bar{Z}_n(\phi^\#_{a,n}, t^\#_{a,n}; \phi_b, t_b), \end{aligned} \qquad (13)$$

where $G_{\Delta,\mathrm{AdS}}(\phi, t)$ is given by (12). The presence of the Φ-factors in summands is related to the change of the causal relation between two points on the boundary under the isometry transformation (9).

Here the subscript l.s. in the LHS means a living space of the boundary of the AdS_3 with static or moving defects and

$$(\phi_a, t_a)^{*n} = (\phi^*_{a,n}, t^*_{a,n}), \quad (\phi_a, t_a)^{\#n} = (\phi^\#_{a,n}, t^\#_{a,n})$$

are coordinates of the image points obtained by n-times applications of the isometry $*$ and #-transformations respectively. We can use the #-transformation as an inverse $*$-transformation as follows:

$$(\phi^\#_a, t^\#_a)^* = (\phi_a, t_a). \qquad (14)$$

In case of deformed AdS spacetime for calculation of correlation function we have to use the extra factors:

$$\Theta_0(\phi^*_{a,n}, t^*_{a,n}; \phi_b, t_b), \quad \bar{\Theta}_0(\phi^\#_{a,n}, t^\#_{a,n}; \phi_b, t_b),$$
$$Z_{\Delta,n}(\phi^*_{a,n}, t^*_{a,n}; \phi_b, t_b), \quad \bar{Z}_{\Delta,n}(\phi^\#_{a,n}, t^\#_{a,n}; \phi_b, t_b),$$

where

$$\Theta_0(\phi_a, t_a, \phi_b, t_b) = \begin{cases} 1, & |\phi_a - \phi_b| > |t_a - t_b|, \\ 0, & |\phi_a - \phi_b| < |t_a - t_b|, \end{cases}$$

Factors Z are related to renormalizations [10].

Geodesic in (13) connecting points a^*, a^{**},... and b are image geodesics. Length of each image geodesic equals to the length of a winding geodesic between a and b points [10].

The images prescription for the correlator can be generalized to the case when the bulk spacetime is deformed by several particles.

4 Conclusion

We have calculated the two point correlation function in holographic approach via geodesic approximation. We have considered the case when geometry of AdS_3 is deformed by moving and static point massive particles. The whole Virasoro symmetry remains unchanged only in case when the mass parameter equals to $2\pi/n$, where n is integer (see Fig.2) but in the general case the symmetry is broken by the conical defect and this is manifest in the fact that correlation function in the geodesic approximation has the singularities in φ dependence (see Fig.3).

DOI: 10.1051/epjconf/201612505026

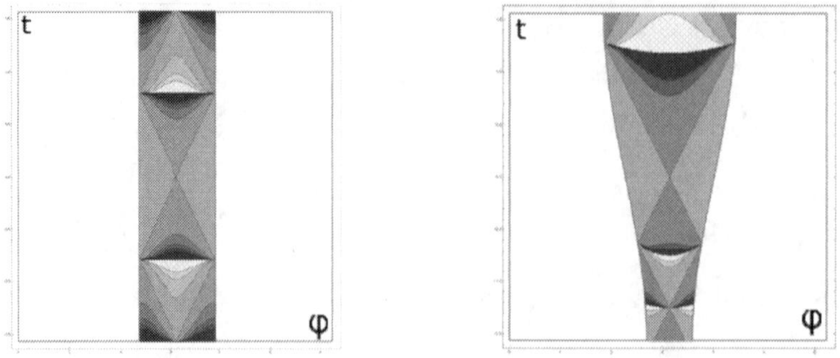

Figure 2. Density plot of inverse correlation function as a function of $G^{-1}(t, \varphi)$ is presented. In both of the cases the massive moving particle with the mass parameter $2\pi/n$ where n is integer deforms the AdS_3 spacetime. Parameter values are: $\xi = 0$, $\alpha = 3\pi/2$. In the left plot the massive particle is static. In the right plot the massive particle is moving. Parameter values are: $\xi = 0.6$, $\alpha = 3\pi/2$

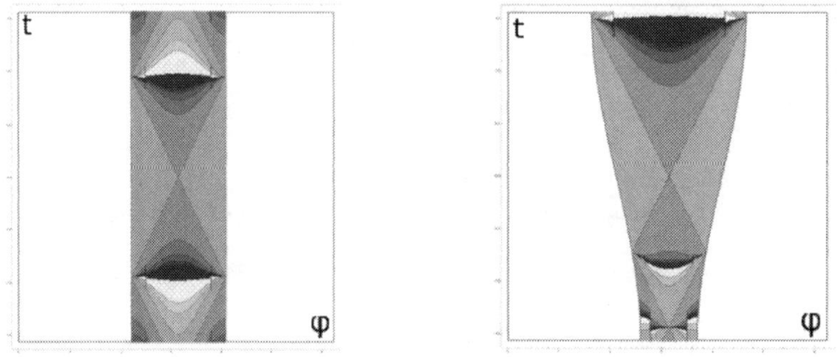

Figure 3. Density plot of inverse correlation function as a function of $G^{-1}(t, \varphi)$ is presented. In the both pictures non orbifold cases are presented. Parameter values are: $\xi = 0$, $\alpha = 4\pi/3 + 0.2$. In the left plot the massive particle is static. In the right plot the massive particle is moving. Parameter values are: $\xi = 0.6$, $\alpha = 4\pi/3 + 0.2$

5 Acknowledgments

This paper is based on a talk by 19th International Seminar on High Energy Physics "Quarks-2016" at Pushkin, Russia on 29 May - 4 July 2016. Author would like to thank the organizers of Quarks-2016 for the support. Author would like to thank Dmitrii Ageev, Irina Arefeva and Mikhail Khramtsov for useful discussions. The author is supported by the Russian Science Foundation (project 14-11-00687).

References

[1] J. M. Maldacena, "The Large N limit of superconformal field theories and supergravity," Adv. Theor. Math. Phys. **2**, 231-252 (1998), [hep-th/9711200].

[2] V. Balasubramanian and S. F. Ross, "Holographic particle detection," Phys. Rev. D **61**, 044007 (2000) [hep-th/9906226].

[3] V. Balasubramanian *et al.*, "Thermalization of the spectral function in strongly coupled two dimensional conformal field theories," JHEP **1304**, 069 (2013) [arXiv:1212.6066 [hep-th]].

[4] I. Arefeva, A. Bagrov, P. Saterskog and K. Schalm, "Holographic dual of a time machine," arXiv:1508.04440 [hep-th].

[5] I. Y. Aref'eva and M. A. Khramtsov, "AdS/CFT prescription for angle-deficit space and winding geodesics," arXiv:1601.02008 [hep-th].

[6] S. S. Gubser, I. R. Klebanov, A. M. Polyakov, "Gauge theory correlators from noncritical string theory," Phys. Lett. **B428**, 105-114 (1998), [hep-th/9802109].

[7] E. Witten, "Anti-de Sitter space and holography," Adv. Theor. Math. Phys. **2**, 253-291 (1998), [hep-th/9802150].

[8] S. Deser, R. Jackiw, and G. 't Hooft, "Three dimensional Einstein gravity: dynamics of flat space", Ann. Phys. **152** (1984) 220

[9] S. Deser and R. Jackiw, "Three-Dimensional Cosmological Gravity: Dynamics of Constant Curvature," Ann. Phys. **153** (1984) 405.

[10] D. S. Ageev, I. Y. Aref'eva and M. D. Tikhanovskaya, "Holographic Dual to Conical Defects: I. Moving Massive Particle," *Theoret. and Math. Phys.*, **188**:1 (2016), 1038–1068 [arXiv:1512.03362 [hep-th]].

[11] I. Ya. Aref'eva, M. A. Khramtsov and M. D. Tikhanovskaya, "Holographic Dual to Conical Defects III: Improved Image Method," [arXiv:1604.08905 [hep-th]]

AUTHOR INDEX

AUTHOR INDEX

AUTHOR INDEX